# CRYOGENIC ENGINEERING

*Second Edition*
*Revised and Expanded*

# CRYOGENIC ENGINEERING

*Second Edition*
*Revised and Expanded*

Thomas M. Flynn
*CRYOCO, Inc.*
*Louisville, Colorado, U.S.A.*

CRC Press is an imprint of the
Taylor & Francis Group, an **informa** business

Published in 2005 by
CRC Press
Taylor & Francis Group
6000 Broken Sound Parkway NW, Suite 300
Boca Raton, FL 33487-2742

First issued in paperback 2020



ISBN 13: 978-0-367-57816-9 (pbk)
ISBN 13: 978-0-8247-5367-2 (hbk)



**Visit the Taylor & Francis Web site at**
**http://www.taylorandfrancis.com**

**and the CRC Press Web site at**
**http://www.crcpress.com**

---

**Library of Congress Cataloging-in-Publication Data**

---

Catalog record is available from the Library of Congress

---

*The places I took him*!
I tried hard to tell
Young Conrad Cornelius o'Donald o'Dell
A few brand new wonderful words he might spell.
I led him around and I tried hard to show
There are things beyond Z that most people don't know.
I took him past Zebra. As far as I could.
And I think, perhaps, maybe I did him some good...

*On Beyond Zebra, by Dr. Seuss.*
*With permission, Random House Inc. 1745 Broadway, New York, NY.*

**To Rita**

# Preface to the Second Edtion

Dr. Flynn's *Cryogenic Engineering 2nd Edition* was written for a specific audience, namely, the professional engineer or physicist who needs to know some cryogenics to get his or her job done, but not necessarily make a career of it. The 2nd Edition was written to follow closely the cryogenic engineering professional course given annually by Dr. Flynn for 25 years, and accordingly has been thoroughly tested to be very practical. This 2nd (and last) edition includes over 125 new literature citations, and features more than 130 new graphs and tables of data, which may no longer be available elsewhere.

# Preface to the First Edition

This book is for the engineer and scientist who work for a living and who need cryogenics to get a job done. It is a deskbook, containing hundreds of tables and chart of cryogenic data that are very hard to come by. Examples and sample calculations of how to use the data are included. It is not a textbook. Instead, it assumes that the reader already has basic engineering and science skills. It is practical, using the measurement units of trade—SI, U.S. customary, and hybrid systems—just as they are commonly used in practice. It is not a design text, but it does contain many useful design guidelines for selecting the right system, either through procurement or in-house construction. In short, it is the cryogenics book I would like to have on my desk.

This book was written to gather into one source much of the technology developed at the National Bureau of Standards (NBS) Cryogenic Engineering Laboratory in Boulder, COL, over the last 40 years.

In the early 1950s, there was a need for the rapid development of a liquid hydrogen technology, and the major responsibility for the progress of this new engineering specialty was entrusted to the Cryogenics Section of the Heat and Power Division of NBS in Washington, D.C. Russell Scott led the work as chief of that section. Scott soon became the individual immediately in charge of the design and construction in Boulder (in March 1952) of the first large-scale liquefier for hydrogen ever built. This was the beginning of the Cryogenic Engineering Laboratory of NBS.

Again, in the late 1950s—when the nation was striving to regain world leadership in the exploration of space—the NBS Cryogenics Engineering Laboratory, which had by then matured under Scott's leadership, assumed a pre-eminent role in the solution of problems important to the nation. Scott, having had the foresight to establish a Cryogenic Data Center within the laboratory, was able to provide a focal point for information on many aspects of cryogenic engineering.

As a result of all this pioneering in the field of low-temperature engineering, a considerable amount of valuable technology was developed that in the course of normal events might have been lost. Scott recognized this, and the result was his book *Cryogenic Engineering*, an important first in its field. Its quality, authority, and completeness constitute a lasting tribute to him.

This present book is a mere shadow of Scott's work but is intended once more to up-date and preserve some of the cryogenics developed at NBS. There is only one author's name on the cover. Nonetheless, this book is a product of the collaboration of hundreds of good men and women of the NBS Cryogenic Engineering Laboratory. It is written to preserve the technology they developed.

When I was about to graduate as a chemical engineer from Rice University in 1955, I proudly told my department chairman, Dr. Arthur J. ("Pappy") Hartsook, that I intended to go to graduate school. Pappy, who knew I was a mediocre student, brightened considerably when I told him I wasn't going to a "good" school, like MIT, Michigan, or Wisconsin. Instead, I was going to the University of Colorado, where I could learn to ski. Dr. Hartsook was so relieved that he gave me a piece of advice pivotal to my career and my life. I will share it with you now. Pappy said that *What* I would work *on* was not nearly as important as *who* I would work *for*.

I took that advice and chose to work for a new professor at the University of Colorado, Dr. Klaus Timmerhaus. Klaus had only been there a year or two; the National Science Foundation and grantsmanship had yet to be invented. I had a teaching assistantship (paper grader) at $150 per month, before taxes. It was the most money I had ever had. To help me get some money for our planned research, Klaus suggested that I work at the Cryogenic Engineering Laboratory of the National Bureau of Standards. And so I did, for the next 28 years. For many years, it was the best of times, a truly nurturing environment for the young engineer scientist, because of the people who either worked there or visited there.

I wish to thank Klaus Timmerhaus, Russell Scott, Bascom Birmingham, Dudley Chelton, Bob Powell, John Dean, Ray Smith, Jo Mandenhall, Jim Draper, Dick Bjorklund, Bill Bulla, M. D. Bunch, Bob Goodwin, Lloyd Weber, Ray Radebaugh, Peter Storch, Larry Sparks, Bob McCarty, Vic Johnson, Bill Little, Bob Paugh, Bob Jacobs, Mike McClintock, Al Schmidt, Pete Vander Arend, Dan Weitzel, Wally Ziegler, John Gardner, Bob Mohling, Bob Neff, Scott Willen, Will Gully, Art Kidnay, Graham Walker, Ralph Surloch, Albert Schuler, Sam Collins, Bill Gifford, Peter Gifford, Ralph Longsworth, Ed Hammel, and Fred Edeskuty. Special thanks are due to Chris Davis and Janet Diaz for manuscript preparation and technical editing. I mention all these names not so much to give credit as to spread the blame.

I apologize in advance to those I have forgotten to mention.

# Contents

# 1
# Cryogenic Engineering Connections

## 1. FOREWARNED

In his masterpiece, *Civilization*, Sir Kenneth Clark writes not a foreword but rather a fore-warned. Clark writes to the effect that he does not know if his recounting is the actual history of civilization or not. He only knows that his story is his own view of the history of civilization. Having said that, Clark found much relief from the tedium of accuracy and was able to tell a more entertaining, and still highly accurate, story.

I wish to do the same. I do not know if the recounting to follow in this chapter is accurate or not. I only know that it is my own view of how it happened. I hope it is accurate—I did not deliberately make any of it up—but who knows?

I am an American and unabashedly proud of that fact. Therefore, let me begin with the American who may have gotten it all going, John Gorrie (1803–1855).

## 2. THE ENTREPRENEURS

John Gorrie was born in Charleston South Carolina in 1803 and graduated from the College of Physicians and Surgeons in New York City in 1833. In the spring of 1833 Gorrie moved to Apalachicola, a small coastal town in Florida situated on the Gulf of Mexico at the mouth of the river after which the town is named. By the time he arrived it was already a thriving cotton port, where ships from the north-east arrived to unload cargoes of supplies and load up with cotton for the northern factories. Within a year Gorrie was involved with town affairs. He served as mayor, city treasurer, council member, bank director, and founder of Trinity Church.

In 1834, he was made postmaster and in 1836 president of the local branch of the Pensacola Bank. In the same year, the Apalachicola Company asked him to report on the effects of the climate on the population, with a view to possible expansion of the town. Gorrie recommended drainage of the marshy, low-lying areas that surrounded the town on the grounds that these places gave off a miasma compounded by heat, damp and rotting vegetation which, according to the Spallanzani theory with which every doctor was intimate, carried disease. He suggested that only brick buildings be erected. In 1837, the area enjoyed a cotton boom, and the town population rose to 1500. Cotton bales lined the streets, and in four months 148 ships arrived to unload bricks from Baltimore, granite from Massachusetts, house framing from New York. Gorrie saw that the town was likely to grow as commerce increased and suggested that there was a need for a hospital. There was already a small medical unit in operation under the auspices of the US Government, and Gorrie was employed there on a part-time basis. Most of his patients were sailors and waterside workers, and most of them had fever, which was endemic in Apalachicola every summer.

Gorrie became obsessed with finding a cure to the disease. As early as 1836, he came close to the answer, over 60 years ahead of the rest of the world. In that year he wrote: "Gauze curtains, though chiefly used to prevent annoyance and suffering from mosquitoes, are thought also to be sifters of the atmosphere and interceptors and decomposers of malaria." The suggestion that the mosquito was the disease carrier was not to be made until 1881, many years after Gorrie's death, and for the moment he presumed that it came in some form of volatile oil, rising from the swamps and marshes.

By 1838, Gorrie had noticed that malaria seemed to be connected with hot, humid weather, and he set about finding ways to lower the temperature of his patients in summer. He began by hanging bowls full of ice in the wards and circulating the cool air above them by means of a fan. The trouble was that in Apalachicola ice was hard to come by. Ever since a Massachusetts merchant named Frederic Tudor had hit on the idea of cutting ice from ponds and rivers in winter and storing it in thick-walled warehouses for export to hot countries, regular ice shipments had left the port of Boston for destinations as far away as Calcutta. But Apalachicola was only a small port which the ships often missed altogether: if the ice crop was poor, the price rose to the exorbitant rate of $1.25 a pound. In 1844, Gorrie found the answer to the problem. It was well known that compressed gases which are rapidly allowed to expand absorb heat from their surroundings, so Gorrie constructed a steam engine to drive a piston back and forward in a cylinder. His machine compressed air, causing it to heat, and then the air through radiant coils where it decompressed and cooled, absorbing heat from a bath of brine. On the next cycle the air remained cool, since the brine had given up most of its heat. This air was then pumped out of the cylinder and allowed to circulate in the ward. Gorrie had invented air-conditioning. By bringing the cold brine into contact with water, Gorrie was then able to draw heat from the water to a point where it froze. Gorrie's application of compressed and decompressed gas as a coolant in radiant coils remains the common method for cooling air in modern refrigeration systems. His first public announcement of this development was made on 14 July 1850 in the Mansion House Hotel, where M. Rosan, the French Consul in Apalachicola, was celebrating Bastille Day with champagne. No ice ship had arrived, so the champagne was to be served warm. At the moment of the toast to the French Republic four servants entered, each carrying a silver tray on which was a block of ice the size of a house-brick, to chill the wine, as one guest put it, "by American genius".

In May of the following year Gorrie obtained a patent for the first ice-making machine (see Fig. 1.1), the first patent ever issued for a refrigeration machine. The patent specified that the water container should be placed in the cylinder, for faster freezing. Gorrie was convinced his idea would be a success. The *New York Times* thought differently: "There is a crank," it said, "down in Apalachicola, Florida, who claims that he can make ice as good as God Almighty!" In spite of this, Gorrie advertised his invention as "the first commercial machine to work for ice making and refrigeration." He must have aroused some interest, for later he was in New Orleans, selling the idea that "a ton of ice can be made on any part of the Earth for less than

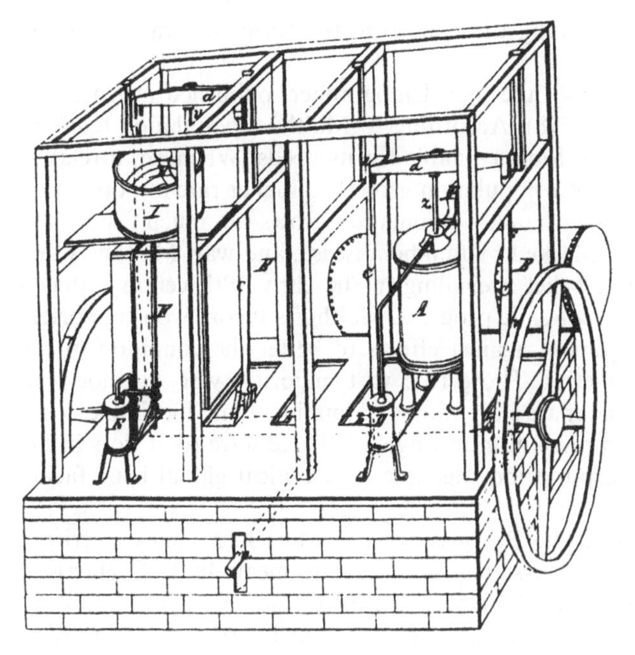

**Figure 1.1** Improved process for the artificial production of ice.

$2.00." But he was unable to find adequate backing, and in 1855 he died, a broken and dispirited man. A statue of John Corrie now stands in Statuary Hall of the US Capitol building, a tribute from the State of Florida to his genius and his importance to the welfare of mankind.

Three years after his death a Frenchman, Ferdinand Carré, produced a compression ice-making system and claimed it for his own, to the world's acclaim. Carré was a close friend of M. Rosan, whose champagne had been chilled by Gorrie's machine eight years before.

Just before he died, Gorrie wrote an article in which he said: "The system is equally applicable to ships as well as buildings . . . and might be instrumental in preserving organic matter an indefinite period of time." The words were prophetic, because 12 years later Dr. Henry P. Howard, a native of San Antonio, used the air-chilling system aboard the steamship *Agnes* to transport a consignment of frozen beef from Indianola, Texas, along the Gulf of Mexico to the very city where Gorrie had tried and failed to get financial backing for his idea. On the morning of Saturday 10 June 1869, the *Agnes* arrived in New Orleans with her frozen cargo. There it was served in hospitals and at celebratory banquets in hotels and restaurants. The New Orleans *Times Picayune* wrote: "[The apparatus] virtually annihilates space and laughs at the lapse of time; for the Boston merchant may have a fresh juicy beefsteak from the rich pastures of Texas for dinner, and for dessert feast on the delicate, luscious but perishable fruits of the Indies."

At the same time that Howard was putting his cooling equipment into the *Agnes*, committees in England were advising the government that mass starvation was likely in Britain because for the first time the country could no longer feed itself. Between 1860 and 1870, consumption of food increased by a staggering 25%. As the population went on rising, desperate speeches were made about the end of democracy and nationwide anarchy if the Australians did not begin immediately to find a way of sending their sheep in the form of meat instead of tallow and wool.

Thomas Malthus (1766–1834), the first demographer, published in An Essay on the Principle of Population (1798). According to Malthus, population tends to increase faster than the supply of food available for its needs. Whenever a relative gain occurs in food production over population growth, a higher rate of population increase is stimulated; on the other hand, if population grows too much faster than food production, the growth is checked by famine, disease, and war. Malthus's theory contradicted the optimistic belief prevailing in the early 19th century, that a society's fertility would lead to economic progress. Malthus's theory won supporters and was often used as an argument against efforts to better the condition of the poor. (the poor should die quietly). Those (the vast majority) who did not read the complete essay, assumed that mass starvation and imminent and inevitable, leading to the so-called Malthusian revolution in England. There were food riots in the streets. Food storage and transportation was seen as a critical global issue facing mankind.

Partially for these reasons, two Britons, Thomas Mort and James Harrison, emigrated to Australia and set up systems to refrigerate meat. In 1873 Harrison gave a public banquet of meat that had been frozen by his ice factory, to celebrate the departure of the S.S. *Norfolk* for England. On board were 20 tons of mutton and beef kept cold by a mixture of ice and salt. On the way, the system developed a leak and the cargo was ruined. Harrison left the freezing business. Mort then tried a different system, using ammonia as the coolant. He too gave a frozen meat lunch, in 1875, to mark the departure for England of the S.S. *Northam*. Another leak ruined this second cargo, and Mort retired from the business. But both men had left behind them working refrigeration plants in Australia. The only problem was to find the right system to survive the long voyage to London. Eventually, the shippers went back to Gorrie's "dry air" system. Aboard ship, it was much easier to replace leaking air than it was to replace leaking ammonia. Even though ammonia was "more efficient", it was sadly lacking while air was ubiquitous. This $NH_3$/air substitution is one early example of a lack of "systems engineering". There are often unintended consequences of technology, and the working together of the system as a whole must be considered. $NH_3$ refrigerators were thermodynamically more efficient, but could not be relished with $NH_3$ readily.

## 3. THE BUTCHERS

The development of the air-cycle refrigerator, patented by the Scottish butchers Bell and Coleman in 1877, made the technical breakthrough. The use of atmospheric air as a refrigerating fluid provided a simple, though inefficient, answer to ship-board refrigeration and led to the British domination of the frozen meat trade thereafter.

The first meat cargo to be chilled in this fashion left Australia aboard the S.S. *Strathleven* on 6 December 1879 to dock in London on 2 February of the following year with her cargo intact (Fig. 1.2). Figure 1.3 shows a similar ship to the S.S. Strathleven being unloaded. The meat sold at Smithfield market for between 5*d* and 6*d* per pound and was an instant success. Queen Victoria, presented with a leg of lamb from the same consignment, pronounced it excellent. England was saved.

**Figure 1.2** The S. S. *Strathleven*, carrying the first successful consignment of chilled beef from Australia to England. Note the cautious mixture of steam and sail, which was to continue into the 20th century.

## 4. THE BREWERS

Harrison's first attempts at refrigeration in Australia had been in a brewery, where he had been trying to chill beer, and although this operation was a moderate success, the profits to be made from cool beer were overshadowed by the immense potential of the frozen meat market. The new refrigeration techniques were to become a boon to German brewers, but in Britain, where people drank their beer almost at room temperature, there was no interest in chilling it. The reason British beer-drinkers take their beer "warm" goes back to the methods used to make the beer. In Britain it is

**Figure 1.3** Unloading frozen meat from Sydney, Australia, at the South West India Dock, London. This shows the hold of the *Catania*, which left port in August 1881 with 120 tons of meat from the same exporters who had filled the *Strathleven*.

produced by a method using a yeast which ferments on the surface of the beer vat over a period of 5–7 days, when the ideal ambient temperature range is from 60°F to 70°F. Beer brewed in this way suffers less from temperature changes while it is being stored, and besides, Britain rarely experiences wide fluctuations in summer temperatures. But in Germany beer is produced by a yeast which ferments on the *bottom* of the vat. This type of fermentation may have been introduced by monks in Bavaria as early as 1420 and initially was an activity limited to the winter months, since bottom fermentation takes place over a period of up to 12 weeks, in an ideal ambient temperature of just above freezing point. During this time the beer was stored in cold cellars, and from this practice came the name of the beer: lager, from the German verb *lagern* (to store). From the beginning there had been legislation in Germany to prevent the brewing of beer in the summer months, since the higher temperatures were likely to cause the production of bad beer. By the middle of the 19th century every medium-sized Bavarian brewery was using steam power, and when the use of the piston to compress gas and cool it became generally known, the president of the German Brewer's Union, Gabriel Sedlmayr of the Munich Spätenbrau brewery, asked a friend of his called Carl Von Linde if he could develop a refrigerating system to keep the beer cool enough to permit brewing all the year round. Von Linde solved Sedlmayr's problem and gave the world affordable mechanical refrigeration, an invention that today is found in almost every kitchen.

Von Linde used ammonia instead of air as his coolant, because ammonia liquefied under pressure, and when the pressure was released it returned to gaseous form, and in so doing drew heat from its surroundings. In order to compress and release the ammonia, he used Gorrie's system of a piston in a cylinder. Von Linde did not invent the ammonia refrigeration system, but he was the first to make it work. In 1879 he set up laboratories in Wiesbaden to continue research and to convert his industrial refrigeration unit into one for the domestic market. By 1891, he had put 12,000 domestic refrigerators into German and American homes. The modern fridge uses essentially the same system as the one with which Von Linde chilled the Spätenbrau cellars.

## 5. THE INDUSTRIALISTS

Interest in refrigeration spread to other industries. The use of limelight, for instance, demanded large amounts of oxygen, which could be more easily handled and transported in liquid form. The new Bessemer steel-making process used oxygen. It may be no coincidence that an ironmaster was involved in the first successful attempt to liquefy the gas. His name was Louis Paul Cailletet, and together with a Swiss engineer, Raoul Pictet, he produced a small amount of liquid oxygen in 1877.

At the meeting of the Académie des Sciences in Paris on 24 December 1887, two announcements were made which may be recognized as the origins of cryogenics as we know it today. The secretary to the Académie spoke of two communications he had received from M. Cailletet working in Paris and from Professor Pictet in Geneva in which both claimed the liquefaction of oxygen, one of the permanent gases.

The term "permanent" had arisen from the experimentally determined fact that such gases could not be liquefied by pressure alone at ambient temperature, in contrast to the nonpermanent or condensable gases like chlorine, nitrous oxide and carbon dioxide, which could be liquefied at quite modest pressures of 30–50 atm. During the previous 50 years or so, in extremely dangerous experiments, a number of workers had discovered by visual observation in thick-walled glass tubes that the permanent gases, including hydrogen, nitrogen, oxygen and carbon monoxide, could not be liquefied at pressures as high as 400 atm. The success of these experimenters marked the end of the idea of permanent gases and established the possibility of liquefying any gas by moderate compression at temperatures below the critical temperature. In 1866, Van der Waals (1837–1923), had published his first paper on "the continuity of liquid and gaseous states" from which the physical understanding of the critical state, and of liquefaction and evaporation, was to grow.

Cailletet had used the apparatus shown in Fig. 1.4 to produce a momentary fog of oxygen droplets in the thick-walled glass tube. The oxygen gas was compressed using the crude Natterer compressor in which pressures up to 200 atmospheres were generated by a hand-operated screw jack. The pressure was transmitted to the oxygen gas in the glass tube by hydraulic transmission using water and mercury. The gas was cooled to −103°C by enclosing the glass tube with liquid ethylene and was then expanded suddenly by releasing the pressure via the hand wheel. A momentary fog was seen, and the procedure could then be repeated for other observers to see the phenomenon.

Figure 1.5 shows the cascade refrigeration system used by Professor Pictet at the University of Geneva in which oxygen was first cooled by sulphur dioxide and then by liquid carbon dioxide in heat exchangers, before being expanded into the atmosphere by opening a valve. The isenthalpic expansion yielded a transitory jet of partially liquefied oxygen, but no liquid could be collected from the high-velocity jet. The figure shows how Pictet used pairs of compressors to drive the $SO_2$ and $CO_2$ refrigerant cycles. This is probably one of the first examples of the cascade refrigeration system invented by Tellier (1866) operating at more than one temperature level. Pictet was a physicist with a mechanical flair, and although he did not

**Figure 1.4** Cailletet's gas compressor and liquefaction apparatus (1877).

**Figure 1.5** Pictet's cascade refrigeration and liquefaction system (1877).

pursue the liquefaction of oxygen (he made a name developing ice-skating rinks), his use of the cascade system inspired others like Kamerlingh Onnes and Dewar.

In the early 1880s one of the first low-temperature physics laboratories, the Cracow University Laboratory in Poland, was established by Szygmunt von Wroblewski and K. Olszewski. They obtained liquid oxygen "boiling quietly in a test tube" in sufficient quantity to study properties in April 1883. A few days later, they also liquefied nitrogen. Having succeeded in obtaining oxygen and nitrogen as true liquids (not just a fog of liquid droplets), Wroblewski and Olszewski, now working separately at Cracow, attempted to liquefy hydrogen by Cailletet's expansion technique. By first cooling hydrogen in a capillary tube to liquid-oxygen temperatures and expanding suddenly from 100 to 1 atm, Wroblewski obtained a fog of liquid-hydrogen droplets in 1884, but he was not able to obtain hydrogen in the completely liquid form.

The Polish scientists at the Cracow University Laboratory were primarily interested in determining the physical properties of liquefied gases. The ever-present problem of heat transfer from ambient plagued these early investigators because the cryogenic fluids could be retained only for a short time before the liquids boiled away. To improve this situation, an ingenious experimental technique was developed at Cracow. The experimental test tube containing a cryogenic fluid was surrounded by a series of concentric tubes, closed at one end. The cold vapor arising from the liquid flowed through the annular spaces between the tubes and intercepted some of the heat traveling toward the cold test tube. This concept of *vapor shielding* is used today in conjunction with high-performance insulations for the long-term storage of liquid helium in bulk quantities.

All over Europe scientists worked to produce a system that would operate to make liquid gas on an industrial scale. The major problem in all this was to prevent the material from drawing heat from its surroundings. In 1882, a French physicist called Jules Violle wrote to the French Academy to say that he had worked out a way of isolating the liquid gas from its surroundings through the use of a vacuum. It had been known for some time that vacua would not transmit heat, and Violle's arrangement was to use a double-walled glass vessel with a vacuum in the space between the walls. Violle has been forgotten, his place taken by a Scotsman who was to do the same thing, much more efficiently, eight years later. His name was Sir James Dewar, and he added to the vessel by silvering it both inside and out (Violle had only silvered the exterior), in order to prevent radiation of heat either into or out of the vessel.

## 6. THE SCIENTISTS

James Dewar was appointed to the Jacksonian Professorship of Natural Philosophy at Cambridge in 1875 and to the Fullerian Professorship of Chemistry at The Royal Institution in 1877, the two appointments being held by him until his death at the age of 81 in 1923. Dewar's research interests ranged widely but his outstanding work was in the field of low temperatures. Within a year of his taking office at The Royal Institution, the successful liquefaction of oxygen by Cailletet and Pictet led Dewar to repeat Cailletet's experiment. He obtained a Cailletet apparatus from Paris, and, within a few months, in the summer of 1878 he demonstrated the formation of a mist of liquid oxygen to an audience at one of his Friday evening discourses at The Royal Institution. This lecture was the first of a long series of demonstrations, extending over more than 30 years and culminating in dramatic and sometimes hazardous demonstrations with liquid hydrogen (Fig. 1.6). In May 1898 Dewar produced $20\,\mathrm{cm}^3$ of liquid hydrogen boiling quietly in a vacuum-insulated tube, instead of a mist.

The use of a vacuum had been used by Dewar and others as early as 1873 and his experiments over several years before 1897 went on to show how he could obtain significant reductions (up to six times) by introducing into the vacuum space powders such as charcoal, lamp black, silica, alumina, and bismuth oxide. For this purpose, he used sets of three double-walled vessels connected to a common vacuum in which one of the set was used as a control. Measurement of the evaporation rate of liquid air in the three vessels then enabled him to make comparative assessments on the test insulations.

In 1910 Smoluchowski demonstrated the significant improvement in insulating quality that could be achieved by using evacuated powders in comparison with unevacuated insulations. In 1937, evacuated-powder insulations were first used in the United States in bulk storage of cryogenic liquids. Two years later, the first vacuum-powder-insulated railway tank car was built for the transport of liquid oxygen.

Following evacuated powders, he made further experiments using metallic and other septa-papers coated with metal powders in imitation of gold and silver, together with lead and aluminum foil and silvering of the inner surfaces of the annular space. He found that three turns of aluminum sheet (not touching) were not as good as silvered surfaces. Had he gone on to apply further turns of aluminum, he would have discovered the principle of multilayer insulation which we now know to be superior to silvering. Nevertheless, his discovery of silvering as an effective means of reducing the radiative heat flux component was a breakthrough. From 1898, the glass Dewar flask quickly became the standard container for cryogenic

**Figure 1.6** Sir James Dewar lecturing at the Royal Institution. Although Violle preceded Dewar in the development of the vacuum flask, there is no evidence that Dewar knew of his work when he presented the details of his new container to the Royal Institution in 1890.

liquids. Meanwhile his work on the absorptive capacity of charcoal at low temperatures paved the way towards the development of all-metal, double-walled vacuum vessels.

Another first for Dewar was his use of mixtures of gases to enhance J-T cooling, a topic revisited in cryogenics in 1995.

Historically, the expansion of a mixture of hydrogen and nitrogen was employed by Dewar in 1894 in attempts to liquefy hydrogen at that time. Dewar wrote: "Expansion into air at one atmosphere pressure of a mixture of 10% nitrogen in hydrogen yielded a much lower temperature than anything that has been recorded up to the present time."

Because his flasks (unlike those of Jules Violle) were silvered inside and out, Dewar's flasks could equally well retain heat as cold. Dewar's vessel became known in scientific circles as the Dewar flask; with it, he was able to use already liquid gases to enhance the chill during the liquefaction of gases whose liquefaction temperature was lower than that of the surrounding liquid. In this way, in 1891, he succeeded for the first time in liquefying hydrogen.

Dewar had considerable difficulty in finding competent glass makers willing to undertake the construction of his double-walled vessels and had been forced to get them made in Germany.

By 1902 a German called Reinhold Burger, whom Dewar had met when visiting Germany to get his vessels made, was marketing them under the name of Thermos. The manufacture of such vessels developed into an important industry in Germany as the Thermos flask, and this monopoly was maintained up to 1914. Dewar never patented his silvered vacuum flask and therefore never benefited financially from his invention.

The word "Cryogenics" was slow in coming.

The word cryogenics is a product of the 20th Century and comes from the Greek—*κρος*—frost and—*γινομαι*—to produce, engender. Etymologically, cryogenics means the science and art of producing cold and this was how Kamerlingh Onnes first used the word in 1894. Looking through the papers and publications of Dewar and Claude, it appears that neither of them ever used the word; indeed a summary of Dewar's achievements by Armstrong in 1916 contains no mention but introduces yet another term, "the abasement of temperature."

In 1882, Kamerlingh Onnes (1853–1926) was appointed to the Chair of Experimental Physics at the University of Leiden in the Netherlands and embarked on building up a low-temperature physics laboratory in the Physics Department. The inspiration for his laboratory was provided by the systematic work of Van der Waals at Amsterdam, and subsequently at Leiden, on the properties of gases and liquids. In 1866, Van der Waals had published his first paper on "the continuity of liquid and gaseous states" from which the physical understanding of the critical state, and of liquefaction and evaporation was to grow.

He operated an open-door policy encouraging visitors from many countries to visit, learn, and discuss. He published all the experimental results of his laboratory, and full details of the experimental apparatus and techniques developed, by introducing in 1885 a new journal "Communications from the Physical Laboratory at the University of Leiden."

As a result, he developed a wide range of contacts and a growing track record of success, so that from the turn of the century his Leiden laboratory held a leading position in cryogenics for almost 50 years, certainly until the mid-1930s. In 1908, for example, he won the race with Dewar and others to liquefy helium and went on to discover superconductivity in 1911.

Onnes' first liquefaction of helium in 1908 was a tribute both to his experimental skill and to his careful planning. He had only 360 L of gaseous helium obtained by heating monazite sand from India. More than 60 $cm^3$ of liquid helium was produced by Onnes in his first attempt. Onnes was able to attain a temperature of 1.04 K in an unsuccessful attempt to solidify helium by lowering the pressure above a container of liquid helium in 1910.

It is interesting to compare Dewar's approach with that of his rival, Kamerlingh Onnes. Having successfully liquefied hydrogen in 1898, Dewar had been able to monopolize the study of the properties of liquid hydrogen and published many papers on this subject. His attempts in 1901 to liquefy helium in a Cailletet tube cooled with liquid hydrogen at 20.5 K, using a single isentropic expansion from pressures up to 100 atm, led at first to a mist, being clearly visible. Dewar was suspicious of a contamination because after several compressions and expansions, the end of the Cailletet tube contained a small amount of solid that sublimed to a gas without passing through the liquid state when the liquid hydrogen was removed. On lowering the temperature of the liquid hydrogen by pumping to 16 K, and repeating the expansions on the gas from which the solid had been separated by the previous expansions at 20.5 K, no mist was seen. From these observations, he concluded that the mist was caused by some material other than helium, in all probability neon, and that the critical temperature and boiling point of helium were below 9 K and about 5 K, respectively.

The Cailletet tube was, of course, limited in its potential and Dewar appreciated that he needed a continuous circulation cascade system employing liquid hydrogen at reduced pressure and a final stage of Joule–Thomson expansion with recuperative cooling. He already had at the Royal Institution a large quantity of hardware, compressors, and pumps for the cascade liquefaction system he had assembled for liquefying hydrogen. From 1901, he joined the race to liquefy helium with competitors like Travers and Ramsay at University College, London, Kamerlingh Onnes and his Leiden team, and Olszewski at Cracow. The race continued for 7 years until 1908 when the Leiden team won. Dewar and the other competitors perhaps failed because they had not appreciated that the magnitude of the effort to win the race required a systems approach to solve the problems of purification, handling small quantities of precious helium gas, maintaining leaky compressors, improving the design of recuperative heat exchangers, and understanding the properties and behavior of fluids and solids at low temperatures, all at the same time.

As a direct result of their success, the Leiden team of Kamerlingh Onnes went on to discover superconductivity in 1911 and thereafter maintained a leading position in low-temperature research for many years.

## 7. THE ENGINEERS

Dr. Hampson was medical officer in charge of the Electrical and X-ray Departments, Queens and St. John's Hospitals, Leicester Square, London. He was a product of the Victorian age, with a classics degree at Oxford in 1878, having subsequently acquired his science as an art and a living; but he possessed an extraordinary mechanical flair. He was completely overshadowed by Dewar at The Royal Institution, Ramsay at University College and Linde at Wiesbaden, each with their considerable laboratory facilities. Indeed, Dewar seems to have been unable to accept Hampson as a fellow experimentalist or to acknowledge his contribution to cryogenics.

And yet, Hampson with his limited facilities was able to invent and develop a compact air liquefier which had a mechanical elegance and simplicity which made Dewar's efforts crude and clumsy in comparison. Indeed, Hampson's design of heat exchanger was so successful that it is still acknowledged today. The Hampson type coiled tube heat exchanger is widely used today. In 1895, when Hampson and Linde independently took out patents on their designs of air liquefier, the Joule–Thomson effect was known and the principle of recuperative cooling by so-called self-intensification had been put forward by Siemens as early as 1857. The step forward in both patents was to break away from the cascade system of cooling and to rely entirely on Joule–Thomson cooling together with efficient heat exchanger designs.

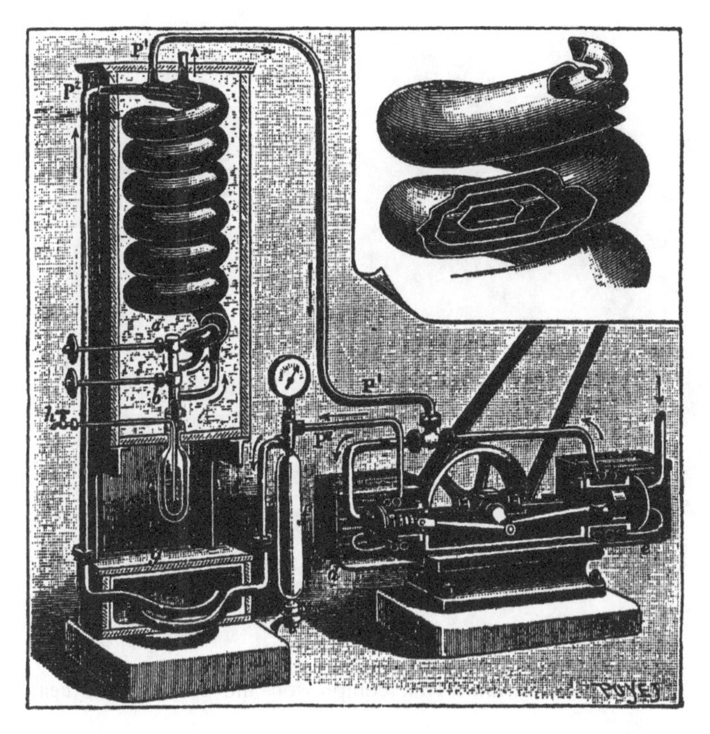

**Figure 1.7** Linde two-stage compressor and 3 L/hr air liquefier (1895).

Linde used concentric tubes of high-pressure line enclosed by a low-pressure return, the two being wound into a single layer spiral to achieve a compact design (Fig. 1.7). "Compact" is a relative term. Linde's first such heat exchanger was made of hammered copper tubing approximately one-quarter of an inch thick. The outer (low pressure) tube was more than 6 in. in diameter. It is said that it took the better part of a month for the heat exchanger to cool down and achieve thermal equilibrium.

After 1895, Linde made rapid progress in developing his Joule–Thomson expansion liquefier making some 3 L of liquid air per hour.

By the end of 1897, Charles Tripler, an engineer in New York had constructed a similar but much larger air liquefier, driven by a 75 kW steam engine, which produced up to 15 L of liquid air per hour (Fig. 1.8). Tripler discovered a market for liquid air as a medium for driving air expansion engines—the internal combustion engine was still unreliable at that time—and he succeeded in raising $10 M on Wall Street to launch his Liquid Air Company.

Using the liquid air he had produced to provide high-pressure air for an air expansion engine to drive his air compressor, he was convinced that he could make more liquid air than he consumed—and coined the word "surplusage." He was of course wrong. In 1902, Tripler was declared bankrupt, and Wall Street and the US lost interest in commercial applications of cryogenics for many years to come, although important cryophysics and cryo-engineering research continued in US universities.

In 1902 Georges Claude, a French engineer, developed a practical system for air liquefaction in which a large portion of the cooling effect of the system was obtained through the use of an expansion engine. The use of an expansion engine

**Figure 1.8** Tripler's air liquefier with steam-driven compressor and tube-in-shell heat exchangers (1898).

caused the gas to cool isentropically rather than isenthalpically, as is the case in Joule–Thomson expansion. Claude's first engines were reciprocating engines using leather seals (actually, the engines were simply modified steam engines). During the same year, Claude established l'Air Liquide to develop and produce his systems. The increase in cooling effect over the Joule–Thomson expansion of the Linde/Hampson/Tripler designs was so large as to constitute another technical breakthrough. Claude went on to develop air liquefiers with piston expanders in the newly formed Société L'Air Liquide.

Although cryogenic engineering is considered a relatively new field in the US, it must be remembered that the use of liquefied gases in US industry began in the early 1900s. Linde installed the first air-liquefaction plant in the United States in 1907, and the first American-made air-liquefaction plant was completed in 1912. The first commercial argon production was put into operation in 1916 by the Linde company in Cleveland, Ohio. In 1917 three experimental plants were built by the Bureau of Mines, with the cooperation of the Linde Company, Air Reduction Company, and the Jefferies-Norton Corporation, to extract helium from natural gas of Clay County, Texas. The helium was intended for use in airships for World War I. Commercial production of neon began in the United States in 1922, although Claude had produced neon in quantity in France since 1920.

Around 1947 Dr. Samuel C. Collins of the department of mechanical engineering at Massachusetts Institute of Technology developed an efficient liquid-helium laboratory facility. This event marked the beginning of the period in which liquid-helium temperatures became feasible and fairly economical. The Collins helium cryostat, marketed by Arthur D. Little, Inc., was a complete system for the safe, economical liquefaction of helium and could be used also to maintain temperatures at any level between ambient temperature and approximately 2 K.

The first buildings for the National Bureau of Standards Cryogenic Engineering Laboratory were completed in 1952. This laboratory was established to provide engineering data on materials of construction, to produce large quantities of liquid hydrogen for the Atomic Energy Commission, and to develop improved processes and equipment for the fast-growing cryogenic field. Annual conferences in cryogenic engineering have been sponsored by the National Bureau of Standards (sometimes sponsored jointly with various universities) from 1954 (with the exception of 1955) to 1973. At the 1972 conference at Georgia Tech in Atlanta, the Conference Board voted to change to a biennial schedule alternating with the Applied Superconductivity Conference. The NBS Cryogenic Engineering Laboratory is now part of history, as is the name "NBS", now the National Institutes for Science and Technology (NIST).

## 8. THE ROCKET SCIENTISTS

The impact of cryogenics was wide and varied. The Dewar flask changed the social habits of the Edwardian well-to-do: picnics became fashionable because of it. In time it changed the working-man's lunch break and accompanied expeditions to the tropics and to the poles, carrying sustenance for the explorers and returning with hot or cold specimens. Later, it saved thousands of lives by keeping insulin and other drugs from going bad. Perhaps its most spectacular impact was made, however, by two men whose work went largely ignored, and by a third who did his work in a way that could not be ignored. The first was a Russian called Konstantin Tsiolkovsky, whose early use of liquid gas at the beginning of the 20th century was to lie buried under governmental lack of interest for decades. The second was an American called Robert Goddard, who did most of his experiments on his aunt's farm in Massachusetts, and whose only reward was lukewarm interest from the weather bureau.

On 16 March 1926, Dr. Robert H. Goddard conducted the world's first successful flight of a rocket powered by liquid-oxygen–gasoline propellant on a farm near Auburn, Massachusetts. This first flight lasted only $2\frac{1}{2}$ sec, and the rocket reached a maximum speed of only 22 m/s (50 mph). Dr. Goddard continued his work during the 1930s, and by 1941 he had brought his cryogenic rockets to a fairly high degree of perfection. In fact, many of the devices used in Dr. Goddard's rocket systems were used later in German V-2 weapons systems (Fig 1.9).

The third was a German, Herman Oberth, and his work was noticed because it aimed at the destruction of London.

His liquid gases were contained in a machine that became known as Vengeance Weapon 2, or V-2, and by the end of the Second World War it had killed or injured thousands of military and civilian personnel. All three men had realized that certain gases burn explosively, in particular hydrogen and oxygen, and that, since in their liquid form they occupy less space than as a gas (hydrogen does this by a factor of 790) they were an ideal fuel. Thanks to the principles of the Dewar flask, they could be stored indefinitely, transported without loss, and contained in a launch vehicle that was essentially a vacuum flask with pumps, navigation systems, a combustion chamber and a warhead.

(One of Oberth's most brilliant assistants was a young man called Werner von Braun, and it was he who brought the use of liquid fuel to its most spectacular expression when his brainchild, the Saturn V, lifted off at Cape Canaveral on 16 July 1969, carrying Armstrong, Aldrin, and Collins to their historic landing on the moon.)

**Figure 1.9** The V-2 liquid-fueled rocket used a mixture of oxygen and kerosene. Originally developed at the experimental rocket base in Peenemunde, on the Baltic, the first V-2 landed in London in 1944. When the war ended, German engineers were working on a V-3 capable of reaching New York.

There is no doubt that the development of today's space technology would have been impossible without cryogenics. The basic reason for this lies with the high specific impulse attainable with kerosene/liquid oxygen (2950 m/s) and liquid hydrogen/liquid oxygen (3840 m/s)—values much higher than with liquid or solid propellants stored at ambient temperatures.

Space cryogenics developed rapidly in the early 1960s for the Apollo rocket series, at the same time as LNG technology. The driving force for space cryogenics development was the competition between the US and Russia in the exploration of Space, and the surface of the Moon in particular, and in the maintenance of detente in the "cold war".

Particular requirements then, and now in the Space Shuttle flight series, include the use of liquid hydrogen as a propellant fuel, liquid oxygen as a propellant oxidizer and for the life support systems, both liquids for the fuel cell electric power supplies and liquid helium to pressurize the propellant tanks. The successful development of the necessary cryogenic technology has provided an extraordinary range of spin-offs and a remarkable level of confidence in the design, construction, and handling of cryogenic systems.

## 9. THE PHYSICISTS AND SUPERCONDUCTIVITY

The phenomenon of zero electrical resistance was discovered in 1911 in mercury by Kamerlingh Onnes and his team at Leiden. Although he realized the great significance of his discovery, Kamerlingh Onnes was totally frustrated by the lack of any practical application because he found that quite small magnetic fields, applied either externally or as self-fields from internal electric currents, destroyed the superconducting state. The material suddenly quenched and acquired a finite electrical resistance.

For the next 50 years, very little progress was made in applying superconductivity, although there was a growing realization that a mixed state of alternate laminae of superconducting and normal phase had a higher critical field which depended on the metallurgical history of the sample being studied. In fact, an impasse had arisen by the late 1950s, and little progress was being achieved on the application of superconductivity.

On the other hand, great progress had been achieved by this time in the UK, USSR, and US in the theoretical description of superconductivity. This progress stemmed largely from two sets of experimental evidence; firstly, demonstrations of the isotope effect, indicating that lattice vibrations must play a central role in the interaction leading to superconductivity; secondly, the accumulation of evidence that an energy gap exists in the spectrum of energy states available to the conduction electrons in a superconductor. In 1956, Bardeen, Cooper, and Schrieffer (see Table 1.1. for a list of the many Nobel Laureates indebted to cryogenics) proposed a successful theory of superconductivity in which conduction electrons are correlated in pairs with the same center of mass momentum via interactions with the lattice. The theory made predictions in remarkably good agreement with experimental data and provided the basis for later developments such as quantum mechanical tunneling, magnetic flux quantization, and the concept of coherence length.

The break out of the impasse in the application of superconductivity came from systematic work at the Bell Telephone Laboratories, led by Matthias and Hulm. In 1961, they published their findings on the brittle compound $Nb_3Sn$, made by the high-temperature treatment of tin powder contained in a Niobium capillary tube. This compound retained its superconductivity in a field of 80 KG (8 Tesla), the highest field available to them, while carrying a current equivalent to a density of 100,000 $A/cm^2$. The critical field at 4.2 K, the boiling point of helium, was much greater than 8 T and was therefore at least 100 times higher than that of any previous superconductor.

This finding was followed by the discovery of a range of compounds and alloys with high critical fields, including NbZr, NbTi, and $V_3Ga$; the race was on to manufacture long lengths of wire or tape and develop superconducting magnets for commercial applications. At first, the early wires were unstable and unreliable and it soon became clear that a major research effort was needed.

### 9.1. High Field, Type 2 Superconductivity (1961)

In any event, NbTi turned out to be a much easier material to develop than $Nb_3Sn$, because it was ductile. However, it took 10 years or more to develop reliable, internally stabilized, multifilamentary composite wires of NbTi and copper. Only then, around 1970, was it possible for high field and large-scale applications of superconductivity to be considered with confidence.

### 9.2. The Ceramic Superconductors

All previous developments in superconductivity were eclipsed in 1986 by the discovery of a new type of superconductor composed of mixtures of ceramic oxides. By the

**Table 1.1** Nobel Laureates Linked with Cryogenics

| | |
|---|---|
| 1902 Pieter Zeeman | Influence of magnetism upon radiation |
| 1904 Lord Rayleigh | Density of gases and his discovery of argon |
| 1910 Johannes Van der Waals | The equation of state of gases and liquids |
| 1913 H. Kamerlingh Onnes | The properties of materials at low temperature, the preparation of liquid helium |
| 1920 Charles Guillaume | Materials for national prototype standards (Ni-Steel) INVAR |
| 1934 H. C. Urey | Discovery of deuterium produced by the distillation of liquid hydrogen |
| 1936 P.J.W. Debye | The behavior of substances at extremely low temperatures, especially heat capacity |
| 1950 Emmanuel Maxwell | "Isotope Effect" in superconductivity |
| 1956 John Bardeen | Semiconductors and superconductivity |
| 1957 Tsung Dao Lee, Chen Ning Yang | Upsetting the principle of conservation of parity as a fundamental law of physics. |
| 1960 D. A. Glaser | Invention of bubble chamber |
| 1961 R. L. Mössbauer | Recoil-less nuclear resonance absorption of gamma radiation |
| 1962 L. D. Landau | Pioneering theories of condensed matter especially liquid $He^3$ |
| 1968 Louis W. Alvarez | Decisive contributions to elementary particle physics, through his development of hydrogen bubble chamber technique and data analysis |
| 1972 John Bardeen, Leon N. Cooper, J. Robert Schrieffer | BCS theory of superconductivity |
| 1973 B. D. Josephson, Ivar Giaever, Leo Esaki | The discovery of tunneling supercurrents (The Josephson Junction) |
| 1978 Peter Kapitza | Methods pf making liquid helium and the characterization of Helium II as a "superfluid". |
| 1987 J. G. Bednorz, K.A. Muller | High-temperature superconductors |
| 1996 D. Lee, D. D. Osheroff, R.C. Richardson | Discovery of superfluidity in liquid helium-3 |
| 1997 Steven Chu, Claude Cohen-Tannoudji | Methods to cool and trap atoms with lasers |
| 1998 R. B. Laughlin, H. L. Stormer, D. C. Tsui | Discovery of a new form of quantum fluid excitations at low temperatures |
| 2001 Eric Cornell, W. Ketterle, Carl Wieman | For achieving the Bose–Einstein condensate at near absolute zero |

end of 1987, superconducting critical temperatures of ceramic materials had risen to 90 K for Y–Ba–Cu–O; 110 K for Bi–Sr–Ga–Cu–O; and 123 K for Tl–Ba–Ca–Cu–O. These temperatures are about ten times higher than the critical temperatures of previous metallic superconductors, thereby allowing liquid nitrogen instead of liquid helium to be used as a cooling medium.

This discovery changes the economics and practicability of engineering applications of superconductivity in a dramatic way. Refrigeration costs with liquid nitrogen are 100 times cheaper than those with liquid helium, and only simple one-stage

**Table 1.2** Notable Events in the History of Cryogenics

| | |
|---|---|
| 1848 | John Gorrie produces first mechanical refrigeration machine |
| 1857 | Siemens suggests recuperative cooling or "self-intensification" |
| 1866 | Van der Waals first explored the critical point, essential to the work of Dewar and Onnes |
| 1867 | Henry P. Howard of San Antonio Texas uses Gorrie's air-chilling system to transport frozen beef from Indianola TX to cities along the Gulf of Mexico |
| 1869 | Malthusian revolution in England, predicting worldwide starvation |
| 1873 | James Harrison attempts to ship frozen beef from Australia to the UK aboard the SS *Norfolk*., the project failed |
| 1875 | Thomas Mort tries again to ship frozen meat from Australia to the UK, this time aboard the SS *Northam*. Another failure. Gorrie's air system eventually produce success by Bell and Coleman 1877 |
| 1877 | Coleman and Bell produce commercial version of Gorrie's system for freezing beef. The frozen meat trade becomes more successful and stems the Malthusian revolution |
| | Cailletet produced a fog of liquid air, and Pictet a jet of liquid oxygen |
| 1878 | James Dewar duplicates the Cailletet/Pictet experiment before the Royal Institution |
| 1879 | The SS. *Strathaven* arrives in London carrying a well preserved cargo of frozen meat from Australia |
| | Linde founded the Linde Eismachinen |
| 1897 | Charles Tripler of NY produces 15 L/hr of liquid air using a 75 kW steam engine power source. Liquid air provides high-pressure gas to drive his air compressor and tripler is convinced he can make more liquid air than he consumes, the "surplusage" effect |
| 1882 | Jules Violle develops the first vacuum insulated flask |
| 1883 | Wroblewski and Olszewski liquefied both nitrogen and oxygen |
| | Vapor cooled shielding developed by Wroblewski and Olszewski |
| 1884 | Wroblewski produced a mist of liquid hydrogen |
| 1891 | Linde had put 12,000 domestic refrigerators in service. |
| | Dewar succeeds for the first time in liquefying hydrogen in a mist |
| 1892 | Dewar developed the silvered, vacuum-insulated flask that bears his name |
| 1894 | Dewar first demonstrated benefit of gas mixtures in J-T expansion |
| | Onnes first uses the word "cryogenics" in a publication |
| 1895 | Kammerlingh Onnes established the University of Leiden cryogenics laboratory |
| | Linde is granted the basic patent on air liquefaction |
| | First Hampson heat exchanger made for air liquefaction plants |
| | Hampson and Linde independently patent air liquefiers using Joule–Thomson expansion and recuperative cooling |
| 1897 | Dewar demonstrates the vacuum powder insulation |
| 1898 | Dewar produced liquid hydrogen in bulk, at the royal Institute of London |
| 1890 | Dewar improves upon the Violle vacuum flask by slivering both surfaces |
| 1902 | Claude developed an air liquefaction system using an expansion engine using leather seals, and established l'Air Liquide |
| | Reinhold Burger markets Dewar vessels under the name Thermos(TM) |
| | Tripler (1897) files for bankruptcy after his "surplusage" is proven false |
| 1907 | Linde installed first air liquefaction plant in the USA |
| | Claude produced neon as a by product of an air plant |
| 1908 | Onnes liquefied helium and received the Noble prize for his accomplishment |
| 1910 | Smoluchowski demonstrated evacuated-powder insulation |
| | Van der Waals receives Nobel prize for work on the critical region |

(*Continued*)

**Table 1.2** *(Continued)*

| | |
|---|---|
| 1911 | Onnes discovered superconductivity |
| 1912 | First American made air liquefaction plant |
| 1913 | Kammerlingh Onnes receives Nobel for work on liquid Helium |
| 1916 | First commercial production of argon in the USA by the Linde Company |
| 1917 | First natural gas plant produces gaseous helium |
| 1920 | Commercial production of neon in France |
| 1922 | First commercial neon production in the USA |
| 1926 | Dr. Goddard fired the first cryogenically propelled rocket |
| | Giauque and Debye independently discover the adiabatic demagnetization principle for producing temperatures much lower than 1 K |
| 1933 | Magnetic cooling first produces temperatures below 1 K |
| 1934 | Kapitza produces first expansion engine for making liquid helium |
| | H. C. Urey receives the Nobel for his discovery of deuterium |
| 1936 | P.J.W. debye receives Nobel prize for discoveries on heat capacity at low temperatures |
| 1937 | Evacuated-powder insulation, originally tested by Dewar, is first used on a commercial scale in cryogenic storage vessels |
| | First vacuum-insulated railway tank car built for liquid oxygen |
| | Hindenburg Zeppelin crashed and burned at Lakehurst, NJ. The hydrogen burned but did not explode. Thirty-seven out of 39 passengers survived, making this disaster eminently survivable as far as air travel goes (63 out of 100 total passengers and crew survived). Nonetheless, hydrogen gains an extremely bad reputation |
| 1942 | V-2 liquid oxygen fueled weapon system fired |
| 1944 | LNG tank in Cleveland OH fails killing 131 persons. LNG industry set back 25 years in the USA |
| 1947 | Collins cryostat developed making liquid helium readily available for the first time |
| 1948 | First 140 ton/day oxygen system built in the USA |
| 1950 | Emmanuel Maxwell receives Nobel prize for discovering the isotope effect in superconductors |
| 1952 | National Bureau of Standards Cryogenic Engineering laboratory built in Boulder, Colorado, including the first large-scale liquid hydrogen plant in the USA |
| 1954 | First Cryogenic Engineering Conference held by NBS at its Boulder Laboratories |
| 1956 | BCS theory of superconductivity proposed |
| 1957 | Atlas ICBM tested, firs use in the USA of liquid oxygen-RP1 propellant |
| | Lee and Yang receive Nobel for upsetting the theory of parity |
| 1958 | Multilayer cryogenic insulation developed |
| 1959 | The USA space agency builds a tonnage liquid hydrogen plant at Torrance, CA |
| 1960 | D.A. Glaser receives Nobel for his invention of the bubble chamber |
| 1961 | Liquid hydrogen fueled Saturn launch vehicle test fired |
| | R. L. Mossbauer receives Nobel for discoveries in radiation absorption |
| 1962 | L. D. Landau awarded the Nobel Prize for discoveries in $He^3$ |
| 1968 | L. W. Alvarez receives Nobel Prize for his work with liquid hydrogen bubble chambers |
| 1969 | Liquid hydrogen fueled Saturn vehicle launched from Cape Canaveral, FL carrying Armstrong, Aldrin, and Collins to the moon |
| 1972 | Bardeen, Cooper, and Schrieffer receive Nobel for the BCS theory of superconductivity |

(*Continued*)

**Table 1.2** (*Continued*)

| | |
|---|---|
| 1973 | B. D. Josephson, I. Giaever, and L. Esaki awarded the Nobel prize for discovery of the Josephson Junction (tunneling supercurrents) |
| 1978 | Peter Kapitza receive the Nobel for the characterization of HeII as a superfluid |
| 1987 | J. G. Bednorz and K. A. Mueller awarded the Nobel Prize for discovering high-temperature superconductors |
| | Y–BA–Cu–O ceramic superconductors found |
| 1996 | D. Lee, D. D. Osheroff, and R. C. Richardson receive the Nobel prize of the discovery of superfluidity in helium-3 |
| 1997 | S. Chu and Claude Cohen-Tannoudji awarded the Nobel for discovering methods to cool and trap atoms with lasers |
| 1998 | R. B. Laughlin, H. L. Stormer, and D. C. Tsui receive the Nobel for discovering a new form of quantum fluid excitations at extremely low temperatures |
| 2001 | E. Cornell, W. Ketterle, and C. Wieman awarded the Nobel for achieving the Bose–Einstein condensate at near absolute zero |

refrigerators are needed to maintain the necessary temperatures. Furthermore, the degree of insulation required is much more simple, and the vacuum requirements of liquid helium disappear.

## 10. SCIENCE MARCHES ON

Many Nobel laureates received their honors either because of work directly in cryogenics, or because of work in which cryogenic fluids were indispensable refrigerants. Table 1.1 provides an imposing list of such Nobel Prize winners.

Two examples from this list of Nobel laureates are considered to illustrate the application of cryogenics to education and basic research.

The first example involves Donald Glaser who invented the bubble chamber. Glaser recognized the principal limitation of the Wilson cloud chamber, namely that the low-density particles in the cloud could not intercept enough high-energy particles that were speeding through it from the beams of powerful accelerators. Glaser's first bubble chamber operated near room temperature and used liquid diethylether. To avoid the technical difficulties presented by such a complex target as diethylether, Prof. Luis Alvarez, of the University of California, devised a bubble chamber charged with liquid hydrogen. Since hydrogen is the simplest atom, consisting only of a proton and an electron, it interferes only slightly with the high-energy processes being studied with the giant accelerators, although it is readily ionized and serves quite well as the detector in the bubble chamber.

The first really large liquid-hydrogen bubble chamber was installed at the Lawrence Radiation Laboratory of the University of California and became operative in March 1959. A remark was made that this 72-in. liquid-hydrogen bubble chamber would be the equivalent of a Wilson cloud chamber one-half mile long. They could not really be equal, however, because the cloud chamber does not have the great advantage of utilizing the simplest molecules as the detector.

The second example chosen from the list of Nobel laureates that serves to link cryogenics to basic science is John Bardeen, who was cited for his work in semiconductors, and more recently has worked in superconductivity. His exposition in 1957, of what is now known as the BCS theory of superconductivity, has helped make this field one of the most active research fields in physics in the last century.

Table 1.2 gives a chronology of some of the major events in the history of cryogenics.

# 2
# Basic Principles

## 1. INTRODUCTION

In the last half of the 19th century, roughly from 1850 to the turn of the century, the principal export of the United States was cotton. The second most important export by dollar volume was *ice*. Americans developed methods for harvesting, storing, and transporting natural ice on a global scale. In 1810, a Maryland farmer, Thomas Moore, developed an ice-box to carry butter to market and to keep it hard until sold. This early refrigerator was an oval cedar tub with an inner sheet of metal serving as a butter container that could be surrounded on four sides by ice. A rabbit skin provided the insulated cover. Moore also developed an insulated box for home use. It featured an ice container attached to the lid and a 6 cu ft storage space below. Ice harvesting was revolutionized in 1825 when Nathaniel J. Wyeth invented a horse-drawn ice cutter (see Fig. 2.1.) It was Wyeth's method of cutting blocks of similar size quickly and cheaply that made the bountiful supplies of natural ice resources of the United States available for food preservation. A steam-powered endless chain was developed in 1855 that could haul 600 tons of ice per hour to be stored. Wyeth and Tudor patented a means to prevent the ice blocks from freezing together by placing saw-dust between the layers. Uniform blocks reduced waste, facilitated transportation, and introduced ice to the consumer level (see Figs. 2.2 and 2.3).

The amount of ice harvested each year was staggering. In the winter of 1879–1880, about 1,300,000 tons of ice was harvested in Maine alone, and in the winter of 1889–1890, the Maine harvest was 3 million tons. The Hudson River supplied about 2 million tons of ice per year to New York City during that period. Even that was not enough, for during those same years, New York City imported about 15,000 tons from Canada and 18,000 tons from Norway annually.

Ice changed the American diet. Fresh food preserved with ice was now preferred to food preserved by smoking or salting. Fresh milk could be widely distributed in the cities. By now, the people were accustomed to having ice on demand, and great disruptions of the marketplace occurred when ice was not available. Ice was, after all, a natural product, and a "bad" winter (a warm one) was disastrous to the marketplace and to the diet. It was time for the invention of machines to make ice on demand and free the market from dependence on the weather.

Machines employing air as a refrigerant, called compressed air or cold air machines, were appearing. They were based on the reverse of the phenomenon of heating that occurred when air was compressed, namely that air cooled as it expanded against resistance. This phenomenon of nature had been observed as early as the middle of the 18th century. Richard Trevithick, who lived until 1833 in

**Figure 2.1** Ice harvesting and storage in the 1850s. Accumulated snow was first removed with a horse-drawn ice plane. Then a single straight groove was made with a hand tool. The ice was marked off from this line into squares 22 in. on a side. Actual cutting was begun with Wyeth's invention and finished with hand saws. The uniform blocks of ice were pulled across the now open water to the lift at the icehouse. Here, they were stored under an insulation of straw or sawdust. The availability of a relatively dependable supply of ice all year, combined with uniformity of size of blocks, did much to create the market for ice.

**Figure 2.2** Delivering ice in New York, 1884. In this neighborhood, apparently, icemen did not enter the kitchen but sold their wares at the curb.

**Figure 2.3** Commercial ice delivery in New York, 1884. We see delivery to a commercial establishment, possibly a saloon. Though ice is available, the butcher a few doors down the street still allows his meat to hang unrefrigerated in the open air.

Cornwall, England, constructed engines in which expanding air was used to convert water to ice. In 1846, an American, John Dutton, obtained a patent for making ice by the expansion of air. The real development of the cold air machine, however, began with the one developed by John Gorrie of Florida and patented in England in 1850 and in the United States in 1851.

Later, we shall see how Gorrie's machine evolved into true cryogenics, not merely ice production. For now, however, in the late 1800s, it was possible to buy an ice-making machine for home use like that shown in Fig. 2.4. This primitive refrigerator had the same components as any refrigerator of today. It also followed exactly the same thermodynamic process as all of today's refrigerators.

All refrigerators involve exchanging energy (work, as the young lady in Fig. 2.4 is doing) to compress a working fluid. The working fluid is later expanded against a resistance, and the fluid cools (usually). Engineers have made it their job to determine how much cooling is produced, how much work is required for that cooling, and what kind of compressor or other equipment is required. More than that, we have made it our business to optimize such a cycle by trading energy for capital until a minimum cost is found. We can make any process more "efficient", that is, less energy-demanding, by increasing the size, complexity, and cost of the capital equipment employed. For instance, an isothermal compressor requires less energy to operate than any other kind of compressor. An isothermal compressor is also the most expensive compressor to build, because an infinity of intercooling stages are required for true isothermal compression.

As another example of trading energy costs for capital costs, consider the Carnot cycle (described later), which is the most thermodynamically efficient heat engine cycle possible. An essential feature of the Carnot engine is that all heat

**Figure 2.4** Home ice machine. Machines such as this were available from catalog centers in the late 1880s. The figure illustrates that energy (work) and capital equipment are at the heart of all cryogenic processes.

transfer takes place with zero $\Delta T$ at the heat exchangers. As a consequence, a Carnot engine must have heat exchangers of infinite size, which can get pretty costly. In addition, the Carnot engine will never "get there" because another consequence of zero $\Delta T$ is that an infinite amount of time will be required for heat transfer to take place.

Someone noted for all posterity:

There once was a young man named Carnot
Whose logic was able to show
For a work source proficient
There is none so efficient
As an engine that simply won't go.

The science that deals with these trade-offs between energy and capital is *thermodynamics.* Thermodynamics is about money. We need to predict how much a refrigerator will cost to run (the energy bill) and how much it will cost to buy (the capital equipment bill). Accordingly, our first job is to define *heat* and *work*, the energy that we must buy. *Work* is fairly easily defined as the product of a force and a displacement. Its symbolic representation takes many different forms, such as $\int P\,\mathrm{d}V$, but it is always the product of an external force and the displacement that goes with that force. Heat is more troublesome to calculate. The driving force for heat, $\Delta T$, seems obvious and analogous to the *force* or $P$ term of the *work* equation. But what is the equivalent *displacement* for heat? What gets moved, if there is such an analogy?

Carnot reasoned, by the flow of water, that there must be a flow of heat. He also reasoned that in the case of heat, something must be conserved, as energy itself is conserved (here he was right). Carnot discovered that the quantity that was conserved, and that was displaced, was not heat but rather the quantity $Q/T$. This led Carnot to the concept of *entropy*. In this book, *entropy* ($S$) is used to calculate heat ($Q$). That calculation is analogous to the one for work. Just as $\mathrm{d}W = P\,\mathrm{d}V$, likewise $\mathrm{d}Q = T\,\mathrm{d}S$. There are certainly more sophisticated implications of *entropy*, but for us, *entropy* is just a way to calculate how much heat energy is involved.

We need these ways to calculate heat and work because it is heat and work (energy) that we pay for to run the equipment. You will never see a work meter or a heat meter on any piece of cryogenic equipment. Instead it is a fact of nature that these two economically important quantities must be calculated from the thermodynamically important quantities, the things that we can measure: pressure ($P$), temperature ($T$), and density (actually, specific volume). Hence, we need two more tools.

One essential tool is an equation of state to tie pressure, volume, and temperature together.

The second essential tool to help us figure out the cost is the thermodynamic network that links not only $P$, $V$, and $T$ together but also links all of the other thermodynamic properties ($H$, $U$, $S$, $G$, $A$, ...—you know what I mean) together in a concise and consistent way. In real life, we usually measure pressure and temperature because it is possible to buy pressure gauges and thermometers and it is not possible to buy entropy or enthalpy gauges. Hence, we need a way to get from $P$ and $T$ to $H$ and $S$. (The beautiful thermodynamic network that gets us around from $P$ to $T$ to heat and work is shown later in Fig. 2.7.) This thermodynamic network is the only reason we delve into partial differential equations, the Maxwell relations, and the like. The payoff is that we can go from easily measured quantities, pressure and temperature, to the hard-to-measure, but ultimately desired, quantities of *heat* and *work*. You might want to look at Fig. 2.7 now to discover a reason to plow through all the arcane stuff between here and there. Once we do get our hands on the thermodynamic network, we can calculate *heat* and *work* for any process, for any fluid, for all time.

We will start our cryogenic analysis with some definitions. Admittedly, this is a boring approach. However, it will ensure that we are all singing from the same sheet of music. We begin with nothing less than the definition of thermodynamics itself.

## 2. THERMODYNAMICS

Thermodynamics is built upon four great postulates (guesses), which are called, rather stuffily, the four laws of thermodynamics.

The second law was discovered first; the first law was discovered second; the third law is called the zeroth law; and the fourth law is called the third law. I am sorry to tell you that things do not get much better.

The *zeroth law* of thermodynamics just says that the idea of temperature makes sense.

The *first law* of thermodynamics is simply the conservation of energy principle. Inelegantly stated, the first law says that what goes into a system must either come out or accumulate. Undergraduates often call this the "checkbook law". If you can balance your checkbook, you can do thermodynamics.

The *second law* is the entropy principle. The second law is also a conservation equation, stating that in the ideal engine, entropy ($S$) is conserved. All real engines make entropy.

The *third law* just says that there is a temperature so low that it can never be reached.

These colloquial expressions of the four great principles of thermodynamics have a calming effect. We will not be able to quantify our analyses unless we use much more precise definitions and, expressly, the mathematical statements of these principles. That is the task we must now turn to.

Thermodynamics is concerned with energy and its transformations into various forms. The laws of thermodynamics are our concepts of the restrictions that nature imposes on such transformations. These laws are *primitive*; they cannot be derived from anything more basic. Unfortunately, the expressions of these laws use words that are also primitive; that is, these words are in general use (e.g., energy) and have no precise definition until we assign one. Accordingly, we begin our discussion of the fundamental concepts of thermodynamics with a few definitions.

**Thermodynamics:** Thermodynamics is concerned with the interaction between a system and its surroundings, the effect of that interaction on the properties of the system, and the flow of heat and work between the system and its surroundings.

**System:** Any portion of the material universe set apart by arbitrarily chosen but specific boundaries. It is essential that the system be clearly defined in any thermodynamic analysis.

**Surroundings:** All parts of the material universe not included in the system.

The definitions of *system* and *surroundings* are coupled. The system is any quantity of matter or region mentally set apart from the rest of the universe, which then becomes the surroundings. The imaginary envelope that distinguishes the system from the surroundings is called the *boundary* of the system.

**Property of the system, or state variable:** Any observable characteristic of the system, such as temperature, pressure, specific volume, entropy, or any other distinguishing characteristic.

**State of the system:** Any specific combination of all the properties of the system such as temperature, pressure, and specific volume. Fixing any two of these three properties automatically fixes all the other properties of a homogeneous pure substance and, therefore, determines the condition or *state* of that substance. In this respect, thermodynamics is a lot like choosing the proper size dress shirt for a gentleman. If you specify the neck size and sleeve length, the shirt size is defined. The size of a man's dress shirt is determined by two variables: neck size and sleeve length. A single-phase homogeneous thermodynamic system is defined by fixing any two thermodynamic variables. If you can pick out shirt size correctly, you can do thermodynamics.

**Energy:** A very general term used to represent heat, work, or the capacity of a system to do heat or work.

**Heat:** Energy in transition between a system and its surroundings that is caused to flow by a temperature difference.

**Work:** Energy in transition between a system and its surroundings. Work is always measured as the product of a force external to the system and a displacement of the system.

If there is no system displacement, there is no work. If there is no energy in transition, there is no work. Consider the following conundrum: It has been raining steadily for the last hour at the rate of 1 in. of rain per hour. How much rain is on the ground? Answer: *none*. Rain is water in transition between heaven and earth. Once the water hits the earth, it is no longer rain. It may be a puddle or a lake, but it is not rain.

Likewise, *heat* and *work* are energy in transition between system and surroundings. Heat and work exist only while in transition. Once transferred, heat and work become internal energy, or increased velocity, or increased kinetic energy, but they are no longer heat and work. Heat and work cannot reside in a system and are not properties of a system.

Because heat and work are in transition, they have direction as well as magnitude, and there is a sign convention to tell us which way they are moving.

The convention among chemists and chemical engineers is:

| Direction of energy flow | Sign convention |
|---|---|
| Heat into the system | (+) Positive |
| Heat out of the system | (−) Negative |
| Work into the system | (−) Negative |
| Work out of the system | (+) Positive |

Thus, the expression $Q - W$ is the algebraic sum of all the energy flowing from the surroundings into the system. $Q - W$ is the net gain of energy of the system.

**Process:** The method or path by which the properties of a system change from one set of values in an initial state to another set of values in a final state. For example, a process may take place at constant temperature, at constant volume, or by any other specified method. A process for which $Q = 0$ is called an adiabatic process, for instance.

**Cyclic process:** A process in which the initial state and final state are identical. The *Carnot cycle*, mentioned earlier, is such a cyclic process. We will now examine the Carnot cycle in the light of the preceeding definitions.

In 1800, there was an extraordinary Frenchman who was a statesman and government minister. His work was so important that he was known as the "organizer of victory of the French Revolution". Lazare Nicholas Marguerite Carnot (1753–1823) was also an outstanding engineer and scientist. A century later, this glorious family tradition was carried on by his grandson, Marie François Sadi Carnot (1837–1894), another gentleman engineer, who served as the President of the Republic of France from 1887 to 1894, when he was assassinated.

The Carnot family member we are interested in is neither of these, but a cousin who lived between them. Nicholas Leonard Sadi Carnot (1796–1832), called "Sadi" by everyone, was very interested in steam engines. He wanted to build the most efficient steam engine of all to assist in military campaigns. Sadi needed the steam engine that required the least fuel for a given amount of work, thereby reducing the logistical nightmare of fueling the war engines.

He began by considering the waterwheel. He reasoned that it should be possible to hook two waterwheels together, one driven by falling water in the conventional way. The other, joined to the first by a common shaft, should be able to pick up the water spent by the first and carry it up to the headgate of the first. In this perfect, frictionless engine, one waterwheel should be able to drive the other forever. We know, as Sadi knew, that this is not possible because of the losses (friction) in the system. However, the idea led Sadi to the conclusion that the most efficient engine would be a *reversible* one, an engine that could run equally well in either direction. The perfect waterwheel could be driven by falling water in one direction or simply run in reverse to carry (pump) the water to the top again. A pair of such *reversible* engines could thus run forever. Anything less than a *reversible* engine would eventually grind to a halt. Hardly anyone paid any attention to Carnot and his seminal concept, and indeed, Carnot's work was further obscured by his early death.

Carnot's perfect abstraction was noticed, however, by the German physicist Rudolph Clausius and the Glaswegian professor of natural philosophy William Thomson. The principle of the conservation of energy was well established by this time. Crudely stated: the energy that goes into a system must either come out or accumulate. In the steady state, exactly the energy that goes into the engine must come out, regardless of the efficiency of the engine. Clausius and Thomson both saw in Carnot's work that a perfect engine would conserve energy like any other engine, but in addition, the quantity $Q/T$ would also be conserved. In the perfect (reversible) engine, $Q_{in}/T_{in} = Q_{out}/T_{out}$. Not only is energy conserved (the first law) but also the quantity $Q/T$ is conserved in the ideal engine.

This astonishing fact leads us to the concept of entropy, $S$. Entropy was defined then as now as $S = Q_{rev}/T$ because of the remarkable discovery that the quantity $Q/T$, *entropy*, is conserved in the ideal (reversible) engine. We can summarize the work of Clausius, Thomson, and Carnot by saying that *in an ideal engine*, $S_{in} = S_{out}$; *in any real engine*, $S_{out} \geq S_{in}$. We now have a concise description of the most efficient engine possible, $S_{in} = S_{out}$. All real engines, regardless of their intended product, produce entropy: $S_{out} \geq S_{in}$ for any real engine.

This is the reason that the concept of reversibility is so important, and it leads to the following definition:

**Reversible process:** A process in which there are no unbalanced driving forces. The system proceeds from the initial state to the final state only because the driving forces in that direction exceed forces opposing the change by an infinitesimal amount. In short, no forces are wasted.

## 2.1. Master Concepts

Thermodynamics is built upon experience and experimentation. A few master concepts have emerged, which may be stated as follows.

*Postulate 1.* There exists a form of energy, known as *internal energy* $U$, that is an intrinsic property of the system, functionally related to pressure, temperature, and the other *measurable* properties of the system. $U$ is a property of the system, but is not *directly measurable*.

The second master concept is simply a conservation equation, known as the *first law of thermodynamics*:

*Postulate 2.* The *total energy* of any system and its surroundings is conserved.

A third postulate qualifies the first law by observing that not all forms of energy have the same quality, or availability for use (the concept of entropy).

*Postulate 3.* There exists a property called *entropy* $S$, which is an intrinsic property of the system that is functionally related to the *measurable* properties that characterize the system. For reversible processes, changes in this property can be calculated by the equation

$$dS = dQ_{rev}/T \tag{2.1}$$

where $T$ is the absolute temperature of the system.

The first law of thermodynamics cannot be properly formulated without the recognition of internal energy $U$ as a property. Internal energy is a concept. Likewise, the second law of thermodynamics can have no complete description without the recognition of entropy as a property. Entropy is also a concept, just as internal energy is a concept, but is usually much harder to grasp.

Abstract quantities such as internal energy $U$ and entropy $S$ are essential to the thermodynamic solution of problems. Accordingly, a necessary first step in thermodynamic analysis is to translate the problem into the terminology of thermodynamics. One of the real systems most often encountered in cryogenic engineering is the one described by pressure, temperature, and specific volume. These systems are called $PVT$ systems and are composed of fluids, either liquids or gases or both. For such systems, yet another postulate is required.

*Postulate 4.* The macroscopic properties of homogeneous $PVT$ systems in equilibrium states can be expressed as functions of pressure, temperature, and composition only.

This postulate is the basis for all the thermodynamic equations that will follow and has enormous utility. It is the basis for assuming the existence of an equation of state relating $P$, $V$, and $T$. Note that it neglects the effects of electric, magnetic, and gravitational fields and also neglects viscous shear and surface effects.

As mentioned, these postulates are *primitive*; they cannot be derived from anything more basic.

## 2.2. The First Law of Thermodynamics

As stated above, the first law is a conservation equation: energy transferred to a system will be conserved as changes in properties of the system or changes in potentials of the system. That is, energy may be transferred or altered in form but never created or destroyed. Energy that enters a system must either come out of the system or accumulate (the checkbook principle):

Heat added to the system + Work done on the system
= Changes in internal energy $U$
+ Changes in potential and kinetic energy

or

$$Q - W = \Delta U + \Delta(\text{potential and kinetic energy}) \tag{2.2}$$

In Eq. (2.2), the term $Q - W$ represents all the energy a system has received from its surroundings. This energy gained by the system is divided into two portions:

1. The increase in internal energy $U$ within the system due to the composite effect of changes in the configuration and motion of all of its ultimate particles. This increment is represented by the term $\Delta U$.
2. The increase in potential and kinetic energy of the entire system due to changes in the position and motion of the system as a whole. This is

written as

$$\Delta(\mathrm{PE} + \mathrm{KE}) = \frac{mg\Delta Z}{g_c} + \frac{m\Delta v^2}{2g_c} \tag{2.3}$$

where $m$ is the mass of the system, $\Delta Z$ is the elevation of the system above a reference plane, $v$ is the velocity of the system, $g$ is the local acceleration of gravity, and $g_c$ is the gravitational constant.

*Example 2.1.* Water is flowing at a velocity of 100 ft/s through a pipe elevated 250 ft above a reference plane. What are the kinetic energy (KE) and the potential energy (PE) of 1 $\mathrm{lb_m}$ of the water?

The work done on a body to increase its velocity leads to the kinetic energy concept. From the first law, the work done on the body to accelerate it is transformed into the kinetic energy of the body.

Work is defined as force times displacement, or

$$W = F \times l$$

and

$$\mathrm{d}W = F\mathrm{d}l \quad \text{for a constant force } F$$

The force may be obtained from Newton's second law of motion,

$$F = \frac{ma}{g_c}$$

where $m$ is the mass of the body, $a$ is the acceleration, and $g_c$ is a proportionality constant called the gravitational constant. In the American engineering system,

$$g_c = 32.174 \frac{(\mathrm{ft})(\mathrm{lb_m})}{(\mathrm{s}^2)(\mathrm{lb_f})}$$

Therefore,

$$\mathrm{d}W = \frac{ma}{g_c}\mathrm{d}l$$

By definition, acceleration is

$$a = \frac{\mathrm{d}v}{\mathrm{d}t}$$

where $v$ is the velocity of the body. Hence,

$$\mathrm{d}W = \frac{m}{g_c}\frac{\mathrm{d}v}{\mathrm{d}t}\mathrm{d}l$$

or

$$\mathrm{d}W = \frac{m}{g_c}\frac{\mathrm{d}l}{\mathrm{d}t}\mathrm{d}v$$

Since the definition of velocity is

$$v = \frac{dl}{dt}$$

the expression for work becomes

$$dW = \frac{m}{g_c} v \, dv$$

This equation may now be integrated for a finite change in velocity from $v_1$ to $v_2$:

$$W = \frac{m}{g_c} \int_{v_1}^{v_2} v \, dv = \frac{m}{g_c} \left( \frac{v_2{}^2}{2} - \frac{v_1{}^2}{2} \right)$$

or

$$W = \frac{mv_2{}^2}{2g_c} - \frac{mv_1{}^2}{2g_c} = \frac{\Delta mv^2}{2g_c}$$

The quantity $mv^2/2g_c$ was defined as the kinetic energy by Lord Kelvin in 1856. Thus,

$$\text{KE} = \frac{mv^2}{2g_c} \tag{2.4}$$

For the case given in the problem statement above,

$$\begin{aligned} \text{Kinetic energy} &= (1 \text{ lb}_\text{m}) \left( 100 \frac{\text{ft}}{\text{s}} \right)^2 \left( \frac{1}{2} \right) \left( \frac{1}{32.174} \frac{\text{s}^2 \text{ lb}_\text{f}}{\text{ft lb}_\text{m}} \right) \\ &= 155 \text{ ft lb}_\text{f} \end{aligned}$$

Similarly, the work done on a body to change its elevation leads to the concept of potential energy.

If a body of mass $m$ is raised from an initial elevation $z_1$ to a final elevation $z_2$, an upward force at least equal to the weight of the body must be exerted on it, and this force must move through the distance $z_2 - z_1$. Since the weight of the body is the force of gravity on it, the minimum force required is given by Newton's second law of motion:

$$F = \frac{1}{g_c} ma = \frac{1}{g_c} mg$$

where $g$ is the local acceleration of gravity and $g_c$ is as before. The minimum work required to raise the body is the product of this force and the change in elevation:

$$W = F(z_2 - z_1) = m \frac{g}{g_c} (z_2 - z_1)$$

or

$$W = mz_2 \frac{g}{g_c} - mz_1 \frac{g}{g_c} = \Delta \left( \frac{mzg}{g_c} \right)$$

The work done on a body in elevating it is said to produce a change in its *potential energy* (PE), or

$$W = \Delta\text{PE} = \Delta\frac{mzg}{g_c}$$

Thus, potential energy is defined as

$$\text{PE} = mz\frac{g}{g_c} \tag{2.5}$$

This term was first proposed in 1853 by the Scottish engineer William Rankine (1820–1872).

For the problem statement above,

$$\text{Potential energy} = (1\,\text{lb}_\text{m})(250\,\text{ft})\left(32.174\frac{\text{ft}}{\text{s}^2}\right)\left(\frac{1}{32.174}\frac{\text{s}^2\,\text{lb}_\text{f}}{\text{ft}\,\text{lb}_\text{m}}\right)$$
$$= 250\,\text{ft}\,\text{lb}_\text{f}$$

Potential energy and kinetic energy are obvious mechanical terms. The energy put into a system may be stored in elevation change or in velocity change as well as in internal energy $U$ by way of property changes.

The internal energy $U$ represents energy stored in the system by changes in its properties $T$, $P$, $V$, etc. $\Delta U$ is a composite effect of changes in the motion and configuration of all of the ultimate particles of a system.

It is essential to note that the left-hand side of Eq. (2.2) describes a system–surroundings interaction, while the right-hand side describes properties of the system alone. This is a very valuable aspect of the first law equation, for it permits the calculation of system–surroundings interactions (heat and work) from changes in system properties only. Some of the consequences of this attribute of the first law equation are discussed below.

As stated earlier, work is a system–surroundings interaction—energy in transit between the system and surroundings, a precise kind of energy that crosses the system boundary. Therefore, work must always have two components: an external (surroundings) force and an internal (system) displacement. Work is always

$$F_\text{external} \times X_\text{system}$$

where $X$ is the system displacement.

For fluids, work is always

$$P_\text{external} \times V_\text{system}$$

Only in the reversible case is $P_\text{external} = P_\text{internal}$. Then, and only then, may the $P$ of the system be substituted for $P_\text{external}$ and equations of state be used to relate $P$ and $V$.

Because the left-hand side of Eq. (2.2) involves a system–surroundings interaction, questions of *path* come into play (e.g., adiabatic, reversible, constant external pressure, etc.). Indeed, the path must always be specified for calculations involving the left-hand side of Eq. (2.2), i.e., heat and work. In thermodynamics, it is essential to distinguish between quantities that depend on path and those that do not. Indeed, we commonly make this distinction in ordinary affairs. For instance, the straight-line distance between Denver and Los Angeles is a fixed quantity, a property. On the other hand, other quantities for an auto-mobile trip between Denver and Los

Angeles—miles traveled, miles per gallon consumed, etc.—depend very much on the path.

As stated, the right-hand side of Eq. (2.2) describes attributes of the system only and contains properties and potentials of the system only. Thus, it can always be evaluated from system properties or potentials alone without considering the surroundings. The question of path is simply irrelevant, because there is no system–surroundings interaction to be found on the right-hand side of the first law statement.

It is a constant goal of thermodynamics to evaluate system–surroundings interactions (e.g., heat and work) in terms of properties and potentials of the system alone. The essence of the first law concept is (a) that it is a conservation equation and (b) that a system–surroundings interaction can be expressed as a function of system parameters only.

The first law is valid for all systems under all cases, as written in Eq. (2.2). There are numerous special cases, and some of the more important will now be considered.

### *2.2.1. Open System—the Flow Case*

One special case of the first law of thermodynamics is the "open" system, or the flow case. For a flow system such as the one shown in Fig. 2.5, the first law can be stated as

$$Q - W = \Delta U + \frac{g}{g_c}\Delta Z + \frac{\Delta v^2}{2g_c}$$

In this case the system is the element of fluid, and

$$W = W_s + W_1 + W_2$$

where $W_s$ is the shaft work by pump, $W_1$ the work done by surroundings on system (element of fluid) pushing it past boundary 1 and $W_2$ the work done by system on surroundings as element of fluid emerges.

Since pressure is the force per unit area, $F_1$ becomes

$$F_1 = P_1 A_1$$

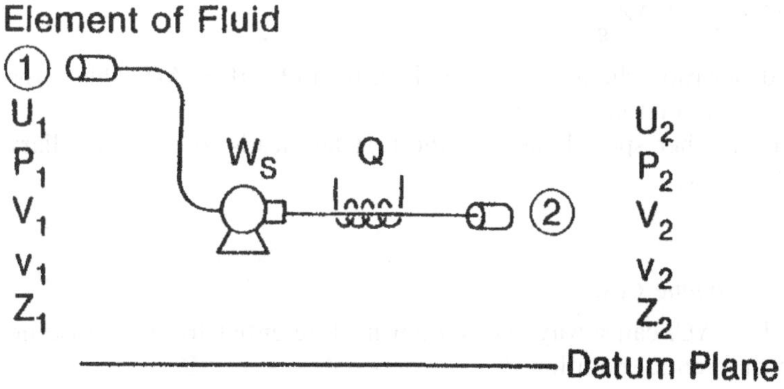

Element of Fluid
1
U1
P1
V1
v1
Z1
Ws
Q
2
U2
P2
V2
v2
Z2
Datum Plane

**Figure 2.5** Open thermodynamic system—the flow case.

and the distance $l_1$ through which $F_1$ acts is

$$l_1 = \frac{V_1}{A_1}$$

where $V_1$ is the specific volume of element fluid and $A_1$ the cross-sectional area of pipe.

Therefore, $W_1 = F_1 P_1$ becomes

$$W_1 = P_1 A_1 \left(\frac{V_1}{A_1}\right) = P_1 V_1$$

Likewise,

$$W_2 = P_2 A_2 \left(\frac{V_2}{A_2}\right) = P_2 V_2$$

and the total work, $W$, is

$$W = W_s + P_2 V_2 - P_1 V_1$$

The first law, as written above, becomes for this special case

$$Q - (W_s + P_2 V_2 - P_1 V_1) = \Delta U + \frac{g}{g_c}\Delta Z + \frac{\Delta v^2}{2g_c}$$

$$Q - W_s = \Delta U = \frac{g}{g_c}\Delta Z + \frac{\Delta v^2}{2g_c} + P_2 V_2 - P_1 V_1$$

$$= U_2 + P_2 U_2 - U_1 - P_1 V_1 + \frac{\Delta v^2}{2g_c} + \Delta Z \frac{g}{g_c}$$

which defines the enthalpy $H$ as

$$H \equiv U + PV \tag{2.6}$$

Then,

$$Q - W_s = H_2 - H_1 + \frac{\Delta v^2}{2g_c} + \Delta Z \frac{g}{g_c}$$

or

$$Q - W_s = \Delta H + \frac{\Delta v^2}{2g_c} + \Delta Z \frac{g}{g_c}$$

If velocity and elevation changes are negligible, then $Q - W_s = \Delta H$. This is the flow system analog of the first law for closed systems.

Consideration of other special cases of the first law leads to other auxiliary functions, discussed below.

### 2.2.2. *The Constant-Volume Case*

The first law, $Q - W = \Delta U$, can always be written in differential form because no path is involved in differential notation:

$$dQ - dW = dU$$

But $dW = P_{ext}dV$, and since $dV = 0$ for constant volume,

$$dQ = (dU)_V$$

Differentiating with respect to temperature suggests the function

$$\left(\frac{dQ}{dT}\right)_V = \left(\frac{dU}{dT}\right)_V \equiv C_V$$

which is the definition of heat capacity at constant volume.

### *2.2.3. The Constant-Pressure Case*

The first law in differential form can be combined with the expression for work under constant pressure, $dW = P_0\,dV$, to yield

$$dQ - dW = dU$$

Thus,

$$dQ = P_0 dV + dU$$

where $P_0$ = the constant system pressure. Rearranging gives

$$\begin{aligned} Q_P &= P_0(V_2 - V_1) + \Delta U \\ &= P_0 V_2 - P_0 V_1 + U_2 - U_1 \\ &= (P_0 V_2 + U_2) - (P_0 V_1 + U_1) \end{aligned}$$

and since $P_0 = P_2 = P_1$ = constant, this may be rewritten as

$$Q_P = (P_2 V_2 + U_2) - (P_1 V_1 = U_1)$$

which, from the definition of enthalpy, $H \equiv U + PV$

$$Q_P = H_2 - H_1 = \Delta H$$

Thus, at constant pressure,

$$Q = (\Delta H)_P$$

Differentiating this expression with respect to temperature suggests the function

$$\left(\frac{dQ}{dT}\right)_P = \left(\frac{dH}{dT}\right)_P \equiv C_P$$

which is the definition of heat capacity at constant pressure.

## 2.3. The Second Law of Thermodynamics

Another master concept, the second law, comes about from observing that $Q$ and $W$ do not have the same "quality". For instance,

1. $W$ is highly ordered and directional.
2. $Q$ is less versatile.
3. $W$ can always be completely converted to $Q$ regardless of the temperature, but not vice versa.
4. $Q$ can flow only from $T_{hot}$ to $T_{cold}$.

$Q$ has a quality as well as a quantity, and that quality depends on the temperature.

Also, the effect of adding $Q$ to a system is always to change the properties of the system, so there must be a *property change* that describes the addition of $Q$.

We postulate a property in thermodynamics to describe these observations that meets the following qualitative requirements:

1. The addition of $Q$ causes a property change in the system.
2. The *way* (path) in which $Q$ is added fixes the magnitude of the property change.
3. Adding $Q$ at a lower temperature causes a greater change than adding the same amount of $Q$ at a higher temperature.

Thus, we invent a property of the system to fit these three criteria, namely entropy, $S$ (see Eq. (2.1)):

$$S = \frac{Q_{rev}}{T}$$

or

$$Q_{rev} = T\Delta S$$

Since $Q$ is not a property, its path must be specified, and $Q_{rev}$ is chosen to complete the definition.

Thus, we have developed qualitatively Postulate 3, which was stated earlier: there exists a property called *entropy S,* which is an intrinsic property of a system that is functionally related to the measurable coordinates that characterize the system. For a reversible process, changes in this property are given by

$$dS = \frac{dQ_{rev}}{T}$$

The definition of entropy, $Q_{rev} = T\Delta S$, can always be written in differential form because neither reversibility nor any other path has any significance in differential notation:

$$dQ = T\,dS$$

and hence the first law,

$$U = Q - W$$

becomes in differential form, for all cases and for all substances,

$$dU = T\,dS - P\,dV \tag{2.7}$$

This is a universal relationship, just like the first law. One can always write the first law in this differential form for all cases. Although derived for a reversible process, Eq. (2.7) relates properties only and is valid for any change between equilibrium states in a closed system. The question of reversibility/irreversibility has no significance to Eq. (2.7) because (a) the concept of path has no bearing on any equation written in differential form and (b) the equation $dU = T\,dS - P\,dV$ contains properties of the system only. *Path* is a concept relevant to system–surroundings interactions, and "surroundings" are completely absent from this description of the system alone.

Entropy is a property of the system that is dependent on the configuration of all the ultimate particles that make up the system. A system in which all the ultimate particles are arranged in the most orderly manner possible has an entropy of zero. As the arrangement within the system is changed from an ordered to a more disordered or more random state, the entropy of the system increases.

The addition of heat to a system causes an increase in scattering and disorder among all the particles within the system, in other words, an increase in entropy.

If heat is added during a reversible process, this increase in entropy is related to the heat added by

$$Q_{\text{rev}} = \int T\,\mathrm{d}S$$

If heat is transferred to the system during an irreversible process, then

$$\int T\,\mathrm{d}S > Q$$

or

$$\int T\,\mathrm{d}S = Q + \text{lost work}$$

The lost work is energy that might have done useful work but is dissipated instead, causing unnecessary randomness and disorder within the system.

The second law can be stated formally as follows.

*Second law of thermodynamics.* The entropy change of any system and its surroundings, considered together, is positive and approaches zero for any process that approaches reversibility.

The second law of thermodynamics is a conservation law only for reversible processes, which are unknown in nature. All natural processes result in an increase in total entropy. The mathematical expression of the second law is simply

$$\Delta S_{\text{total}} \geq 0$$

where the label "total" indicates that both the system and its surroundings are included. The equality applies only to the limiting case of a reversible process.

## 2.4. Clausius Inequality and Entropy

In 1824, Nicholas Leonard Sadi Carnot proposed a heat power cycle composed of the following reversible processes shown in Fig. 2.6.

1. Isothermal reversible expansion and absorption of heat $Q_1$ at temperature $T_1$ (a–b).
2. Isentropic (adiabatic, reversible) expansion to $T_2$ (b–c).
3. Isothermal reversible compression and rejection of heat $Q_2$ at $T_2$ (c–d).
4. Isentropic compression back to temperature $T_1$ (d–a).

Since the energy change of the system for the complete cycle must be zero,

$$\Delta U_{\text{cycle}} = 0 = Q_1 - Q_2 - W$$

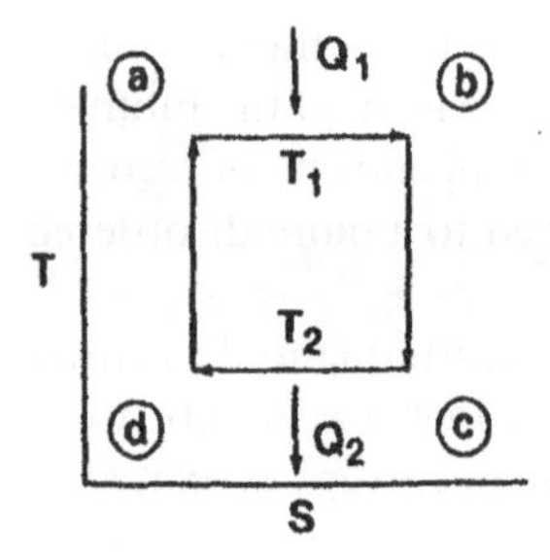


**Figure 2.6** Carnot's reversible heat power cycle.

or

$$Q_1 - Q_2 = W$$

where $W$ is the net work, which is the difference between the work produced in the expansion process (b–c) and the work required in the compression process (d–a). Also, the absolute values of $Q_1$ and $Q_2$ are used, temporarily abandoning the usual sign convention.

The efficiency of this reversible cycle is

$$\eta = \frac{W}{Q_1} = \frac{Q_1 - Q_2}{Q_1}$$

since $Q_1 = T_1 \Delta S$ and $Q_2 = T_2 \Delta S$

The efficiency of the Carnot engine can be expressed in terms of the temperature of the heat reservoirs alone:

$$\eta = \frac{T_1 - T_2}{T_1}$$

Carnot stated (and proved) the following principles, which are known today as the *Carnot principles*:

1. No engine can be more efficient than a reversible engine operating between the same high-temperature and low-temperature reservoirs. Here the term "heat reservoir" is taken to mean either a heat source or a heat receiver or sink.
2. The efficiencies of all reversible engines operating between the same constant-temperature reservoirs are the same.
3. The efficiency of a reversible engine depends only on the temperatures of the heat source and heat receiver.

The two equations for the efficiency of the Carnot cycle can be combined to yield

$$\eta = \frac{W}{Q_1} = \frac{Q_1 - Q_2}{Q_1} = \frac{T_1 - T_2}{T_1} = \left(1 - \frac{T_2}{T_1}\right)$$

or

$$W = Q_1\left(1 - \frac{T_2}{T_1}\right)$$

But

$$W = Q_1 - Q_2$$

Therefore,

$$Q_1 - Q_2 = Q_1\left(1 - \frac{T_2}{T_1}\right)$$

which reduces to

$$\frac{Q_1}{T_1} = \frac{Q_2}{T_2}$$

or $S_1 = S_2$, which is to repeat that entropy is *conserved* in a reversible engine.

In 1848, Lord Kelvin (William Thomson of Glasgow) derived a temperature scale based on the above equation, independent of the thermodynamic substance employed in the cycle. He defined the scale such that

$$\frac{T_1}{T_2} = \frac{Q_1}{Q_2}$$

where $Q_1$ is the heat received by a Carnot engine from a source at temperature $T_1$ and $Q_2$ is the heat rejected to a receiver at temperature $T_2$. One advantage of the absolute scale is the absence of negative temperatures.

Clausius noted that the Kelvin equation could be rearranged to give

$$\frac{Q_1}{T_1} = \frac{-Q_2}{T_2} \quad \text{or} \quad \frac{Q_1}{T_1} + \frac{Q_2}{T_2} = 0$$

Here, the usual sign convention is used and the minus sign with $Q_2$ denotes that heat is leaving the system. This equation indicates that the change in the quantity $Q/T$ around a reversible or Carnot cycle is zero. Clausius extended this relationship to embrace all cyclic processes by stating that the cyclic integral of $\delta Q/T$ is less than zero or in the limit equal to zero; thus,

$$\oint \frac{\delta Q}{T} \leq 0$$

This equation is now known as the *inequality of Clausius.*

It can be proved that the cyclic integral of $\delta Q/T$ is equal to zero for all reversible cycles and less than zero for all irreversible cycles. Since the summation of the quantity $\delta Q/T$ for a reversible cycle is equal to zero, it follows that the value of the integral of $\delta Q/T$ *is the same for any reversible process between states* 1 *and* 2 *of a system and thus is a property of the system.* It is this proof that gives credence to Postulate 3, that there is a property called entropy that is defined as the ratio of heat transferred during a reversible process to the absolute temperature of the system, or mathematically,

$$\mathrm{d}s = \left(\frac{\delta Q}{T}\right)_{\text{rev}}$$

Entropy, since it is a property, is especially useful as one of the coordinates when representing a reversible process graphically.

## 2.5. The Maxwell Relations

We are now in a position to construct a complete thermodynamic property network and develop the Maxwell relations. The purpose is to develop a set of property functions (also known as *state functions* or thermodynamic functions) that will enable us to calculate any thermodynamic property change (e.g., $\Delta S$, $\Delta H$, $\Delta U$) for any substance for any process.

By definition, $H = U + PV$ and

$$\mathrm{d}H = \mathrm{d}U + P\,\mathrm{d}V + V\,\mathrm{d}P$$

or

$$\mathrm{d}U = \mathrm{d}H = P\,\mathrm{d}V - V\,\mathrm{d}P$$

This form of $U$ may be substituted into the first law in differential form:

$$\mathrm{d}U = T\,\mathrm{d}S + P\,\mathrm{d}V \tag{2.8}$$

to yield

$$\mathrm{d}H = T\,\mathrm{d}S + V\,\mathrm{d}P \tag{2.9}$$

It can easily be shown that for a reversible flow process,

$$-W_{\mathrm{max}} = H - Q_{\mathrm{rev}}$$

or

$$-\mathrm{d}W = \mathrm{d}H - T\,\mathrm{d}S$$

This form of the work available from a reversible flow process suggests the function

$$G = H - TS \tag{2.10}$$

which is the definition of *Gibbs free energy*. This can likewise be substituted into the differential form of the first law to yield

$$\mathrm{d}G = -S\,\mathrm{d}T + V\,\mathrm{d}P \tag{2.11}$$

Similarly, the maximum work for a closed reversible process is

$$-W_{\mathrm{max}} = \Delta U - Q_{\mathrm{rev}}$$

or

$$-\mathrm{d}W = \mathrm{d}U - T\,\mathrm{d}S$$

which suggests another new function,

$$A = U - TS \tag{2.12}$$

which is known as the *Helmholtz free energy*. If this is likewise substituted into the first law in differential form, we have

$$\mathrm{d}A = S\,\mathrm{d}T - P\,\mathrm{d}V \tag{2.13}$$

We have thus defined the *state functions,* functions of the properties of a system that, under various conditions, can relate changes in the properties of the system to the heat and work transferred between it and the surroundings. Since these functions or potentials are functions only of the properties of the system, they can be expressed as exact differentials. Thus, changes in these functions depend only on the initial and final states of the system and not on the method or path by which the change takes place. These functions are

| | |
|---|---|
| $U = Q - W$ | Internal energy |
| $H = U + PV$ | Enthalpy |
| $G = H - TS$ | Gibbs free energy |
| $A = U - TS$ | Helmholtz free energy |

The four state functions written in differential form are

$$\mathrm{d}E = T\,\mathrm{d}S - P\,\mathrm{d}V \qquad \text{(Eq. 2.8)}$$

$$\mathrm{d}H = T\,\mathrm{d}S + V\,\mathrm{d}P \qquad \text{(Eq. 2.9)}$$

$$\mathrm{d}G = -S\,\mathrm{d}T + V\,\mathrm{d}P \qquad \text{(Eq. 2.11)}$$

$$\mathrm{d}A = S\,\mathrm{d}T - P\,\mathrm{d}V \qquad \text{(Eq. 2.13)}$$

It would be imprecise to stop with the impression that these four state functions are derived from considering the maximum possible work connected with a reversible process. Such a consideration only suggests the function. New thermodynamic functions such as $H$, $G$, and $A$ cannot in general be defined by the random combination of variables. As a minimum, dimensional consistency is required. Instead, a standard mathematical method exists for the systematic definition of the desired functions; this is the *Legendre transformation*. The functions for $G$, $H$, and $A$ are in fact Legendre transformations of the fundamental property relation of a closed $PVT$ system, $\mathrm{d}U = T\mathrm{d}S - P\mathrm{d}V$.

A consequence of the Legendre transformation is that the resulting expressions for $G$, $H$, and $A$ are exact differential equations with the following special property:

If $z = f(x,y)$ and $\mathrm{d}z = M\mathrm{d}x + N\mathrm{d}y$, then the property of exactness is that

$$\left(\frac{\partial M}{\partial y}\right)_x = \left(\frac{\partial N}{\partial x}\right)_y$$

Applying this property of exact differential equations to the four equations above yields the following Maxwell relations:

$$\left(\frac{\partial T}{\partial V}\right)_S = -\left(\frac{\partial P}{\partial S}\right)_V \qquad (2.14)$$

$$\left(\frac{\partial T}{\partial P}\right)_S = \left(\frac{\partial V}{\partial S}\right)_P \qquad (2.15)$$

$$\left(\frac{\partial S}{\partial P}\right)_T = -\left(\frac{\partial V}{\partial T}\right)_P \qquad (2.16)$$

$$\left(\frac{\partial S}{\partial V}\right)_T = \left(\frac{\partial P}{\partial T}\right)_V \qquad (2.17)$$

The last two are among the most useful, because they relate entropy changes to an equation of state.

The development of many of these relations is shown in Fig. 2.7, as proposed by Dr. A. J. Kidnay of the Colorado School of Mines.

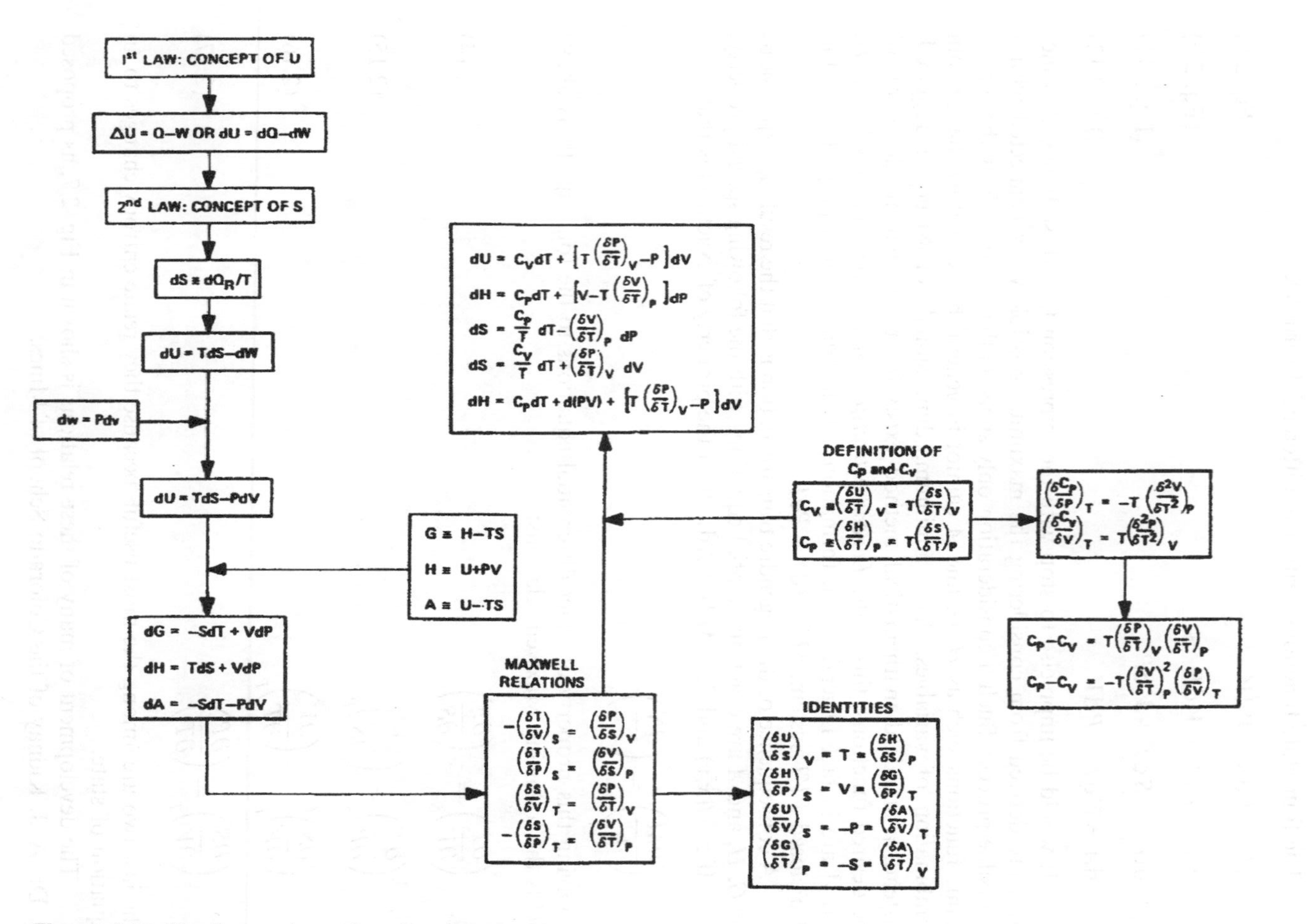

1st LAW: CONCEPT OF U
$\Delta U = Q - W$ OR $dU = dQ - dW$
2nd LAW: CONCEPT OF S
$dS \equiv dQ_R/T$
$dU = TdS - dW$
$dw = Pdv$
$dU = TdS - PdV$
$G \equiv H - TS$
$H \equiv U + PV$
$A \equiv U - TS$
$dG = -SdT + VdP$
$dH = TdS + VdP$
$dA = -SdT - PdV$
$dU = C_V dT + \left[T\left(\frac{\delta P}{\delta T}\right)_V - P\right] dV$
$dH = C_P dT + \left[V - T\left(\frac{\delta V}{\delta T}\right)_P\right] dP$
$dS = \frac{C_P}{T} dT - \left(\frac{\delta V}{\delta T}\right)_P dP$
$dS = \frac{C_V}{T} dT + \left(\frac{\delta P}{\delta T}\right)_V dV$
$dH = C_P dT + d(PV) + \left[T\left(\frac{\delta P}{\delta T}\right)_V - P\right] dV$
MAXWELL RELATIONS
$-\left(\frac{\delta T}{\delta V}\right)_S = \left(\frac{\delta P}{\delta S}\right)_V$
$\left(\frac{\delta T}{\delta P}\right)_S = \left(\frac{\delta V}{\delta S}\right)_P$
$\left(\frac{\delta S}{\delta V}\right)_T = \left(\frac{\delta P}{\delta T}\right)_V$
$-\left(\frac{\delta S}{\delta P}\right)_T = \left(\frac{\delta V}{\delta T}\right)_P$
DEFINITION OF $C_p$ and $C_v$
$C_V \equiv \left(\frac{\delta U}{\delta T}\right)_V = T\left(\frac{\delta S}{\delta T}\right)_V$
$C_P \equiv \left(\frac{\delta H}{\delta T}\right)_P = T\left(\frac{\delta S}{\delta T}\right)_P$
$\left(\frac{\delta C_P}{\delta P}\right)_T = -T\left(\frac{\delta^2 V}{\delta T^2}\right)_P$
$\left(\frac{\delta C_V}{\delta V}\right)_T = T\left(\frac{\delta^2 P}{\delta T^2}\right)_V$
$C_P - C_V = T\left(\frac{\delta P}{\delta T}\right)_V\left(\frac{\delta V}{\delta T}\right)_P$
$C_P - C_V = -T\left(\frac{\delta V}{\delta T}\right)_P^2\left(\frac{\delta P}{\delta V}\right)_T$
IDENTITIES
$\left(\frac{\delta U}{\delta S}\right)_V = T = \left(\frac{\delta H}{\delta S}\right)_P$
$\left(\frac{\delta H}{\delta P}\right)_S = V = \left(\frac{\delta G}{\delta P}\right)_T$
$\left(\frac{\delta U}{\delta V}\right)_S = -P = \left(\frac{\delta A}{\delta V}\right)_T$
$\left(\frac{\delta G}{\delta T}\right)_P = -S = \left(\frac{\delta A}{\delta T}\right)_V$


**Figure 2.7** The thermodynamic network.

## 2.6. Equations of State

It is evident from the preceding discussion that thermodynamics provides a multitude of equations inter-relating the properties of substances. Given appropriate data, thermodynamics allows the development of a complete set of thermodynamic property values from which one can subsequently calculate the heat and work effects of various processes. *PVT* data are among the most important and are concisely given by equations of state.

The simplest equation of state is the *ideal gas law*:

$$\frac{PV}{RT} = 1 \tag{2.18}$$

The ideal gas law is useful for calculating limiting, low-pressure, cases; for defining the ideal gas thermodynamic temperature scale; and for calculating so-called residual properties; but it fails to predict the existence of the liquid state.

The ideal gas law, with all of its shortcomings, is nonetheless very useful and is found at the core of all other equations of state. It is fun to develop from mechanics. Doing so shows us that temperature is a direct measure of kinetic energy and that pressure is kinetic energy per unit volume.

### *2.6.1. Deriving the Ideal Gas Law*

The requisites for an ideal gas are that the molecules occupy zero volume (or negligible volume compared to the total available) and that there be no interactions (forces) between molecules.

Consider $n$ molecules of an ideal gas confined in a volume $V^{\mathrm{t}}$. We want to find the pressure, i.e., the force per unit area, exerted on a confining wall. Force is the time rate of change of momentum, so we want to find that rate

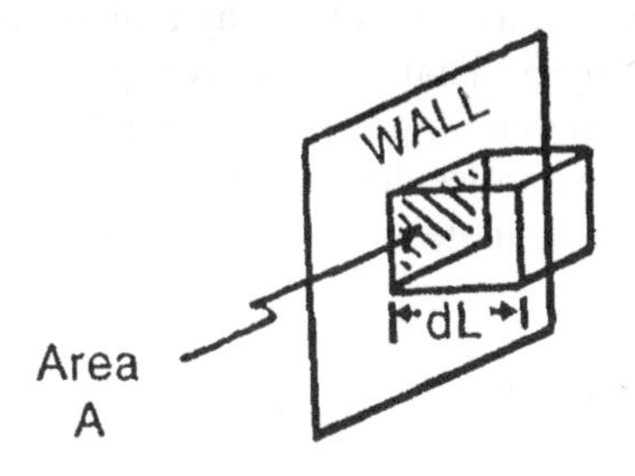


The molecules are in random motion. Their velocity, a vector, can be resolved into three components. Take one component normal to the wall. Since the motion is random, on the average, one-third of the velocities are normal to the wall and half of those, one-sixth of the total, are moving toward the wall at any time (the other half are moving away from the wall).

The total number of molecules in the small box shown in the figure is the number per unit volume, $n/V^{\mathrm{t}}$, times the volume $A\,\mathrm{d}L$. One-sixth of that number is moving toward the wall and will strike the wall in the time $\mathrm{d}t$ that it takes to travel the distance $\mathrm{d}L$ at the average velocity, $u = \mathrm{d}L/\mathrm{d}t$. Thus, the rate at which molecules strike the wall is

$$\frac{1}{6}\left(\frac{n}{V^{\mathrm{t}}}\right)A\,\mathrm{d}L\left(\frac{1}{\mathrm{d}t}\right) = \frac{1}{6}\left(\frac{n}{V^{\mathrm{t}}}\right)A\,\mathrm{d}L\left(\frac{u}{\mathrm{d}L}\right) = \frac{1}{6}\left(\frac{n}{V^{\mathrm{t}}}\right)Au$$

When molecules strike the wall, their velocity is reversed. So the total change in velocity is $2u$. The rate of change of momentum is the rate of striking the wall times the mass per molecule (i.e., molecular weight times the change in velocity, $2u$). Thus,

$$F = \frac{1}{6}\left(\frac{n}{V^{t}}\right)Au(M)(2u) = \frac{1}{3}\left(\frac{Mn}{V^{t}}\right)Au^2$$

and the pressure is

$$P = \frac{F}{A} = \frac{1}{3}\left(\frac{Mn}{V^{t}}\right)u^2$$

The ideal gas law is $P = (n/V^{t})\,RT$; thus,

$$\frac{1}{3}Mu^2 = RT \quad \text{or} \quad \frac{Mu^2}{2} = \frac{3}{2}RT$$

where $Mu^2/2$ is the average kinetic energy per mole. Thus, we have shown that for an ideal gas the kinetic energy is directly proportional to the absolute temperature. Or, put another way, temperature is a direct measure of kinetic energy. Notice also that pressure is equal to kinetic energy per unit volume.

This short derivation also provides some insight into the universal gas constant $R$. $R$ is the chemist's version of Boltzmann's constant $k$. $R$ is in units per mole, and $k$ is in units per molecule. Both $R$ and $k$ are scaling factors between temperature and energy. For physicists, $E = kT$. For chemists, it is $PV = RT$, where the product $PV$ is energy. Several values of $R$ are given in Table 2.1.

We have already defined the heat capacities $C_v$ and $C_p$. The ideal gas derivation can also breathe some life into these two important quantities. Heat capacity describes the ability of some substance to "soak up" energy.

Consider the molecules of a monatomic ideal gas. The only way a monatomic gas can soak up energy is to increase in kinetic energy. That is, the only way a monatomic ideal gas can "contain" energy (short of energies large enough to affect electronic or nuclear changes) is by the motion of its molecules—translational kinetic energy. Thus, if energy is added to a box of such molecules at constant volume, the change in their internal energy will be

$$\mathrm{d}U = \mathrm{d}(\text{kinetic energy}) = \mathrm{d}((3/2)RT) = (3/2)R\,\mathrm{d}T$$

At constant volume, $\mathrm{d}U = C_V\,\mathrm{d}T$, therefore $C_V = (3/2)\,R$ for a monatomic ideal gas. If energy is added at constant pressure, the volume will change. Work is done when the volume changes. The work is $P\,\mathrm{d}V$, due to pushing back the surroundings at the same pressure, $P$, as the system. The total energy added is the change in internal energy plus the work ($\mathrm{d}U + P\,\mathrm{d}V$). This quantity equals $\mathrm{d}H$ (at constant $P$), and $\mathrm{d}H = C_P\,\mathrm{d}T$ (at constant $P$). So,

$$\mathrm{d}H = C_P\,\mathrm{d}T = \mathrm{d}U + P\,\mathrm{d}V = C_V\,\mathrm{d}T + R\,\mathrm{d}T = (C_V + R)\,\mathrm{d}T$$

Thus, $C_P = C_V + R$. This is true for any ideal gas (monatomic or not, because the work expression is always the same).

For polyatomic molecules, the internal energy includes the translational kinetic energy, but the molecules also can "soak up" energy in the form of rotation and vibration about their bonds. Polyatomic molecules thus have greater $C_V$ (and $C_P$)

**Table 2.1** Values of the Gas Constant $R$ for 1 mol of Ideal Gas

| Temp. unit | Pressure unit | Volume unit | Energy unit | $R$ per gram mol |
|---|---|---|---|---|
| K | | | Calorie | 1.9872 |
| K | atm | $cm^3$ | | 82.057 |
| K | atm | L | | 0.082054 |
| K | mmHg | L | | 62.361 |
| K | bar | L | | 0.08314 |
| K | $kg/cm^2$ | L | | 0.08478 |
| | | | | $R$ per lb mol |
| °R | | | Btu (IT) | 1.986 |
| °R | | | hp hr | 0.0007805 |
| °R | | | kW hr | 0.0005819 |
| °R | atm | cu ft | | 0.7302 |
| °R | in. Hg | cu ft | | 21.85 |
| °R | mmHg | cu ft | | 555.0 |
| °R | psia | cu ft | | 10.73 |
| | psfa | cu ft | ft lb | 1545.0 |
| K | atm | cu ft | | 1.314 |
| K | mmHg | cu ft | | 998.9 |

Selected values from API Research Project 44:
Ice point = 0°C = 273.16 K = 491.69°R.
Gram molar volume of ideal gas at 1 atm and 0°C = 22,414.6 $cm^3$ = 22.4140 L.
Pound molar volume of ideal gas at 1 atm and 32°F = 359.05 cu ft.
1 atm = 760 mmHg = 29.921 in. Hg = 14.696 psia.
1 hp = 550 ft lb/s = 745.575 W (IT).
1 cal = 41833 J (IT).
1.8 Btu/lb = 1.0 IT cal/g = 1.000657 cal/g.

than do monatomic molecules. As a first approximation, $C_V$ increases by $R$ for each additional atom in the molecule. This is a consequence of the principle of equipartition of energy: the value of $C_V$ increases by $(1/2)R$ for each additional mode by which energy can be soaked up (three translational modes plus one rotational and one vibrational per bond formed). This holds well for diatomic molecules [$C_V = (5/2)R$, $C_P = (7/2)R$], but as more atoms are added it does not work so well. The vibrational modes and rotational modes are not completely independent in more complex molecules.

The molecules of real gases do occupy a volume of their own, and there is an attraction among them, the intermolecular force. The equation of state for a real gas will therefore contain the ideal gas law as a first approximation and add other terms to account for the volume of the molecules and the potential between the molecules. These *other terms* express the nonideality of real gases. We need these terms, not only for accuracy but also to explain why some gases cool as they expand and others do not. Without the nonideality expressed by these terms, there would be no change in temperature upon expansion. Ideal gases have a zero Joule–Thomson coefficient.

It is easy to find the ideal gas law in the virial equation of state.

The *virial equation of state* is extremely powerful and can be written in two forms.

1. As an expansion in volume ($V$) or density ($\rho$):

$$\frac{PV}{RT} = 1 + \frac{B}{V} + \frac{C}{V^2} + \cdots \tag{2.19}$$

or

$$\frac{PV}{RT} = 1 + B\rho + C\rho^2 + \cdots \tag{2.20}$$

2. As an expansion in pressure:

$$\frac{PV}{RT} = 1 + B'P + C'P^2 + \cdots \tag{2.21}$$

The virial coefficients ($B$, $C$, ...) can be calculated from first principles (two-body interactions, three-body interactions, etc.) and are functions of temperature only.

*Generalized or corresponding state* forms of the equation of state are also useful:

$$\frac{PV}{RT} = Z, \qquad Z = Z(T_r, P_r) \tag{2.22}$$

where $Z$ is the compressibility factor and $T_r$ and $P_r$ are the reduced temperature and pressure, e.g., $T_r = T/T_c$. $T$ is the system temperature, and $T_c$ is the critical temperature of the system component.

Another useful generalized equation of state is the three-parameter corresponding states form employing the acentric factor $\omega$,

$$Z = f(T_r, P_r, \omega) \tag{2.23}$$

where $\omega$ is a tabulated parameter for a specific substance.

Several of the more common equations of state are shown in Table 2.2.

The values of $\Delta S$, $\Delta H$, $\Delta E$, etc., can now be calculated for any process for any substance according to the following scheme.

**Step 1.** Choose a pair of coordinates from $P$, $T$, and $V$, for example, $P$–$T$, $T$–$V$, or $P$–$V$. The problem statement is useful; use what is given. If the process is isothermal or isobaric, choose $T$ (or $P$) as one of the pair, because it is constant.

The $P$–$T$ pair is commonly chosen (but any two will do for homogeneous systems of a pure component), as $P$ and $T$ are the most commonly measured variables.

**Step 2.** Write the total differential in terms of the pair chosen, e.g.,

$$S = f(P, T)$$
$$\mathrm{d}S = \left(\frac{\partial S}{\partial T}\right)_P \mathrm{d}T + \left(\frac{\partial S}{\partial P}\right)_T \mathrm{d}P \tag{2.24}$$

**Step 3.** Evaluate the partial differential coefficients from the basic Maxwell relations. For example, evaluate the first coefficient, $(\partial S/\partial T)_P$. Because this coefficient involves an entropy term ($S$) that changes with $T$ at constant $P$, it is likely that $H$ will be involved.

From the definition of $\mathrm{d}H$,

$$\mathrm{d}H = T\,\mathrm{d}S + V\,\mathrm{d}P$$

**Table 2.2** Equations of State (for 1 mol)

Van der Waals:

$$\left(p+\frac{a}{\widehat{V}^2}\right)(\widehat{V}-b)=RT$$

Lorenz:

$$p=\frac{RT}{\widehat{V}^2}(\widehat{V}+b)-\frac{a}{\widehat{V}^2}$$

Dieterici:

$$p=\frac{RT}{\widehat{V}-b}e^{-a/\widehat{V}RT}$$

Berthelot:

$$p=\frac{RT}{\widehat{V}-b}-\frac{a}{T\widehat{V}^2}$$

Redlich–Kwong:

$$\left[p+\frac{a}{T^{1/2}\widehat{V}(\widehat{V}+b)}\right](\widehat{V}-b)=RT$$

$$a=0.4278\frac{R^2T_c^{2.5}}{p_c}$$

$$b=0.0867\frac{RT_c}{p_c}$$

Onnes:

$$p\widehat{V}=RT\left(1+\frac{B}{\widehat{V}}+\frac{C}{\widehat{V}^2}+\cdots\right)$$

Holborn:

$$p\widehat{V}=RT(1+B'p+C'p^2+\cdots)$$

Beattie–Bridgeman:

$$p\widehat{V}=RT+\frac{\beta}{\widehat{V}}+\frac{\gamma}{\widehat{V}^2}+\frac{\delta}{\widehat{V}^3}$$

$$\beta=RTB_o-A_o-\frac{Rc}{T^2}$$

$$\gamma=-RTB_ob+aA_o-\frac{RB_oc}{T^2}$$

$$\delta=\frac{RB_obc}{T^2}$$

Benedict–Webb–Rubin:

$$p\widehat{V}=RT+\frac{\beta}{\widehat{V}}+\frac{\sigma}{\widehat{V}^2}+\frac{\eta}{\widehat{V}^4}+\frac{\omega}{\widehat{V}^5}$$

$$\beta=RTB_o-A_0-\frac{C_o}{T^2}$$

$$\sigma=bRT-a+\tfrac{c}{T^2}\exp\left(-\tfrac{\gamma}{\widehat{V}^2}\right)$$

$$\eta=c\gamma\exp\left(-\tfrac{\gamma}{\widehat{V}^2}\right)$$

$$\omega=a\alpha$$

Peng–Robinson:

$$p=\frac{RT}{\widehat{V}-b}-\frac{a\alpha}{\widehat{V}(\widehat{V}+b)+b(\widehat{V}-b)}$$

$$a=0.45724\left(\frac{R^2T_c^2}{p_c}\right)$$

$$b=0.07780\left(\frac{RT_c}{p_c}\right)$$

$$\alpha=[1+\kappa(1-T_r^{1/2})]^2$$

$$\kappa=0.37464+1.5422\omega-0.26992\omega^2$$

$\omega=$ acentric factor

we have

$$\left(\frac{\partial H}{\partial T}\right)_P=T\left(\frac{\partial S}{\partial T}\right)_P+0$$

Since at constant pressure $dP=0$, and since

$$\left(\frac{\partial H}{\partial T}\right)_P\equiv C_P$$

$$C_P=T\left(\frac{\partial S}{\partial T}\right)_P$$

therefore,

$$\left(\frac{\partial S}{\partial T}\right)_P=\frac{1}{T}C_P \tag{2.25}$$

So far, there are no restrictions on this equation.

**Step 4.** Now evaluate the second coefficient, $(\partial S/\partial P)_T$. From the relationship

$$dG = -S\,dT + V\,dP$$

and the property of exactness described before, we have

$$\left(\frac{\partial S}{\partial P}\right)_T = -\left(\frac{\partial V}{\partial T}\right)_P \tag{2.26}$$

**Step 5.** The total differential, $dS$, is now complete as

$$dS = \frac{C_P}{T}dT - \left(\frac{\partial V}{\partial T}\right)_P dP \tag{2.27}$$

There are still no restrictions on this equation.
**Step 6.** Now evaluate $(\partial V/\partial T)_P$ from an equation of state or numerically. If we choose $PV = RT$ as the arbitrarily selected equation of state, then

$$V = RT/P \quad \text{and} \quad \left(\frac{\partial V}{\partial T}\right)_P = \frac{R}{P}$$

and

$$dS = C_P\frac{dT}{T} - R\frac{dP}{P}$$

We now have the total derivative of $S$ as a function of properties only for any process. The integration to yield $\Delta S$ is straightforward. This equation is now restricted to the ideal gas state, since that was the equation of state chosen to evaluate $\partial V/\partial T_P$.

This is a perfectly general method for any thermodynamic function: $\Delta S$, $\Delta H$, $\Delta U$, $\Delta G$, etc. Since such equations contain only *properties of the system*, they are independent of path and can be tabulated for various fluids. Such tabulations form the basis for the various thermodynamic charts: $T$–$S$, $H$–$S$, etc.

*Example 2.2.* Calculate internal energy ($U$) and entropy ($S$) as a function of $T$ and $V$.

It is sometimes more convenient to take $T$ and $V$ as independent variables rather than $T$ and $P$. The method described above is perfectly general and can be applied here. Writing $U$ and $S$ as total differentials:

$$dU = \left(\frac{\partial U}{\partial T}\right)_V dT + \left(\frac{\partial U}{\partial V}\right)_T dV \tag{2.28}$$

$$dS = \left(\frac{\partial S}{\partial T}\right)_V dT + \left(\frac{\partial S}{\partial V}\right)_T dV \tag{2.29}$$

Recall that

$$C_V = \left(\frac{\partial U}{\partial T}\right)_V \tag{2.30}$$

Two relations follow immediately from differentiating Eq. (2.8):

$$\left(\frac{\partial U}{\partial T}\right)_V = T\left(\frac{\partial S}{\partial T}\right)_V \tag{2.31}$$

and

$$\left(\frac{\partial U}{\partial V}\right)_T = T\left(\frac{\partial S}{\partial V}\right)_T - P \tag{2.32}$$

As a result of Eq. (2.30), Eq. (2.31) becomes

$$\left(\frac{\partial S}{\partial T}\right)_V = \frac{C_V}{T} \tag{2.33}$$

and as a result of Eq. (2.17), Eq. (2.32) becomes

$$\left(\frac{\partial U}{\partial V}\right)_V = T\left(\frac{\partial P}{\partial T}\right)_V - P \tag{2.34}$$

Combining Eqs. (2.28), (2.30), and (2.34) gives

$$\mathrm{d}U = C_V\,\mathrm{d}T + \left[T\left(\frac{\partial P}{\partial T}\right)_V - P\right]\mathrm{d}V \tag{2.35}$$

and combining Eqs. (2.29), (2.17), and (2.33) gives

$$\mathrm{d}S = \frac{C_V}{T}\,\mathrm{d}T + \left(\frac{\partial P}{\partial T}\right)_V\mathrm{d}V \tag{2.36}$$

Equations (2.35) and (2.36) are general equations that express the internal energy and entropy of homogeneous fluids *at constant composition* as functions of temperature and molar volume. The coefficients of $\mathrm{d}T$ and $\mathrm{d}V$ are expressed in terms of measurable quantities.

### *2.6.2. Useful Thermodynamic Relations*

With these basic definitions and concepts we can now obtain a whole host of useful thermodynamic relations. Table 2.3 lists these relationships as functions of the process for both real systems and ideal gas systems. A distinction is also made, where necessary, between nonflow and flow systems, which are sometimes designated as closed and open systems, respectively.

## 2.7. Thermodynamic Analysis of Low-Temperature Systems

The fundamental tools for a low-temperature analysis are the first and second laws of thermodynamics. Assuming that the kinetic and potential energy terms in the first law can be neglected (frequently a valid assumption), the first law for a steady-state process such as a cryogenic refrigerator can be simplified to

$$\Delta H = Q - W_s \tag{2.37}$$

where $\Delta H$ is the summation of the enthalpy differences of all the fluids entering and leaving the system (or piece of equipment) being analyzed, $Q$ is the summation of all heat exchanges between the system and its surroundings, and $W_s$ is the net shaft work. There can be as many independent first law analyses as there are independent systems or pieces of equipment in the refrigeration and liquefaction systems. These can involve numerous unknowns, some of which may have to be fixed as parameters.

**Table 2.3** Useful Thermodynamic Relationships

| Process | $W$ | $Q$ | $\Delta U$ | $\Delta H$ |
|---|---|---|---|---|
| Isobaric (constant $P$) | | | | |
| Real | $P(V_2 - V_1)$ | $nC_P(T_2 - T_1)$ | $Q - W$ | $nC_P(T_2 - T_1)$ |
| Ideal gas | $nR(T_2 - T_1)$ | | $nC_V(T_2 - T_1)$ | |
| Isometric (constant $V$) | | | | |
| Real | 0 | $nC_V(T_2 - T_1)$ | $Q$ | $\Delta U + \Delta(PV)$ |
| Ideal gas | | | $nC_V(T_2 - T_1)$ | $nC_P(T_2 - T_1)$ |
| Isothermal (constant $T$) | $Q - \Delta U$ | $\Delta U + W$ | $Q - W$ | $\Delta U + \Delta(PV)$ |
| Real | $nRT\ln(V_2/V_1)$ | $nRT\ln(P_1/P_2)$ | 0 | 0 |
| Ideal gas | $nRT\ln(P_1/P_2)$ | $nRT\ln(V_2/V_1)$ | | |
| Isentropic (constant $S$) | | | | |
| Real (nonflow) | $-\Delta U$ | 0 | $-W$ | $\Delta U + \Delta(PV)$ |
| Ideal gas (nonflow) | $(P_2V_2 - P_1V_1)/(1 - \gamma)$ | | $nC_V(T_2 - T_1)$ | $nC_P(T_2 - T_1)$ |
| | $nC_V(T_2 - T_1)$ | | | |
| Real (flow) | $-\Delta H$ | 0 | $\Delta H - \Delta(PV)$ | $-W$ |
| Ideal (flow) | $\gamma(P_2V_2 - P_1V_1)/(1 - \gamma)$ | | | |
| Polytropic | | | | |
| Real (nonflow) | $Q - \Delta U$ | $\Delta U + W$ | $Q - W$ | $\Delta U + \Delta(PV)$ |
| | $(P_2V_2 - P_1V_1)/(1 - n)$ | | | |
| Ideal (nonflow) | $Q - nC_V(T_2 - T_1)$ | | $nC_V(T_2 - T_1)$ | $nC_P(T_2 - T_1)$ |
| Real (flow) | $Q - \Delta H$ | $\Delta H + W$ | $\Delta H - \Delta(PV)$ | $Q - W$ |
| Ideal (flow) | $n(P_2V_2 - P_1V_1)/(1 - n)$ | | | |
| Cyclic | $Q_{net}$ | $W_{net}$ | 0 | 0 |

The choice of these parameters will generally be made on the basis of experience or convenience.

Application of the first law to the ideal work of isothermal compression in a refrigerator cycle with appropriate substitution for the heat summation reduces to

$$-W_s/\dot{m} = T_1(S_1 - S_2) - (H_1 - H_2) \tag{2.38}$$

where $\dot{m}$ is the mass rate of flow through the compressor, $S$ is the specific entropy, $H$ is the specific enthalpy, and the subscripts 1 and 2 refer to the inlet and exit streams, respectively, of the isothermal compressor. In a real system, an overall compressor efficiency factor will have to be incorporated in this equation to obtain meaningful results.

Evaluation of the refrigeration duty in a refrigerator cycle is dependent on the process chosen. For example, if the refrigeration process is isothermal, the refrigeration duty can be evaluated using the relation

$$Q = \dot{m}T(S_2 - S_1) \tag{2.39}$$

where $T$ is the temperature at which the refrigeration takes place and the subscripts 1 and 2 refer to the inlet and exit streams of the process. On the other hand, if the refrigeration process involves only the vaporization of the saturated liquid refrigerant at constant pressure, evaluation of the duty reduces to

$$Q = \dot{m}h_{fg} \tag{2.40}$$

where $h_{fg}$ is the specific heat of vaporization of the refrigerant. If the process is performed at constant pressure with the absorption of only sensible heat by the refrigerant (e.g., warming a cold gas), the duty can be evaluated from

$$Q = \dot{m}(H_2 - H_1) \tag{2.41}$$

Similar first law balances could be written for other units of equipment used at low temperatures.

## 2.8. Joule–Thomson Coefficient

The slope at any point of an isenthalpic curve on a temperature–pressure diagram is called the Joule–Thomson (or J–T) coefficient. The J–T coefficient is usually denoted by $\mu$ and is given by

$$\mu = \left(\frac{\partial T}{\partial P}\right)_H$$

The locus of all points at which the J–T coefficient is zero, i.e., the locus of the maxima of the isenthalpic curves, is known as the inversion curve and is shown as a dotted curve in Fig. 2.8. The region to the left and inside the inversion curve, where the J–T coefficient is positive, is the region of cooling; the region outside, where the J–T coefficient is negative, is the region of heating.

Hydrogen, helium, and neon have negative J–T coefficients at ambient temperature. Consequently, when used as refrigerants in a throttling process, they must be cooled either by a separate precoolant fluid or by a work-producing expansion to a temperature below which the J–T coefficient is positive. Only then will throttling cause a further cooling rather than heating.

Nitrogen, methane, and certain other fluids, on the other hand, have positive J–T co-efficients at ambient temperatures and hence produce cooling when expanded across a valve. Accordingly, any of these fluids can be used directly as the refrigerant in a throttling process without the necessity of a precooling step or expansion through a work-extracting device. Maximum inversion temperatures for some of the more common cyrogens are given in Table 2.4.

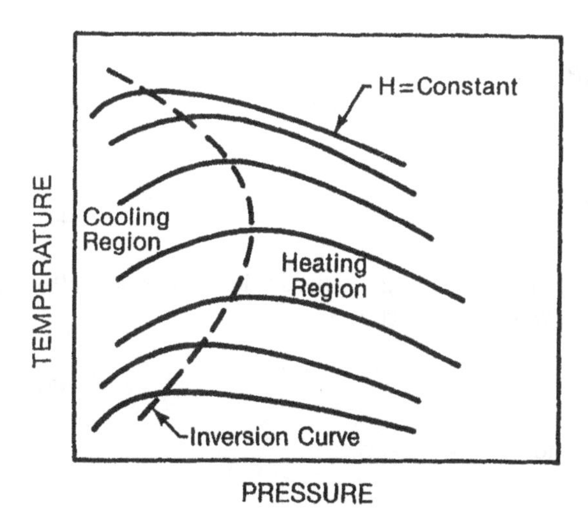


**Figure 2.8** Joule–Thomson inversion curve.

**Table 2.4** Maximum Inversion Temperature

| Fluid | Maximum inversion temperature | |
|---|---|---|
| | K | °R |
| Oxygen | 761 | 1370 |
| Argon | 722 | 1300 |
| Nitrogen | 622 | 1120 |
| Air | 603 | 1085 |
| Neon | 250 | 450 |
| Hydrogen | 202 | 364 |
| Helium | 40 | 72 |

*Example 2.3.* Show that the Joule–Thomson coefficient for a perfect gas is always zero.

The Joule–Thomson coefficient has been defined as

$$\mu = \left(\frac{\partial T}{\partial P}\right)_H \tag{2.42}$$

From calculus it can be shown that

$$\mu_{\mathrm{J-T}} = \left(\frac{\partial T}{\partial P}\right)_H = -\left(\frac{\partial T}{\partial H}\right)_P \left(\frac{\partial H}{\partial P}\right)_T \tag{2.43}$$

The expression $(\partial H/\partial T)_P$ has previously been identified as the heat capacity, $C_p$. The expression $(\partial H/\partial P)_T$ can be found by differentiating Eq. (2.9 ) and combining the result with Eq. (2.16 ) to yield

$$\left(\frac{\partial H}{\partial T}\right)_P = \left[V - T\left(\frac{\partial V}{\partial T}\right)_P\right]$$

The total differential for $H$,

$$\mathrm{d}H = \left(\frac{\partial H}{\partial T}\right)_P \mathrm{d}T + \left(\frac{\partial H}{\partial P}\right)_T \mathrm{d}P$$

becomes

$$\mathrm{d}H = C_P\,\mathrm{d}T + \left[V - T\left(\frac{\partial V}{\partial T}\right)_P\right]\mathrm{d}P \tag{2.44}$$

Setting $\mathrm{d}H = 0$ (constant enthalpy), Eq. (2.45 ) gives the Joule–Thomson coefficient in terms of measurable thermodynamic properties as

$$\mu_{\mathrm{J-T}} = \frac{1}{C_P}\left[T\left(\frac{\partial V}{\partial T}\right)_P - V\right] \tag{2.45}$$

For a perfect gas, $V = RT/P$, and

$$\left(\frac{\partial V}{\partial T}\right)_P = \frac{R}{P} = \frac{V}{T}$$

Therefore, for a perfect gas,

$$\mu_{\mathrm{J-T}} = \frac{1}{C_P}\left[T\left(\frac{V}{T}\right) - V\right] = 0 \tag{2.46}$$

A perfect gas would not experience any temperature change upon expansion.

## 3. HEAT TRANSFER

All cryogenic systems operate, that is to say complete their required tasks, at very low temperatures. This fact requires a reasonably strong understanding of heat transfer and all associated technologies. As stated earlier, the best that one can hope for in some cases (e.g., cryogenic fluid storage) is to severely limit the energy transfer across boundaries. Since heat always flows from hot to cold and cannot be stopped, how it is controlled is the design aspect relating to heat transfer. In other cases (like gasification of cryogenic fluids) the ability to maximize the amount of heat introduced into the fluid is the desired design state. Due to the nature of the temperature difference between the everyday world we live in and the extremes of operating thermal conditions of cryogenic systems, a number of unique problems present themselves. These include, but not limited to: (1) variability of material properties, (2) thermal insulation, (3) fluid property differences for near-critical-point convection, (4) effects of thermal radiation, (5) equipment design characteristics and operation, (6) selection of instrumentation and control sensors, (7) transfer of cryogenic fluids, and (8) disposal and/or control of cryogenic boil-off. Each of these problems will be addressed in the remaining chapters of this book. Heat transfer at low temperatures is governed by the same three mechanisms present at ambient and elevated temperatures: conduction, convection, and radiation. It is not surprising, therefore, that all the general equations are appropriate for low-temperature applications as long as they are adjusted for the property changes in both materials and fluids.

### 3.1 Conductive Heat Transfer

Another method of determining the heat transfer rate is in the expression:

$$Q = S(K_H - K_C)$$

where $K_H$ and $K_C$ are the thermal conductivity integrals at the evaluated temperature $T_H$ and $T_C$, respectively. These two terms are defined as:

$$K_H = \int_0^{T_H} (k_\mathrm{t}\mathrm{d}t) \text{ and } K_C = \int_0^{T_C} (k_\mathrm{t}\mathrm{d}T)$$

and $S$ is defined as a conduction shape factor as

$$1/S = \int_{x_1}^{x_2} (\mathrm{d}x/A(x))$$

where $k_\mathrm{t}$ is the material thermal conductivity at some temperature within the temperature field expressed by the integral; d$x$ is the change in length along the axis of heat transfer; and $A$ is the cross-sectional area of heat transfer.

The quantities $K_H$ and $K_C$ represent the integrated thermal conductivity for any solid material that has had its thermal conductivity variation with temperature determined. The values of $K_H$ and $K_C$ are also useful when their difference is divided by the difference of $T_H$ and $T_C$ and becomes the apparent thermal conductivity of the material, expressed as

$$K_M = (K_H - K_C)/(T_H - T_C)$$

Figures 2.9 and 2.10 are graphical depictions of thermal conductivity integrals for some metals and insulators, respectively.

The conduction shape factor $S$ is derived from the Fourier equation and simply places all the physical parameters of the shape of the material into a single integral. This allows for various geometric shapes to be evaluated and conduction shape factors to be determined based on the desired physical dimension of interest, such as wall thickness or sphere diameters. Table 2.5 gives some representative conduction shape factors for various geometries of interest. These geometries are pictured in Fig. 2.11.

With this simple expression it is possible to get a good engineering estimate on the heat balance around various cryogenic components under steady-state conditions

$$Q = -kA\frac{dT}{dx} \tag{2.47}$$

where $Q$ is the heat transferred, $k$ the material thermal conductivity (normally a function of temperature), $A$ the area of the material through which the heat is transferred and $dT/dx$ the change in temperature observed in the material in the $x$ direction due to the heat transfer.

If $Q$, $k$, and $A$ are constant, the relation becomes

$$Q = -kA\frac{T_2 - T_1}{\Delta_x} \tag{2.48}$$

This relation is useful when determining the heat transfer through a layer of insulation or a series of insulation materials. If the conduction is through insulation wrapped around a pipe, the relationship is modified to

$$Q = -kA\frac{dT}{dr} = 2\,\pi\,Lk\frac{T_1 - T_2}{\ln(r_2/r_1)} \tag{2.49}$$

where $L$ is the length of the pipe, $r_1$, $r_2$ the inside and outside radii of the insulation and $T_1$, $T_2$ the corresponding temperatures at these locations.

For nonsteady-state conductive heat transfer, a heat balance may be made around a volume element of material of dimensions $dx$, $dy$, and $dz$. In the $x$ direction, the heat transferred into the volume element is given by

$$dQ_x = -k\,dy\,dz\frac{\partial T}{\partial x} \tag{2.50}$$

The heat leaving the volume element is

$$dQ'_x = dQ_x + \frac{\partial dQ_x}{\partial x}\,dx \tag{2.51}$$

If $k$ is constant, then

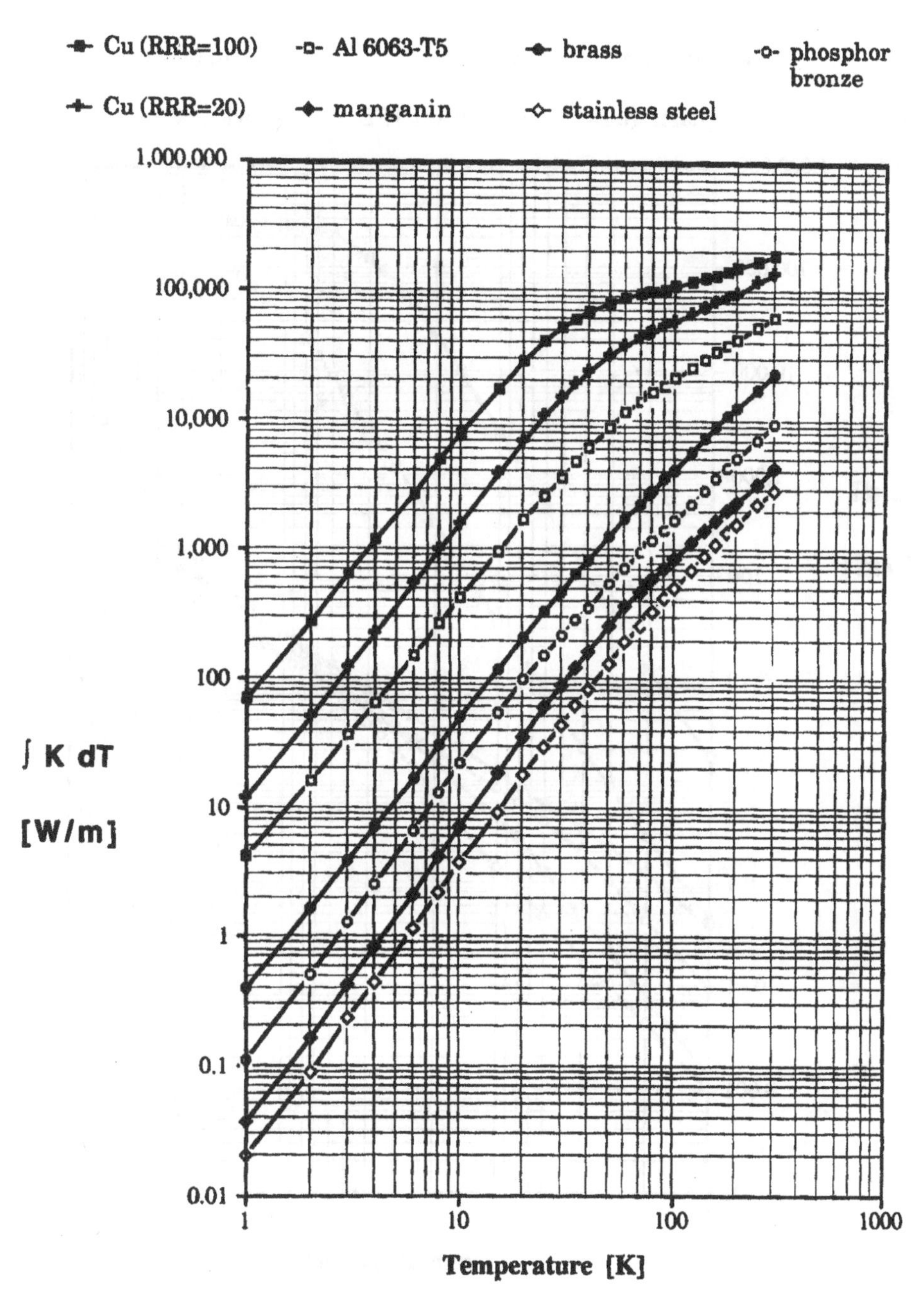


**Figure 2.9** Thermal conductivity integrals of metals. Data provided from S. Courts, Lakeshore Cryotronics.

$$dQ'_x - dQ_x = k\,dy\,dz\,dx\frac{\partial^2 T}{\partial x^2} \tag{2.52}$$

Similar expressions are valid for the other two directions. If heat is stored in the volume element, then

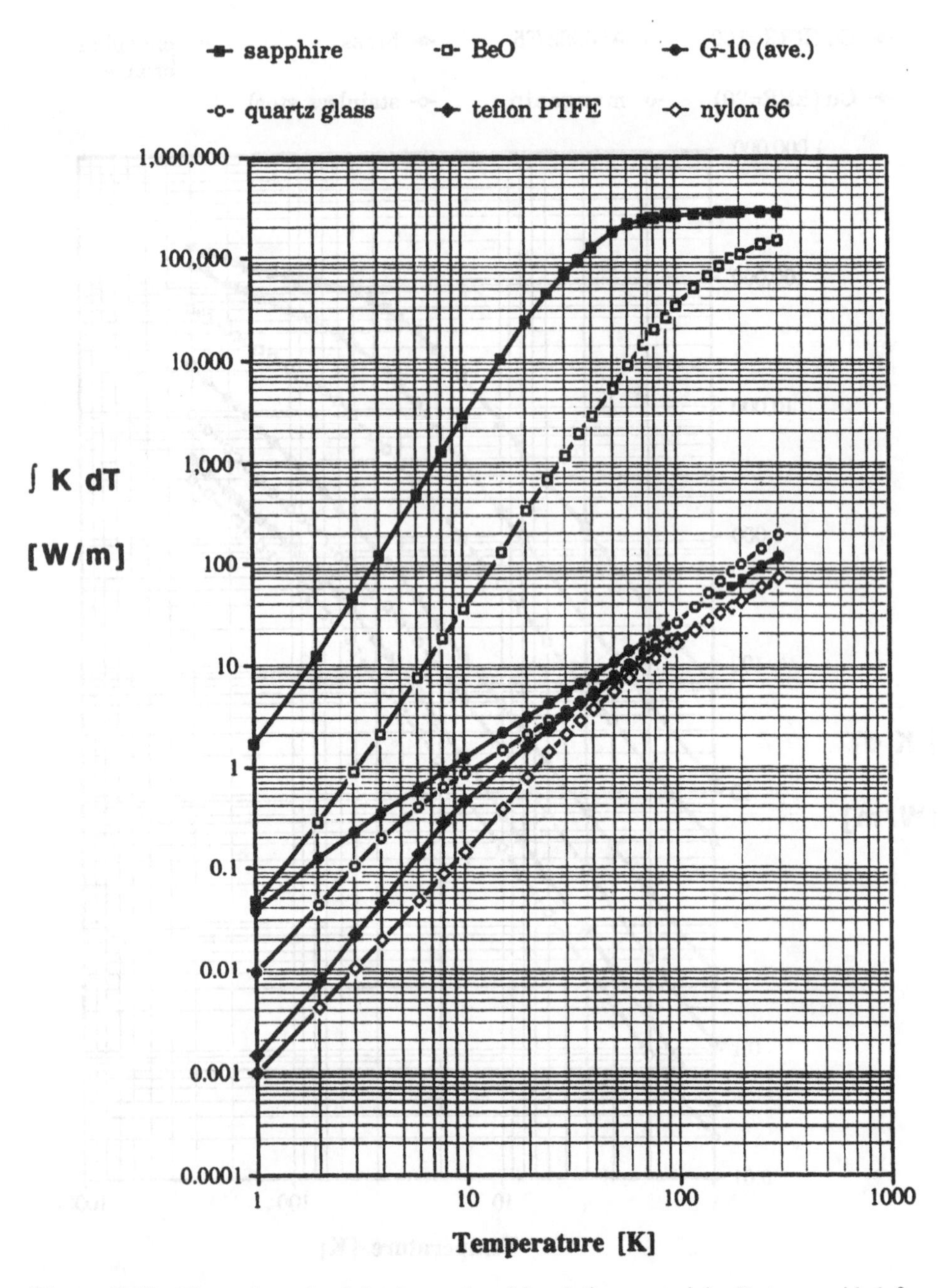


**Figure 2.10** Thermal conductivity integrals of insulating materials. Data provided from S. Courts, Lakeshore Cryotronics.

$$dQ_s = dQ_x - dQ'_x + dQ_y + dQ_z - dQ'_z \tag{2.53}$$

but

$$dQ_s = C_P\rho \, dx \, dy \, dz \frac{\partial T}{\partial \theta} \tag{2.54}$$

**Table 2.5** Example Conduction Shape Factors[a]

| Geometry | Shape factor |
|---|---|
| [1] Plane wall or slab | $S = A/\Delta x$ |
| [2] Hollow cylinder | $S = \frac{2\pi L}{\ln(D_0/D_1)}$ |
| [3] Hollow sphere | $S = \frac{2\pi D_0 D_i}{(D_0 - D_i)}$ |
| [4] Isothermal cylinder placed vertically in a semi-infinite medium | $S = \frac{2\pi L}{\ln(4L/D)}$ |
| [5] Isothermal cylinder buried in a semi-infinite medium | $S = \frac{2\pi L}{\cosh^{-1}(2H/D)}$ |
| [6] Isothermal sphere buried in a semi-infinite medium | $S = \frac{8\pi D}{1 - (D/4H)}$ |
| [7] Cylinder centered in a large plate | $S = \frac{2\pi L}{\ln(4H/D)}$ |

[a] See Fig. 2.11 for correlated sketch of referenced geometry.

so

$$\left(\frac{\partial T}{\partial \theta}\right) = \alpha\left[\left(\frac{\partial^2 T}{\partial x^2}\right) + \left(\frac{\partial^2 T}{\partial y^2}\right) + \left(\frac{\partial^2 T}{\partial z^2}\right)\right] \tag{2.55}$$

where $\theta$ is the time, $C_P$ and $\rho$ are the heat capacity and density of the material,

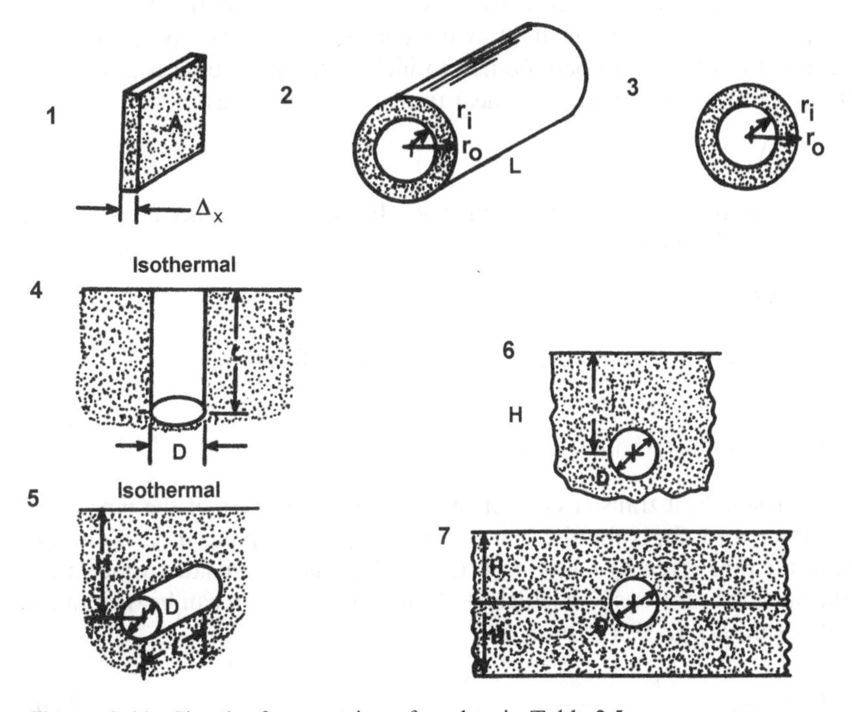


**Figure 2.11** Sketch of geometries referred to in Table 2.5.

respectively, and $\alpha$ is defined as $k/\rho C_P$. At a steady state, $\partial T/\partial\theta = 0$. The solution of Eq. (2.55 ) involves either the separation of variables or the use of integral transforms (usually Laplace transforms). For regular geometries, plots have been developed that permit rapid and reasonably reliable solutions even at low temperatures for heating or cooling times.

### 3.2. Convective Heat Transfer

Thermal convection is the transfer of heat from a fluid to a colder surface by means of fluid particle motion. There are two types of *convective heat transfer. Forced convection* occurs when a fluid is forced or pumped past a surface; *free convection* or natural convection occurs when fluid motion is caused by density differences within the fluid. In either case, the basic equation is

$$Q = h_c A(T_s - T_b) \tag{2.56}$$

where $h_c$ is the individual heat transfer coefficient, $T_s$ is the surface temperature, and $T_b$ is the bulk fluid temperature. The individual heat transfer coefficient is critical in the solution of Eq. (2.56) and must generally be obtained with the aid of some empirical relationship that has been found to correlate the convective heat transfer reasonably well for a specific geometry or flow pattern. The empirical relationships used to obtain the individual heat transfer coefficients at low temperatures are very similar to the relationships used at ambient temperatures.

Tables 2.6–2.8 provide some useful heat transfer relationships for low-temperature heat exchangers for a variety of operating conditions. The units for the relationships in these tables are in consistent American engineering units. Conversion factors to other systems of measurement are found in the Appendix.

If heat is transferred from one fluid to another separated by a pipe as in a heat exchanger, then Eq. (2.56 ) can be modified to include the individual heat transfer coefficients for both fluids in an overall heat transfer coefficient as

$$Q = U_o A_o \Delta T_{\text{overall}} \tag{2.57}$$

where $U_o$ is based on the outside surface area $A_o$. If the inside surface area is used, the relationship can be modified to

$$Q = U_i A_i \Delta T_{\text{overall}} \tag{2.58}$$

If the heat exchange is across a pipe, $U_i$ can be obtained from

$$U_i = \frac{1}{1/h_i + \Delta r_p A_i / k_p A_{p,\text{lm}} + A_i/h_o A_o} \tag{2.59}$$

where $h_i$ is the inside heat transfer coefficient, $\Delta r_p$ is the thickness of the pipe, $A_i$ is the inside surface area, $k_p$ is the thermal conductivity of the pipe, $A_{p,\text{lm}}$ is the log mean area of the pipe given by $(A_o - A_i)[\ln(A_o/A_i)]$, $A_o$ is the outside surface area, and $h_o$ is the outside heat transfer coefficient. A similar expression can be written for $U_o$. For a flat wall, Eq. (2.59 ) reduces to

$$U_i = U_o = \frac{1}{1/h_i + \Delta x/k + 1/h_o} \tag{2.60}$$

**Table 2.6** Empirical Heat Transfer and Corresponding Pressure Drop Relationships for Concentric Tube and Extended Surface Heat Exchangers

| Type | Flow conditions | Empirical heat transfer equation | Empirical pressure drop/length equation |
|---|---|---|---|
| Straight tubular pipe | Flow inside, Re < 2100; no phase change (gas) | $h=\left[3.658+\dfrac{0.0668(D_e/L)\mathrm{Re\,Pr}}{1+0.04(D_e/L)\mathrm{Re\,Pr}^{2/3}}\right]\dfrac{k}{D_e}$ | $\dfrac{\Delta P}{\Delta L}=\dfrac{32\,G^2}{\mathrm{Re}\,g_c D_e \rho}$ |
| Straight tubular pipe | Flow inside, 2100 < Re < 10,000; no phase change (gas) | $h=2.439\times10^{-6}(\mathrm{Re}^{2/3}\times125)\left[1+\left(\dfrac{D_e}{L}\right)^{2/3}\right]\left(\dfrac{C_P}{D_e}\right)\left(\dfrac{\mu_f}{\mu_b}\right)^{-0.14}$ | $\dfrac{\Delta P}{\Delta L}=\dfrac{0.158\,G^2}{\mathrm{Re}^{0.2}\,g_c D_e \rho}$ |
| Straight tubular pipe | Flow inside, Re > 10,000; no phase change (gas) | $h=0.023\,C_P G\,\mathrm{Re}^{-0.2}\,\mathrm{Pr}^{-2/3}$ | $\dfrac{\Delta P}{\Delta L}=\dfrac{0.092\,G^2}{\mathrm{Re}^{0.2}\,g_c D_e \rho}$ |
| Helical tubular pipe | Flow inside, Re > 10,000; no phase change (gas) | $h=0.023\,C_P G\,\mathrm{Re}^{-0.2}\,\mathrm{Pr}^{-2/3}[1+3.5(D_e/D_h)]$ | $\dfrac{\Delta P}{\Delta L}=\dfrac{0.092\,G^2[1+3.5(D_e/D_h)]}{\mathrm{Re}^{0.2}\,g_c D_e \rho}$ |
| Collins tubing | Flow in annulus, 400 < Re < 10,000; no phase change (gas) | $h=0.118\,C_P G\,\mathrm{Re}^{-0.2}\,\mathrm{Pr}^{-2/3}$ | $\dfrac{\Delta P}{\Delta L}=\dfrac{0.952\,G^2_{max}}{\mathrm{Re}^{0.2}\,g_c D_e \rho}$ |

For tubular pipe: $D_e$ = inside pipe diameter; for noncircular tubes, $D_e = 4(A_c L/A_w)$, where $A_c$ = inside free-flow cross-sectional area, $L$ = length of pipe, and $A_w$ = heat transfer or wetted tube surface area; for annulus, $D_e = D_2 - D_1$, where $D_2$ = inside diameter of outer pipe and $D_1$ = outside diameter of inner pipe; $D_h$ = diameter of helix.
For Collins tubing: $D_e = 4(V_a/A_w)$, where $V_a$ = active flow volume in annulus, $A_w$ = wetted area in annulus, including fins.
Evaluate fluid properties for all relations at the mean film temperature, $T_m = 0.5(T_w + T_b)$, where $T_w$ = tube wall temperature and $T_b$ = bulk temperature of fluid. Constants in these equations are for use with 51 units.

**Table 2.7** Empirical Heat Transfer and Corresponding Pressure Drop Relationships for Flows Normal to Outside of Tubes on Shell Side of Coiled Tube Heat Exchangers

| Type | Flow conditions | Empirical heat transfer equation | Empirical pressure drop/tube equation |
|---|---|---|---|
| Banks of staggered tubes | Flow normal to outside of tubes, $2000 < \mathrm{Re} < 3.2 \times 10^4$; no phase change (gas) | $h = 0.33\ C_P G\, \mathrm{Re}^{-0.4}\ \mathrm{Pr}^{-2/3}$ | $\frac{\Delta P}{\Delta N} = \left[0.5 + 0.235(X_T - 1)^{-1.08}\right] \mathrm{Re}^{-0.15}\ G_{max}^2/g_c\rho$ |
| Banks of in-line tubes | Flow normal to outside of tubes, $2000 < \mathrm{Re} < 3.2 \times 10^4$; no phase change (gas) | $h = 0.26\ C_P G\, \mathrm{Re}^{-0.4}\ \mathrm{Pr}^{-2/3}$ | $\frac{\Delta P}{\Delta N} = [0.088 + 0.16 X_L (X_T - 1)^{-n}] \mathrm{Re}^{-0.15}\ G_{max}^2/g_c\rho$ |

$\mathrm{Re} = D_o G_{max}/\mu$, where $D_o$ = outside diameter of tubes; $G_{max} = \dot{m}/A_{min}$, with $A_{min}$ = minimum open free-flow area between tubes; $X_T$ = transverse pitch/tube outside diameter; $N$ = total number of tubes in line of flow; $X_L$ = longitudinal pitch/tube outside diameter; $n = 0.43 + (1.13/X_L)$. Fluid properties for all relations are evaluated at the mean film temperature, $T_m = 0.5(T_w + T_b)$, where $T_w$ = tube wall temperature and $T_b$ = bulk temperature of fluid.

**Table 2.8** Empirical Heat Transfer and Corresponding Pressure Drop Correlations for Brazed Aluminum Plate-Fin Heat Exchangers[a]

| Type and flow conditions | Empirical heat transfer equation | Empirical pressure drop/length equation |
|---|---|---|
| Straight fins, $500 < \mathrm{Re} < 10^4$; $h = 7.87$ mm (0.310 in.); $w = 0.15$ mm (0.006 in.); 492 fins/m (12.5 fins/in.); no phase change (gas) | $h = 0.0291\,\mathrm{Re}^{-0.24}\,\mathrm{Pr}^{-2/3}\,C_P G$ | $\frac{\Delta P}{\Delta L} = \frac{(0.0099 + 40.8\,\mathrm{Re}^{-1.033})G^2}{g_c \rho D_e}$ |
| Wavy fins, $300 < \mathrm{Re} < 10^4$; $h = 9.53$ mm (0.375 in.); $w = 0.20$ mm (0.008 in.); 591 fins/m (15 fins/in.); no phase change (gas) | $h = 0.085\,\mathrm{Re}^{-0.265}\,\mathrm{Pr}^{-2/3}\,C_P G$ | $\frac{\Delta P}{\Delta L} = \frac{(0.0834 + 23.6\,\mathrm{Re}^{-0.062})G^2}{g_c \rho D_e}$ |
| Herringbone fins, $400 < \mathrm{Re} < 10^4$; $h = 10.82$ mm (0.426 in.); $w = 0.15$ mm (0.006 in.); 472 fins/m (12 fins/in.); no phase change (gas) | $h = 0.555(\mathrm{Re} + 500)^{-0.482}\,\mathrm{Pr}^{-2/3}\,C_P G$ | $\frac{\Delta P}{\Delta L} = \frac{7.04 G^2\,\mathrm{Re}^{-0.547}}{g_c \rho D_e}$ |

[a] To match present vendor specifications, dimensions are also given in customary English units. $D_e = 4(A_c L/A_w)$, where $A_c$ = inside free-flow cross-sectional area, $L$ = length of pipe, and $A_w$ = heat transfer or wetted tube surface area. Evaluate fluid properties for all relations at mean film temperature, $T_m = 0.5(T_w + T_b)$, where $T_w$ = tube wall temperature and $T_b$ = bulk temperature of fluid.

Heat transfer from a condensing vapor to a cooler surface may be considered as a special case of convective heat transfer. Heat transfer calculations with condensing vapors are generally made using a relation similar to Eq. (2.56 ), where $T_b$ is the temperature of the condensing vapor and $T_s$ is the temperature of the surface upon which condensation occurs. For filmwise condensation on the outside of horizontal tubes, the convective heat transfer co-efficient can be obtained from

$$h_c = 0.725\left[\frac{k^3 \rho^2 g h_{fg}}{\mu_f D_o N \Delta T}\right]^{1/4} \tag{2.61}$$

where $h_{fg}$ is the latent heat of vaporization, $N$ is the number of rows of tubes in the vertical plane, $D_o$ is the outside diameter of the tubes, and $\Delta T$ is the temperature difference between the saturated vapor and the outside tube surface. For filmwise condensation on the outside of vertical tubes, the convective heat transfer coefficient relationship is modified to

$$h_c = 0.942\left[\frac{k^3 \rho^2 g h_{fg}}{\mu_f L \Delta T}\right]^{1/4} \tag{2.62}$$

Heat transfer from a surface to a boiling liquid can also be considered a special case of convective heat transfer. There are commonly seen two types of boiling: nucleate and film boiling. Within the nucleate boiling category, there are two distinct classifications: natural convection (also known as pool boiling) or forced convection

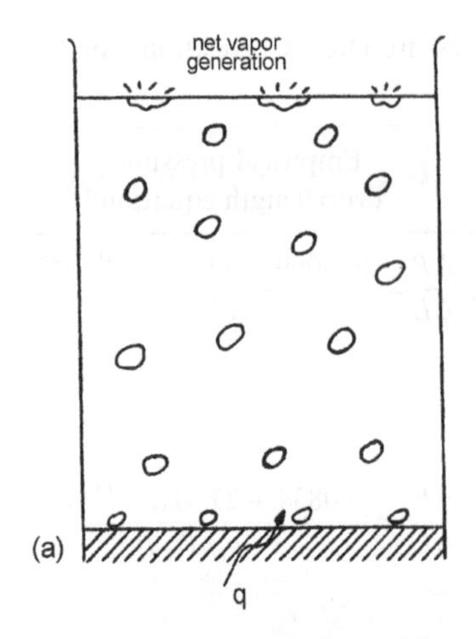


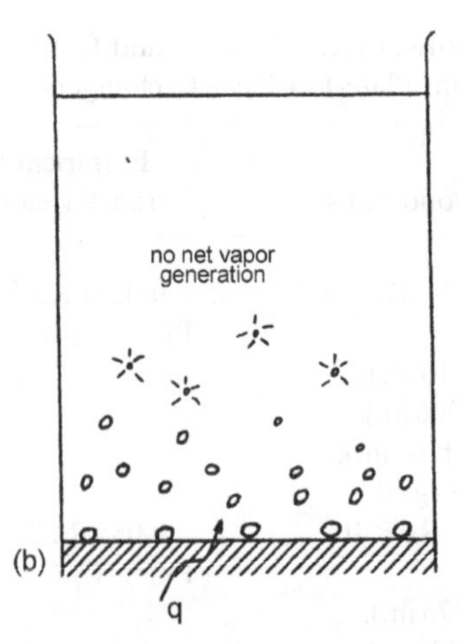


**Figure 2.12** Pooling boiling depictions: (a) saturated conditions generate vapor and raises the bulk fluid temperature while (b) subcooled conditions shown allows for no vapor generation but does raise in the bulk fluid temperature, as well.

boiling. There is a further subdivision of two primary boiling types (nucleate and film) based upon whether the boiling fluid is under saturated or sub-cooled conditions. The nucleate boiling conditions are depicted in Fig. 2.12 and 2.13, while 2.14 illustrates the filming boiling process. Most experimental work at low temperatures has been conducted on pool boiling. Such studies have shown that the boiling action at these temperatures is also largely dependent on the temperature difference between the submerged surface and the boiling liquid. Unfortunately, the available information on boiling liquids at low temperatures is limited and also shows wide scatter in the experimental data. Figure 2.15 shows typical experimental nucleate and film pool boiling data for nitrogen at 1atm compared with predictive correlations. Figures 2.16–2.18 give similar data for oxygen, hydrogen, and helium. See "Suggested Reading" at the end of the book for the original data sources called out on Figs. 2.15–2.18.

### 3.3. Radiative Heat Transfer

Thermal radiation involves the transfer of heat from one body to another at a lower temperature by electromagnetic waves passing through the intervening medium. Radiant energy striking a material may be partly absorbed, transmitted, or reflected.

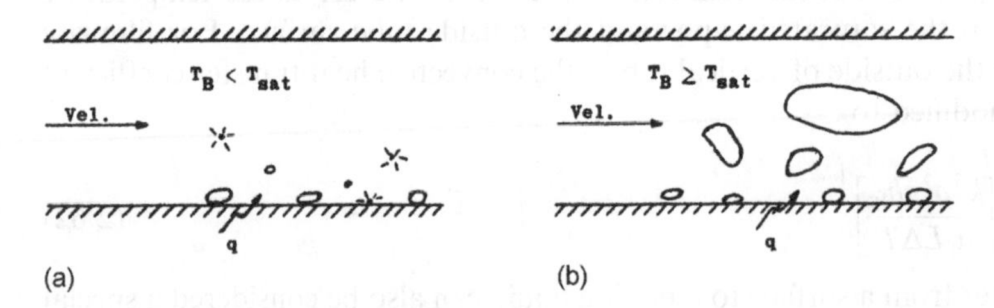


**Figure 2.13** Forced convection boiling: (a) subcooled conditions again allow for no vapor generation while (b) saturated or bulk boiling generates vapor and any number of different vapor-liquid flow conditions.

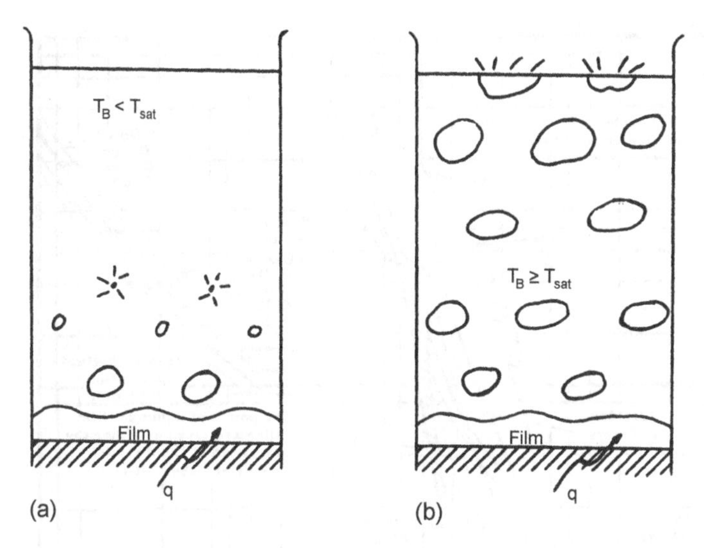


**Figure 2.14** Film boiling occurs when a vapor covers the super-heated bottom surface: (a) if conditions are subcooled, then no vapor will be generated, and (b) at saturated conditions large amounts of vapor can be formed.

For some materials, for example glass or plastic, this can be written as

$$\alpha + \tau + \rho = 1 \tag{2.63}$$

where $\alpha$ is the absorptivity, $\tau$ is the transmissivity, and $\rho$ is the reflectivity. For an opaque material, $\alpha + \rho = 1$, and for a blackbody, $\alpha = 1$. The emissive power $E$ of a unit surface is the amount of heat radiated by the surface per unit of time. The

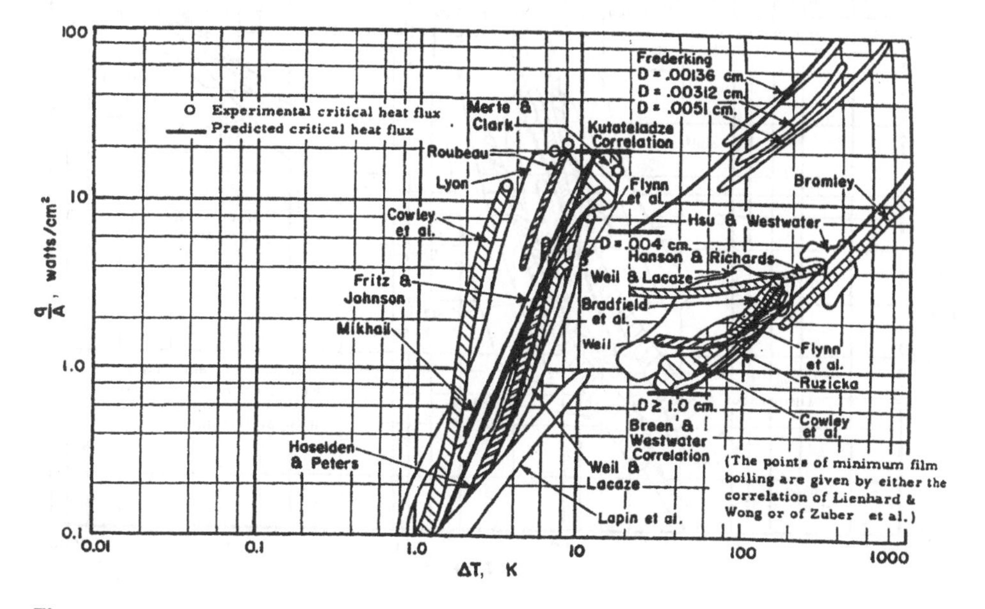


**Figure 2.15** Experimental and predictive nucleate and film boiling of nitrogen at 1 atm.

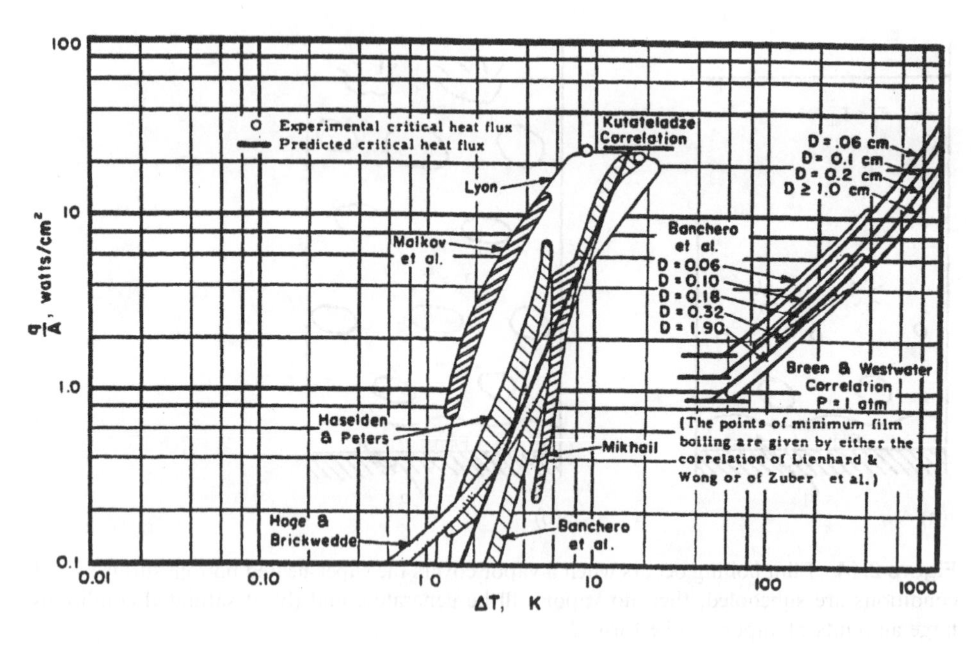


**Figure 2.16** Experimental and predictive nucleate and film boiling of oxygen at 1 atm.

emissivity $\epsilon$ of a surface is defined by

$$\epsilon = E/E_b \tag{2.64}$$

where $E_b$ is the emissive power of a blackbody at the same temperature. The emissive power for a blackbody is given by

$$E_b = \sigma T^4 \tag{2.65}$$

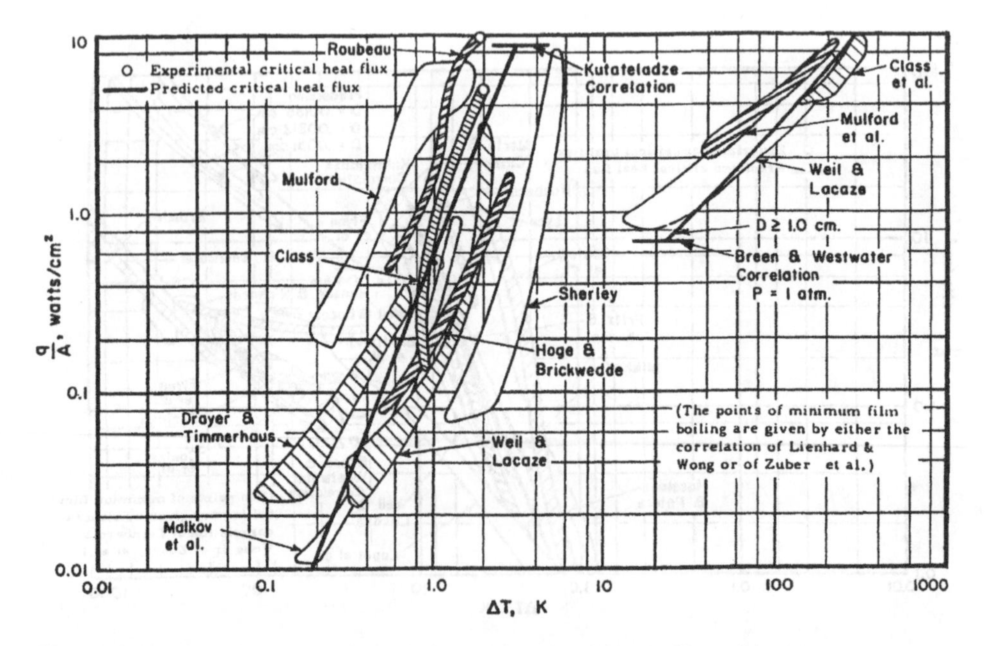


**Figure 2.17** Experimental and predictive nucleate and film boiling of hydrogen at 1 atm.

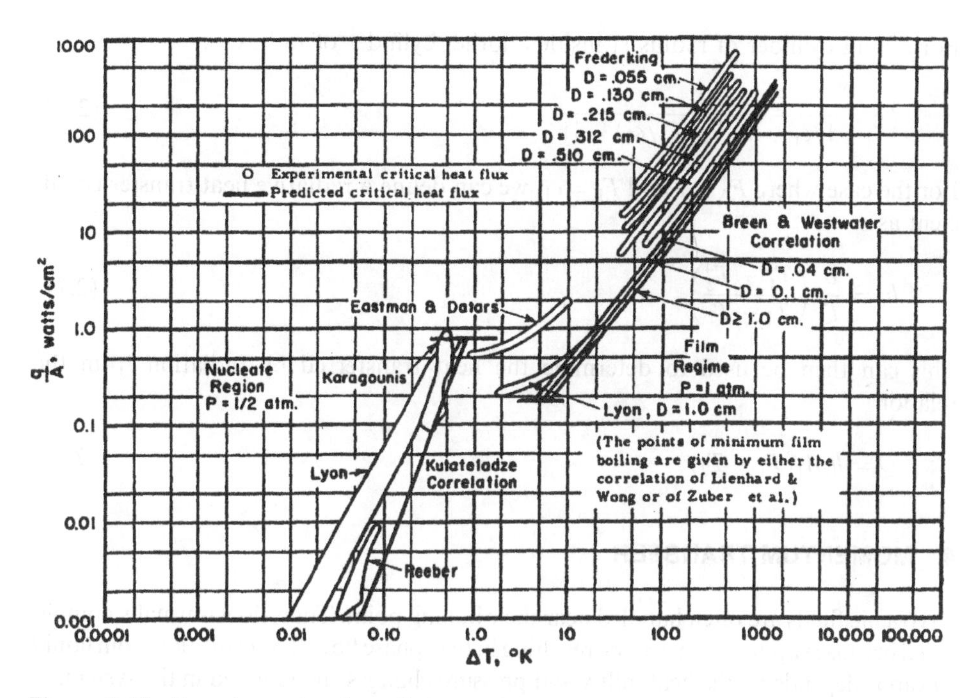


**Figure 2.18** Experimental and predictive nucleate and film boiling of helium at 1 atm.

where $\sigma$ is the Stefan–Boltzmann constant. For a real surface, the relation is modified to

$$E = \epsilon \sigma T^4 \tag{2.66}$$

When two bodies within visual range of each other are separated by a medium that does not absorb radiation, an energy exchange occurs as a result of reciprocal processes of emission and absorption. The net rate of heat transfer from one surface at $T_1$ to another surface at $T_2$ can be calculated from

$$Q = \sigma A_1 F_A F_E (T_1^4 - T_2^4) \tag{2.67}$$

where $A_1$ is the surface area of one of the bodies, $F_A$ is a shape and orientation factor for the two bodies relative to area $A_1$, and $F_E$ is the emission and absorptance factor for the two bodies. If the surface of one body is small or enclosed by the surface of the other body, $F_A = 1$; otherwise, $F_A$ must be evaluated. For a surface $A_1$ radiating to a large area or space, $F_E = \epsilon_1$. For two large parallel plates with a small space between them,

$$F_E = \frac{1}{1/\epsilon_1 + 1/\epsilon_2 - 1} \tag{2.68}$$

For a sphere of radius $r_1$ inside another sphere of radius $r_2$,

$$F_E = \frac{1}{1/\epsilon_1 + (r_1/r_2)^2[(1/\epsilon_2) - 1]} \tag{2.69}$$

For a long cylinder of radius $r_1$ inside another cylinder of radius $r_2$,

$$F_E = \frac{1}{1/\epsilon_1 + (r_1/r_2)[(1/\epsilon_2) - 1]} \tag{2.70}$$

For the case where $F_A = 1$ and $F_E = \epsilon_1$, we can define a radiative heat transfer coefficient as

$$h_r = \frac{\sigma}{T_1}\left(\frac{T_1^4 - T_2^4}{T_1 - T_2}\right) \tag{2.71}$$

This can then be used to determine the heat transferred by radiation from the relation

$$Q = h_r A_1 (T_1 - T_2) \tag{2.72}$$

## 4. MOMENTUM TRANSFER

Cryogenic fluids are often handled near their boiling point. Thus, flow is usually complicated because vaporization occurs, resulting in two-phase flow. The transition from liquid to vapor depends on the heat influx and pressure changes encountered in the system.

For single-phase flow, the most important point to recognize is the distinction between laminar and turbulent flow. For low temperatures, the same concepts that have been developed for ambient flow apply. For flows with Reynolds numbers less than 2100, laminar flow is assumed to occur; and for flows with Reynolds numbers greater than 3000, turbulent flow is assumed to occur. The range in between 2100 and 3000 is normally considered the transition region between laminar and turbulent flow. Thus, the empirical relationships that have been used successfully in calculating, for example, the pressure drop in a pipe at ambient conditions are also valid for systems using cryogenic fluids.

Tables 2.5–2.7 provide some useful relationships to determine the empirical heat transfer coefficient and pressure drop per unit length of many common heat exchangers.

In cryogenic systems, two-phase flow cannot be avoided, particularly during the cooldown of transfer lines. If a cooldown situation is encountered frequently, complete system performance reliability requires design for this condition. It takes only a slight localized heat influx or pressure drop to cause the evaporation of part of the fluid. Vapor in such lines can significantly reduce the liquid transfer capacity. This transition from liquid to vapor is usually an unequilibrated condition and presents an undefined problem for design purposes.

The Martinelli–Nelson correlation has been found to be useful for estimating pressure drop for a single-phase stream under steady-state flow conditions. However, this correlation, developed for ambient conditions, has a tendency to underestimate the pressure drop by as much as 10–30% because of the high tendency of cryogenic fluids to vaporize.

Another fluid problem encountered rather frequently is that of *cavitation* in pumps. Cavitation begins at a pressure slightly below where vaporization just begins. The difference in pressure between cavitation and normal vaporization is equal to the net positive suction head minus the drop in head pressure in the pump. Normally, this must be determined experimentally and is a characteristic of the pump and the fluid.

A critical flow can be achieved in the transfer of a cryogenic fluid. In such a situation, the flow rate of the fluid cannot be altered by lowering the downstream pressure as long as the supply pressure remains constant. Such a situation is due to the increased down-stream vaporization, which reduces the cross-sectional area available for liquid transfer.

Cooldown of cryogenic systems involves a period of surging flow and pressure fluctuations with accompanying sound effects. The primary factors that affect surging and maximum pressure include inlet pressure and temperature, pipe length and diameter, pipe precooling, throttling, and liquid properties. Other factors such as heat capacity, external heat leak, and pipe wall thickness have a minor effect on surging, because the pipe temperature undergoes little change during such a surge.

## 5. HEAT LEAK AND PRESSURE DROP IN CRYOGENIC TRANSFER LINES

Cryogenic fluid transfer presents a number of challenges to any designer but two problem areas, in particular, have wide ranging impacts on the overall system design: (1) heat leak in cryogenic transfer, and (2) pressure drop during cryogenic transfer of fluids.

The transfer piping design should accommodate the desired heat transfer requirements as well as the fluid transfer conditions necessary to accommodate the overall system design. A comparison of the heat leak into a bare pipe vs. those with various insulations techniques is given in Table 2.9 for liquid nitrogen. The effect of improved insulation techniques, such as vacuum jacketed pipe (VJP), is obvious and can result in reduced heat loads of several orders of magnitude over bare pipes. VJP comes in rigid and flexible configurations to accommodate the designer's need for making bends, etc. The flexible configuration requires a more complex and slightly less efficient insulation scheme and therefore do not offer the same performance as the rigid versions but is still very good vs. the nonvacuum jacketed alternatives. Both rigid and flexible transfer lines come in various sizes and a number of commercial vendors are available to supply these types of units.

Table 2.10 gives estimated heat leak for liquid nitrogen, oxygen, and hydrogen with various line sizes. An additional consideration of VJP is the required cooldown heat removal necessary to allow the line to efficiently transfer cryogenic fluids. Note that flow is not a parameter. This is the case because fully developed turbulent flow is assumed. For laminar flow to exist, the pipe diameter would have to be extremely large, resulting in excessive costs and unnecessarily high heat leak into the system. At turbulent flow conditions, the controlling insulating parameter to heat leak into the fluid is the vacuum jacket around the transfer pipe. Solid thermal conductivity through the pipe walls, as well as the heat transfer coefficient to turbulent cryogenic

**Table 2.9** Comparison of Insulation Techniques for Fluid Transfer Piping Versus Bare or Non-insulated Piping Illustrates Needs for Insulation (Courtesy Technifab Products, Inc.)

| Pipe comparison | Heat leak Btu/ft/hr |
| --- | --- |
| Bare copper pipe | 200.0 |
| Foam insulated pipe | 20.0 |
| Bare vacuum pipe | 4.0 |
| Vacuum jacketed pipe from TPI | 0.47 |

**Table 2.10** Estimate Heat Leak for Vacuum Jacketed Piping, Both Rigid and Flexible. (Courtesy of PHPK Technologies, Inc.)

| Vacuum jacketed piping heat leak in Btu/hr/ft and (Watts/ft) | | | | | | |
|---|---|---|---|---|---|---|
| | $LN_2$ | | $LO_2$ | | $LH_2$ | |
| Line Size | Rigid | Flex | Rigid | Flex | Rigid | Flex |
| $\frac{3}{4}$" OD × $1\frac{1}{4}$" NPS | 0.37 (0.11) | 0.97 (0.28) | 0.37 (0.11) | 0.96 (0.28) | 0.40 (0.12) | 1.05 (0. 31) |
| $\frac{3}{4}$" NPS × 2" NPS | 0.43 (0.13) | 1.21 (0.25) | 0.42 (0.12) | 1.19 (0.35) | 0.46 (0.13) | 1.29 (0.38) |
| 1" NPS × $2\frac{1}{2}$" NPS | 0.47 (0.14) | 1.43 (0.42) | 0.47 (0.14) | 1.41 (0.41) | 0.51 (0.15) | 1.54 (0.45) |
| $1\frac{1}{2}$" NPS × 3" NPS | 0.58 (0.17) | 1.74 (0.51) | 0.57 (0.17) | 1.71 (0.50) | 0.63 (0.18) | 1.89 (0.55) |
| 2" NPS × 4" NPS | 0.79 (0.23) | 2.37 (0.70) | 0.65 (0.19) | 1.95 (0.57) | 0.85 (0.25) | 2.56 (0.75) |
| 3" NPS × 5" NPS | 0.98 (0.29) | 2.95 (0.86) | 0.84 (0.25) | 2.52 (0.74) | 1.08 (0.32) | 3.24 (0.95) |
| 4" NPS × 6" NPS | 1.28 (0.38) | 3.85 (1.13) | 1.01 (0.30) | 3.03 (0.89) | 1.40 (0.41) | 4.22 (1.24) |
| 6" NPS × 8" NPS | 1.65 (0.48) | 4.97 (.146) | 1.36 (0.40) | 4.10 (1.20) | 1.83 (0.54) | 5.50 (1.61) |

flow are both negligible to that of the radiation heat load through the vacuum space per length of pipe. This radiation heat load is a first order effect relating to surface areas, the larger the surface area, the larger the heat load. The internal and external surface areas of pipes are driven by their respective diameters.

Table 2.10 also shows very little increase in heat leak in changing the fluid transferred. All three fluids have heat leak conditions within about 10% of each other even though the temperature changed from 90 to 20 K. From simple radiation heat transfer calculations alone, one might expect a heat leak increase by a factor of over 400 ($=(90/20)^4$). However, the actual case is that the hydrogen temperatures do a much better job of cryopumping than do oxygen or nitrogen cooled surfaces. Thus at 90 K, there is still some small residual gas conduction, which is not present at 20 K. Therefore, the apparent thermal conductivities are not that much different.

Table 2.11 illustrates using vacuum jacket valves the cooldown loss in Btu for various sizes at two temperatures, liquid nitrogen and liquid hydrogen, and then compares the cooldown loss to the steady-state heat leaks of these units. The heat leak into vacuum jacketed (VJ) valves is expected to be higher due to the (1) increase of the penetrations into the VJ around the valve body, (2) the required increased

**Table 2.11** Estimated Performance Data for Different Size Vacuum Jacketed Valves. (Courtesy of PHPK Technologies, Inc.

| Cool-down and heat leak performance information (jacketed) | | | | |
|---|---|---|---|---|
| | Cool-down loss Btu | | Steady state heat leak Btu/hr (W) | |
| Valve Size | 80 K | 20 K | 80 K | 20 K |
| $\frac{1}{2}$" | 81 | 86 | 10.5 (3.1) | 11.6 (3.4) |
| $\frac{3}{4}$" | 81 | 86 | 10.5 (3.1) | 11.6 (3.4) |
| 1" | 99 | 105 | 10.5 (3.1) | 11.6 (3.4) |
| $1\frac{1}{2}$" | 189 | 200 | 13.7 (4.0) | 15.0 (4.4) |
| 2" | 324 | 342 | 18.1 (5.3) | 20.0 (5s.9) |
| 3" | 990 | 1045 | 39.7 (11.6) | 44.1 (12.9) |
| 4" | 1260 | 1330 | 47.9 (14.0) | 53.1 (15.6) |

support for the internal valve components, (3) heat from the power source to drive the valve, (4) instrumentation required to operate and monitor the valve, etc. Additional heat leak considerations should be included for the mechanical connections between VJP segments and the components included in the fluid systems. Each individual connection can be roughly estimated to be about 1.24 times a one-foot segment of the same size line.

The second design problem is pressure drop in cryogenic transfer lines. Two figures (Figs. 2.19 and 2.20) are provided to show the expected pressure drop per foot of pipe vs. flow in gallons per minute for nitrogen and hydrogen. Inside pipe diameter is the parameter for the data lines. Additionally, Fig. 2.21 is provided and illustrates the pressure drop differences of rigid pipe vs. flexible pipe for flowing liquid nitrogen. The increase in pressure drop can be directly attributed to the internal design use of welded bellows to form the interior pipe walls. This is true in all, not just cryogenic, fluid transfer involving flexible lines.

The values shown in Table 2.10 and the graphs of Figs. 2.19 and 2.20 can be expected from currently available, commercially fabricated transfer lines using current vacuum-jacketing technology. These values have been confirmed by the two largest US manufacturers of this equipment. At their request, no credit is given, as they have no control over how the data will be used and accordingly wish to assume no responsibility for potential misuse by others.

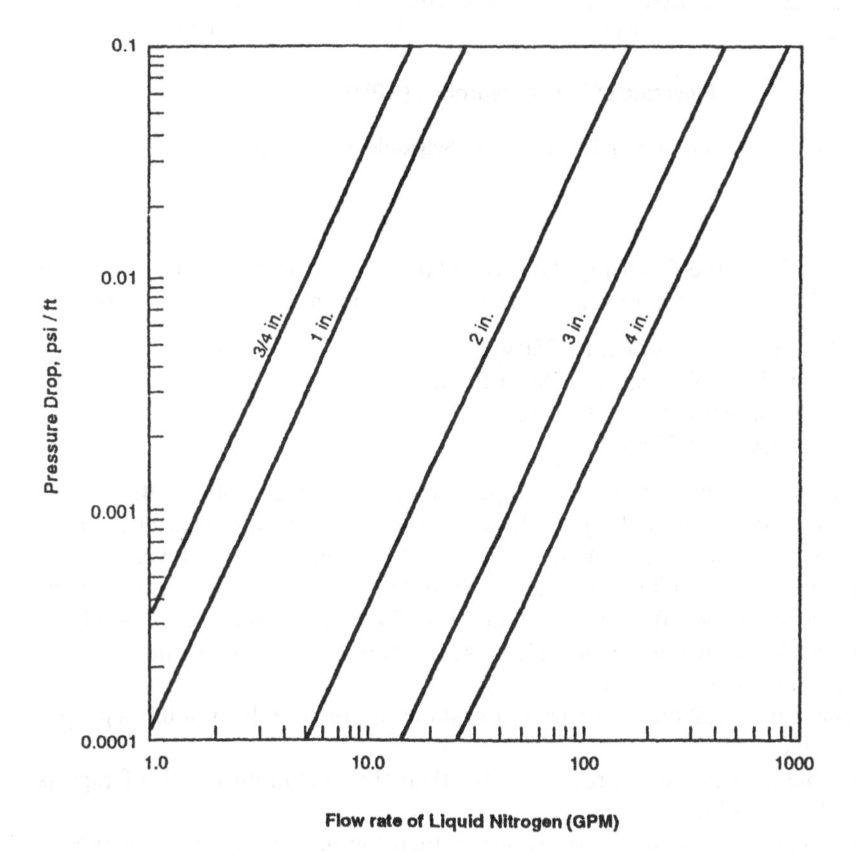


**Figure 2.19** Pressure drop of liquid nitrogen in Schedule 5 rigid pipe.

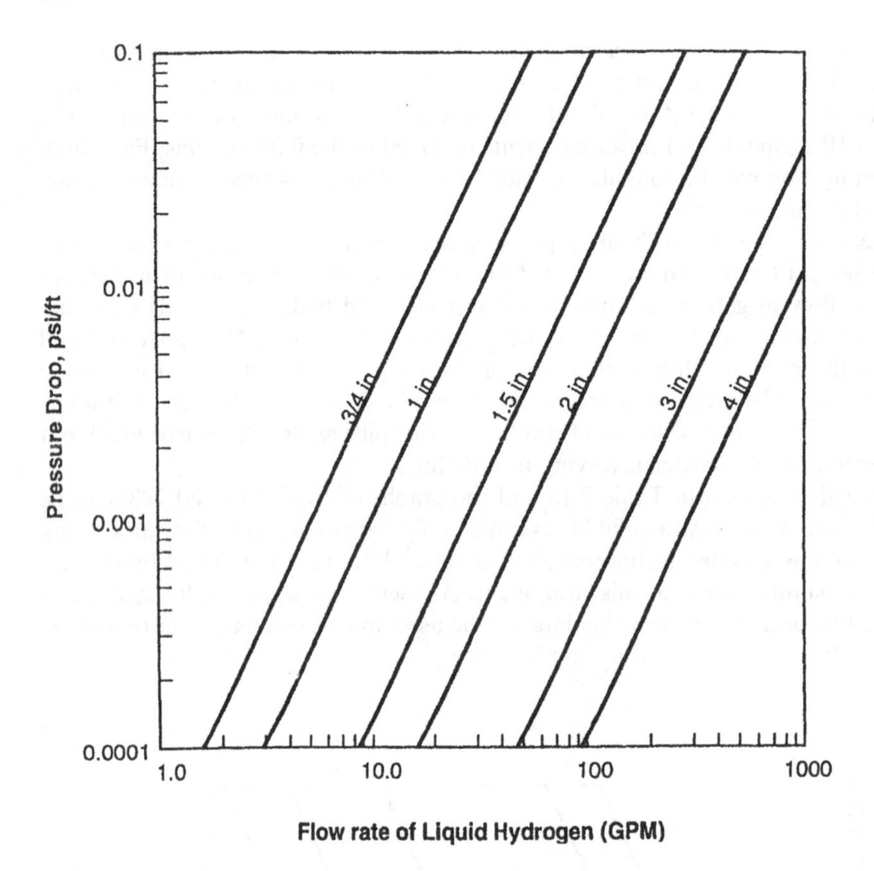


**Figure 2.20** Pressure drop of liquid hydrogen in Schedule 5 rigid pipe.

*Example 2.4.* Given the following data, calculate the heat leak and pressure drop for liquid hydrogen in a commercially available vacuum-insulated transfer line:

Liquid hydrogen transfer rate: 250 gpm
Maximum allowable heat leak: 800 Btu/hr
Acceptable pressure drop: 20–30 psi
Desirable pipe size: 2 in. i.d.

**Heat leak**. Table 2.10 shows that a 2 in. pipe will have a heat leak of 0.85 Btu/hr per foot of pipe in which liquid hydrogen is flowing. The various fittings (spacers, valves, bayonet connectors, etc.) will each amount to 1.24 equivalent feet of pipe.

Therefore, if 800 Btu/hr can be accommodated, the maximum run of equivalent pipe is $800/0.85 = 941$ ft of pipe. If there were 24 components (a large number), their equivalent length would be $24 \times 1.24 = 41$ ft. Thus, the pipe run could comprise 900 ft of pipe and 24 components.

**Pressure drop.** Figure 2.20 for hydrogen flow shows a pressure drop of 0.034 psi per foot of 2 in. pipe.

If the tolerated pressure drop is 20 psi, then the maximum length of pipe is $20/0.034 = 588$ ft of pipe.

If the acceptable pressure drop is 30 psi, then the maximum length of pipe is $30/0.034 = 882$ ft of pipe.

**PRESSURE DROP IN LIQUID NITROGEN FLUID FLOW**

Tube/Schedule 5 Pipe - psi/100ft Flex Hose - psi/10ft

| Flow | 3/4" | 3/4 inch | | 1 inch | | 1 1/2 inch | | 2 inch | | 3 inch | | 4 inch | |
|---|---|---|---|---|---|---|---|---|---|---|---|---|---|
| (gpm) | Tube | Pipe | Flex | Pipe | Flex | Pipe | Flex | Pipe | Flex | Pipe | Flex | Pipe | Flex |
| 0.4 | 0.052 | - | - | - | - | - | - | - | - | - | - | - | - |
| 0.6 | 0.109 | - | 0.033 | - | - | - | - | - | - | - | - | - | - |
| 0.8 | 0.185 | 0.033 | 0.058 | - | - | - | - | - | - | - | - | - | - |
| 1 | 0.281 | 0.050 | 0.091 | - | - | - | - | - | - | - | - | - | - |
| 2 | 1.047 | 0.182 | 0.363 | 0.051 | 0.082 | - | - | - | - | - | - | - | - |
| 3 | 2.291 | 0.392 | 0.816 | 0.109 | 0.184 | - | - | - | - | - | - | - | - |
| 4 | 4.009 | 0.681 | 1.451 | 0.188 | 0.326 | - | 0.040 | - | - | - | - | - | - |
| 5 | 6.202 | 1.046 | 2.267 | 0.287 | 0.510 | 0.038 | 0.062 | - | - | - | - | - | - |
| 6 | 8.868 | 1.490 | 3.264 | 0.407 | 0.734 | 0.054 | 0.089 | - | - | - | - | - | - |
| 8 | 15.622 | 2.608 | 5.803 | 0.708 | 1.305 | 0.092 | 0.159 | - | 0.035 | - | - | - | - |
| 10 | - | 4.035 | 9.067 | 1.091 | 2.040 | 0.141 | 0.248 | 0.043 | 0.055 | - | - | - | - |
| 15 | - | 8.95 | - | 2.406 | 4.589 | 0.306 | 0.558 | 0.092 | 0.124 | - | - | - | - |
| 20 | - | 15.79 | - | 4.231 | 8.159 | 0.534 | 0.991 | 0.159 | 0.221 | - | - | - | - |
| 25 | - | - | - | 6.56 | 12.748 | 0.824 | 1.549 | 0.244 | 0.346 | 0.033 | 0.041 | - | - |
| 30 | - | - | - | 9.40 | 18.358 | 1.176 | 2.231 | 0.347 | 0.498 | 0.047 | 0.060 | - | - |
| 35 | - | - | - | 12.75 | - | 1.590 | 3.036 | 0.468 | 0.678 | 0.063 | 0.081 | - | - |
| 40 | - | - | - | 16.616 | - | 2.066 | 3.966 | 0.607 | 0.885 | 0.081 | 0.106 | - | - |
| 45 | - | - | - | - | - | 2.604 | 5.019 | 0.763 | 1.120 | 0.101 | 0.134 | - | - |
| 50 | - | - | - | - | - | 3.203 | 6.196 | 0.938 | 1.383 | 0.124 | 0.166 | 0.033 | 0.037 |
| 60 | - | - | - | - | - | 4.589 | 8.922 | 1.340 | 1.991 | 0.176 | 0.239 | 0.047 | 0.053 |
| 70 | - | - | - | - | - | 6.221 | 12.144 | 1.814 | 2.711 | 0.238 | 0.325 | 0.063 | 0.072 |
| 80 | - | - | - | - | - | 8.102 | 15.862 | 2.359 | 3.540 | 0.308 | 0.425 | 0.081 | 0.094 |
| 90 | - | - | - | - | - | 10.229 | - | 2.976 | 4.481 | 0.387 | 0.538 | 0.102 | 0.119 |
| 100 | - | - | - | - | - | 12.604 | - | 3.663 | 5.532 | 0.476 | 0.664 | 0.124 | 0.147 |
| 125 | - | - | - | - | - | 19.624 | - | 5.694 | 8.644 | 0.736 | 1.037 | 0.192 | 0.229 |
| 150 | - | - | - | - | - | - | - | 8.169 | 12.447 | 1.053 | 1.493 | 0.273 | 0.330 |
| 175 | - | - | - | - | - | - | - | 11.089 | 16.941 | 1.426 | 2.032 | 0.369 | 0.449 |
| 200 | - | - | - | - | - | - | - | 14.454 | - | 1.855 | 2.654 | 0.480 | 0.586 |
| 250 | - | - | - | - | - | - | - | - | - | 2.882 | 4.147 | 0.743 | 0.916 |
| 300 | - | - | - | - | - | - | - | - | - | 4.135 | 5.972 | 1.064 | 1.319 |
| 350 | - | - | - | - | - | - | - | - | - | 5.611 | 8.129 | 1.442 | 1.795 |
| 400 | - | - | - | - | - | - | - | - | - | 7.313 | 10.618 | 1.877 | 2.345 |
| 450 | - | - | - | - | - | - | - | - | - | 9.239 | 13.438 | 2.369 | 2.968 |
| 500 | - | - | - | - | - | - | - | - | - | 11.390 | 16.590 | 2.919 | 3.664 |
| 600 | - | - | - | - | - | - | - | - | - | 16.366 | - | 4.188 | 5.276 |
| 700 | - | - | - | - | - | - | - | - | - | - | - | 5.687 | 7.181 |
| 800 | - | - | - | - | - | - | - | - | - | - | - | 7.413 | 9.379 |
| 900 | - | - | - | - | - | - | - | - | - | - | - | 9.368 | 11.870 |
| 1000 | - | - | - | - | - | - | - | - | - | - | - | 11.552 | 14.655 |

**Figure 2.21** Pressure drop estimates for rigid and flexible transfer lines with liquid nitrogen. (Courtesy of PHPK Technologies, Inc.).

Since heat transfer and pressure drop follow Rayleigh's equivalence, a safe assumption is that the pressure drop caused by a component is about 1.24 times that of an equivalent length of pipe. Let us pick a pipe run of 500 ft of actual pipe with 24 components in it. The 24 components represent $24 \times 1.24 = 30$ equivalent feet of pipe. The heat leak for this case is then

$$(530\,\text{ft of equivalent pipe}) \times [0.85\text{Btu/hr ft}]) = 450.5\,\text{Btu total heat leak}$$

The pressure drop for this case is

$$(530\,\text{ft equivalent pipe}) \times (0.034\,\text{psi/ft}) = 18\,\text{psi}$$

**Summary.** For this sample case, 500 ft of 2 in. I.D. pipe carrying liquid hydrogen with 24 fittings will have a heat leak of 450 Btu and a pressure drop of 18 psi, which is well within our stated parameters. Other cases are readily calculated in a similar manner.

## 6. COOLDOWN

The amount of liquid required for cooldown can be estimated rapidly by equating the sensible heat of a pipeline and its insulation to the heat of vaporization of the fluid. This, of course, does not consider heat leaking into the system, frictional losses, or changing temperature of the gas discharge from the system. Should any one of these items contribute a sizable heat load to the system, then the amount of liquid required for cooldown would obviously be low. In the final analysis, economics will normally be the most important factor in the selection of a piping system. Such economics will depend largely on the duration of the transfer and the time between transfers. These times must be evaluated rather closely if the economic evaluation is to be valid.

In Chapter 4, we will see that the heat capacity of all crystalline materials goes to zero as the absolute temperature goes to zero. Accordingly, the amount of cryogenic fluid needed for cooldown also goes to zero as the initial temperature of the solid goes to zero. In other words, prechilling a liquid hydrogen system with liquid nitrogen will save a lot of liquid hydrogen.

Figures 2.22–2.25 give cooldown values for nitrogen, oxygen, hydrogen, and helium. They are based on the heat capacity of aluminum (Al); stainless steel (SS), typically a 304 series; and copper. They all show that the amount of coolant per mass of metal decreases rapidly as the initial cooldown temperature decreases. For each metal, there is a pair of MAX and MIM curves. The MAX curve is the maximum amount of coolant required if only the heat of vaporization of the liquid is available to cool the metal. This is the worst case. The MIN curve is the minimum amount of coolant required if both the heat of vaporization of the respective liquid *and* all of the sensible heat capacity of the liquid turned gas are available to cool the metal. This is the best case and will occur only if all of the boil-off gas is somehow in perfect heat exchange with the metal being cooled, such that the boil-off gas vents at the same temperature as the warm metal. The vent gases must move slowly past the metal in order to achieve this "best case" cooling. In other words, all of the possible refrigeration is captured; none is wasted. Only by very careful and expensive design, such as in the cooling of some space satellites, is the minimum ever achieved.

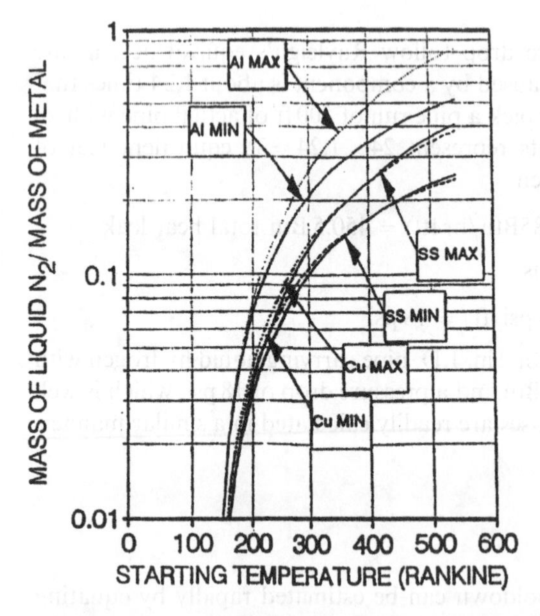


**Figure 2.22** Cooldown to liquid nitrogen temperatures. Mass of liquid nitrogen required per mass of metal vs. starting temperature.

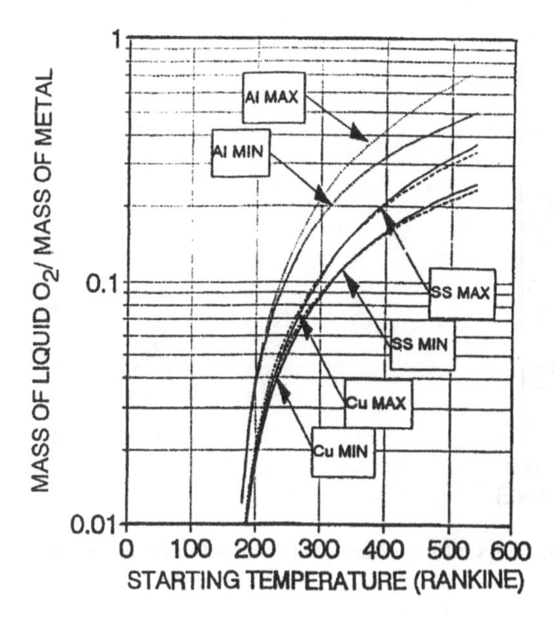


**Figure 2.23** Cooldown to liquid oxygen temperatures. Mass of liquid oxygen required per mass of metal vs. starting temperature.

It is instructive to see just how much precooling a liquid hydrogen system with liquid nitrogen can help. Referring to Fig. 2.2 for hydrogen, we see that starting at 540°R (80°F), it will take (MAX case) about 0.2 lb of liquid hydrogen to cool 1 lb of stainless steel to 36°R (20 K). Now suppose that the stainless steel system is precooled with liquid nitrogen to 139°R (77 K). The same MAX curve now predicts that only 0.01 lb of liquid hydrogen will be needed per pound of steel to cool it the rest of the way to 36°R (77 K). The liquid hydrogen use has been cut by a factor of 20. The price we pay (see Fig. 2.19) is that 0.4 lb of liquid nitrogen is used per pound of metal

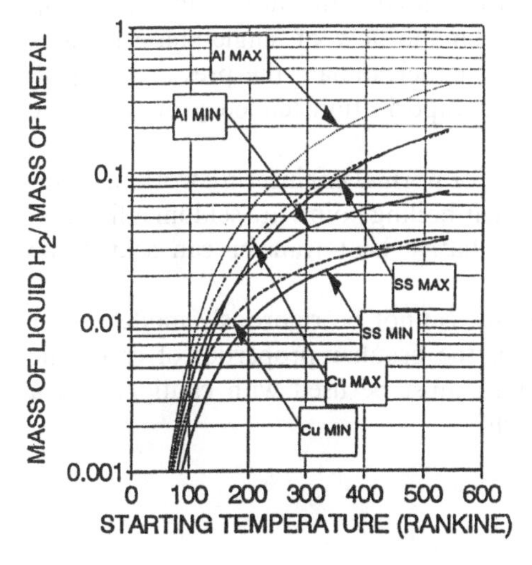


**Figure 2.24** Cooldown to liquid hydrogen temperatures. Mass of liquid hydrogen required per mass of metal vs. starting temperature.

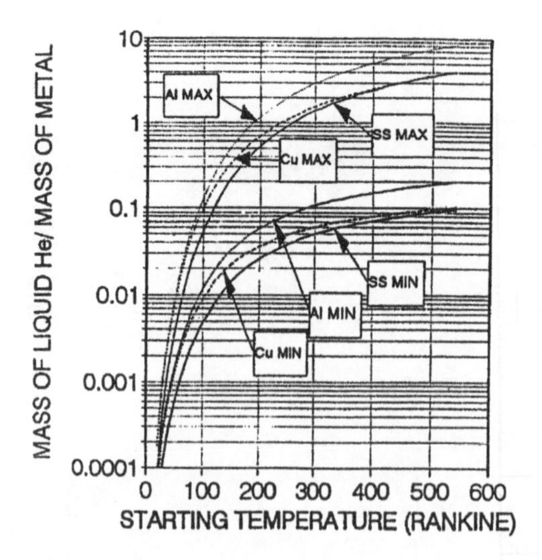


**Figure 2.25** Cooldown to liquid helium temperatures. Mass of liquid helium required per mass of metal vs. starting temperature.

for precooling from 540°R (300 K) to 139°R (77 K). Such a trade is very economical indeed.

## 7. SUMMARY

As one would expect, the basic principles of thermodynamics, heat transfer, and momentum transfer apply equally as well to cryogenic engineering as they do to all other fields of engineering. Appropriate use must be made of the sometimes unusual low-temperature properties of fluids and materials (see Chapter 3 and 4).

Simple engineering correlations, usually based on the behavior of a large number of near-room-temperature fluids, must be used with caution. Pressure drop correlations, for instance, are usually based on the flow of single-phase streams under steady-state conditions. Thus they may underestimate the pressure drop in cryogenic systems by 10–30% due to the high vaporization tendency of cryogenic fluids.

Finally, the usual modeling assumptions for normal engineering work may not apply. Cryogenic fluids are nearly always at their boiling point, are seldom adiabatic, and are often composed of two phases. Radiative heat transfer can seldom be neglected.

This is not to say that cryogenic systems cannot be rigorously modeled, for they can be, if one takes into account all of the peculiar properties of cryogenic materials and properties. Rather, correlations must be used with caution, and a return to basics is usually the safest approach.

# 3

# Cryogenic Fluids

## 1. PVT BEHAVIOR OF A PURE SUBSTANCE

The relationship of specific or molar volume to temperature and pressure for a pure substance in an equilibrium state can be represented by a surface in three dimensions, as shown in Fig. 3.1. The shaded surfaces marked S, L, and G in Fig. 3.1 represent, respectively, the solid, liquid, and gas regions of the diagram. The unshaded surfaces are the regions of co-existence of two phases in equilibrium, and there are three such regions: solid–gas (S–G), solid–liquid (S–L), and liquid–gas (L–G).

Heavy lines separate the various regions and form boundaries of the surfaces representing the individual phases. Because of the difficulty of visualizing and drawing these surfaces, it is customary to depict the data on projections such as the *PT* and *PV* planes. Figure 3.2 is an example of a *PT* plane for water; other materials have similar diagrams.

The heavy line passing through points A and B marks the intersections of the two-phase regions and is the three-phase line, along which solid, liquid, and gas phases exist in three-phase equilibrium. According to the phase rule, such three-phase, single-component systems have zero degrees of freedom; they exist for a given pure substance at only one temperature and one pressure. For this reason, the projection of this line on the *PT* plane of Fig. 3.2 (shown to the left of the main diagram) is a point. This is known as a triple point.

The triple point is merely the point of intersection of the sublimation and vaporization curves. It must be understood that only on a *P–T* diagram is the triple point represented by a point. (On a *P–V* diagram it is a line.)

The phase rule also requires that systems made up of two phases in equilibrium have just one degree of freedom, and therefore the two-phase regions must project as lines on the *PT* plane, forming a *P–T* diagram of three lines—fusion (or melting), sublimation, and vaporization—which meet at the triple point.

The *P–T* diagram is divided into areas by three lines: the solid–vapor, solid–liquid, and liquid–vapor lines. The *triple point* is a region in which liquid, solid, and vapor can exist in equilibrium simultaneously.

The *P–T* projection of Fig. 3.2 is shown to a larger scale in Fig. 3.3.

Upon projecting the surface on the *PT* plane, the whole solid–vapor region projects into the sublimation curve, the whole liquid–vapor region projects into the vaporization curve, the whole solid–liquid region projects into the fusion curve, and finally the "triple point line" projects into the triple point.

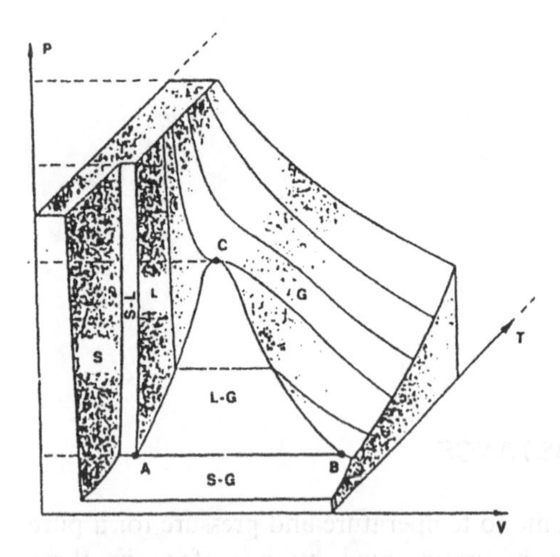


**Figure 3.1** *P–V–T* surface of a pure substance.

The solid lines of Fig. 3.3 clearly represent phase boundaries. The fusion curve (line 2–3) normally has a positive slope, but for a few substances (water is the best known) it has a negative slope. This line is believed to continue upward indefinitely.

The two curves 1–2 and 2–C represent the vapor pressures of solid and liquid, respectively.

The terminal point C is the critical point, which represents the highest pressure and highest temperature at which liquid and gas can coexist in equilibrium. The termination of the vapor pressure curve at point C means that at higher temperatures and pressures no clear distinction can be drawn between what is called liquid and what is called gas. At the critical point, there is no distinction between the liquid and vapor phases.

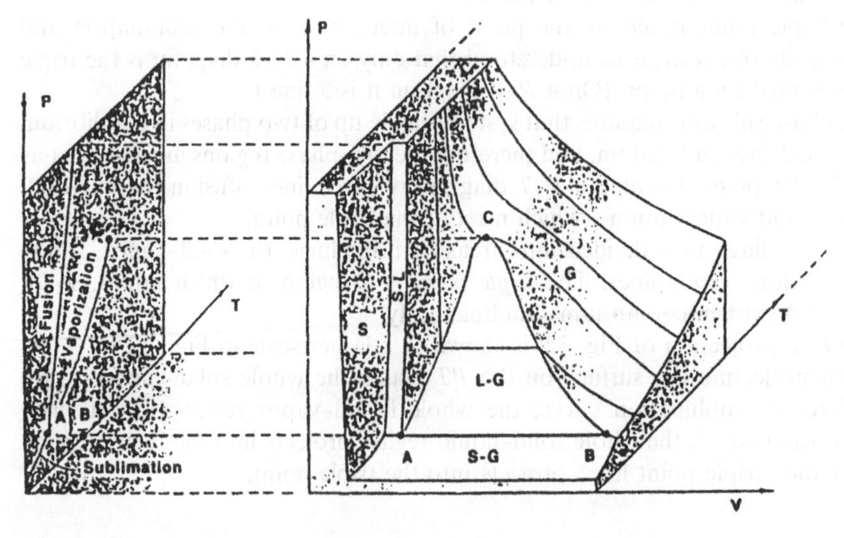


**Figure 3.2** *PT* projection of the *P–V–T* surface of a pure fluid.

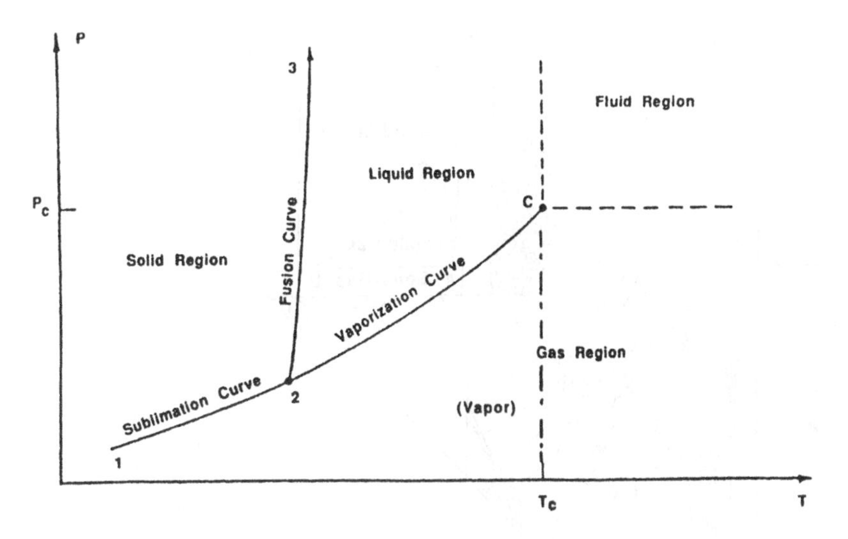


**Figure 3.3** *P–T* plot for a pure substance.

Homogeneous fluids are normally divided into two classes, liquids and gases. However, the distinction cannot always be sharply drawn, because the two phases become indistinguishable at the critical point.

Thus, there is a region extending indefinitely upward from and indefinitely to the right of the critical point that is called simply the fluid region. The fluid region, existing at higher temperatures and pressures, is marked off by dashed lines that do not represent phase changes but depend on arbitrary definitions of what constitutes liquid and gas phases.

The region designated as liquid in Fig. 3.3 lies above the vaporization curve. Thus, a liquid can always be vaporized by a sufficient reduction in pressure at constant temperature. The gas region of Fig. 3.3 lies to the right of the sublimation and vaporization curves. Thus, a gas can always be condensed by a sufficient reduction in temperature at constant pressure. Because the fluid region fits neither of these definitions, it can be called neither a gas nor a liquid.

The gas region is sometimes considered to be divided in two parts, as shown by the dash-dot vertical line of Fig. 3.3. A gas to the left of this line, which can be condensed either by compression at constant temperature or by cooling at constant pressure, is frequently called a vapor. A *vapor* is a gas existing at temperatures below $T_c$, and it may therefore be condensed either by reduction of temperature at constant $P$ or by increase of pressure at constant $T$.

The $P$–$T$ projection of Fig. 3.3 provides no information about the volumes of the systems represented. This is, however, given explicitly by the projection of Fig. 3.4, where all surfaces of the three-dimensional diagram show up as areas on the $PV$ plane.

The liquid and gas regions of the $PV$ projection are shown in more detail by Fig. 3.5.

Since the $PV$ plane is perpendicular to the $PT$ plane in the $P,V,T,$ coordinate system, the solid–vapor, liquid–vapor, and solid–liquid areas of the $PV$ plane appear as lines on the $PT$ plane or phase diagram.

The primary curves of Fig. 3.5 give the pressure–volume relations for *saturated liquid* (A–C) and for *saturated vapor* (C–B). The area lying below the curve ACB

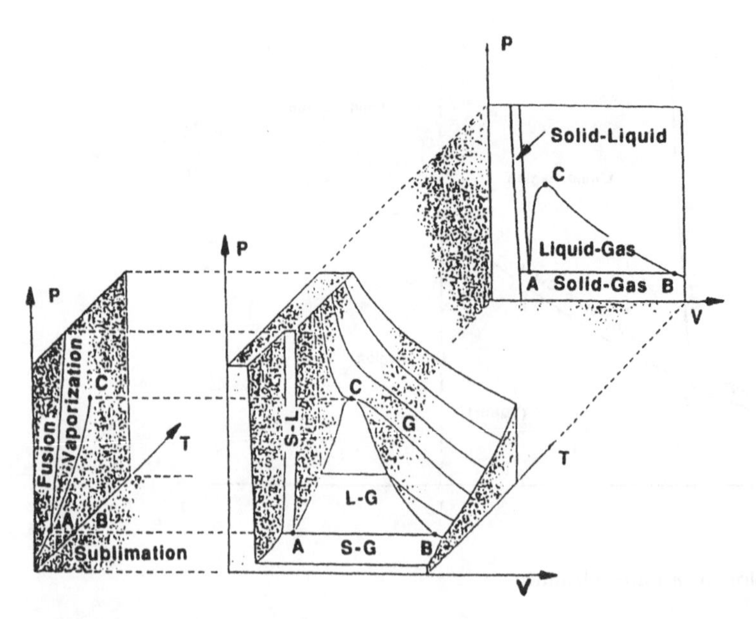


**Figure 3.4** Projection of the *P-V-T* surface of a pure fluid on *PV* and *PT* planes.

represents the two-phase region where saturated liquid and saturated vapor coexist at equilibrium. Point C is the critical point, for which the coordinates are $P_c$ and $V_c$.

A number of constant-temperature lines or *isotherms* are included on Fig. 3.5. The critical isotherm at temperature $T_c$ passes through the critical point. Isotherms

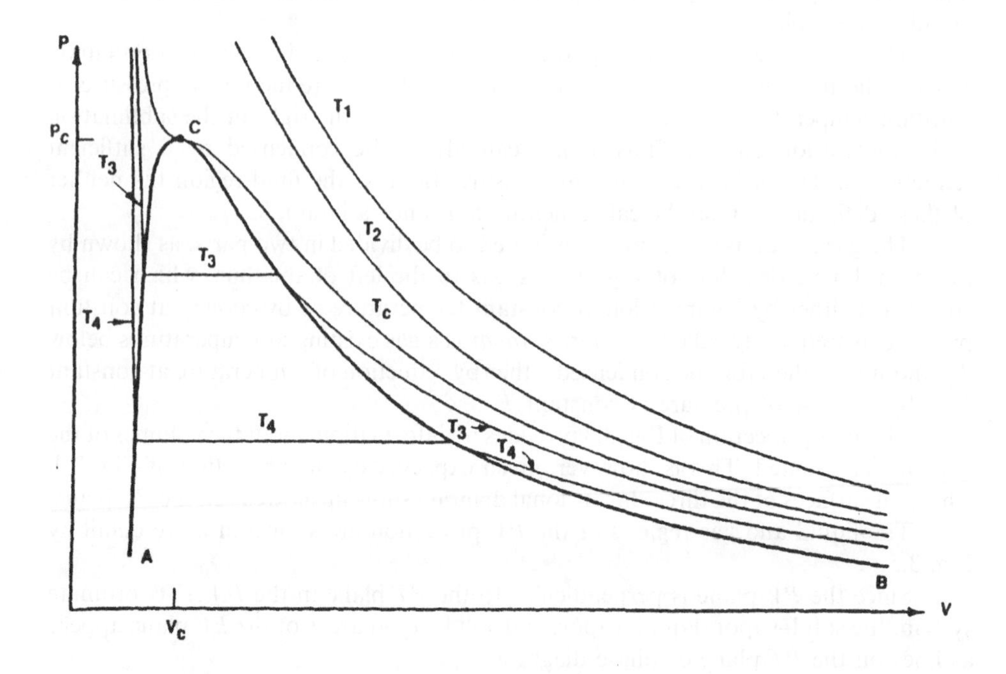


**Figure 3.5** *P–V* plot for a pure substance.

for higher temperatures $T_1$ and $T_2$ lie above the critical isotherm. Such isotherms do not cross a phase boundary and are therefore smooth.

Lower temperature isotherms $T_3$ and $T_4$ lie below the critical isotherm and are made up of three sections. The middle section traverses the two-phase region and is horizontal, because equilibrium mixtures of liquid and vapor at a fixed temperature have a fixed pressure, the *vapor pressure*, regardless of the proportions of liquid and vapor present. Points along this horizontal line represent different proportions of liquid and vapor, ranging from all liquid at the left end to all vapor at the right.

The locus of these end points is represented by the dome-shaped curve labeled ACB, the left half of which (from A to C) represents saturated liquid, and the right half (from C to B), saturated vapor.

The horizontal line segments become progressively shorter with increasing temperature, until at the critical point the isotherm exhibits a horizontal inflection. As this point is approached, the liquid and vapor phases become more and more nearly alike, eventually becoming unable to sustain a meniscus between them and ultimately becoming indistinguishable.

The sections of isotherms to the left of line AC traverse the liquid region and are very steep, because liquids are not very compressible; that is, it takes large changes in pressure to effect a small volume change. A liquid that is not saturated (not at its boiling point, along line AC) is often called subcooled liquid or compressed liquid.

The sections of isotherms to the right of line CB lie in the vapor region. Vapor in this region is called superheated vapor to distinguish it from saturated vapor as represented by the line CB.

## 2. TEMPERATURE–ENTHALPY AND TEMPERATURE–ENTROPY DIAGRAMS OF PURE SUBSTANCES

Consider 1 lb of solid nitrogen encased in a cylinder fitted with a weighted piston such that the pressure is always constant at $P_1$.

Let the solid nitrogen be heated from a very low temperature (40 or 50 K). The temperature of the solid nitrogen ice will rise as the nitrogen ice absorbs the energy until the melting point (B) is reached. This process is represented on the *T–H* and *T–S* diagrams of Figs. 3.6 and 3.7 by line AB. The energy added is represented on the *T–H* diagram by the change in enthalpy (*H*) between A and B and by the area under the line AB on the *T–S* diagram, since $Q = \int T\,ds$.

As more energy is added to the nitrogen ice, it will melt at constant pressure and form a subcooled liquid at point C. Additional heat causes the liquid to increase in temperature until it reaches the saturation temperature at point D. As vapor is formed, the amount of liquid decreases until all liquid disappears and saturated vapor fills the cylinder, corresponding to point F. At some intermediate point such as E, the mixture of liquid and vapor is referred to as wet vapor having a quality *X*.

If the heating process described in the preceding paragraphs is repeated at various pressures, a series of constant-pressure lines similar to line $P_1$ may be obtained. The locus of all B points forms the saturated solid curve, while the locus of all D and F points forms the saturated liquid and saturated vapor curves, respectively.

Figure 3.8 gives the generalized layout of a typical *T–S* chart for any fluid except helium. The solid lines are phase boundaries. The region labeled "Liquid I" is the homogeneous liquid region. This region is bounded on the left by the

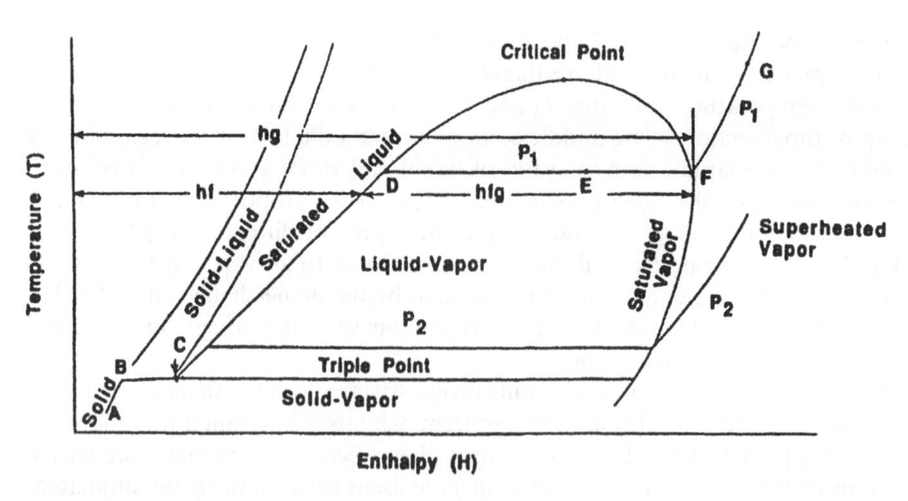


**Figure 3.6** *T–H* diagram for a pure substance.

solid/liquid melting curve, which extends indefinitely, almost vertically. Liquid I is bounded on the right by the saturated liquid/vapor line of the liquid–vapor II region. The horizontal phase boundary at the bottom is the solid locus. The three phase boundaries meet at the triple point, which is a single fixed temperature. Since this projection shows specific volume information, the triple point shows as the line that it really is, the triple point line.

Every fluid except helium has a characteristic *T–S* diagram showing the regions where the three phases exist. For many fluids, the solid region is either not of interest or not of enough interest to pay for measuring its boundaries. In this, the usual case, only *T–S* information above the triple point will be displayed. Figure 3.9 shows the complete *T–S* region for nitrogen, with all three phases. It is a bit unusual to find a thermodynamic diagram with all three phases present, but that third phase and the triple point line exist nonetheless.

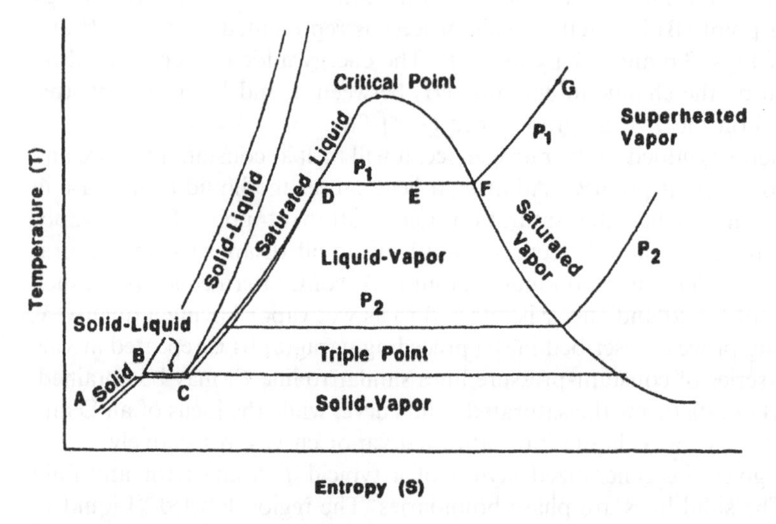


**Figure 3.7** *T–S* diagram for a pure substance.

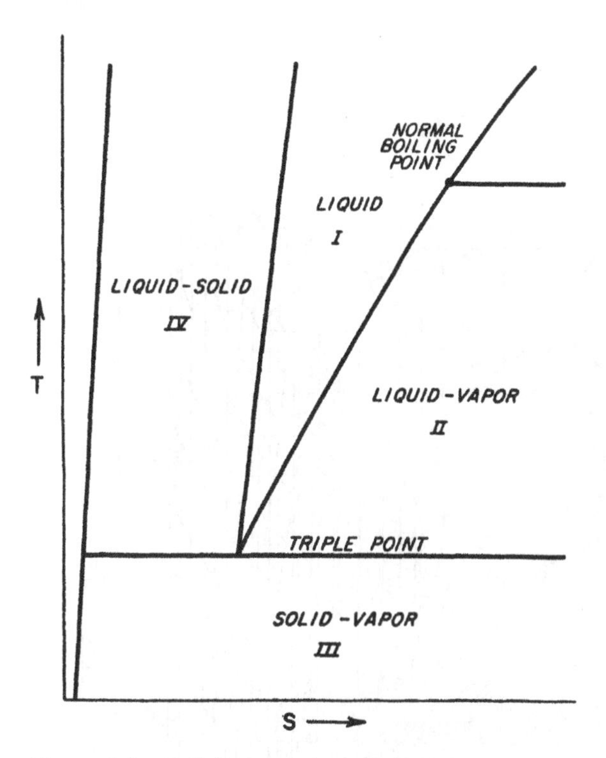


**Figure 3.8** *T–S* diagram phase boundaries near the triple point.

## 3. PROPERTIES AND USES OF CRYOGENIC FLUIDS

The cryogenic region of most interest is characterized principally by five fluids: oxygen, nitrogen, neon, hydrogen, and helium. We do, in fact, speak of the "oxygen range" or the "hydrogen range." Table 3.1 gives the normal (0.987 bar, or 1 atm) boiling temperature, the normal melting temperature, the critical temperature and pressure, and the normal latent heat of vaporization for these five cryogenic fluids and several other common cryogens (Tables 3.2–3.6).

The liquid temperature ranges of several cryogenic fluids are shown in Fig. 3.10. The bottom point (○) is the triple point; the midpoint (+) is the normal boiling point (NBP), and the topmost point (▲) is the critical point. Note that there is no convenient fluid to provide refrigeration between the critical point of neon, 44.5 K, and the freezing point of nitrogen, 63.5 K. The chart shows that, in principle, subatmospheric pressure oxygen could be used. In reality, liquid oxygen is considered to be much too hazardous for use in refrigeration. In addition, the subatmospheric pressure invites undetected leaks into the dewar, a most dangerous situation indeed.

Figure 3.10 also shows that, in general, the liquid range increases as the NBP increases. Oxygen and propylene ($C_3^=$), are anomalous in this respect. Their liquid ranges are much greater than the liquid ranges of their neighbors, no doubt because of their chemical bond hyperactivity. The freezing point of methane is tantalizingly close to the NBP of oxygen, indicating that methane might be soluble in liquid oxygen. This is indeed the case, as proved by McKinley and again recently by Flynn. In addition, Flynn has shown that such mixtures are not overly sensitive to shock and can be handled safely with about the same precautions as any other rocket fuel.

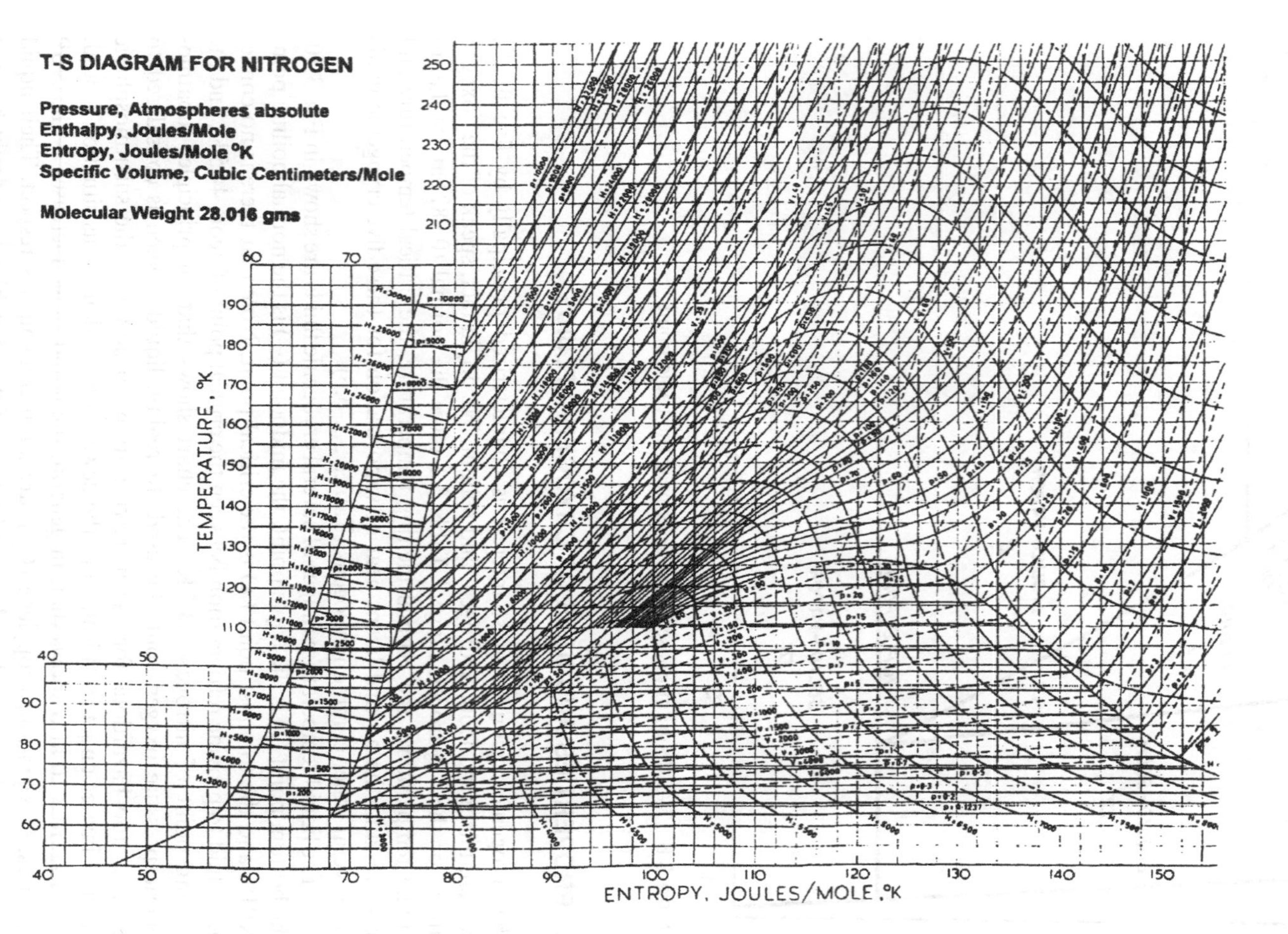


**Figure 3.9** *T–S* diagram for nitrogen near the triple point.

**Table 3.1** Selected Properties of Cryogenic Liquids at Their Normal Boiling Points

| Property | Oxygen | Nitrogen | Neon | e-Hydrogen[a] | Helium-4 | Air | Fluorine | Argon | Methane |
|---|---|---|---|---|---|---|---|---|---|
| Normal boiling point (K) | 90.18 | 77.347 | 27.09 | 20.268 | 4.224 | 78.9 | 85.24 | 87.28 | 111.7 |
| Density ($kg/m^3$) | 1141 | 808.9 | 1204 | 70.78 | 124.96 | 874 | 1506.8 | 1403 | 425.0 |
| Heat of vaporization (kJ/kg) | 212.9 | 198.3 | 86.6 | 445.6 | 20.73 | 205.1 | 166.3 | 161.6 | 511.5 |
| Specific heat (kJ/(kg K)) | 1.70 | 2.04 | 1.84 | 9.78 | 4.56 | 1.97 | 1.536 | 1.14 | 3.45 |
| Viscosity ($kg/(m\,s) \times 10^6$) | 188.0 | 157.9 | 124.0 | 13.06 | 3.57 | 168 | 244.7 | 252.1 | 118.6 |
| Thermal conductivity(mW/(m K)) | 151.4 | 139.6 | 113 | 118.5 | 27.2 | 141 | 148.0 | 123.2 | 193.1 |
| Dielectric constant | 1.4837 | 1.434 | 1.188 | 1.226 | 1.0492 | 1.445 | 1.43 | 1.52 | 1.6758 |
| Critical temperature (K) | 154.576 | 126.20 | 44.4 | 32.976 | 5.201 | 133.3 | 144.0 | 150.7 | 190.7 |
| Critical pressure (MPa) | 5.04 | 3.399 | 2.71 | 1.293 | 0.227 | 3.90 | 5.57 | 4.87 | 4.63 |
| Temperature at triple point (K) | 54.35 | 63.148 | 24.56 | 13.803 | – | – | 53.5 | 83.8 | 88.7 |
| Pressure at triple point ($MPa \times 10^3$) | 0.151 | 12.53 | 43.0 | 7.042 | – | – | 0.22 | 68.6 | 10.1 |

[a] For a discussion of forms of hydrogen, see Sec. 3.7.

**Table 3.2** Properties of Cryogenic Liquids at Their NBP, Critical and Triple Points

| Property | Helium | Hydrogen | Neon | Nitrogen | Air | Fluorine | Argon | Oxygen | Methane |
|---|---|---|---|---|---|---|---|---|---|
| Molecular weight | 4.0026 | 2.01594 | 20.1790 | 28.0134 | 28.975 | 37.9968 | 39.9480 | 31.9988 | 16.0420 |
| Normal boiling point, K | 4.2 | 20.4 | 27.1 | 77.3 | 78.9 | 84.95 | 87.3 | 90.2 | 111.7 |
| Normal boiling point, R | 7.57 | 36.7 | 48.8 | 139.2 | 141.8 | 152.9 | 157.1 | 162.4 | 201.1 |
| NBP Liquid density, $kg/m^3$ | 124.9 | 70.9 | 1204 | 810.8 | 874.0 | 1502 | 1403 | 1134 | 425.0 |
| NBP Liquid density, $lb_m/ft^3$ | 7.80 | 4.43 | 75.17 | 50.61 | 54.56 | 93.7 | 87.56 | 70.8 | 26.53 |
| NBP Vapor density, $kg/m^3$ | 16.845 | 1.3395 | 9.5924 | 4.624 | 4.487 | 5.629 | 5.7765 | 4.4668 | 1.8176 |
| NBP Vapor density, $lb_m/ft^3$ | 1.0516 | 0.08362 | 0.59883 | 0.28866 | 0.28011 | 0.3513 | 0.36062 | 0.27885 | 0.11346 |
| NBP Heat of vaporization, kJ/kg | 20.7 | 446.3 | 85.71 | 198.8 | 205.1 | 157.4 | 161.6 | 213.1 | 510.33 |
| NBP Heat of vaporization, $Btu/lb_m$ | 8.92 | 191.9 | 36.849 | 85.32 | 88.2 | 67.8 | 69.5 | 91.63 | 219.2 |
| Critical temperature, K | 5.2 | 33.2 | 44.4 | 126.12 | 133.3 | 144.31 | 150.9 | 154.6 | 190.7 |
| Critical temperature, R | 9.36 | 59.74 | 79.9 | 227.0 | 240 | 259.8 | 271.2 | 278.3 | 343.3 |
| Critical pressure, MPa | 0.23 | 1.31 | 2.71 | 3.38 | 3.90 | 5.22 | 4.87 | 5.06 | 4.63 |
| Critical pressure, Atm | 2.26 | 12.98 | 26.84 | 33.5 | 38.7 | 51.47 | 48.3 | 50.1 | 45.8 |
| Triple point, K | 2.177($\lambda$) | 13.9 | 24.56 | 63.2 | 60.6 | 53.5 | 83.8 | 54.4 | 88.7 |
| Triple point, R | 3.92($\lambda$) | 25.1 | 44.2 | 113.7 | 109.1 | 96.37 | 150.8 | 98.0 | 159.7 |
| Pressure at triple point (MPa)$10^3$ | 4.73($\lambda$) | 7.2 | 43.0 | 12.8 | 7.035 | 0.25 | 68.6 | 0.15 | 10.1 |
| Pressure at triple point, atm | 0.0469($\lambda$) | 0.0711 | 0.426 | 0.1268 | 0.0697 | 0.00249 | 0.679 | 0.00150 | 0.099 |
| Gas density NTP 1 atm, 70°F | | | | | | | | | |
| $\rho$ $lb/f^3$ | 0.010343 | 0.0052089 | 0.052148 | 0.072445 | 0.07493 | 0.09803 | 0.10355 | 0.082787 | 0.041552 |
| $\rho$ $kg/m^3$ (g/L) | 0.16569 | 0.083438 | 0.83532 | 1.1604 | 1.2000 | 1.5704 | 1.6555 | 1.3261 | 0.66559 |
| Volume ratio | | | | | | | | | |
| $\frac{\text{GAS NTP, 1 atm 70°F}}{\text{Liquid NBP}}$ | 754.2 | 848.5 | 1433.7 | 695.3 | 728.1 | 956.7 | 842.0 | 860.6 | 634.7 |
| $\frac{\text{Ft}^3\text{ GAS NTP, 1 atm 70°F}}{\text{Liter Liq. NBP}}$ | 26.63 | 29.96 | 50.9 | 24.55 | 25.71 | 33.78 | 29.70 | 30.39 | 22.41 |

**Table 3.3** Enthalpy of Phase Changes and Density of Five Common Cryogenic Fluids

| Fluid<br>Molecular weight | He<br>4.0026 | P-$H_2$<br>2.01594 | $N_2$<br>28.0134 | $O_2$<br>31.9988 | $C_1$<br>16.0420 |
|---|---|---|---|---|---|
| Triple point | ($\lambda$) | | | | |
| $T$, K | 2,176.8 | 13.800 | 63.148 | 54.359 | 90.685 |
| $T$, °F | −455.75 | −434.83 | −346.00 | −361.82 | −296.44 |
| $P$, atm | 0.04976 | 0.06950 | 0.12356 | 0.001461 | 0.11543 |
| $P$, psi | 0.73125 | 1.0214 | 1.8159 | 0.021465 | 1.6964 |
| $\rho_l$, lb/f$^3$ | 9.1239 | 4.8093 | 54.294 | 81.536 | 28.187 |
| $\rho_l$, kg/m$^3$ (g/l) | 146.15 | 77.038 | 869.70 | 1306.1 | 451.51 |
| $\rho_v$, lb/ft$^3$ | – | 0.0079284 | 0.00421 | 0.0006715 | 0.015732 |
| $\rho_v$, kg/m$^3$ (g/l) | – | 0.127 | 0.672 | 0.010756 | 0.252 |
| $\rho_s$, lb/ft$^3$ | – | 5.43 | (62.4) | 84.82 | 30.41 |
| $\rho_s$, kg/m$^3$ (g/l) | – | 86.5 | (1000) | 1358.7 | 487.1 |
| $\Delta H_{melt}$, BTU/lb | – | 25.03 | (11.0) | 5.98 | 27.04 |
| $\Delta H_{melt}$, J/gm | – | 58.23 | (25.6) | 13.908 | 62.9 |
| $\Delta H_{subl}$, BTU/lb | – | 218.14 | 103.6 | 111.5 | 260.1 |
| $\Delta H_{subl}$, J/gm | – | 507.39 | 241.0 | 259.3 | 605 |
| $\Delta H_{vap}$, BTU/lb | – | 192.97 | 92.602 | 104.35 | 233.52 |
| $\Delta H_{vap}$, J/gm | – | 448.85 | 215.39 | 242.72 | 543.16 |
| Normal boiling point | | | | | |
| $T$, K | 4.2219 | 20.277 | 77.313 | 90.191 | 111.69 |
| $T$, °F | −452.07 | −423.17 | −320.51 | −297.33 | −258.63 |
| $\rho_l$, lb/f$^3$ | 7.8004 | 4.4197 | 50.367 | 71.240 | 26.372 |
| $\rho_l$, kg/m$^3$ (g/l) | 124.95 | 70.797 | 806.79 | 1141.2 | 422.44 |
| $\rho_v$, lb/ft$^3$ | 1.0516 | 0.08362 | 0.28866 | 0.27885 | 0.11347 |
| $\rho_v$, kg/m$^3$ (g/l) | 16,845 | 1.3395 | 4.6240 | 4.4668 | 1.8176 |
| $\Delta H_v$ BTU/lb | 8.9081 | 191.51 | 85.496 | 91.597 | 219.40 |
| $\Delta H_v$ J/gm | 20.720 | 445.44 | 198.86 | 213.05 | 510.33 |
| Critical point | | | | | |
| $T$, K | 5.1953 | 32.938 | 126.19 | 154.58 | 190.55 |
| $T$, °F | −450.32 | −400.38 | −232.52 | −181.42 | −116.68 |
| $P$, atm | 2.2453 | 12.670 | 33.534 | 49.771 | 45.391 |
| $P$, psia | 32.996 | 186.20 | 492.81 | 731.43 | 667.06 |
| $\rho$, lb/f$^3$ | 4.3476 | 1.9577 | 19.547 | 27.228 | 10.154 |
| $\rho$, kg/m$^3$ (g/l) | 69.641 | 31.360 | 313.11 | 436.14 | 162.65 |
| NTP 1 atm, 70°F | | | | | |
| $\rho$, lb/f$^3$ | 0.010343 | 0.0052089 | 0.072445 | 0.082787 | 0.041552 |
| $\rho$, kg/m$^3$ (g/l) | 0.16569 | 0.083438 | 1.1604 | 1.3261 | 0.66559 |
| Volume ratio | | | | | |
| $\frac{\text{GAS NTP, 1 atm 70°F}}{\text{Liquid NBP}}$ | 754.2 | 848.5 | 695.3 | 860.6 | 634.7 |
| $\frac{\text{Ft}^3\text{ GAS NTP, 1 atm 70°F}}{\text{Liter Liq. NBP}}$ | 26.63 | 29.96 | 24.55 | 30.39 | 22.41 |

Figure 3.11 shows the vapor pressure of several cryogenic liquids. Again, the unusual liquid range of oxygen is shown.

Figure 3.12 gives the unusual correlation between the NBP and critical temperature for several cryogenic fluids. In the absence of data, a rule of thumb is that $T_c$ is about 1.6 times $T_{NBP}$ or that the NBP is about 60% of the critical temperature. Helium, of course, marches to its own drummer. It is surprising that the other quantum fluids, hydrogen and neon, are so well behaved.

**Table 3.4** Triple Point, Normal Boiling Point and Critical Point Properties of Five Common Cryogenic Fluids

| | | T | | | P | | | Density | | | | | |
|---|---|---|---|---|---|---|---|---|---|---|---|---|---|
| | | | | | | | | Liquid | | Vapor | | Solid | |
| | MW | K | °F | °C | atm | psi | mm | lb/ft$^3$ | g/L | lb/ft$^3$ | g/L | lb/ft$^3$ | g/L |
| *Triple point (λ He)* | | | | | | | | | | | | | |
| He | 4.0026 | 2.1768 | −455.75 | −270.97 | 0.04976 | 0.73125 | 37.82 | 9.1239 | 146.15 | | | – | – |
| p-$H_2$ | 2.01594 | 13.80 | −434.38 | −254.35 | 0.06950 | 1.0214 | 52.822 | 4.8093 | −77.038 | 0.0079284 | 0.127 | 5.43 | 86.5 |
| $N_2$ | 28.0134 | 63.148 | −346.00 | −210.00 | 0.12356 | 1.8159 | 93.909 | 54.294 | 869.70 | 0.00421 | 0.672 | 56.19 | 900 |
| $O_2$ | 31.9988 | 54.359 | −361.82 | −218.77 | 0.001461 | 0.021465 | 1.1101 | 81.536 | 1306.1 | 0.0006715 | 0.010756 | 84.82 | 1358.7 |
| $CH_4$ | 16.0420 | 90.685 | −296.44 | −182.47 | 0.11543 | 1.6964 | 87.729 | 28.187 | 451.51 | 0.015732 | 0.252 | 30.75 | 492 |

| | | T | | | Density | | | | Heat of vaporization | |
|---|---|---|---|---|---|---|---|---|---|---|
| | | | | | Liquid | | Vapor | | | |
| | MW | K | °F | °C | lb/ft$^3$ | g/L | lb/ft$^3$ | g/L | BTU/lb | J/g |
| *Normal boiling point* | | | | | | | | | | |
| He | 4.0026 | −4.2219 | −452.07 | −268.93 | 7.8004 | 124.95 | 1.0516 | 16.845 | 8.9081 | 20.720 |
| p-$H_2$ | 2.01594 | 20.277 | −423.17 | −252.87 | 4.4197 | −70.797 | 0.08362 | 1.3395 | 191.51 | 445.44 |
| $N_2$ | 28.0134 | 77.313 | −320.51 | −195.84 | 50.367 | 806.79 | 0.28866 | 4.6240 | 85.496 | 198.86 |
| $O_2$ | 31.9988 | 90.191 | −297.33 | −182.76 | 71.240 | 1141.2 | 0.27885 | 4.4668 | 91.597 | 213.05 |
| $CH_4$ | 16.0420 | 111.69 | −258.63 | −161.46 | 26.372 | 422.44 | 0.11347 | 1.8176 | 219.40 | 510.33 |

| | MW | T | | | P | | Density | |
|---|---|---|---|---|---|---|---|---|
| | | K | °F | °C | atm | psia | lbs/ft$^3$ | g/l |
| *Critical point* | | | | | | | | |
| He | 4.0026 | −5.1953 | −450.32 | −267.95 | 2.2453 | 32.996 | 4.3476 | 69.641 |
| p-$H_2$ | 2.01594 | 32.938 | −400.38 | −240.21 | 12.670 | 186.20 | 1.9577 | 31.360 |
| $N_2$ | 28.0134 | 126.19 | −232.52 | −146.96 | 33.534 | 492.81 | 19.547 | 313.11 |
| $O_2$ | 31.9988 | 154.58 | −181.42 | −118.57 | 49.771 | 731.43 | 27.228 | 436.14 |
| $CH_4$ | 16.0420 | 190.55 | −116.68 | −82.6 | 45.391 | 667.06 | 10.154 | 162.65 |

**Table 3.5** Weight Volume Equivalents for Air, Argon, Nitrogen, Oxygen, Helium and Hydrogen

| | | Volume of liquid at normal boiling point | | Volume of gas at 70°F and 14.696 PSIA | | |
|---|---|---|---|---|---|---|
| | Weight lb | Gallons | L | Cu ft | Cu M | L |
| Air | 1.000 | 0.1371 | 0.5190 | 13.35 | 0.3779 | 377.9 |
| | 7.294 | 1.000 | 3.785 | 97.33 | 2.756 | 2756. |
| | 1.927 | 0.2642 | 1.000 | 25.71 | 0.7281 | 728.1 |
| | 7.493 | 1.027 | 3.889 | 100.0 | 2.832 | 2832. |
| | 2.646 | 0.3628 | 1.373 | 35.31 | 1.000 | 1000. |
| Argon | 1.000 | 0.08600 | 0.3255 | −9.671 | 0.2739 | 273.9 |
| | 11.63 | 1.000 | 3.785 | 112.5 | 3.184 | 3184. |
| | 3.072 | 0.2642 | 1.000 | 29.71 | 0.8412 | 841.2 |
| | 10.34 | 0.8893 | 3.366 | 100.0 | 2.832 | 2832. |
| | 3.652 | 0.3141 | 1.189 | 35.31 | 1.000 | 1000. |
| Nitrogen | 1.000 | 0.1482 | 0.5612 | 13.80 | 0.3908 | 390.8 |
| | 6.745 | 1.000 | 3.785 | 93.11 | 2.637 | 2637. |
| | 1.782 | 0.2642 | 1.000 | 24.60 | 0.6965 | 696.5 |
| | 7.245 | 1.074 | 4.066 | 100.0 | 2.832 | 2832. |
| | 2.559 | 0.3792 | 1.436 | 35.31 | 1.000 | 1000. |
| Oxygen | 1.000 | 0.1050 | 0.3973 | 12.08 | 0.3420 | 342.0 |
| | 9.527 | 1.000 | 3.785 | 115.0 | 3.258 | 3258. |
| | 2.517 | 0.2642 | 1.000 | 30.39 | 0.8606 | 860.6 |
| | 8.281 | 0.8692 | 3.290 | 100.0 | 2.832 | 2832. |
| | 2.924 | 0.3070 | 1.162 | 35.31 | 1.000 | 1000. |
| Helium | 1.000 | 0.9593 | 3.361 | 96.71 | 2.739 | 2739. |
| | 1.042 | 1.000 | 3.785 | 100.8 | 2.855 | 2855. |
| | 0.2754 | 0.2642 | 1.000 | 26.63 | 0.7542 | 754.2 |
| | 1.034 | 0.9919 | 3.755 | 100.0 | 2.832 | 2832. |
| | 0.3652 | 0.3503 | 1.326 | 35.31 | 1.000 | 1000. |
| Hydrogen[a] | 1.000 | 1.693 | 6.409 | 192.0 | 5.436 | 5436. |
| | 0.5906 | 1.000 | 3.785 | 113.4 | 3.210 | 3210. |
| | 0.1560 | 0.2642 | 1.000 | 29.95 | 0.8481 | 848.1 |
| | 0.5209 | 0.8821 | 3.339 | 100.0 | 2.832 | 2832. |
| | 0.1840 | 0.3115 | 1.179 | 35.31 | 1.000 | 1000. |

[a] Hydrogen data are for equilibrium hydrogen (99.79% para, 0.21% ortho).

Volume Equivalents

| | | Density | | Volume ratio | |
|---|---|---|---|---|---|
| | | 1 atm, 70°F | | | |
| | MW | lb/ft$^3$ | g/L | Gas, 1 atm 70°F Liquid NBP | Ft$^3$ Gas 1 atm, 70°F Liter liquid NBP |
| He | 4.0026 | 0.010343 | 0.16569 | 754.2 | 26.63 |
| p-$H_2$ | 2.01594 | 0.0052089 | 0.083438 | 848.5 | 29.96 |
| $O_2$ | 28.0134 | 0.07244 | 1.1604 | 695.3 | 24.55 |
| N | 31.9988 | 0.082787 | 1.3261 | 860.6 | 30.39 |
| $CH_4$ | 16.0420 | 0.041552 | 0.66559 | 634.7 | 22.41 |

**Table 3.6** Weight Volume Equivalents of Commercially Important Cryogenic Fluids (courtesy of PHPK Technologies)

| Weight of liquid or gas | | Volume of liquid or normal boiling point | | | | Volume of gas at 70°F | |
|---|---|---|---|---|---|---|---|
| lb | kg | cu ft | l | quarts | gallons | cu ft | cu m |
| Oxygen | | | | | | | |
| 1.000 | 0.454 | 0.0140 | 0.396 | 0.419 | 0.105 | 12.08 | 0.342 |
| 2.205 | 1.000 | 0.0308 | 0.873 | 0.923 | 0.231 | 26.63 | 0.754 |
| 71.5 | 32.43 | 1.000 | 28.32 | 29.92 | 7.481 | 863.5 | 24.45 |
| 2.524 | 1.145 | 0.0353 | 1.000 | 1.056 | 0.264 | 30.48 | 0.863 |
| 2.388 | 1.083 | 0.0334 | 0.946 | 1.000 | 0.250 | 28.84 | 0.817 |
| 9.560 | 4.336 | 0.134 | 3.786 | 4.000 | 1.000 | 115.5 | 3.271 |
| 8.28 | 3.756 | 0.116 | 3.285 | 3.472 | 0.868 | 100.0 | 2.832 |
| 2.92 | 1.324 | 0.0408 | 1.155 | 1.221 | 0.305 | 35.31 | 1.000 |
| Air | | | | | | | |
| 1.000 | 0.454 | 0.0183 | 0.518 | 0.548 | 0.137 | 13.35 | 0.379 |
| 2.205 | 1.000 | 0.0404 | 1.144 | 1.209 | 0.302 | 29.44 | 0.834 |
| 54.57 | 24.77 | 1.000 | 28.32 | 29.92 | 7.481 | 728.5 | 20.63 |
| 1.927 | 0.875 | 0.0353 | 1.000 | 1.056 | 0.264 | 25.73 | 0.729 |
| 1.824 | 0.828 | 0.0334 | 0.946 | 1.000 | 0.250 | 24.35 | 0.690 |
| 7.294 | 3.311 | 0.134 | 3.786 | 4.000 | 1.000 | 97.37 | 2.758 |
| 7.493 | 3.402 | 0.137 | 3.880 | 4.099 | 1.025 | 100.0 | 2.832 |
| 2.646 | 1.201 | 0.0485 | 1.374 | 1.449 | 0.363 | 35.31 | 1.000 |
| Argon | | | | | | | |
| 1.000 | 0.454 | 0.0114 | 0.323 | 0.341 | 0.0853 | 9.681 | 0.274 |
| 2.205 | 1.000 | 0.0252 | 0.714 | 0.755 | 0.189 | 21.35 | 0.605 |
| 87.4 | 39.6 | 1.000 | 28.32 | 29.92 | 7.481 | 846.1 | 23.96 |
| 3.085 | 1.399 | 0.0353 | 1.000 | 1.056 | 0.264 | 29.86 | 0.846 |
| 2.919 | 1.324 | 0.0334 | 0.946 | 1.000 | 0.250 | 28.26 | 0.800 |
| 11.71 | 5.310 | 0.134 | 3.786 | 4.000 | 1.000 | 113.4 | 3.212 |
| 10.33 | 4.686 | 0.118 | 3.342 | 3.533 | 0.883 | 100.0 | 2.832 |
| 3.648 | 1.655 | 0.0417 | 1.181 | 1.248 | 0.312 | 35.31 | 1.000 |
| Nitrogen | | | | | | | |
| 1.000 | 0.454 | 0.0198 | 0.561 | 0.593 | 0.148 | 13.79 | 0.391 |
| 2.205 | 1.000 | 0.0437 | 1.239 | 1.310 | 0.327 | 30.41 | 0.861 |
| 50.4 | 22.86 | 1.000 | 28.32 | 29.92 | 7.481 | 695.2 | 19.69 |
| 1.78 | 0.807 | 0.0353 | 1.000 | 1.056 | 0.264 | 24.55 | 0.695 |
| 1.68 | 0.762 | 0.0334 | 0.946 | 1.000 | 0.250 | 23.17 | 0.656 |
| 6.75 | 3.062 | 0.134 | 3.786 | 4.000 | 1.000 | 93.10 | 2.637 |
| 7.25 | 3.289 | 0.144 | 4.078 | 4.310 | 1.077 | 100.0 | 2.832 |
| 2.56 | 1.161 | 0.0508 | 1.439 | 1.521 | 0.380 | 35.31 | 1.000 |
| Helium | | | | | | | |
| 1.000 | 0.454 | 0.128 | 3.625 | 3.832 | 0.958 | 97.08 | 2.749 |
| 2.205 | 1.000 | 0.283 | 8.00 | 8.468 | 2.117 | 214.1 | 6.063 |
| 7.80 | 3.541 | 1.000 | 28.32 | 29.92 | 7.481 | 757.3 | 21.45 |
| 0.2753 | 0.125 | 0.0353 | 1.000 | 1.056 | 0.264 | 26.73 | 0.757 |
| 0.2605 | 0.118 | 0.0334 | 0.946 | 1.000 | 0.250 | 25.29 | 0.716 |
| 1.045 | 0.474 | 0.134 | 3.786 | 4.000 | 1.000 | 101.5 | 2.875 |
| 1.033 | 0.469 | 0.132 | 3.738 | 3.948 | 0.987 | 100.0 | 2.832 |
| 0.3637 | 0.165 | 0.0466 | 1.320 | 1.396 | 0.349 | 35.31 | 1.000 |

*(Continued)*

**Table 3.6** (*Continued*)

| Weight of liquid or gas | | Volume of liquid or normal boiling point | | | | Volume of gas at 70°F | |
|---|---|---|---|---|---|---|---|
| lb | kg | cu ft | l | quarts | gallons | cu ft | cu m |
| Hydrogen | | | | | | | |
| 1.000 | 0.454 | 0.226 | 6.400 | 6.764 | 1.691 | 191.6 | 5.426 |
| 2.205 | 1.000 | 0.498 | 14.10 | 14.90 | 3.726 | 422.4 | 11.96 |
| 4.43 | 2.011 | 1.000 | 28.32 | 29.92 | 7.481 | 848.7 | 24.04 |
| 0.156 | 0.0708 | 0.0353 | 1.000 | 1.056 | 0.264 | 29.89 | 0.846 |
| 0.148 | 0.0672 | 0.0334 | 0.946 | 1.000 | 0.250 | 28.35 | 0.803 |
| 0.594 | 0.270 | 0.134 | 3.786 | 4.000 | 1.000 | 113.8 | 3.223 |
| 0.522 | 0.237 | 0.118 | 3.342 | 3.532 | 0.883 | 100.0 | 2.832 |
| 0.184 | 0.0835 | 0.0416 | 1.178 | 1.244 | 0.311 | 35.31 | 1.000 |

It is also possible to estimate the Joule–Thomson inversion temperature, as shown in Fig. 3.13. Ignore the quantum fluids for a moment, and we see that the J–T inversion temperature is about eight times the NBP. Including helium, hydrogen, and neon, the approximation is about 9 or 10 times the NBP.

Some of the important characteristics of the most widely used cryogenic liquids are discussed more specifically in the following sections. For a comprehensive listing of fluid properties, see Johnson (1960).

## 3.1. Oxygen

Oxygen was the base used for chemical atomic weights, being assigned the atomic weight 16.000, until 1961, when the International Union of Pure and Applied Chemistry (IUPAC) adopted carbon-12 as the new basis. Oxygen has eight isotopes. Naturally occurring oxygen consists of three stable isotopes of atomic mass numbers 16, 17, and 18, having abundances in the proportion 10,000:4:20. Oxygen condenses into a light blue liquid whose density, 1134.2 kg/m$^3$ (70.8 lb$_m$/ft$^3$) at its boiling temperature, is slightly greater than that of water at room temperature. At 0.987 bar (1 atm) pressure, liquid oxygen boils at 90.2 K (162.4°R) and freezes at 54.4 K (97.9°R).

In the liquid state, there is thought to be some weak transient association of oxygen molecules forming $O_4$, which is said to be responsible for the blue color of both liquid and solid oxygen. Ozone ($O_3$), a highly active allotropic from of oxygen, is formed by the action of an electrical discharge or ultraviolet light on oxygen. Ozone's presence in the atmosphere (amounting to the equivalent of a layer 3 mm thick at ordinary pressures and temperatures) is of vital importance in preventing harmful ultraviolet rays of the sun from reaching the earth's surface.

Table 3.7 lists the temperature, pressure, and specific volume of each of the various phases of oxygen at several important points. Tables 3.8–3.10 give properties of oxygen in metric and English units at three of these fixed points.

Figures 3.14 and 3.15 and *T–S* diagrams for oxygen. A Mollier chart (pressure vs. heat content) is given in Fig. 3.16, and the Joule–Thomson inversion chart in Fig. 3.17. Table 3.11 presents data for the J–T curve.

### 3.1.1. *Properties of Oxygen*

**a. Magnetism.** Liquid oxygen is slightly magnetic, in contrast to the other cryogenic fluids, which are nonmagnetic. Its paramagnetic susceptibility is 1.003 at

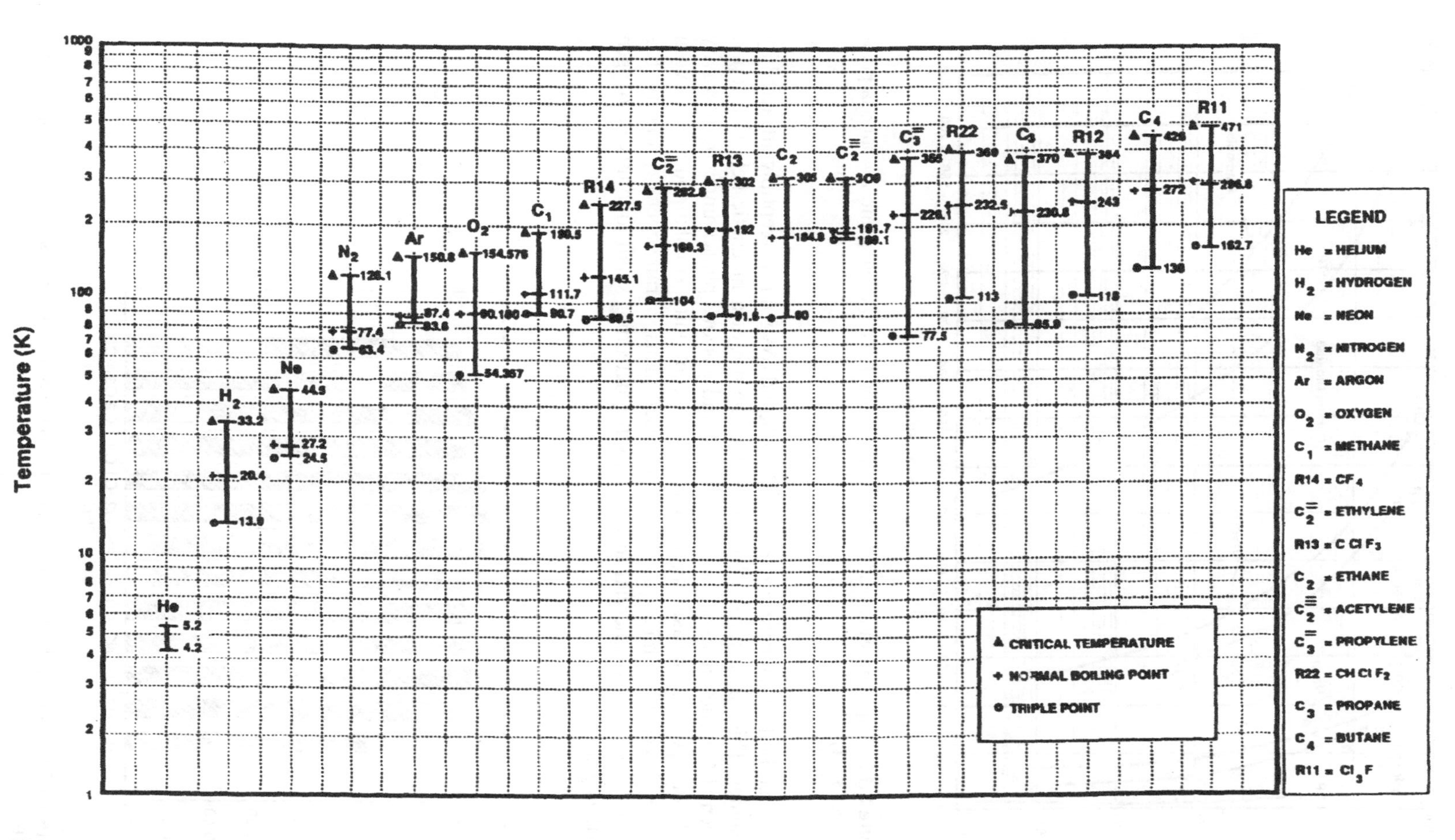


**Figure 3.10** Liquid temperature range of different cryogens.

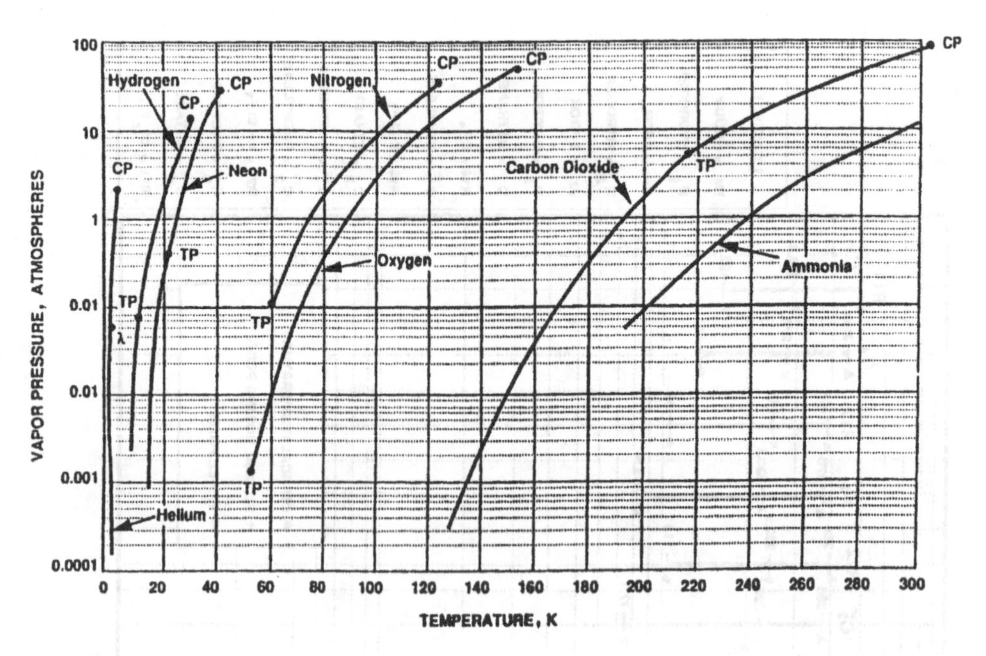


**Figure 3.11** Vapor pressure curves. CP = critical point; TP = triple point; $\lambda$ = lambda point (see Sec. 3.8.1).

its normal boiling temperature, so a bath of liquid oxygen is perceptibly attracted by a magnet. This characteristic has prompted the use of a magnetic field in a liquid oxygen (LOX) dewar to separate the liquid and gaseous phases under zero gravity conditions. Also, by measuring the magnetic susceptibility, small amounts of oxygen can be detected in mixtures of other gases. This paramagnetic property has been exploited in commercial instruments for detecting small amounts of oxygen in other gases.

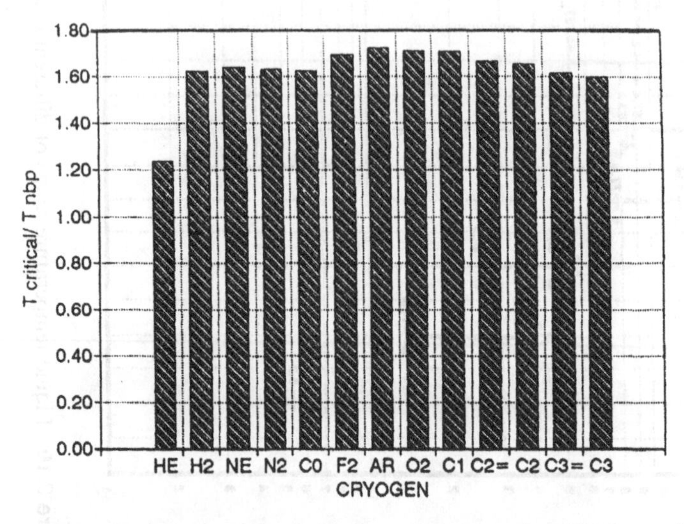


**Figure 3.12** Ratio of the critical temperature to the NBP for various cryogens. For $C$ terms, see Fig. 3.11.

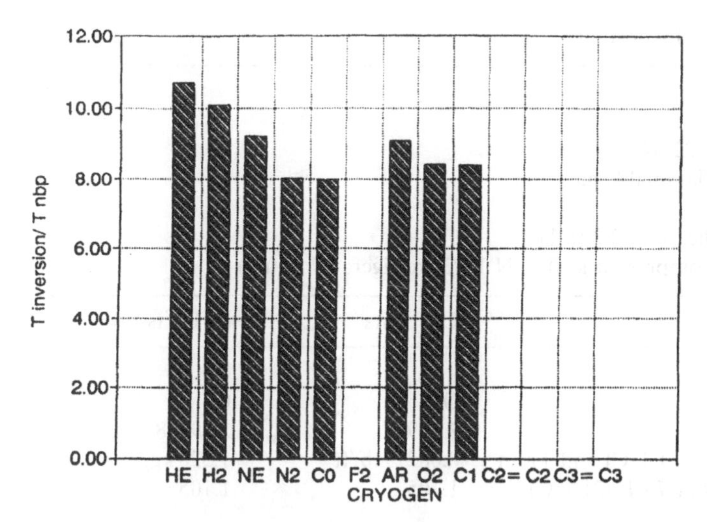


**Figure 3.13** Ratio of the inversion temperature to the NBP.

**b. Chemical Reactivity.** Both gaseous and liquid oxygen are chemically reactive, especially with hydrocarbon materials. Because of its chemical activity, oxygen presents a serious safety problem. Several explosions have resulted from the combination of oxygen and hydrocarbon lubricants. Although at 90 K most chemical reaction rates are negligibly slow, a small amount of energy added under the right conditions can cause an explosion in a system containing liquid oxygen and a substance with which it combines chemically.

**Table 3.7** Fixed Points for Oxygen

*Critical point*
$T = 154.576 \pm 0.010$ K $T = 154.581 \pm 278.237$°R
$P = 49.76 \pm 0.02$ atm (731.4 psia; 5.043 MPa)
$V = 73.37 \pm 0.10\ \mathrm{cm^3/mol}$ ($0.03673\ \mathrm{ft^3/lb_m}$; $436.2\ \mathrm{kg/m^3}$)
*Normal boiling point*
$T = 90.180 \pm 0.01$ K $T = 90.188 \pm 162.324$°R
$P = 1$ atm (14.696 psia; 101.3 kPa)
$V$ (liquid) $= 28.05 \pm 0.028\ \mathrm{cm^3/mol}$ ($0.01404\ \mathrm{ft^3/lb_m}$; $1141\ \mathrm{kg/m^3}$)
$V$ (vapor) $= 7150 \pm 7\ \mathrm{cm^3/mol}$ ($3.579\ \mathrm{ft^3/lb_m}$; $4.48\ \mathrm{kg/m^3}$)
*Normal melting point*
$T = 54.362 \pm 0.001$ K $T = 54.372 \pm 97.852$°R
$P = 1$ atm (14.696 psia; 101.3 kPa)
$V$ (liquid) $= 24.49 \pm 0.024\ \mathrm{cm^3/mol}$ ($0.01226\ \mathrm{ft^3/lb_m}$; $1307\ \mathrm{kg/m^3}$)
*Triple point*
$T = 54.351 \pm 0.001$ K $T = 54.361 \pm 97.832$°R
$P = (1.50 \pm 0.06) \times 10^{-3}$ atm (0.0220 psia; 151.7 Pa)
$V$ (solid) $= 23.55 \pm 0.03\ \mathrm{cm^3/mol}$ ($0.01179\ \mathrm{ft^3/lb_m}$; $1359\ \mathrm{kg/m^3}$)
$V$ (liquid) $= 24.49 \pm 0.024\ \mathrm{cm^3/mol}$ ($0.01226\ \mathrm{ft^3/lb_m}$; $1307\ \mathrm{kg/m^3}$)
$V$ (vapor) $= 2.97 \pm 0.01 \times 10^6\ \mathrm{cm^3/mol}$ ($1487\ \mathrm{ft^3/lb_m}$; $10.8\ \mathrm{g/m^3}$)
*Solid-solid transitions.* There are two such transitions at atmospheric pressure: at 43.8 K and at 23.9 K

*Source*: H. M. Roder and L. A. Weber (1972).

**Table 3.8** Oxygen Physical Constants

*Oxygen*
Chemical Symbol: $O_2$
CAS Registry Number: 7782-44-7
DOT Classification: Nonflammable gas
DOT Label: Oxidizer
Transport Canada Classification: 2.2 (5.1)
UN Number: UN 1072 (compressed gas); UN 1073 (refrigerated liquid)

| | US Units | SI Units |
|---|---|---|
| *Physical constants* | | |
| Chemical formula | $O_2$ | $O_2$ |
| Molecular weight | 31.9988 | 31.9988 |
| Density of the gas at 70°F (21.1°C) and 1 atm | 0.08279 lb/ft$^3$ | 1.326 kg/m$^3$ |
| Specific gravity of the gas at 70°F (21.1°C) and 1 atm (air = 1) | 1.105 | 1.105 |
| Specific volume of the gas at 70°F (21.1°C) and 1 atm | 12.08 ft$^3$/lb | 0.7541 m$^3$/kg |
| Boiling point at 1 atm | −297.33°F | -182.96°C |
| Freezing point at 1 atm | −361.80°F | −218.78°C |
| Critical temperature | −181.43°F | −118.57°C |
| Critical pressure | 731.4 psia | 5043 kPa abs |
| Critical density | 27.22 lb/ft$^3$ | 436.1 kg/m$^3$ |
| Triple point | −361.82°F at 0.02147 psia | −218.79°C at 0.1480 kPa abs |
| Latent heat of vaporization at boiling point | 91.7 Btu/lb | 213 kJ/kg |
| Latent heat of fusion at −361.1°F (−218.4°C) melting point | 5.959 Btu/lb | 13.86 kJ/kg |
| Specific heat of the gas at 70°F (21.1°C) and 1 atm | | |
| $C_P$ | 0.2197 Btu/(lb)(°F) | 0.9191 kJ/(kg)(°C) |
| $C_V$ | 0.1572 Btu/(lb)(°F) | 0.6578 kJ/(kg)(°C) |
| Ratio of specific heats ($C_P/C_v$) | 1.40 | 1.40 |
| Solubility in water, vol/vol at 32°F (0°C) | 0.0491 | 0.0491 |
| Density of the liquid at boiling point | 9.52 lb/gal or 71.23 lb/ft$^3$ | 1141 kg/m$^3$ |
| Density of the gas at boiling point | 0.2799 lb/ft$^3$ | 4.483 kg/m$^3$ |
| Gas/liquid ratio (gas at 70°F (21.1°C) and 1 atm, liquid at boiling point, vol/vol) | 860.5 | 860.5 |

To ensure against such unwanted chemical reactions, systems using liquid oxygen must be maintained scrupulously clean of any foreign matter. The phrase "LOX clean" in the space industry has come to be associated with a set of elaborate cleaning and inspection specifications representing a near ultimate in large-scale equipment cleanliness. (See Chapter 10.) Ordinary hydrocarbon lubricants are dangerous to use in oxygen compressors and vacuum pumps exhausting oxygen. Also valves, fittings, and lines used with oil-pumped gases should never be used with oxygen. Serious explosions have resulted from the combination of oxygen with a lubricant. In fact, combustible materials soaked in liquid oxygen are used as inexpensive commercial explosives.

**Table 3.9** Fixed Point Properties of Oxygen (Metric Units)

| Property | Triple point | | | Normal boiling point | | | Standard conditions | |
|---|---|---|---|---|---|---|---|---|
| | Solid | Liquid | Vapor | Liquid | Vapor | Critical point | STP(0°C) | NTP(20°C) |
| Temperature (K) | 54.351 | 54.351 | 54.351 | 90.180 | 90.180 | 154.576 | 273.15 | 293.15 |
| Pressure (mmHg) | 1.138 | 1.138 | 1.138 | 760 | 760 | 37,823 | 760 | 760 |
| Density $(mol/cm^3) \times 10^3$ | 42.46 | 40.83 | 0.000336 | 35.65 | 0.1399 | 13.63 | 0.04466 | 0.04160 |
| Specific volume $(cm^3/mol) \times 10^{-3}$ | 0.02355 | 0.02449 | 2975 | 0.028047 | 7.1501 | 0.07337 | 22.392 | 24.038 |
| Compressibility factor, $Z = PV/RT$ | — | 0.0000082 | 0.9986 | 0.00379 | 0.9662 | 0.2879 | 0.9990 | 0.9992 |
| Heats of fusion and vaporization (J/mol) | 444.8 | — | 7761.4 | 6812.3 | — | 0 | — | — |
| Specific heat (J/(mol K)) | | | | | | | | |
| $C_g$, at saturation | 46.07 | 53.313 | −108.7 | 54.14 | −53.2 | (very large) | — | — |
| $C_P$, at constant pressure | — | 53.27 | 29.13 | 54.28 | 30.77 | (very large) | 29.33 | 29.40 |
| $C_V$ at constant volume | — | 35.65 | 20.81 | 29.64 | 21.28 | (38.7) | 20.96 | 21.04 |
| Specific heat ratio, $\gamma = C_P/C_V$ | — | 1.494 | 1.400 | 1.832 | 1.446 | (large) | 1.40 | 1.40 |
| Enthalpy (J/mol) | −6634.46 | −6189.6 | 1571.8 | −4270.3 | 2542.0 | 1032.2 | 7937.8 | 8525.1 |
| Internal energy (J/mol) | −6634.4 | −6189.6 | 1120.0 | −4273.1 | 1817.5 | 662.3 | 5668.9 | 6089.5 |
| Entropy (J/(mol K)) | 58.92 | 67.11 | 209.54 | 94.17 | 169.68 | 134.42 | 202.4 | 204.5 |
| Velocity of sound (m/s) | — | 1159 | 141 | 903 | 178 | 164 | 315 | 326 |
| Viscosity | | | | | | | | |
| $(N\,sec/m^3) \times 10^3$ | — | 0.6194 | 0.003914 | 0.1958 | 0.00685 | (0.031) | 0.01924 | 0.02036 |
| Centipoises | — | 0.6194 | 0.003914 | 0.1958 | 0.00685 | (0.031) | 0.01924 | 0.02036 |
| Thermal conductivity, k(mW/(cm K)) | — | 1.929 | 0.04826 | 1.515 | 0.08544 | —[b] | 0.2428 | 0.2575 |
| Prandtl number, $N_{pr} = uC_{p/k}$ | — | 5.344 | 0.7392 | 2.193 | 0.7714 | — | 0.7259 | 0.7265 |
| Dielectric constant, $\varepsilon$ | (1.614) | 1.5687 | 1.000004 | 1.4870 | 1.00166 | 1.17082 | 1.00053 | 1.00049 |
| Index of refraction, $n = \sqrt{\varepsilon^a}$ | (1.271) | 1.2525 | 1.000002 | 1.219 | 1.00083 | 1.0820 | 1.00027 | 1.00025 |
| Surface tension $(N/m) \times 10^3$ | — | 22.65 | — | 13.20 | — | 0 | — | — |
| Equiv. vol./vol. liquid at NBP | 0.8397 | 0.8732 | 106,068 | 1 | 254.9 | 2.616 | 798.4 | 857.1 |

[a] Long wavelengths.
[b] Anomalously large.
Gas constant: $R = 62{,}365.4\,cm^3\,mmHg/(mol\,K)$.
Values in parentheses are estimates based on the NBS-1955 temperature scale using 90.180 K as a fixed point for the normal boiling temperature of oxygen. (The IPTS-1968 temperature scale uses 90.188 K as the normal boiling temperature of oxygen, but the reported data used in these tables have not yet been converted to the new temperature scale.) 273.15 K = 0°C = 32°F = 491.67°R base point (zero values) for enthalpy, internal energy, and entropy are 0 K for the ideal gas at 1 atm pressure.

**Table 3.10** Fixed Point Properties of Oxygen (English Units)

| Property | Triple point | | | Normal boiling point | | Critical point | Standard conditions | |
|---|---|---|---|---|---|---|---|---|
| | Solid | Liquid | Vapor | Liquid | Vapor | | STP(32°F) | NTP(68°F) |
| Temperature (°F) | −361.84 | −361.84 | −361.84 | −297.35 | −297.35 | −181.43 | 32.0 | 68.0 |
| Pressure (psia) | 0.0220 | 0.0220 | 0.0220 | 14.696 | 14.696 | 731.4 | 14.696 | 14.696 |
| Density (lb/ft$^2$) | 84.82 | 81.57 | 0.0006715 | 71.23 | 0.2794 | 27.23 | 0.0892 | 0.0831 |
| Specific volume (ft$^3$/lb) | 0.01179 | 0.01226 | 1489.2 | 0.01404 | 3.5793 | 0.03673 | 11.21 | 12.03 |
| Compressibility factor, $Z = PV/RT$ | — | 0.0000082 | 0.9985 | 0.00379 | 0.9662 | 0.2879 | 0.9990 | 0.9992 |
| Heats of fusion and vaporization (Btu/lb) | 5.976 | — | 104.348 | 91.588 | 91.588 | 0 | — | — |
| Specific heat (Btu/lb °R) | | | | | | | | |
| $C_s$, at saturation | 0.345 | 0.398 | −0.812 | 0.404 | −0.397 | (very large) | — | — |
| $C_P$, at constant pressure | — | 0.398 | 0.218 | 0.405 | 0.230 | (very large) | 0.219 | 0.220 |
| $C_V$, at constant volume | — | 0.266 | 0.155 | 0.221 | 0.159 | (0.289) | 0.157 | 0.157 |
| Specific heat ratio, $\gamma = C_P/C_V$ | — | 1.496 | 1.406 | 1.833 | 1.447 | (large) | 1.40 | 1.40 |
| Enthalpy (Btu/lb) | −89.192 | −83.216 | 21.132 | −57.412 | 34.176 | 13.88 | 106.72 | 114.62 |
| Internal energy (Btu/lb) | −89.192 | −83.216 | 15.057 | −57.450 | 24.435 | 8.90 | 76.22 | 81.875 |
| Entropy (Btu/lb °R) | 0.4401 | 0.50122 | 1.5651 | 0.70339 | 1.2674 | 1.004 | 1.5123 | 1.5278 |
| Velocity of sound (ft/sec) | — | 3804 | 461 | 2963 | 583 | 537 | 1033 | 1070 |
| Visosity, $\mu$ | | | | | | | | |
| (lb/sec-ft) × 10$^5$ | — | 41.62 | 0.263 | 13.16 | 0.460 | (2.1) | 1.293 | 1.368 |
| (centipoise) | — | 0.6194 | 0.003914 | 0.1958 | 0.00685 | (0.031) | 0.01924 | 0.02036 |
| Thermal conductivity (Btu/hr ft °R), $k$ | — | 0.11156 | 0.00279 | 0.08758 | 0.00494 | —[b] | 0.01404 | 0.01489 |
| Prandtl number, $N_{pr} = \mu C_P/k$ | — | 5.3437 | 0.7392 | 2.1929 | 0.7714 | — | 0.7259 | 0.7265 |
| Dielectric constant, $\varepsilon$ | (1.614) | 1.5687 | 1.000004 | 1.4870 | 1.00166 | 1.17082 | 1.00053 | 1.00049 |
| Index of refraction, $n = n = \sqrt{\varepsilon^a}$ | (1.1271) | 1.2525 | 1.000002 | 1.219 | 1.00083 | 1.0820 | 1.00027 | 1.00025 |
| Surface tension (lb/ft) × 10$^3$ | — | 1.552 | — | 0.9046 | — | 0 | — | — |
| Equiv. vol./vol. liquid at NBP | 0.8397 | 0.8732 | 106,068 | 1 | 254.9 | 2.616 | 798.4 | 857.1 |

[a] Long wavelengths.
[b] Anomalously large.
Gas constant: $R = 0.335385\,ft^3\,psi/(lb\,°R)$.
Values in parentheses are estimates based on the NBS-1955 temperature scale using 90.180 K as a fixed point for the normal boiling temperature of oxygen. (The IPTS-1968 temperature scale used 90.188 K as the normal boiling temperature of oxygen, but the reported data used in these tables have not yet been converted to the new temperature scale.) 273.15 K = 0°C = 32°F = 491.67°R. Calculated from property values given in this table. Base point (zero values) for enthalpy, internal energy, and entropy are 0 K for the ideal gas at 1 atm pressure.

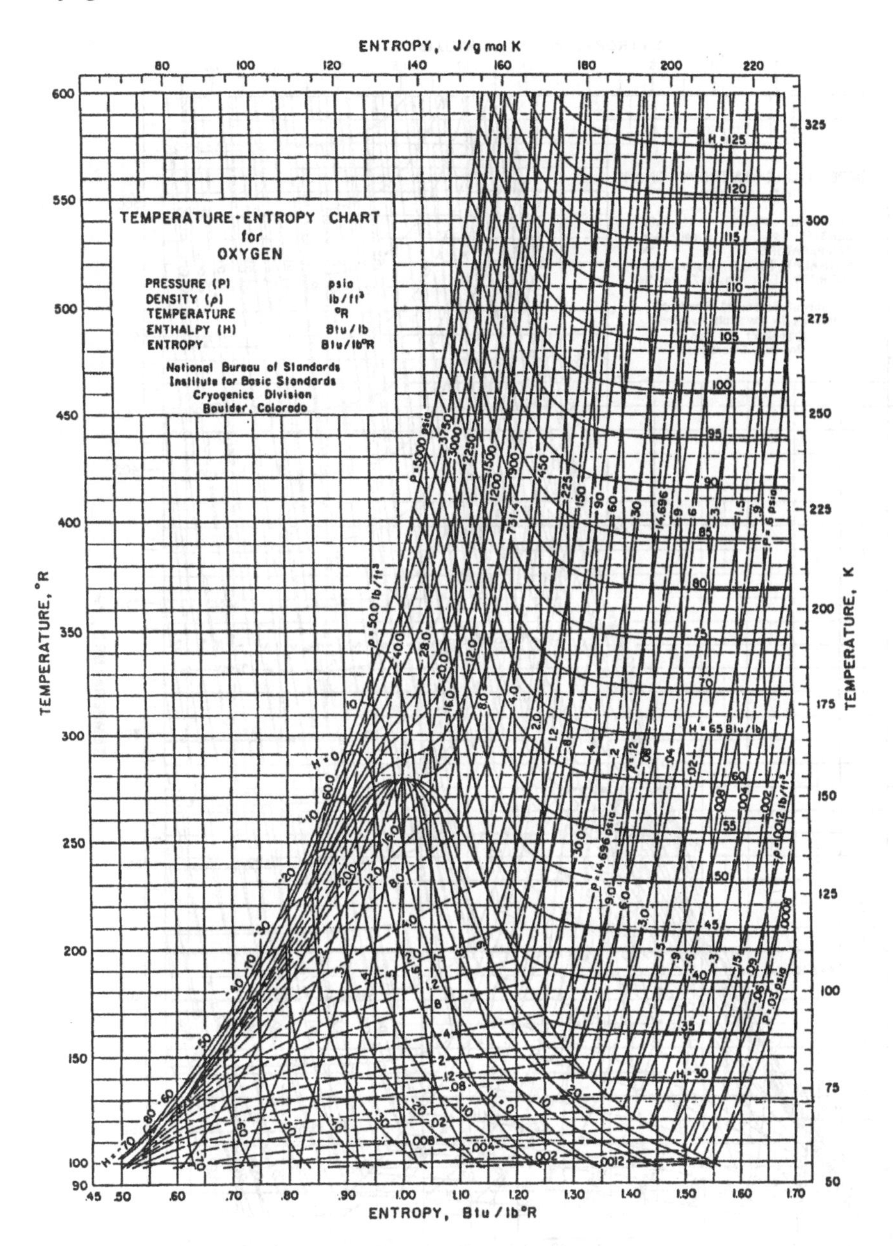


**Figure 3.14** Temperature–entropy chart for oxygen, English units.

Liquid oxygen equipment must also be designed of construction materials incapable of initiating or sustaining a reaction. Only a few polymeric materials (plastics), for example, can be used in the design of such equipment, as most will react violently with oxygen under mechanical impact. Also, reactive metals such as titanium or aluminum must be used cautiously, because they are potentially hazardous. Once the reaction is started, for instance, an aluminum pipe containing oxygen burns rapidly and intensely. With proper design and care, however, liquid oxygen systems can be operated safely. For a complete discussion of this important topic, please see the comprehensive work of Schmidt and Forney (1975).

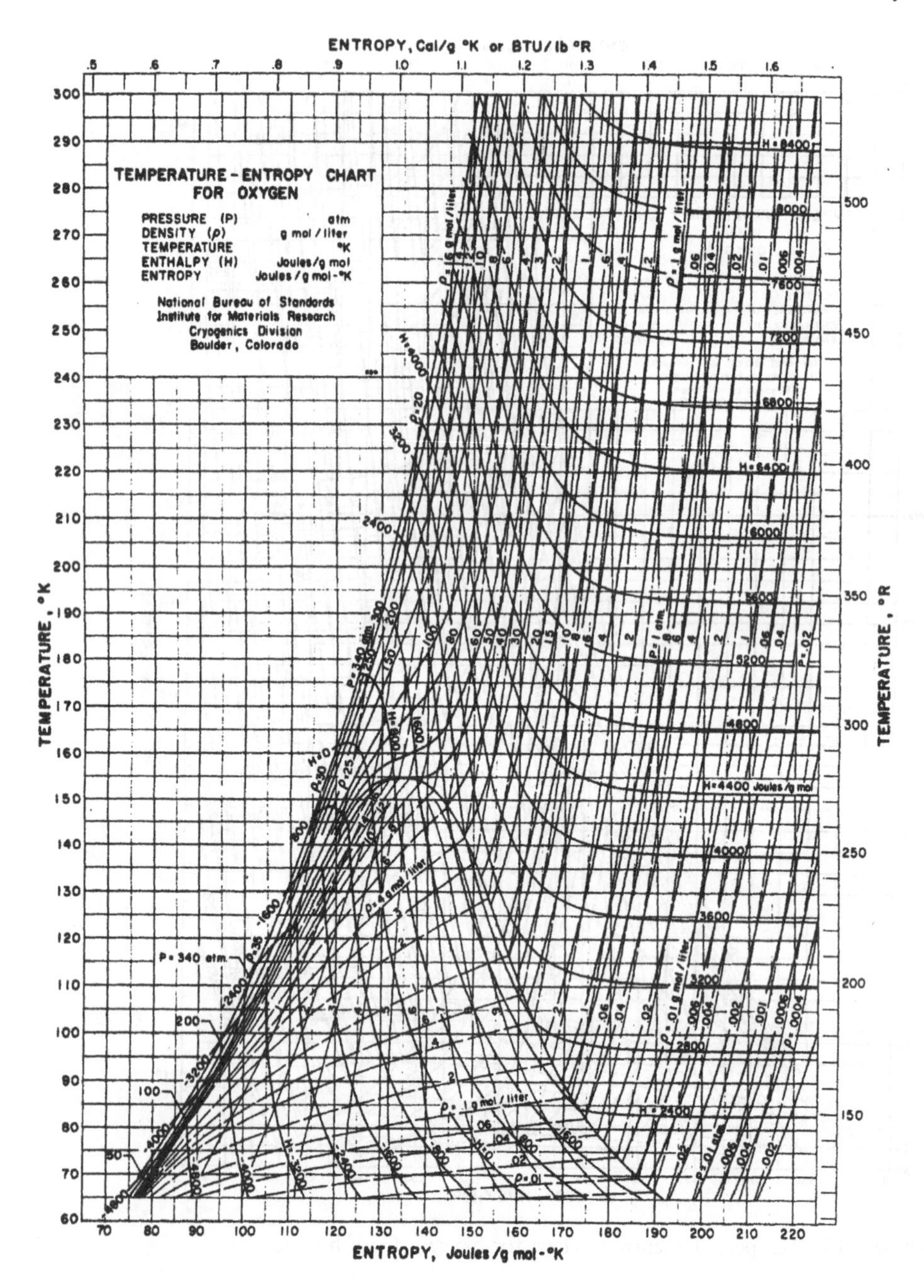


**Figure 3.15** Temperature–entropy chart for oxygen, metric units.

Oxygen is manufactured in large quantities by distillation of liquid air because oxygen is the second most abundant substance in air (20.95% by volume or 23.2% by weight). In comparison, the atmosphere of Mars contains about 0.15% oxygen. Oxygen under excited conditions is responsible for the bright red and yellow-green colors of the aurora borealis. The element and its compounds make up 49.2% by weight of the earth's crust.

Property data for oxygen are available from Roder and Weber (1972) unless otherwise noted.

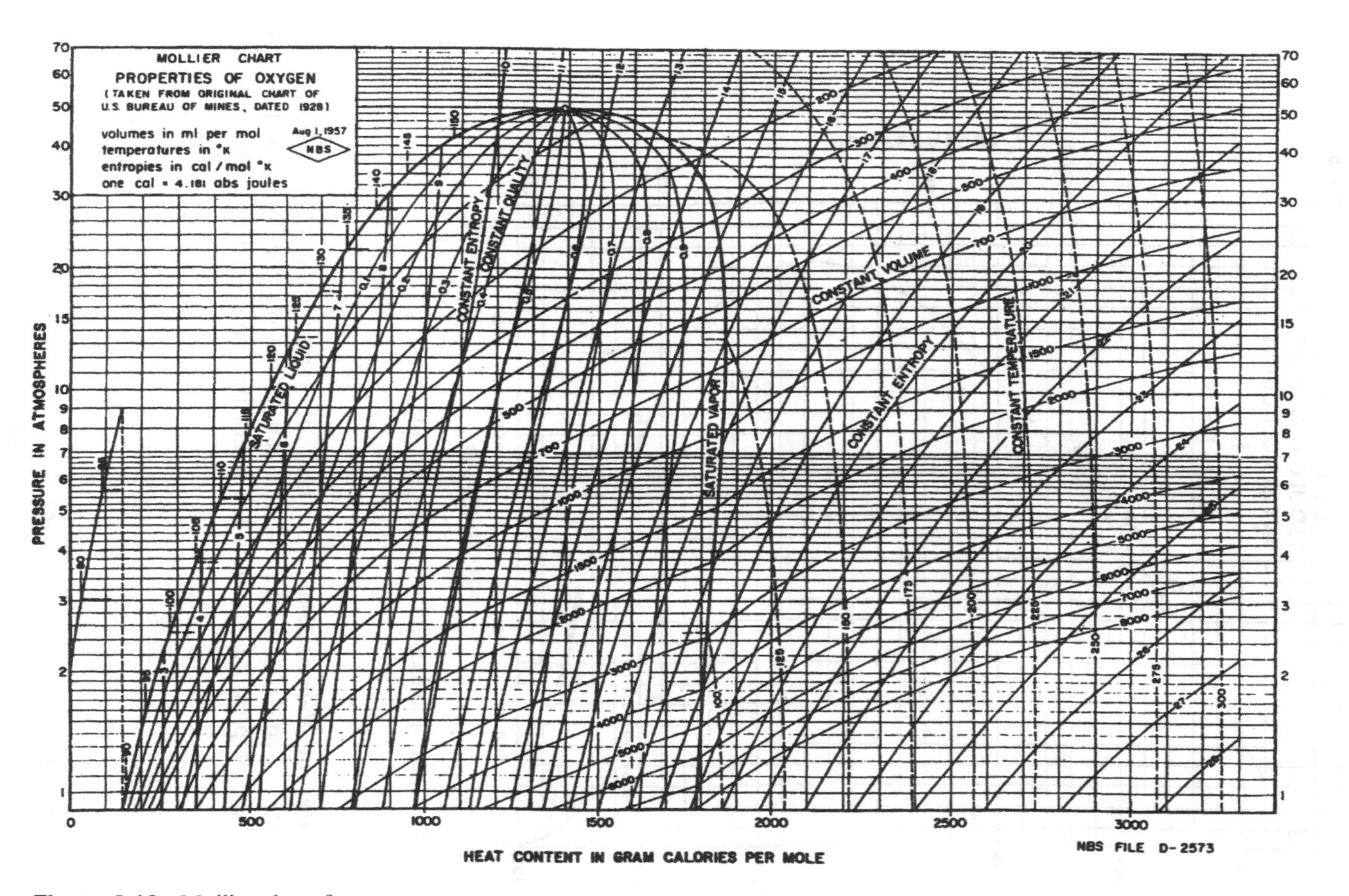


**Figure 3.16** Mollier chart for oxygen.

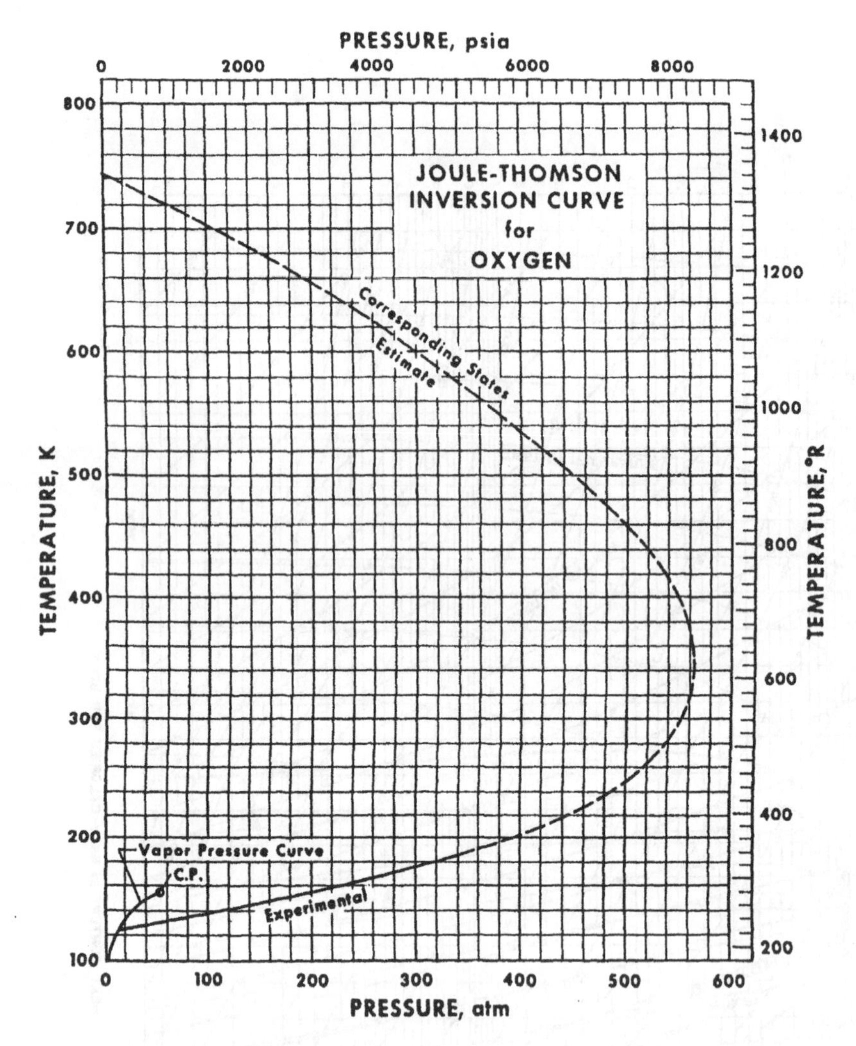


**Figure 3.17** Joule–Thomson inversion curve for oxygen.

**c. Heat of Vaporization.** The heat of vaporization (Fig. 3.18) is the heat required to convert a unit mass of a substance from the liquid to the vapor state at constant pressure. Its units are joules per mole or British thermal units per pound:

| Unit | Triple point | Boiling point |
|---|---|---|
| J/mol | 7761 | 6812 |
| Btu/lb | 104.3 | 91.588 |

The uncertainty is estimated to be $\pm 10\,\text{J/mol}$ (0.13 Btu/lb).

**d. Heat of Sublimation.** The heat of sublimation is the heat required to vaporize a unit mass of solid.

**Table 3.11** Joule–Thomson Inversion Curve for Oxygen

| $T$ (K) | $P$ (atm) | Density (mol/cm$^3$) | $\Delta P^a$ (atm) |
|---|---|---|---|
| 125 | 15.60 | 0.02942 | 1.83 |
| 130 | 52.71 | 0.02907 | 1.98 |
| 135 | 86.45 | 0.02869 | 2.13 |
| 140 | 119.17 | 0.02835 | 2.32 |
| 145 | 150.87 | 0.02802 | 2.58 |
| 150 | 178.30 | 0.02765 | 2.61 |
| 155 | 204.56 | 0.02729 | 2.91 |
| 160 | 232.57 | 0.02700 | 2.92 |
| 165 | 256.29 | 0.02666 | 3.13 |
| 170 | 278.96 | 0.02633 | 3.23 |
| 175 | 300.60 | 0.02600 | 3.43 |
| 180 | 321.28 | 0.02569 | 3.49 |
| 185 | 340.06 | 0.02537 | 3.81 |
| 190 | 357.65 | 0.02504 | 3.80 |

[a] Estimated uncertainty.

Range of values:

| Units | 20.0–23.8 K (gas–$\alpha$-solid) | 23.8–43.8 K (gas–$\beta$-solid) | 43.8–54.35 K (gas–$\gamma$-solid) |
|---|---|---|---|
| J/mol | 9309–9265 | 9216–9131 | 8389–8207 |
| Btu/lb | 125.1–124.5 | 123.8–122.7 | 112.7–110.3 |

The uncertainty is estimated to be 10 J/mol (0.13 Btu/lb), with the uncertainty in temperature as large as 0.1 K.

**e. Vapor Pressure.** The vapor pressure is the pressure, $P(T)$, of a liquid and its vapor in equilibrium. For $P$ in atmospheres and $T$ in kelvins, the equation is

$$\ln \mathrm{P} = A_1 + A_2T + A_3T^2 + A_4T^3 + A_5T^4 + A_6T^5 + A_7T^6 + A_8T^7 \tag{3.1}$$

where

$$A_1 = -62.5967185 \qquad A_5 = -4.09349868 \times 10^{-6}$$

$$A_2 = 2.47450429 \qquad A_6 = 1.91471914 \times 10^{-8}$$

$$A_3 = -4.68973315 \times 10^{-2} \qquad A_7 = -5.13113688 \times 10^{-11}$$

$$A_4 = 5.48202337 \times 10^{-4} \qquad A_8 = 6.0265693 \times 10^{-14}$$

The range of values for the vapor pressure of oxygen is:

| Unit | Triple point | Critical point |
|---|---|---|
| atm | 0.0015 | 49.77 |
| psia | 0.022 | 731 |

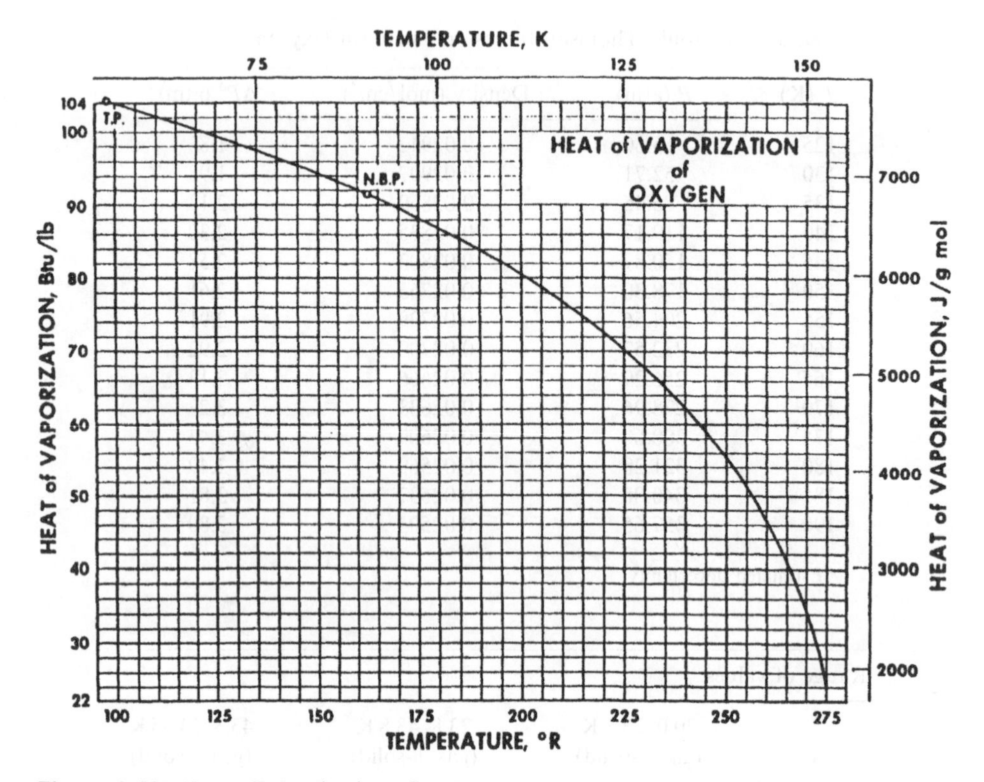


**Figure 3.18** Heat of vaporization of oxygen.

The estimated uncertainty is 0.02% from the normal boiling point to the critical point. Below the normal boiling point the uncertainty increases, reaching about 3% at the triple point. The equivalent uncertainty is 0.02 K in the range of 60–154.58 K.

Figures 3.19 and 3.20 are graphs of the vapor pressure of oxygen.

**f. Virial Coefficients.** The virial coefficients (Table 3.12) are usually defined from the virial equation in density,

$$P = RT\rho[1 + B(T)\rho + C(T)\rho^2 + \cdots] \tag{3.2}$$

The virial coefficients are functions of temperature only.

Two coefficients, *B(T)* and *C(T)*, are adequate to describe the *PVT* surface accurately up to a density of about one-half critical. The uncertainty for *B* varies from $\pm 30\,\text{cm}^3/\text{mol}$ at the boiling point to $\pm 0.25\,\text{cm}^3/\text{mol}$ for temperatures greater than 150 K; for *C*, from $\pm 10{,}000(\text{cm}^3/\text{mol})^2$ at the boiling point to $\pm 30(\text{cm}^3/\text{mol})^2$ above 150 K.

The *Boyle point* ($B = 0$) for oxygen is at 405.88 K.

**g. Dielectric Constant.** The dielectric constant can be calculated from an extension of the Clausius–Mossotti relationship

$$\frac{\varepsilon - 1}{\varepsilon + 2} = A\rho + B\rho^2 + C\rho^3 \tag{3.3}$$

where $A = 0.12361$, $B = 3.2 \times 10^{-4}$, $C = -1.21 \times 10^{-3}$, and $\rho$ is in $\text{g/cm}^3$. The equation is valid over the range 54.35–340 K, 0.2–340 atm, and will yield reasonable values upon extrapolation.

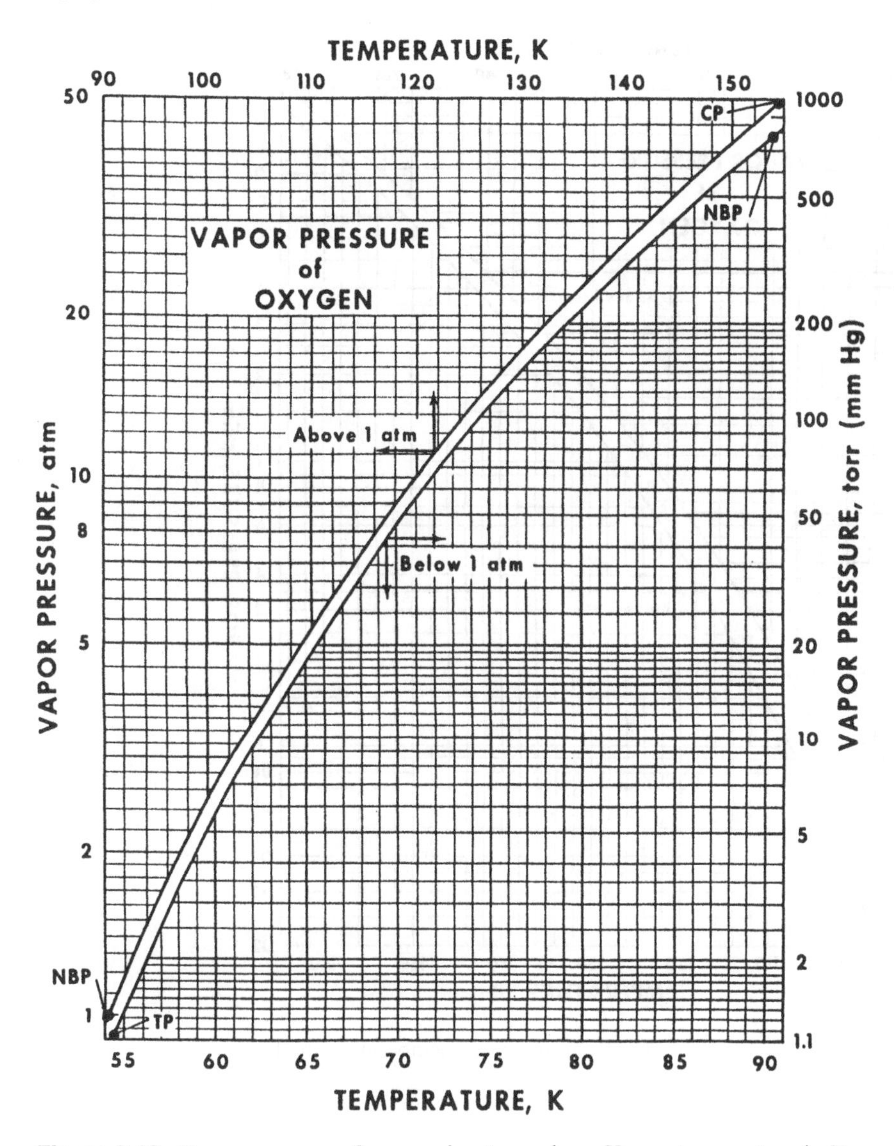


**Figure 3.19** Vapor pressure of oxygen in atm and mmHg vs. temperature in K.

The uncertainty in the term $\varepsilon - 1$ varies from 0.15% at low densities to less than 0.05% at high densities.

**h. Other Properties.** Tables 3.13–3.15 give values for the viscosity of gaseous and liquid oxygen as well as the density of saturated liquid oxygen at various temperatures. Figures 3.21–3.29 were compiled by the National Bureau of Standards, Cryogenic Engineering Laboratory for the US Air Force, Wright Air Development Division in WADD Technical Report 60–56, V. J. Johnson ed., October 1960. They are commonly referred to as the WADD data. In Figs. 3.21–3.29 you will find best values of the melting curve, dielectric constant, density, specific heats, and viscosity, for oxygen.

### *3.1.2. Uses of Oxygen*

Commercial oxygen demand in the United States is about $45 \times 10^9$ kg (50 million tons) per year, at an average growth rate of 5.1% annually. For the most part,

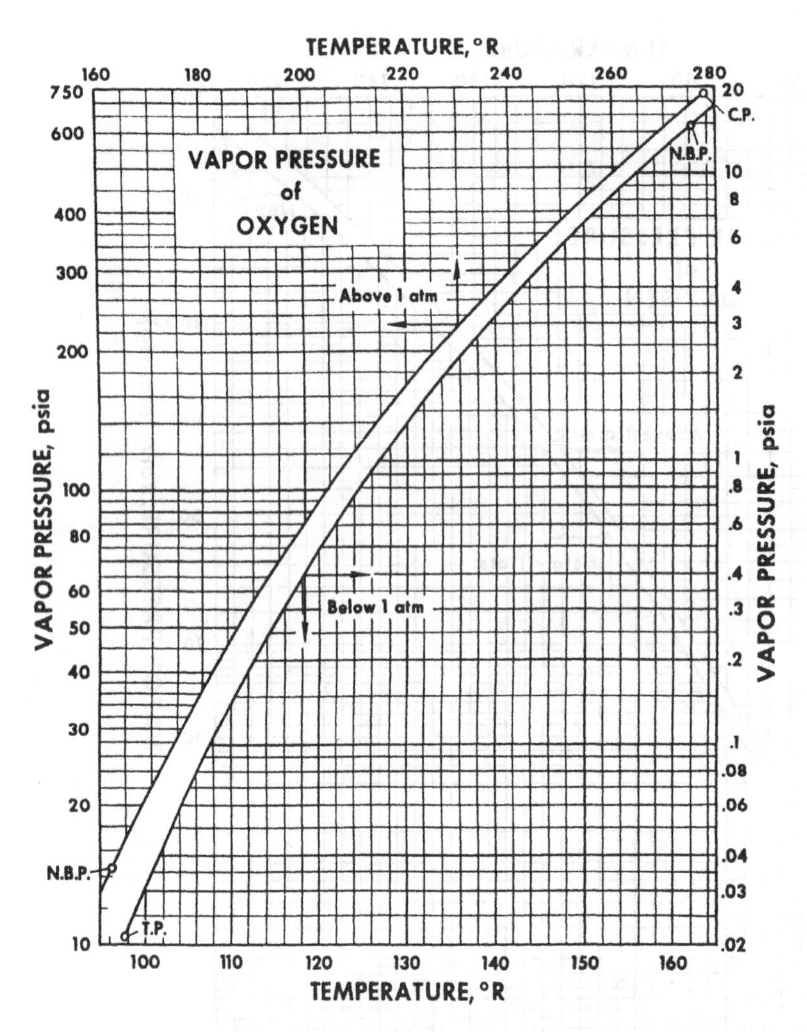


**Figure 3.20** Vapor pressure of oxygen in psia vs. temperature in °R.

tonnage oxygen (that delivered via pipeline) is consumed in steel plants. Present consumption in other industries has seldom reached levels requiring individual oxygen plants; those industries are usually supplied with liquid "merchant oxygen", which is the name given to liquid oxygen delivered to the consumer by truck. The tonnage oxygen plants currently fill about 80% of the oxygen demand.

The first continuous machine for liquefying air was built by Karl von Linde of Munich, Germany, in 1895. The first liquid air installation of commercial size in the United States was built for the American Linde Co., now Linde Division of Union Carbide, in Niagara Falls, New York, in 1907. It had a capacity of 21,240 $m^3$ (750,000 cu ft) per month of oxygen, equivalent to about 907 kg (1 ton) per day.

There is no substitute for oxygen in any of its uses, and it is not recycled or reclaimed.

**a. Steel.** About 60–70% of the oxygen produced in the United States is consumed in the manufacture of iron and steel. Concurrently with the development of pelletizing and other improvements in blast furnace practice, low-cost tonnage

**Table 3.12** Second and Third Virial Coefficients for Oxygen

| $T$ (K) | $B$ (cm$^3$/mol) | $C$(cm$^3$/mol)$^2$ | $T$ (K) | $B$ (cm$^3$/mol) | $C$ (cm$^3$/mol)$^2$ |
|---|---|---|---|---|---|
| 85 | −267.78 | −21,462 | 210 | −44.66 | 1580 |
| 90 | −240.67 | −12,764 | 215 | −42.47 | 1537 |
| 95 | −217.51 | −7058 | 220 | −40.02 | 1498 |
| 100 | −197.54 | −3326 | 225 | −37.90 | 1461 |
| 105 | −180.20 | −904 | 230 | −35.89 | 1428 |
| 110 | −165.05 | 644 | 235 | −33.98 | 1397 |
| 115 | −151.71 | 1609 | 240 | −32.17 | 1368 |
| 120 | −139.91 | 2187 | 245 | −30.45 | 1342 |
| 125 | −129.41 | 2507 | 250 | −28.81 | 1317 |
| 130 | −120.02 | 2659 | 255 | −27.25 | 1294 |
| 135 | −111.59 | 2702 | 260 | −25.77 | 1273 |
| 140 | −103.98 | 2677 | 265 | −24.34 | −1253 |
| 145 | −97.08 | 2611 | 270 | −22.98 | 1234 |
| 150 | −90.81 | 2522 | 275 | −21.68 | 1217 |
| 155 | −85.09 | 2423 | 280 | −20.44 | 1201 |
| 160 | −79.84 | 2320 | 285 | −19.24 | 1186 |
| 165 | −75.02 | 2219 | 290 | −18.09 | 1172 |
| 170 | −70.58 | 2122 | 295 | −16.98 | 1160 |
| 175 | −66.48 | 2031 | 300 | −15.92 | 1149 |
| 180 | −62.67 | 1948 | | | |
| 185 | −59.14 | 1871 | | | |
| 190 | −55.85 | 1801 | | | |
| 195 | −52.77 | 1738 | | | |
| 200 | −49.89 | 1680 | | | |
| 205 | −47.20 | 1628 | | | |

oxygen became available in the 1950s and was instrumental in the development of the basic oxygen furnace (BOF), which is a converter using oxygen instead of air.

**b. Chemical.** The chemical industry uses about 12% of the total oxygen produced. Major uses are in partial oxidation of methane to produce acetylene; in the oxidation of ethylene (the chief hydrocarbon produced in the United States) to

**Table 3.13** Viscosity of Gaseous Oxygen at 1 atm

| Temp. (K) | Viscosity (cP) × $10^{-3}$ | Temp. (K) | Viscosity (cP) × $10^{-3}$ |
|---|---|---|---|
| 100 | 7.715 | 200 | 14.775 |
| 110 | 8.500 | 210 | 15.400 |
| 120 | 9.225 | 220 | 16.000 |
| 130 | 10.000 | 230 | 16.620 |
| 140 | 10.725 | 240 | 17.200 |
| 150 | 11.440 | 250 | 17.790 |
| 160 | 12.135 | 260 | 18.350 |
| 170 | 12.820 | 270 | 19.000 |
| 180 | 13.490 | 280 | 19.450 |
| 190 | 14.140 | 290 | 20.000 |
| | | 300 | 20.645 |

*Source:* Hilsenrath et al. (1955).

**Table 3.14** Viscosity of Liquid Oxygen

| Rudenko and Shubnikow (1934) | | Rudenko (1934) | |
|---|---|---|---|
| Temp. (K) | Viscosity (cP) | Temp. (K) | Viscosity (cP) |
| 54.4 | 0.873 | 90 | 0.190 |
| 54.5 | 0.863 | 111 | 0.123 |
| 54.6 | 0.821 | 112 | 0.121 |
| 54.9 | 0.772 | | |
| 56.4 | 0.717 | 125.8 | 0.110 |
| 57.1 | 0.638 | 138.2 | 0.100 |
| 57.4 | 0.648 | | |
| 59.7 | 0.631 | 145.5 | 0.098 |
| 61.7 | 0.521 | 154.1 | 0.090 |
| 63.5 | 0.476 | | |
| 65.4 | 0.435 | | |
| 68.9 | 0.377 | | |
| 72.3 | 0.323 | | |
| 77.4 | 0.273 | | |
| 80.0 | 0.250 | | |
| 90.1 | 0.190 | | |

*Source:* Rudenko (1939), Rudenko and Shubnikow (1934).

produce ethylene oxide, an important chemical intermediate; and in various processes to produce hydrogen from hydrocarbons for the manufacture of chemicals such as ammonia.

**c. Nonferrous Metals.** About 6% of the oxygen produced is used by the nonferrous metals industry. It is used to enrich air in primary and secondary lead blast furnaces and in copper reverberatory furnaces, converters, flash smelters, and autogenous and continuous copper smelting processes.

*Example 3-1 (Oxygen Sample Problem).* What is the boiling point of oxygen at 150 psia? What is the heat of vaporization at this pressure?

**Table 3.15** Density of Saturated Liquid Oxygen

| Van Itterbeek (1955) | | Mathias and Onnes (1911) | |
|---|---|---|---|
| Temp. (K) | Density (g/cm$^3$) | Temp. (K) | Density (g/cm$^3$) |
| 61 | 1.282 | 62.7 | 1.2746 |
| 65 | 1.263 | 91.1 | 1.1415 |
| 70 | 1.239 | 118.6 | 0.9758 |
| 75 | 1.215 | 132.9 | 0.8742 |
| 80 | 1.191 | 143.2 | 0.7781 |
| 85 | 1.167 | 149.8 | 0.6779 |
| 90 | 1.142 | 152.7 | 0.6032 |
| | | 154.3[a] | 0.4299 |

[a] Critical point (current accepted value is 154.77 K based on present international scale of 0°C = 273.15 K instead of the Leiden temperature scale of 0°C = 273.09 K).

*Source:* Van Itterbeek (1955), Mathias and Onnes (1911).

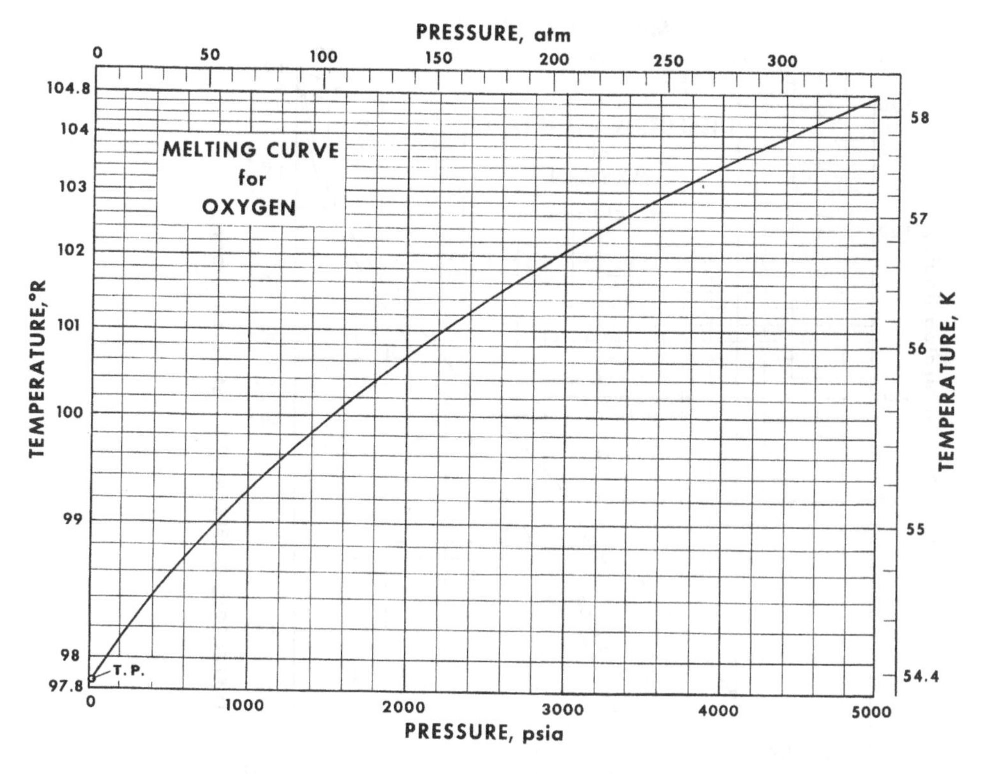


**Figure 3.21** Melting curve for oxygen (NASA SP3071).

The *T–S* diagram for oxygen (Fig. 3.30) contains much more information than first meets the eye. The answers to the above questions are readily found on the oxygen *T–S* chart.

First, find the 150 psia isobar for oxygen and follow it to the two–phase region. The isobars and isotherms are coincident horizontal lines under the two–phase dome. Therefore, the boiling point of oxygen is read directly from the temperature scale to be about 216°R. The published value is 216.38°R at 150 psia.

Likewise, the heat of vaporization can be found in two independent ways. First, it is only necessary to read the enthalpy values for saturated vapor ($H_g$) and saturated liquid ($H_f$). Along the 150 psia isobar we find $H_g = 40\,\text{Btu/lb}_m$ and $H_f = -34\,\text{Btu/lb}_m$. The difference is $H_{fg} = H_g - H_f = 40 - (-34) = 74\,\text{Btu/lb}_m$. The published value is $74.39\,\text{Btu/lb}_m$.

Heat of vaporization is also $T\Delta S$. From the $T-S$ chart we find $S_g = 1.167\,\text{Btu/(lb}_m\,°\text{R)}$ and $S_f = 0.824\,\text{Btu/(lb}_m\,°\text{R)}$. Accordingly, $\Delta S = 1.167 - 0.824 = 0.343\,\text{Btu/(lb}_m\,°\text{R)}$. $T\Delta S$ in appropriate units is $(216.4°\text{R}) \times [0.343\,\text{Btu/(lb}_m\,°\text{R)}] = 74.2\,\text{Btu/lb}_m$, which compares well with the published value of $74.4\,\text{Btu/lb}_m$.

## 3.2. Nitrogen

Nitrogen (mol wt. 28.0134) has two stable isotopes of mass numbers 14 and 15 with a relative abundance of 10,000:38. Liquid nitrogen is of considerable importance to the cryogenic engineer because it is a safe refrigerant. Because it is rather inactive

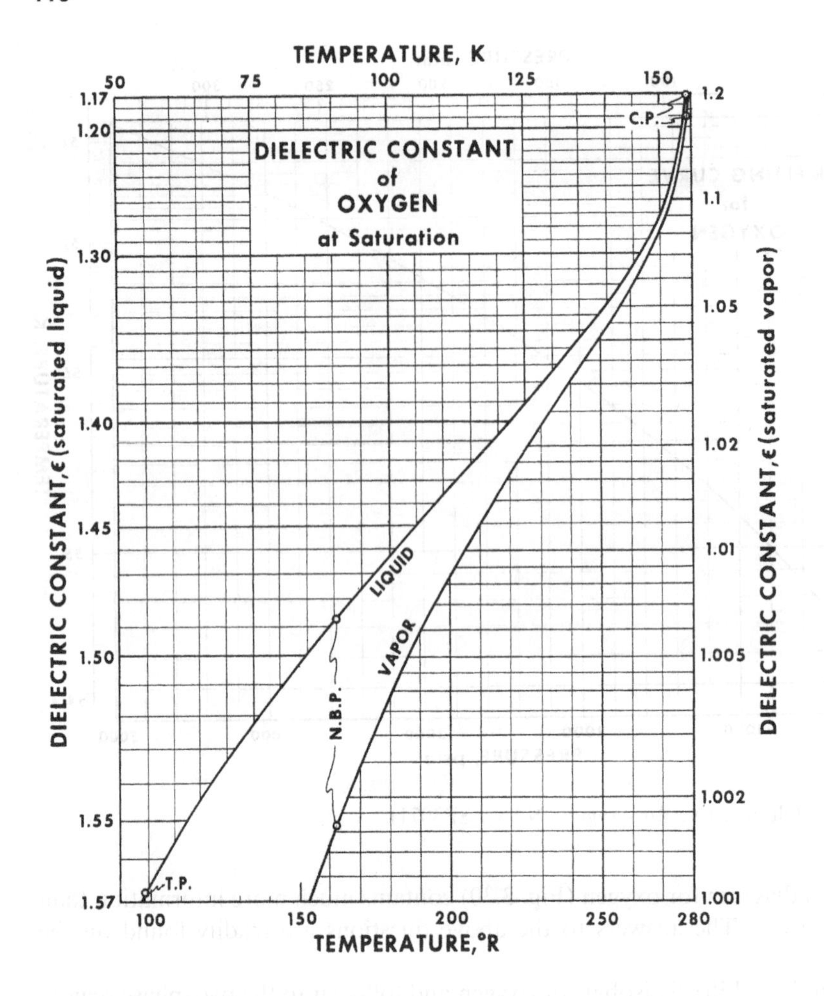


**Figure 3.22** Dielectric constant of oxygen (NASA SP3071).

chemically and is neither explosive nor toxic, liquid nitrogen is commonly used in hydrogen and helium liquefaction cycles as a precoolant.

Nitrogen is the major constituent of air (78.09% by volume or 75.45% by weight). The atmosphere of Mars, by comparison, is 2.6% nitrogen. The estimated amount of this element in out atmosphere is more than 4000 billion tons. From this inexhaustible source, it can be obtained by liquefaction and fractional distillation.

Liquid nitrogen is a clear, colorless fluid that resembles water in appearance. At 1 bar pressure, liquid nitrogen boils at 77.3 K (139.2°R) and freezes at 63.2 K (113.8°R). Saturated liquid nitrogen at the normal boiling point has a density of 808.9 kg/m$^3$ (50.5 lb$_m$/ft$^3$) in comparison with water at 520°R (60°F), which has a density of 998.0 kg/m$^3$ (62.3 lb$_m$/ft$^3$). One of the significant differences between the properties of liquid nitrogen and those of water (apart from the differences in normal boiling points) is that the heat of vaporization of nitrogen (Table 3.16) is more than an order of magnitude less than that of water. At its normal boiling point, liquid nitrogen has a heat of vaporization of 198.3 kJ/kg (85.32 Btu/lb$_m$), while water has a heat of vaporization of 2255 kJ/kg (970.3 Btu/lb$_m$).

Table 3.17 gives the physical constants of nitrogen.

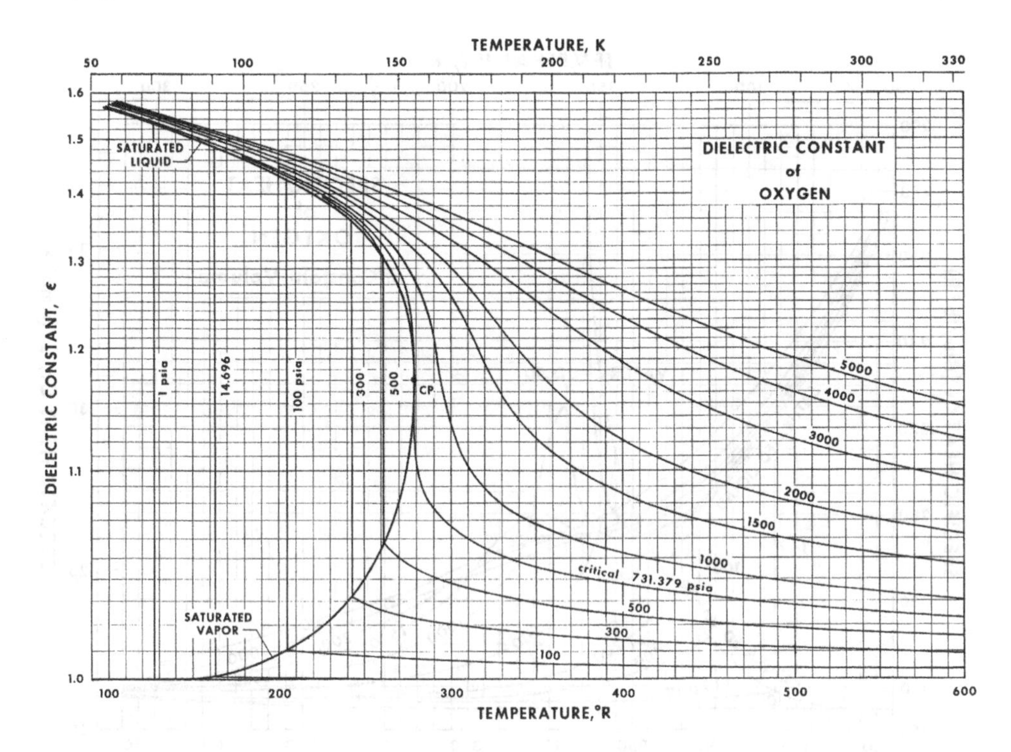


Figure 3.23 Dielectric constant of oxygen, extended range (NASA SP3071).

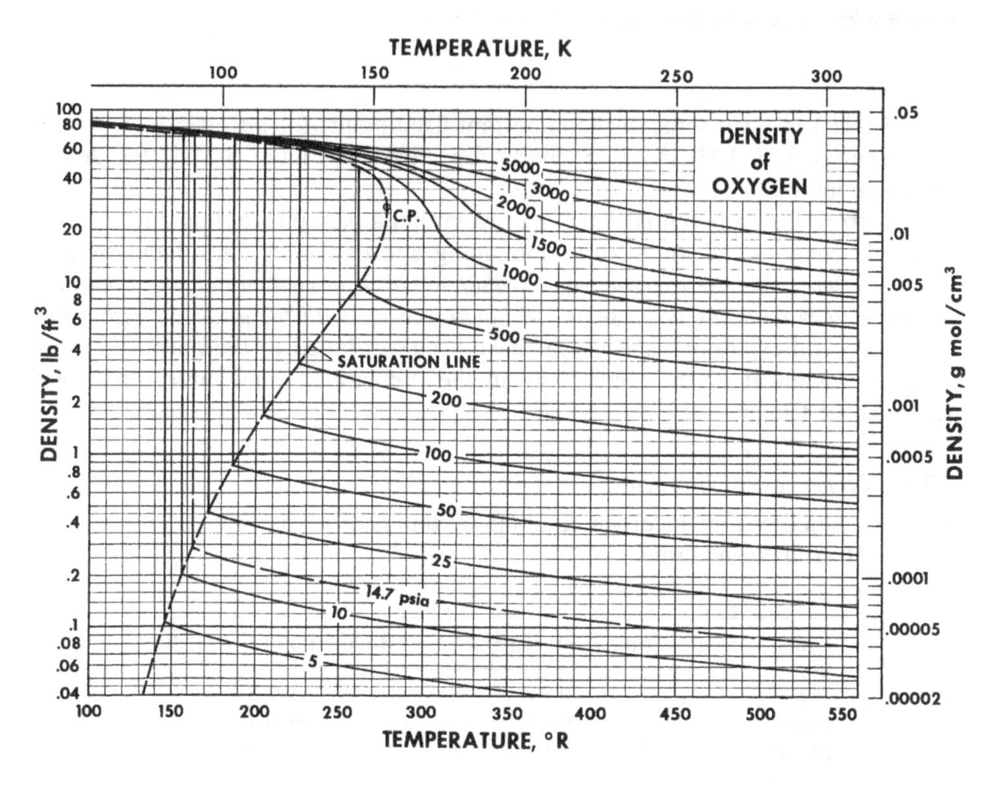


Figure 3.24 Density of oxygen (NASA SP3071).

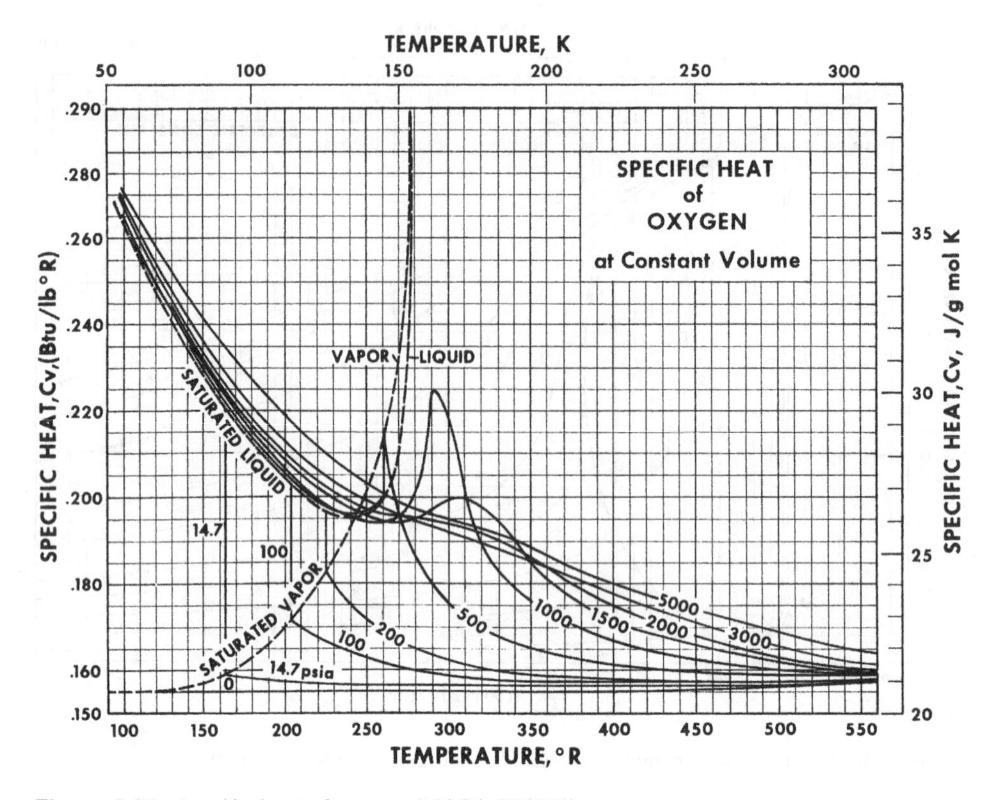


**Figure 3.25** Specific heat of oxygen (NASA SP3071).

Table 3.18 gives some of the important fixed points for nitrogen.

Figures 3.31–3.33 are *T–S* charts for nitrogen. Figure 3.34 is the Joule–Thomson inversion curve.

Nitrogen has a potential safety hazard in that a bare pipe of liquid nitrogen at 77 K will condense an air mixture containing approximately 50% liquid oxygen. *Thus, one must exercise extreme caution that so-called inert liquid nitrogen does not in fact become an unsuspected source of liquid oxygen, with the attendant risks cited earlier.* Several explosions and deaths have been attributed to the phenomenon of oxygen enrichment of the atmosphere in the presence of liquid nitrogen cooled surfaces. See Chapter 10.

### 3.2.1. *Properties of Nitrogen*

Property data are from Jacobsen et al. (1973) unless otherwise noted.

**a. Heat of Vaporization.**

| Units | Triple point (63.148 K) | Boiling point (77.313 K) |
|---|---|---|
| J/g | 215.1 | 198.8 |
| J/mol | 6025.7 | 5569.1 |
| Btu/lb | 92.57 | 85.55 |
| Btu/(lb mol) | 2593.2 | 2396.5 |

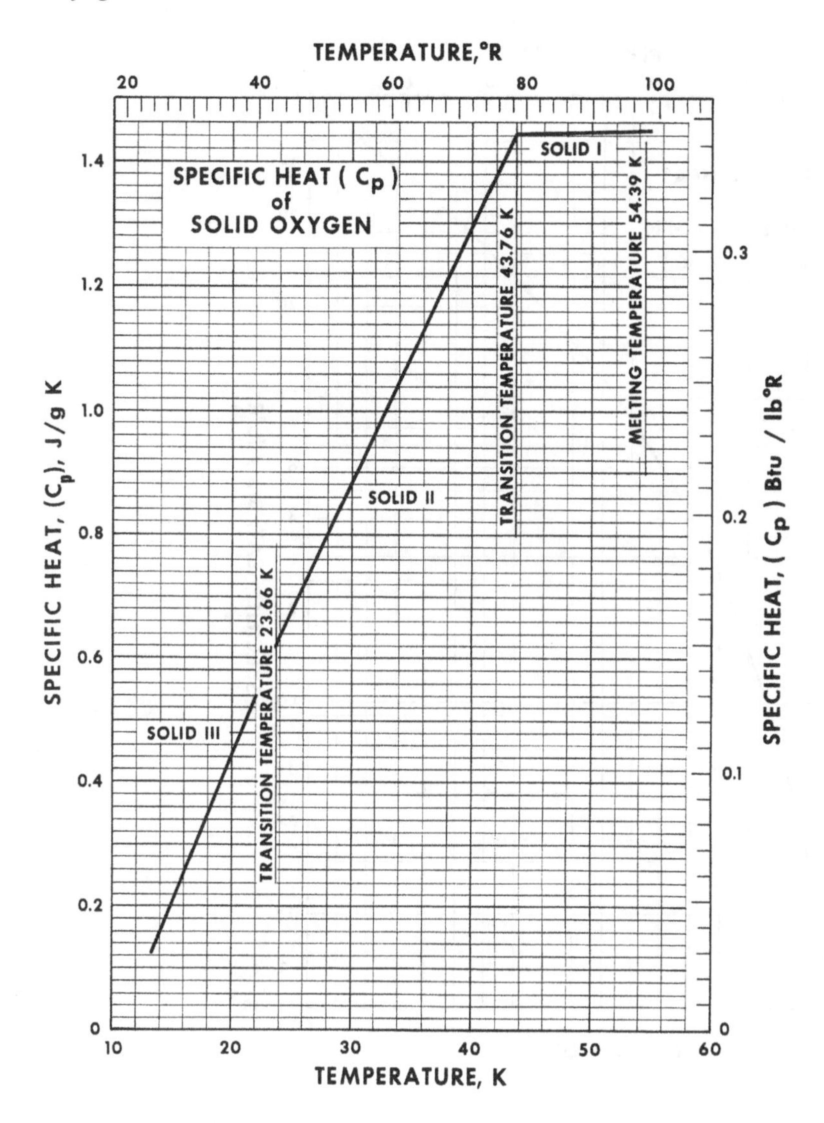


**Figure 3.26** Specific heat ($C_P$) of solid oxygen (NASA SP3071).

**b. Vapor Pressure.** For solid nitrogen.

$$\log p = a - \frac{359.093}{T} \tag{3.4}$$

where $T$ is in $K$, $a = 7.65894$ for p in mmHg, $a = 4.77813$ for p in atm and $a = 5.94532$ for p in psia.

or, for $p$ below 760 mmHg,

$$\log p = -\frac{334.64}{T} + 7.577 - 0.00476\,T \tag{3.5}$$

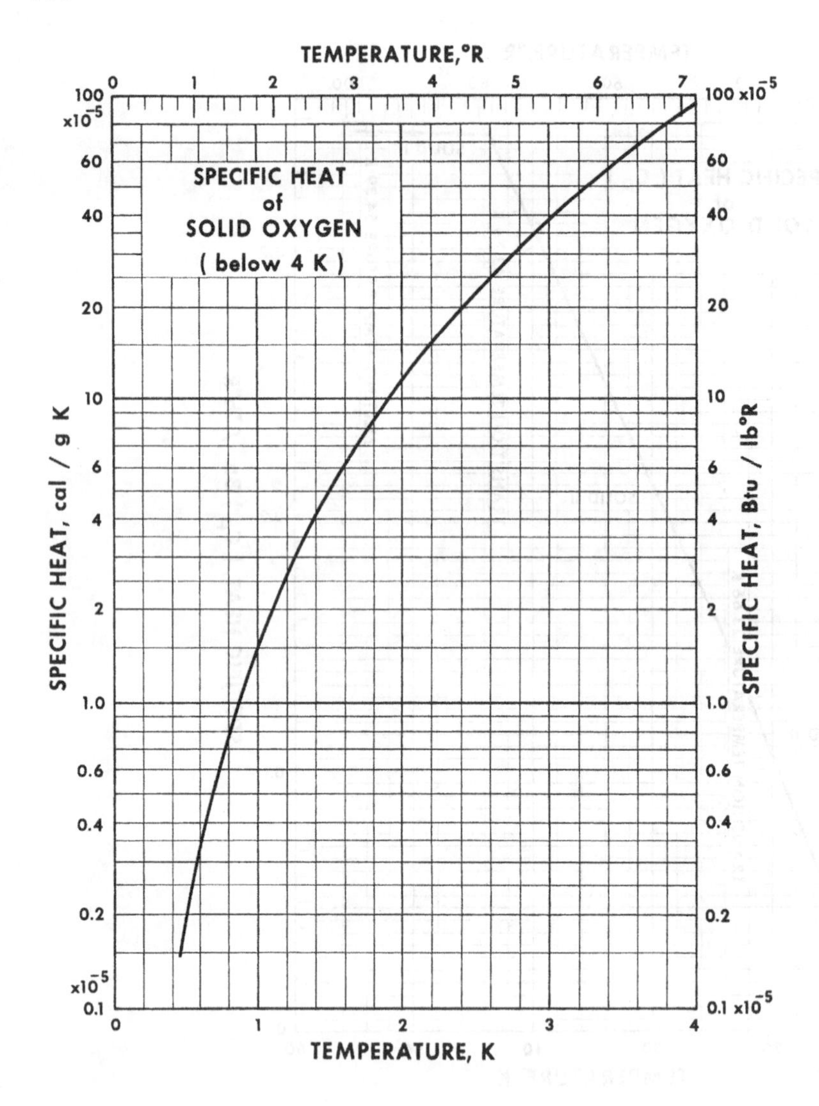


**Figure 3.27** Specific heat ($C_P$) of solid oxygen below 4 K (NASA SP3071).

where $p$ is in mmHg and $T$ is in K. (This equation is also valid for liquid nitrogen from 63.14 to 77.35 K; 96.4–760 mmHg.)

For liquid nitrogen,

$$\log p = 6.49594 - \frac{255.821}{T - 6.600} \tag{3.6}$$

where $p$ is in mm Hg and $T$ is in K for the range 64–78 K

or

$$\log p = -\frac{316.824}{T} + 4.47582 - 0.0071701T + 2.940 \times 10^{-5}T^2 \tag{3.7}$$

where, $p$ is in atm, $T$ is in K and 1–32 atm.

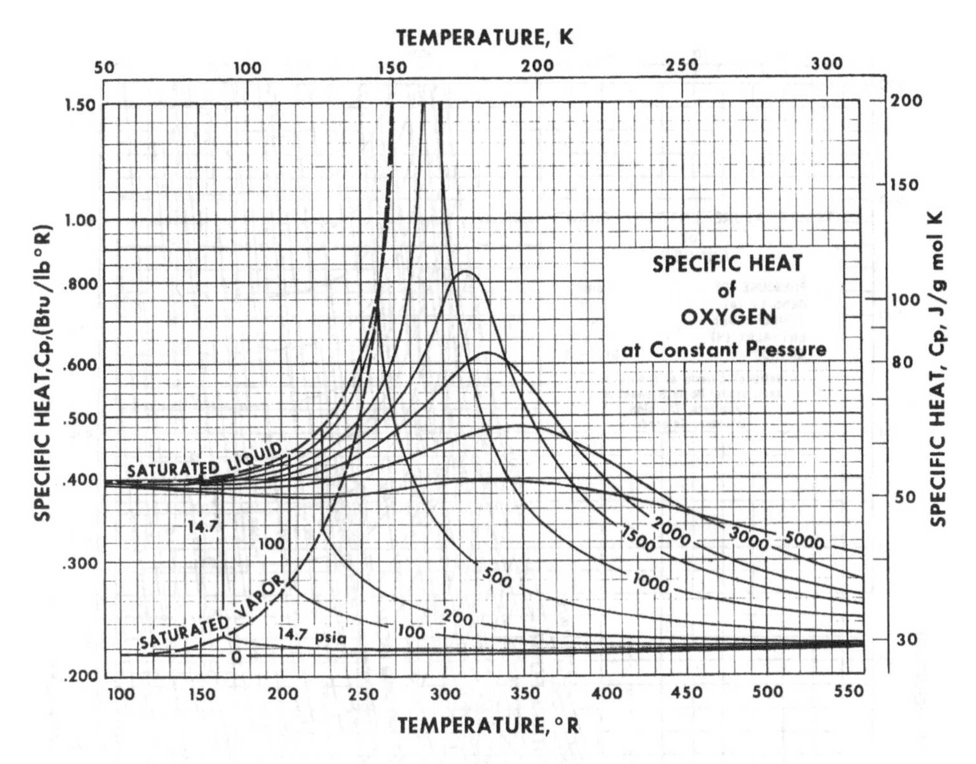


**Figure 3.28** Specific heat of oxygen at constant pressure (NASA SP3071).

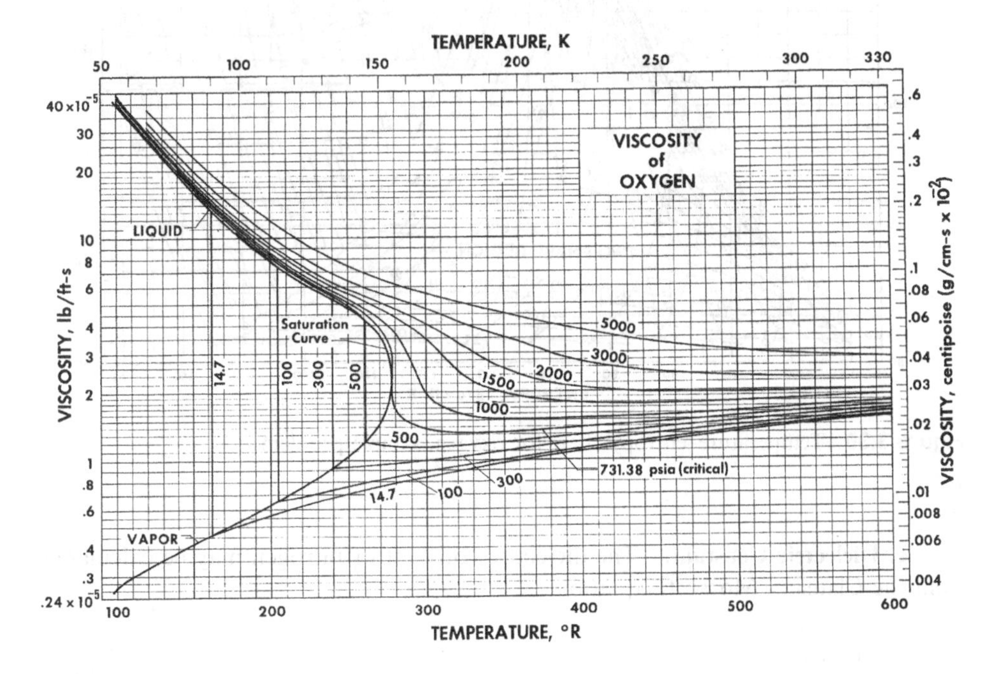


**Figure 3.29** Viscosity of oxygen.

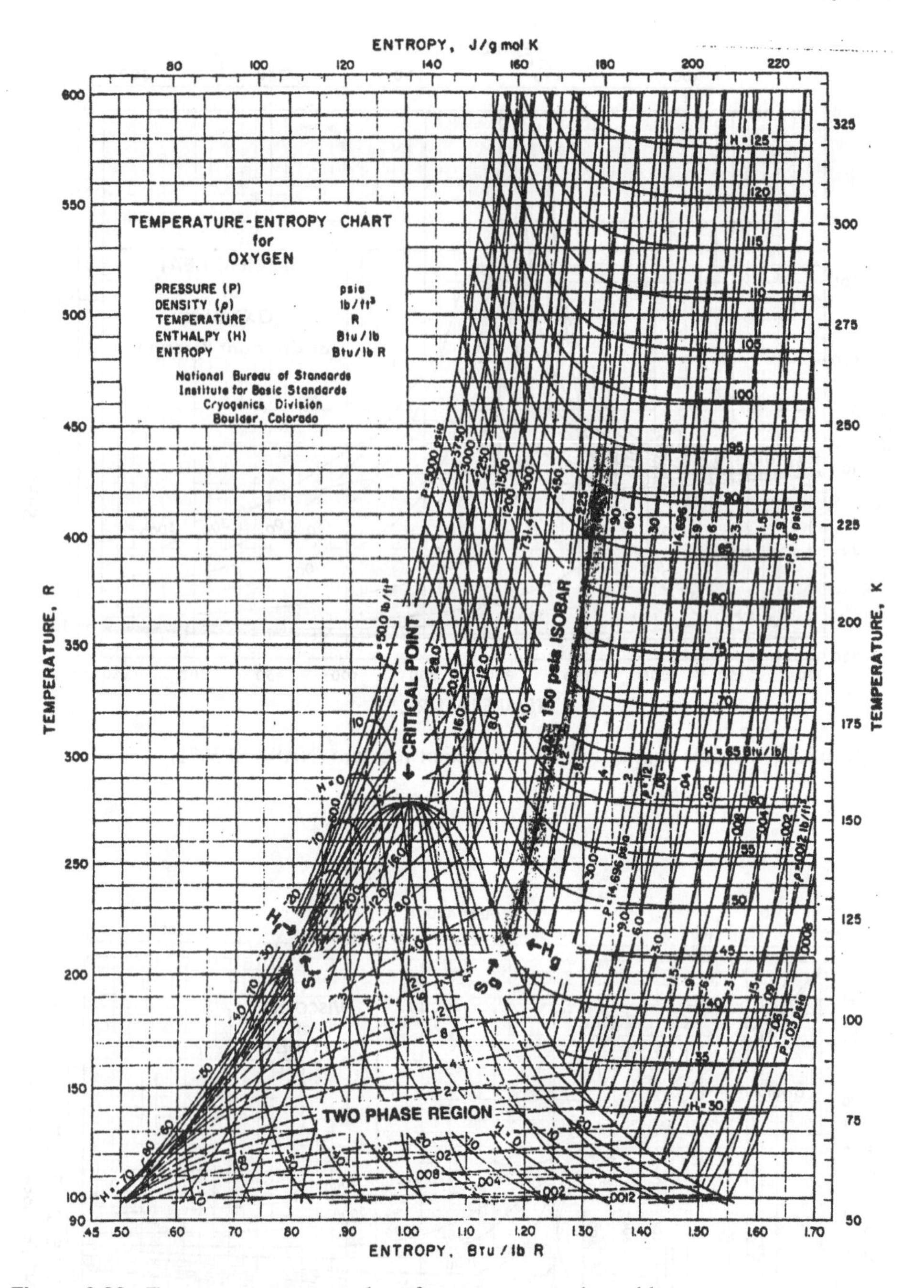


Figure 3.30 Temperature–entropy chart for oxygen, sample problem.

**c. Dielectric Constant.** The dielectric constant of nitrogen may be calculated from the Clausius–Mossotti equation,

$$\frac{\varepsilon - 1}{\varepsilon + 2}\left(\frac{1}{\rho}\right) = p \tag{3.8}$$

**Table 3.16** Heat of Vaporization of Nitrogen

| Temp. (K) | Heat of vaporization, $hv$ | |
|---|---|---|
| | J/mol | cal/mol |
| *Solid nitrogen* | | |
| 62.00 | 6775.0 | |
| 62.0018 | 6787.4 | |
| 62.0172 | 6762.4 | |
| *Liquid nitrogen* | | |
| 67.9588 | 5901.6 | |
| 67.9620 | 5899.0 | |
| 68.00 | 5899.0 | |
| 73.0913 | 5739.1 | |
| 73.0887 | 5732.1 | |
| 73.10 | 5735.2 | |
| 77.395 | 5735.2 | |
| 78.00 | 5579.4 | |
| 78.0147 | 5563.1 | |
| 78.0153 | 5571.8 | |
| 80 | | 1313 |
| 85 | | 1266 |
| 90 | | 1213 |
| 95 | | 1155 |
| 100 | | 1086 |
| 105 | | 1010 |
| 110 | | 918 |
| 115 | | 803 |
| 120 | | 643 |
| 125 | | 328 |
| 126.1[a] | | 0 |

[a] Critical point.
*Source*: Furukawa and McCoskey (1953), Millar and Sullivan (1928).

where $\varepsilon$ is the dielectric constant, $\rho$ is the density, and $p$ is the specific polarization, a property of the substance having dimensions of specific volume:

$$p = A_n + B_n\rho + C_n\rho^2 \tag{3.9}$$

where $p$ is the specific polarization in $cm^3/mol$ and $\rho$ is in units of $mol/cm^3$.

The parameters $A_n$, $B_n$, and $C_n$ are

$$A_n = 4.389\,cm^3/mol, \quad B_n = 2.2(cm^3/mol)^2, \quad C_n = -114.0(cm^3/mol)^3$$

**d. Other Properties.** Tables 3.19–3.23 present data on the solid and liquid vapor pressure, density, viscosity of nitrogen. Figures 3.35–3.53, taken from the WADD report mentioned before, show the heat of vaporization, vapor pressure, melting curve, density, thermal expansivity, thermal conductivity, specific heats, dielectric constant, surface tension, and viscosity of nitrogen.

**Table 3.17** Nitrogen Physical Constants

*Nitrogen*
Chemical Symbol: $N_2$
Synonyms: LIN (liquid only)
CAS Registry Number: 7727-37-9
DOT Classification: Nonflammable gas
DOT Label: Nonflammable gas
Transport Canada Classification: 2.2
UN Number: UN 1066 (compressed gas); UN 1977 (refrigerated liquid)

| | US Units | SI Units |
|---|---|---|
| *Physical Constants* | | |
| Chemical formula | $N_2$ | $N_2$ |
| Molecular weight | 28.01 | 28.01 |
| Density of the gas at 70°F (21.1°C) and 1 atm | 0.072 lb/ft$^3$ | 1.153 kg/m$^3$ |
| Specific gravity of the gas at 70°F (21.1°C) and 1 atm (air = 1) | 0.967 | 0.967 |
| Specific volume of the gas at 70°F (21.1°C) and 1 atm | 13.89 ft$^3$/lb | 0.867 m$^3$/kg |
| Density of the liquid at boiling point and 1 atm | 50.47 lb/ft$^3$ | 808.5 kg/m$^3$ |
| Boiling point at 1 atm | −320.4°F | −195.8°C |
| Melting point at 1 atm | −345.8°F | −209.9°C |
| Critical temperature | −232.4°F | −146.9°C |
| Critical pressure | 493 psia | 3399 kPa abs |
| Critical density | 19.60 lb/ft$^3$ | 314.9 kg/m$^3$ |
| Triple point at 1.81 psia (12.5 kPa abs) | −346.0°F | −210.0°C |
| Latent heat of vaporization at boiling point | 85.6 Btu/lb | 199.1 kJ/kg |
| Latent heat of fusion at melting point | 11.1 Btu/lb | 25.1 kJ/kg |
| Specific heat of the gas at 70°F (21.1°C) and 1 atm | | |
| $C_P$ | 0.249 Btu/(lb)(°F) | 1.04 kJ/(kg)(°C) |
| $C_V$ | 0.177 Btu/(lb)(°F) | 0.741 kJ/(kg)(°C) |
| Ratio of specific heats, $C_P/C_V$ | 1.41 | 1.41 |
| Solubility in water vol/vol at 32°F (0°C) | 0.023 | 0.023 |
| Weight of liquid at boiling point | 6.747 lb/gal | 808.5 kg/m$^3$ |
| Gas/liquid ratio (gas at 70°F (21.1°C) and 1 atm, liquid at boiling point, vol/vol) | 696.5 | 696.5 |

### 3.2.2. *Uses of Nitrogen*

Nitrogen is used in two distinct ways: (1) as an element, which may be used either as a gas to exclude air from industrial processes or as a liquid to provide refrigeration, and (2) as fixed nitrogen, in compounds with other elements, in which form it is an essential plant nutrient and a constituent of many important chemical products.

Elemental nitrogen is produced by cryogenic means and is used primarily for two purposes. Most is used as a gas to provide an inert atmosphere in a variety of chemical and metallurgical processes, electronic devices, and packages; about 10% is used as a liquid to provide refrigeration. The major consuming areas are discussed below.

**a. Chemicals.** Numerous chemical reactions require an inert atmosphere to prevent oxidation, fire, or explosion, for drying or purging equipment, for

**Table 3.18** Fixed Points for Nitrogen

| |
|---|
| *Critical point* |
| $T = 126.20\,K$ (227.16°R) |
| $P = 33.55$ atm (493.0 psia;3.399 MPa) |
| $\rho = 0.01121\,mol/cm^3$ ($19.60\,lb_m/ft^3$; $314\,kg/m^3$) |
| *Normal boiling point* |
| $T = 77.347\,K$ (139.225°R) |
| $P = 1$ atm (14.696 psia; 101.3 kPa) |
| $\rho_{gas} = 0.0001647\,g\,mol/cm^3$ ($0.2880\,lb_m/ft^3$; $4.614\,kg/m^3$) |
| $\rho_{liquid} = 0.02887\,g\,mol/cm^3$ ($50.49\,lb_m/ft^3$; $808.9\,kg/m^3$) |
| *Triple point* |
| $T = 63.148\,K$ (113.666°R) |
| $P = 0.1237$ atm (1.818 psia; 12.53 kPa) |
| $\rho_{gas} = 2.41 \times 10^{-5}\,g \cdot mol/cm^3$ ($4.21 \times 10^{-2}\,lb_m/ft^3$; $674\,g/m^3$) |
| $\rho_{liquid} = 0.03098\,g \cdot mol/cm^3$ ($54.18\,lb_m/ft^3$; $867.96\,kg/m^3$) |

*Source:* Jacobsen et al. (1973).

pressurizing, or as a dilutent to control gas reaction rates. About 25% of elemental nitrogen is used to provide blanketing atmospheres for chemical processes.

**b. Electronic Components.** Gaseous nitrogen is used to pressurize telephone and electric cables and to flush air out of and dry electronic tubes. Liquid nitrogen is used as a coolant for lasers, masers, and infrared detectors. Use in electronic components constitutes about 15% of the demand for elemental nitrogen.

**c. Food Products.** The food industry is the largest consumer of liquid nitrogen. Evaporation of liquid nitrogen is a fast, uncomplicated, reliable, and inexpensive method of providing refrigeration. Although it is not competitive for small installations requiring moderately low temperatures, such as household refrigerators, it provides a much lower temperature than competitive large refrigeration units and is particularly useful for fast freezing, as in the meat backing industry. Liquid nitrogen is especially convenient for refrigeration during transportation of foodstuffs because it is relatively simple to install compared to mechanical refrigerators and has the additional advantage of naturally providing an inert atmosphere. This use is slowing down in its rate of growth and now accounts for about 5% of the nitrogen market.

**d. Iron and Steel.** A large use of elemental nitrogen, about 15% of the total, is in annealing stainless steel; the nitrogen may be mixed with hydrogen, which acts as a deoxidizing agent. Annealing and heat treating of other steel mill products such as tinplate also are conducted in a nitrogen atmosphere.

**e. Nonferrous Metal Production.** Nitrogen gas is used for preparing reactive metals, such as the rare earths and sodium and potassium, and to degas molten aluminum. Liquid nitrogen is used in shrink-fitting metals and in the grinding of heat-sensitive materials. About 2% of elemental nitrogen production goes into all these uses.

**f. Other Applications.** Nitrogen is used in various aerospace applications (about 5% of the market) such as for transferring propellants and in wind tunnels. Nitrogen appears to have developed its own big new use over the past few years—enhanced oil recovery (EOR), used to boost the output from old oilfields. This use now accounts for 15% of all nitrogen consumed in the United States.

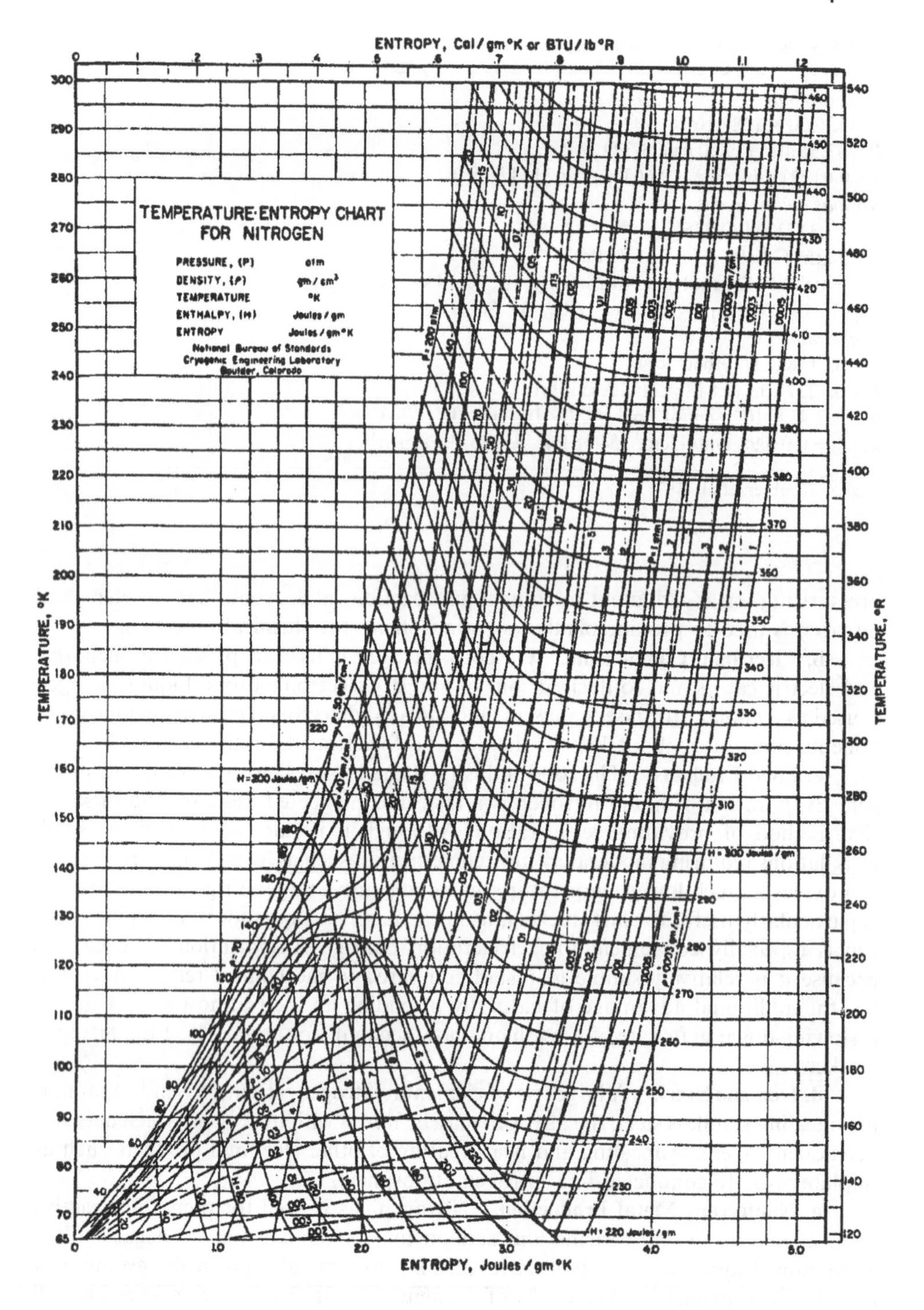


**Figure 3.31** Temperature–entropy chart for nitrogen.

The diversity of nitrogen markets resulted from producers' efforts over many years to find some way to sell this formerly discarded coproduct of many air separation plants. Oddly, producers succeeded so well that they now have the

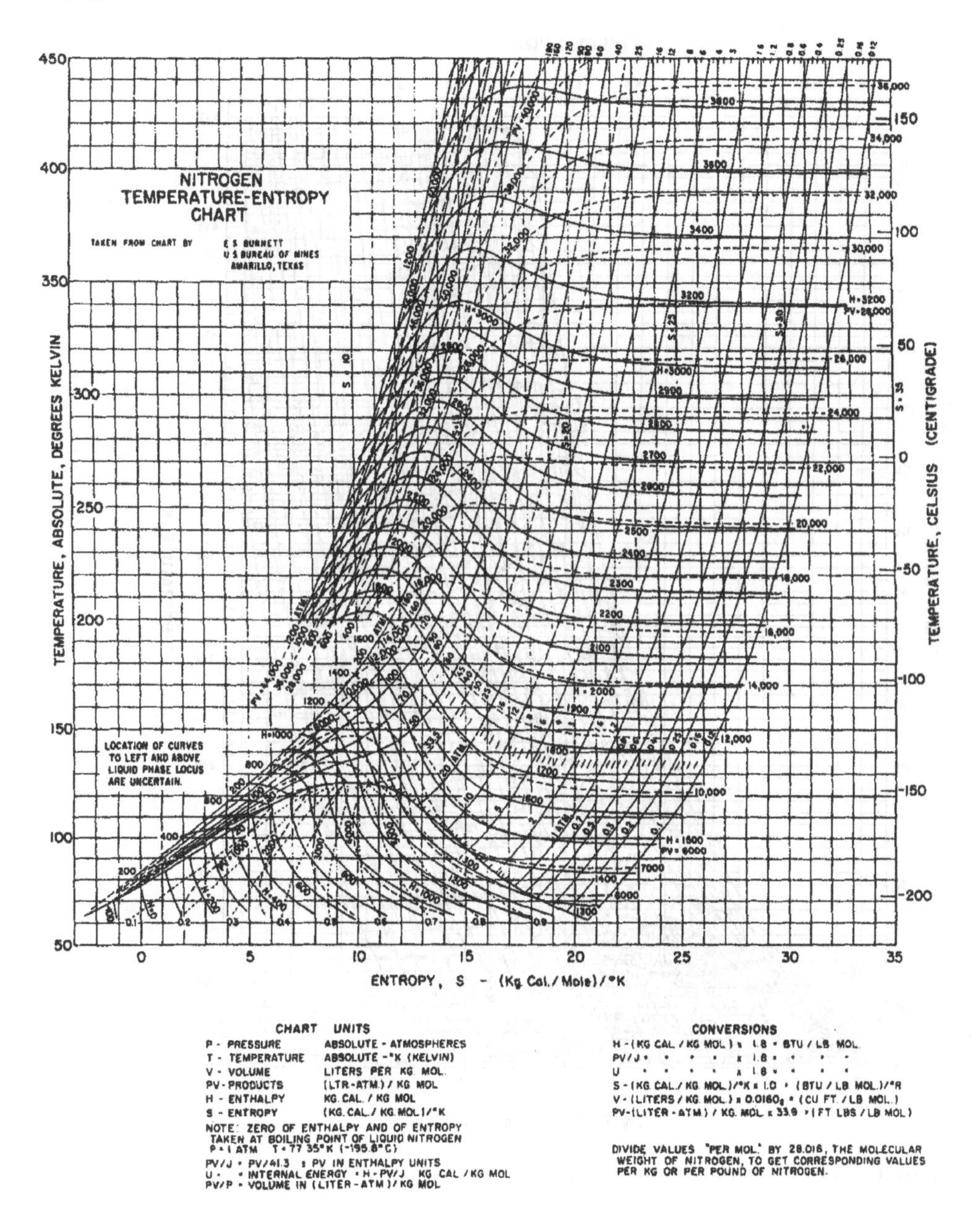


**Figure 3.32** Temperature–entropy chart for nitrogen.

reverse problem at many plants; they are looking for new oxygen uses to balance the demand for nitrogen.

*Example 3–2 Nitrogen Sample Problem.* Nitrogen gas at 138 K and 220 atm pressure is expanded through a throttling valve. What is the final temperature? What is the fraction liquefied?

This problem statement is typical of the conditions found in a Joule–Thomson nitrogen liquefier. The nitrogen gas has been compressed to 200 atm and precooled to 138 K in a countercurrent heat exchanger just prior to expansion through the J–T

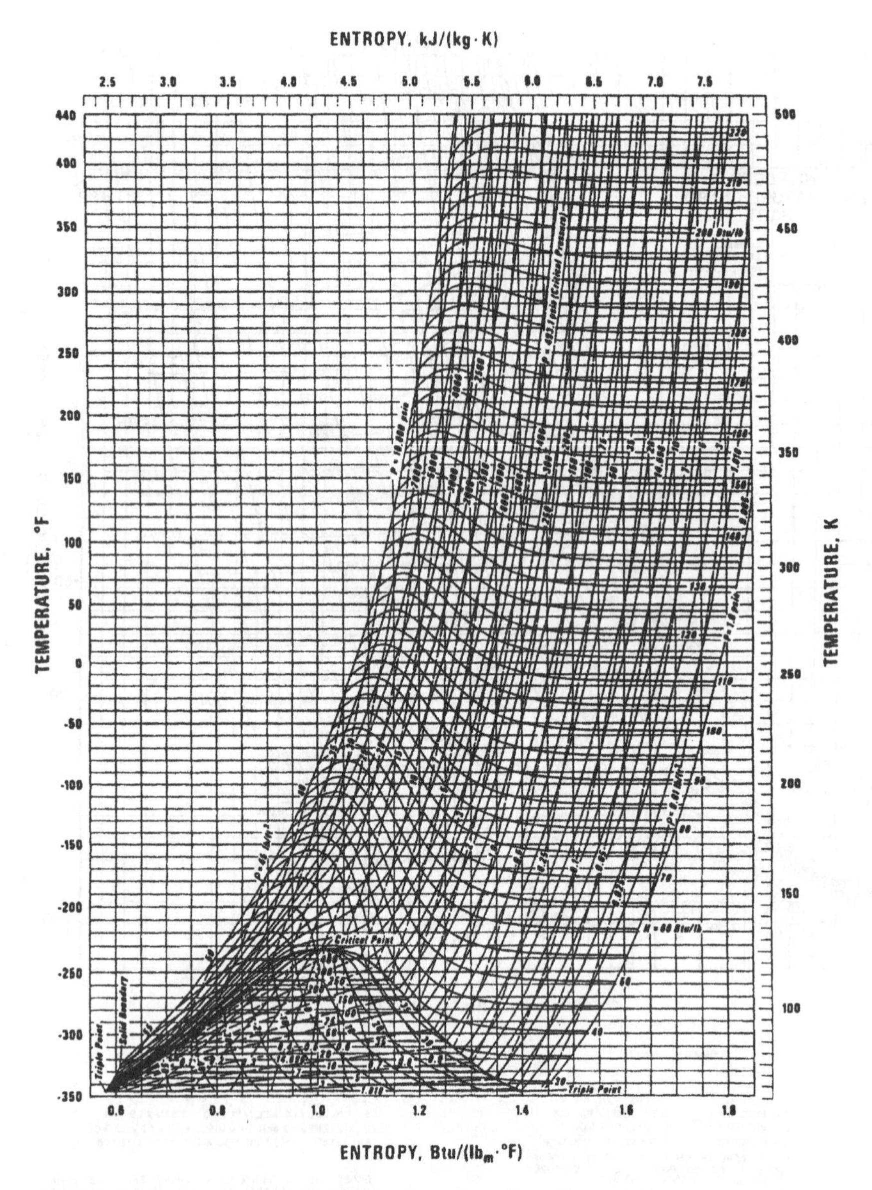


**Figure 3.33** Temperature–entropy chart for nitrogen.

valve. We know the starting conditions, 138 K, 200 atm. The thermodynamic process is expansion through a J–T valve, which is isenthalpic.

The nitrogen *T–S* diagram of Fig. 3.54 contains all the information we need to solve this problem.

The starting point is 138 K, 200 atm, which also fixes the enthalpy at 160 J/gK. Since the process is isenthalpic, we follow the 160 J/gK isenthalp to the final pressure of 1 atm. Note that temperature and pressure lines are coincident and horizontal in the two-phase liquid–vapor region. The final temperature is easily found, namely the NBP of nitrogen, 77 K. The fraction liquefied is given by the lines of constant quality ($X$) in the two-phase region. Our final quality falls between $X = 60\%$ vapor and $X = 70\%$ vapor. We estimate the final quality to be about 63% vapor.

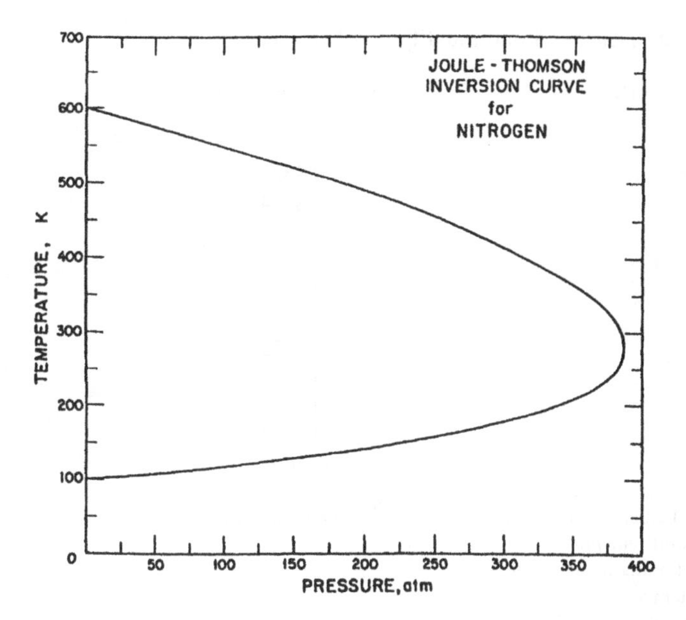


**Figure 3.34** Joule–Thomson inversion curve for nitrogen.

**Table 3.19** Vapor Pressure of Solid and Liquid Nitrogen

| Temp. (K) | Vapor pressure (atm) | Temp. (K) | Vapor pressure (atm) | Temp.(K) | Vapor pressure (atm) |
|---|---|---|---|---|---|
| 52 | 0.0075 | 76 | 0.8461 | 104 | 10.07 |
| 54 | 0.0134 | 77.35 ± .02[b] | 1.000 | 105.0 | 10.71 |
| 56 | 0.0232 | 78 | 1.073 | 106 | 11.42 |
| 57.9 | 0.0379 | 80.0 | 1.36 | 108 | 12.91 |
| 58 | 0.0387 | 80 | 1.341 | 110.0 | 14.54 |
| 59.0 | 0.0496 | 82 | 1.657 | 110 | 14.52 |
| 60 | 0.0621 | 84 | 2.026 | 112 | 16.26 |
| 61.0 | 0.0787 | 85.0 | 2.25 | 114 | 18.15 |
| 62 | 0.0968 | 86 | 2.460 | 115.0 | 19.28 |
| 63.14[a] | 0.1268 | 88 | 2.967 | 116 | 20.20 |
| 63.156 | 0.1237 | 90.0 | 3.54 | 118 | 22.41 |
| 64 | 0.1439 | 90 | 3.548 | 120.0 | 25.04 |
| 64.55 | 0.1591 | 92 | 4.203 | 120 | 24.81 |
| 66 | 0.2028 | 94 | 4.937 | 122 | 27.40 |
| 68 | 0.2797 | 95.0 | 5.31 | 124 | 30.21 |
| 68.4 | 0.3005 | 96 | 5.76 | 125.0 | 31.94 |
| 70 | 0.3784 | 98 | 6.68 | 126 | 33.27 |
| 72 | 0.5033 | 100.0 | 7.67 | 126.1[c] | 33.49 |
| 74 | 0.6579 | 100 | 7.70 | 126.1 ± 0.1[c] | 33.5 |
| 74.9 | 0.7386 | 102 | 8.83 | | |

[a] Melting point.
[b] Normal boiling point.
[c] Critical point.
*Source*: Hoge (1950); Hoge and King (1950); Keesom (1922, 1923); Mathias and Crommelin (1924).

**Table 3.20** Density of Saturated Liquid Nitrogen

| Temperature (K) | Density ($g/cm^3$) |
|---|---|
| 63.14[a] | |
| 64.73 | 0.8622 |
| 77.31 | 0.8084 |
| 77.32[b] | |
| 77.5 | |
| 78.00 | 0.8043 |
| 90.58 | 0.7433 |
| 99.36 | 0.6922 |
| 111.89 | 0.6071 |
| 119.44 | 0.5332 |
| 125.01 | 0.4314 |
| 125.96[c] | 0.31096 |

*Note*: The Leiden temperature scale (0°C = 273.09 K) was used.
[a] Triple point temperature.
[b] Normal boiling temperature.
[c] Critical temperature.
*Source*: Gerold (1921); Mathias and Crommelin (1924).

## 3.3. Air

Dry air (avg mol wt. approximately 29.00) is a mixture consisting principally of nitrogen, oxygen, and argon with traces of other gases as shown in Table 3.25.

When air is liquefied, the carbon dioxide is usually removed; thus, for practical purposes, liquid air can be considered to consist of 78% nitrogen, 21% oxygen, and 1% argon, the other constituents being present in negligible amounts. Sometimes the presence of argon is ignored and liquid air is considered to be a binary mixture of 21% oxygen and 79% nitrogen. Since argon has a vapor pressure between those of oxygen and nitrogen, this assumption is a rather good approximation for some purposes.

Liquid air has a density of 874 g/L at its normal boiling point of 78.9 K.

Figure 3.55 is a *T–S* diagram for air.

**Table 3.21** Viscosity of Liquid Nitrogen

| Temp. (K) | Viscosity (cP) |
|---|---|
| 63.9 | 0.292 |
| 64.3 | 0.290 |
| 64.8 | 0.284 |
| 69.1 | 0.231 |
| 69.25 | 0.228 |
| 71.4 | 0.209 |
| 76.1 | 0.165 |
| 77.33 | 0.158 |
| 111.7 | 0.074 |

*Source:* Rudenko (1939); Rudenko and Shubnikow (1934).

**Table 3.22** Viscosity of Gaseous Nitrogen above 1 atm at $T = 298$ K

| Pressure (atm) | Viscosity (cP) × $10^2$ |
|---|---|
| 1 | 1.778 |
| 2 | 1.780 |
| 5 | 1.784 |
| 10 | 1.793 |
| 20 | 1.809 |
| 30 | 1.827 |
| 40 | 1.846 |
| 50 | 1.867 |
| 60 | 1.890 |
| 70 | 1.916 |
| 80 | 1.946[a] |
| 90 | 1.978[a] |
| 100 | 2.015[a] |

[a] Extrapolated.
*Source:* Kestin and Wang (1958).

### *3.3.1. Properties of Air*

All data are from Johnson (1960) and Din (1956) unless otherwise noted.

Table 3.24 gives the physical constants for air.

**a. Heat of Vaporization.** The heat of vaporization is usually defined for a single component and is the heat required to completely change a given quantity of liquid at its bubble point (boiling point) to vapor at its dew point at constant temperature and pressure. However, for air, which is a multicomponent system, only the pressure remains constant while the temperature rises throughout the process. The data given in Table 3.26 are for air at the temperature and pressure of the saturated liquid (bubble point). Table 3.27 presents data on the latent heat of vaporization of an oxygen–nitrogen mixture at 1 atm.

**Table 3.23** Viscosity of Gaseous Nitrogen at 1 atm

| Temp. (K) | Viscosity (cP) | Temp. (K) | Viscosity (cP) |
|---|---|---|---|
| 90 | 0.006 298 | 210 | 0.013 499 |
| 100 | 0.006 975 | 220 | 0.014 029 |
| 110 | 0.007 631 | 230 | 0.014 547 |
| 120 | 0.008 264 | 240 | 0.015 052 |
| 130 | 0.008 876 | 250 | 0.015 547 |
| 140 | 0.009 484 | 260 | 0.016 031 |
| 150 | 0.010 083 | 270 | 0.016 502 |
| 160 | 0.010 676 | 280 | 0.016 960 |
| 170 | 0.011 253 | 290 | 0.017 410 |
| 180 | 0.011 829 | 298.1 | 0.017 777 |
| 190 | 0.012 394 | 300 | 0.017 857 |
| 200 | 0.012 954 | | |

*Source:* Johnston and McCloskey (1940).

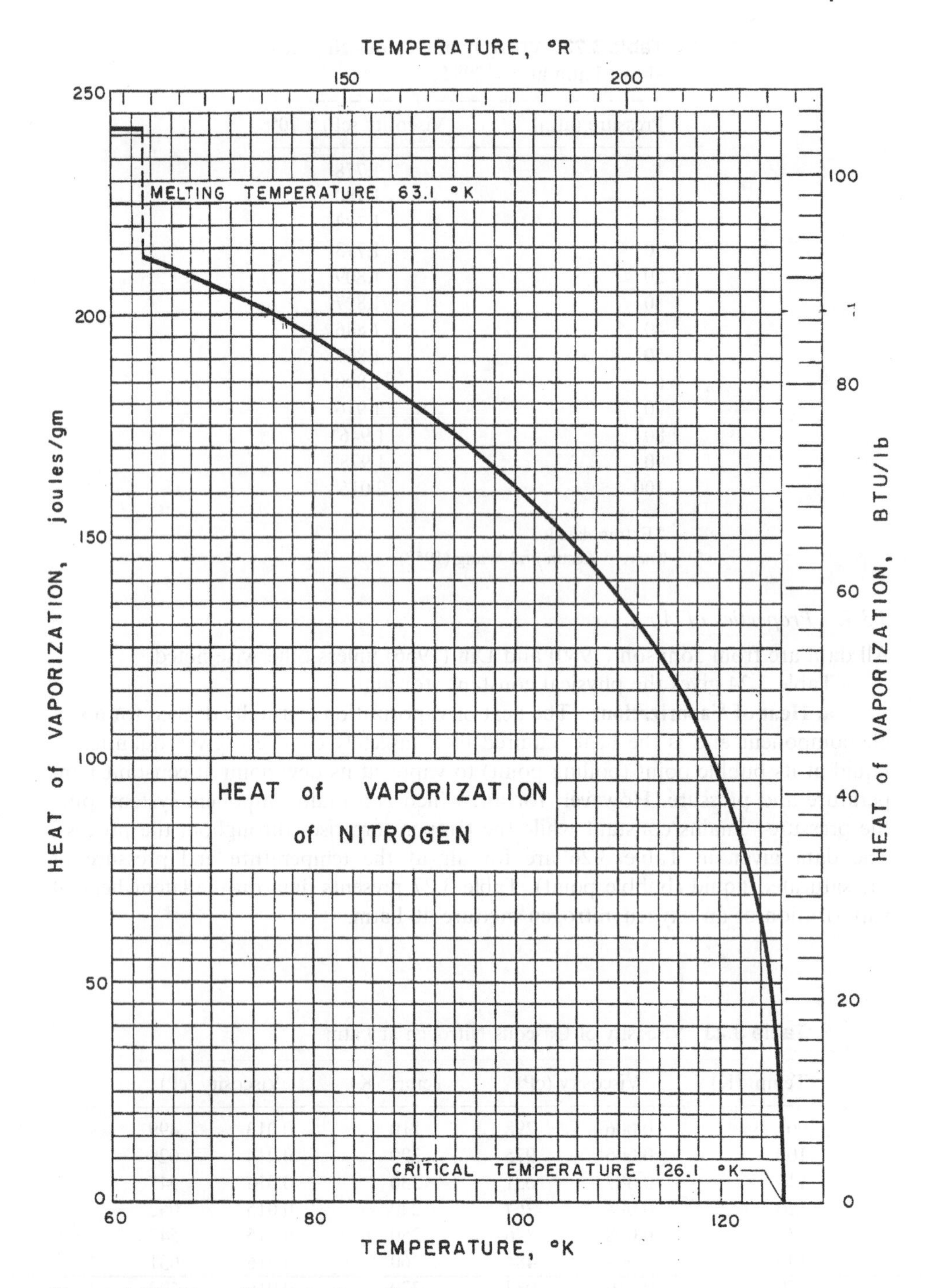


**Figure 3.35** Heat of vaporization of nitrogen.

**b. Vapor Pressure.** It should be noted that there are separate curves for the vapor pressure of saturated liquid (bubble) and saturated vapor (dew). Air is a multi-component mixture, and at a given pressure the saturated liquid is at a different temperature than the saturated vapor (see Table 3.28 and Figure 3.56).

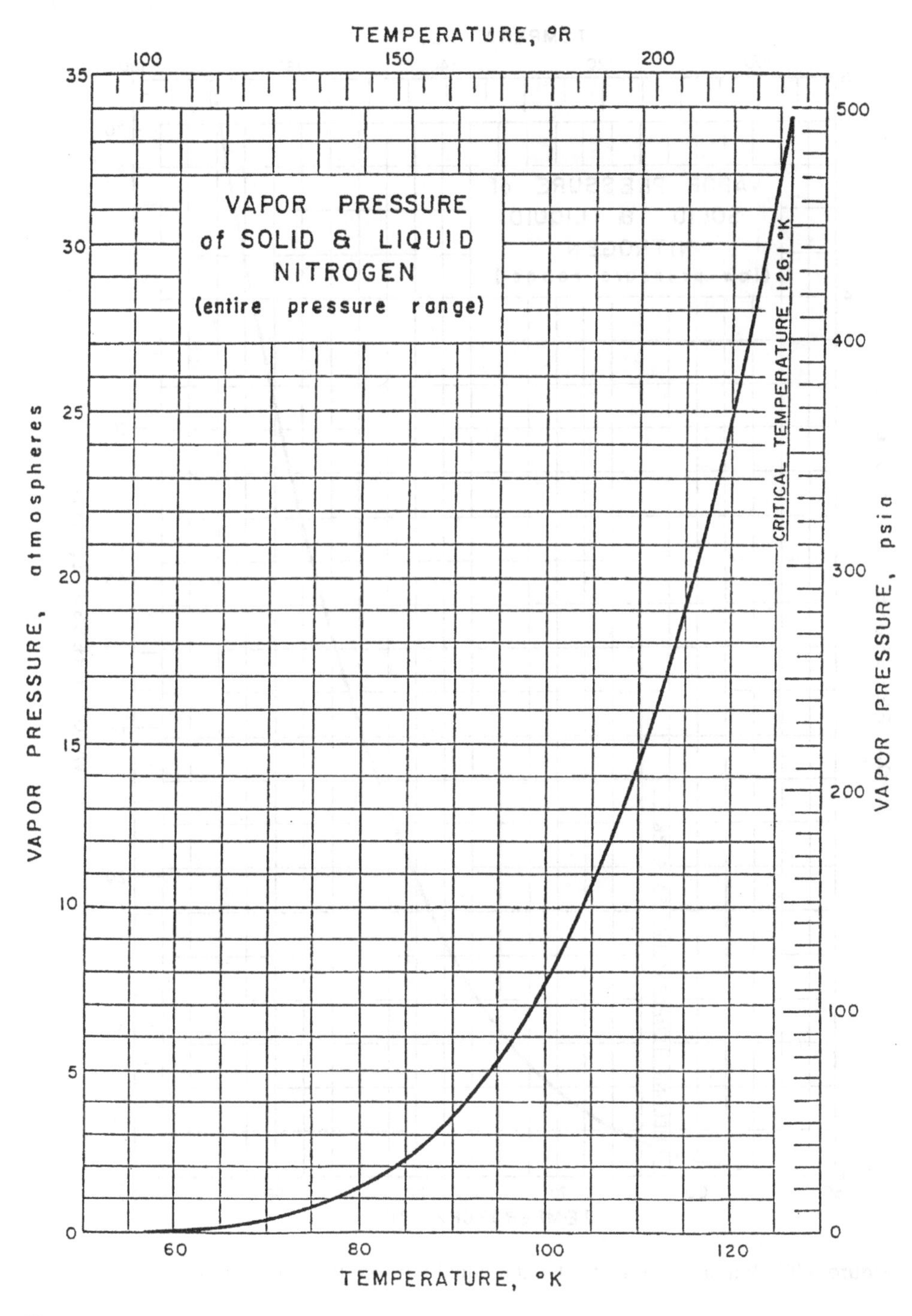


**Figure 3.36** Vapor pressure of solid and liquid nitrogen (entire pressure range).

Bubble curve:

$$\log P = 3.5713 - \frac{290.70}{T} + 0.001494T \tag{3.10}$$

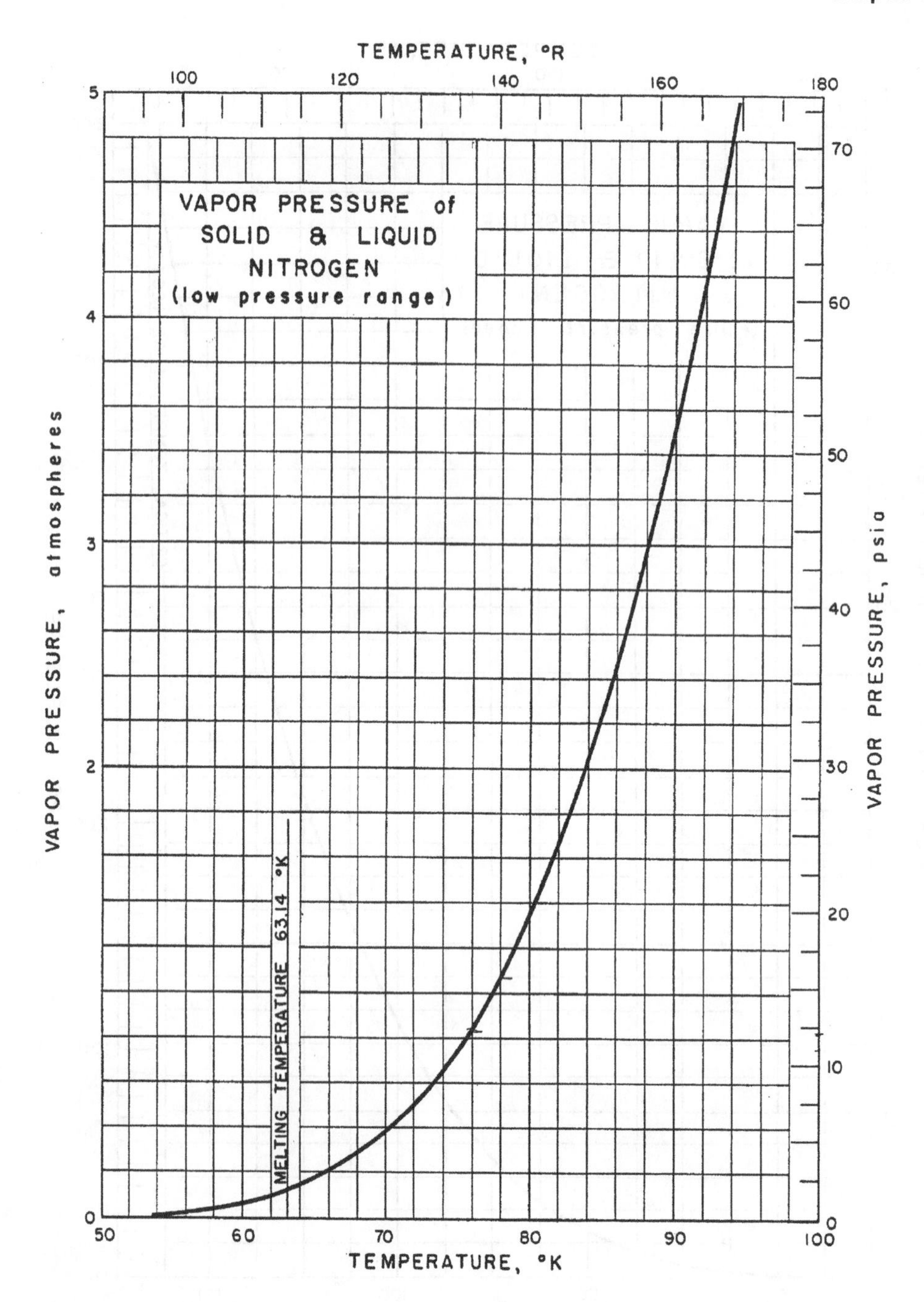


**Figure 3.37** Vapor pressure of solid and liquid nitrogen (low pressure range).

Dew curve:

$$\log P = 4.0816 - \frac{333.88}{T} \tag{3.11}$$

where $P$ is in atm and $T$ is kelvin.

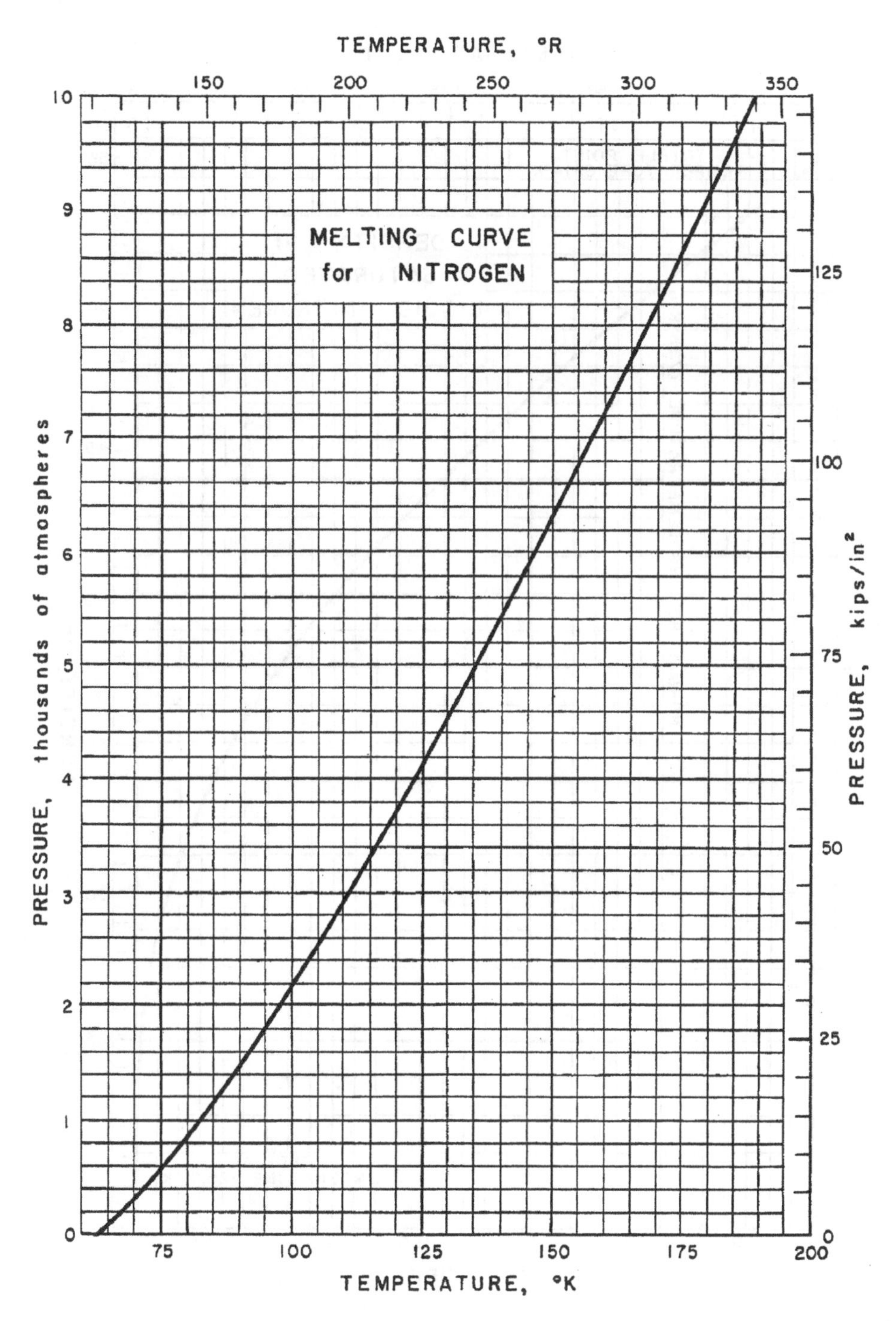


**Figure 3.38** Melting curve for nitrogen.

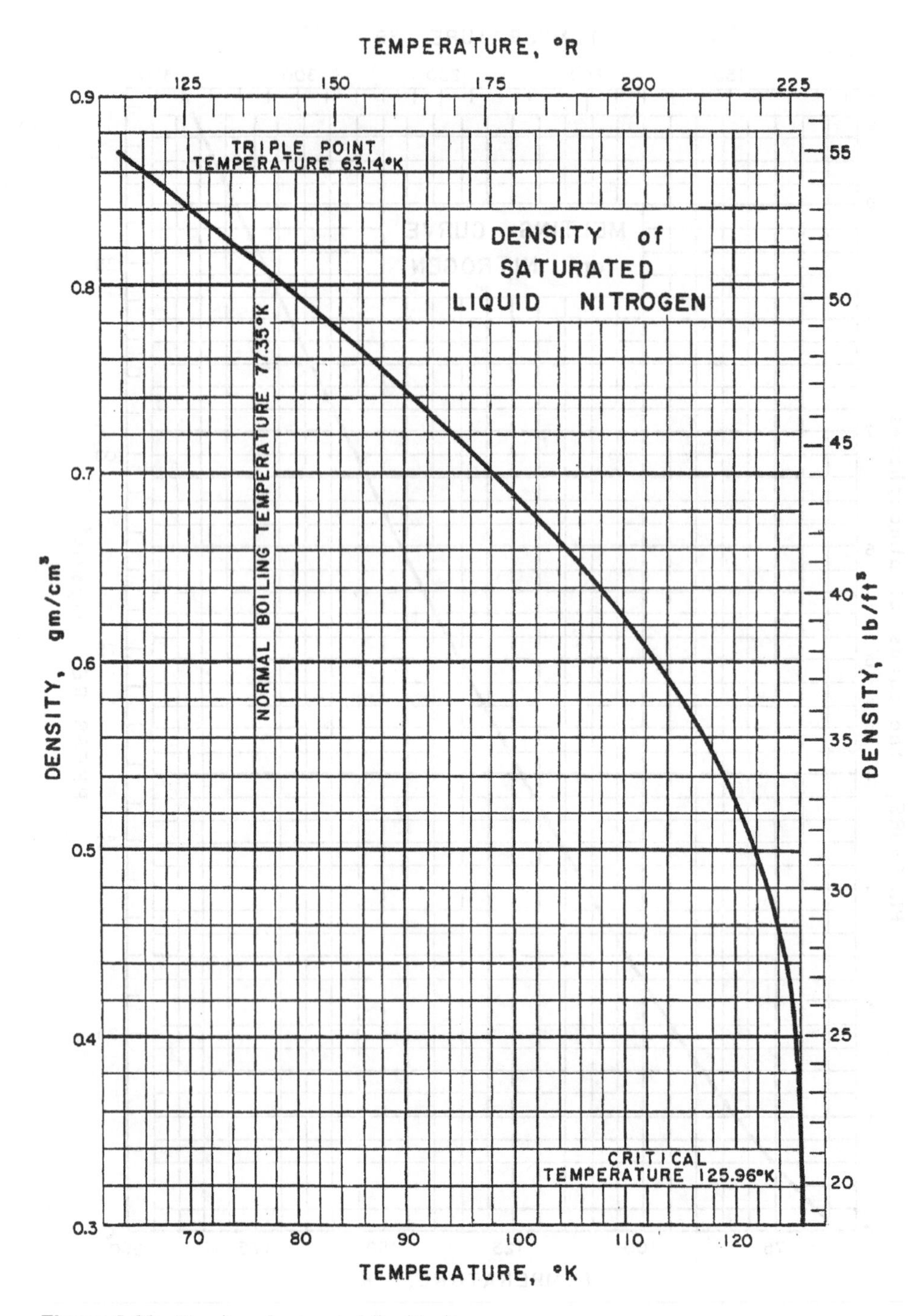


**Figure 3.39** Density of saturated liquid nitrogen.

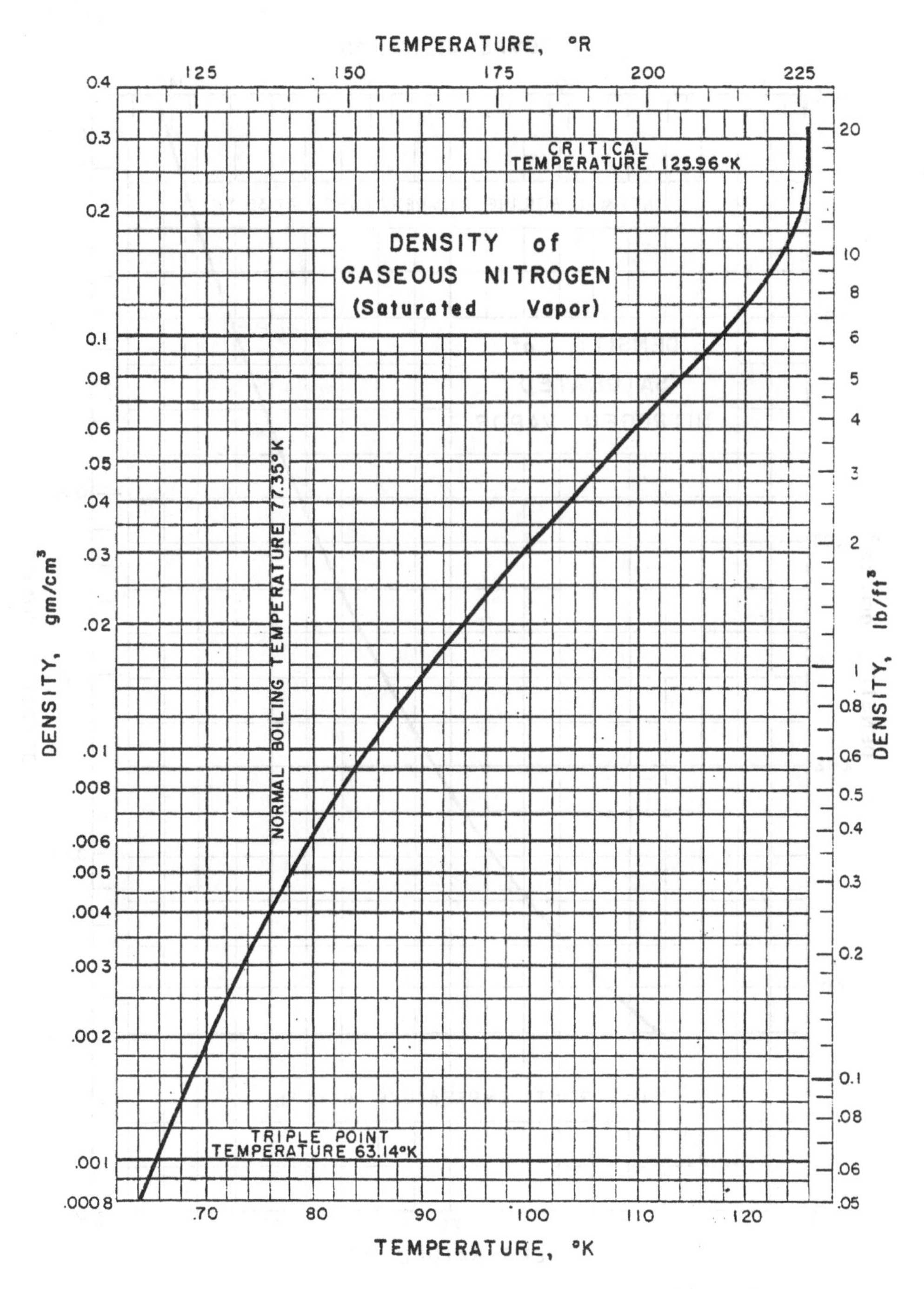


**Figure 3.40** Density of gaseous nitrogen (saturated vapor) to the critical point.

**c. Thermal Conductivity.** See Table 3.29

**d. Density.** See Table 3.30

**e. Vapor–Liquid Equilibrium Constants for the Major Components of Air.** The relationship between the mole fraction composition of a component ($i$) in the liquid

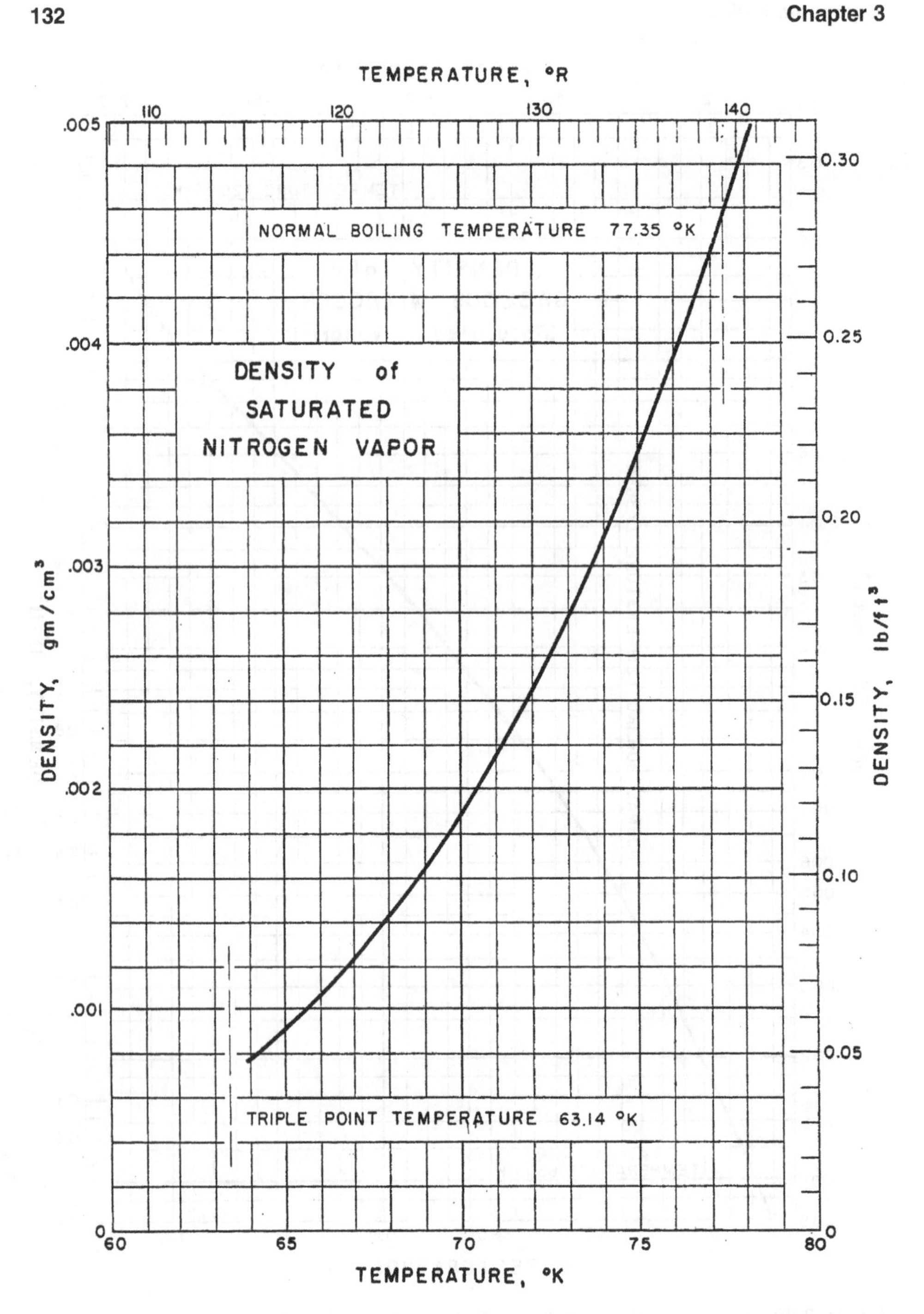


**Figure 3.41** Density of gaseous nitrogen (saturated vapor) triple point to the NBP.

phase ($x_i$) and the vapor phase ($y_i$) at equilibrium is given by the equilibrium constant ($K_i$) :

$$K_i = y_i/x_i \tag{3.12}$$

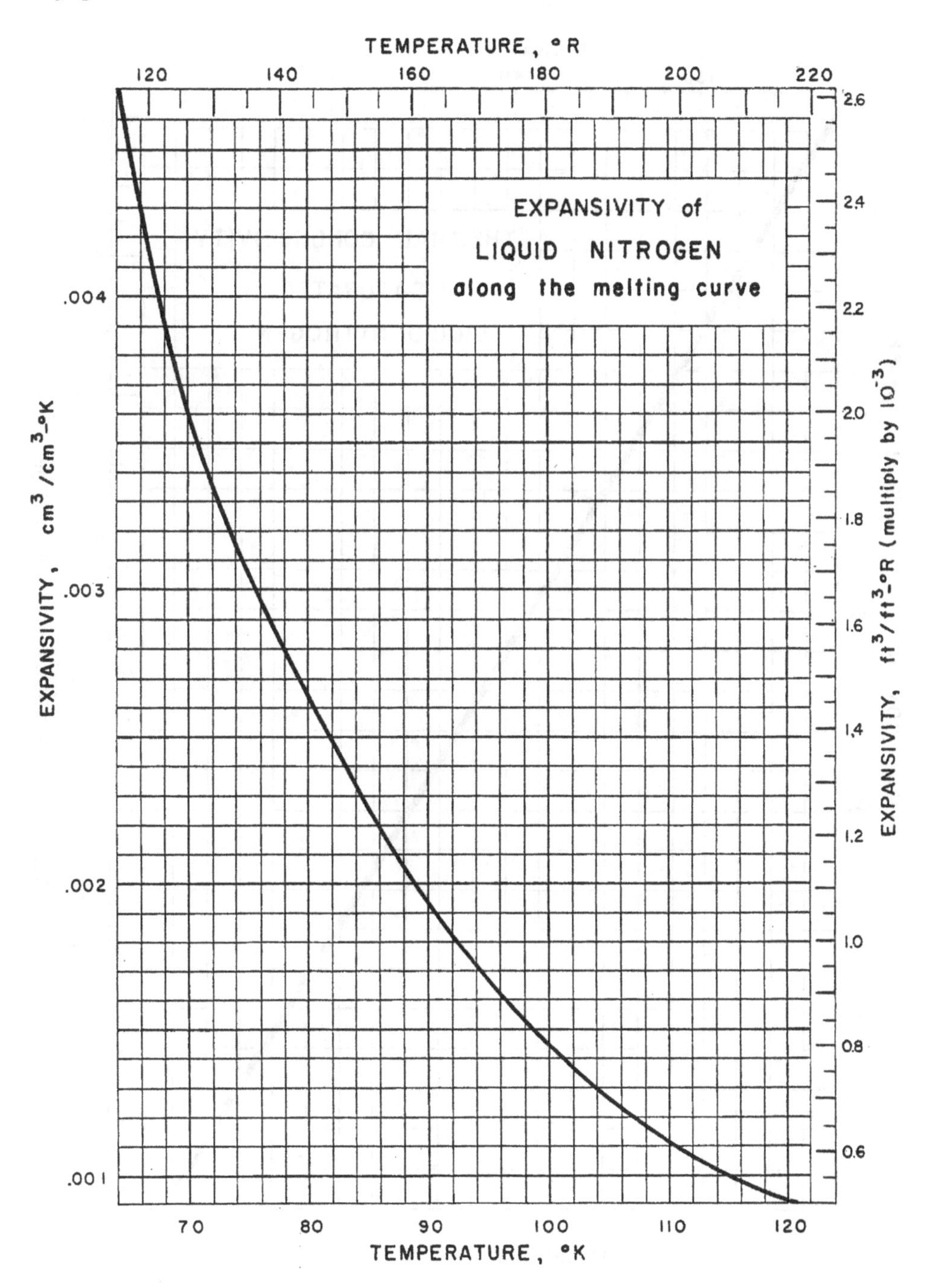


**Figure 3.42** Expansivity of liquid nitrogen along the melting curve.

If both the liquid and vapor phases are ideal mixtures, it can be shown from Raoult's and Dalton's laws that

$$K_i = P_i/P \tag{3.13}$$

where $P_i$ is the vapor pressure of component $i$ and $P$ are total system pressure.

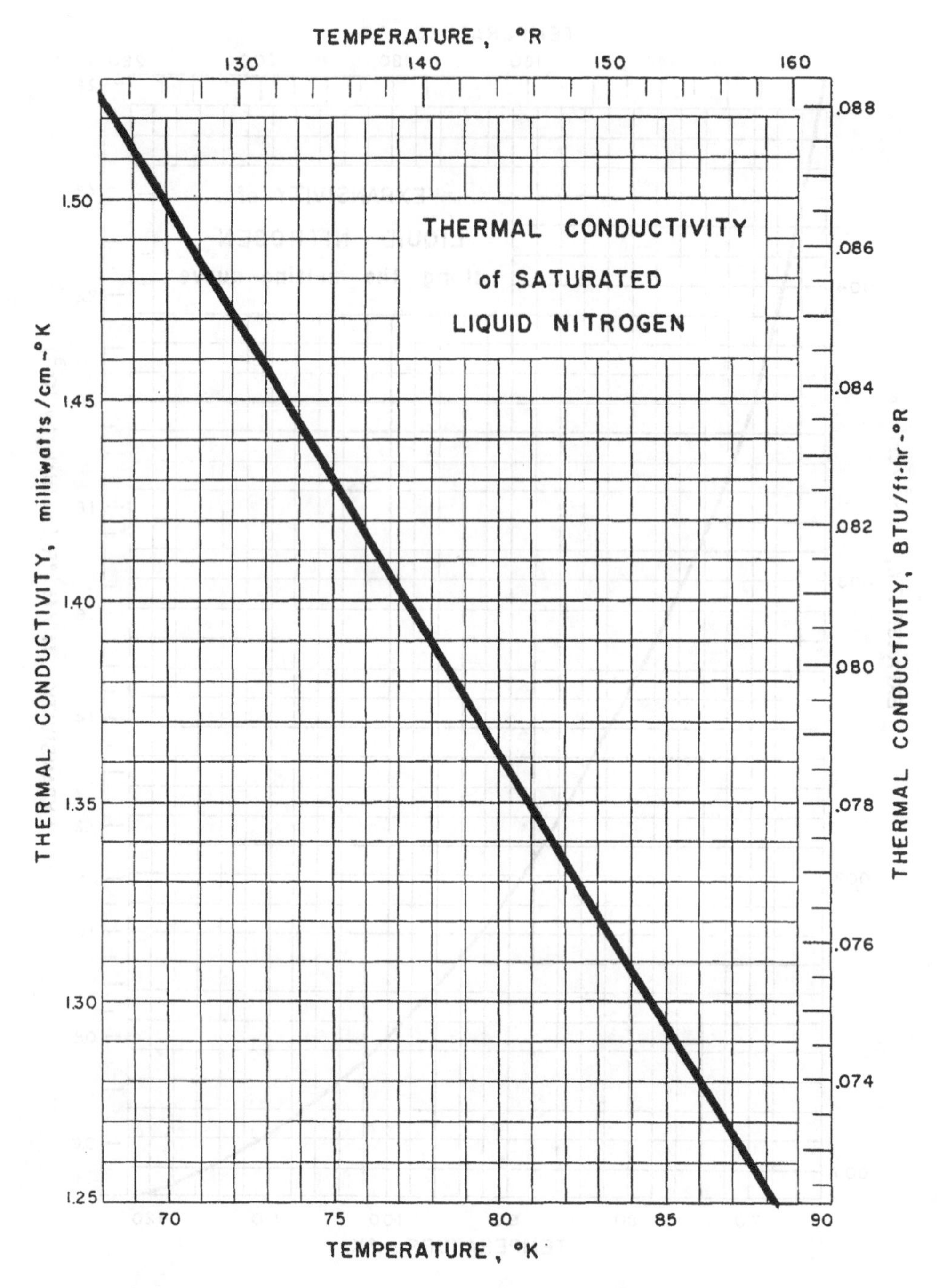


**Figure 3.43** Thermal conductivity of saturated liquid nitrogen.

For real systems, values of $K$ are determined experimentally and are often tabulated as the *distribution coefficient function*, ln $(K_p/p_o)$, where $p_o$ is some reference pressure, such as 101.3 kPa (1 atm). Values for the distribution coefficient function for nitrogen, oxygen, and argon are given in Table 3.31.

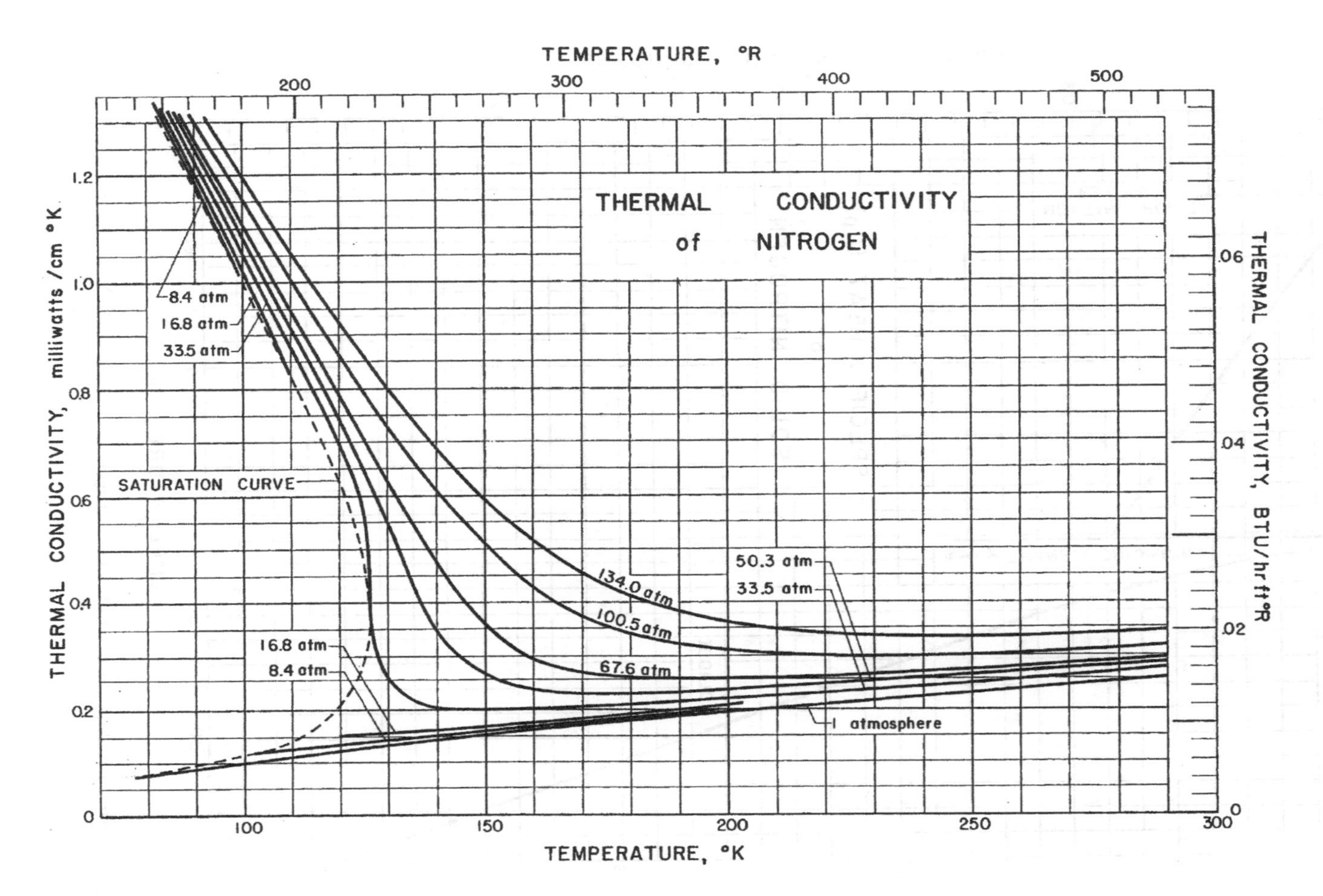


**Figure 3.44** Thermal conductivity of nitrogen.

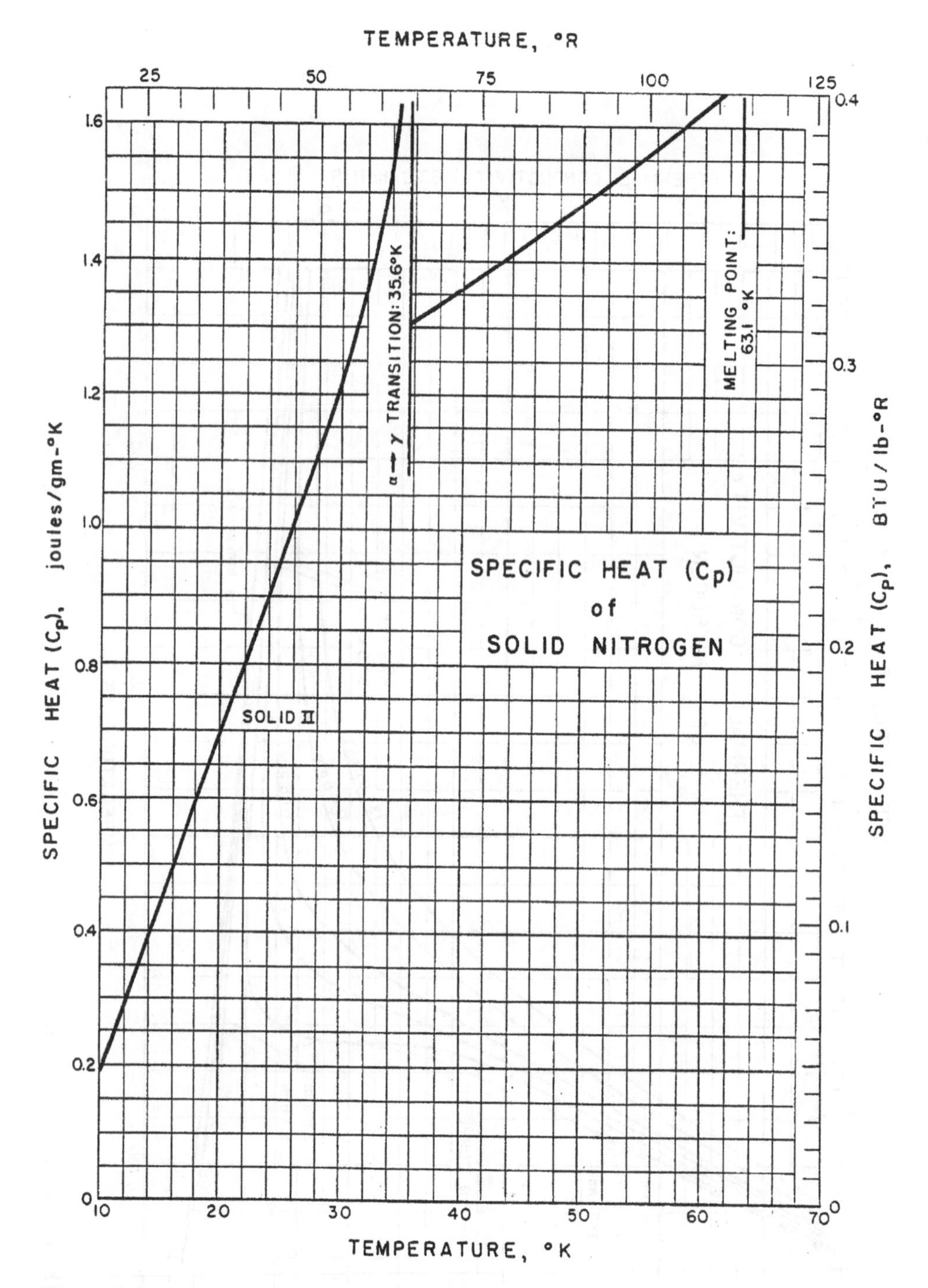


**Figure 3.45** Specific heat ($C_P$) of solid nitrogen.

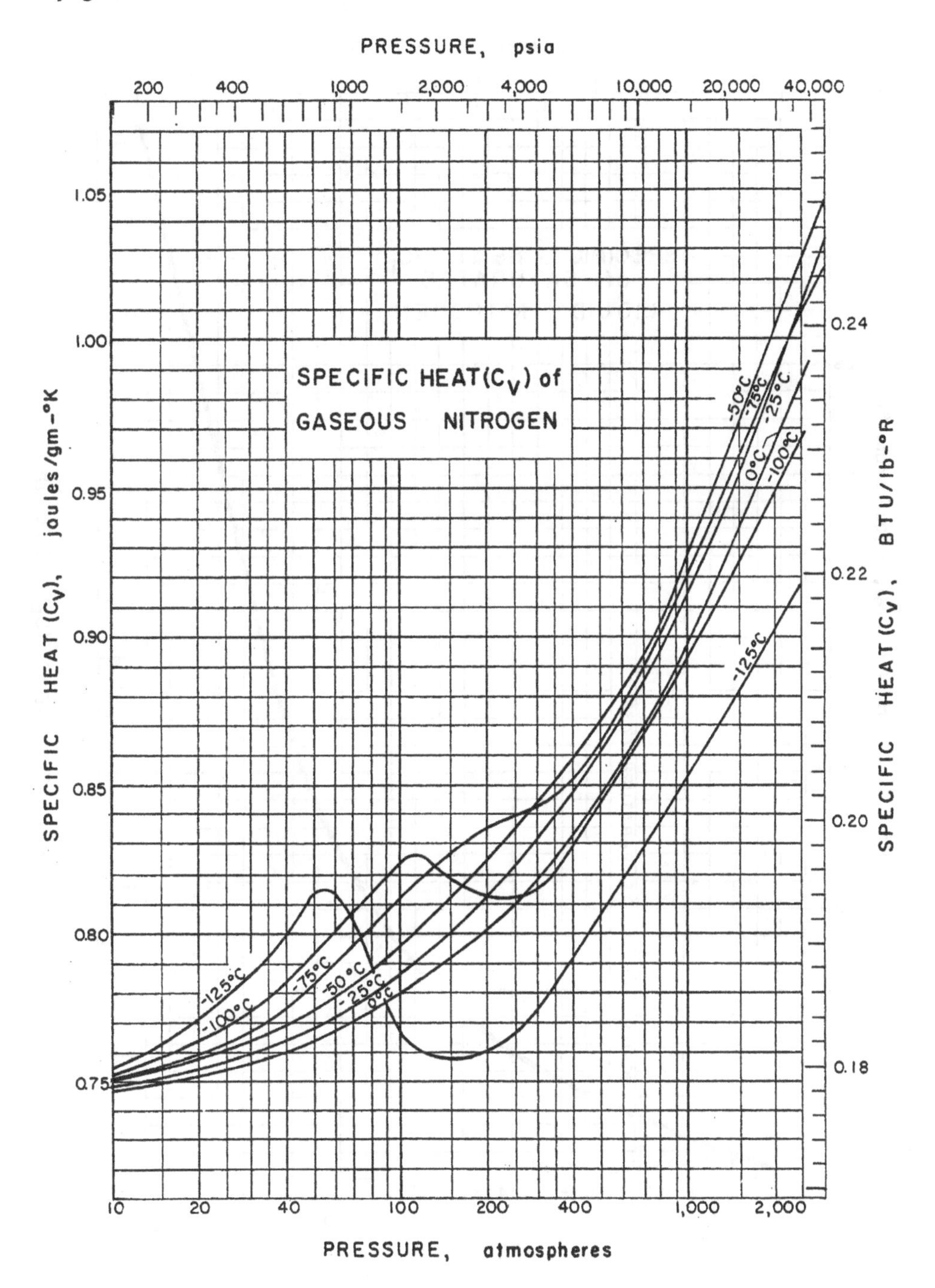


**Figure 3.46** Specific heat ($C_V$) of gaseous nitrogen.

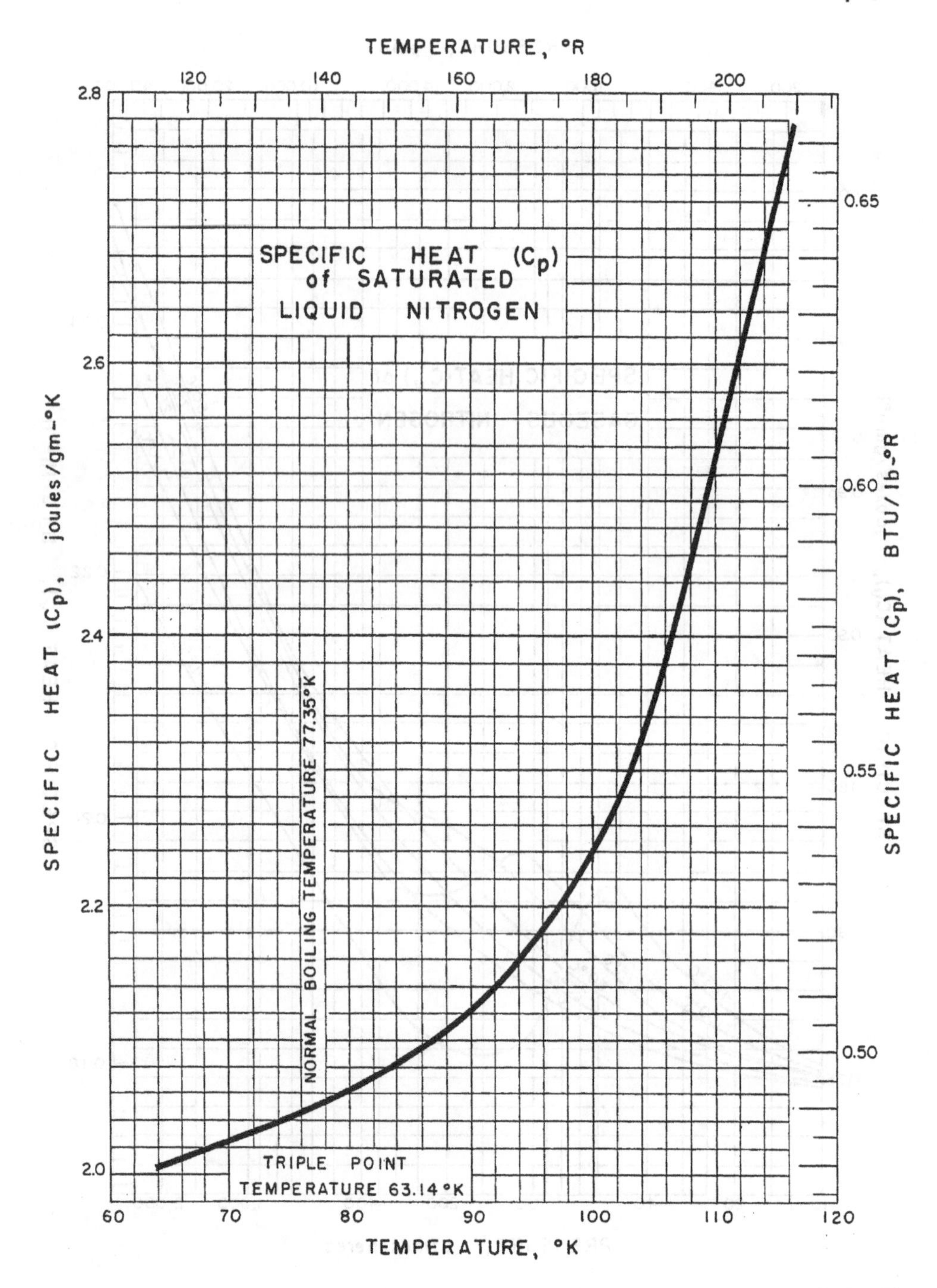


**Figure 3.47** Specific heat ($C_P$) of saturated liquid nitrogen.

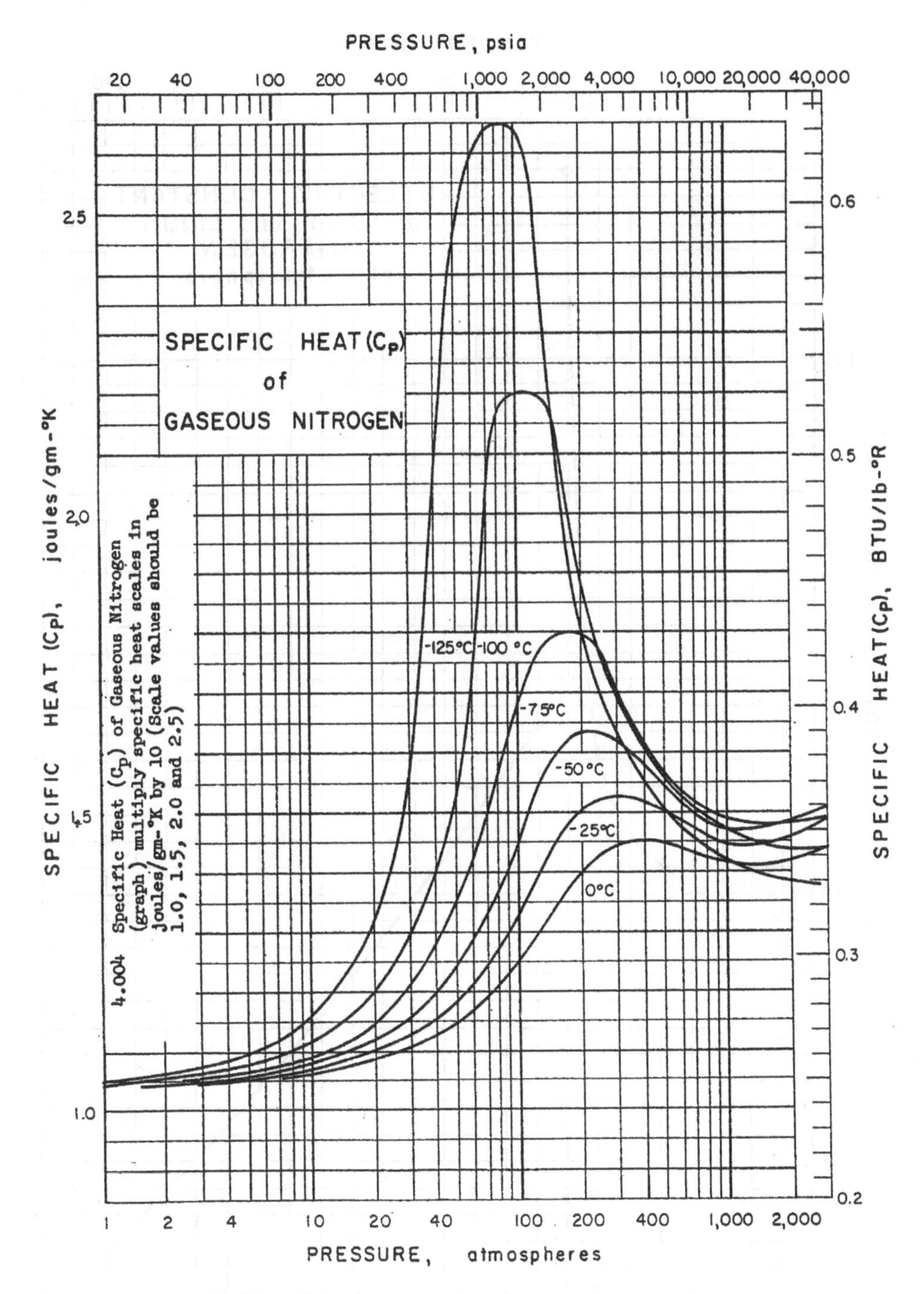


**Figure 3.48** Specific heat ($C_P$) of gaseous nitrogen.

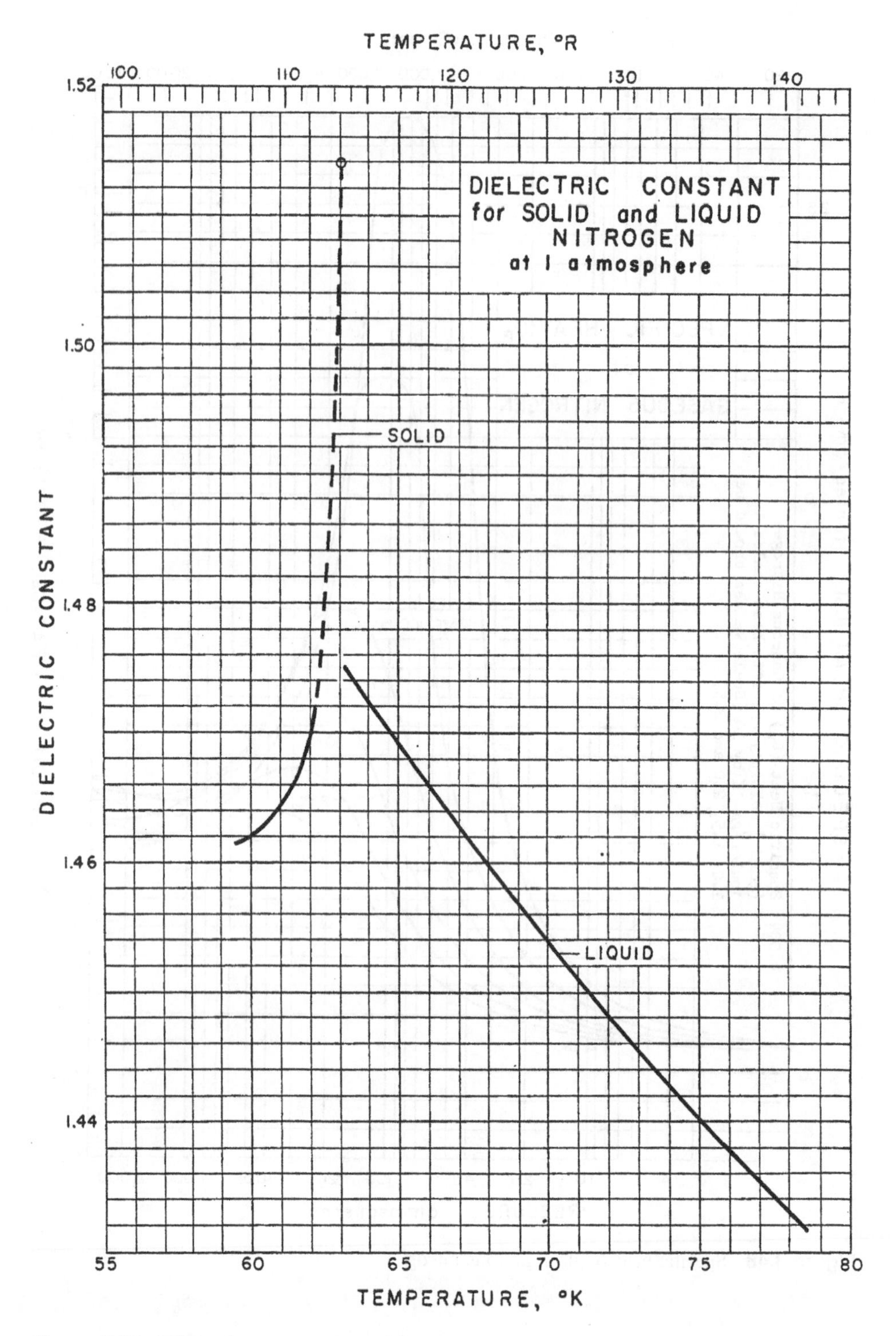


**Figure 3.49** Dielectric constant for solid and liquid nitrogen at 1 atm.

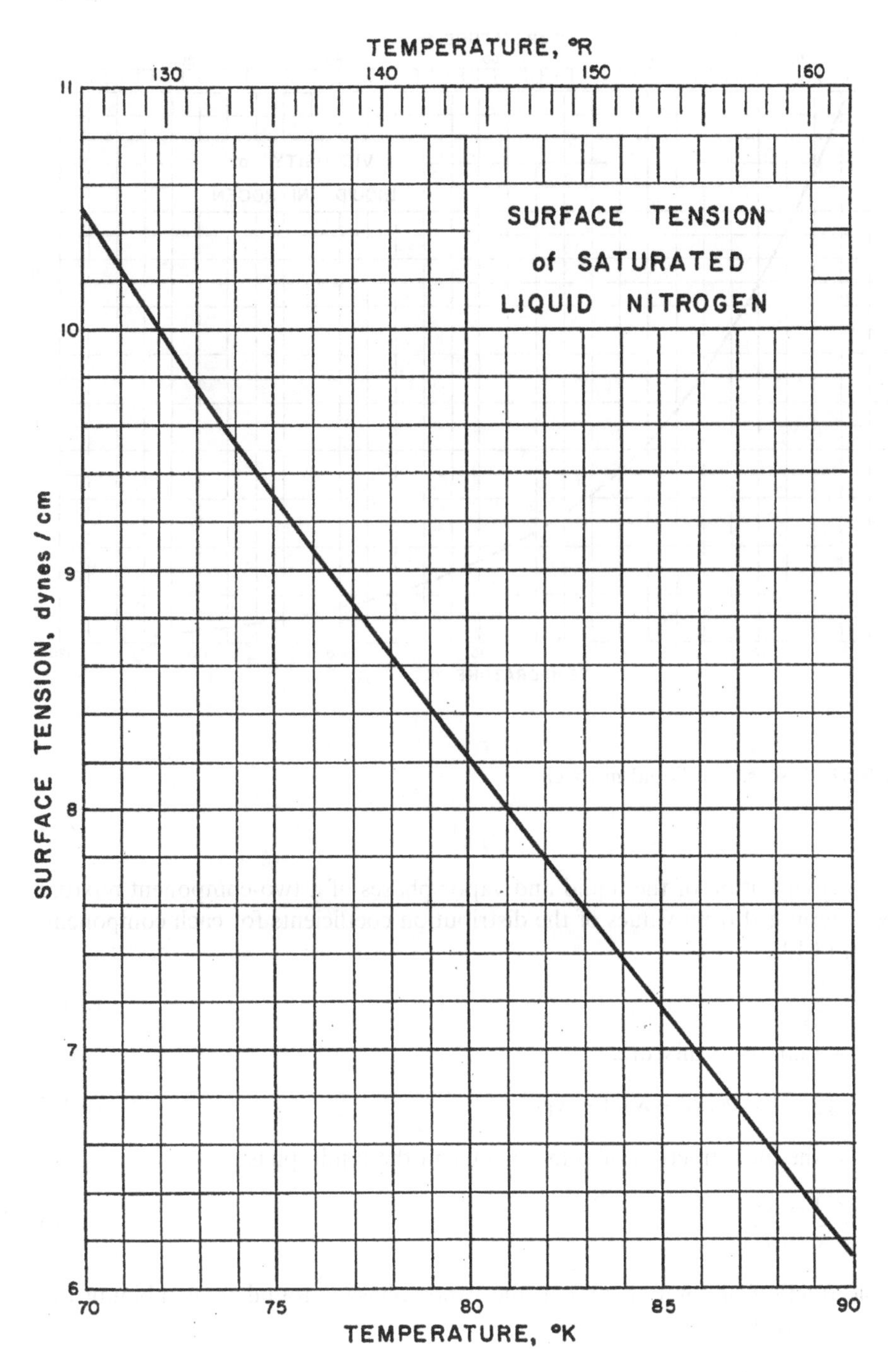


**Figure 3.50** Surface tension of saturated liquid nitrogen.

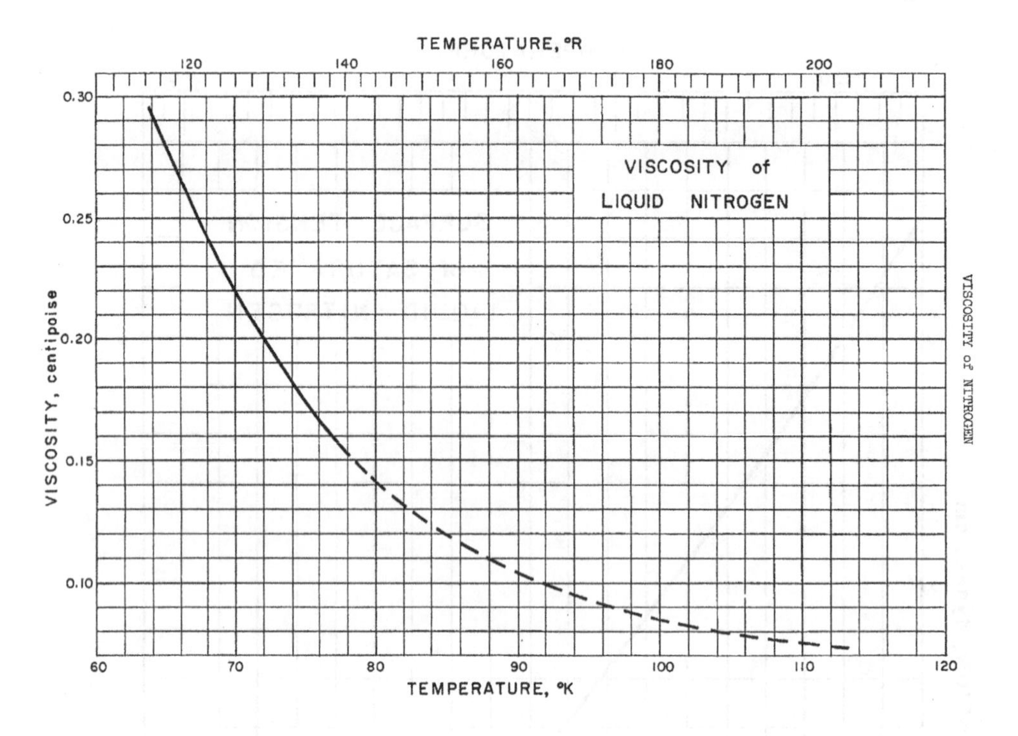


**Figure 3.51** Viscosity of liquid nitrogen.

The composition of the liquid and vapor phases of a two-component mixture may be determined from values of the distribution coefficients for each component. From Eq. (3.12),

$$\begin{aligned} y_1 &= K_1 x_1 \\ y_2 &= K_2 x_2 = K_2(1 - x_1) \end{aligned} \tag{3.14}$$

For a two-component mixture.

$$y_1 + y_2 = 1 = K_1 x_1 + K_2(1 - x_1) \tag{3.15}$$

Solving for the mole fraction of component 1 in the liquid phase,

$$x_1 = \frac{1 - K_2}{K_1 - K_2} \tag{3.16}$$

The mole fraction of component 1 in the vapor phase is found by combining Eq. (3.13 ) and (3.16 ) :

$$y_1 = k_1 x_1 = \frac{K_1(1 - K_2)}{K_1 - K_2} \tag{3.17}$$

### *3.3.2. Uses of Liquid Air*

Liquid air was once widely used as a refrigerant for low-temperature investigations, but the relative simplicity of producing liquid nitrogen from liquid air by distillation

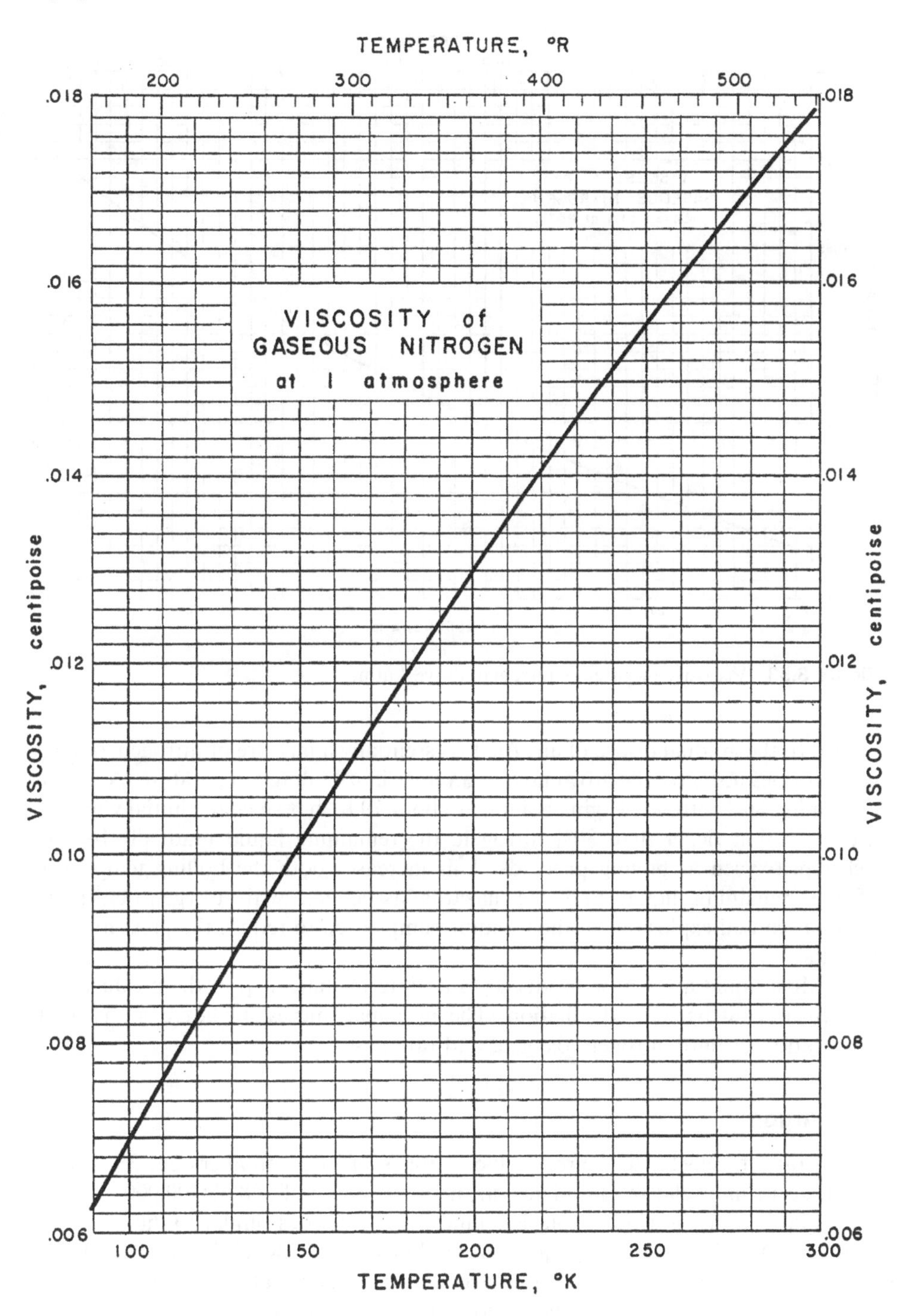


**Figure 3.52** Viscosity of gaseous nitrogen at 1 atm.

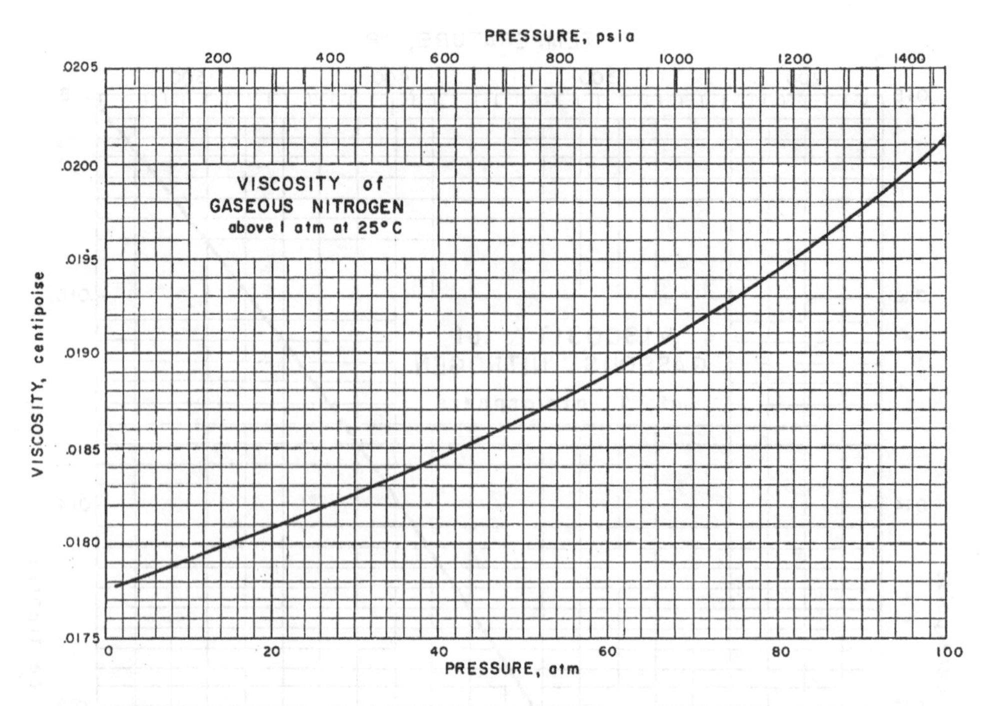


**Figure 3.53** Viscosity of gaseous nitrogen above 1 atm.

has led to the gradual disuse of air. As it is essentially a mixture of nitrogen and oxygen, air does not have the advantage of an invariant boiling temperature. A mixture containing 80% nitrogen begins boiling at about 79 K, but since the initial vapor has a composition richer than 80% nitrogen, the remaining liquid gradually becomes richer in oxygen as boiling proceeds. This increases both the boiling temperature of the remaining liquid and the hazards associated with oxygen-rich liquid. This has caused explosions in the vacuum pumps used to reduce the pressure above liquid air.

Most liquid air is now produced as an intermediate step in the production of oxygen and nitrogen by distillation. The principal interest in liquid air is in the preparation of pure nitrogen, oxygen, and rare gases.

## 3.4. Argon

Argon (at wt. 39.948) has three stable isotopes of mass numbers 36, 38, and 40, which occur in a relative abundance in the atmosphere of 338:63:100,000.

Liquid argon is a clear, colorless fluid with properties similar to those of liquid nitrogen. At 1 bar pressure, liquid argon boils at 87.3 K (157.1°R) and freezes at 83.8 K (151.0°R). Saturated liquid argon at 0.987 bar (1 atm) is more dense than oxygen, as one would expect, since argon has a larger molecular weight than oxygen [argon density = 1403 kg/m$^3$ (87.5 lb$_m$/ft$^3$) for saturated liquid at 0.987 bar (1 atm) ].

See Fig. 3.57 for a *T–S* diagram of argon.

Argon is present in atmospheric air in a concentration of 0.934% by volume or 1.25% by weight. The atmosphere of Mars contains 1.6% $^{40}$Ar and 5 ppm $^{36}$Ar, for comparison. Since the boiling point of argon lies between that of liquid oxygen and that of liquid nitrogen (slightly closer to that of liquid oxygen), a crude grade

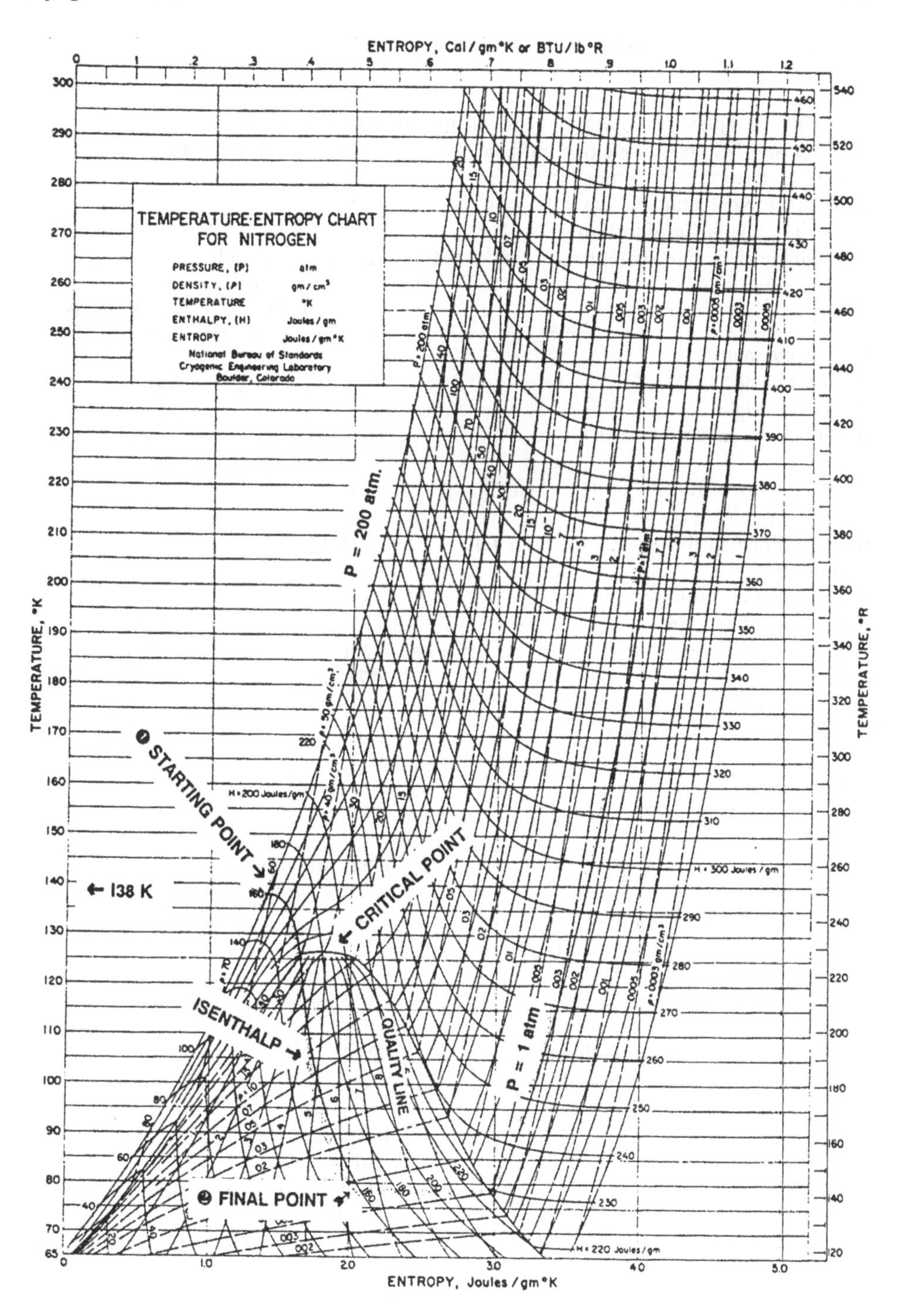


**Figure 3.54** Temperature–entropy chart for nitrogen, sample problem.

of argon (90–95% pure) can be obtained by adding a small auxiliary argon recovery column in an air separation plant. About 900 kg (1 ton) of argon is recovered for every $36 \times 10^3$ kg (40 tons) of oxygen. This is a little less than 50% of the argon passing through the oxygen column and represents the point of present maximum

**Table 3.24** Physical Constants for Air

*Air*
Synonyms: Compressed air, atmospheric air, the atmosphere (of the earth)
CAS Registry Number: None (for nitrogen, 7727–37-9; for oxygen, 7782–44-7)
DOT Classification: Nonflammable gas
DOT Label: Nonflammable gas
Transport Canada Classification: 2.2
UN Number: UN 1002 (compressed gas); UN 1003 (refrigerated liquid)

| | U.S. Units | SI Units |
|---|---|---|
| *Physical constants* | | |
| Chemical name | Air | Air |
| Molecular weight | 28.975 | 28.975 |
| Density of the gas at 70°F (21.1°C) and 1 atm | 0.07493 lb/ft$^3$ | 1.2000 kg/m$^3$ |
| Specific gravity of the gas at 70°F (21.1°C) and 1 atm (air = 1) | 1.00 | 1.00 |
| Specific volume of the gas at 70°F (21.1°C) and 1 atm | 13.346 ft$^3$/lb | 0.8333 m$^3$/kg |
| Boiling point at 1 atm | −317.8°F | −194.3°C |
| Freezing point at 1 atm | −357.2°F | −216.2°C |
| Critical temperature | −221.1°F | −140.6°C |
| Critical pressure | 547 psia | 3771 kPa abs |
| Critical density | 21.9 lb/ft$^3$ | 351 kg/m$^3$ |
| Latent heat of vaporization at normal boiling point | 88.2 Btu/lb | 205 kJ/kg |
| Specific heat of gas at 70°F (21.1°C) and 1 atm | | |
| $C_P$ | 0.241 Btu/(lb)(°F) | 1.01 kJ/(kg)(°C) |
| $C_V$ | 0.172 Btu/(lb)(°F) | 0.720 kJ/(kg)(°C) |
| Ratio of specific heats ($C_P/C_V$) | 1.40 | 1.40 |
| Solubility in water, vol/vol at 32°F (0°C) | 0.0292 | 0.0292 |
| Weight of liquid at normal boiling point | 7.29 lb/gal | 874 kg/m$^3$ |
| Density of liquid at boiling point and 1 atm | 54.56 lb/ft$^3$ | 874.0 kg/m$^3$ |
| Gas/liquid ratio (liquid at boiling point, gas at 70°F and 1 atm), vol/vol | 728.1 | 728.1 |
| Thermal conductivity | | |
| at −148°F (−100°C) | 0.0095 Btu/(hr)(ft)(°F/ft) | 0.0164 W/(m)(°C) |
| at 32°F (0°C) | 0.0140 Btu/(hr)(ft)(°F/ft) | 0.0242 W/(m)(°C) |
| at 212°F (100°C) | 0.0183 Btu/(hr)(ft)(°F/ft) | 0.0317 W/(m)(°C) |

economic recovery. Some oxygen plants do not recover argon, and US argon production is thus about a quarter of the theoretical maximum.

The first commercial production, amounting to about 227 m$^3$ (8000 ft$^3$), or 363 kg (800 lb), was carried out by the American Cyanamid Co., Niagara Falls, Ontario, Canada, in 1914–1915. The argon was produced, not by the air liquefaction

**Table 3.25** Composition of the Air Near Ground Level

| Gas | Mol wt. | Percent by volume |
|---|---|---|
| Nitrogen | 28 | 78.09 |
| Oxygen | 32 | 20.95 |
| Argon | 40 | 0.93 |
| Carbon dioxide | 44 | 0.02–0.04 |
| Neon | 20.2 | $18 \times 10^{-4}$ |
| Helium | 4 | $5.3 \times 10^{-4}$ |
| Krypton | 83.7 | $1.1 \times 10^{-4}$ |
| Hydrogen | 2 | $0.5 \times 10^{-4}$ |
| Xenon | 131.3 | $0.08 \times 10^{-4}$ |
| Ozone | 48 | $0.02 \times 10^{-4}$ |
| Radon | 222 | $7 \times 10^{-18}$ |

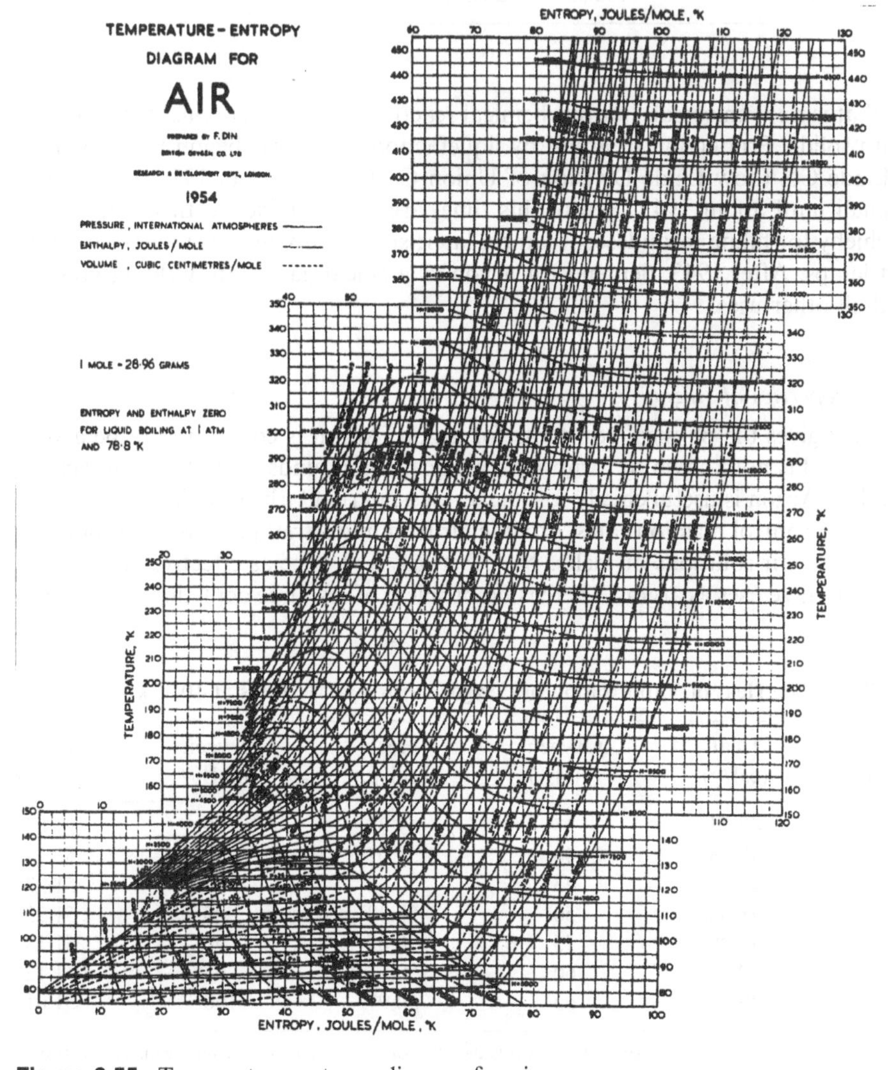


**Figure 3.55** Temperature–entropy diagram for air.

**Table 3.26** Heat of Vaporization of Air at the Bubble Point

| | Temp. (sat. liq.) | | Heat of vaporization | |
|---|---|---|---|---|
| Pressure (atm) | K | °R | J/g | Btu/lb |
| 1 | 78.8 | 142.0 | 205.2 | 88.22 |
| 2 | 85.55 | 154.0 | 197.4 | 84.87 |
| 3 | 90.94 | 163.7 | 191.4 | 80.86 |
| 5 | 96.38 | 173.5 | 183.9 | 79.06 |
| 7 | 101.04 | 181.87 | 175.0 | 75.24 |
| 10 | 106.47 | 191.65 | 163.5 | 70.29 |
| 15 | 113.35 | 204.03 | 143.5 | 61.69 |
| 20 | 118.77 | 213.79 | 124.1 | 53.35 |
| 25 | 123.30 | 221.94 | 103.2 | 44.37 |
| 30 | 127.26 | 229.07 | 80.39 | 34.56 |
| 35 | 130.91 | 235.64 | 48.45 | 20.83 |
| 37.17 | 132.52 | 238.50 | 0 | 0 |

route that is now universal, but by chemical reaction of the other constituents of the air with solids. Production of argon by liquefaction of air was begun by the American Linde Co. (now Linde Division of Union Carbide Corp.) in 1915. Production was small until about 1943. Before this time, argon was mixed with nitrogen and used chiefly for filling incandescent lamps; later, similar mixtures were used in fluorescent lamps. After 1943, demand for argon as a shield gas in arc welding grew to exceed all other uses.

### 3.4.1. *Properties of Argon*

Table 3.32 gives the physical constants for argon. Argon is considered to be a very inert gas and is not known to form true chemical compounds as do krypton, xenon, and radon. Argon is useful chiefly for its inertness, particularly at high temperatures; it is used to control reaction rates, exclude air from industrial processes and devices, and mix and stir solutions or melts. No other gas is at once as effective and as economical in these uses as argon. There is no problem of resource depletion because the

**Table 3.27** The Heat of Vaporization of Oxygen–Nitrogen Mixtures at 101.3 kPa (1 atm)

| Oxygen (mol T) | Heat (cal/g) | Oxygen (mol%) | Heat (cal/g) | Oxygen (mol%) | Heat (cal/g) |
|---|---|---|---|---|---|
| 0 | 47.74 | 35 | 49.83 | 70 | 51.08 |
| 5 | 48.06 | 40 | 50.07 | 75 | 51.17 |
| 10 | 48.37 | 45 | 50.28 | 80 | 51.23 |
| 15 | 48.68 | 50 | 50.48 | 85 | 51.23 |
| 20 | 48.98 | 55 | 50.67 | 90 | 51.18 |
| 25 | 49.28 | 60 | 50.83 | 95 | 51.12 |
| 30 | 49.57 | 65 | 50.97 | 100 | 51.01 |

*Note:* There is a maximum at 82% oxygen. Argon is usually considered part of the nitrogen. The heat of a 21% oxygen mixture is 49.04 cal/g.

**Table 3.28** Air Vapor–Liquid Equilibria—Selected Values

| Pressure | | Sat. liquid (bubble) temperature | | Sat. vapor (dew) temperature | |
|---|---|---|---|---|---|
| atm | psi | K | °R | K | °R |
| 1 | 14.70 | 78.8 | 141.84 | 81.8 | 147.24 |
| 2 | 29.40 | 85.55 | 153.99 | 88.31 | 158.96 |
| 3 | 44.10 | 90.94 | 163.69 | 92.63 | 166.73 |
| 5 | 73.50 | 96.38 | 173.48 | 98.71 | 177.68 |
| 7 | 102.90 | 101.04 | 181.87 | 103.16 | 185.69 |
| 10 | 147.00 | 106.47 | 191.65 | 108.35 | 195.03 |
| 15 | 220.50 | 113.35 | 204.03 | 114.91 | 206.84 |
| 20 | 294.00 | 118.77 | 213.79 | 120.07 | 216.13 |
| 25 | 367.50 | 123.30 | 221.94 | 124.41 | 223.94 |
| 30 | 441.00 | 127.26 | 229.07 | 128.12 | 230.62 |
| 35 | 514.50 | 130.91 | 235.64 | 131.42 | 236.56 |
| 37.17 | 546.40 | 132.52 | 238.54 | 132.52 | 238.54 |
| 37.25 | 547.58 | 132.42 | 238.35 | 132.42 | 238.36 |

amount of argon available in the atmosphere is inexhaustible; it is not consumed in use but returns to the atmosphere.

**a. Heats of Vaporization and Fusion.**

| Units | Triple point | Boiling point |
|---|---|---|
| K | 83.8 | 87.3 |
| kJ/kg | 27.78 | 161.6 |

**b. Vapor Pressure.** The vapor pressure of argon is given to a high degree of accuracy ($\pm 0.02\%$) by the equation

$$\log p(mm) = A - B/T \tag{3.18}$$

where $T$ is in kelvins. for liquid argon,

$$A = 6.9224 \text{ (for } 83.77 - 88.2\,\text{K)} \quad \text{and} \quad \text{B} = 352.8$$

For solid argon,

$$A = 7.7353 \text{ (for } 82 - 83.77\,\text{K)} \quad \text{and} \quad \text{B} = 420.9$$

These data are from Johnson (1960).

**c. Dielectric Constant.** The Clausius–Mosotti function

$$\frac{\varepsilon - 1}{\varepsilon + 2}\left(\frac{1}{\rho}\right) \tag{3.19}$$

should be a constant and independent of temperature provided the molecules of the liquid studied have no permanent dipole moment. For liquid argon this quantity

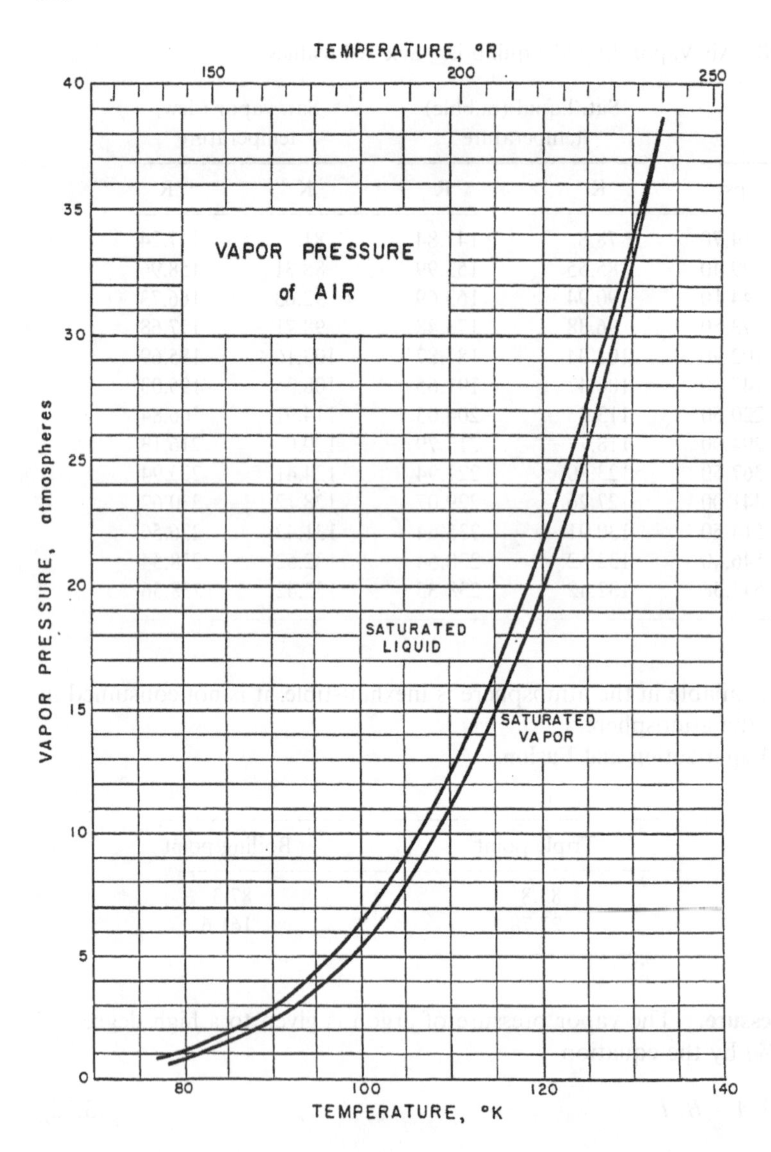


**Figure 3.56** Vapor pressure of air.

is constant within the accuracy of the data. $\varepsilon$ is the dielectric constant and $\rho$ is the density.

For liquid argon, values for the dielectric constant are given in Table 3.33 For gaseous argon at STP, according to Jelatis (1948),

$$\varepsilon = 1.000554$$

**d. Density.** The data tabulated below are from the work of Baly and Donnan (1902). These results may be expressed by the formula

$$d = 1.42333 - 0.006467(T - 84)$$

**Table 3.29** Thermal Conductivity of Gaseous Air

| Temperature | | Thermal conductivity | |
|---|---|---|---|
| K | °R | Watt/(cm K) | Btu/(ft hr °R) |
| 80 | 144 | $7.464 \times 10^{-2}$ | $4.313 \times 10^{-3}$ |
| 90 | 162 | $8.350 \times 10^{-2}$ | $4.825 \times 10^{-3}$ |
| 100 | 180 | $9.248 \times 10^{-2}$ | $5.344 \times 10^{-3}$ |
| 110 | 198 | $1.015 \times 10^{-1}$ | $5.863 \times 10^{-3}$ |
| 120 | 216 | $1.105 \times 10^{-1}$ | $6.383 \times 10^{-3}$ |
| 130 | 234 | $1.194 \times 10^{-1}$ | $6.902 \times 10^{-3}$ |
| 140 | 252 | $1.284 \times 10^{-1}$ | $7.419 \times 10^{-3}$ |
| 150 | 270 | $1.373 \times 10^{-1}$ | $7.933 \times 10^{-3}$ |
| 160 | 288 | $1.461 \times 10^{-1}$ | $8.442 \times 10^{-3}$ |
| 170 | 306 | $1.549 \times 10^{-1}$ | $8.953 \times 10^{-3}$ |
| 180 | 324 | $1.637 \times 10^{-1}$ | $9.458 \times 10^{-3}$ |
| 190 | 342 | $1.723 \times 10^{-1}$ | $9.957 \times 10^{-3}$ |
| 200 | 360 | $1.809 \times 10^{-1}$ | $1.045 \times 10^{-2}$ |
| 210 | 378 | $1.894 \times 10^{-1}$ | $1.094 \times 10^{-2}$ |
| 220 | 396 | $1.978 \times 10^{-1}$ | $1.143 \times 10^{-2}$ |
| 230 | 414 | $2.062 \times 10^{-1}$ | $1.192 \times 10^{-2}$ |
| 240 | 432 | $2.145 \times 10^{-1}$ | $1.239 \times 10^{-2}$ |
| 250 | 450 | $2.227 \times 10^{-1}$ | $1.287 \times 10^{-2}$ |
| 260 | 468 | $2.308 \times 10^{-1}$ | $1.334 \times 10^{-2}$ |
| 270 | 486 | $2.388 \times 10^{-1}$ | $1.380 \times 10^{-2}$ |
| 280 | 504 | $2.467 \times 10^{-1}$ | $1.426 \times 10^{-2}$ |
| 290 | 522 | $2.547 \times 10^{-1}$ | $1.472 \times 10^{-2}$ |
| 300 | 540 | $2.624 \times 10^{-1}$ | $1.516 \times 10^{-2}$ |

*Source*: Hilsenrath et al. (1955).

**Table 3.30** Density of Saturated Liquid Air

| | | | Density | |
|---|---|---|---|---|
| Pressure (atm) | Temp. (K) | Volume (cm$^3$/mol) | g/cm$^3$ | lb/ft$^3$ |
| 1 | 78.8 | 33.14 | 0.8739 | 54.56 |
| 2 | 85.55 | 34.39 | 0.8421 | 52.57 |
| 3 | 90.94 | 35.40 | 0.8181 | 51.07 |
| 5 | 96.38 | 36.94 | 0.7840 | 48.94 |
| 7 | 101.04 | 38.21 | 0.7579 | 47.31 |
| 10 | 106.47 | 40.00 | 0.7240 | 45.20 |
| 15 | 113.35 | 43.21 | 0.6702 | 41.84 |
| 20 | 118.77 | 46.63 | 0.6211 | 38.77 |
| 25 | 123.30 | 50.37 | 0.5749 | 35.89 |
| 30 | 127.26 | 55.69 | 0.5200 | 32.46 |
| 35 | 130.91 | 64.90 | 0.4462 | 27.86 |
| 37.17 | 132.52 | 90.52 | 0.3199 | 19.97 |
| 37.25 | 132.42 | 88.28 | 0.3280 | 20.48 |

**Table 3.31** Values of Distribution Coefficient Functions, ln $(K_p/p_o)$, for Nitrogen, Oxygen, and Argon. The Value of the Reference Pressure is $p_o = 101.3$ kPa (1 atm)

| | Nitrogen | | | Oxygen | | | Argon | | |
|---|---|---|---|---|---|---|---|---|---|
| Temp. (K) | 101.3 kPa | 202.6 kPa | 506.6 kPa | 101.3 kPa | 202.6 kPa | 506.6 kPa | 101.3 kPa | 202.6 kPa | 506.6 kPa |
| 78 | 0.0798 | – | – | –1.3368 | – | – | –0.9075 | – | – |
| 80 | 0.3040 | – | – | –1.1164 | – | – | –0.7157 | – | – |
| 82 | 0.5282 | – | – | –0.8959 | – | – | –0.5238 | – | – |
| 84 | 0.7582 | 0.7048 | – | –0.6755 | –0.4573 | – | –0.3319 | –0.2516 | – |
| 86 | 0.9766 | 0.9030 | – | –0.4550 | –0.3016 | – | –0.1400 | –0.0717 | – |
| 88 | 1.2008 | 1.1012 | – | –0.2346 | –0.1459 | – | –0.0519 | +0.1082 | – |
| 90 | 1.4249 | 1.2994 | – | –0.0141 | +0.0098 | – | +0.1400 | 0.2880 | – |
| 92 | – | 1.4976 | – | – | 0.1655 | – | – | 0.4679 | – |
| 94 | – | 1.6958 | 1.5503 | – | 0.3211 | 0.6605 | – | 0.6477 | 0.5518 |
| 96 | – | 1.8939 | 1.7017 | – | 0.4768 | 0.7877 | – | 0.8276 | 0.7323 |
| 98 | – | – | 1.8531 | – | – | 0.9148 | – | – | 0.8629 |
| 100 | – | – | 2.0045 | – | – | 1.0420 | – | – | 1.0434 |
| 102 | – | – | 2.1559 | – | – | 1.1692 | – | – | 1.2240 |
| 104 | – | – | 2.3073 | – | – | 1.2963 | – | – | 1.4045 |
| 106 | – | – | 2.4588 | – | – | 1.4235 | – | – | 1.5851 |
| 108 | – | – | 2.6102 | – | – | 1.5506 | – | – | 1.7656 |

where $d$ is the density in g/cm$^3$ and $T$ is the absolute temperature in kelvins:

| Temperature (K) | Density (g/cm$^3$) |
|---|---|
| 84.0 | 1.4233 |
| 84.5 | 1.4201 |
| 85.0 | 1.4169 |
| 85.5 | 1.4136 |
| 86.0 | 1.4104 |
| 86.5 | 1.4072 |
| 87.0 | 1.4039 |
| 87.5 | 1.4007 |
| 88.0 | 1.3975 |
| 88.5 | 1.3942 |
| 89.0 | 1.3910 |
| 89.5 | 1.3878 |
| 90.0 | 1.3845 |

**e. Other Properties.** Figures 3.58–3.66 give the density, specific heats, melting curve, surface tension, and viscosity of argon, taken from the WADD report mentioned before (Johnson, 1960)

### *3.4.2. Uses of Argon*

Demand for argon has grown rapidly as new applications have developed. Gas-shielded arc welding was once the major consumer. The principal use of argon now is in the argon–oxygen decarburization (AOD) process in the stainless steel industry. Argon is also used in metal working and in electrical and electronic applications. In 1983, US argon production was estimated to be 1.8 Mg (1800 metric tons) per day. The US industry is currently operating at 100% of practical capacity, primarily due to the demand of the AOD process.

The uses of argon result largely from its property of inertness in the presence of reactive substances but also from its low thermal conductivity, low ionization potential, and good electrical conductivity.

**a. Fabricated Metal Products.** Argon is used in electric arc welding as an inert gas shield, alone or mixed with other gases. It is used both with consumable electrodes, most often similar in composition to the material being welded, or with nonconsumable tungsten electrodes in order to protect the weld from the atmosphere. It also acts as a conductor to carry the arc welding current. It is especially suitable for metals such as aluminum, stainless steel, titanium, and copper that are difficult to weld.

Approximately 26% of the argon produced is used as a shielding gas in welding fabricated metal products.

**b. Iron and Steel Production.** Argon is used to mix and degas carbon steel and steel castings, to purge molds, and to provide neutral atmospheres in heat-treating furnaces. As mentioned, the AOD process in the stainless steel industry is now the largest single user of argon.

**c. Nonferrous Metals Production.** The major use of argon in nonferrous metals production is to furnish an inert atmosphere in vessels in which titanium tetrachloride is reduced to metal sponge by the Kroll process. Argon is also used

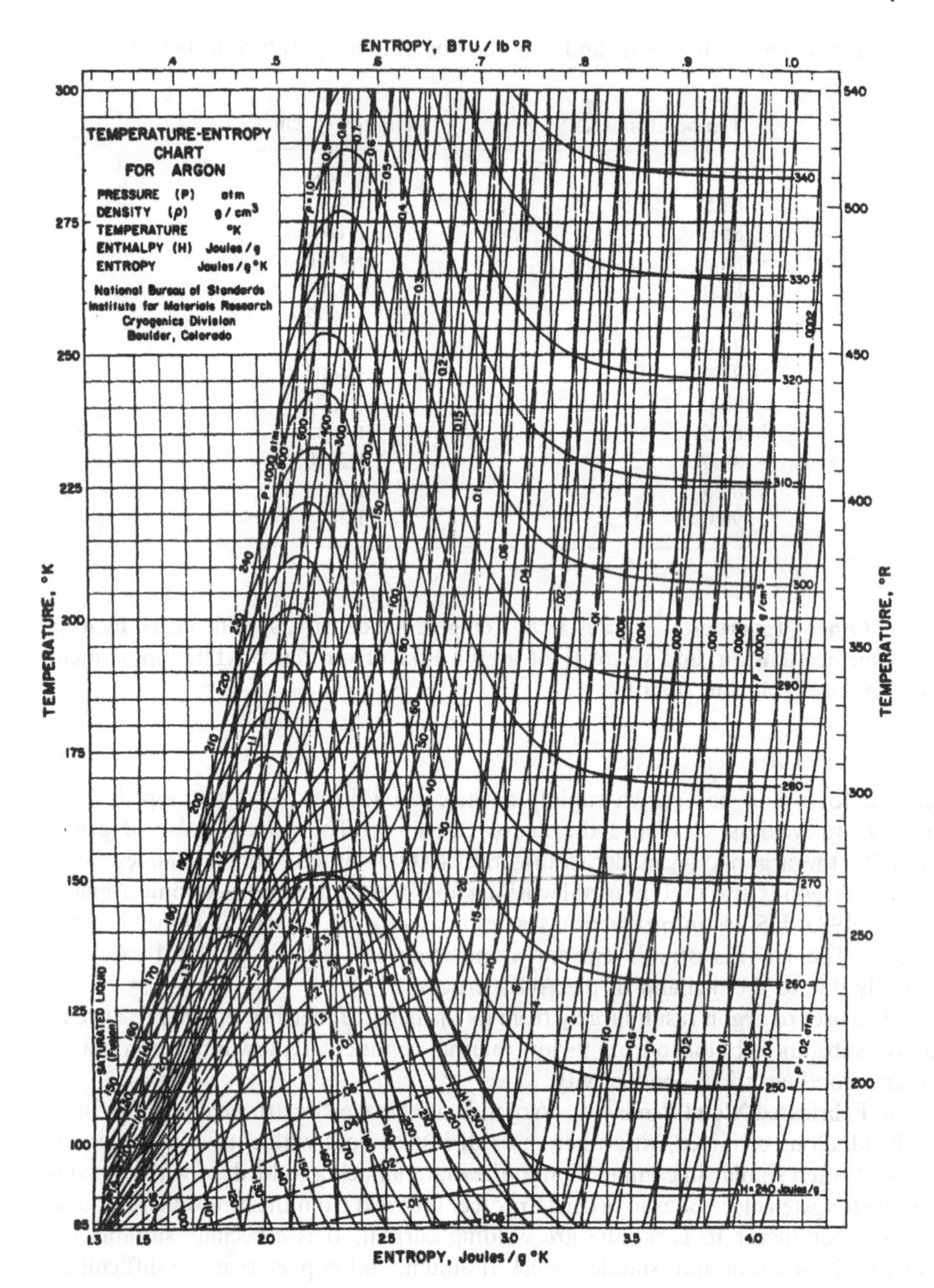


**Figure 3.57** Temperature–entropy chart for argon.

in similar treatment of other metals, such as molybdenum, vanadium, and cesium, that are reactive at the temperatures of preparation.

**d. Chemicals.** Argon is used for blanketing and atmosphere control in the production of chemicals. It prevents reaction with elements in the air and, used as a dilutent, reduces in a controlled fashion the rate of reactions involving gases.

**e. Electronic Components.** Argon atmospheres are used in integrated circuit manufacturing, in single-crystal-growing furnaces, and as a carrier gas in constructing the circuits.

**Table 3.32** Physical Constants of Argon

*Argon*
Chemical Symbol: Ar
Synonym: LAR (liquid only)
CAS Registry Number: 7440–37–1
DOT Classification: Nonflammable gas
DOT Label: Nonflammable gas
Transport Canada Classification: 2.2
UN Number: UN 1006 (compressed gas); UN 1951 (refrigerated liquid)

| | US Units | SI Units |
|---|---|---|
| *Physical constants* | | |
| Chemical formula | Ar | Ar |
| Molecular weight | 39.95 | 39.95 |
| Density of the gas at 70°F (21.1°C) and 1 atm | 0.103 lb/ft$^3$ | 1.650 kg/m$^3$ |
| Specific gravity of the gas at 70°F (21.1°C) and 1 atm | 1.38 | 1.38 |
| Specific volume of the gas at 70°F (21.1°C) and 1 atm | 9.71 ft$^3$/lb | 0.606 m$^3$/kg |
| Density of the liquid at boiling point and 1 atm | 87.02 lb/ft$^3$ | 1394 kg/m$^3$ |
| Boiling point at 1 atm | −302.6°F | −185.9°C |
| Melting point at 1 atm | −308.6°F | −189.2°C |
| Critical temperature | −188.1°F | −122.3°C |
| Critical pressure | 711.5 psia | 4905 kPa abs |
| Critical density | 33.44 lb/ft$^3$ | 535.6 kg/m$^3$ |
| Triple point | −308.8°F at 9.99 psia | −199.3°C at 68.9 kPa abs |
| Latent heat of vaporization at boiling point and 1 atm | 69.8 Btu/lb | 162.3 kJ/kg |
| Latent heat of fusion at triple point | 12.8 Btu/lb | 29.6 kJ/kg |
| Specific heat of the gas at 70°F (21.1°C) and 1 atm | | |
| $C_P$ | 0.125 Btu/(lb)(°F) | 0.523 kJ/(kg)(°C) |
| $C_V$ | 0.075 Btu/(lb)(°F) | 0.314 kJ/(kg)(°C) |
| Ratio of specific heats | 1.67 | 1.67 |
| Solubility in water at 32°F (0°C) vol/vol | 0.056 | 0.056 |
| Weight of the liquid at boiling point | 11.63 lb/gal | 1394.0 kg/m$^3$ |

**Table 3.33** Dielectric Constant for Argon—Selected Values

| Temp. (K) | Density (g/cm$^3$) | Dielectric constant (ref. to vacuum) |
|---|---|---|
| 88.8 | 1.393 | 1.516 |
| 88.5 | 1.395 | 1.518 |
| 87.1 | 1.404 | 1.520 |
| 85.8 | 1.414 | 1.525 |
| 84.3 | 1.422 | 1.530 |
| 82.4[a] | 1.434 | 1.537 |

[a] Supercooled liquid.
*Source:* McLennan et al. (1930).

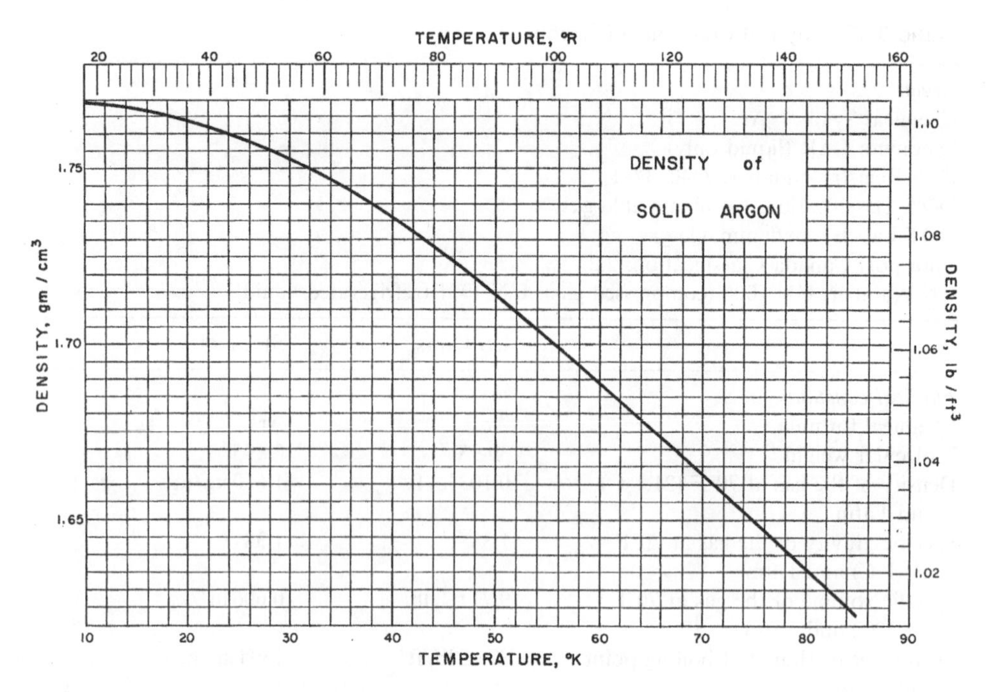


**Figure 3.58** Density of solid argon.

**f. Other Uses.** About 1% of the demand for argon results from its use in lamps. Infrared and photoflood lamps are filled with a mixture of about 88% argon and 12% nitrogen. The high density and low reactivity of argon help retard the rate of evaporation of the tungsten filament, thus prolonging the life of the lamp; the nitrogen prevents arcing.

## 3.5. Neon

Neon (at wt. 20.183) has three stable isotopes of mass numbers 20, 21, and 22, which occur in atmosphere air in the relative abundance 10,000:28:971. Five unstable isotopes are also known.

Neon is a rare gas element present in the atmosphere to the extent of 1 part in 65,000 of air. It is obtained by liquefaction of air and seperated from the other gases by fractional distillation. It is a very insert element; however, it is said to form a compound with fluorine. Of all the rare gases, the discharge of neon is the most intense at ordinary voltages are currents.

Temperature entropy diagrams for neon are shown in Fig 3.67 and 3.68.

### *3.5.1. Properties of Neon*

Liquid neon is a clear, colorless liquid that boils at 0.987 bar (1 atm) and 27.1 K (48.8°R) and has a triple point at 24.56 K.

Table 3.34 gives the physical constants for neon, as well as krypton and xenon.

**a. Phase Transition.** See Table 3.35.

**b. Heat of Vaporization and Heat of Fusion.** See Table 3.36. The heat of fusion of neon at the melting point is 76.67 kJ/kg (3.969 cal/g; 80.1 cal/mol).

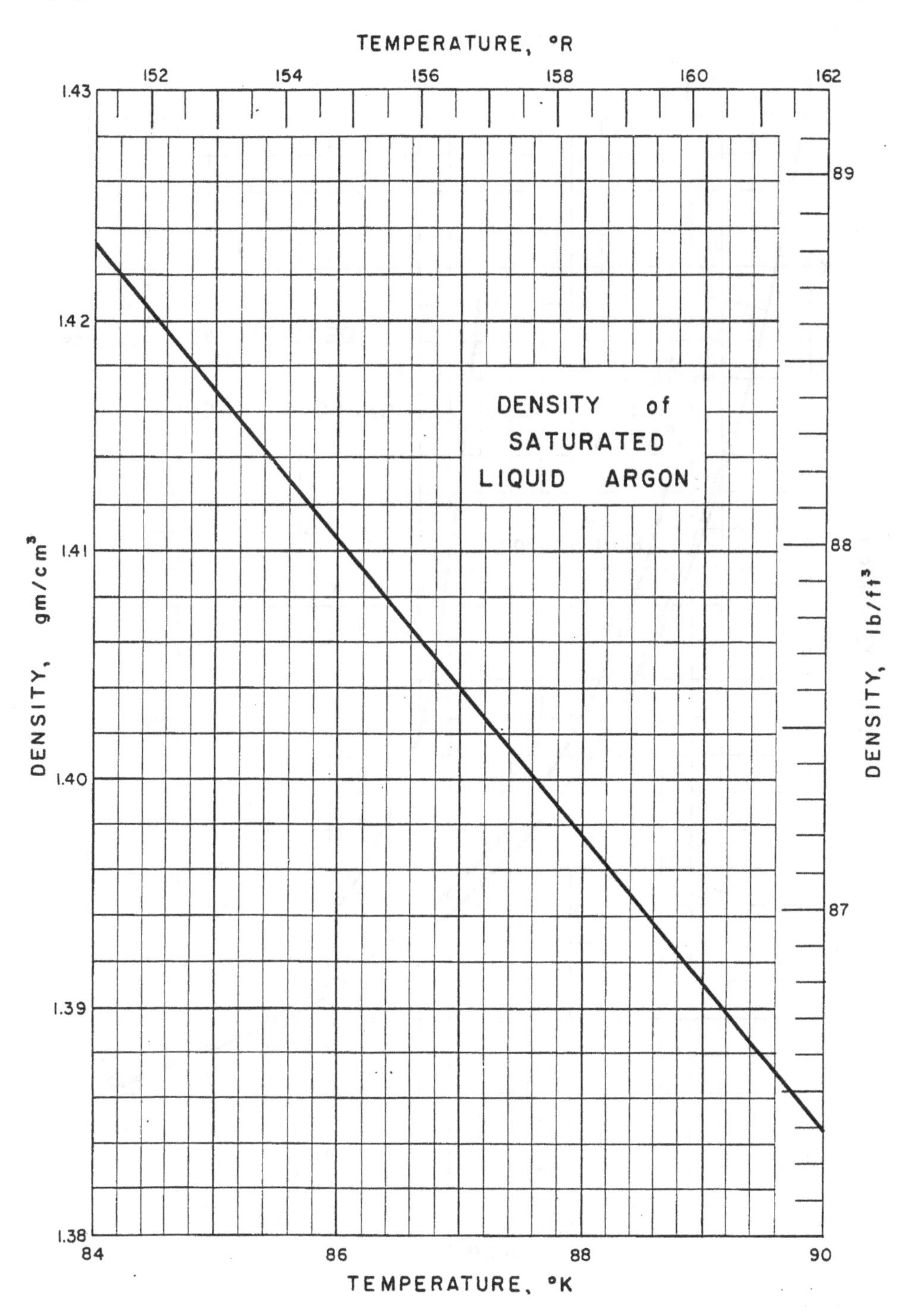


**Figure 3.59** Density of saturated liquid argon.

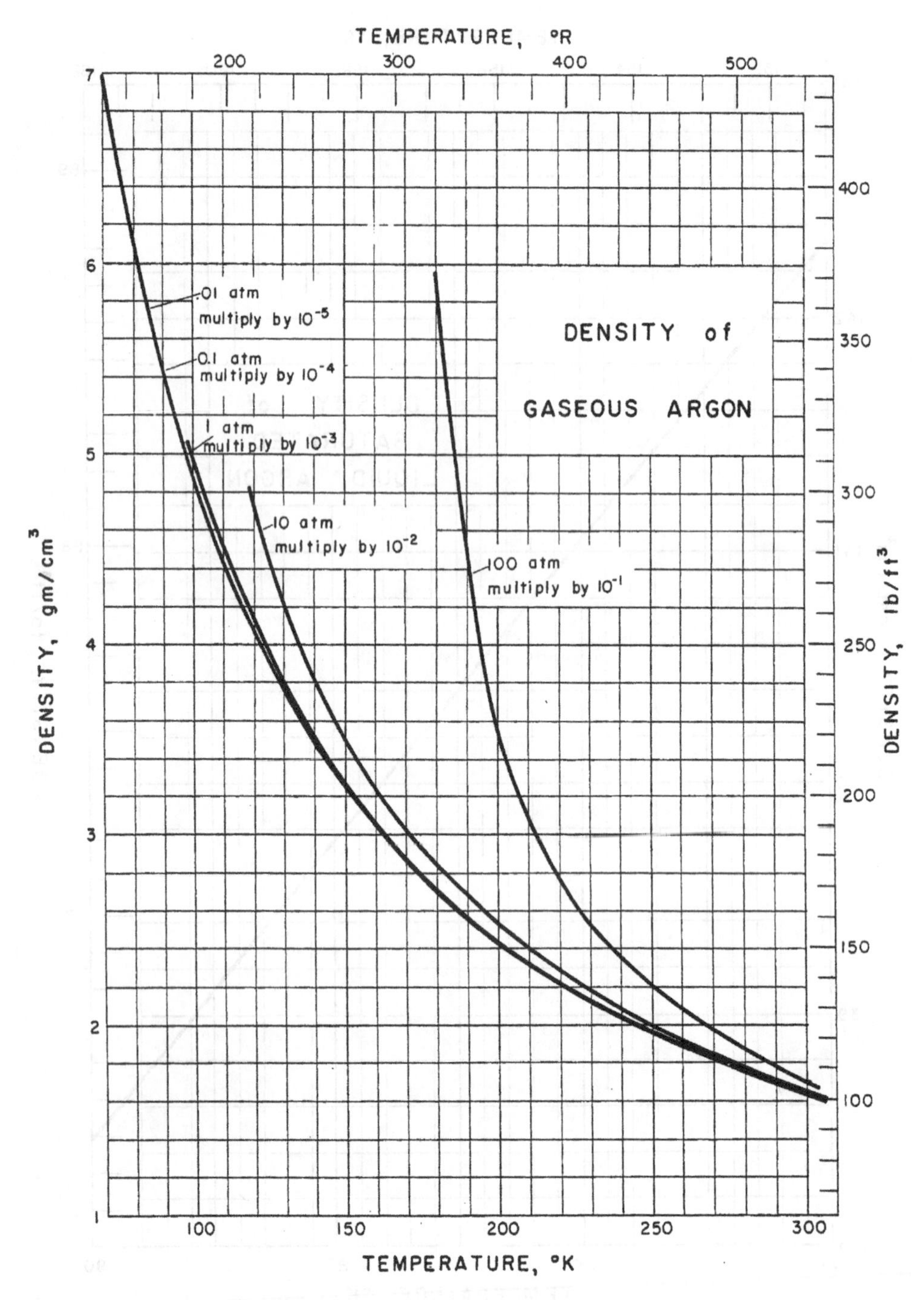


Figure 3.60 Density of gaseous argon.

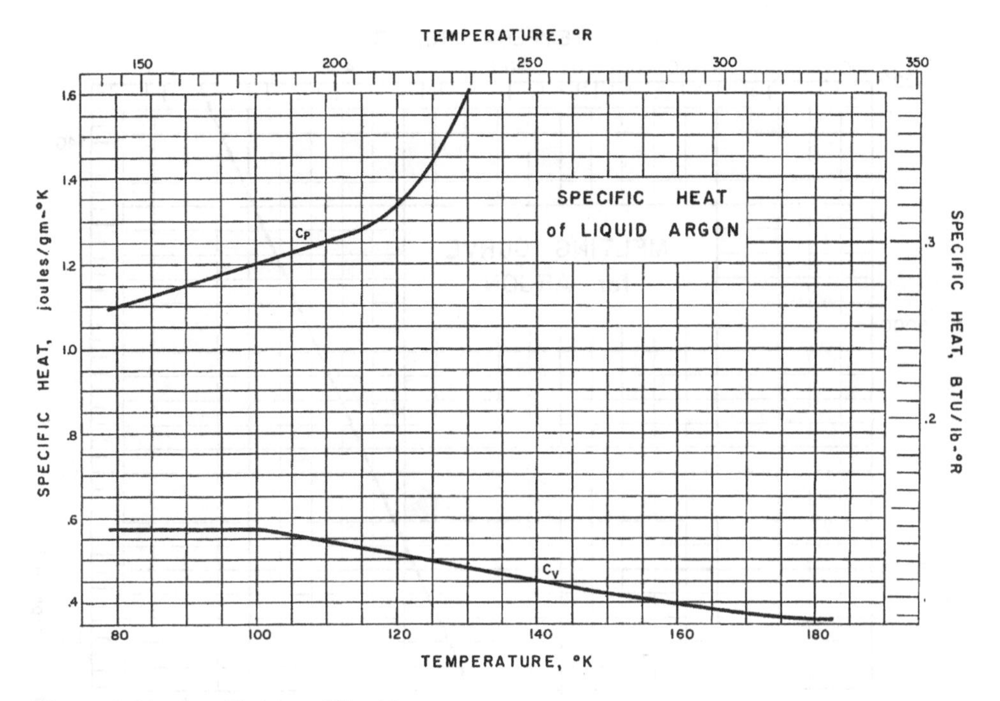


**Figure 3.61** Specific heat of liquid argon.

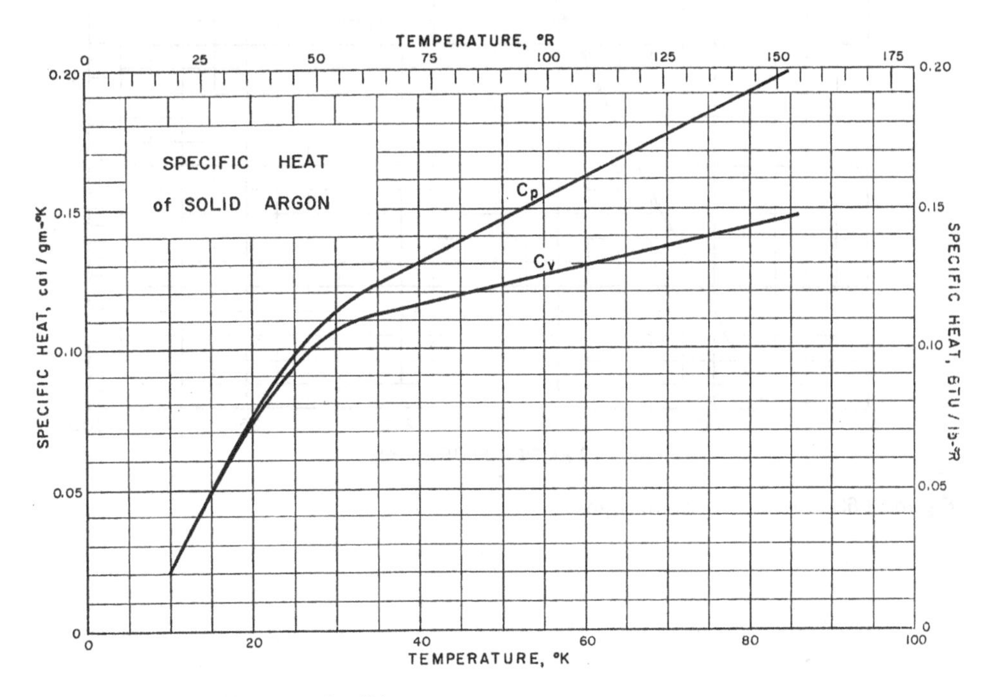


**Figure 3.62** Specific heat of solid argon

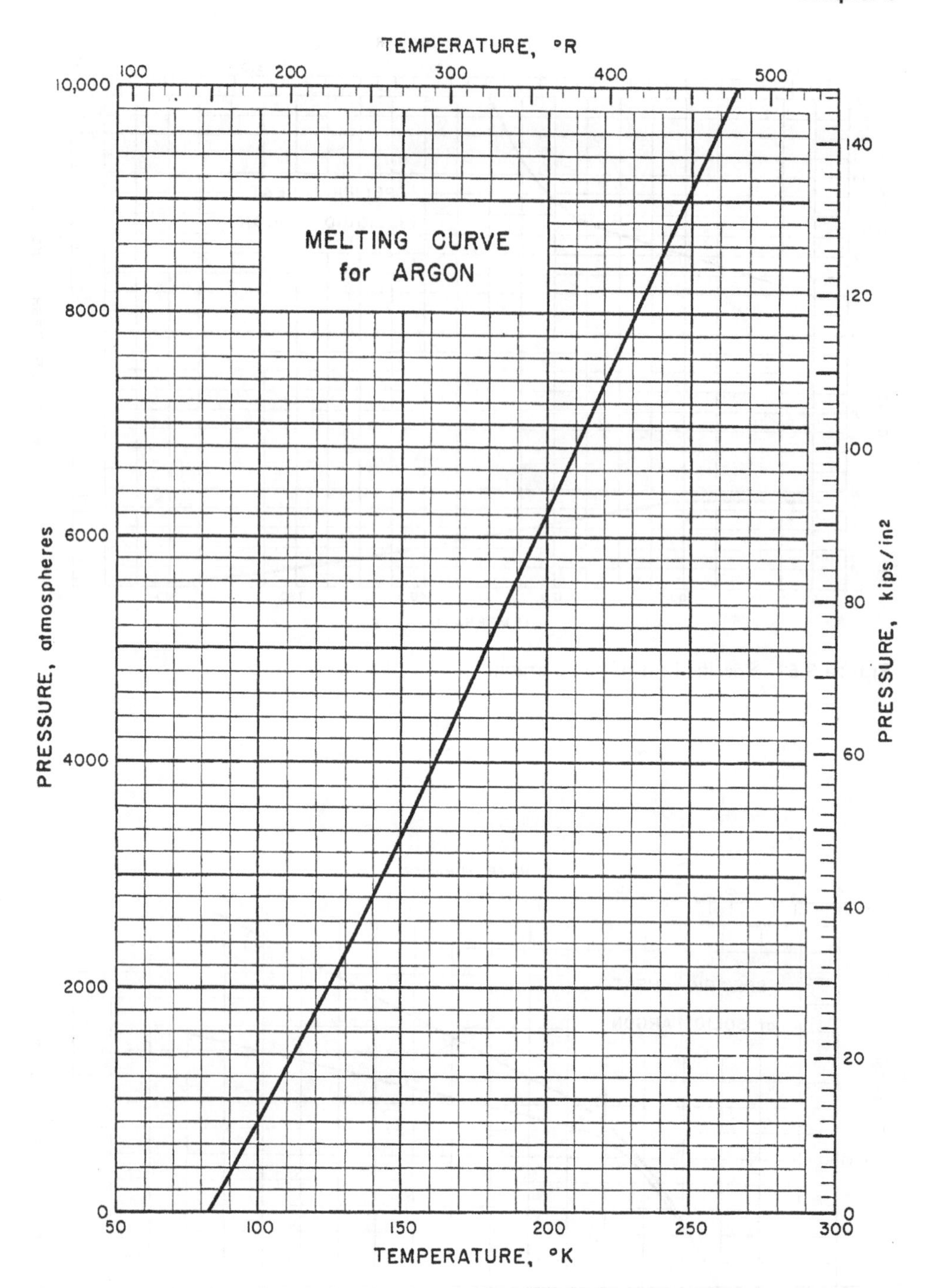


**Figure 3.63** Melting curve for argon.

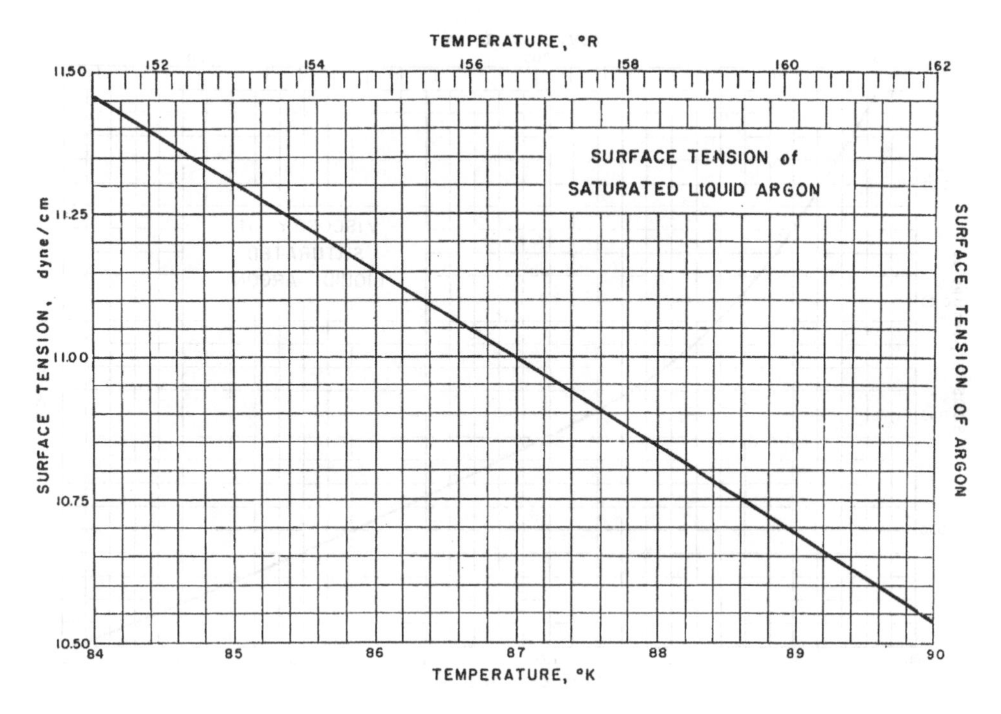


Figure 3.64 Surface tension of saturated liquid argon.

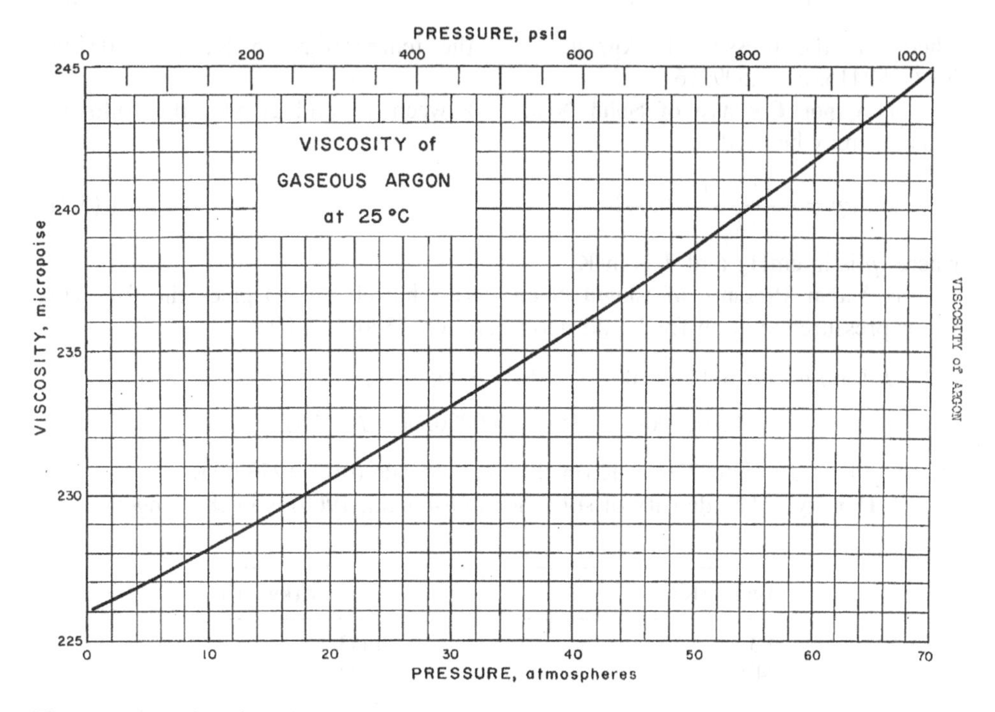


Figure 3.65 Viscosity of gaseous argon at 25°C

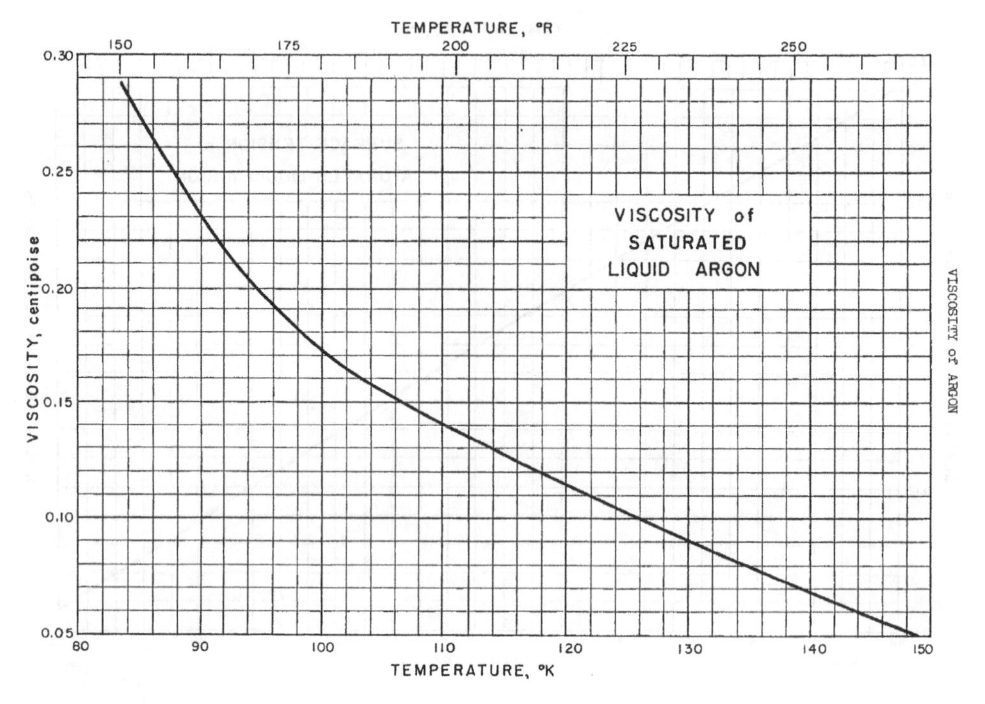


**Figure 3.66** Viscosity of saturated liquid argon.

**c. Melting Point Temperature of Neon.** Mills and Grilley (1955) give the melting curve of neon as

$$p = a + bT^c \tag{3.20}$$

where $p$ is the pressure in kg/cm$^2$, $T$ is the temperature in K, $a = -1057.99$, $b = 6.289415$, $c = 1.599916$.

**d. Vapor Pressure of Solid Neon.** Between 15 and 20.5 K, the following equation may be used:

$$\log \nu p = -\frac{111.76}{T} + 6.0424 \tag{3.21}$$

where $vp$ is in cmHg and $T$ is in K.

**e. Vapor Pressure of Liquid Neon.** Verschaffelt (1929) gives the following vapor pressure equation for liquid neon between 25 and 35 K:

$$\begin{aligned} T \log \nu p = & 28.100 + 36.00(T - 35) - 0.003333(T - 35)^2 \\ & + 0.000400(T - 35)^3 + 0.00002667(T - 35)^4 \end{aligned} \tag{3.22}$$

where $vp$ is in atm and $T$ is in K on the Leiden scale (0°C = 273.09 K).

**f. Density.** The density of solid neon (Johnson, 1960) is in the range:

| Temp. (K) | Density (kg/m$^3$) |
|---|---|
| 4.3 | 1443 |
| 24.57[a] | 1444 |

[a] Normal melting point value.

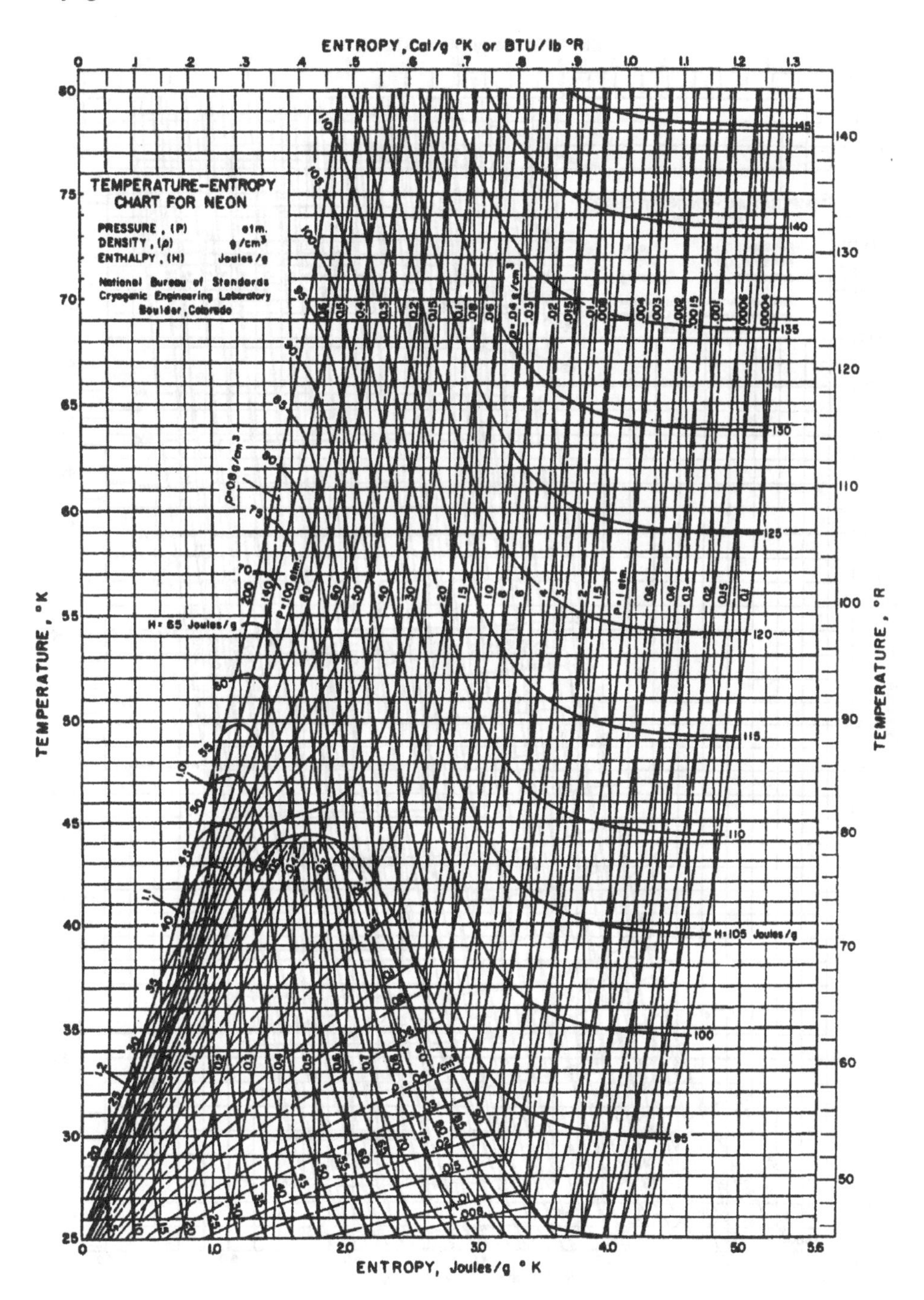


**Figure 3.67** Temperature–entropy chart for neon, 25–80 K.

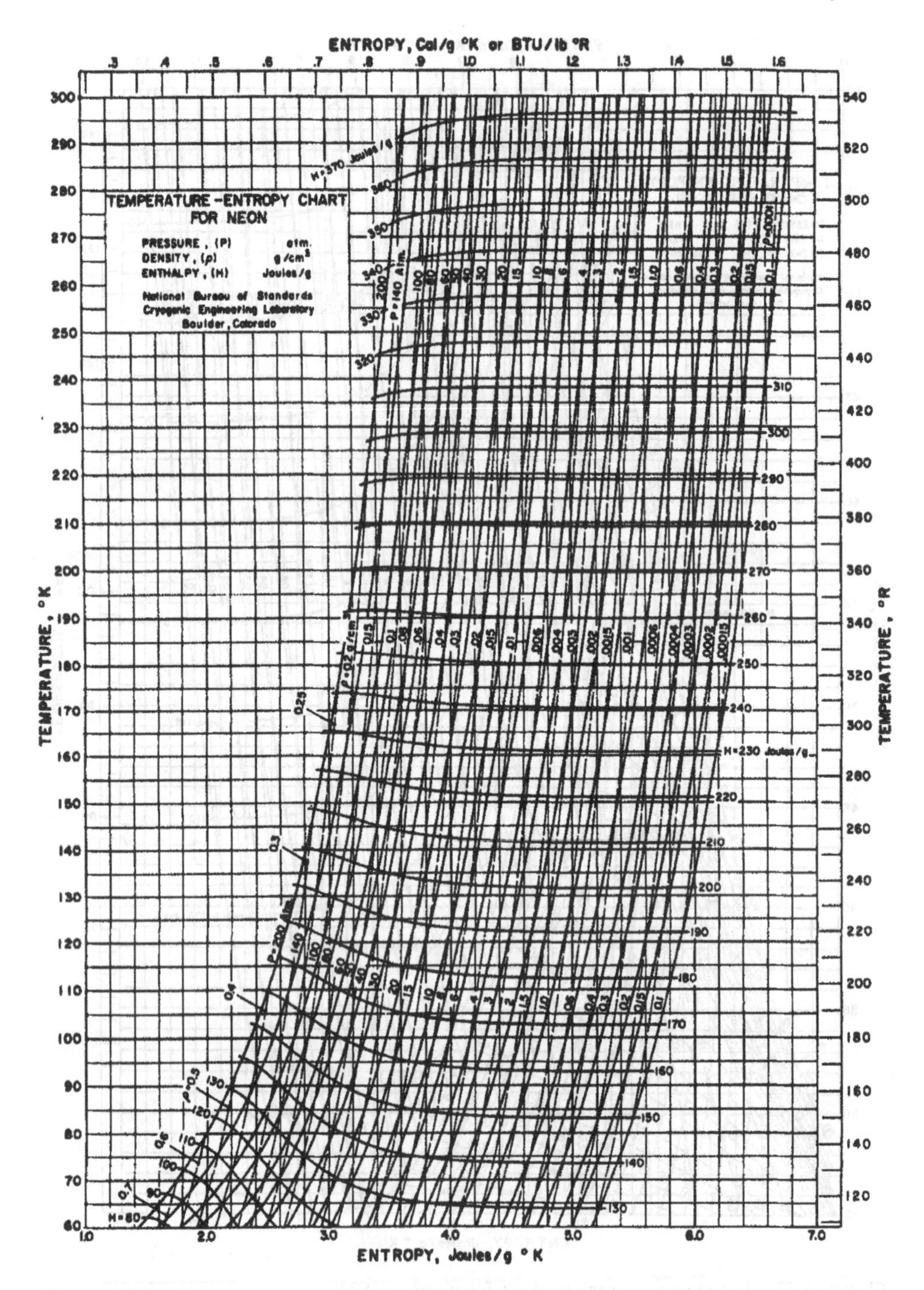


**Figure 3.68** Temperature–entropy chart for neon, 60–300 K.

**Table 3.34** Physical Constants of Krypton, Neon, and Xenon

| *Rare Gases: Krypton, Neon, Xenon* | Krypton | Neon | Xenon |
|---|---|---|---|
| Chemical Symbol: | Kr | Ne | Xe |
| CAS Registry Number: | 7439-90-9 | 7440-01-9 | 7440-63-3 |
| DOT Classification: | Nonflammable gas | Nonflammable gas | Nonflammable gas |
| DOT Label: | Nonflammable gas | Nonflammable gas | Nonflammable gas |
| Transport Canada Classification: | 2.2 | 2.2 | 2.2 |
| UN Number: | | | |
| (compressed gas): | UN 1056 | UN 1065 | UN 2036 |
| (refrigerated liquid): | UN 1970 | UN 1913 | UN 2591 |

| | US Units | SI Units |
|---|---|---|
| *Krypton* | | |
| Chemical formula | Kr | Kr |
| Molecular weight | 83.80 | 83.80 |
| Density of the gas at 70°F (21.1°C) and 1 atm | 0.2172 lb/ft$^3$ | 3.479 kg/m$^3$ |
| Specific gravity of the gas at 70°F (21.1°C) and 1 atm (air = 1) | 2.899 | 2.899 |
| Specific volume of the gas at 70°F (21.1°C) and 1 atm | 4.604 ft$^3$/lb | 0.287 m$^3$/kg |
| Boiling point at 1 atm | −244.0°F | −153.4°C |
| Melting point at 1 atm | −251°F | −157°C |
| Critical temperature | −82.8°F | −63.8°C |
| Critical pressure | 798.0 psia | 5502 kPa abs |
| Critical density | 56.7 lb/ft$^3$ | 908 kg/m$^3$ |
| Triple point | −251.3°F at 10.6 psia | −157.4°C at 73.2 kPa abs |
| Latent heat of vaporization at boiling point | 46.2 Btu/lb | 107.5 kJ/kg |
| Latent heat of fusion at triple point | 8.41 Btu/lb | 19.57 kJ/kg |
| Specific heat of the gas at 70°F (21.1°C) and 1 atm | | |
| $C_P$ | 0.060 Btu/(lb)(°F) | 0.251 kJ/(kg)(°C) |
| $C_V$ | 0.035 Btu/(lb)(°F) | 0.146 kJ/(kg)(°C) |
| Ratio of specific heats, $C_P/C_V$ | 1.69 | 1.69 |
| Solubility in water, vol/vol at 68°F (20°C) | 0.0594 | 0.0594 |
| Weight of the liquid at boiling point | 20.15 lb/gal | 2415 kg/m$^3$ |
| Density of the liquid at boiling point | 150.6 lb/ft$^3$ | 2412.38 kg/m$^3$ |
| Gas/liquid volume ratio (liquid at boiling point, gas at 70°F and 1 atm, vol/vol) | 693.4 | 693.4 |
| *Neon* | | |
| *Physical Constants* | | |
| Chemical formula | Ne | Ne |
| Molecular weight | 20.183 | 20.183 |
| Density of the gas at 70°F (21.1°C) and 1 atm | 0.05215 lb/ft$^3$ | 0.83536 kg/m$^3$ |
| Specific gravity of the gas at 70°F (21.1°C) and 1 atm (air = 1) | 0.696 | 0.696 |

(*Continued*)

**Table 3.34** (*Continued*)

| | US Units | SI Units |
|---|---|---|
| Specific volume of the gas at 70°F (21.1°C) and 1 atm | 19.18 ft$^3$/lb | 1.197 m$^3$/kg |
| Boiling point at 1 atm | –410.9°F | –246.0°C |
| Melting point at 1 atm | –415.6°F | –248.7°C |
| Critical temperature | –379.8°F | –228.8°C |
| Critical pressure | 384.9 psia | 2654 kPa abs |
| Critical density | 30.15 lb/ft$^3$ | 483 kg/m$^3$ |
| Triple point | –415.4°F at 6.29 psia | –248.6°C at 43.4 kPa abs |
| Latent heat of vaporization at boiling point | 37.08 Btu/lb | 86.3 kJ/kg |
| Latent heat of fusion at triple point | 7.14 Btu/lb | 16.6 kJ/kg |
| Specific heat of the gas at 70°F (21.1°C) and 1 atm | | |
| $C_P$ | 0.25 Btu/(lb)(°F) | 1.05 kJ/(kg)(°C) |
| $C_V$ | 0.152 Btu/(lb)(°F) | 0.636 kJ/(kg)(°C) |
| Ratio of specific heats, $C_P/C_V$ | 1.64 | 1.64 |
| Solubility in water, vol/vol at 68°F (20°C) | 0.0105 | 0.0105 |
| Weight of the liquid at boiling point | 10.07 lb/gal | 1207 kg/m$^3$ |
| Density of saturated vapor at 1 atm | 0.5862 lb/ft$^3$ | 9.390 kg/m$^3$ |
| Density of the gas at boiling point | 0.6068 lb/ft$^3$ | 9.7200 kg/m$^3$ |
| Density of the liquid at boiling point | 75.35 lb/ft$^3$ | 1207 kg/m$^3$ |
| Gas/liquid volume ratio (liquid at boiling point, gas at 70°F and 1 atm, vol/vol) | 1445 | 1445 |
| *Xenon* | | |
| *Physical Constants* | | |
| Chemical formula | Xe | Xe |
| Molecular weight | 131.3 | 131.3 |
| Density of the gas at 70°F (21.1°C) and 1 atm | 0.3416 lb/ft$^3$ | 5.472 kg/m$^3$ |
| Specific gravity of the gas at 70°F and 1 atm (air = 1) | 4.560 | 4.560 |
| Specific volume of the gas at 70°F (21.1°C) and 1 atm | 2.927 ft$^3$/lb | 0.183 m$^3$/kg |
| Boiling point at 1 atm | −162.6°F | −108.2°C |
| Melting point at 1 atm | −168°F | −111°C |
| Critical temperature | 61.9°F | 16.6°C |
| Critical pressure | 847.0 psia | 5840 kPa abs |
| Critical density | 68.67 lb/ft$^3$ | 1100 kg/m$^3$ |
| Triple point | −169.2°F at 11.84 psia | −111.8°C at 81.6 kPa abs |
| Latent heat of vaporization at boiling point | 41.4 Btu/lb | 96.3 kJ/kg |
| Latent heat of fusion at triple point | 7.57 Btu/lb | 17.6 kJ/kg |
| Specific heat of the gas at 70°F (21.1°C) and 1 atm | | |
| $C_P$ | 0.038 Btu/(lb)(°F) | 0.269 kJ/(kg)(°C) |
| $C_V$ | 0.023 Btu/(lb)(°F) | 0.096 kJ/(kg)(°C) |
| Ratio of specific heats, $C_P/C_V$ | 1.667 | 1.667 |
| Solubility in water, vol/vol at 68°F (20°C) | 0.108 | 0.108 |
| Weight of the liquid at boiling point | 25.51 lb/gal | 3057 kg/m$^3$ |
| Density of the liquid at boiling point | 190.8 lb/ft$^3$ | 3057 kg/m$^3$ |
| Gas/liquid volume ratio (liquid at boiling point, gas at 70°F and 1 atm, vol/vol) | 558.5 | 558.5 |

**Table 3.35** Phase Transition Temperatures of Neon

| | Pressure | | | Temperature | |
|---|---|---|---|---|---|
| Property | kPa | atm | psia | K | °R |
| Critical point[a] | 2719 | 26.84 | 394.5 | 44.38 | 79.88 |
| Boiling point | 101.3 | 1 | 14.696 | 27.17 | 48.91 |
| Melting point | 101.3 | 1 | 14.696 | 24.57 | 44.23 |
| Triple point | 43.1 | 0.4257 | 6.256 | 24.57 | 44.23 |

[a] Critical volume = 42 $cm^3$/mol.

**Table 3.36** Latent Heat of Vaporization of Neon

| Temperature | | Heat of vaporization | |
|---|---|---|---|
| K | °R | cal/g | kJ/kg |
| 25.17 | 45.31 | 21.36 | 89.37 |
| 26.15 | 47.07 | 20.96 | 87.70 |
| 27.15 | 48.87 | 20.56 | 86.02 |
| 27.17 | 48.91 | 20.6 | 86.19 |
| 30.13 | 54.23 | 19.34 | 80.92 |
| 33.09 | 59.56 | 17.97 | 75.19 |
| 36.05 | 64.89 | 16.23 | 67.91 |
| 37.83 | 68.09 | 14.87 | 62.22 |
| 39.08 | 70.34 | 13.69 | 57.28 |
| 41.065 | 73.92 | 11.26 | 47.11 |
| 43.02 | 77.44 | 7.491 | 31.34 |

The density of liquid neon is tabulated below.

| Temp. (K) | Density (kg/$m^3$) |
|---|---|
| 24.57[a] | 1248 |
| 25.17 | 1238 |
| 27.17[b] | 1204 |
| 30.13 | 1149 |
| 36.05 | 1017 |
| 39.08 | 928 |
| 43.02 | 749 |
| 44.38[c] | 483 |

[a] Triple point temperature.
[b] Normal boiling temperature.
[c] Critical temperature.

*Symbols:* $\rho_{sat}$ = saturation pressure or vapor pressure, $\rho$ = density, $Cz_P$ = speci-specific heat at constant pressure, $\mu$ = viscosity, $k$ = thermal conductivity, $h_{fg}$ = heat of vaporization, $N_{pr} = \mu C_p/K$ = Prandtl number, $\sigma_L$ = surface tension.

**g. Dielectric Constant of Gaseous Neon.** One value has been published at 70°F and 1 atm. The Clausius–Mosotti function may be used to estimate values at other conditions with $\varepsilon = 1.0001274$.

**h. Other Properties.** Figures 3.69–3.76 give the density, specific heats, thermal conductivity, vapor pressure, and melting curve of neon, taken from the WADD report mentioned before (Johnson 1960).

### *3.5.2. Uses of Neon*

Neon is used in making the common neon advertising signs, which accounts for its largest use. It is also used to make high-voltage indicators, lightning arrestors, wave-meter tubes, and TV tubes.

Liquid neon is now commercially available and is finding important application as an economical cryogenic refrigerant. It has over 40 times more refrigerating capacity per unit volume than liquid helium and more than three times that of liquid hydrogen. It is compact, inert, and less expensive than helium when it meets refrigeration requirements.

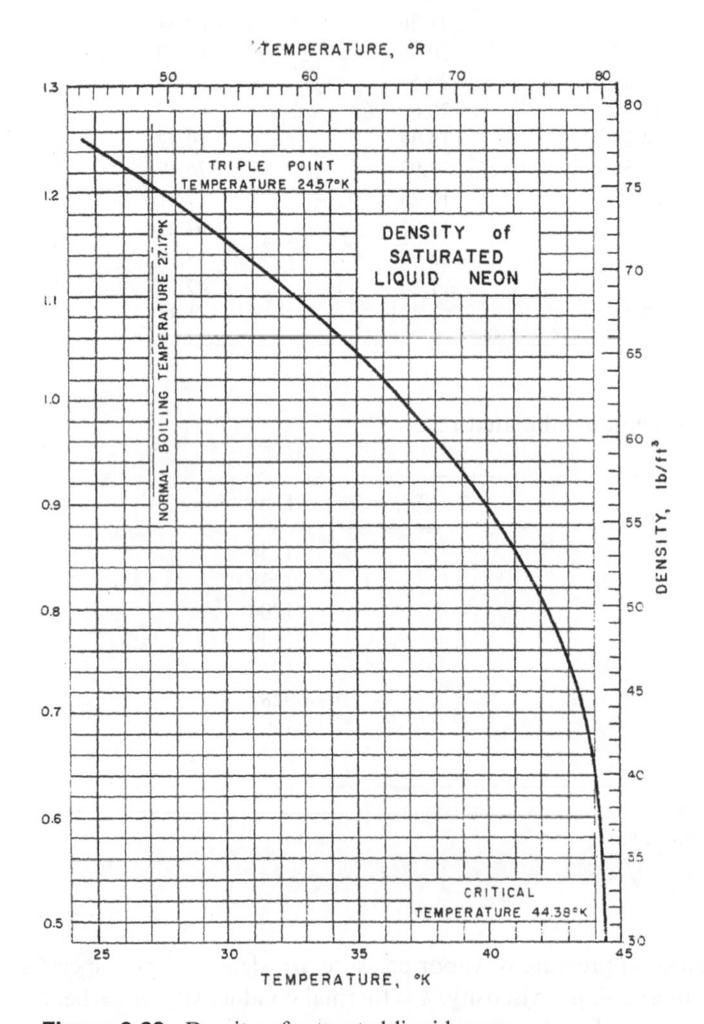


**Figure 3.69** Density of saturated liquid neon.

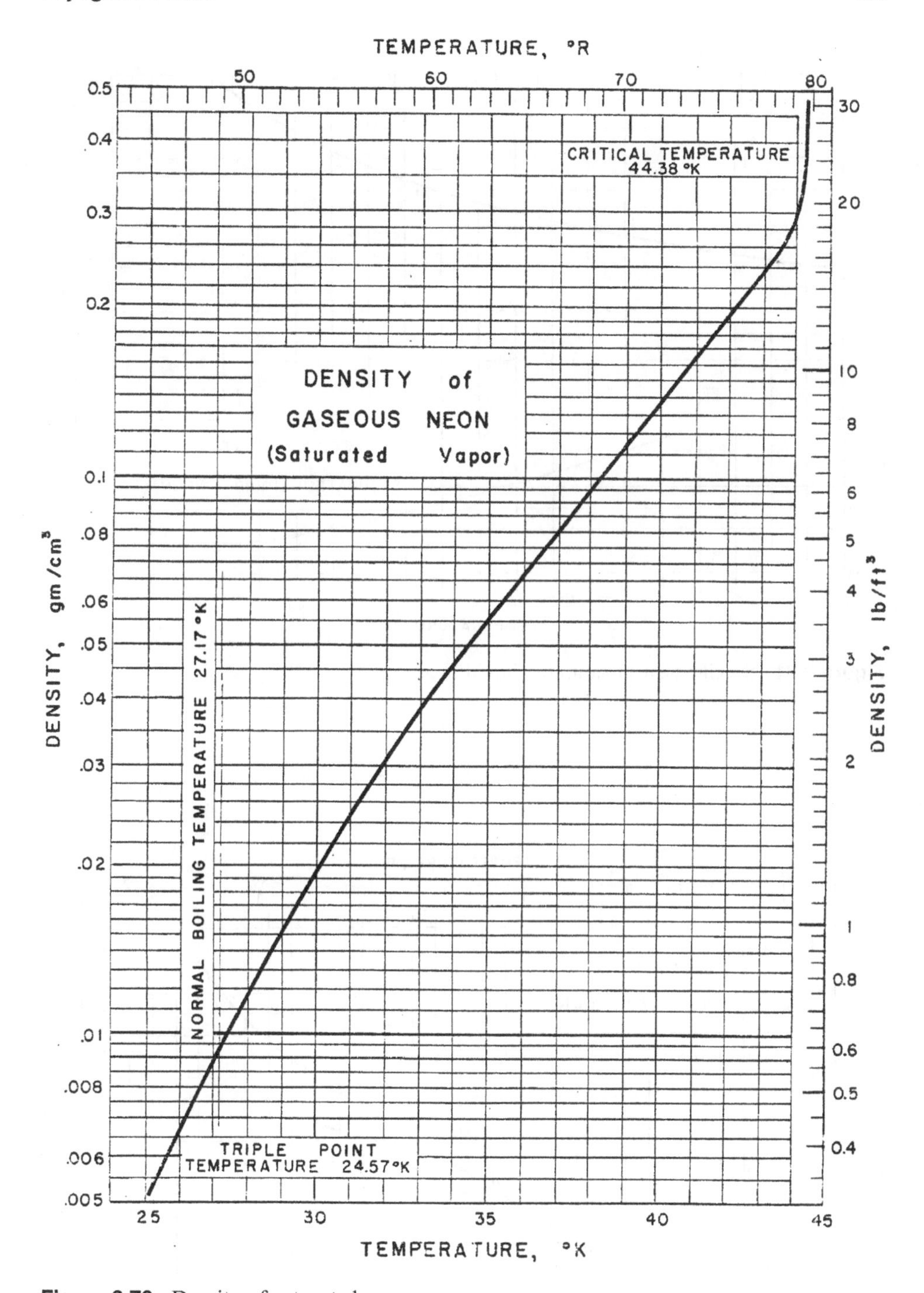


**Figure 3.70** Density of saturated vapor neon.

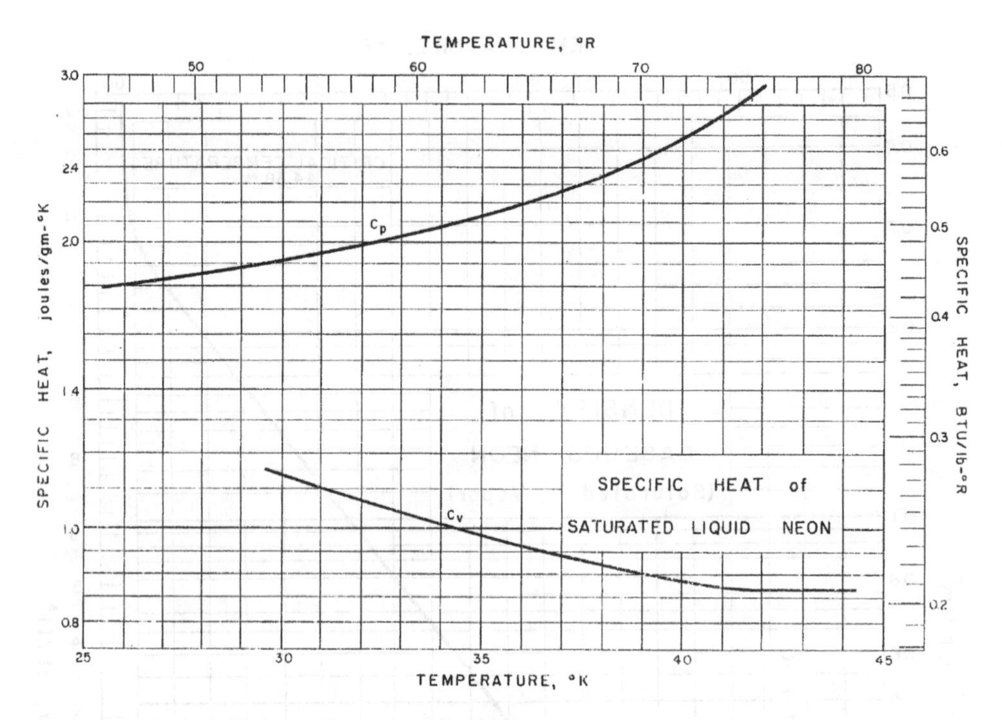


**Figure 3.71** Specific heat of saturated liquid neon.

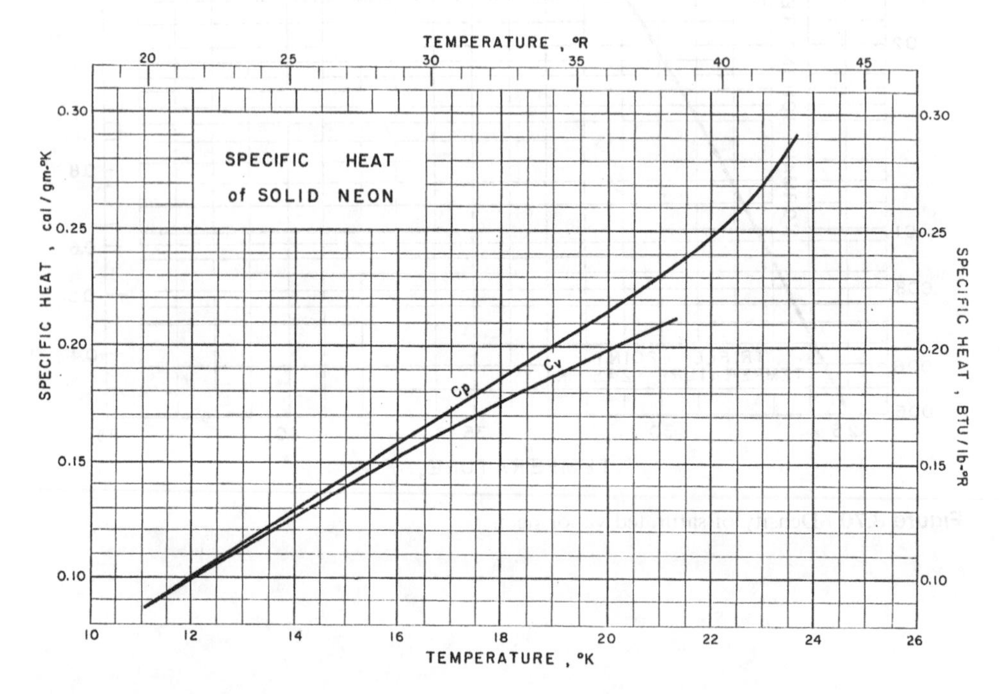


**Figure 3.72** Specific heat of solid neon.

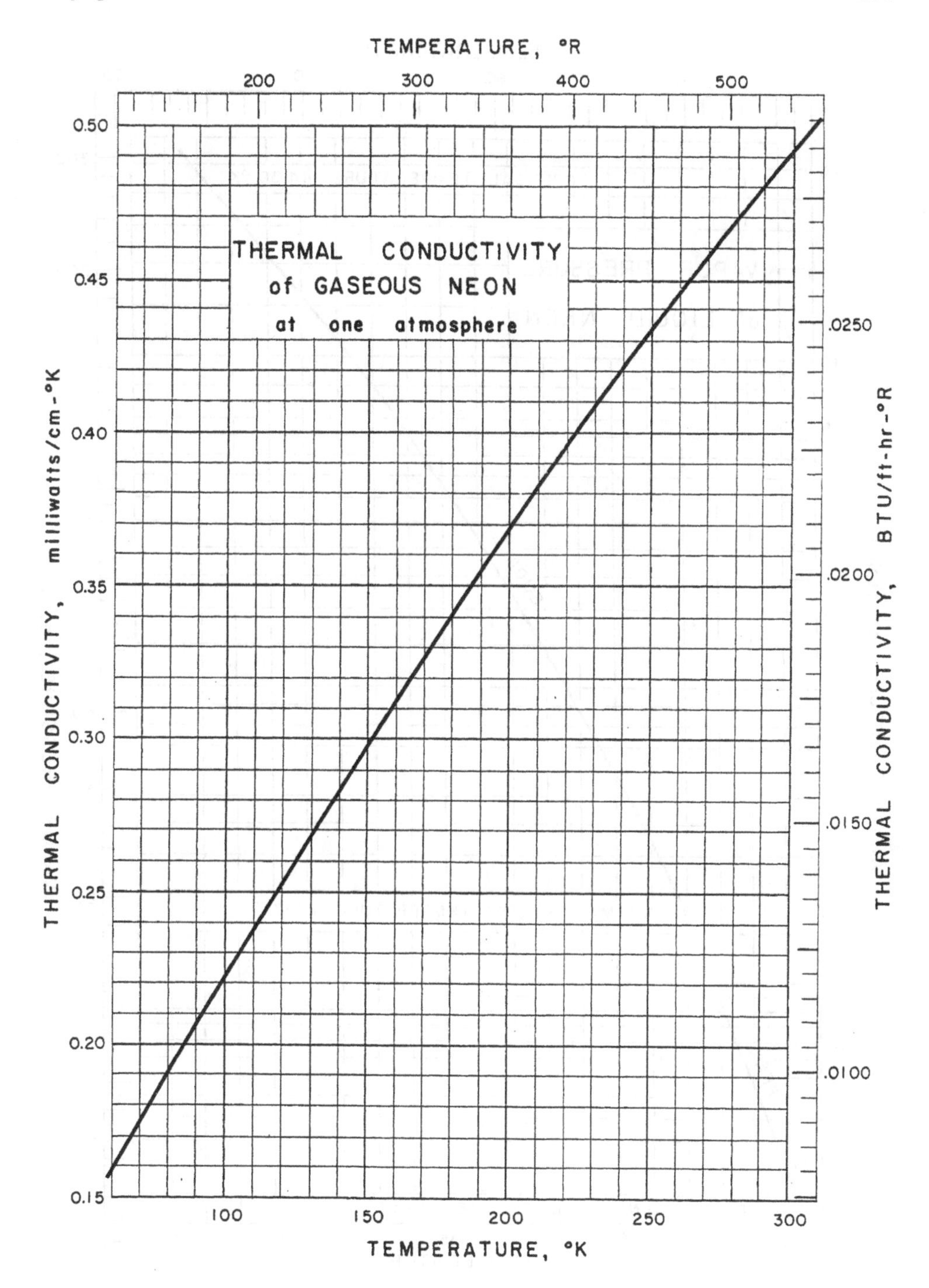


**Figure 3.73** Thermal conductivity of gaseous neon at 1 atm.

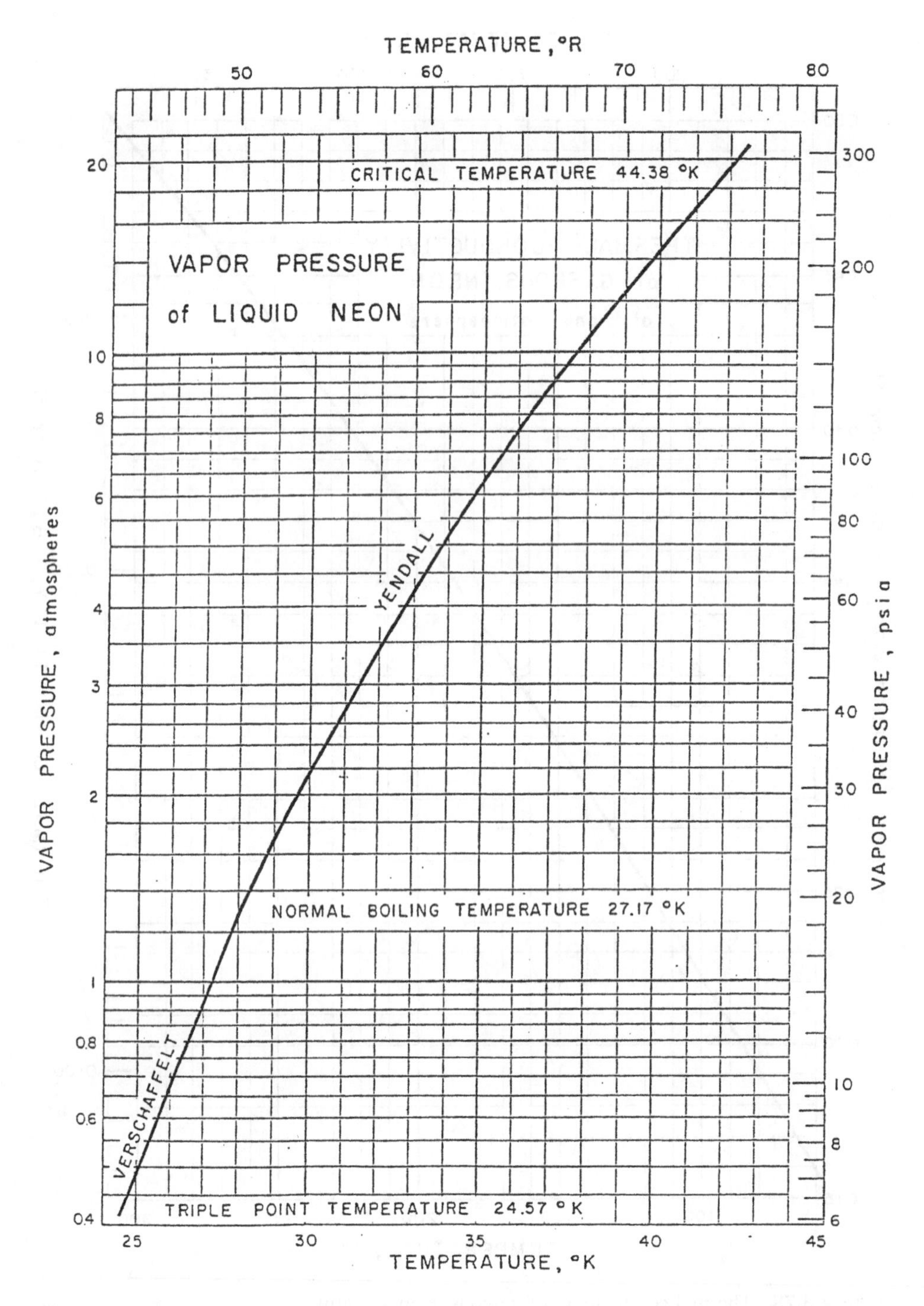


**Figure 3.74** Vapor pressure of liquid neon.

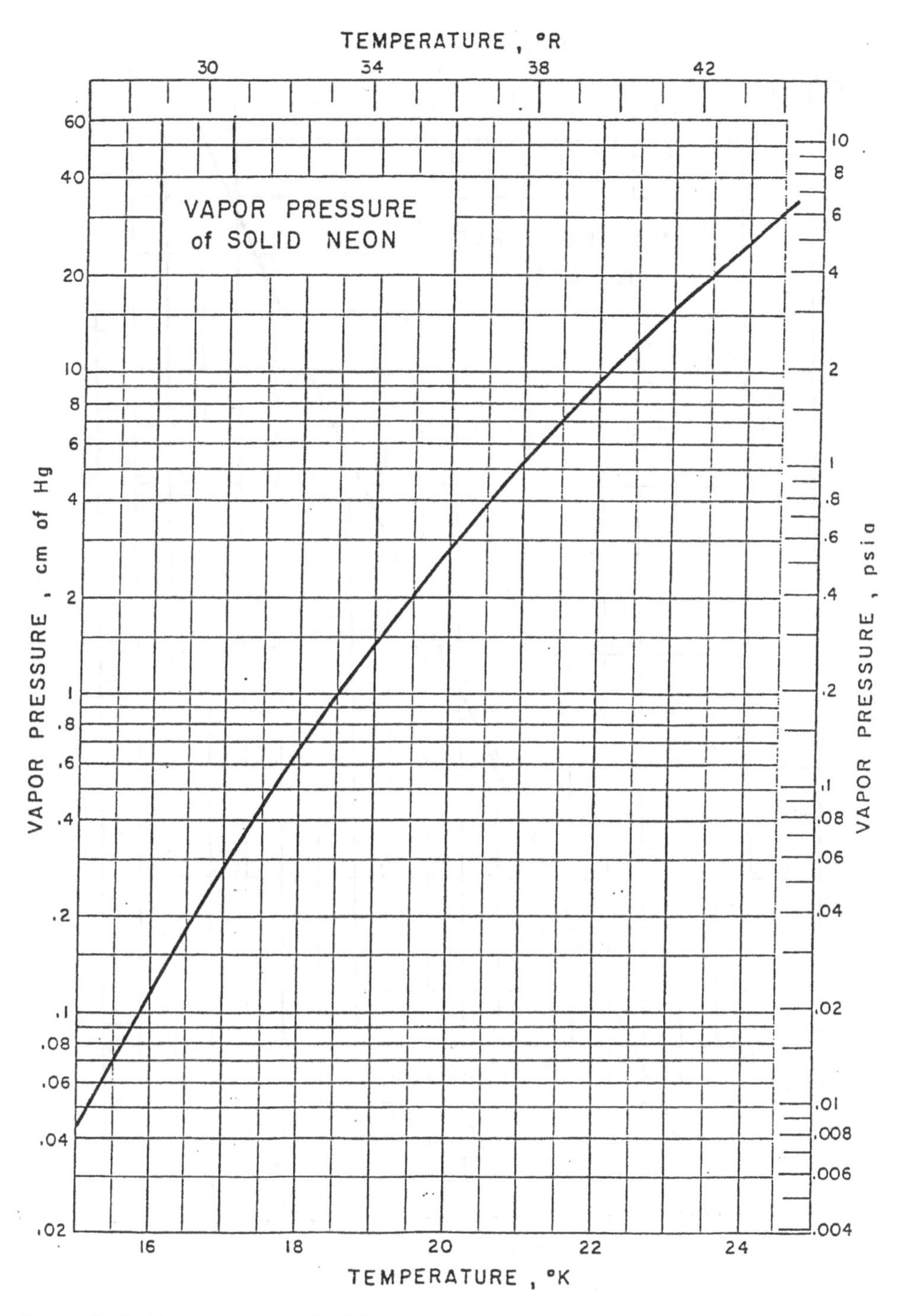


**Figure 3.75** Vapor pressure of solid neon.

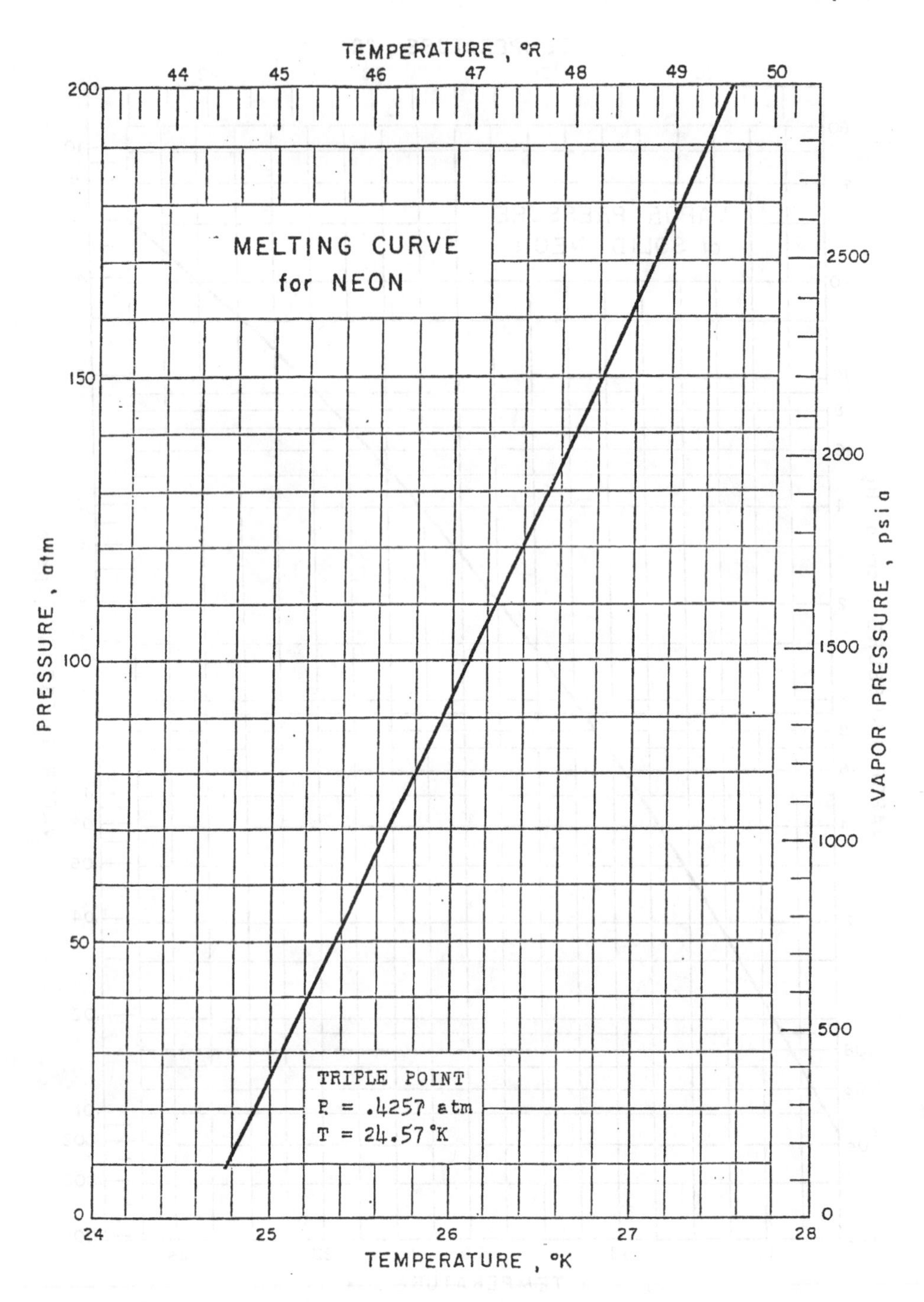


**Figure 3.76** Melting curve for neon.

## 3.6. Fluorine

Liquid fluorine (at wt. 18.9984) is a light yellow liquid having a normal boiling point of 85.3 K (153.6°R). At 53.5 K (96.5°R) and 0.987 bar (1 atm), liquid fluorine freezes as a yellow solid, but upon subcooling to 45.6 K (82°R), it transforms to a white solid. Liquid fluorine is one of the most dense cryogenic liquids [density at normal boiling point = 1504.3 $kg/m^3$ (93.9 $lb_m/ft^3$)].

Fluorine will react with almost all substances. If it comes in contact with hydrocarbons, it will react hypergolically with a high heat of reaction, which is sometimes sufficiently high that the metal container for the fluorine is ignited. Satisfactory handling systems for liquid fluorine have been operated, but this substance presents even greater difficulty than oxygen because of its higher reactivity. The flow of liquid fluorine through a system is preceded by a process of "passivation" in which fluorine gas is admitted slowly to the system in order to build up a layer of passive fluorides on its surfaces. This layer then presents a chemically inert barrier to the large quantities of fluorine that follow.

Fluorine is highly toxic. The fatal concentration range for animals is 200 ppm hr; i.e., for an exposure time of 1 hr, 200 ppm of fluorine is fatal; for an exposure time of 15 min, 800 ppm is fatal; and for an exposure time of 4 hr, 50 ppm is fatal. The maximum allowable concentration for human exposure is usually considered to be approximately 1 ppm hr. The presence of fluorine in air may be detected by its sharp, pungent odor for concentrations as low as 1–3 ppm.

Fluorine was first isolated in 1886 by the French chemist Henri Moissan after nearly 75 years of unsuccessful attempts by several others. For many years after its isolation, fluorine remained little more than a scientific curiosity, to be handled with extreme caution because of its toxicity. Commercial production of fluorine began during World War II when large quantities were required in the fluorination of uranium tetrafluoride ($UF_4$) to produce uranium hexafluoride ($UF_6$) for the isotopic separation of uranium-235 by gaseous diffusion in the development of the atomic bomb. Today, commercial production methods are essentially variations of the Moisson process, and safe techniques have been developed for the bulk handling of liquid fluorine.

The major use of fluorine is in the form of fluorspar. Although not a major commodity in terms of total quantity produced, fluorspar is a critical raw material for the aluminum, chemical, and steel industries of the world. There is at present no significant use of cryogenic liquid fluorine per se.

Table 3.37 gives the physical constants for fluorine.

### *3.6.1. Phase Transition*

The phase transition heats of fluorine are listed in Table 3.38.

### *3.6.2. Heat of Vaporization*

See Table 3.39

### *3.6.3. Vapor Pressure*

Measurements of the vapor pressure have been made by many authors; agreement is good. Hu et al. (1953), whose data extend over a wide range give

$$\log P = 7.08718 - \frac{357.258}{T} - \frac{1.3155 \times 10^{13}}{T^3} \tag{3.23}$$

where $P$ is in mmHg and $T$ is in K.

**Table 3.37** Physical Constants of Fluorine

Chemical Symbol: $F_2$
CAS Registry Number: 7782-41-4
DOT Classification: Nonflammable gas
DOT Label: Poison and Oxidizer
Transport Canada Classification: 2.3 (5.1)
UN Number: UN 1045

| | U.S.Units | SI Units |
|---|---|---|
| *Physical constants* | | |
| Chemical formula | $F_2$ | $F_2$ |
| Molecular weight | 38.00 | 38.00 |
| Density of the gas | | |
| at 32°F (0.0°C) and 1 atm | 0.106 lb/ft$^3$ | 1.70 kg/m$^3$ |
| at 70°F (21.1°C) and 1 atm | 0.098 lb/ft$^3$ | 1.57 kg/m$^3$ |
| Specific gravity of the gas | | |
| at 70°F (21.1°C) and 1 atm | 1.312 | 1.312 |
| Specific volume of the gas | | |
| at 70°F (21.1°C) and 1 atm | 10.17 | 0.635 |
| Density of the liquid | | |
| at −306.8°F (−188.2°C) | 94.2 lb/ft$^3$ | 1509 kg/m$^3$ |
| at −320.4°F (−195.8°C) | 97.9 lb/ft$^3$ | 1568 kg/m$^3$ |
| Boiling point at 1 atm | −306.8°F | −188.2°C |
| Melting point at 1 atm | −363.4°F | −219.6°C |
| Critical temperature | −199.9°F | −128.8°C |
| Critical pressure | 756.4 psia | 5215 kPa abs |
| Critical density | 35.8 lb/ft$^3$ | 573.6 kg/m$^3$ |
| Triple point | −363.4°F at 0.324 psia | −219.6°C at 0.223 kPa abs |
| Latent heat of vaporization | | |
| at −306.8°F (−188.2°C) | 74.8 Btu/lb | 173.5 kJ/kg |
| Latent heat of fusion | | |
| at −363.4°F (−219.6°C) | 5.8 Btu/lb | 13.4 kJ/kg |
| Weight of liquid per gallon | | |
| at −306.8°F (−188.1°C) | 12.6 lb/gal | 1509.8 kg/m$^3$ |
| at −320.4°F (−195.8°C) | 13.1 lb/gal | 1569.7 kg/m$^3$ |
| Heat capacity of the gas, $C_P$ | | |
| at 32°F (0°C) | 0.198 Btu/(lb)(°F) | 0.828 kJ/(kg)(°C) |
| at 70°F (21.1°C) | 0.197 Btu/(lb)(°F) | 0.825 kJ/(kg)(°C) |
| Heat capacity of the gas, $C_V$ | | |
| at 70°F (21.1°C) | 0.146 Btu/(lb)(°F) | 0.610 kJ/(kg)(°C) |
| Ratio of specific heats, $C_P/C_V$ | | |
| at 70°F (21.1°C) | 1.353 | 1.353 |
| Thermal conductivity of gas | | |
| at 32°F (0°C) and 1 atm | 0.172 Btu.in./hr°F | 0.248 W/m.K |
| Viscosity | | |
| of gas at 32°F (0°C) and 1 atm | 0.0218 cP | 0.0218 mPa s |
| of liquid at −314°F (−192.2°C) | 0.275 cP | 0.275 mPa s |

**Table 3.38** Phase Transition Heats of Fluorine

| Transition | Temp. (K) | Transition heats (kJ/kg) |
| --- | --- | --- |
| Solid transition | 45.55 | 19.14 |
| Fusion | 53.51 | 13.43 |
| Vaporization | 85.21 | 166.22 |

*Source:* Hu et al. (1953); Rossini et al. (1952).

**Table 3.39** Heat of Vaporization of Fluorine

| Temp. (K) | Heat of vaporization (kJ/kg) |
| --- | --- |
| 85.25 | 166.2 |
| 90 | 160.9 |
| 95 | 154.9 |
| 100 | 148.6 |
| 105 | 142.0 |
| 110 | 134.6 |
| 115 | 126.6 |
| 120 | 117.7 |
| 125 | 107.6 |
| 130 | 95.65 |
| 135 | 80.81 |
| 140 | 59.12 |
| 144 | 14.85 |

*Source:* Fricke (1948).

### *3.6.4. Density of Liquid Fluorine*

The data of Jarry and Miller (1956) are represented by the following equation and are tabulated below.

$$\rho = 1.907 - 2.201 \times 10^{-3}T - 2.948 \times 10^{-5}T^2 \tag{3.24}$$

where $\rho$ (density) is in g/cm$^3$ and $T$ is in K.

Density of Fluorine—Selected Values

| Temp. (K) | Density (kg/m$^3$) | Temp. (K) | Density (kg/m$^3$) |
| --- | --- | --- | --- |
| 65.78 | 1638 | 86.91 | 1496 |
| 71.76 | 1594 | 88.26 | 1481 |
| 74.93 | 1578 | 88.50 | 1484 |
| 78.59 | 1550 | 90.08 | 1472 |
| 78.62 | 1553 | 91.55 | 1458 |
| 81.72 | 1532 | 91.75 | 1460 |
| 81.73 | 1528 | 94.73 | 1434 |
| 84.34 | 1514 | 97.56 | 1412 |
| 85.05 | 1505 | 100.21 | 1391 |
| 85.67 | 1505 | 102.75 | 1370 |

*Source:* Jarry and Miller (1956).

**Table 3.40** Viscosity of Gaseous Fluorine

| Temp. (K) | Viscosity (cP) |
|---|---|
| 90.0 | 0.767 |
| 169.3 | 1.424 |
| 200.0 | 1.680 |
| 289.1 | 2.345 |
| 327.1 | 2.547 |

*Source:* Franck and Stober (1952).

### 3.6.5. *Viscosity*

Tables 3.40 and 3.41 present data on the viscosity of gaseous and liquid fluorine, respectively.

### 3.6.6. *Thermal Conductivity*

For data on the thermal conductivity of gaseous fluorine, see Table 3.42.

*Symbols:* $\rho_{sat}$ = saturation pressure or vapor pressure, $\rho$ = density, $C_P$ = specific heat at constant pressure, $\mu$ = viscosity, $k$ = thermal conductivity, $n_{fg}$ = heat of vaporization, $N_{pr} = \mu C_P/K$ = Prandtl number, $\sigma_L$ = surface tension.

### 3.6.7. *Other Properties*

Figures 3.77–3.82 give the specific heats, density, heat of vaporization, vapor pressure, surface tension, and viscosity of fluorine, taken from the WADD report mentioned before (Johnson, 1960).

**Table 3.41** Viscosity of Liquid Fluorine

| Temp. (K) | Viscosity (cP) | Temp. (K) | Viscosity (cP) |
|---|---|---|---|
| 69.2 | 0.414 | 78.2 | 0.299 |
| 73.2 | 0.349 | 80.9 | 0.275 |
| 75.3 | 0.328 | 83.2 | 0.257 |

*Source:* Elverum and Doescher (1952).

**Table 3.42** Thermal Conductivity of Gaseous Fluorine

| Temperature (K) | Thermal conductivity (mW/(cm K)) |
|---|---|
| 100 | 0.0862 |
| 150 | 0.134 |
| 200 | 0.183 |
| 250 | 0.228 |
| 273 | 0.247 |
| 300 | 0.269 |
| 350 | 0.308 |

*Source:* Franck and Wicke (1951).

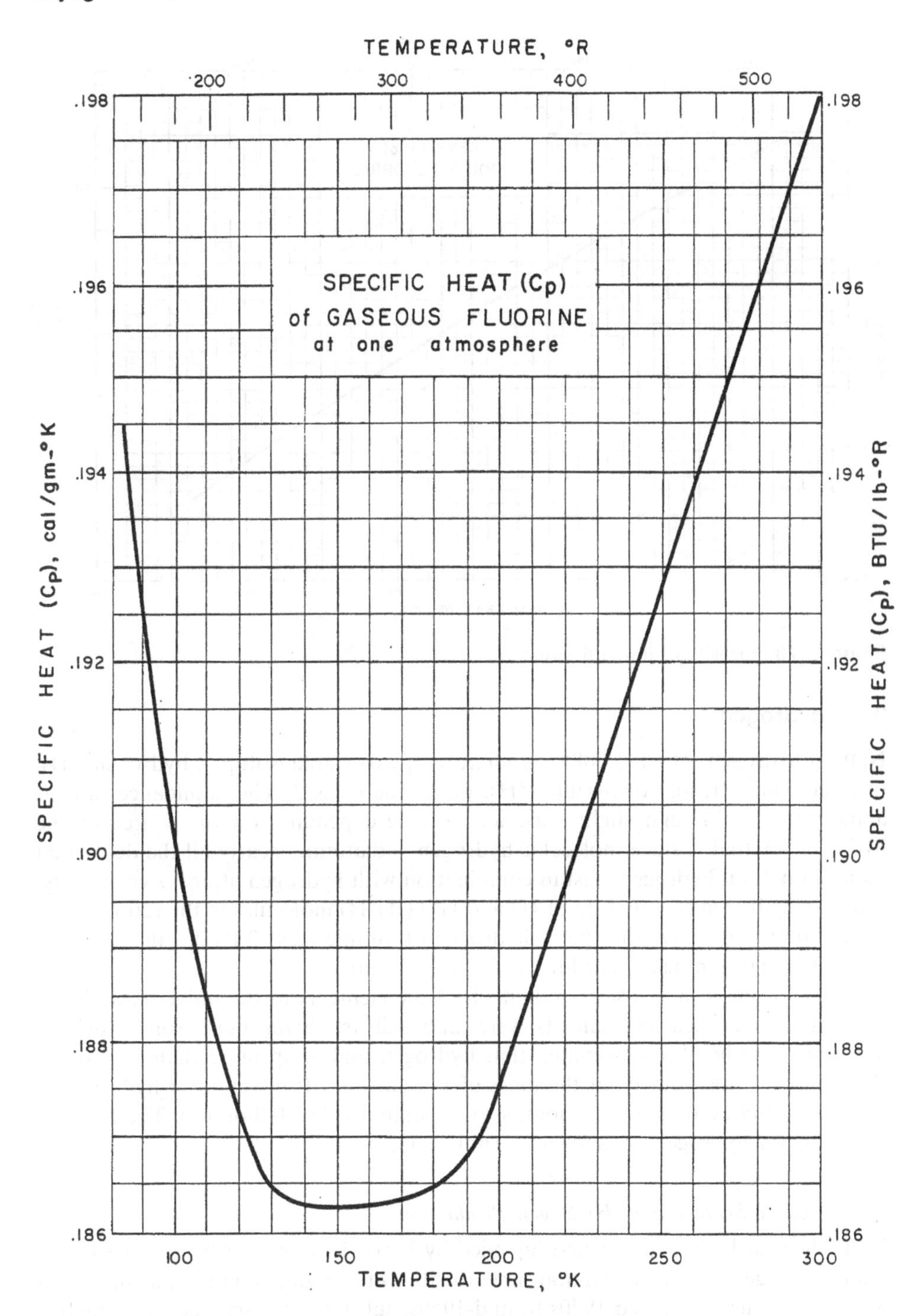


**Figure 3.77** Specific heat ($C_P$) of gaseous fluorine at 1 atm.

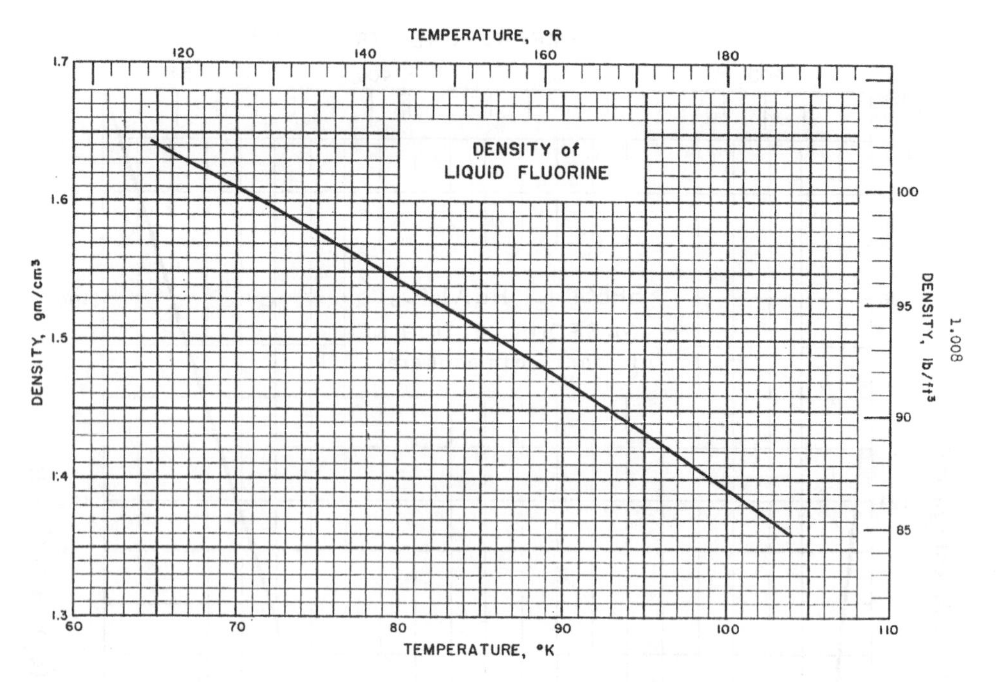


**Figure 3.78** Density of liquid fluorine.

## 3.7. Hydrogen

Natural hydrogen (at wt. 1.008) is a mixture of two stable isotopes; hydrogen ($^1H$), of atomic mass 1, and deuterium ($^2H$), of atomic mass 2. The abundance ratio is about 6400:1, although slight variations occur, depending on the source of the hydrogen. Actually, since molecular hydrogen is diatomic, nearly all the deuterium atoms in natural hydrogen exist in combination with hydrogen atoms. Accordingly, ordinary hydrogen is a mixture of $H_2$ and HD ($^1H^2H$) molecules in the ratio 3200:1. A very rare radioactive isotope of hydrogen of atomic mass 3 ($^3H$), called tritium, also exists. Tritium has a half-life of about 12.5 years.

Hydrogen is the most abundant of all the elements in the universe, and it is thought that the heavier elements were, and still are, being built from hydrogen and helium. It has been estimated that hydrogen makes up more than 90% of all the atoms or three-fourths of the mass of the universe. Production of hydrogen in the United States alone now amounts to 85 million $m^3$ (3 billion $ft^3$). The physical constants for hydrogen are shown in Table 3.43.

### *3.7.1. Large-Scale Liquid Hydrogen Production*

From the time hydrogen was first liquefied by James Dewar in 1898 until the 1940s, liquid hydrogen remained a laboratory curiosity, with production rates measured in a few grams. During the mid-1940s to mid-1950s, substantial-scale laboratory hydrogen liquefiers were built to meet the needs of the US government's nuclear and aerospace programs, with the largest such liquefier being a 240 L/hr (0.41 metric ton/day) unit at the National Bureau of Standards in Boulder, Colorado. The first commercial-scale liquid hydrogen production plants were built in the late 1950s under a then secret US. Air Force program. Under contract to the Air Force, three successively larger liquid hydrogen plants code named Baby Bear (0.75 US tons/day or 0.68

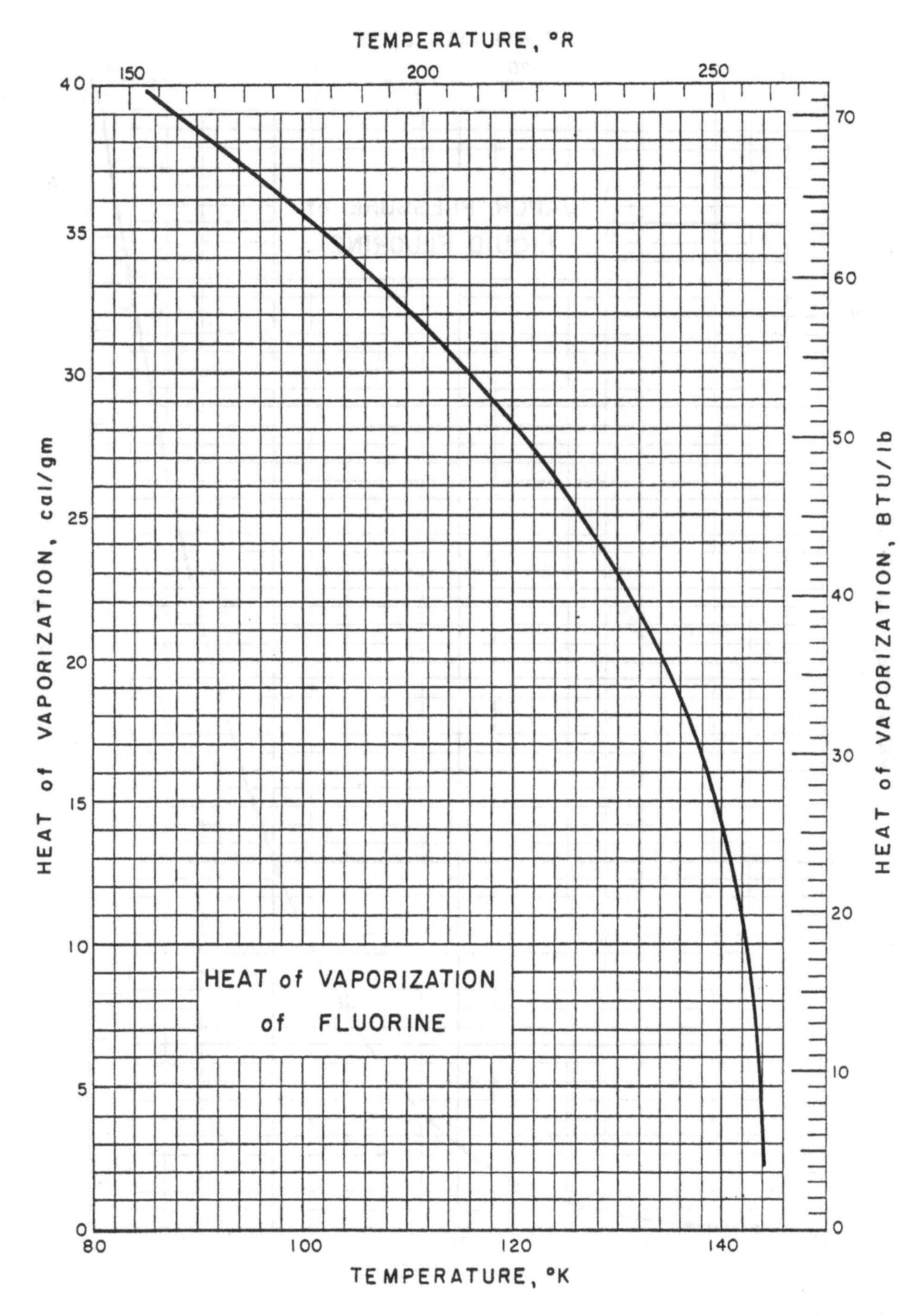


**Figure 3.79** Heat of vaporization of fluorine.

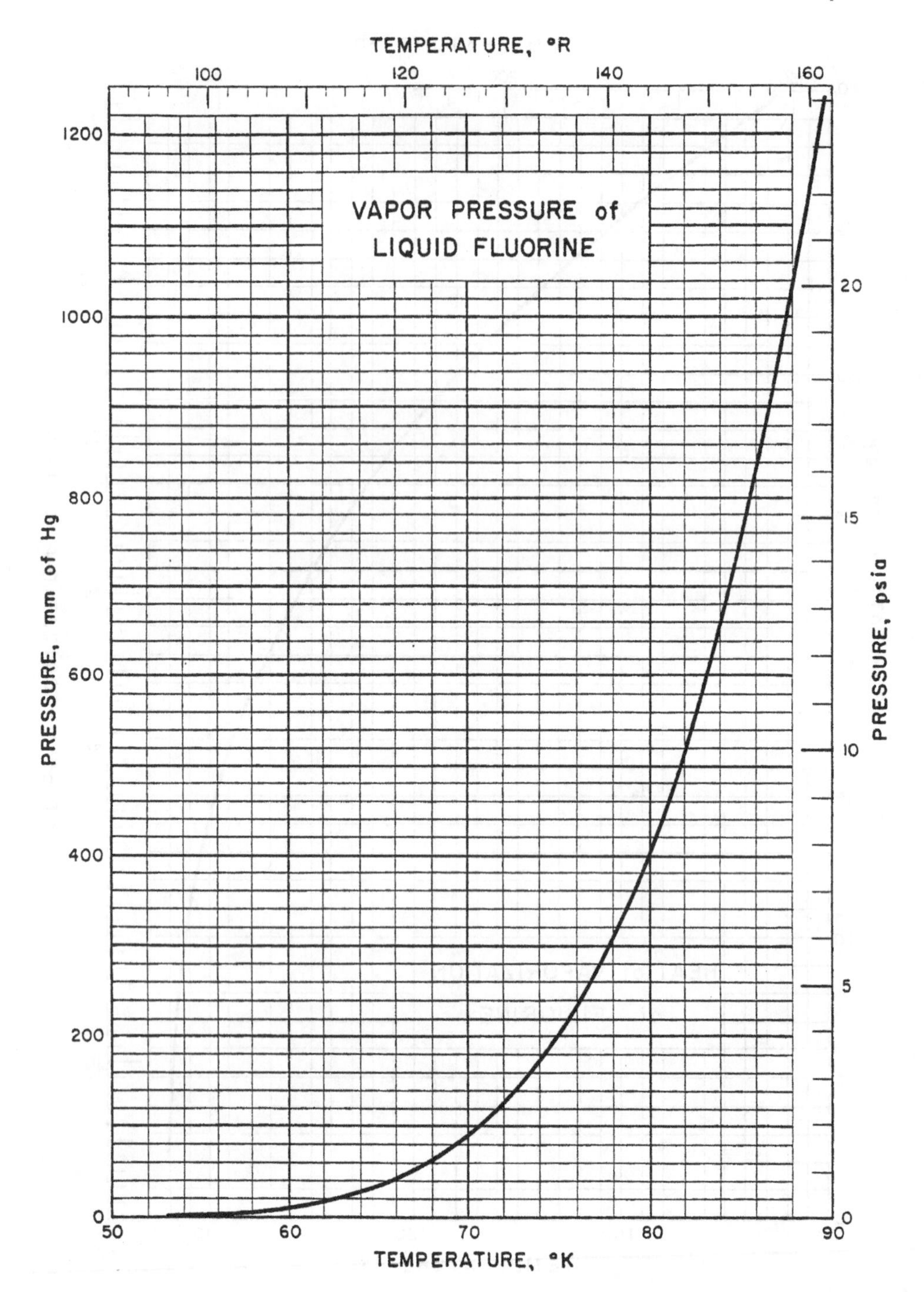


**Figure 3.80** Vapor pressure of liquid fluorine.

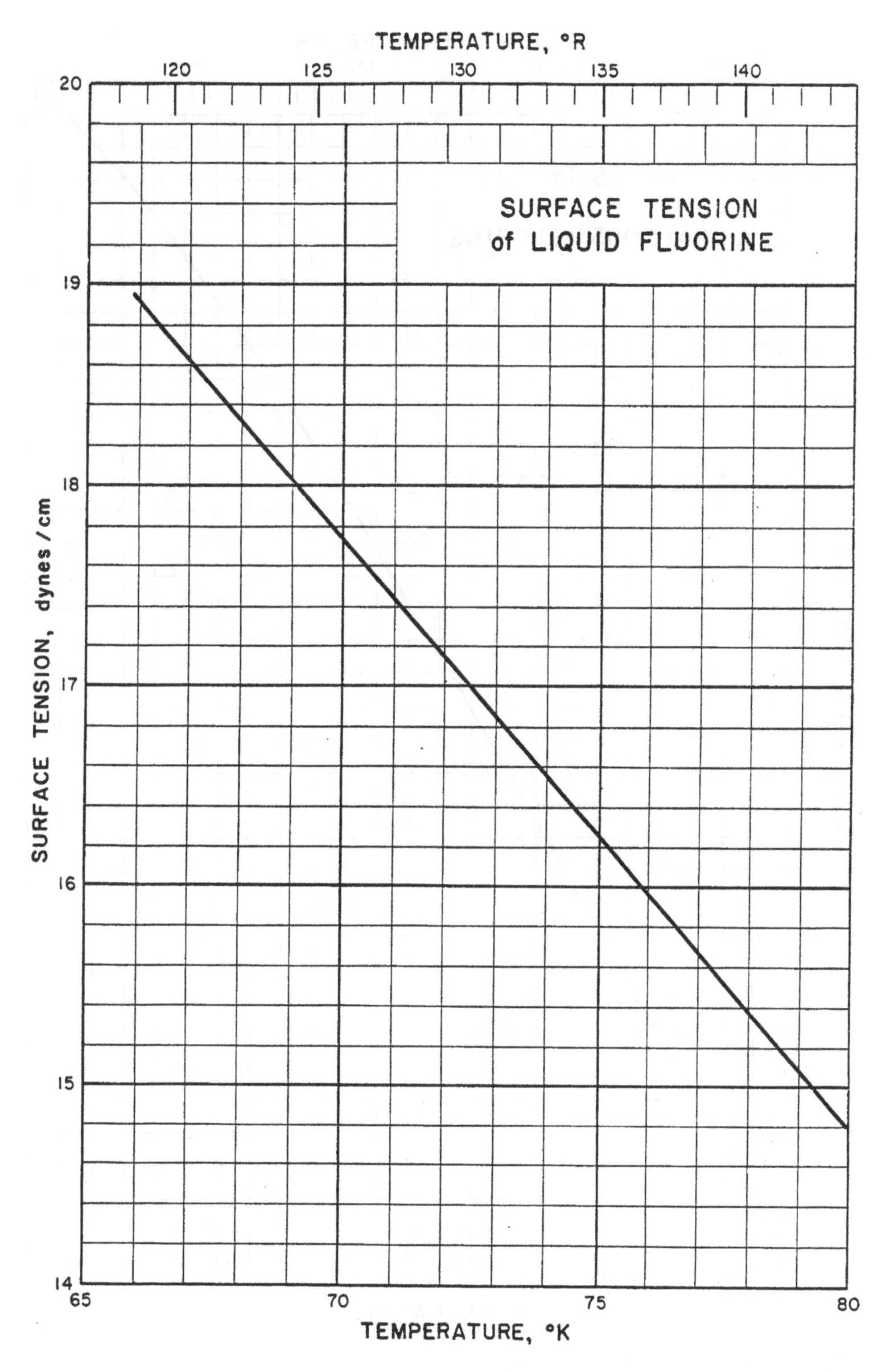


**Figure 3.81** Surface tension of liquid fluorine.

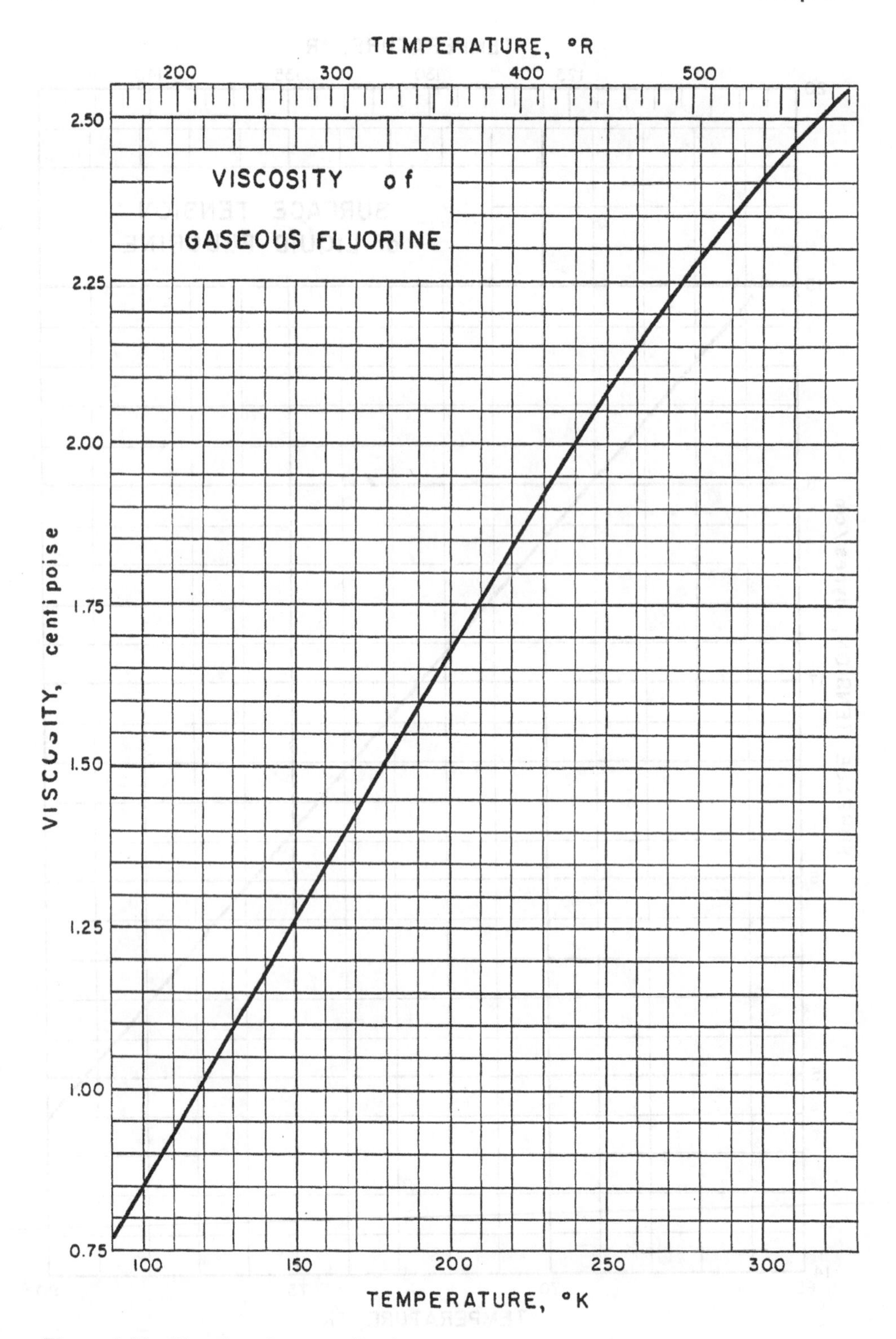


**Figure 3.82** Viscosity of gaseous fluorine.

**Table 3.43** Physical Constants of Hydrogen

*Hydrogen*
Chemical Symbol: $H_2$
CAS Registry Number: 1333-74-0
DOT Classification: Flammable gas
DOT Label: Flammable gas
Transport Canada Classification: 2.1
UN Number: UN 1049 (compressed gas); UN 1966 (refrigerated liquid)

| | U.S. Units | SI Units |
|---|---|---|
| *Normal Hydrogen* | | |
| Chemical formula | $H_2$ | $H_2$ |
| Molecular weight | 2.016 | 2.016 |
| Density of the gas at 70°F (21.1°C) and 1 atm | 0.00521 lb/ft$^3$ | 0.08342 kg/m$^3$ |
| Specific gravity of the gas at 32°F (0°C) and 1 atm (air = 1) | 0.06960 | 0.06960 |
| Specific volume of the gas at 70°F (21.1°C) and 1 atm | 192.0 ft$^3$/lb | 11.99 m$^3$/kg |
| Boiling point at 1 atm | −423.0°F | −252.8°C |
| Melting point at 1 atm | −434.55°F | −259.2°C |
| Critical temperature | −399.93°F | −239.96°C |
| Critical pressure | 190.8 psia | 1315 kPa abs |
| Critical density | 1.88 lb/ft$^3$ | 30.12 kg/m$^3$ |
| Triple point | −434.55°F at 1.045 psia | −259.2°C at 7.205 kPa abs |
| Latent heat of vaporization at boiling point | 191.7 Btu/lb | 446.0 kJ/kg |
| Latent heat of fusion at triple point | 24.97 Btu/lb | 58.09 kJ/kg |
| Specific heat of the gas at 70°F (21.1°C) and 1 atm | | |
| $C_P$ | 3.425 Btu/(lb)(°F) | 14.34 kJ/(kg)(°C) |
| $C_V$ | 2.418 Btu/(lb)(°F) | 10.12 kJ/(kg)(°C) |
| Ratio of specific heats | | |
| $(C_P/C_V)$ | 1.42 | 1.42 |
| Solubility in water, vol/vol at 60°F (15.6°C) | 0.019 | 0.019 |
| Density of the gas at boiling point and 1 atm | 0.083 lb/ft$^3$ | 1.331 kg/m$^3$ |
| Density of the liquid at boiling point and 1 atm | 4.43 lb/ft$^3$ (0.5922 lb/gal) | 70.96 kg/m$^3$ |
| Gas/liquid ratio (liquid at boiling point, gas at 70°F (21.1°C) and 1 atm), vol/vol | 850.3 | 850.3 |
| Heat of combustion at 70°F (21.1°C) and 1 atm | | |
| Gross | 318.1 Btu/ft$^3$ | 11,852 kJ/m$^3$ |
| Net | 268.6 Btu/ft$^3$ | 10,009 kJ/m$^3$ |
| *Para Hydrogen* | | |
| (The following constants are different from normal hydrogen) | | |
| Boiling point at 1 atm | −423.2°F | −252.9°C |
| Melting point | −434.8°F | −259.3°C |

(*Continued*)

**Table 3.43** (*Continued*)

| | | |
|---|---|---|
| Critical temperature | −400.31°F | −240.17°C |
| Critical pressure | 187.5 psia | 1293 kPa abs |
| Critical density | 1.96 lb/ft$^3$ | 31.43 kg/m$^3$ |
| Triple point | −434.8°F at 1.021 psia | −259.3°C at 7.042 kPa abs |
| Latent heat of vaporization at boiling point | 191.6 Btu/lb | 445.6 kJ/kg |
| Latent heat of fusion at triple point | 25.06 Btu/lb | 58.29 kJ/kg |
| Specific heat of the gas at 70°F (21.1°C) and 1 atm | | |
| $C_P$ | 3.555 Btu/(lb)(°F) | 14.88 kJ/(kg)(°C) |
| $C_V$ | 2.570 Btu/(lb)(°F) | 10.76 kJ/(kg)(°C) |
| Ratio of specific heats ($C_P/C_V$) | 1.38 | 1.38 |
| Density of the gas at boiling point and 1 atm | 0.084 lb/ft$^3$ | 1.338 kg/m$^3$ |
| Density of the liquid at boiling point and 1 atm | 4.42 lb/ft$^3$ (0.5907 lb/gal) | 70.78 kg/m$^3$ |
| Gas/liquid ratio (liquid at boiling point, gas at 70°F (21.1°C) and 1 atm, vol/vol | 848.3 | 848.3 |

metric tons/day), Mama Bear (3.5 US tons/day or 3.2 metric tons/day), and Papa Bear (30 tons/day or 27 metric tons/day) were designed and built by Air Products and Chemicals, Inc. Papa Bear, which went on-stream in West Palm Beach, Florida, in February 1959, was the world's first truly large-scale liquid hydrogen plant.

From 1960 to the present, several additional large-scale commercial liquid hydrogen plants were built in North America. For example, in 1964 the Linde Division of Union Carbide built the largest single plant to date, a 60 US ton/day (54 metric tons/day) plant in Sacramento, California. Between 1965 and 1976, Air Products constructed a 66 US tons/day (60 metric tons/day) two-plant liquid hydrogen production complex in New Orleans, Louisiana, and between 1982 and 1990 three liquid hydrogen plants were built in Canada by Air Products, Liquid Air, and BOC.

Today, nine large-scale liquid hydrogen plants are in operation in the United States and Canada with a combined capacity of about 190 US tons/day (172 metric tons/day). Some older plants like Papa Bear and the Linde Sacramento plant no longer exist. Table 3.44 lists some key milestones in the creation of large-scale commercial liquid hydrogen plants, and Fig. 3.83 shows the location, size, and ownership of liquid hydrogen plants in North America in 1990. The aerospace program, which created the initial demand for large quantities of liquid hydrogen, accounted for only 20% of liquid hydrogen demand in 1990. A variety of commercial applications account for the remaining demand.

The diversity of commercial liquid hydrogen containers for both stationary storage and transportation are listed below. Three insulation technologies are employed. In increasing order of thermal efficiency and cost, they are evacuated perlite insulation, evacuated multilayer insulation, and liquid nitrogen shielded multilayer insulation. Large stationary storage containers are reasonably efficient, using just an evacuated annulus filled with perlite, due to their relatively low surface-to-volume ratio. Smaller stationary containers and all transportation containers require multilayer insulation for acceptable efficiency (i.e., low vent rates). Liquid nitrogen shielded multilayer insulation is widely used for liquid helium storage and transport,

**Table 3.44** Milestones in the Development of Large-Scale Commercial Liquid Hydrogen Plants[a]

| Date | Event |
|---|---|
| 1898 | Hydrogen first liquefied by James Dewar |
| 1898–mid-40s | Laboratory curiosity |
| Mid 1940s–1956 | Laboratory and pilot-scale liquefiers built to support US nuclear weapons and aerospace programs. Largest unit: 0.45 ton/day at National Bureau of Standards |
| 1956–1959 | US Air Force and Air Products—"Bear" program: |
| 1957 | Baby Bear—Painsville, OH (0.75 ton/day) |
| 1957 | Mama Bear—West Palm Beach, FL (3.5 tons/day) |
| 1959 | Papa Bear—West Palm Beach, FL (30 tons/day) |
| 1960–present | Several large-scale plants constructed in North America through private funding and operation. Examples: |
| 1964 | Linde 60 tons/day $LH_2$ plant in Sacramento, CA; largest ever built |
| 1965, 1976 | Air Products builds two-plant, 66 tons/day $LH_2$ complex in New Orleans, LA |
| 1982–1990 | Three $LH_2$ plants built in Canada—Air Products, Liquid Air, Airco |
| 1980s–present | $LH_2$ plants built in Europe and Japan |

[a] All plant capacities given in US tons/day.

but for liquid hydrogen it is typically limited to portable containers that must be shipped long distances (often overseas) without venting for several days.

Boiloff from product liquid hydrogen storage tanks at the liquid hydrogen plant site is usually redirected back to the liquefier plant for reliquefaction.

### *3.7.2. Molecular Forms of Hydrogen and Ortho–Para Conversion*

One of the properties of hydrogen that sets it apart from other substances is the fact that it can exist in two different molecular forms: orthohydrogen ($o$-$H_2$) and parahydrogen ($p$-$H_2$).

Liquid Hydrogen Transportation and Storage[a]

| Container | Capacity (US gal)[b] |
|---|---|
| *Transportation* | |
| Over-the-road tankers | 13,000–15,000 |
| Seagoing containers | 11,000 |
| Inland waterway barges | 250,000 |
| Railcars | 28,000–34,000 |
| *Storage* | |
| Large tanks | |
| Spheres at Kennedy Space Center | 830,000 |
| Spheres at Air Products' New Orleans plant | 500,000 |
| Horizontal cylinders at Air Products' Sacramento plant | 70,000 |
| *Customer tanks* | |
| Horizontal/vertical cylinder | 1500–9000 |
| Horizontal cylinder | 9000–20,000 |

[a] Insulation types include perlite, multilayer, and liquid nitrogen shielded multilayer.
[b] 1000 US gal = 3.785 L.

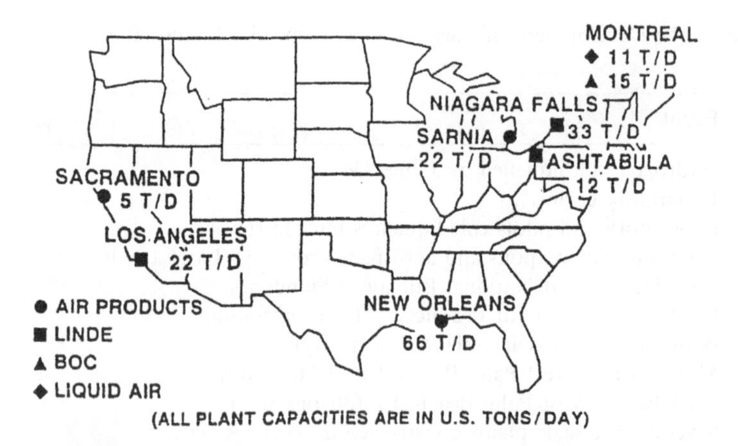


**Figure 3.83** Liquid hydrogen plants in the United States and Canada, 1990.

Figure 3.84 illustrates the two states of the hydrogen molecule: the higher energy ortho state in which the protons in the two nuclei spin in the same direction, and the lower energy para state in which the protons spin in opposite directions. Temperature determines the relative abundance of each state, as shown in Table 3.45 and by the ortho–para thermodynamic equilibrium diagram in Fig. 3.85. At ambient and higher temperatures, hydrogen consists of 25% para and 75% ortho and is referred to as "normal" hydrogen ($n$-$H_2$). At the normal boiling point of liquid hydrogen (20.4 K), equilibrium shifts to almost 100% para. Conversion from ortho to para is exothermic, with the heat of conversion of 302 Btu/lb (702 J/g) substantially exceeding the latent heat of vaporization of normal boiling point hydrogen of 195 Btu/lb (454 J/g).

The equilibrium mixture of $o$-$H_2$ and $p$-$H_2$ at any given temperature is called equilibrium hydrogen ($e$-$H_2$). At the normal boiling point of hydrogen, 20.4 K (36.7°R), equilibrium hydrogen has a composition of 0.21% $o$-$H_2$ and 99.79% $p$-$H_2$, or practically all parahydrogen.

Two deuterium atoms can combine to form a deuterium molecule, often denoted $D_2$ and sometimes called heavy hydrogen. As in the case of "light" hydrogen, the nuclear spins of the atoms in a deuterium molecule can assume different

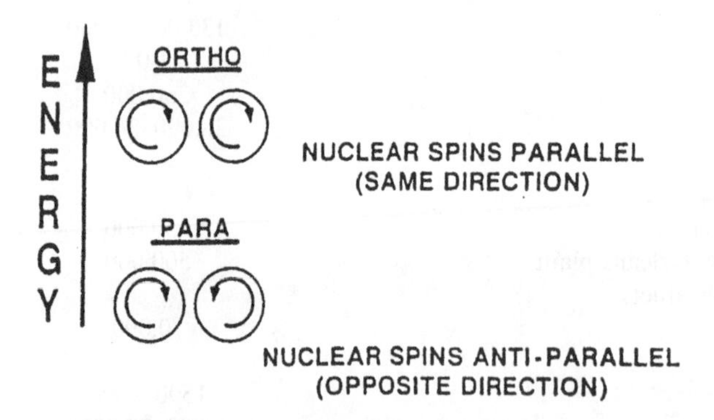


**Figure 3.84** Ortho- and parahydrogen differ in the manner in which their two protons spin in the nucleus of the hydrogen atom.

**Table 3.45** Ortho–Para Hydrogen Data

| Temperature (K) | $H_2$, % in para form | $D_2$, % in para form | Heat of conversion[a] (cal/(g mol)) |
|---|---|---|---|
| 10 | 99.9999 | 0.0277 | 338.648 |
| 20 | 99.821 | 1.998 | 338.649 |
| 20.39 | 99.789 | – | 338.648 |
| 23.57 | – | 3.761 | – |
| 30 | 97.021 | 7.864 | 338.648 |
| 33.1 | 95.034 | – | 338.648 |
| 40 | 88.727 | 14.784 | 338.634 |
| 50 | 77.054 | 20.718 | 338.460 |
| 60 | 65.569 | 25.131 | 337.616 |
| 70 | 55.991 | 28.162 | 335.200 |
| 80 | 48.537 | 30.141 | 330.164 |
| 90 | 42.882 | 31.395 | 321.700 |
| 100 | 38.620 | 32.164 | 309.440 |
| 120 | 32.99 | 32.916 | 274.475 |
| 150 | 28.603 | 33.246 | 207.175 |
| 200 | 25.974 | 33.327 | 105.20 |
| 250 | 25.264 | – | 45.31 |
| 298.16 | 25.075 | 33.333 | 18.35 |
| 300 | 25.072 | 33.333 | 17.71 |
| 350 | 25.019 | – | – |
| 400 | 25.005 | – | – |
| 500 | 25.000 | – | – |

[a] For conversion of $o$-$H_2$ to $p$-$H_2$.

spatial orientations with respect to each other, although the orientations are a bit more complicated. *Ortho*- and *para deuterium* have different *sets* of possible spatial orientations, and the ortho–para equilibrium concentrations are temperature-dependent, as in the case of hydrogen. The nucleus of the deuterium atom consists

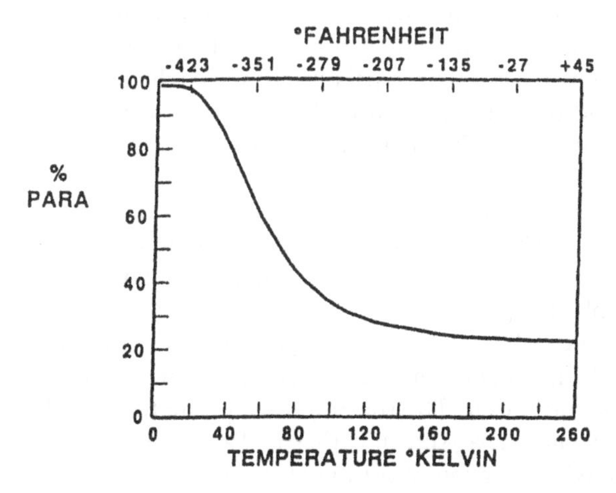


**Figure 3.85** Equilibrium para content of hydrogen vs. temperature.

of one proton and one neutron, so that the high-temperature composition (composition of normal deuterium) is two-thirds orthodeuterium and one-third paradeuterium. In the case of deuterium, $p$-$D_2$ converts to $o$-$D_2$ as the temperature is decreased, in contrast to hydrogen, in which $o$-$H_2$ converts to $p$-$H_2$ upon a decrease in temperature. (The prefix ortho- simply means *common* and does not inherently refer to the more energetic form.) The hydrogen deuteride molecule does not have the symmetry that hydrogen and deuterium molecules possess; therefore, HD exists in only one form.

Temperature–entropy diagrams for parahydrogen and deuterium are presented in Figs. 3.86–3.91

Table 3.46 lists the heat of conversion for normal hydrogen to parahydrogen. Figure 3.92 presents similar data graphically.

Table 3.47 gives the fixed points for hydrogen, and Tables 3.48 and 3.49 summarize the fixed point properties of normal hydrogen and parahydrogen.

A catalyst is normally required to convert between the ortho and para forms of hydrogen. Without a catalyst the ortho to para reaction proceeds only very slowly in the liquid phase and not at all in the vapor phase. In the liquid phase, unconverted normal liquid hydrogen will slowly convert to $p$-$H_2$, releasing sufficient heat to vaporize most of the liquid. Thus, ortho-to-para conversion must be incorporated into the hydrogen liquefier so that the product liquid will consist almost entirely of stable *para*-hydrogen.

To minimize the impact of heat removal due to ortho-to-para conversion on the thermodynamic efficiency of the liquefier cycle, it is important to accomplish ortho-to-para conversion at the highest possible temperature levels consistent with thermodynamic equilibrium. Continuous conversion would be the most thermodynamically efficient. In continuous conversion, ortho-to-para conversion occurs simultaneously with cooling so that the hydrogen cooling path closely follows the equilibrium curve. (Some offset from full equilibrium is necessary to provide some driving force for conversion). Continuous conversion, however, requires a complex integration of catalyst and heat exchange surface that is difficult to design into practical systems. A common design approach is to approximate continuous conversion by using several simple adiabatic catalyst converters interdispersed with heat exchange. Starting at around liquid nitrogen temperatures, hydrogen is partially cooled, removed from the heat exchanger, and sent through an adiabatic converter where the hydrogen partially rewarms due to the heat of conversion. Conversion to parahydrogen is accomplished to the extent permitted by approaching near equilibrium at the converter outlet temperature. The partially converted hydrogen is returned to the heat exchanger, and the stepwise heat exchange-conversion process is repeated until the product hydrogen from the final converter is at least 95% parahydrogen.

Two types of ortho–para catalysts are commercially available: an iron oxide catalyst and a nickel silicate catalyst. The nickel silicate catalyst is the more active of the two. Anywhere from 50 to a few hundred pounds of catalyst would be used for each ton per day of liquefier plant capacity.

Most of the physical properties of hydrogen and deuterium, such as vapor pressure, density of the liquid, and triple point temperature and pressure, are mildly dependent upon the ortho–para composition. The transition from one form to the other involves a substantial energy change, as shown earlier in Table 3.45.

Unfortunately, the uncatalyzed conversion of orthohydrogen to parahydrogen is relatively slow, which leads to an important consideration in the liquefaction and

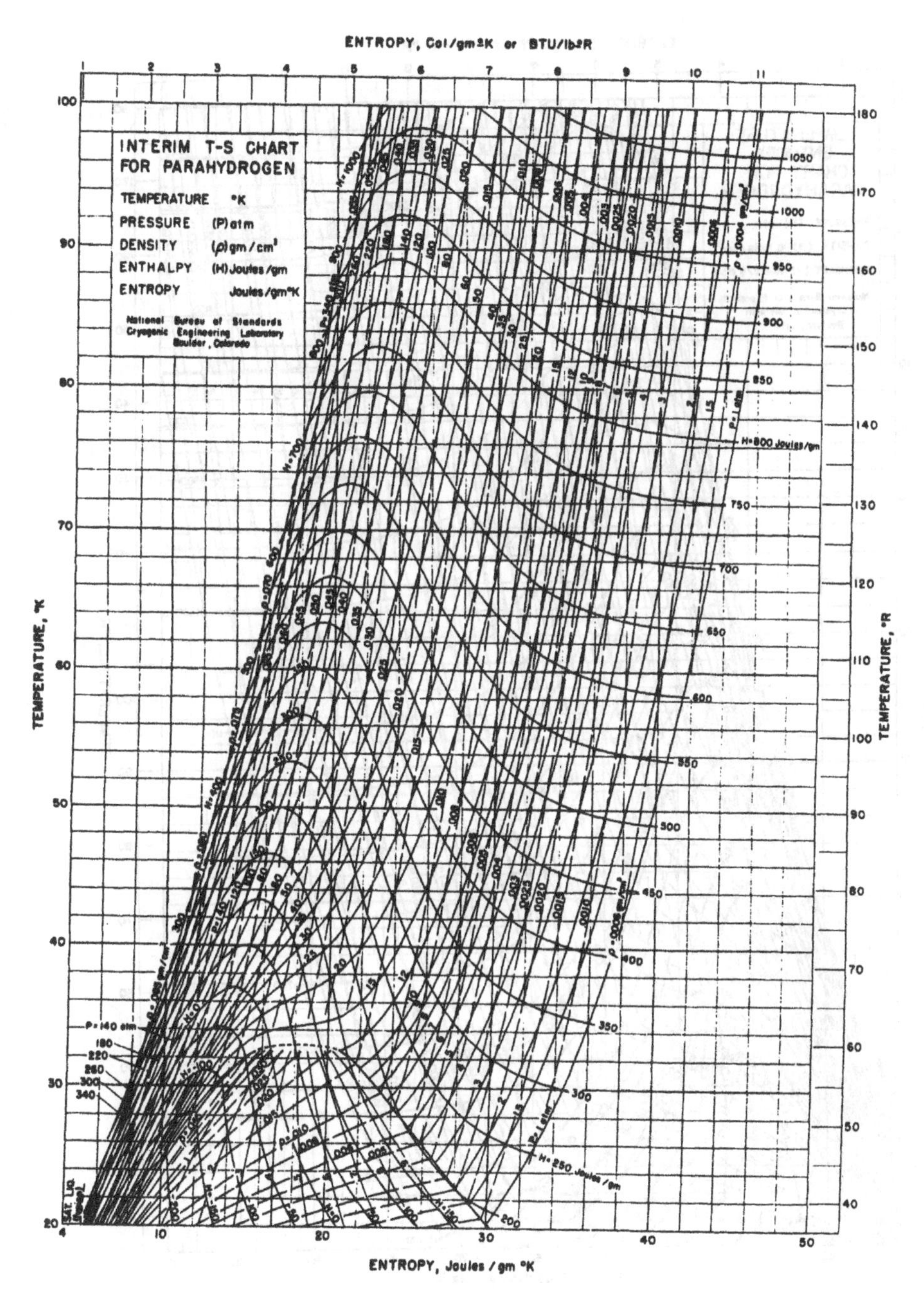


**Figure 3.86** Interim *T–S* chart for parahydrogen.

storage of hydrogen that does not exist for other gases. As mentioned, equilibrium hydrogen at room temperature consists of 75% orthohydrogen and 25% parahydrogen. At 20.39 K, however, the approximate temperature at which hydrogen boils at atmospheric pressure, the equilibrium concentration, is 99.8% parahydrogen. If one starts with equilibrium or normal hydrogen at room temperature and liquefies it, the liquid orthohydrogen will slowly convert to parahydrogen. Since parahydrogen is

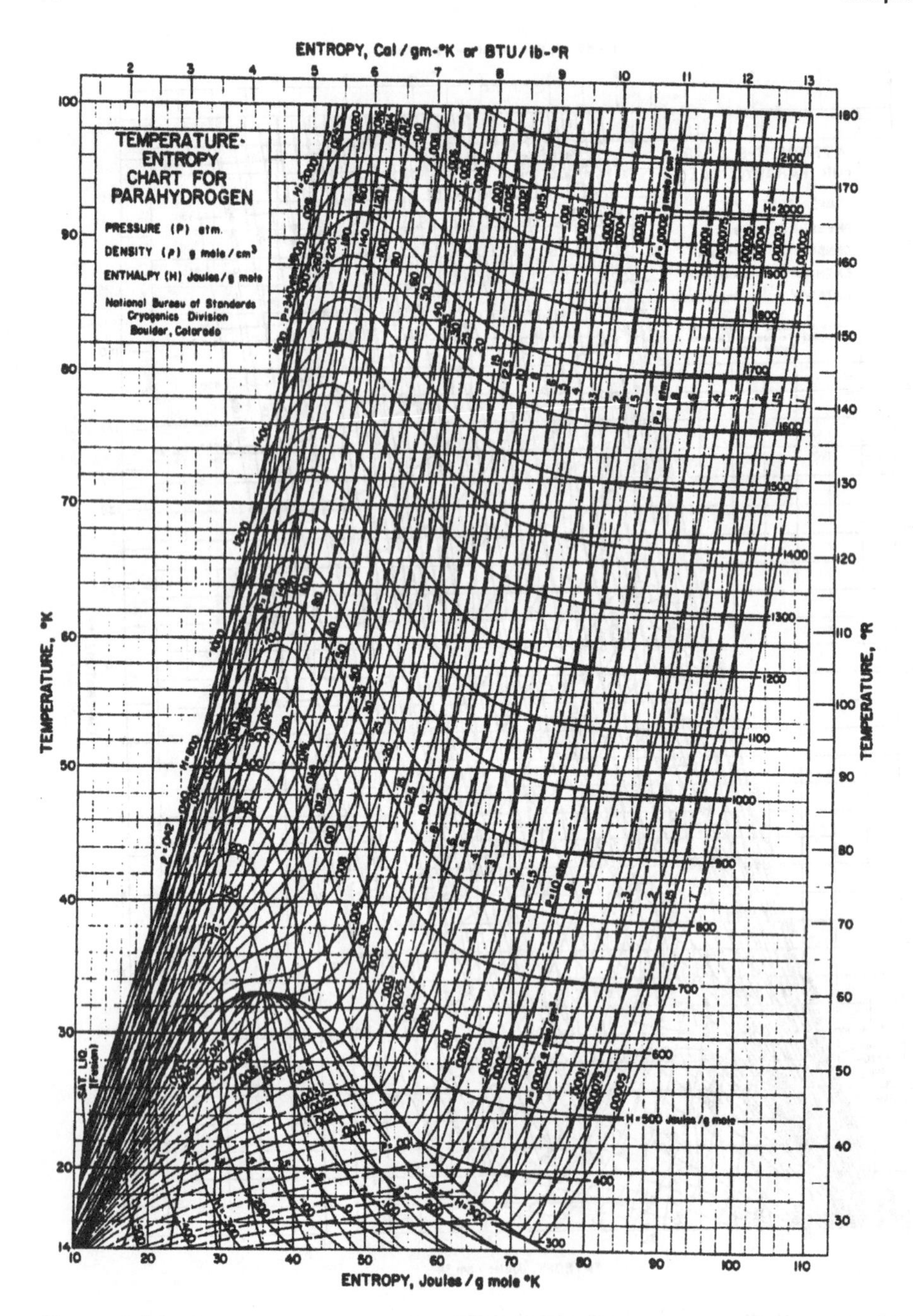


Figure 3.87 Temperature–entropy chart for parahydrogen.

the lower energy form, heat is liberated in the conversion process, and this is sufficient to cause evaporation of nearly 70% of the hydrogen originally liquefied. Liquefaction cycles, therefore, usually incorporate a catalyst through which the newly liquefied hydrogen must pass. This causes the conversion to take place in the liquefier, where the heat can be absorbed conveniently, and eliminates the otherwise gradual evaporation of the stored liquid from this cause.

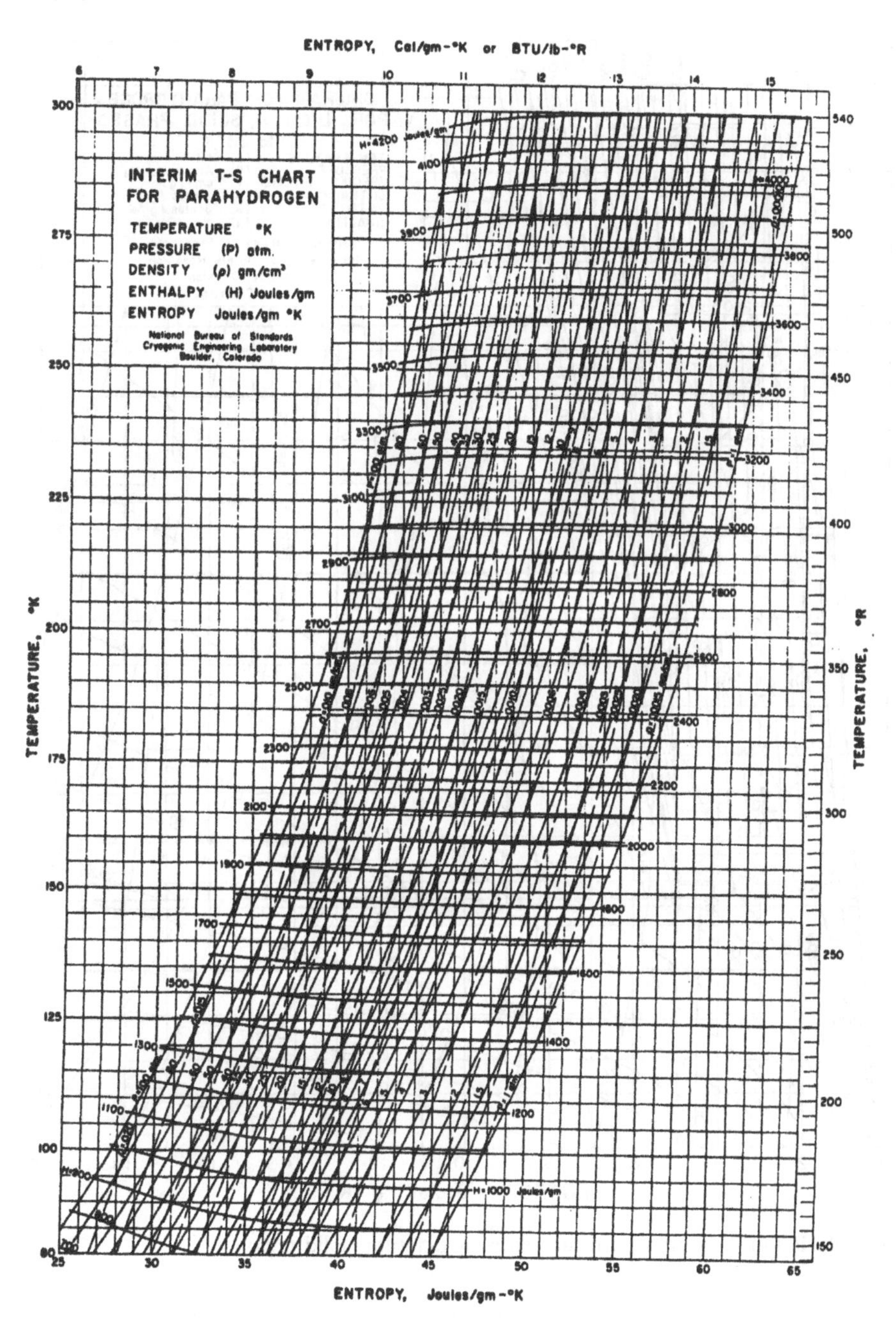


**Figure 3.88** Interim *T–S* chart for parahydrogen, upper temperatures.

If an ortho–para catalyst is not employed in the liquefier, it is possible to produce liquid normal hydrogen. The conversion from normal to equilibrium hydrogen will then occur in the storage container. The heat of conversion is then unavoidably present to cause evaporation of the stored product, even in a perfectly insulated vessel. The simple derivation below gives the rate of this conversion and the effect it has on the stored contents.

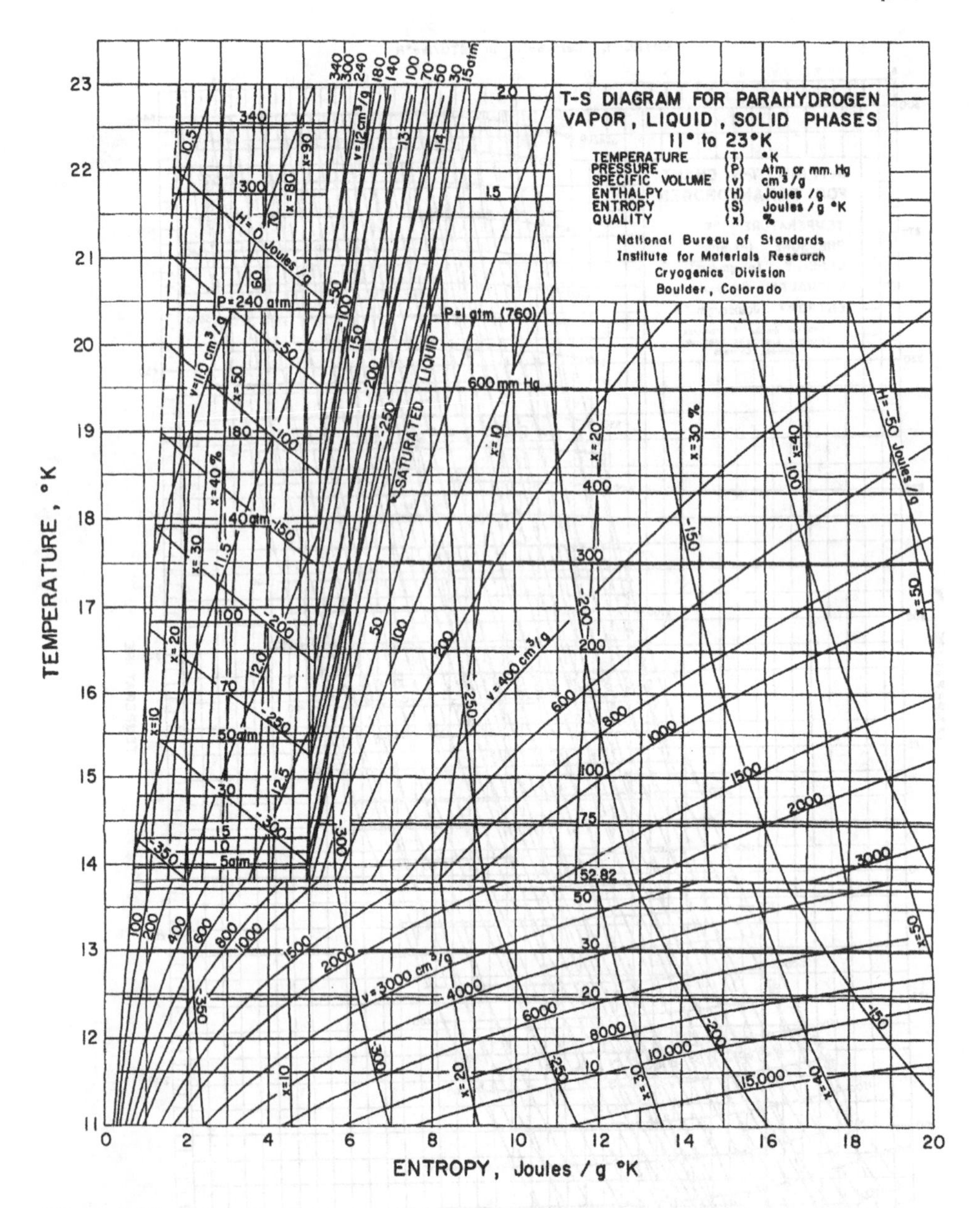


**Figure 3.89** Parahydrogen temperature–entropy diagram, solid region.

The fractional rate of conversion is given by

$$\frac{\mathrm{d}x}{\mathrm{d}t} = -kx^2 \tag{3.25}$$

where $x$ is the ortho fraction at time $t$.

The mass rate of conversion is given by

$$m\frac{\mathrm{d}x}{\mathrm{d}t} = -kmx^2 \tag{3.26}$$

where $m$ is the mass of liquid remaining in the container at time $t$.

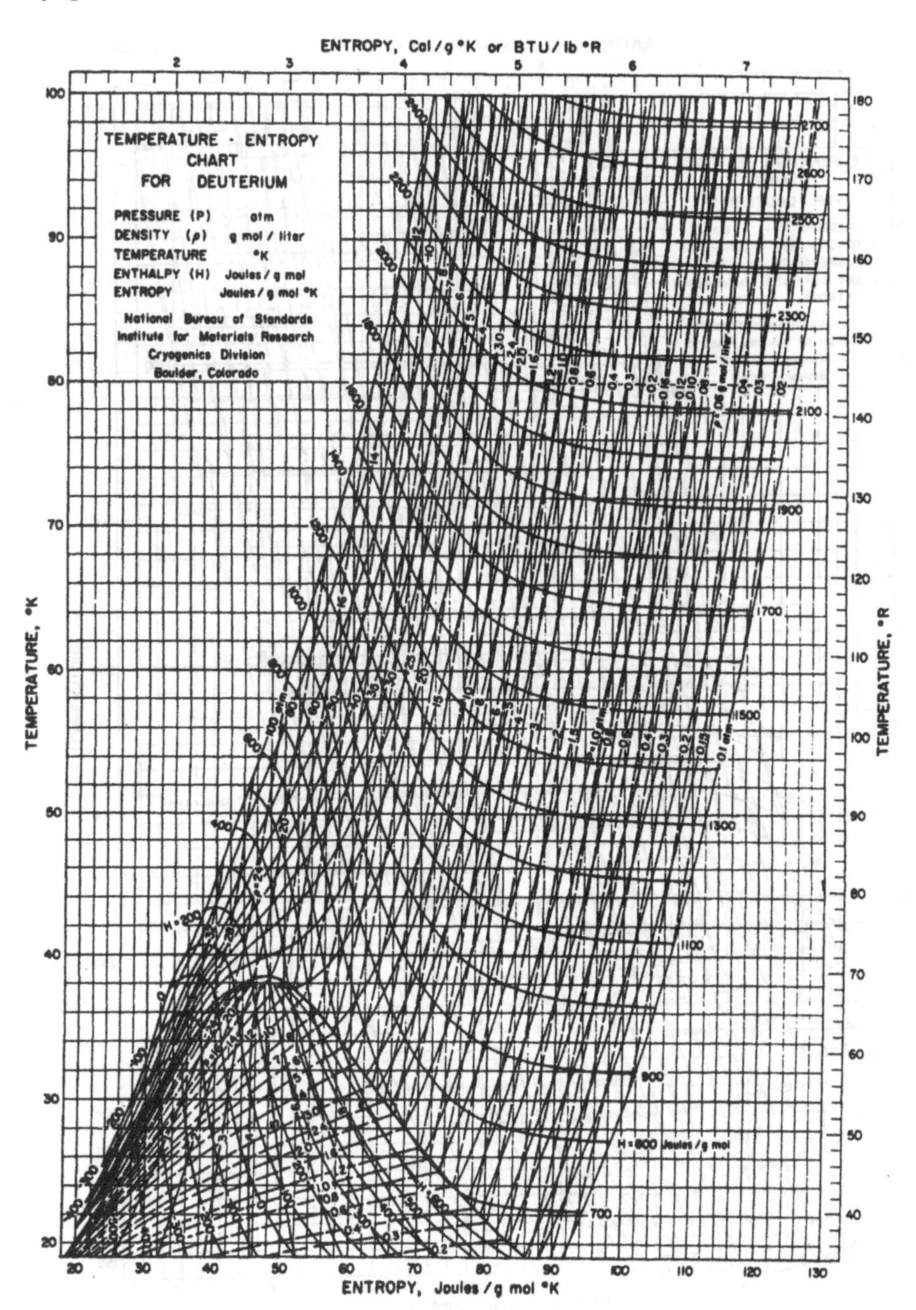


**Figure 3.90** Temperature–entropy chart for deuterium.

The rate of generation of heat is

$$\frac{dq}{dt} = Hm\frac{dx}{dt} = -Hkmx^2 \tag{3.27}$$

where $H$ is heat generated when converting unit mass from ortho- to parahydrogen.

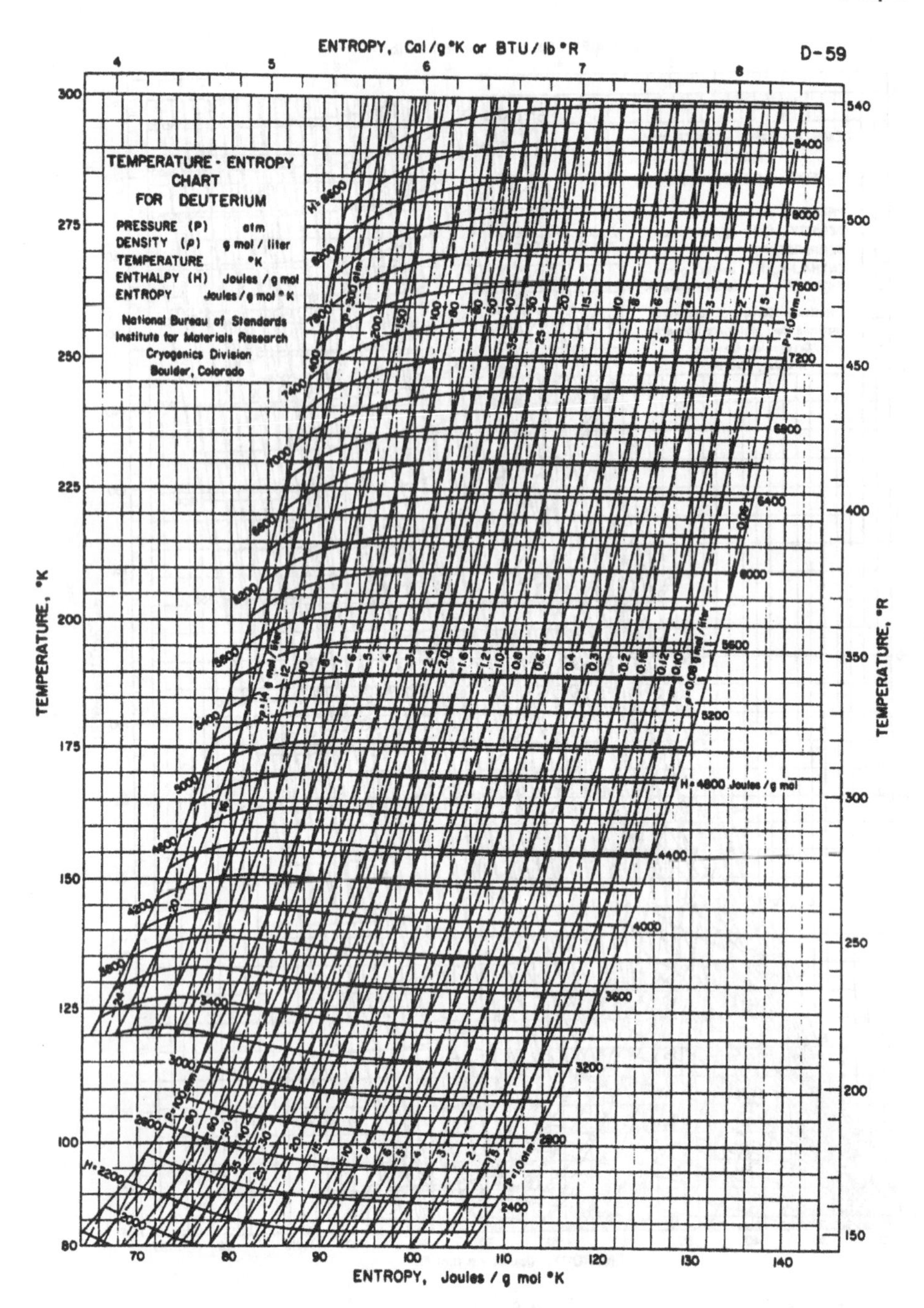


Figure 3.91 Temperature–entropy chart for deuterium.

The rate of evaporation of liquid is

$$\frac{\mathrm{d}m}{\mathrm{d}t} = \frac{1}{L}\frac{\mathrm{d}q}{\mathrm{d}t} = -\frac{Hk}{L}mx^2 \tag{3.28}$$

where $L$ is the latent heat of vaporization per unit mass.

**Table 3.46** Heat of Conversion from Normal Hydrogen to Parahydrogen

| | Heat of conversion | |
|---|---|---|
| Temp.(K) | J/g | cal/mol |
| 10 | 527.139 | 253.9865 |
| 20 | 527.140 | 253.987 |
| 20.39 | 527.138 | 253.986 |
| 30 | 527.138 | 253.986 |
| 33.1 | 527.138 | 253.986 |
| 40 | 527.117 | 253.976 |
| 50 | 526.845 | 253.845 |
| 60 | 525.531 | 253.212 |
| 70 | 521.770 | 251.400 |
| 80 | 513.932 | 247.623 |
| 90 | 500.757 | 241.275 |
| 100 | 481.671 | 232.079 |
| 120 | 427.248 | 205.857 |
| 150 | 322.495 | 155.385 |
| 200 | 163.774 | 78.91 |
| 250 | 70.524 | 33.98 |
| 298.16 | 28.558 | 13.76 |
| 300 | 27.562 | 13.28 |

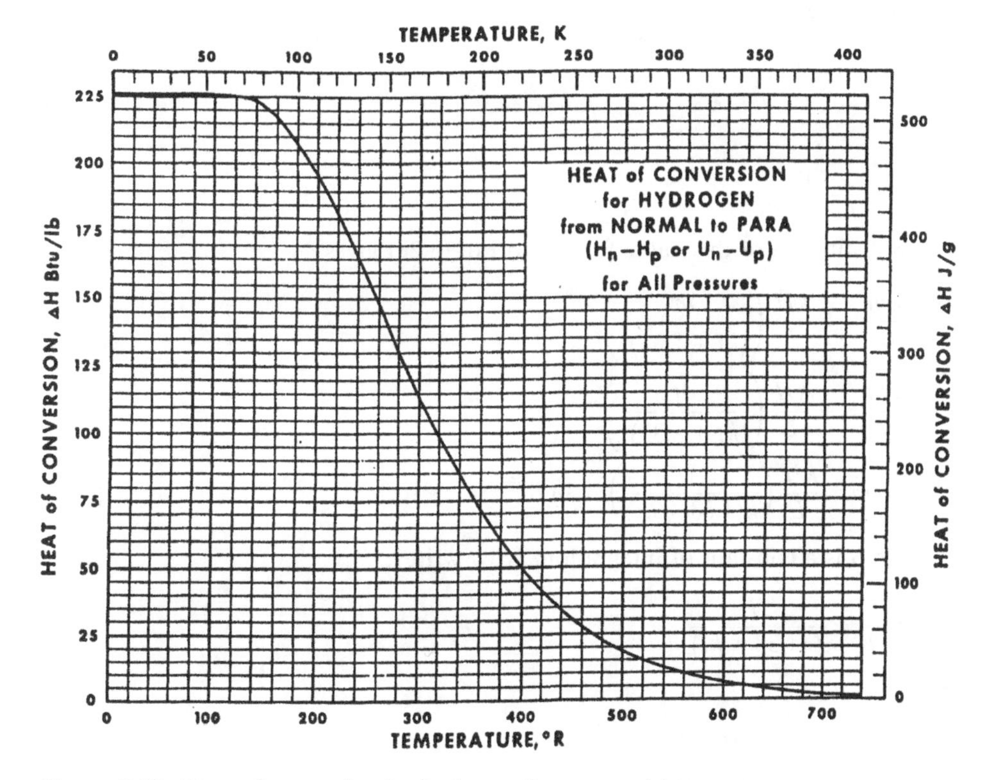


**Figure 3.92** Heat of conversion for hydrogen from normal to para.

**Table 3.47** Fixed Points for Hydrogen

| $p$-$H_2$ or $e$-$H_2$ | $n$-$H_2$ |
|---|---|
| *Critical point* | |
| $T = 32.976 \pm 0.05$ K (59.357°R) | $T = 33.19$ K (59.742°R) |
| $P = 12.759$ atm (1.2928 MPa) | $P = 12.98$ atm (1.315 MPa) |
| $\rho = 15.59$ mol/L (31.43 kg/m$^3$) | $\rho = 14.94$ mol/L (30.12 kg/m$^3$) |
| *Normal boiling point* | |
| $T = 20.268$ K (36.482°R) | $T = 20.39$ K (36.702°R) |
| $>P = 1$ atm (0.101325 MPa) | $P = 1$ atm (0.101325 MPa) |
| $\rho_{liq} = 35.11$ mol/L (70.78 kg/m$^3$) | $\rho_{liq} = 35.2$ mol/L (71.0 kg/m$^3$) |
| $\rho_{vap} = 0.6636$ mol/L (1.338 kg/m$^3$) | $\rho_{vap} = 0.6604$ mol/L (1.331 kg/m$^3$) |
| *Triple point* | |
| $T = 13.803$ K (24.845°R) | $T = 13.957$ K (25.123°R) |
| $P = 0.0695$ atm (0.00704 MPa) | $P = 0.0711$ atm (0.00720 MPa) |
| $\rho_{solid} = 42.91$ mol/L (86.50 kg/m$^3$) | $\rho_{solid} = 43.01$ mol/L (86.71 kg/m$^3$) |
| $\rho_{liq} = 38.21$ mol/L (77.03 kg/m$^3$) | $\rho_{liq} = 38.3$ mol/L (77.2 kg/m$^3$) |
| $\rho_{vap} = 0.0623$ mol/L (0.126 kg/m$^3$) | $\rho_{vap} = 0.0644$ mol/L (0.130 kg/m$^3$) |

*Source:* McCarty and Roder (1981).

Equation (3.28) gives the rate of variation of $m$ with $t$. To solve this equation, it is convenient to first express $x$ as an explicit function of $t$, i.e., solve Eq. (3.25). The general solution is $-1/x = -kt + A$, where $A$ is the constant of integration.

Starting with normal liquid in the dewar, $x = 0.75$ when $t = 0$, from which $A = -1.33$. then

$$\frac{1}{x} = kt + 1.33$$

or

$$x = \frac{1}{kt + 1.33} \tag{3.29}$$

Substitution of this value for $x$ into Eq. (3.28) gives

$$\frac{dm}{dt} = -\frac{kH}{L}\left(\frac{m}{(kt + 1.33)^2}\right)$$

$$\frac{dm}{m} = -\frac{kH}{L}\left(\frac{dt}{(kt + 1.33)^2}\right)$$

$$\ln m = \frac{kH}{L}\left(\frac{1}{k(1.33 + kt)}\right) + \ln B \tag{3.30}$$

where $\ln B$ is a constant of integration. Now write

$$\ln \frac{m}{B} = \frac{H}{L(1.33 + kt)} \tag{3.31}$$

or

$$\frac{m}{B} = \exp\left[\frac{H}{L(1.33 + kt)}\right] \tag{3.31a}$$

**Table 3.48** Fixed Point Properties of Normal Hydrogen

| | Triple point | | | Normal boiling point | | | Standard conditions | |
|---|---|---|---|---|---|---|---|---|
| Property | Solid | Liquid | Vapor | Liquid | Vapor | Critical point[a] | STP(0°C) | NTP(20°C) |
| Temperature (K) | 13.957[b] | 13.957[b] | 13.957[b] | 20.390[b] | 20.390[b] | 33.19[b] | 273.15 | 293.15 |
| Pressure (mmHg) | 54.04 | 54.04 | 54.04 | 760 | 760 | 9865 | 760 | 760 |
| Density (mol/cm$^3$) × $10^3$ | 43.01 | 38.30 | 0.0644 | 35.20 | 0.6604 | 14.94 | 0.04460 | 0.04155 |
| Specific volume (cm$^3$/mol) × $10^{-3}$ | 0.02325 | 0.026108 | 15.519 | 0.028409 | 1.5143 | 0.066949 | 22.423 | 24.066 |
| Compressibility factor, $Z = PV/RT$ | – | 0.001621 | 0.9635 | 0.01698 | 0.9051 | 0.3191 | 1.00042 | 1.00049 |
| Heats of fusion and vaporization[c] (J/mol) | 117.1 | 911.3 | – | 899.1 | – | 0 | – | – |
| Specific heat (J/(mol K) | | | | | | | | |
| $C_\sigma$, at saturation | 5.73 | 13.85 | –46.94 | 18.91 | –33.28 | (Very large) | – | – |
| $C_P$, at constant pressure | – | 13.23 | 21.22 | 19.70 | 24.60 | (Very large) | 28.59 | 28.89 |
| $C_V$, at constant volume | – | 9.53 | 12.52 | 11.60 | 13.2 | (19.7) | 20.30 | 20.40 |
| Specific heat ratio[d], $\gamma = C_P/C_V$ | – | 1.388 | 1.695 | 1.698 | 1.863 | (Large) | 1.408 | 1.416 |
| Enthalpy[e] (J/mol) | 321.6 | 438.7 | 1350.0 | 548.3 | 1447.4 | 1164 | 7749.2 | 8324.1 |
| Internal energy[e] (J/mol) | 317.9[d] | 435.0 | 1234.8 | 545.7 | 1294.0 | – | 5477.1 | 5885.4 |
| Entropy[e] (J/mol k) | 20.3 | 28.7 | 93.6 | 34.92 | 78.94 | 54.57 | 139.59 | 141.62 |
| Velocity of sound (m/s) | – | 1282 | 307 | 1101 | 357 | – | 1246 | 1294 |
| Viscosity, $\mu$ | – | 0.026 | 0.00074 | 0.0132 | 0.0011 | (0.0035) | 0.00839 | 0.00881 |
| N s/m$^2$ × $10^3$ | | | | | | | | |
| centipoise | – | 0.026 | 0.00074 | 0.0132 | 0.0011 | (0.0035) | 0.00839 | 0.00881 |
| Thermal conductivity, k(mW(cm K)) | 9.0 | 0.73 | 0.124 | 0.99 | 0.169 | –[f] | 1.740 | 1.838 |
| Prandtl number, $N_{pr} = \mu C_p/k$ | – | 2.24 | 0.623 | 1.29 | 0.809 | – | 0.6849 0.6848 | |
| Dielectric constant, $\varepsilon$ | 1.287 | 1.253 | 1.00039 | 1.231 | 1.0040 | 1.0937 | 1.000271 | 1.000253 |
| Index of refraction[g,h], $n = \sqrt{\varepsilon}$ | 1.134 | 1.119 | 1.000196 | 1.1093 | 1.0020 | 1.0458 | 1.000136 | 1.000126 |
| Surface tension (N/m) × $10^3$ | – | 3.00 | – | 1.94 | – | 0 | – | – |
| Equiv. vol./vol. liquid at NBP | 0.8184 | 0.9190 | 546.3 | 1 | 53.30 | 2.357 | 789.3 | 847.1 |

Gas constant: $R = 8.31434$ J(mol K); molecular weight = 2.01594; mole = gram mole.

[a] Values in parentheses are estimates.

[b] These temperatures are based on the IPTS-1968 temperature scale. To compare with the corresponding temperatures based on the NBS-1955 temperature scale, 0.01 K should be subtracted; i.e., use 13.947, 20.380, and 33.18, respectively, for the triple point, normal boiling point, and critical point when determining property differences between normal and parahydrogen.

[c] Heats of fusion and vaporization calculated from enthalpy differences.

[d] Calculated from property values given in this table.

[e] Base point (zero values) for enthalpy, internal energy, and entropy are 0 K for the ideal gas at 0.101325 Mpa (1 atm) pressure.

[f] Anomalously large.

[g] Long wavelengths.

[h] Index of refraction calculated from dielectric constant data.

**Table 3.49** Fixed Point Properties of Parahydrogen

| | Triple point | | | Normal boiling point | | | Standard conditions | |
|---|---|---|---|---|---|---|---|---|
| Property | Solid | Liquid | Vapor | Liquid | Vapor | Critical point[a] | STP(0°C) | NTP(20°C) |
| Temperature (K) | 13.803 | 13.803 | 13.803 | 20.268 | 20.268 | 32.976 | 273.15 | 293.15 |
| Pressure (mmHg) | 52.82 | 52.82 | 52.85 | 760 | 760 | 9696.8 | 760 | 760 |
| Density (mol/cm$^3$) × $10^3$ | 42.91 | 38.207 | 0.0623 | 35.11 | 0.6636 | 15.59 | 0.5459 | 0.04155 |
| Specific volume (cm$^3$/mol) × $10^{-3}$ | 0.02330 | 0.026173 | 16.057 | 0.028482 | 1.5069 | 0.064144 | 22.425 | 24.069 |
| Compressibility factor, $Z = PV/RT$ | – | 0.001606 | 0.9850 | 0.01712 | 0.9061 | 0.3025 | 1.0005 | 1.0006 |
| Heats of fusion and vaporization[b] (J/mol) | 117.5 | 905.5 | – | 898.3 | – | 0 | – | – |
| Specific heat (J/(mol K) | | | | | | | | |
| $C_\sigma$ at saturation | 5.73 | 13.85 | −46.94 | 18.91 | −33.28 | (Very large) | – | – |
| $C_p$, at constant pressure | – | 13.13 | 21.20 | 19.53 | 24.50 | (Very large) | 30.35 | 30.02 |
| $C_V$, at constant volume | – | 9.50 | 12.52 | 11.57 | 13.11 | 19.7 | 21.87 | 21.70 |
| Specific heat ratio[c] $\gamma = C_P/C_V$ | – | 1.382 | 1.693 | 1.688 | 1.869 | (Large) | 1.388 | 1.383 |
| Enthalpy[d] (J/mol) | −740.2 | −622.7 | 282.8 | −516.6 | 381.7 | 77.6 | 7656.6 | 8260.6 |
| Internal energy[d] (J/mol) | −740.4[c] | −622.9 | 169.8 | −519.5 | 229.0 | 5.7 | 5384.5 | 5822.0 |
| Entropy[d] (J/(mol K) | 1.49 | 10.00 | 75.63 | 16.08 | 60.41 | 35.4 | 127.77 | 129.90 |
| Velocity of sound (m/s) | – | 1273 | 305 | 1093 | 355 | 350 | 1246 | 1294 |
| Viscosity, $\mu$ N s/m$^2$ × $10^3$ | – | 0.026 | 0.00074 | 0.0132 | 0.0011 | 0.0035 | 0.00839 | 0.00881 |
| centipoise | – | 0.026 | 0.00074 | 0.0132 | 0.0011 | 0.0035 | 0.00839 | 0.00881 |
| Thermal conductivity, k(mW(cm K)) | 9.0 | 0.73 | 0.124 | 0.99 | 0.169 | –[e] | 1.841 | 1.914 |
| Prandtl number, $N_{pr} = \mu C_p/k$ | – | 2.24 | 0.623 | 1.29 | 0.809 | | 0.6866 | 0.6855 |
| Dielectric constant, $\varepsilon$ | 1.286 | 1.252 | 1.00038 | 1.230 | 1.0040 | 1.098 | 1.00027 | 1.00026 |
| Index of refraction[f,g], $n = \sqrt{\varepsilon}$ | 1.134 | 1.119 | 1.00019 | 1.109 | 1.0020 | 1.048 | 1.00013 | 1.00012 |
| Surface tension (N/m) × $10^3$ | – | 2.99 | – | 1.93 | – | 0 | – | – |
| Equiv. vol./vol. liquid at NBP | 0.8181 | 0.9190 | 563.8 | 1 | 52.91 | 2.252 | 787.4 | 845.1 |

Gas constant: $R = 8.31434$ J(mol K); molecular weight = 2.01594; mole = gram mole.
[a] Values in parentheses are estimates.
[b] Heats of vaporization calculated from enthalpy differences.
[c] Calculated from property values given in this table.
[d] Base point (zero values) for enthalpy, internal energy, and entropy are 0 K for the ideal gas at 0.101325 MPa (1 atm) pressure.
[e] Anomalously large.
[f] Long wavelengths.
[g] Index of refraction calculated from dielectric constant data.

when $t = 0$, $m = m_o$, so

$$\frac{m_o}{B} = \exp\left(\frac{3H}{4L}\right) \quad \text{or} \quad B = \frac{m_o}{\exp(3H/4L)} \tag{3.32}$$

The solution then becomes

$$\frac{m}{m_o}\exp\left(\frac{3H}{4L}\right) = \exp\left[\frac{H}{L(1.33 + kt)}\right] \tag{3.33}$$

or

$$\frac{m}{m_o} = \exp\left[\frac{H}{L(1.33 + kt)} - \frac{3H}{4L}\right] \tag{3.33a}$$

$L = 906.09$ J/mol (216.56 cal/mol) for $n$-$H_2$ at 20 K (36°R) and 894.87 J/mol (213.88 cal/mol) for 99.8% $p$-$H_2$ at 20 K. 890 J/mol (215 cal/mol) is a reasonable average value. The value for $H$ is 1416.19 J/mol (338.65 cal/mol), and the value for the rate constant $k$ is 0.0114/hr$^{-1}$.

These numerical substitutions reduce Eq. (3.33) to the form

$$\frac{m}{m_o} = \exp\left[\frac{1.57}{(1.33 + 0.0114t)} - 1.18\right] \tag{3.34}$$

or

$$\ln\frac{m}{m_o} = \frac{1.57}{1.33 + 0.0114t} - 1.18 \tag{3.34a}$$

If the original composition of the liquid is not 0.75 orthohydrogen at $t = 0$, then a new constant of integration based on the starting composition can be evaluated for Eq. (3.29). The new expression for $m/m_o$ then follows directly as outlined above, and Fig. 3.93 shows the results.

### 3.7.3. *Properties of Hydrogen*

Physical constants for five isotopic forms of hydrogen are presented in Table 3.50.

**a. Melting Curve.** The melting curve is the boundary between the solid and liquid regions in a phase diagram. Range of values (kg/m$^3$) for parahydrogen:

| Triple point | | 10.1325 Mpa (100 atm) | | 30.3975 MPa (300 atm) | |
|---|---|---|---|---|---|
| Liquid | Solid | Fluid | Solid | Fluid | Solid |
| 77.0 | 86.9 | 81.8 | 89.9 | 88.7 | 96.0 |

Uncertainty is estimated to be 0.1% for the liquid phase and 0.5% for the solid phase.

**b. Heat of Vaporization.** Range of values (J/g):

| | Triple point | Boiling point |
|---|---|---|
| Para | 448.2 | 445.5 |
| Normal | 449.1 | 445.6 |

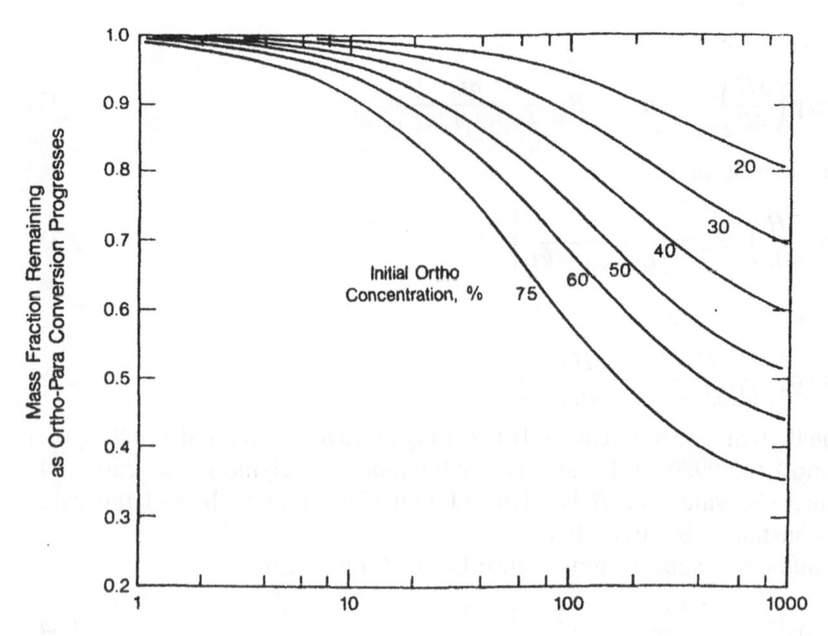


**Figure 3.93** Fraction of liquid hydrogen evaporated due to ortho–para conversion as a function of time.

Uncertainty is estimated to be 1.24 J/g for $p$-$H_2$ and 2.5 J/g for $n$-$H_2$.

The heat of vaporization for parahydrogen and normal hydrogen is tabulated in Table 3.51 and graphed in Figs. 3.94 and 3.95.

**c. Vapor Pressure.** The vapor pressure is the pressure, as a function of temperature, of a liquid in equilibrium with its own vapor.

**Table 3.50** Physical Constants of the Isotopic Forms of Hydrogen

| | Normal hydrogen, 75% $o$-$H_2$ | 20.39 Equilibrium hydrogen, 0.21% $o$-$H_2$ | Normal deuterium, 66.67% $o$-$D_2$ | 20.39 Equilibrium deuterium, 97.8% $o$-$D_2$ | Hydrogen deuteride |
|---|---|---|---|---|---|
| Triple point temp., K | 13.96 | 13.81 | 18.72 | 18.69 | 16.60 |
| Triple point pressure | | | | | |
| atm | 0.07105 | 0.0695 | 0.1691 | 0.1691 | 0.122 |
| kPa | 0.07197 | 0.0704 | 0.1713 | 0.1713 | 0.124 |
| Normal boiling point, K | 20.39 | 20.27 | 23.57 | 23.53 | 22.13 |
| Critical temp., K | 33.19 | 33.1 | | 38.3 | 35.9 |
| Critical pressure | | | | | |
| atm | 12.98 | 12.8 | | 16.3 | 14.6 |
| kPa | 13.15 | 13.1 | | 16.5 | 14.8 |
| Critical volume, $cm^3/mol$ | 66.95 | 65.5 | | 60.3 | 62.8 |

**Table 3.51** Heat of Vaporization ($hv$) of Normal Hydrogen

| Pressure (atm) | Temperature (K) | $hv$ (cal/g) |
|---|---|---|
| 0.12 | 15.0 | 108.6 |
| 0.6 | 17.8 | 106.7 |
| 0.8 | 19.7 | 106.0 |
| 1.0 | 20.5 | 105.5 |
| 1.5 | 22.0 | 103.3 |
| 2.0 | 23.0 | 102.0 |
| 3.0 | 24.75 | 98.0 |
| 4.0 | 26.2 | 94.3 |
| 5.0 | 27.3 | 89.0 |
| 6.0 | 28.3 | 84.0 |
| 8.0 | 30.0 | 71.2 |
| 10.0 | 31.3 | 57.1 |
| 12.0 | 32.6 | 33.0 |
| 12.98 | 33.19 | 0 |

*Source:* Wooley et al. (1948); Simon and Lange (1923).

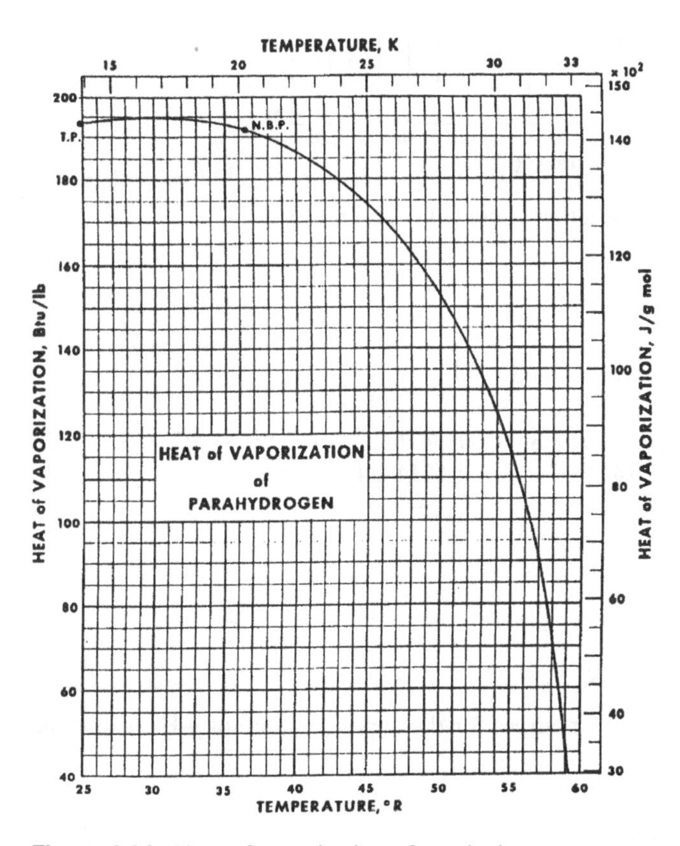


**Figure 3.94** Heat of vaporization of parahydrogen.

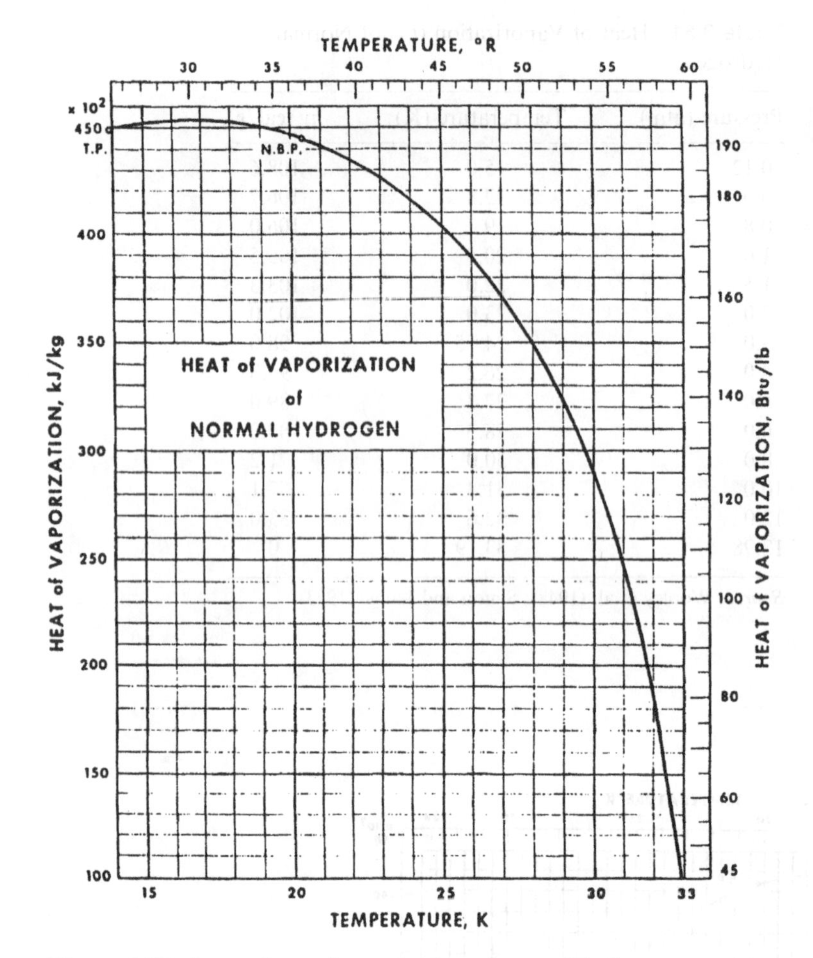


**Figure 3.95** Latent heat of vaporization of normal hydrogen.

Parahydrogen:

$$\ln\frac{(\nu p)}{(\nu p_t)} = n_1 x + n_2 x^2 + n_3 x^3 + n_4 (1-x)^{n_5} \tag{3.35}$$

where $vp_t = 0.007042$ Mpa, $n_1 = 3.05300134164$, $n_2 = 2.80810925813$, $n_3 = -0.655461216567$, $n_4 = 1.59514439374$, $n_5 = 1.5814454428$, and

$$x = \frac{1 - T_t/T}{1 - T_t/T_c}$$

with

$$T_t = 13.8\,\mathrm{K},\ \ T_C = 32.938\,\mathrm{K}$$

Figures 3.96 and 3.97 present curves for the vapor pressure of $p$-$H_2$ below and above $p = 1$ atm, respectively.

Normal hydrogen:

$$\log P(\mathrm{kg/cm^2}) = A + B/T + C \log T \tag{3.36}$$

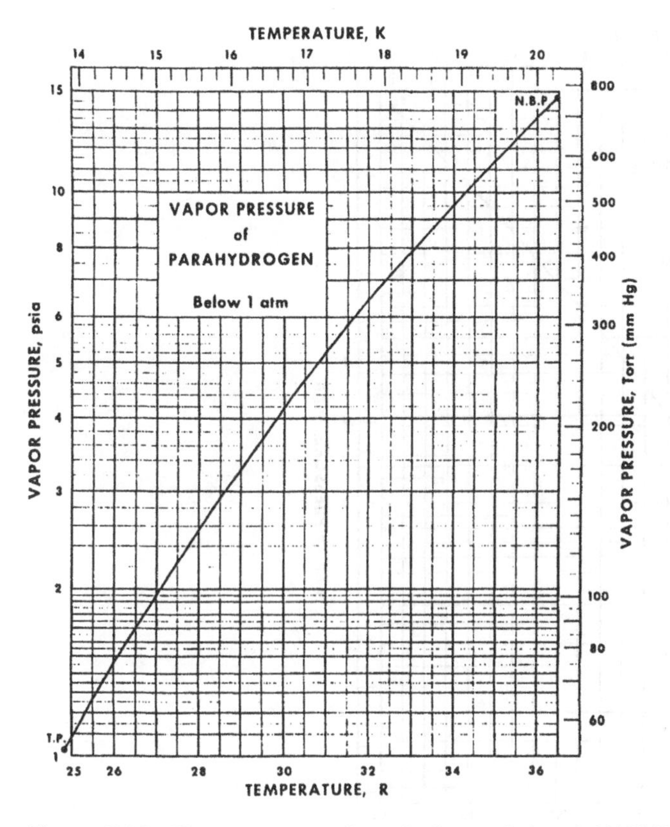


**Figure 3.96** Vapor pressure of parahydrogen below 0.101325 MPa (1 atm).

where

$$A = -1.55447, \ B = -31.875, \ C = 2.39188$$

and $T$ is in kelvins.

Table 3.52 presents vapor pressure values for the isotopic forms of hydrogen.

The vapor pressure equations of Table 3.52 agree with the experimental data for hydrogen to within the experimental error over most of the liquid range to the critical point. The values for deuterium and hydrogen deuteride should not be used at pressures much above 1 atm.

Figures 3.98 and 3.99 present vapor pressure curves for $n$-$H_2$.

**d. Virial Coefficients.** The virial coefficients are commonly defined in two ways as follows:

$$P = RT\rho[1 + B(T)\rho + C(T)\rho^2 + \cdots]$$

and

$$PV = RT + B'(T)P + C'(T)P^2$$

The first two terms of these series may be inverted by the following:

$$B(T) = B'(T) \qquad \text{and} \qquad C'(T) = \frac{C(T) - B(T)^2}{RT} \tag{3.37}$$

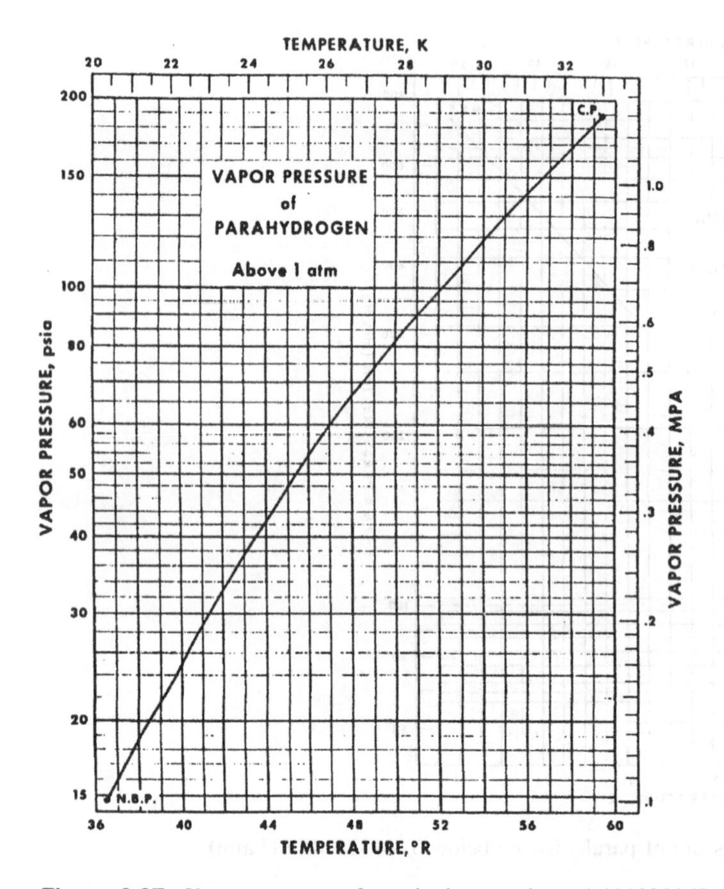


**Figure 3.97** Vapor pressure of parahydrogen above 0.101325 MPa (1 atm).

where $P$ is the pressure, $T$ is the temperature, $\rho$ is the density, $V$ is the $1/\rho$, and $R$ is the gas constant. $B(T)$, $B'(T)$ and $C(T)$, $C'(T)$ are virial coefficients of a power series expansion in density and pressure, respectively.

Both series are theoretically infinite in length; however, the coefficients beyond the first two [$B(T)$ and $C(T)$ ] are of less interest because of their complexity. The temperature at which $B(T) = 0$ is called the *Boyle point*. Either of the above two

**Table 3.52** Vapor Pressure of the Isotopic Forms of Hydrogen

| Material | State | $A$ | $B$ | $C$ |
|---|---|---|---|---|
| Normal hydrogen, 75% $o$-$H_2$ | Liquid | 4.66687 | −44.9569 | 0.020537 |
| | Solid | 4.56488 | −47.2059 | 0.03939 |
| 20.39 K Equilibrium hydrogen, 0.21% $o$-$D_2$ | Liquid | 4.64392 | −44.3450 | 0.02093 |
| | Solid | 4.62438 | −47.0172 | 0.03635 |
| Normal deuterium, 66.67% $o$-$D_2$ | Liquid | 4.7312 | −58.4619 | 0.02671 |
| | Solid | 5.1626 | −68.0782 | 0.03110 |
| 20.39 K Equilibrium deuterium, 97.8% $o$-$D_2$ | Liquid | 4.7367 | −58.4440 | 0.02670 |
| | Solid | 5.1625 | −67.9119 | 0.03102 |
| Hydrogen deuteride | Solid | 5.04964 | −55.2495 | 0.01479 |
| | Liquid | 4.70260 | −56.7154 | 0.04101 |

equations is adequate to describe the *PVT* surface for densities up to about one-half the critical density.

Range of values:

| | Unit | 20 K | 100 K | 200 K | 300 K |
|---|---|---|---|---|---|
| $B(T)$ | $cm^3/mol$ | 146.7 | −2.51 | 10.73 | 14.48 |
| $C(T)$ | $(cm^3/mol)^2$ | −1503 | 608.6 | 421.7 | 343.8 |

Boyle point ($B = 0$):112.4 K.

The uncertainty of $B$ is probably a maximum of 5% at the highest and lowest temperatures. The uncertainty of $C$ is a minimum of 5% between 55 and 100 K and as much as 20% below critical temperature.

Values for $B$ and $C$ for parahydrogen are listed in Table 3.53.

Table 3.54 gives values for converting between specific volume and density.

**e. Joule–Thomson Coefficient.** The Joule–Thomson coefficient μ is defined as

$$\mu = -C_p - 1\left(\frac{\partial H}{\partial P}\right)_T = \left(\frac{\partial T}{\partial P}\right)_H \tag{3.38}$$

The sign of $\mu$ indicates whether the expansion of a gas will cause an increase or decrease in its temperature. If μ is positive, the expanding gas will be cooled. The locus of points where $\mu = 0$ is called the Joule–Thomson (J–T) inversion curve. Data

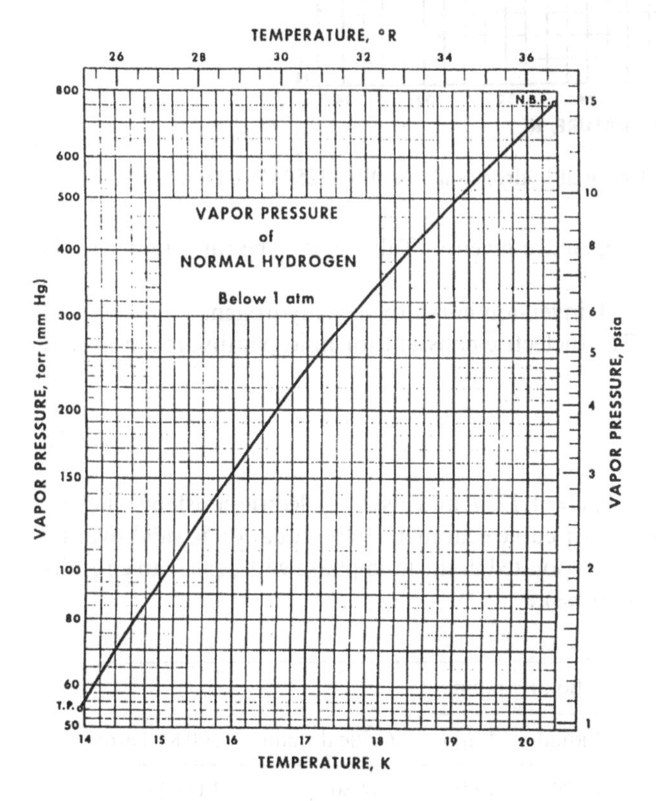


**Figure 3.98** Vapor pressure of normal hydrogen below 0.101325 MPa (1 atm).

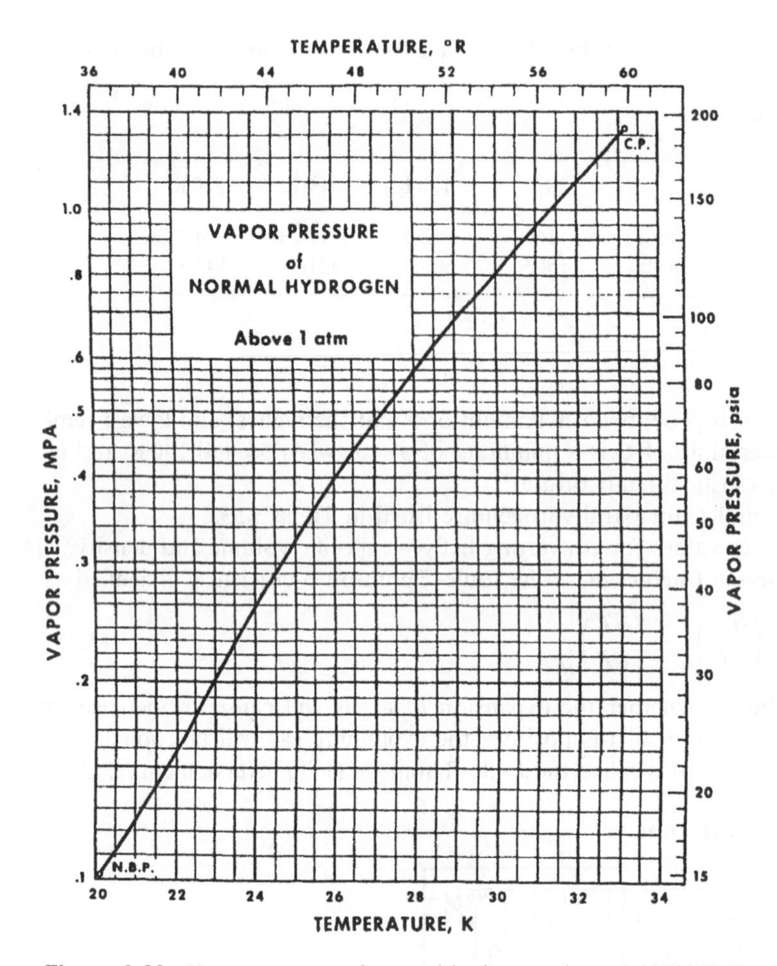


**Figure 3.99** Vapor pressure of normal hydrogen above 0.101325 MPa (1 atm).

for the J–T curve for parahydrogen are given in Table 3.55. The uncertainty in the data is estimated to be 5%.

Figure 3.100 shows the J–T inversion curve for parahydrogen.

**f. Dielectric Constant.** The dielectric constant for fluid parahydrogen may be calculated from

$$\frac{\varepsilon - 1}{\varepsilon + 2} = A\rho + B\rho^2 + C\rho^3 \tag{3.39}$$

where $A = 0.99575$, $B = -0.09069$, $C = 1.1227$, and $\varepsilon$ is the dielectric constant. The density, $\rho$, must be in units of grams per cubic centimeter. The equation is valid over the range of the tables and may be extrapolated in the fluid phase with reasonable results.

Range of values (dimensionless) :

| Triple point | | | Boiling point | | | |
|---|---|---|---|---|---|---|
| Liquid | Vapor | Solid | Liquid | Vapor | Critical point | 300 K, 1 atm |
| 1.252 | 1.0004 | 1.285 | 1.229 | 1.004 | 1.0980 | 1.00025 |

**Table 3.53** Second and Third Virial Coefficients for Parahydrogen

| Temp. (K) | $B$ (cm$^3$/*mol*) | $C$ (cm$^3$/mol)$^2$ |
|---|---|---|
| 14 | $-2.372 \times 10^2$ | $-6.713 \times 10^4$ |
| 16 | $-1.992 \times 10^2$ | $-2.265 \times 10^4$ |
| 18 | $-1.698 \times 10^2$ | $-7.121 \times 10^3$ |
| 20 | $-1.467 \times 10^2$ | $-1.503 \times 10^3$ |
| 22 | $-1.281 \times 10^2$ | $5.656 \times 10^2$ |
| 24 | $-1.129 \times 10^2$ | $1.323 \times 10^3$ |
| 26 | $-1.003 \times 10^2$ | $1.580 \times 10^3$ |
| 28 | $-8.969 \times 10^1$ | $1.640 \times 10^3$ |
| 30 | $-8.066 \times 10^1$ | $1.615 \times 10^3$ |
| 32 | $-7.290 \times 10^1$ | $1.550 \times 10^3$ |
| 34 | $-6.615 \times 10^1$ | $1.467 \times 10^3$ |
| 36 | $-6.025 \times 10^1$ | $1.376 \times 10^3$ |
| 38 | $-5.504 \times 10^1$ | $1.290 \times 10^3$ |
| 40 | $-5.042 \times 10^1$ | $1.214 \times 10^3$ |
| 45 | $-4.086 \times 10^1$ | $1.066 \times 10^3$ |
| 50 | $-3.343 \times 10^1$ | $9.638 \times 10^2$ |
| 60 | $-2.264 \times 10^1$ | $8.351 \times 10^2$ |
| 70 | $-1.522 \times 10^1$ | $7.510 \times 10^2$ |
| 80 | $-9.824 \times 10^1$ | $6.878 \times 10^2$ |
| 90 | $-5.725 \times 10^1$ | $6.405 \times 10^2$ |
| 100 | $-2.510 \times 10^1$ | $6.085 \times 10^2$ |
| 120 | $2.144 \times 10^1$ | $5.491 \times 10^2$ |
| 140 | $5.342 \times 10^1$ | $5.063 \times 10^2$ |
| 160 | $7.654 \times 10^1$ | $4.726 \times 10^2$ |
| 180 | $9.388 \times 10^1$ | $4.449 \times 10^2$ |
| 200 | $1.073 \times 10^1$ | $4.217 \times 10^2$ |
| 250 | $1.300 \times 10^1$ | $3.768 \times 10^2$ |
| 300 | $1.438 \times 10^1$ | $3.438 \times 10^2$ |
| 400 | $1.589 \times 10^1$ | $2.976 \times 10^2$ |
| 500 | $1.660 \times 10^1$ | $2.661 \times 10^2$ |

The uncertainty of $\varepsilon - 1$ is estimated to be no greater than 0.1% for the fluid phase and 0.2% for the solid phase. Values of the dielectric constant are listed in Table 3.56.

*Symbols:* $\rho_{sat}$ = saturation pressure or vapor pressure, $\rho$ = density, $C_P$ = specific heat at constant pressure, $\mu$ = viscosity, $k$ = thermal conductivity, $h_{fg}$ = heat of vaporization, $N_{pr} = \mu C_P / K$ = Prandtl number, $\sigma_L$ = surface tension, $\varepsilon$ = dielectric constant.

### *3.7.4. The Hydrogen Density Problem*

Hydrogen has the lowest liquid density of any known substance. This is an inescapable fact of liquid hydrogen behaving as a quantum fluid. This is true because liquid hydrogen possesses a high quantum mechanical zero-point energy relative to its classical thermal energy. According to classical theory, at absolute zero temperature the particles of matter should be in static equilibrium with one another. Since they then have no thermal energy, perfect static balance is supposed to exist between the electromagnetic attractive and repulsive forces of the atoms.

**Table 3.54** Conversion Table for Specific Volumes to Density, for Hydrogen

| Specific volume | | |
| --- | --- | --- |
| $cm^3/mol$ | $ft^3/lb$ | Density ($lb/ft^3$) |
| 26 | 0.207 | 4.84 |
| 26.5 | 0.211 | 4.75 |
| 27 | 0.215 | 4.66 |
| 27.5 | 0.219 | 4.58 |
| 28 | 0.222 | 4.49 |
| 28.5 | 0.226 | 4.42 |
| 29 | 0.230 | 4.34 |
| 29.5 | 0.234 | 4.27 |
| 30 | 0.238 | 4.20 |
| 31 | 0.246 | 4.06 |
| 32 | 0.254 | 3.93 |
| 33 | 0.262 | 3.81 |
| 34 | 0.270 | 3.70 |
| 35 | 0.278 | 3.60 |
| 36 | 0.286 | 3.50 |
| 37 | 0.294 | 3.40 |
| 38 | 0.302 | 3.31 |
| 39 | 0.310 | 3.23 |
| 40 | 0.318 | 3.15 |
| 42 | 0.334 | 3.00 |
| 44 | 0.350 | 2.86 |
| 46 | 0.365 | 2.74 |
| 48 | 0.381 | 2.62 |
| 50 | 0.397 | 2.52 |
| 55 | 0.437 | 2.29 |
| 60 | 0.477 | 2.10 |
| 65 | 0.516 | 1.94 |
| 70 | 0.556 | 1.80 |

Above 0 K, however, all matter has thermal energy in the form of rapid random motion of its atoms, and the balance of forces among the particles becomes dynamic rather than static. In cooling matter slowly to 0 K, then, classical theory would predict the loss of thermal energy of the material through the loss of the kinetic energy of its atoms until at 0 K there would exist perfect motionless order.

Quantum theory, on the other hand, shows that each atom has an irreducible minimum of kinetic energy amounting to $(1/2)\,h\nu$, where $h$ is Planck's constant, $6.6 \times 10^{-27}$ erg sec, and $\nu$ is the frequency of oscillation of the atom. Even at 0 K, when a substance has lost all its thermal energy, it will still have this zero-point energy.

This amount is not large, and in most cases it is effectively inundated by the thermal energy of matter at higher temperatures. At very low temperatures, however, the zero-point energy becomes a significant fraction of the total energy of some substances.

Solid hydrogen, for example, has a zero-point energy of about 200 cal/mol, which counteracts about 50% of its computed lattice energy of 400 cal/mol.

**Table 3.55** Joule–Thomson Inversion Curve Data for Parahydrogen

| Temperature | | Pressure | | | Density | |
|---|---|---|---|---|---|---|
| K | °R | MPa | atm | psia | mol/cm$^3$ × 10$^3$ | lb$_m$/ft$^3$ |
| 28 | 50.4 | 1.000 | 9.87 | 145.1 | 30.06 | 3.783 |
| 29 | 52.2 | 1.525 | 15.05 | 221.2 | 29.90 | 3.763 |
| 30 | 54.0 | 2.035 | 20.08 | 295.1 | 29.73 | 3.742 |
| 31 | 55.8 | 2.534 | 25.01 | 367.6 | 29.56 | 3.720 |
| 32 | 57.6 | 3.025 | 29.85 | 438.7 | 29.40 | 3.700 |
| 34 | 61.2 | 3.968 | 39.16 | 575.5 | 29.05 | 3.656 |
| 36 | 64.8 | 4.870 | 48.06 | 706.3 | 28.70 | 3.612 |
| 40 | 72.0 | 6.545 | 64.59 | 949.2 | 27.99 | 3.523 |
| 50 | 90.0 | 10.02 | 98.93 | 1454 | 26.16 | 3.292 |
| 60 | 108.0 | 12.60 | 124.4 | 1828 | 24.30 | 3.058 |
| 80 | 144.0 | 15.55 | 153.5 | 2256 | 20.58 | 2.590 |
| 100 | 180.0 | 16.35 | 161.4 | 2372 | 17.04 | 2.145 |
| 120 | 216.0 | 16.42 | 162.1 | 2353 | 14.12 | 1.777 |
| 140 | 252.0 | 14.24 | 140.5 | 2064 | 10.86 | 1.367 |
| 160 | 288.0 | 10.36 | 102.2 | 1502 | 7.176 | 0.9031 |
| 180 | 324.0 | 5.165 | 50.97 | 749.1 | 3.321 | 0.4179 |
| 200 | 360.0 | 0.0547 | 0.54 | 8.6 | 0.036 | 0.0045 |

The measured heat of sublimation of hydrogen is therefore only 400−200 = 200 cal/mol. The zero-point energy acts as though it were additional thermal energy and effectively counteracts part of the attractive force between molecules of hydrogen. The practical result is that solid hydrogen melts very easily. Indeed, the thermal properties of solid hydrogen resemble those of liquid helium more than they do those of liquid hydrogen. Solid hydrogen melts readily, and liquid hydrogen vaporizes very easily.

Figure 3.101 is the familiar plot of intermolecular potential energy vs. interatomic separation distance. Above the horizontal axis, the coulombic forces dominate, causing a repulsion among the molecules. Below the abscissa, the gravitational attractive forces between the molecules dominate. The lower dashed curve is the classical energy curve representing the balance between repulsive and attractive forces among molecules vs. their separation. The upper dashed curve shows the zero-point energy predicted by quantum mechanics. The solid curve is the sum of the classical energy and quantum energy forces vs. molecular separation. The average interatomic spacing is the sum of the classical and quantum forces and is considerably larger than if the zero-point energy were negligible.

The effect of the zero-point energy is to shift the curve to the right in Fig. 3.101 $R'_0$ is the molecular separation for a classical fluid. $R_0$ is the molecular separation that results from adding the zero-point energy to the classical energy. Shifting $R_0$ to the right inflates the distance between the molecules and therefore decreases the density of the liquid. This is an inescapable fact of life for liquid hydrogen and accounts for the extremely low liquid density of 70 g/L or a specific gravity of only 7%.

The atoms in liquid hydrogen do not pack well. Indeed, there is more hydrogen in a gallon of water than there is in a gallon of pure liquid hydrogen. Methane is a much better carrier of hydrogen than is hydrogen, as shown in Table 3.57.

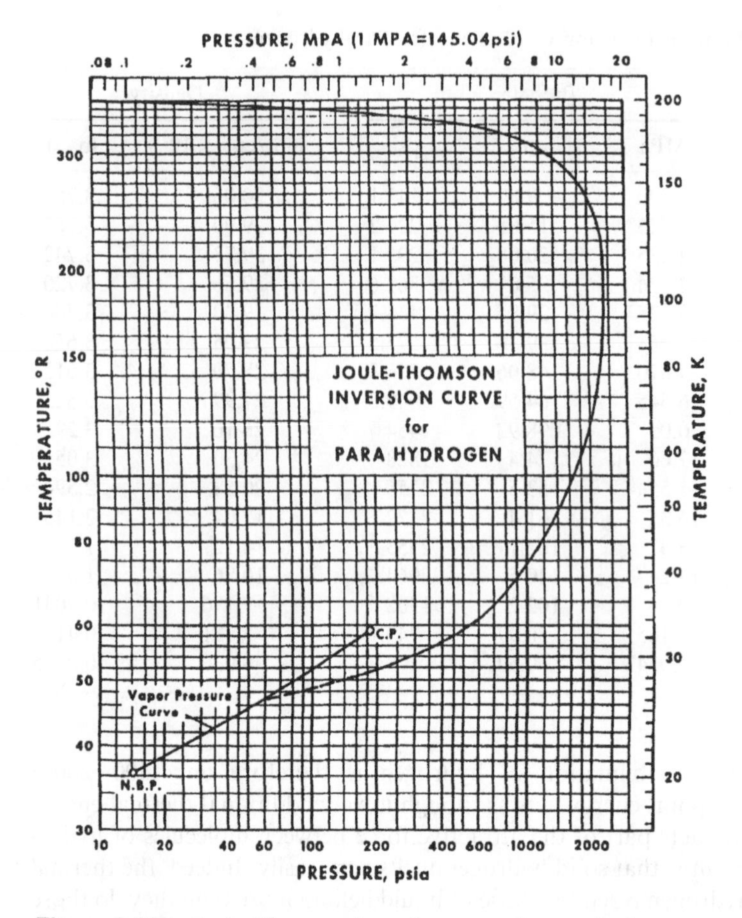


**Figure 3.100** Joule–Thomson inversion curve for parahydrogen.

**Table 3.56** Dielectric Constant for Parahydrogen

| Density (g/cm$^3$) | Dielectric constant, $\varepsilon$ |
|---|---|
| 0.005 | 1.01515 |
| 0.010 | 1.03046 |
| 0.015 | 1.04594 |
| 0.020 | 1.06158 |
| 0.025 | 1.07739 |
| 0.030 | 1.09336 |
| 0.035 | 1.10950 |
| 0.040 | 1.12580 |
| 0.045 | 1.14226 |
| 0.050 | 1.15889 |
| 0.055 | 1.17569 |
| 0.060 | 1.19265 |
| 0.065 | 1.20977 |
| 0.070 | 1.22705 |
| 0.075 | 1.24449 |
| 0.080 | 1.26210 |

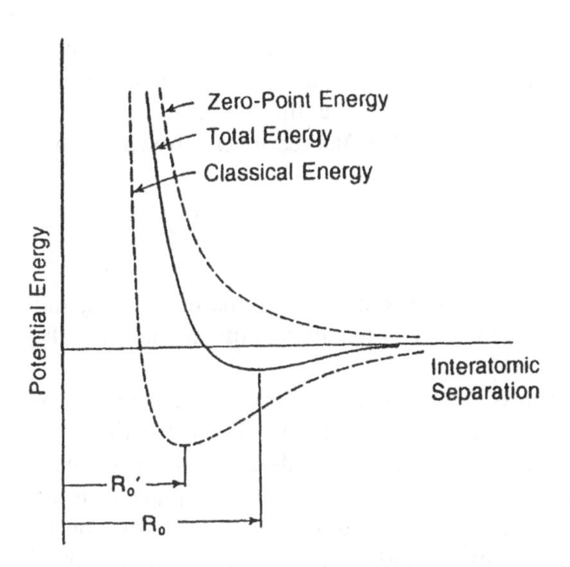


**Figure 3.101** The effect of zero-point energy on atomic separation.

Another effect of the zero—point energy is to reduce the depth of the potential well in Fig. 3.101. The "well" region below the horizontal axis actually represents the liquid phase of a fluid. It is only here that the fluid can exist in two different densities (interatomic separations) at the same potential energy (temperature). The potential well therefore represents the liquid–vapor equilibrium region of a fluid. Reducing the depth of the potential well reduces the temperature range over which vapor–liquid equilibrium is possible. Therefore, the depth of the potential well is proportional to the heat of vaporization (or sublimation) of the fluid. Hence, the low heat of vaporization of liquid hydrogen is a macroscopic manifestation of the zero-point energy of liquid hydrogen.

We shall see shortly that the zero-point energy is manifested even more prominently in liquid helium. It should be noted that neon also exhibits strong quantum mechanical effects. That is why these three fluids—hydrogen, helium, and neon—are referred to as the quantum fluids. This is also the reason the *PVT* properties of these three fluids are omitted from the usual generalized correlations of state. These equations depend upon the similarity among fluids, especially the rule that fluids equally distant from their critical points behave similarly. That is to say, at the same reduced temperature, $T_R = T/T_c$, and the same reduced pressure, $P_R = P/P_C$, the fluids will have the same compressibility factor, $Z = PV/RT$. Helium, hydrogen, and neon do not follow the rules of normal fluids and do not fit well the generalized equations of state. They are quantum fluids instead.

**Table 3.57** How Much Hydrogen is There in a Gallon of Liquid Hydrogen?

| Fluid | Density at NBP (g/cm$^3$) | mol/cm$^3$ at NBP | moles $H_2$/cm$^3$ | ratio to $H_2$ |
|---|---|---|---|---|
| $H_2$ | 0.0707 | 0.03507 | 0.03507 | 1.00 |
| $CH_4$ | 0.5110 | 0.03194 | 0.063875 | 1.82 |
| $H_2O$ | 1.0000 | 0.05560 | 0.05560 | 1.58 |

**Other Properties.** Figures 3.102–3.113 give the dielectric constant, specific heats, density, melting curve, surface tension, vapor pressure, differences in specific heat between normal and para-differences between the thermal conductivity of normal and para- of hydrogen taken from the WADD report mentioned before (Johnson, 1960).

### 3.7.5. *Uses of Hydrogen*

More than 97% of current domestic hydrogen use is in chemical manufacturing and petroleum refining. The other hydrogen uses are often important in their specific industries.

**a. Petroleum Refining.** Hydrogen is used to treat and upgrade oils in hydrotreating and hydrocracking processes in petroleum refining. Hydrotreating, the reacting of hydrogen with refinery fractions to desulfurize feedstocks, to hydrogenate olefins, and to treat lube oils and kerosene-type jet fuels, has grown rapidly with the introduction of catalytic reforming and the resulting supply of low-cost hydrogen. Catalytic reforming produces around 125 $m^3$ of hydrogen for every cubic meter of oil processed (700 $ft^3$/bl). The amount of hydrogen required for hydrotreating ranges from 9 $m^3$ per cubic meter of straight-run naphthas (50 $ft^3$/bl) to more than 15 times as much for some coker distillates. The average is estimated to be around 44.5 $m^3$ per cubic meter of feed (250 $ft^3$/bl).

**b. Chemicals.** Between 1982 and 2266 $m^3$ (70,000–80,000 $ft^3$) of hydrogen is required to produce 907 kg (1 ton) of synthetic anhydrous ammonia, including the amount that does not react to form ammonia.

The making of synthetic methanol requires about 2.2 $m^3$ of hydrogen per kilogram (36 $ft^3$/lb).

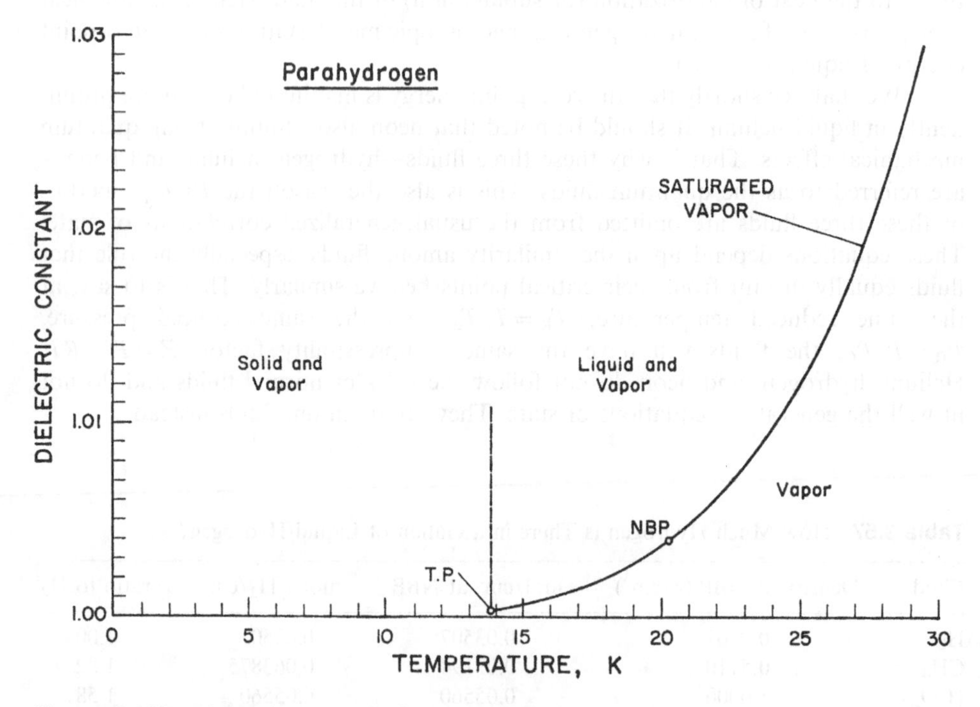


**Figure 3.102** Dielectric constant of saturated parahydrogen vapor.

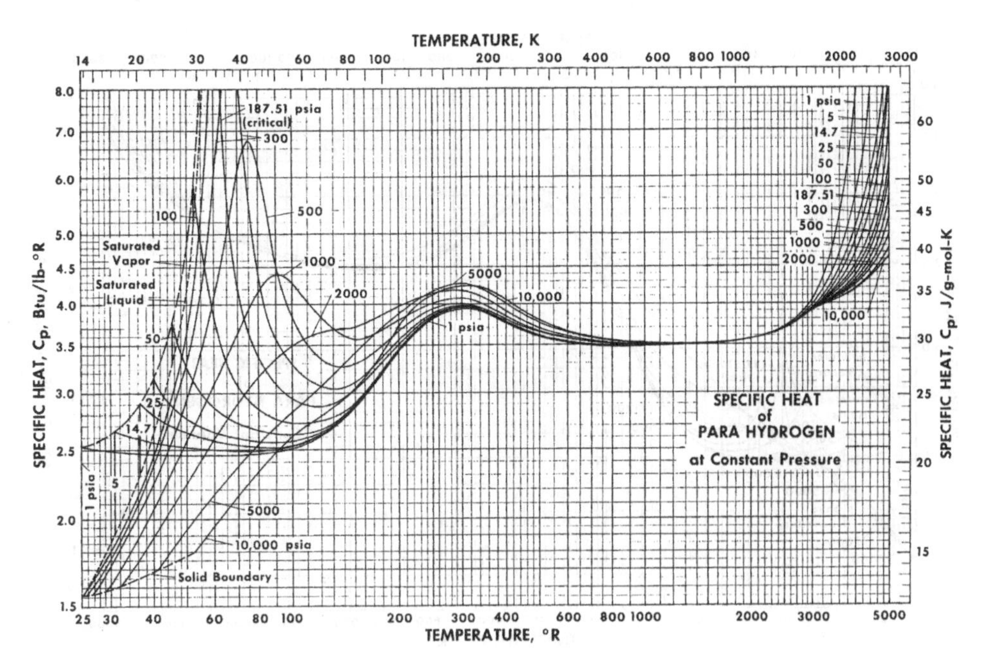


**Figure 3.103** Specific heat ($C_P$) of parahydrogen at constant pressure.

Cyclohexane is directly recoverable from petroleum and is also made by the catalytic hydrogenation of benzene. The benzene route for cyclohexane production, representing 96% of cyclohexane capacity in 1972, uses a minimum of 814 $m^3$ (28,750 $ft^3$) of hydrogen per 907 kg (ton) of cyclohexane produced.

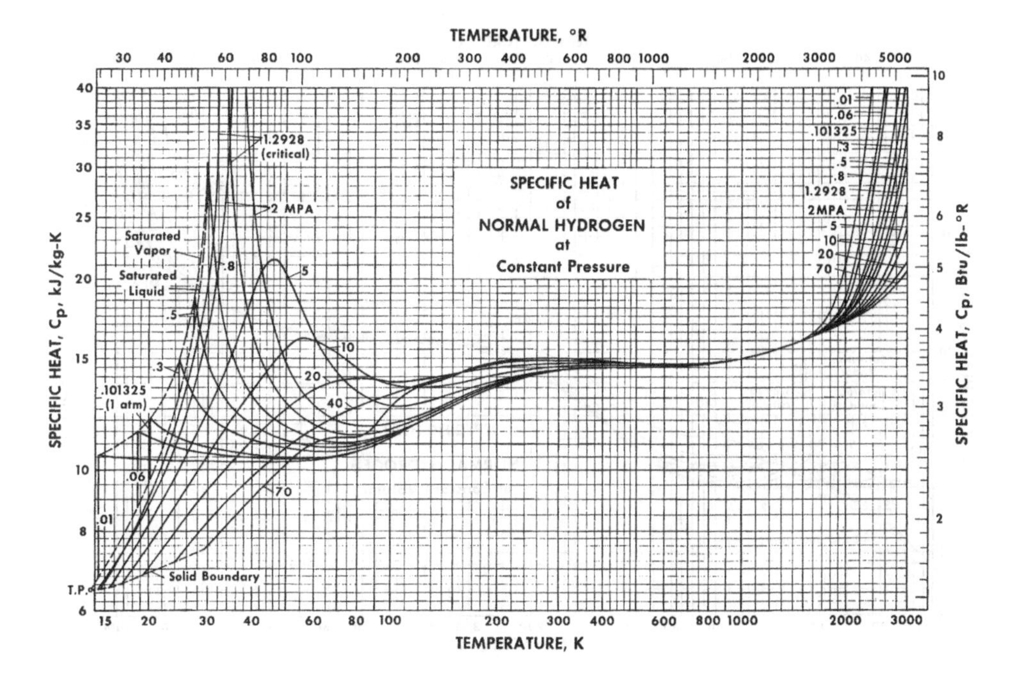


**Figure 3.104** Specific heat $C_P$ of normal hydrogen at constant pressure.

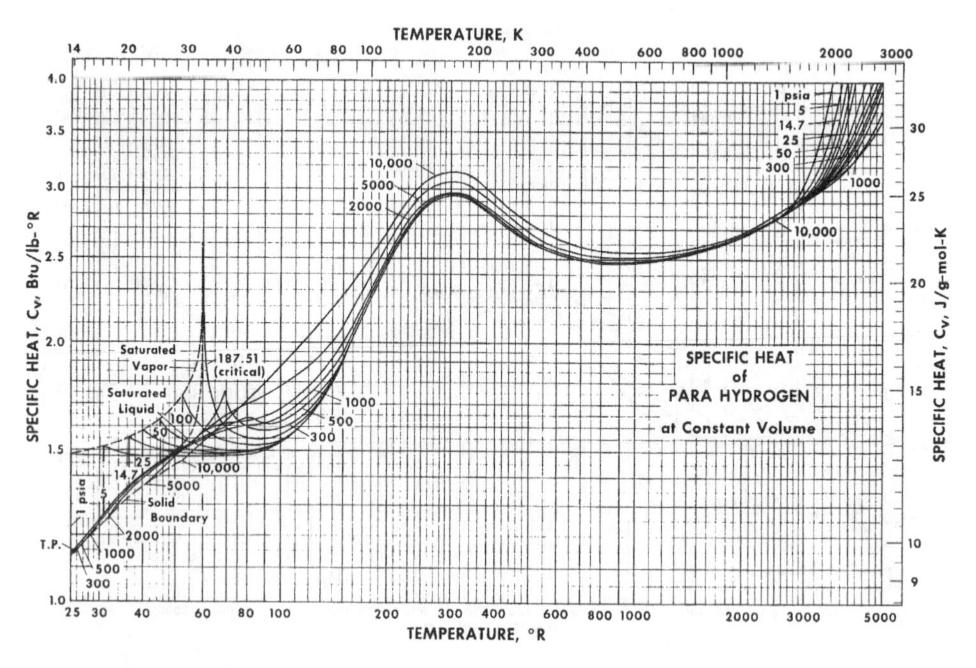


Figure 3.105 Specific heat ($C_V$) of parahydrogen at constant volume.

Aniline is produced from benzene by nitration and then catalytic hydrogenation requiring a minimum of 0.81 $m^3$ of hydrogen per kilogram of product (13 $ft^3$/lb).

Naphthalene production was coal-based until the early 1960s when petroleum-based processes were developed. Today, around 40% of naphthalene production is

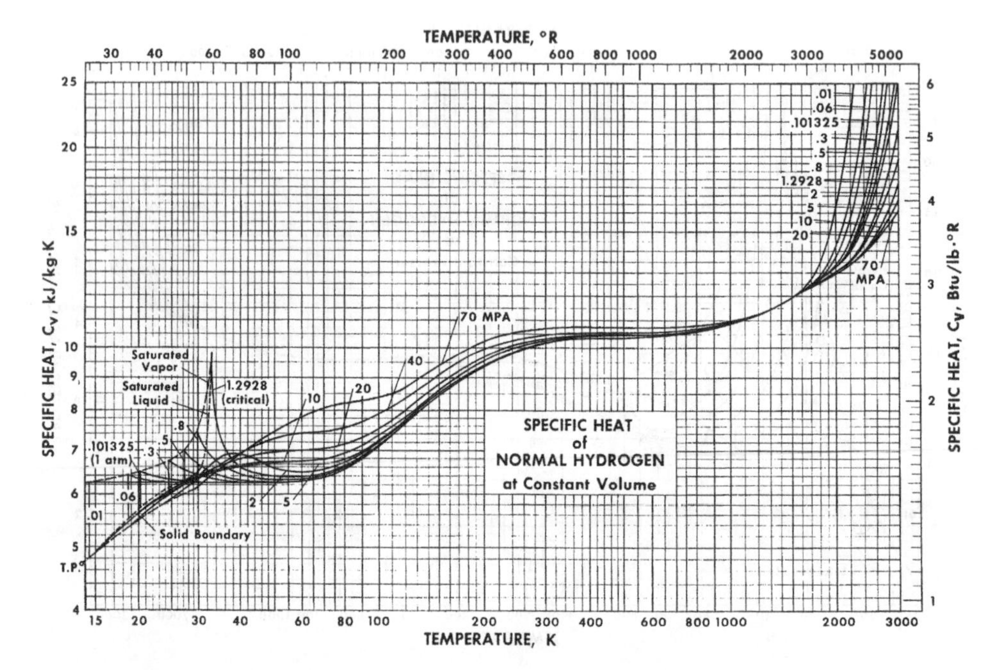


Figure 3.106 Specific heat $C_V$ of normal hydrogen at constant volume.

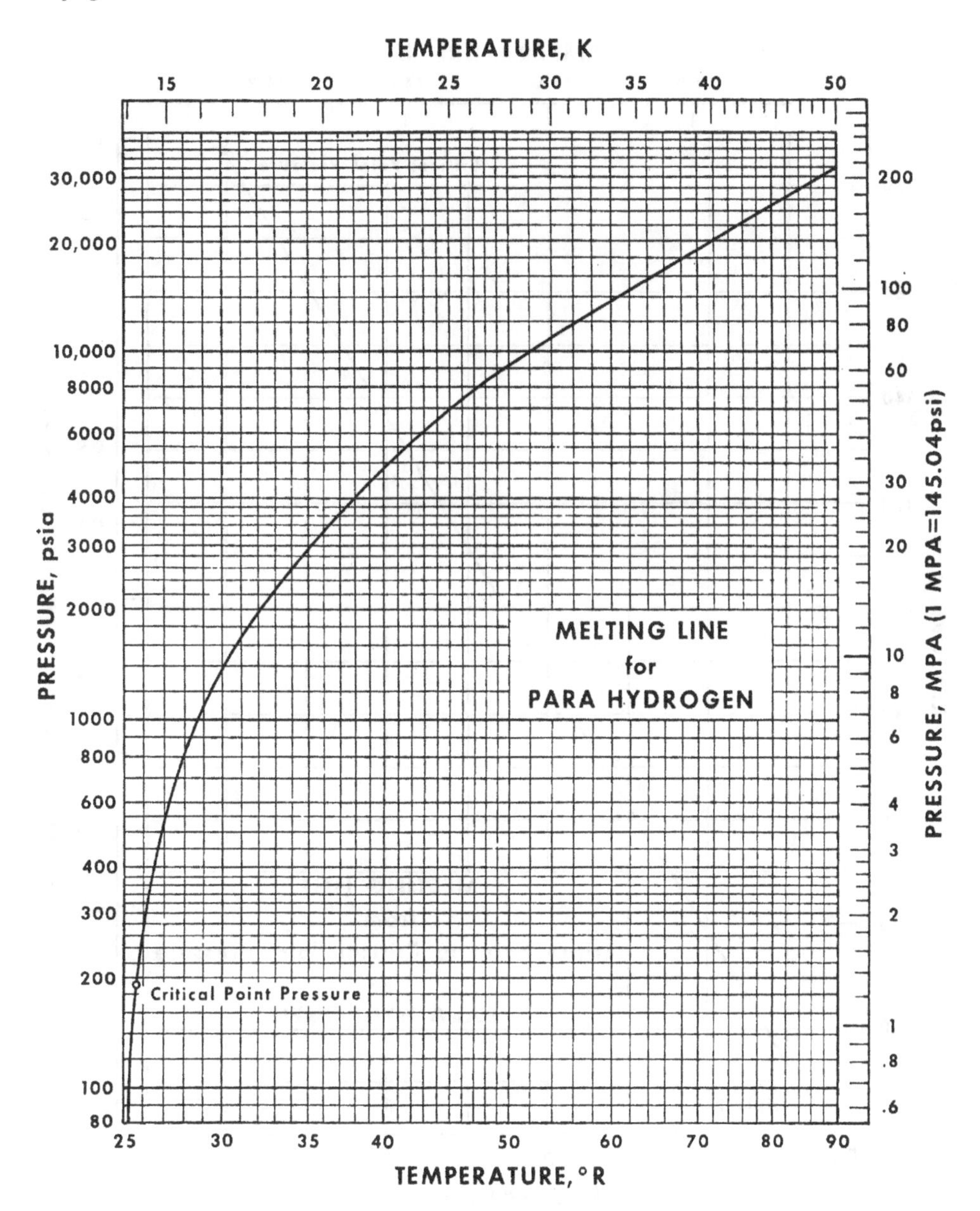


**Figure 3.107** Melting line for parahydrogen, critical point and above.

from petroleum. Hydrogen is used for desulfurization and dealkylation in a hydrogen atmosphere. Total hydrogen use in naphthalene production is estimated to be around 1.87 $m^3$ per kilogram of naphthalene (30 $ft^3$/lb).

Other hydrogen uses in the chemical industry include the production of a variety of chemicals such as hydrogen chloride, aldehyde, caprolactam, ethylene, hydrogen peroxide, and styrene.

**c. Hydrogenation of Fats and Oils.** Hydrogen plays a major role in the hardening of vegetable and fish oils used in the production of margarine, lard, shortening, and cooking oils.

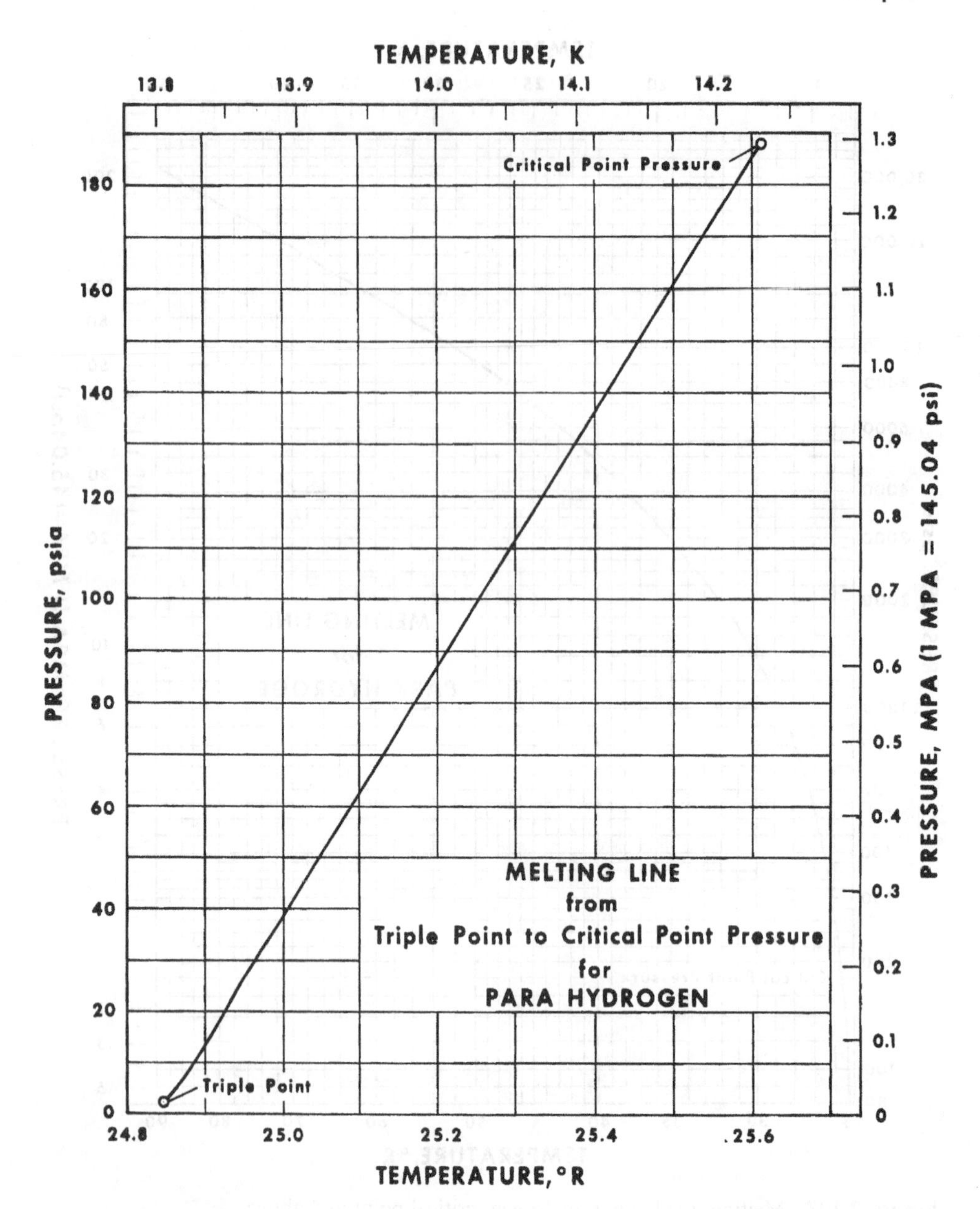


**Figure 3.108** Melting line from triple point to critical point pressure for parahydrogen.

**d. Aerospace.** A significant amount of the liquid hydrogen produced, perhaps half, is used by the aerospace industry for rocket propellant and rocket engine component testing. Hydrogen is also used in spacecraft fuel cells.

**e. Metallurgy.** Hydrogen is used for direct reduction of ores, heat treatments, welding, and in the production of pure metals. The major present use is in the direct reduction of iron oxide ores to sponge iron, which uses around 566 $m^3$ of hydrogen per 907 kg iron produced (20,000 $ft^3$/ton).

**f. Electricity and Electronics.** Some public electric utilities require hydrogen gas as a coolant for large generators, motors, and frequency changers. In electronics,

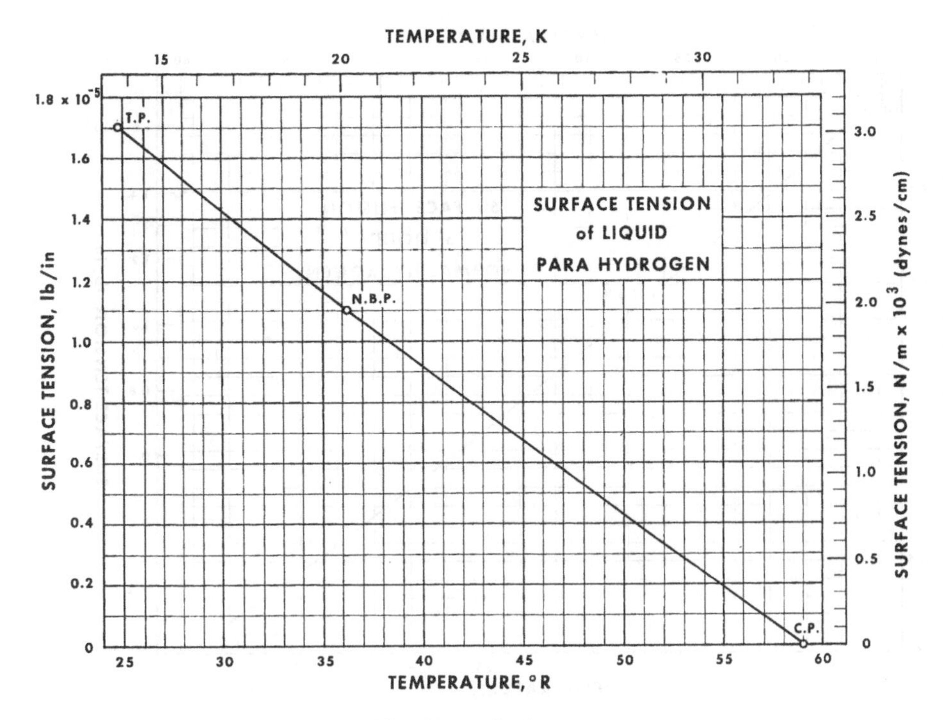


**Figure 3.109** Surface tension of liquid parahydrogen.

hydrogen is used as a reducing atmosphere and for brazing of metals and silicon compounds for solid-state components, vacuum tubes, light bulbs, and crystal growing.

**g. Other Uses.** Hydrogen use in other areas is difficult to quantify. These use areas include glass manufacture via the float glass process, oxygen-hydrogen glass cutting, artificial gem production, refrigeration, pharmaceuticals, nuclear fuel processing, and nuclear research using liquid hydrogen bubble chambers.

## 3.8. Helium

Evidence of the existence of helium (He) was first obtained by Janssen during the solar eclipse of 1868 when he detected a new line in the solar spectrum; Lockyer and Frankland suggested the name *helium* for the new element. In 1895, Ramsay discovered helium in the uranium mineral clevite, and it was independently discovered in clevite by the Swedish chemists Cleve and Langlet about the same time.

Onnes first liquefied helium in 1908 in his laboratory at Leiden, Netherlands, by using liquid hydrogen precooling in a Joule–Thomson liquefier. Liquid helium was not available outside of Leiden until 1923.

In 1920, Aston discovered that, in addition to the common isotope helium-4, there existed a very rare isotope helium-3. The helium-4 atom consists of two electrons orbiting a nucleus of two protons plus two neutrons. The helium-4 nucleus is the alpha particle associated with radioactive decay and other atomic processes. The helium-3 atom consists of two electrons orbiting a nucleus of two protons plus one neutron.

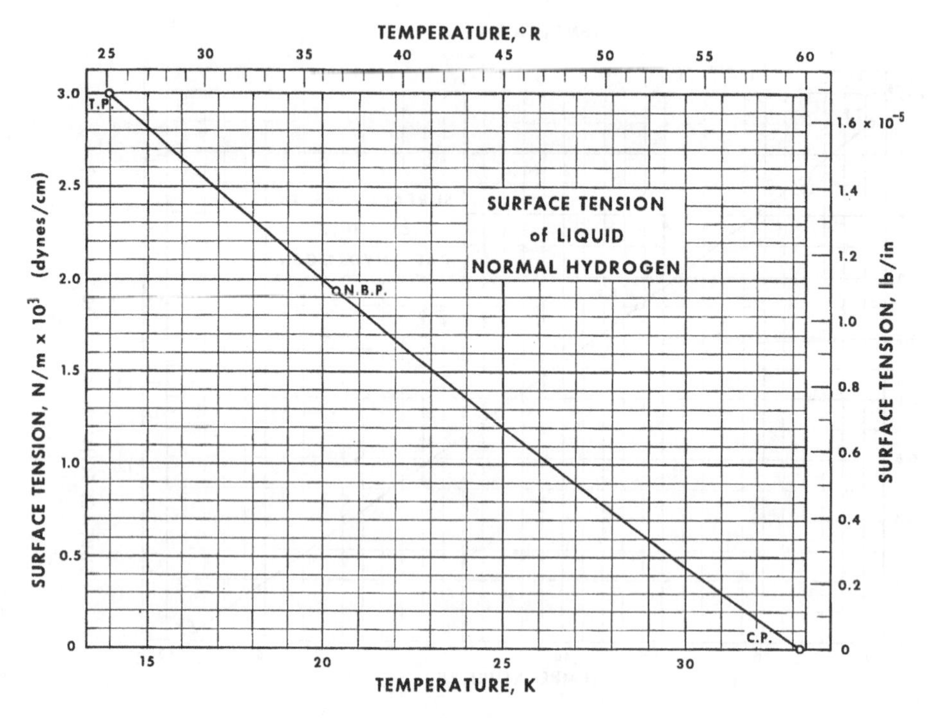


**Figure 3.110** Surface tension of liquid normal hydrogen.

Helium-4 is by far the more common of the two isotopes. Ordinary helium gas contains about $1.3 \times 10^{-4}\%$ helium-3, so that when we speak of helium or liquid helium, we normally are referring to helium-4 (molecular weight 4.0026). Liquid helium-4 has a normal boiling point of 4.216 K (7.57°R) and a density at the normal boiling point of 124.96 kg/m$^3$ (7.8 lb$_m$/ft$^3$), or about one-eighth that of water.

The physical constants for helium are shown in Table 3.58.

### *3.8.1. Helium Phase Diagram*

Helium, like hydrogen, possesses a high quantum mechanical zero-point energy relative to its classical thermal energy. As we mentioned earlier, solid hydrogen has a zero-point energy of about 200 cal, which counteracts about 50% of its computed lattice energy of 400 cal/mol. The zero-point energy contribution in helium is even more dramatic.

The zero-point energy of solid helium near 0 K, about 50 cal/mol, counteracts about 80% of the calculated lattice energy of 62 cal/mol, and the zero point energy of *liquid* helium is about 80% of the cohesive energy.

The average interatomic spacing of liquid helium is determined by a balance of the zero-point energy and the classical energy forces. Hence, the average interatomic spacing of helium is also greatly affected by the zero-point energy. The density of liquid helium is more gas-like than liquid-like due to the high zero-point energy. One particularly gas-like property is the high compressibility of liquid helium. At a pressure of 20,000 atm, helium-4 occupies only two-fifths of its volume at 1 atm.

The zero-point energy also reduces the binding energy between atoms in the liquid state and hence reduces the heat of vaporization of the liquid. In most liquids,

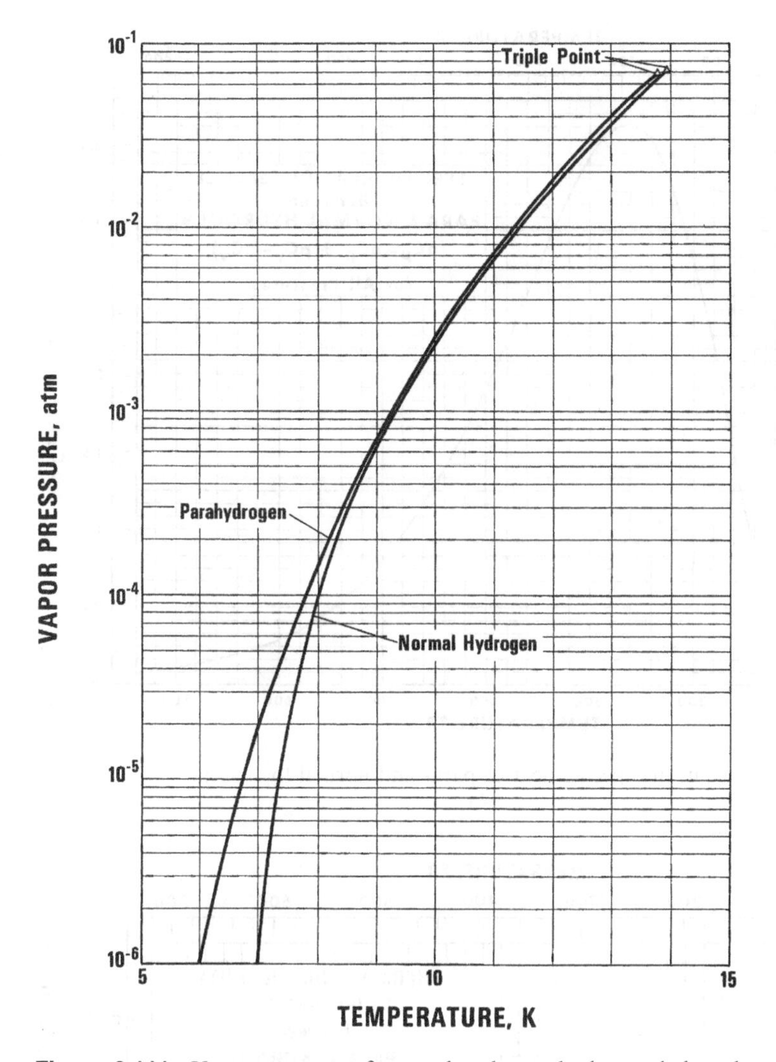


**Figure 3.111** Vapor pressure of normal and para-hydrogen below the triple point.

the entropy difference between the liquid and gaseous states (equal to the heat of vaporization divided by the temperature) is about 80 cal/(mol K) at the normal boiling point (Trouton's rule). In liquid helium, the influence of zero-point energy causes this number to drop to about 20 cal/mol. This zero-point energy accounts for the very low heat of vaporization of liquid helium. It is as if the zero-point energy had already started the pot to boil.

The large zero-point energy of liquid helium has a profound and totally unexpected effect on its phase diagram.

Figure 3.114 shows the phase diagram of a normal fluid on the left and the phase diagram of helium on the right. The normal diagram shows that there are three phases, solid, liquid, and vapor; that these are separated by the melting, vapor pressure, and sublimation curves; that there is a critical point where the vapor and liquid phases are no longer distinguishable; and that there is a solid–liquid–vapor coexistence point, the triple point.

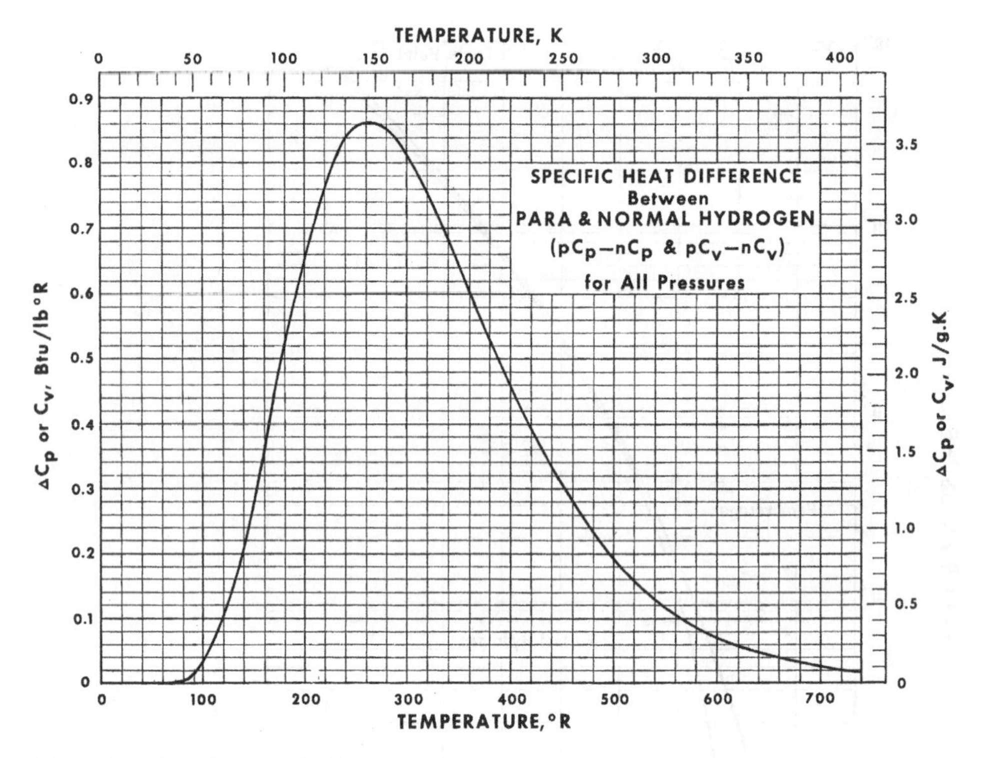


Figure 3.112 Difference in the specific heat of para- and normal hydrogen.

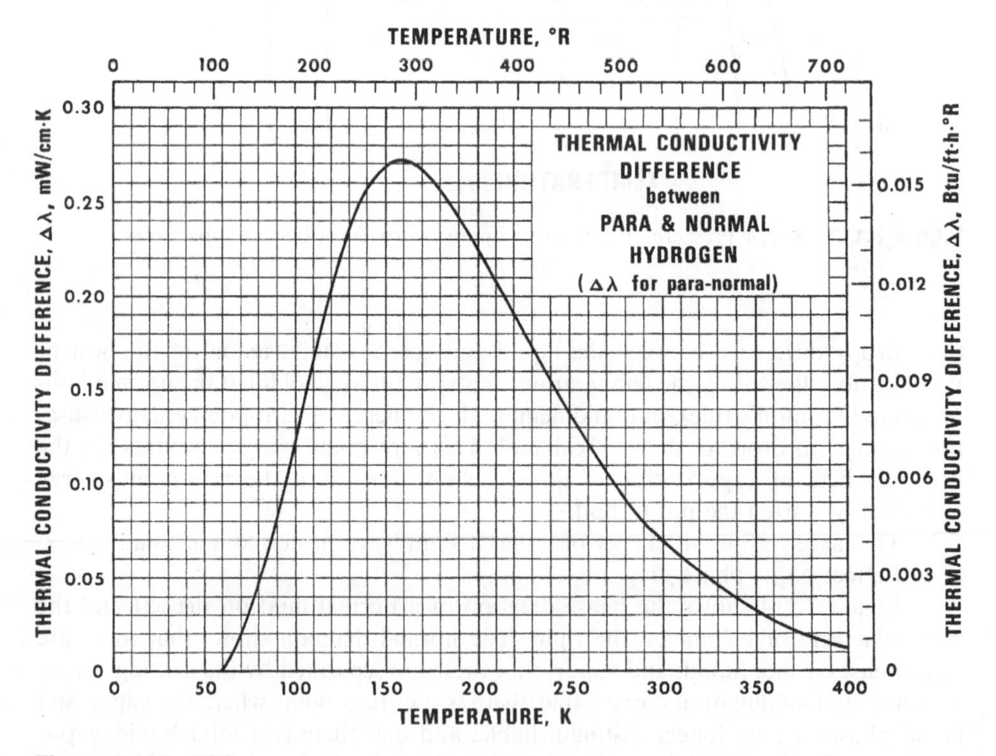


Figure 3.113 Difference in the thermal conductivity of para- and normal hydrogen.

**Table 3.58** Physical Constants for Helium

*Helium*
Chemical Symbol: He
CAS Registry Number: 7440-59-7
DOT Classification: Nonflammable gas
DOT Label: Nonflammable gas
Transport Canada Classification: 2.2
UN Number: UN 1046 (compressed gas); UN 1963 (refrigerated liquid)

| | US Units | SI Units |
|---|---|---|
| *Physical constants* | | |
| Chemical formula: | He | He |
| Molecular weight | 4.00 | 4.00 |
| Density of the gas at 70°F (21.1°C) and 1 atm | 0.0103 lb/ft$^3$ | 0.165 kg/m$^3$ |
| Specific gravity of the gas at 70°F (21.1°C) and 1 atm | 0.138 | 0.138 |
| Specific volume of the gas at 70°F (21.1°C) and 1 atm | 97.09 ft$^3$/lb | 6.061 m$^3$/kg |
| Density of the liquid at boiling point and 1 atm | 7.802 lb/ft$^3$ | 124.98 kg/m$^3$ |
| Boiling point at 1 atm | –452.1°F | –268.9°C |
| Melting point at 1 atm | None | None |
| Critical temperature | –450.3°F | –267.9°C |
| Critical pressure | 33.0 psia | 227 kPa abs |
| Critical density | 4.347 lb/ft$^3$ | 69.64 kg/m$^3$ |
| Triple point | None | None |
| Latent heat of vaporization at boiling point and 1 atm | 8.72 Btu/lb | 20.28 kJ/kg |
| Latent heat of fusion at triple point | No T.P. | No T.P. |
| Specific heat of the gas at 70°F (21.1°C) and 1 atm | | |
| $C_P$ | 1.24 Btu/(lb)(°F) | 5.19 kJ/(kg)(°C) |
| $C_V$ | 0.745 Btu/(lb)(°F) | 3.121 kJ/(kg)(°C) |
| Ratio of specific heats ($C_P/C_V$) | 1.66 | 1.66 |
| Solubility in water, vol/vol at 32°F (0°C) | 0.0094 | 0.0094 |
| Weight of the liquid at boiling point | 1.043 lb/gal | 125.0 kg/m$^3$ |

The helium phase diagram (Fig. 3.114) exhibits many of the characteristics displayed by normal fluids: the existence of a critical point on the liquid–vapor curve, for instance. One difference between helium and all other substances is the shape of the solid–liquid (melting) curve. For other substances, this line intersects the liquid–vapor line at some point, the triple point. For helium, there is no temperature and pressure at which the solid, liquid, and vapor phases of helium can exist in equilibrium. Helium has no triple point. Linear extrapolation of the solid–liquid line would place a triple point at about 1 K. However, from about 2 K down, this extrapolated line deviates from its higher temperature linear behavior and approaches 0 K horizontally at a value of 25 atm. This means that helium can never be solidified at atmospheric pressure merely by reducing its temperature, as is the case with all other substances. In addition, it is necessary to subject the helium to a pressure of 25 atm. This is a direct result of the quantum mechanical zero-point energy of helium.

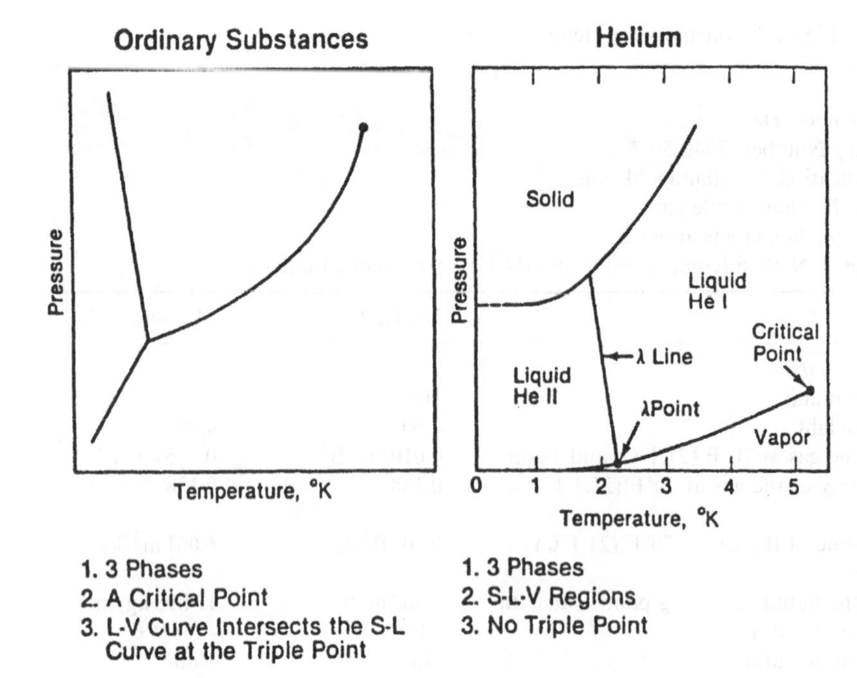


**Figure 3.114** Vapor pressure curve for ordinary liquids and helium.

Consider the gedanken experiment of sliding down the helium vapor pressure curve, starting at the critical point, 5.2 K and 2.24 atm. As the pressure is reduced, the vapor pressure is reduced correspondingly, reaching the normal boiling point of 4.2 K and 1 atm. Our progress down the vapor pressure curve continues uneventfully until at about 2.17 K there is a totally unexpected event. A totally new liquid phase of helium-4 is formed that has the properties of zero viscosity, anomalously large thermal conductivity, zero absolute temperature, and zero viscosity. This is called the superfluid.

Classical theory predicts a state of perfect order, and hence zero entropy, at absolute zero. Liquids become solids (freeze) as the temperature is decreased, to achieve the most orderly, crystalline state. The large zero-point energy of helium, however, keeps the interatomic separation too large for the atoms to fall into a lattice, that is, to freeze into a solid. Nevertheless, helium must still approach zero entropy, so it does the next best thing. Instead of forming a solid under its own vapor pressure, helium creates a second liquid phase of zero entropy, called He II. This superfluid is helium's answer to remaining a liquid (the atoms are too far apart for a solid) and at the same time heading for perfect order. Helium II plays something of the role that solids do for normal fluids.

Liquid helium-4 is odorless and colorless and somewhat difficult to see in a container because its index of refraction is so near that of the gas ($n_r = 1.02$ for liquid He). The heat of vaporization of liquid He at the normal boiling point is 20.73 kJ/kg (8.92 Btu/$lb_m$), which is only 1/110 that of water.

Table 3.59 gives some of the important fixed points of helium.

Figures 3.115–3.117 are *T–S* charts for helium-4. The Joule–Thomson inversion curve is given in Fig. 3.118 and data for the J-T curve are listed in Table 3.60.

**Table 3.59** Fixed Points for Helium

*Critical point*
$T = 5.2014$ K (9.3625°R)
$P = 2.245$ atm (32.99 psia; 2.275 kPa)
$\rho = 0.017399$ mol/cm$^3$ (4.348 lb$_m$/ft$^3$)
*Normal boiling point*
$T = 4.224$ K (7.604°R)
$P = 1$ atm (14.696 psia; 1.01325 kPa)
$\rho_{gas} = 0.004220$ mol/cm$^3$ (1.054 lb$_m$/ft$^3$)
$\rho_{liq} = 0.03122$ mol/cm$^3$ (7.802 lb$_m$/ft$^3$)
*Lower lambda point*
$T = 2.177$ K (3.919°R)
$P = 0.0497$ atm (0.730 psia; 5036 Pa)
$\rho_{liq} = 0.03653$ mol/cm$^3$ (9.127 lb$_m$/ft$^3$
*Upper lambda point*
$T = 1.763$ K (3.174°R)
$P = 29.74$ atm (437.1 psia; 3013 M Pa
$\rho_{liq} = 0.04507$ mol/cm$^3$ (11.26 lb$_m$/ft$^3$)

*Source:* McCarty (1972).

Helium exhibits striking properties at temperatures below 2.17 K (3.9°R). As the liquid is cooled, instead of solidifying it changes to a new liquid phase in the neighborhood of 2 K (3.6°R). The phase diagram of helium thus takes on an additional line separating the two phases into liquid He I at temperatures above the line and liquid He II at lower temperatures (see Fig. 3.119). The low-temperature liquid phase, called liquid He II, has properties exhibited by no other liquid. Helium II expands on cooling, its conductivity for heat is enormous, and neither its heat conduction nor its viscosity obeys normal rules (see below). The phase transition between the two liquid phases is identified as the *lambda line*, and the intersection of the latter with the vapor pressure curve is known as the *lambda point*. The transition between the two forms of liquid helium, I and II, is called the lambda-point or lambda-line because of the resemblance of the specific heat curve to the Greek letter lambda ($\lambda$) (Fig. 3.120).

Heat transfer in helium II is spectacular. When a container of liquid helium I is pumped to reduce the pressure above the liquid, the fluid boils vigorously as the pressure on the liquid decreases. During the pumping operation, the temperature of the liquid decreases as the pressure is decreased and part of the liquid is boiled away. When the temperature reaches the lambda point and the fluid becomes liquid helium II, all apparent boiling suddenly stops. The liquid becomes clear and quiet, although it is vaporizing quite rapidly at the surface. The thermal conductivity of liquid helium II is so large that vapor bubbles do not have time to form within the body of the fluid before the heat is quickly conducted to the surface of the liquid. Liquid He I has a thermal conductivity of approximately 0.024 W/(m K) at 3.3 K [0.014 Btu/(hr ft °F) at 6°R], whereas liquid He II can have an apparent thermal conductivity as large as 86,500( W/(m K) [50,000 But/hr ft °F) ]—much higher than that of pure copper at room temperature.

Not all the physical properties of helium-4 undergo such dramatic changes at the lambda transition. For instance, there is no latent heat involved in crossing the lambda line and no discontinuous change in volume. The lambda transition is

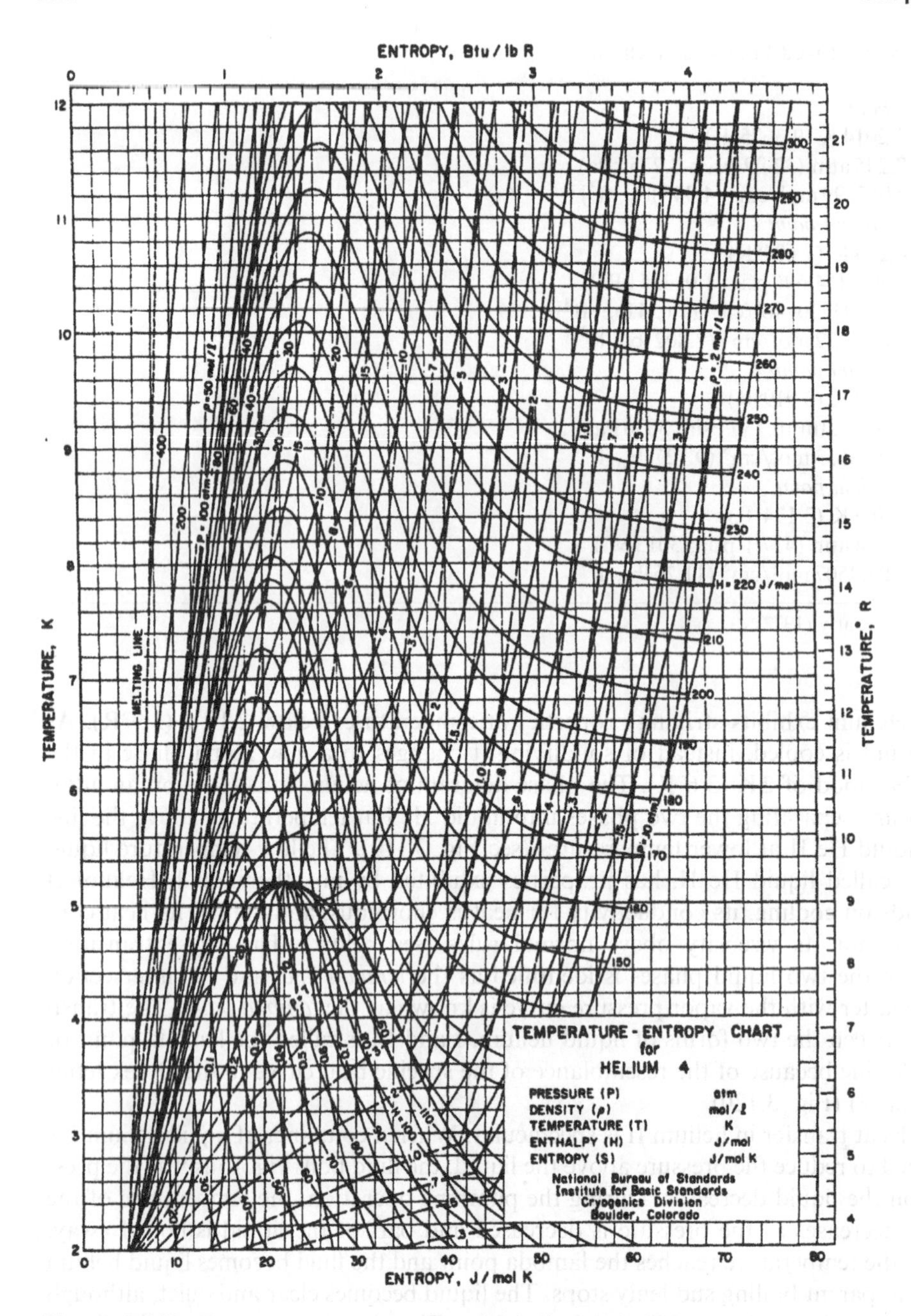


**Figure 3.115** Temperature–entropy chart for helium-4, 2–12 K.

usually considered a transition of the "second order," i.e., the Gibbs energy has a discontinuity in the second derivative.

As mentioned earlier, He II is often referred to as a *superfluid*. Immediately below the lambda point, the flow of liquid through narrow slits or channels becomes very rapid. Figure 3.121 shows the viscosity of liquid helium measured by an oscillating disk. Helium I has a viscosity of about $3 \times 10^{-6}$ (Pa s), whereas this experiment showed that He II has a viscosity of about $1 \times 10^{-7}$ Pa s at 1.3 K. However, Kapitza

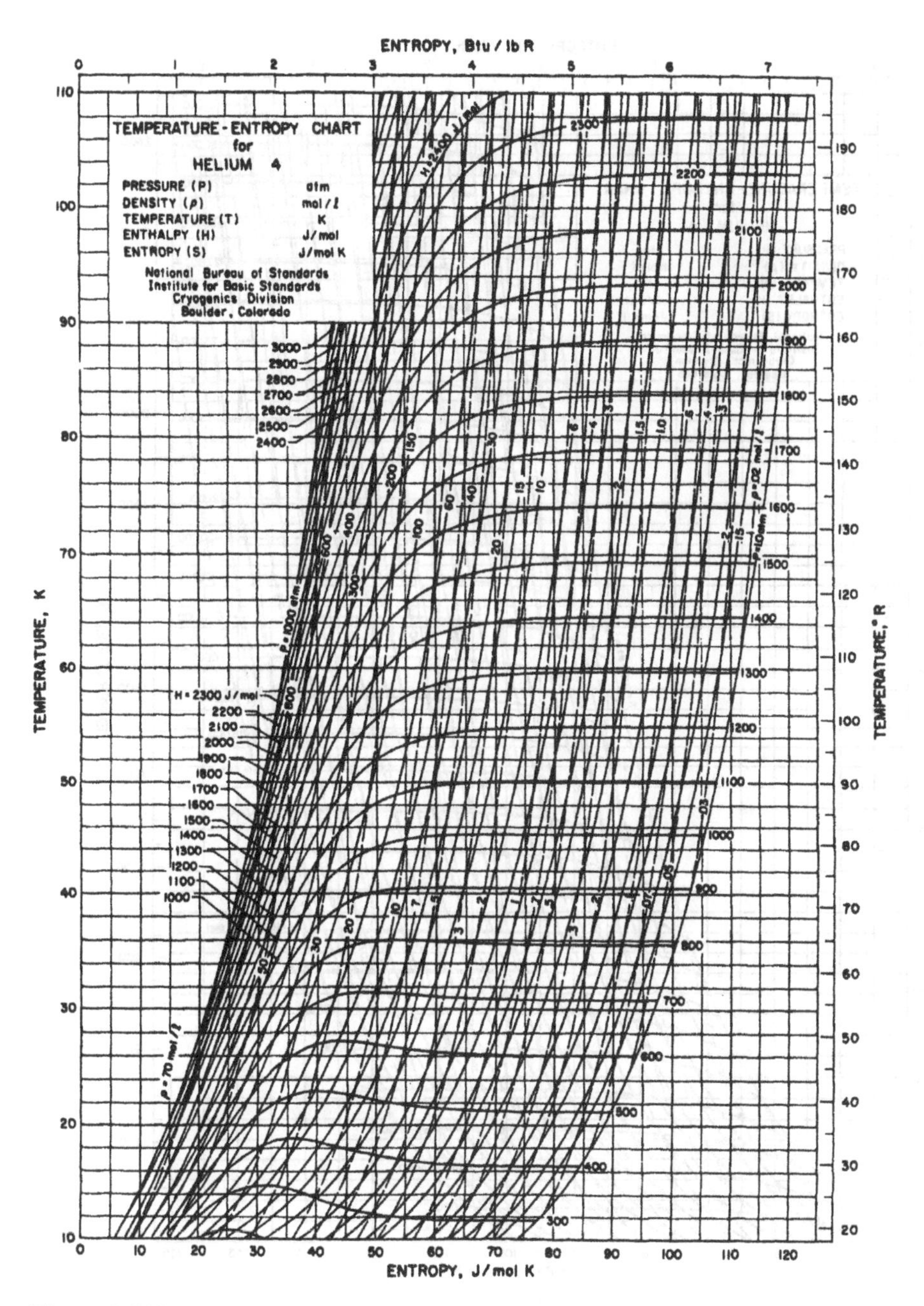


**Figure 3.116** Temperature–entropy chart for helium-4, 10–110 K.

and Allen and Misener showed that the viscosity for flow through thin channels ($10^{-4}$–$10^{-5}$ cm) was about $10^{-12}$ Pa s and was independent of the pressure drop or length of channel and dependent only on the temperature. Thus, the viscosity of helium below the lambda point is different for bulk flow and for flow through very thin channels.

To explain these viscosity effects, a "two-fluid" model is used. The liquid is considered to be composed of two fluids where the total density $\rho$ is made up of a normal density $\rho_n$ and a superfluid density component $\rho_s$ such that $\rho = \rho_n + \rho_s$.

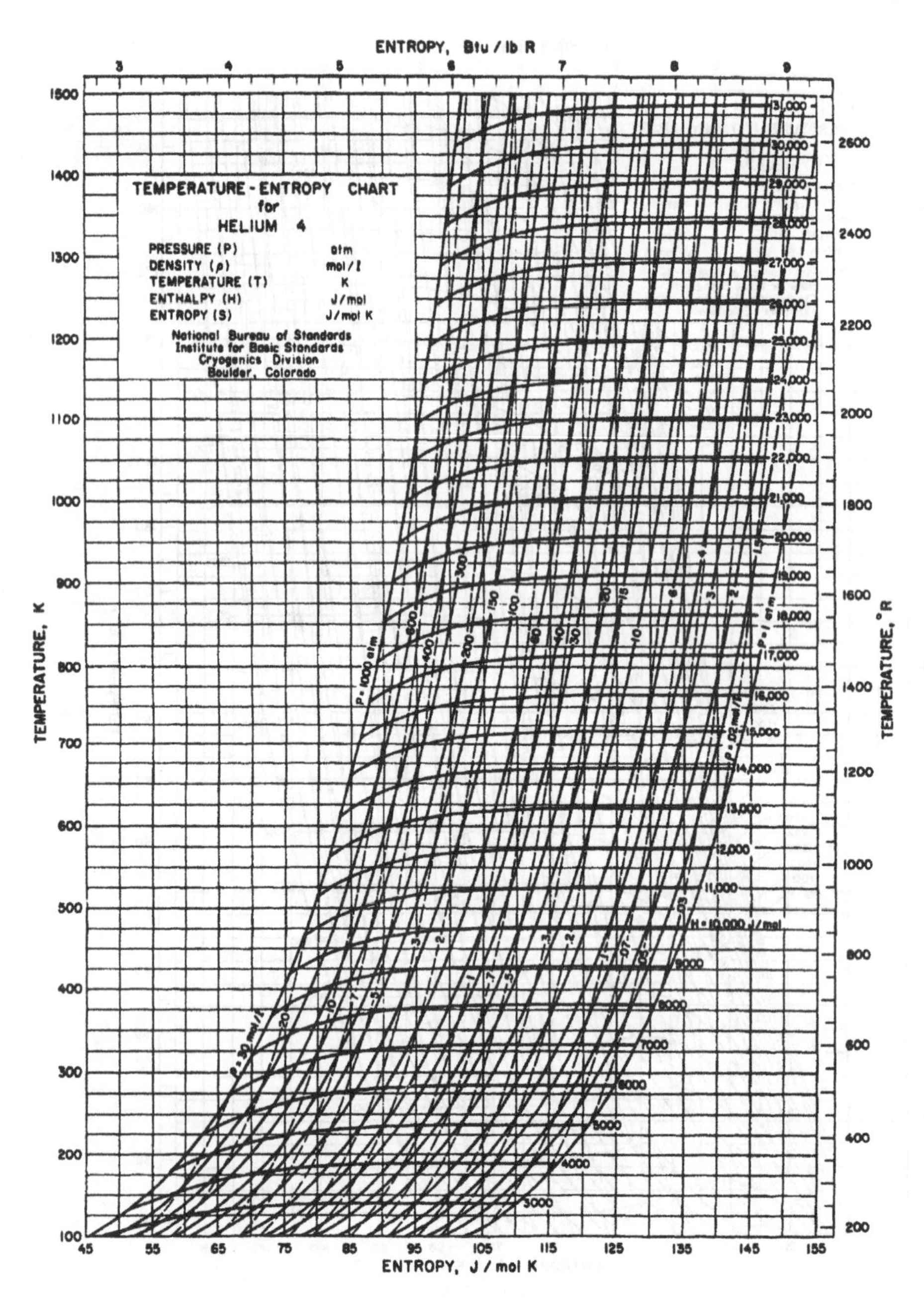


**Figure 3.117** Temperature–entropy chart for helium-4, 100–1500 K.

The former is termed the normal fluid, while the latter has been designated the superfluid because under certain conditions the fluid acts as if it had no viscosity.

Figure 3.122 shows the temperature dependence of $\rho_n$ and $\rho_s$ below the lambda transition. This picture is very useful in explaining many of the physical observations of He II. However, the detailed theoretical description of the helium problem is much more complicated, with the two-fluid model being only a first approximation.

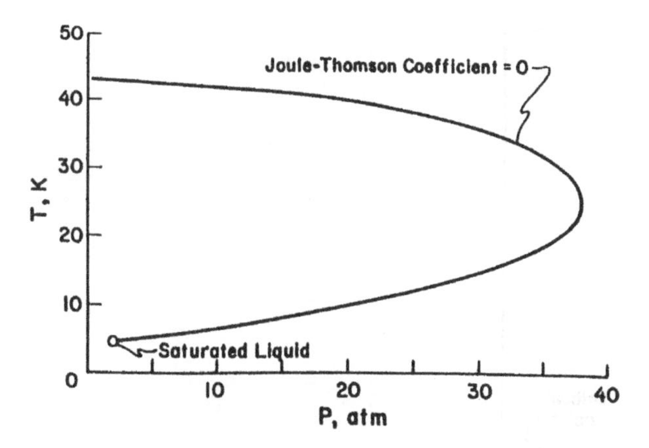


**Figure 3.118** Joule–Thomson inversion curve for helium.

To resolve the viscosity paradox, assume that at the lambda point all the fluid is normal fluid with a normal viscosity and that at absolute zero all the fluid is superfluid with zero viscosity. In the thin-channel experiment described above, only the superfluid atoms, which have zero entropy and do not interact, can flow through

**Table 3.60** Joule–Thomson Inversion Curve Data for Helium

| Temperature | | Pressure | | Density | |
|---|---|---|---|---|---|
| K | °R | atm | psia | mol/L | lb/ft$^3$ |
| 4.5 | 8.1 | 1.821 | 26.76 | 30.83 | 7.703 |
| 5 | 9.0 | 3.768 | 55.37 | 30.68 | 7.667 |
| 6 | 10.8 | 7.266 | 106.8 | 30.03 | 7.504 |
| 7 | 12.6 | 10.74 | 157.8 | 29.53 | 7.378 |
| 8 | 12.4 | 14.10 | 207.2 | 28.99 | 7.245 |
| 9 | 16.2 | 17.31 | 254.4 | 28.43 | 7.106 |
| 10 | 18.0 | 20.36 | 299.2 | 27.86 | 6.962 |
| 12 | 21.6 | 25.57 | 375.8 | 26.42 | 6.602 |
| 14 | 25.2 | 29.29 | 430.5 | 24.72 | 6.177 |
| 16 | 28.8 | 32.07 | 471.3 | 23.07 | 5.764 |
| 18 | 32.4 | 34.44 | 506.2 | 21.61 | 5.400 |
| 20 | 36.0 | 36.18 | 531.7 | 20.20 | 5.046 |
| 22 | 39.6 | 37.33 | 548.6 | 18.82 | 4.703 |
| 24 | 43.2 | 37.93 | 557.4 | 17.48 | 4.367 |
| 26 | 46.8 | 37.98 | 558.2 | 16.15 | 4.035 |
| 28 | 50.4 | 37.48 | 550.8 | 14.83 | 3.705 |
| 30 | 54.0 | 36.40 | 535.0 | 13.49 | 3.372 |
| 32 | 57.6 | 34.71 | 510.1 | 12.13 | 3.030 |
| 34 | 61.2 | 32.32 | 475.0 | 10.71 | 2.675 |
| 36 | 64.8 | 29.13 | 428.0 | 9.194 | 2.297 |
| 38 | 68.4 | 24.89 | 365.8 | 7.527 | 1.881 |
| 40 | 72.0 | 19.11 | 280.8 | 5.567 | 1.391 |
| 42 | 75.6 | 9.80 | 144.0 | 2.780 | 0.695 |
| 43 | 77.4 | 0.03 | 0.5 | 0.009 | 0.002 |

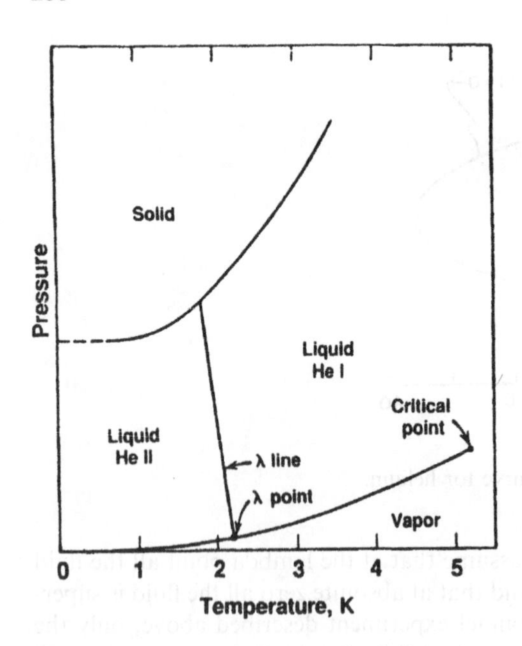


**Figure 3.119** Phase diagram for helium-4.

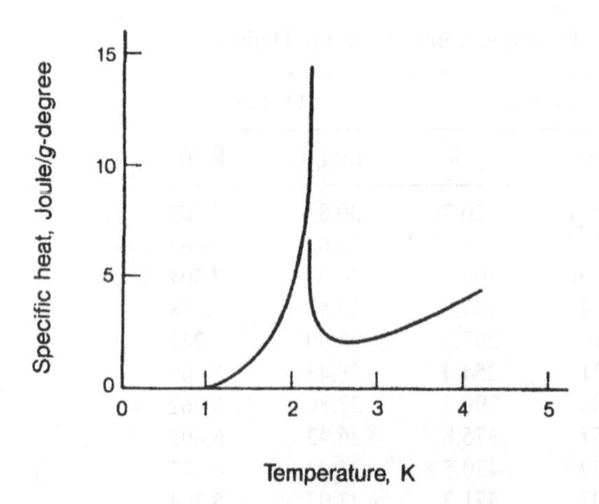


**Figure 3.120** The specific heat of saturated liquid helium.

the slit. On the other hand, the oscillating disk is damped by the normal fluid and thus accounts for the shape of the viscosity curve below the lambda point.

The flow of He II through very thin channels is accompanied by two very interesting thermal effects called the thermomechanical effect and the mechanocaloric effect.

The thermomechanical effect was discovered in 1938 and is illustrated in Fig. 3.123. When heat is applied to the fluid in the inner container, the superfluid component (being cold) tends to move toward a region of higher temperature. The superfluid can flow very rapidly through the narrow channel, whereas the flow of the normal component is inhibited by the channel resistance. Thus, a

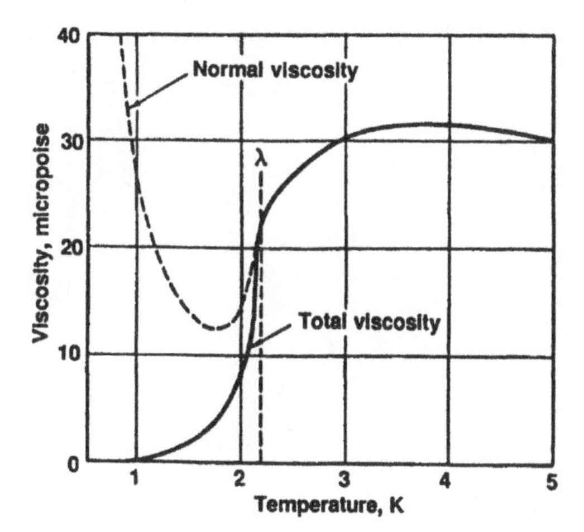


**Figure 3.121** The total viscosity and normal viscosity of helium II.

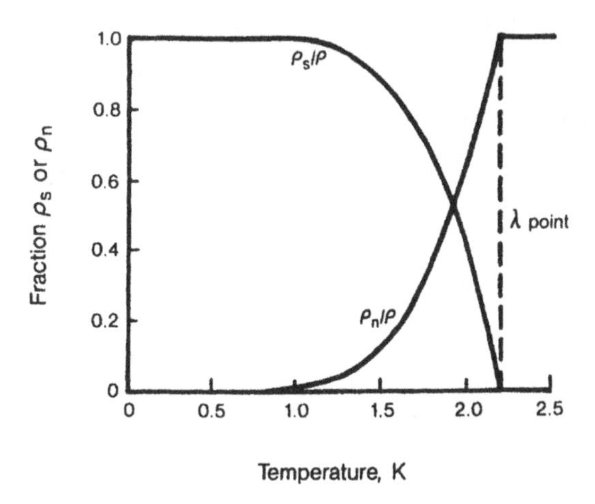


**Figure 3.122** Density vs. temperature of helium II.

thermomechanical pressure head $\Delta h$ commensurate with the temperature rise $\Delta T$ builds up in the inner container.

The mechanocaloric effect, Fig. 3.124, was observed in 1939. When liquid He II is allowed to flow through a fine powder, the remaining liquid is observed to rise in temperature. This effect can be explained by assuming that nearly all of the fluid flowing out of the container is superfluid, thus carrying zero entropy. The concentration of normal fluid in the liquid above the powder increases with time, resulting in a rise in temperature.

Yet another interesting phenomenon, that of the helium film, is shown in Fig. 3.125. A surface in contact with He II will be covered with a film of liquid. If a beaker is lowered into a flask of He II, it will gradually fill, as shown in Fig. 3.125a. When raised above the liquid, it will empty, as shown in Fig. 3.125b. The rate of flow proceeds at a constant rate (at constant temperature) and is independent of the difference in level, the length of the path, and the height of the barrier over which the flow takes place. The rate of film transfer increases with decreasing temperature,

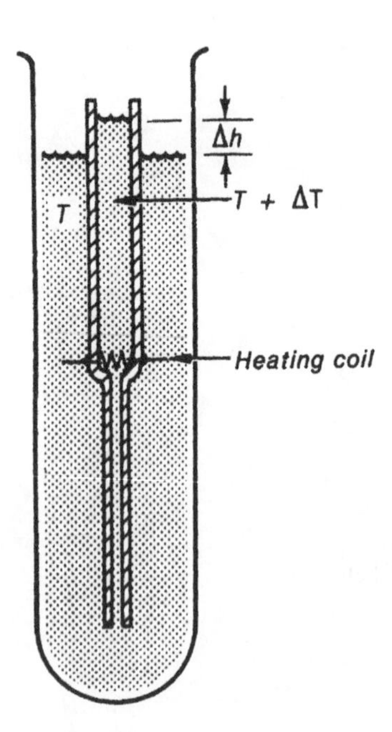


**Figure 3.123** The thermomechanical effect.

### *Cryogenic Fluids*

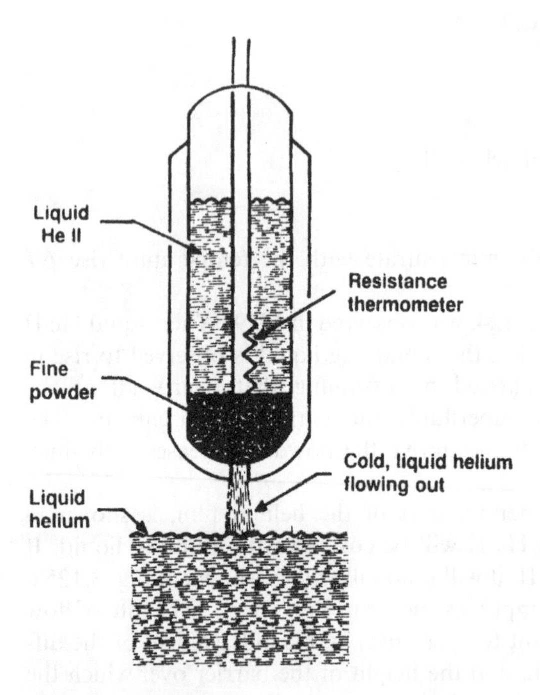


**Figure 3.124** The mechanocaloric effect.

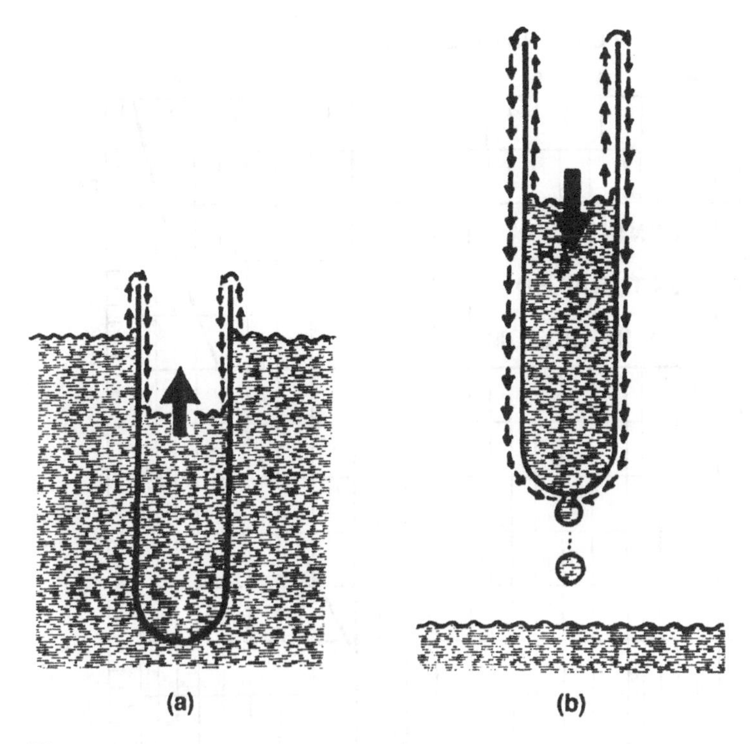


**Figure 3.125** The helium film effect.

being zero at the lambda point and becoming nearly temperature-independent below 1.5 K.

Sound can be propagated in liquid He II by at least three mechanisms. First, or ordinary, sound is the transfer of energy by a pressure wave. Second sound is a temperature wave caused by out-of-phase oscillations of superfluid and normal components of the He II. The velocity of second sound is zero at the lambda point and rises to a value of about 20 m/s between 1 and 2 K.

Third sound is a wave motion present in helium films in which the superfluid component oscillates but the normal component remains fixed to the walls. The wave motion appears as an oscillation in the thickness of the film. The velocity if propagation of third sound is approximately 0.5 m/s.

Yet another form of wave propagation has been postulated to exist in helium-3 at very low temperatures; this form has been termed zero sound. The attenuation of ordinary sound waves (first sound) in helium-3 becomes infinite as the temperature goes to zero. In this limit, zero sound replaces first sound. Zero sound is described as an oscillation in the shape of the Fermi surface. In the limit of $T=0$, the velocity of zero sound will be $\sqrt{3}$ times the velocity of first sound (at finite temperature). It is not at all obvious what type of experiments should be tried to detect zero sound. The situation is analogous to that which occurred after second sound was predicted to exist in helium-4: Experiments to detect it failed until it was realized that thermal rather than acoustical techniques were required.

*Other Properties.* Figures 3.126–3.142 give the density, thermal conductivity, specific heats, heat of fusion, heat of vaporization, vapore pressure, surface tension, viscosity, of helium, taken from the WADD report mentioned before (Johnson, 1960).

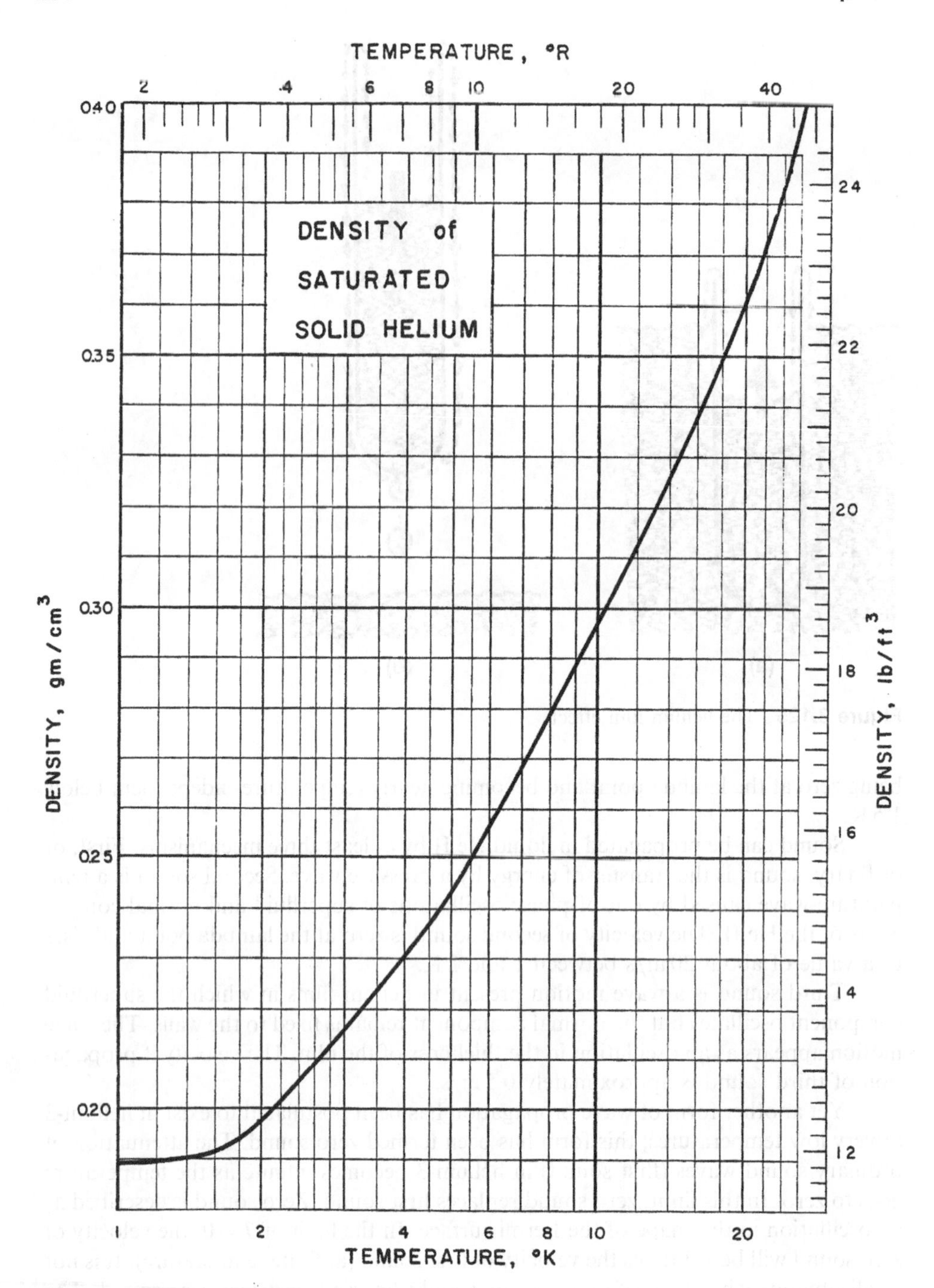


**Figure 3.126** Density of saturated solid helium.

*3.8.2. Helium-3*

The isotope helium-3 is very rare and is difficult to isolate from the isotope helium-4. Interest in helium-3 is twofold. First, its use in low-temperature refrigerators extends the minimum attainable temperature (outside of using magnetic cooling techniques)

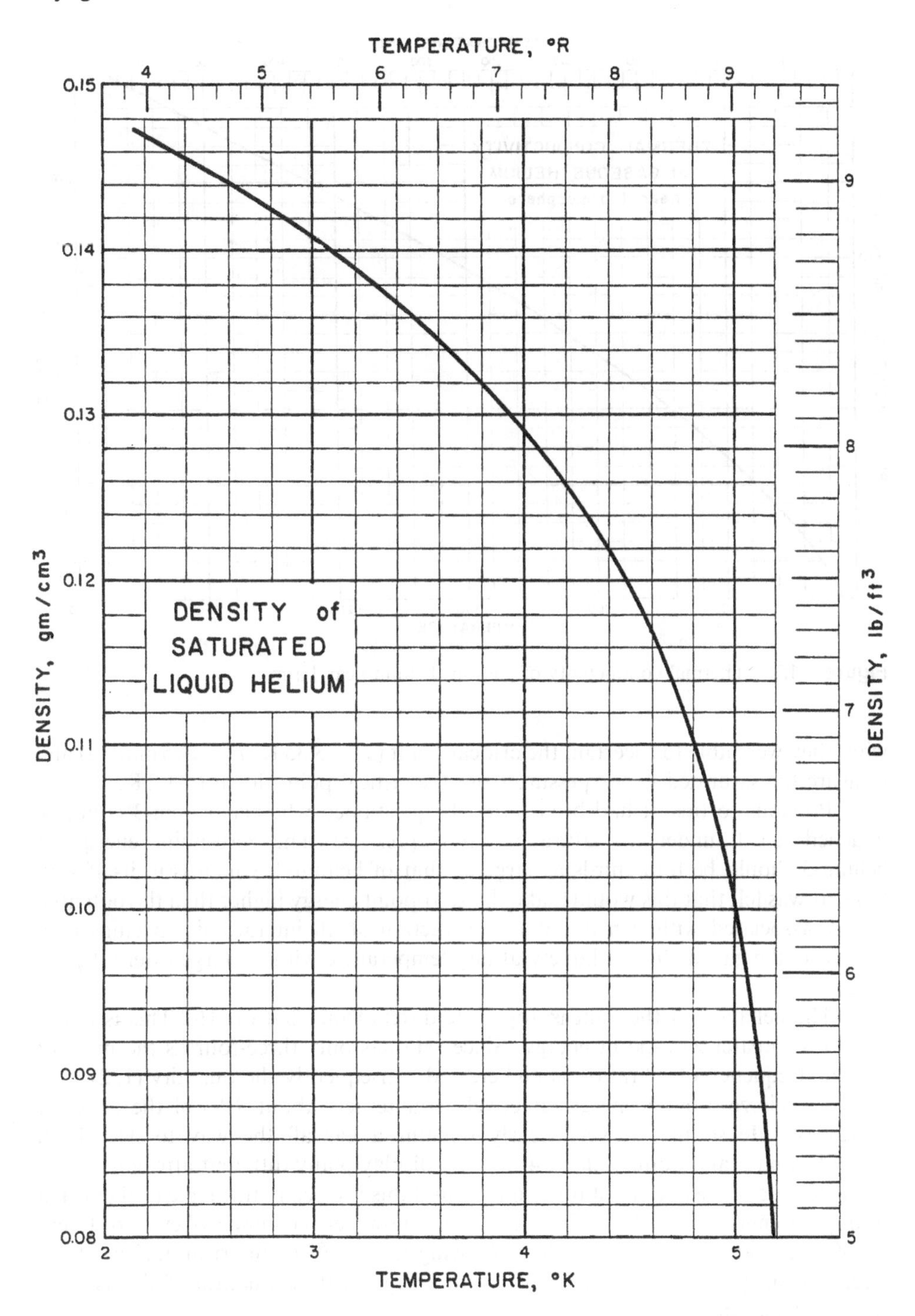


**Figure 3.127** Density of saturated liquid helium.

to about 0.3 K. Second, its properties are of fundamental interest in relation to theories of quantum statistical mechanics.

The first liquefaction of helium-3 was reported by Sydoriak, Grilly, and Hammel in 1948. Although they had only a total of 20 $cm^3$ of gas at STP to work

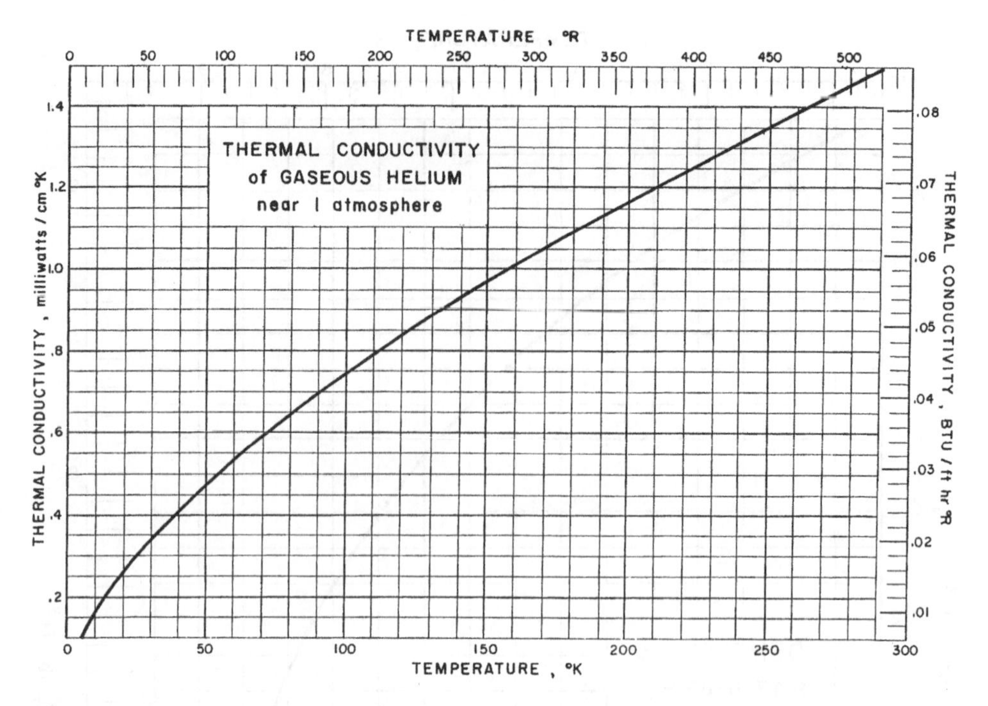


**Figure 3.128** Thermal conductivity of gaseous helium near 1 atm.

with, they were able to ascertain the critical point ($T_c = 3.35$ K, $P_c = 875$ mmHg) and measure the saturated vapor pessure from the critical point down to 1.2 K.

Previous to this, it had been seriously questioned whether helium-3 could be liquefied at all under its saturated vapor pressure. The zero-point energy of helium-3 should be four-thirds as large as that of helium-4, which is in itself very large. It was felt that this would make the zero-point energy higher than the potential energy associated with van der Waals attraction at all interatomic spacings and, therefore, the gas might not liquefy at any temperature without large external pressures.

The behavior of the light isotope $^3$He differs from that of $^4$He. This is partly due to its greater zero-point energy. Since $^3$He has only three-fourths the mass of $^4$He, its frequency of vibration is higher, and consequently the quantity (1/2) $h\nu$ is greater. This zero-point energy effectively counteracts about 95% of the cohesive energy of $^3$He so that it very nearly remains a gas all the way to 0 K. Both liquid helium-3 and liquid helium-4, in fact, display many different attributes more characteristic of gases than of other liquid, and this is directly traceable to their high zero-point energy. This is not the whole story, however. Helium-3 obeys a different set of quantum statistics than helium-4 owing to its antisymmetrical nuclear structure. For our purposes, the high zero-point energy of these quantum fluids is sufficient explanation.

Liquid helium-3 is a clear, colorless substance having a normal boiling point of 3.19 K (5.74°R) and a density at the normal boiling point of 58.95 kg/m$^3$ (3.68 lb$_m$/ft$^3$). The heat of vaporization of liquid helium-3 at the normal boiling point is only 8.48 kJ/kg (3.65 Btu/lb$_m$)—so small that there was some doubt in the minds of early investigators that helium-3 could be liquefied at normal atmospheric pressure. As in the case of liquid helium-4, liquid helium-3 remains in the liquid state under

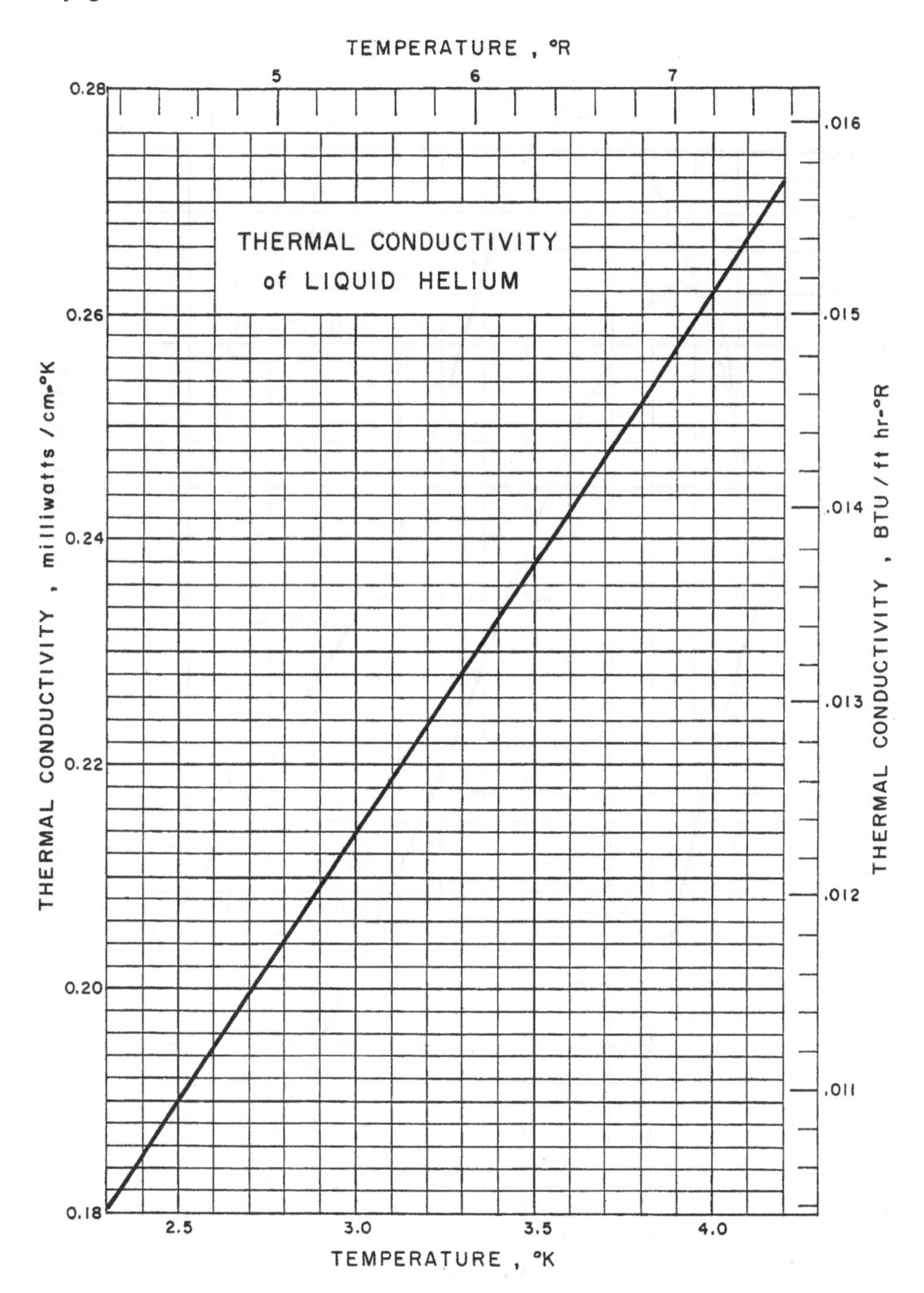


**Figure 3.129** Thermal conductivity of liquid helium.

its own vapor pressure all the way down to absolute zero. Helium-3 must be compressed to 28.9 bar (29.3 atm) at 0.32 K (0.576°R) before it will solidify.

An approximate pressure–temperature diagram for helium-3 is shown in Fig. 3.140 These data were used by Arp and Kropschot (1962) to prepare Fig. 3.141 From this figure the practical interest in helium-3 becomes clear. Liquid helium-3

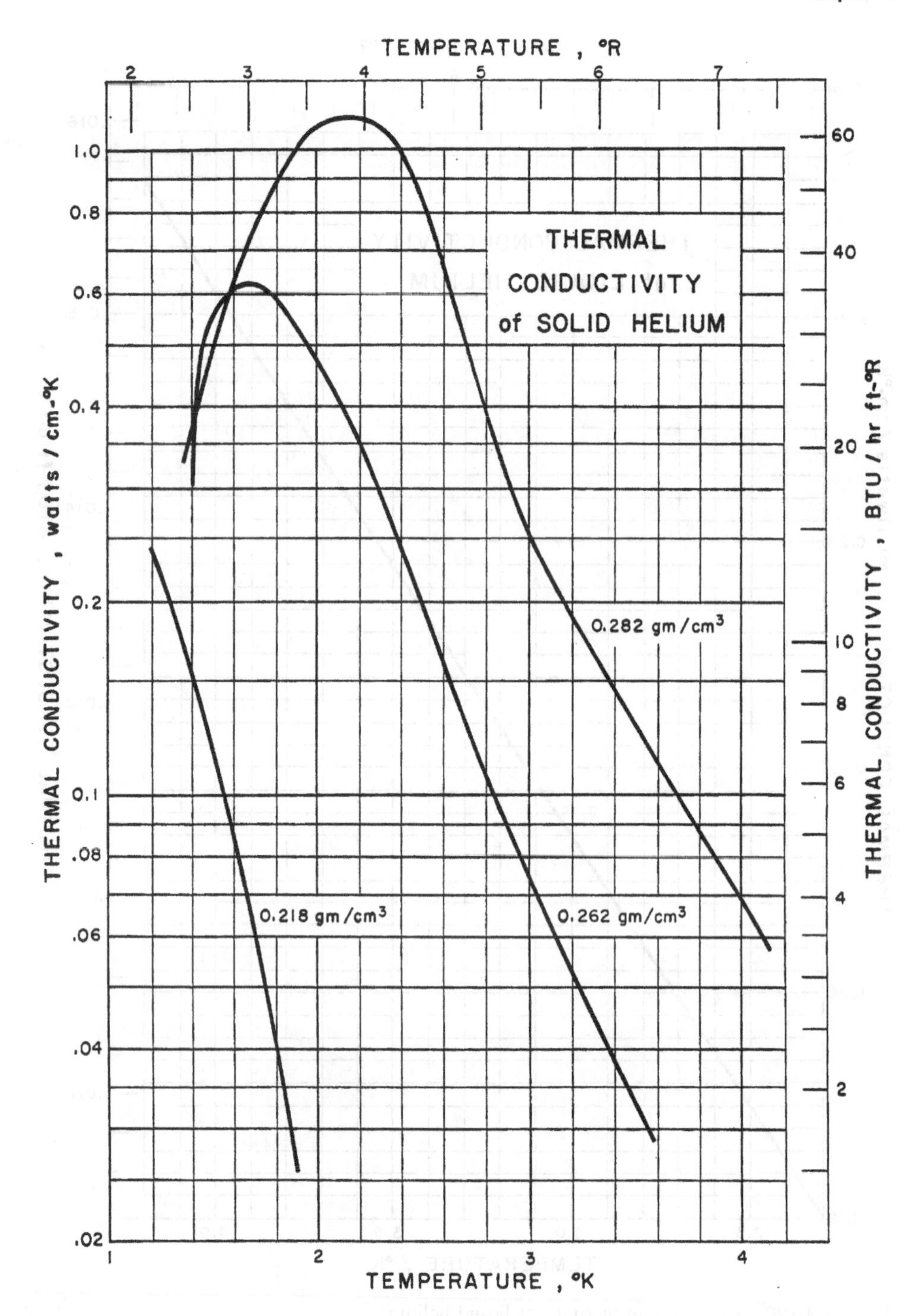


**Figure 3.130** Thermal conductivity of solid helium.

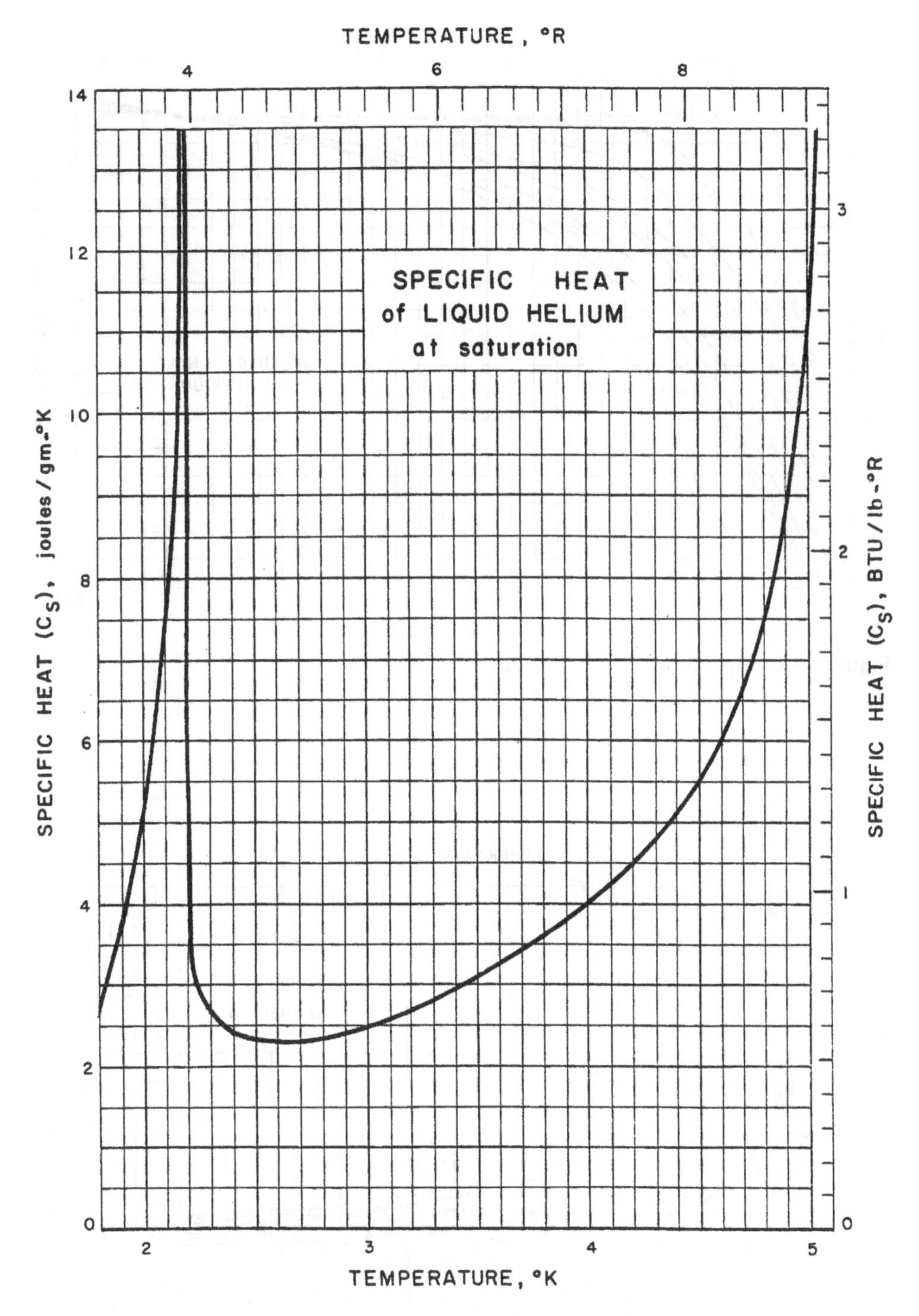


**Figure 3.131** Specific heat of saturated liquid helium.

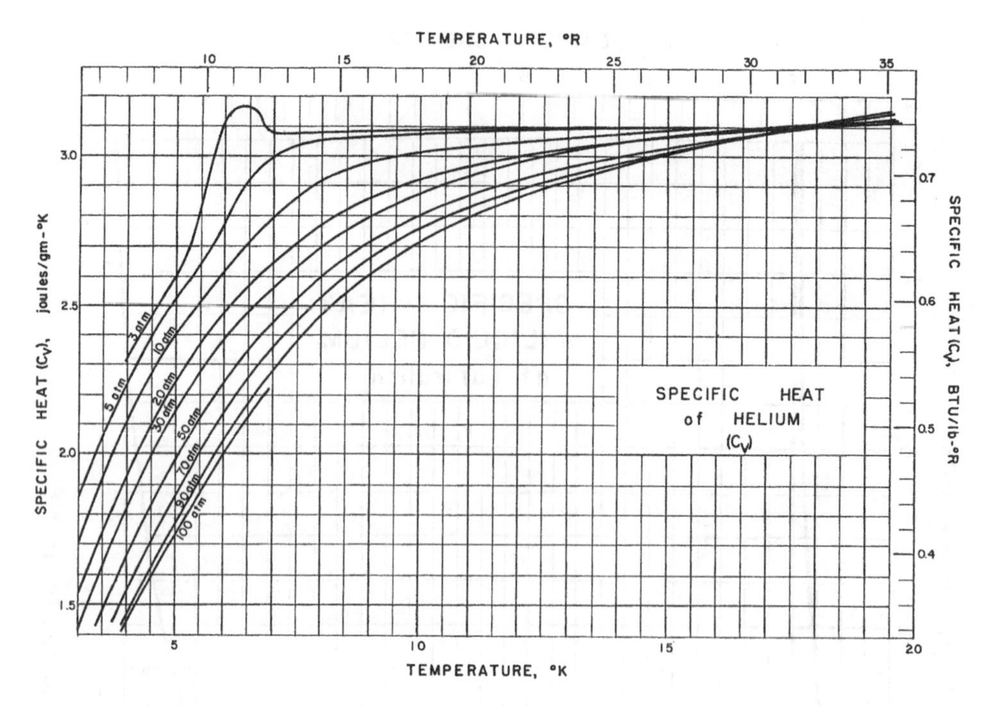


**Figure 3.132** Specific heat ($C_V$) of helium.

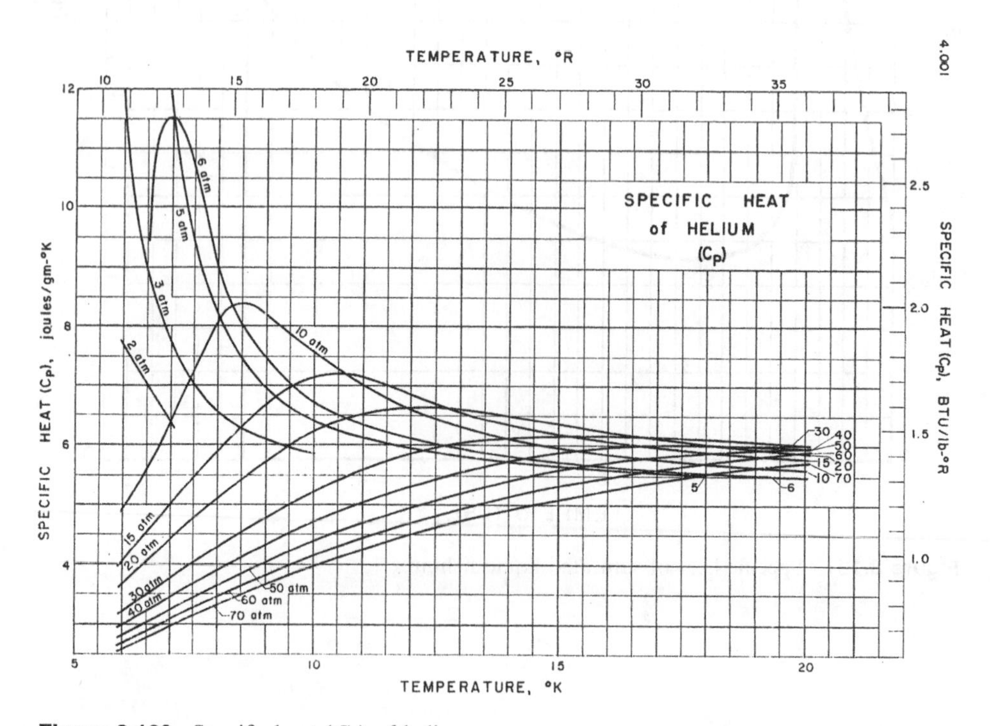


**Figure 3.133** Specific heat ($C_P$) of helium.

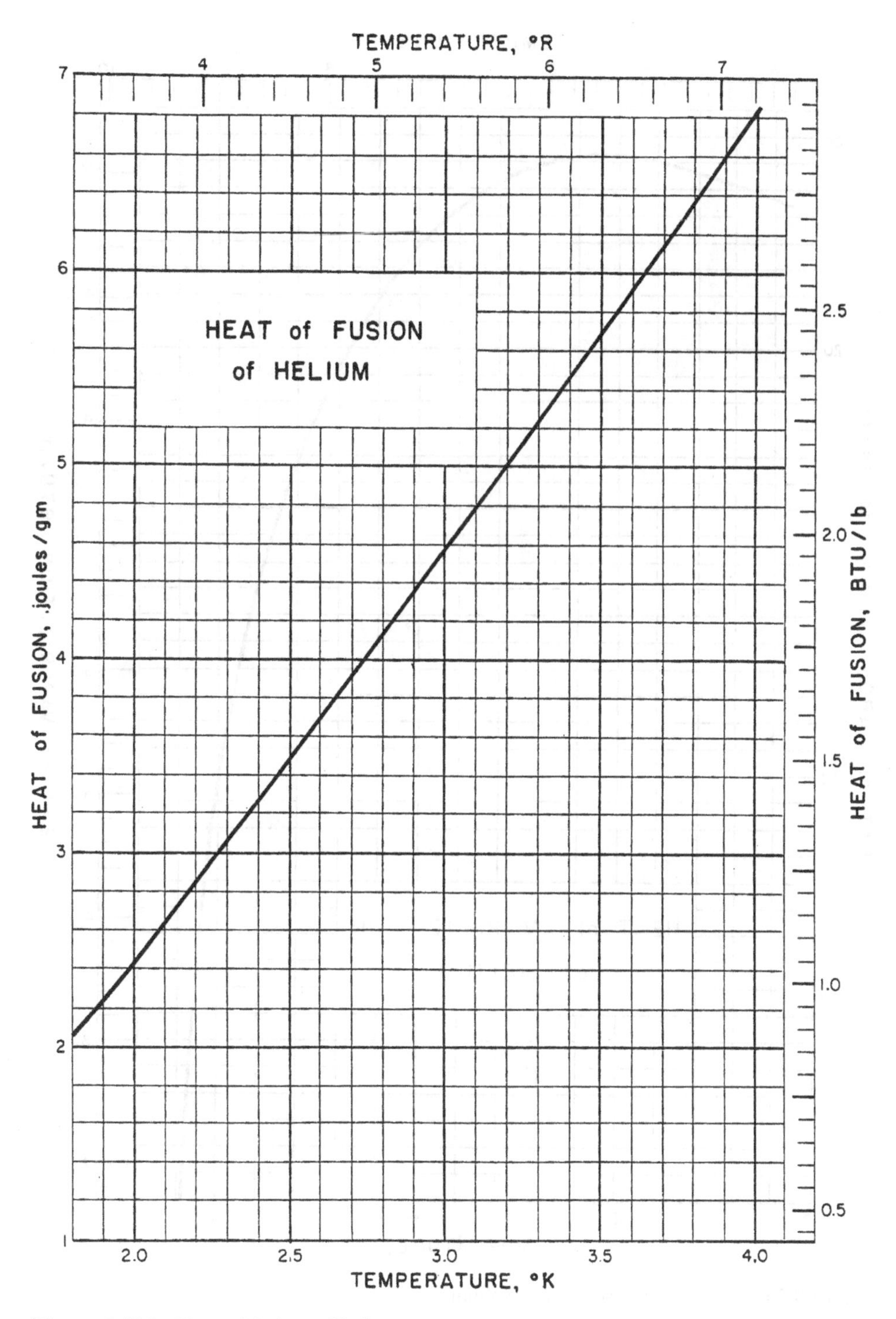


**Figure 3.134** Heat of fusion of helium.

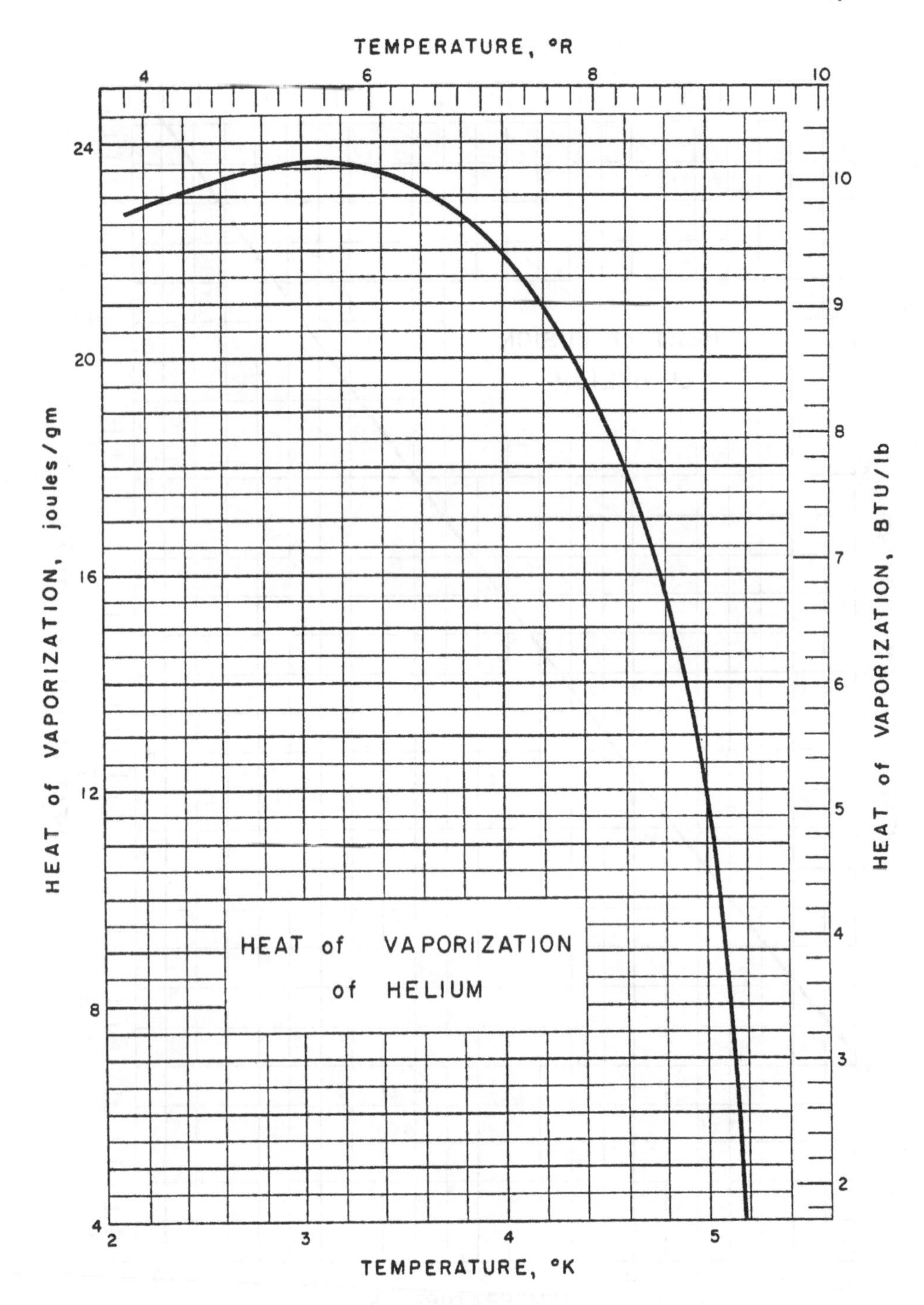


**Figure 3.135** Heat of vaporization of helium.

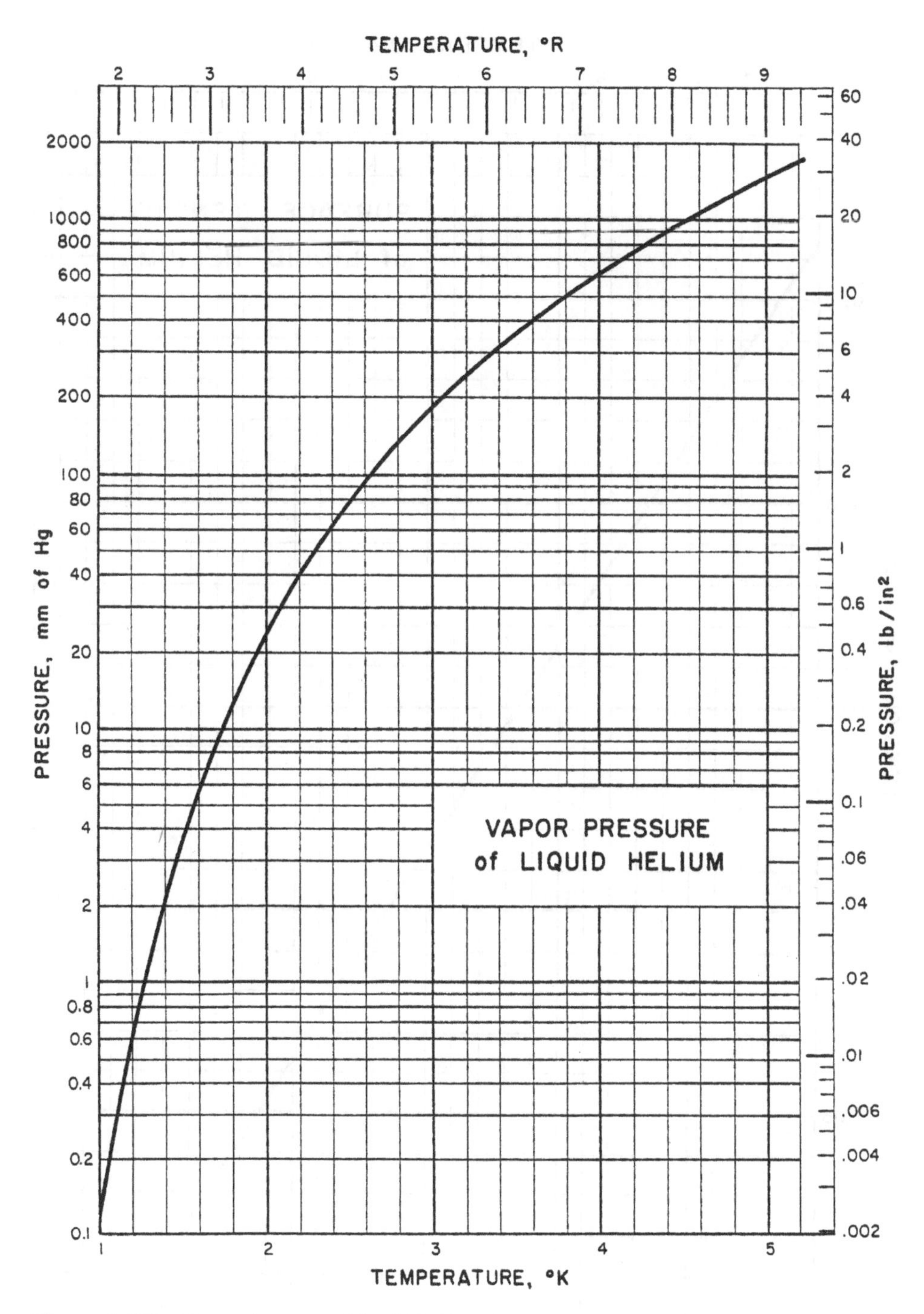


**Figure 3.136** Vapor pressure of liquid helium.

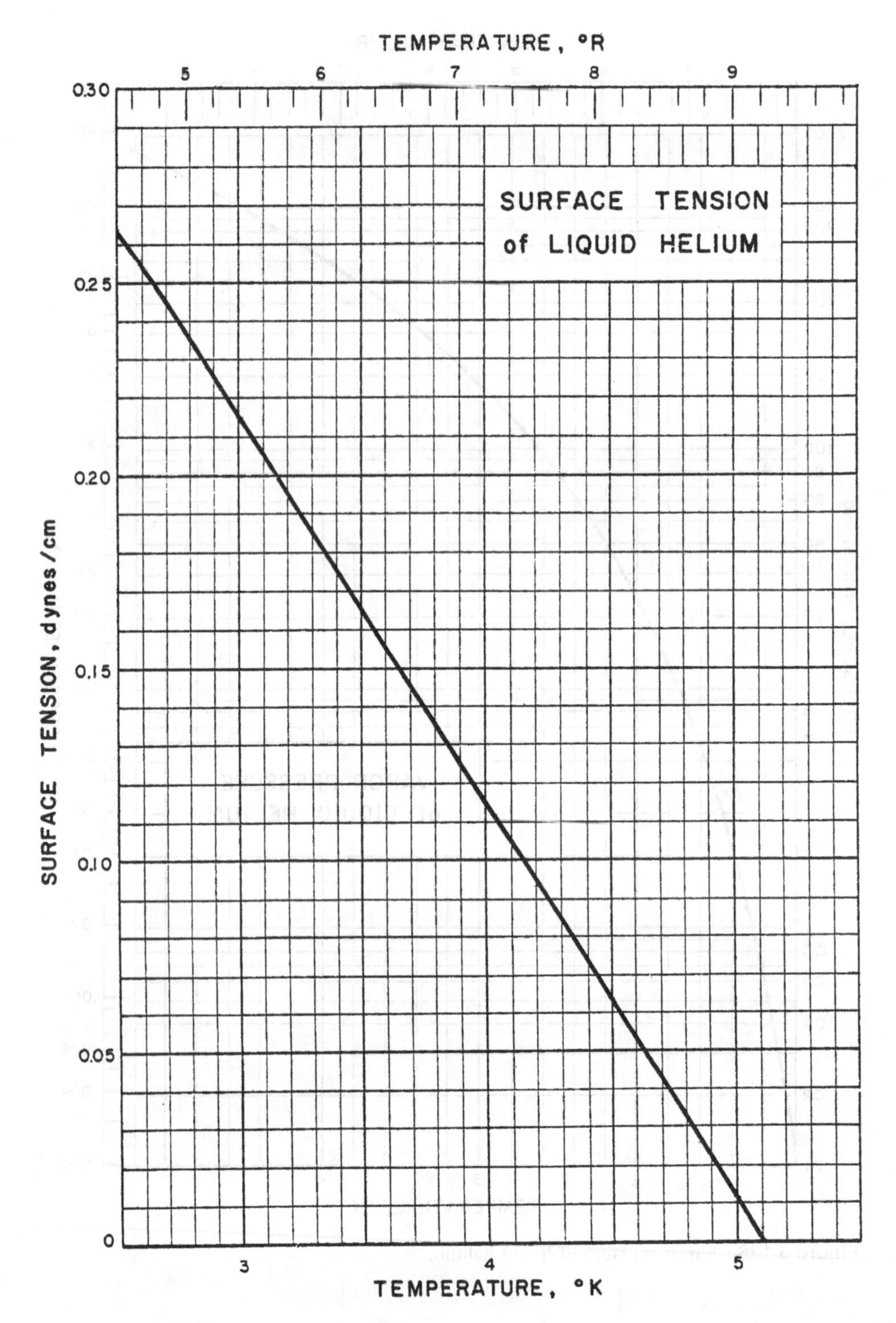


**Figure 3.137** Surface tension of liquid helium.

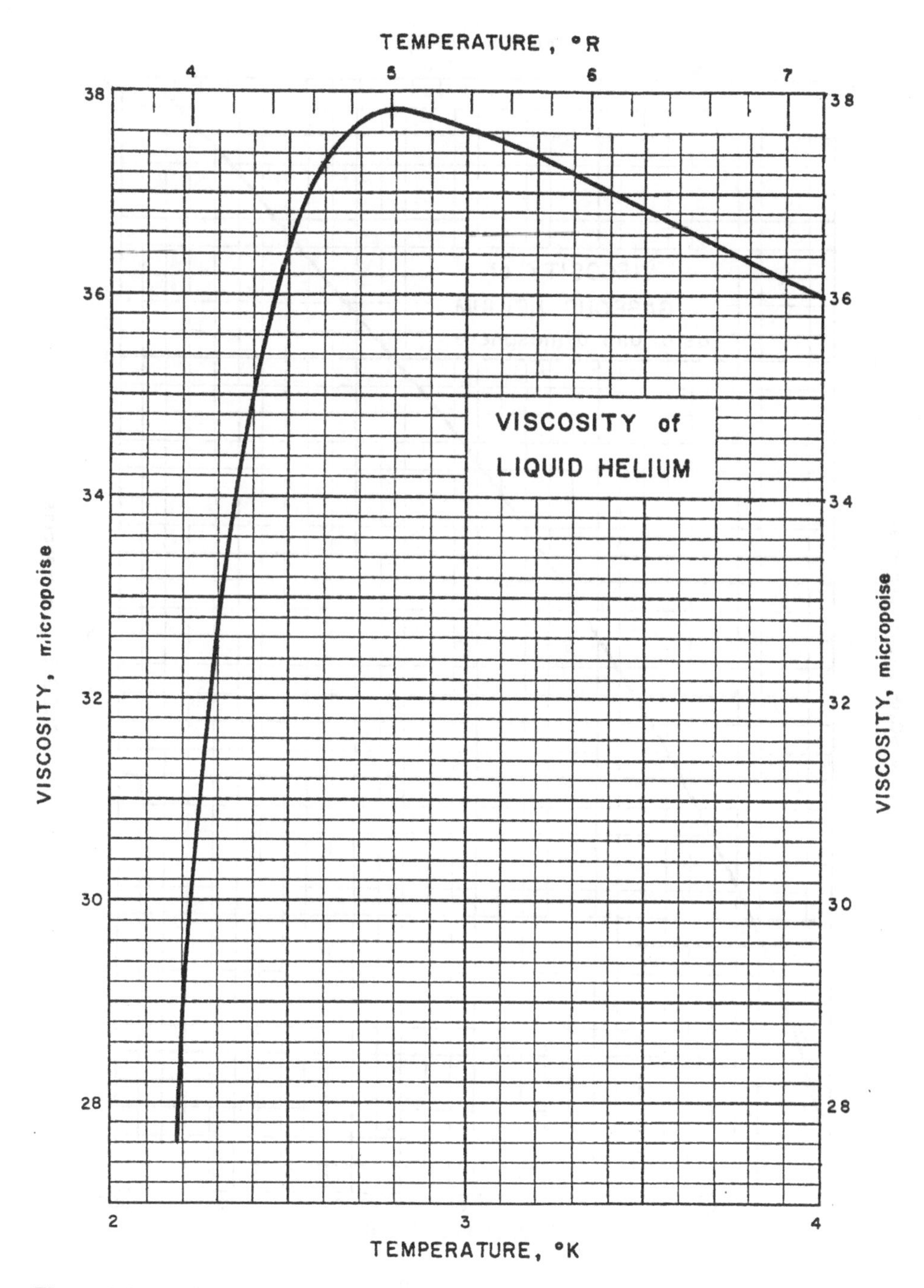


**Figure 3.138** Viscosity of liquid helium.

in equilibrium with its vapor at a given pressure is significantly colder than liquid helium-4 at the same pressure. In an ordinary dewar system, it is difficult to reduce the helium vapor pressure to less than a corresponding temperature of about 0.8 K. With a modest (closed cycle) pumping system, it is not difficult to reach 0.3 K by

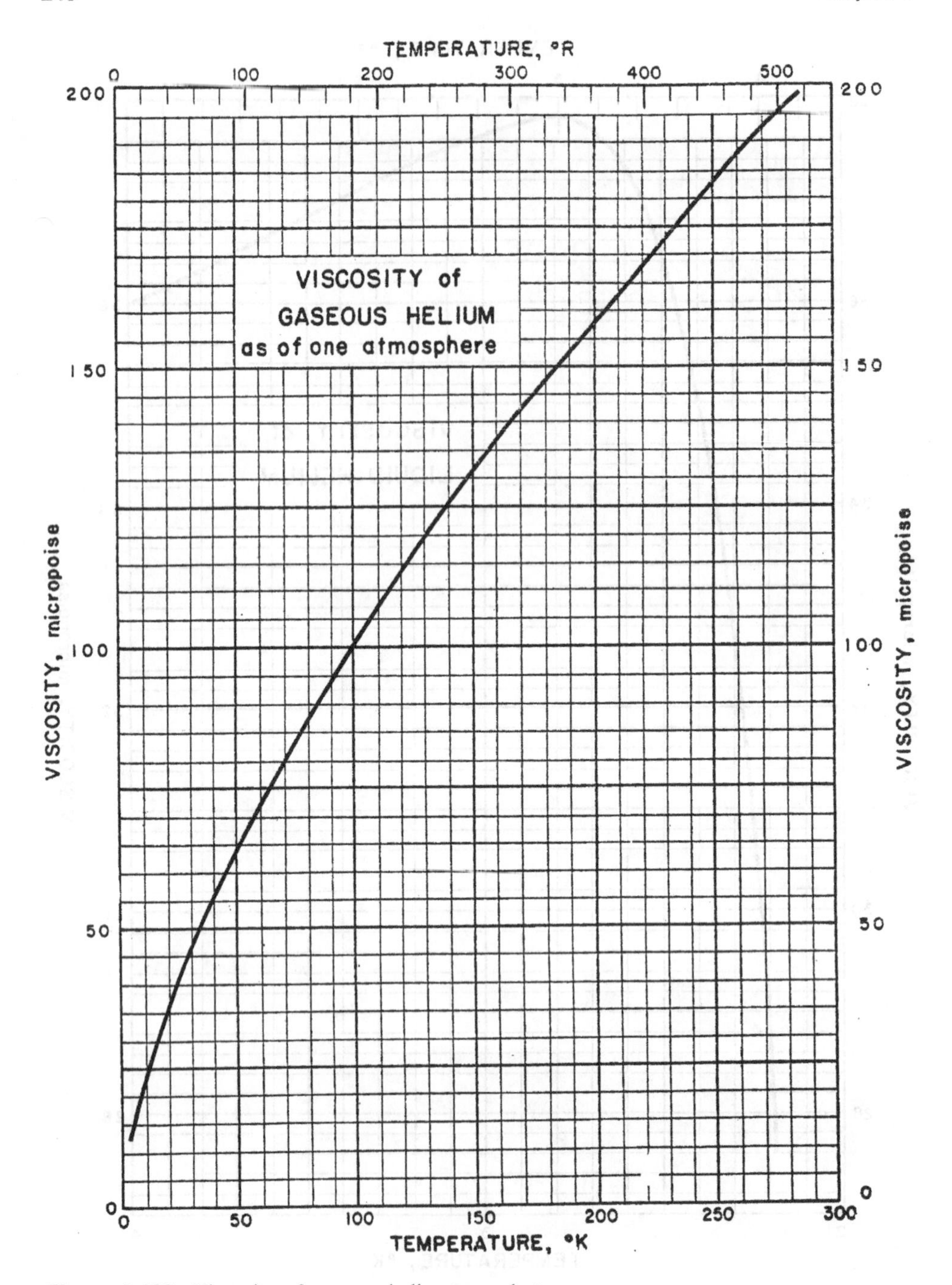


**Figure 3.139** Viscosity of gaseous helium near 1 atm.

using helium-3. This decrease of about 0.5 K may seem small, but in reality it is more like a gain of a factor of about 3 in temperature.

Mixtures of helium-3 and helium-4 are not completely miscible at very low temperatures, as shown in Fig. 3.142. In fact, as absolute zero is approached, helium-3 appears to be completely insoluble in helium-4. At other temperatures, a phase separation occurs in any mixture whose average concentration, at the given

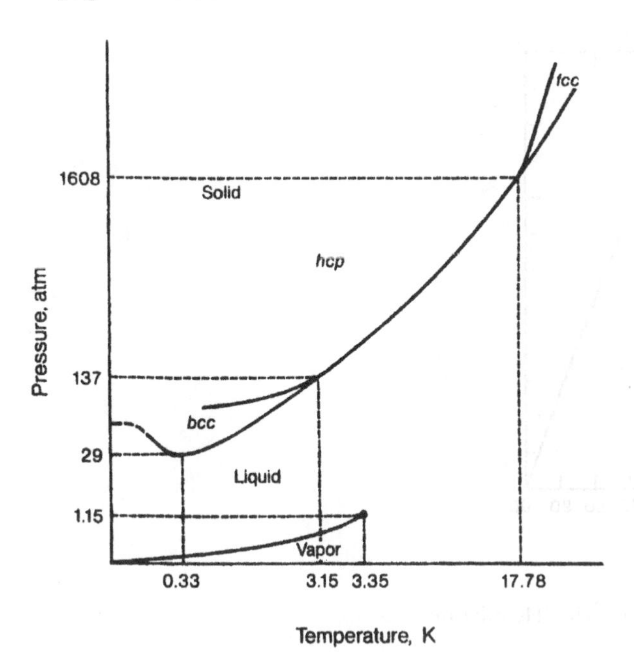


**Figure 3.140** Pressure–temperature diagram for helium-3.

temperature, falls beneath this curve. The concentrations of the separated phases are given by the two intersections of this curve with the (horizontal) line of constant temperature. This separation into two liquid phases, and the difference in vapor pressures, forms the basis for the helium-3/helium-4 dilution refrigerator used to obtain temperatures close to absolute zero.

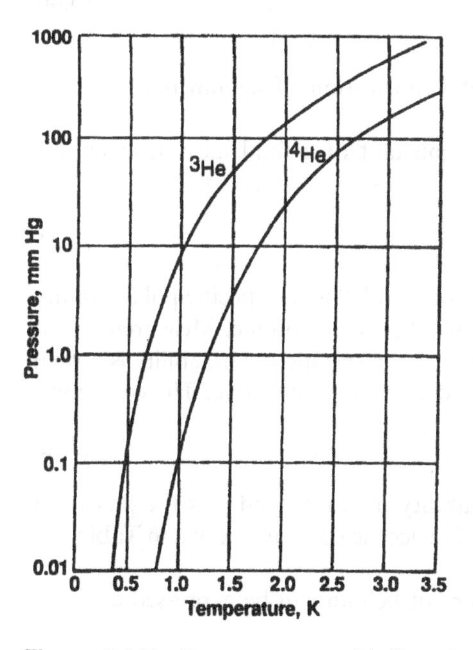


**Figure 3.141** Vapor pressure of helium-3 and helium-4.

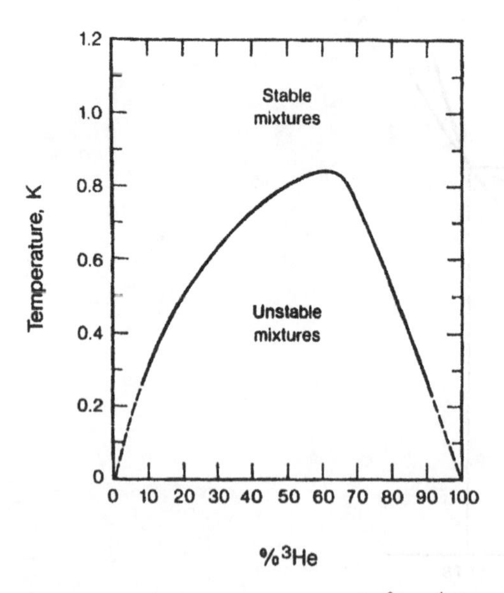


**Figure 3.142** Phase separation in $^3$He–$^4$He mixtures.

### *3.8.3. Properties of Helium*

All helium property data are from McCarty (1972) unless otherwise noted.

**a. Saturation Properties.** The saturation properties of liquid helium-4 are given in Table 3.61 for gaseous and liquid helium. Table 3.62 gives the thermal conductivity of saturated liquid helium for 2.3–2.8 K.

**b. Surface Tension.** The surface tension for helium-4 can be calculated using the equation

$$\gamma = \gamma_0(1 - T/T_c) \tag{3.40}$$

where $\gamma_0 = 0.5308$ dyn/cm and $T_c = 5.2014$.

**c. Heat of Vaporization.** The heat of vaporization of helium is listed for a range of temperatures in Table 3.63.

**d. Dielectric Constant.** The dielectric constant of a fluid may be calculated from the Clausius–Mossotti equation

$$\frac{\varepsilon - 1}{\varepsilon + 2}\left(\frac{3M}{4\pi\rho}\right) = P \tag{3.41}$$

where $\varepsilon$ is the dielectric constant, $\rho$ is the density, and $P$ is the specific polarizability, a property of the substance having dimensions of specific volume. Measurements of the dielectric constant indicated that for helium-4 the specific polarizability is a weak function of density and that the first density correction is negative. The equation

$$p = 0.123396 - 0.0014\rho \tag{3.42}$$

can be used, where $p$ is the specific polarizability in cm$^3$/g and $\rho$ is the density in g/cm$^3$. The uncertainty of tabulated values of dielectric constant shown in Table 3.61 is estimated to be 0.01%.

**e. Melting Curve.** The melting pressure of helium can be expressed as

$$P = 25.00 + 0.053T^8 \tag{3.43}$$

**Table 3.61** Thermodynamic Properties of Coexisting Liquid and Gaseous Helium

| Temp. (K) | Pressure (atm) | Density[a] $(g/cm^3) \times 100$ | Enthalpy[a] (J/g) | Entropy[a] (J/(g.K)) | Spec. heat (J/(g K)) | | Thermal conductivity (mW/cm. K) | Viscosity[a] $(g/(cm\ s)) \times 10^6$ | Dielectric constant $\varepsilon$ | Prandtl number $N_{pr}$ |
|---|---|---|---|---|---|---|---|---|---|---|
| | | | | | $C_V$ | $C_P$ | | | | |
| 2.177[b] | 0.04969 | 14.62 | | | | | | | | |
| | | 0.1177 | 25.41 | 11.92 | 3.20 | 5.61 | 0.0447 | 5.38 | 1.00044 | 0.675 |
| 2.20 | 0.05256 | 14.61 | 3.276 | 1.671 | 3.10 | 3.16 | 0.148 | 36.1 | 1.05500 | 0.778 |
| | | 0.1235 | 25.51 | 11.85 | 3.20 | 5.63 | 0.0453 | 5.45 | 1.00046 | 0.677 |
| 2.30 | 0.06629 | 14.58 | 3.567 | 1.796 | 2.46 | 2.56 | 0.155 | 36.7 | 1.05486 | 0.605 |
| | | 0.1503 | 25.91 | 11.57 | 3.21 | 5.69 | 0.0480 | 5.76 | 1.00056 | 0.683 |
| 2.40 | 0.08228 | 14.53 | 3.816 | 1.898 | 2.10 | 2.25 | 0.160 | 37.1 | 1.05468 | 0.521 |
| | | 0.1805 | 26.30 | 11.32 | 3.22 | 5.77 | 0.0508 | 6.07 | 1.00067 | 0.690 |
| 2.50 | 0.1008 | 14.48 | 4.045 | 1.986 | 1.93 | 2.13 | 0.165 | 37.3 | 1.05448 | 0.0482 |
| | | 0.2144 | 26.88 | 11.09 | 3.22 | 5.85 | 0.0535 | 6.38 | 1.00079 | 0.698 |
| 2.60 | 0.1219 | 14.42 | 4.269 | 2.068 | 1.85 | 2.10 | 0.168 | 37.3 | 1.05424 | 0.468 |
| | | 0.2521 | 27.04 | 10.87 | 3.23 | 5.93 | 0.0562 | 6.69 | 1.00093 | 0.707 |
| 2.70 | 0.1460 | 14.35 | 4.496 | 2.147 | 1.84 | 2.14 | 0.171 | 37.3 | 1.05399 | 0.468 |
| | | 0.2939 | 27.38 | 10.66 | 3.24 | 6.02 | 0.0589 | 7.01 | 1.00109 | 0.717 |
| 2.80 | 0.1730 | 14.28 | 4.731 | 2.225 | 1.85 | 2.22 | 0.174 | 37.2 | 1.05371 | 0.475 |
| | | 0.3401 | 27.71 | 10.47 | 3.25 | 6.13 | 0.0616 | 7.32 | 1.00126 | 0.728 |
| 2.90 | 0.2033 | 14.20 | 4.975 | 2.304 | 1.88 | 2.31 | 0.176 | 37.0 | 1.05341 | 0.487 |
| | | 0.3908 | 28.03 | 10.28 | 3.26 | 6.24 | 0.0643 | 7.64 | 1.00145 | 0.741 |
| 3.00 | 0.2371 | 14.11 | 5.231 | 2.382 | 1.92 | 2.42 | 0.178 | 36.8 | 1.05309 | 0.500 |
| | | 0.4463 | 28.33 | 10.11 | 3.26 | 6.36 | 0.0671 | 7.96 | 1.00165 | 0.755 |
| 3.10 | 0.2744 | 14.02 | 5.501 | 2.462 | 1.96 | 2.54 | 0.180 | 36.5 | 1.05274 | 0.514 |
| | | 0.5070 | 28.61 | 9.937 | 3.27 | 6.49 | 0.0699 | 8.29 | 1.00188 | 0.770 |
| 3.20 | 0.3156 | 13.93 | 5.783 | 2.542 | 2.00 | 2.67 | 0.182 | 36.2 | 1.05238 | 0.529 |
| | | 0.5731 | 28.87 | 9.774 | 3.28 | 6.64 | 0.0726 | 8.62 | 1.00212 | 0.787 |
| 3.30 | 0.3607 | 13.83 | 6.081 | 2.624 | 2.04 | 2.80 | 0.184 | 35.8 | 1.05199 | 0.545 |
| | | 0.6452 | 29.11 | 9.616 | 3.28 | 6.80 | 0.0755 | 8.95 | 1.00239 | 0.807 |

*(Continued)*

**Table 3.61** (*Continued*)

| Temp. (K) | Pressure (atm) | Density[a] $(g/cm^3) \times 100$ | Enthalpy[a] (J/g) | Entropy[a] (J/(g.K)) | Spec. heat (J/(g K)) | | Thermal conductivity (mW/cm. K) | Viscosity[a] $(g/(cm\,s)) \times 10^6$ | Dielectric constant $\varepsilon$ | Prandtl number $N_{pr}$ |
|---|---|---|---|---|---|---|---|---|---|---|
| | | | | | $C_V$ | $C_P$ | | | | |
| 3.40 | 0.4100 | 13.72 | 6.393 | 2.706 | 2.07 | 2.95 | 0.186 | 35.5 | 1.05157 | 0.562 |
| | | 0.7235 | 29.32 | 9.463 | 3.29 | 6.98 | 0.0784 | 9.30 | 1.00268 | 0.828 |
| 3.50 | 0.4637 | 13.60 | 6.722 | 2.790 | 2.11 | 3.10 | 0.188 | 35.1 | 1.05113 | 0.580 |
| | | 0.8085 | 29.52 | 9.314 | 3.30 | 7.18 | 0.0813 | 9.65 | 1.00300 | 0.852 |
| 3.60 | 0.5220 | 13.48 | 7.068 | 2.875 | 2.14 | 3.28 | 0.190 | 34.7 | 1.05066 | 0.599 |
| | | 0.9008 | 29.69 | 9.169 | 3.30 | 7.40 | 0.0843 | 10.0 | 1.00334 | 0.878 |
| 3.70 | 0.5849 | 13.35 | 7.432 | 2.962 | 2.18 | 3.47 | 0.191 | 34.2 | 1.05016 | 0.621 |
| | | 1.001 | 29.84 | 9.025 | 3.31 | 7.66 | 0.0874 | 10.4 | 1.00371 | 0.908 |
| 3.80 | 0.6528 | 13.21 | 7.816 | 3.050 | 2.21 | 3.68 | 0.192 | 33.8 | 1.04963 | 0.646 |
| | | 1.109 | 29.96 | 8.884 | 3.32 | 7.94 | 0.0906 | 10.7 | 1.00411 | 0.942 |
| 3.90 | 0.7257 | 13.06 | 8.221 | 3.141 | 2.25 | 3.92 | 0.194 | 33.3 | 1.04906 | 0.674 |
| | | 1.228 | 30.06 | 8.743 | 3.33 | 8.27 | 0.0939 | 11.1 | 1.00455 | 0.930 |
| 4.00 | 0.8040 | 12.90 | 8.650 | 3.234 | 2.28 | 4.19 | 0.195 | 32.8 | 1.04845 | 0.707 |
| | | 1.356 | 30.12 | 8.603 | 3.33 | 8.65 | 0.0974 | 11.5 | 1.00503 | 1.02 |
| 4.10 | 0.8878 | 12.73 | 9.105 | 3.330 | 2.32 | 4.51 | 0.196 | 32.3 | 1.04780 | 0.745 |
| | | 1.496 | 30.15 | 8.463 | 3.34 | 9.10 | 0.101 | 11.9 | 1.00555 | 1.07 |
| 4.20 | 0.9772 | 12.54 | 9.508 | 3.429 | 2.36 | 4.88 | 0.196 | 31.8 | 1.04710 | 0.791 |
| | | 1.649 | 30.14 | 8.322 | 3.35 | 9.63 | 0.105 | 12.4 | 1.00612 | 1.13 |
| 4.224 | 1.000 | 12.50 | 9.711 | 3.454 | 2.37 | 4.98 | 0.196 | 31.7 | 1.04692 | 0.884 |
| | | 1.689 | 30.13 | 8.287 | 3.35 | 9.78 | 0.106 | 12.5 | 1.00626 | 1.15 |
| 4.30 | 1.073 | 12.35 | 10.10 | 3.532 | 2.40 | 5.32 | 0.197 | 31.3 | 1.04635 | 0.847 |
| | | 1.818 | 30.08 | 8.177 | 3.35 | 10.3 | 0.110 | 12.8 | 1.00674 | 1.20 |
| 4.40 | 1.174 | 12.13 | 10.66 | 3.640 | 2.44 | 5.86 | 0.197 | 30.7 | 1.04553 | 0.916 |
| | | 2.005 | 29.98 | 8.029 | 3.36 | 11.1 | 0.114 | 13.3 | 1.00744 | 1.28 |
| 4.50 | 1.282 | 11.89 | 11.25 | 3.753 | 2.49 | 6.55 | 0.199 | 30.1 | 1.04463 | 0.994 |
| | | 2.213 | 29.81 | 7.876 | 3.37 | 12.1 | 0.128 | 13.8 | 1.00821 | 1.39 |
| 4.60 | 1.397 | 11.63 | 11.90 | 3.873 | 2.54 | 7.44 | 0.200 | 29.5 | 1.04364 | 1.10 |

| | | | | | | | | | | |
|---|---|---|---|---|---|---|---|---|---|---|
| | | 2.449 | 29.58 | 7.714 | 3.37 | 13.5 | 0.127 | 14.3 | 1.00909 | 1.52 |
| 4.70 | 1.519 | 11.34 | 12.61 | 4.002 | 2.59 | 8.68 | 0.202 | 28.9 | 1.04252 | 1.24 |
| | | 2.179 | 29.25 | 7.540 | 3.38 | 15.5 | 0.135 | 14.8 | 1.01010 | 1.70 |
| 4.80 | 1.648 | 11.01 | 13.40 | 4.144 | 2.66 | 10.5 | 0.205 | 28.2 | 1.04125 | 1.45 |
| | | 3.037 | 28.80 | 7.350 | 3.38 | 18.5 | 0.147 | 15.5 | 1.01128 | 1.95 |
| 4.90 | 1.784 | 10.61 | 14.30 | 4.304 | 2.73 | 13.6 | 0.210 | 27.4 | 1.03974 | 1.78 |
| | | 3.425 | 28.18 | 7.133 | 3.39 | 23.6 | 0.164 | 16.2 | 1.01273 | 2.32 |
| 5.00 | 1.929 | 10.11 | 15.39 | 4.495 | 2.81 | 19.9 | 0.222 | 26.4 | 1.03785 | 2.37 |
| | | 3.930 | 27.28 | 6.871 | 3.39 | 34.6 | 0.198 | 17.0 | 1.01461 | 2.97 |
| 5.10 | 2.082 | 9.489 | 16.82 | 4.747 | 2.92 | 38.5 | 0.263 | 25.1 | 1.03520 | 3.68 |
| | | 4.680 | 25.83 | 6.513 | 3.37 | 71.5 | 0.301 | 18.1 | 1.01742 | 4.30 |
| 5.201[b] | 2.245 | 6.964 | 21.36 | 5.589 | | | | | | |
| | | 6.964 | 21.36 | 5.589 | | | | | | |

[a] Top line of data = saturated liquid properties; bottom line data = saturated vapor properties.
[b] Lambda point and critical point
*Source:* McCarty (1972).

**Table 3.62** Thermal Conductivity of Saturated Liquid Helium

| Temp. (K) | Thermal conductivity (mW/(cm K)) | Temp. (K) | Thermal conductivity (mW/(cm K)) |
|---|---|---|---|
| 2.3 | 0.181 | 3.0 | 0.214 |
| 2.4 | 0.185 | 3.5 | 0.238 |
| 2.6 | 0.195 | 4.0 | 0.262 |
| 2.8 | 0.205 | 4.2 | 0.271 |

*Source:* Grenier (1951).
*Note:* Thermal conductivity is a linear function of temperature between 2.5 and 4.5 K.

**Table 3.63** Heat of Vaporization of Helium

| Temp. (K) | $hv$(J/g) | Temp. (K) | $hv$(J/g) |
|---|---|---|---|
| 2.20 | 22.8 | 4.00 | 21.9 |
| 2.40 | 23.1 | 4.20 | 20.9 |
| 2.60 | 23.3 | 4.40 | 19.7 |
| 2.80 | 23.5 | 4.60 | 18.0 |
| 3.00 | 23.7 | 4.80 | 15.6 |
| 3.20 | 23.6 | 5.00 | 12.0 |
| 3.40 | 23.5 | 5.10 | 8.99 |
| 3.60 | 23.2 | 5.15 | 6.70 |
| 3.80 | 22.7 | 5.18 | 4.00 |

*Source*: Berman and Mate (1958).

from 1.0 to 1.4 K and as

$$\frac{P}{16.45} = \left(\frac{T}{0.992}\right)^{1.554} - 1 \tag{3.44}$$

above 4 K, where $P$ is expressed in atmospheres.

**f. Vapor Pressure.** The saturated vapor curve of helium is used as the international temperature scale from 0.5 to 5.2 K. The 1958 international scale is reproduced in part in Table 3.64.

**g. Specific Heat.** The specific heat of helium at constant pressure ($C_P$) and at constant volume ($C_V$) is listed in Tables 3.65 and 3.66, respectively.

*Symbols:*, $\rho_{\text{sat}}$ = saturation pressure or vapor pressure, $\rho$ = density, $C_P$ = speci-specific heat at constant pressure, $\mu$ = viscosity, $k$ = thermal conductivity, $h_{fg}$ = heat of vaporization, $N_{pr} = \mu C_P/k$ = Prandtl number, $\sigma_L$ = surface tension, $\varepsilon$ = dielectric constant.

### *3.8.4. Uses of Helium*

Except for hydrogen, helium is the most abundant element found throughout the universe. The helium content of the earth's atmosphere is about 1 part in 200,000. While it is present in various radioactive minerals as a decay product, the bulk of the Western world's supply is obtained from wells in Texas, Oklahoma, and Kansas.

**Table 3.64** Vapor Pressure of Helium-4 P in $10^{-3}$ mmHg (Hg at 0°C) $g = 980.665\,cm/sec^2$

| $T$ (K) | 0.00 | 0.01 | 0.02 | 0.03 | 0.04 | 0.05 | 0.06 | 0.07 | 0.08 | 0.09 |
|---|---|---|---|---|---|---|---|---|---|---|
| 0.5 | 0.016342 | 0.022745 | 0.031287 | 0.042561 | 0.057292 | 0.076356 | 0.10081 | 0.13190 | 0.17112 | 0.22021 |
| | 0.28121 | 0.35649 | 0.44877 | 0.56118 | 0.69729 | 0.86116 | 1.0574 | 1.2911 | 1.5682 | 1.8949 |
| | 2.2787 | 2.7272 | 3.2494 | 3.8549 | 4.5543 | 5.3591 | 6.2820 | 7.3365 | 8.5376 | 9.9013 |
| | 11.445 | 13.187 | 15.147 | 17.348 | 19.811 | 22.561 | 25.624 | 29.027 | 32.800 | 36.974 |
| | 41.581 | 46.656 | 52.234 | 58.355 | 85.059 | 72.386 | 80.382 | 89.093 | 98.567 | 108.853 |
| 1.0 | 120.000 | 132.070 | 145.116 | 159.198 | 174.375 | 190.711 | 208.274 | 227.132 | 247.350 | 269.006 |
| | 292.169 | 316.923 | 343.341 | 371.512 | 401.514 | 433.437 | 467.365 | 503.396 | 541.617 | 582.129 |
| | 625.025 | 670.411 | 718.386 | 769.057 | 822.527 | 878.916 | 938.330 | 1000.87 | 1066.67 | 1135.85 |
| | 1208.51 | 1284.81 | 1364.83 | 1448.73 | 1536.61 | 1628.62 | 1724.91 | 1825.58 | 1930.79 | 2040.67 |
| | 2155.35 | 2274.99 | 2399.73 | 2529.72 | 2665.08 | 2805.99 | 2952.60 | 3105.04 | 3263.48 | 3428.07 |
| 1.5 | 3598.97 | 3776.32 | 3960.32 | 4151.07 | 4348.79 | 4553.58 | 4765.68 | 4985.18 | 5212.26 | 5447.11 |
| | 5689.88 | 5940.76 | 6199.90 | 6467.42 | 6743.57 | 7028.47 | 7322.31 | 7625.21 | 7937.40 | 8259.02 |
| | 8590.22 | 8931.18 | 9282.06 | 9643.02 | 10014.3 | 10395.9 | 10788.2 | 11191.2 | 11605.1 | 12030.1 |
| | 12466.1 | 12913.7 | 13372.8 | 13843.6 | 14326.1 | 14820.7 | 15327.3 | 15846.3 | 16377.7 | 16921.7 |
| | 17478.2 | 18047.7 | 18630.1 | 19225.5 | 19834.1 | 20455.9 | 21091.1 | 21739.7 | 22402.0 | 23077.9 |
| 2.0 | 23767.4 | 24470.9 | 25188.1 | 25919.2 | 26664.2 | 27423.3 | 28196.3 | 28983.2 | 29784.2 | 30599.1 |
| | 31428.1 | 32271.1 | 33128.0 | 33998.6 | 34882.8 | 35780.3 | 36690.9 | 37614.3 | 38550.2 | 39500.3 |
| | 40465.6 | 41446.6 | 42443.5 | 43456.5 | 44485.7 | 45531.3 | 46593.5 | 47672.5 | 48768.6 | 49881.8 |
| | 51012.3 | 52160.2 | 53325.8 | 54509.2 | 55710.5 | 56930.0 | 58167.8 | 59423.8 | 60698.8 | 61992.0 |
| | 63304.3 | 64635.2 | 65985.4 | 67354.8 | 68743.5 | 70152.0 | 71580.2 | 73028.1 | 74496.0 | 75984.2 |
| 2.5 | 77493.1 | 79022.2 | 80572.2 | 82142.9 | 83734.6 | 85347.2 | 86981.2 | 88636.7 | 90313.8 | 92012.6 |
| | 93733.1 | 95476.0 | 97240.8 | 99028.2 | 100838 | 102669 | 104525 | 106403 | 108304 | 110228 |
| | 112175 | 114145 | 116139 | 118156 | 120198 | 122263 | 124353 | 126465 | 128603 | 130765 |
| | 132952 | 135164 | 137401 | 139663 | 141949 | 144260 | 146597 | 148961 | 151349 | 153763 |
| | 156204 | 158671 | 161164 | 163684 | 166230 | 168802 | 171402 | 174028 | 176682 | 179364 |
| 3.0 | 182073 | 184810 | 187574 | 190366 | 193187 | 196037 | 198914 | 201820 | 204755 | 207719 |
| | 210711 | 213732 | 216783 | 219864 | 222975 | 226115 | 229285 | 232484 | 235714 | 238974 |

(*Continued*)

**Table 3.64** (*Continued*)

| $T$ (K) | 0.00 | 0.01 | 0.02 | 0.03 | 0.04 | 0.05 | 0.06 | 0.07 | 0.08 | 0.09 |
|---|---|---|---|---|---|---|---|---|---|---|
| | 242266 | 245587 | 248939 | 252322 | 255736 | 259182 | 262658 | 266166 | 269706 | 273278 |

(*Continued*)

| $T$ (K) | 0.00 | 0.01 | 0.02 | 0.03 | 0.04 | 0.05 | 0.06 | 0.07 | 0.08 | 0.09 |
|---|---|---|---|---|---|---|---|---|---|---|
| | 276880 | 280516 | 284183 | 287883 | 291615 | 295380 | 299178 | 303008 | 306871 | 310768 |
| | 314697 | 318659 | 322654 | 326684 | 330747 | 334845 | 338976 | 343141 | 374341 | 351575 |
| 3.5 | 355844 | 360147 | 364485 | 368860 | 373269 | 377714 | 382194 | 386710 | 391262 | 395849 |
| | 400471 | 405130 | 409825 | 414556 | 419324 | 424128 | 428968 | 433846 | 438760 | 443713 |
| | 448702 | 453729 | 458794 | 463897 | 469038 | 474218 | 479435 | 484691 | 489985 | 495317 |
| | 500688 | 506098 | 511547 | 517036 | 522564 | 528132 | 533739 | 539387 | 545075 | 550805 |
| | 556574 | 562383 | 568234 | 574126 | 580059 | 586034 | 592051 | 598110 | 604210 | 610352 |
| 4.0 | 616537 | 622764 | 629033 | 635345 | 641700 | 648099 | 654541 | 661026 | 667554 | 674125 |
| | 680740 | 687399 | 694103 | 700851 | 707643 | 714479 | 721360 | 728285 | 735255 | 742269 |
| | 749328 | 756431 | 763579 | 770772 | 778010 | 785294 | 792623 | 799999 | 807422 | 814893 |
| | 822411 | 829978 | 837592 | 845255 | 852966 | 860725 | 868533 | 876390 | 884296 | 892252 |
| | 900258 | 908313 | 916418 | 924573 | 932778 | 941033 | 949338 | 957693 | 966099 | 974556 |
| 4.5 | 983066 | 991628 | 1000239 | 1008905 | 1017621 | 1026390 | 1035213 | 1044087 | 1053014 | 1061995 |
| | 1071029 | 1080144 | 1089254 | 1098449 | 1107699 | 1117002 | 1126359 | 1135772 | 1145239 | 1154761 |
| | 1164339 | 1173972 | 1183662 | 1193407 | 1203209 | 1213066 | 1222981 | 1232955 | 1242983 | 1253069 |
| | 1263212 | 1273414 | 1283673 | 1293991 | 1304367 | 1314802 | 1325297 | 1335850 | 1346462 | 1357136 |
| | 1367870 | 1378662 | 1389516 | 1400429 | 1411404 | 1422438 | 1433533 | 1444690 | 1455911 | 1467191 |
| 5.0 | 1478535 | 1489940 | 1501409 | 1512940 | 1524535 | 1536192 | 1547912 | 1559698 | 1571546 | 1583458 |
| | 1595437 | 1607481 | 1619589 | 1631761 | 1644000 | 1656305 | 1668673 | 1681108 | 1693612 | 1706180 |
| | 1718817 | 1731521 | 1744290 | | | | | | | |

*Source*: Brickwedde et al. (1960).

**Table 3.65** Specific Heat ($C_P$) of Helium (cal/g K)

| Temp. (K) | 3 atm | 5 atm | 6 atm | 10 atm | 15 atm | 30 atm | 50 atm | 70 atm |
|---|---|---|---|---|---|---|---|---|
| 6 | 2.91 | | | 1.18 | 0.97 | 0.77 | 0.67 | 0.615 |
| 6.5 | 2.14 | 3.40 | 2.25 | 1.35 | 1.09 | | | |
| 7 | 1.84 | 2.83 | 2.84 | 1.53 | 1.20 | 0.90 | 0.74 | 0.71 |
| 8 | 1.55 | 1.96 | 2.17 | 1.93 | 1.41 | 1.02 | 0.87 | 0.79 |
| 9 | 1.46 | 1.67 | 1.78 | 1.96 | 1.60 | 1.14 | 0.91 | 0.87 |
| 10 | 1.40 | 1.53 | 1.61 | 1.81 | 1.71 | 1.25 | 1.05 | 0.95 |
| 12 | | 1.42 | 1.46 | 1.59 | 1.64 | 1.40 | 1.18 | 1.08 |
| 14 | | 1.37 | 1.39 | 1.47 | 1.52 | 1.47 | 1.28 | 1.18 |
| 16 | | 1.34 | 1.35 | 1.40 | 1.45 | 1.48 | 1.37 | 1.22 |
| 18 | | 1.32 | 1.33 | 1.36 | 1.41 | 1.46 | 1.41 | 1.32 |
| 20 | | | 1.31 | 1.34 | 1.38 | 1.44 | 1.43 | 1.37 |

*Source*: Lounasmaa (1958).

The only helium plant in the western world outside the United States is near Swift Current, Saskatchewan.

Helium is a low-density inert gas with a combination of unique properties that make its availability essential for a wide range of industrial and scientific endeavors. These include applications not only in cryogenics, but also in special welding technology, space exploration, chromatography, heat transfer, and controlled atmosphere.

**a. Pressurizing and Purging.** Most of the helium for pressurizing and purging is used in the space program. Helium is used to pressurize liquid oxygen, liquid hydrogen, kerosene, and the hypergolic propellants. Helium is used also to pressurize pneumatic systems on boosters and spacecraft. A Saturn booster such as that used on the Apollo lunar missions requires about $13 \times 10^6\ ft^3$ of helium for a firing plus more for checkouts. At the height of the Apollo program, almost one-third of all domestic helium production was used for this purpose.

**Table 3.66** Specific Heat ($C_V$) of Helium (cal/(g K))

| Temp. (K) | 3 atm | 5 atm | 10 atm | 20 atm | 40 atm | 60 atm | 80 atm | 100 atm |
|---|---|---|---|---|---|---|---|---|
| 3 | | 0.444 | 0.408 | 0.365 | 0.318 | | | |
| 4 | 0.555 | 0.534 | 0.504 | 0.464 | 0.415 | 0.382 | 0.358 | 0.345 |
| 5 | 0.621 | 0.604 | 0.575 | 0.538 | 0.492 | 0.459 | 0.434 | 0.414 |
| 6 | 0.746 | 0.664 | 0.626 | 0.593 | 0.552 | 0.523 | 0.501 | 0.480 |
| 6.5 | 0.751 | | | | | | | |
| 7 | 0.737 | 0.719 | 0.669 | 0.633 | 0.600 | 0.574 | .0555 | |
| 8 | 0.738 | 0.730 | 0.699 | 0.664 | 0.634 | 0.613 | 0.596 | |
| 10 | 0.739 | 0.736 | 0.722 | 0.701 | 0.680 | 0.667 | 0.654 | |
| 12 | 0.740 | 0.740 | 0.732 | 0.721 | 0.712 | 0.700 | 0.796 | |
| 16 | 0.741 | | 0.742 | 0.740 | 0.736 | 0.730 | 0.730 | |
| 20 | | | 0.750 | 0.752 | 0.755 | 0.758 | 0.760 | |

*Source*: Lounasmaa (1958).

**b. Controlled Atmospheres.** Helium is used to maintain a controlled atmosphere for cooling vacuum furnaces in processing fuel elements for nuclear reactors and for producing germanium and crystal growth.

**c. Research.** Uses include aerodynamic research in wind tunnels and shock tubes to stimulate velocities up to about 50 times the speed of sound, the development of improved seals and valves for positive shutoff and leak proof operation, pharmaceutical and biological research, particle investigations in physics, development of improved light sources, radiation detection devices, plasma arc studies, mass transfer studies, tensile and impact tests of materials, solid-state physics, and the development of new analytical techniques.

**d. Welding.** Helium alone or mixed with other gases, usually argon, is used as a shield in the high-speed mass production welding of tubes from strip material; ship, aircraft, spacecraft, and rocket structures; food handling equipment; diesel engine parts; hardware; electrical devices; storage tanks; vessels and piping; and other products made from stainless steel, 9% nickel steel, aluminum, copper, titanium, zirconium, and other metals.

**e. Lifting Gas.** Until after World War II, the principal use of helium was for the inflation of military lighter-than-air craft. Although such craft are no longer in operation, helium continues to be used as a lifting gas because it is much safer than hydrogen. Although its density is almost twice that of hydrogen, it has about 98% of the lifting power of hydrogen. At sea level, $28\,m^3$ ($1000\,ft^3$) of helium lifts 31 kg ($68.5\,lb_m$) while the same volume of hydrogen lifts 35 kg ($76\,lb_m$). The Weather Bureau uses helium in about half of its weather balloon operations. Commercial uses of helium include its use in blimps for advertising purposes and in special balloons used to remove logs from mountains in inaccessible locations.

# 4

# Mechanical Properties of Solids

## 1. INTRODUCTION

A knowledge of the properties and behavior of materials used in any cryogenic system is essential for proper design considerations. Often the choice of materials for the construction of cryogenic equipment will be dictated by consideration of mechanical and physical properties such as thermal conductivity (heat transfer along a structural member), thermal expansivity (expansion and contraction during cycling between ambient and low temperatures), and density (weight of the system relative to its volume). Since properties at low temperatures are often significantly different from those at ambient temperature, there is no substitute for test data, and fortunately there are now several excellent data compilations (McClintock and Gibbons, 1960; Durham et al., 1962; Campbell, 1974; Johnson, 1960). To help make sense out of all of the data that do exist, and to help estimate properties when no data are present, it is useful to have certain general rules in mind. Such is the purpose of the following discussion.

In any branch of engineering design, the choice of material is dictated by questions of safety and economy. The range of choice in cryogenic design is perhaps limited by the issue of low-temperature embrittlement. The purpose of this section is to show which materials that are ductile at ordinary temperatures retain their ductility at low temperatures. This determination requires an explanation of why ductility exists at all in some materials and not others and a knowledge of the temperature-related mechanisms that govern ductility. Common indices of strength of materials are also explored along with the temperature dependence of this important property.

## 2. STRENGTH, DUCTILITY, AND ELASTIC MODULUS

When a bar of a structural solid is subjected to an elongating force, the first response is a slight stretching that is directly proportional to the force producing it. If the elongating force is released during this initial period, the material will return to very nearly its original length. This is called elastic behavior, and all solid materials exhibit it in some measure. At some higher value of stress, however, the material will no longer behave elastically. It may break without further deformation (brittle behavior), or it may take on a permanent deformation (ductile behavior).

The ductile or brittle behavior of structural materials is usually determined by the familiar stress–strain test shown in Fig. 4.1. The figure represents a steel bar of cross-section $a$, on which a gauge length $l$ has been marked. Let $P$ be the value of any

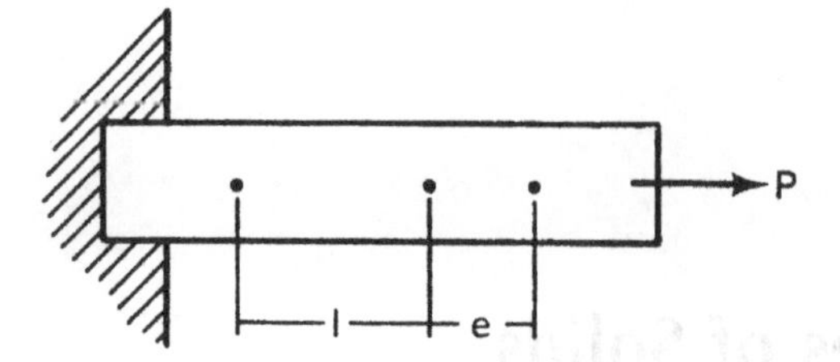


**Figure 4.1** Tensile test of a structural material.

axial tensile load that gradually increases from zero value until the steel bar breaks. As the value of $P$ increases, values of the elongation $e$ are taken.

Stress:

$$S = P/A$$

Strain:

$$\epsilon = e/l$$

The relationship between $s$ and $\epsilon$ is the stress–strain curve found experimentally and is shown in Fig. 4.2.

Another class of materials (nonmetals) are capable of extreme great elastic deformations. These are called elastomeric materials.

Usually, materials exhibiting plastic deformation under stress are the more desirable for structures. Ductility is desirable so that accidental stresses beyond design values can be redistributed to safer levels by means of plastic flow.

Brittle materials have no such mechanism to protect them from excessive stress. When a local overstress occurs in a brittle material, the result is often failure of the piece rather than a deformation. It is useful to know, therefore, which structural materials that are acceptable (ductile) at normal temperatures will remain ductile at low temperatures. This requires that we know the mechanisms of ductility in solids and how those mechanisms change with temperature. To begin, let us examine the structure of solids.

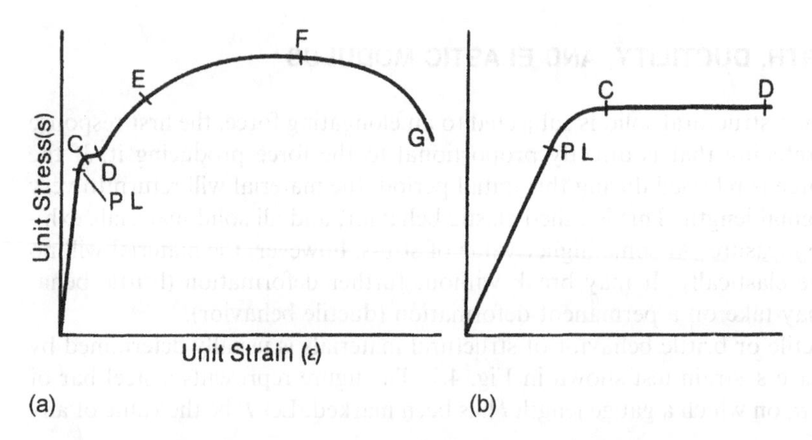


**Figure 4.2** The stress–strain relationship of a ductile material (a) and a brittle material (b).

## 3. THE STRUCTURE OF SOLIDS

The theory of elasticity treats solids as continuous elastic media. Solids are not continuous media. They are composed of atoms bound together in more or less regular arrays. Essentially, three different types of solid structures exist, represented by glasses, plastics, and metals.

### 3.1. Glasses

The structure of an idealized glass is shown in Fig. 4.3 (Jones, 1956). The chemical composition of this glass is $G_2O_3$, where G represents a metallic element and O represents oxygen. The principal characteristic of the glass to be noted is the lack of any spatial order, even though the stoichiometric relationship between G and O is observed. Liquids also show such lack of order, and glasses can be described to a first approximation as extremely viscous liquids.

### 3.2. Plastics

Plastics, or polymers, are composed of giant long-chain molecules. Polymer molecules may have from tens to thousands of atoms each and are essentially linear. They lie in tangled disarray; a segment of a typical polymer molecule is shown in Fig. 4.4. The intermolecular force that binds the polymer molecules to one another is the rather weak van der Waals force. Some cross-linking among the polymer molecules may also be present. Cross-linking is actual chemical bonding between adjacent molecules and is much stronger than the van der Waals forces. The more highly cross-linked (thermoset) polymers form much more rigid solids than those that are only sparingly cross-linked (thermoplastics).

### 3.3. Metals

Compared to glasses and polymers, metals have a highly ordered structure. Metal atoms are arranged in symmetrical crystal lattices. The three most common metal crystal lattices are shown in Fig. 4.5 (Quarrel, 1959). The face-centered cubic (fcc)

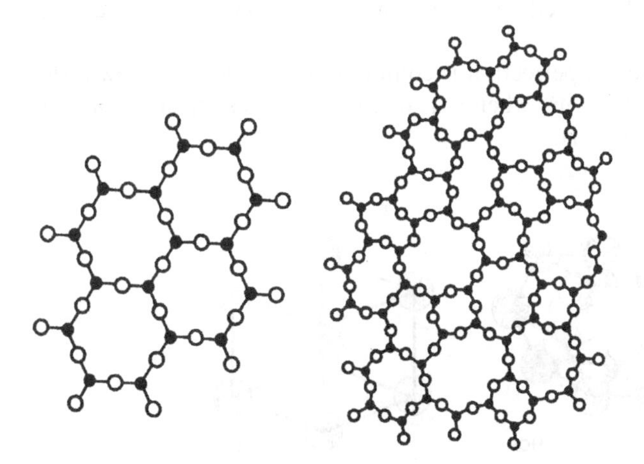

**Figure 4.3** Idealized structure of glass.

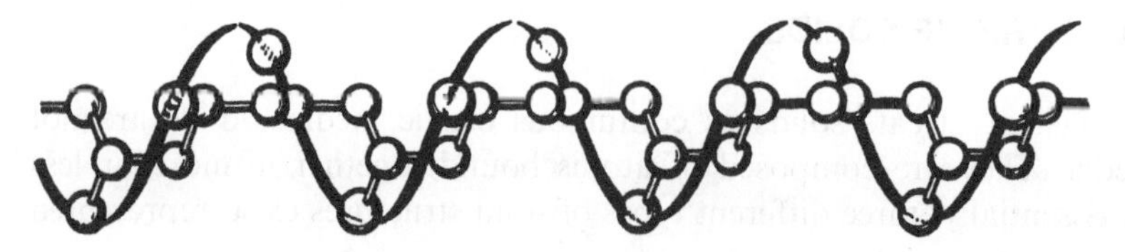

**Figure 4.4** Structure of a representative polymer.

lattice consists of a cube that has an atom at each of its eight corners and one at the center of each of the six crystal faces. The hexagonal close-packed (hcp) lattice consists of a right hexagonal prism with an atom at each of its 12 vertices, an atom at the center of each of its two regular hexagonal ends, and three more atoms located midway between the ends of the prism, each on a line parallel with the vertical axis of the prism and running through the center of the equilateral triangle formed by three neighboring atoms in the end faces. The body-centered cubic (bcc) lattice is a cube with an atom at each of its eight corners and an atom at the center of the body of the cube. Other more complicated metal structures exist, but only these three basic types will be considered here.

The drawings of Fig. 4.5 are useful for visualizing the slip planes that exist in various metal crystal systems. For example, from the fcc lattice shown in Fig. 4.5a, it is easy to see how planes of atoms might slide rather easily over one another from upper left to lower right. This lattice also has several other such slip planes. Although it is not readily apparent from the figure, a three-dimensional model of these crystal systems would show that the bcc lattice offers the least number of slip planes of the three shown, and the hcp system falls in between the fcc and bcc systems.

Real crystals do not have the perfect geometric arrangement of atoms shown in Fig. 4.5 (Van Buren, 1960). For example, there are always some impurity atoms present. These impurity atoms cause irregularities in the crystal lattice. Real crystals may also have locations where atoms are simply out of place. An extra atom in the interstices between regular crystal sites is called an interstitial atom. A missing atom is called a vacancy. Interstitial atoms and vacancies are examples of "point defects" in the crystal.

Extended imperfections can also occur. An extra plane of atoms can be present in a crystal lattice. Extended imperfections of this type are called edge dislocations.

Another type of extended imperfection is common in crystals: the screw dislocation. This type of dislocation occurs when part of a crystal has slipped one atomic distance relative to its adjacent part.

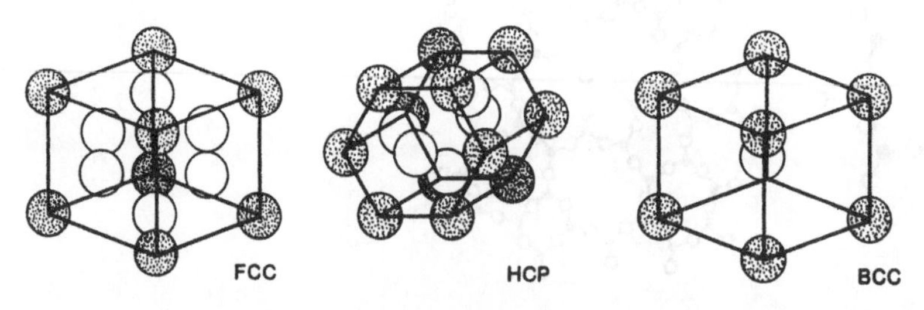


**Figure 4.5** The three most common crystal structures of metals.

Both dislocations and point defects are nearly always present to some degree in real metallic crystals (Cottrell, 1953). The concept of a geometrically perfect crystal is useful in explaining certain properties of crystals, but there are other properties that can be adequately explained only by consideration of the defects. The concept of dislocations, for example, helps explain the yield stress and ductility of metals. The regular arrangement of atoms in the crystal (and not so much its defects) helps explain the modulus of elasticity.

## 4. DUCTILITY

As mentioned, the role of dislocations is useful in explaining the yield strength and ductility of certain crystalline solids.

One could account for elasticity by imagining a regular crystal structure without dislocations such as those of Fig. 4.5. If a certain horizontal shearing stress were applied to the top of the crystal lattice, then the structure would bend, but it would snap back to its original shape when the stress was removed. Above a certain stress, however, plastic shear would suddenly take place along one of the planes of atoms, resulting in a permanent deformation.

The "perfect crystal" was the model originally used to explain the plastic deformation of metals. However, when the stress necessary to shear a perfect crystal was calculated on the basis of interatomic forces, the results were a thousand times greater than the experimentally observed stresses. Taking the presence of dislocations into account gives better results, because the stress required to force a dislocation through a crystal is of the required magnitude (Barrett, 1957). The movement of one plane of atoms over another in a crystal actually occurs progressively as the dislocation moves through the crystal. The resulting crystal deformation is the same as if the process had taken place suddenly. Shearing a perfect crystal lattice structure is like moving a heavy rug all at once; forcing a dislocation through the crystal is like moving a wrinkle through the rug. The displacement of the rug is the same in both cases.

The ability of crystalline materials to deform plastically (their ductility) depends primarily on the mobility of dislocations within the crystal (Fisher, 1957). The mobility of dislocations depends on temperature, as one might expect, and on several other things as well. The number of slip systems available in a crystal influences dislocation mobility. Slip systems are directions within the crystal in which the planes can slip easily over one another. Dislocations can move most easily through a crystal in the direction of slip planes.

The movement of dislocations can also be impeded in many ways. Impurity atoms at lattice sites can lock dislocations in place. Vacancies and interstitial atoms also provide obstacles to the movement of dislocations. The existence of hard foreign particles in the crystal can also impede the mobility of dislocations.

Grain boundaries, which are discontinuities in the orientation of neighboring crystals, also are obstacles. A metal having a small grain size will have a higher yield stress than one of large grain size. One reason annealed metals are more ductile than work-hardened ones is that work hardening reduces the size of grains. Finally, dislocations themselves provide obstacles to the progress of other dislocations.

Glasses have no slip systems and are therefore very brittle. When stress is applied to a piece of glass, the atomic bonds rupture at some point on the surface. The crack then propagates, causing a fracture. The final stage of rupture of a metal is glass-like. After plastic deformation, dislocation motion is blocked by tangles of the great number of dislocations produced by work hardening. Greater stress then ruptures atomic bonds, much like a glass, at some location where the stress is too great for the deformed crystal. Further stress then propagates the crack through the material (fracture).

Plastics and elastomers that are not highly cross-linked can yield considerably to tension stress by merely uncoiling their long molecules. As only van der Waals bonds are ruptured in this process, elongation takes place easily. The long molecules merely slide over each other. When this sliding process is completed, the force of the chemical bonds comes into play.

The resistance of chemical bonds in cross-linked plastics or elastomers comes into play early, and these materials elongate less under stress. Highly cross-linked plastics, in fact, are so brittle that they resemble glass.

## 4.1. Low-Temperature Ductility

The effect of temperature on ductility is, to a first approximation, the effect of temperature on the response mechanisms discussed above—the crystal system (slip planes) and dislocation motion. This will be discussed first for metals, then for elastomers. Embrittlement of these materials at low temperatures is summarized in Table 4.1.

### *4.1.1. Metals*

Since the number of slip systems is not usually a function of temperature, the ductility of fcc metals is relatively insensitive to a decrease in temperature (Brick, 1953). Metals of other crystal lattice types tend to become brittle at low temperatures. Crystal structure and ductility are related because the fcc lattice has more slip systems than the other crystal structures. In addition, the slip planes of bcc and hcp crystals tend to change at low temperature, which is not the case for fcc metals. Therefore, copper, nickel, all of the copper–nickel alloys, aluminum and its alloys, and the

**Table 4.1** Embrittlement of Structural Materials at Low Temperatures

| Materials that remain ductile at low temperatures, if ductile at room temperature | Materials that become brittle at low temperatures |
|---|---|
| Copper | Iron |
| Nickel | Carbon and low alloy steels |
| All copper–nickel alloys | |
| Aluminum and all its alloys | Molybdenum |
| Austenitic stainless steels containing more than 7% nickel | Niobium |
| | Zinc |
| | Most bcc metals |
| Zirconium | Most plastics |
| Titanium | |
| Most fcc metals | |
| Polytetrafluoroethylene | |

austenitic stainless steels that contain more than approximately 7% nickel, which are all face-centered cubic, remain ductile down to low temperatures if they are ductile at room temperature. Iron, carbon and low-alloy steels, molybdenum, and niobium, which are all body-centered cubic, become brittle at low temperatures. The hexagonal close-packed metals occupy an intermediate place between fcc and bcc behavior. Zinc undergoes a transition to brittle behavior in tension, whereas zirconium and pure titanium remain ductile (Corruccini, 1957).

The thermal vibration of atoms in the crystal lattice is strongly temperature-dependent and is less effective in assisting dislocation motion at low temperatures. The interaction of dislocations with thermal vibrations is complicated, but it is nonetheless satisfying to find that ductility usually decreases somewhat with a decrease in temperature.

The complete situation regarding brittleness in metals at low temperature depends on more than just crystal structure. For instance, bcc potassium and beta brass remain ductile down to 4.2 K, and the bcc metals lithium and sodium show no signs of brittleness down to 4.2 K. In general, however, brittleness will not occur in face-centered cubic metals in which dislocations cannot be firmly locked by impurity atoms.

#### *4.1.2. Elastomers*

As mentioned, the tendency to show a ductile–brittle transition is correlated with the lattice type. Thus, the fcc metals show only a few cases of this effect and for structural purposes may be regarded as almost uniformly well behaved. These include copper, nickel, aluminum, the solid solution alloys of each of these, and the austenitic stainless steels. In contrast, bcc metals for the most part show brittle behavior (though the transition zones of some can be depressed to low temperatures). The ferritic steels are by far the most prominent of these. Prominent among the established structural metals with a different lattice are the hexagonal close-packed metals magnesium and titanium. The impact strength of magnesium is low at all subambient temperatures because its brittle transition zone is above room temperature. Tests on commercial titaniums indicate that ductility is retained in tension to low temperatures if the amounts of the interstitial solutes carbon, oxygen, nitrogen, and hydrogen are small. However, notched-bar impact tests show a transition above ambient temperature.

For temperatures much below 200 K, it is common practice to use the face-centered cubic metals almost exclusively, especially where shock and vibration are encountered. However, less expensive steels can be used in many less critical applications, especially for temperatures above 150–200 K.

All but one plastic or elastomer become brittle at low temperatures. Polytetrafluoroethylene (PTFE) is unique in that it can still be deformed plastically to a small degree at 4 K.

Plastics and elastomers do not respond to stress as metals do. The less cross-linked elastomers yield by uncoiling their long-chain molecules and by sliding over one another. The thermal energy of the material at room temperature facilitates this motion.

At low temperatures, however, the attractive intermolecular forces are more effective than the thermal energy "lubricant," and the material deforms less readily. This effect is especially pronounced through its "glass transition" in the temperature range at the onset of brittleness.

## 4.2. Ductility as a Function of Temperature

The curve of ductility (elongation in a tensile test, for example) vs. temperature generally has an S shape similar to that shown in Fig. 4.6. The material shown in Fig. 4.6 has considerable ductility at higher temperatures but is brittle at low temperatures. All materials have a ductility temperature dependence represented by at least some portion of Fig. 4.6. Between temperatures $T_1$ and $T_2$, there is a transition from ductile to brittle behavior. This brittle transition may occur in any temperature range, wide or narrow, or, for some materials, may not occur at all. If the material is either ductile or brittle over the entire temperature range, then the curve consists only of the portion above $T_2$ or below $T_1$. For example, face-centered cubic metals show only the curve above $T_2$, and brittle materials such as glass show only the curve below $T_2$.

Some hard copper alloys appear to undergo a gradual transition over most of the temperature range, and a similar graph for these materials looks like a portion of the region between $T_1$ and $T_2$ but spread out over a considerably greater temperature range.

Ductility is well represented by the entire curve in Fig. 4.6 for steels except for the fcc austenitic types. These austenitic steels display ductility near room temperature but undergo a transition to brittle behavior at some temperature below room temperature.

The brittle transition region is that range of temperatures where the material's important mechanisms of response to stress are becoming inactive. These response mechanisms include the mobility of dislocations for a metal and the ability of the molecules to slide over one another in a polymer.

A particular case of brittle behavior is shown by tensile data on a low-carbon steel in Fig. 4.7 (Corruccini, 1957). Two features are characteristic of these materials:

1. The large decrease in elongation and reduction of area that occur in a relatively narrow region of temperature, the brittle transition zone
2. The rapid rise in yield strength, which approaches the material's tensile strength as the temperature is lowered through the transition zone

In contrast, Fig. 4.8 shows a similar set of properties (Corruccini, 1957) for Type 347 stainless steel, an alloy that is apparently not brittle at any low temperature. The two main characteristics of these alloys are (1) the relative constancy of the yield strength and the increased capacity for work hardening at lower temperatures as shown by the steep rise of the ultimate strength, and (2) the

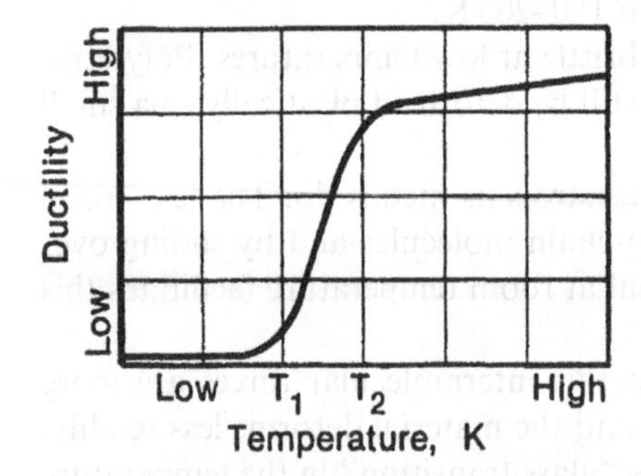


**Figure 4.6** The general curve of ductility vs. temperature.

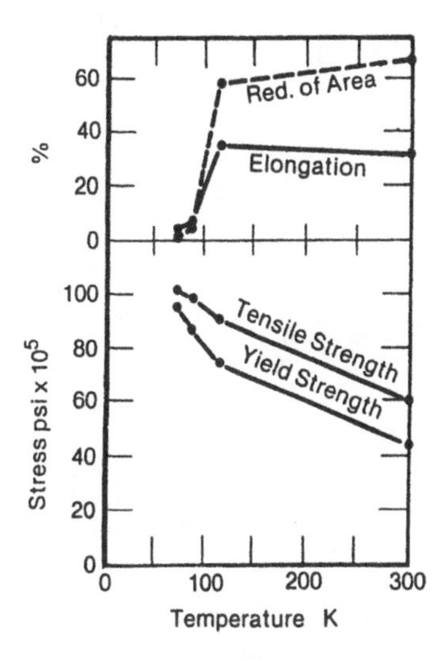


**Figure 4.7** Brittle behavior of a low-carbon steel.

maintenance of ductility at all temperatures as shown by the high values of elongation and reduction of area.

The curves of ductility vs. temperature for several important structural materials are shown in Fig. 4.9 (Durham et al., 1962).

## 4.3. Ductility as a Function of Strain Rate and Stress Complexity

Increasing strain rate and increasing complexity of the stress system have the effect of decreasing ductility. Two tests of mechanical properties, the tensile and impact tests, are used to measure the effect of strain rate and stress complexity on ductility. The tensile test involves unidirectional stresses applied at comparatively slow rates. The impact test applies stresses in several directions at rapid rates. At a given

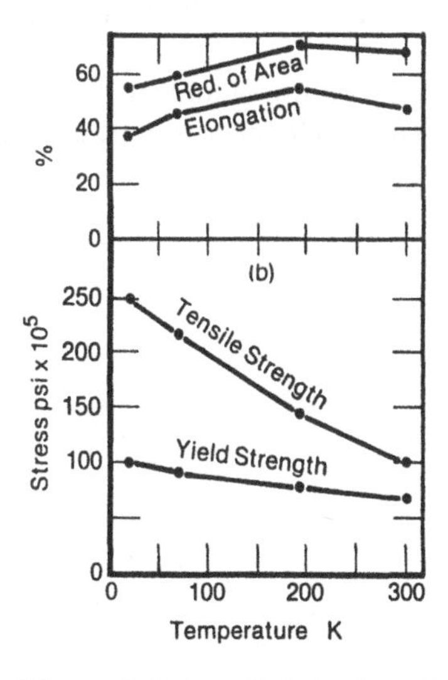


**Figure 4.8** Ductile behavior of a Type 347 stainless steel.

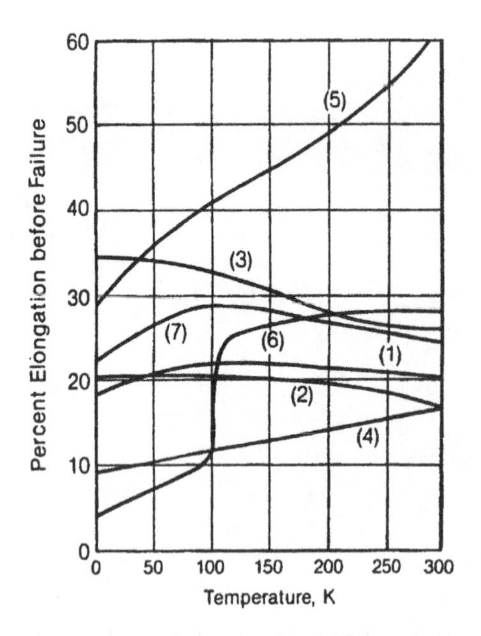


**Figure 4.9** The curve of ductility vs. temperature for several important structural materials. (1) 2024 T4 aluminum; (2) beryllium copper; (3) K Monel; (4) titanium; (5) 304 stainless steel; (6) C1020 carbon steel; (7) 9% nickel steel.

temperature, a material may exhibit considerable ductility in the tensile test but not in the impact test.

Figure 4.10 shows this effect for an ordinary carbon steel. This figure illustrates that the ductility of a material is affected by the type of stress system and the rate of application of this stress system. Between $T_2$ and $T_3$, for example, the carbon steel displays ductile behavior in a simple uniaxial stress system (tensile test) and displays brittle characteristics at high rates of loading (impact test). Increasing either the strain rate or the complexity of the stress system moves the curves to the right. This amounts to an increase in the brittle transition temperature. Uniformly ductile or brittle behavior is observed above $T_4$ and below $T_1$.

The tensile elongation and the impact energy vs. temperature of annealed oxygen-free copper are shown in Fig. 4.11. The tensile elongation of copper is quite high

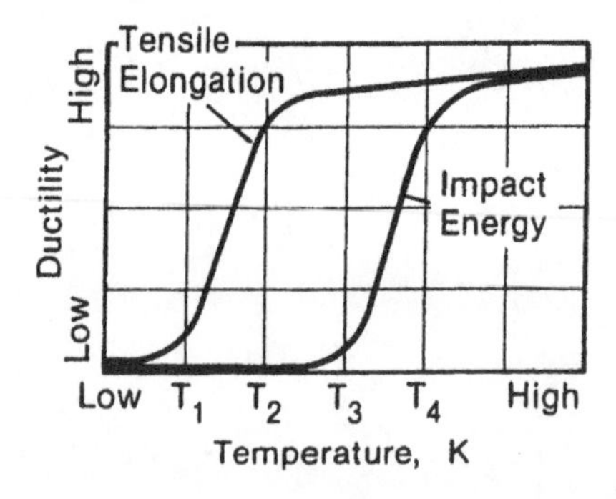


**Figure 4.10** Effect of strain rate and stress complexity.

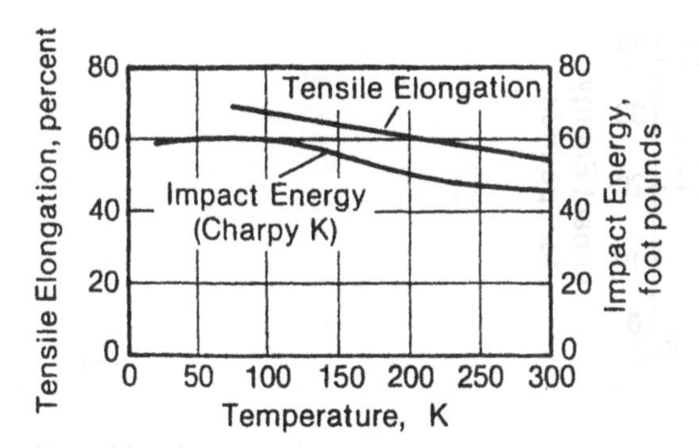


**Figure 4.11** Tensile elongation and impact energy of annealed oxygen-free copper.

at room temperature, and it even increases somewhat at low temperatures. The impact energy curve follows the same trend. Therefore, copper will exhibit no brittle behavior at low temperatures for the usual applications. All of the common copper alloys investigated so far are ductile at low temperatures if they are ductile at higher temperatures (Teed, 1950). Copper alloys that have a predominantly face-centered cubic crystal structure have a ductility largely independent of temperature.

The behavior of another fcc alloy, one of the chromium–nickel stainless steels, is shown in Fig. 4.12. There is a small decrease in tensile elongation and impact energy with decreasing temperature. The ductility of this group of alloys is largely independent of temperature. When the nickel is removed from these stainless steel alloys, they crystallize into the body-centered cubic lattice. Type 430 stainless steel is typical of this group, and its low-temperature behavior is shown in Fig. 4.13. At about 200 K, this alloy has become brittle in the impact test. Below about 75 K, it displays very little ductility even in the tensile test.

These transitions can become very sharp, as shown in Fig. 4.14. The ductility of plain carbon steels can drop precipitously at the transition temperature.

Figure 4.15 shows the Charpy impact strength at low temperatures for several common structural materials. Note that in general the fcc and hcp materials retain their resistance to impact, whereas the bcc materials tend to become brittle. Curve 6 of Fig. 4.15 shows the effect of a solid–solid phase transition around 140 K.

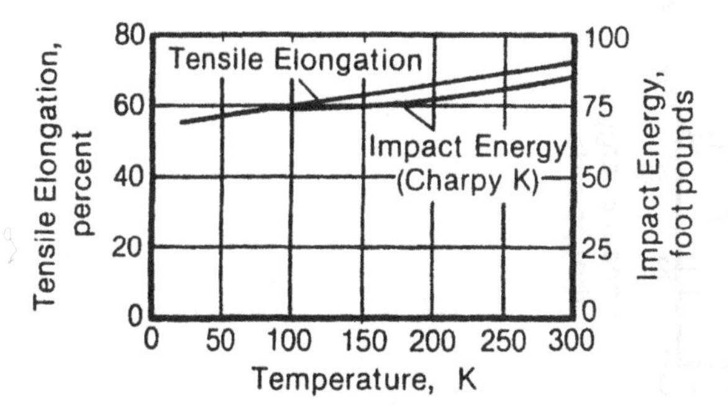


**Figure 4.12** Tensile elongation and impact energy of a chromium–nickel stainless steel (AISI 316, annealed).

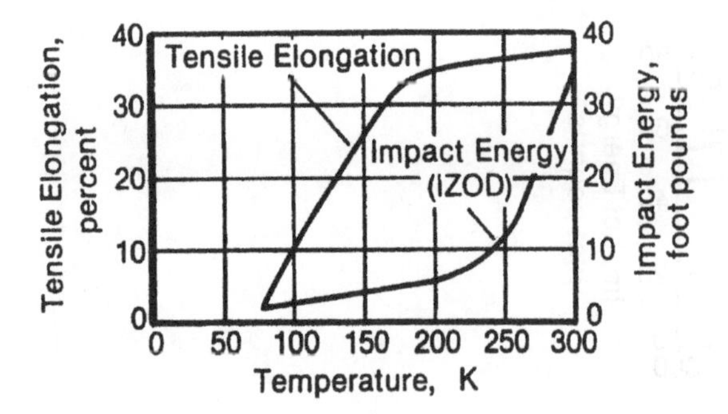


**Figure 4.13** Tensile elongation and impact energy of a chromium stainless steel (AISI 430, annealed).

## 5. LOW-TEMPERATURE STRENGTH OF SOLIDS

In general, the *ultimate tensile strength* of a solid material is greater at low temperatures than it is at ordinary temperatures. This is true for both crystalline and noncrystalline solids and for many heterogeneous materials as well (e.g., glass-reinforced plastics). For metals, the strength at 4 K may be 2–5 times that at room temperature. For plastics, the strength at 76 K may be 1.5–8 times greater than the room-temperature value. Glasses show less change in strength at low temperatures; at 76 K, glasses have between 1.5 and 2 times their room-temperature strength.

Reduction of the thermal energy of the metal lattice at lower temperatures is responsible for part of the increase in the strength of metals. The decreased thermal vibration at low temperatures also results in stronger plastics, because it becomes more difficult for the long-chain molecules to slide over one another.

The change in strength of glass at low temperatures is not as great as that of metals and plastics. The important cohesive forces of glasses are the chemical bonds between atoms (rather than the intermolecular forces of plastics). Such chemical bond forces are less affected by temperature.

There are exceptions to the above general rule. The thermal and mechanical history of a material, various solid-state phase changes, the form of stress system, and the rate of application of this stress system can also affect mechanical properties.

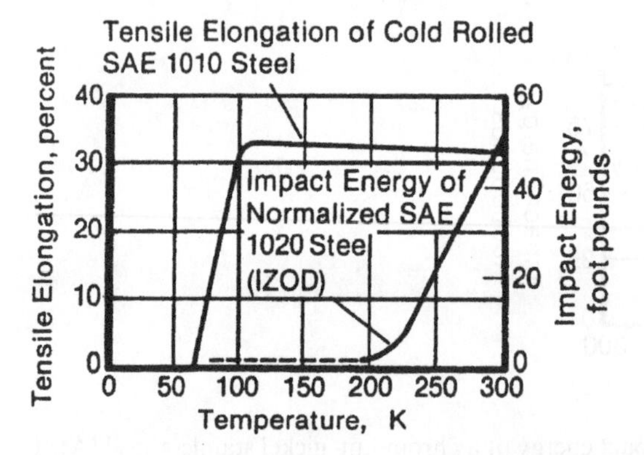


**Figure 4.14** Tensile elongation and impact energy of plain carbon steels.

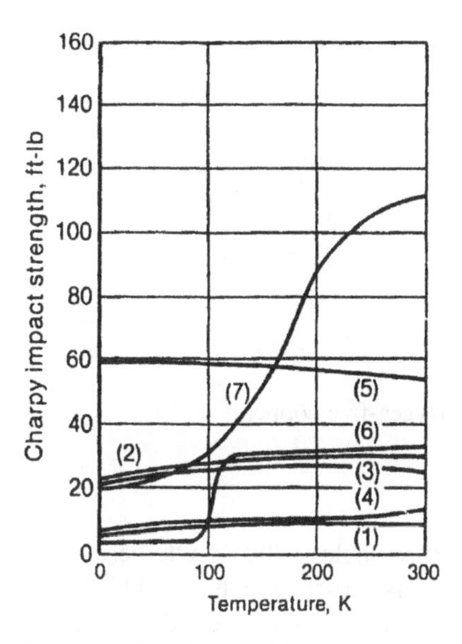


**Figure 4.15** The Charpy impact strength at low temperatures for several common structural materials. (1) 2024 T4 aluminum; (2) beryllium copper; (3) K Monel; (4) titanium; (5) 304 stainless steel; (6) C1020 carbon steel; (7) 9% nickel steel.

Some of the exceptions to the general rule of increasing strength with decreasing temperature can be explained qualitatively by phase changes during deformation. Some austenitic stainless steels undergo a partial transformation under strain from a face-centered cubic to an intermediate hexagonal close-packed phase and then to a body-centered cubic lattice (martensite). The martensite form is stronger than the parent structure and contributes to the high strength of these materials. The tensile strength of austenitic steels shows the usual increase with decreasing temperature to some low temperature and then a decrease, reflecting a maximum in the phase transformation curve.

### 5.1. Metals

The *yield strength* of ductile metals increases with decreasing temperature. The yield strength at 20 K is between one and three times the room-temperature yield strength for fcc and hcp metals. An even greater tendency toward increases in yield strength at low temperatures is shown for bcc metals. In fact, for body-centered cubic steels, the yield strength increases so rapidly with decreasing temperature that it becomes greater than the fracture strength. Thus, below a certain temperature, these materials fracture before they reach their yield strengths.

The *yield* and *ultimate tensile strengths* of annealed copper are shown in Fig. 4.16. This figure shows that the yield strength of a material can be insensitive to temperature while its tensile strength is increasing by more than a factor of 2. The material has "work hardened." That is, the material has hardened itself by generating obstacles to dislocation motion.

The behavior of a plain carbon (bcc) steel is shown in Fig. 4.17 for contrast. This behavior is typical of the metals that become brittle at low temperatures. The yield and tensile strength curves approach one another as temperature is reduced. When the two curves converge, the material is brittle in the tensile test.

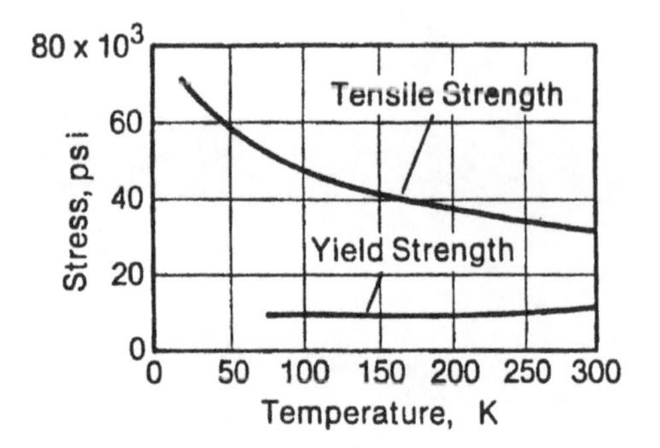


**Figure 4.16** Tensile and yield strength of annealed oxygen-free copper.

## 5.2. Plastics

Only a few plastics have been tested at temperatures below 200 K. Of these, only Teflon showed ductility down to the lowest test temperature, which was 4 K. However, reinforced plastics such as the glass fiber laminates can have good properties, the tensile strength parallel to the laminations increasing at low temperatures and the modulus being approximately constant. Mylar breaks with fragmentation in a tensile test and so is obviously brittle. Yet in films about $2.540 \times 10^{-5}$ m (0.001 in.) or less in thickness, Mylar shows remarkable flexibility in bending tests at temperatures as low as 20 K.

Figure 4.18 shows the tensile strength of several plastics. The increase in strength as the temperature is depressed is accompanied by a rapid decrease in elongation and impact resistance. The glass-reinforced plastics are the only plastics that retain appreciable impact resistance as the temperature is lowered. Figure 4.19 shows the tensile strength and impact energy of a glass fiber-reinforced epoxy resin sample. These glass-reinforced epoxies are very useful for cryogenic applications because of their high strength-to-weight ratios and high strength-to-thermal conductivity ratios (Weitzel et al., 1960).

## 5.3. Glass

The strength of glass at room temperature varies inversely with load duration, is sensitive to atmospheric water vapor, and is also sensitive to rather minute surface

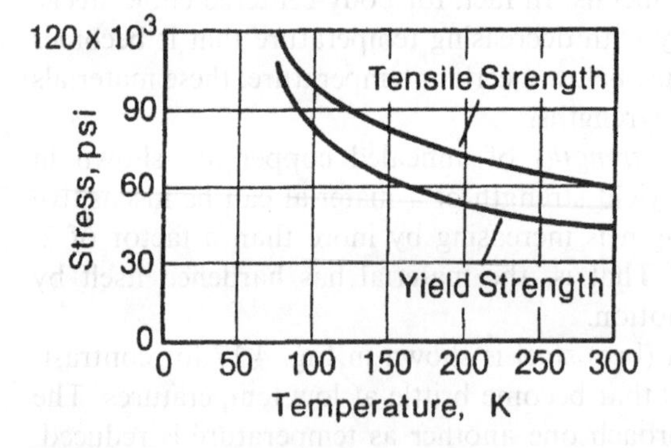


**Figure 4.17** Tensile and yield strength of a plain carbon steel (SAE 1010 m, cold-rolled).

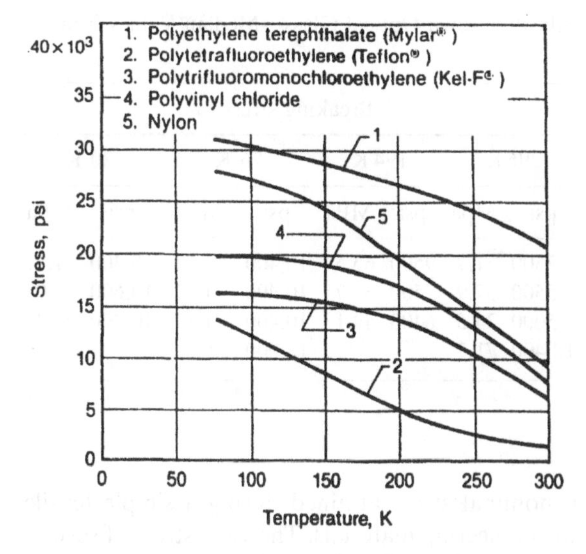


**Figure 4.18** Tensile strength of selected plastics. (1) Polyethylene terephthalate (Mylar); (2) polytetrafluoroethylene (Teflon); (3) polytrifluoromonochloroethylene (Kel-F); (4) polyvinyl chloride; (5) nylon.

defects (Kropschot and Mikesell, 1957). The dependence of strength on load duration (fatigue) has been found to decrease as the temperature is lowered below ambient, but for a soda lime glass it was still appreciable at 83 K.

Table 4.2 shows the median breaking stress of borosilicate crown optical glass (BSC-2) as a function of temperature. Owing to the brittle nature of glass, there is large scatter in the breaking stress data; therefore, these data should not be used for design purposes. For example, the median value at room temperature (296 K) is 34.47 MPa (5000 psi) at a stress rate of increase of 1 psi/s. One percent of the samples will fail at a stress of approximately 27.58 MPa (4000 psi) or below. The strength properties of glass can be improved by tempering the surface, i.e., placing the surface layer in compression. Large optical glass (BSC-2) windows have been successfully tempered for use in liquid hydrogen bubble chambers.

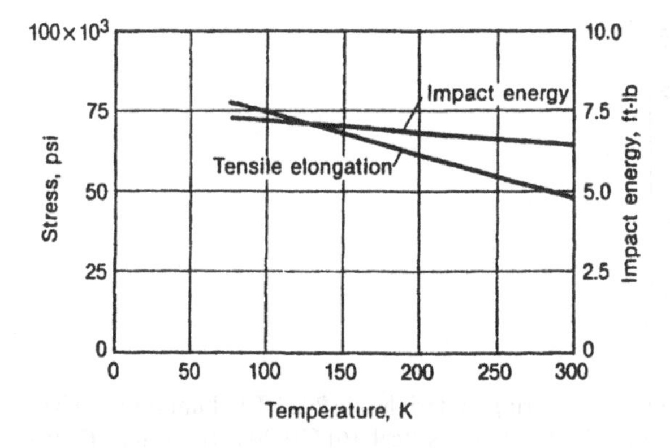


**Figure 4.19** Tensile strength and impact energy of a glass fiber-reinforced epoxy resin.

**Table 4.2** Breaking Stress of a Borosilicate Crown Optical Glass—Median Values from Probability Plots

| | Rate of stress increase | | Breaking stress | | | | | | | |
|---|---|---|---|---|---|---|---|---|---|---|
| | | | 296 K | | 194 K | | 76 K | | 20 K | |
| Condition | psi/s | MPa/s | psi | Mpa | psi | MPa | psi | MPa | psi | MPa |
| Abraded | 800 | 5.52 | 7500 | 51.7 | 9500 | 65.5 | 10,400 | 71.71 | 10,400 | 71.71 |
| Abraded | 10 | 0.0689 | 5500 | 37.9 | 7500 | 51.71 | 10,400 | 71.17 | 10,600 | 73.08 |
| Abraded | 1 | 0.00689 | 5000 | 34.5 | 6400 | 44.13 | 10,400 | 71.17 | 10,200 | 70.33 |
| Unabraded | 800 | | 10,400 | 71.7 | | | 18,000 | 124.11 | | |

## 5.4. Ultimate Stress

The ultimate stress (the maximum nominal stress attained during a simple tensile test) is shown in Fig. 4.20 for several engineering materials. The yield stress of several materials is shown in Fig. 4.21 (Durham et al., 1962). (The *yield stress* is the value of stress at which the strain of the material in a simple tensile test begins to increase rapidly with increase in stress or cause a permanent set of 0.1–0.2%.) At low temperatures, less thermal agitation is available to assist dislocation motion, and hence the yield stress usually increases.

*Example 4.1.* In many design situations, the yield strength-to-density ratio is an important parameter. Determine the strength-to-density ratio $S_y\rho$ in Nm/kg for the following materials at 22.2 K (density given in parenthesis):

a. 2024 T4 aluminum (2740 kg/m$^3$)
b. 304 stainless steel (7506 kg/m$^3$)
c. Monel (8885 kg/m$^3$)

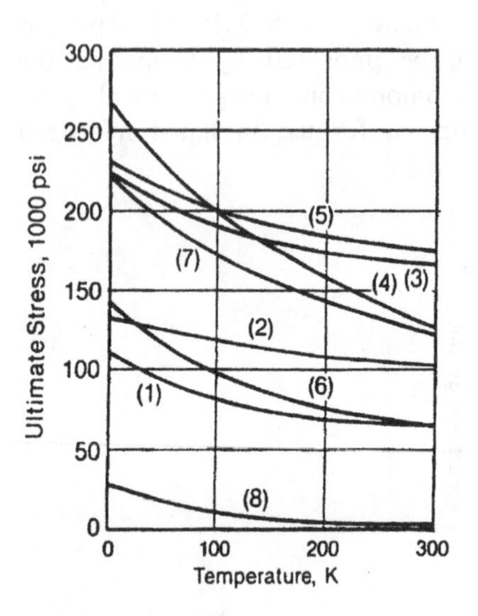


**Figure 4.20** Ultimate stress for several engineering materials. (1) 2024 T4 aluminum; (2) beryllium copper; (3) K Monel; (4) titanium; (5) 304 stainless steel; (6) C1020 carbon steel; (7) 9% nickel steel; (8) Teflon.

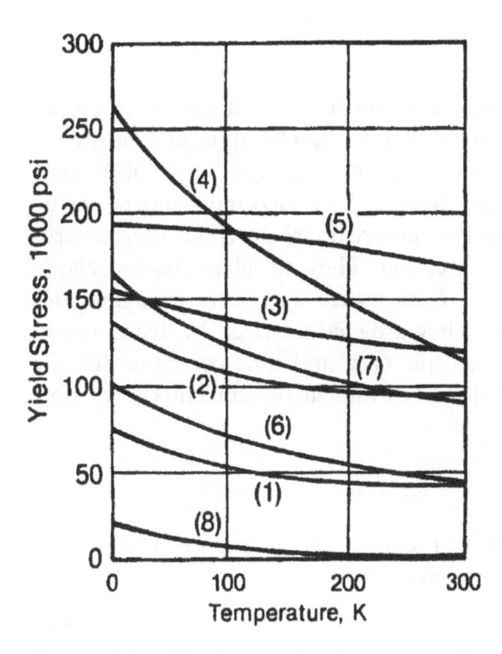


**Figure 4.21** Yield stress for the same engineering materials as in Fig. 4.20.

d. Beryllium copper ($8304\,kg/m^3$)
e. Teflon ($2297\,kg/m^3$)

From Fig. 4.21:

a. 2024 T4 aluminum

$$S_y = 7.24 \times 10^8\,N/m^2$$
$$S_y/\rho = 2.64 \times N\,m/kg$$

b. 304 stainless steel

$$S_y = 1.32 \times 10^9\,N/m^2$$
$$S_y/\rho = 1.76 \times 10^5\,N\,m/kg$$

c. Monel

$$S_y = 1.03 \times 10^9\,N/m^2$$
$$S_y/\rho = 1.16 \times 10^5\,N\,m/kg$$

d. Beryllium copper

$$S_y = 8.96 \times 10^8\,N/m^2$$
$$S_y/\rho = 1.08 \times 10^5\,N\,m/kg$$

e. Teflon

$$S_y = 1.38 \times 10^8\,N/m^2$$
$$S_y/\rho = 6.00 \times 10^4\,N\,m/kg$$

## 6. ELASTIC CONSTANTS

Elastic constants are physical properties, which relate stress to strain, or force per unit area to relative length change. Either stress or strain can be induced by any elastically coupled force—mechanical, thermal, magnetic, or electrical. Since solids resist both volume change and shape change, they have at least two independent elastic constants. Most materials considered in this chapter are quasi-isotropic; their macroscopic elastic constants do not depend on direction. They are characterized elastically by two independent elastic constants. Composites and textured aggregates are anisotropic rather than quasi-isotropic. They are characterized by five independent elastic constants for the transverse-isotropic case and nine constants for the orthotropic case. This section defines and describes the elastic constants used most often to characterize polycrystals.

### 6.1. Compressibility, $K_T$

All matter—gas, liquid, or solid—responds to pressure change and exhibits at least one elastic constant, the compressibility. It is given by

$$K_T = -(1/V)(\partial V/\partial P)_T \tag{4.1}$$

where $P$, $T$, and $V$ denote pressure, temperature, and volume. Figure 4.22 illustrates the mechanical deformations with which compressibility is associated. $K_T$ is always positive and has units of reciprocal pressure.

$$k_T = (A/\ell)(\mathrm{d}\ell/\mathrm{d}L)_T \tag{4.2}$$

where $A$ denotes the cross-sectional area of a rod-shaped, hydrostatically loaded specimen, $\ell$ denotes length, and $L$ denotes load. For isotropic solids, linear and bulk compressibilities relate according to

$$3k = K_T \tag{4.3}$$

### 6.2. Bulk Modulus, $B$

The bulk modulus, $B$, is used to describe a solid's resistance to volume change rather than compressibility. For isotropic materials, $B$ and $K$ relate reciprocally:

$$B_T = 1/K_T = -V(\partial P/\partial V)_T \tag{4.4}$$

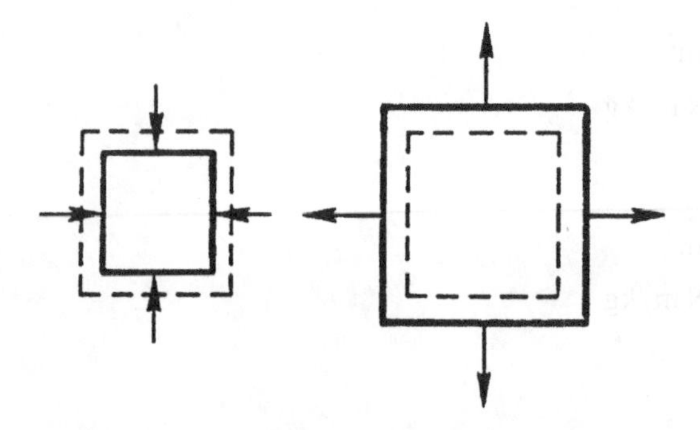

**Figure 4.22** Mechanical deformations with which compressibility or bulk modulus is associated.

$B$ has pressure units, and, herein, elastic constants with pressure units are called elastic moduli. For describing solids, hydrostatic stress, $\sigma$, is often preferred over pressure, which is a negative stress. Thus,

$$B_T = V(\partial\sigma/\partial V)_T \tag{4.5}$$

A high bulk modulus usually implies strong interatomic forces, high cohesive energy, high melting point, and high elastic-stiffness moduli.

## 6.3. Young's Modulus, *E*

Engineers find Young's modulus the most familiar elastic constant, as evidenced by its many pseudonyms: extension modulus, tensile modulus, tension modulus, elastic modulus, and modulus. Young's moduli in tension or compression are identical, theoretically and experimentally.

Usually defined in terms of a uniaxially stressed rod where both stress, $\sigma$, and strain, $\varepsilon$, are measured along the rod axis, $z$. Young's modulus, $E$, becomes (see Fig. 4.1)

$$E = \sigma_z/\varepsilon_z \tag{4.6}$$

Figure 4.23 illustrates the mechanical deformations with which Young's modulus is associated.

As shown in Fig. 4.23, rod bending also involves Young's modulus. The stress along a bent rod is

$$\sigma_z = E, \qquad \varepsilon_z = Ex/r \tag{4.7}$$

where $x$ is a coordinate perpendicular to $z$, and $r$ is the curvature radius of the neutral surface near the origin.

## 6.4. Shear Modulus, *G*

Sometimes called the torsional modulus, rigidity modulus, or transverse modulus, the shear modulus, $G$, relates the shear stress, $\tau$ in simple shear to the shear strain, $\gamma$. Thus,

$$\tau = G\gamma \tag{4.8}$$

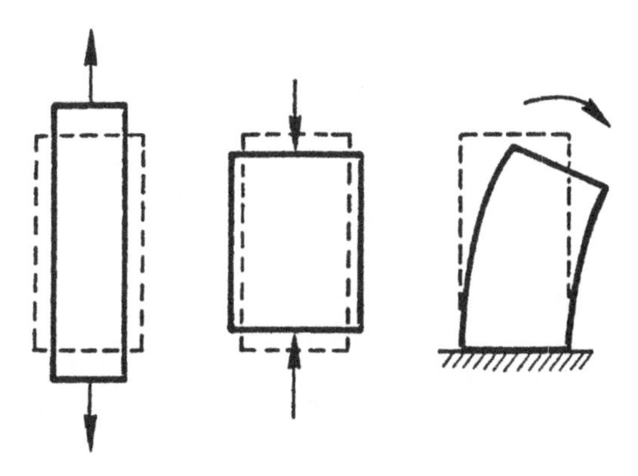

**Figure 4.23** Mechanical deformations with which Young's modulus is associated.

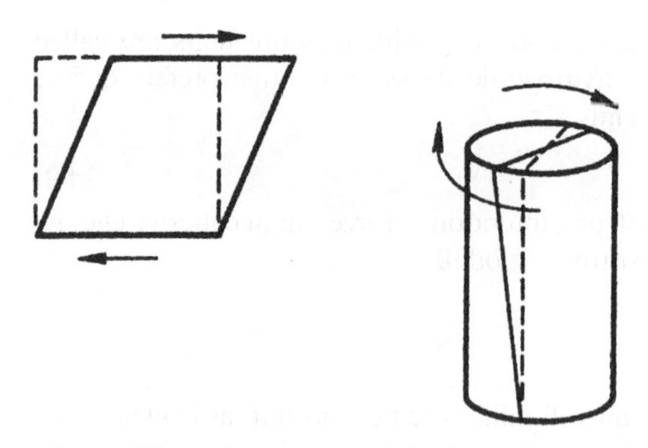

**Figure 4.24** Mechanical deformations with which shear modulus and torsional modulus are associated.

Figure 4.24 illustrates the mechanical deformations with which the shear modulus is associated. A cube sheared on one face has a shear strain related to the length change of the face diagonal by

$$\gamma = 2\Delta\ell/\ell \tag{4.9}$$

In torsion (see Fig. 4.24), $\tau$ is the torsional stress. For isotropic materials, simple-shear and torsional $G$'s are identical. Both shear and torsion conserve volume but change shape. The bulk modulus, as described above, describes volume change without shape change.

### 6.5. Poisson's Ratio, $\nu$

Poisson's ratio, $\nu$, is not an elastic modulus, but a dimensionless ratio of two elastic compliances. Usually defined by a uniaxially stressed rod (Young's-modulus) experiment, Poisson's ratio is the negative ratio of transverse ($x$) and longitudinal ($z$) strains:

$$\nu = -\varepsilon_x/\varepsilon_z \tag{4.10}$$

Figure 4.25 illustrates the deformation and strains involved. A typical value for metals is 1/3, 0.28–0.42 being the observed range for most materials. Such $\nu$ values

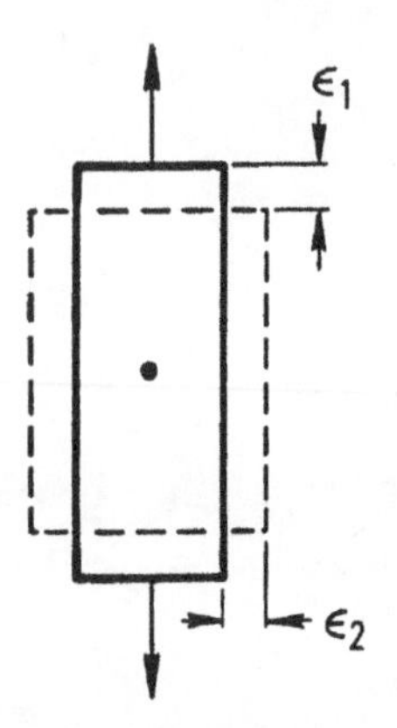


**Figure 4.25** Mechanical deformation and strains with which Poisson's ratio is associated.

mean that a material *tends* to maintain constant volume during uniaxial deformation. Liquids exhibit constant volume during deformation, and some solids (such as rubber) tend to.

The four elastic constants just described—$B$, $E$, $G$, and $\nu$—are sometimes referred to as the bulk, engineering, and macroscopic, polycrystalline, practical, or technological elastic constants. These four constants are those used most frequently to describe materials such as polycrystalline aggregates.

## 6.6. Hooke's Law

Hooke's law states a linear proportionality between strain-response intensity, $\varepsilon$, and imposed stress, $\sigma$

$$\sigma = C\varepsilon \tag{4.11}$$

where $C$ denotes an elastic-stiffness modulus. For small strains, Eq. 4.11 is an example of Hooke's law written inversely:

$$\varepsilon = S\sigma \tag{4.12}$$

where the $S$'s are elastic-compliance moduli, and for isotropic media

$$C = 1/S \tag{4.13}$$

This chapter considers mainly Hooke's-law, or linear-elastic, solids. Nonlinear cases involve elastic constants higher than second order and are not treated in this book.

## 6.7. Elastic-Constant Inter-relationships

Although for complete characterization isotropic solids require only two independent elastic constants, practice requires many more. This multiplicity arises because particular elastic constants best describe particular mechanical deformations. As already discussed, $B$, $E$, and $G$ best describe dilatation, extension, and shear, respectively. Table 4.3 summarizes inter-relationships among various elastic constants.

## 6.8. Typical Elastic-Constant Values

Tables 4.4–4.7 give elastic constants at various temperatures for six metals—nickel, iron, copper, titanium, aluminum, and magnesium—that form the bases for most

**Table 4.3** Connecting Identities Among Isotropic-Solid Elastic Constants

| | $B$ | $E$ | $G$ | $\nu$ |
|---|---|---|---|---|
| $B$, $E$ | – | – | $\frac{3BE}{9B-E}$ | $\frac{1}{2}-\frac{E}{6B}$ |
| $B$, $G$ | – | $\frac{9BG}{3B+G}$ | – | $\frac{1}{2}-\frac{3B-2G}{3B+G}$ |
| $B$, $\nu$ | – | $3B\,(1-2\nu)$ | $\frac{3B}{2}-\frac{1-2\nu}{1+\nu}$ | – |
| $E$, $G$ | $\frac{GE}{3(3G-E)}$ | – | – | $\frac{E}{2G}-1$ |
| $E$, $\nu$ | $\frac{E}{3(1-2\nu)}$ | – | $\frac{E}{2(1+\nu)}$ | – |
| $G$, $\nu$ | $\frac{2G}{3}\frac{1+\nu}{1-2\nu}$ | $2G\,(1+\nu)$ | – | – |

**Table 4.4** Bulk Modulus (Reciprocal Compressibility) of Six Metals at Selected Temperatures in Units of GPa

| T (K) | Ni | Fe | Cu | Ti | Al | Mg |
|---|---|---|---|---|---|---|
| 0 | 187.6 | 187.6 | 142.0 | 110.0 | 79.4 | 36.9 |
| 20 | 187.6 | 173.1 | 142.0 | 109.8 | 79.4 | 36.9 |
| 40 | 187.5 | 172.9 | 141.9 | | 79.4 | 36.8 |
| 60 | 187.3 | 172.7 | 141.7 | | 79.3 | 36.8 |
| 80 | 187.2 | 172.1 | 141.4 | 109.5 | 79.2 | 36.7 |
| 100 | 187.0 | 171.6 | 141.1 | | 79.0 | 36.6 |
| 120 | 186.6 | 171.1 | 140.8 | 109.0 | 78.8 | 36.5 |
| 140 | 186.2 | 170.7 | 140.5 | | 78.6 | 36.4 |
| 160 | 185.8 | 170.3 | 140.1 | | 78.3 | 36.3 |
| 180 | 185.5 | 169.9 | 139.7 | 108.5 | 78.0 | 36.1 |
| 200 | 185.2 | 169.5 | 139.3 | | 77.7 | 36.0 |
| 220 | 184.9 | 169.2 | 138.8 | 108.0 | 77.4 | 35.9 |
| 240 | 184.6 | 168.9 | 138.4 | | 77.1 | 35.7 |
| 260 | 184.2 | 168.6 | 138.0 | | 76.7 | 35.6 |
| 280 | 183.9 | 168.2 | 137.5 | 107.5 | 76.4 | 35.4 |
| 300 | 183.6 | 168.0 | 137.1 | 107.3 | 76.1 | 35.3 |

technological alloys. These data are taken from "Materials at Low Temperatures," Am. Soc. Metals (1983).

Table 4.8 gives ambient-temperature elastic constants for 24 cubic elements for which reliable data exist. This table also includes mass density, $\rho$.

Table 4.9 gives zero-temperature values extrapolated from 4 K data. Since the values shown in Tables 4.4–4.9 come from many different sources, some small differences in values will occur among the tables.

The data in Tables 4.4–4.9 are all taken from "Materials at Low Temperatures," Am. Soc. Metals, 1983.

**Table 4.5** Young's Modulus of Six Metals at Selected Temperatures in Units of GPa

| *T* (K) | Ni | Fe | Cu | Ti | Al | Mg |
|---|---|---|---|---|---|---|
| 0 | 240.1 | 224.1 | 138.6 | 130.6 | 78.4 | 49.4 |
| 20 | 240.1 | 224.1 | 138.6 | 130.6 | 78.3 | 49.3 |
| 40 | 239.8 | 223.9 | 138.4 | | 78.2 | 49.2 |
| 60 | 239.2 | 223.5 | 138.0 | | 77.9 | 49.1 |
| 80 | 238.5 | 222.9 | 137.5 | 129.4 | 77.6 | 48.9 |
| 100 | 237.5 | 222.3 | 136.9 | | 77.1 | 48.6 |
| 120 | 236.4 | 221.7 | 136.2 | 127.0 | 76.5 | 48.3 |
| 140 | 235.1 | 220.9 | 135.4 | | 75.8 | 48.0 |
| 160 | 233.8 | 220.1 | 134.6 | | 75.2 | 47.6 |
| 180 | 232.7 | 219.3 | 133.7 | 123.7 | 74.5 | 47.2 |
| 200 | 231.3 | 218.4 | 132.9 | | 73.8 | 46.8 |
| 220 | 230.0 | 217.7 | 132.0 | 120.0 | 73.1 | 46.4 |
| 240 | 228.6 | 216.8 | 131.1 | | 72.3 | 46.0 |
| 260 | 227.2 | 215.8 | 130.1 | | 71.6 | 45.6 |
| 280 | 225.8 | 214.9 | 129.2 | 116.6 | 70.8 | 45.1 |
| 300 | 224.5 | 214.0 | 128.2 | 114.6 | 70.1 | 44.7 |

**Table 4.6** Shear Modules of Six Metals at Selected Temperatures in Units of GPa

| $T$ (K) | Ni | Fe | Cu | Ti | Al | Mg |
|---|---|---|---|---|---|---|
| 0 | 93.3 | 87.3 | 51.9 | 50.2 | 29.3 | 19.4 |
| 20 | 93.3 | 87.2 | 51.8 | 50.2 | 29.3 | 19.3 |
| 40 | 93.2 | 87.2 | 51.8 | | 29.3 | 19.3 |
| 60 | 93.0 | 87.0 | 51.6 | | 29.2 | 19.1 |
| 80 | 92.6 | 86.8 | 51.4 | 49.7 | 29.0 | 19.1 |
| 100 | 92.2 | 86.6 | 51.2 | | 28.8 | 19.0 |
| 120 | 91.8 | 86.4 | 50.9 | 48.8 | 28.6 | 18.9 |
| 140 | 91.2 | 86.1 | 50.5 | | 28.3 | 18.8 |
| 160 | 90.6 | 85.7 | 50.2 | | 28.1 | 18.6 |
| 180 | 90.1 | 85.4 | 49.9 | 47.2 | 27.8 | 18.4 |
| 200 | 89.6 | 85.0 | 49.6 | | 27.5 | 18.3 |
| 220 | 89.0 | 84.7 | 49.2 | 45.7 | 27.2 | 18.1 |
| 240 | 88.4 | 84.3 | 48.8 | | 26.9 | 17.9 |
| 260 | 87.8 | 83.9 | 48.5 | | 26.6 | 17.7 |
| 280 | 87.2 | 83.5 | 48.1 | 44.2 | 26.3 | 17.6 |
| 300 | 86.6 | 83.1 | 47.7 | 43.4 | 26.0 | 17.4 |

## 7. MODULUS OF ELASTICITY

The constant of proportionality relating tensile stress and strain in the range of elastic response of a material is the modulus of elasticity, or Young's modulus, $E$. Two other commonly used elastic moduli are (1) the shear modulus $G$, which is the rate of change of shear stress with respect to shear strain at constant temperature in the elastic region, and (2) the bulk modulus $B$, which is the rate of change of pressure (corresponding to a uniform three-dimensional stress) with respect to volumetric strain (change in volume per unit volume) at constant temperature.

**Table 4.7** Poisson's Ratio of Six metals at Selected Temperatures

| $T$ (K) | Ni | Fe | Cu | Ti | Al | Mg |
|---|---|---|---|---|---|---|
| 0 | 0.287 | 0.285 | 0.338 | 0.302 | 0.336 | 0.277 |
| 20 | 0.287 | 0.285 | 0.338 | 0.302 | 0.336 | 0.278 |
| 40 | 0.287 | 0.285 | 0.338 | | 0.336 | 0.278 |
| 60 | 0.287 | 0.285 | 0.338 | | 0.336 | 0.278 |
| 80 | 0.288 | 0.285 | 0.338 | 0.303 | 0.337 | 0.279 |
| 100 | 0.288 | 0.284 | 0.338 | | 0.337 | 0.279 |
| 120 | 0.289 | 0.284 | 0.339 | 0.306 | 0.338 | 0.280 |
| 140 | 0.290 | 0.285 | 0.339 | | 0.339 | 0.281 |
| 160 | 0.290 | 0.285 | 0.340 | | 0.340 | 0.282 |
| 180 | 0.291 | 0.285 | 0.341 | 0.311 | 0.341 | 0.283 |
| 200 | 0.292 | 0.285 | 0.341 | | 0.342 | 0.284 |
| 220 | 0.293 | 0.286 | 0.342 | 0.315 | 0.343 | 0.285 |
| 240 | 0.294 | 0.286 | 0.342 | | 0.344 | 0.286 |
| 260 | 0.294 | 0.287 | 0.343 | | 0.345 | 0.287 |
| 280 | 0.295 | 0.287 | 0.343 | 0.321 | 0.346 | 0.288 |
| 300 | 0.296 | 0.288 | 0.344 | 0.322 | 0.347 | 0.289 |

**Table 4.8** Room-Temperature Elastic Properties of 24 Cubic Elements

| | $\rho$ (g/cm$^3$) | $A$ | $B$ (GPa) | $E$ (GPa) | $G$ (Gpa) | $\nu$ |
|---|---|---|---|---|---|---|
| Ag | 10.500 | 2.8825 | 101.20 | 78.89 | 28.79 | 0.3701 |
| Al | 2.697 | 1.2231 | 75.86 | 70.27 | 26.11 | 0.3456 |
| Au | 19.300 | 2.8522 | 173.50 | 76.12 | 26.68 | 0.4269 |
| C | 3.512 | 1.2109 | 442.00 | 1139.37 | 532.23 | 0.0704 |
| Cr | 7.200 | 0.7144 | 161.87 | 278.38 | 114.72 | 0.2134 |
| Cu | 8.930 | 3.2727 | 135.30 | 121.69 | 45.07 | 0.3501 |
| Fe | 7.872 | 2.4074 | 166.93 | 205.87 | 79.52 | 0.2945 |
| Ge | 5.323 | 1.6644 | 75.02 | 130.30 | 53.82 | 0.2105 |
| Ir | 22.520 | 1.5645 | 370.37 | 556.74 | 222.79 | 0.2495 |
| K | 0.851 | 6.7143 | 3.33 | 2.25 | 0.81 | 0.3874 |
| Li | 0.532 | 8.5243 | 12.13 | 9.21 | 3.35 | 0.3734 |
| Mo | 10.228 | 0.7140 | 259.77 | 321.83 | 124.40 | 0.2935 |
| Na | 0.971 | 7.1624 | 6.51 | 4.81 | 1.74 | 0.3788 |
| Nb | 8.570 | 0.4925 | 163.78 | 103.07 | 36.94 | 0.3951 |
| Ni | 8.910 | 2.6652 | 185.97 | 212.27 | 81.03 | 0.3098 |
| Pb | 11.344 | 4.0743 | 44.76 | 23.17 | 8.19 | 0.4136 |
| Pd | 12.038 | 2.8096 | 193.06 | 128.23 | 46.15 | 0.3893 |
| Pt | 21.500 | 1.5938 | 282.70 | 176.23 | 63.12 | 0.3961 |
| Rb | 1.560 | 7.4545 | 2.60 | 1.85 | 0.67 | 0.3811 |
| Si | 2.331 | 1.5636 | 97.89 | 161.61 | 65.97 | 0.2248 |
| Ta | 16.626 | 1.5609 | 189.72 | 183.69 | 68.61 | 0.3386 |
| Th | 11.694 | 3.6212 | 57.70 | 69.83 | 26.89 | 0.2983 |
| V | 6.022 | 0.7867 | 155.57 | 129.11 | 47.41 | 0.3617 |
| W | 19.257 | 1.0114 | 310.38 | 409.83 | 160.10 | 0.2799 |

If the material is isotropic (many polycrystalline materials can be considered isotropic for engineering purposes), these three moduli are related through Poisson's ratio $\mu$, the ratio of strain in one direction due to a stress applied perpendicular to that direction to the strain parallel to the applied stress (see also Table 4.3):

$$B = \frac{E}{3(1-2\mu)} \tag{4.14}$$

(in some texts, the symbol $K$ is used for the bulk modulus),

$$G = \frac{E}{2(1+\mu)} \tag{4.15}$$

$$E = \frac{2G(1+\mu)}{1+\mu} \tag{4.16}$$

where

$$\mu = \frac{E-2G}{2G}$$

**Table 4.9** Zero-Temperature Elastic Properties of 24 Cubic Elements

| | $\rho$ (g/cm$^3$) | $A$ | $B$ (GPa) | $E$ (GPa) | $G$ (GPa) | $\nu$ | $v_m$ (cm $\mu$s$^{-1}$) | $v_l$ (cm $\mu$s$^{-1}$) | $v_t$ (cm $\mu$s$^{-1}$) | $\theta$ (K) |
|---|---|---|---|---|---|---|---|---|---|---|
| Ag | 10.634 | 2.9912 | 108.72 | 87.02 | 31.84 | 0.3666 | 0.1950 | 0.3770 | 0.1730 | 226.46 |
| Al | 2.732 | 1.2073 | 79.38 | 78.24 | 29.29 | 0.3357 | 0.3674 | 0.6583 | 0.3274 | 430.56 |
| Au | 19.478 | 2.8436 | 180.32 | 83.19 | 29.23 | 0.4231 | 0.1391 | 0.3355 | 0.1225 | 161.74 |
| C | 3.516 | 1.2109 | 442.00 | 1139.37 | 532.23 | 0.0704 | 1.3416 | 1.8099 | 1.2304 | 2239.62 |
| Cr | 7.231 | 0.6865 | 201.00 | 300.58 | 120.16 | 0.2508 | 0.4526 | 0.7068 | 0.4076 | 589.57 |
| Cu | 9.024 | 3.1904 | 142.03 | 132.51 | 49.28 | 0.3445 | 0.2625 | 0.4798 | 0.2337 | 344.40 |
| Fe | 7.922 | 2.3431 | 169.00 | 213.17 | 82.64 | 0.2898 | 0.3603 | 0.5936 | 0.3230 | 472.44 |
| Ge | 5.340 | 1.6651 | 76.52 | 132.92 | 54.90 | 0.2105 | 0.3544 | 0.5295 | 0.3207 | 373.36 |
| Ir | 22.659 | 1.5698 | 366.67 | 557.12 | 223.43 | 0.2468 | 0.3485 | 0.5416 | 0.3140 | 429.62 |
| K | 0.910 | 7.6267 | 3.66 | 3.10 | 1.14 | 0.3587 | 0.1261 | 0.2386 | 0.1120 | 90.52 |
| Li | 0.5375 | 9.2400 | 13.30 | 10.70 | 3.90 | 0.3650 | 0.3049 | 0.5868 | 0.2706 | 326.74 |
| Mo | 10.223 | 0.9024 | 265.29 | 335.72 | 130.22 | 0.2891 | 0.3481 | 0.6552 | 0.3569 | 474.53 |
| Na | 1.005 | 8.5000 | 7.24 | 5.98 | 2.20 | 0.3623 | 0.1664 | 0.3181 | 0.1478 | 147.43 |
| Nb | 8.616 | 0.5183 | 173.03 | 110.87 | 39.79 | 0.3932 | 0.2430 | 0.5122 | 0.2149 | 276.60 |
| Ni | 8.972 | 2.3859 | 187.60 | 233.52 | 90.33 | 0.2925 | 0.3541 | 0.5860 | 0.3173 | 475.98 |
| Pb | 11.593 | 3.8379 | 48.79 | 30.34 | 10.86 | 0.3964 | 0.1095 | 0.2336 | 0.0968 | 105.33 |
| Pd | 12.092 | 2.4552 | 195.43 | 134.72 | 48.63 | 0.3851 | 0.2266 | 0.4640 | 0.2005 | 275.92 |
| Pt | 21.578 | 1.4828 | 288.40 | 183.54 | 65.83 | 0.3939 | 0.1976 | 0.4175 | 0.1747 | 238.44 |
| Rb | 1.629 | 8.0657 | 3.21 | 2.39 | 0.87 | 0.3761 | 0.0824 | 0.1638 | 0.0730 | 55.31 |
| Si | 2.331 | 1.5640 | 99.22 | 163.21 | 66.57 | 0.2258 | 0.5916 | 0.8980 | 0.5344 | 648.87 |
| Ta | 16.754 | 1.6154 | 194.21 | 191.15 | 71.54 | 0.3360 | 0.2319 | 0.4158 | 0.2066 | 263.77 |
| Th | 11.888 | 3.4545 | 58.10 | 75.58 | 29.45 | 0.2832 | 0.1754 | 0.2862 | 0.1574 | 163.83 |
| V | 6.050 | 0.8130 | 157.04 | 135.23 | 49.85 | 0.3565 | 0.3230 | 0.6078 | 0.2870 | 399.19 |
| W | 19.313 | 0.9959 | 314.15 | 417.76 | 163.40 | 0.2784 | 0.3240 | 0.5248 | 0.2909 | 384.39 |

or

$$\mu = \frac{3K - E}{6K}$$

It has been found that Poisson's ratio for isotropic materials does not change appreciably with change in temperature in the cryogenic range. Therefore, all three of the elastic moduli above vary in the same manner with temperature. Accordingly, in the discussion that follows, "modulus" shall mean Young's modulus $E$, which is the slope of the initial portion of the stress–strain curve of Fig. 4.1.

We expect materials to get stiffer at low temperatures, and that is the usual case. However, the change in modulus of elasticity with temperature is not very great in metals compared to the change in strength over the same temperature range. Some general statements can be made regarding the temperature dependence of the elastic modulus.

The elastic moduli of most polycrystalline metals increase by about 10% in going from room temperature to about 20 K, and below this there is little change (Campbell, 1974). For some alloys, the modulus of elasticity may either decrease or increase in value or even exhibit maxima or minima as the temperature is lowered, depending upon their composition. The moduli of noncrystalline plastics are 2–20 times their room temperature values at 4 K, whereas the moduli of glasses, also

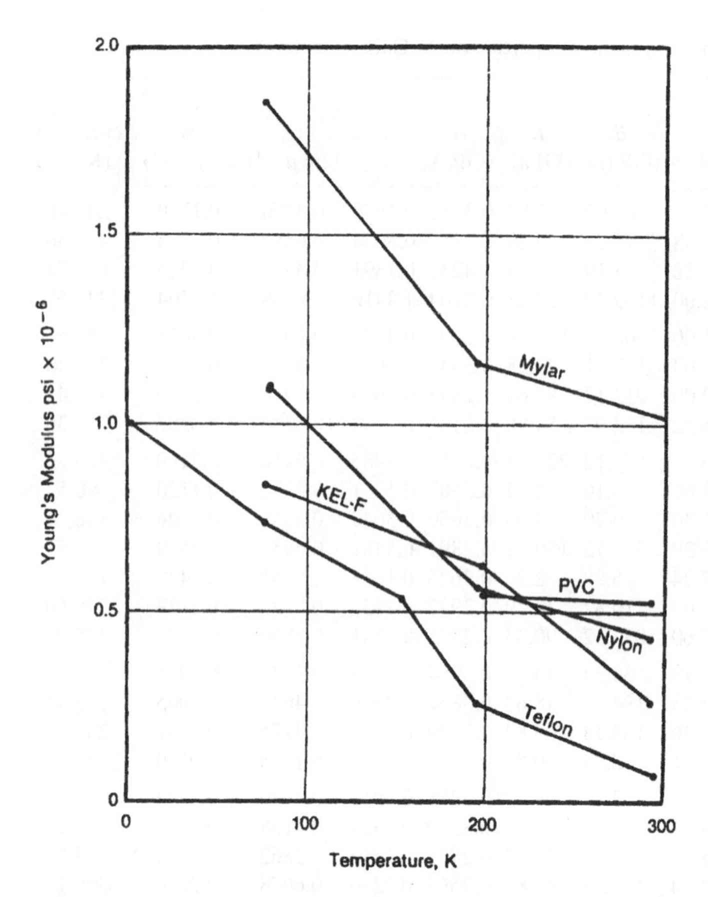


**Figure 4.26** Young's modulus of fine common plastics.

noncrystalline, either increase or decrease by about 3% between room temperature and 76 K, depending upon composition. Thermodynamic arguments predict that the elastic constants of all solids will be independent of temperature at absolute zero, so no great changes are to be expected in the moduli below about 20 K.

Nonmetallic materials exhibit a much greater change in the modulus of elasticity at low temperatures. For polyethylene terephthalate, $E$ increases by a factor of nearly 2 as the temperature is decreased from 300 to 76 K, and the modulus of polytetrafluorethylene at 4 K is about 20 times that at room temperature (Fig. 4.26) (Corruccini, 1957).

Curves of the modulus of elasticity of representative engineering materials are shown in Fig. 4.27 (Durham et al., 1962). Similar information is summarized in Table 4.10 (McClintock and Gibbons, 1960).

## 8. FATIGUE STRENGTH

A simple reverse bending test is the usual method used to measure fatigue strength. *Fatigue strength* is defined as the stress required to cause failure after a certain number of bending cycles and is given as $s_f$.

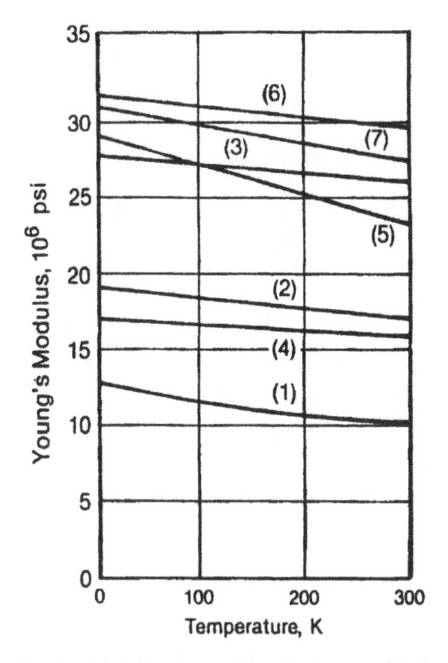


**Figure 4.27** Young's modulus of several metals at low temperatures. (1) 2024 T4 aluminum; (2) beryllium copper; (3) K Monel; (4) titanium; (5) 304 stainless steel; (6) C1020 carbon steel; (7) 9% nickel steel; (8) Teflon.

**Table 4.10** Temperature Dependence of Modulus of Elasticity of Some Common Metals

| Metals exhibiting very small change in modulus of elasticity from 300 to 20 K | Metals exhibiting very large change in modulus of elasticity from 300 to 20 K |
|---|---|
| Aluminum | Aluminum |
| 356 | 1100 |
| 7075 | 2024 |
| Cobalt | |
| Elgiloy | |
| Copper | |
| Berylco 25 | |
| Pure copper[a] | |
| Iron | Iron |
| Vascojet 1000[a] | 17–7 pH |
| 2800 (9% Ni) | 304, 310 |
| 4340[a] | 321, 347 |
| Nickel | Nickel |
| Inconel | K Monel |
| Inconel X[a] | Pure nickel |
| Titanium | Titanium |
| A-110-AT[a] | C-210-AV |
| B-120-VCA[a] | |

[a] No significant temperature dependence.

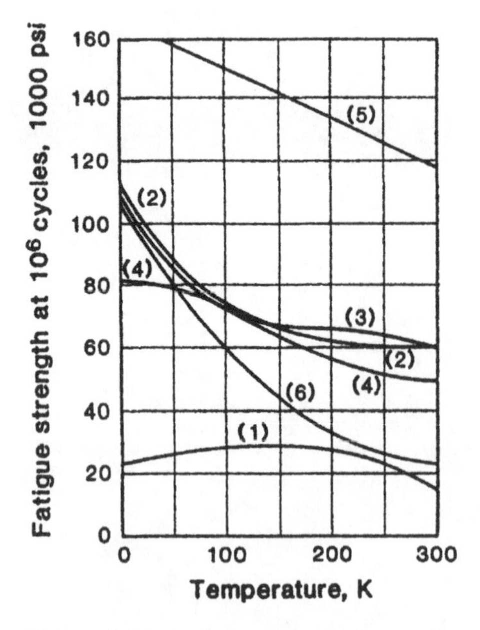


**Figure 4.28** Fatigue strength of several materials at $10^6$ cycles. (1) 2024 T4 aluminum; (2) beryllium copper; (3) K Monel; (4) titanium; (5) 304 stainless steel; (6) C1020 carbon steel; (7) 9% nickel steel; (8) Teflon.

Some materials, such as carbon steel, have the property that fatigue failure will not occur if the stress is maintained below a certain value, called the endurance limit $s_e$ no matter how many cycles have elapsed.

Fatigue strength data at low temperatures are not as common as data on yield strength and ultimate tensile strength, because the fatigue tests are more time-consuming and hence more expensive to perform. As one would expect, however, the data that have been reported all show that fatigue strength increases as temperature decreases, in the same manner as the yield and ultimate tensile strength.

Fortunately, for aluminum alloys at least, it has been found that the ratio of fatigue strength to ultimate strength remains fairly constant as the temperature is lowered. Therefore, the fatigue strength of these alloys varies with temperature in the same manner that the ultimate strength varies with temperature. This fact may be used in estimating the fatigue strength for nonferrous materials at cryogenic temperatures if no fatigue data are available.

Representative data of fatigue strength of several materials at low temperature are shown in Fig. 4.28.

## 9. MECHANICAL PROPERTIES SUMMARY

It is convenient to classify metals by their lattice structure for low-temperature mechanical properties. The face-centered cubic (fcc) metals and their alloys are most often used in the construction of cryogenic equipment. Aluminum, copper, nickel, their alloys, and the austenitic stainless steels of the 18–8 type are fcc and do not exhibit an impact ductile-to-brittle transition at low temperatures. As a general rule, the structural properties of these metals improve as the temperature is reduced. The yield strength at 22.2 K (40°R) is quite a bit greater than at ambient temperature;

Young's modulus is 5–20% greater at the lower temperatures, and fatigue properties are also improved at the lower temperatures (with the exception of 2024 T4 aluminum). Annealing of these metals and alloys can affect both the ultimate and yield strengths.

The body-centered cubic (bcc) metals and alloys are usually undesirable for low-temperature construction. This class of metals includes iron, the martensitic steels (low-carbon and the 400 series of stainless steels), molybdenum, and niobium. If not already brittle at room temperature, these materials exhibit a ductile-to-brittle transition at low temperatures. Working of some steels, in particular, can induce the austenite-to-martensite transition.

The hexagonal close-packed (hcp) metals exhibit structural properties intermediate between those of the fcc and bcc metals. For example, zinc suffers a ductile-to-brittle transition, whereas zirconium and pure titanium do not. Titanium and its alloys, having an hcp structure, remain reasonably ductile at low temperatures and have been used for many applications where weight reduction and reduced heat leakage through the material have been important. Small impurities of O, N, H, and C can have a detrimental effect on the low-temperature ductility properties of Ti and its alloys.

Plastics also increase in strength as the temperature is decreased, but this is accompanied by a rapid decrease in elongation in a tensile test and a decrease in impact resistance as the temperature is lowered. The glass-reinforced plastics also have high strength-to-weight and strength-to-normal conductivity ratios. Conversely, all elastomers become brittle at low temperatures. Nevertheless, many of these materials, including rubber, Mylar, and nylon, can be used for static seal gaskets provided they are highly compressed at room temperature prior to cooling.

The strength of glass under constant loading also increases with decreases in temperature. Since failure occurs at a lower stress when the glass has surface defects, the strength can be improved by tempering the surface.

See Figs. 4.29–4.32 for the temperature dependence of several important module: $E$.

## 10. DESIGN CONSIDERATIONS

A cryogenic storage container must be designed to withstand forces resulting from the internal pressure, the weight of the contents, and bending stresses. Material compatibility with low temperatures, which has already been discussed, results in choosing among the fcc metals (copper, nickel, aluminum, stainless steels, etc.). Because these materials are more expensive than ordinary carbon steels, a design goal is to make the inner vessel as thin as possible. This constraint also reduces cooldown time and the amount of cryogenic liquid required for cooldown. Thus, the inner vessel of cryogenic containers is nearly always thin walled.

A thin-walled cylinder has a wall thickness such that the assumption of constant stress across the wall results in negligible error. Cylinders having internal diameter-to-thickness ($D/t$) ratios greater than 10 are usually considered thin-walled. An important stress in thin-walled cylinders is the circumferential, or hoop, stress (see Fig. 4.33), which is

$$s = pr/t \tag{4.17}$$

where, $p$ is the internal pressure, $r$ the radius and, $t$ the thickness of cylinder.

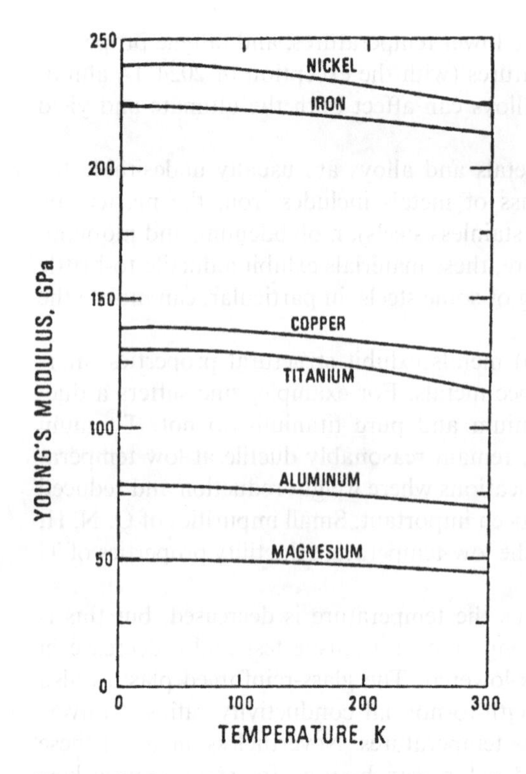


**Figure 4.29** Temperature variation of ($E$) Young's modulus for six metals.

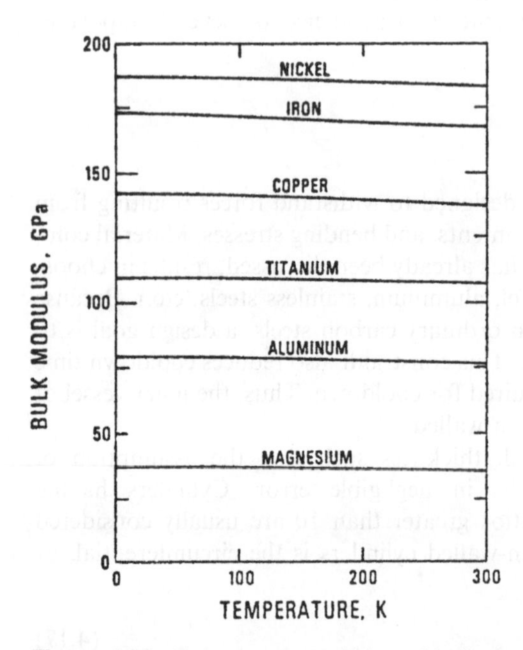


**Figure 4.30** Temperature variation of bulk modulus ($B$)(reciprocal compressibility) for six metals.

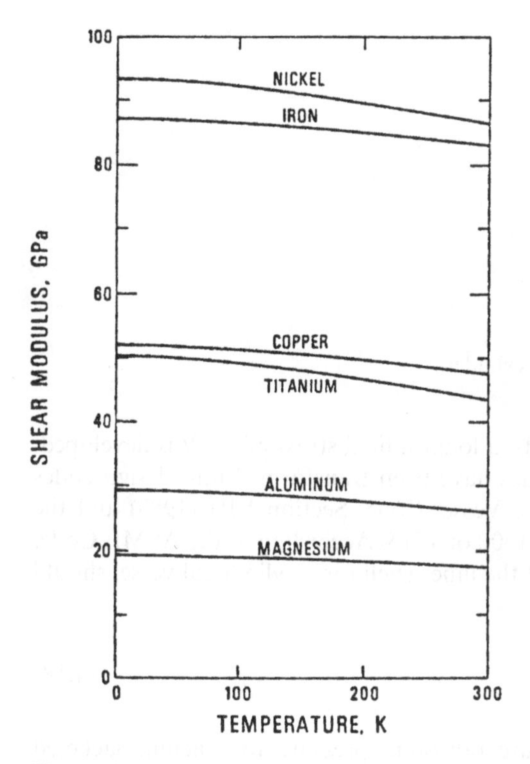


**Figure 4.31** Temperature variation of shear modulus ($G$) for six metals.

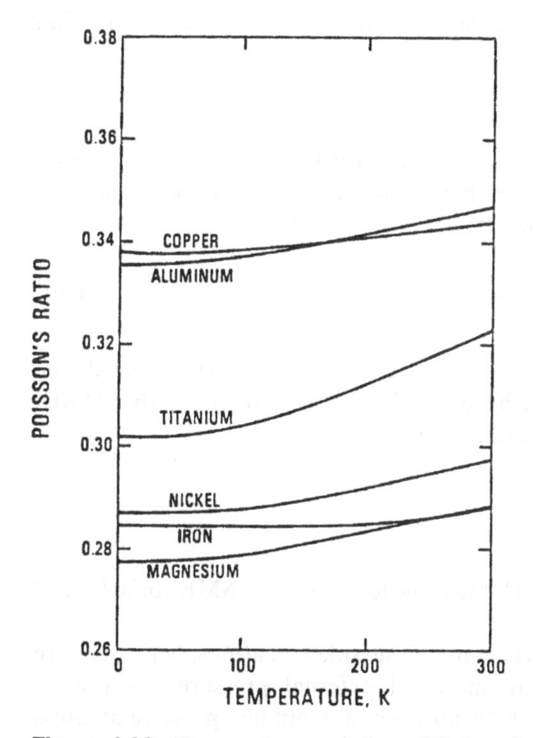


**Figure 4.32** Temperature variation of Poisson's ratio ($\nu$) for six metals.

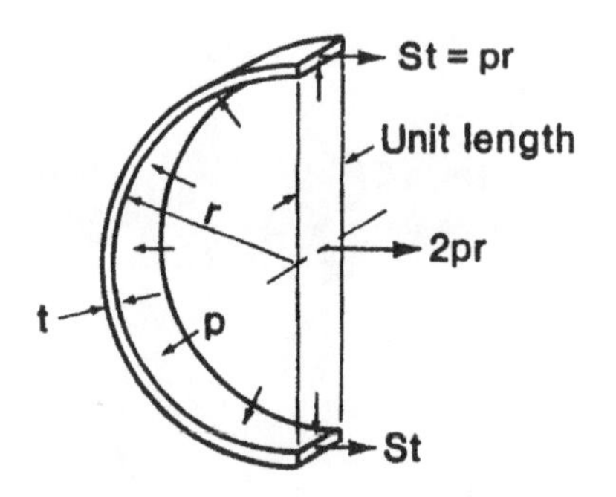


**Figure 4.33** Hoop stress in a thin-walled cylinder.

If the cylinder is closed at the ends, a longitudinal stress of $pr/2t$ is developed.

These basic equations of mechanics have been transformed into design codes such as the ASME Boiler and Pressure Vessel Code, Section VIII (1983) and the British Standards Institution Standard 1500 or 1515. According to the ASME Code, Section VIII, the minimum thickness of the inner shell for a cylindrical vessel should be determined from

$$t = \frac{PD}{2s_a e_w - 1.2P} = \frac{PD_o}{ws_a e_w + 0.8P} \tag{4.18}$$

where $P$ is the design internal pressure (absolute pressure for vacuum-jacketed vessels), $D$ the inside diameter of shell, $D_o$ the outside diameter of shell, $s_a$ the allowable stress (approximately one-fourth minimum ultimate strength of material) and $e_w$ is the weld efficiency.

Referring to Fig. 4.33, it can be seen that the tensile stress developed in a thin hollow sphere is

$$s = Pr/2t$$

The ASME Code that transforms this equation for practical use requires that the minimum thickness for spherical shells, hemispherical heads, elliptical heads, or ASME torispherical heads be determined from the equation

$$t_h = \frac{PDK}{2s_a e_w - 0.2P} = \frac{PD_o K}{2s_a e_w + 2P(K - 0.1)} \tag{4.19}$$

where $D$ is the inside diameter of the spherical vessel or hemispherical head, the inside major diameter for an elliptical head, or 2 (crown radius) for the ASME torispherical head. The value of the constant $K$ is given by

$$K = \frac{1}{6[2 + (D/D_1)^2]}$$

where $D_1$ is the minor diameter of the elliptical head. For the ASME torispherical head, $K = 0.885$.

The outer shell, which has a high vacuum on one side and atmospheric pressure on the other, is subject to the failure mode of external pressure collapse. A thin-walled external shell will collapse under an externally applied pressure at stress values much lower than the yield strength for the material.

To determine the collapsing pressure, the outer shell is assumed to be perfectly round and of uniform thickness, the material obeys Hook's law, the radial stress is negligible, and the normal stress distribution is linear. Other assumptions are also made. Using the theory of elasticity, the *collapsing pressure* is

$$P_c = KE\left(\frac{t}{D}\right)^3 \tag{4.20}$$

The factor $K$, a numerical coefficient, depends upon the $L/D$ and $D/t$ ratios ($D$ is the outer shell diameter), the kind of end support, and whether pressure is applied radially only or at the ends as well. This failure mode is covered by the ASME Code, in which design charts are presented for the design of cylinders and spheres subjected to external pressure.

One simple embodiment of the collapsing pressure for a long cylinder exposed to external pressure is given by

$$P_c = \frac{2E(t/D_o)^3}{1-\mu^2} \tag{4.21}$$

where $E$ is the Young's modulus of shell material, $t$ the shell thickness $D_o$ the outside diameter of shell and $\mu$ the Poisson's ratio for shell material.

A long cylinder in this analysis has an $L/D$ such that

$$L/D_o > 1.140(1-\mu^2)^{1/4}(D_o/t)^{1/2} \tag{4.22}$$

where $L$ is the unsupported length of the cylinder (distance between stiffening rings for the outer shell). When the cylinder is stiffened with rings, the shell may be assumed to be divided into a series of shorter shells equal in length to the ring spacing.

For short cylinders, the collapsing pressure may be estimated from

$$P_c = \frac{2.42E(t/D_o)^{5/2}}{(1-\mu^2)^{3/4}[L/D_o - 0.45(t/D_o)^{1/2}]} \tag{4.23}$$

where $L$ is the distance between stiffening rings for the outer shell.

The heads that enclose the ends of the outer shell cylinder are also subject to collapsing pressure. The critical pressure for a hemispherical, elliptical, or torispherical head (or for a spherical vessel) is given by

$$P_c = \frac{0.5E(t_h/R_o)^2}{[3(1-\mu^2)]^{1/2}} \tag{4.24}$$

where $t_h$ is the thickness of the head and $R_o$ is the outside radius of the spherical head, or the equivalent radius for the elliptical head, or the crown radius of the torispherical head. The equivalent radius for elliptical heads is given by $R_o = K_1 D$, where $D$ is the major diameter and $K_1$ is the shape factor.

Stresses on the piping needed to fill and empty cryogenic containers must also be taken into account.

The minimum wall thickness for piping under internal pressure is given by the ASA Code as

$$t = \frac{PD_o}{2s_a + 0.8P}$$

where $P$ is the design pressure, $D_o$ the outside diameter of pipe, and $s_a$ the allowable stress of pipe material.

For piping subjected to external pressure, Eq. (4.21) may be used.

In addition to pressure-induced stress, thermal stresses on the piping are often a serious consideration. When the deformation arising from change of temperature is prevented, temperature stresses arise that are proportional to the amount of deformation that is prevented. Let $\alpha$ be the coefficient of expansion per degree of temperature, $l_1$ the length of bar at temperature $T_1$, and $l_2$ the length at temperature $T_2$. Then

$$l_2 = l_1[1 + \alpha(T_2 - T_1)]$$

If, subsequently, the structured member is cooled to a temperature $T_1$, the proportionate deformation is $s = \alpha(T_2 - T_1)$, and the corresponding unit stress is

$$s = E\,\alpha\,x(T_2 - T_1).$$

This discussion of design considerations is by no means complete. Load-induced bending stresses, flexure stresses, and support systems are not mentioned at all. The intent here is merely to show the practical application of the low-temperature properties discussed in this section. For complete design considerations, one should consult *Mark's Standard Handbook for Mechanical Engineers* (1978), the appropriate codes, and such classic strength of materials books as that of Timoshenko and Gere (1961).

*Example 4.2.* A food freezing firm is currently using an ammonia refrigeration system to maintain a food warehouse temperature of 255.6 K. Plans are being drawn to change to a spray liquid nitrogen system in which controlled amounts of liquid nitrogen are sprayed directly into the cold storage room. Because of this, it must be assumed that at times parts of the warehouse shelving will drop to 77.8 K.

You must determine whether or not the existing structures can be safely used. Assume that the shelving was assembled at 300 K and that

Coefficient of linear expansion $\beta = 1.656 \times 10^{-7}/^\circ C$
Modulus of elasticity $E = 1.86 \times 10^{13}$ Pa
Yield strength $= 2.76 \times 10^8$ Pa
Ultimate tensile strength $= 1.17 \times 10^9$ Pa
Working stress $= 1/5 \times$ ultimate strength

a. Will the shelves rupture if exposed to liquid nitrogen?
b. Will permanent set occur?
c. From the working stress viewpoint, is the design satisfactory?
d. For the next generation plant, a quick choice must be made between aluminum and brass shelving. Is there any obvious low-temperature material reason (not considering cost) to prefer one over the other at this stage of the design?

By definition,

$$E = \frac{s}{\epsilon} = \frac{\text{stress}}{\text{strain}} = 1.86 \times 10^{13}\ \text{Pa} \tag{4.25}$$

$$\beta = \frac{\Delta l}{l \Delta T}, \quad \epsilon = \frac{\Delta l}{l} = \beta = \frac{\epsilon}{\Delta T} \quad \text{or } \epsilon = \beta \Delta T$$

Stress:

$$\begin{aligned} s = \epsilon E &= \beta \Delta T E \\ &= (1.656 \times 10^{-7})(255.6 - 77.8)(1.86 \times 10^{13}) \\ &= 5.48 \times 10^{8}\ \text{Pa} \end{aligned} \tag{4.26}$$

a. Stress < ultimate strength of $1.17 \times 10^9$ Pa. Therefore, material will not rupture.
b. Stress > yield strength of $2.76 \times 10^8$ Pa. Therefore, permanent set will occur. Not satisfactory.
c. Working stress = $(1/5)(1.17 \times 10^9\ \text{Pa}) = 2.34 \times 10^8$ Pa.
Stress $5.48 \times 10^8$ Pa > $2.34 \times 10^8$ Pa.
Stress exceeds working stress. Not satisfactory.
d. No. Aluminum and brass are both fcc metals and should serve equally well, all other things being the same.

## 11. MATERIAL SELECTION CRITERIA FOR CRYOGENIC TANKS

### 11.1. General Design Considerations

High-performance cryogenic storage vessels employ the concept of the dewar flask design principle—a double-walled container with the space between the two vessels evacuated and filled with an insulation. A fill line and a drain line are essential elements for larger vessels, and these may be one and the same. A vapor vent line must also be provided to allow the vapor formed as a result of heat inleak to escape and prevent overpressurization of the vessel.

Cryogenic fluid storage vessels are not designed to be filled completely for two major reasons. First, heat leak into the container is always present; therefore, the vessel pressure would rise quite rapidly if little or no vapor space were allowed. Second, inadequate cooldown of the inner vessel during a rapid filling operation would result in excessive boil-off, and the liquid would be percolated through the vent tube and wasted if no ullage space were provided. A 10% ullage volume is commonly used for large storage vessels.

The details of conventional cryogenic fluid storage vessel design are covered in such standards as the ASME Boiler and Pressure Vessel Code, Section VIII (1983) for unfired vessels. Most users require that the vessels be designed, fabricated, and tested according to this code.

The inner vessel must be constructed of a material compatible with the cryogenic fluid. Therefore 18-8 stainless steel, aluminum, Monel, Inconel, and titanium are commonly used for the inner shell. These materials are much more expensive than ordinary carbon steel, so the designer would like to make the inner vessel wall as thin as possible to hold the cost within reason. In addition, a thick-walled vessel requires a longer cooldown time, wastes more liquid in cooldown, and introduces the possibility of thermal stresses in the vessel wall during cooldown.

Table 4.11 lists the allowable stresses for some of these materials.

As discussed earlier, an important property of metals chosen for cryogenic service is their grain size. Grain boundaries, which are discontinuities in the orientation of neighboring crystals, are obstacles to slip-plane movement. A metal having a small grain size will have a higher yield stress than one with a larger grain size.

**Table 4.11** Allowable Stress for Materials at Room Temperature or Lower

| Material | | Allowable Stress | |
|---|---|---|---|
| | | MPa | psi |
| Carbon steel (for outer shell only) | SA-285 Grade C | 94.8 | 13,750 |
| | SA-299 | 129.2 | 18,750 |
| | SA-442 Grade 55 | 94.8 | 13,750 |
| | SA-516 Grade 60 | 103.4 | 15,000 |
| Low-alloy steel | SA-202 Grade B | 146.5 | 21,250 |
| | SA-353-B (9% Ni) | 163.7 | 23,750 |
| | SA-203 Grade E | 120.6 | 17,500 |
| | SA-410 | 103.4 | 15,000 |
| Stainless steel | SA-240 (304) | 129.2 | 18,750 |
| | SA-240 (304L) | 120.6 | 17,500 |
| | SA-240 (316) | 129.2 | 18,750 |
| | SA-240 (410) | 112.0 | 16,250 |
| Aluminum | SB-209 (1100-0) | 16.2 | 2,350 |
| | SB-209 (3004-0) | 37.9 | 5,500 |
| | SB-209 (5083-0) | 68.9 | 10,000 |
| | SB-209 (6061-T4) | 41.4 | 6,000 |
| Copper | SB-11 | 46.2 | 6,700 |
| | SB-169 (annealed) | 86.2 | 12,500 |
| Nickel alloys (annealed) | SB-127 (Monel) | 128.2 | 18,600 |
| | SB-168 | 137.9 | 20,000 |

*Source*: ASME Code, Sec. VIII (1983).

One reason annealed metals are more ductile than work-hardened ones is that work hardening increases the grain size.

For large storage vessels, 9% nickel steels are commonly used with higher boiling cryogens ($T > 75$ K), whereas aluminum alloys and austenitic steels can generally be used over the entire range of liquid temperatures. To speed up cooldown time and lower the cost of the cooldown of the container, the thickness of the inner shell is kept as small as possible and is only designed to withstand the maximum internal pressure and the bending forces. Section VIII of the ASME Boiler and Pressure Vessel Code provides all the necessary design equations and should be adhered to for safety reasons when designing vessels of capacities greater than 0.2 $m^3$.

## 11.2. Constitution and Structure of Steel

As a result of the methods of production, the following elements are always present in steel: carbon, manganese, phosphorus, sulfur, silicon, and traces of oxygen, nitrogen, and aluminum. Various alloying elements are frequently added, such as nickel, chromium, copper, molybdenum, and vanadium. The most important constituent is carbon, and it is necessary to understand the effect of carbon on the internal structure of steel.

When pure iron is heated to 910°C, its internal crystalline structure changes from a body-centered cubic arrangement of atoms (alpha-iron) to face-centered cubic (gamma-iron). The alpha-iron containing carbon, or any other element in solid solution, is called ferrite; and the gamma-iron containing other elements in solid solution is called austenite.

When the iron solution is cooled rapidly, and the carbon does not have time to separate out in the form of carbide, the austenite transforms to a highly stressed structure supersaturated with carbon. This structure is called martensite. Martensite is exceedingly hard, but it is brittle and requires tempering to increase the ductility. Therefore, martensite should be avoided for cryogenic service.

### 11.3. Nickel Steel and Nickel Alloys

Nickel steel is steel that contains nickel as the predominant alloying element. Some nickel steels are the toughest structural materials known. Compared to steel, other nickel steels have ultrahigh strength, high proportional limits, and high moduli of elasticity. At cryogenic temperatures, nickel alloys are strong and ductile.

Nickel added to carbon steel increases the steel's ultimate strength, elastic limit, hardness, and toughness. It narrows the hardening range but lowers the critical range of steel, reducing the danger of warping and cracking, and balances the intensive deep-hardening effect of chromium. Nickel steels are also of finer structure than ordinary steels, and the nickel retards grain growth.

When the percentage of nickel is high, the steel is very resistant to corrosion. Above 20% nickel, the steel becomes a single-phase austenitic structure. The steel is nonmagnetic above 29% nickel, and the maximum value of magnetic permeability is at about 78% nickel. The lowest coefficient of thermal expansion value of the steel is at 36% nickel.

The fracture and toughness properties of steels have been extensively studied. The results consistently indicate that the 5–9% nickel steels retain satisfactory properties down to −260°F. As the temperature is reduced from −260°F to −320°F, the fatigue and fracture resistance may be degraded for the 5% nickel steel. Therefore, for temperatures below −320°F, such as liquid hydrogen temperatures, a 9% nickel steel is a minimum requirement. The data in Table 4.12 give some

**Table 4.12** Relative Values for the Properties of Nickel Alloys

| Temperature (°F) | Density ($lb/in^3$) | Young's modulus ($10^4$ psi) | Shear modulus ($10^6$ psi) | Poisson's ratio | Thermal conductivity (Btu/ft/h/°F) | Thermal expansion (mean) ($10^{-6}$/°F) | Specific heat ($10^{-1}$Btu/lb/°F) |
|---|---|---|---|---|---|---|---|
| *3.5Ni* | | | | | | | |
| 70 | 0.284 | 29.6 | 11.5 | 0.282 | 20 | 6.6 | 108 |
| −150 | – | 30.5 | 11.9 | 0.281 | 17 | 5.7 | 84 |
| *5Ni* | | | | | | | |
| 70 | 0.282 | 28.7 | 11.2 | 0.283 | 18 | 6.6 | 108 |
| −260 | – | 30.1 | 11.8 | 0.277 | 12 | 5.2 | 60 |
| −320 | – | 30.3 | 11.8 | 0.277 | 9.2 | 4.9 | 36 |
| *9Ni* | | | | | | | |
| 70 | 0.283 | 28.3 | 10.7 | 0.286 | 16 | 6.6 | 108 |
| −260 | – | 29.6 | 11.2 | 0.281 | 10 | 5.2 | 60 |
| −320 | – | 29.8 | 11.3 | 0.280 | 7.5 | 4.9 | 36 |
| *36Ni (Invar)* | | | | | | | |
| 70 | 0.282 | 22.1 | 8.08 | 0.284 | 8.0 | 0.65 | |
| −320 | – | 20.4 | 7.37 | 0.307 | 3.6 | 1.00 | |
| −452 | – | 20.6 | 7.33 | 0.332 | 0.18 | 0.77 | |

relative values for the properties of nickel alloys at both room and cryogenic temperatures.

Commercial nickel and nickel alloys are available in a wide range of wrought and cast grades; however, considerably fewer casting grades are available. Wrought alloys tend to be better known by trade names such as Monel, Hastelloy, Inconel, and Incoloy. There is also a class of superalloys that are nickel-based, strengthened by intermetallic compound precipitation in a face-centered cubic matrix.

Most wrought nickel alloys can be hot- and cold-worked, machined, and welded successfully. The casting alloys can be machined or ground, and many can be welded or brazed. Nearly any shape that can be forged in steel can be forged in nickel and nickel alloys. However, because nickel work-hardens easily, severe cold-forming operations require frequent intermediate annealing to restore soft temper. Annealed cold-rolled sheet is best for spinning and other manual work.

### *11.3.1. Behavior in Welding*

Nickel alloys can be joined by shielded metal arc, gas tungsten arc, gas metal arc, plasma arc, electron beam, oxyacetylene, and resistance welding; by silver and bronze brazing; and by soft soldering. Resistance welding methods include spot, seam, protection, and flash welding.

Special nickel alloys, including superalloys, are best worked at about 800–2200°F. In the annealed condition, these alloys can be cold-worked by all standard methods. Required forces and rate of work hardening are intermediate between those of mild steel and Type 304 stainless steel. These alloys work-harden to a greater extent than the austenitic stainless steels, so they require more intermediate annealing steps.

## 11.4. Stainless Steels

The principal reason for the existence of stainless steels is their outstanding resistance to corrosion. However, stainless steels also exhibit excellent mechanical properties, and in some applications these mechanical properties are as important as corrosion resistance in determining the life or utility of a given structure.

Within the stainless steel family, there are martensitic, ferritic, austenitic, and precipitation-hardenable steels. These steels exhibit a wide variety of mechanical properties at room and cryogenic temperatures. The martenstic and precipitation-hardenable steels can be heat-treated to high strengths at room temperature. Thus, these steels are frequently used for structural applications. The ferritic stainless steels exhibit better resistance to corrosion than the martensitic stainless steels and can be polished to a high luster. However, they cannot be treated to high strengths. Thus, they are used in applications where appearance is more important, such as in automotive and appliance trim.

## 11.5. Austenitic Stainless Steels

The austenitic stainless steels exhibit a wide range of mechanical properties. They are soft and ductile in the annealed condition and can be easily formed. They can also be appreciably strengthened by cold-working, but even at high strength levels they exhibit good ductility and toughness. This good ductility and toughness are retained to cryogenic temperature. Thus, austenitic steels are used in the annealed condition in applications requiring good formability. In the cold-worked condition, they are frequently used in structural applications. For cryogenic temperature applications, the austenitic steels are unsurpassed because they are relatively easy to fabricate and weld, do not require heat treatment after fabrication, and exhibit relatively high strength with excellent stability and toughness at very low temperatures.

Austenitic chromium–nickel stainless steels are the most widely used stainless steels. While resistance to corrosion is their chief attribute, they also display low magnetic permeability, good high-temperature strength, and excellent toughness at low temperatures.

The variations in composition among the standard austenitic stainless steels are important, both in the performance of the steel in service and in its behavior in welding. As examples, Types 302, 304, and 304L represent the so-called 18-8 stainless steels. They differ chiefly in carbon content, which determines the amount of carbide precipitation that can occur in the base metal heat-affected zones near a weld. Types 316 and 316L contain an addition of molybdenum for improved corrosion resistance. The presence of molybdenum will have both advantages and disadvantages in welding.

The austenitic steels are very well suited to cryogenic service. Consider the effect of cryogenic temperatures on the tensile properties of the austenitic stainless steels shown in Table 4.13 (ASM, 1983). This table lists the tensile properties of four austenitic stainless steels used in cryogenic service at room temperature, −320°F, and −425°F: Types 304, 304L, 310, and 347. Note that the high ductility (elongation and reduction of area) of the austenitic stainless steels is retained at cryogenic temperatures. Note also that the yield and tensile strengths of these steels increase as the temperature decreases, with the largest increase occurring in the tensile strength. The one exception is Type 310, which exhibits almost the same increase in tensile strength as in yield strength. This behavior is probably due to the slightly higher amount of carbon in Type 310 compared to the other steels listed. Interstitial elements, such as carbon, are known to have marked effects on the yield strength exhibited at low temperatures.

As indicated by the ductility exhibited by the austenitic steels, the toughness of these steels is excellent at cryogenic temperatures. The results of Charpy V-notch impact tests on Types 304, 304L, 310, and 347 conducted at room temperature,

**Table 4.13** Tensile Properties of Some Austenitic Stainless Steels

| AISI type | Testing temp. (°F) | Yield strength (0.2 % offset) (psi) | Tensile strength (psi) | Elongation in 2 in. (%) | Reduction in area (%) |
|---|---|---|---|---|---|
| 304 | 75 | 33,000 | 85,000 | 60 | 70 |
| 304 | −320 | 57,100 | 205,500 | 43 | 45 |
| 304 | −425 | 63,700 | 244,500 | 48 | 43 |
| 304L | 75 | 28,000 | 85,000 | 60 | 60 |
| 304L | −320 | 35,000 | 194,500 | 42 | 50 |
| 304L | −425 | 33,900 | 220,000 | 41 | 57 |
| 310 | 75 | 45,000 | 95,000 | 60 | 65 |
| 310 | −320 | 84,900 | 157,500 | 54 | 54 |
| 310 | −425 | 115,500 | 177,500 | 56 | 61 |
| 347 | 75 | 35,000 | 90,000 | 50 | 60 |
| 347 | −320 | 41,200 | 186,000 | 40 | 32 |
| 347 | −425 | 45,500 | 210,500 | 41 | 50 |

*Source:* ASM (1983).

**Table 4.14** Transverse Charpy V-Notch Impact Strength of Some Austenitic Stainless Steels

| AISI type | Energy absorbed (ft lb) | | |
|---|---|---|---|
| | 80°F | −320°F | −425°F |
| 304 | 154 | 87 | 90 |
| 304L | 118 | 67 | 67 |
| 310 | 142 | 89 | 86 |
| 347 | 120 | 66 | 57 |

*Source*: ASM (1983).

−320°F, and −425°F are shown in Table 4.14. The toughness of these four steels decreases somewhat as the temperature is decreased from room temperature to −320°F, but on the further decrease to −425°F the toughness remains about the same. It should be noted that the toughness of these steels is excellent at −425°F and that the toughness of Types 304 and 310 is significantly higher than those of Types 304L and 347. Shown in Table 4.15 are the results of impact tests on the same four steels at −320 and −425°F for various thickness of plate. These results demonstrate that the toughness of the austenitic stainless steels is not markedly affected by plate thickness.

Table 4.16 shows the results of impact tests done on Type 304 stainless steel after up to 1 year of exposure at −320°F. These results indicate that Type 304 is very

**Table 4.15** Low-Temperature Impact Strength of Several Annealed Austenitic Stainless Steels

| AISI type | Testing temp (°F) | Specimen orientation | Type of notch | Product size | Energy absorbed (ft lb) |
|---|---|---|---|---|---|
| 304 | -320 | Longitudinal | Keyhole | 3 in. plate | 80 |
| 304 | -320 | Transverse | Keyhole | 3 in. plate | 80 |
| 304 | -320 | Transverse | Keyhole | $2^1/_2$ in. plate | 70 |
| 304 | -425 | Longitudinal | Keyhole | $^1/_2$ in. plate | 80 |
| 304 | -425 | Longitudinal | V-notch | $3^1/_2$ in. plate | 91.5 |
| 304 | -425 | Transverse | V-notch | $3^1/_2$ in. plate | 85 |
| 304L | −320 | Longitudinal | Keyhole | $^1/_2$ in. plate | 73 |
| 304L | −320 | Transverse | Keyhole | $^1/_2$ in. plate | 43 |
| 304L | −320 | Longitudinal | V-notch | $3^1/_2$ in. plate | 67 |
| 304L | −425 | Longitudinal | V-notch | $3^1/_2$ in. plate | 66 |
| 310 | -320 | Longitudinal | V-notch | $3^1/_2$ in. plate | 90 |
| 310 | -320 | Transverse | V-notch | $3^1/_2$ in. plate | 87 |
| 310 | -425 | Longitudinal | V-notch | $3^1/_2$ in. plate | 86.5 |
| 310 | -425 | Transverse | V-notch | $3^1/_2$ in. plate | 85 |
| 347 | -320 | Longitudinal | Keyhole | $^1/_2$ in. plate | 60 |
| 347 | -320 | Transverse | Keyhole | $^1/_2$ in. plate | 47 |
| 347 | -425 | Longitudinal | V-notch | $3^1/_2$ in. plate | 59 |
| 347 | -425 | Transverse | V-notch | $3^1/_2$ in. plate | 53 |
| 347 | -300 | Longitudinal | V-notch | $6^1/_2$ in. plate | 77 |
| 347 | -300 | Transverse | V-notch | $6^1/_2$ in. plate | 58 |

**Table 4.16** Charpy Keyhole Impact Values at −320°F on Type 304 After Prolonged Exposure at −320°F

| Time of exposure | Energy absorbed (ft lb) |
|---|---|
| 0 | 85.0 |
| 30 min | 72.5 |
| 6 months | 73.0 |
| 12 months | 75.0 |

stable and does not exhibit any marked degradation of toughness due to long-time exposure at −320°F.

### *11.5.1. Behavior in Welding*

The austenitic stainless steels are considered the most weldable of the high alloy steels. This favorable appraisal is mainly based on the great toughness of welded joints, even in the as-welded condition. Comparatively little trouble is experienced in making satisfactory welded joints if the inherent physical characteristics and mechanical properties of an austenitic steel are given proper consideration. For example, the austenitic stainless steels have a coefficient of expansion approximately 50% greater than that of plain steel, while the thermal conductivity is about one-third as great. This calls for more attention to factors that control warpage and distortion and to thermally induced stresses.

An important part of the successful welding of austenitic stainless steels is the control of compositions and microstructures through selection of type, welding procedure, and postweld treatment. As the steels become more complex in composition or heavier in section or the service conditions become more demanding, a greater degree of knowledge of stainless steel metallurgy is needed.

For instance, in the heat-affected zone adjacent to the welds in austenitic stainless steels, chromium carbides will precipitate to varying degrees. This precipitation of carbides, known as sensitization, depends on the carbon content of the stainless steel and the exposure time in the range of 800–1600°F. Because these carbides precipitate in the grain boundaries, they may be detrimental to the toughness of the steels. However, the toughness of the lower carbon Type 304 and 304L steels remains at relatively high levels after sensitization. Thus, for cryogenic equipment that is welded, low-carbon austenitic stainless steels such as Type 304 appear to be better materials of construction than the high-carbon austenitic stainless steels such as Type 302.

## 11.6. Titanium Alloys

Depending on the predominant phase or phases in their microstructure, titanium alloys are categorized as alpha, beta, and alpha–beta. The alpha phase in pure titanium is characterized by a hexagonal close-packed crystalline structure that remains stable from room temperature to approximately 1620°F. The beta phase in pure titanium has a body-centered cubic structure and is stable from 1620°F to the melting point, 3040°F.

The addition of alloying elements to titanium provides a wide range of physical and mechanical properties. Some elements, notably tin and zirconium, behave as neutral solutes and have little effect except to strengthen the alpha phase.

Like stainless steel, titanium sheet and plate work-harden significantly during forming. Minimum bend-radius rules are nearly the same for both, although spring-back is greater for titanium. Commercially pure grades of heavy plate are cold-formed or, for more severe shapes, warm-formed at temperatures of about 800°F. Alloy grades can be formed at temperatures as high as 1400°F in inert gas atmospheres.

Titanium plates can be sheared, punched, or perforated on standard equipment. The harder alloys are more difficult to shear, so thickness limitations are generally about two-thirds of those for stainless steel. Titanium and its alloys can be machined and abrasive-ground; however, sharp tools and continuous feed are required to prevent work hardening. Tapping is difficult because the metal galls. Coarse threads are used whenever possible.

### *11.6.1. Welding Titanium Alloys*

Generally, titanium is welded by gas tungsten arc (GTA) or plasma arc techniques. Inert gas processes can be used under certain conditions. In all aspects, GTA welding of titanium is similar to that of stainless steel. Normally, a sound weld appears bright silver with no discoloration on the surface or along the heat-affected zone.

## 11.7. Aluminum and Aluminum Alloys

Nonferrous metals such as aluminum offer a wide variety of mechanical properties and material characteristics. In addition to property variations, nonferrous metals and alloys differ in cost, based on availability, abundance, and the ease with which the metal can be converted into useful forms.

When selecting any material for a mechanical or structural application, some important considerations include how easily the material can be shaped into a finished product and how its properties change or are altered, either intentionally or inadvertently, in the shaping process. Like steels, nonferrous metals can be cast, rolled, formed, forged, extruded, or worked by other deforming processes. Although the same operations are used with ferrous and nonferrous metals and alloys, the reaction of nonferrous metals to these forming processes is often more severe. Consequently, properties may differ considerably between cast and wrought forms of the same metal or alloy. Each forming method imparts unique physical and mechanical characteristics to the final component.

Aluminum and its alloys, numbering in the hundreds, are available in all common commercial forms. Aluminum alloy sheet can be formed, drawn, stamped, or spun. Many wrought or cast aluminum alloys can be welded, brazed, or soldered, and aluminum surfaces readily accept a wide variety of finishes. Aluminum reflects radiant energy throughout the entire spectrum, is nonsparking, and is nonmagnetic.

Though light in weight, commercially pure aluminum has a tensile strength of about 13,000 psi. Cold working the metal approximately doubles it strength. In other attempts to increase strength, aluminum is alloyed with elements such as manganese, silicon, copper, magnesium, or zinc. The alloys can also be strengthened by cold working. Some alloys are further strengthened and hardened by heat treatments.

At subzero temperatures, aluminum is stronger than at room temperature and is no less ductile. However, most aluminum alloys lose their strength at elevated temperatures.

Wrought aluminum alloys are the most useful for low-temperature service. A four-digit number that corresponds to a specific alloying element usually designates wrought aluminum alloys (Guyer, 1989).

To develop strength, heat-treatable wrought aluminum alloys are solution heat-treated, then quenched and precipitation-hardened. Solution heat treating consists of heating the metal, then holding it at temperature to bring the hardening constituents into solution, then cooling to retain those constituents in solution. Precipitation hardening after solution heat treatment increases the strength and hardness of these alloys.

Wrought aluminum alloys are also strengthened by cold working. The high-strength alloys, whether heat-treatable or not, work-harden more rapidly than the softer, lower strength alloys and so may require annealing after cold working. Because hot forming does not always work-harden aluminum alloys, this method is used to avoid annealing and straightening operations. However, hot forming fully heat-treated materials is difficult. Generally, aluminum formability increases with temperature.

# 5

# Transport Properties of Solids

## 1. THERMAL PROPERTIES

The thermal properties of materials at low temperatures of most interest to the process engineer are specific heat, thermal conductivity, and thermal expansivity. (We shall use the term "specific heat" to refer to a material property and the term "heat capacity" to refer to the atomic scale contributions to specific heat.) It will be shown that each of these properties depends on the intermolecular potential of the lattice and accordingly that they are interrelated.

### 1.1. Specific Heat

#### *1.1.1. Lattice Heat Capacity*

Nearly all the physical properties of a solid (e.g., specific heat, thermal expansion) depend on the vibration or motion of the atoms in the solid. Specific heat is often measured at low temperatures for design purposes. However, specific heat measurements are important in their own right, because the variation of specific heat with temperature shows how energy is distributed among the various energy-absorbing modes of the solid. Thus specific heat measurements give important clues to the structure of the solid. Finally, because other properties also depend on the lattice structure and its vibration, specific heat measurements are used to predict or correlate other properties, such as thermal expansion. Therefore, an understanding of the temperature dependence of specific heat not only gives useful design information but also is helpful in predicting other thermal properties.

The specific heat of any material is defined from thermodynamics as

$$C_V = \left(\frac{\partial U}{\partial T}\right)_V \tag{5.1}$$

where $U$ is the internal energy, $T$ is the absolute temperature, and $V$ is the volume.

$C_V$ is the property more useful to theory than $C_P$, because it directly relates internal energy, and hence the microscopic structure of the solid, to temperature. However, it must be remembered that most solids expand when they are heated at constant pressure. As a result, the solid does work against both internal and external forces. The specific heat measured at constant pressure, $C_P$, then, includes some extra heat to provide this work. Under ordinary circumstances, $C_P$ is the specific heat observed. Therefore, $C_V$ must be calculated from $C_P$. Fortunately, using thermodynamic theory, this is not difficult, and the relation between $C_P$ and $C_V$ will be shown later. In addition, for solids (and liquids), the difference between $C_P$ and $C_V$ is usually less than 5% at room temperature. Accordingly, we shall limit the present discussion to $C_V$, to lattice effects only, and to electrical insulators. The specific heat contribution of free electrons in a metal will be taken up after the lattice effects have been explored.

**(a) Models of Lattice Specific Heat.** The various models of lattice specific heat will be discussed in the order of their historical development, which is also more or less the order of their exactness.

**Dulong and Petit.** In 1911, Dulong and Petit observed that the heat capacity of many solids, both metals and nonmetals, was very nearly independent of temperature near room temperature, and furthermore that it had a value of 6 cal/(mol K) regardless of the substance. This experimentally observed fact can be explained if it is assumed that each lattice point can absorb energy in the same way as every other lattice point (the equipartition of energy). In this model, the atoms are assumed to oscillate as *independent classical* particles. Thus the total thermal energy of 1 mol of material would be

$$U = (3)(2)N_o\left(\frac{kT}{2}\right) = 3N_okT \tag{5.2}$$

where 3 is the number of degrees of freedom, 2 arises from the fact that the bound oscillator has equal amounts of both kinetic and potential energy, $N_o$ is Avogadro's number, the number of particles per mole, and $(kT/2)$ is the thermal energy of each degree of freedom.

From Eq. (5.1), then,

$$C_V = \frac{\partial U}{\partial T} = 3N_ok = 6\,\text{cal/(mol K)}$$

which is the Dulong–Petit value.

This brief derivation indicates that $C_V$ is not a function of temperature, but this is not the case. Instead, heat capacity varies with temperature as shown in Fig. 5.1 and goes to zero as the absolute temperature goes to zero.

**Einstein.** The theory of lattice specific heat was basically solved by Einstein, who introduced the idea of quantized oscillation of the atoms. He pointed out that, owing to the quantization of energy, the law of equipartition must break down at low temperatures. Improvements have since been made on this model, but they all include the quantization. Einstein treated the solid as a system of simple harmonic oscillators of the same frequency. He assumed that each oscillator is independent. This is not really the case, but the results, even with this assumption, were remarkably good. All the atoms are assumed to vibrate, due to their thermal motions, with a frequency $\nu$, and according to quantum theory each of the three degrees of freedom has an associated energy $E = h\nu/(e^{h\nu/kT} - 1)$ in place of the $kT$ postulated by classical mechanics. The lattice specific heat, $C_{V1}$, thus becomes temperature-dependent. Einstein deduced for the specific heat at constant volume the formula

$$C_{V1} = \left(\frac{\mathrm{d}U}{\mathrm{d}T}\right)_V = 3R\frac{\mathrm{e}^{\theta_\mathrm{E}/T}(\theta_\mathrm{E}/T)^2}{(\mathrm{e}^{\theta_\mathrm{E}/T} - 1)^2} \tag{5.3}$$

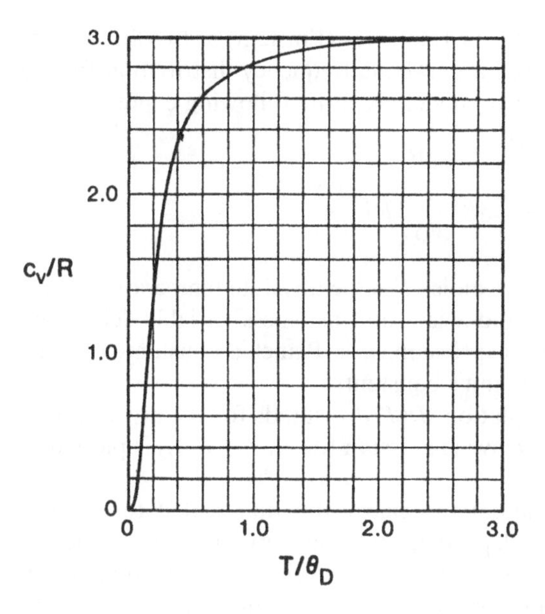


**Figure 5.1** Specific heat curve typical of simple isotropic solids.

or

$$C_{V1} = 3R\left(\frac{\theta_E}{T}\right)^2\left(\frac{e^{\theta_E/T}}{(e^{\theta_E/T}-1)^2}\right) \tag{5.4}$$

where $N_o k$ has been replaced by its equivalent $R$, the general gas constant, and where $\theta_E = h\nu/k$, which has the dimensions of temperature and is the Einstein characteristic temperature. The characteristic temperature $\theta_E$ is the parameter that permits variation from one material to another. It permits plotting all materials on one curve by replacing $T$ by $T/\theta_E$ as the abscissa. Expression (5.4) fits experimental data quite well for all materials except that at low temperatures it drops below the experimental data. At intermediate and high temperatures, the fit is good, and Eq. (5.4) provides the approach to a limiting Dulong–Petit value of $3R$ at high $T$. Although now superseded by more exact models, the Einstein model included the most important effect, the quantization.

**Nernst and Lindemann**. The Einstein model was a considerable improvement over that of Dulong and Petit, but it still did not have the correct low-temperature shape. Real atoms in the lattice do not vibrate at only one frequency, but are coupled to one another in a complex way that depends upon both the lattice type and the nature of the interatomic forces.

Nernst and Lindemann improved upon the Einstein model by assuming that the atoms vibrate at two frequencies, one equal to the Einstein frequency $\nu$, and one at half that value, $\nu/2$. This treatment, while an improvement, was still a considerable oversimplification and was on shaky theoretical grounds because it required the introduction of a half quantum of energy.

**Debye**. Debye made a major advance in the theory of heat capacity at low temperatures by treating a solid as an infinite elastic continuum and considering the excitement of all possible standing waves in the material. These range from the acoustic vibrations up to a limiting frequency, $\nu_m$. Instead of a system of simple harmonic oscillators all vibrating at only one frequency (Einstein) or two frequencies (Nernst and Lindemann), Debye derived a parabolic frequency distribution for the atoms vibrating in the lattice. The Debye model gives the following expression for the lattice heat capacity per mole ($C_{V1}$):

$$C_{V1} = 9R\left(\frac{T}{\theta_D}\right)^3 \int_0^{\theta_D/T} \frac{x^4 e^x \, dx}{(e^x - 1)^2} = 3R\left(\frac{T}{\theta_D}\right) D\left(\frac{T}{\theta_D}\right) \tag{5.5}$$

where $R$ is the universal gas constant per mole, $T$ is the absolute temperature, $\theta_D$ is the Debye characteristic temperature, and $x$ is a dimensionless variable defined by the expression $x = h\nu/kT$. In this last expression, $h$ is Planck's constant, $\nu$ is the frequency of vibration, and $k$ is Boltzmann's constant.

In the abbreviated form of the equation, $D(T/\theta_D)$ stands for the expression in the integral, the *Debye function*. The Debye expression for $C_V$ vs. $T/\theta_D$ is plotted in Fig. 5.1.

The important thing to notice about the Debye function is that for a given substance, the lattice heat capacity is dependent only on a mathematical function of the ratio of the Debye characteristic temperature $\theta_D$ to the absolute temperature. This mathematical function is the same for all materials, and only $\theta_D$ changes from material to material. $\theta_D$ thus appears as a constant characteristic of the material at hand.

The value of $\theta_D$ is an adjustable parameter for the temperature scale, similar to $\theta_E$ for the Einstein model. Its value usually lies between 200 and 400 K, although values occur far outside this range. Values of $\theta_D$ are given in Table 5.1.

The Debye function predicts heat capacity surprisingly well, especially considering that Debye merely assumed a parabolic form for the vibration spectrum of every solid without regard for the individual differences to be expected among different solids.

At high temperatures ($T > 2\theta_D$), the specific heat given by Eq. (5.5) approaches a constant value of $3R$, which is the *Dulong–Petit value*, as it should be. At low temperatures ($T < \theta_D/12$), the Debye function approaches a constant value of $D(0) = 4\pi^4/5$; thus, the lattice specific heat at temperatures less than $\theta_D/12$ may be expressed as

$$C_{V1} = (12\pi^4 R/5)(T/\theta_D)^3 = 464.5(T/\theta_D)^3 = AT^3 \tag{5.6}$$

or

$$C_V = AT^3 \tag{5.7}$$

where $A$ is a constant.

From Eq. (5.6), we see that the lattice contribution to the specific heat of solids varies as the third power of the absolute temperature at very low temperatures. This behavior is useful in separating the effects of electrons from those of the lattice for electrical conductors, and this will be used later.

The Debye temperature, $\theta_D$, is not only important to estimating heat capacity but is also a useful property in itself. $\theta_D$ can be calculated directly from first principles by the theoretical expression

$$\theta_D = \frac{h\nu_a}{k}\left(\frac{3N}{4\pi V}\right)^{1/3} \tag{5.8}$$

**Table 5.1** Debye Characteristic Temperatures ($\theta_D$) of Some Representative Elements and Compounds at $T \sim \theta_D/2$

| | $\theta_D$ | | | $\theta_D$ | | | $\theta_D$ | | | $\theta_D$ | |
|---|---|---|---|---|---|---|---|---|---|---|---|
| Element | °R | K | Element | °R | K | Element | °R | K | Compound | °R | K |
| Ac | 180 | 100 | Ge | 666 | 370 | Pb | 153 | 85 | AgCl | 324 | 180 |
| Ag | 396 | 220 | H (para) | 207 | 115 | Pd | 495 | 275 | Alums | 144 | 80 |
| Al | 693 | 385 | H (ortho) | 189 | 105 | Pr | 216 | 120 | $As_2O_3$ | 252 | 140 |
| Ar | 162 | 90 | H ($n$-$D_2$) | 189 | 105 | Pt | 405 | 225 | $As_2O_5$ | 432 | 240 |
| As | 495 | 275 | He | 54 | 30 | Rb | 108 | 60 | $AuCu_3$ (ord.) | 360 | 200 |
| Au | 324 | 180 | Hf | 351 | 195 | Re | 540 | 300 | $AuCu_3$ (disord.) | 324 | 180 |
| B | 2196 | 1220 | Hg | 180 | 100 | Rh | 630 | 350 | BN | 1080 | 600 |
| Be | 1692 | 940 | I | 189 | 105 | Rn | 720 | 400 | $CaF_2$ | 846 | 470 |
| Bi | 216 | 120 | In | 252 | 140 | Sb | 252 | 140 | $Cr_2O_3$ | 648 | 360 |
| C (diamond) | 3650 | 2028 | Ir | 522 | 290 | Se | 270 | 150 | FeS | 1134 | 630 |
| C (graphite) | 1500 | 2700 | K | 180 | 100 | Si | 1134 | 630 | KBr | 324 | 180 |
| Ca | 414 | 230 | Kr | 108 | 60 | Sn (fcc) | 432 | 240 | KCl | 414 | 230 |
| Cd (hcp) | 504 | 280 | La | 234 | 130 | Sn (tetra) | 252 | 140 | KI | 351 | 195 |
| Cd (bcc) | 306 | 170 | Li | 756 | 420 | Sr | 306 | 170 | LiF | 1224 | 680 |
| Ce | 198 | 110 | Mg | 594 | 330 | Ta | 414 | 230 | MgO | 1440 | 800 |
| Cl | 297 | 165 | Mn | 756 | 420 | Tb | 315 | 175 | $MoS_2$ | 522 | 290 |
| Co | 792 | 440 | Mo | 675 | 375 | Te | 234 | 130 | NaCT | 504 | 280 |
| Cr | 774 | 430 | N | 126 | 70 | Th | 252 | 140 | RbBr | 234 | 130 |
| Cs | 81 | 45 | Na | 270 | 150 | Ti | 639 | 355 | RbI | 207 | 115 |
| Cu | 558 | 310 | Nb | 477 | 265 | Tl | 162 | 90 | $SiO_2$ (quartz) | 459 | 255 |
| Dy | 279 | 155 | Nd | 270 | 150 | V | 504 | 280 | $TiO_2$ (rutile) | 810 | 450 |
| Er | 297 | 165 | Ne | 108 | 60 | W | 567 | 315 | ZnS | 468 | 260 |
| Fe | 828 | 460 | Ni | 792 | 440 | Y | 414 | 230 | | | |
| Ga (rhomb) | 432 | 240 | O | 162 | 90 | Zn | 450 | 250 | | | |
| Ga (tetra) | 225 | 125 | Os | 450 | 250 | Zr | 432 | 240 | | | |
| Gd | 288 | 160 | Pa | 270 | 150 | AgBr | 252 | 140 | | | |

where $h$ is the Planck's constant; $\nu_a$ the speed of sound in the solid; $k$ the Boltzmann's constant and $N/V$ is the number of atoms per unit volume for the solid.

In practice, the Debye temperature is usually determined by selecting the value that makes the theoretical specific heat curve fit the experimental specific heat curve as closely as possible. In such cases, $\theta_D$ has lost its physical significance as a measure of limiting frequency and has become an "effective" characteristic temperature.

A modification of the Debye theory is sometimes used to calculate heat capacity in a semiempirical way. Essentially, $\theta_D$ is treated as a function of temperature rather than as a characteristic constant. A few heat capacities are measured, and $\theta_D$ is allowed to vary with temperature to get the best fit between the Debye function and experiment. Usually, the resulting variation of $\theta_D$ with temperature is not too great for metals. Plastics, however, are characterized by large changes in their interatomic force constants with temperature (equivalent to large changes in their elastic moduli). As a result, $\theta_D$ for these materials increases greatly with decreasing temperature.

The situation for compounds and alloys is more complicated. Near room temperature, the specific heat of a compound or alloy can be approximated quite well by linear combination of the specific heats of the constituent elements (the Kopp-Joule rule). This is because the temperature is sufficiently high that all modes of lattice vibration are fully excited (except for a few hard substances of low atomic weight such as diamond, BeO, $B_2O_3$), and the gram atomic heat capacity is about the same for all the constituent elements (the Dulong-Petit value).

This procedure is not proper at low temperatures because there the specific heat is determined by the arrangement and bonding of the lattice, and for a compound this bears no relation to the lattice characteristics of the component elements. This is seen at once by comparing the solid properties of an alkali halide with those of the alkali metal and the halogen as separate entities. Nevertheless the Debye function is found to represent simple compounds that form lattices with only one type of bonding (such as alkali halides, other binary salts, and some oxides). The $\theta_D$ value in this case bears no relation to the $\theta_D$ value of the constituent elements.

**Other Models.** Although the model of Debye gives good agreement with observation in many cases, deviations are observed. The cause seems to be that the detailed crystal structure is not introduced. Much further work has been done, beginning with Born and von Karman and Blackman and continued by many others. These models attempt to express the detailed situation of actual crystal lattices. This leads immediately to dispersion; that is, the velocity of sound depends upon the frequency. In all these models, the phonon (quantized sound wave or lattice vibration) picture is retained, and it has become one of the well-established particles. The frequency spectra differ from model to model, and each crystal leads to a different frequency spectrum. These other models are very important to achieving a detailed understanding of the crystal lattice. For process design purposes, however, the Debye function and $\theta_D$ yield adequate estimations. As usual, though, good experimental data are much to be preferred.

The specific heats of several materials used in low-temperature construction are shown in Fig. 5.2 as a function of temperature. Values for other materials are tabulated in several excellent references (Corruccini and Gniewek, 1960; Johnson, 1960; Landolt-Bornstein, 1961), and a few values are abstracted in Table 5.2 (Gopal, 1966).

*Example 5.1.* Determine the lattice specific heat of chromium at 200 K as given by the Debye function.

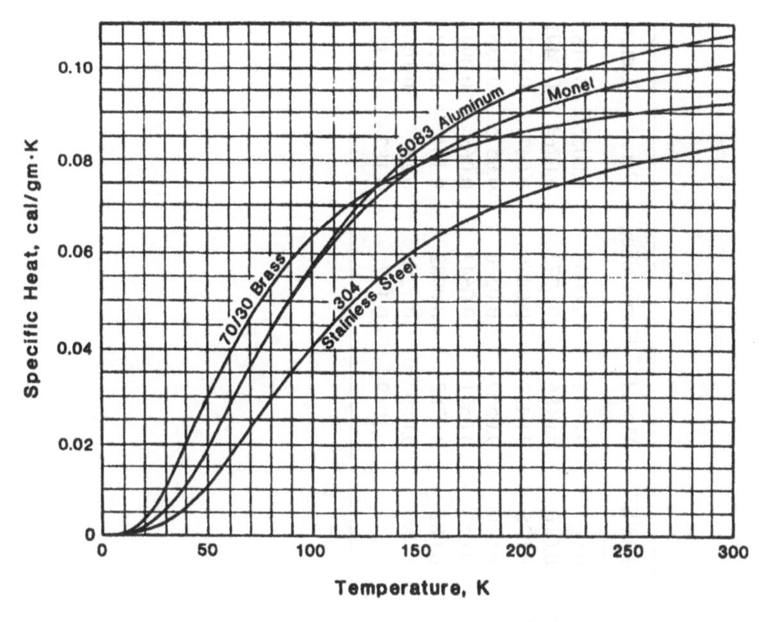


**Figure 5.2** Specific heat of several materials of construction.

For chromium, $\theta_D = 430$ K from Table 5.1 and molecular weight = 52
$T/\theta_D = 200/430 = 0.046$
From Fig. 5.1, $C_V/R = 2.5$ and $R = 1.987$ cal/(mol K)

$$C_V = (2.5)[1.987\,\text{cal/(mol K)}]/(52\,\text{g/mol})$$
$$= 0.096\,\text{cal/(g K)}\ \text{(compares to published value of 0.099 cal/(g K), Table 5.2)}$$

**(b) Estimation Techniques.** In the absence of experimental data, it may be necessary to estimate heat capacity and hence $\theta_D$. This may be done with modest success for elements and simpler compounds.

Procedures for calculating values of $\theta_D$, using either elastic constants, compressibility, melting point, or the temperature dependence of the expansion coefficient, and other properties, are outlined by Dillard and Timmerhaus (1968) and are summarized briefly here.

Of all the properties that do depend on the lattice constants and hence reflect $\theta_D$, usually only the melting point and thermal expansion coefficient are likely to be known better than the heat capacity itself. The formula relating melting point $T_m$ with $\theta_D$ is due to Lindemann:

$$\theta_D = 140\left(\frac{n}{V}\right)^{1/3}\left(\frac{nT_m}{M}\right)^{1/2} \tag{5.9}$$

where $V$ and $M$ are molar volume ($cm^3$) and molecular weight, respectively, and $n$ is the number of atoms in the molecule. Equation (5.9) has been shown to be a good approximation for common metals and cubic binary salts.

An experimental study of alloys showed that the specific heat of the copper–nickel alloys can be closely represented by using either the Kopp–Joule additive rule or a weighted mean $\theta_D$ value, and it seems probable that generally this will be true for simple alloy systems in which all compositions and the parent metals are of the same lattice type and in which the $\theta_D$ values of the parent metals are not widely

**Table 5.2** Specific Heats ($C_P$) of Some Selected Substances [cal/(g K)]

| Temperature (K) | Al | Mg | Cu | Ni | α-Mn | α-Fe | γ-Fe | Cr | 18–18 SS[a] | Monel | Fused silica | Pyrex | Teflon |
|---|---|---|---|---|---|---|---|---|---|---|---|---|---|
| 20 | 0.0024 | 0.0040 | 0.0019 | 0.0012 | 0.0025 | 0.0011 | 0.0014 | 0.0006 | 0.0011 | 0.0014 | 0.006 | 0.0055 | 0.0183 |
| 50 | 0.0337 | 0.0580 | 0.0236 | 0.0164 | 0.0211 | 0.0129 | 0.0218 | 0.0090 | 0.016 | 0.0186 | 0.0272 | 0.0264 | 0.0491 |
| 77 | 0.0815 | 0.119 | 0.0471 | 0.0392 | 0.0473 | 0.0343 | 0.0487 | 0.0277 | 0.038 | 0.0417 | 0.0470 | 0.047 | 0.0739 |
| 90 | 0.102 | 0.141 | 0.0554 | 0.0488 | 0.0574 | 0.0441 | 0.0604 | 0.0381 | 0.050 | 0.0509 | 0.0570 | 0.0575 | 0.0851 |
| 100 | 0.116 | 0.155 | 0.0607 | 0.0555 | 0.0641 | 0.0516 | 0.0684 | 0.0459 | 0.057 | 0.0571 | 0.0643 | 0.065 | 0.0931 |
| 150 | 0.164 | 0.202 | 0.0774 | 0.0785 | 0.0872 | 0.0775 | 0.0975 | 0.0757 | 0.085 | 0.0782 | 0.0982 | 0.101 | 0.132 |
| 200 | 0.191 | 0.221 | 0.0854 | 0.0915 | 0.1003 | 0.0918 | 0.118 | 0.0925 | 0.099 | 0.0897 | 0.129 | 0.132 | 0.166 |
| 298 | 0.215 | 0.235 | 0.0924 | 0.1060 | 0.1146 | 0.1070 | 0.1251 | 0.1073 | 0.114 | 0.1019 | 0.177 | 0.182 | 0.248[b] |

[a] SS = stainless steel.
[b] At 280 K.

different. The additive rule will fail for alloys having crystal structures and elastic constants widely different from those of either of the major components.

It was noted that the Debye and other functions give $C_V$, whereas one desires $C_P$ for most purposes. The difference between $C_P$ and $C_V$ is given by several thermodynamic expressions, one of which is

$$C_P - C_V = \alpha^2 VT/\beta_T \tag{5.10}$$

in which $V$ is the molar volume, $\alpha$ is $(1/V)(\mathrm{d}V/\mathrm{d}T)_P$, which is equal to three times the coefficient of linear expansion, and $\beta_T$ is $-(1/V)(\mathrm{d}V/\mathrm{d}P)_T$. However, these coefficients are not always known. Since the difference between the two specific heats is small, 1–10% of $C_P$, it can often be neglected or calculated approximately from the expression

$$C_P - C_V = AC_{V^2}T \tag{5.11}$$

in which $A = \alpha^2 V/\beta_T C_{V^3}$ is regarded as a constant and is obtained from data at normal temperatures.

If, to go one step further, the data are not available for calculating $A$ at any temperature from the above formula, $A$ can be obtained to a less satisfactory degree of accuracy from a formula by Nernst and Lindemann

$$A = 0.0214/T_\mathrm{m} \tag{5.12}$$

for specific heat in cal/(mol K) and $T_\mathrm{m}$ in K. Finally, it may be admitted that if $C_V$ itself has to be estimated by an indirect method, the accuracy of the resulting heat capacities will probably not be such as to justify taking $C_P - C_V$ into account at all.

Often the property actually desired (as in heat balances) is the enthalpy difference over a large temperature interval, and this difference is fortunately much less sensitive to estimation errors than is the specific heat.

### *1.1.2. Electron Heat Capacity*

We have seen that the heat capacities of both metals and nonmetals can be well predicted by considering only the energy of the lattice. The Dulong–Petit high-temperature value is the same for both. This is paradoxical because the free electrons in a metal should be capable of accepting some of the thermal energy of the metal, and this demands that metals have a significantly larger heat capacity than nonmetallic crystals. This is, in fact, not the case, because only a small fraction of the conduction electrons can accept thermal energy. The basic reason for this lies in the Pauli exclusion principle, which limits the number of electrons that can occupy the same energy state. All the possible low-lying energy states of electrons are already occupied. Therefore, energy can be accepted only by those electrons that are in states below the highest occupied energy level by an amount that corresponds to the average energy of a thermal vibration. This energy is of the order of $kT$, where $k$ is Boltzmann's constant. Therefore, only electrons within about $kT$ of the highest energy level occupied, the Fermi level, can be excited by thermal processes out of the sea of already occupied states to a higher energy state. All others, since they have less energy, cannot be excited to an unoccupied level by this amount of energy and so are unable to accept it. This characteristic of the element is termed *degeneracy*. Because the conduction electrons are degenerate until thermal energies characteristic of temperatures on the order of 10,000 K are available, the contribution of conduction electrons to the specific heat of solids is small at ordinary temperatures.

Oddly enough, it is possible to observe the electron heat capacity of metals at very low temperatures. Since theory predicts that only the electrons within an energy range $kT$ of the Fermi level of electron energy states will accept thermal energy from the lattice, the electronic heat capacity will be linearly proportional to the absolute temperature.

Meanwhile, at sufficiently low temperatures, Debye's equation reduces to

$$C_V = 464.5\left(\frac{T}{\theta_D}\right)^3 = \text{constant} \times T^3$$

See Eq. (5.6). This fits well the measured specific heat curves of many substances, provided the temperature is below $\theta_D/12$; for copper and aluminum this is at about 30 K. On further reduction of the temperature to the boiling point of helium, departures from the $T^3$ law become evident and the specific heat is given by an equation of the form

$$C_V = A\left(\frac{T}{\theta_D}\right)^3 + \gamma T \tag{5.13}$$

The first term on the right is, as before, the contribution arising from the lattice vibrations, and the second has been shown to be due to the specific heat of the conduction electrons, which is negligible at high temperatures in comparison with the lattice contribution and proportional to temperature at low temperatures. For a typical metal, copper for example,

$$A = 464.5, \quad \gamma = 0.000177, \quad \theta_D = 310\,\text{K}$$

so that at 26 K, where $T = \theta_D/12$, the lattice specific heat is 0.275 cal/(g-atom) and the electronic specific heat is only 0.0046 cal/(g-atom). At 4 K, the two contributions to the specific heat are 0.001 and 0.007 cal/(g-atom), respectively, and at 1 K they are 0.0000155 and 0.000178 cal/(g-atom). Thus the electronic contribution, which is of considerable theoretical interest, can most accurately be determined by measurements at temperatures below about 4 K. This is a typical example of the contributions to physical theory that can be made as a result of low-temperature research. The coefficient $\gamma$ in Eq. (5.13 ) is given by

$$C_{Ve} = \frac{4\pi^4 a m_e M R^2 T}{h^2 N_o (3\pi^2 N/V)^{2/3}} = \gamma_e T \tag{5.14}$$

where $a$ is the number of valence electrons per atom; $m_e$ the electron mass; $M$ the atomic weight of material; $R$ the specific gas constant for material; $T$ the absolute temperature; $h$ the Planck's constant; $N_o$ the Avogadro's number and $N/V$ is the number of free electrons per unit volume.

*Example 5.2.* Determine the ratio of the electronic contribution to the specific heat, $C_{ve}$, to the lattice specific heat (as determined by the Debye function) for copper at (a) 5.6 K, (b) 0.56 K, and (c) 0.056 K.

$$\theta_D = 309.4\,\text{K}, \qquad C_{Ve} = \frac{4\pi^4 a m_e M R^2 T}{h^2 N_o (3\pi^2 N/V)^{2/3}} = \gamma_e T$$

$$\gamma_e = 1.13 \times 10^{-2}\ \text{J/(kg K}^2\text{) for copper}, \qquad R = 1.987\ \text{cal/(mol K)}$$

a. At 5.6 K:

$$C_{Ve} = (1.13 \times 10^{-2}) \frac{\text{J}}{\text{kg K}^2} (5.6\,\text{K}) = 0.0645\,\text{J/(kg K)}$$

$$T/\theta_D = 5.6/309.4 = 0.0181$$

$$C_V = 233.8\,R \left( \frac{T}{\theta_D} \right)^3 = \left( 2.75 \times 10^{-3} \frac{\text{cal}}{\text{g K}} \right) (4184) = 11.5\,\text{J/(kg K)}$$

$$C_{Ve}/C_V = 5.61 \times 10^{-3}$$

b. At 0.56 K, by similar calculations,

$$C_{Ve}/C_V = 0.561$$

c. At 0.056 K,

$$C_{Ve}/C_V = 56.1$$

#### *1.1.3. Anomalies in the Heat Capacity Curve*

The Debye theory used to present the heat capacity–temperature relationship shown in Fig. 5.1 depends on the way in which heat is accepted by the lattice and by conduction electrons in metals. This model gives smooth functions of heat capacity vs. temperature. Heat capacity measurements on many materials, however, do not show such regular behavior. Sharp spikes are sometimes apparent in the heat capacity curves. Other mechanisms, besides those of conduction electrons and the lattice, must come into play. Obviously, other mechanisms must exist to accept thermal energy as the temperature is raised. These other energy-absorbing mechanisms include phase transformations, Curie points, order–disorder transformations, rotational transitions of molecules, transitions between spin states in paramagnetic salts, and electronic excitations.

If these other mechanisms can accept energy, they will become active when the thermal energy of the lattice approaches the activation energy of the new process. That is, when $kT$ approaches the order of the energy needed for the new process, then the process can begin to accept additional heat. These phenomena can lead to the peaks and bumps found in some materials.

### 1.2. Thermal Conductivity

The thermal conductivity $K_t$ of any material is defined such that the heat transferred per unit time $dQ/d\theta$ is given by

$$\frac{dQ}{d\theta} = K_t A \frac{dT}{dx} \tag{5.15}$$

where $A$ is the cross-sectional area and $dT/dx$ is the thermal gradient.

To understand how thermal conductivity depends on temperature, especially at low temperatures, it is useful to understand the basic mechanisms for energy transport through materials.

There are three basic energy transports, and hence heat conduction, through a solid:

1. By lattice vibrational energy transport, also called phonon conduction, which occurs in all solids—dielectrics and metals. In nonmetallic crystals and some intermetallic compounds, the principal mechanism of heat

conduction is by this lattice vibration mode or the mechanical interaction between molecules. For single crystals at quite low temperatures, this mode of heat conduction can be very effective, equaling or exceeding the conduction by pure metals.

2. By electron motion, as in metals. Metals, of course, also have a lattice structure and hence a lattice contribution to their thermal conductivity. However, thermal conductivity in pure metals (particularly at low temperatures) is due principally to the "free" conduction electrons, those that are so loosely bound to the atoms that they wander readily throughout the crystal lattice and thus transfer thermal energy.
3. By molecular motion, such as in organic solids and gases. This characteristic disorder and the lattice imperfections of these materials introduce resistance to heat flow. Accordingly, the disordered dielectrics such as glass and polymeric plastics are the poorest solid conductors of heat.

For the present, we are concerned with structural materials, so this overview is limited to dielectrics (insulators) and metals. Furthermore, since dielectrics have only one heat transport mechanism (phonons) and metals have two (phonons plus electrons), we shall begin by examining heat transport in dielectrics.

### *1.2.1. Dielectric Heat Conduction*

We have already used the term *phonon* several times. In the discussion of thermal conduction by the thermal vibrations of the crystal lattice, it has been found convenient to treat the quantized vibrational modes as quasi-particles. Since these wave packets have many of the properties of particles, they have been named *phonons* in analogy with the photon, the "particle" of light. The phonon then is a "particle" of thermal energy. These quasi-particles undergo collisions with each other, just as the molecules of a gas do, and the transfer of thermal energy in a perfect crystal becomes closely analogous to the momentum transfer between gas molecules in thermal agitation. A thermally excited solid is thus treated as a gas composed of phonons.

Since the thermal energy of a solid can be treated in terms of phonons, the solid becomes comparable to a box full of a phonon gas. In this model, the thermal conductivity of a solid can be calculated in terms of the thermal conductivity of a gas. For the latter, we can go to the classical kinetic theory of gases. This predicts that heat flows across an imaginary surface, drawn perpendicular to a temperature gradient, by hot (excess velocity) atoms passing through the surface plane from one side to the other and by cold (low-velocity) atoms passing through from the other side. This leads to the following expression for the thermal conductivity:

$$K_t = (1/3)C_V UL \tag{5.16}$$

where $C_V$ is the specific heat of the phonons, the lattice specific heat; $U$ the velocity of propagation of the phonons, which travel at the speed of sound; and $L$ is the mean free path of the phonon between collisions.

The 1/3 is included because the phonons are in random motion and are free to move in any of the three spatial directions.

This general expression for lattice thermal conductivity results directly from the adoption of a kinetic model in which "particles" carry the heat. It can be imagined that the capacity of a phonon to carry heat is $C_V$; that, on the average, one-third of the phonons are traveling in the desired direction, they travel at a velocity $U$, and they travel a distance $L$ before being diverted. To arrive at a knowledge of how the thermal conductivity should vary with temperature, it is necessary to consider the various mechanisms by which phonon scattering takes place and the dependence upon temperature of each mechanism. This amounts to examining the temperature dependence $C_V$, $U$, and $L$ (Eq. (5.16)).

The specific heat $C_V$ is the lattice specific heat of the solid, as already discussed. Its variation with $T$ is plotted in Fig. 5.3b. The average velocity $U$ of the phonons is the mean of the velocity of sound. $U$ varies only slightly with temperature as plotted in Fig. 5.3c. The phonon gas differs from a real gas in that the number of particles varies with the temperature, increasing in number as temperature is increased. At high temperatures, the large number of phonons leads to collisions between phonons. Then, as $T$ increases, $L$ decreases, as shown in Fig. 5.3d.

At low temperatures, phonon–phonon collisions disappear and $L$ is determined by the distance between imperfections in the crystal. The phonons collide with imperfections such as impurity atoms, dislocations, intercrystalline boundaries, or finally, the specimen boundaries. The distance between these does not depend on temperature, so $L$ becomes constant at low $T$.

The general shape of the curve of $K$ vs. $T$ can now be readily expressed in terms of the other curves. $U$ is almost constant for all $T$, so it has negligible effect on the shape. At low $T$, $L$ is constant. Thus the $K$ curve has the shape of the one remaining variable, $C_V$. At high $T$, $C_V$ is constant, so the $K$ curve has the shape of the $L$ curve, which drops off with temperature. At intermediate temperature, a broad maximum occurs in the transition between the high- and low-temperature behavior.

These considerations lead to the general temperature dependence of $K_t$ shown in Fig. 5.3a, which is experimentally observed also. For all materials, it starts at zero for absolute zero temperature, rises to a maximum, and then falls asymptotically toward zero as the temperature becomes very high.

Materials differ in the height of the maximum and its position in temperature, as can be seen in Fig. 5.4, which shows the thermal conductivity curves for several samples. The variability of thermal conductivity with impurity and cold work and the suppression of the maximum in $K_t$ is shown in Fig. 5.3. Note especially the thermal conductivity of the dielectric, corundum. Its peak conductivity, around 37 K, is almost 15 times that of copper at room temperature, making it an excellent thermal

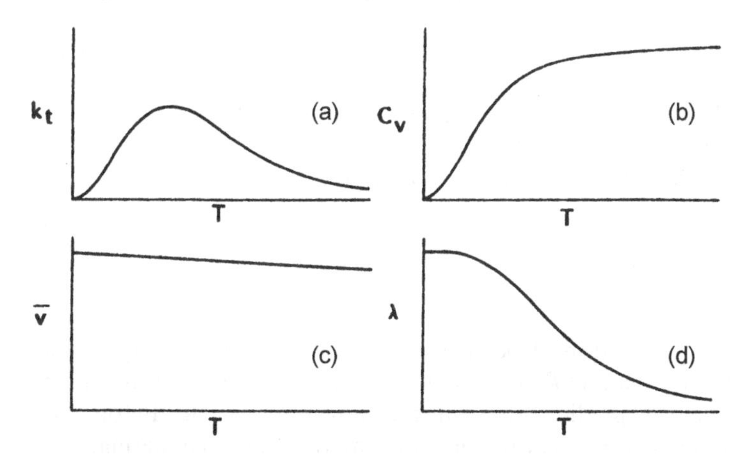


**Figure 5.3** Thermal conductivity mechanisms of dielectrics vs. temperature.

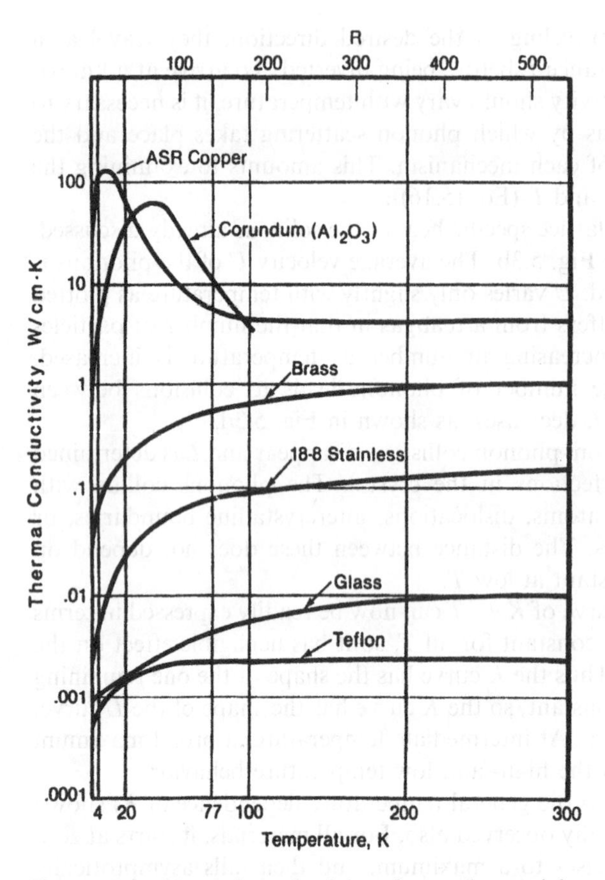


**Figure 5.4** Thermal conductivity of several materials.

conductor. Glass is an extreme example of a disordered material and is shown for comparison. In some cases, thermal conductivity becomes a function of specimen dimensions.

It has been shown that thermal conductivity can be expressed in terms of the Debye temperature $\theta_D$ and the thermal conductivity at the Debye temperature, $K_0$:

$$K_t = K_0(\theta_D/T) \tag{5.17}$$

This expression holds when $T$ is equal to or greater than $\theta_D$ and gives a $1/T$ dependence on temperature.

When $T$ is considerably less than $\theta_D$, i.e., below about $\theta_D/10$, then

$$K_t = K_0\left(\frac{T}{\theta_D}\right)^3 \exp\left(\frac{\theta_D}{bT}\right)$$

which is an exponential temperature dependence. The thermal conductivity should therefore rise more rapidly than $1/T$ as the temperature is reduced in this region. In fact, this exponential rise predicts an infinite conductivity at 0 K. Before this happens, however, surface reflections of the phonons interfere, and the thermal conductivity is held in check. As the absolute temperature approaches 0, the mean free path increases until it reaches the order of the dimensions of the crystal. Phonons are then reflected from its surfaces. With $L$ and $U$ now constant, $K_t$ depends on the dimensions of the crystal and varies as $C_V$ with temperature.

We have already seen that the heat capacity of a dielectric varies as $T^3$ at very low temperatures, and so we also expect a $T^3$ temperature dependence for $K_t$ at the lowest temperatures. This dependence has in fact been observed experimentally.

The type of thermal conductivity curve suggested by all of these considerations is exemplified by that of corundum in Fig. 5.4. Above temperatures corresponding to about $\theta_D/10$, phonon–phonon collisions govern. These are decreased at lower temperatures to cause a maximum near $\theta_D/20$, and below this the size of the crystal limits the mean free path and the $T^3$ dependence causes a drop to zero at 0 K.

The phonon interaction and boundary scattering process establish the general shape of the conductivity curve for a perfect crystalline dielectric. Further scattering may occur at internal lattice imperfections such as vacancies, interstitial atoms, impurity atoms (among which may be included isotopes of the primary constituents), and dislocations. These added resistances have the effect of lowering and flattening the conductivity curve and, in the limit of a glassy or disordered solid, lead to a curve without a maximum.

### *1.2.2. Metallic Heat Conduction*

Metals, having a lattice structure, conduct heat by phonons as do dielectrics. In addition, heat may be conducted by the free electrons. These two mechanisms are additive:

$$K_t = K_e + K_g \tag{5.18}$$

where $K_e$ is the electron conduction; and $K_g$ is the phonon conduction.

In metals having a large number of free electrons (e.g., Na, Ag, Au, Cu), roughly one conduction electron per atom, $K_e \gg K_g$ and $K_g$ may be ignored.

Why do free electrons contribute so little to heat capacity but so much to heat transport? Consider the following equation for electronic heat conduction, analogous to the expression (Eq. (5.16)) for lattice heat conduction:

$$K_e = (1/3) C_e U_e L_e \tag{5.19}$$

where $C_e$ is the electronic heat capacity; $U_e$ the mean velocity of the electron and $L_e$ is the mean free path of electrons between collisions with obstacles.

Although $C_e$ is very small, $U_e$ is the speed of *light*, not the speed of sound, as is the case with phonons. Furthermore, electrons effectively scatter phonons, reducing the contribution of $K_g$ still further. In fact, the striking difference between the very high phonon conduction of dielectrics and its very low value in metals is due to the effectiveness of the free electrons in scattering phonons. Accordingly, $K_e$ is much greater than $K_g$ for most pure metals.

The details of thermal conductivity in metals can be seen by examining the temperature dependence of the electron heat capacity $C_e$ and the lattice heat capacity $C_l$. The former is given by

$$C_e = \pi^2 \frac{k^2}{m^* v^2} Z N_o T \tag{5.20}$$

where $m^*$ is the effective electron mass: $\nu$ is the electron velocity at the energy of the Fermi face: $Z$ the number of free electrons per atom and $N_o$ is Avogadro's number (number of atoms per mole). The coefficient of $T$ in the preceding equation is a constant of the material, and it is convenient to give it the symbol $\gamma$. Then

$$C_e = \gamma T \tag{5.21}$$

that is, the electron specific heat varies linearly with $T$. Ordinarily, $\gamma T$ is small compared to the lattice specific heat. (See Sec. 1.1.1.) However, it becomes appreciable at *very high T*, where the lattice $C_V$ remains constant at a value $3R$. The electron $C_V$ also becomes appreciable at *low T*, where the Debye model predicts that the lattice $C_V$ varies as $(T/\theta_D)^3$. Then the total $C_V$ is given by

$$C_V = \gamma T + 234R\left(\frac{T}{\theta_D}\right)^3 \tag{5.22}$$

For copper, at 1 K, the first term is roughly four times the magnitude of the second term in this equation.

In dealing with the thermal conductivity due to the electrons, $C'_V$, the specific heat per unit volume, is needed. This differs from $C_V$, the specific heat per mole, by having $N_o$ replaced by $n$, the number of electrons per unit volume.

For metals, the thermal conductivity is almost entirely the result of energy carried by the electrons rather than by the phonons. The two heat fluxes are additive, but the phonon conductivity becomes negligible because collisions of phonons with electrons drastically reduce the phonon mean free path in metals.

In treating the thermal conductivity due to electrons, the solid is considered to contain an electron gas. The gas thermal conductivity expression may again be applied, that is

$$K = (1/3)C'_V\, UL \tag{5.23}$$

The manner in which $K$ varies with $T$ for an electrical conductor is indicated in Fig. 5.5a. The causes of this variation may be seen by looking at how the terms of Eq. (5.23 ) vary with temperature. $C'_V = \gamma T$ is plotted in Fig. 5.5b. The electron velocity $V$ is independent of $T$ as indicated in Fig. 5.5c. The mean free path varies with $T$

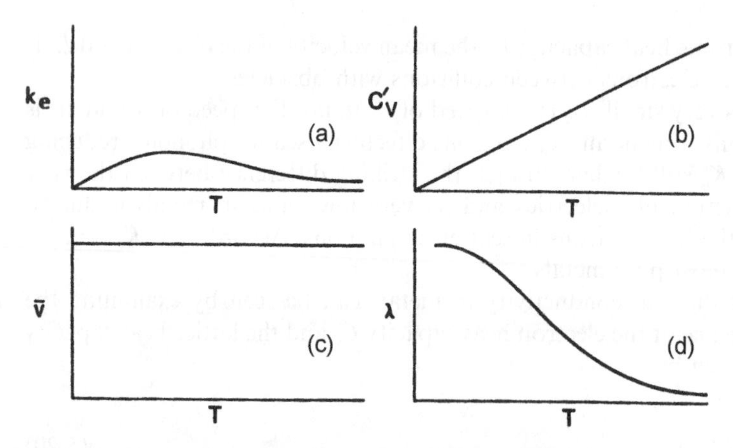


**Figure 5.5** Thermal conductivity by electron mechanisms.

in much the same way as for phonons. At high $T$, the electrons collide with the phonons, so $L$ decreases as the number of phonons increases. At low temperature, the number of phonons is small and the electron mean free path is limited by distances between crystal imperfections. These are impurities, dislocations, intercrystalline boundaries, or, finally, the specimen walls. Looking again at the product of $C_V'$, $V$, and $L$, the quantity $V$ is constant, so $K$ depends on $C_V'$ and $L$. At low $T$, $L$ is constant, so $K$ follows $C_V'$ and varies linearly with $T$. At high $T$, $C_V'$ still varies linearly with $T$, but $L$ begins to fall off so rapidly that the product begins to drop off. At intermediate $T$, a broad maximum occurs.

Figures 5.6 and 5.7 show the thermal conductivity curves for several solids, with the highest curve having the highest purity. It will be noted that the metals of high purity exhibit a maximum of conductivity at low temperatures that in some cases is many times the room temperature value. Moreover, the conductivities of these metals approach a room temperature value that is almost temperature-independent. The height (or existence) of the maximum is quite sensitive to certain slight impurities, although it will be noted that some of the commercial coppers exhibit maxima.

Table 5.3 gives the thermal conductivity and thermal conductivity integral for a few common materials of construction (Johnson, 1960).

The variability of thermal conductivity with alloying and cold working and the suppression of the maximum are shown clearly in Figs. 5.5 and 5.6. Inconel, Monel,

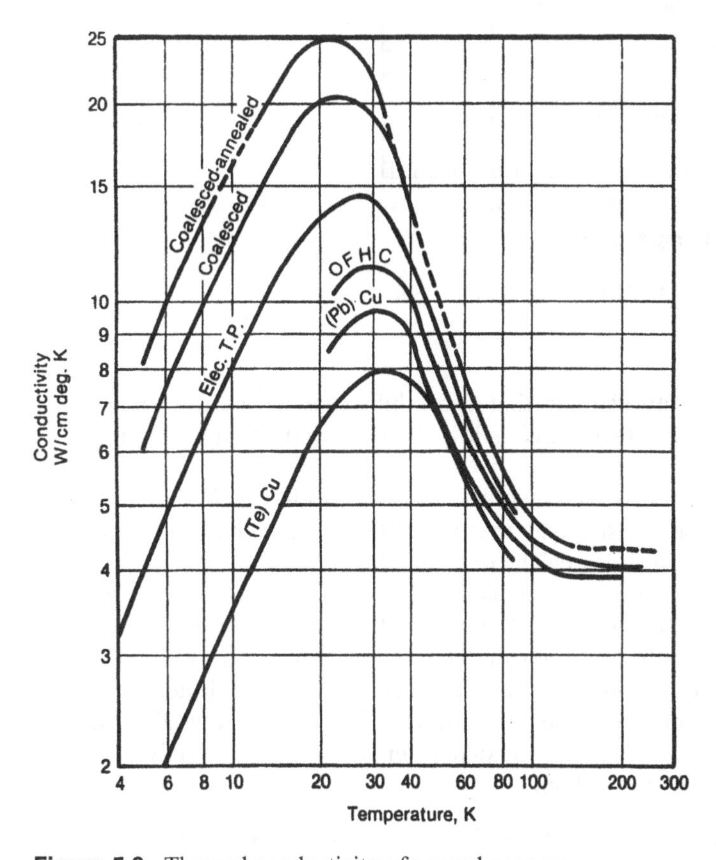


**Figure 5.6** Thermal conductivity of several coppers.

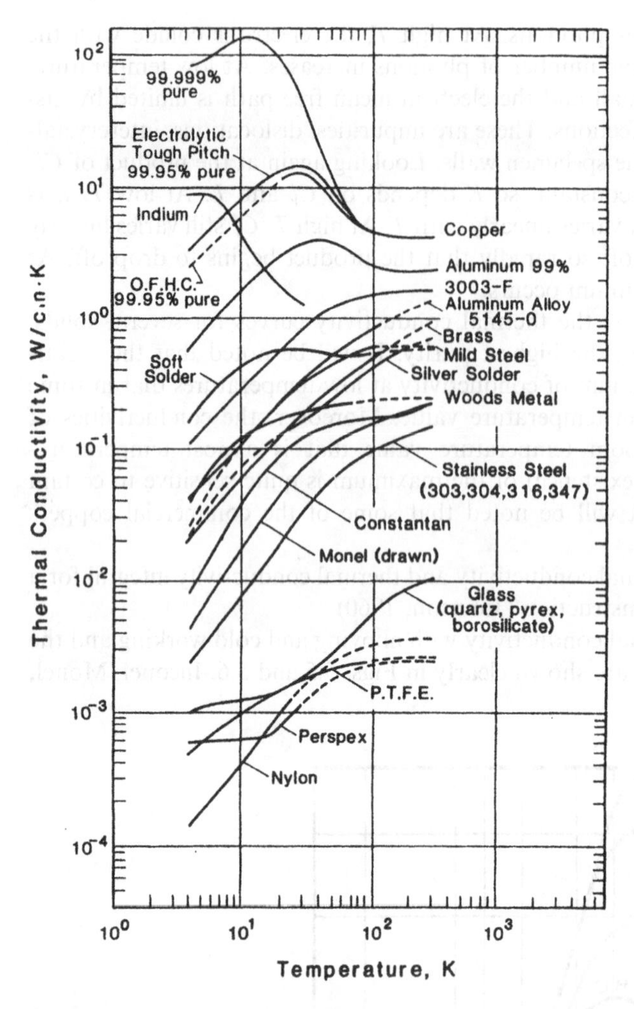


**Figure 5.7** Thermal conductivity of some common solids.

and stainless steel are structural alloys that also exhibit these properties and are thus useful in cryogenic service, which requires low thermal conductivity over the entire temperature range.

Alloys and highly cold-worked metals represent cases in which scattering from imperfections in the lattice is so great that this effect predominates over phonon scattering even at room temperature. The resulting flattening of the thermal conductivity curve is illustrated in Fig. 5.4 for nickel–chromium stainless steel and brass. Furthermore, pure metals having a small number of conduction electrons per atom (e.g., antimony and bismuth) and semiconductors (e.g., germanium) are all metals for which phonon thermal conduction can no longer be safely neglected.

Experimental thermal conductivity values for most materials used in low-temperature design are readily available in the literature (Powell, 1963; Touloukian, 1964; Powell et al., 1966).

*Example 5.3.* In support members for cryogenic storage vessels and other equipment, it is desirable to minimize the heat transfer through the members but maximize

**Table 5.3** Thermal Conductivities and Thermal Conductivity Integrals for Several Materials[a]

| Temperature (K) | Cu (beryllium tough pitch) | | Aluminium (6063 T5) | | Low-carbon steel (C1020) | | Stainless steel (304) | | Monel (drawn) | | Teflon | |
|---|---|---|---|---|---|---|---|---|---|---|---|---|
| | $K_t$ | $\int_{4.2\,K}^{T} K_t\,dT$ | $K_t$ | $\int_{4.2\,K}^{T} K_t\,dT$ | $K_t$ | $\int_{4.2\,K}^{T} K_t\,dT$ | $K_t$ | $\int_{4.2\,K}^{T} K_t\,dT$ | $K_t$ | $\int_{4.2\,K}^{T} K_t\,dT$ | $K_t$ | $\int_{4.2\,K}^{T} K_t\,dT$ |
| 4.2 | 1.9 | 0 | 34 | 0 | 3.0 | 0 | 0.24 | 0 | 0.43 | 0 | 0.046 | 0.0 |
| 10 | 4.8 | 19 | 86 | 360 | 11.5 | 43 | 0.77 | 2.9 | 1.74 | 6.3 | 0.096 | 0.44 |
| 20 | 10.6 | 95 | 170 | 1650 | 24.0 | 222 | 1.95 | 16.3 | 4.30 | 36.4 | 0.141 | 1.64 |
| 30 | 16.2 | 229 | 230 | 3650 | 32.0 | 502 | 3.30 | 42.4 | 6.90 | 92.9 | 0.174 | 3.23 |
| 40 | 21.0 | 415 | 270 | 6200 | 38.6 | 867 | 4.70 | 82.4 | 9.00 | 173 | 0.193 | 5.08 |
| 50 | 26.1 | 650 | 280 | 8950 | 47.6 | 1310 | 5.80 | 135 | 10.95 | 273 | 0.208 | 7.16 |
| 60 | 30.0 | 930 | 270 | 11700 | 53.6 | 1810 | 6.80 | 198 | 12.09 | 368 | 0.219 | 9.36 |
| 70 | 33.7 | 1250 | 248 | 14300 | 57.5 | 2360 | 7.60 | 270 | 13.06 | 513 | 0.228 | 11.6 |
| 80 | 37.0 | 1600 | 230 | 16700 | 60.0 | 2950 | 8.26 | 349 | 13.90 | 647 | 0.235 | 13.9 |
| 90 | 40.1 | 1990 | 222 | 19000 | 61.8 | 3350 | 8.86 | 436 | 14.63 | 791 | 0.241 | 16.3 |
| 100 | 43.0 | 2400 | 216 | 21100 | 62.9 | 4170 | 9.40 | 528 | 15.27 | 940 | 0.245 | 18.7 |
| 120 | 48.4 | 3300 | 207 | 25300 | 64.1 | 5450 | 10.36 | 726 | 16.26 | 1260 | 0.251 | 23.7 |
| 140 | 53.3 | 4320 | 201 | 29300 | 64.6 | 6750 | 11.17 | 939 | 17.34 | 1590 | 0.255 | 28.7 |
| 160 | 57.6 | 5440 | 200 | 33300 | 64.8 | 8050 | 11.86 | 1170 | 18.25 | 1950 | 0.257 | 33.8 |
| 180 | 61.5 | 6640 | 200 | 37300 | 64.9 | 9350 | 12.47 | 1410 | 19.02 | 2320 | 0.258 | 39.0 |
| 200 | 65.0 | 7910 | 200 | 41300 | 65.0 | 10700 | 13.00 | 1660 | 19.69 | 2710 | 0.259 | 44.2 |
| 250 | 72.4 | 11300 | 200 | 51300 | 65.0 | 13900 | 14.07 | 2340 | 21.02 | 3730 | 0.260 | 57.2 |
| 300 | 78.5 | 15000 | 200 | 61300 | 65.0 | 17200 | 14.90 | 3060 | 22.00 | 4800 | 0.260 | 70.2 |

[a] The thermal conductivity ($K_t$) is expressed in dimensions of W/(m K), and the thermal conductivity integral ($\int_{4.2\,K}^{T} K_t\,dT$) is given in dimensions of W/m.

the strength of the members at the same time. An important parameter in this situation is the strength-to-conductivity ratio. Determine this ratio, $S_y/K_t$ [in $(N\ K)/(W\ m)$] at 22.2 K for

a. 2024 T4 aluminum
b. 304 stainless steel
c. Monel
d. Beryllium copper
e. Teflon

Using Fig. 4.21 for $S_y$ values, and Fig. 5.7 for $K_t$ values:

a. 2024 T4 aluminum

$S_y = 7.24 \times 10^8\ \mathrm{N/m^2}$
$K_t = 17.3\ \mathrm{W/(m\,K)}$
$S_y/K_t = 4.18 \times 107\ \mathrm{N\,K/(W\,m)}$

b. 304 stainless steel

$S_y = 1.32 \times 10^9\ \mathrm{N/m^2}$
$K_t = 2.77\ \mathrm{W/(m\ K)}$
$S_y/K_t = 4.77 \times 10^8\ \mathrm{N\,K/(W\,m)}$

c. Monel

$S_y = 1.03 \times 10^9\ \mathrm{N/m^2}$
$K_t = 7.61\ \mathrm{W/(m\,K)}$
$S_y/K_t = 1.35 \times 10^8\ \mathrm{N\,K/(W\,m)}$

d. Beryllium copper

$S_y = 8.96 \times 10^8\ \mathrm{N/m^2}$
$K_t = 9.52\ \mathrm{W/(m\,K)}$
$S_y/K_t = 9.41 \times 10^7\ \mathrm{N\,K/(W\,m)}$

e. Teflon

$S_y = 1.38 \times 10^8\ \mathrm{N/m^2}$
$K_t = 0.138\ \mathrm{W/(m\,K)}$
$S_y/K_t = 1.0 \times 10^9\ \mathrm{N\,K/(W\,m)}$

*1.2.3. Design Considerations*

It is difficult to predict thermal conductivities of nearly pure metals on the basis of their chemical analyses because (1) the effect of each kind of impurity is specific and depends on its electronic band structure and (2) the effect of a given impurity is much greater if it is in solid solution than if it is segregated at grain boundaries. Thus a gross or overall chemical analysis provides insufficient information as shown by the bands of Fig. 5.8.

It is believed that the relatively poor thermal conductivity of OFHC (oxygen-free, high-conductivity) copper is due to the latter effect (solution vs. inclusion). The removal of the oxygen may have permitted metallic impurities that were previously segregated as oxides at the grain boundaries to enter the lattice. On the other hand, the good conductivities of the free-machining lead and tellurium coppers may exist in spite of the additives because the impurities in this case are soluble only slightly (if at

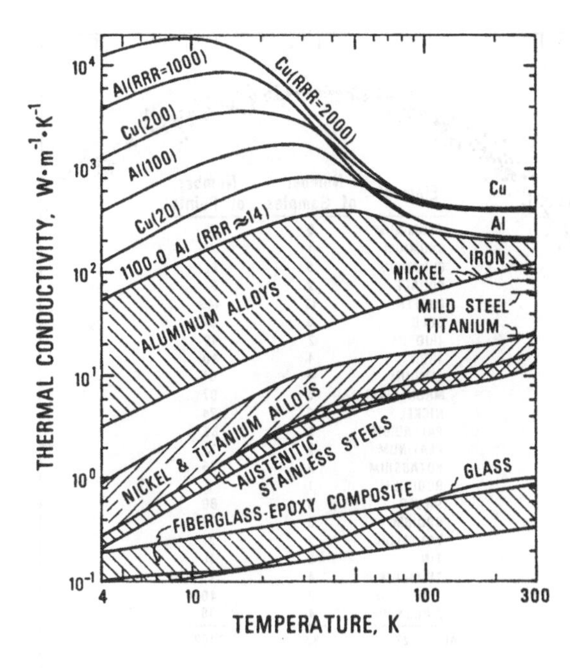


**Figure 5.8** Thermal conductivity of selected classes of materials.

all) in the copper. Thus they are localized at grain boundaries. This condition is also responsible for the free-machining characteristic.

The extreme sensitivity of the low-temperature conductivity of nearly pure metals to impurities and cold working is in strong contrast to the specific heat and expansivity, both of which are somewhat sensitive to the type of lattice structure and to the interatomic forces but relatively insensitive to local imperfections of the lattice. The variability shown by coppers is demonstrated by other metals. This serves to illustrate that where conductivity is concerned, there are no "pure" metals but only alloys of various degrees of dilution. Fortunately for the cryogenics designer, the only other widely used cryogenic metal (aluminum) is available in only one nominally pure commercial grade.

As noted earlier, the lattice conduction of pure metals and dilute metal alloys is generally small, so $k_e$ can be taken as the total conduction. This principle has been used as the basis for a generalized low temperature equation for thermal conductivity as represented in Fig. 5.9, where $k^*$ is equal to $k/k_m$, $T^*$ is equal to $T/T_m$, and $T_m$ is the temperature at which $k$ is a maximum, $k_m$. In analogy to reduced equations of state for thermodynamic properties of fluids, this correlation is referred to as the principle of corresponding states. The reducing parameters chosen are the temperature and thermal conductivity at the maximum in conductivity. The results of this correlation are illustrated in Fig. 5.9. The figure represents 1002 data points for a total of 22 metals (83 specimens) in varying states of impurity and imperfection content. The standard deviation of the data is only 3.2%.

## 1.3. Thermal Expansivity

In our discussion of thermal properties so far, it has been assumed that the atoms of a solid vibrate symmetrically about some equilibrium position in the solid. If this

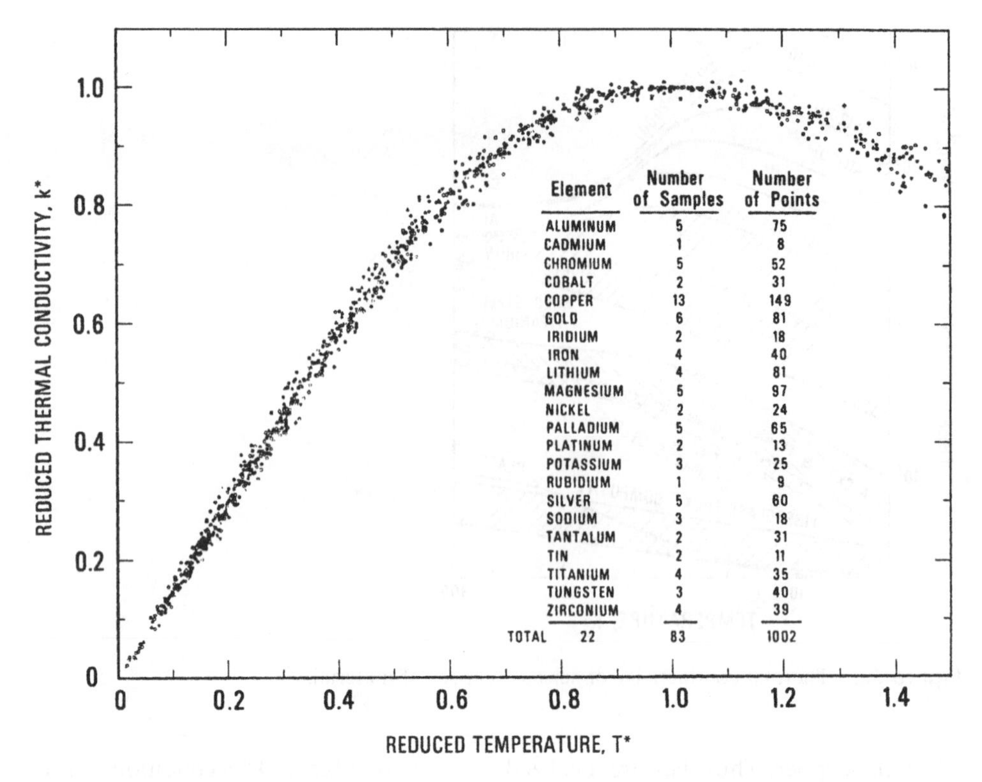


**Figure 5.9** Reduced thermal conductivity vs. reduced temperature of 22 metallic elements. The $k^*$ is $k/k_m$ where $k_m$ is the maximum value of $k$ and $T^*$ is $T/T_m$ where $T_m$ is the temperature corresponding to $k_m$.

were the case, there would be no thermal expansion. Instead, the intermolecular potential energy curve (forces of attraction and repulsion) is not symmetrical. The general shape of the intermolecular potential, as illustrated in Fig. 5.10, is asymmetric. This results from a repulsive force that increases very rapidly as the two atoms approach each other closely (one atom is considered at rest at the origin in Fig. 5.10) plus an attractive force that is not such a strong function of distance. Their algebraic sum provides a curve having a minimum at $r_0$, the classical equilibrium separation of the atoms at 0 K. Increasing the temperature from 0 K corresponds to raising the energy of the system slightly, and now thermal vibrations take place between two higher values of $r$. But because of the asymmetry of the potential well, the average distance no longer corresponds to $r_0$; it is slightly greater and results in the observed thermal expansion of materials. The mean spacing of the atoms increases with temperature as the energy (temperature) of the material increases. Thus, the coefficient of thermal expansion increases as temperature is increased.

Both the specific heat and the coefficient of thermal expansion arise from the intermolecular potential, and accordingly these two properties are related. Our present understanding of this relation is largely due to Gruneisen (1926), who developed an equation of state for solids based on the lattice dynamics of Debye. He showed that the dimensionless ratio $\gamma$ should be a universal constant:

$$\gamma = \alpha V/\beta_T C_V \tag{5.24}$$

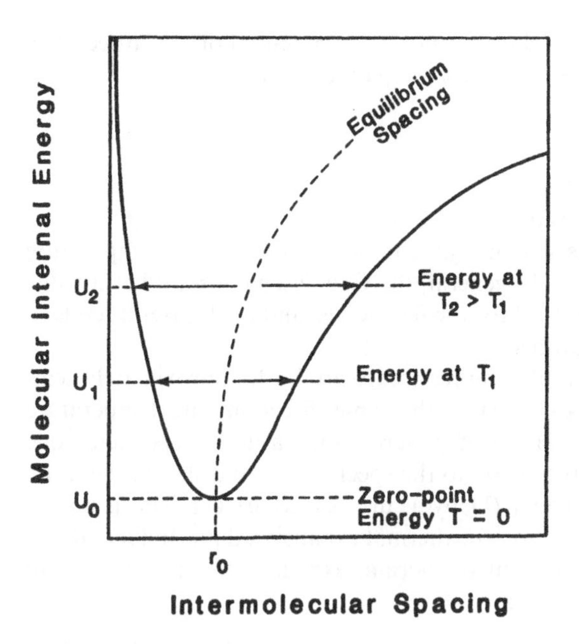


**Figure 5.10** Intermolecular potential curve.

or

$$\gamma = \alpha V / \beta_S C_V \tag{5.25}$$

where

$$\alpha = \frac{1}{V}\left(\frac{\partial V}{\partial T}\right)_P \quad \text{volume coefficient of thermal expansion}$$

$$\beta_T = -\frac{1}{V}\left(\frac{\partial V}{\partial P}\right)_T \quad \text{isothermal compressibility}$$

$$\beta_S = -\frac{1}{V}\left(\frac{\partial V}{\partial T}\right)_S \quad \text{adiabatic compressibility}$$

Here the volumetric coefficient of thermal expansion $\alpha$ is defined as the fractional change in volume per unit change in temperature as the pressure on the material remains constant.

The linear coefficient of thermal expansion $\lambda_t$ is defined as the fractional change in length (or any linear dimension) per unit change in temperature while the stress on the material remains constant. For isotropic materials, $\alpha = 3\lambda_t$.

The Gruneisen relation can also be expressed in terms of the more commonly measured mechanical properties as

$$\alpha = \frac{\gamma C_V \rho}{B} \tag{5.26}$$

where $\rho$ is the density of the material; $B$ the bulk modulus (see Eq. (4.1)); $\gamma$ the Gruneisen constant and $C_V$ is the specific heat and

$$\alpha = (1/3)\lambda_t \quad \text{for isotropic materials}$$

where $\lambda_t$ = linear coefficient of thermal expansion.

Since both $C_V$ and $\alpha$ depend on the intermolecular potential of the lattice, they are related to a first approximation by the Debye temperature $\theta_D$

$$\gamma = -\frac{V}{\theta_D}\frac{d\theta_D}{dV} \tag{5.27}$$

where $V$ is the volume of the material.

Some values of the Gruneisen constant computed from room temperature property data are given in Table 5.4. In general, the Gruneisen constant has a value close to 1.5 for the bcc metals, about 2.3 for the fcc metals, and in all cases it tends to increase with increasing atomic number.

For most materials, volume and compressibility are both approximately temperature-independent, and since both vary in the same direction with temperature their ratio is even more constant. Hence if $\gamma$ were temperature-independent, one could hold that $\alpha$ should vary in proportion to the specific heat and should therefore have a temperature variation like the $C_V$ (Debye) curve shown earlier. This is indeed the case and provides a useful correlation for thermal expansion data. In fact, at low temperatures ($T < \theta_D/12$), the coefficient of thermal expansion is proportional to $T^3$, in accord with the Debye expression.

*Example 5.4.* Determine the Gruneisen constant for beryllium copper at 22.2 K if copper has a density of 8304 kg/m$^3$ and a Poisson ratio of 0.335. Assume that the Debye expression is valid at this temperature

$$\theta_D = 309, \quad MW = 63.54$$

$$T/\theta_D = 22/309 = 0.0712 < 1/12$$

$$C_V = 233.8R\left(\frac{T}{\theta_D}\right)^3$$

$$C_V = (233.8)\left(\frac{1.987}{63.54}\right)(0.0712)^3 = 0.00264 \text{ cal/(g K)} = 11.04 \text{ J/(kg K)}$$

**Table 5.4** Values of the Gruneisen Constant of Selected Cryogenic Materials Computed from Room Temperature Properties

| Material | $\gamma$ | Material | $\gamma$ |
|---|---|---|---|
| Li | 1.17 | Mo | 1.57 |
| Na | 1.25 | W | 1.62 |
| K | 1.34 | Fe | 1.60 |
| Mg | 1.51 | Ni | 1.88 |
| Al | 2.17 | Co | 1.87 |
| Sn | 2.14 | Pd | 2.23 |
| Pb | 2.73 | Pt | 2.54 |
| Sb | 0.92 | NaCl | 1.63 |
| Bi | 1.14 | KCl | 1.60 |
| Cu | 1.96 | KBr | 1.68 |
| Au | 3.03 | $FeS_2$ | 1.47 |
| Zn | 2.01 | PbS | 1.94 |
| Cd | 2.19 | Ta | 1.75 |

For isotropic materials, $\beta = 3\lambda_t$

$$\lambda_t = 0.1 \times 10^{-6}/^\circ\text{R} = 1.8 \times 10^{-7}/\text{K}$$

$$\beta = 5.4 \times 10^{-7}/\text{K}$$

$$B = \frac{E}{3(1-2\mu)}$$

$$E = 1.31 \times 10^{11}\ \text{N/m}^3$$

$$\mu = 0.335$$

$$B = \frac{1.31 \times 10^{11}\ \text{N/m}^2}{3[1-2(0.225)]} = 1.32 \times 10^{11}\ \text{N/m}^2$$

$$\gamma = \frac{\beta B}{C_V \rho} = \frac{(5.4 \times 10^{-7}\ \text{K}^{-1})(1.32 \times 10^{11}\ \text{N/m}^2)}{[11.04\ \text{J/(kg K)}]\ (8304\ \text{kg/m}^3)}$$

$$\gamma = 0.78$$

Figures 5.11 and 5.12 illustrate the expansivity of several metals as a function of temperature and show that the behavior predicted by Eq. (5.26) is actually observed experimentally. Namely, the coefficient of thermal expansion $\beta$ shows the same temperature dependence as that of the heat capacity $C_V$. Two sources (Corruccini and Gniewek, 1961; White, 1959) may be consulted for tabulations of the mean linear thermal expansion of other selected construction materials.

In general, organic polymers have expansivities that are considerably larger than those of metals. Expansivities of such materials may be reduced considerably

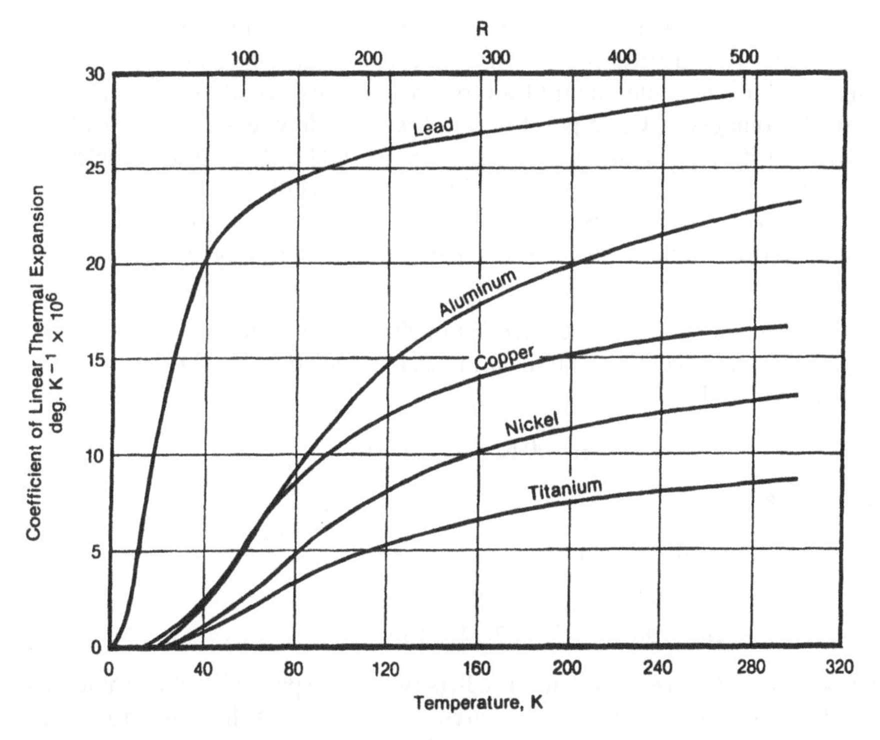


**Figure 5.11** Coefficient of linear expansion for several metals as a function of temperature in K.

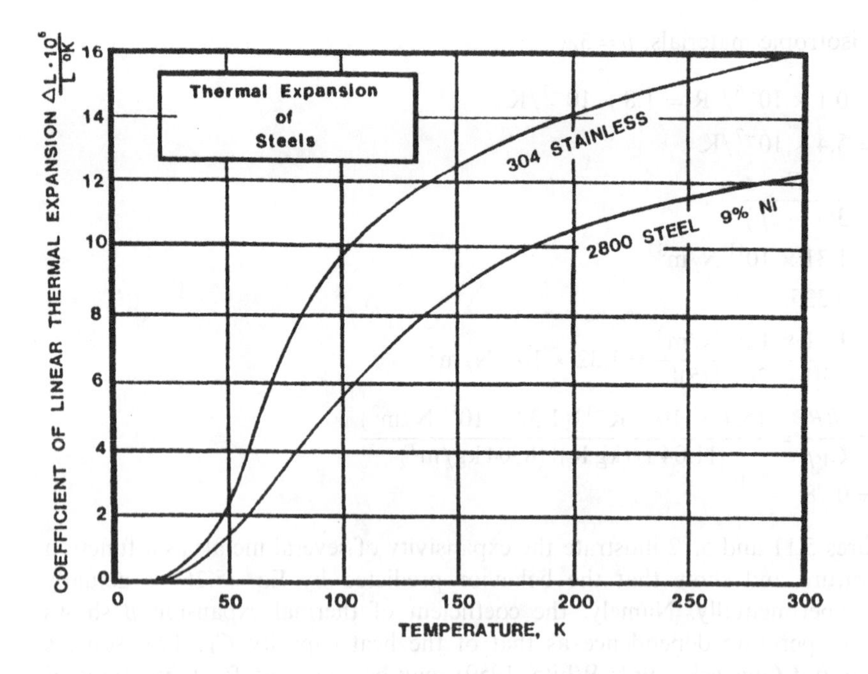


**Figure 5.12** Thermal expansion of steels vs. temperature in °R.

by the addition of filler material of low expansivity such as glass fiber, silica, alumina, and asbestos.

As in the case of heat capacity, anomalies can also occur in the thermal expansivity of materials. Various transformations that occur in the solid state can alter the interatomic forces of the material and lead to changes in its overall dimensions. For example, the ferromagnetic Curie point of dysprosium at low temperatures leads to an irregularity in its expansivity. Other examples are $KH_2PO_4$ with a ferroelectric Curie point at 122 K, and the "glass" transitions of soft rubbers. In fact, the measurement of thermal expansion is a useful method by which solid-state transitions can be investigated. Tables 5.5–5.7 give values of thermal expansion for some structural materials at low temperatures.

*Example 5.5.* An aluminum distillation column 20 m high is placed in an air separation service. How much will the column shrink as it is cooled from 300 K to an average temperature of 89 K?

$$\alpha \text{ at } 89\,\text{K} = 35 \times 10^{-5}, \quad \alpha \text{ at } 300\,\text{K} = 431 \times 10^{-5}$$

From 89 to 300 K

$$\bar{\alpha} = (431 - 35) \times 10^{-5} = 396 \times 10^{-5}, \quad \bar{\alpha} = \frac{\Delta L}{L_0}$$

or

$$\Delta L = L_0\bar{\alpha} = (20\,\text{m})(396 \times 10^{-5}) = 7.92 \times 10^{-2}\,\text{m}$$

The third law of thermodynamics predicts that the expansivity of all materials must go to 0 at 0 K. At very low temperatures, a few materials show an anomalous expansion with decreasing temperature. In addition to 304 stainless steel and Pyrex glass, uranium, fused silica, and the intermetallic compound indium antimonide have

**Table 5.5** Mean Linear Thermal Expansion of Various Metals at Selected Temperatures[a]

| Temperature (K) | Ni | Mg | Zn | Cu | Al | 1020 Low-carbon steel | Stainless steel | Inconel | Monel | Yellow brass |
|---|---|---|---|---|---|---|---|---|---|---|
| 0 | 0 | 0 | 0 | 0 | 0 | 0 | 0 | 0 | 0 | 0 |
| 20 | 0 | 1 | 1 | 1 | 0 | 0 | 0 | 0 | 0 | 1 |
| 40 | 1 | 5 | 9 | 4 | 3 | 1 | 0 | 1 | 1 | 4 |
| 60 | 4 | 12 | 28 | 12 | 9 | 4 | 4 | 5 | 6 | 16 |
| 80 | 12 | 29 | 57 | 26 | 24 | 10 | 13 | 12 | 15 | 34 |
| 100 | 23 | 55 | 93 | 45 | 46 | 20 | 32 | 24 | 28 | 58 |
| 120 | 38 | 87 | 133 | 67 | 71 | 32 | 49 | 38 | 45 | 85 |
| 140 | 55 | 124 | 176 | 92 | 104 | 47 | 72 | 55 | 64 | 115 |
| 160 | 74 | 164 | 221 | 120 | 138 | 63 | 97 | 74 | 84 | 147 |
| 180 | 95 | 208 | 267 | 149 | 175 | 81 | 122 | 95 | 107 | 180 |
| 200 | 117 | 254 | 314 | 179 | 214 | 101 | 150 | 117 | 130 | 215 |
| 220 | 140 | 303 | 363 | 210 | 254 | 121 | 179 | 140 | 154 | 249 |
| 240 | 164 | 353 | 413 | 241 | 295 | 142 | 208 | 163 | 179 | 283 |
| 260 | 188 | 403 | 465 | 273 | 339 | 164 | 240 | 187 | 207 | 320 |
| 280 | 213 | 453 | 518 | 305 | 384 | 187 | 272 | 212 | 233 | 357 |
| 300 | 239 | 503 | 572 | 337 | 418 | 210 | 304 | 238 | 261 | 397 |

[a] Net expansion from 0 K to any other temperature in the table, $\Delta L/L_0 \times 10^5$. $\Delta L$ is the change in length from 0 K to $T$. $L_0$ is the length at 0 K.
*Source:* Corruccini and Gniewek (1961).

been reported to exhibit this characteristic. (This effect for Pyrex glass between 40°C and 0°C can be seen in Table 5.6.) Some anisotropic materials such as zinc show a negative expansivity in one crystallographic direction although they have an average positive expansivity, but for the materials just mentioned, it is the average linear expansivity that has been reported negative. In 304 stainless steel, this occurrence is ascribed to the partial transformation of the material at low temperatures to a second phase, martensite.

Figure 5.13 shows the range of Thermal expansion values for various categories of materials.

## 2. EMISSIVITY, ABSORPTIVITY, AND REFLECTIVITY

Estimating radiant heat transfer across an insulating vacuum requires data on the emissivity, absorptivity, and reflectivity of the surfaces involved. The common definitions for these terms are as follows.

*Total normal emissivity* ($\epsilon_n$) is the rate of radiant energy emission normal to the surface, divided by the corresponding rate from a blackbody.

*Total hemispherical emissivity* ($\varepsilon_h$) is the rate of radiant energy emission into a hemisphere (centered on the normal to the emitting surface), divided by the corresponding rate from a blackbody.

*Total absorptivity* ($\alpha$) is the fraction of energy incident upon a surface that is absorbed. Quantities $\alpha_n$ and $\alpha_h$ analogous to $\varepsilon_n$ and $\varepsilon_h$ may be distinguished.

*Reflectivity* ($r$) is the fraction of incident energy that is reflected; $r = 1 - \alpha$. The value of the reflectivity also depends on the angle of incidence.

**Table 5.6** Mean Linear Thermal Expansion of Various Nonmetals at Selected Temperatures[a]

| Temperature (K) | Pyrex | Teflon | Molded polyester rod reinforced with glass fiber | Cast phenolic rod | Cast epoxy polymer | Nylon rod | Fluoroethylene | Polystyrene |
|---|---|---|---|---|---|---|---|---|
| 0 | 0 | 0 | 0 | 0 | 0 | 0 | 0 | 0 |
| 20 | −1 | 30 | 3 | 14 | 10 | 10 | 21 | 27 |
| 40 | −2 | 90 | 11 | 38 | 39 | 37 | 65 | 82 |
| 60 | −1.5 | 160 | 21 | 70 | 78 | 81 | 116 | 152 |
| 80 | +1 | 245 | 34 | 109 | 126 | 142 | 173 | 235 |
| 100 | 4.5 | 335 | 49 | 154 | 181 | 217 | 235 | 329 |
| 120 | 8.5 | 430 | 67 | 205 | 242 | 301 | 301 | 432 |
| 140 | 13 | 540 | 88 | 261 | 310 | 393 | 372 | 542 |
| 160 | 17.5 | 660 | 110 | 321 | 385 | 493 | 444 | 658 |
| 180 | 22.5 | 805 | 134 | 385 | 467 | 600 | 531 | 778 |
| 200 | 27.5 | 995 | 159 | 452 | 556 | 716 | 618 | 900 |
| 220 | 33 | 1215 | 184 | 524 | 651 | 841 | 711 | 1024 |
| 240 | 39 | 1440 | 210 | 602 | 753 | 977 | 811 | 1152 |
| 260 | 44.5 | 1670 | 237 | 688 | 862 | 1124 | 921 | 1284 |
| 280 | 50.5 | 1900 | 264 | 782 | 980 | 1282 | 1045 | 1422 |
| 300 | 57 | 2460 | 291 | 889 | 1107 | 1450 | 1187 | 1566 |

[a] Net expansion from 0 K to any other temperature in the table, $\Delta L/L_0 \times 10^5$.
*Source:* Corruccini and Gniewek (1961).

**Table 5.7** Thermal Expansion of Selected Metals[a]

| Temperature | Aluminum 2024 | | Aluminum 5083 | | Aluminum 7039 | | Titanium 6A1-4V | | Copper ETP & OFHC | | Brass 70/30 | | Monel K & S | | Stainless steel 304 | | 2800 Steel (9% Ni) | |
|---|---|---|---|---|---|---|---|---|---|---|---|---|---|---|---|---|---|---|
| (°R) | *A* | *B* | *A* | *B* | *A* | *B* | *A* | *B* | *A* | *B* | *A* | *B* | *A* | *B* | *A* | *B* | *A* | *B* |
| 0 | 434 | 0 | 455 | 0 | 462 | 0 | 187 | 0 | 356 | 0 | 410 | 0 | 277 | 0 | 326 | 0 | 232 | 0 |
| 20 | 434 | 0.03 | 455 | 0.03 | 462 | 0.05 | 187 | 0.01 | 356 | 0.03 | 410 | 0.07 | 277 | 0.02 | 326 | 0.02 | 232 | 0.02 |
| 40 | 434 | 0.27 | 455 | 0.27 | 462 | 0.32 | 187 | 0.08 | 356 | 0.23 | 409 | 0.40 | 277 | 0.14 | 326 | 0.15 | 232 | 0.09 |
| 60 | 433 | 0.78 | 454 | 0.80 | 461 | 0.95 | 187 | 0.24 | 355 | 0.73 | 408 | 1.31 | 276 | 0.44 | 325 | 0.40 | 231 | 0.29 |
| 80 | 431 | 1.58 | 452 | 1.63 | 458 | 1.82 | 186 | 0.51 | 353 | 1.62 | 404 | 2.47 | 275 | 0.98 | 324 | 0.85 | 230 | 0.62 |
| 100 | 427 | 2.56 | 447 | 2.63 | 453 | 3.02 | 184 | 1.00 | 348 | 2.64 | 398 | 3.59 | 272 | 1.61 | 322 | 1.83 | 229 | 1.08 |
| 120 | 420 | 3.65 | 441 | 3.74 | 446 | 4.05 | 182 | 1.56 | 342 | 3.61 | 389 | 4.60 | 268 | 2.33 | 317 | 3.05 | 226 | 1.60 |
| 140 | 412 | 4.70 | 432 | 4.82 | 438 | 5.10 | 178 | 2.01 | 334 | 4.50 | 379 | 5.56 | 263 | 3.07 | 310 | 4.00 | 222 | 2.13 |
| 160 | 402 | 5.68 | 422 | 5.82 | 426 | 6.06 | 174 | 2.45 | 324 | 5.21 | 367 | 6.36 | 256 | 3.66 | 301 | 4.78 | 218 | 2.62 |
| 180 | 389 | 6.55 | 409 | 6.72 | 413 | 6.89 | 168 | 2.81 | 313 | 5.84 | 354 | 7.06 | 248 | 4.16 | 291 | 5.33 | 212 | 3.07 |
| 200 | 376 | 7.26 | 395 | 7.50 | 398 | 7.60 | 162 | 3.09 | 301 | 6.33 | 339 | 7.54 | 239 | 4.63 | 280 | 5.75 | 205 | 3.50 |
| 220 | 360 | 7.94 | 379 | 8.18 | 382 | 8.26 | 156 | 3.37 | 288 | 6.74 | 324 | 7.93 | 230 | 5.00 | 268 | 6.13 | 198 | 3.92 |
| 240 | 344 | 8.47 | 362 | 8.77 | 365 | 8.81 | 149 | 3.61 | 274 | 7.13 | 308 | 8.21 | 219 | 5.33 | 255 | 6.48 | 190 | 4.32 |
| 260 | 327 | 8.95 | 344 | 9.29 | 347 | 9.30 | 142 | 3.82 | 259 | 7.46 | 291 | 8.48 | 208 | 5.62 | 242 | 6.78 | 181 | 4.67 |
| 280 | 308 | 9.38 | 325 | 9.70 | 328 | 9.73 | 134 | 4.00 | 244 | 7.75 | 274 | 8.74 | 197 | 5.89 | 228 | 7.02 | 171 | 4.99 |
| 300 | 289 | 9.74 | 305 | 10.10 | 308 | 10.13 | 126 | 4.15 | 228 | 7.96 | 256 | 8.96 | 185 | 6.14 | 214 | 7.24 | 161 | 5.24 |
| 320 | 269 | 10.03 | 285 | 10.49 | 288 | 10.49 | 117 | 4.28 | 212 | 8.15 | 238 | 9.14 | 172 | 6.34 | 199 | 7.43 | 150 | 5.47 |
| 340 | 249 | 10.30 | 264 | 10.80 | 266 | 10.80 | 108 | 4.42 | 196 | 8.30 | 219 | 9.31 | 159 | 6.51 | 184 | 7.59 | 139 | 5.65 |

(*Continued*)

**Table 5.7** (*Continued*)

| Temperature | Aluminum 2024 | | Aluminum 5083 | | Aluminum 7039 | | Titanium 6A1-4V | | Copper ETP & OFHC | | Brass 70/30 | | Monel K & S | | Stainless steel 304 | | 2800 Steel (9% Ni) | |
|---|---|---|---|---|---|---|---|---|---|---|---|---|---|---|---|---|---|---|
| (°R) | *A* | *B* | *A* | *B* | *A* | *B* | *A* | *B* | *A* | *B* | *A* | *B* | *A* | *B* | *A* | *B* | *A* | *B* |
| 360 | 228 | 10.56 | 242 | 11.06 | 244 | 11.10 | 99 | 4.55 | 179 | 8.44 | 201 | 9.46 | 146 | 6.67 | 169 | 7.78 | 128 | 5.81 |
| 380 | 207 | 10.79 | 219 | 11.31 | 222 | 11.37 | 90 | 4.64 | 162 | 8.56 | 182 | 9.60 | 133 | 6.82 | 153 | 7.92 | 116 | 5.96 |
| 400 | 185 | 10.98 | 196 | 11.55 | 199 | 11.62 | 81 | 4.73 | 145 | 8.68 | 162 | 9.72 | 119 | 6.97 | 137 | 8.07 | 104 | 6.09 |
| 420 | 163 | 11.15 | 173 | 11.75 | 175 | 11.85 | 71 | 4.81 | 127 | 8.78 | 143 | 9.84 | 105 | 7.10 | 121 | 8.19 | 92 | 6.21 |
| 440 | 141 | 11.30 | 149 | 11.94 | 151 | 12.06 | 61 | 4.89 | 110 | 8.87 | 123 | 9.95 | 91 | 7.22 | 104 | 8.32 | 79 | 6.32 |
| 460 | 118 | 11.47 | 125 | 12.13 | 127 | 12.27 | 51 | 4.97 | 92 | 8.96 | 103 | 10.05 | 76 | 7.33 | 87 | 8.45 | 66 | 6.42 |
| 480 | 95 | 11.60 | 101 | 12.31 | 102 | 12.45 | 41 | 5.04 | 74 | 9.05 | 83 | 10.15 | 61 | 7.44 | 70 | 8.56 | 53 | 6.51 |
| 500 | 72 | 11.73 | 76 | 12.47 | 77 | 12.62 | 31 | 5.11 | 56 | 9.15 | 62 | 10.25 | 46 | 7.54 | 53 | 8.67 | 40 | 6.60 |
| 520 | 48 | 11.85 | 51 | 12.62 | 52 | 12.79 | 21 | 5.17 | 38 | 9.25 | 42 | 10.3 | 31 | 7.63 | 36 | 8.78 | 27 | 6.68 |
| 540 | 24 | 11.97 | 26 | 12.77 | 26 | 12.96 | 11 | 5.23 | 19 | 9.33 | 21 | 10.4 | 16 | 7.72 | 18 | 8.89 | 14 | 6.76 |
| 560 | 0 | 12.08 | 0 | 12.90 | 0 | 13.12 | 0 | 5.28 | 0 | 9.40 | 0 | 10.5 | 0 | 7.82 | 0 | 9.00 | 0 | 6.84 |

[a] *A* is the net linear contraction coefficient, the change in length from the base temperature of 560°R to any temperature in the table. $A \times 10^{-5} = (L_{560} - L)/L_{560}$, in m/m, in./in., etc. *B* is the coefficient of linear expansion at the temperature shown. $B \times 10^{-6} = (\Delta L/\Delta T)(1/L_{560})$, per degree Rankine in m/(m °R) in./(in. °R). etc.

*Source*: Schmidt (1968).

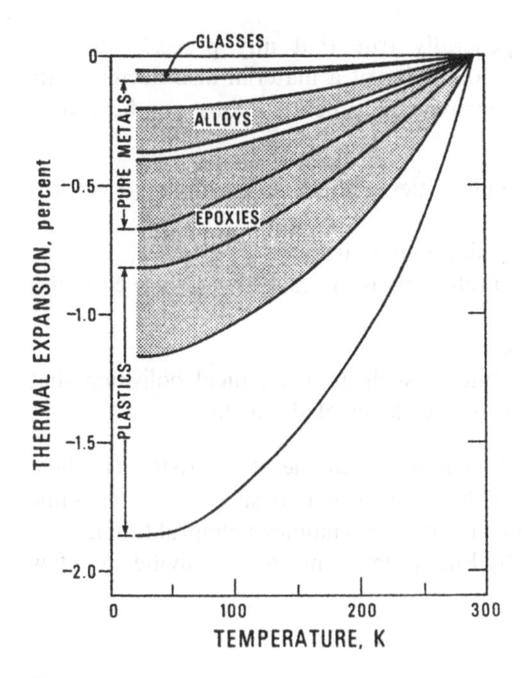


**Figure 5.13** Range of thermal expansion values for various categories of materials.

The rate of energy emission from a blackbody at temperature $T$ is

$$Q = \sigma T^4$$

where $\sigma$ is the Stefan–Boltzmann blackbody radiation constant, and $T$ is the Kelvin temperature.

The emissivities of electrical conductors are obtained theoretically from electromagnetic theory provided the surfaces are smooth and oxide-free and the wavelength of radiation is not too small. For these types of pure materials, the emissivity is directly proportional to the absolute temperature of the material

$$n = \frac{T/51.3}{k_{e,492^{1/2}}} \tag{5.28}$$

where $k_{e,492}$ is the electrical conductivity at 492°R in mho/cm and $T$ is the absolute temperature in degrees Rankine.

Equation (5.28) shows that the emissivity of a pure electrical conductor is directly proportional to its absolute temperature. Emissivity is also shown to be related to the electrical conductivity for electrical conductors, as both are electromagnetic in origin.

For the same reason, the emissivity of electrical insulators can be related to the index of refraction $n_r$ of the material through electromagnetic theory. For electrical insulators,

$$\epsilon = \frac{4n_r}{(n_r + 1)^2} \tag{5.29}$$

for normal incidence of radiation. Since the index of refraction for most electrical insulators is less than 2, the emissivity of insulators is near unity. [For $n_r = 2$, $\epsilon = 8/9 = 0.889$ from Eq. (5.29).]

For electrical conductors, it is generally true that any process, such as annealing, that increases the electrical conductivity of a material also decreases its emissivity. In addition, the following approximate generalizations can be drawn from experimental observations:

1. Materials having the lowest emissivities also have the lowest electrical resistances.
2. Emissivity decreases with decreasing temperature.
3. The apparent emissivity of good reflectors is increased by surface contamination.
4. Alloying a metal increases its emissivity.
5. Emissivity is increased by treatments such as mechanical polishing that result in work hardening of the surface layer of the metal.

At a fixed temperature, the emissivity must equal the absorptivity. (If these differed, there could be a net transfer of heat between two surfaces at the same temperature—a violation of the second law of thermodynamics.) Helpful tabulations of emissivity values have been prepared by Fulk (1959), and the emissivities of a few selected materials are shown in Table 5.8.

## 3. ELECTRICAL PROPERTIES

In metals, the primary mechanisms of thermal conductivity and electrical conductivity are the same. Accordingly, the mechanisms of electron–lattice interaction described in Sec. 1 for thermal conductivity apply equally well to electrical conductivity. This fact becomes the starting point for the present discussion.

Superconductivity, to be discussed in Sec. 4, also depends on the electron–lattice interaction, albeit in a much more subtle way.

### 3.1. Electrical Resistivity and Conductivity

#### *3.1.1. Metals*

The most striking property of metals, in fact the one used to define this class of materials, is electrical conductivity.

An electrical field applied to a metal produces a drift of the free electrons. The resistance that this flow encounters is due to the scattering of electrons by obstacles in their path. This scattering is provided (1) by the thermal vibrations of the crystal lattice, i.e., the phonons, and (2) by lattice imperfections, which we will consider in this order. The effect of temperature on resistance can be deduced by considering the effect of temperature on these two resistance mechanisms, that is, the collision of electrons with phonons and with imperfections in the crystal lattice. The first is a dynamic scattering mechanism; the second, static. As in the case of thermal conductivity, these scattering mechanisms can be considered independent to a very good approximation, and we can write for the electrical resistivity, $\rho$,

$$\rho = \rho_{\mathrm{i}} + \rho_{\mathrm{r}} \tag{5.30}$$

where $\rho_{\mathrm{i}}$ is the phonon scattering resistivity, the resistivity of an ideal crystal; $\rho_{\mathrm{r}}$ is the imperfection scattering, the "residual" resistivity of a real crystal at 0 K, when phonon scattering is inoperative. This is *Matthiessen's rule*, observed empirically on metals of different purity.

**Table 5.8** Total Normal Emissivity $\epsilon$ of Various Materials[a]

| Material | Blackbody source temperature (K) | $\epsilon_n$ at absorptivity temperature $T$ | | |
|---|---|---|---|---|
| | | Same as source | 77 K | 4 K |
| Aluminum, annealed (electropolished) | 300 | 0.03 | 0.018 | 0.011 |
| Aluminum with oxide layer | | | | |
| 0.25 μm thick | 311 | 0.06 | | |
| 1.0 μm thick | 311 | 0.30 | | |
| 7.0 μm thick | 311 | 0.75 | | |
| Aluminum with lacquer layer | | | | |
| 0.5 μm thick | 311 | 0.05 | | |
| 2.0 μm thick | 311 | 0.30 | | |
| 8.0 μm thick | 311 | 0.57 | | |
| Aluminum vaporized on 0.5 mil Mylar (both sides) | 300 | | 0.04 | |
| Brass | | | | |
| Polished | 373 | 0.03 | | |
| Rolled plate | 300 | 0.06 | | |
| Shim stock, 65 Cu/35 Zn | 295 | 0.35 | 0.029 | 0.018 |
| Oxidized | 373 | 0.60 | | |
| Cadmium, electroplate (mossy) | 295 | | 0.03 | |
| Chromium plated on copper | 300 | 0.08 | 0.08 | |
| Copper | | | | |
| Black oxidized | 300 | 0.78 | | |
| Scraped | 300 | 0.07 | | |
| Commercial polish | 300 | 0.03 | | |
| Electrolytic, careful polish | 295 | | 0.015 | |
| Chromic acid dip | 295 | | 0.017 | |
| Polished | 295 | | 0.019 | |
| Liquid honed[b] | 295 | | 0.088 | |
| Electrolytic polish | 295 | | | 0.0062 |
| Carefully prepared surface of pure Cu | 295 | 0.018 | 0.0082 | 0.0050 |
| Gold | | | | |
| 0.0015 in. foil (on glass or Lucite) | 295 | 0.02 | 0.01 | |
| 0.0005 in. foil (on glass or Lucite) | 295 | | 0.016 | |
| 0.000,010 in. leaf (on glass or Lucite) | 295 | | 0.063 | |
| Vaporized onto both sides of 0.5 mil Mylar | 295 | | 0.02 | |
| Plate, 0.0002 in. on stainless steel (1% Ag in Au) | 295 | | 0.025 | |
| Plate, 0.00005 in. on stainless steel (1% Ag in Au) | 295 | | 0.027 | |
| Iron | | | | |
| electrolytic | 450–500 | 0.05–0.65 | | |
| Cast, polished | 311 | 0.21 | | |

(*Continued*)

**Table 5.8** (*Continued*)

| Material | Blackbody source temperature (K) | $\epsilon_n$ at absorptivity temperature $T$ | | |
|---|---|---|---|---|
| | | Same as source | 77 K | 4 K |
| Cast, oxidized | 311 | 0.63 | | |
| galvanized | 365 | 0.07 | | |
| stainless steel, polished | 373 | 0.08 | | |
| stainless steel, type 302 | 300 | | 0.048 | |
| Lead | | | | |
| 0.004 in. foil | 295 | 0.05 | 0.036 | 0.011 |
| Gray oxidized | 295 | 0.28 | | |
| Magnesium | 295 | 0.07 | | |
| Mercury | 273–373 | 0.09–0.12 | | |
| Nickel | | | | |
| Electrolytic | 295 | 0.04 | | |
| 0.004 in. foil | 295 | | 0.022 | |
| Electroplated on Cu | 300 | | 0.03 | |
| Platinum | 290 | | 0.016 | |
| Rhodium, plated on stainless steel | 295 | | 0.078 | |
| Silver | 295 | 0.022 | 0.008 | 0.044 |
| Plate on Cu | 300 | | 0.017 | |
| Stainless steel, 18–8 | 300 | 0.08 | 0.048 | |
| Tin, 0.001 in. foil | 295 | 0.05 | 0.013 | 0.014 |
| Tin with 1% In | 295 | | | 0.0125 |
| Tin with 5% In | 295 | | | 0.0174 |
| Solder, 50–50 on Cu | 295 | | 0.032 | |
| Stellite | 293 | 0.11 | | |
| Monel, smooth, not polished | 366 | 0.16 | | |
| Copper–nickel | 373 | 0.059 | | |
| Ice, smooth, $H_2O$ | 273 | 0.96 | | |
| Glass | 293 | 0.94 | | |
| Lacquer, white | 373 | 0.925 | | |
| Lacquer, black matte | 373 | 0.97 | | |
| Oil paints, all colors | 273–373 | 0.92–0.96 | | |
| Candle soot | 273–373 | 0.952 | | |
| Paper | 373 | 0.92 | | |
| Quartz (fused) | 295 | 0.932 | | |

[a] Values are actually absorptivities from a source at the temperature shown in the first column. Normal and hemispherical values are included indiscriminately.
[b] A commercial liquid "sandblast".

Since the lattice vibrations (phonons) are temperature dependent, $\rho_i$ will have a temperature dependence. Since the numbers of defects and impurities are not functions of temperature, $\rho_r$ will have no temperature dependence.

The temperature dependence of $\rho_i$ can be defined from what we already know about the effect of temperature on phonon activity. Let use thermal conductivity as an index of the phonon effect and recall Eq. (5.17):

$$K_t = K_o(\theta_D/T)$$

This equation holds well at the higher temperatures (approximately $T > \theta_D$) and gives a $1/T$ dependence for thermal conductivity. Because thermal conductivity and electrical conductivity in metals are due to the same carriers (electrons), it follows that electrical conductivity will also have a $1/T$ dependence. Since resistivity, $\rho$, is the reciprocal of electrical conductivity, it follows that $\rho_i$ will have a linear dependence on $T$, or $\rho_i \propto T$.

At lower temperatures ($T < \theta_D$), we have shown that phonon scattering is less and less important as the lattice vibrations quiet down. In this low-temperature region, imperfection scattering becomes more important. Since the number of imperfections is not temperature dependent, the total resistance levels off at a constant value, $\rho_r$.

The resulting curve from these two effects has a steep slope at higher temperatures but levels off rapidly around 40–80 K and then becomes constant at lower temperatures, the constant or residual value depending on the degree of perfection of the lattice (see Fig. 5.14).

From a theoretical analysis by Bloch and Gruneisen, the following expression for electrical resistivity due to phonon scattering was evolved:

$$\rho_i = K\frac{T^5}{\theta_{R^6}}\int_0^{\theta_D/T}\frac{x^5 e^x \, dx}{(e^x - 1)^2} = K\frac{T^5}{\theta_{R^6}} J_5\left(\frac{\theta_R}{T}\right) \tag{5.31}$$

where $K$ is constant for a given metal: $\theta_R$ is a characteristic temperature for electrical resistivity that compares conceptually with $\theta_D$ for thermal processes, and the other symbols are clear from Eq. (5.31). Similar to our analysis of the Debye expression, we can conclude that for $T/\theta_R$ greater than about 0.5, Eq. (5.31) becomes very nearly

$$\rho_i = AT/\theta_{R^2} \tag{5.32}$$

which is the required linear relationship to temperature at high temperatures.

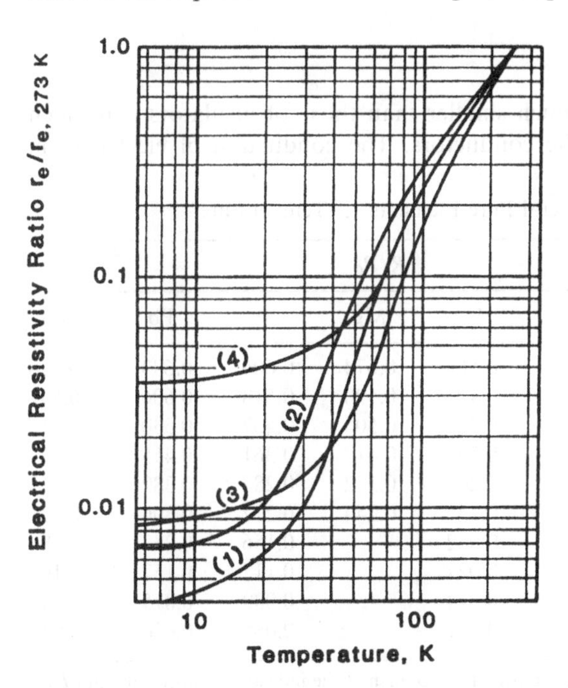


**Figure 5.14** Electrical resistivity ratio for several materials as a function of temperature.

For $T/\theta_R$ less than about 0.1, Eq. (5.31 ) becomes approximately

$$\rho = BT^5/\theta_{R^6} \tag{5.33}$$

and predicts that the resistivity will have a $T^5$ dependence at very low temperatures. These relations have been observed to hold best for monovalent metals such as sodium, lithium, potassium, copper, silver, and gold. To fit the data properly, $\theta_R$ must be considered a function of temperature rather than a constant. This situation is similar to the procedure found necessary for $\theta_D$ when considering thermal processes on the basis of the Debye model. Other metals show a less satisfactory agreement with the Bloch–Gruneisen theory, and in fact some of the simpler metals in slightly impure from even show a resistivity minimum at low temperatures.

### 3.1.2. *Alloys*

Alloys show a much different resistivity temperature dependence than that of relatively pure metals. Alloys and highly cold-worked metals are materials in which the disorder of the lattice is so great that the scattering of electrons by imperfections overshadows that by phonons, even at room temperature. Alloys, as a rule, have resistivities much higher than those of their constituent elements and resistivity–temperature coefficients that are quite low. For example, the alloy that is 60 parts copper and 40 parts nickel (constantan) has a room temperature resistivity of about 44 μohm cm, although copper and nickel separately have resistivities of 1.7 and 6 μohm cm, respectively. Also, while the residual resistivities of the pure metallic elements at very low temperatures are very small, that of constantan is about 95% of the room temperature value.

Table 5.9 and 5.10 show the temperature dependence of resistivity of several common elements and commercial alloys.

### 3.1.3. *Semiconductors*

Semiconductors have resistivities much smaller than those of insulators and much greater than those of most metallic conductors. The conduction of electricity in

**Table 5.9** Electrical Resistivity of Several Pure Elements vs. Temperature[a]

| | $R/R_0$ | | | | | | |
|---|---|---|---|---|---|---|---|
| Temperature (K) | Al | Cu | Fe | Mg | Ni | Pb | Zn |
| 193 | 0.641 | 0.649 | 0.569 | 0.674 | 0.605 | 0.683 | 0.678 |
| 173 | 0.552 | 0.557 | 0.473 | 0.590 | 0.518 | 0.606 | 0.597 |
| 153 | 0.464 | 0.465 | 0.381 | 0.505 | 0.437 | 0.530 | 0.516 |
| 133 | 0.377 | 0.373 | 0.292 | 0.419 | 0.361 | 0.455 | 0.435 |
| 113 | 0.289 | 0.286 | 0.207 | 0.332 | 0.287 | 0.380 | 0.353 |
| 93 | 0.202 | 0.201 | 0.131 | 0.244 | 0.217 | 0.306 | 0.271 |
| 73 | 0.120 | 0.117 | 0.062 | | 0.156 | 0.232 | 0.188 |
| 53 | 0.071 | 0.047 | 0.027 | | 0.112 | 0.157 | 0.108 |
| 33 | 0.049 | 0.012 | 0.014 | | 0.089 | 0.075 | 0.041 |
| 20 | 0.0427 | 0.00629 | 0.011 | | 0.085 | 0.0303 | 0.014 |

[a] Values are given as $R/R_0$, where $R$ is the resistance of a specimen at the indicated temperature and $R_0$ is its resistance at 273 K.

**Table 5.10** Electrical Resistivity of Some Commercial Alloys vs. Temperature[a]

| | Resistivity (ohms circ mil/ft) | | | |
|---|---|---|---|---|
| Material | 20 K | 76 K | 195 K | 273.2 K |
| Manganin | 255 | 267 | 282 | 284.8 |
| Cupron | 259 | 275 | 283 | 285.2 |
| Advance | 262 | 278 | 284.5 | 286.5 |
| Stainless steel type 304 | 332 | 337 | 425 | 616.5 |
| Chromel A | 613 | 617 | 627.5 | 632.5 |
| Trophet A | 653 | 655 | 665.5 | 670.5 |
| Evanohm | 797 | | 803 | 803.0 |

[a] Values are in ohms per foot for a cross-sectional area of 1 circular mil.

semiconductors is achieved by carriers, which may be either occasional excess electrons loosely bonded to lattice positions or holes, which are scattered lattice spaces in which there are no electrons. In the presence of an electric field, these electrons or holes can migrate through the lattice and conduct an electric current. The temperature coefficient of resistance of semiconductors is negative (the resistance decreasing with increasing temperature) because the increased thermal vibrations accompanying an increase in temperature promote the transition of a carrier from one lattice point to the next.

Semiconductors with excess electrons are called *n*-type (negative), and those with holes are called *p*-type (positive). Among the semiconductors that have been studied are germanium and silicon, with minute traces of impurities to provide carriers, and several metallic oxides and mixtures of oxides in which carriers are present because the oxides are not prepared with exact stoichiometric proportions.

The electrons and holes that are the charge carriers responsible for the electrical conductivity of semiconductors are produced by thermal excitation. For intrinsic semiconductors, the carriers occur by thermal excitation of electrons across the forbidden gap, $\epsilon_g$, from the lower filled valence band to the upper, empty, conduction band. This results in an electrical conductivity for pure semiconductors that is proportional to $\exp(-\epsilon_g/kT)$. This temperature variation is exhibited at temperatures high enough that the effects of impurities can be neglected. For impurity semiconductors, the carriers are excited from impurity states. In either case, lowering the temperature will ultimately reduce the number of carriers toward zero and the resistivity of the material will rise toward infinity. In the temperature region from 1 to 10 K, the electrical conductivity decreases quite rapidly with a small decrease in temperature.

Ordinary carbon and pure polycrystalline graphite, which have small crystallites, have negative temperature coefficients of resistance (they are semiconductors), whereas single crystals of graphite have positive temperature coefficients. The resistance–temperature relations for ordinary carbon resistors widely used in radio circuitry are of particular interest to the low-temperature worker because of their usefulness as thermometers and liquid level sensing devices (described in Chapter 8, Sec. 6). In fact, in the very low-temperature range, the electrical resistivity of many semiconductors increases quite rapidly with a small decrease in temperature and forms the basis for the development of numerous sensitive semiconductor resistance thermometers for very low-temperature measurements.

### *3.1.4. Insulators*

Insulators are that class of solids that offer very high resistance to the flow of current. Solid electrical insulators usually improve in insulating quality at low temperatures, as no mechanism can come into play to change this property. The observed improvement may merely be due to the fact that accidental surface films of water become less conducting at low temperatures.

## 3.2. Relationship Between Electrical and Thermal Conductivity

As mentioned, it is difficult to estimate or predict thermal conductivities of even nearly pure metals on the basis of their chemical composition alone. It is possible, however, to make adequate predictions of this property if there are data on the electrical conductivity of these good conductors.

This correlation is possible because the thermal and electrical conductivity of metals are closely related. Both thermal and electrical conductivity of metals are due to transport of electrons. As mentioned previously, the heat flux carried by the phonons is very small because of phonon–electron collisions. As a result, the mean free path of the electrons is approximately the same for both thermal and electrical conductivity. Taking them to be the same and dividing the thermal conductivity $K$ by the electrical conductivity $\sigma$ and simplifying yields the Lorenz ratio $L$

$$L = \frac{K}{\sigma} = \frac{\pi^2}{3}\left(\frac{k}{e}\right)^2 T \tag{5.34}$$

or

$$\begin{aligned} \frac{K}{\sigma T} = \frac{\pi^2}{3}\left(\frac{k}{e}\right)^2 = \frac{\pi^2}{3}\left(\frac{R}{F}\right)^2 &= 24.45\,\mathrm{N\,W\,ohm/K^2} \\ &= 2.58 \times 10^{-8}\,\mathrm{Btu\,ohm/(h\,{}^\circ R^2)} \end{aligned} \tag{5.35}$$

where $K$ is the thermal conductivity; $\sigma$ the electrical conductivity ($\sigma = 1/\rho$); $k$ the Boltzmann's constant; $e$ the charge on an electron; $T$ the absolute temperature; $R$ the universal gas constant and $F$ is the Faraday's constant. This relationship is known as the Wiedemann–Franz–Lorenz rule. It is usually valid for temperatures above 100–200 K. It is a useful working rule because the coefficient of $T$ is an absolute constant that does not depend on the material. It is helpful in obtaining a value of $K$, which is usually rather difficult to measure, from an easily measured quantity.

This rule fails below 100–200 K because the mean free path $\lambda$ actually has different meanings for thermal conductivity ($K$) and electrical conductivity ($\sigma$). For $\sigma$, it is the mean distance of travel of electrons before the first power of $v$ is brought back to equilibrium. For $K$, it is the mean distance of travel before $(1/2)\, m^* v^2$ is brought back to equilibrium, where $v$ is the electron velocity at the energy of the Fermi surface and $m^*$ is the effective electron mass (Crittenden, 1963).

Table 5.11 gives some values of the constant $JK/\sigma T$ for several metals in the range of temperature 103–373 K, where the approximate validity of the Wiedemann–Franz–Lorenz law is apparent from their agreement and $J$ is the mechanical equivalent of heat.

## 3.3. Summary of Electrical Properties

The electrical resistivity of most pure metallic elements at ambient and moderately low temperatures is approximately proportional to the absolute temperature. At very

**Table 5.11** Values of $JK/\sigma T$ in $10^{-8}$ $V^2/deg^2$ [a]

| Metal | 103 K | 173 K | 273 K | 291 K | 373 K |
|---|---|---|---|---|---|
| Copper | 1.85 | 2.17 | 2.30 | 2.32 | 2.32 |
| Silver | 2.04 | 2.29 | 2.33 | 2.33 | 2.37 |
| Zinc | 2.20 | 2.39 | 2.45 | 2.43 | 2.33 |
| Cadmium | 2.39 | 2.43 | 2.40 | 2.39 | 2.44 |
| Tin | 2.48 | 2.51 | 2.49 | 2.47 | 2.49 |
| Lead | 2.55 | 2.54 | 2.53 | 2.51 | 2.51 |

[a] $J$ is the mechanical equivalent of heat.
*Source*: Zemansky (1957).

low temperatures, however, the resistivity (with the exception of superconductors) approaches a residual value almost independent of temperature. Alloys, on the other hand, have resistivities much higher than those of their constituent elements and resistance–temperature coefficients that are quite low. Their electrical resistivity as a consequence is largely independent of temperature and may often be of the same magnitude as the room temperature value. The quality of an insulator usually improves as the temperature is lowered. All the common cryogenic fluids are good electrical insulators.

## 4. SUPERCONDUCTIVITY

### 4.1. Introduction

The phenomenon of superconductivity is described as the absence of electrical resistance in certain materials that are maintained below a certain temperature, electrical current, and magnetic field.

Superconductivity is surely the most distinguishing characteristic of cryogenic materials. Cryogenic engineering is, for the most part, the extension of common engineering practices into the extreme of one thermodynamic variable, temperature. Superconductivity, however, is unique to cryogenics and had to await the development of cryogenic techniques to be discovered, as the following historical development shows.

By 1835 researchers had reached temperatures as low as 163 K. Although this temperature is low enough to liquefy many gases, common gases such as helium, hydrogen, oxygen, and nitrogen still defied liquefaction and became referred to as "permanent" gases. It was not until 1877 that oxygen and nitrogen were liquefied, and it was 1899 before hydrogen was liquefied. Helium remained a stubborn holdout to attempts at liquefaction. Finally, in 1908 Professor Kamerlingh Onnes of the Leiden Laboratories in Holland succeeded in liquefying helium. The stage was set for the discovery of superconductivity.

In 1911 Onnes was investigating the electrical resistance of mercury at the very low temperatures now available to him. The theory of electrical resistance was then in a rather rudimentary state, and three postulates had been offered. If electrical resistance were due only to the scattering of electrons by phonons, then the resistance should go to 0 at absolute zero, as the number of phonons decreased. If, in addition to phonons, scattering were caused by impurities and structural defects in the lattice, then the resistance should level off at some residual value. A third theory advanced the possibility that the free conduction electrons might become bound to the atoms of the metal at low temperatures by a sort of condensation. This would lead to a low-temperature minimum resistance. Below the minimum, the resistance would rise again as the conduction electrons in the metal dropped out of circulation.

We have seen (Sec. 3.1) that the second postulate represents the actual behavior, the value of the residual resistance depending upon the degree of structural and chemical purity of the metal. Onnes suspected that impurities in metals might affect the resistance at low temperatures. He therefore chose mercury, the only metal that one could hope to get into a higher state of purity than platinum and gold. He found that mercury behaved as a normal metal, having a residual resistance that leveled off at about 14 K. However, at 4.2 K, the temperature of liquid helium, the resistance vanished (Fig. 5.15). On this occasion, he noted that mercury had passed into a new state, which, because of its extraordinary properties, he called the *superconductive state* (Onnes, 1911).

> The value of the mercury resistance used was 172.7 ohm in the liquid condition at 273 K; extrapolation from the melting point to 273 K by means of the temperature coefficient of solid mercury gives a resistance corresponding to this of 39.7 ohm in the solid state. At 4.3 K this had sunk to 0.084 ohm, that is, to 0.0021 times the resistance which the solid mercury would have at 273 K. At 3 K the resistance was found to have fallen below $3 \times 10^{-6}$ ohm, that is to one ten-millionth of the value which it would have at 273 K. As the temperature sank further to 1.5 K, this value remained the upper limit of the resistance.

In the superconducting state, the dc electrical resistivity is zero. Persistent electric currents have been observed to flow without attenuation in superconducting

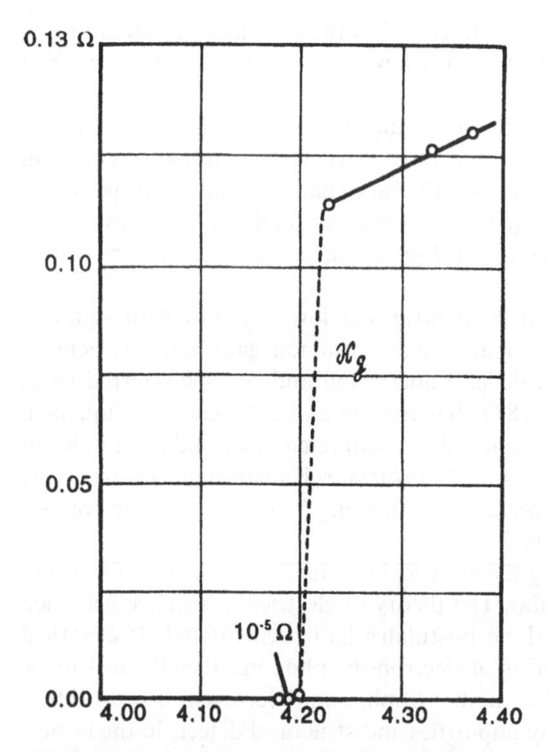


**Figure 5.15** Resistance of mercury vs. temperature, the original sketch by Onnes.

rings for more than a year, until at last the experimentalist wearied of the experiment. The record appears to be $2\frac{1}{2}$ years (S.C. Collins, quoted in Lynton, 1969). The decay of supercurrents in a solenoid was studied by File and Mills (1963), who used precision nuclear magnetic resonance methods to measure the magnetic field associated with the supercurrent. They concluded that the decay time of the supercurrent is not less than 100,000 years.

All attempts to measure a resistance for a superconductor below $T_c$ have failed. Only an upper limit can be placed, this limit depending on the sensitivity of the measurement. The present limit is

$$\rho < 10^{-22} \text{ ohm cm}$$

By comparison, the resistivity of pure copper at low temperatures is

$$\rho_{Cu} \sim 10^{-9} \text{ ohm cm}$$

Superconductivity is not peculiar to a few metals. More than 20 metallic elements can become superconductors (see Table 5.12). Even certain semiconductors can be made superconducting under suitable conditions, and the list of alloys whose superconducting properties have been measured stretches into the thousands. A few

**Table 5.12** Properties of Selected Superconducting Elements

| Element | Transition temperature in zero field, $T_0$ (K) | Critical field at absolute zero, $H_{c2}$ ($T$) (T) |
|---|---|---|
| Aluminum | 1.19 | 0.0102 |
| Cadmium | 0.55 | 0.00288 |
| Gallium | 1.09 | 0.0055 |
| Indium | 3.4 | 0.0285 |
| Iridium | 0.14 | 0.0019 |
| Lanthanum ($\beta$) | 6.10 | 0.160 |
| Lead | 7.19 | 0.0803 |
| Mercury ($\alpha$) | 4.15 | 0.0415 |
| Molybdenum | 0.92 | 0.0098 |
| Niobium | 9.2 | 0.1390 ($H_{c1}$); 0.2680 ($H_{c2}$) |
| Osmium | 0.66 | 0.0065 |
| Rhenium | 1.70 | 0.0193 |
| Ruthenium | 0.49 | 0.0066 |
| Tantalum | 4.48 | 0.0805 |
| Technetium | 7.77 | 0.1410 |
| Thallium | 2.39 | 0.0170 |
| Thorium | 1.37 | 0.0150 |
| Tin | 3.72 | 0.0305 |
| Titanium | 0.39 | 0.0100 |
| Uranium | 0.68 | |
| Vanadium | 5.41 | 0.0430 ($H_{c1}$); 0.0820 ($H_{c2}$) |
| Zinc | 0.85 | 0.0053 |
| Zirconium | 0.55 | 0.0047 |

**Table 5.13** Properties of Superconducting Compounds and Alloys[a]

| Compound | Transition temperature in zero field, $T_0$ (K) | Critical field at absolute zero, $H_{c2}$ (T) (T) |
|---|---|---|
| $Au_2Bi$ | 1.7 | 0.0085 |
| Cu–S | 1.6 | 0.0104 |
| Hg–Cd | 4.1 | 0.041 |
| $KBi_2$ | 3.6 | 0.0234 |
| Na–Bi | 2.2 | 0.0198 |
| $Nb_3Al$ | 17.1 | 25.0 ($H_{c2}$) |
| NbTi | 9.5 | 12.2 |
| $Nb_3Sn$ | 18.1 | 25 |
| $Nb_3Ge$ | 23.2 | 38 |
| $Nb_3Al_{0.7}Ge_{0.3}$ | 20.7 | 44 |
| Pb–In | 7.3 | 0.359 |
| Pb–Tl | 7.3 | 0.448 |
| Sn–Bi | 3.8 | 0.0807 |
| Sn–Cd | 3.6 | 0.0381 |
| $Sn_3$–$Sb_2$ | 4.0 | 0.0673 |
| Sn–Zn | 3.7 | 0.0305 |
| $Tl_2Bi_5$ | 6.4 | 0.0702 |
| Tl–Mg | 2.75 | 0.0220 |
| $Ta_3Sn$ | 8.35 | 24.5 ($H_{c2}$) |
| $V_3Ga$ | 16.5 | 35 |
| $YBa_2\ Cu_3\ O_7$ (bulk) | 95.0 | 18 ($H_{c2}$) |

[a] For type II superconductors, the zero temperature critical field quoted is obtained from an equal-area construction. The low-field ($H_{c1}$) magnetization is extrapolated linearly to a field $H_c$ chosen to give an enclosed area equal to the area under the actual magnetization curve.

of the more technically important compounds and alloys are listed in Table 5.13. For pure, unstrained metals, the transition is very sharp, ~0.001 K. For alloys, the transition takes place over a small range of temperature. A superconducting thermometer, phosphor bronze +0.1% lead, becomes superconducting nearly linearly over the temperature range of 1–7 K. This is an extreme case caused by the combination of alloy, impurity, and severe strain (from being drawn into a wire).

The range of transition temperatures at present extends from 23.2 K for the alloy $Nb_3Ge$ to 0.01 K for some semiconductors. In many metals, superconductivity has not been found down to the lowest temperatures at which the metal was examined, usually well below 1 K. Thus Li, Na, and K have been investigated for superconductivity down to 0.08, 0.09, and 0.08 K, respectively, where they were still normal conductors. Similarly, Cu, Ag, and Au have been investigated down to 0.05, 0.35, and 0.05 K, where they are still normal conductors. It has been predicted that if sodium and potassium superconduct at all, the transition temperature will be much lower than $10^{-5}$ K.

These values of the transition temperature refer to normal (atmospheric) pressure. Since electrical conductivity depends on at least an electron–lattice interaction (see Sec. 3.1), it is not surprising that superconductivity also depends on such an interaction (although in a much more subtle way). Applying pressure to a specimen causes its crystal lattice to be changed, so it is to be expected that transition temperatures will also be altered. In general, it is found that at high pressures superconducting transitions occur at lower temperatures than they do at 1 atm for the same material. For example, at 1 atm the transition temperature of indium is about 3.41 K. At 1370 atm, it is 3.34 K, and at 1730 atm it is about 3.325 K. Cesium becomes a superconductor ($T_c = 1.5$ K) at a pressure of 110 kbar, after several phase transformations. Pure bismuth, not a superconductor at 1 atm, has been found to become superconducting at pressures in the region of 20,000 atm with a transition temperature of about 7 K. There are, however, some metals such as thallium for which the transition temperature increases slightly with moderate pressure, although at still higher pressures, around 13,000 atm, it falls again.

It should be noted that some of the best electrical conductors—Cu, Ag, Au—do not become superconducting; none of the elements of group I superconduct either. Something more than normal electronic conduction is going on. Superconductors are more than merely *perfect conductors*, which may be defined as a conductor in which the electrons find no impurity or lattice obstacles.

## 4.2. Magnetic Properties of Superconductors

The magnetic properties exhibited by superconductors are as dramatic as their electrical properties. They cannot be accounted for by the assumption that the superconducting state is characterized properly by zero electrical resistivity.

Onnes quickly realized the possibility of using his discovery to produce electromagnets with zero power requirement, because, with superconductors, there would be no joule heat to be dissipated. Unfortunately, he found a threshold value of current in the superconducting materials that were available at that time that caused them to revert to the normal resistive stage (quench) in magnetic fields of less than 1000 gauss.

In 1933, Meissner in Germany discovered that superconductors expel magnetic flux regardless of whether the field was applied before or after the specimen was cooled. This is known as perfect diamagnetism or the Meissner effect (see Fig. 5.16). The unique magnetic properties of superconductors are of central importance to the characterization of the superconducting state.

For a given material, the value of magnetic field that quenches superconductivity is called the critical field or threshold field. At the critical temperature, the critical field is 0. $H_c(T_c) = 0$. The variation of the critical field with temperature for several superconducting elements is shown in Fig. 5.17. The result is a nearly parabolic

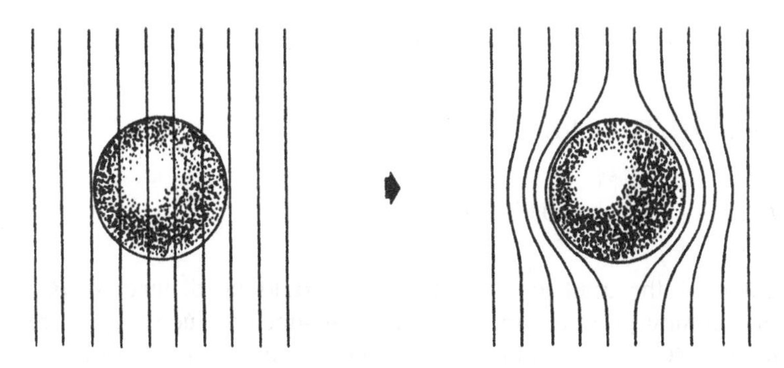

**Figure 5.16** The Meissner effect.

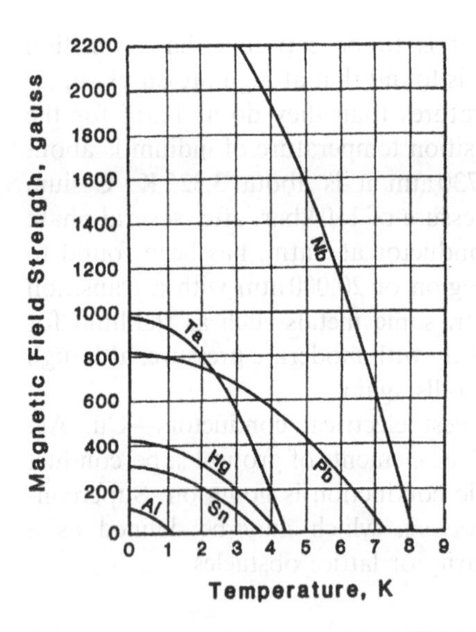


**Figure 5.17** Variation of critical field with temperature.

curve, called the threshold curve. Each threshold curve divides the realm of magnetic field strength and temperature into two areas. The superconducting state for each metal is in the lower left of the figure.

For a specimen in the shape of a long cylinder or wire placed parallel to the applied field, the threshold field is well defined at every temperature and is a function of temperature. The threshold field is related approximately to the absolute temperature $T$ by the expression

$$H_T = H_0\left[1 - \left(\frac{T}{T_0}\right)^2\right] \tag{5.36}$$

Although this parabolic relationship is not exactly true, it is adequate for many purposes.

*Example 5.6.* Determine the threshold field strength $H_T$ for vanadium at 3.3 K, assuming that the parabolic rule for the transition curve is valid for vanadium.

$$H_0 = 1340\,\text{G} \quad \text{and} \quad T_0 = 9.18°\text{R} = 5.1\,\text{K}$$

$$H_T = H_0\left[1 - \left(\frac{T}{T_0}\right)^2\right] = 1340\,\text{G}\left[1 - \left(\frac{3.3}{5.1}\right)^2\right] = 779\,\text{G}$$

In ISO units, $H_0 = 0.1340T$; $T_0 = 5.1$ K.

$$H_T = (0.134T)\left[1 - \left(\frac{3.3}{5.1}\right)^2\right] = 0.0779\,T$$

The critical value of the applied magnetic field should be defined as $B_{ac}$. However, this is not common practice among workers in superconductivity, where the external field in gauss (G) or tesla (T) is used. One should note that in the cgs system, $H_c = B_{ac}$, and in SI, $H_c = B_{ac}/\mu_0$. The symbol $B_a$ denotes the applied magnetic field. The symbol $\mu_0$ is the magnetic induction, and the subscript "c" denotes the critical conditions.

The Meissner effect clarified theoretical thought on the possibility of applying thermodynamics to superconductivity and opened the door to further understanding. The *HT* plane is divided into two distinct regions, much like a vapor pressure curve separates the liquid and gas regions of a fluid. It is useful, therefore, to ask whether crossing the *H–T* curve represents a phase transition in the material analogous to the more familiar types of thermodynamic phase transitions. This is the same as asking whether the superconducting transition is thermodynamically reversible. If so, thermodynamic relationships could be used to obtain additional information about superconductivity. Early attempts to answer this question started from the assumption that the condition of a superconductor is fully described by the property of perfect electrical conductivity. Ohm's law states that

$$E = j/\sigma$$

where $E$ is the electric field producing the current; $j$ the current density; and $\sigma$ is the electrical conductivity. As $\sigma \to \infty$, then, $\mathrm{E} \to 0$. But if $E = 0$, Maxwell's statement of Faraday's law is

$$\frac{\partial B}{\partial t} = 0$$

where $B$ is the magnetic induction and $t$ the time. That is, the magnetic induction $B$ is constant with time.

To test the thermodynamic reversibility of crossing the *HT* plane, let us imagine two different paths between identical initial and final conditions.

If the initial and final states of the superconductor are the same, then the dependence of transition temperature on magnetic field is reversible, and we can proceed with other ordinary thermodynamic arguments to understand superconductivity.

Imagine a metal cylinder cooled to a temperature below $T_c$ and then placed in a transverse magnetic field. The requirement developed above that $B$ remains constant in the superconductor requires the behavior shown in Fig. 5.18a. The superconductor is perfectly diamagnetic in the presence of the field, with no penetration of magnetic flux into its interior.

Figure 5.18b shows another path between the same initial and final states. In this case, the specimen is placed in a magnetic field while still a normal conductor, and flux penetrates it. It is then cooled below $T_c$. Again the requirement of constant $B$ applies, so in this case the flux lines remain constant in the specimen in the final state. In both cases, the temperature and applied magnetic field are the same at the end of the process, but the magnetic state of the superconductor is obviously different. Since a thermodynamically reversible process does not depend on the process, the conclusion from this analysis would be that the superconducting transition is *not* thermodynamically reversible.

In spite of these early conclusions that the transition temperature–magnetic field dependence was irreversible, application of reversible thermodynamic principles gave results in good agreement with experiment.

In 1933, Meissner and Ochsenfeld performed the experiment described above and found that the superconductor *expelled* the magnetic field. Not only is $\mathrm{d}B/\mathrm{d}t = 0$ (and constant), but also $B = 0$! Thus the superconducting state exhibits not only zero

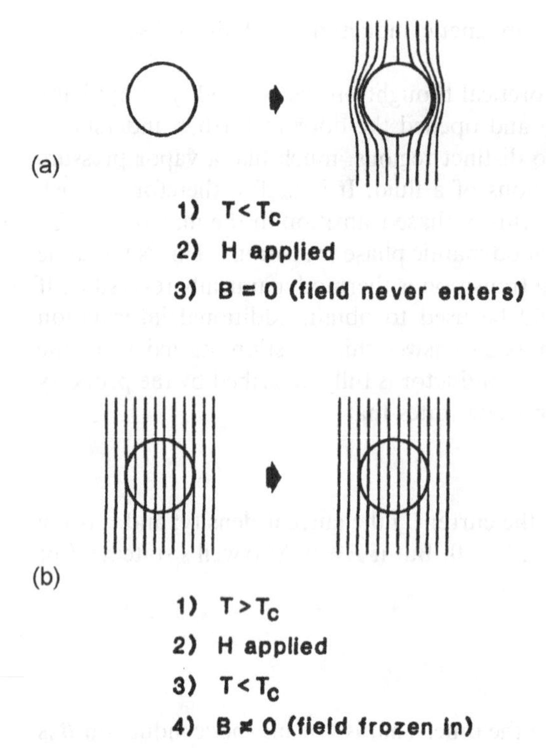


**Figure 5.18** Behavior of a metal of infinite conductivity in a magnetic field.

resistivity and zero electric field but also zero magnetic induction (it is perfectly diamagnetic).

The discovery of the Meissner effect settled the question of whether or not the transition was reversible (the transition is thermodynamically reversible). Instead of Fig. 5.15b, therefore, we have the following situation as we cool the specimen to a temperature below its crucial temperature. As the transition to the superconducting state takes place, the magnetic field is expelled by the metal. It is seen that this state corresponds to what has been called the perfect diamagnetism of a superconductor. In terms of the normal diamagnetic effect, which is quite small, the diamagnetism of a superconductor is enormous and represents an effect as surprising as perfect conductivity.

This result is not implied by perfect conductivity (i.e., $\sigma = \infty$) alone, even though perfect conductivity does imply a somewhat related property. If a perfect conductor, initially in zero magnetic field, is moved into a region of nonzero field (or if a field is turned on), then Faraday's law of induction gives rise to eddy currents that cancel the magnetic field in the interior. It turns out that a perfect conductor placed in a magnetic field cannot produce a permanent eddy current screen; the field will penetrate about 1 cm in an hour.

If, however, a magnetic field were established in a perfect conductor, its expulsion would be equally resisted. Eddy currents would be induced to maintain the field if the sample were moved into a field-free region (or if the applied field were turned off). Thus perfect conductivity implies a time-independent magnetic field in the interior but is noncommittal as to the value that field must have. In a superconductor, the field is not only independent of time; it is zero.

The superconductor is a perfect diamagnet, and the magnetic field is expelled by shielding currents that flow on the surface of the superconductor and penetrate only a small distance ($10^{-5}$ cm) into the bulk of the specimen. The surface currents are just high enough to cancel the magnetic field in the interior. The distance the field penetrates is called the penetration depth $\lambda$. As the field is increased, it will begin to penetrate the sample.

The existence of the critical magnetic field also explains why a superconductor has a critical current. The current in the superconductor will increase until the magnetic field in the surface of the conductor (generated by the self-current) reaches $H_c$. At this point, complete flux penetration occurs and superconductivity is destroyed.

This second method of destroying superconductivity is the Silsbee effect, which was suggested by Silsbee (1916). The Silsbee hypothesis is that when the electric current flowing in a superconductor produces a magnetic field at the surface of the material that equals or exceeds the threshold field, the normal state is restored. The current corresponding to the threshold field is called the threshold current. For example, for a long wire of diameter $d$, the magnetic field produced at the surface of the wire by an electric current $I$ is $H = I/\pi d$, so the critical current in this case would be

$$I_c = \pi H_c d \tag{5.37}$$

where the critical field is expressed in units of amperes per meter. (The units conversion factor between tesla and A/m is the permeability of free space, $\mu_0 = 4\pi \times 10^{-7}$ T m/A.)

The threshold current, as defined by the Silsbee hypothesis, is the largest current that a superconductor can carry and still remain superconducting. The critical current relationship for type II superconductors is more complicated and must generally be determined experimentally.

The magnetization curve expected for a superconductor under the conditions of the Meissner–Ochsenfeld experiment is sketched in Fig. 5.19a. This applies quantitatively to a specimen in the form of a long solid cylinder placed in a longitudinal magnetic field. Pure specimens of many materials exhibit this behavior; they are called type I superconductors (formerly, soft superconductors). The values of $H_c$ are always too low for type I superconductors to have any useful technical application in coils for superconducting magnets.

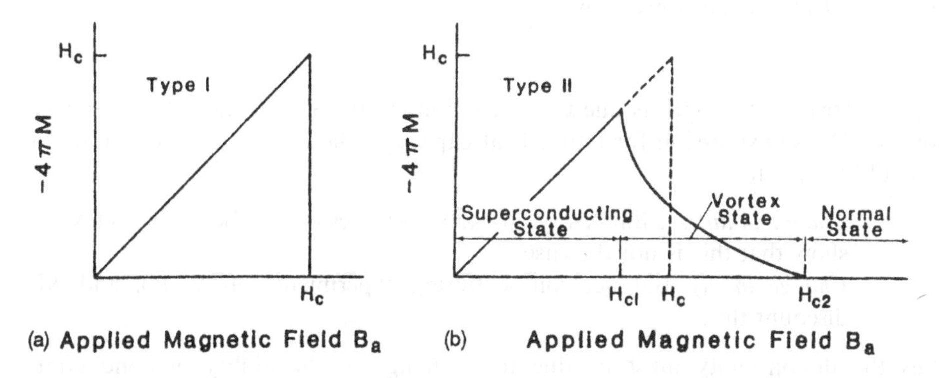


**Figure 5.19** Magnetization vs. applied magnetic field for type I and type II superconductors.

*Example 5.7.* Determine the threshold current predicted by Silsbee's rule for tantalum wire at 3.33 K, assuming that the parabolic threshold rule is valid. The wire diameter is 0.00127 m (0.127 cm).

First, it is necessary to find the critical field for tantalum at 3.3 K. From Table 5.12, $H_0 = 0.0805\,\text{T}$ at $T_0 = 4.48\,\text{K}$.

From Eq. (5.20),

$$H_c = H_0\left[1 - \left(\frac{T}{T_0}\right)^2\right] = 0.0805[1 - (3.33/4.48)^2] = 0.0360\,\text{T}$$

or

$$H_c = (0.0360)/(4\pi)(10^{-7}) = 2685\,\text{A/m}$$

From Eq. (5.21 ), the critical current for a long wire is

$$I_c = \pi H_c d = \pi(2685)(1.27 \times 10^{-3}) = 10.7\,\text{A}$$

Other materials exhibit a magnetization curve of the form shown in Fig. 5.19b and are known as type II superconductors. They tend to be alloys or transition metals with high values of electrical resistivity in the normal state; that is, the electronic means free path in the normal state is short. Type II superconductors have superconducting electrical properties up to a field denoted by $H_{c2}$. Between the lower critical field $H_{c1}$ and the upper critical field $H_{c2}$, the flux density $B \neq 0$, and the Meissner effect are said to be incomplete. The value of $H_{c2}$ may be 100 times or more higher than the value of the critical field $H_c$ calculated from the thermodynamics of the transition. In the region between $H_{c1}$ and $H_{c2}$, the superconductor is threaded by flux lines called fluxoids and is said to be in the vortex state. A field $H_{c2}$ of 410 kG (41 T) has been attained in an alloy of Nb, Al, and Ge at the boiling point of helium, and 510 kG has been reported for $Pb_1Mo_{5.1}S_6$. Commercial solenoids with a hard superconductor would produce high steady fields, some over 100 kG.

## 4.3. Thermodynamic Properties of Superconductors

### *4.3.1. Specific Heat*

Experimental measurements on superconducting materials such as tin in both the normal and superconducting states show a discontinuity in heat capacity at the transition temperature. (The specific heat of the normal metal is obtained by making measurements in a magnetic field large enough to prevent superconductivity in the specimen.) To interpret these observations, we recall that at low temperatures, the specific heat of a metal is given by

$$C = \gamma T + AT^3 \tag{5.38}$$

The first term on the right is due to conduction electrons, and the $T^3$ term comes from the Debye expression for lattice heat capacity. The observed discontinuity in $C_V$ could be due to

1. *Change in lattice.* But X-ray diffraction studies of Pb above and below $T_c$ show that this is not the case.
2. *Change in* $\theta_D$. But neutron scattering experiments on V, Pb, and Nb discount this.

Thus the discontinuity must be due to a change in the ability of conduction electrons to accept heat. This observation led to an early phenomenological theory of superconductivity, the "two-fluid" model of superconductivity. This model is not unlike the two-fluid model used to explain certain properties of helium II (see Sec. 3.8 of Chapter 3). In this model, the superconductor is said to be composed of two fluids: (1) normal electrons, and (2) superconducting electrons.

Below $T_c$, both normal and superconducting electrons exist; above $T_c$, only normal electrons exist. As $T$ is decreased below $T_c$, more and more normal electrons condense into the superconducting state until $T = 0\,\text{K}$ only superconducting electrons exist.

The "superfluid" is responsible for all superconducting effects such as $\rho = 0$, $E = 0$, and the Meissner effect.

The ratio of normal electrons to superconducting electrons is governed by temperature. As heat is added, part of the energy goes to the lattice and normal electrons and part goes to promote some superconducting electrons to the normal state. (The superconducting state is a lower energy state.) Thus, the discontinuity in the heat capacity curve is due to the normal/superconducting electron spectrum, because additional energy is required to promote some superconducting electrons to the normal state. If the lattice contribution to $C_V$ ($\sim T$) is subtracted, the remaining portion is due only to the electron spectrum. This residual fits the exponential form

$$C_e = A \exp(-D/T) \tag{5.39}$$

which is strongly suggestive of an energy gap.

This says that there is a region of energies forbidden to the conduction electrons. Superconducting electrons lie in a low-energy state below this gap and must be excited across it to become normal electrons. The number excited across the gap varies exponentially with the reciprocal of the temperature.

*4.3.2. Thermal Conductivity*

The thermal conductivity of superconductors is not nearly as their electrical conductivity. Nothing like the Lorenz relation holds. From the explanation of thermal conductivity of normal metals (Sec. 1.2), We can show that the thermal conductivity of superconductors can either rise or fall. Superconducting electrons can neither accept heat nor scatter phonons. (Heat cannot be accepted in amounts less than the energy gap amounts. Greater amounts demote superconducting electrons to the normal state.)

For pure metals, in which thermal conduction is carried out largely by electrons, the thermal conductivity is lowered in the superconducting state as normal electrons "condense" into the superconducting state and become unavailable to conduct heat. Thus the thermal conductivity decreases. In impure or highly disordered metals, which conduct heat largely by phonons, the thermal conductivity increases below $T_c$ because fewer normal electrons are present to scatter the phonons. A degree of impurity can exist between these extremes such that the thermal conductivity is unaffected by the transition to the superconducting state, and hence the thermal conductivity remains relatively constant as the transition temperature is passed.

## 4.4. The Isotope Effect

It has been observed that the critical temperature of superconductors varies with isotopic mass. In mercury, $T_c$ varies from 4.185 to 4.146 K as the average atomic mass

$M$ varies from 199.5 to 203.4 atomic mass units (amu). The transition temperature changes smoothly when we mix different isotopes of the same element. The experimental results within each series of isotopes may be fitted by a relation of the form

$$T_c = \text{constant}/\sqrt{M} \tag{5.40}$$

The constant is the Debye temperature, and the $-1/2$ power of $M$ is only approximate. The important thing to notice is that this relation predicts that $T_c$ would be 0 K if the mass $M$ were infinite, that is, if the ions were *immovably* fixed in the lattice. Thus this isotope effect suggests that (since $T_c > 0$ K) superconductivity depends not on the lattice array but on its dynamic motion. Like normal conductivity, superconductivity depends on an electron–phonon interaction, but in a more complicated fashion than in normal metals.

## 4.5. A Classical Representation of the BCS Theory of Superconductivity

The BCS theory of superconductivity is due to Bardeen et al. (1957). Superconductivity appears in a variety of metals that have widely differing lattices, purities, alloys, etc., but a striking similarity or uniformity in their superconducting properties. Accordingly, for a first approximation, construct a hypothesis that ignores the difference among metals and concentrates on the similarities, i.e., the basic interactions between their ions and conduction electrons. Clearly this will not explain the differences among superconductors, but it will give a qualitative understanding of the basic aspects.

In the electrical conductivity of normal metals, we considered electrons colliding with phonons and imperfections and ignored electron–electron collisions. This was possible because the Coulomb forces between any two electrons are diluted by intermediate electrons. They experience a "screened" coulombic force that makes this repulsion (scattering) a very short-range force and insignificant compared to scattering by phonons and impurities.

But for superconductors, one must consider every force. The Meissner effect allows a calculation of the quenching energy, which is $10^{-8}$ eV per atom. The theory of *normal* metals ignores effects smaller than 1 eV.

Basic to the BCS theory is an attractive interaction between pairs of electrons in the metal. This interaction results from an interaction between each electron and the lattice of the metal.

As an electron passes through the lattice, the positive ions of the lattice are momentarily attracted to it. Therefore, in its wake the electron leaves a ripple in the lattice. Because the lattice is composed of charged ions, this can be thought of as a vibrating polarization of the medium through which the electron has passed. The second electron sees this deformed lattice and is attracted to it, thereby creating an attraction between the electrons even though each has a negative charge. The pair of electrons bound in this manner is referred to as a Cooper pair. These electrons orbit around each other, each "surfing" on the virtual lattice wave the other leaves in its wake. The distance by which they are separated is determined by the relative strengths of the forces between them. The binding is very weak, which explains why the thermal energy ($kT$) is enough to separate the pairs at temperatures higher than $T_c$.

The attractive force of the electron–phonon interaction must balance the repulsive force of the screened coulombic interaction, and this occurs at a distance of about $10^{-4}$ cm. This result would obviously disappear if the mass of the ions were infinite; furthermore, the effect leads to the proper square-root dependence of the transition temperatures on the mass as observed in the isotope effect.

The BCS theory is successful in accounting for most of the basic features of the superconducting state. Its basic elements are the following:

1. An attractive interaction between electrons can lead to a ground state separated from excited states by an energy gap. The critical field, the thermal properties, and most of the electromagnetic properties are consequences of the energy gap.
2. The electron–lattice–electron interaction leads to an energy gap of the observed magnitude. The indirect interaction proceeds when one electron interacts with the lattice and deforms it; a second electron sees the deformed lattice and adjusts itself to take advantage of the deformation to lower its energy. Thus the second electron interacts with the first electron via the lattice deformation.

The BCS theory also helps explain the paradox that the best electrical conductors (e.g., gold, silver, copper) do not become superconductors. Good conductivity in a normal metal depends on poor interaction between electrons and the lattice (little scattering of electrons by phonons), whereas superconductivity depends on strong interactions of this type.

A point to be noted in connection with the BCS theory is its inability to predict which elements of the periodic table will become superconductors at low temperatures. It has nothing to say on this subject because it disregards differences among various metals from the start. The transition temperature $T_c$ depends on (1) an average phonon frequency that is proportional to the Debye temperature; (2) the density of states in the normal form of the metal at the Fermi level, which is proportional to the constant $\gamma$ in the expression for electronic specific heat; and (3) an electron–phonon interaction parameter, which is greatly oversimplified to allow mathematical solution of the problem. As a result of the oversimplification of these parameters in the BCS approach, the transition temperature cannot be derived from the BCS theory but is treated as an empirical parameter. Progress still has to be made on such important aspects of the subject as distinguishing between superconductors and nonsuperconductors and predicting transition temperatures from first principles. These things can be done to some degree, but it is necessary to rely on empirical rules that have been formulated out of a great deal of experience with superconducting alloys and compounds.

The major contribution of the BCS theory lies in its formulation of the pertinent interaction responsible for superconductivity and the assumption of a sufficiently simple mathematical expression for this interaction to allow an initial solution of the problem. Even with an oversimplified expression for the interaction, remarkable agreement is obtained with some of the observed properties of superconductors.

## 4.6. Superconducting Materials

Three important characteristics of the superconducting state are the critical temperature (the temperature below which the superconducting state occurs), the critical magnetic field, and the critical current. All these are parameters that can be varied and sometimes improved by using new materials and giving them special metallurgical treatments. The science and technology involved in producing useful characteristics has now progressed to the level at which materials (superconductors) are on the shelf in forms that can support this new technology.

The crystal structure that stands out as being the most favorable to superconductivity is the beta-tungsten structure. The spatial arrangement evidently helps. Compounds with this structure are of the form $A_3B$, and the atoms of the transition element A are placed so as to form linear chains running along the three cubic directions. An unusual feature is that along the chains the atoms come about 10% closer to one another than they do in their elemental forms. At least 20 superconductors have been found in the A & B system since Hulm and Hardy discovered superconductivity in the compound $V_3Si$ at 17 K in 1952 at the University of Chicago. Shortly thereafter, Matthias found $Nb_3Sn$ to be superconducting at 18 K; this compound can be fabricated into wire that has very useful high-field properties.

Three conditions appear to be necessary to obtain high transition temperatures in the beta-tungsten structure: (1) the chains of A atoms must be long and unbroken and must consist of either niobium or vanadium, whereas the B atoms must be nontransition elements such as tin or aluminum; (2) the compound must be well ordered (i.e., niobium atoms must sit on A sites, B atoms on B sites); and (3) the compound must be formed in the ideal 3:1 chemical ratio.

It is quite reasonable to assume that technologically important new binary or ternary systems of compounds with the beta-tungsten structure will be discovered in the next few years with increasing transition temperatures.

In parallel with the experimental search for higher temperature superconductors in favorable structures (such as beta-tungsten or rock salt), there is always the possibility of finding superconductivity in new and unusual compounds. One expects to find unusual properties associated with unusual structures. Gamble and coworkers discovered an entirely new class of highly anisotropic superconductors by using a chemical reaction that inserts organic molecules between sheets of conducting layers of metallic $TaS_2$. In these substances, the molecular $TaS_2$ sheets remain intact, but their stacking sequence along the perpendicular axis frequently changes and their separation can be varied from 3 to 57 Å. These structures have been confirmed by Humberto Fernandez-Moran, Mitsuo Ohtsuki, and Akemi Hibino of the University of Chicago with an electron microscope.

Along with an increase of about 3 K in the transition temperature of $Nb_3$ (AlGe) over that of $Nb_3Sn$, Foner at the National Magnet Laboratory at MIT found that $Nb_3$ (AlGe) retains superconductivity in magnetic fields up to 400 kG, almost twice that of $Nb_3Sn$. However, it is very difficult to make $Nb_3$ (AlGe) in suitable wire from with useful critical currents.

$Nb_3Sn$ remains a likely material for use in high-field applications in the near future, although NbTi and $V_3Ga$ are strong rivals. $Nb_3Sn$, $V_3Ga$, NbTi, and NbZr are commercially available, whereas $Nb_3Ge$ and $Nb_3$ (AlGe) are materials for which the development is still in progress.

The alloy niobium titanium (NbTi) and the intermetallic compound of niobium and tin ($Nb_3Sn$) are the most technologically advanced, with NbTi being used more. Even though NbTi has the lower critical field and critical current density of the two, it is often favored because of its more convenient metallurgical properties. NbTi is a ductile alloy that can be easily drawn into a wire. It can also be extruded from a billet composed of NbTi along with other metals such as copper or aluminum. The wire can then be fabricated into devices by using conventional wire-handling machinery with no damage to the superconducting properties. In contrast, $Nb_3Sn$

and $V_3Ga$ are very brittle materials. They are compounds of crystalline structure designated as the A-15 type, which have to be formed under very well-defined temperature conditions.

$Nb_3Ge$ is also of the A-15 structure. However, it is even more difficult to fabricate because the high-$T_c$ phase is not the equilibrium phase. That is, if Nb and Ge are put in contact with each other at the required temperature, a low-$T_c$ phase is formed. The material must be formed under nonequilibrium conditions by sputtering, chemical vapor deposition, or other evaporation technique.

We have a good theoretical understanding of the properties of type II superconductors in high magnetic fields thanks primarily to the Russian theorists Ginsburg, Landau, Abrikosov, and Gorkov (GLAG theory). In the presence of a high magnetic field, the superconducting electrons form a lattice of vortices through which run single magnetic flux lines. The flux lines move in response to magnetic forces when the critical electric current is passed through the superconductor. In doing so, they generate voltage, resistance, and heat and destroy the superconductivity. It is this macroscopic interaction that limits the critical current to much lower values than that thermodynamically expected. The ability of the flux lines to resist the magnetic forces is caused by interactions that pin them to the lattice. The pinning forces depend very much on the metallurgical preparation.

For dc applications, the pinning of the flux lines keeps the resistance to 0. For ac applications, however, the flux lines seem to be able to flop around between pinning sites, which results in effective resistance. A better understanding of the ac loss mechanism and the pinning interactions could easily result in a 10-fold increase in critical current, which, for example, might be even more useful to immediate and future technology than the raising of the critical temperature by 5 or even 10 K.

Two mechanisms that result in premature quenching are the flux jumps and internal mechanical motion in the conductors. The effects of mechanical motion are reduced by designing adequate support for each conductor, and flux jumps can be minimized by training the magnet over many charging cycles. In many situations, the small thermal energies resulting from conductor motion are enough to quench magnets.

Flux jumps result from a momentary breakdown of the pinning force holding the flux lines to the lattice. This phenomenon introduces a spike of thermal energy to the system, which can raise the temperature of the wire to a value greater than $T_c$, thus causing the wire to lose its superconductivity and "go normal". The energy released in a flux jump is proportional to the diameter of the superconducting wire. Therefore, instead of a superconducting wire being made from a single filament, conductors are now made of many small superconducting strands embedded in a matrix of high-conductivity material such as copper or aluminum. A typical size for a superconducting filament in a cable is around 10 μm.

The fluxoid motion can be reduced by introducing structural defects in the crystal structure. Defects such as dislocations, voids, precipitates, and grain boundaries all play a role in enhancing the critical current by "pinning" the fluxoids to the defect sites, thus making it possible to resist the force arising from the superconducting transport current. Only when the pinning force is exceeded by the Lorenz force (the critical current is reached) will the vortices move. Thus the problem of producing high-$T_c$ materials reduces to that of producing many pinning sites in the superconductor.

Economic considerations come into play at this point. Both the cost of the materials and the cost of fabrication are appreciable. The transition metal alloys such as NbTi are much more flexible and hence easier to fabricate and handle than the beta-tungsten compounds such as $Nb_3Sn$, which have higher critical currents and fields. For example, one method of fabricating an $Nb_3Sn$ wire is to form a billet of Nb and Sn by drilling holes in a tin cylinder and inserting niobium rods into the holes. Copper is added to the billet for stability. The billet is then extruded. The number of Nb strands in the final wire can be increased by adding more billets, stacking, and continuing the extruding process. $Nb_3Sn$ cables with over 500,000 strands have been built up in this manner. The final geometrical configuration of the wire is not yet superconducting because it contains no $Nb_3Sn$. The wire must be heated to allow the Nb and Sn to react to form the superconducting material. The process is normally performed in an inert atmosphere at a temperature of around 700°C. The $Nb_3Sn$ compound forms at the interface between the Nb and Sn. Once the wire has been reacted, it must be handled very carefully because of the brittleness of the A-15 structure.

## 4.7. The High-Temperature Superconductors

Over the past few years, there has been a renewed experimental search for high-temperature superconductors. This effort has been augmented by the unexpected discovery of a superconducting copper oxide-based ceramic by Bednorz and Müller. This flurry of activity, shown in Fig. 5.20 has culminated in the development of a completely new class of mixed oxide materials that are superconducting to temperatures above the temperature of liquid nitrogen. The bulk superconducting behavior of these mixed oxide materials has been established not only by the disappearance of electrical resistance, magnetic shielding, and expulsion of flux (Meissner effect) but also through related studies covering the tunneling, acoustic, thermodynamic, and electromagnetic properties of the materials. Additionally, flux quantization studies

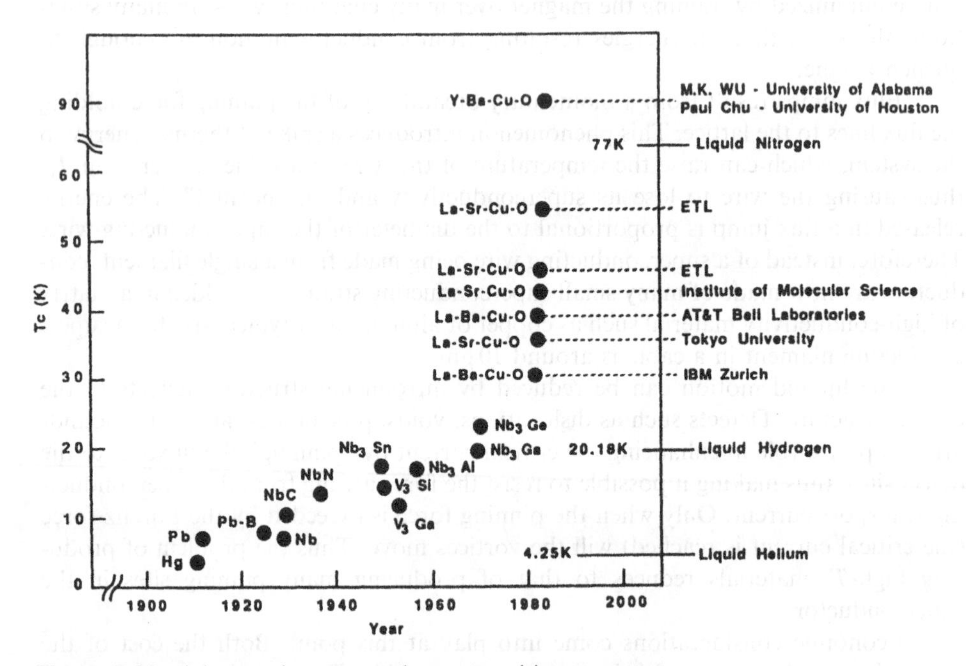


**Figure 5.20** Advances in superconductor transition temperatures.

have demonstrated the existence of electron pairs in these materials, as in the BCS theory.

The new high-temperature superconductors occur in at least two crystalline forms, both of which are closely related to the perovskite structure. The common feature of these superconductors is the sheets of Cu–O bonds, which are critical to both the conducting and superconducting modes of operation. One of the new superconductors, $La_{1.85}Sr_{0.15}CuO_4$ ($T_c \sim 40\,K$) has been obtained by doping $La_2CuO_4$. The other class of oxide superconductors, typified by $YBa_2Cu_3O_7$ ($T_c \sim 95\,K$), is not derived by doping but is a compound with separate sites for the yttrium, barium, copper, and oxygen atoms as shown in Fig. 5.21 and has an intrinsic metallic conductivity. The flow of electrons in this compound in the superconducting and normal states is sketched in Fig. 5.22. An interesting property of the latter compound is that the yttrium can be replaced by nearly any other member of the rare earth series of

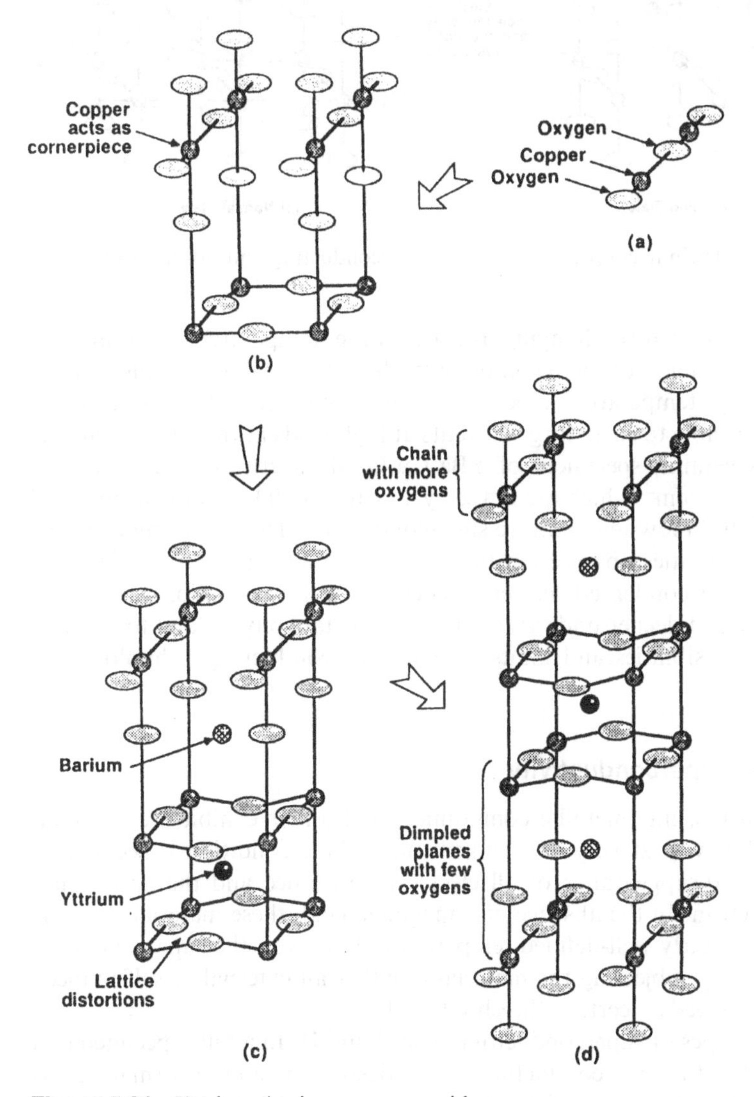


**Figure 5.21** Yttrium–barium–copper oxide.

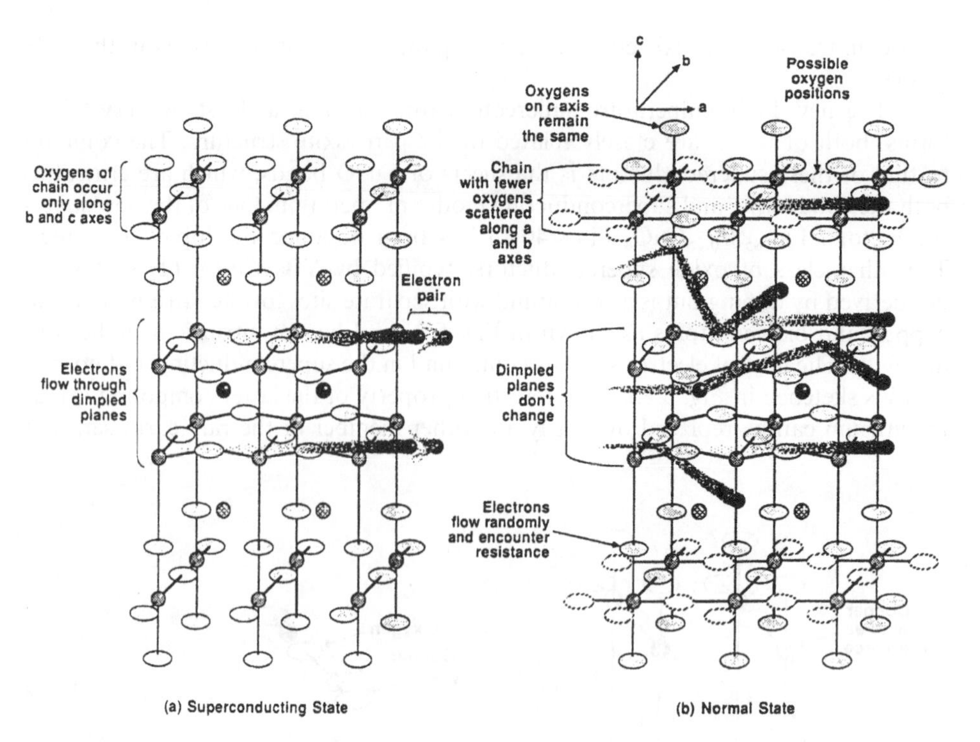


**Figure 5.22** Yttrium–barium–copper oxide in the superconducting and normal states.

elements without significantly changing the transition temperature. However, if oxygen is removed from any of these compounds, the superconductivity disappears.

For the new high-temperature superconductors to be used in large-scale applications, they must be able to carry large currents at high fields at the newly available temperatures. Bulk ceramic specimens of $YBa_2Cu_3O_7$ show critical current density values of $10^2$ and $10^3$ $A/cm^2$, which are lower by a factor of 100 than those observed for the NbTi and $Nb_3Sn$ low-temperature superconductors. The exact origins of the low critical current densities observed in these new materials are possibly the most challenging problem in condensed matter physics today. Its resolution is expected to lead to a broader and deeper understanding of solid-state physics and to be helpful in defining the possibilities and limits of new superconducting technologies as well.

## 4.8. Summary of Superconductivity

Some metallic elements, intermetallic compounds, and alloys exhibit the phenomenon of superconductivity at very low temperatures. This condition is recognized by the simultaneous disappearance of all electrical resistance and the appearance of perfect diamagnetism. In the absence of a magnetic field, these materials become superconducting at a fairly well-defined temperature. However, the superconductivity can be destroyed by subjecting the material to either an external or self-induced magnetic field that exceeds a certain threshold field.

There are two types of superconductors, types I and II. In a bulk specimen of a type I superconductor, the superconducting state is destroyed and the normal state is restored when an external magnetic field in excess of a critical value $H_c$ is applied. A type II superconductor has two critical fields, $H_{c1} < H_c < H_{c2}$; a vortex state exists in the range between $H_{c1}$ and $H_{c2}$.

During the transition to the superconducting state

1. The X-ray diffraction pattern remains constant and indicates that no change in the crystal lattice is involved.
2. The reflectivity in both the optical range and the infrared range remains unchanged.
3. Neutron absorption and photoelectric properties are unchanged.
4. There is no latent heat exhibited in zero field. This indicates a second-order transformation. If the magnetic field is not 0 however, heat is absorbed during the transition from the superconducting to the normal state.

In addition to the obvious electrical properties, properties that do change during the transition include:

1. The specific heat, which rises abruptly when a material becomes superconducting.
2. Thermoelectric effects. All the thermoelectric effects (Peltier, Thomson, and Seebeck effects) vanish when a material becomes superconducting.

The thermal conductivity may increase, decrease, or remain nearly constant. (See Sec. 4.3.)

Several of the low-temperature superconducting metals, such as lead, brass, and some solders (particularly lead-tin alloys), experience property changes when they become superconducting. Properties that may change include specific heat, thermal conductivity, electrical resistance, magnetic permeability, and thermoelectric resistance. The use of these metals may at times have to be avoided in the construction of low-temperature equipment if operational problems are to be minimized.

# 6

# Refrigeration and Liquefaction

## 1. INTRODUCTION

Why is it so easy to make something hotter and so much harder to make the same thing colder? You can rub your hands together and make them warmer. Why isn't there a way to rub your hands together backwards, somehow, and make them get colder? Put some soup in a microwave oven, push a button, and the soup gets warmer. Will there ever be a cryogenic analog to the microwave oven? Will we ever be able to put the soda pop in a box, push a button, and in a few seconds, the pop gets colder? Such an idea is so foreign to us that not even the best science fiction writers have ever proposed such a device.

Will there ever be such a machine? No. Here's why.

## 2. REFRIGERATION AND LIQUEFACTION

We understand intuitively that a refrigerator is a machine for making cold. A liquefier is a refrigerator that produces a cold liquid that is then drawn off. The thermodynamics is the same for both a refrigerator and a liquefier. The engineering is quite a bit different.

1. A refrigerator operates as a closed loop, constantly circulating the same working fluid, or refrigerant. There is no accumulation or withdrawal of product in a refrigerator. The mass flow rate of the refrigerant is the same at all points of the cycle. We say that there is *balanced flow* of the refrigerant. A refrigerator can certainly make a liquid coolant. This liquid refrigerant will be evaporated continuously by the heat added at the low-temperature sink and the vapor will be returned to complete the cycle.
2. A liquefier is an open system. A liquid product is removed, and an equivalent make-up stream must be added. The mass flow rate of the stream being cooled is greater than the countercurrent return stream by just the amount of liquid removed. Liquefier design differs from that of a refrigerator because the mass flow rates of these two streams are not equal. There is *unbalanced flow*. The return stream is smaller than the stream being cooled. The countercurrent refrigerant effect of the product withdrawn has been lost.

Since the thermodynamic cycle is the same for both refrigerators and liquefiers, we shall begin by looking at their similarities and use the broader term *refrigerator* to describe them both.

Most refrigerators use a gas as the working fluid. Dr. Will Gulley of Ball Aerospace Corporation says that there are only four things you can do to a gas. You can only (1) compress a gas or (2) expand a gas and these two steps can each be done in two ways: (1) adiabatically (in thermal isolation from the surroundings) or (2) isothermally (in thermal contact with the surroundings). See Fig. 6.1.

The basis of a gas refrigerator, then, is

1. Do work on a gas to compress it. Do this while the gas is thermally connected to a heat sink so that the gas is compressed isothermally. The gas is now compressed to the compressor outlet pressure and at its original temperature.
2. Move the gas to a different location either in space or time. No change in temperature has occurred.
3. Expand the gas while it is thermally isolated, causing it to cool. This is the essential refrigeration step. The gas temperature is now lower than when it started, and its pressure is back to the starting pressure of the compressor inlet.
4. Move the gas to a different location in space or time. The gas remains at the new, lower temperature.
5. Thermally connect the cold gas to the object to be refrigerated. Because the gas is colder than its surroundings, it will inevitably absorb heat from the surroundings.
6. Move the warmed gas to a different location in space or time; specifically, remove it from the object being cooled.
7. Warm the gas to its original temperature, and return it to the inlet of the compressor.

All refrigerators (gas or otherwise) have these essential elements in common. Note that it is essential that these steps somehow be out of phase with one another. If the gas did not physically move between the steps, the net effect would be no

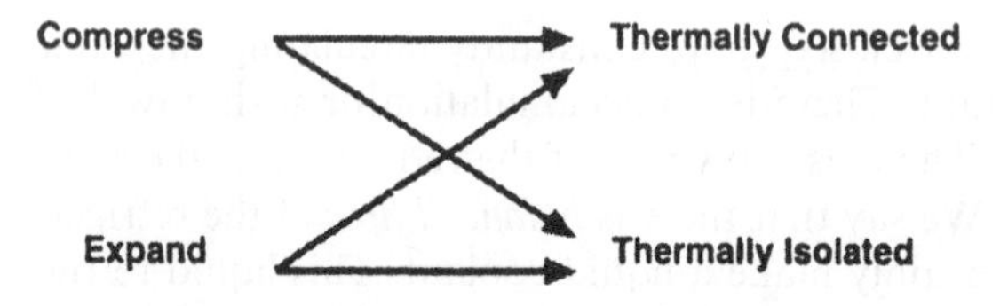


- **Compressing thermally connected to a reservoir**
  **- Rejects heat to the reservoir at constant temperature**
- **Compressing thermally isolated**
  **- Causes the gas to heat**
- **Expanding thermally connected to a reservoir**
  **- Absorbs heat from the reservoir at constant temperature**
- **Expanding thermally isolated**
  **- Causes the gas to cool**

**Figure 6.1** Four things you can do to a gas.

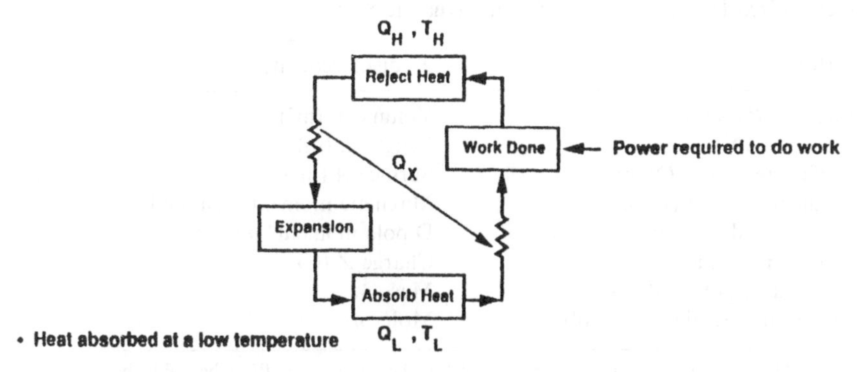


**Figure 6.2** Closed cycle cryogenic refrigerator.

refrigeration and the work done would be totally wasted. Such phasing is obvious in common refrigerators where the refrigerant is moved to the cold point to absorb heat and then moved back to the compressor inlet. The out-of-phase character of refrigeration is perhaps not immediately evident in Stirling-type coolers and pulse tube refrigerators but is nonetheless present and essential to their operation.

The purpose of these steps is to absorb heat at a low temperature and reject it at a higher temperature. Therefore, all refrigerators have a low-temperature heat absorption step and a high-temperature heat rejection step (Fig. 6.2).

In addition, all refrigerators have

1. A step where work is done on the system. To generalize, this is where the entropy of the refrigerant is reduced. A gas compressor is a common example.
2. A step where there is an order–disorder transition (the entropy of the system increases). A liquid evaporator and an expansion engine are common examples.

There may be heat exchange within the system itself for two reasons:

1. To permit the "compressor" to work at room temperature
2. To conserve, or recuperate, the "cold" produced

In every closed-cycle refrigerator, external power must be supplied to do work on the system. Typically, work is done on the working substance to move it to different physical locations in the refrigerator and also to move it thermodynamically to different places on the *T–S* diagram. A gas compressor is an easily understood component used for this inherently entropy-reducing step. However, the necessary work may, in fact, be done by the product of any generalized force and displacement. For examples, see Table 6.1.

In the most common case, work is done on the refrigerant fluid by compressing a gas. Work in the most general sense is the product of a force and a displacement. Therefore, refrigeration work can be done by the product of any force in Table 6.1 and its general displacement. For instance, a magnetic field may "compress" (that is, increase the order of) a paramagnetic salt in a magnetic refrigerator. Figure 6.3 shows the generalized nature of the refrigerator "compressor."

**Table 6.1** Examples of Generalized Forces and their Associated Generalized Displacements in Thermodynamic Systems

| $F$ (force) | $X$ (displacement) |
|---|---|
| Pressure $P$ (N/m$^2$) | Volume $V$ (m$^3$) |
| Tension $\tau$ (N) | Length $L$ (m) |
| Surface tension $\sigma$ (N/m) | Surface $A$ (m$^2$) |
| Magnetic field $H$ (A/m) | Magnetic moment[a] $\mu_o m$ (Wb m) |
| Electric field $E$ (V/m) | Dipole moment[a] $p$ (cm) |
| Electric potential $\varepsilon$ (V) | Charge $Z$ (C) |
| Centrifugal potential $r^2\omega^2/2$(m$^2$/s$^2$) | Mass $M$ |
| Chemical potential $\mu$ (J/mol) | Moles $n$ |

[a] $m = MV$, where $M$ is the magnetization and $V$ is the volume; $p = PV$, where $P$ is the polarization.

In the expansion step (Fig. 6.4), the working substance is expanded while it is thermally isolated from its surroundings. Since "internal work" is done as the molecules separate from one another, the temperature of the isolated system will drop. This is the essential refrigeration step. For fluids, this step may be the isoenthalpic expansion across a valve ($\Delta H = 0$) or the isentropic expansion through an engine ($\Delta S = 0$). In either case, the system is thermally isolated (adiabatic) with respect to its surroundings. In this step for a magnetic refrigerator, the field is turned off while the sample is thermally isolated.

In all cases, the generalized force on the system is reduced adiabatically. This causes the system temperature to drop.

The heat absorption step might typically be the evaporation of a liquid at a constant temperature. In general, this step depends on ordered and disordered states being in equilibrium with each other at some temperature.

The heat absorbed at temperature $T$ is related to the entropy change by

$$\mathrm{d}Q \leq T\,\mathrm{d}S$$

If the process is performed reversibly, the heat absorbed is

$$\mathrm{d}Q_{\mathrm{rev}} - T\,\mathrm{d}S$$

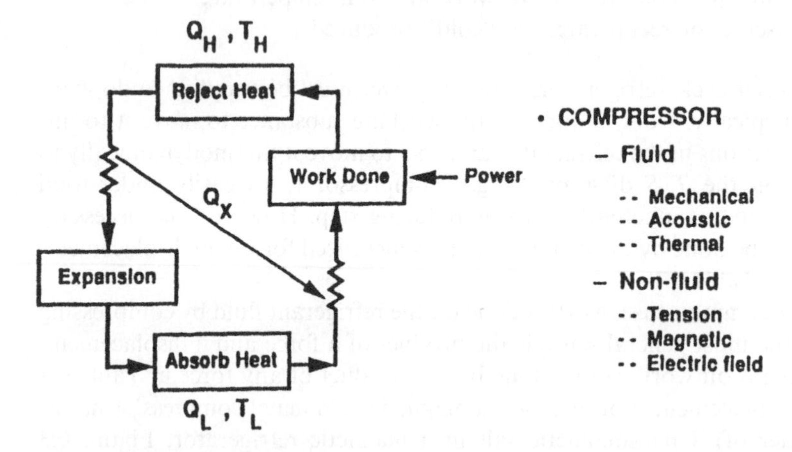


**Figure 6.3** Refrigeration steps—the compressor.

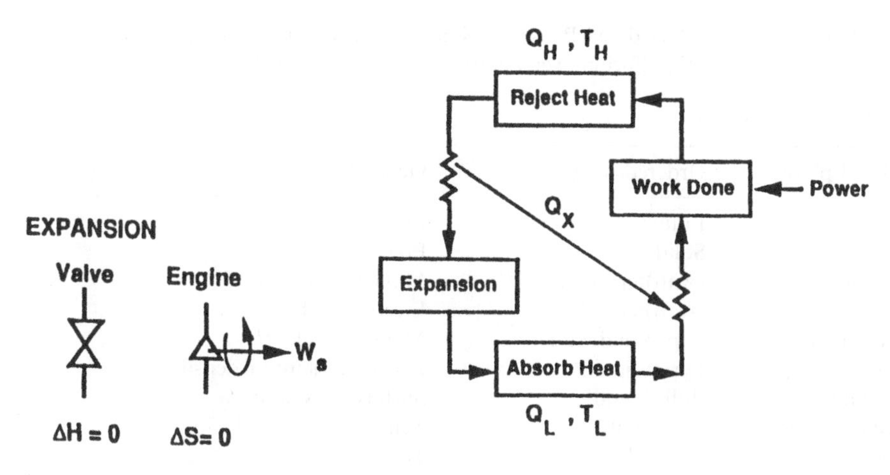


**Figure 6.4** Refrigeration steps—expansion.

Any material whose entropy can be changed can serve as the working substance (see Fig. 6.5).

Table 6.2 lists several systems that have ordered and disordered phases in equilibrium. The use of these two phases permits refrigeration. The temperature of the phase equilibrium can be altered by the force shown in the third column in Table 6.2. Some of the systems, e.g., liquid–gas, have sharp transitions, whereas other system transitions are broader. The question mark in the force column for some systems means that the best generalized force to act on such a system has not been determined.

There are as many ways to combine these few components as there are engineers to combine them. A relatively few "unit operations" of chemical engineering are strung together on the chemical plant design necklace to produce a seemingly infinite number of chemical processes and products. Likewise, these relatively few refrigeration operations can be combined in different ways to yield a large number of refrigeration cycles that have names like Joule–Thomson, Stirling, Brayton, Claude, Linde, Hampson, Postle, Ericsson, Gifford–McMahon, and

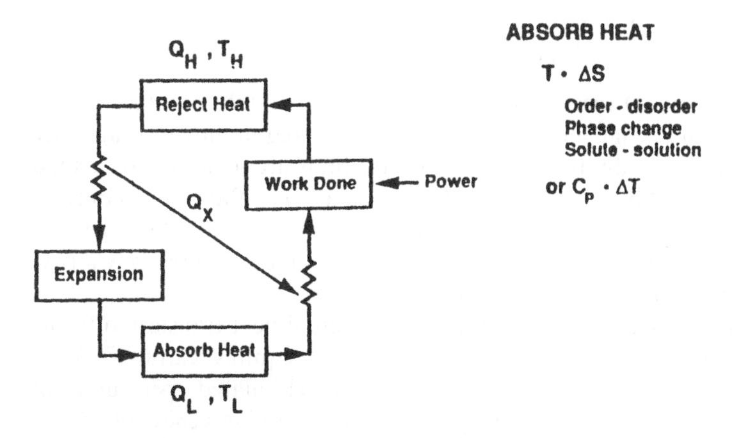


**Figure 6.5** Refrigeration steps—heat absorption.

**Table 6.2** Ordered and Disordered Phases of Some Systems and the Generalized Force Necessary to Change the Equilibrium Temperatures of the Entropy $S_d$

| System | | |
|---|---|---|
| Disordered phase | Ordered phase | Generalized force |
| Gas | Liquid | Pressure $P$ |
| Liquid | Solid | Pressure $P$ |
| Paramagnetic | (Anti)ferromagnetic | Magnetic field $H$ |
| Paraelectric | (Anti)ferroelectric | Electric field $E$ |
| Normal metal | Superconductor | Magnetic field $H$ |
| Fermi gas | Fermi liquid | Fermi pressure or density $n$ |
| Surface film | Bulk liquid | Surface tension $\sigma$ |
| Interstitial atoms | Martensite | Tension $\tau$ |
| Flexible polymer | Crystalline polymer | Tension $\tau$ |
| Disordered $\beta$-brass | Ordered ß-brass | ? |
| Uniform solution | Precipitate | Pressure $P$ or concentration $x$ |
| Molecular rotation | Rotational order | ? |
| $Z_n + CuSO_4$ | $Cu + ZnSO_4$ | Electric potential $E$ |
| Dilute solution | Concentrated solution | Chemical potential $\mu$ or concentration $x$ |

Vuilleumier. The essence of all these cycles is essentially the same, as described above. A classification system proposed by Graham Walker helps to demystify these names.

Walker proposed a classification system based on the kind of heat exchanger used in the cycle.

The purpose of the heat exchanger is to:

1. Warm the working substance to a higher temperature in order to
   a. Reject the heat absorbed at the higher temperature—the very essence of refrigeration
   b. Let a warm "compressor" be used
2. Recuperate or recover the "cold" produced in the cycle to
   a. Promote cycle efficiency
   b. Often, "bootstrap" the cycle

Heat exchangers are either (1) recuperative or (2) regenerative (see Fig. 6.6).

A recuperative heat exchanger is a device in which separate flow passages are provided for the hot and cold fluids (see Fig. 6.7). The fluids are separated by a solid wall, and heat is transferred by conduction across the wall. The fluids flow continuously, always in the same direction. Recuperative heat exchangers are "dc" devices. Cryocoolers with recuperative heat exchangers include the Linde, Hampson, Claude, Collins, and Joule–Thomson cycles.

Heat exchanger effectiveness $\varepsilon$ is the ratio of the actual heat transferred to the maximum possible energy transfer. Poor effectiveness requires extra work of compression because more working fluid must be circulated per unit of refrigeration. It also cuts into the "bootstrapping" process. Effectiveness of 95% is virtually a minimum requirement.

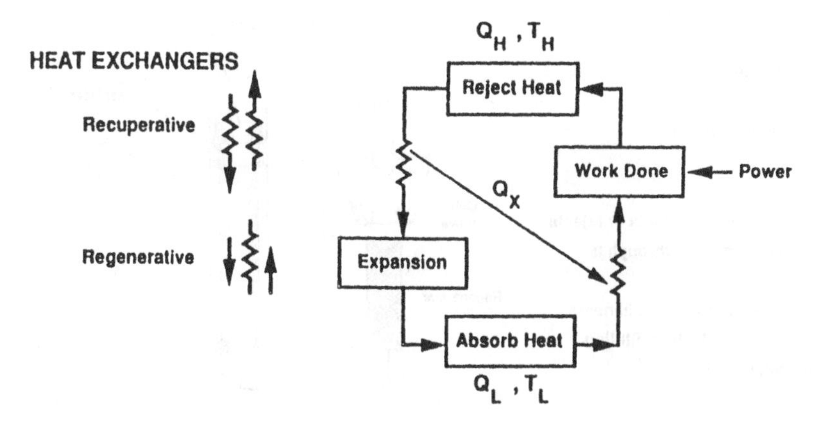


**Figure 6.6** Refrigeration steps—heat exchange.

A regenerative heat exchanger (Fig. 6.8) has a single set of flow passages through which the hot and cold fluids flow alternately and periodically. Recuperative heat exchangers are "ac" devices. The regenerative matrix, often a porous, finely divided mass (of metal wire or spheres), may be thought of as a thermodynamic sponge alternately accepting and rejecting heat as the hot or cold fluid flows through it. Cryocoolers with regenerative heat exchangers include Solvay and Gifford–McMahon, Stirling, Ericsson, and Vuilleumier (VM) cycles.

There is a fundamental low-temperature limit to the use of regenerators. The regenerator must be able to store heat on alternate cycles, although the heat capacity of most crystalline solids approaches zero as $T \rightarrow 0$. Regenerative heat exchanger effectiveness $\varepsilon$ must be 99% or higher.

Classifying mechanical refrigerators according to the kind of heat exchanger used leads to the tree shown in Fig. 6.9. One can see that regenerative heat

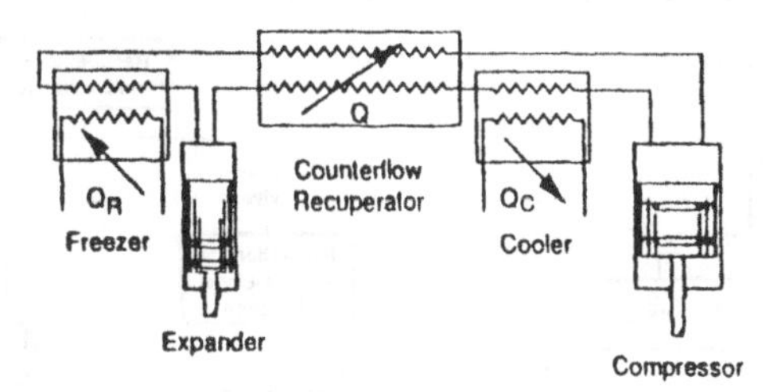


- **Separate flow passages for the hot and cold fluids**
- **Fluids are separated by a solid wall**
- **Fluids flow continuously (dc)**
- **Linde, Hampson, Claude, Collins, and Joule-Thomson cycles use recuperative heat exchangers**
- **Good heat exchanger effectiveness required**

**Figure 6.7** Recuperative heat exchanger.

- Single set of flow passages
- The hot and cold fluids flow alternately and periodically
- Regenerative matrix alternately accepts or rejects heat as the hot or cold fluid flows through it
- Cryocoolers with regenerative heat exchangers include Solvay and Gifford-McMahon, Stirling, Ericsson, and Vuilleumier cycles

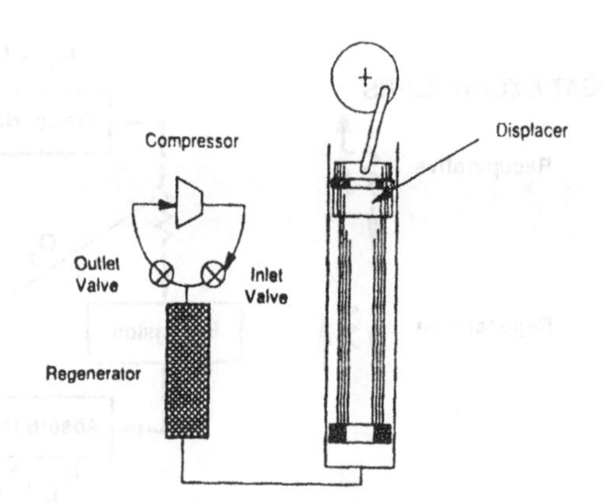


- Fundamental low temperature limit heat capacity of crystalline solids → 0 as T → 0

**Figure 6.8** Regenerative heat exchanger.

exchangers (Solvay, Postle, Gifford–McMahon, Ericsson, Stirling, VM) have many features in common:

1. Not dependent (significantly) on the operating fluid.
2. Cyclic (ac) operation and hence fluctuating cold temperatures.
3. Moving parts at low temperatures.
4. Fundamental low-temperature limit imposed by vanishing heat capacity at low temperatures.
5. Relatively low operating pressures, allowing them to be smaller and lighter than recuperative heat exchangers.

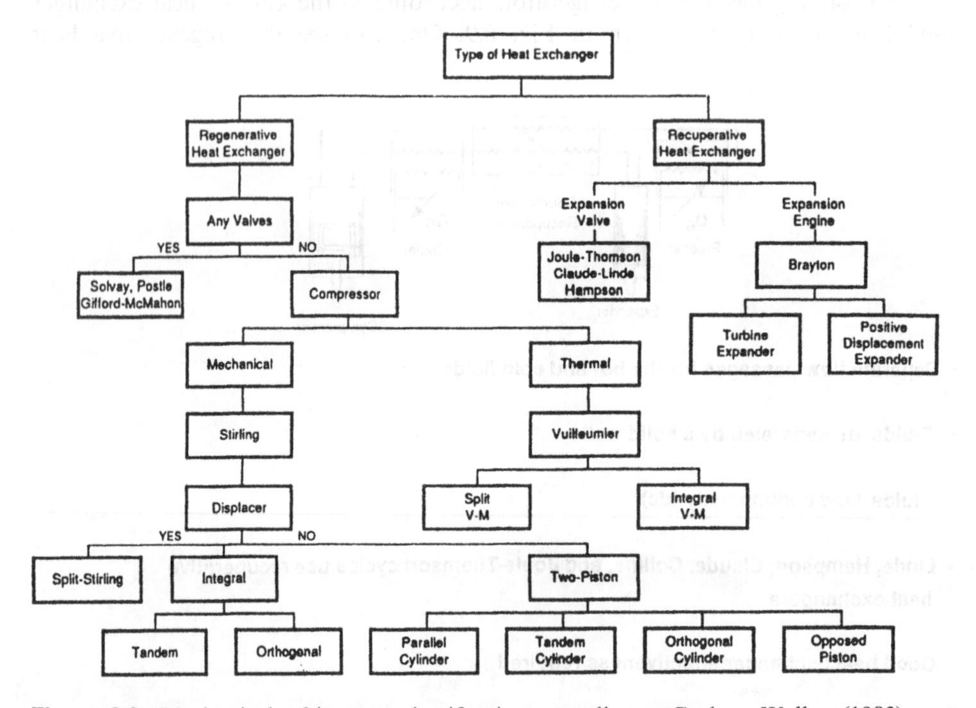


**Figure 6.9** Mechanical refrigerator classification according to Graham Walker (1983).

Likewise, the recuperative cycles (Joule–Thomson, Claude, Brayton, etc.) will have many features in common:

1. Very dependent upon the operating fluid.
2. Produce even, not cyclic, cooling (dc type operation).
3. Can have no moving parts at low temperatures.
4. No fundamental low-temperature limit as long as a fluid exists.
5. Higher pressures typically required means they are bigger and weigh more than their regenerative counterparts.
6. May require precooling because of Joule–Thomson inversion temperature consideration.

Recuperative cycles tend to be used for larger scale operations—tens of watts to megawatts at low temperatures—while the regenerative cycles tend to be used for the smaller applications—a few watts to milliwatts at low temperatures. There is no fundamental reason for this apparent division, and crossovers are common. Nonetheless, we shall organize the following discussion roughly along these historic lines, dealing first with the larger scale recuperative cycles and then with the smaller scale regenerative cycles.

## 3. RECUPERATIVE CYCLES

John Gorrie, a physician in Apalachicola, Florida in the 1800s, had observed that malaria was associated with warm, humid air. In the 1840s, Gorrie developed an air expansion engine to produce ice and air conditioning for his patients. Similar steam-driven air expansion machines were in use in the 1860s and 1870s to produce frozen beef for shipment from Australia to England and, not incidentally, for the refrigeration of Australian breweries. In the middle of the 19th century, the president of the German Brewers' Union, Gabriel Sedlmayr, asked a friend, Carl von Linde, to develop a refrigeration system for the Munich Spatenbrau brewery. Von Linde at first used ammonia instead of air in Gorrie's expansion system of a piston in a cylinder. Von Linde also included a recuperative single-tube heat exchanger. This heat exchanger consisted of a smaller inner tube in which the warm high-pressure gas flowed concentric to a larger tube in which the cold low-pressure gas returned.

Interest in refrigeration spread to other industries, such as steelmaking, where the newly invented Bessemer process used large quantities of oxygen. It may be no coincidence that an ironmaster, Louis Paul Cailletet (Fig. 6.10), together with a Swiss engineer, Raoul Pictet (Fig. 6.11), first produced a small amount of liquid oxygen in 1877.

This pattern has persisted. Most large-scale gas liquefaction systems today use the same principle of expanding air through a work-producing device such as an expansion engine or expansion turbine in order to extract energy from the gas so that it can be liquefied. The liquefied gases so produced continue to be used for steelmaking, chemical production, accelerated sewage treatment by aerobic decay, food freezing and refrigerated transportation, fuels and oxidizer for space propulsion, breathing gases, and as refrigerants to enable the various applications of superconductivity.

**Figure 6.10** Louis Paul Cailletet.

**Figure 6.11** Raoul Pictet.

## 4. LIQUEFACTION OF GASES

Liquefaction of gases is always accomplished by refrigerating the gas to some temperature below its critical temperature so that liquid can be formed at some suitable pressure below the critical pressure. Thus, gas liquefaction is a special case of gas refrigeration and cannot be separated from it. In both cases, the gas is first compressed to an elevated pressure in an ambient temperature compressor. This high-pressure gas is passed through a countercurrent recuperative heat exchanger to a throttling valve or expansion engine (see Fig. 6.2). Upon expanding to the lower pressure, cooling may take place, and some liquid may be formed. The cool, low-pressure gas returns to the compressor inlet to repeat the cycle. The purpose of the countercurrent heat exchanger is to warm the low-pressure gas prior to recompression and simultaneously to cool the high-pressure gas to the lowest temperature possible prior to expansion. Both refrigerators and liquefiers operate on this basic principle.

There is nonetheless an important distinction between refrigerators and liquefiers. In a continuous refrigeration process, there is no accumulation of refrigerant in any part of the system. This contrasts with a gas liquefying system, where liquid accumulates and is withdrawn. Thus, in a liquefying system, the total mass of gas that is warmed in the counter-current heat exchanger is less than that of the gas to be cooled by the amount liquefied, creating an imbalance of flow in the heat exchanger. In a refrigerator, the warm and cool gas flows are equal in the heat exchanger. This results in what is usually referred to as a "balanced flow condition" in a refrigerator heat exchanger. The thermodynamic principles of refrigeration and liquefaction are identical. However, the analysis and design of the two systems are quite different because of the condition of balanced flow in the refrigerator and unbalanced flow in liquefier systems.

### 4.1. Liquefaction Principles

The prerequisite refrigeration for gas liquefaction is accomplished in a thermodynamic process when the process gas absorbs heat at temperatures below that of the environment. As mentioned, a process for producing refrigeration at liquefied gas temperatures always involves some equipment at ambient temperature in which the gas is compressed and heat is rejected to a coolant. During the ambient temperature compression process, the enthalpy and entropy, but not usually the temperature of the gas, are decreased. The reduction in temperature of the gas is usually accomplished by recuperative heat exchange between the cooling and warming gas streams followed by an expansion of the high-pressure stream. This expansion may take place through either a throttling device (isenthalpic expansion) where there is a reduction in temperature only (when the Joule–Thomson coefficient is positive; see below) or in a work-producing device (isentropic expansion) where both temperature and enthalpy are decreased.

For reasons of efficiency, isentropic expansion devices might seem to be preferred for gas liquefaction. However, expansion engines have yet to be developed that can operate with more than a few percent of liquid present, due to erosion. Therefore, simple throttling devices are employed at some point in nearly all gas liquefaction systems. Throttling is an isenthalpic process, and the slope at any point on an isenthalpic curve on a pressure–temperature diagram is a quantitative measure of how the temperature will change with pressure. This slope is called the

**Table 6.3** Maximum Inversion Temperature

| Fluid | Maximum inversion temperature K | °R |
|---|---|---|
| Oxygen | 761 | 1370 |
| Argon | 722 | 1300 |
| Nitrogen | 622 | 1120 |
| Air | 603 | 1085 |
| Neon | 250 | 450 |
| Hydrogen | 202 | 364 |
| Helium | 40 | 72 |

Joule–Thomson (or J–T) coefficient and is usually denoted by $\mu_{J\text{–}T}$

$$\mu_{J\text{–}T} = \left(\frac{\partial T}{\partial P}\right)_H$$

The Joule–Thomson coefficient is a property of each specific gas, is a function of temperature and pressure, and may be positive, negative, or zero. For instance, hydrogen, helium, and neon have negative J–T coefficients at ambient temperature. Consequently, to be used as refrigerants in a throttling process they must first be cooled either by a separate precoolant fluid or by a work-producing expansion engine to a temperature below which the J–T coefficient is positive. Only then will throttling cause a further cooling rather than a heating of these gases.

Nitrogen, methane, and other gases, on the other hand, have positive J–T coefficients at ambient temperatures and hence produce cooling when expanded across a valve. Accordingly, they may be liquefied directly in a throttling process without the necessity of a precooling step or expansion through a work-extracting device. Maximum inversion temperatures for some of the more common liquefied gases are given in Table 6.3.

The complete inversion curves for several fluids are shown in Fig. 6.12. The maximum inversion temperature is seen to occur at zero pressure.

### 4.2. Adiabatic Expansion

Another method of producing low temperatures is the adiabatic expansion of the gas through a work-producing device such as an expansion engine. In the ideal case, the expansion would be reversible and adiabatic and therefore isentropic. In this case, we can define the *isentropic expansion coefficient* $\mu_S$, which expresses the temperature change due to a pressure change at constant entropy:

$$\mu_S = \left(\frac{\partial T}{\partial P}\right)_S$$

An isentropic expansion through an expander always results in a temperature decrease, whereas an expansion through an expansion valve may or may not result in a temperature decrease. The isentropic expansion process removes energy from the gas in the form of external work, so this method of low-temperature production is sometimes called the *external work method.* Expansion through an expansion valve

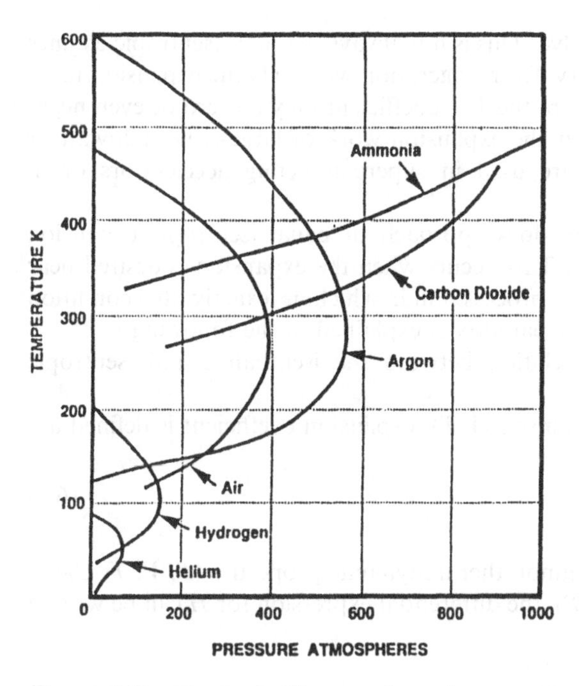


**Figure 6.12** The Joule–Thomson inversion curve for some common gases.

does not remove any energy from the gas but moves the molecules farther apart under the influence of intermolecular forces, so this method is called the *internal work method.*

It appears that expanding the gas through an expander would always be the most effective means of lowering the temperature of a gas, and this is the case as far as the thermodynamics of the two processes is concerned. Between any two given pressures, an isentropic expansion will always result in a lower final temperature than will an isenthalpic expansion from the same initial temperature. In reality, the difference between isenthalpic and isentropic expansion may be minimal, as explained below.

## 4.3. Comparison of Isenthalpic and Isentropic Expansion Processes

All refrigeration systems, regardless of the cycle chosen, use the same components, namely a compressor, heat exchangers, and an expansion device. The first differentiation among refrigerator cycles we have made concerns the type of heat exchanger, which can be regenerative or recuperative. The regenerative cycles are represented by the Vuilleumier (VM), Gifford–McMahon (G–M), and linear Stirling machines. The recuperative cycles are represented by the rotary reciprocating refrigeration ($R^3$) turbo Brayton and Joule–Thomson (J–T) machines. Hybrid cycles use the J–T process for the coldest stages and any of the other devices for the warmer stages. Another basic differentiation among refrigeration cycles lies in the expansion device. This may be either an expansion engine (expansion turbine or reciprocating expansion engine) or a throttling valve. The expansion engine approaches an isentropic process, and the valve, an isenthalpic process.

Isentropic expansion implies an adiabatic reversible process, while isenthalpic expansions are irreversible. For this reason, it appears that expansion engines are preferable to the isenthalpic J–T valve. This is not always the case. Isentropic engines are preferable, and indeed necessary, for refrigeration when: (1) the expansion takes place at elevated temperatures, where the J–T coefficient may be zero or even negative, or (2) it is desirable to recover the expansion work of large-scale (kilowatt or megawatt) refrigerators such as are used in superconducting accelerators or in tonnage capacity air plants.

However, the J–T expansion does approach or equal isentropic expansion efficiency under certain conditions. This occurs when the expansion is desired near or below the critical point of the process fluid, which is exactly the condition desired for liquefiers. This apparent paradox is explained in the following.

We shall first develop the relation between the isenthalpic and isentropic expansion coefficients.

The isenthalpic, or Joule–Thomson (J–T), expansion coefficient is defined as

$$\mu_H = \left(\frac{\partial T}{\partial P}\right)_H \tag{6.1}$$

$\mu_H$ may be derived in terms of common thermodynamic properties ($P$, $V$, $T$, $C_P$) in the following way. Since $H = f(T, P)$, the differential expression for $H$ can be written for a pure fluid as

$$\mathrm{d}H = \left(\frac{\partial H}{\partial P}\right)_T \mathrm{d}P + \left(\frac{\partial H}{\partial T}\right)_P \mathrm{d}T$$

Since the J–T process is isenthalpic, $\mathrm{d}H = 0$. By the rules of calculus,

$$\left(\frac{\partial T}{\partial P}\right)_H = \mu_H = -\frac{(\partial H/\partial P)_T}{(\partial H/\partial T)_P}$$

but

$$\left(\frac{\partial H}{\partial T}\right)_P = C_P$$

Let us call

$$\left(\frac{\partial H}{\partial P}\right)_T = \phi$$

and therefore,

$$\mu_H = -\frac{1}{C_P}\phi$$

To find $\mu_H = \mu_H(C_P, V, T)$, we shall evaluate $\phi$. Since $\mathrm{d}H = T\,\mathrm{d}S + V\,\mathrm{d}P$,

$$\left(\frac{\partial H}{\partial P}\right)_T = \phi = T\left(\frac{\partial S}{\partial P}\right)_T + V\left(\frac{\partial P}{\partial P}\right)_T$$

or

$$\phi = T\left(\frac{\partial S}{\partial P}\right)_T + V$$

From the Maxwell relation,

$$\left(\frac{\partial S}{\partial P}\right)_T = -\left(\frac{\partial V}{\partial T}\right)_P$$

Therefore,

$$\phi = \left(\frac{\partial H}{\partial P}\right)_T = \left[V - T\left(\frac{\partial V}{\partial T}\right)_P\right]$$

Since $\mu_H = -(1/C_P)\phi$,

$$\mu_H = \frac{1}{C_P}\left[T\left(\frac{\partial V}{\partial T}\right)_P - V\right]$$

For a perfect gas, $V = RT/P$, and

$$\left(\frac{\partial V}{\partial T}\right)_P = \frac{R}{P} = \frac{V}{T}$$

Therefore, for a perfect gas

$$\mu_{J-T} = \frac{1}{C_p}\left[T\left(\frac{V}{T}\right) - V\right] = 0$$

A perfect gas would not experience any temperature change upon expansion through an expansion valve.

It follows that the more imperfect the gas, the better. What is the most imperfect gas known? A liquid. This is the reason we try to achieve the liquid state before the J–T expansion (see the discussion of the vapor compression cycle that follows). The working fluid for this cycle is always chosen to be a liquid under ordinary temperatures; that is, the critical temperature of the fluid is less than ambient temperature.

Among liquids, which are the most nonideal? The highly polar fluids. This explains why ammonia, sulfur dioxide, and the fluorocarbons (the R-series refrigerants) are chosen. We can also see why replacing the fluorocarbons with less reactive (supposedly more environmentally benign) fluids will not be so easy.

One equation of state for real gases is the van der Waals equation of state,

$$\left(P + \frac{a}{V^2}\right)(V - b) = RT \tag{6.2}$$

where $a$ is a measure of the intermolecular forces and $b$ is a measure of the finite size of the molecules. For a perfect gas, $a = b = 0$, because a perfect gas has no intermolecular forces and the molecules are considered to be mass points with no volume. A van der Waals gas has molecules that are considered to be weakly attracting rigid spheres. If we calculate the Joule–Thomson coefficient for a van der Waals gas, we obtain

$$\mu_{J-T} = \frac{(2a/RT)(1 - b/V)^2 - b}{C_P[1 - (2a/V)(1 - b/V^2)]}$$

For large values of the specific volume, this equation reduces to

$$\mu_{\text{J-T}} = \frac{1}{C_P}\left(\frac{2a}{RT} - b\right)$$

which shows that at some temperatures $\mu_{\text{J-T}}$ is positive for ($T < 2a/bR$) and at other temperatures $\mu_{\text{J-T}}$ is negative (for $T > 2a/bR$).

The inversion curve is represented by all points at which the Joule–Thomson coefficient is zero. For a van der Waals gas, this happens when $(2a/RT)(1 - b/V)^2 = b$. Solving for the inversion temperature $T_{\text{i}}$ for a van der Waals gas,

$$T_{\text{i}} = \frac{2a}{bR}\left(1 - \frac{b}{V}\right)^2$$

which can be related to the pressure through the equation of state. In the higher temperature range, the inversion curve for a van der Waals gas can be approximated by

$$P_{\text{i}} = \frac{2R}{3b}T_{\text{i}} - \frac{R^2}{3a}T_{\text{i}}^2$$

The maximum inversion temperature for a van der Waals gas is the temperature on the inversion curve at $P = 0$ or $T_{\text{i,max}} = 2a/bR$.

The isentropic expansion coefficient, $\mu_S$, is defined as

$$\mu_S = \left(\frac{\partial T}{\partial P}\right)_S \tag{6.3}$$

(All properties per mole, $V = \hat{V}$, $S = \hat{S}$, etc.)

Since $S = f(T, P)$, the differential expression for $S$ can be written as

$$\text{d}S = \left(\frac{\partial S}{\partial T}\right)_P \text{d}T + \left(\frac{\partial S}{\partial P}\right)_T \text{d}P$$

Multiply both sides of this equation by $T$ to obtain

$$T\,\text{d}S = T\left(\frac{\partial S}{\partial T}\right)_P \text{d}T + T\left(\frac{\partial S}{\partial P}\right)_T \text{d}P \tag{6.4}$$

From $\text{d}H = T\text{d}S + V\text{d}P$,

$$= T\left(\frac{\partial S}{\partial T}\right)_P + 0$$

Therefore,

$$C_P = T\left(\frac{\partial S}{\partial T}\right)_P$$

which is the first coefficient in Eq. (6.4).

From $\text{d}G = -S\text{d}T - V\text{d}P$,

$$-\left(\frac{\partial S}{\partial P}\right)_T = \left(\frac{\partial V}{\partial T}\right)_P$$

which gives the second coefficient in Eq. (6.4).

Therefore,

$$T\,\mathrm{d}S = C_P\,\mathrm{d}T - T\left(\frac{\partial V}{\partial T}\right)_P \mathrm{d}P$$

For an isentropic expansion, $\mathrm{d}S = 0$, and therefore

$$0 = \left[C_P\,\mathrm{d}T - T\left(\frac{\partial V}{\partial T}\right)_P \mathrm{d}P\right]_S$$

Hence,

$$\left(\frac{\partial T}{\partial P}\right)_S = \frac{1}{C_P}\left[T\left(\frac{\partial V}{\partial T}\right)_P\right] = \mu_S$$

For a perfect gas, $(\partial V/\partial T)_P = R/P = V/T$, so

$$\mu_S = V/C_P$$

For a van der Waals gas, one can show that

$$\mu_S = \frac{V}{C_P}\left(\frac{(1 - b/V)}{1 - (2a/VRT)(1 - b/V)^2}\right)$$

which is positive, since $V > b$.

We are now in a position to compare $\mu_H$ and $\mu_s$. Dr. Arthur J. Kidnay of the Colorado School of Mines developed it this way. Recall that

$$\mu_H = \frac{1}{C_P}\left[T\left(\frac{\partial V}{\partial T}\right)_P - V\right]$$

and

$$\mu_S = \frac{1}{C_P}\left[T\left(\frac{\partial V}{\partial T}\right)_P\right]$$

Therefore,

$$\begin{aligned}
\mu_H - \mu_S &= \frac{1}{C_P}\left[T\left(\frac{\partial V}{\partial T}\right)_P - V\right] - \frac{1}{C_P}\left[T\left(\frac{\partial V}{\partial T}\right)_P\right] \\
&= \frac{1}{C_P}\left[T\left(\frac{\partial V}{\partial T}\right)_P - V - T\left(\frac{\partial V}{\partial T}\right)_P\right] \\
&= \frac{1}{C_P}(-V) \\
\mu_H - \mu_S &= -\frac{V}{C_P} = -\frac{RT}{PC_P}
\end{aligned}$$

and $\mu_s = \mu_H$ when either $\hat{V} = 0$ or $C_P = \infty$.

$\hat{V}$ does not ever become zero but does become increasingly smaller at low temperatures and elevated pressures. Therefore, at low temperatures and high pressures, $\mu_S \rightarrow \mu_H$, so they are the preferred conditions of the J–T expansion.

Furthermore, $C_P \rightarrow \infty$ near the critical point for all fluids (see Figs. 6.13–6.16). Therefore, $\mu_H \rightarrow \mu_s$ as the critical point of the gas is approached, which is exactly the condition desired for J–T expansion. Hence, the conditions for J–T expansion are

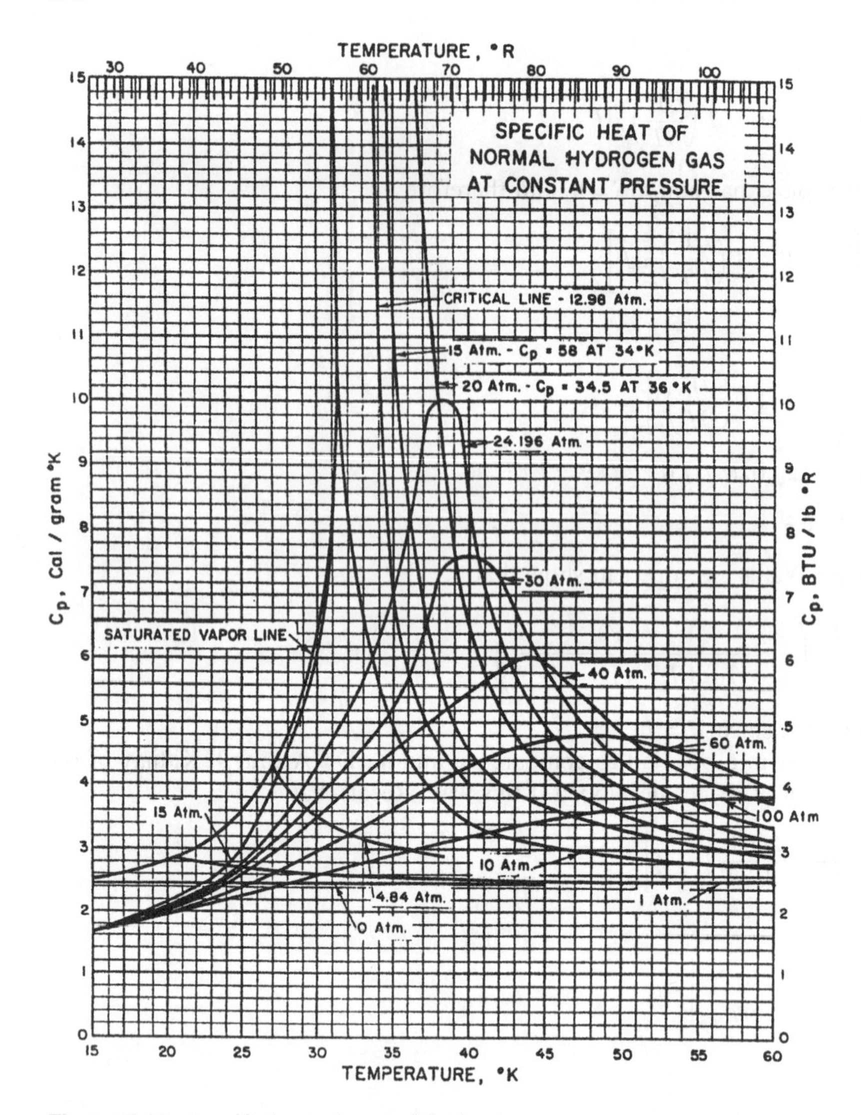


**Figure 6.13** Specific heat of normal hydrogen gas at constant pressure.

often the same as the conditions for $\hat{V} \to \min$ and $C_P \to \infty$, and $\mu_H \to \mu_S$ under real conditions.

Therefore, when the expansion is meant to produce liquid or occurs near or below the critical point, J–T expansion may be indistinguishable thermodynamically from isentropic expansion.

Furthermore, real expansion engines have difficulty operating in the two-phase region. In addition, the expansion may be only approximately 85% or so of isentropic. Thus, the J–T expansion approaches or equals the efficiency of a "real" isentropic expansion engine.

Thus, there are no thermodynamic arguments for preferring an expansion engine over a J–T valve. In addition, small scale (1 lb m/hr or so) expansion engines have yet to be developed, due to the high speed of rotation (500,000 rpm), lubrication-free bearings, and seals required. Joule–Thomson valves, in some instances, are nothing more than a crimp in the line.

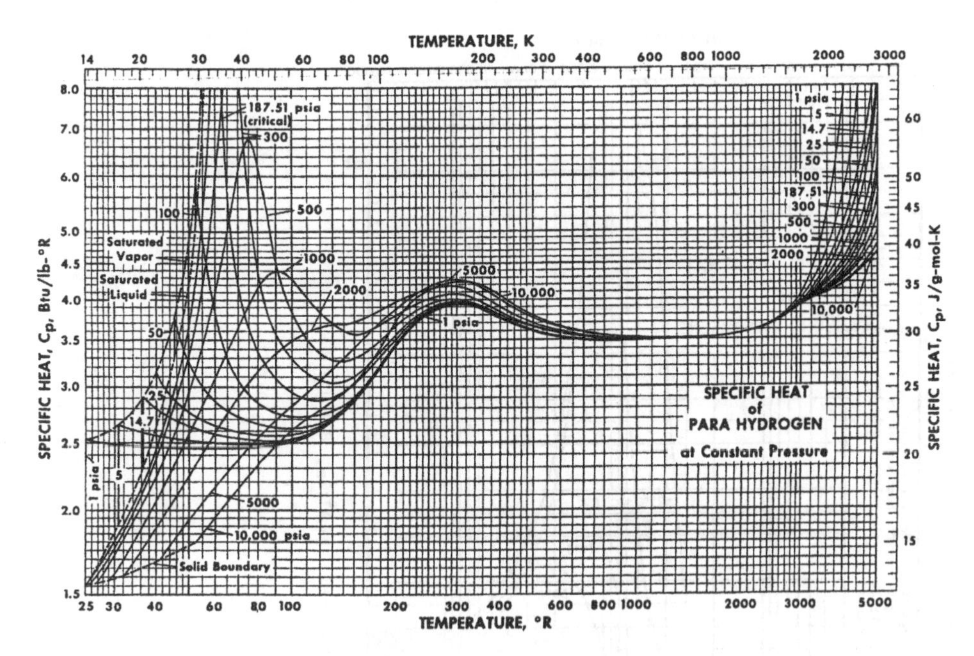


**Figure 6.14** Specific heat (heat capacity) of parahydrogen at constant pressure.

Accordingly,

1. There are no thermodynamic reasons to prefer expansion engines over expansion valves in all cases.
2. There are numerous hardware reasons to prefer valves over engines.

## 5. REFRIGERATOR EFFICIENCY

Before discussing various refrigeration methods, it is desirable to have a method of comparing real refrigerators with the ideal refrigerator. It is of interest to know the maximum efficiency that can be achieved by such an engine operating between two reservoirs at different temperatures. The French engineer Carnot described an engine operating in a particularly simple cycle known as the Carnot cycle. This cycle consists of two isothermal (constant-temperature) processes and two isentropic (constant-entropy) processes. A machine made to operate on this cycle so that heat is absorbed at the lower temperature is known as a refrigerator. A convenient measure of the effectiveness of a Carnot refrigerator is the work of compression $W$ required per unit of heat $Q$ removed at the low temperature. For a Carnot cycle, this is

$$\frac{W}{Q} = \frac{T_h - T_c}{T_c}$$

where $T_h$ and $T_c$ refer to the upper and lower temperature isotherms in the cycle.

The performance of a real refrigerator is measured by the coefficient of performance (COP), which is defined as the ratio of the refrigeration effect to the work input, the inverse of the efficiency term above.

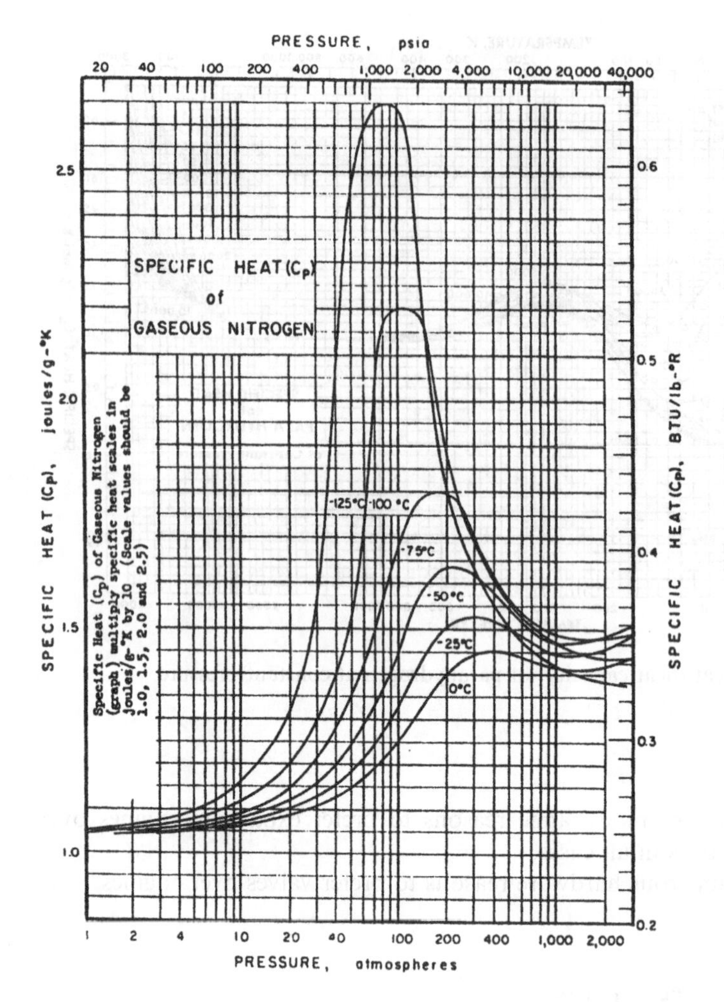


**Figure 6.15** Specific heat ($C_p$) of gaseous nitrogen.

Thus,

$$\text{COP} = \frac{\text{heat absorbed from low-temperature source}}{\text{net work input}} = \frac{Q}{W} \tag{6.5}$$

Figure 6.17 shows the change in coefficient of performance for a Carnot refrigerator as a function of refrigerator temperature for a fixed upper temperature of 530°R.

The figure of merit (FOM) is still another means of comparing the performance of practical refrigeration and is defined as

$$\text{FOM} = \frac{\text{COP}}{\text{COP}_{\text{ideal}}} = \frac{\text{COP}}{\text{COP}_{\text{Carnot}}}$$

where COP is the coefficient of performance of the actual refrigerator system and $\text{COP}_{\text{ideal}}$ and $\text{COP}_{\text{Carnot}}$ are the coefficient of performance of the thermodynamically ideal system and the Carnot refrigerator, respectively. This figure of merit for a

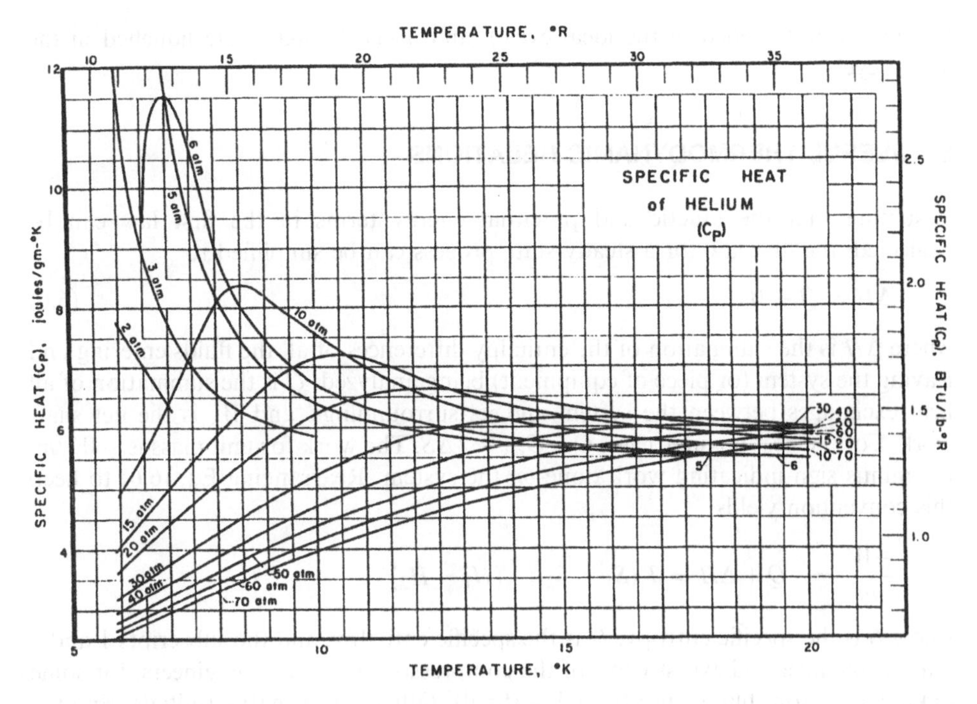


**Figure 6.16** Specific heat ($C_P$) of helium.

liquefier is generally rewritten as

$$\text{FOM} = \frac{W_{\text{i}}/\dot{m}}{W/\dot{m}_{\text{f}}}$$

where $W_{\text{i}}$ is the work of compression for the ideal cycle, $W$ is the work of compression for the actual cycle, $\dot{m}$ is the mass flow rate through the compressor (and is also

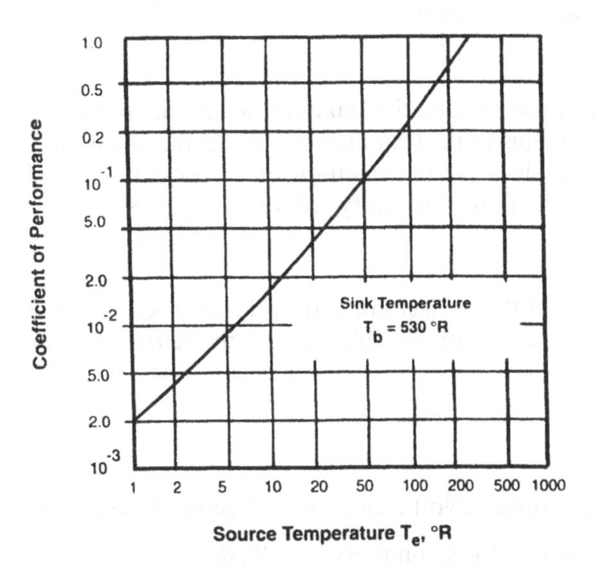


**Figure 6.17** Coefficient of performance for a Carnot refrigerator as a function of the source temperature for a sink temperature of 530°R.

the mass rate liquefied in the ideal cycle), and $\dot{m}_f$ is the mass rate liquefied in the actual cycle.

## 6. USEFUL THERMODYNAMIC RELATIONS

Assuming that the kinetic and potential energy terms in the first law can be neglected, the first law for a steady-state process can be simplified to

$$\Delta H = Q - W_s \tag{6.6}$$

where $\Delta H$ is the summation of the enthalpy differences of all the fluids entering and leaving the system (or piece of equipment) being analyzed, $Q$ is the summation of all heat exchanges between the system and its surroundings, and $W_s$ is the net shaft work. For the best possible case, $Q = Q_{rev} = T\Delta S$. The work for unit mass is $-W_s/m$, the minus sign indicating work done *on* the system. Rearranging Eq. (6.6) to keep this convention yields

$$\frac{-W_s}{\dot{m}} = -Q + \Delta H = T_i(S_1 - S_2) - (H_1 - H_2)$$

where $S$ is the specific entropy, $H$ is the specific enthalpy, and the subscripts 1 and 2 refer to the inlet and exit streams of the compressor. (Cryogenic engineers, for some unknown reason, like to use (initial) – (final). Others use (final) – (initial). Enough said.)

Application of the first law to the ideal work of isothermal compression with appropriate substitution for the heat summation reduces to

$$-W_s/\dot{m} = T_i(S_1 - S_2) - (H_1 - H_2) \tag{6.7}$$

where $\dot{m}$ is the mass rate of the flow through the compressor.

Evaluation of the refrigeration duty ($Q_{ref}$) in a refrigerator depends on the process chosen. For example, if the refrigeration process is an isothermal one, the refrigeration duty can be evaluated using the relation

$$Q_{ref} = \dot{m}T(S_2 - S_1)$$

Now $T$ is the temperature at which the refrigeration takes place and the subscripts 1 and 2 refer to the inlet and exit streams of the heat absorption. On the other hand, if the refrigeration process only involves the vaporization of the saturated liquid refrigerant at constant pressure, evaluation of the duty reduces to

$$Q_{ref} = \dot{m}h_{fg}$$

where $h_{fg}$ is the heat of vaporization of the refrigerant. If the process is performed at constant pressure with the absorption of only sensible heat by the refrigerant, the duty can be evaluated from

$$Q_{ref} = \dot{m}(H_2 - H_1) = \dot{m}C_P\Delta T$$

if $C_P$ is constant.

Randall Barron says there are three payoff functions to describe liquefaction:

1. Work required per unit mass of gas compressed, $-\dot{W}/\dot{m}$.
2. Work required per unit mass of gas liquefied, $-\dot{W}/\dot{m}_f$.
3. Fraction of the total flow of gas that is liquefied, $y = \dot{m}_f/\dot{m}$.

The last two payoff functions are related to the first one by

$$\frac{-\dot{W}}{\dot{m}} = \frac{-\dot{W}}{\dot{m}_f} y$$

In any liquefaction system, we should want to minimize the work requirements and maximize the fraction of gas that is liquefied. We shall use each of these indices of performance in the analysis that follows.

## 7. REFRIGERATION AND LIQUEFACTION METHODS

This section presents the general principles of the more useful refrigeration and liquefaction methods. Recall that although a refrigerator may cause liquefaction to provide an isothermal source of refrigeration, that does not make it a liquefier. A liquefier is a refrigerator that makes a liquid that is to be withdrawn. A make-up stream is required by a liquefier but not by a refrigerator.

### 7.1. Vapor Compression

A schematic diagram of a typical single-stage compressed vapor refrigerator is shown in Fig. 6.18a. Heat is absorbed by evaporation of a suitable liquid at $T_1$ and $P_1$ in an evaporator. In a dry compression cycle, saturated vapor is compressed to pressure $P_2$, and superheated vapor at the compressor outlet is condensed in a condenser at $T_2$ by exchanging heat with a suitable coolant. The vapor may be fully condensed into its liquid at point 3 simply because the critical temperature of the vapor is less than the temperature of the external coolant (air, water). For this reason, a vapor compression refrigerator does not necessarily require a true recuperative heat exchanger. A throttling valve then allows the saturated liquid formed in the condenser to expand approximately isenthalpically from point 3 to point 4 to the evaporator, completing the cycle. Figure 6.18b shows the states of the refrigerant in this compressed vapor process on a temperature–entropy diagram.

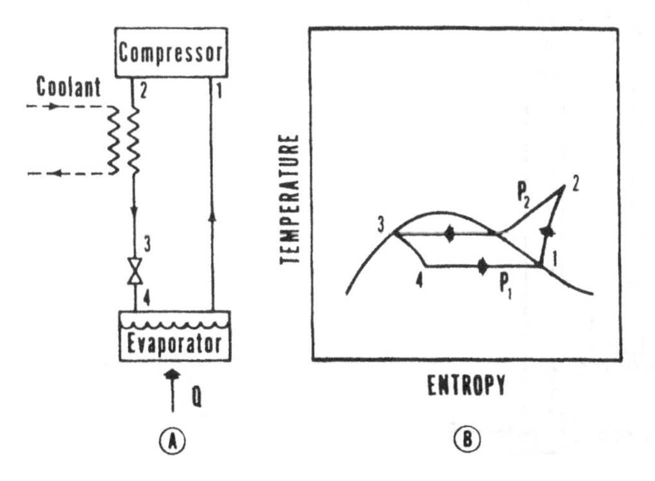


**Figure 6.18** Compressed vapor refrigerator.

Certain criteria can be established to aid in choosing a gas for this type of liquefier.

1. The temperature of the evaporator is usually near the normal boiling point of the gas, so the pressure of the evaporator may be approximately atmospheric. Temperatures lower than the normal boiling temperature can be used; however, care must be taken to prevent air from leaking into the subatmospheric parts of the cycle.
2. The gas must be chosen with a small compression ratio. This is necessary because the work of compression increases rapidly with the compression ratio. High discharge pressures require massive equipment for proper containment of the refrigerant and result in costly and heavy refrigerators.
3. Ambient temperature must be less than the critical temperature of the gas to effect condensation using the environment as a coolant.

The efficiency of the compressed vapor refrigerator is generally higher than that of other methods discussed in this chapter because heat is transferred nearly isothermally, and the compression process can be made nearly reversible and isothermal.

## 7.2. Cascade Vapor Compression

It is possible to cascade several vapor compression systems, each containing different refrigerants, to reach lower temperatures. Such a system is shown schematically in Fig. 6.19. Cascading is necessary because the gases with the lower boiling points have critical temperatures well below ambient. Thus, it is necessary for condenser temperatures for these vapor compression cycles to be provided by the evaporators of other vapor compression cycles, which reject heat at higher temperatures. The efficiency of cascade systems is greatly improved by the use of heat exchangers that cool the incoming warm gases. The exchangers warm the cool gases prior to reaching the compressor intake. Strictly speaking, a recuperative heat exchanger is not necessary.

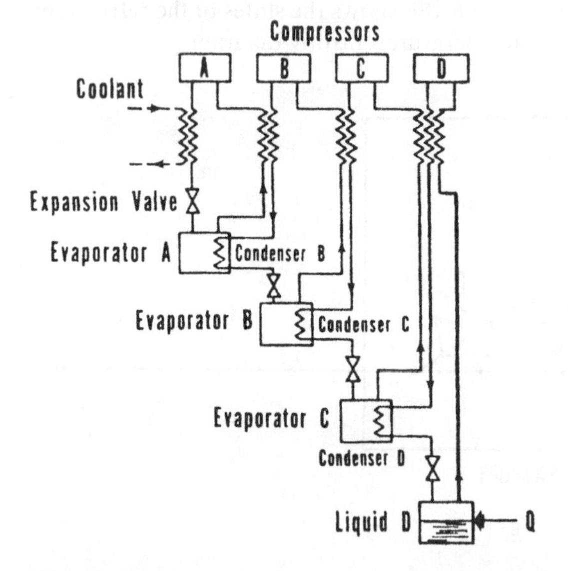


**Figure 6.19** Cascade compressed vapor refrigerator.

Instead, each gas is cooled by a different gas return stream in the next warmer refrigerator to the left, as drawn.

## 7.3. Mixed Refrigerant Cascade Cycle

Very large natural gas liquefaction plants often use the mixed refrigerant cascade (MRC) cycle, which is a variation of the cascade cycle just described. It involves the circulation of a single refrigerant stream carefully prepared from a mixture of refrigerants. The simplification of the compression and heat exchange processes in such a cycle may, under certain circumstances, offer potential for reduced capital expenditure over the conventional cascade cycle.

Figure 6.20 shows one version of the MRC cycle. The refrigeration required for the cooling and liquefaction of the natural gas is provided by a circulating mixed refrigerant stream containing components such as butane, propane, ethane, methane, and nitrogen. This stream is compressed by a two-stage compressor, and the heat of compression is removed in the intercooler and aftercooler with cooling water. The mixed refrigerant stream, as a result of being compressed and cooled, is partially condensed, and the two phases are separated in the first separator. The liquid and overhead vapor flow as separate streams to the first exchanger, where heat is transferred to the returning refrigerant stream. The vapor is thereby partially condensed and flows from this heat exchanger to the second separator, where the phases are again separated.

The liquid phase from the first separator is subcooled in flowing through the first exchanger, then reduced in pressure before being returned through the first exchanger to the warm end to provide refrigeration to the three cooling streams (natural gas feed and vapor and liquid refrigerant streams from the first separator). The overhead vapors (B) and liquid bottoms (C) from the second separator are sent as separate streams to the second exchanger for further cooling. The liquid from the

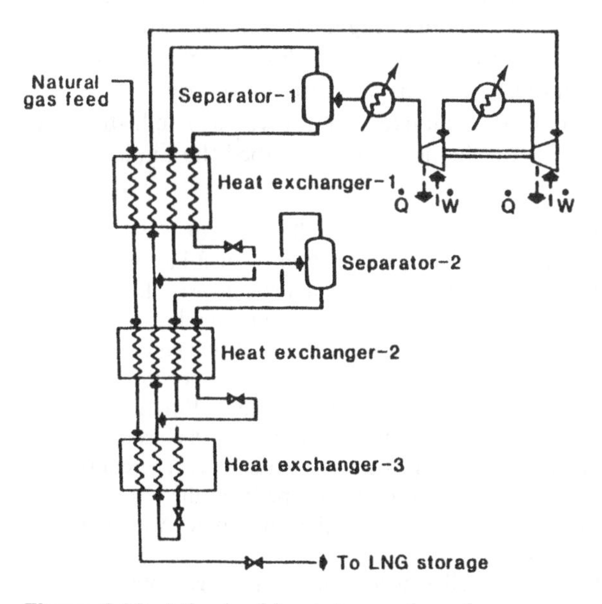


**Figure 6.20** Mixed refrigerant cascade cycle.

second separator (C) is again subcooled in the second exchanger and reduced in pressure before being returned to the second exchanger as a refrigerant stream (D).

The vapor overhead from the second separator is cooled and totally condensed in the second and third exchangers and is then reduced in pressure to supply the final level of refrigeration in flowing back through the third exchanger. This returning stream joins with the other liquid streams injected at the cold end of the second and first exchangers to provide the refrigeration to (1) cool and liquefy the natural gas feed, (2) partially condense the refrigerant vapor streams, and (3) subcool the liquid refrigerant streams. The total combined refrigerant stream from the first exchanger flows to the two-stage compressor, where it is recompressed for recirculation.

All of the mixed refrigerant processes use a multicomponent refrigerant mix that is repeatedly partially condensed, separated, cooled, expanded, and warmed as the refrigerant moves through the natural gas liquefaction process. Thus, these processes require more sophisticated design methods as well as a more complete knowledge of the thermodynamic properties of gaseous mixtures than is required for the simple expander or cascade cycles. Also, these processes must handle two-phase mixtures in all heat exchangers.

## 7.4. Isenthalpic Expansion—The Linde Apparatus

Linde's first successful experiments at air liquefaction were performed in May 1895. Fifteen hours of pumping was required to liquefy air, and then he collected some three quarts of liquid air per hour, containing about 70% oxygen. He used an "interchanger" (recuperative heat exchanger) made of iron tubes over 300 ft long and 1.2 and 2.4 in. in internal diameter. His pump was a carbon dioxide gas compressor that raised the air pressure from 22 to 65 atm. He reported that the liquid air was crystal clear and bluish. This early description of liquid air was both accurate and definitive. First, the blue, or any color for that matter, was unexpected. Second, other, earlier workers had reported that their "liquid air" closely resembled water. It was.

Von Linde described his apparatus as eliminating heat from gas "exclusively by expenditure of internal work." This internal work Linde held to be "the work of separating the gas's own sluggish molecules from each others' vicinity."

The basic Linde cycle has remained unchanged for 100 years. The historian of science and technology Thomas O'Conor Sloane (1900) described the Linde apparatus as follows:

> Linde's apparatus, which is described as utilizing this small increment of cold, if the expression may be allowed, and by constant summation of such increments bringing about a high degree of refrigeration, caused much interest when its supposed principles were first stated and its operations were first disclosed. The term self-intensive has been aptly coined to describe machines of this type.
>
> What the apparatus of the original Linde type does is this: air is pumped through a circuit of pipes; the pipe from the outlet of the pump, after going through the given circuit, returns to the inlet, so that the air under treatment goes constantly around the same circuit. When a gas is pumped against resistance, it is compressed or diminished in volume and heated. The outlet pipe from the pump is kept at a uniform temperature by cold water circulating in contact with the outside of the pipe, like a surface condenser.
>
> The air thus cooled is forced through a small aperture, and the passage from high to low pressure, with consequent expansion, causes cooling. Between the water

> cooling apparatus and the aperture a long length of pipe intervenes. The cooled air is carried back to the pump so as to circulate around this pipe on its way back, and it abstracts heat from the air already cooled [J–T valve] by the water [recuperative heat exchanger]. Hence the air reaches the aperture constantly at a lower temperature but leaves the water condenser always at a uniform temperature. The real cold production is done after the air leave the water condenser. The degree of cold keeps increasing until liquid air drops from the aperture and lies in the bottom of the apparatus. By a cock it can be drawn therefrom like water.
> It seems at first sight impossible that the small decrease of temperature, due to the imperfection of the gaseous state as it exists in air, should be able to produce such refrigeration. What Hampson calls thermal advantages are to be aimed at. The surface on which the cooled air acts on its return must be large, the material of the pipes thin. These elements provide for a rapid cooling by the returning air of the counter-stream on its way to the aperture. The entire mass to be cooled must be as light as possible. The action of the pump is constantly heating the gas by compression, and this heat is removed by the water. The atmosphere surrounding the apparatus constantly heats the portions colder than itself by contact. The colder portions, therefore, must be protected from this action by thick jacketing or other means. Concentric air spaces produce a good effect, and doubtless if it were practicable Dewar's vacuum heat insulation might be applied with excellent effect.

Linde made quite a sensation by his description of his apparatus, which, by purely mechanical means, liquefied air, although his first results were far from encouraging. Sloane continues his description as follows:

> What is called Linde's simplest form of apparatus is illustrated in the cut (Fig. 6.21) and will be readily understood, especially if the reader has grasped the very simple general theory on which its operation depends. It will be understood that the drawing is not a reproduction of the exact apparatus but is diagrammatic, being purposely made as clear as possible without permitting detail to interfere with intelligibility.

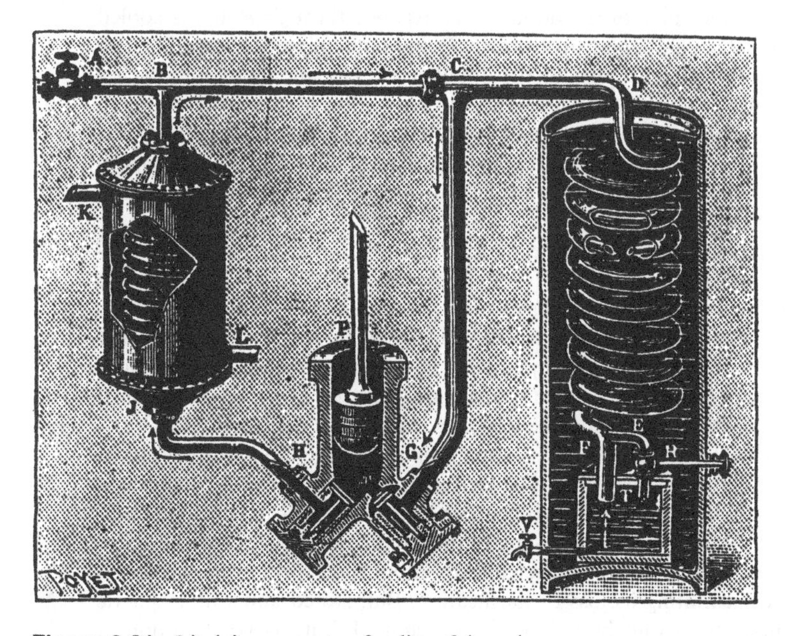


**Figure 6.21** Linde's apparatus for liquefying air.

*P* represents a pump which aspirates air from the pipe, *G*, and forces it out, under pressure, through *H*. The air forced out through *H* goes through a complete circuit of pipes and returns through *G*, thus constantly and repeatedly going around the circuit.

*J* is a water condenser or more properly a cooling apparatus. It is a cylindrical vessel, and the air pipe goes through it in a coil. Water enters at *K* and emerges at *L*, so that as the gas leaves the vessel it is always at the temperature of the inflowing water. The arrows show the direction of the current of gas, and all is perfectly clear to the point *C*. The arrows might be taken to indicate that the gas, on reaching *C*, goes down directly to *G*, but they do not indicate this. The pipe, *B*, is of small diameter, and, without any opening or break, runs straight on to *D*, is bent into a coil, and descends to *E* and *T*. But from *C* to *F* it is surrounded by a second pipe concentric with it, and it is this outer pipe which is connected to the pump suction by the vertical pipe extending downward from *C* and ending in *G*.

The cylindrical vessel on the right is simply a nonconducting casing or jacket to protect the pipes from the heating effect of the outer air. In the illustration the interior of the coil is shown, a part of the pipe being supposed to be broken away to show this.

In the course of the air in the pipes to the right of the point, *C*, lies the soul of the apparatus. The small pipe running down through the protecting vessel terminates in the chamber, *T*. A valve, *R*, is provided which may be opened or shut so as to regulate the pressure drop, and this valve constitutes the aperture through which the gas passes and expands with attendant cooling.

The end of the pipe, *E*, enters the small airtight box of chamber, *T*. From the chamber rises a larger pipe, *F*, which, just above the top of the chamber, receives within it the smaller inlet pipe, *E*, and winds up through the protecting vessel concentric with the smaller pipe. On the second and third turns from the top the interior arrangement of the pipes is shown very clearly.

The operation is now clear. The air enters the pump at *G*, is forced through *H* and compressed, thereby being heated. The heat is removed in the cooling apparatus, *J*, and the compressed air, at the temperature of the water, goes on to *D*. There it descends in the inner pipe of the double coil, expands through *R* and is cooled thereby, passes through *T* and up through *F*, the outer pipe of the coil. There it cools the air in the inner pipe of the double coil. The air, therefore, reaches the valve, *R*, at a lower temperature than before, so that it is constantly falling in temperature, reaches *R* at lower and lower temperatures, and eventually the critical temperature of liquid air is reached and passed, and liquid air begins to collect in the chamber, *F*, as shown in the cut. By the faucet, *V*, can be drawn therefrom as required.

If air is liquefied in the apparatus, every cubic inch of liquid represents about one-half a cubic foot of air withdrawn from circulation in the apparatus. Once the apparatus begins to liquefy air, it has to have new material supplied it, just as a grist mill needs a supply of grain to keep the stones in operation. A pipe at *A* connects with a second pump which pumps in new air as required, so as to maintain an advantageous pressure in the system—one which will give an economical relation between the pressures on the opposite sides of the aperture.

A minor yet important feature of this apparatus is that the liquid air collects at atmospheric pressure, or thereabout. The effect is twofold. It can be withdrawn much more easily than when it has to be taken from a receiver in which it is subjected to 50 or 100 atmospheres pressure. In the latter case it rushes out, only controllable by the faucet, and the mechanically atomizing effect plays a part in wasting it and facilitating its loss by gasification. But, stored under atmospheric pressure, it not only is quietly withdrawn, as required, but, by volatilization, it

> keeps its own temperature down. Maintaining it in a quiet state and in bulk operates to make it evaporate more slowly, the battle of the squares and the cubes, as it has aptly been termed, being involved.

Sloane is apparently referring to a fundamental issue of cryogenics today: larger storage tanks have a better thermal performance than smaller tanks. Heat transfer increases as the square; volume stored increases as a cubic function.

A thermodynamic process of the simple Linde cycle is shown schematically in Fig. 6.22a. The gas is compressed at ambient temperature approximately isothermally from 1 to 2, rejecting heat to a coolant. The compressed gas is cooled in a true recuperative heat exchanger by the stream returning to the compressor intake until it reaches the throttling valve. Joule–Thomson cooling upon expansion from 3 to 4 further reduces the temperature until, in the steady state, a portion of the gas is liquefied. Figure 6.22b shows this process on a temperature–entropy diagram.

For the simple Linde refrigerator shown, the liquefied portion is continuously evaporated from the reservoir by the refrigerator head load $Q$. All of the return gas is warmed in the recuperative counter-current heat exchanger and returned to the compressor. The fraction $y$ that is liquefied is obtained by applying the first law to the heat exchanger, J–T valve, and liquid reservoir. This results in

$$y = \frac{H_1 - H_2}{H_1 - H_f} \tag{6.8}$$

where $H_f$ is the specific enthalpy of the liquid being produced. Note that maximum liquefaction occurs when $H_1$ and $H_2$ refer to the same temperature. That is, the warm end of the heat exchanger has no $\Delta T$ across it: $T_1 = T_2$, which is never possible.

Gases used in this process have a critical temperature well below ambient; consequently, liquefaction by direct compression (such as in the vapor compression cycles) is not possible. In addition, the inversion temperature must be *above* ambient temperature to provide cooling as the process is started. Auxiliary refrigeration is required if the simple Linde cycle is to be used to liquefy gases whose inversion temperature is below ambient. Liquid nitrogen is the optimum refrigerant for

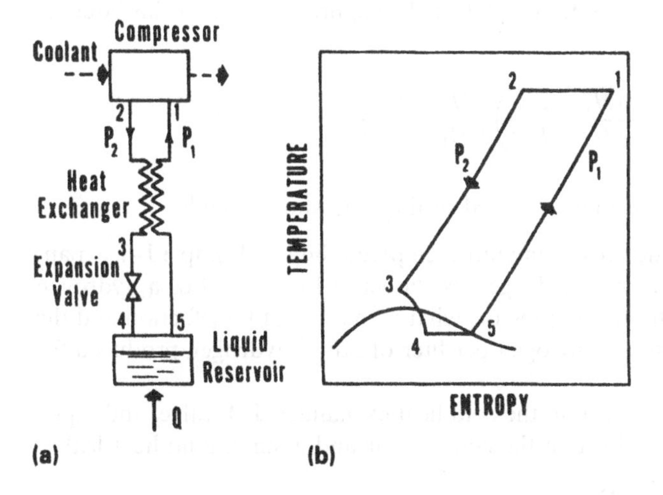


**Figure 6.22** Refrigerator using simple Linde cycle.

hydrogen and neon liquefaction systems, whereas liquid hydrogen is the normal refrigerant for helium liquefaction systems.

It is always a surprise to find that the Linde cycle refrigeration is totally dependent on what is going on at the top end or compressor. An overall energy balance just shows

Inputs = outputs

$$Q_{ref} + mH_1 = mH_2, \qquad Q_{ref} = m(H_2 - H_1)$$

This may well be the most profound statement of refrigeration: the compressor does it all. However, 99.999% of all refrigeration research has historically been done on the cold end. Something is out of kilter here.

Thus, the coefficient of performance for the ideal simple Linde refrigerator is given by

$$\text{COP} = \frac{H_1 - H_2}{T_1(S_1 - S_2) - (H_1 - H_2)}$$

For a simple Linde liquefier, the liquefied portion is continuously withdrawn from the reservoir and only the unliquefied portion of the fluid is warmed in the countercurrent heat exchanger and returned to the compressor. The fraction $y$ that is liquefied is obtained by applying the first law to the heat exchanger, J–T valve, and liquid reservoir. This results in

$$y = \frac{H_1 - H_2}{H_1 - H_f}$$

where $H_f$ is the specific enthalpy of the liquid being withdrawn. Note that maximum liquefaction occurs when $H_1$ and $H_2$ refer to the same temperature. To account for the heat inleak, $Q_L$, the relation needs to be modified to

$$y = \frac{H_1 - H_2 - Q_L}{H_1 - H_f}$$

with a resultant decrease in the fraction liquefied. The figure of merit for this liquefier is

$$\text{FOM} = \left(\frac{T_1(S_1 - S_f) - (H_1 - H_f)}{T_1(S_1 - S_2) - (H_1 - H_2)}\right)\left(\frac{H_1 - H_2}{H_1 - H_f}\right)$$

Evaluation of such a system is demonstrated in the following example.

*Example 6.1.* An ideal Linde cycle using nitrogen precooling and simple J–T expansion is used for liquefying normal hydrogen as shown in Fig. 6.23. For a hydrogen flow rate of 1 mol/s through the compressor, what is the rate of liquefaction and the nitrogen consumption in liters of nitrogen per liter of liquid hydrogen produced for the conditions shown?

Make a first law balance around the cold heat exchanger, J–T valve, and liquefier pot. For 1 mol/s of $n$-$H_2$ through the compressor and assuming no heat leak,

$$\dot{m}H_3 = (\dot{m} - \dot{m}_f)H_4 + \dot{m}_f H_f$$

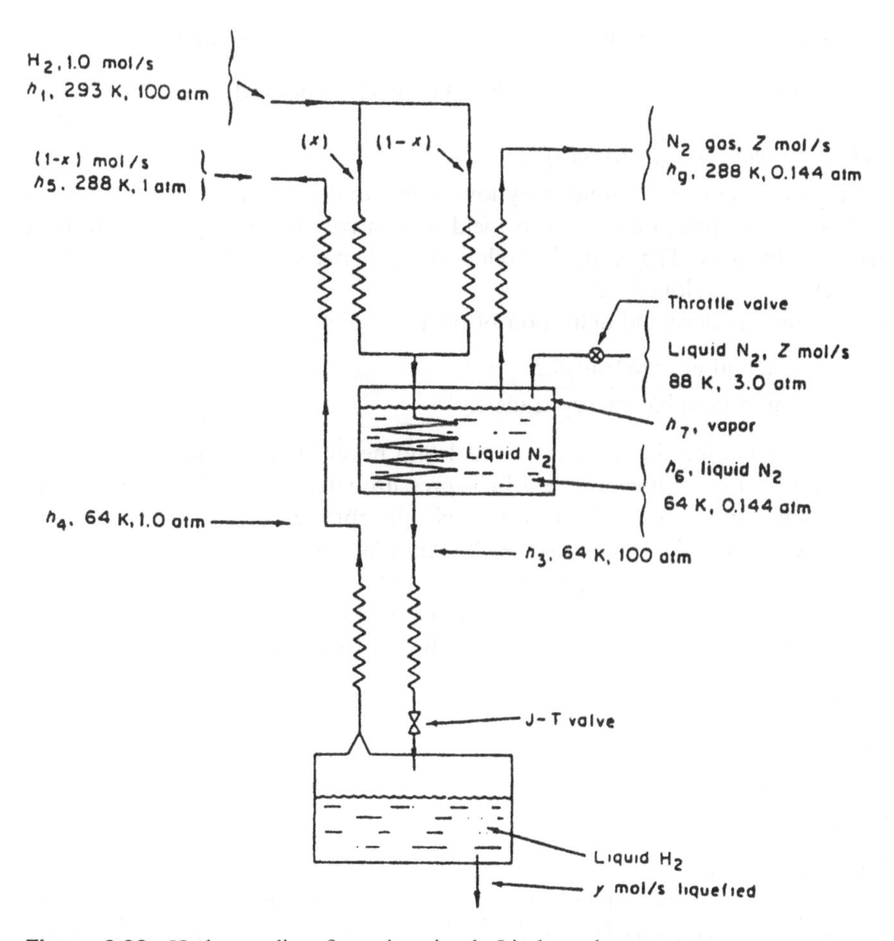


**Figure 6.23** Hydrogen liquefier using simple Linde cycle.

or

$$y = \frac{H_4 - H_3}{H_4 - H_f}$$

where $y = \dot{m}_f/\dot{m}$

Using enthalpy values from Chapter 3,

$$y = \frac{566 - 451}{566 - 130} = 0.263 \text{ mol liquid } H_2 \text{ produced/mol compressed}$$

To find the liters of nitrogen consumed per liter of liquid hydrogen produced, make a first law balance around the two warm heat exchangers and the liquid nitrogen ($LN_2$) precooling bath:

$$H_1 + zH_{LN2} + (1 - y)H_4 = (1 - y)H_5 + zH_8 + H_3$$

$$z = \frac{(H_1 - H_3) - (H_5 - H_4)(1 - \dot{m}_y)}{H_8 - H_{LN_2}}$$

$$= \frac{(1994 - 451) - (1938 - 566)(1 - 0.263)}{(3009 - 345)}$$

$$= 0.2 \text{ mol } N_2/\text{mol } H_2 \text{ gas compressed}$$

$$= 0.2/0.263 = 0.76 \text{ mol } N_2/\text{mol } H_2 \text{ liquid produced}$$

Molar volume of $H_2 = 28.4$ L/mol. Molar volume of $N_2 = 37.4$ L/mol.

$$z = (0.76)(37.4)(28.4) = 1.0 \text{ L } N_2/\text{liter liquid } H_2 \text{ produced}$$

### *7.4.1. Heat Exchanger Effectiveness*

Joule–Thomson systems have inherently low yields, that is, only 15–20% of the gas circulated becomes liquid. Furthermore, yield is sensitive (in a negative way) to heat exchanger effectiveness. The analytical relationship between yield and heat exchanger effectiveness is developed below.

The commonly accepted definition of heat exchanger effectiveness $\varepsilon$ is

$$\varepsilon = \frac{\text{actual energy transfer}}{\text{maximum possible energy transfer}}$$

Effectiveness is a number between zero and 1 and never greater than 1.

Liquefier heat exchangers have a high-pressure stream (stream 2) and a low-pressure stream (stream 1). Which stream is appropriate for calculating $\varepsilon$? The answer depends on which stream has the "maximum possible energy transfer."

Let us take a simple J–T system like that in Fig. 6.22, where a liquid product $\dot{m}_f$ is produced from a total compressed stream of $\dot{m}$. Making an energy balance around the heat exchanger, J–T valve, and liquid product withdrawal point, we have

$$0 = (\dot{m} - \dot{m}_f)H_1 + \dot{m}_f H_f - \dot{m} H_2$$

The yield $y$ is

$$y = \frac{\dot{m}_f}{\dot{m}} = \frac{H_1 - H_2}{H_1 - H_f}$$

The performance of a liquefier is determined by the conditions at the top of the warm heat exchanger. Yield depends on (1) the pressure and temperature at ambient conditions (point 1), which fix $H_1$ and $H_f$, and (2) the pressure after isothermal compression, which solely determines $H_2$ (since $T_1 = T_2$ for isothermal compression). Since in a perfect gas the effect of pressure on enthalpy is zero, it follows immediately that $H_1 = H_2$ and $y = 0$ for a perfect gas.

For a real gas, let us define two average specific heats of the two streams: (see Fig. 6.22)

$$\bar{C}_{P,h} = \frac{H_2 - H_3}{T_2 - T_3}$$

and

$$\bar{C}_{P,c} = \frac{H_1 - H_g}{T_1 - T_5}$$

where the subscripts h and c refer to the hot and cold streams, and $C_P$ is consistent with the definition

$$\bar{C}_P = \frac{\Delta H}{\Delta T_P}$$

The rate of heat transfer in each stream is (assuming that the $\Delta T$ of both streams is equal)

$$Q_h = \dot{m} \bar{C}_{P,h}$$

and

$$Q_c = \dot{m}(1 - y)\bar{C}_{P,c}$$

As pressure is increased on a real gas, the constant-enthalpy lines on a $T$–$S$ diagram become closer together. This means that for a given temperature change, the corresponding enthalpy change at constant pressure becomes larger as pressure is increased. Therefore,

$$\bar{C}_{P,h} > \bar{C}_{P,c}$$

Since $\dot{m} > \dot{m}(1 - y)$, it follows that $Q_h > Q_c$. Therefore, the cold stream is always the limiting stream for heat transfer. Hence, the cold stream is the appropriate stream for calculation of $\varepsilon$.

Figure 6.24 shows the real case for a J–T cycle where $\varepsilon < 1.0$. The cold stream will leave the heat exchanger at point 1′, at a lower temperature than the ideal case, point 1. Hence the effectiveness is

$$\varepsilon = \frac{Q_{real}}{Q_{ideal}} = \frac{H_1' - H_g}{H_1 - H_g}$$

To see this relation rigorously, consider that

$$\varepsilon = \frac{C_c(T_1 - T_g)}{C_i(T_2 - T_g)}$$

where $C_c$ is defined above based on the actual flow rate $\dot{m}(1 - y)$ and the actual average heat capacity $C_{P,c}$. $C_{ideal}$ is the ideal capacity representing the amount of heat that could be transferred by the limiting stream. Since $C_{ideal} = C_c$ and $T_1 = T_2$, we

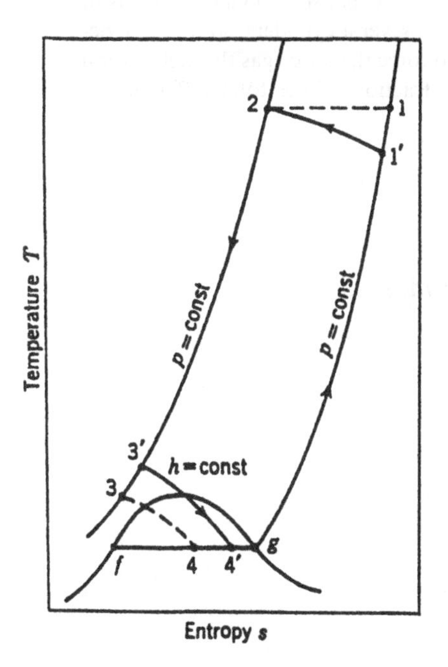


**Figure 6.24** Joule–Thomson liquefication system with heat exchanger effectiveness. Points 1, 2, 3, and 4 for $\varepsilon = 1.0$ (ideal system). Points 1′, 2′, 3′, and 4′ for $\varepsilon < 1.0$.

get another form for ε, namely

$$\varepsilon = \frac{H_{\mathrm{i}} - H_{\mathrm{g}}}{H_{\mathrm{l}} - H_{\mathrm{g}}}$$

A material and energy balance around all the components (except the compressor) gives

$$y = \frac{H_1' - H_2}{H_1 - H_{\mathrm{f}}}$$

Combining the last two equations results in

$$y = \frac{(H_1 - H_2) - (1 - \varepsilon)(H_1 - H_{\mathrm{g}})}{(H_1 - H_{\mathrm{f}})}$$

This equation shows analytically the direct effect of heat exchanger effectiveness upon yield $y$ for the simple Linde cycle.

### *7.4.2. Dual-Pressure Process*

To reduce the work of compression, a two-stage or dual-pressure process may be used whereby the pressure is reduced by two successive isenthalpic expansions as shown in Fig. 6.25.

Sloane (1900) described the Linde dual-pressure system as follows:

> It is evident that to make the difference of pressure $P^2 - P^1$ large, recourse may be had to the expedient adopted in steam engineering for expansion engines of high initial pressure. These are constructed with two cylinders (compound engines) or with three or more cylinders working in series, the steam passing seriatim from one cylinder into the next (triple, quadruple, etc., expansion engines). Just as in these engines the expansion is divided between several cylinders, so it is practicable in self-intensive refrigerating machines to force the air or gas through several apertures, letting each one take care of its fraction of the total difference of pressures, $P^2 - P^1$.

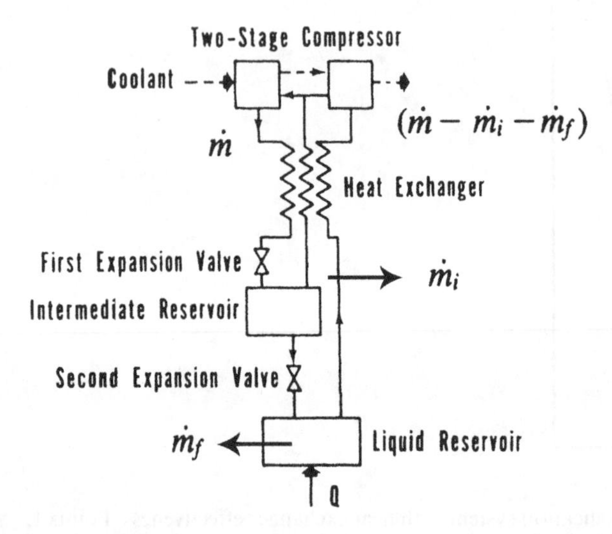


**Figure 6.25** Linde dual-pressure refrigerator cycle.

> Linde has done this in a partial way in his laboratory apparatus, and the cut (Fig. 6.26) shows the modification in question. If the description of the simple apparatus has been understood, the drawing alone will be almost self-explanatory.

Since the work of compression is approximately proportional to the logarithm of the pressure ratio, and the Joule–Thomson cooling is roughly proportional to the pressure difference, there is a much greater reduction in compressor work than in refrigerating performance. Hence the dual-pressure process produces a given amount of refrigeration with less energy input than the simple Linde process.

*Example 6.2.* What is the theoretical liquid yield for a dual-pressure cycle as shown by Fig. 6.25?

An energy balance around the cold heat exchanger, J–T valve, and liquid reservoir results in

$$(\dot{m} - \dot{m}_f - \dot{m}_i)H_1 + \dot{m}_i H_2 + \dot{m}_f H_f - \dot{m}H_3 = 0$$

Collecting terms gives

$$\dot{m}_i(H_2 - H_1) + \dot{m}_f(H_f - H_1) = \dot{m}(H_3 - H_1)$$

Let $\dot{m}_i/\dot{m} = i$ and $\dot{m}_f/\dot{m} = y$; then

$$i(H_2 - H_1) + y(H_f - H_1) = H_3 - H_1$$

or

$$y = \frac{H_1 - H_3}{H_1 - H_f} - i\left[\frac{H_1 - H_2}{H_1 - H_f}\right]$$

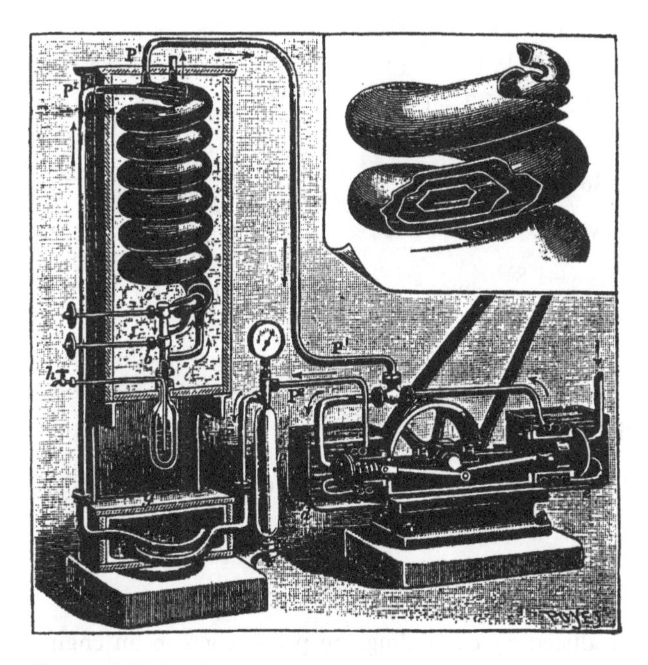

**Figure 6.26** Laboratory apparatus.

*Example 6.3.* Find (a) the theoretical liquid yield, (b) work per unit mass liquefied, and (c) the figure of merit for a dual-pressure liquefaction system using air as the working fluid. The liquefier operates between 1 and 200 atm. The inlet and exit temperatures to the compressor are 80°F. The intermediate pressure is 50 atm, and the intermediate pressure/flow rate ratio is 0.80. What is the work per unit mass liquefied if the simple Linde cycle has been used to accomplish the liquefaction?

a. From Example 6.2, the amount liquefied is given by

$$y = \frac{H_1 - H_3}{H_1 - H_f} - i\left(\frac{H_1 - H_2}{H_2 - H_f}\right)$$

Substituting enthalpy values gives

$$y = \frac{6.7 - (-1.8)}{6.7 - (-97)} - 0.8\left(\frac{6.7 - 3.8}{6.7 - (-97)}\right)$$
$$= 0.059, \text{ or } 5.9\% \text{ liquid yield}$$

Work per unit mass liquefied is obtained by applying the first law around the two compressors. For steady flow this results in

$$-W/\dot{m} = [T_1(S_1 - S_3) - (H_1 - H_3)] - i[T_1(S_1 - S_2) - (H_1 - H_2)]$$

Substituting appropriate thermodynamic values gives

$$\begin{aligned} -W/\dot{m} &= \{300[0.023 - (-0.367)] - [6.7 - (-1.8)]\} \\ &\quad - 0.8\{300[0.023 - (-0.253)] - [6.7 - 3.8]\} \\ &= 44.6\,\text{cal/g} = 80.3\,\text{Btu/lb} \\ &= 80.3/0.0597 = 1345\,\text{Btu/lb liquefied} \end{aligned}$$

b. Figure of merit:

$$\begin{aligned} \text{FOM} &= (W_i\dot{m})/(W/\dot{m}_f) \\ -W_i/\dot{m} &= T_1(S_t - S_f) - (H_1 - H_f) \\ &= 300[0.023 - (-0.912)] - [6.7 - (-97)] \\ &= 176.8\,\text{cal/g} = 318\,\text{Btu/lb} \\ \text{FOM} &= 318/1345 = 0.236 \end{aligned}$$

c. For the simple Linde cycle,

$$y = \frac{H_1 - H_3}{H_1 - H_f} = \frac{6.7 - (-1.8)}{6.7 - (-97)} = 0.082 \text{ or } 8.2\% \text{ liquid yield}$$

$$\begin{aligned} -W/\dot{m}_f &= [T_1(S_1 - S_3) - (H_1 - H_3)]/y \\ &= \{300[0.023 - (-0.367)] - (6.7 - 1.8)\}(1.8)/0.082 \\ &= 2380\,\text{Btu/lb liquefied} \end{aligned}$$

which compares to 1345 Btu/lb liquefied for the dual-pressure cycle, above.

## 7.5. Isentropic Expansion

Refrigeration can always be produced by expanding the process gas in an engine (instead of a valve) and causing it to do work. Recuperative heat exchangers are employed here also. A schematic of a simple refrigerator using this principle and the corresponding temperature–entropy diagram are shown in Fig. 6.27. Gas compressed isothermally from point 1 to point 2 at ambient temperature is cooled in a recuperative heat exchanger by gas being warmed on its way to the compressor intake. Further cooling takes place during the engine expansion from 3 to 4. In practice this expansion is never truly isentropic, and this is reflected by the *curved* path 3–4 on the temperature–entropy diagram. (Isentropic expansion would be a vertical straight line.)

In any work-producing expansion, the temperature of the gas is always reduced; hence, cooling does not depend on being below the inversion temperature prior to expansion. In large machines, the work produced during expansion is conserved. In small liquefiers, the energy from the expansion is usually expended in a gas or hydraulic pump and is not recovered.

The refrigerator in Fig. 6.27 produces a cold gas that absorbs heat from 4 to 5 and provides a method of refrigeration that can be used to obtain temperatures between those of the boiling points of the lower boiling cryogenic fluids.

*Example 6.4.* A cold gas refrigeration system using helium as the working fluid operates from 1 atm and 540°R to 15 atm. The maximum temperature in the cooling load is 36°R. The gas is expanded reversibly and adiabatically from 15 atm and 36°R down to 1 atm. This refrigeration system does not use a precooler purifier but consists of a 75% efficient compressor, a 95% effective warm heat exchanger, a reversible adiabatic expander, and the cooling load heat exchanger, which is isobaric. Determine (a) the refrigeration effect, (b) the coefficient of performance (assuming that the expander work is not used to aid in the compression process), and (c) the figure of merit for the refrigerator.

a. Figure 6.27a provides a schematic for the refrigeration system described. The refrigeration effect is given by

$$Q = \dot{m}(H_5 - H_4)$$

For an isentropic expansion from 15 atm and 36°R to 1 atm, $S_3 = S_4 = 11.39\,\text{J/(g K)}$. By interpolation, $H_4 = 11.1\,\text{cal/g}$ at an equivalent temperature of

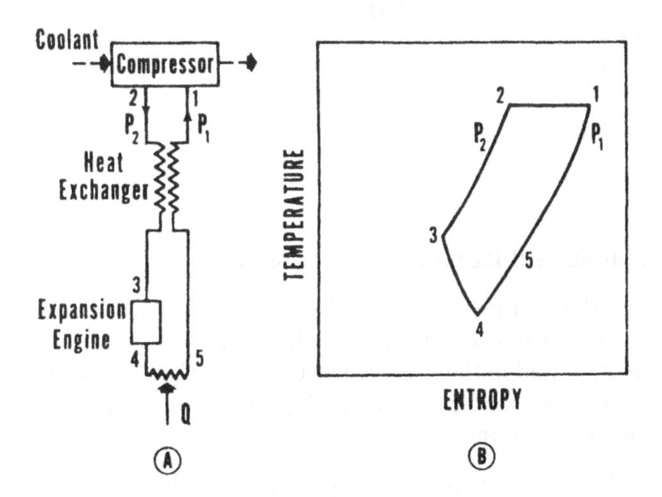


**Figure 6.27** Isentropic expansion refrigerator.

12°R. Thus,

$$Q = 28.1 - 11.1 = 17\ \text{cal/g} = 30.6\ \text{Btu/lb}$$

b. The coefficient of performance is obtained from Eq.(6.5) where the net work input is obtained from a modification of Eq.(6.7) assuming that the expander work is not utilized. $T_2$ is determined from the effectiveness of the warm heat exchanger, defined as

$$\varepsilon = \frac{T_1 - T_5}{T_2 - T_5}$$

or

$$T_2 = T_5 + \frac{(T_1 - T_5)}{\varepsilon}$$

$$= 36 + \frac{540 - 36}{0.95} = 566°\text{R}$$

Since the compression is not isothermal, assume an average temperature for this step or

$$-\frac{W}{\dot{m}} = \frac{T_{\text{avc}}(S_1 - S_2) - (H_1 - H_2)}{\eta}$$

where $\eta$ is the efficiency of the compressor. Substituting appropriate thermodynamic values results in

$$-\frac{W}{\dot{m}} = \left[\frac{553}{1.8}(7.50 - 6.21) - (375 - 394)\right]/0.75$$

$$= 553\ \text{cal/g} = 995\ \text{Btu/lb}$$

$$\text{COP} = Q/W = 17/553 = 0.031$$

The coefficient of performance of an ideal gas refrigerator with varying refrigerator temperature can be related to the temperature using the expression

$$\text{COP}_1 = \frac{T_5 - T_4}{T_{\text{ave}} \ln(T_5/T_4) - (T_5 - T_4)]}$$

$$= \frac{36 - 12}{(553/1.8) \ln[(36/12) - (36 - 12)]} = 0.076$$

c. The figure of merit is thus

$$\text{FOM} = \frac{\text{COP}}{\text{COP}_i} = \frac{0.031}{0.076} = 0.408$$

## 7.6. Combination of Isenthalpic and Isentropic Expansion

It is not uncommon to combine the isentropic and isenthalpic expansions to allow for the formation of liquid in the refrigerator. This is done because of the technical difficulties associated with forming liquid in the engine. The Claude cycle is an example of a combination of these methods and is shown in Fig. 6.28 along with the corresponding temperature–entropy diagram.

*Example 6.5.* Hydrogen gas is liquefied in a system using the ideal Claude concept. The gas initially enters a reversible isothermal compressor at 70°F and 1 atm and is

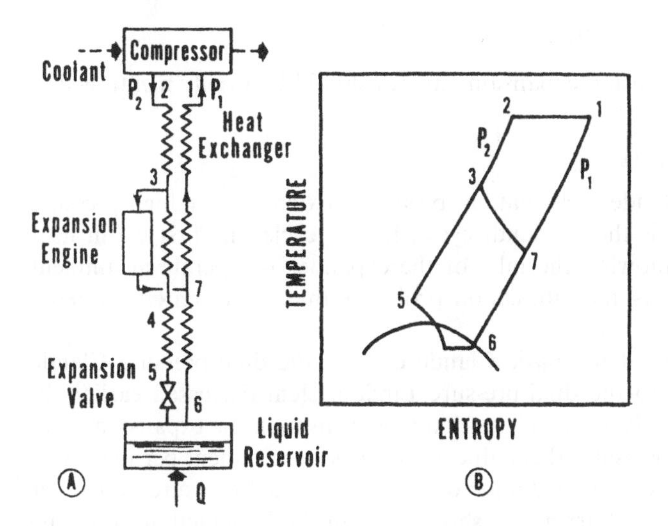


**Figure 6.28** The Claude cycle: a combination of Joule–Thomson and adiabatic expansion.

compressed to 40 atm. The high-pressure gas then is cooled in a heat exchanger to 324°R. At this point 50% of the total flow is diverted and expands through the expander to 1 atm, reversibly and adiabatically. The remainder of the gas continues through two more heat exchangers and expands through the expansion valve to 1 atm. Determine the liquid yield and the work per unit mass liquefied (assuming that the expander work is used in the compression).

Figure 6.28 can be used in this solution by removing liquid $m_f$ from the liquid reservoir rather than absorbing a refrigeration load $Q$. For simplicity assume that the fluid is $n$-$H_2$. Thermodynamic data are available in Chapter 3. A first law balance around the heat exchangers, expansion valve, and liquid reservoir provides the fraction liquefied:

$$y = \frac{H_1 - H_2}{H_1 - H_f} + x\frac{H_3 - H_e}{H_1 - H_f}$$

where $x$ is the fraction diverted through the expansion engine and $H_c$ is the specific enthalpy at the outlet of the expansion engine. Substituting appropriate enthalpy values gives

$$\begin{aligned} y &= \frac{990.6 - 994.8}{990.6 - 65.0} + 0.5\left(\frac{610.0 - 237.0}{990.6 - 65.0}\right) \\ &= 0.197 \end{aligned}$$

Another first law balance provides the relation for the work per unit mass liquefied if the expander work is used in the compression as

$$\begin{aligned} -\frac{W}{\dot{m}} &= T_1(S_1 - S_2) - (H_1 - H_2) - x(H_3 - H_c) \\ &= (530/1.8)(16.86 - 13.21) - (990.6 - 994.8) - 0.5(610.0 - 237.0) \\ &= 892.4\,\text{cal/g} \end{aligned}$$

$$-\frac{W}{\dot{m}_f} = \frac{892.4}{0.197} = 4530\,\text{cal/g} = 8155\,\text{Btu/lb}$$

## 7.7. Other Liquefaction Systems

Other refrigeration systems using expansion engines should be briefly mentioned.

### *7.7.1. Modified Claude Cycle*

One modification of the Claude cycle that has been used extensively in high-pressure liquefaction plants for air is the Heylandt cycle. In this cycle, the first warm heat exchanger is eliminated, allowing the inlet of the expander to operate at ambient temperatures and minimizing the lubrication problems that are often encountered at lower temperatures.

Another modification of the basic Claude cycle is the dual-pressure Claude cycle, similar in principle to the dual-pressure Linde system discussed earlier. In the dual-pressure Claude cycle, only the gas that is sent through the expansion valve is compressed to the high pressure; this reduces the work requirement per unit mass of gas liquefied. To illustrate the advantage of the Claude dual-pressure cycle over the Linde dual-pressure cycle, Barron has shown that in the liquefaction of air the liquid yield can be doubled while the work per unit mass liquefied can be halved when the Claude dual-pressure cycle is selected.

Still another extension of the Claude cycle is the Collins helium liquefier. Depending upon the helium inlet pressure, from two to five expansion engines are used in this system. The addition of a liquid nitrogen precooling bath to this system results in a two- to threefold increase in liquefaction performance.

### *7.7.2. Mixed Refrigerant Cycle*

The mixed refrigerant cycle is a variation of the mixed refrigerant cascade cycle described in Sec. 7.3. The simplification of the compression and heat exchange services in such a cycle may, under certain circumstances, offer the potential for reduced capital expenditure in comparison with the conventional cascade cycle. This cycle is widely used to liquefy natural gas.

A simplified flow sheet of the classical cascade process was shown in Fig. 6.20. After purification, the natural gas stream is cooled successively by vaporization of propane, ethylene, and methane. These gases have each been liquefied in a conventional refrigeration loop. Each refrigerant may be vaporized at two or three pressure levels to increase the natural gas cooling efficiency, but at the cost of a considerable increase in process complexity.

Cooling curves for natural gas liquefaction by the cascade process are shown in Figs. 6.29a and 6.29b. It is evident that the cascade cycle efficiency can be improved considerably by increasing the number of refrigerants employed. The actual work required for the nine-level cascade cycle depicted in Fig. 6.29b is approximately 80% of that required by the three-level cascade cycle depicted in Fig. 6.29a, for the same throughput. This increase in efficiency is achieved by minimizing the temperature difference throughout each increment of the cooling curve.

The cascade system can be adapted to any cooling curve; i.e., the quantity of refrigeration supplied at the various temperature levels can be chosen so that the temperature differences in the evaporators and heat exchangers approach a practical minimum (small temperature differences mean low irreversibility and therefore lower power consumption).

The mixed refrigerant cycle is a variation of the cascade cycle described above and involves the circulation of a single refrigerant stream, which may be a mixture of

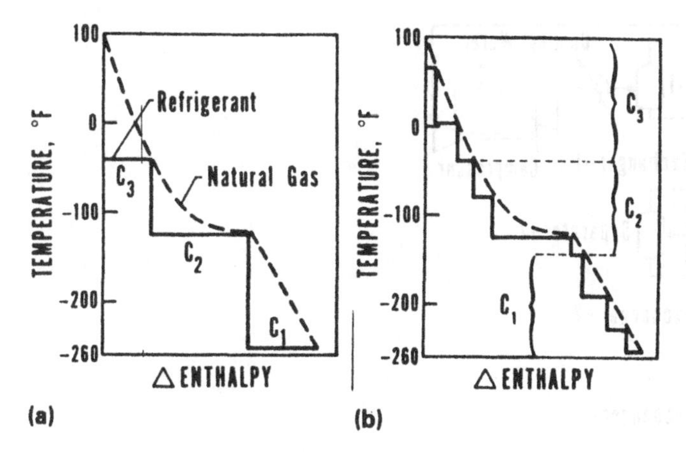


**Figure 6.29** Cascade cycle cooling curves for natural gas. (a) three-level; (b) nine-level.

refrigerants. The simplification of the compression and heat exchange services in such a cycle may, under certain circumstances, offer potential for reduced capital expenditure over the conventional cascade cycle. Note the similarity of the mixed refrigerant cycle temperature vs. enthalpy diagram in Fig. 6.30 with the corresponding diagram for the nine-level cascade cycle in Fig. 6.29b.

Figure 6.31 shows a version of the mixed refrigerant cycle. The refrigeration required for the cooling and liquefaction of the natural gas is provided by a circulating mixed refrigerant stream containing components such as butane, propane, ethane, methane, and nitrogen. This stream is compressed by the compressor, and the heat of compression is removed by the intercooler and aftercooler by means of heat transfer with cooling water. The mixed refrigerant stream, as a result of being compressed and cooled, is partially condensed, and the two phases are separated in separator 1. The liquid and overhead vapor flow as separate streams to exchanger 1, where heat is transferred to the returning refrigerant stream. The vapor is thereby

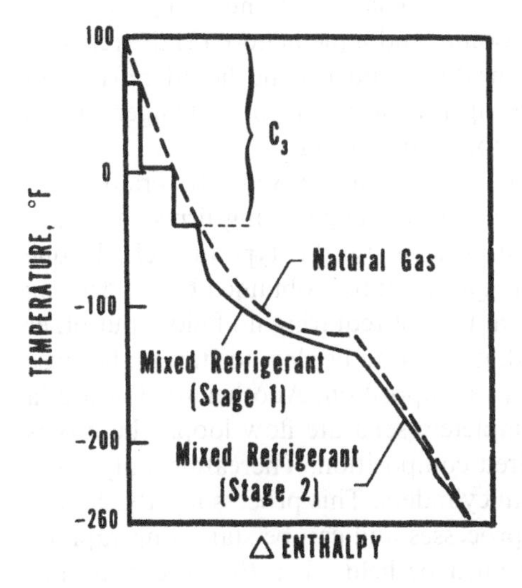


**Figure 6.30** Propane precooled mixed refrigerant cycle cooling curve for natural gas.

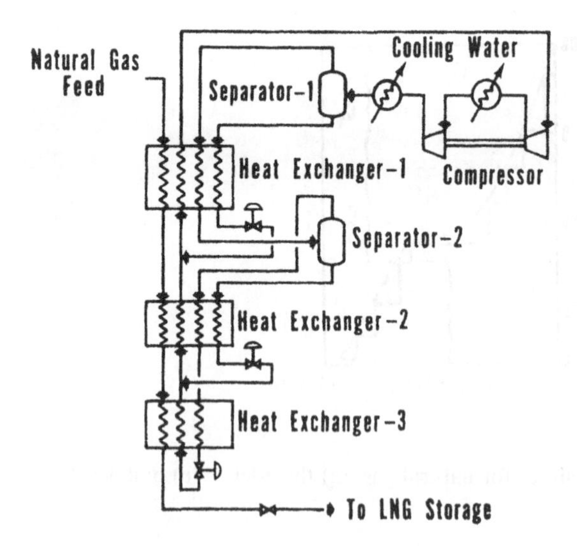


**Figure 6.31** Mixed refrigerant cycle.

partially condensed and flows from exchanger 1 to separator 2, where the phases are again separated.

The liquid phase from separator 1, in flowing through exchanger 1, is subcooled and then reduced in pressure and allowed to flow back through that exchanger to the warm end to provide refrigeration to the three cooling streams (natural gas feed and vapor and liquid refrigerant streams from separator 1). The overhead vapor and liquid bottoms from separator 2 flow into exchanger 2 to be further cooled. The liquid from separator 2 is further subcooled in exchanger 2 and reduced in pressure before flowing back to that exchanger.

The vapor overhead from separator 2 is further cooled and totally condensed in exchangers 2 and 3 and is then reduced in pressure to supply the final level of refrigeration in flowing back through exchanger 3. This returning stream joins with the other liquid streams injected at the cold end of exchangers 1 and 2 to provide the refrigeration to (1) the natural gas feed for cooling and liquefaction, (2) the refrigerant vapor stream for further partial condensation, and (3) the liquid refrigerant stream for subcooling. The total combined refrigerant stream from exchanger 1 flows to the compressor, where it is recompressed for recirculation.

The above description covers the basic concepts of a mixed refrigerant cycle. Variations of the cycle are proprietary with cryogenic engineering firms that have developed the technology. Many commercial projects use this type of cycle. In one commercial cycle, for example, the refrigerant gas mixture is obtained by condensing part of the feed natural gas. This has the advantage of requiring no fluid input other than the natural gas itself. On the other hand, it is said to be slow to start up because of the need to collect refrigerant and adjust its composition. Another version uses a mixed refrigerant that is contained in a completely separate flow loop. The mix is made up of purchased gases to get the desired composition. Thereafter, only occasional make-up gases are added, usually from cylinders. This process has the advantage of simplicity compared to some other processes and can be started up rapidly. On the other hand, the refrigerant mixture must be held when the process is shut down. Therefore, suction and surge drums are needed.

All of the mixed refrigerant processes use a multicomponent refrigerant that is repeatedly condensed, vaporized, separated, and expanded. Thus, these processes require more sophisticated design methods and more complete knowledge of the thermodynamic properties of gaseous mixtures than do the expander or cascade cycles. Also, these processes must handle two-phase mixtures in heat exchangers.

## 8. LARGE SYSTEMS

Industrial uses of cryogenics are spectacular and commonplace and most frequently at the 77 K (nitrogen) or 90 K (oxygen) level. Uses include food freezing (McDonalds restaurants alone use $50 million worth of liquid nitrogen a year), sewage treatment (many cities use oxygen produced on site from liquid air to speed up treatment), breathing oxygen for hospitals obtained from liquid oxygen storage, and the production of chemicals (antifreeze) and steel by processes that require liquid oxygen.

Hydrogen for industrial and aerospace use is routinely stored as a liquid at 20 K, strictly as a convenience. Texas Instruments' Stafford, Texas plant uses the country's largest commercial liquid hydrogen storage tank to supply hydrogen gas for semiconductor processing. NASA and the US Air Force both have 1 million gallon liquid hydrogen storage tanks supplied by a network of hydrogen liquefiers. Liquid hydrogen (20 K) technology is available "off the shelf" from such firms as Air Products and Linde.

The status of large-scale liquid helium facilities is perhaps even more surprising. Most large high-energy accelerators use helium-cooled superconducting magnets (1) simply because it is cheaper to do so, compared to the electric power required for normal conductor magnets; and (2) to achieve higher levels of magnet performance. Table 6.4 shows a partial list of these facilities. At present, these facilities have a combined capacity of 82.5 kW at 4 K. It is a fact that the large-scale production of helium temperatures is a routine, commercially available technology.

## 9. REGENERATIVE CYCLES

A regenerative heat exchanger has only a single set of flow passages through which the hot and cold fluids flow alternately and continuously. This gives rise to a periodic or ac type of cooling cycle. Refrigerators with regenerative heat exchangers include the Stirling, Philips, Solvay, Gifford–McMahon, Vuillemier, Ericsson, and Postle cycles. Although it is misleading to do so, all of these cycles are often lumped under the name Stirling cycle, although regenerative cycles would be more apt.

The regenerative matrix is called upon to store and release heat when the cycle alternates the flow of the hot and cold streams. This requires that the matrix have some useful heat storage ability at the temperature of interest and leads to a fundamental limitation of these cycles. Namely, the lattice heat capacity of all crystalline solids tends to zero as the absolute temperature tends to zero. To produce very low temperatures, then, materials must be used that have some energy storage mechanism other than lattice heat capacity, such as magnetic phase transitions. Hence, materials such as erbium and gadolinium compounds are used as regenerator materials at very low temperatures.

There are system design considerations between recuperative and regenerative coolers. Recuperative (e.g., J–T) coolers will nearly always require higher pressures

**Table 6.4** Some Large-Scale Cryogenic Facilities with Low-Temperature Superconductors

| Facility | Location | Operational | Cryogenic capacity | Comments |
|---|---|---|---|---|
| Fermilab Tevatron | Batavia, IL | 1983 | 24 kW at 4 K; 5000 L/hr liquid He; 63,000 L Liquid He storage; 254,000 L liquid $N_2$ storage | |
| Mirror Fusion Test Facility (MFTF) | Livermore, CA | MFTF-A 1982 | 10.5 kW at 4 K | MFTF-B: $1.05 \times 10^6$ kg magnets at $4_2$ 35 K; |
| | | MFTF-B 1985 | 500 kW at 77 K | 900 $m^2$ of cryopanels at 4.35 K; 10 days to cool system to 4 K |
| Joint European Torus (JET) | Oxon, UK | Yes | 500 W at 3.8 K; 20 kW at 77 K | |
| Tore-Supra Experimental European Fusion Tokomak | France | 1986 | 300 W at 1.75 K; 700 W at 4.0 K; 10.7 kW at 80 K | 50,000 kg at 1.8 K; 120,000 kg at 4.5 K; 20,000 kg at 80 K |
| International Fusion Superconducting Magnet Test Facility (IFSMTF) | Oak Ridge, TN | 1983 | 1.4 kW at 4 K | 360,000 kg at 3.8 K; 20 days to cool system to 4 K |
| Brookhaven National Lab (BNL) | Upton, NY | 1985 | 24.8 kW at 3.8 K | |
| Electron–Proton Collider | Hamburg, Germany | | 20.3 kW at 4 K; 20 kW at 40–80 K | |
| Superconducting Super Collider (SSC) | | Cancelled | 31.5 kW at 4.15 K 48.2 kW at 20 K 390 kW at 84 K | $5 \times 10^7$ kg mass cooled to 4 K |

than regenerative coolers. Hence tubing and compressor walls will always be thicker, and these systems will weigh more than the regenerative coolers. There is no fundamental limit to the low-temperature level that can be produced by J–T cycles as long as there is a liquid to expand. Hydrogen and helium working fluids must be cooled below a certain inversion temperature before they produce cooling on J–T expansion. For helium, this temperature is about 24 K or less, and the colder the better. Joule–Thomson type cycles produce continuous ("dc") cooling. The major components of a J–T system can be separated from one another by practical distances. Thus, the compressor can be isolated from the expansion valve to assist in noise reduction. Furthermore, the cooling may be distributed over the system. There is no "cold head" per se that must be rigidly attached to a displacer/compressor system. The J–T cold head will always be simpler and much less expensive than any regenerative cycle cold head.

Regenerative coolers will always be lighter in weight than the J–T coolers and can be much smaller in volume as well, because the compressor duty is much less. Regenerative cycles do not depend significantly upon the operating fluid used, although helium is the fluid of choice. These coolers generally have some moving mechanical components at the lower temperatures. Cooling will always be periodic, or of the "ac" type.

A common misconception is that Stirling cycles are always more efficient than J–T cycles. This simply is not true. In fact, every liquefier of consequence ever made always uses a J–T expansion in the last step to produce liquid. (See the discussion of isenthalpic vs. isentropic expansion in Sec. 4.3.)

There is no "silver bullet" thermodynamic cycle that determines which cooler to use. The choice is very application-dependent. For instance, is the cooling load concentrated at one physical point and hence amenable to Stirling cycle machines (e.g., LWIR weapon systems)? Or is the cooling load physically distributed, as in a telescopic optics system? The system requirements dictate the choice.

## 9.1. The Stirling Cycle

The Stirling cycle was invented in 1816 by a Scottish minister, Robert Stirling, for use in a hot air engine. Stirling had an interest in some Scottish coal mines. The coal was plentiful, but the mines often filled with water because many were located along the coast of Scotland. Stirling's financial investments would be more valuable if an economical system for pumping out the water could be devised. Steam power was available, but the fuel was somewhat expensive. Stirling sought a more economical engine that could use any waste material as fuel. The Stirling cycle was the result.

As early as 1834, John Herschel suggested that this engine could be used as a refrigerator. The first Stirling cycle refrigerator was constructed by Alexander Kirk around 1864. At the end of World War II, the famous Philips Co. of Eindhoven, The Netherlands, repeated Stirling's quest for a cheap heat engine that could be fueled by any waste material. The need was for a driver to generate electricity in remote or war-devastated regions. Hence, Philips turned first to a Stirling cycle heat engine. Perhaps a decade later, J. W. L. Kohler and others at the N.V. Philips Gloeilampenfabrieken research laboratories had an industrial scale Stirling refrigerator for sale. Kohler introduced the Philips refrigerator on September 5, 1956 at the Cryogenic Engineering Conference held at the National Bureau of Standards in Boulder, Colorado. Kohler was an exceptional showman as well as a brilliant scientist and engineer. He first ran the machine in one direction until liquid air condensed and dripped off the cold head. He then reversed the direction of the machine until it literally glowed red hot in the darkened NBS auditorium. Kohler leaned over his machine and used the incandescent heat to light a cigarette. The audience cheered. It was a cryogenic Woodstock. I was there.

The Stirling machine is manufactured today as small, hand-held, ground-based I-R coolers about the size of a pop can. Perhaps 10,000 coolers are produced per year by the Hughes Aircraft Co., Texas Instruments, and other suppliers of so-called common module coolers for the US Defense Department.

Kohler (1956) described the Philips refrigerator in this way:

> Reviewing present-day means of producing cold, one is struck by the fact that refrigerators, that is, machines for producing cold, operate over only the temperature range between the ambient or room temperature and −80°C. A second point is that in most refrigerators the cold is produced by evaporating a liquid. It is amazing that of the many possibilities offered by Nature, only one is actually utilized. In this article, we give a survey of another method, which may provide a welcome extension of refrigerating technique.
>
> The cold is produced in this machine by the reversible expansion of a gas. The gas performs a cycle, during which it is compressed at ambient and expanded at the desired low temperature, the transition between these temperatures being accomplished by heat exchange between the gas going in the one direction and the other. The gas refrigerating machine is fundamentally a special case of the Stirling cycle (Robert Stirling, a Scottish minister, invented the air engine in 1816; the first actual refrigerating machine, based on the Stirling cycle, was described by A.G. Kirk [1873–1874] whereby the compression and the expansion are alternating processes and the heat exchange already mentioned takes place in a regenerator). The *p*–*V* diagram of the ideal cycle consists of two isotherms and two isochors, as shown in (Fig. 6.32); it is to be noted, however, that the isochors may be replaced by isobars, or in general by any pair of lines which permit heat exchange by regeneration.

Figure 6.33 shows stages in carrying out the ideal cycle. In this diagram, *1* is a cylinder, closed by the piston *2*, and containing a nearly perfect gas. A second piston,

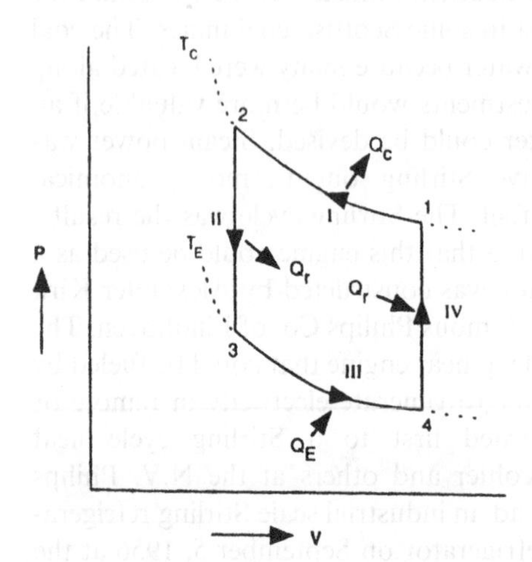


**Figure 6.32** *P*–*V* Diagram of the ideal Stirling cycle between temperatures $T_c$ and $T_E$. At $T_E$ the amount of heat $Q_E$ is absorbed during phase III; at $T_c$ the heat $Q_c$ is rejected during phase I. The heat $Q_r$, rejected in phase II, is stored in the regenerator and absorbed in phase IV.

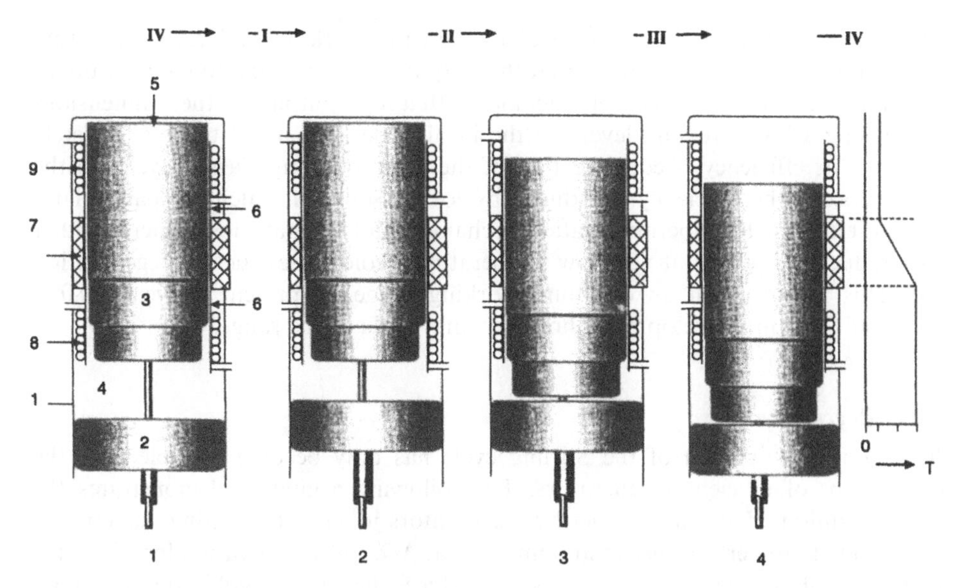


**Figure 6.33** The ideal Stirling cycle. The figure shows points 1–4 of Fig. 6.32. I–IV refer to the four phases of the cycle. The graph on the right shows the temperature distribution in each machine.

the displacer *3*, divides the cylinder into two spaces, *4* at room temperature and *5* at the low temperature, connected by the narrow annular passage *6*. This passage contains the regenerator *7*, a porous mass with a high heat capacity; the temperature in the passage is shown on the far right of Fig. 6.33. The cycle, consisting of four phases, runs as follows:

1. Compression in space *4* by the piston *2*; the heat of compression is discharged through the cooler *8*.
2. Transfer of the gas through the regenerator to space *5* by movement of the displacer. The gas is reversibly cooled down in the regenerator, the heat of the gas being stored in the regenerator mass.
3. Expansion in the cold space by the combined movement of the piston and the displacer; the cold produced is discharged into the refrigerant in *9* and utilized.
4. Return of the gas to space *4*; thereby the gas is reheated, the heat stored in the regenerator being restored to the gas.

Actually, the discontinuous motion of the pistons implied here is not very practicable. Fortunately, little change is required if the pistons are actuated by a crank mechanism, with the condition that space *5* is leading in phase with respect to space *4*. In this harmonic cycle the four phases are of course fused, so that the basic processes are more difficult to distinguish. The harmonic motion allows the high speed of the machine, causing the compression and the expansion to take place adiabatically; this entails only a small loss, as the compression ratio is rather low (approximately 2), due to the volume of the heat exchangers (dead space).

### *9.1.1. General Characteristics*

The properties of the ideal cycle are easily derived from the *P–V* diagram. We find that, in contrast to the evaporation process, the output drops only gradually with temperature, so that the cycle is particularly suited to work over a large temperature range. Another important result is that the output is proportional to the pressure of the gas; this enables the design of machines with a high output for their dimensions by choosing a high pressure level. As the ideal cycle is assumed to be completely reversible, its efficiency is equal to that of the corresponding Carnot cycle. In the actual process, the efficiency is reduced by losses, some of which are particularly important at a small temperature ratio (mechanical and adiabatic loss) whereas other losses reduce the output at very low temperatures (cold losses such as regeneration loss). This brings about an optimum working range, at present between −80°C and −200°C; future development, however, may extend this range.

### *9.1.2. Regeneration*

The practical realization of the Stirling cycle has only become possible with the development of efficient regenerators. The following argument demonstrates the severe conditions that must be met by regenerators in gas refrigerating machines.

Owing to imperfections, an amount of heat $\Delta Q_r$, the regeneration loss, is transmitted through the regenerator. As this loss reduces the effective cold production, we may compare it with the ideal production $Q_E$. For well-designed machines we find

$$\frac{\Delta Q_r}{Q_E} \approx 7(1-\eta_r)\frac{T_C - T_E}{T_E}$$

$T_C$ being the compression temperature (300 K), $T_E$ the expansion temperature, and $\eta_r$ the efficiency of the regenerator. Taking $T_E = 75$ K, this reduces to

$$\frac{\Delta Q_r}{Q_E} \approx 21(1-\eta_r)$$

showing that a regeneration loss of 1% entails a loss of 21% in output (at 20 K the result would be 98%). Thus, the lowest temperature that can be obtained is mainly determined by the quality of the regenerator.

It has been found that with a light, feltlike mass of very fine wire, regenerators can be prepared with efficiency higher than 99% (at frequencies as high as 50 Hz). It should be borne in mind, however, that the regenerator acts as a resistance to the gas flow as well, so that in designing such machines there must be a compromise between regeneration loss, flow loss, and heat loss due to dead space. Finally, the heat capacity of the regenerator must lie above a certain value and the heat capacity of the gas must be independent of pressure (nearly a perfect gas), requirements that are difficult to meet at very low temperatures.

### *9.1.3. Applications*

A first application of the new process is a machine working at −194°C to liquefy atmospheric air. The fact that the air to be liquefied need not be compressed makes this unconventional air liquefier very simple and easy to handle. Its efficiency compares favorably with that of conventional types.

Kohler (1959) made an illuminating comparison between the Claude cycle (recuperator/expansion) and the Stirling cycle (regenerator/expansion):

> Liquefaction and refrigeration at one temperature are different processes and therefore cannot be compared with each other.

It is well known that the gas liquefaction process can be divided into two distinct operations, namely the precooling to the condensation point and the actual liquefaction at this temperature. Both operations may be performed in one stage by an ideal refrigerator removing heat at the condensation temperature; the precooling is then performed with the Carnot efficiency for that temperature $\eta_c$. But it is more advantageous to precool the gas reversibly, the efficiency of this process $\eta_{rev}$ being higher than $\eta_c$. This is shown in Fig. 6.34, where the ratio $\eta_c/\eta_{rev}$ is plotted against the temperature ratio $\tau = T_0/T_1$, $T_0$ being ambient temperature and $T_1$ the lowest temperature, according to the expression

$$\frac{\eta_c}{\eta_{rev}} = \frac{[\tau/(\tau-1)]\ln(\tau-1)}{\tau-1}$$

It is assumed that the specific heat of the gas to be cooled is independent of temperature. For air, with $\tau \approx 4$, $\eta_c \approx 0.28\ \eta_{rev}$, so that only a nonideal precooling process whose efficiency is lower than $0.28\ \eta_{rev}$ requires more power than a Carnot process producing all the refrigeration at the condensation point. Since with air the heat absorbed for precooling is approximately equal to the heat of condensation, a great reduction in power consumption can be obtained by precooling even if this process has a poor efficiency.

As a result, precooling in some form or another is generally applied in conventional gas liquefaction techniques. Precooling is accomplished in the Linde-type liquefier with an ammonia cycle and/or with a second Joule–Thomson process. In the Claude liquefier, precooling is accomplished with adiabatic expansion (employed twice in the Collins helium liquefier). The total efficiency of these processes is roughly the same, although they consist of different steps with different efficiencies. As mentioned earlier, the best value obtained is a power consumption of 0.80 kW hr/kg for air liquefaction. This figure has to be compared with that for the ideal process with

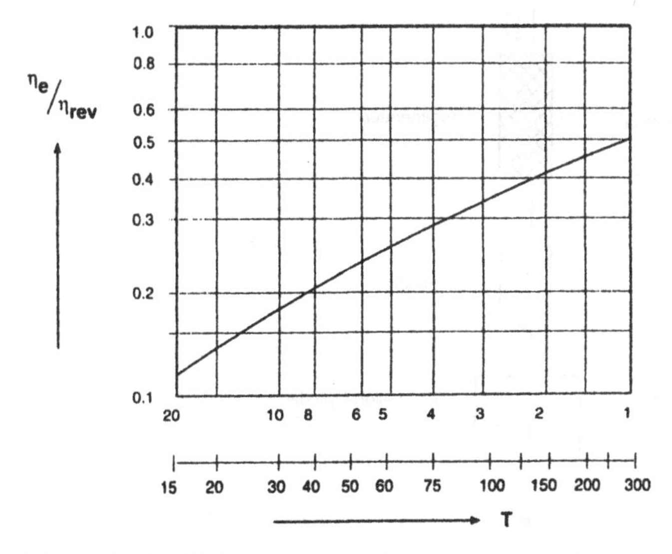


**Figure 6.34** Efficiency ratio of single-stage precooling to that for reversible continuous precooling.

reversible precooling, 0.194 kW hr/kg. The best process therefore exhibits a figure of merit of 0.24.

The Philips air liquefier, based on the Stirling cycle, is an attempt to provide a very simple and small installation. This simplification is obtained by removing all the heat at the lowest temperature only, in contrast with conventional techniques. Even with this unfavorable procedure the efficiency is rather high in view of the small size of the refrigerator, i.e., 1.0 kW hr/kg. This figure has to be compared with that of the Carnot process of 0.30 kW hr/kg, giving us a figure of merit of 0.30. One precooling stage would reduce the power consumption down to 75% of the present value; this may easily be achieved with multicylinder machines, e.g., by using one cylinder for precooling at an intermediate temperature and three cylinders for final precooling and condensation.

From this discussion it is apparent that conventional liquefaction installations cannot readily be converted into refrigeration units because precooling cannot be used. Since these installations employ a combination of two different processes, it is practically impossible to obtain information about the quality of one of the refrigeration processes from data on the efficiency of the liquefaction process as a whole.

For the Claude and Stirling processes, shown schematically in Fig. 6.35, one has to look deeper to find important differences. In both, gas is compressed at ambient temperature and expanded at the desired low temperature, and in both processes the temperature difference is bridged by heat exchange between the outgoing and returning gas flows. This is accomplished in the Claude process in a counterflow heat

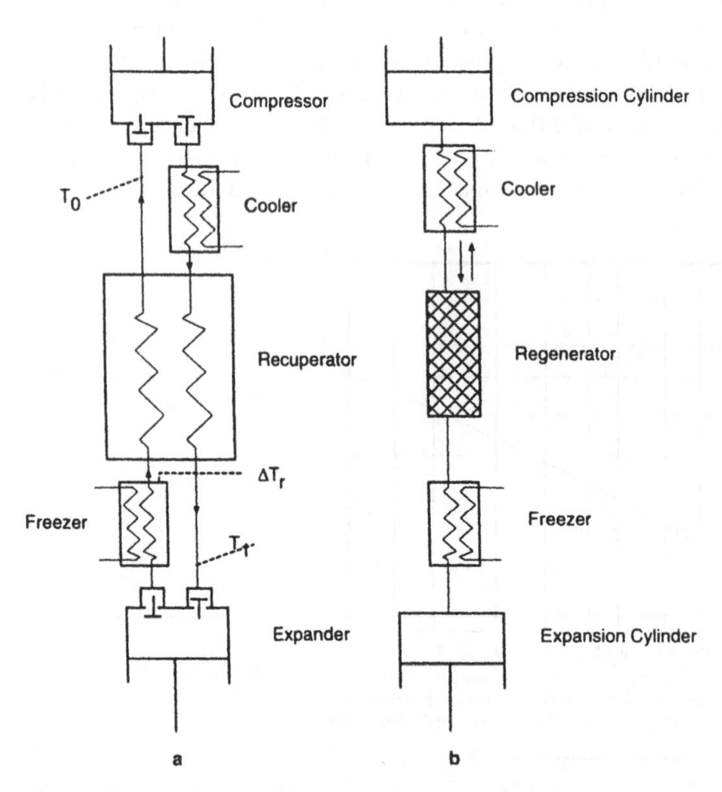


**Figure 6.35** Schematic diagrams of (a) the Claude process and (b) the Philips–Stirling process.

exchanger or recuperator with two channels and in the Stirling process in a regenerator through which the gas flows alternately in opposite directions.

It has been intimated in the literature that the difference between the two processes is essentially that the transfer of the gas between the limiting temperatures is performed with constant pressure in the Claude process, whereas in the Stirling cycle it is performed with constant volume. Actually any manner of transfer is permissible that allows regeneration, and constant-volume transfer is only one example of this; another example, which more closely approaches the actual Stirling process with harmonically moving pistons, is the transfer with constant pressure. In this respect, therefore, the difference between the Claude cycle and the Stirling cycle is very small indeed.

The Claude process is quite sensitive to the efficiency of the compressor and the expander. This becomes apparent when one tries to obtain the total power consumption of the process. Here the power output of the expander has to be subtracted from the power input of the compressor, the two have different efficiencies and are of a comparable order of magnitude. (With piston expanders the mechanical loss should also be included.) If the coupling of the two component processes can be achieved only via the electric mains, it is usually not worthwhile to try to regain the expansion power.

In contrast with this, the Stirling process with displacer performs this subtraction in a very direct manner with ideal efficiency through the influence of both processes on the pressure of the gas; thus, only the transfer of the power difference from outside to inside is subjected to a loss. This is enhanced by the fact that the efficiencies of the driver for the Stirling process can be rather favorable, i.e., between 0.8 and 0.9. It is obvious that this conclusion is greatly affected by the assumed values for the efficiencies. We have the impression that the assumption used is quite realistic, taking into account that the compressor efficiency is defined as the departure from isothermal compression and the omission of the mechanical losses of the expander.

With respect to adiabatic compression vs. isothermal compression, the Stirling process is in a better position for another reason, namely that it is only partially adiabatic. The reason is that the gas is not only compressed and expanded in the cylinders, but also in the rest of the working circuit, the so-called dead space. While the temperature behaves nearly adiabatically in the cylinders, it is practically constant in the regenerator. The reduction in the adiabatic loss due to this effect is quite appreciable; a rough estimate indicates a reduction down to one-half of the full effect. In the Claude process, the adiabatic loss may be reduced in principle by using intermediate cooling; with the low compression ratios considered, its effect is questionable, however.

Although the losses in the heat exchanger are not unduly high, they should not be underestimated. The use of a regenerator in the Stirling cycle permits this loss to be cut down quite effectively, because a regenerator can be subdivided more readily than a counterflow heat exchanger. This effect may be counteracted somewhat by the fact that in the Stirling process the dead space of the regenerator should be kept within limits in order to provide a reasonable pressure ratio. This effect is absent in the Claude process. On the whole, the application of a regenerator seems to be advantageous. Apart from this, the regenerator enhances the compactness of the installation, with lower insulation losses.

In conclusion, it seems difficult to find points where the Claude process can be improved. Therefore, it is to be concluded that if refrigeration at a temperature below $-100°C$ is desired, the Stirling process can supply it with the highest efficiency.

### 9.2. The Philips Refrigerator

The Philips refrigerator operates on the Stirling cycle. A schematic of the sequence of operations of the Philips refrigerator is shown in Fig. 6.36, and the cycle is shown on the temperature–entropy plane at the left in Fig. 6.37.

The Philips refrigerator consists of a cylinder enclosing a piston, a displacer, and a regenerator. The piston compresses the gas, while the displacer simply moves the gas from one chamber to another without changing the gas volume, in the ideal case. The heat exchange during the constant-volume process is carried out in the regenerator. The sequence of operations for the system is as follows.

**Process 1 → 2.** The gas is compressed isothermally while rejecting heat to the high-temperature sink (surroundings).

**Process 2 → 3.** The gas is forced through the regenerator by the motion of the displacer. The gas is cooled at constant volume during this process. The energy removed from the gas is not transferred to the surroundings but is stored in the regenerator matrix.

**Process 3 → 4.** The gas is expanded isothermally while absorbing heat from the low-temperature source.

**Process 4 → 1.** The cold gas is forced through the regenerator by the motion of the displacer; the gas is heated during this process. The energy stored during process 2 → 3 is transferred back to the gas. In the ideal case (no heat inleaks), heat is transferred to the refrigerator only during process 3 → 4, and heat is rejected from the refrigerator only during process 1 → 2.

Assuming that reversible heat transfers occur in the first and third process steps (1 → 2 and 3 → 4) of Fig. 6.38, the second law of thermodynamics yields the expressions

$$\text{Heat rejected} = \dot{Q}_h = \dot{m}T_1(S_2 - S_1)$$

$$\text{Heat absorbed} = \dot{Q}_c = \dot{m}T_3(S_4 - S_3)$$

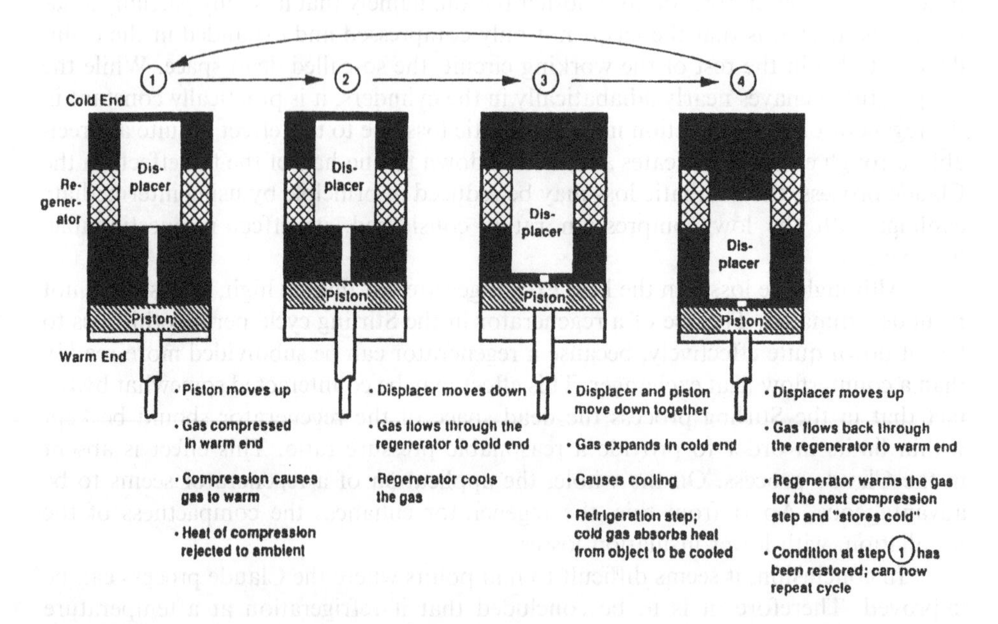


**Figure 6.36** Stirling cycle integral type cryocooler.

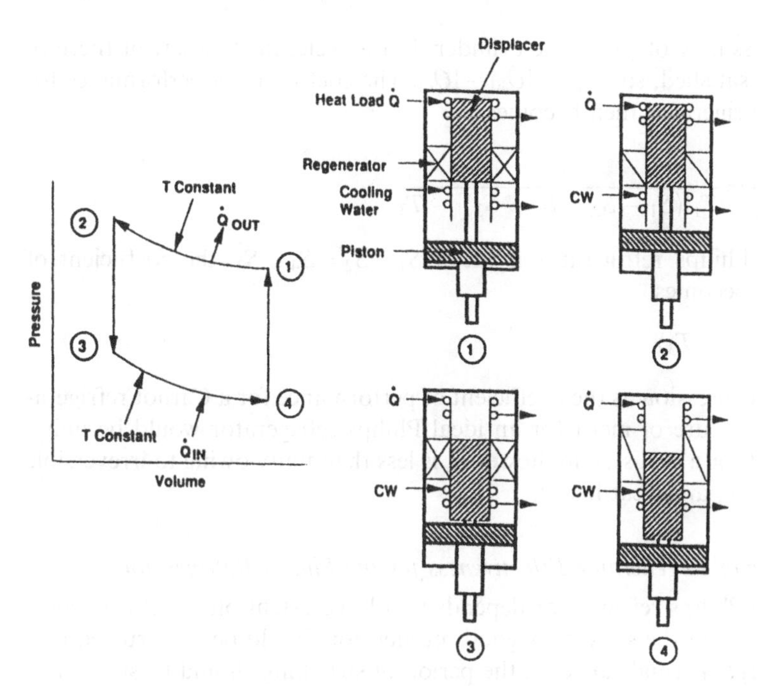


Figure 6.37 Philips refrigerator using Stirling cycle.

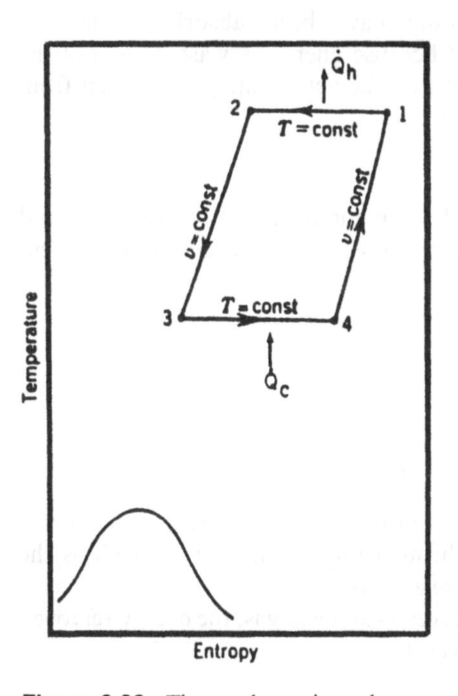


Figure 6.38 Thermodynamic cycle representing the Philips refrigerator.

where $\dot{m}$ is the mass flow of gas in the cylinder. For a cycle, the first law of thermodynamics must be satisfied, so $\dot{W}_{net} = |\dot{Q}_h| - |\dot{Q}_c|$. The coefficient of performance for an ideal Philips refrigerator then becomes

$$\mathrm{COP} = \frac{\dot{Q}_c}{\dot{W}_{net}} = \frac{T_3}{T_1[(S_1 - S_2)/(S_4 - S_3)] - T_3}$$

If the gas in the Philips refrigerator is ideal, $S_1 - S_2 = S_4 - S_3$, the coefficient of performance then becomes

$$\mathrm{COP} = T_3/(T_1 - T_3)$$

which is the same expression as the coefficient of performance for a Carnot refrigerator. Therefore, the figure of merit for an ideal Philips refrigerator would be unity. The figure of merit for a real system, however, is less than unity owing to irreversible mechanical, pressure, and heat losses.

### *9.2.1. Importance of Regenerator Effectiveness for the Philips Refrigerator*

The success of the Philips refrigerator depends to a large extent on the effectiveness of the generator used in the system. A good regenerator should be constructed of a material with a large thermal capacity, the period of switching should be small (i.e., the frequency of cycling the fluid through the regenerator should be large), the heat transfer coefficient and surface area should be large, and the loss flow rate of gas through the regenerator should be small. In the Philips refrigerator, a light feltlike mass of fine wire is used as the regenerator matrix material to attain the large thermal capacity with large heat transfer area needed for good regenerator performance.

If the regenerator were less than 100% effective, the temperature of the gas leaving the regenerator would be somewhat higher than the source temperature. This means that some of the energy that could have been absorbed from the low-temperature region cannot be absorbed because energy is wasted to cool the refrigerator gas down to the source temperature. The actual energy absorbed from the low-temperature source becomes

$$Q_a = Q_{a,ideal} - \Delta Q$$

where $Q_{a,\ ideal}$ is the ideal heat absorbed and $\Delta Q$ is the heat that must be removed from the working fluid to overcome the ineffectiveness of the regenerator. The regenerator effectiveness is defined as

$$\varepsilon = \frac{Q_{actual}}{Q_{ideal}} = \frac{Q_{2-3,ideal} - \Delta Q}{Q_{2-3,ideal}}$$

then

$$\Delta Q = (1 - \varepsilon)Q_{2-3,ideal} = (1 - \varepsilon)mC_V(T_2 - T_3)$$

where $Q_{2\text{-}3,ideal}$ is the ideal heat transferred from the gas to the regenerator during process $2 \rightarrow 3$, $m$ the mass of gas flowing through the regenerator, and $C_V$ is the specific heat of the gas flowing through the regenerator.

If we assume that the working fluid behaves as an ideal gas, the energy removed from the cooling load in the ideal case is given by

$$Q_{a,ideal} = mT_3(S_4 - S_3) = mRT_3 \ln(V_4/V_3)$$

or

$$Q_{\text{a,ideal}} = (\gamma - 1)mC_V T_3 \ln(V_4/V_3)$$

where $\gamma = C_P/C_V$ = specific heat ratio for the gas. We obtain the fraction of the ideal refrigeration effect that is wasted because of the ineffectiveness of the regenerator:

$$\frac{\Delta Q}{Q_{\text{a,ideal}}} = \left(\frac{1-\varepsilon}{\gamma - 1}\right)\left(\frac{T_2/T_3 - 1}{\ln(V_4/V_3)}\right)$$

As a numerical example, suppose that we have helium as the working fluid, for which $\gamma = 1.67$, and suppose that $V_4/V_3 = V_1/V_2 = 1.24$. For refrigeration between 540°R (300 K) and 140°R (77.8 K), the loss in refrigeration effect becomes

$$\frac{\Delta Q}{Q_{\text{a,ideal}}} = \frac{1-\varepsilon}{0.67}\left[\frac{(540/140) - 1}{\ln 1.24}\right] = 19.87(1-\varepsilon)$$

If the regenerator has an effectiveness $\varepsilon$ of 99%, then $\Delta Q/Q_{\text{a,ideal}} = 19.87\%$, or almost 20% of the refrigeration effect is wasted because of a 1% decrease in the effectiveness from the ideal or 100% effectiveness. We see that in this example all the refrigeration would be wasted if the regenerator effectiveness were

$$\varepsilon_{\min} = 1 - \frac{1}{19.87} = 1 - 0.0503 = 0.9497$$

or about 95%. This example illustrates how critical the regenerator is in the performance of the whole refrigeration system.

*Example 6.6.* A Philips refrigerator is used to maintain a refrigeration temperature of 100 K while operating from an ambient temperature of 295 K. Helium is used as the working fluid. The system operates from a low pressure of 0.101 MPa to a maximum pressure of 1.01 MPa. What are (a) the refrigeration per kilogram of helium compressed, (b) the coefficient of performance, and (c) the figure of merit if all of the process steps in the system are assumed to be ideal and the helium behaves as an ideal gas at these operating conditions?

a. If we use the state conditions of Fig. 6.38, the pressures at points 2 and 4 are 1.01 and 0.101 MPa, respectively. The missing pressures can be obtained from a ratio of the temperatures (because of ideal gas conditions) or

$$P_1 = P_4(T_1/T_4) = (0.101)(295/100) = 0.295\ \text{MPa}$$
$$P_3 = P_2(T_3/T_2) = (1.01)(100/295) = 0.339\ \text{MPa}$$
$$\dot{Q}_c = \dot{m}RT_3 \ln\left(\frac{P_3}{P_4}\right) = (1)\left(\frac{8.314}{4}\right)(100)\ln\left(\frac{0.339}{0.101}\right)$$
$$= 254\ \text{kJ/kg of helium compressed}$$

b. From the definition of COP:

$$\text{COP} = \dot{Q}_c/\dot{W}_{\text{net}}$$
$$W_{\text{net}} = |\dot{Q}_h| - |\dot{Q}_c|$$
$$\dot{Q}_h = \dot{m}RT_1 \ln\left(\frac{P_1}{P_2}\right) = (1)\left(\frac{8.314}{4}\right)(295)\ln\left(\frac{0.295}{1.01}\right) = -750\text{kJ/kg}$$
$$\text{COP} = 254/(|750| - |254|) = 0.512$$

which yields $COP_i$:

$$COP_i = T_c/(T_h - T_c) = 100/(295 - 100) = 0.512$$

c. The figure of merit is

$$FOM = COP/COP_i = 0.512/0.512 = 1.0$$

## 9.3. The Solvay Refrigerator

The Solvay refrigerator was invented in Germany about 1887 (Solvay, 1887) and was the first system planned for air liquefaction using an expansion engine. Solvay's prototype apparatus was able to achieve a low temperature of only 178 K (320°R or −140°F), so the system was not considered for cryogenic refrigeration until the late 1950s when Gifford and McMahon of A.D. Little, Inc. described the use of the refrigerator in a miniature infrared cooler. This refrigerator diagrammed in Fig. 6.39 consists of a compressor, a regenerator, and a piston within a cylinder. The features of the process can be described by following the temperature–entropy diagram shown in Fig. 6.40:

1. The piston is initially at the bottom of the cylinder. The inlet valve in process $1 \rightarrow 2$ is opened, which allows high-pressure gas to enter the system.
2. The high-pressure gas in process $2 \rightarrow 3$ is directed through the regenerator, where it is cooled, and the energy is stored in the regenerator. With the expander piston raised, the three-way valve permits the cold gas to flow into the cylinder.
3. The inlet valve in process $3 \rightarrow 4$ is closed, and the gas in the cylinder is isentropically expanded (ideally) to the initial system pressure. Because

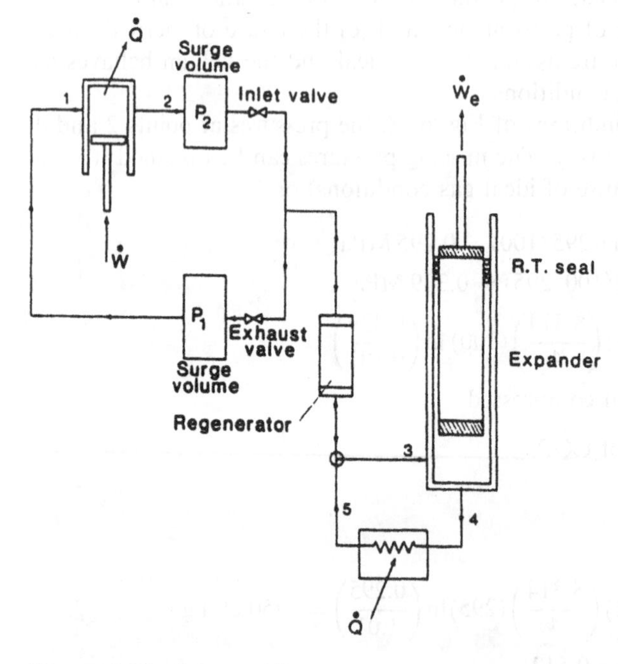


**Figure 6.39** Solvay refrigerator.

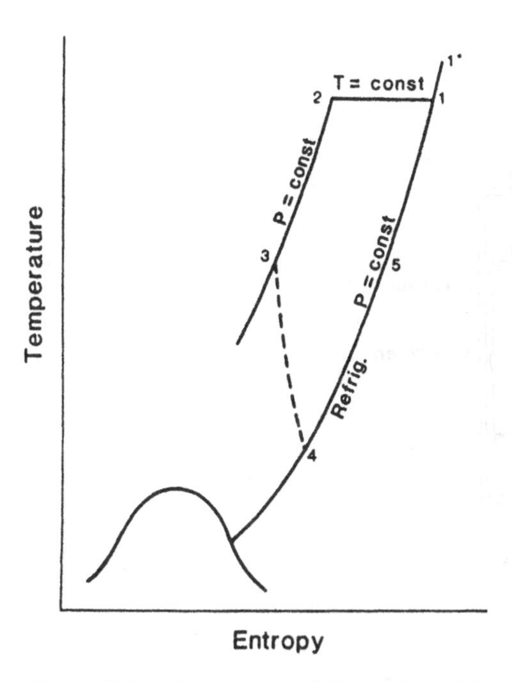


**Figure 6.40** Process steps followed by refrigent gas is the Solray and A. D. Little refrigerators.

work is performed by the gas in raising the piston, energy is removed from the gas and the temperature of the gas decreases.

4. The exhaust valve in process $4 \rightarrow 5$ is opened, and the cold gas is forced out of the cylinder by the downward motion of the piston. This cold gas is diverted by the three-way valve through a heat exchanger, where heat is absorbed from the region to be cooled.
5. The gas in process $5 \rightarrow 1$ is then warmed back to room temperature by passing it through the regenerator where the gas absorbs the energy stored in process $2 \rightarrow 3$.

## 9.4. The Gifford–McMahon Refrigerator

The Gifford–McMahon (G–M) refrigerator is similar to the Solvay refrigerator, except that the cylinder is closed at both ends with the piston inside the cylinder now serving as a displacer (see Fig. 6.41). In this system, no work is transferred from the system because the displacer moves the gas from one expansion space to another. The process can be described by following the same temperature–entropy diagram used to highlight the process for the Solvay refrigerator:

1. Process $1 \rightarrow 2$ is identical to that for the Solvay refrigerator. The displacer is initially at the bottom of the cylinder, the inlet valve is opened, and the high-pressure gas flows to the regenerator.
2. In process $2 \rightarrow 3$, the displacer is raised to the top of the cylinder; this moves the gas originally in the upper expansion space through the three-way valve into the lower expansion space. Because the volume of the gas decreases as it is cooled through the regenerator, the inlet valve remains open to maintain constant pressure throughout the system.

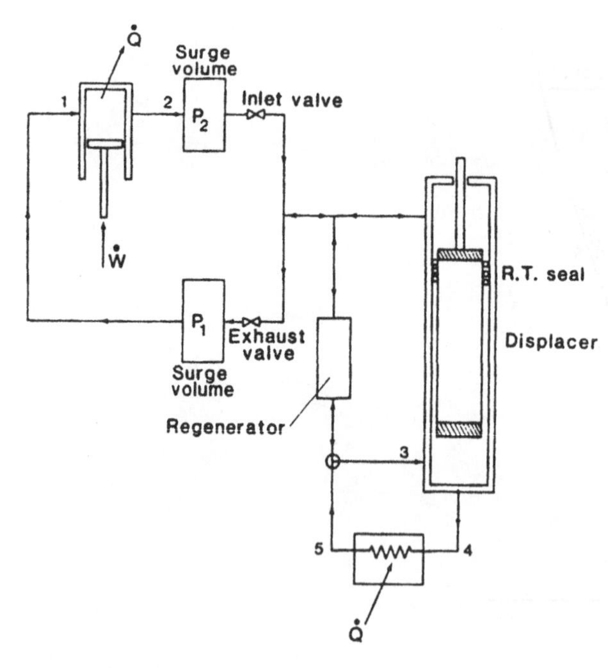


**Figure 6.41** G–M refrigerator.

3. The gas within the lower expansion space is allowed to expand in process $3 \rightarrow 4$ to the initial system pressure by closing the inlet valve, redirecting the three-way valve, and opening the exhaust valve. During the expansion, the gas experiences a temperature drop.
4. The displacer in process $4 \rightarrow 5$ moves downward, forcing the remaining gas out of the bottom of the cylinder and through a heat exchanger, where the gas absorbs heat from the region to be cooled.
5. Finally, in process $5 \rightarrow 1^*$, the gas is warmed back to near room temperature by sending it back through the regenerator.

In both the Solvay and G–M refrigerators, regenerators are used instead of ordinary heat exchangers, which eliminates problems caused by the use of slightly impure gas. As the gas passes back and forth through the regenerator, impurities are deposited during the inlet phase and removed during the exhaust phase. Such deposits would quickly clog ordinary heat exchangers. Low-temperature sealing problems also are eliminated in both refrigerators, because the flow valves and displacer piston seals are at room temperature.

The Solvay refrigerator provides a slightly higher coefficient of performance than the G–M because more energy is removed from the working fluid during the expansion process. In the G–M refrigerator, a small motor is required to move the displacer back and forth, whereas the expanding gas moves the piston in the Solvay arrangement. On the other hand, the leakage past the displacer in the G–M arrangement is considerably less than across the expander piston because of the smaller pressure difference across the displacer seals.

A major advantage of the G–M system is the ease with which several units may be multistaged to achieve temperatures as low as 15–20 K with helium as the working fluid. In fact, with a three-stage refrigerator, refrigeration may be obtained at three different temperature levels using a common piston with room temperature seals for all three displacers.

*Example 6.7.* A G–M refrigerator operates between the pressure limits of 0.101 and 2.02 MPa using helium gas as the working fluid. The maximum temperature in the space to be cooled is 90 K, and the temperature of the gas leaving the compressor is 300 K. Assume that the regenerator is 100% effective and the compressor has an efficiency of 100%. The expansion process in the expansion space is assumed to be isentropic. Determine (a) the refrigeration effect, (b) the COP, and (c) the FOM for the system.

The operation of the G–M refrigerator is depicted in Fig. 6.40 except for a slightly higher temperature observed for the low-pressure exhaust stream designated by 1*. From the temperature–entropy property tabulation for helium as listed by Mann (1962), we find

$$H_2(2.02\,\mathrm{MPa}, 300\,\mathrm{K}) = 1579\,\mathrm{kJ/kg}$$
$$S_2(2.02\,\mathrm{MPa}, 300\,\mathrm{K}) = 25.2\,\mathrm{kJ/(kg\,K)}$$
$$H_5(0.101\,\mathrm{MPa}, 90\,\mathrm{K}) = 482\,\mathrm{kJ/kg}$$

Since the regenerator effectiveness is 100%, the temperature of the gas leaving the regenerator at point 3 must be the same as that of the gas entering the regenerator at point 5. The thermodynamic values at point 3 are then

$$H_3(2.02\,\mathrm{MPa}, 90\,\mathrm{K}) = 486\,\mathrm{kJ/kg}$$
$$S_3(2.02\,\mathrm{MPa}, 90\,\mathrm{K}) = 18.9\,\mathrm{kJ/(kg\,K)}$$

Process $3 \rightarrow 4$ is assumed to be an isentropic expansion; therefore, point 4 is found by following a constant entropy line from point 3 to a pressure of 0.101 MPa. Interpolation gives a temperature of 27.1 K at point 4. Thus,

$$H_4(0.101\,\mathrm{MPa}, 27.1\,\mathrm{K}) = 155\,\mathrm{kJ/kg}$$

a. The refrigeration effect can be found once the mass ratio $\dot{m}_e/\dot{m}$ is determined. As the volume of the expansion space remains constant during the expansion process, the mass ratio can be related to the density ratio by the relationship

$$\frac{\dot{m}_e}{\dot{m}} = \frac{\rho_4}{\rho_3} = \frac{V_3}{V_4}$$

where $\rho$ is the density and $V$ is the specific volume at the given point. From the temperature–entropy diagram for helium,

$$V_3 = 95\,\mathrm{cm^3/g} \quad \text{and} \quad V_4 = 555\,\mathrm{cm^3/g}$$

The mass ratio is then

$$\dot{m}_e/\dot{m} = V_3/V_4 = 95/555 = 0.171$$

The refrigeration effect is given by

$$\dot{Q}/\dot{m} = (\dot{m}_e/\dot{m})(H_5 - H_4) = 0.171(482 - 155) = 56\,\mathrm{kJ/kg}$$

b. From an enthalpy balance around the system without the compressor, we have

$$\dot{Q}/\dot{m} = H_1 - H_2 = (\dot{m}_e/\dot{m})(H_5 - H_4)$$

The condition of the gas leaving the regenerator at the warm end can now be determined from

$$H_1 = H_2 + (\dot{Q}/\dot{m}) = 1579 + 56 = 1635\,\text{kJ/kg}$$

At 0.101 MPa this value for $H_1$ is equivalent to 312 K with $S_1 = 31.6\,\text{kJ/(kg K)}$. This indicates that the gas leaving the regenerator must either be cooled in the exit surge volume to 300 K before entering the compressor or isothermal compression will occur at 312 K, and cooling to 300 K will have to be accomplished in the compressor intercoolers.

The work requirement for the compressor, assuming an inlet condition of 312 K, is given by

$$-\dot{W}/\dot{m} = T_1(S_1 - S_2) - (H_1 - H_2) = 312(31.6 - 25.2) - (1635 - 1579)$$
$$= 1941\,\text{kJ/kg}$$

The COP for this system is then

$$\text{COP} = \frac{\dot{Q}_c/\dot{m}}{\dot{W}/\dot{m}} = 56/1941 = 0.029$$

Assuming that the helium gas behaves as an ideal gas in the temperature and pressure ranges under consideration, the ideal COP for an isobaric source refrigerator can be found as:

$$\text{COP}_i = \frac{T_5 - T_4}{T_h \ln(T_5/T_4) - (T_5 - T_4)}$$

where $T_h$ is the sink temperature of 312 K, $T_5$ is the maximum refrigerator temperature of 90 K, and $T_4$ is the minimum refrigeration temperature of 27.1 K. Therefore,

$$\text{COP}_i = \frac{90 - 27.1}{312 \ln(90/27.1) - (90 - 27.1)} = 0.202$$

c. The figure of merit for this G–M system is

$$\text{FOM} = \text{COP}/\text{COP}_i = 0.029/0.202 = 0.143$$

## 9.5. Vuilleumier Refrigerator

The Vuilleumier refrigeration cycle has frequently been described as a Stirling cycle with a thermal compressor instead of a mechanical compressor. Recently, it has generated much interest in the area of spacecraft applications, where the advantages of long-life operation, low acoustical noise, and minimal wear of moving parts have been extensively examined.

The steady-state operation of Vuilleumier refrigeration cycle is most easily followed with reference to Fig. 6.42, a schematic of four crank positions encountered

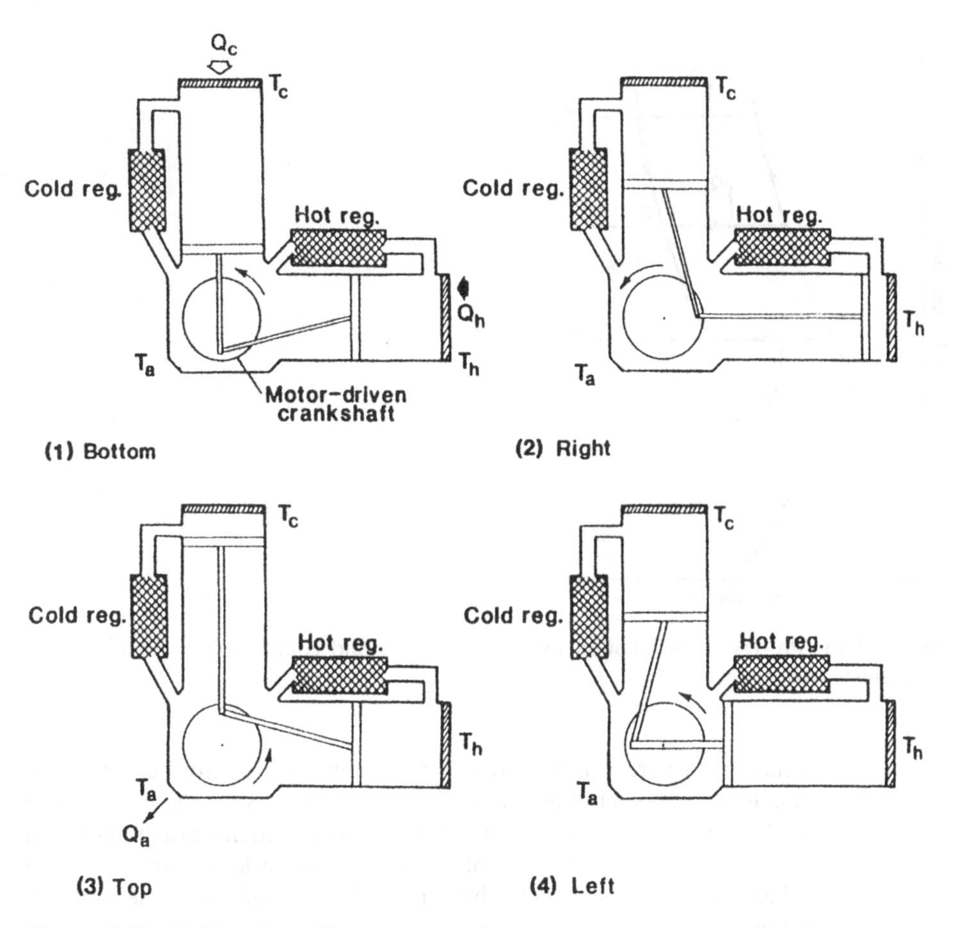


**Figure 6.42** Vuilleumier refrigerator operation.

during the cycle, and Fig. 6.43, a temperature–entropy diagram showing the process steps for the cycle. The process is as follows.

With the crank in the bottom position, the cold displacer is at its maximum displacement and the hot displacer is at half its maximum displacement. Most of the helium gas is in the cold cylinder volume. Therefore, the average gas temperature and pressure are relatively low.

**Process 1 → 2.** As the crank moves in a counter current direction to a position on the right, both the hot and cold cylinder volumes decrease. Part of the cold gas is forced through the cold regenerator, where it is warmed close to $T_a$ (ambient temperature) before entering the ambient section. Similarly, most of the gas in the hot cylinder is forced through the hot regenerator, where it is cooled to essentially ambient temperature before entering the ambient section. Heat is thus stored in the hot regenerator. Since both cylinder volumes have decreased, the average gas temperature and pressure change very little.

**Process 2 → 3.** With continued countercurrent movement of the crank to the top position, the hot cylinder volume increases. The rest of the gas in the cold cylinder is forced through the cold regenerator and warmed to near ambient temperature while part of the ambient gas flows through the hot

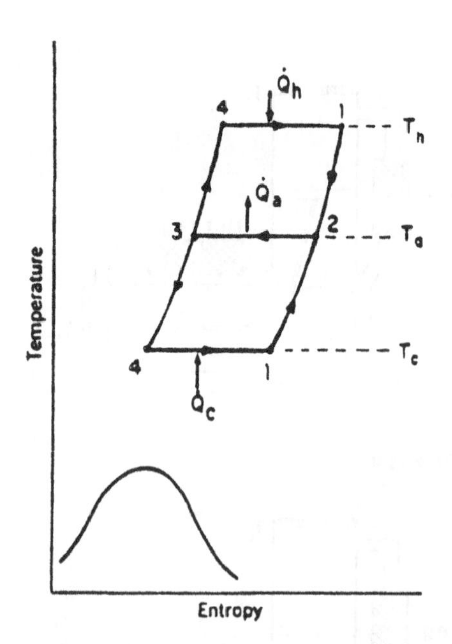


**Figure 6.43** Process steps followed by gas in the hot and cold cylinder volumes in the Vuilleumier refrigerator.

regenerator where it is heated to approximately $T_h$ before entering the hot cylinder. The net effect of the volumetric increase in the hot cylinder and the volumetric decrease in the cold cylinder is an increase in the mean gas temperature and pressure of the system. In order for this to be an isothermal compression step, heat must be rejected from the ambient section.

**Process 3 → 4.** As the crank turns from the top position to the position on the left, the volumes of both cylinders increase. Part of the ambient gas moves back through the cold regenerator, releasing heat, and enters the cold volume with a temperature close to $T_c$. On the hot side, part of the ambient gas moves through the hot regenerator, absorbs heat, and enters the hot volume with a temperature near $T_h$. Since both the cold and hot volumes increase, the system pressure does not change greatly.

**Process 4 → 1.** Once the crank turns from the left position to the bottom position, the hot cylinder volume once again decreases while the cold cylinder volume increases. Part of the hot cylinder gas is forced through the hot regenerator, where it is cooled to a temperature near $T_a$, while part of the ambient gas is forced through the cold regenerator, where it is cooled to a temperature near $T_c$. The mean gas temperature and the pressure of the system decrease, resulting in an expansion of the gas and consequent heat absorption by the system.

The net effect of this entire cycle is that heat is absorbed at the hot and cold cylinder ends and is rejected from the ambient section. The coefficient of performance for this refrigerator in terms of the three different temperature levels, $T_c$, $T_h$, and $T_a$, can be shown to be

$$\mathrm{COP} = T_c(T_h - T_a)/T_h(T_a - T_c)$$

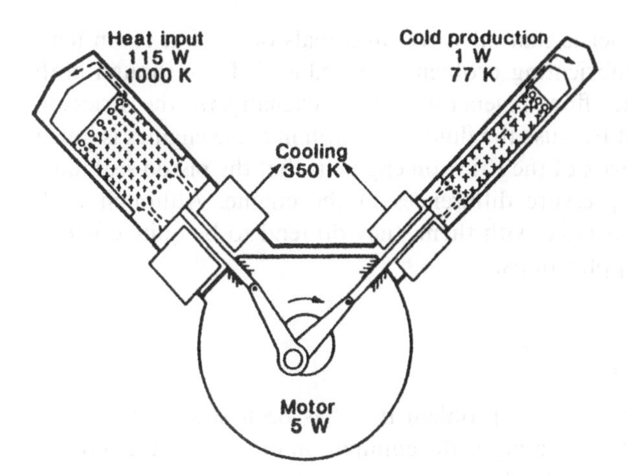


**Figure 6.44** Schematic of miniature Vuilleumier refrigerator.

when the effects of harmonic motion of the displacers, ineffectiveness of the regenerators, and the thermodynamic losses are neglected.

A schematic of an actual miniature Vuilleumier refrigerator is shown in Fig. 6.44. The refrigerator consists of a hot cylinder, a cold cylinder, and a connecting space. Helium gas successively enters the hollow displacers through short channels in the bottom, passes through the regenerator, and exits through small openings above the regenerator. The gas then flows through a slit to the expansion volume above each cylinder. Helium gas flowing in the space between the two displacers allows for heat exchange to the ambient. The displacers operate about 90° out of phase, with the low-temperature displacer leading the high-temperature displacer.

By its very design, the Vuilleumier refrigerator is essentially a one-component system because the compressor and expander are in the same housing. Since pressure differences among the three sections of the refrigerator are very small, piston pressure seals are not required as in many other refrigeration concepts such as the Stirling refrigerator. In fact, a feasible seal design has been effected by simply using the clearance between the displacers and the cylinder walls. Thus, failure due to pressure seal wear can essentially be eliminated, thereby extending the operating life between instances of mechanical failure. However, since a fraction of the heat added at the high-temperature source of the Vuilleumier refrigerator has to be rejected, whereas in the Stirling refrigerator all of the mechanical work can be fully utilized, the Vuilleumier refrigerator cannot achieve a comparable coefficient of performance.

## 9.6. Ericsson and Postle Refrigerators

The Ericsson and Postle refrigerators are quite similar. They differ from the Stirling refrigerator in that they control fluid flows by pressure changes instead of volume changes. The only difference between the two is in the means of expansion. The Ericsson refrigerator (which includes the Solvay system) uses a reciprocating piston connected to a crankshaft to expand the working fluid, whereas the Postle cycle uses a displacer that slides in a cylinder.

No working refrigerators employing the Postle cycle are presently in operation, but the G–M refrigerators described earlier use a modified Postle cycle. This permits some general comparisons between Ericsson and G–M cryocoolers. One important advantage of the G–M cryocooler is that all valves and seals operate at room temperature; thus, operating and lubricating problems are reduced. Even though both cycles use very efficient counter-flow regenerative heat exchangers, the Ericsson engines tend to be more efficient because the fluid expansion in these engines is much more complete. The only drawback of the Ericsson engine is that the piston seal must be able to withstand the full pressure differences in the engine. Although both refrigerators have problems associated with them, they do tend to be quite efficient and well suited for cryogenic applications.

## 9.7. Pulse Tube Refrigerator

High reliability in small cryocoolers is a problem that has been studied for many years. One approach to increased reliability is the elimination of some of the moving parts in a mechanical refrigerator. Stirling refrigerators have only two moving parts—the compressor piston and the displacer. In 1963 Gifford and Longsworth discovered a refrigeration technique that eliminates the displacer from the Stirling refrigerator (Gifford and Longsworth, 1964). They called this new technique pulse tube refrigeration. Under Gifford's direction, the pulse tube refrigerator was advanced to the point where temperatures achieved were 124 K in one stage and 79 K in two stages (Longsworth, 1967). A single-stage unit operating from 65 K achieved 30 K. Figure 6.45 illustrates the development of pulse tube refrigerators.

Radebaugh (1986) described three types of pulse tubes:

> A schematic of the basic pulse tube refrigerator is shown in (Fig. 6.46). The principle of operation, as given by Gifford and coworkers and by Lechner and Ackermann, is qualitatively simple. The pulse tube is closed at the top end where a good heat transfer surface exists between the working gas (helium is best) and

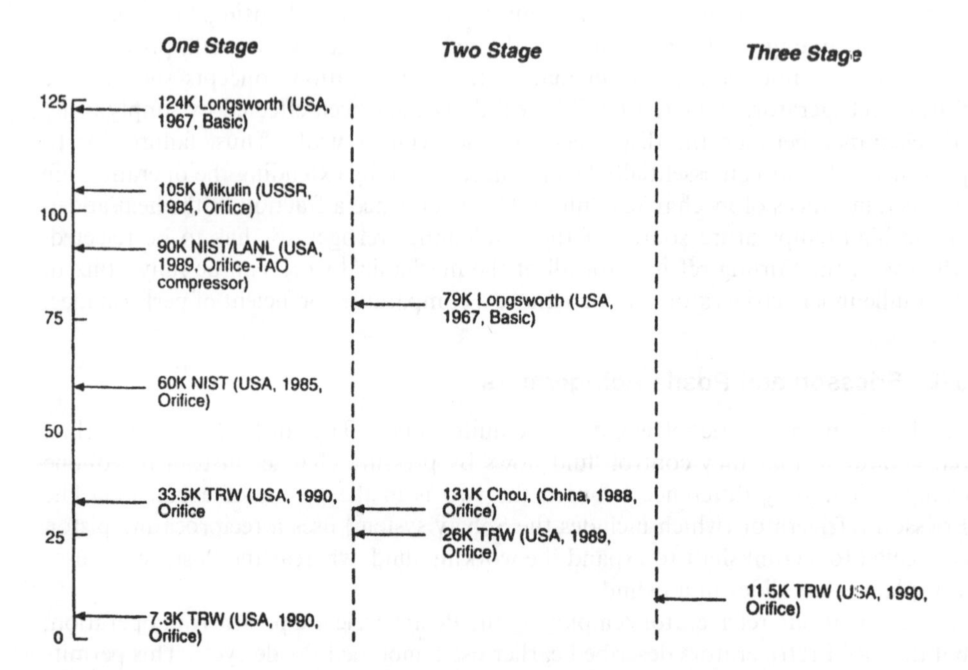


**Figure 6.45** Pulse tube refrigerator developments.

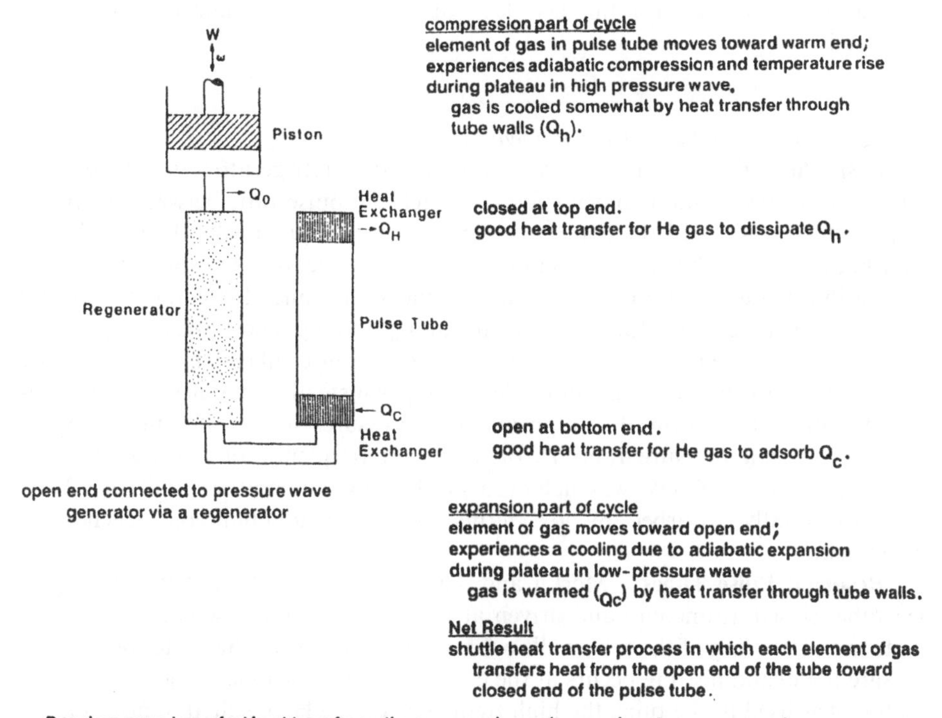

Basic Pulse Tube
W
Piston
Q0
Heat Exchanger
QH
Regenerator
Pulse Tube
QC
Heat Exchanger
compression part of cycle
element of gas in pulse tube moves toward warm end;
experiences adiabatic compression and temperature rise
during plateau in high pressure wave.
gas is cooled somewhat by heat transfer through
tube walls (Qh).
closed at top end.
good heat transfer for He gas to dissipate Qh.
open at bottom end.
good heat transfer for He gas to adsorb Qc.
open end connected to pressure wave
generator via a regenerator
expansion part of cycle
element of gas moves toward open end;
experiences a cooling due to adiabatic expansion
during plateau in low-pressure wave
gas is warmed (Qc) by heat transfer through tube walls.
Net Result
shuttle heat transfer process in which each element of gas
transfers heat from the open end of the tube toward
closed end of the pulse tube.
Requires some imperfect heat transfer so the compression and expansion steps are somewhere
between isothermal and adiabatic- out of phase


**Figure 6.46** Basic pulse tube.

the surroundings in order to dissipate heat $Q_H$. The bottom end is open but also has a good heat transfer surface to absorb refrigeration power $Q_C$. The open end is connected to a pressure-wave generator via a regenerator. During the compression part of the cycle any element of gas in the pulse tube moves toward the closed end and at the same time experiences a temperature rise due to the adiabatic compression. At that time the pressure is at its highest value $P_h$. During the plateau in the pressure wave, the gas is cooled somewhat by heat transfer to the tube walls. In the expansion part of the cycle the same element of gas moves toward the open end of the pulse tube and experiences a cooling due to the adiabatic expansion. During the plateau at $P_l$ the gas is warmed through heat transfer from the tube walls. The net result of cycling the pressure in this manner is a shuttle heat transfer process in which each element of gas transfers heat toward the closed end of the pulse tube. Note that the heat pumping mechanism described here requires that the thermal contact between the gas and the tube be imperfect, so the compression and expansion processes are somewhere between isothermal and adiabatic. The best intermediate heat transfer generally occurs when the thermal relaxation time $\tau_t$ between the gas and the tube walls is $\omega\tau_t \approx 1$, where $\omega$ is the frequency of operation in rad/s.

The work of Gifford and coworkers in the mid-1960s on the basic pulse tube led to some semiempirical expressions for the performance. Overall efficiencies were estimated to be about comparable to a Joule–Thomson refrigerator. The low temperature of 124 K using a pressure ratio of 4.25/1 was certainly interesting, but most practical applications begin at about 80 K or below. No further work on pulse tube refrigeration occurred until the last few years when some new concepts were introduced.

### *9.7.1. Four Types of Pulse Tube Refrigerators*

**Basic Pulse Tube.** The early work on pulse tube refrigerators was done using a valved compressor and a rotary valve to switch the pulse tube between the high- and low-pressure sides of the compressor. Such a technique lowers the overall efficiency because no work is recovered in the expansion process even though the actual refrigeration process is improved because of the short times for compression and expansion and long times for heat transfer at $P_h$ and $P_l$. The lowest temperature of 124 K was achieved in 1967 by Longsworth on a pulse tube 19 mm diameter by 318 mm long with the top 31.8 mm made of copper instead of stainless steel to provide an isothermal section. The high and low pressures were 2.38 and 0.56 MPa, respectively, for a pressure ratio $P_r$ of 4.25 at a frequency of 0.67 Hz. A gross heat pumping rate of 5 W was achieved at the low-temperature limit of 124 K, which was totally absorbed by loss terms such as conduction and regenerator ineffectiveness

**Resonant Pulse Tubes.** Merkli and Thomann (1975) showed that in a gas-filled tube, closed at one end and driven at resonance with an oscillating piston at the other end, cooling of the tube walls occurs at positions near the center of the tube (pressure node) and heating occurs at the ends (pressure antinode). Since no internal structure was used in the tube, the high frequency of 122 Hz resulted in nearly adiabatic conditions ($\omega\tau_t \gg 1$) so little refrigeration was produced. Wheatley et al. (1985) improved the resonant pulse tube or thermoacoustic refrigerator by adding internal structure at the appropriate position along the tube to get $\omega\tau_t \approx 1$ (see Fig. 6.47). The internal structure of the 39 mm diameter tube consisted of cloth epoxy plates 67 μm thick spaced 380-μm apart. A frequency of 516 Hz was used with a mean pressure $P_m$. The lowest temperature achieved with this type of device has been 195 K. Note that this device has no regenerator and the thermodynamically active region in the gaps between plates is located differently from that of the basic pulse tube. The reason for this difference will be explained later, but the resonant pulse tube can use the same geometry, including regenerator, as in the basic pulse tube (Wheatley et al., 1983).

**Orifice Pulse Tube.** Mikulin et al. (1984) inserted an orifice at the top of the pulse tube to allow some gas to pass through to a large reservoir volume. Figure 6.48 shows the NBS version of this modification, although the original work of Mikulin et al. (1984) placed the orifice just below the isothermal section instead of above it as in the NBS work. They applied the pressure wave by the use of valves and used air as the working gas even though the Joule–Thomson effect is not used in the orifice. They were able to achieve a low temperature of 105 K by using a pulse tube (see Fig. 6.49) 10 mm in diameter by 450 mm long with pressures of $P_h = 0.4$ MPa and $P_l \approx 0.2$ MPa at a frequency of 15 Hz. The net refrigeration capacity at about 120 K was 10 W. It is not clear from their work what fraction of the gas passes through the orifice.

Optimum adjustments gave a low temperature of 60 K with $P_h = 1.24$ MPa and $P_l = 0.71$ MPa ($P_h/P_l = 1.75$) and a frequency of 9 Hz. The orifice adjustment that gave 60 K was such that the pressure variation in the reservoir volume was about 0.9% of the average pressure. It was then calculated that mass flowing in and out

**Driven at resonance by piston (100-500 Hz)**
**Cooling near center-pressure node**
**Heating near ends-pressure antinode**
**Little useful cooling**

**Add plates for phase shift**

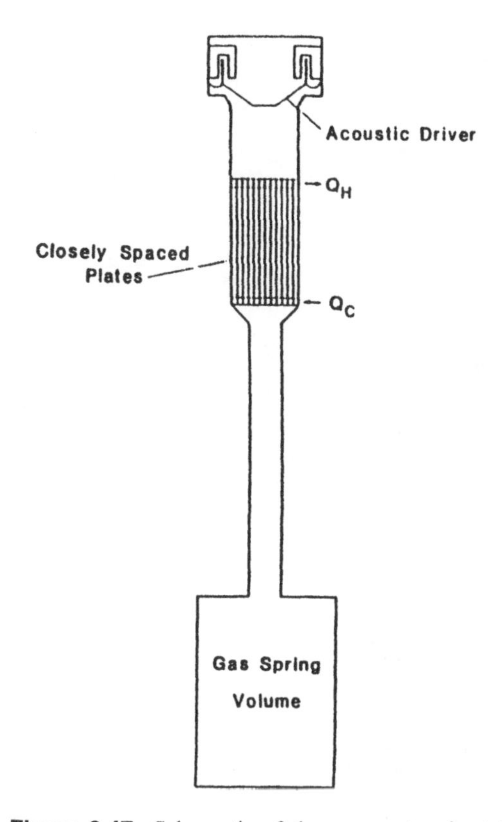


**Figure 6.47** Schematic of the resonant pulse tube (thermoacoustic) refrigerator.

of the reservoir volume was about half the total mass flow at the cold end of the pulse tube. Gross heat pumping values from the cold to hot ends were about 18 W at 60 K with those conditions.

**MultiStage Pulse Tubes.** It is easy to add any number of stages, still using one pressure wave generator, for the basic and orifice types of pulse tubes. The hot end of the second-stage pulse tube is attached to the cold end of the first stage. A second regenerator is added below the first, with some gas passing on to the second stage. (It is not entirely clear how a multistage resonant pulse tube would be best configured.) The only experiments done to date on multiple stages have been the work of Longsworth. He reached 79 K with two stages of a valved basic pulse tube and 32 K with four stages using $P_h \approx 2.0$ MPa and $P_l \approx 0.68$ MPa (Longsworth, 1967). It is interesting to speculate on the low-temperature limit with about four stages of the orifice pulse tubes. It may be possible to reach 10 K if sufficiently good regenerators can be made.

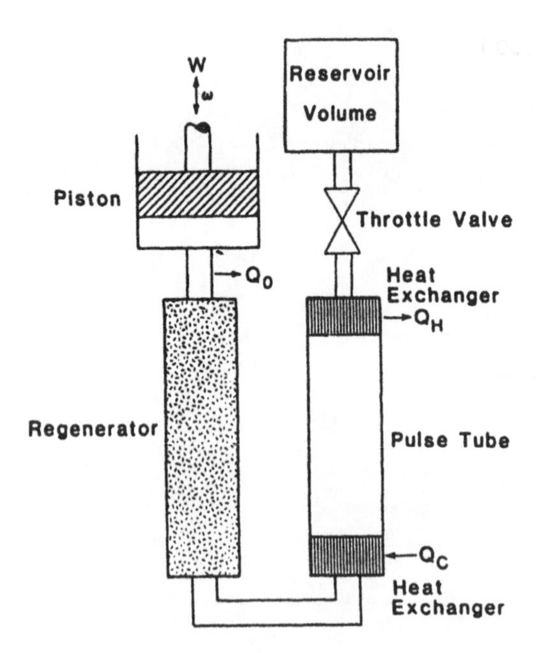


**Figure 6.48** Schematic of the orifice pulse tube refrigerator.

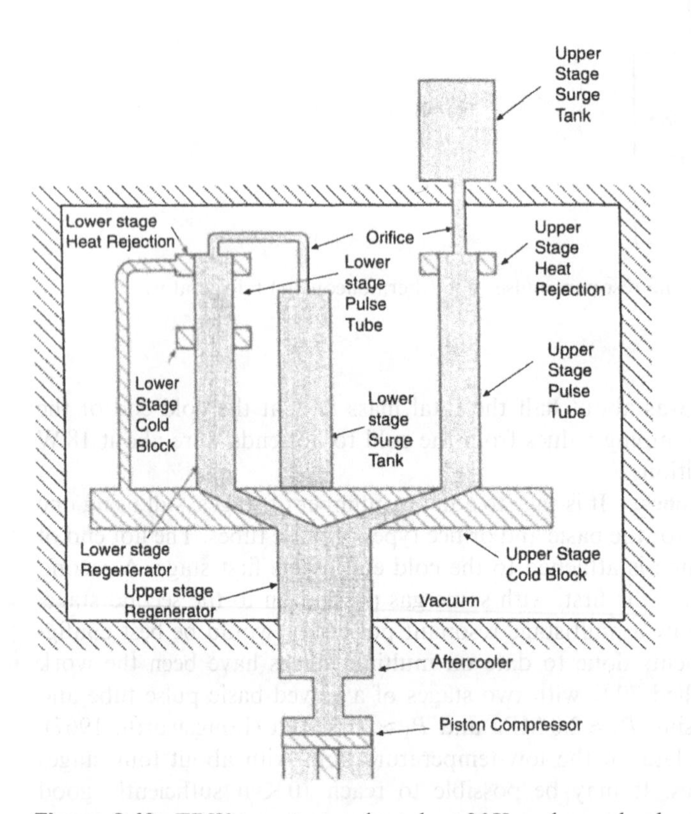


**Figure 6.49** TRW two-stage pulse tube—26 K under no load.

## 9.8. Summary

The orifice pulse tubes can now reach 60 K in a single stage, and they offer a viable alternative to Stirling and Joule–Thomson refrigerators for situations where high reliability is needed. Of the three types of pulse tube refrigerators, the orifice type is the most promising. Advantages over other cycles are:

1. Only one moving part, and that is at room temperature.
2. Uses modest pressure and pressure ratios.
3. Possible to achieve temperature ratios of greater than 4 in one stage.
4. Works on ideal gas, which implies one fluid for all temperatures.
5. Large orifice will not collect impurities at high-temperature part of cycle.
6. Good intrinsic efficiency.
7. Several stages can be operated from the same pressure wave generator.

The one disadvantage is the low refrigeration rate per unit mass flow, which means that better regenerators are required. Further studies of intrinsic efficiencies, refrigeration capacities, regenerator performance, and compressor designs are needed to better understand and improve the overall system performance.

## 9.9. Comparison with Other Refrigerators

Radebaugh (1990) wrote in *Advances in Cryogenic Engineering*, Vol. 35:

> Because it has only one moving part, the pulse tube refrigerator offers the potential for greater reliability than the Stirling cryocooler. Vibration caused by a moving displacer is also eliminated. Joule–Thomson refrigerators also have no low-temperature moving parts, but the mechanical compressor is notoriously unreliable because it must have inlet and outlet valves; it [J–T] must work with high absolute pressures of 10–20 MPa, and it must work with pressure ratios of 100–200. The intrinsic efficiency (see Fig. 6.50) is better than that of a Joule–Thomson refrigerator using a single-component fluid.

It appears to be about a factor of 1.5–2 less efficient than the Stirling expansion process. The estimated specific power input of about 100 W/W at 80 K for a 75% efficient compressor is greater than a typical value of about 40 W/W for a Stirling cryocooler with the same compressor. However, the pulse tube results were obtained with systems where heat rejection to the ambient was poor. The hot end temperatures were usually about 45°C during those measurements. Further efforts to reduce the specific power input are under way and are of much interest to system managers and designers.

# 10. MAGNETOCALORIC REFRIGERATION

Figure 6.51 shows other sources of entropy useful for refrigeration, particularly magnetic systems. The entropy of the disordered paramagnetic phase can be quite high. In addition, this disordered phase in some cases can exist down to very low temperatures before ordering to a ferromagnetic or antiferromagnetic phase takes place and the entropy is removed. The nucleus has a small magnetic moment and hence a low ordering temperature. The entropy of the copper nucleus is shown in Fig. 6.51 for the cases of applied magnetic fields of 0 and 10 T. Strictly speaking, the applied magnetic field $H$ has SI units of amperes per meter (A/m). In this book we are using the

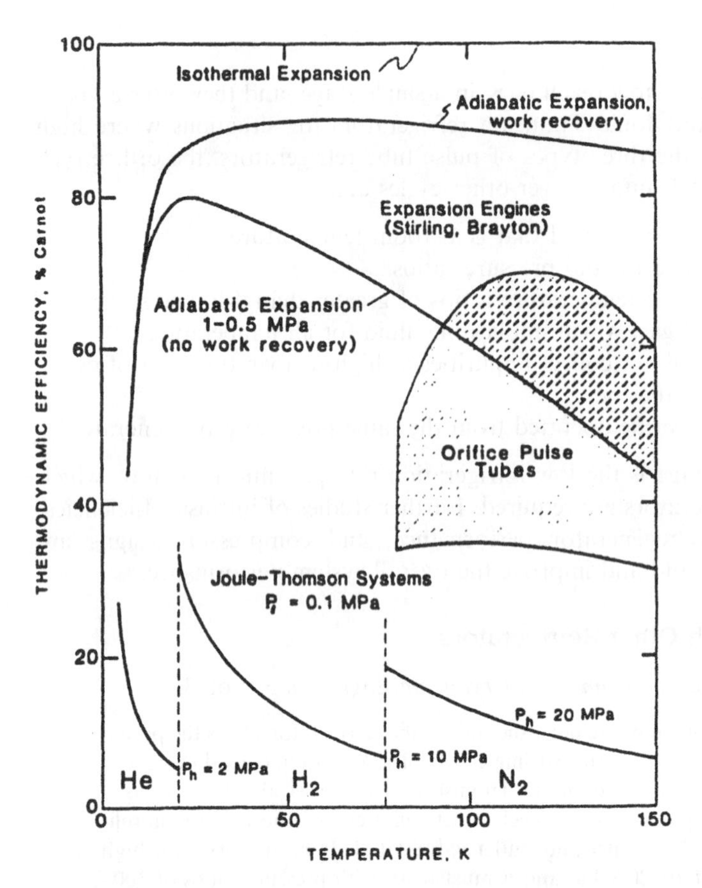


**Figure 6.50** Comparison of intrinsic cooling efficiency for various refrigerators.

magnetic flux density $B$ with SI units of tesla ($1\,T = 10^4\,G$) as the applied field in free space, i.e., $B = \mu_0 H$, where $\mu_0$ is the permeability of free space. Ordering for $B = 0$ takes place at about $10^{-7}$ K. Thus, adiabatic demagnetization from about 5 mK and 10 T will result in a nuclear temperature of about $10^{-6}$ K or less.

Magnetic moments of electron spin systems are much higher, which means that ordering temperatures are higher. For a magnetic system, the entropy in the disordered phase is given by

$$S/R = \ln(2J + 1)$$

where $2J + 1$ is the number of possible orientations of the magnetic dipole of angular momentum $J$. The three electronic magnetic systems are shown in Fig. 6.51 at $Ce_2Mg_3(NO_3)_{12}{\cdot}24H_2O$ (CMN), $Gd_2(SO_4)_3 \cdot 8H_2O$, and pure Gd. Values of $J$ for these systems are 1/2, 7/2, and 7/2, respectively. Their ordering temperatures cover the range from about $10^{-3}$ K to room temperature. Cerium magnesium nitrate (CMN) has been used for refrigeration down to 1 mK for many years. The material $Gd_2(SO_4)_3 \cdot 8H_2O$ has an entropy of $S/R = 2.1$ that is easily removed by a field of 10 T at 4 K. Because $Gd_2(SO_4)_3{\cdot}8H_2O$ is a solid with a resulting low molar volume, it may prove to be competitive with helium for refrigeration in the vicinity of 4 K. The system could use the Carnot cycle with time domain operation by using heat switches such as magnetoresistive beryllium (Radebaugh, 1977). Such

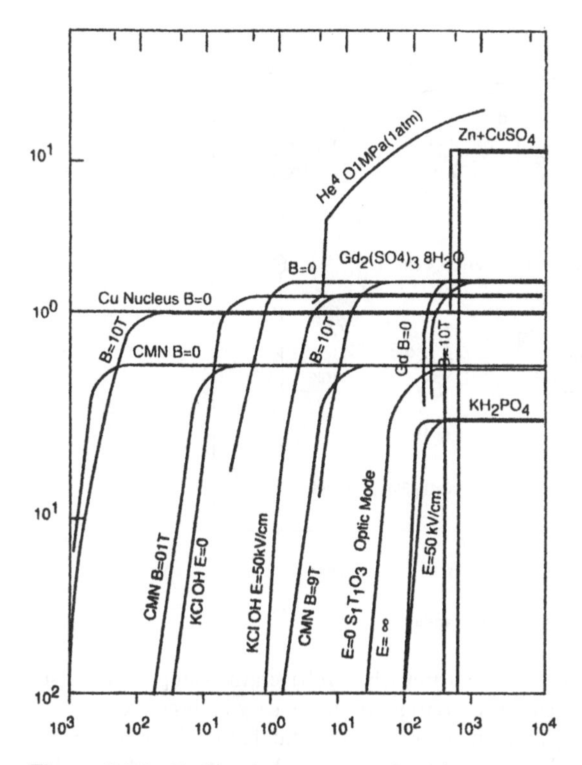


**Figure 6.51** Reduced entropy as a function of temperature for magnetic, dielectric, and electrolytic systems.

a refrigerator would have no moving mechanical parts. Stirling or Ericsson cycle operation may also be used by placing the material on a rotating wheel. Steyert (1978a) discussed designs for Carnot cycle refrigerators using $Gd_2(SO_4)_3 \cdot 8H_2O$ to reach temperatures of about 2 K with the upper temperature at 10 K. One of these uses time domain cycling with beryllium heat switches, and the other uses a rotating wheel for space domain cycling. The wheel is made porous to allow helium gas at 10 K and liquid helium at 2 K to pass through the wheel for good heat exchange. A schematic diagram of the apparatus is shown in Fig. 6.52. In this diagram, the high-field region is at the top of the wheel and the low-field region is at the bottom. Calculated efficiencies are also given in Fig. 6.52 as a function of refrigeration capacity.

A prototype magnetic refrigerator using a Carnot wheel of $Gd_2(SO_4)_3 \cdot 8H_2O$ was built and demonstrated by the Los Alamos group (Pratt et al., 1977). This refrigerator (Fig. 6.53) uses 4 K liquid helium as the upper bath and cools the lower superfluid bath to 2.1 K. One of the major difficulties with it concerns the seal between the two helium baths. The wheel rotates at about 0.5 Hz in a field of about 3 T. Even though it is a magnetic refrigerator, the weak link with respect to reliability is still in the mechanical parts because it uses mechanical work input. The speeds, however, are quite slow. In time domain operation the work input could be in the form of electrical energy for cycling the magnetic field, but large electrical losses may occur.

There are many other magnetic materials suitable for refrigeration at various temperatures. Magnetic refrigeration around room temperature utilizing gadolinium

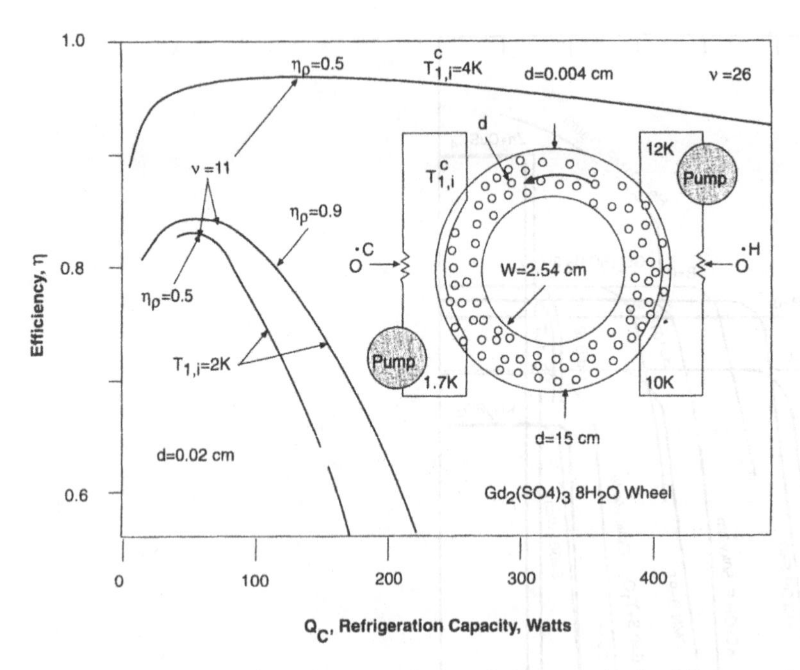


**Figure 6.52** Schematic of a $Gd_2(SO_4)_3 \cdot 8H_2O$ Carnot wheel refrigerator along with its estimated efficiency. Fraction of Carnot efficiency vs. refrigeration capacity is shown for various assumed values of the pump efficiency $\eta_p$ and temperature of cold fluid at the inlet, $T_{1,i}$. Some values of rotation rate $\nu$ (in Hz) required are shown on the curves. A maximum field of 5 T is required for this refrigerator (after Steyert, 1978a).

metal was proposed two decades ago by Brown (1976) of the NASA Lewis Research Center.

The entropy can be changed by about $\Delta S/R = 0.3$ with a 10 T field at room temperature. Note that the entropies shown in Fig. 6.51 do not include the phonon entropies of these solids. The specific heat associated with this additional entropy can be great at room temperature and act as a heat load on the refrigeration cycle. Figure 6.51 shows how a field of 10 T has less and less effect on the magnetic entropy of a system as the temperature increases. This is due to the tendency of the phonons to disorder the magnetic spins. Carnot cycle operation with gadolinium would be limited to temperature differences of less than 14 K. Ericsson cycle operation would follow the $B = 0$ and 10 T curves down to much lower temperatures, so larger temperature differences can be provided with that cycle or with the Stirling cycle. The high phonon specific heat places a severe requirement on the heat exchangers for such cycles. The design of a room temperature refrigerator using a rotating wheel of Gd was described by Steyert (1978b). In this case, the cycle followed is the Ericsson cycle (constant field lines) although Steyert refers to it as the Stirling cycle. A schematic of the device is shown in Fig. 6.54. In the Carnot version in Fig. 6.53, the fluids pass through the wheel in a direction parallel to the axis of rotation. In the Ericsson version in Fig. 6.54, the fluid passes through the wheel in a direction opposite to the rotation. A working model of the device was described by Barclay et al. (1979). With a 3 T magnetic field and a rotational speed of 0.1 Hz, the refrigerator maintained a temperature difference of 7 K with 500 W heat input to the cold side. In this model, water is the heat exchange fluid and the 15 cm diameter wheel contains 2.3 kg of Gd.

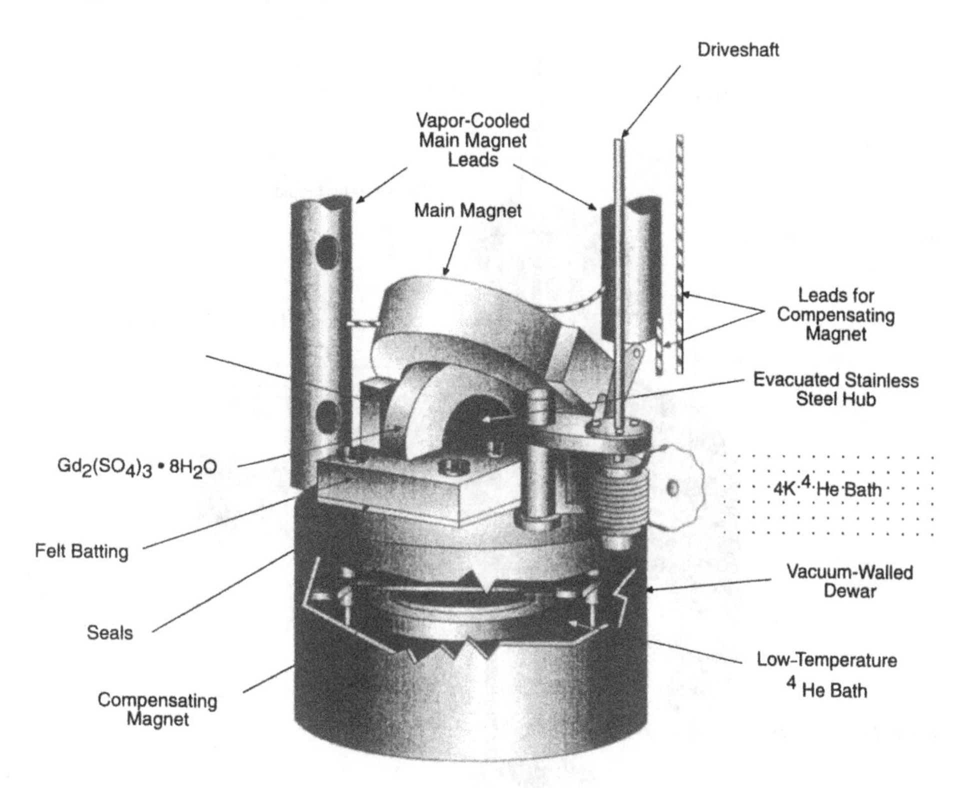


**Figure 6.53** Drawing of a $Gd_2((SO)_4)_3 \cdot 8H_2O$ Carnot wheel refrigerator operating between 2 and 4 K (after Pratt et al., 1977).

Brown (1979) described a laboratory model of a reciprocating magnetic refrigerator for room temperature use that is smaller (0.9 kg Gd) than the Los Alamos rotating model, but achieves a temperature span of 80 K with 6 W heat input to the cold end. The field used is 7 T. Figure 6.55 shows the general layout for an engineering prototype of the refrigerator, which is similar to that originally proposed by

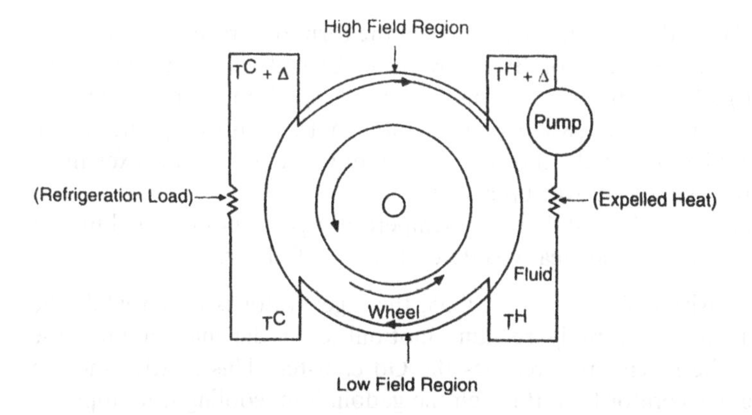


**Figure 6.54** Schematic of a magnetic Stirling or Ericsson cycle refrigerator using a rotating wheel. $\Delta$ is the inherent temperature change of the magnetic material upon entering and leaving the magnetic field (after Steyert, 1978b).

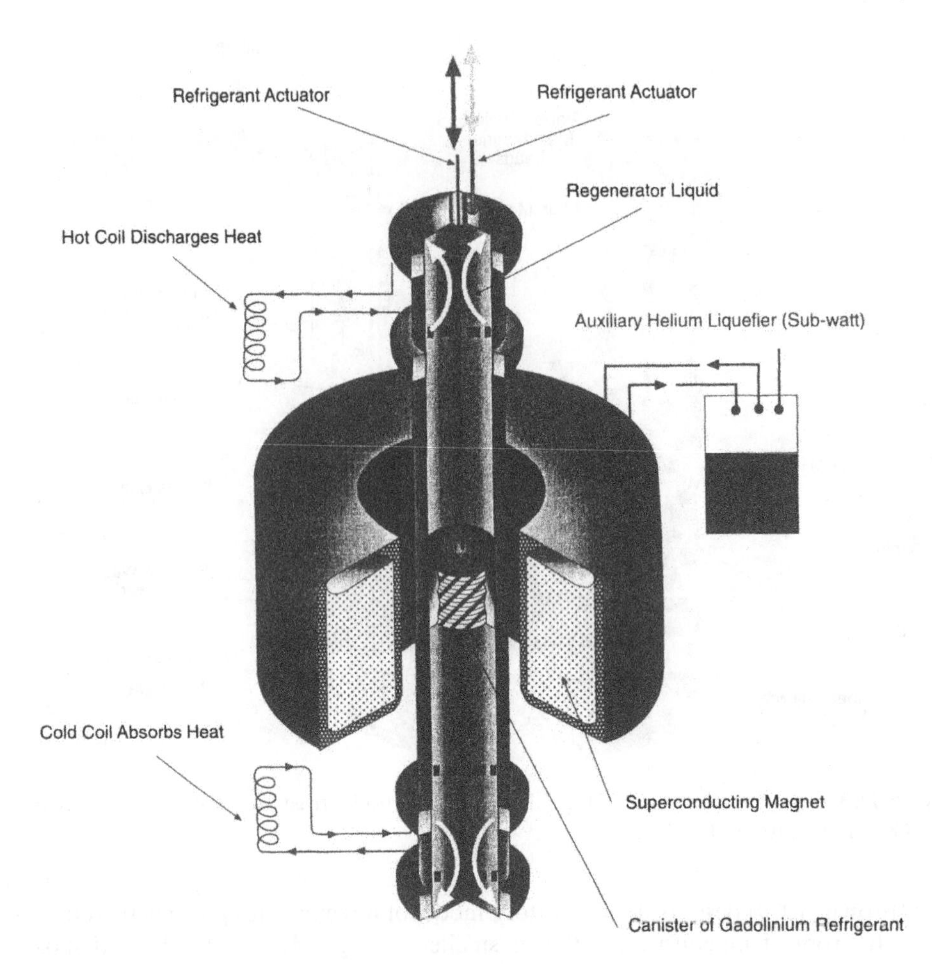


**Figure 6.55** Reciprocating magnetic refrigerator using pure gadolinium around room temperature. Either Stirling or Ericsson cycle operation could be followed in this configuration (after Brown, 1979).

Van Geuns (1966). The Gd shown in Fig. 6.55 is in the form of screens to give better heat transfer, but the experiments of Brown were performed with Gd plates. The porous assembly of gadolinium plates fits inside a 1 m long, 5 cm diameter cylinder containing the regenerator liquid (water plus ethanol) that can move up and down. The regenerator liquid flows freely through the gadolinium plates and exchanges heat with them as the cylinder moves up and down.

After a start-up phase that establishes a temperature gradient in the column of regenerator fluid, the steps in the steady-state cycle are as follows:

1. From the position shown in Fig. 6.55, the regenerator is raised while the gadolinium (Gd) assembly remains motionless in the magnet until the bottom of the regenerator reaches the Gd canister. This relative motion passes the regenerator fluid through the gadolinium, cooling it to approximately the temperature of the bottom (cold end) of the regenerator.
2. The regenerator continues upward, carrying the Gd out of the magnet bore. The Gd experiences a decreasing magnetic field strength and

therefore becomes cooler. In this process, it absorbs heat from the cold end of the regenerator and from the external load because of the fluid circulating in the lower heat transfer loop.

3. When the gadolinium and the regenerator have been completely withdrawn from the magnet, they come to rest. The Gd remains at rest while the regenerator is lowered until its top contacts the Gd assembly. The regenerator fluid thus passes down through the Gd, warming it to approximately the temperature at the hot end of the regenerator.
4. The regenerator continues down, and the gadolinium now moves with it. As the gadolinium enters the magnetic field it is warmed by the "magnetocaloric effect" (not by eddy currents) and discharges heat into the hot end of the regenerator and to the hot external heat exchanger through the upper heat transfer loop.
5. With the gadolinium at rest, the regenerator is raised to repeat the cycle.

The thermodynamic cycle followed by this device is the Ericsson cycle, although, like the Los Alamos group, Brown also refers to it as a Stirling cycle device. Strictly speaking, in a Stirling cycle device the isothermal paths should be connected by constant-magnetization paths instead of constant-field paths.

Though Brown's laboratory refrigerator used a water-cooled magnet that consumed an enormous amount of power, any commercial device would use a superconducting magnet. It seems odd that a superconducting magnet at 4 K would be wed with a refrigerator operating at near room temperature. However, in large sizes the unit could be economically feasible because the power requirement of the 4 K refrigerator would be small compared with that of the overall refrigerator. In some applications, such as air conditioning for large ships, there may already be a helium liquefier on board for a superconducting motor and generator. The principle used here for a room temperature refrigerator could be used at any temperature with the proper choice of refrigerants and regenerator materials.

As mentioned earlier, a magnetic field will increase the entropy of an antiferromagnet until a transition to the paramagnetic phase occurs at some high field. Thus, at low fields, cooling takes place during adiabatic magnetization. The maximum entropy change will always be less than the entropy of the paramagnetic phase in zero field. Since paramagnetic systems exist that have transitions at nearly any desired temperature, it is usually best to use them just above the transition instead of an antiferromagnet below the transition.

## 11. ULTRA LOW-TEMPERATURE REFRIGERATORS

### 11.1. Dilution Refrigerator Fundamentals

Figure 6.56 illustrates a dilution refrigeration process for making ice cream. Ice ($H_2O$) a pure single component, is first made in the ice plant. As in all refrigerators, Fig. 6.56 shows that work ($\dot{W}$) is put into the system and hear ($\dot{Q}$) is rejected. Ice is made at 0°C. Temperatures of −21°C are reached when salt is added to the ice at 0°C. Since pure ice and pure salt are not in equilibrium, the ice melts and dissolves the salt to form a salt solution. The ice cannot absorb enough heat to melt immediately, and so it cools and stays on the equilibrium curve until it reaches the point where all of the salt is dissolved and forms a solution of 29% or less. If enough salt is added the cooling stops at the eutectic point F in Fig. 6.57.

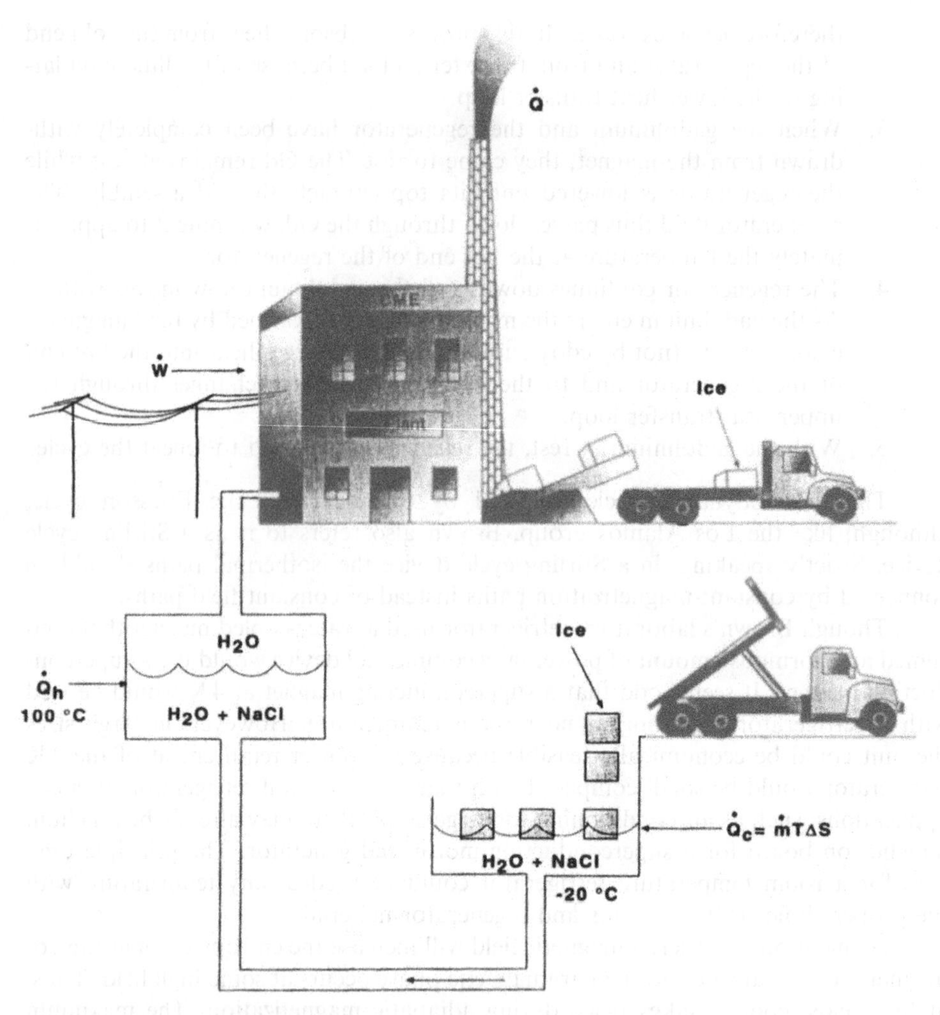


**Figure 6.56** Salt dilution refrigerator.

In a similar manner we can start with pure ice at some temperature between 0 and −21°C and add salt isothermally. The ice is then converted to liquid and the heat of fusion is still more related to the entropy change because the addition of pure salt not in equilibrium phase, causes an irreversible mixing. That irreversible process would significantly decrease the efficiency of a complete refrigeration cycle. In any case we have seen an example of the salt concentration being the generalized force controlling the phase equilibrium temperature and the system entropy. The irreversible processes are not necessarily a problem if an isolated and portable cooling system is needed for a short period of time (refrigerator "battery") and the energy required to complete the cycle is of little concern.

To achieve a reversible cycle with this system, pure salt cannot be used to control the salt concentration because it is not in equilibrium with the system. One reversible scheme that can be used with the salt–water system is like that of the dilution refrigerator. Let the ice and salt solution be contained in a low-temperature container (mixing chamber). The heavier salt solution on the bottom can pass through a heat exchanger to a still, where heat causes the water to vaporize. The water vapor

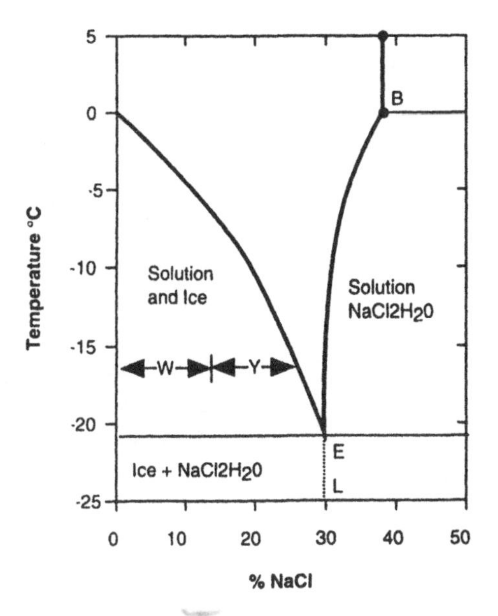


**Figure 6.57** Phase diagram for the mixture NaCl and $H_2O$ (after Zemansky, 1957).

is condensed and frozen. The ice then passes to the mixing chamber. In this entire process, the salt remains stationary and water diffuses through the solution from the mixing chamber to the still, just as $^3He$ does in a $^3He$–$^4He$ dilution refrigerator (see below).

Since the process is reversible, the heat absorbed in the mixing chamber is simply $T\Delta S$, where $\Delta S$ is the entropy change of water between the solid state and the solution state. The entropy of the water in solution is known as the partial entropy. Though the system is certainly not going to make any impact on the household refrigerator market, it does illustrate the concept of refrigeration with mixtures or solutions.

### 11.2. $^3He$–$^4He$ Dilution Refrigerator

London (1951) suggested the use of a solution of helium-3 in helium-4 in a dilution refrigerator operating on the principle of an ice cream maker (Fig. 6.56). In any dilute solution, the solute molecules can be considered to behave like a gas whose pressure and volume correspond to the osmotic pressure and volume of the solution. Dilution of the solution by adding more solvent causes an "expansion" of the solute "gas," and cooling should result. A practical dilution cycle was first developed in 1962.

The density of $^3He$ is less than that of $^4He$. Therefore, at temperatures below 0.87 K, solutions of $^3He$ and $^4He$ exist in two liquid phases, with the $^3He$ phase residing above the $^4He$ phase. The migration of $^3He$ atoms across the liquid/liquid interface is similar to evaporation, like water evaporating from the brine mixture in Fig. 6.56. During the phase transition of $^3He$ into the $^4He$ solution at constant temperature, the entropy increases and heat is absorbed by the $^3He$ to increase its enthalpy. This driving force behind the dilution cycle is analogous to the movement of pure $H_2O$ as ice into the salt–water mixture, increasing the entropy of the $H_2O$.

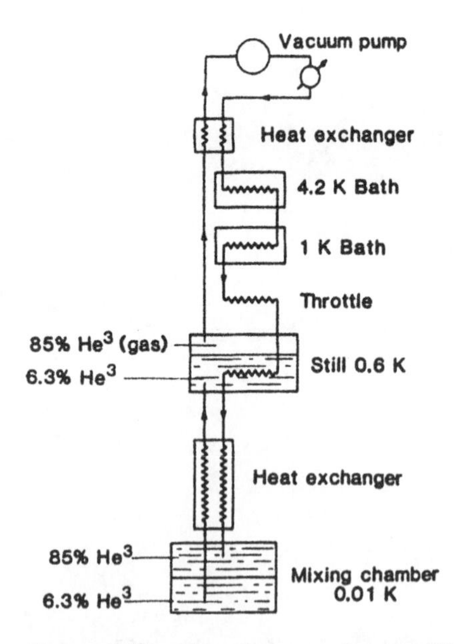


**Figure 6.58** Flow diagram of a simplified helium dilution cycle.

In a continuously operating dilution cycle (see Fig. 6.58), helium gas composed of about 85% $^3$He and 15% $^4$He is compressed with the aid of a vacuum pump, cooled, and sent through a heat exchanger followed by heat exchange in two helium baths, the first at 4.2 K and the second at 1 K. A throttling device in the line provides the necessary temperature adjustment to permit the $^3$He–$^4$He mixture to supply the necessary energy to operate the still. After further cooling in another heat exchanger, the $^3$He–$^4$He mixture is admitted to the mixing chamber. Two liquid phases appear as the temperature drops below 0.87 K with the less dense $^3$He-rich mixture on the top. As this is a continuous process, $^3$He diffuses into the $^4$He phase in the mixing chamber and must be replenished to maintain equilibrium. The lower part of the mixing chamber contains $^4$He in the superfluid form (He II) through which $^3$He easily diffuses. This expansion of the $^3$He into the more dense $^4$He phase provides the refrigeration effect. The $^3$He–$^4$He mixture is returned through the heat exchanger to the still, where the $^3$He is evaporated from the mixture. This vapor is warmed and returned to the vacuum pump, where recompression of the $^3$He-rich mixture completes the cycle. This compressed $^3$He replaces the $^3$He originally dissolved in the $^4$He phase. Temperatures down to about 15 mK have been achieved and maintained for hundreds of hours using this method.

The refrigeration effect for the dilution refrigerator can be determined by an energy balance around the mixing chamber,

$$\dot{Q} = \dot{n}_{He3}(h_{out} - h_{in})$$

where $\dot{n}_{He3}$ is the molar flow of $^3$He in the refrigerator, $h_{out}$ is the molar enthalpy of the $^3$He in the more dense phase leaving the mixing chamber, and $h_{in}$ is the molar enthalpy of the $^3$He in the less dense phase entering the mixing chamber. Below a temperature of 40 mK, enthalpies in units of J/mol can be approximated by

$$h_{out} = 94\,T_{out}^2, \quad h_{in} = 12\,T_{in}^2$$

where $T_{out}$ and $T_{in}$ are the temperatures of the streams leaving and entering the mixing chamber, respectively. Typical refrigeration effects are on the order of 2–10 μW.

The simplest dilution refrigeration can be operated as a single cycle. The $^3He$–$^4He$ mixture is condensed into the mixing chamber, precooled, and then circulated. Temperatures of 4–5 mK have been obtained with such a device because heat is not added continuously with the liquid feed.

## 11.3. Pomeranchuk Cooling

A method for achieving temperatures around 2 mK by using the properties of solid helium-3 was originally proposed by Pomeranchuk and subsequently developed by Wheatley. Pure liquid $^3He$ will not solidify at 0 K unless a pressure of about 3.5 MPa is applied to the liquid. The melting curve has a minimum at 0.3 K, for which the required pressure is close to 3 MPa. Below 0.3 K, the solid has a higher molar entropy than the liquid. In fact, at 20 mK the molar entropy of the liquid is only one-seventh that of the solid. This implies that a substantial cooling effect could be gained by adiabatic solidification under compression at these low temperatures. However, for this process to work, the compression has to be performed without frictional heating. This frictional effect is so significant that at 8 mK 1% conversion of mechanical work into heat reduces the available refrigerating capacity to zero.

A schematic of a Pomeranchuk cooling machine proposed by Sato (1968) is shown in Fig. 6.59. A major problem in the operation of such cooling devices on a large scale is the method of precooling helium-3. In Sato's device, two Leiden dilution refrigerators (LDRs) use $^4He$ to precool the $^3He$. After cooling the liquid below 300 mK, the liquid $^3He$ is sealed into the compressor cell and the bellows are used to mechanically and adiabatically compress the cell.

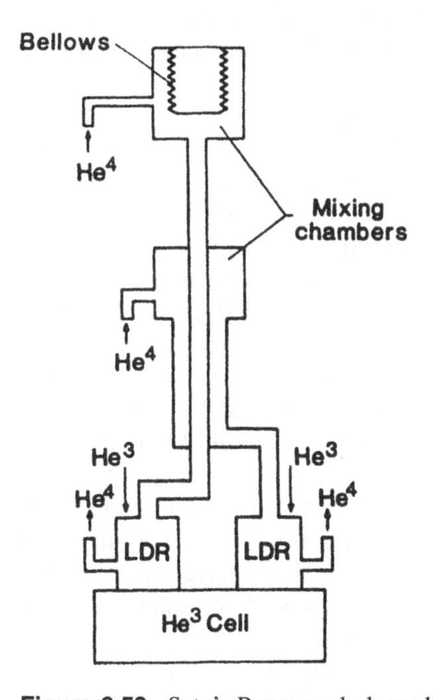


**Figure 6.59** Sato's Pomeranchuk cooling machine.

## 12. VERY SMALL COOLERS

The first major requirement for small ground-based cryocoolers was brought about by the need to refrigerate ground-based parametric amplifiers to 4 K for use in the satellite communications network. Several units meeting this requirement were developed based on the G–M cycle. These units produced a few watts at 3.8–4 K. Further development in amplifier performance led to an amplifier that would perform satisfactorily at 20 K. As a result, a major market for the G–M closed cycle, 20 K cooler evolved. Cryogenic Technology, Inc. has produced thousands of these units, which are in continuous operation in the satellite communications network. This basic G–M cooler is also produced by Cryomech, Inc. and Air Products, Inc. These two companies, as well as Cryogenic Technology, Inc. have built a number of G–M units for specific applications with the addition of a J–T loop to provide a final stage of refrigeration at 4 K. Applications include cooling of computer systems and cooling of superconducting magnets for magnetic separation processing and NMR experiments.

The next major use to evolve was that of cooling infrared (IR) detectors. The initial cooling requirements for IR detectors were 0.25–2 W at 80 K. A number of manufacturers become involved in producing refrigerators for this level of refrigeration. Thousands of open cycle J–T units have been produced as well as several thousand integral Stirling and split Stirling refrigerators. These units are used for cooling military IR detectors and are produced both in the United States and abroad. For instance, each Bradley Fighting Vehicle of the US Army uses eight separate cryocoolers for IR systems.

This requirement for large numbers of coolers for military infrared detectors led to the development of a common module cryocooler (CMC) meeting specific size, weight, and performance requirements. These units are manufactured by a number of US companies and by companies abroad. These companies, in addition to CTI Cryogenics and Air Products, Inc., include Hughes Aircraft, Texas Instruments, H. R. Textron, and Magnavox in the United States; Telefunken in Germany; L'Air Liquide and A. B. G. Semca in France; Hymatic in England; Philips in Holland; Ricor in Israel; and Galileo in Italy.

The third major commercial use is that of cryopumping. Cryopumping is a means of producing a high vacuum by condensing residual gases on cryogenically cooled panels. A number of G–M and Stirling cycle refrigerators were installed on cryopumping systems in the early 1970s. However, the market did not develop fully until coolers were required for semiconductor production.

The general range of cooling required for cryopumping systems is 50–65 W at 80 K and 5 W at 12–15 K. Closed cycle refrigerators for cryopumps in this range are produced by CTI Cryogenics, Air Products, and CVI. In addition, the major vacuum equipment companies produce their own refrigerator systems. These include Balzers High Vacuum, Varian, and Sargent Welch in the United States; L'Air Liquide in France; Leybold-Heraeus in Germany; and Osaka Oxygen, Suzuki Shokan, Ulvac Cryogenics, and Toshiba in Japan.

Although no companies produce many refrigerators that meet the requirement of 1 W at 4 W with a reasonable efficiency and of a size suitable for cooling small superconductive devices, the major manufacturers listed above have the capability to develop such systems. Ground-based cryocoolers, large-scale or small-scale, for use at 4, 20, or 77 K, are virtually off-the-shelf systems.

Cryogenic cooling and storage have been used in space instruments for over 20 years. The cooling needed is typically for temperatures below the boiling point of liquid nitrogen (77 K) and for heat loads of 10 W or less. The primary need for cooling is for IR detectors for astronomy and for surveillance; X-ray and gamma-ray detectors have also been cooled. The storage of cryogenic fluids for the atmosphere of manned spacecraft and for power production in fuel cells is a major use dating back to the late 1950s.

Three basic refrigeration methods are used to meet these cooling requirements: (1) passive thermal radiation to space, (2) storage of cryogenic fluids or solids, and (3) active refrigerators. Passive radiation to space is a simple and reliable method of producing small amounts of cooling. This method has limited applications because the amount of cooling obtainable is very small at cryogenic temperatures—typically fractions of a watt. The practical limit is that the radiator becomes very large and heavy for larger loads.

Storage of fluids or solids has been the mainstay for cryogenic cooling in space. This method employs highly insulated storage tanks. The cooling temperatures obtained range all the way down to less than 2 K in the case of superfluid helium storage. Many cryogenic materials have been used to produce cooling by using the heat of vaporization to absorb heat loads. The practical limitation of cryogenic storage is that the size and weight become large for long-duration missions and for high heat loads. The development of a long-life space cryocooler has been an elusive goal because of fundamental problems related to contamination and wear.

Stirling coolers were used in space to cool gamma ray detectors to about 80 K in the P78 satellite. These coolers were built by the Philips Corporation and employed a linear drive mechanism to achieve several years of operation. Degradation in performance occurred during the mission in the form of steadily rising cooler temperature.

A very promising development of a Stirling-type refrigerator is now taking place following an innovation in Great Britain at Oxford University and the Rutherford Appleton Laboratories in conjunction with British Aerospace Corporation. Operational times of about 20,000 hr have been achieved, and this machine has been slated for use in several space systems. The limitations of Stirling coolers are that temperatures below about 20 K are difficult to achieve with acceptable power input requirements.

Additional benefits accrue to the J–T approach in comparison to the Stirling and Brayton machines due to the fact that J–T coolers produce liquid cryogens. High, short-term peak heating loads and variable heating loads can be accommodated at the constant temperature of the boiling liquid refrigerant. This is not possible with other systems that produce only a cold gas and is sometimes very important to space instrument cooling.

Some estimates of the size and weight of spaceborne cryocoolers are shown in Figs. 6.60 and 6.61.

Sloane (1900) reminded us of the "battle of the cubes vs. the squares" to point out that volume increases as the cube and surface area for heat loss as the square. Thus, it is inevitable that small coolers have more heat loss per mass than their larger counterparts. Figure 6.62 shows how small the net refrigeration capacity can actually be.

## 13. SUPERCONDUCTORS AND THEIR COOLING REQUIREMENTS

The three important characteristics of the superconducting state are the critical temperature ($T_c$) (that is, the temperature below which the superconducting state

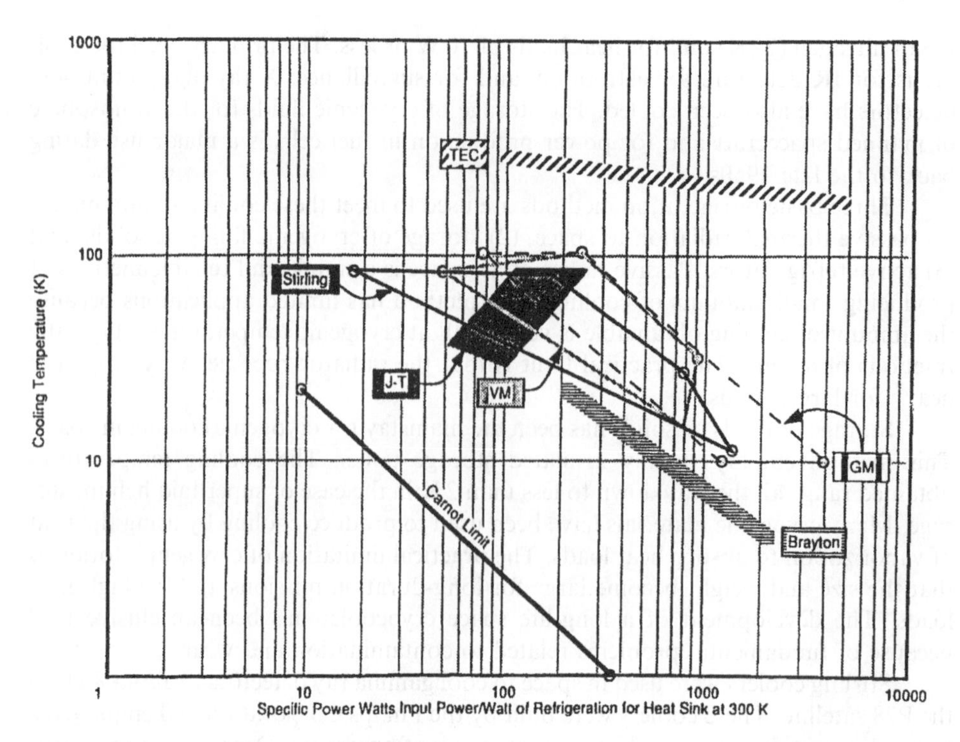


**Figure 6.60** Specific power of cryocoolers.

occurs), the critical magnetic field ($H_c$), and the critical current ($I_c$). It is well known that temperatures above $T_c$ quench superconductivity. Perhaps, it is less well known that the superconducting state is also quenched by an external magnetic field, $H > H_c$, or an electric current, $I > I_c$. The highest temperature possible for the transition to the superconducting state, $T_c$, occurs when both $H$ and $I$ are zero. For useful values of $H$ and $I$, the operating temperature of the superconductor must be less than $T_c$. A useful guideline is that the absolute operating temperature $T$ should be about one-half the transition temperature. This temperature level provides useful values of $H$ and $I$ as well as a margin below $T_c$ to accommodate any localized transient heating of the superconductor that might occur. (The actual relationships among $T$, $H$, and $I$ are complex, but the above guideline is nonetheless handy.)

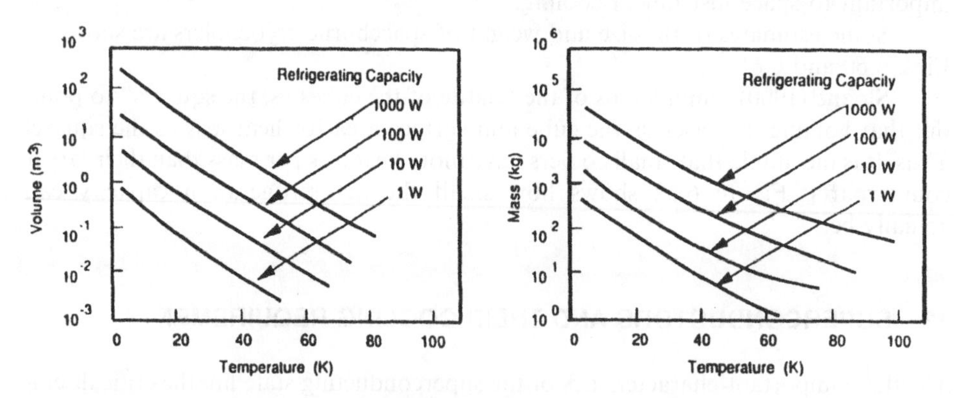


**Figure 6.61** Volume and mass of cryocoolers (complete operational system).

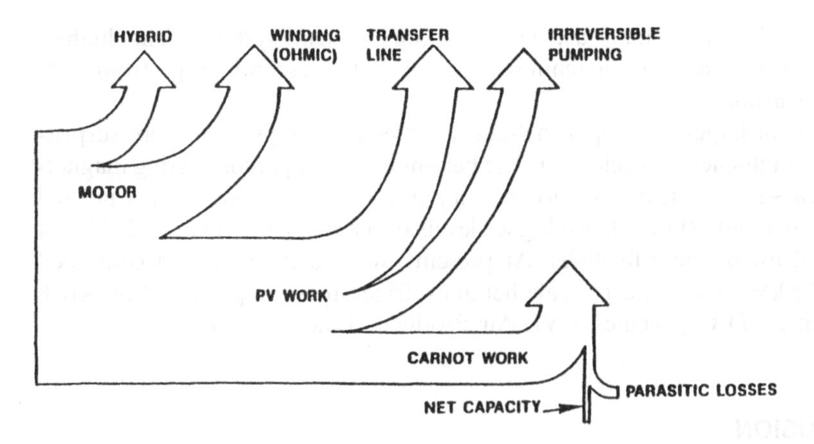


**Figure 6.62** Stirling coolers.

Accordingly, the ordinary (type II) superconductors with $T_c$ around 20 K could typically be usefully operated at 10 K. Since no stable, liquid refrigerant exists at 10 K, the operating temperature is usually lowered to 4 K, the temperature of liquid helium. By the same rule, the new class of superconducting compounds, with $T_c \sim 90$ K, would be useful if cooled to 45 K. Cooling to the temperature of liquid hydrogen, 20 K may not be unreasonable because hydrogen is much less expensive than the 40 K liquid refrigerant, neon. Similarly, to accommodate a useful magnetic field and electric current, the yet undiscovered room temperature (300 K) superconductors would be cooled to 150 K. For practical reasons, the temperature of liquid nitrogen, 77 K, would probably be chosen. Therefore, superconductors, old, new, and yet undiscovered, require cryogenic temperatures (4, 20, and 77 K, respectively) to have useful properties.

## 14. CRYOCOOLERS

Cryocoolers are often rated by the available refrigeration capacity measured in watts. To be meaningful, however, it is necessary to specify not only the refrigeration capacity but also the temperature at which the refrigeration is available. A cryocooler with a capacity of 1 W at 4 K (liquid helium temperature) is very different from a cryocooler having a capacity of 1 W at 77 K (liquid nitrogen temperature). Thus, the level of refrigeration (4 K, helium; 20 K, hydrogen; 77 K, nitrogen) is specified as well as the capacity at that level (usually in watts).

Another important parameter is the power input or work required to achieve refrigeration, usually measured in watts of input power per watt of useful refrigeration ($W_{input}/W_{cooling}$). This figure is closely related to the efficiency of the refrigerator. The efficiency of presently available cryocoolers ranges from a minimum of less than 1% to a maximum of nearly 50%. Efficiency depends more on the scale of the machine than on the temperature level or thermodynamic cycle employed, and therefore the size of the cryocooler is important to this discussion.

The small machines used for electronic applications have the lowest efficiencies. This is because nearly all the refrigeration generated is consumed in cooling the low-temperature parts of the machine itself. The surplus or useful refrigeration available from these units is very small, only fractions of a watt. Applications requiring a larger useful refrigeration load use larger, more efficient systems. The highest efficiencies are found in large machines used for liquefiers and range from 20% to 50% of the Carnot value.

The status of large-scale liquid helium facilities is perhaps even more surprising. Most large high-energy accelerators use helium-cooled superconducting magnets simply (1) because it is cheaper to do so compared to electric power or a normal conductor magnet and (2) to achieve higher levels of magnet performance. Table 6.4 shows a partial list of these facilities. At present, these facilities have a combined capacity of 82.5 kW at 4 K. Large-scale helium refrigerators are produced by Koch Process Systems, CTI Cryogenics, CVI, Air Products, Linde, and others.

## 15. CONCLUSION

An earlier comparison of low-temperature refrigeration systems examined the ratio of $Q/W$ for a Carnot refrigerator. A comparison of the inverse of this ratio and the $W/Q$ quantity for an actual refrigerator indicates the extent to which an actual refrigerator approaches ideal performance. In Fig. 6.63, this approach to ideal performance is plotted as a function of refrigeration capacity for actual refrigerators and liquefiers. The capacity of the liquefiers included has been converted to equivalent refrigeration capacity by determining the percent of Carnot performance these units achieved as liquefiers and then calculating the refrigeration output of a refrigerator operating at the same efficiency with the same input power. In some instances, equivalent refrigeration capacity was given by the manufacturers and was used directly. In all instances, the input power was taken as the installed drive power, not the power measured at the input to the drive motor.

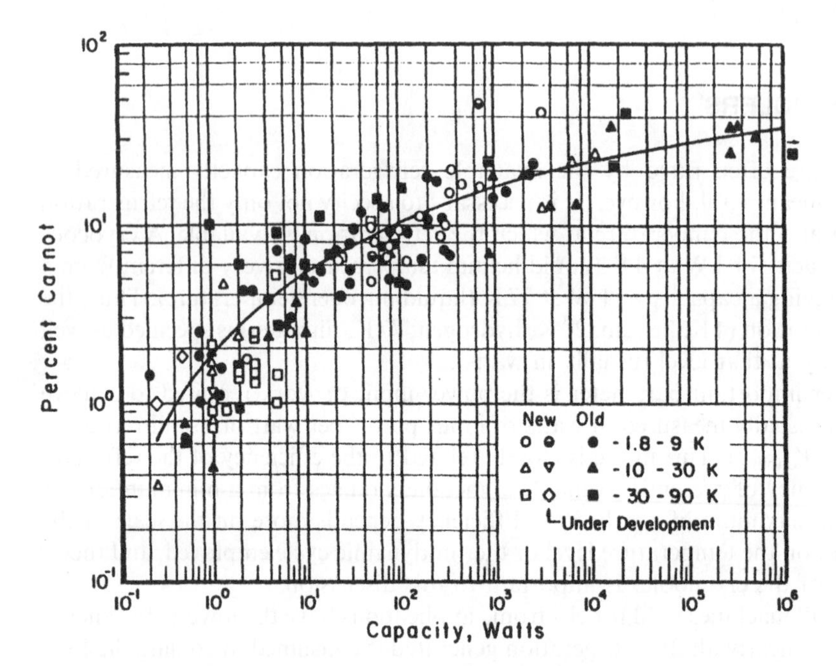


**Figure 6.63** Efficiency of low-temperature refrigerators and liquefiers as a function of refrigeration capacity (Strobridge, 1974).

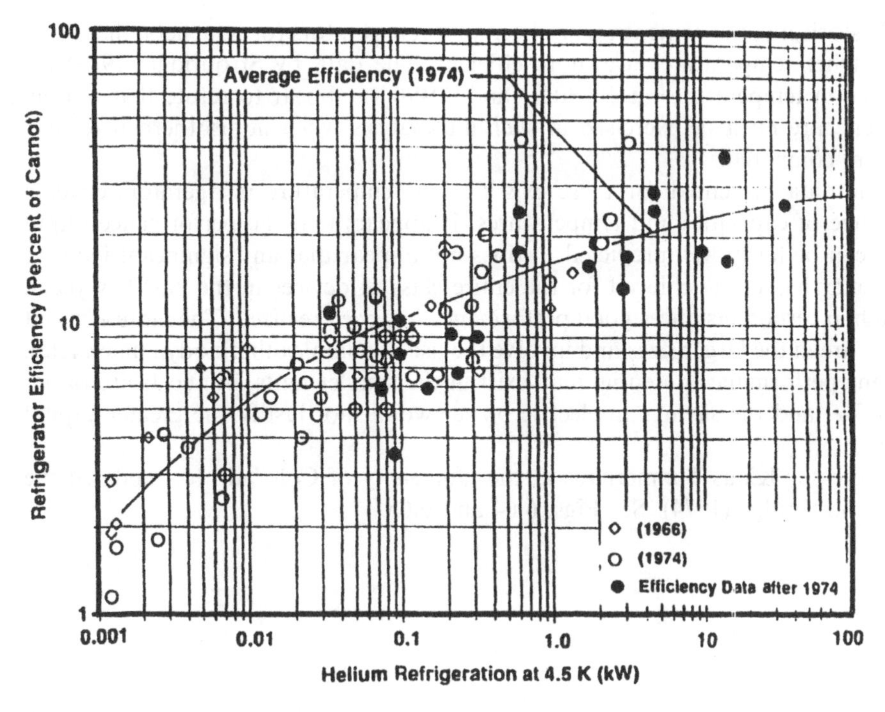


**Figure 6.64** Average efficiency (Strobridge, 1974).

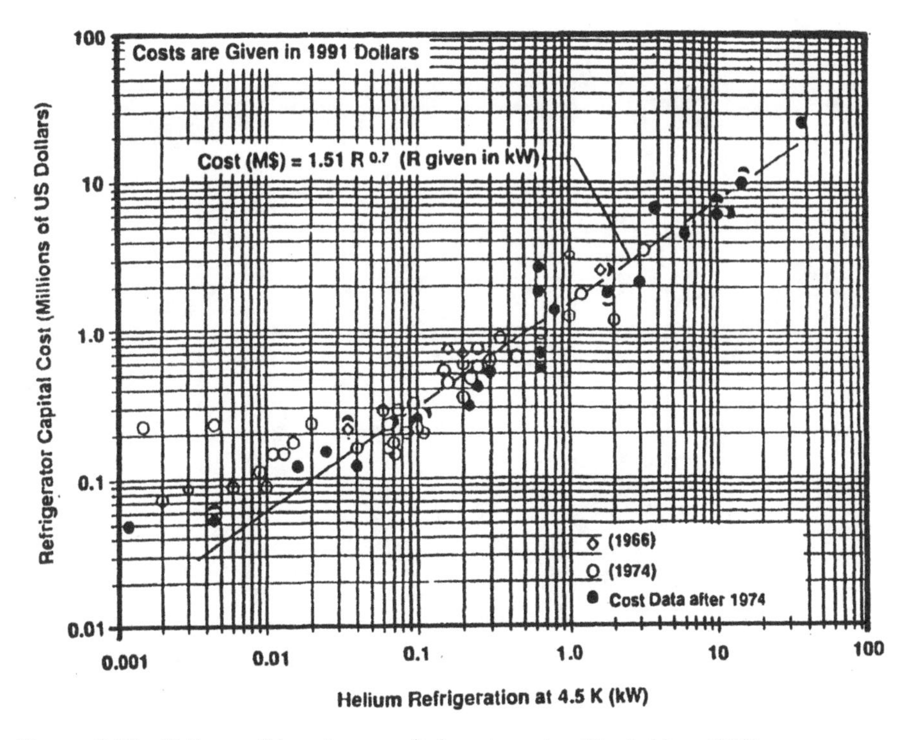


**Figure 6.65** Helium refrigerator—capital cost vs. size (Strobridge, 1974).

Historically, the contention has been that refrigerators (or liquefiers) that operate at higher temperatures are more efficient. The data by Strobridge (1974) for refrigeration temperatures at 10–30 K (and 30–90 K) appear to refute that notion. However, care must be exercised in such an analysis when nonisothermal refrigerators are compared.

The data presented for these existing low-temperature refrigerators cover a wide range of capacities and temperatures. It appears that the performance (input power) can be fairly well predicted. It does not appear that any significant increase in efficiency has been achieved for the entire class of devices in the past few years, although in certain instances good performance has been realized. The more efficient facilities are in the larger sizes and involve the use of complex thermodynamic cycles. The same performance potential may exist in the smaller units, but costs are rather prohibitive, and the savings in electric power would not justify the greater capital outlay.

Professor Klaus Timmerhaus at the University of Colorado has updated the work of Strobridge (1974). See Figs. 6.64 and 6.65.

# 7
# Insulation

## 1. INTRODUCTION

The great American cryogenic engineer Russel B. Scott once noted that many cryogenic applications require a perfection of thermal insulation unapproached in any other field. For instance, in some experiments, in which temperatures of 0.01 K were produced by adiabatic demagnetization, a 2.5 g specimen was so well insulated from its surroundings that it received heat at a rate of only about 3.6 ergs/min. This may be better appreciated when the energy is expressed in more easily visualized terms: the specimen would produce more energy upon being dropped 1 mm than it would receive by heat transfer in an hour.

Even in relatively large-scale equipment, the heat flow must be kept very small to conserve refrigeration or to preserve liquids having small heats of vaporization. For example, there is a commercial storage vessel holding 50 L of liquid helium with an evaporation loss of only 0.25 L/day. Translated into terms of energy flow, this is less than 0.01 W.

The choice of insulation for a particular application is a compromise among economy, convenience, weight, ruggedness, volume, and insulation effectiveness. Thus, although the major part of this chapter is devoted to the principles and techniques of high quality insulation, some attention is given to less perfect insulation methods that may be desirable for special applications.

The effectiveness of a thermal insulation is judged on the basis of thermal conductivity. *Thermal conductivity* is defined as the property of a homogeneous body measured by the ratio of steady-state heat flux (time rate of heat flow per unit area) to the temperature gradient (temperature difference per unit length of heat flow path) in the direction perpendicular to the area.

To be meaningful, thermal conductivity must be identified with respect to the mean temperature, and it is usually measured with a material exposed to a definite temperature difference. Because of the complex interactions of several heat transfer mechanisms, the term "effective" or "apparent thermal conductivity" is employed to distinguish this value from an *ideal* thermal conductivity corresponding to a very small temperature difference of a pure material.

In general, the performance of a thermal insulation depends on the temperature and emittance of the boundary surfaces, its density, the type and pressure of gas contained within it, its moisture content, its thermal shock resistance, the compressive loads applied to it, and the effects of mechanical shocks and vibrations.

*Emissivity* is the ratio of emission of radiant energy by an opaque material to the emission of a perfect emitter ( a blackbody) at the same temperature. Emittance, a practical measure of a material's radiant emission, is distinct from *emissivity*, a more nearly ideal property obtained by measuring a material whose surface has been carefully prepared to be optically flat.

## 2. HEAT TRANSFER

During cooldown, the heat capacity of the insulation must be considered. Heat can flow through an insulation by the simultaneous action of several different mechanisms:

- Solid conduction through the materials making up the insulation and conduction between individual components of the insulation across areas of contact.
- Gas conduction in void spaces contained within the insulation material.
- Radiation across the void spaces and through the components of the insulation.

Because these heat transfer mechanisms operate simultaneously and interact with each other, it is not possible to superimpose the separate mechanisms to obtain an overall thermal conductivity. The thermal conductivity of an insulation is not strictly definable analytically in terms of variables such as temperature, density, and physical properties of the component materials. It is therefore useful to refer to *apparent* thermal conductivity, which is measured experimentally during steady-state heat transfer and evaluated from the basic Fourier equation:

$$\dot{Q} = kA\frac{T_1 - T_2}{L} \tag{7.1}$$

where $\dot{Q}$ is the rate of heat flow through the material, $k$ the apparent thermal conductivity, $A$ the area, $T_1$, $T_2$ the boundary temperatures and $L$ the thickness of the insulation.

### 2.1. Solid Conduction

The contribution of solid conduction can be reduced by breaking up the heat flow paths within an insulation. This is accomplished by using either finely divided particles or fine fibers so that a resistance to heat flow is formed at the surface of each insulation component. In foams, small pore sizes and thin wall structures are desirable for the same reason. To maintain these circuitous heat flow paths, the contract area between individual particles or fibers must be reduced to point contacts, whose resistance depends upon the deformation caused by the compressive load imposed on the insulation. Hollow particles or those with a structured interior tend to exhibit a lower thermal conductivity than solid particles. Fine screens separating radiation shields in multilayer insulations, small-diameter powders and fibers, or disks of metal foils dusted with fine powder also exhibit markedly lower thermal conductivity because of increased contact resistance.

### 2.2. Gas Conduction

The low thermal conductivity of evacuated thermal insulation can be largely attributed to the removal of gas from the void spaces within the insulation. The gas itself is a dominant contributor to heat flow through the insulation because it provides good thermal contact between the components of the insulation.

The degree of vacuum necessary to achieve the desired insulating effectiveness can be found by considering the mechanism by which heat flows through a gas bounded by the surfaces of the insulation components. According to the kinetic theory of gases, the thermal conductivity of a gas is proportional to the mean free path of the molecules and the gas density.

The effect of gas conduction can be divided into two separate regions:

1. The region ranging from atmospheric pressure down to a few torr (1 torr = 1 mm Hg), in which gas conduction is independent of pressure.
2. The region at pressures below a few torr, in which gas conduction depends on pressure.

The transition from one type of gas conduction region to the other depends on the dimensions of the individual components of the insulation. For example, the diameter of a particle, the spacing between radiation shields, and the diameter and arrangement of pores determine the pressure required to obtain the mean free path of the gas molecules within the desired gas conduction region. A *decrease* in gas pressure results in an *increase* of the mean free path until the mean free path is on the order of the distance between insulation components. A further decrease in pressure can no longer influence the mean free path of the gas molecules being constrained by the insulation components. However, the density of the gas is directly proportional to the pressure and will continue to decrease as gas molecule, which then transfer heat directly between adjacent components without suffering collision, are removed.

The larger the voids between the components of the insulation, the lower the pressure required to approach the pressure-dependent gas conduction region. Because the degree of vacuum required to obtain a high insulating effectiveness in evacuated powder, fiber, or multilayer insulations is difficult to achieve, voids between components of the insulation should be small enough that gas conduction can be greatly reduced. For example, it is desirable to use a range of sizes of powder so that large voids will be filled by smaller particles and small voids in turn by still smaller particles. In fiber insulations, submicrometer sized fibers will, for the same reasons, provide a more effective insulation performance at high gas pressures.

In the pressure-dependent gas conduction region, the thermal conductivity of the insulation still has a finite value because heat can be transferred by residual gas conduction, by solid conduction within and through the components of the insulation, and by radiation across the voids and through these components. The contribution to the heat flow by gas conduction in an insulation and the mechanisms for transferring heat between adjacent particles have been the subject of many detailed investigations. These investigations have shown that for fine particles up to 0.004 in. in diameter, a reduction in gas pressure to about $10^{-2}$ mm Hg is sufficient to decrease gas conduction to a negligible value.

Cryogenic insulation can be conveniently subdivided into five categories: (1) vacuum, (2) multilayer insulation, (3) powder and fibrous insulation, (4) foam insulation, and (5) special-purpose insulations. The boundaries between these general categories are by no means distinct but offer a framework within which the widely varying types of cryogenic insulation can be discussed. We begin with vacuum insulation.

## 3. VACUUM INSULATION

James Dewar, the first to liquefy hydrogen, was also the first to use vacuum insulation. His invention, the double-walled glass vessel with a high vacuum in the space between the walls, is a common household article in addition to being extensively used in research laboratories. The inventor is honored by the name "Dewar vessel" or sometimes simply "dewar". The importance of vacuum insulation is evident when it is realized that it can almost completely eliminate two of the principal modes of heat transfer, gaseous conduction and convection. When appropriate measures are also taken to minimize heat transfer by radiation and conduction by solid structural members, vacuum insulation is by far the most effective known. Two common types of vacuum-insulated vessels (dewars) are illustrated in Fig. 7.1. Because of its importance, a large part of this chapter is devoted to a discussion of high-vacuum insulation. First let us examine the principal modes by which heat enters a vacuum-insulated low-temperature vessel and show how the heat flow can be computed.

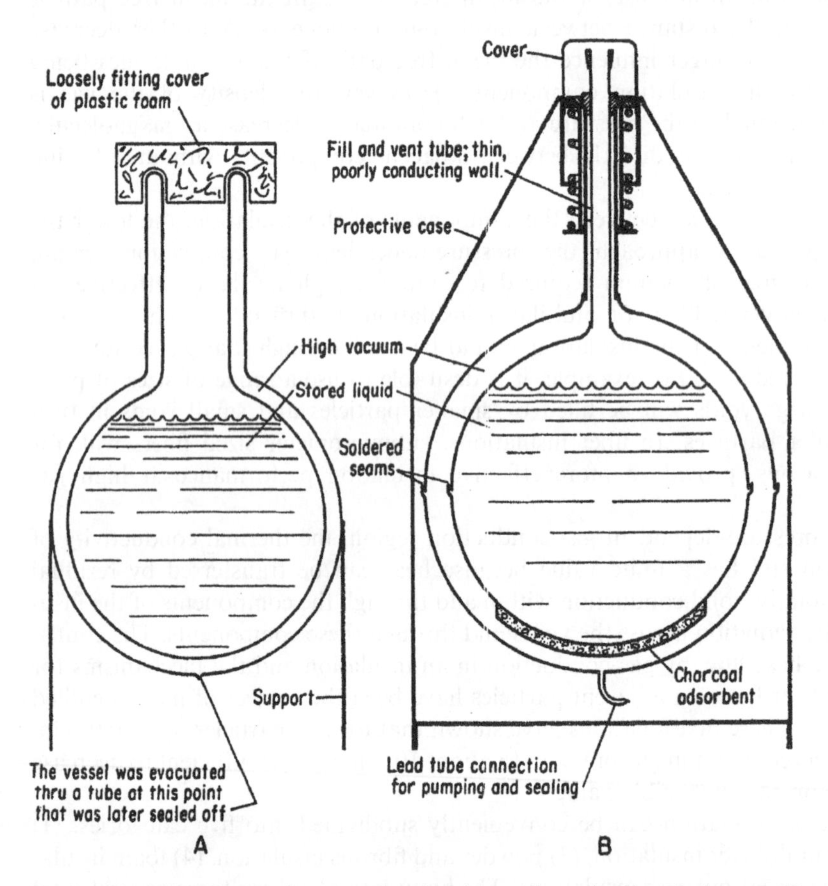


**Figure 7.1** Two types of vacuum-insulated containers for liquefied gases. Vessel A is Pyrex glass. Surfaces of the glass facing the vacuum space are silvered to reduce heat transfer by radiation. Vessel B is metal. Spheres are copper, with surfaces facing the vacuum space cleaned to achieve the high intrinsic reflectivity of copper.

### 3.1. Heat Transfer by Radiation

The rate at which a surface emits thermal radiation is given by the Stefan–Boltzmann equation

$$W = \sigma e A T^4 \tag{7.2}$$

where $e$ is the total emissivity at temperature $T$; $A$ the area and $\sigma$ is the constant having the value $5.67 \times 10^{-12}\,\mathrm{W/(cm^2\,K^4)}$

The net exchange of radiant energy between two surfaces is given by the expression

$$W = \sigma E A(T_2^4 - T_1^4)$$

where subscripts 1 and 2 refer to the cold and warm surfaces, respectively. A is an area factor. In the case of cylinders or spheres, it will be taken as the area of the enclosed (inner) surface; in the case of parallel plates, it is obviously the area of either surface. $E$ is a factor involving the two emissivities. For spheres and cylinders, its value depends on whether the reflections at the enclosing surface are specular (mirrorlike) or diffuse (i.e., with intensity proportional to the cosine of the angle between the direction of emission and the surface). The mode of reflection at the enclosed surface is immaterial, and for parallel plates the mode of reflection at both surfaces is immaterial. Table 7.1 gives values of $E$ in terms of individual emissivities and geometries for gray surfaces, that is, surfaces for which the emissivity is independent of wavelength. The assumption of grayness is not strictly correct but can be shown to introduce negligible error in practical situations.

The formula giving the value of $E$ for long cylinders and concentric spheres (diffuse reflection) is a good approximation for practical vessels such as short cylinders with elliptical or flat ends or for other conservative shapes if $A_1$ and $A_2$ are taken as the total areas, facing the insulating space, of the inner and outer surfaces, respectively. The mode of reflection at practical surfaces is usually not known. It is often assumed that the reflection is diffuse rather than specular, but this assumption may not be justified for the good reflecting surfaces and long wavelengths involved in cryogenic applications. In the absence of information on this point, the formulas serve only to define the limits within which $E$ must lie. For the common case, $e_1 \cong e_2 \ll 1$, the ratio $E$ (specular)/$E$ (diffuse) equals $(1/2)(1 + A_1/A_2)$ and obviously ranges from 1/2 to 1. In general, the heat transfer where specular reflection is involved is equal to or less than that for diffuse reflection.

Table 7.2 also gives the minimum recorded emissivities for some selected cryogenic materials but at different temperatures (Corruccini, 1957; Kropschot, 1959,

**Table 7.1** Values of the Factor $e$ for Gray Surfaces[a]

| Geometry | Specular reflection | Diffuse reflection |
|---|---|---|
| Parallel plates | $\dfrac{e_1 e_2}{e_2 + (1 - e_2)e_1}$ | $\dfrac{e_1 e_2}{e_2 + (1 - e_2)e_1}$ |
| Long coaxial cylinders ($L \gg r$) | $\dfrac{e_1 e_2}{e_2 + (1 - e_2)e_1}$ | $\dfrac{e_1 e_2}{e_2 + \frac{A_1}{A_2}(1 - e_2)e_1}$ |
| Concentric spheres | $\dfrac{e_1 e_2}{e_2 + (1 - e_2)e_1}$ | $\dfrac{e_1 e_2}{e_2 + \frac{A_1}{A_2}(1 - e_2)e_1}$ |

[a] $A_1$ and $A_2$ are the respective areas of the inner and outer surfaces. See text.

**Table 7.2** Selected Minimum Total Emissivities[a]

| Surface | Surface temperature, K | | | |
|---|---|---|---|---|
| | 4 | 20 | 77 | 300 |
| Copper | 0.0050 | | 0.008 | 0.018 |
| Gold | | | 0.01 | 0.02 |
| Silver | 0.0044 | | 0.008 | 0.02 |
| Aluminum | 0.011 | | 0.018 | 0.03 |
| Magnesium | | | | 0.07 |
| Chromium | | | 0.08 | 0.08 |
| Nickel | | | 0.022 | 0.04 |
| Rhodium | | | 0.078 | |
| Lead | 0.012 | | 0.036 | 0.05 |
| Tin | 0.012 | | 0.013 | 0.05 |
| Zinc | | | 0.026 | 0.05 |
| Brass | 0.018 | | | 0.035 |
| Stainless steel, 18–8 | | | 0.048 | 0.08 |
| 50 Pb 50 Sn solder | | | 0.032 | |
| Glass, paints, carbon | | | | >0.9 |
| Silver plate on copper | | 0.013 | 0.017 | |
| Nickel plate on copper | | 0.027 | 0.033 | |

[a] These are actually absorptivities for radiation from a source at 300 K. Normal and hemispherical values are included indiscriminately.
*Source*: Kropschot (1959,1963).

1963). These data include normal and hemispherical emissivities and absorptivities because their differences can usually be neglected for engineering calculations.

So far we have been discussing total emissivities without bothering to prefix the qualifying adjective. Occasionally total emissivity data will be lacking for a substance for which "spectral" emissivities have been determined. The latter are simply emissivities determined using monochromatic radiation. Now the total emissivity is the integral over all wavelengths of the spectral emissivity weighted according to the blackbody energy–wavelength distribution function. For a gray body the total and spectral emissivities are equal. Fortunately, metals approximate gray bodies reasonably well at the wavelengths that are important for thermal radiators at temperatures of 300 K or less, that is, wavelengths of a few micrometers or greater. This means that spectral emissivity values in the infrared will be reasonably good approximations to the total emissivity, especially if they are available at wavelengths equal to or greater than $\lambda_{max}$. The "greater than" is specified because more energy is produced at $\lambda > \lambda_{max}$ than at $\lambda < \lambda_{max}$. It is the same fact that metals approximate gray bodies in the infrared that permits us to use the expressions "emissivity of the cold surface" and "absorptivity of the cold surface for radiation characteristics of the warm surface" as if they were interchangeable.

Examination of the available data on low-temperature emissivities discloses certain generalizations:

1. The best reflectors are also the best electrical conductors (copper, silver, gold, aluminum).
2. The emissivity decreases with decreasing temperature.
3. The emissivity of good reflectors is increased by surface contamination.

4. Alloying a metal with good reflectivity increases its emissivity.
5. The emissivity is increased by treatments such as mechanical polishing that result in work hardening of the surface layer of metal.
6. Visual appearance (i.e., brightness) is not a reliable criterion of reflecting power at long wavelengths.

Items 2, 4, and 5 are closely related to item 1 in that the electrical resistivity usually increases with temperature, impurity, and strain.

The classical free-electron theory leads to an expression for reflectivity at long wavelengths in terms of resistivity. This expression is approximately correct for good conductors at room temperature and for poor conductors such as alloys at all temperatures, but it gives reflectivities that are much too high for good conductors at low temperatures where the electronic mean free path becomes large compared with the depth of penetration of the electromagnetic wave into the metal. However, the strong dependence of the reflectivity on surface contamination makes this a field in which precise agreement between theory and experiment is hardly to be expected. We may note parenthetically that a superconductor does not have perfect reflectivity but rather shows little or no difference from the normal state.

As a converse of the "don'ts" implied on the above list of rules, we may state that the best reflecting surface will be a pure good conducting metal that has been annealed and cleaned in some manner that avoids strain, such as by an acid or residue-free organic solvent.

As noted, the net heat transfer by radiation between two surfaces depends on two quantities: (1) the emissivity of the warm surface, and (2) the absorptivity of the cold surface for radiation having an energy–wavelength distribution characteristic of the warm surface. For the sake of conciseness, both of these quantities are often designated as emissivities. Figure 7.2 shows graphically the energy–wavelength distribution for some representative cryogenic temperatures. Of course, the radiant energy decreases very rapidly with falling temperature, and in addition, the wavelength for maximum energy becomes greater with decreasing temperature. This wavelength is given by Wien's displacement formula

$$\lambda_{\max} T = \text{constant}$$

The constant is 2898 μm K.

For completeness, it should be pointed out that the literature contains two principal classes of emissivities: (1) those measured at normal incidence, and (2) those averaged over all angles between normal and grazing with the application of a weighting factor of the co-sine of the angle made with the surface. These are called normal emissivities and hemi-spherical emissivities, respectively. In heat transfer between extended surfaces, it is the hemispherical emissivity that is applicable. However, the difference between the two kinds can be at most 30%, and it is usually ignored in engineering calculations because of the large uncertainties that can exist due to surface contamination, the presence of piping or other complicated geometrical factors, etc. Unfortunately, the difference between normal and hemispherical emissivities is greatest for the case of greatest practical interest, metals for which $e \ll 1$.

Figures 7.3 and 7.4 are useful in making rapid approximate computations of heat transfer by radiation for two situations frequently encountered by low-temperature workers. Figure 7.3 gives the net radiant heat transfer from a surface at room temperature to one at 77 K or lower, and Fig. 7.4 gives the net radiant heat transfer from a surface at 77 K to one at 20 K or lower.

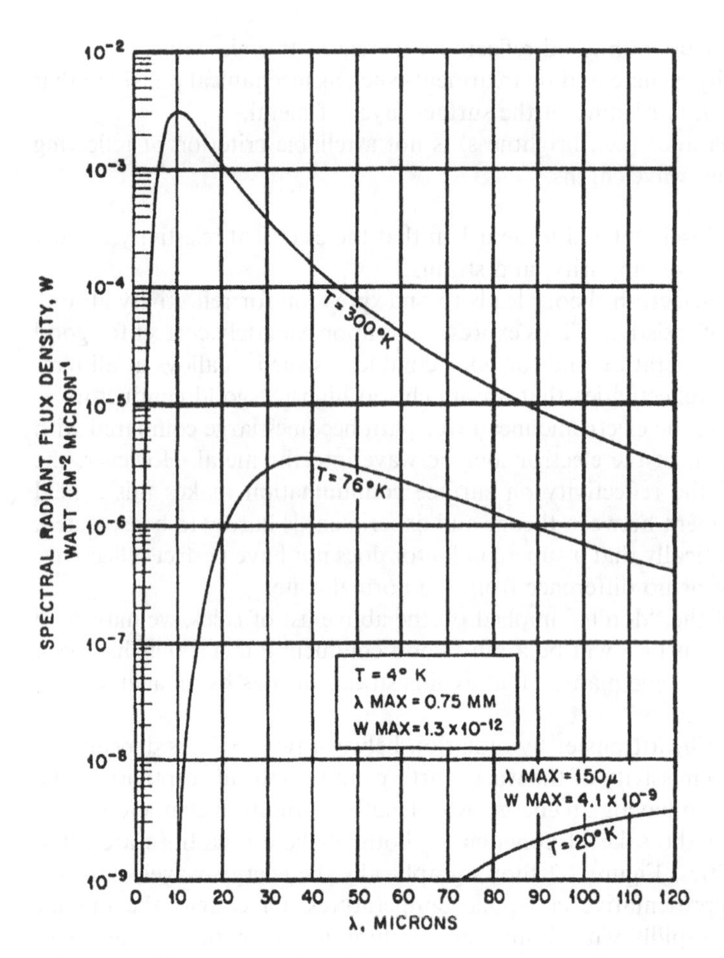


**Figure 7.2** Energy–wavelength graph (blackbody radiation).

Radiation transfer can be reduced by interposing thermally isolated shields parallel to the radiating surfaces. Here we will limit ourselves for simplicity to the special case of two parallel planes between which are $n$ shields. We will further simplify by distinguishing only two kinds of emissivity, that of the two bounding surfaces, $e_0$, assumed to be the same, and that of the shields, $e_s$, assumed to be all equal to each other. Then it can be shown that the heat transfer for $n \geq 1$ is

$$\dot{Q} = (I_2 - I_1)\frac{E_0 E_s}{(n-1)E_0 + 2E_s} \tag{7.3}$$

where $I = \sigma A T^4$ the blackbody emission. $E_0$ is the emissivity factor applying between either boundary and the adjacent shield. $E_s$ is the emissivity factor applying between any two adjacent shields:

$$E_0 = \frac{e_0 e_s}{e_s + e_0 - e_0 e_s}, \quad E_s = \frac{e_s}{2 - e_s}$$

*Case A.* $e_0 = e_s$ and consequently

$$E_0 = E_s = E$$

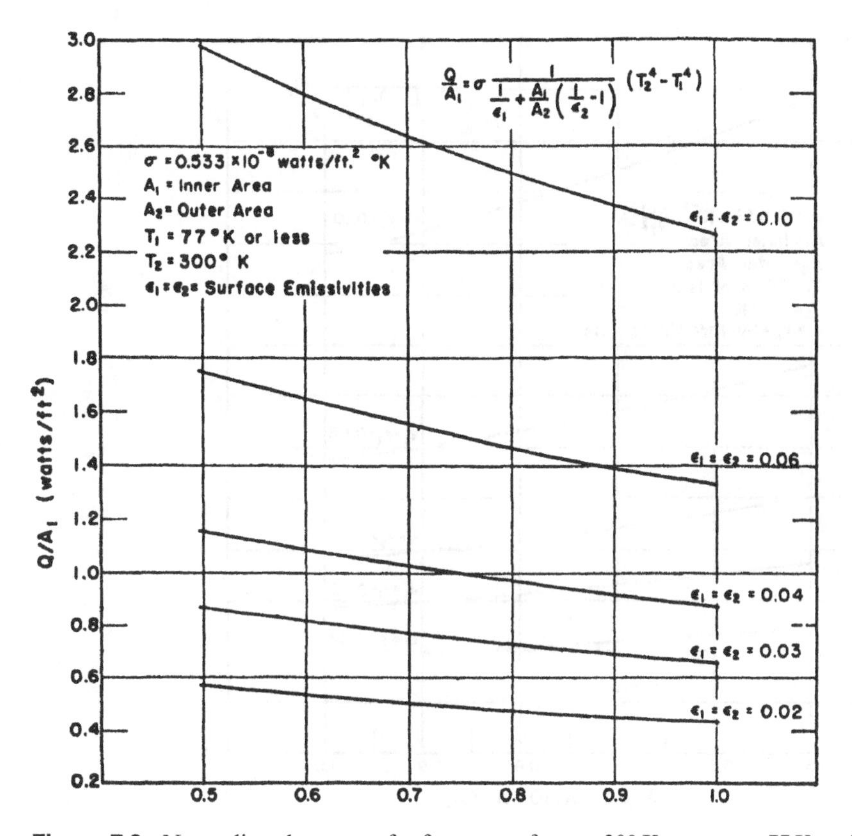


**Figure 7.3** Net radiant heat transfer from a surface at 300 K to one at 77 K or lower.

Here

$$\dot{Q} = (I_2 - I_1)\left(\frac{E}{n+1}\right)$$

This case represents the use of floating metallic radiation shields in a dewar. In an actual installation it may prove difficult to realize the reduction in heat transfer indicated by the above formula because of necessary openings, irregularities, and connecting lines that behave like areas of high emissivity.

*Case B.* $e_0 \ll 1$, $e_s \cong 1$, i.e., bright boundaries, black shields. Here

$$\dot{Q} = (I_2 - I_1)\frac{E_0}{(n-1)E_0 + 2}$$

For small values of $n$, this becomes

$$\dot{Q} \cong (I_2 - I_1)\frac{E_0}{2}$$

But here $E_0 \cong e_0$. If $n$ were zero, we would have

$$\dot{Q} = (I_2 - I_1)\frac{e_0}{2 - e_0} \cong (I_2 - I_1)\frac{e_0}{2}$$

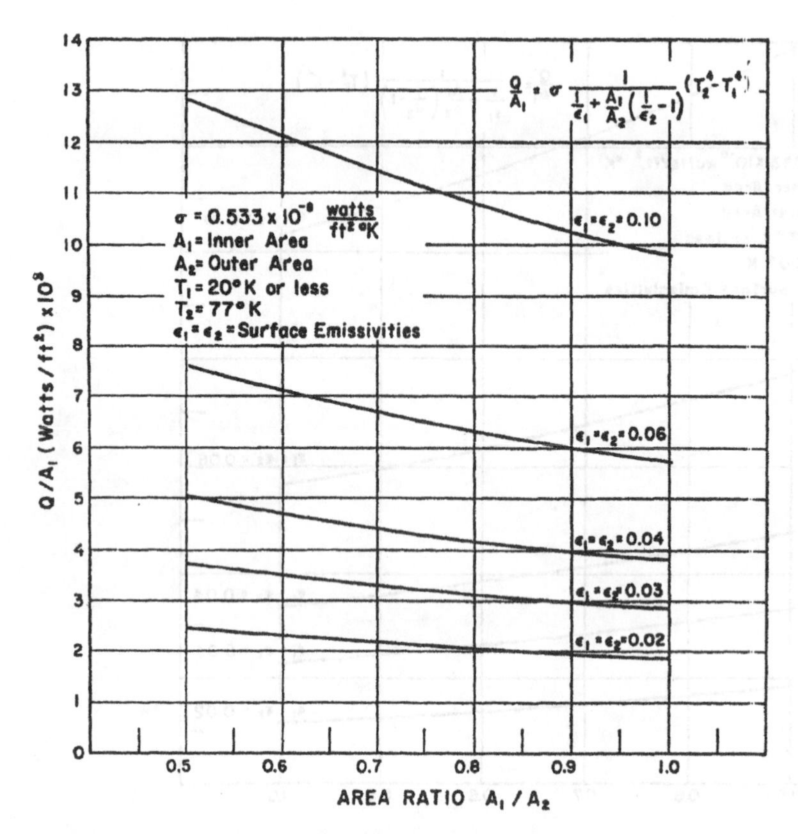


**Figure 7.4** Net radiant heat transfer from a surface at 77 K to one at 20 K or lower for inner/outer surface area ratios of 0.5-1.0.

which is the same result. Hence a small number of such shields is virtually without effect. As $n$ is made very large, the thermal resistance of the shields, though individually small, takes effect through sheer numbers and $W$ approaches the same value as in case A. Thus the boundary emissivity is of importance only for small numbers of shields. The case of numerous black shields is simulated by opaque nonmetallic powders in the vacuum space.

*Example 7.1.* If 10 floating shields are added to a high-vacuum insulation, by how much is the radiative heat transfer reduced?

Radiation from the warm surface to the cold surface is the dominating mode of heat transfer in high-vacuum insulation and can be approximated by the modified Stefan–Boltzmann equation

$$\dot{Q}_r/A_1 = \sigma F_e F_{1-2}(T_2^4 - T_1^4) \tag{7.4}$$

where $\dot{Q}_r/A_1$ is the heat transfer rate by radiation per unit area of inner surface, $\sigma$ the Stefan–Boltzmann constant, $F_e$ the emissivity factor and $F_{1-2}$ a configuration factor relating the two surfaces whose temperatures are $T_1$ and $T_2$. Here the subscripts 1 and 2 refer to the inner (colder) and outer (warmer) surfaces, respectively. When the inner surface is completely enclosed by the outer surface, as is the case with cryogenic storage vessels, the configuration factor ($F_{1-2}$) is unity. For diffuse radiation between concentric spheres or cylinders (with length much greater than diameter), the emissivity factor is given by

$$F_e = \left[\frac{1}{e_1} + \frac{A_1}{A_2}\left(\frac{1}{e_2} - 1\right)\right]^{-1} \tag{7.5}$$

where $A$ is the surface area and $e$ is the emissivity of the surfaces. The emissivity is a non-dimensional parameter defined as the ratio of the amount of energy a surface actually emits to that which it could emit if it were a blackbody radiator at the same temperature. Some typical emissivities for metals used in container construction are given in Table 7.3.

In order to reduce the emissivity factor and thus the heat transfer, 10 floating shields with low-emissivity surfaces have been inserted between the hot and cold surfaces. In general, for $n_s$ shields or $n_s + 2$ surfaces where the emissivity of the inner and outer surfaces is $e_0$ and that of the shields is $e_s$, the emissivity factor becomes

$$F_{e_{(n=n_s)}} = \left[2\left(\frac{1}{e_0} + \frac{1}{e_s} - 1\right) + \frac{(n_s - 1)(2 - e_s)}{e_s}\right]^{-1} \tag{7.6}$$

when the ratio of the inner to the outer surface is unity. In the absence of floating shields where $e_s = 1$ and $n = 0$, the emissivity factor reduces to

$$F_{e_{(n=0)}} = \frac{e_0}{2 - e_0} \tag{7.7}$$

If we arbitrarily select $e_0 = 0.90$, $e_s = 0.05$, we see that 10 radiation shields ($n_s = 10$) reduce the radiant heat transfer rate by a factor of 1/348.

### 3.2. Heat Transfer by Residual Gas

It is predicted by the kinetic theory of gases and confirmed by experiment that the thermal conductivity of a gas in which the mean free path is small in relation to the distances between the surfaces constituting the heat source and the heat sink is independent of the gas pressure. Consider measuring the rate of heat flow through air at 300°C between surfaces of slightly different temperatures with a separation of about 1 cm as the pressure is reduced from a starting pressure of 1 atm. It is found that the rate is essentially unchanged until the pressure becomes quite small. However, as the pressure approaches the micrometer range (~10–100 μm Hg), there is a marked decrease in the rate of heat transfer. At pressures below 1 μm Hg, the rate is nearly proportional to the gas pressure. There is a constant small amount of heat transferred by radiation, and if this is corrected for, the transfer by the residual gas is quite precisely proportional to the pressure. Figure 7.5 illustrates the variation in heat flux with changes in pressure and the significance of the mean free path $\bar{\lambda}$ with respect to the distance $L$ between the surfaces.

This result is also very simply explained by kinetic theory. At sufficiently low pressures, the molecules of a gas collide with the walls of the enclosure much more frequently than with each other. Thus, each molecule becomes a vehicle that travels without interruption from one boundary to the opposite boundary and transports energy from the warm to the cold surface on each trip. It is this region of low pressure and long mean free path ("free-molecule" conduction) that is of interest in vacuum insulation. It is also worthy of mention that under these conditions the rate of heat transfer is independent of the separation between the two surfaces. Increasing the separation lengthens the time required for an individual molecule to make a trip;

**Table 7.3** Emissivities of Various Metals

| Metal | Surface preparation | Surface temperature (K) | Emissivity, total normal |
|---|---|---|---|
| Aluminum (annealed) | Electropolished | 300 | 0.03 |
| | Electropolished | 300 | 0.018 (77 K) |
| | Electropolished | 300 | 0.011 (4.2 K) |
| Aluminum vaporized on 12.7 μm Mylar (both sides) | – | 300 | 0.04 (77 K) |
| Aluminum foil (Kaiser annealed 25.4 μm) | – | 300 | 0.018 (77 K) |
| Aluminum foil (household) | – | 273 | 0.043 (77 K) |
| Aluminum (Alcoa No. 2, 510 μm) | Hot acid cleaned | | |
| | Alcoa process | 300 | 0.029 (77 K) |
| | Alkali cleaned | 300 | 0.035 (77 K) |
| Brass (rolled plate) | Natural surface | 295 | 0.06 |
| Brass | Clean, some scratches | 273 | 0.10 (77 K) |
| Brass (73.2% Cu, 26.72% Zn) | Highly polished | 520 | 0.028 |
| Brass shim stock (65% Cu, 35% Zn) | Highly polished | 295 | 0.029 (77 K) |
| Brass shim stock (65% Cu, 35% Zn) | Highly polished | 295 | 0.018 (4.2 K) |
| Copper | Commercial emery polish | 292 | 0.03 |
| Copper foil (127 μm) | Dilute chromic acid | 300 | 0.017 (77 K) |
| 301 Stainless | Cleaned with toluene and methanol | 297 | 0.021 |
| 316 Stainless | Cleaned with toluene and methanol | 297 | 0.028 |
| 347 Stainless | Cleaned with toluene and methanol | 297 | 0.039 |
| Electroplate silver (silver polish) | Commercially supplied | 300 | 0.017 |
| | | 300 | 0.0083 (77 K) |

*Source*: Kropschot (1959,1963).

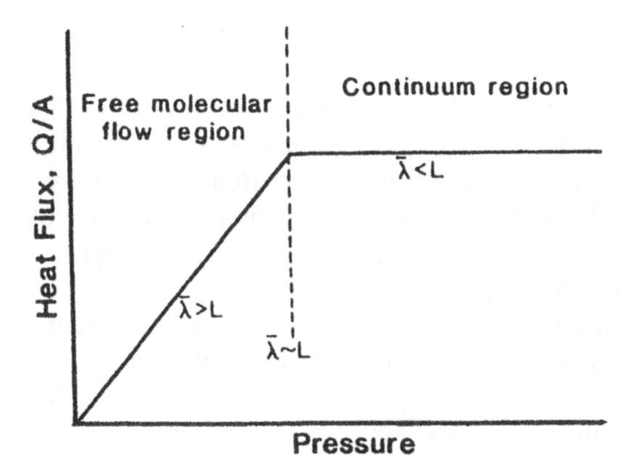


**Figure 7.5** Heat transfer vs. pressure for vacuum insulation.

but if the pressure (molecular concentration) is unchanged, the total number of molecules available as heat transfer vehicles is increased in the same proportion. If each gas molecule achieved thermal equilibrium with the wall as it collided and rebounded, the rate of heat transfer would be uniquely determined by the molecular weight, specific heat, and pressure of the residual gas. However, such thermal equilibrium is seldom if ever realized. The approach to equilibrium is a property of the wall–gas combination and has been treated quantitatively by Knudsen (1910,1930) by the concept of the accommodation coefficient $\alpha$, which is defined by

$$\alpha = \frac{T_i - T_e}{T_i - T_w}$$

where $T_i$ is the effective temperature of the incident molecules, $T_e$ the effective temperature of the emitted (reflected) molecules, and $T_w$ is the temperature of the wall. If complete equilibrium is established between the wall temperature and that of the molecules striking it, $\alpha$ will be unity. If the molecules rebound from the wall without change of kinetic energy, $\alpha$ will be zero. In practice neither of these extremes is encountered; $\alpha$ is found to have values between 0 and 1, although for rough surfaces at temperatures close to the condensing temperature of the molecule the value of $\alpha$ approaches 1.

The residual gas in the vacuum space insulating a cryogenic vessel presents a curious situation when one considers the "temperature" of the gas. In the simple case of parallel surfaces with accommodation coefficients of unity at each wall, it is apparent that there are two distinct sets of molecules. The set leaving the warm surface has a velocity distribution characteristic of the temperature of the warm surface, and each molecule has a velocity component in the direction of the cold surface. The other set of molecules has a speed distribution bestowed by the cold surface and a component of velocity toward the warm surface. Knudsen has investigated this problem and obtained the following relation for the heat transfer between long coaxial cylinders:

$$\dot{Q} = \frac{A_1}{2}\left(\frac{\alpha_1\alpha_2}{\alpha_2 + (r_1/r_2)(1-\alpha_2)\alpha_1}\right)\left(\frac{\gamma+1}{\gamma-1}\right)\left(\frac{R}{2\pi}\right)^{1/2}\left(\frac{P}{\sqrt{TM}}\right)(T_2 - T_1)$$

where $\dot{Q}$ is the rate of heat transfer, $A_1$ the area of the inner cylinder, $\alpha$ is the accommodation coefficient, $r$ the radius of a cylinder, $\gamma$ is the ratio of specific heats, $C_P/C_V$,

$R$ is the universal gas constant, $P$ the pressure, $T$ the absolute temperature and $M$ the molecular weight of the residual gas.

The subscripts 1 and 2 refer to the inner and outer cylinders, respectively. The temperature $T$ without subscript should be measured at the gauge that measures the pressure $P$. In a system in which the mean free paths of molecules are large compared to the dimensions of the spaces, $P/\sqrt{T}$ has a constant value. This phenomenon of proportionality between $P$ and $\sqrt{T}$ has been called thermal transpiration, thermo-molecular pressure, or the Knudsen effect.

This behavior can be derived quite simply from the elementary kinetic theory of gases. Consider a chamber containing a gas at low density divided by an insulating partition in which there is an aperture of area $A$. The temperatures on the two sides of the partition are maintained at $T_1$ and $T_2$, respectively.

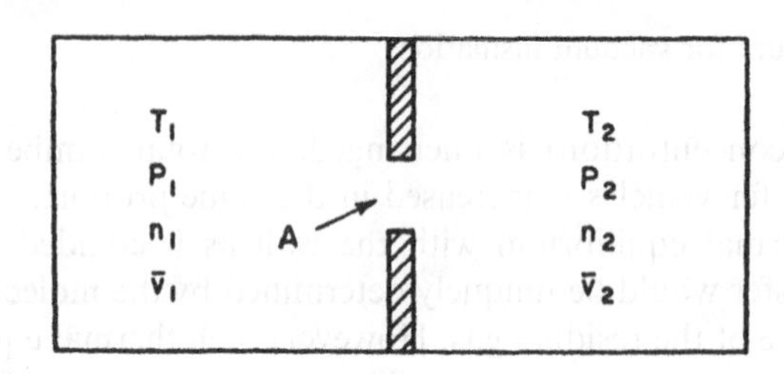


Now the number of molecules entering the aperture from the left in unit time is proportional to $n_1\bar{v}_1$, where $n_1$ is the number of molecules in a unit volume and $\bar{v}_1$ is their mean root square velocity. Likewise, the number entering the aperture from the right in unit time is proportional to $n_2\bar{v}_2$. Moreover, $n$ and $\bar{v}$ are related to the pressure by

$$P = \frac{1}{3}nM\bar{v}^2$$

where $M$ is the molecular mass. Also, $\bar{v}^2$ is proportional to the absolute temperature $T$. At equilibrium, of course, the rate at which molecules move through the aperture from left to right must equal the number moving through from right to left:

$$n_1\bar{v}_1 = n_2\bar{v}_2$$

Then by simple substitution it is seen that

$$P_1/P_2 = \sqrt{T_1/T_2}$$

For practical computations of heat transfer in systems other than long coaxial cylinders, it is sufficiently accurate to substitute the ratio of the areas of the inner and outer surfaces, $A_1/A_2$, for the ratio of the radii, $r_1/r_2$. This will allow the computation of the heat transfer between surfaces of various geometries: short cylinders, concentric spheres, etc.

The following is a general equation that can be used with any consistent system of units. For units commonly used in vacuum work and cryogenics, it can be stated as

$$\dot{Q} = 2.426 \times 10^{-4} A_1 \left(\frac{\alpha_1\alpha_2}{\alpha_2 + (A_1/A_2)(1-\alpha_2)\alpha_1}\right) \times \left(\frac{\gamma+1}{\gamma-1}\right)\left(\frac{P}{\sqrt{MT}}\right)(T_2 - T_1) \quad (7.8)$$

where $\dot{Q}$ is the rate of heat transfer in W, $A_1$ and $A_2$ are the areas of the inner and outer walls in cm$^2$, $P$ the pressure in μm Hg and $T$, $T_1$ and $T_2$ are temperatures in kelvin.

Equation (7.8) has been made more tractable by Corruccini (1957,1958): At low pressures, where the mean free path of the gas molecules is large compared to the dimension of the container, the heat transport $\dot{Q}_{gc}$, for concentric spheres, coaxial cylinders, and parallel plates due to free-molecule gas conduction is

$$\dot{Q}_{gc} = \frac{\gamma + 1}{\gamma - 1}\alpha\left(\frac{R}{8\pi MT}\right)^{1/2} P(T_2 - T_1) \tag{7.9}$$

where $\alpha = \alpha_1\alpha_2/(\alpha_2 + \alpha_1(1 - \alpha_2)(A_1/A_2))$, $\dot{Q}_{gc}$ is the net energy transfer per unit time per unit area of inner surface, $\gamma$ the $C_P/C_V$, the specific heat ratio of the gas, assumed constant, $R$ the molar gas constant, $P$ the pressure, $M$ the molecular weight of the gas, $T$ the absolute temperature at the point where $P$ is measured, $\alpha$ the overall accommodation coefficient, and $A$ is the area and subscripts 1 and 2 refer to the inner and outer surfaces, respectively.

This expression reduces to

$$\dot{Q}_{gc} = \text{constant} \times \alpha P(T_2 - T_1)$$

where the constant is

$$\frac{\gamma + 1}{\gamma - 1}\left(\frac{R}{8\pi MT}\right)^{1/2}$$

Values of this constant for cryogenic applications have been calculated by Corruccini (1957,1958) and are shown in Table 7.4. The temperature at the pressure gauge is assumed to be 300 K, $\dot{Q}_{gc}$ is given in W/cm$^2$ of inner surface, $P$ is in mm Hg, and the temperature is in K.

Table 7.5, which lists the mean free paths of some gases, may be helpful in deciding whether or not free-molecule conduction exists in a given situation.

**Table 7.4** Constants for the Gas Conduction Equation

| Gas | $T_2$ and $T_1$ (K) | Constant |
|---|---|---|
| $N_2$ | $<400$ | 0.0159 |
| $O_2$ | $<300$ | 0.0149 |
| $H_2$ | 300 and 77 or 300 and 90 | 0.0528 |
| $H_2$ | 77 and 20 | 0.0398 |
| He | Any | 0.0280 |

*Source*: Corruccini (1957,1958).

**Table 7.5** Mean Free Paths (cm) of Air, Hydrogen, and Helium at a Pressure of 1 μm Hg

| Temperature (K) | Air | Hydrogen | Helium |
|---|---|---|---|
| 4 | – | – | 0.11 |
| 20 | – | 0.30 | 0.67 |
| 76 | 0.87 | 1.8 | 3.2 |
| 300 | 5.1 | 9.5 | 15.0 |

The mean free path (mfp) is given by

$$L = 8.6 \times 10^3 \frac{\eta}{P} \sqrt{\frac{T}{M}} \tag{7.10}$$

where $L$ is the mfp in cm, $\eta$ the viscosity of the gas in poises, $P$ the pressure in μm, $T$ the Kelvin temperature and $M$ the molecular weight. Thus, $L$ varies inversely with pressure and approximately as $T^{3/2}$ because $\eta$ is roughly proportional to $T$.

The limited literature on accommodation coefficients has been summarized by Partington (1952). Except on surfaces that have had unusual preparation, the common gases have accommodation coefficients between 0.7 and 1 at or below room temperature. The light gases, hydrogen and helium, may have much lower values. Table 7.6 gives the approximate value of $\alpha$ for He, H, and air at 20, 76, and 300 K. $\alpha$ will tend toward unity as the temperature approaches the critical temperature of the gas. However, the value of the accommodation coefficient $\alpha$ is sensitive to the condition of the surface.

Because of the great variations in accommodation coefficients, it is difficult to make accurate estimates of heat transfer by residual gas. This is usually not a serious lack, however, because as a rule the objective is to obtain a vacuum of such quality that the heat transfer by residual gas does not contribute seriously to the overall heat transfer. It is interesting to note that in some cases residual helium in a vacuum will transfer less heat than the same pressure of air, even though the thermal conductivity of helium is much greater than that of air. The values of accommodation coefficients given in Table 7.6 for helium and air at 300 and 76 K will lead to this result when used in conjunction with Eq. (7.9). The lower accommodation coefficient of helium more than compensates for its higher intrinsic thermal conductivity, which is a consequence of its lower molecular weight.

*Example 7.2.* Liquid oxygen is stored in a vacuum-insulated spherical container with an outer vessel of 1.6 m i.d. and an inner vessel of 1.2 m o.d. The space between the two vessels is evacuated to a pressure of 1.5 mPa. The emissivity of the inner vessel is 0.04, and the emissivity of the outer vessel is 0.09. If the temperature of the outer vessel is 300 K and the temperature of the inner vessel is 90 K, determine (a) the heat transfer rate by radiation and (b) the total heat transfer rate by radiation and molecular conduction through the evacuated space.

(a) The emissivity factor for the spherical container is obtained from the equation

$$F_e = \left[\frac{1}{e_1} + \left(\frac{A_1}{A_2}\right)\left(\frac{1}{e_2} - 1\right)\right]^{-1} = \left[\frac{1}{e_1} + \left(\frac{D_1}{D_2}\right)^2 \left(\frac{1}{e_2} - 1\right)\right]^{-1}$$

$$= \left[\frac{1}{0.04} + \left(\frac{1.2}{1.6}\right)^2 \left(\frac{1 - 0.09}{0.09}\right)\right]^{-1} = 0.03259 \tag{7.11}$$

**Table 7.6** Approximate Accommodation Coefficients

| T (K) | Helium | Hydrogen | Air |
|---|---|---|---|
| 300 | 0.3 | 0.3 | 0.8–0.9 |
| 76 | 0.4 | 0.5 | 1 |
| 20 | 0.6 | 1.0 | – |

The radiant heat transfer is given by

$$\dot{Q}_r = \pi D_1^2 \sigma F_e (T_2^4 - T_1^4)$$
$$= \pi(1.2)^2(5.669 \times 10^{-8})(0.03259)[(300)^4 - (90)^4] = 67.16\,\text{W} \qquad (7.12)$$

where $F_{1\text{-}2} = 1$ and $A_1 = \pi D_1^2 A_1 = \pi D_{1\text{-}2}$.

(b) To evaluate the heat transfer by molecular conduction, we will first need to determine the mean free path for the remaining gas in the evacuated space from Eq. (7.10). Assume that the gas in the evacuated space is air. The viscosity of air at 300 K is 18.47 μ Pa s,

$$L = \frac{\mu}{P}\left(\frac{\pi RT}{2g_c}\right)^{1/2} = \left(\frac{18.47 \times 10^{-6}}{1.5 \times 10^{-3}}\right)\left(\frac{\pi(8314/28.97)(300)}{(2)(1)}\right)^{1/2} = 4.53\,\text{m} \qquad (7.13)$$

Since the mean free path of the molecules is much larger than the space between the inner and outer vessels, free molecular conduction occurs. The accommodation coefficients from Table 7.6 for air at 300 and 90 K are 0.85 and 1.0, respectively. The accommodation coefficient factor is obtained with the aid of the equation

$$F_a = \left[\frac{1}{\alpha_1} + \left(\frac{A_1}{A_2}\right)\left(\frac{1}{\alpha_2} - 1\right)\right]^{-1} = \left[\frac{1}{1} + \left(\frac{1.2}{1.6}\right)^2\left(\frac{1 - 0.85}{0.85}\right)\right]^{-1} = 0.9097 \qquad (7.14)$$

The free molecular gas conduction is given by

$$\dot{Q}_g = \left(\frac{\gamma + 1}{\gamma - 1}\right)\left[\frac{g_c\ R}{8\pi\ T}\right]^{1/2} A_1 F_a p (T_2 - T_1) = \left(\frac{1.4 + 1}{1.4 - 1}\right)\left[\frac{(1)(8314/28.97)}{(8\pi)(300)}\right]^{1/2}$$
$$\times\, \pi(1.2)^2(0.9097)(1.5 \times 10^{-3})(300 - 90) = 1.52\,\text{W} \qquad (7.15)$$

The total heat transfer rate through the evacuated space to the cold surface is

$$\dot{Q}_t = \dot{Q}_r + \dot{Q}_g = 67.16 + 1.52 = 68.68\,\text{W} \qquad (7.16)$$

### *3.2.1. Outgassing*

When a vacuum enclosure is pumped, the surfaces gradually release sizable quantities of gas, and accordingly the attainment of a good vacuum is greatly retarded. Some of the gas being liberated is bound to the surfaces by physical adsorption, some is dissolved in the solid, and some exists in chemical combination—for example, in oxide on the surfaces. The most effective means of removing these gases and thus speeding up the evacuation process is by heating during evacuation. When this is not feasible, heating the components in a vacuum oven just prior to assembly may accomplish nearly the same result. Heating in a hydrogen atmosphere will remove surface oxides, adsorbed air, and water vapor, but it is believed that for components to be used for a cryogenic insulating vacuum, heating in vacuum is greatly preferable because this will also remove some of the dissolved hydrogen. Residual hydrogen in an insulating vacuum is particularly objectionable because of its high thermal conductivity and its resistance to condensation on cold surfaces. In some metals large quantities of hydrogen trapped in pores during the smelting process may slowly diffuse to the surface and spoil the vacuum. Diffusion is more rapid in steel than in copper or aluminum.

It is advisable to avoid, insofar as possible, organic materials in a vacuum system. These as a rule contain appreciable quantities of air in solution and also may contain entrapped or dissolved solvent or plasticizer. Particularly bad are rubber and phenolic plastics. Teflon, Kel-F, polyethylene, polystyrene, and epoxy resins are much better.

It is very important that the interior of a vacuum system be scrupulously clean. A film of oil on the pipes or soldering flux at the joints may make it almost impossible to achieve a satisfactory vacuum. In cleaning a vacuum system, it is good practice to use hot water to dissolve the water-soluble dirt, dry, and clean a second time with an organic solvent. If the principal contaminant is not water-soluble, the order of washing should be reversed. Of course, the presence of soluble plastics may proscribe the use of organic solvents. In this case, a detergent in hot water may be satisfactory.

Because of the difficulty of outgassing metals, many large vacuum-insulated metal containers are continuously pumped with a diffusion pump and forepump permanently attached. Smaller metal vacuum vessels are usually provided with adsorbents that take up the residual gas when cold.

### *3.2.2. Getters*

A *getter* is a substance that takes up gas at very low pressures and so is used to improve or maintain a vacuum in a closed system. The processes responsible for the action of various getters include physical adsorption of the gas on an extended surface adsorbent, chemical combination with suitable active materials, and the solution of gases in certain metals.

**(a) Chemical Getters.** The term "getter" originated with substances of this type used to maintain vacuum in electronic tubes. The one having the most general utility is barium, but other alkali and alkaline earth metals have also been used. To be effective at ambient temperatures, the substance must be finely dispersed so as to present a large surface area. This is usually done by subliming it onto the walls of the vacuum space.

**(b) Solution Getters.** Solution getters consist of certain transition metals of which titanium, zirconium, and thorium are the most widely used. These take common gases such as $N_2$, $O_2$, and $H_2$ into solid solution either as dissociated atoms or possibly by dissolving compounds formed by the gas and the metal. Stoichiometric proportions are not maintained overall.

The more active of these metals may absorb very large volumes of gas. The system palladium–hydrogen is one of the most striking, especially at ordinary temperatures. Such systems show thermochemical and electrical properties that indicate them to be more properly considered as a special class of alloys than as ordinary solutions.

**(c) Adsorbents.** Adsorbents such as activated coconut charcoal have long been used to maintain good insulating vacua in vessels used to store cryogenic liquids. This is a particularly apt utilization because the effectiveness of an adsorbent is greatly enhanced by lowering the pressure of the liquid being stored. Figure 7.1, presented earlier, shows a cross-section of a typical commercial vessel used for storing and transporting liquid oxygen and liquid nitrogen. The adsorbent is activated charcoal contained so as to be very nearly at the temperature of the liquid being stored. Although the vacuum between the walls of such a vessel may not be particularly good when it is empty (warm), as soon as the cryogenic liquid is introduced the adsorbent will take up gas and greatly improve the insulating vacuum.

### 3.3. Heat Transfer by Supports

It is a simple matter to compute the amount of heat conducted through ordinary solid supports, such as rods or cables, which may be used to bear the weight of the inner container of a dewar. The rate of heat transfer is given by

$$\dot{Q} = -Ak\frac{\mathrm{d}T}{\mathrm{d}x} \tag{7.17}$$

where $A$ is the cross-sectional area of the support ($A$ = constant), $k$ the thermal conductivity and $\mathrm{d}T/\mathrm{d}x$ the temperature gradient. The negative sign merely indicates that heat flows in the direction of lower temperatures. The thermal conductivity usually varies markedly with temperature at low temperatures, so that an integration is required to obtain the heat flow. Thus

$$\dot{Q} = \frac{A}{L}\int_{T_1}^{T_2} k\ \mathrm{d}T \tag{7.18}$$

where $L$ is the length of the support with one end at the higher temperature $T_2$ and the other end at $T_1$. This expression may be evaluated by inserting an analytic function for $k = f(T)$, but as a rule there is no such simple function; $k = f(T)$ is commonly represented by a graph in which $k$ is plotted as ordinate with $T$ as abscissa. In this case, Eq. (7.18) may be evaluated by numerical integration.

$$\dot{Q} = \frac{A}{L}\sum_{T_1}^{T_2} k_{\mathrm{m}}\Delta T \tag{7.19}$$

where $k_{\mathrm{m}}$ is the mean thermal conductivity in the temperature interval $\Delta T$. By taking sufficiently small intervals of $\Delta T$, any desired accuracy can be achieved. If the thermal conductivity is a linear function of the temperature, it is correct to use an average value of $k$ midway between $T_1$ and $T_2$.

Most convenient are thermal conductivity integrals that show directly the conduction of a unit area between any two temperatures.

When the supporting member is the tube that carries the evaporating vapor, as in the dewars shown in Fig. 7.1, there is a circumstance that reduces the heat transfer to the inner container. Some of the heat that starts down the tube is intercepted by the issuing vapor and never reaches the inner container. This effect will be analyzed later.

There is a type of support that is of value when available space or other considerations make it difficult to use long rods or cables. This is the laminated support, which consists of many layers of thin stainless steel or other poorly conducting metal. These supports, of course, are used only for compressive loads, but for such use their strength is almost the same as that of solid metal although the heat flow may be only a small fraction (e.g., 1%) of that of a similar solid support. These insulating supports owe their effectiveness to the high thermal resistance of the contact between two metal surfaces in a vacuum. Two forms have been devised; one consists of a stack of disks or washers, and the other is a coiled strip like a tightly wound clock spring. It has been found that the thermal resistance of these multiple-contact supports can be increased considerably by lightly dusting the surfaces with manganese dioxide. Figure 7.6 gives data on the thermal resistance of some laminated stainless steel supports.

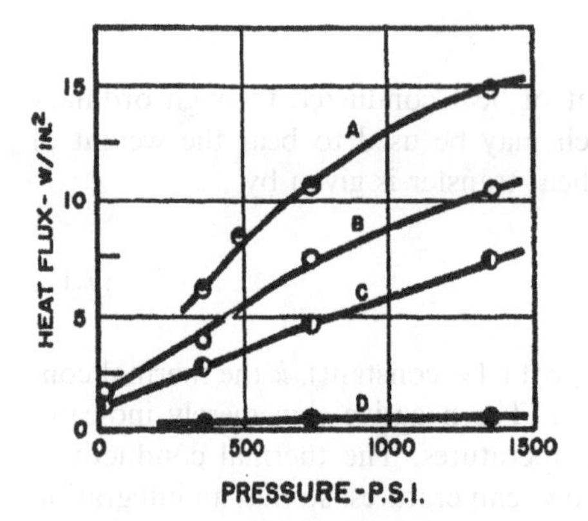


**Figure 7.6** Heat conducted as a function of mechanical pressure by stacks of Type 302 round stainless steel plates, each plate 0.00008 in. thick: (A) 148 plates, (B) 209, (C) 313, (D) 315. Boundary temperatures for A, B, and C are 76 and 296 K; for D, 20 and 76 K.

### *3.3.1. Support Materials*

Thermal conductivity, tensile strength, and modulus of elasticity are the most significant design properties for a support system. A number of suitable structural materials are listed in Table 7.7 along with values for each property.

The heat leak per unit of load for a support member in pure tension is proportional to $k/\sigma$, where $k$ is the thermal conductivity and $\sigma$ is the maximum working stress of the member. Using room temperature values of the conductivity and stress, Table 7.7 shows the $k/\sigma$ ratio for several materials. Among these materials, the industrial yarns—Dacron, Nomex, and glass fibers—have significantly lower values of $k/\sigma$ than do stainless steels and the titanium alloys. Thus, on this basis alone, the yarns are promising for support materials. However, other factors, such as the percent elongation at the maximum loads and the temperature coefficient of expansion, must also be considered in the design of support systems.

The natural frequency of the support systems must be greater than the frequency of any excitations that result from the vibrations and accelerations induced by handling or transportation. The criteria for determining whether a material will give a minimum conductance support for a given system's natural frequency (tension support) is to have a minimum ratio of the thermal conductivity of the support to its modulus of elasticity, $k/E$. Table 7.7 shows the value of this ratio for various materials; again, it is apparent that the synthetic yarns—Dacron, Nomex, and glass fibers—are among the best.

### *3.3.2. Tension Supports*

Two examples of the tension type of support are shown in Fig. 7.7. Each system requires six tension members. In the system of Fig. 7.7a, each set of three structural members connects to a rigid or cantilevered support that penetrates the insulation. In the other system (Fig. 7.7b), each structural member connects to the tank at a different location, and therefore each penetrates the insulation separately. Analysis of the effects of vibration shows that the natural frequency of the tension members is given by

$$f_n = \frac{1}{2\pi}\left(\frac{6k\cos^2\alpha}{M}\right)^{1/2}$$

where $k$ is the spring constant, $\alpha$ is the angle of the tension member with the vertical, and $M$ is the mass of the container.

**Table 7.7** Thermal and Mechanical Properties of Support Materials

| Material | Modulus of elasticity | | Thermal conductivity | | Tensile strength | | Conductivity/strength ratio | | Conductivity/modulus of elasticity ratio | |
|---|---|---|---|---|---|---|---|---|---|---|
| | psi ($\times 10^6$) | kg/cm$^2$ ($\times 10^6$) | Btu in./(h ft$^2$ °F) | mW/(cm °C) | psi ($\times 10^3$) | kg/cm$^2$ ($\times 10^3$) | Btu in.$^3$ (h °F ft$^2$ lb) ($\times 10^{-4}$) | mW $\times$cm (kg.°C) ($\times 10^{-4}$) | Btu in.$^3$ (h °F ft$_2$ lb) ($\times 10^{-6}$) | mW cm (kg °C) ($\times 10^{-6}$) |
| Stainless steel | 29 | 2.0 | 98 | 140 | 90–280 | 6.1–19.4 | 3.4–11 | 70–230 | 3.4 | 72 |
| Titanium alloy (7% Al, 4% Mo) | 17 | 1.2 | 49 | 70 | 60–170 | 4.1–11.6 | 2.9–8.2 | 60–170 | 2.9 | 61 |
| Pyroceram (9608) | 12 | 0.85 | 14 | 20 | | | | | 1.12 | 24 |
| Borosilicate glass | 9.1 | 0.62 | 7.8 | 11 | | | | | 8.6 | 18 |
| Wood | | | | | | | | | | |
| Balsa | 0.5 | 0.03 | 0.36 | 0.52 | | | | | 0.75 | 17 |
| Maple | 1.8 | 0.12 | 1.3 | 1.9 | | | | | 7.2 | 160 |
| White pine | 1.2 | 0.08 | 1.0 | 1.4 | | | | | 0.83 | 1200 |
| Industrial yarn | | | | | | | | | | |
| Dacron | 2.0 | 0.14 | 1.1 | 1.6 | 80–140 | 5.4–9.9 | 0.07–0.14 | 1.6–2.9 | 5.5 | 11 |
| Nylon | 0.95 | 0.06 | 1.7 | 2.5 | | | | | 1.79 | 41 |
| Nomex | 2.8 | 0.19 | 0.9 | 1.3 | 70–90 | 4.8–6.1 | 0.10–0.13 | 2.1–2.7 | 0.32 | 6 |
| Epoxies, cast | | | | | | | | | | |
| Unfilled | 0.35 | 0.02 | 1.3 | 1.9 | | | | | 3.7 | 93 |
| Glass-filled | 3.0 | 0.20 | 2.4 | 3.5 | | | | | 0.8 | 17 |
| Glass fibers | 17 | 1.1 | 7.2 | 10 | 200–220 | 13.6–15.0 | 0.33–0.36 | 6.9–7.6 | 0.43 | 9 |

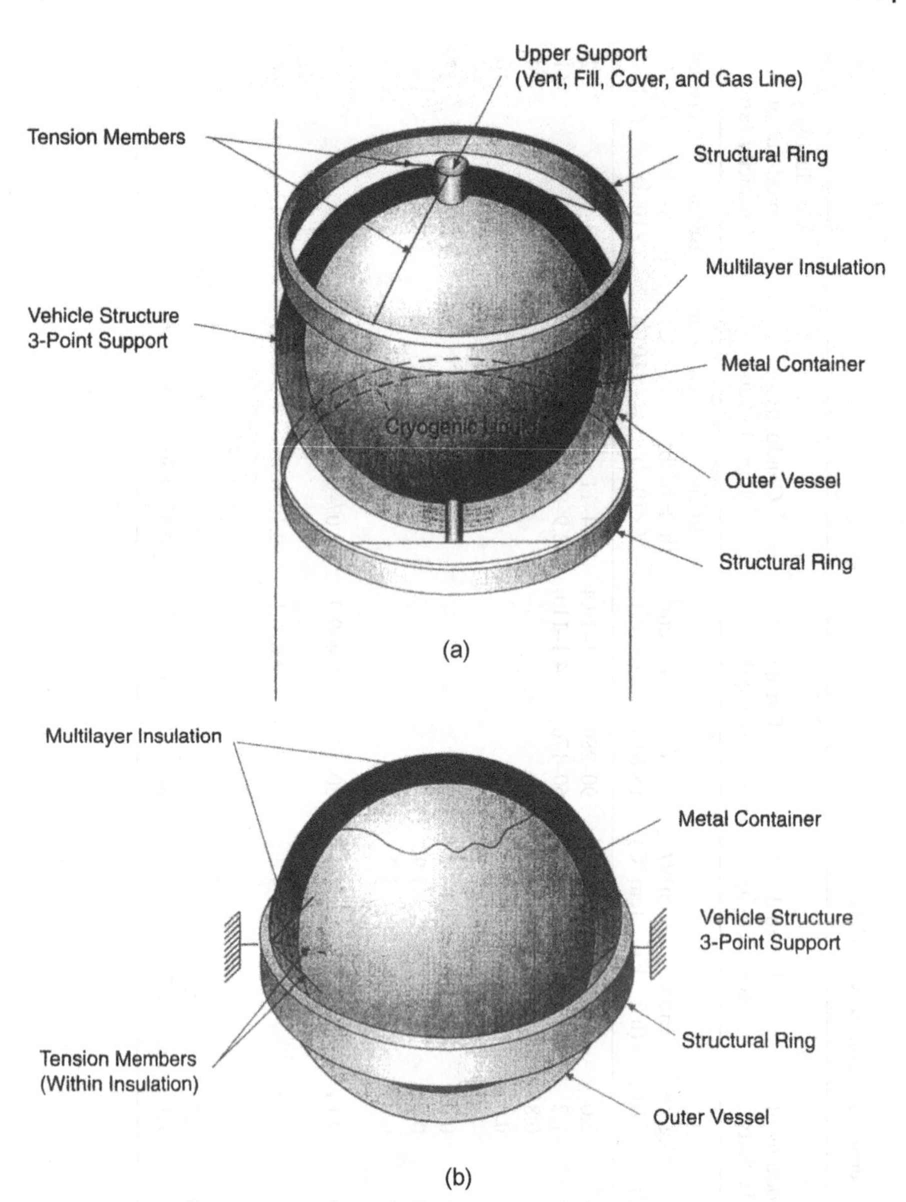


**Figure 7.7** Tension support systems for cryogenic containers.

The strength-to-conductivity ratios for several materials employed in support members are listed in Table 7.8. The austenitic steels are most suitable for larger vessels, while Dacron webbing has been successfully used to support smaller inner vessels.

As mentioned, the best materials have a high design stress and a low thermal conductivity. The ratio of design stress to thermal conductivity is a good parameter to use in the selection of materials. Table 7.9 shows a comparison of some selected support materials (Kropschot, 1963).

Since the design safety factor varies with each application, the yield strength $\sigma_y$ has been used for these comparisons. The second column of Table 7.9 shows the

**Table 7.8** Strength-to-Conductivity Ratios for Typical Materials Used in Support Members

| | Yield strength (MPa) | Thermal conductivity[a] [W/(m K)] | Strength-to-conductivity ratio |
|---|---|---|---|
| Teflon | 20.7 | 1.522 | 13.6 |
| Nylon | 75.1 | 1.552 | 48.4 |
| Mylar | 275.7 | 0.962 | 286.6 |
| Dacron fibers | 606.7 | 0.962 | 630.7 |
| Kel-F oriented fibers | 206.8 | 0.381 | 542.8 |
| Glass fibers | 896.1 | 4.88 | 183.6 |
| 304 stainless steel[b] | 627.3 | 12.41 | 50.5 |
| 316 stainless steel[b] | 820.3 | 12.41 | 66.1 |
| 347 stainless steel[b] | 882.3 | 12.41 | 71.1 |
| 1100-H16 aluminum | 135.8 | 260.1 | 0.52 |
| 2024-0 aluminum | 75.8 | 88.4 | 0.86 |
| 5056-0 aluminum | 137.9 | 114.9 | 1.20 |
| K Monel (45% cold-drawn) | 641.0 | 18.63 | 34.4 |
| Hastelloy C (annealed) | 379.1 | 12.36 | 30.7 |
| Inconel[b] | 310.2 | 13.35 | 23.2 |

[a] Values of thermal conductivity are mean values between 300 and 90 K.
[b] Cold-drawn.

average thermal conductivity $k$ between 300 and 20 K, except as noted; the third column shows the ratio $\sigma_y/k$; and the fourth column shows $\sigma_y/k$ divided by $\sigma_y/k$ for annealed stainless steel. The English units for $\sigma$ and $k$ are used to make the ratio $\sigma_y/k$ near unity. When compared in this manner, the nonmetallic materials appear to be very good for supports.

**Table 7.9** Comparison of Materials Used for Support Members[a]

| Material | $\sigma_y$ (1000 psi) | $K$ [Btu/(h ft °R)] | $\sigma_y/K$ | $F^+$ |
|---|---|---|---|---|
| Aluminum 2024 | 55 | 47 | 1.17 | 0.20 |
| Aluminum 7075 | 70 | 50 | 1.4 | 0.24 |
| Copper, annealed | 12 | 274 | 0.044 | 0.007 |
| Hastelloy B | 65 | 5.4 | 12 | 2.0 |
| Hastelloy C | 48 | 5.9 | 8.1 | 1.4 |
| "K" Monel | 100 | 9.9 | 10.1 | 1.7 |
| Stainless steel 304, annealed | 35 | 5.9 | 5.9 | 1 |
| Stainless steel (drawn, 210,000 psi) | 150 | 5.2 | 29 | 4.9 |
| Titanium, pure | 85 | 21 | 4.0 | 0.68 |
| Titanium, alloy (4Al–4Mn) | 145 | 3.5 | 41.4 | 7.0 |
| Dacron | 20 | 0.088[b] | 227 | 38.5 |
| Mylar | 10 | 0.088[b] | 113 | 19.1 |
| Nylon | 20 | 0.18 | 111 | 18.8 |
| Teflon | 2 | 0.14 | 14.3 | 2.4 |

[a] $\sigma_y$ is the yield stress, $k$ is the average thermal conductivity between 20 and 300 K. $F^+$ is $\sigma_y/k$ divided by $\sigma_y/k$ for stainless steel (annealed).
[b] Room temperature value.
*Source*: Fulk (1959).

## 4. EVACUATED POROUS INSULATION

When an insulating space is filled with a powder having a low gross density (a large ratio of volume of gas-filled voids to volume of solid material), it is found that the apparent thermal conductivity is approximately that of the gas. It appears that the amount of heat transferred by solid conduction through the powder is relatively small. Also the presence of the powder inhibits, to some extent, heat transfer by convection and radiation. Now if the gas pressure in the interstices is reduced by pumping the enclosure, the rate of heat transfer is little affected at first because the thermal conductivity of the gas is almost independent of pressure in this higher pressure region. However, when the gas pressure approaches the value at which the mean free paths of the molecules are comparable with interstitial distances, there is a marked reduction in the apparent thermal conductivity. This is illustrated in Fig. 7.8. For most powders used with nitrogen as the interstitial gas, it appears that the condition of free-molecule conduction sets in at pressures on the order 1.0–0.1 mm Hg. If the only mechanism for heat transfer were that of gaseous conduction, the rate of heat transfer at all lower pressures would be proportional to the pressure of the gas. However, this is found not to be the case; at quite low pressures the rate of heat transfer becomes almost independent of pressure, showing that heat is being transmitted by other means. This residual heat transfer can be by solid conduction through the powder or by radiation or by both. Figure 7.8 is somewhat idealized to represent the behavior of a powder with voids of uniform size and therefore having a rather sharp transition into the range of free-molecule conduction. For most actual powders, this transition is more diffuse.

### 4.1. Powders and Fibers

Expanded perlite is produced from a volcanic glassy rock that occurs in widely distributed deposits. This rock is ground to a fine powder and then expanded by heating to above 1500°F (800°C). Microscopic voids forming in the expanding, heat-softened

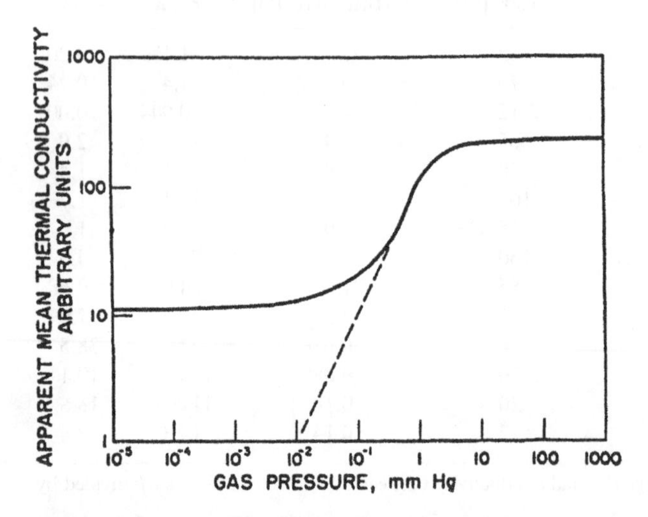


**Figure 7.8** Variation of the apparent thermal conductivity of an insulating powder as the pressure of the interstitial gas is changed.

glass cause a porous structure and account for the expanded perlite's low density. Perlite is available in various grades containing particles ranging in size from 100 to 1600 μm ($4 \times 10^{-3}$ to $6 \times 10^{-3}$ in.).

Colloidal silica is a white, fluffy, submicroscopic particulate silica prepared in a hot gaseous environment (2000°F; 1100°C) by the vapor-phase hydrolysis of silicon tetrachloride. The average size of primary particles of colloidal silica ranges between 150 and 200 Å.

Silica aerogel is prepared from a gel in which water is replaced with an alcohol. The alcohol is subsequently removed, leaving a submicroscopic skeleton structure that is retained as the solid is ground to powder. These materials have been mixed with fine metallic powders or flakes (such as copper and aluminum) and carbon.

Preheating permits the efficient removal of adsorbed gases and particularly moisture. After preheating, fine powders are effective absorbers of reduced residual gases; therefore, their exposure to the atmosphere should be minimized prior to use.

For any given particle shape, a specific packing configuration will yield a maximum density. For spheres, closest packing occurs with a rhombohedral arrangement in which each sphere has 12 points of contact. In actual lattices obtained by prolonged shaking of spherical particles, the voids account for about 39.5% of the volume, characteristic of orthorhombic or eight-point packing, which yields the highest density that can be practically achieved.

When powder is used at low pressures (10 μm Hg or less), gas conduction is negligible and heat transport is chiefly by radiation and solid conduction. For some powders, radiation adsorption reduces heat transfer more than solid conduction increases it, and these powders thus reduce the overall heat transport. Figure 7.9 shows the apparent mean thermal conductivity of several powders as a function of interstitial gas pressure (Fulk, 1959).

The radiant heat transmission through a partially transparent powder is not simple. Some of the heat is radiated directly from the warm surface to the cold surface, some is absorbed by the powder and reradiated, and some is reflected by the powder. For the layers of silica aerogel (less than 1 in. thick), the apparent thermal conductivity depends on the thickness and on the emissivities of the boundaries. Thus the concept of thermal conductivity cannot be properly applied to thin layers. However, for layers 1 in. or more in thickness, the apparent thermal conductivities given in Fig. 7.9 may be employed in the conventional way with errors not exceeding 10%. It is seen in Fig. 7.9 that when the cold surface is at 76 K, the apparent thermal conductivity is reduced only slightly by lowering the pressure below 10 μm Hg. The mechanism responsible for the residual heat transfer has been investigated by varying the temperature of the outer, warm surface of the calorimeter. Since heat transferred by radiation is proportional to the differences of the fourth powers of the temperatures of the two surfaces, while that transferred by solid conduction is approximately proportional to the difference of the first powers of the temperatures, a moderate variation of the temperature of the warm surface will distinguish the two modes. With silica aerogel as the insulating medium, the temperature of the outer surface was varied from 273 to 329 K while the inner surface was kept at 76 K, and it was found that the residual heat transfer was nearly proportional to the difference of the fourth powers of the temperatures. From this, it was concluded that solid conduction is unimportant for this temperature span. However, for heat transfer from a surface of 76 K to one at 20 K, there is evidence that solid conduction through the powder is a major factor. Since the energy radiated from a surface at 76 K is only about 0.004 of that radiated at room temperature, it is not surprising

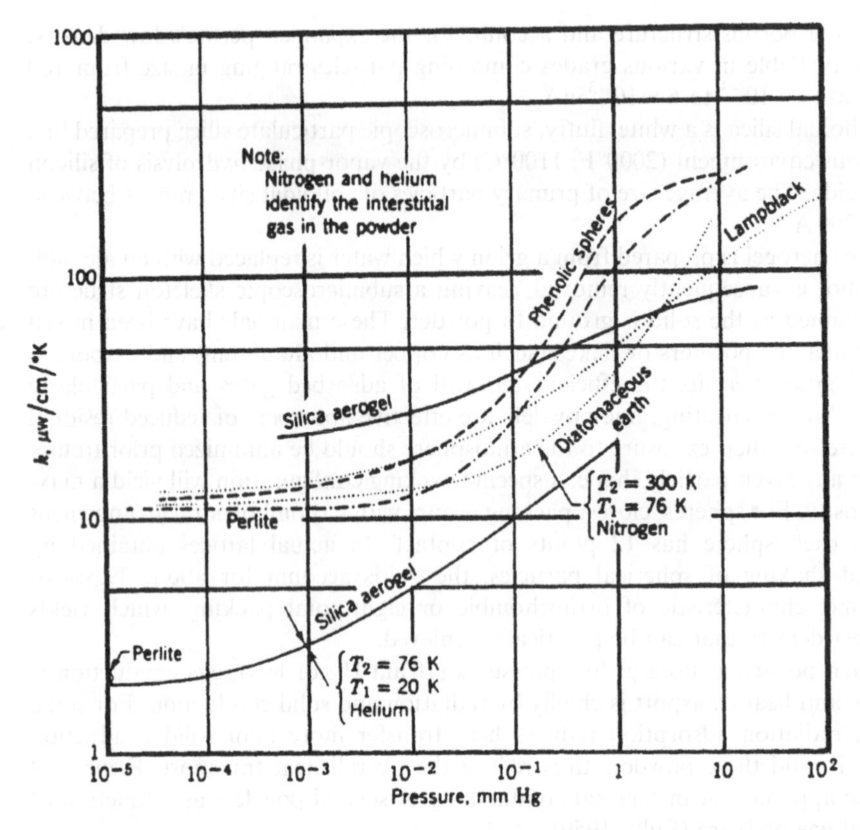


**Figure 7.9** Apparent mean thermal conductivities of several powders as a function of interstitial gas pressure.

that the relative importance of the two modes of heat transfer can be reversed by such a temperature change.

Table 7.10 shows the apparent mean thermal conductivity between 76 and 300 K of some selected powders (Fulk, 1959). The effects of density and particle size on the conductivity of perlite are shown in Figs. 7.10 and 7.11.

The amount of heat transport due to radiation through these dielectric powders can be reduced by adding metallic powders. Figure 7.12 shows the effect of adding aluminum powder to the dielectric powders Cab-O-Sil, Santocel, and perlite (Fulk, 1959; Hunter et al., 1960). The conductivity is dependent upon the amount of metal added, and the effect is much more pronounced for Cab-O-Sil than for perlite.

The vacuum required with powder insulations is much less extreme than with high-vacuum or multilayer insulations. However, most powders can adsorb considerable moisture and therefore must be thoroughly dried before the required vacuum can be attained. In addition, fine mesh filters must be used to protect the vacuum pumps from the highly abrasive particles.

Generally, for highly evacuated powders operating between room temperature and liquid nitrogen temperature, the heat transfer by radiation is greater than that by solid conduction. Therefore, between these temperatures, evacuated powders can be superior to vacuum alone because the powders limit radiative heat transfer. On the

**Table 7.10** Apparent Mean Thermal Conductivity of Several Evacuated Powders[a]

| Powder | Remarks | Density ($g/cm^3$) | Gas pressure (mm Hg) | Conductivity [μW/(cm K)] |
|---|---|---|---|---|
| Silica aerogel | Chemically prepared | 0.10 | $< 10^{-4}$ | 21 ± 1 |
| | 250 A | 0.10 | 628 ($N_2$) | 196 ± 5 |
| | | 0.10 | 628 (He) | 620 ± 10 |
| | | 0.10 | 628 ($H_2$) | 800 ± 10 |
| | +10% free silicon dust, by weight | 0.11 | $< 10^{-4}$ | 18 ± 1 |
| Silica | flame prepared, | 0.06 | $< 10^{-4}$ | 21 ± 1 |
| | 150–200 A | 0.06 | 630 ($N_2$) | 185 ± 5 |
| Perlite, expanded | +30 mesh | 0.06 | $< 10^{-4}$ | 21 ± 0.5 |
| | +30 mesh | 0.10 | $< 10^{-4}$ | 18 ± 0.5 |
| | +30 mesh | 0.10 | 628 ($N_2$) | 332 ± 5 |
| | +30 mesh | 0.10 | 628 (He) | 1270 ± 10 |
| | +30 mesh | 0.10 | 628 ($H_2$) | 1460 ± 20 |
| | −30 + 80 mesh | 0.13 | $< 10^{-4}$ | 12 ± 0.5 |
| | −30 + 80 mesh | 0.13 | 628 ($N_2$) | 325 ± 5 |
| | −30 + 80 mesh | 0.13 | 628 (He) | 1260 ± 20 |
| | | 0.13 | 628 ($H_2$) | 1450 ± 20 |
| | −80 mesh | 0.14 | $< 10^{-4}$ | 10 ± 0.5 |
| | −80 mesh | 0.14 | 628 ($N_2$) | 350 ± 5 |
| | −80 mesh | 0.14 | 628 (He) | 1350 ± 10 |
| | −80 mesh | 0.14 | 628 ($H_2$) | 1453 ± 20 |
| | −30 mesh | 0.14 | $< 10^{-4}$ | 10 ± 0.5 |
| Diatomaceous earth | 1–100 μm | 0.24 | $< 10^{-4}$ | 16 ± 0.5 |
| | | 0.25 | $< 10^{-4}$ | 14 ± 0.5 |
| | | 0.29 | $< 10^{-4}$ | 10 ± 0.5 |
| Alumina, fused | −50+100 mesh | 2.0 | $< 10^{-4}$ | 18 ± 0.5 |
| Alumina, laminar | 0.1–10 μm | 0.07 | $< 10^{-4}$ | 23 ± 0.5 |
| Mica, expanded | −20 + 30 mesh 30% | 0.15 | $< 10^{-4}$ | 18 ± 0.5 |
| | −30 + 15 mesh 70% | 0.15 | 628 ($N_2$) | 503 ± 10 |
| | | 0.15 | 628 (He) | 1500 ± 20 |
| Lampblack | | 0.2 | $< 10^{-4}$ | 12.5 ± 0.5 |
| Charcoal peach pits | 20–30 mesh | 0.49 | $< 10^{-4}$ | 18 ± 1 |
| Carbon + 7% ash[b] | 30% < 44 μm | 0.2 | $< 10^{-4}$ | 6 ± 0.5 |
| Talc | | 1.2 | $< 10^{-4}$ | 16 ± 0.5 |
| Phenolic spheres | 25–100 μm | 0.2 | $< 10^{-3}$ | 13 ± 1 |
| Calcium silicate | 0.02 μm | 0.17 | $< 10^{-5}$ | 7.5 ± 0.5 |
| (synthetic) | 0.02–0.07 μm | 0.36 | $< 10^{-5}$ | 5.5 ± 0.5 |
| | 0.02–0.07 μm | 0.36 | 628 ($N_2$) | 455 ± 5 |
| Titanium dioxide | 110 Å | 0.35 | $< 10^{-3}$ | 16 ± 1 |
| Iron oxide ($Fe_2O_3$) | 200 Å | 0.19 | $10^{-3}$ | 14 ± 0.5 |

[b] Sample thickness, 2.54 cm; boundary temperatures, 304 and 76 K; wall emissivities greater than 0.8. The variations denoted by ± represent estimate of accuracy based on a combination of probable errors of measurement and the observed differences of the thermal conductivity of samples that were nominally the same.

[b] Mostly $SiO_2$ and $Al_2O_3$.

*Source*: Fulk (1959) and Kropśchot (1959,1963).

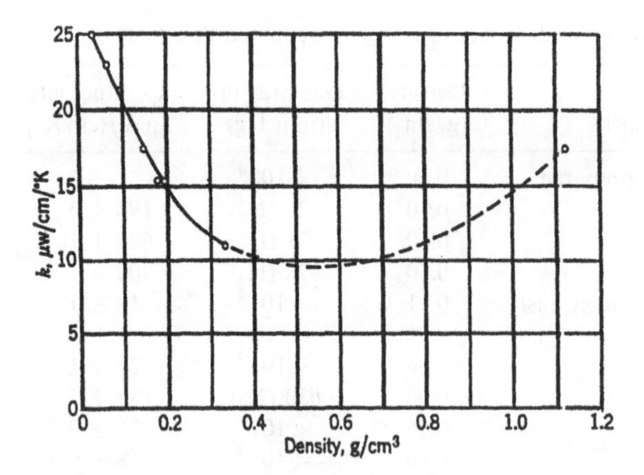


**Figure 7.10** Thermal conductivity vs. density for evacuated perlite. (Over 80% of the particles were $450 \pm 150\,\mu m$; boundary temperatures 300 and 76 K; pressure less than $10^{-4}$ mm Hg).

other hand, however, heat transfer by solid conduction becomes greater than that for radiation below liquid nitrogen temperatures; therefore, vacuum insulation alone is frequently better in this temperature range.

### *4.1.1. Convenient Formula for Porous Insulation*

A good approximation for the heat transfer through a porous insulation of constant thickness in a vessel of conservative shape is given by

$$W = \frac{\bar{k}(T_2 - T_1)}{t} \times \sqrt{A_1 A_2}$$

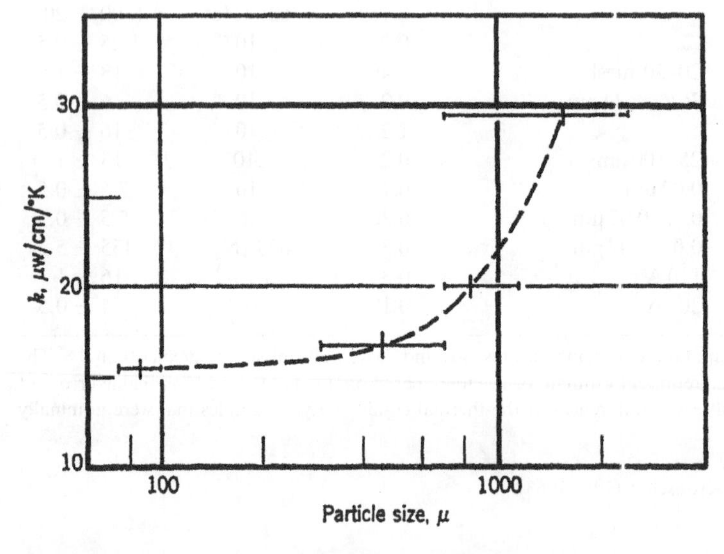


**Figure 7.11** Thermal conductivity vs. particle size for evacuated perlite (density $0.19\,g/cm^3$; boundary temperatures 300 and 77 K).

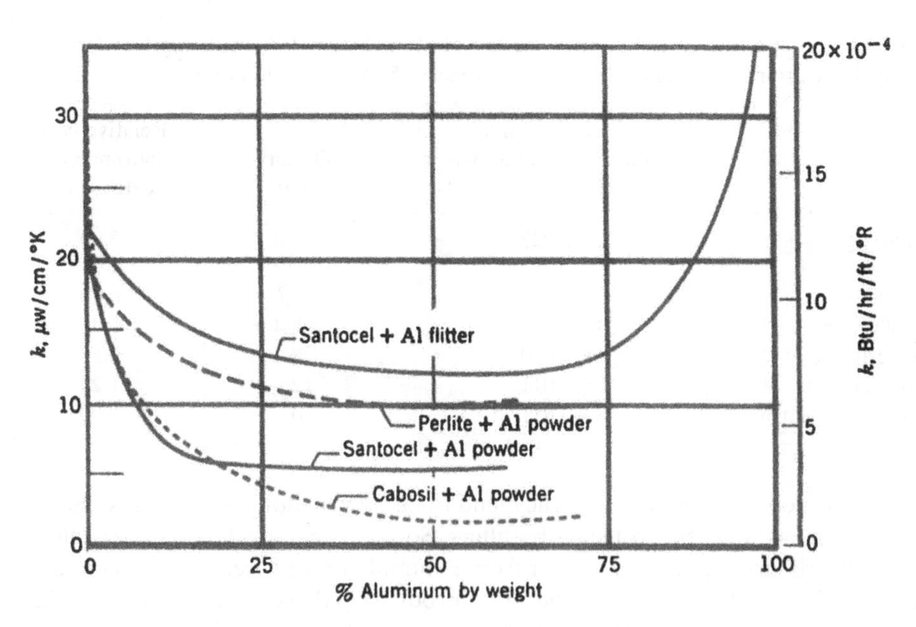


**Figure 7.12** Thermal conductivity of Santocel, perlite, and Cab-O-Sil with added aluminum between surfaces at 300 and 76 K.

where $\bar{k}$ is the mean effective thermal conductivity between the temperatures $T_2$ and $T_1$, $T_2$ the temperature of the outer surface, $T_1$ the temperature of the inner surface, $A_1$ and $A_2$ are the areas of the inner and outer surfaces, respectively and $t$ is the thickness of the insulation. This formula is exact for concentric spheres. For cylinders with spherical, dished, or elliptical heads, it has been estimated that it will give results within 5% if the thickness of the insulating space is not more than 50% of the radius of the inner cylinder. It is correspondingly reliable for other shapes in which the thickness of the insulation is a small fraction of the minor perimeter of the vessel.

In reporting values of the thermal properties of insulators, the mean effective thermal conductivity is often given between commonly encountered temperatures. If the available data are presented in the more general form of instantaneous conductivities as a function of temperature, the mean effective thermal conductivity can be obtained from the relation

$$\bar{k} = \frac{\int_{T_1}^{T_2} k \, \mathrm{d}T}{T_2 - T_1}$$

where $k$ is the thermal conductivity at temperature $T$. If $k$ is a linear function of $T$, then $\bar{k}$ is the arithmetic mean conductivity between $T_1$ and $T_2$.

### *4.1.2. Other Relevant Properties of Insulating Powders*

There are some properties of insulating powders other than thermal conductivity that should also be taken into consideration. First is the difficulty of evacuation. The gas being removed usually has to filter through the powder for rather long distances, and the resistance to flow is serious. Evacuation time can be reduced by providing a number of pumping taps or, better still, a number of pumping channels with many openings distributed so that the paths of the gas through the powder are short.

**Table 7.11** Comparison of Relevant Properties of Some Evacuated Insulating Powders Tested at the National Bureau of Standards Cryogenic Engineering Laboratory

| Material | Mean apparent thermal conductivity (300–76 K) [mW/(cm K)] | Density (lb/ft$^3$) | Relative cost, diatomaceous earth = 1.0 |
|---|---|---|---|
| Silica aerogel | 0.022 | 5.0 | 5.00 |
| Perlite | | | |
| 30 mesh | 0.017 | 6.6 | 0.40 |
| 30–80 mesh | 0.012 | 8.4 | 0.40 |
| 80 mesh | 0.010 | 8.7 | 0.50 |
| Lampblack | 0.011 | 12 | 1.20 |
| Diatomaceous earth | 0.011 | 20 | 1.00 |

Since the powders are light and fine, they tend to be carried along with the gas; therefore the openings must be protected by filters. Some of the powders, notably silica aerogel, absorb large amounts of water from a humid atmosphere. It saves pumping time to dry them by heating before they are introduced. Also evacuation time can be reduced by heating to drive off adsorbed gases while the vessel is being pumped.

The cost of the powder is important when it is used to insulate a large vessel. For transportable equipment, the density may also be of some concern. Table 7.11 lists properties of several insulating powders tested at the National Bureau of Standards.

## 4.2. Microsphere Insulation

Good thermal insulation performance has also been achieved by using packed, hollow glass spheres, typically of a size ranging from 15 to 150 μm in diameter and coated on the exterior with a film of low-emittance material such as aluminum. These hollow spheres, which generally have a wall thickness of 0.5–2.0 μm, substantially increase the conduction thermal resistance but markedly reduce the heat capacity and the mass relative to solid particles. Further, hollow microspheres offer a lightweight and low heat capacity alternative to multilayer insulation (MLI). A comparison of these two insulations is presented in Table 7.12.

## 4.3. Opacified-Powder Insulation

By adding copper or aluminum flakes to evacuated powders, the radiant heat transfer rate can be significantly reduced. The insulating ability of these opacified powder

**Table 7.12** Comparison of Microsphere Insulation with MLI

| Characteristic | Microsphere | Multilayer |
|---|---|---|
| Predictability of performance | ± 10% | ± 100% (installed) |
| Anisotropy, ratio of normal to parallel conductivity | 1 | $10^{-1}$–$10^{-4}$ |
| Bulk density, kg/m$^3$ | 70–80 | 50–60 |
| Apparent thermal conductivity, W/(m K) (installed) | 2.2–4.0 × $10^{-4}$ | 5.0–10.0 × $10^{-5}$ |
| Compressive strength, N/m$^2$ (compressive vacuum jacket application) | > 3 × $10^6$ | Nil |

*Source*: Fulk [1959] and Hunter et al. [1960].

insulations approaches an optimum when 35–50 wt% of the powder is the metal powder. Since aluminum has a large heat of combustion with oxygen, copper flakes are generally preferred.

## 5. GAS-FILLED POWDERS AND FIBROUS MATERIALS

Permeable materials of low density, such as powders and fibers with gas at atmospheric pressure in the interstices, have been used to insulate equipment in which the temperature does not fall below the boiling point of nitrogen. In this type of insulating material, the volume of the gas in the voids may be 10–100 times the volume of the solid material.

Perlite, colloidal silica, and silica aerogel have been widely used as powder insulations as well as mica, diatomaceous earth, carbon, calcium, silicate, and plastic microspheres. All are finely divided particulate materials that make solid conduction paths disjointed and discontinuous.

Figure 7.13 shows the thermal conductivities of some of the more common permeable insulations. It is seen that the best of these approach the conductivity of air, indicating that the air that occupies the voids is the agent responsible for most of the heat conduction. This suggests the principle of gas-filled insulation. The solid portion of such insulation stops radiation and gas convection. In the ideal case the conduction through the solid would be negligible and the gas in the voids would be responsible for the heat flow. In actual insulations, the solid particles of powder or the fibers short-circuit the heat flow through the gas to some extent and the resultant

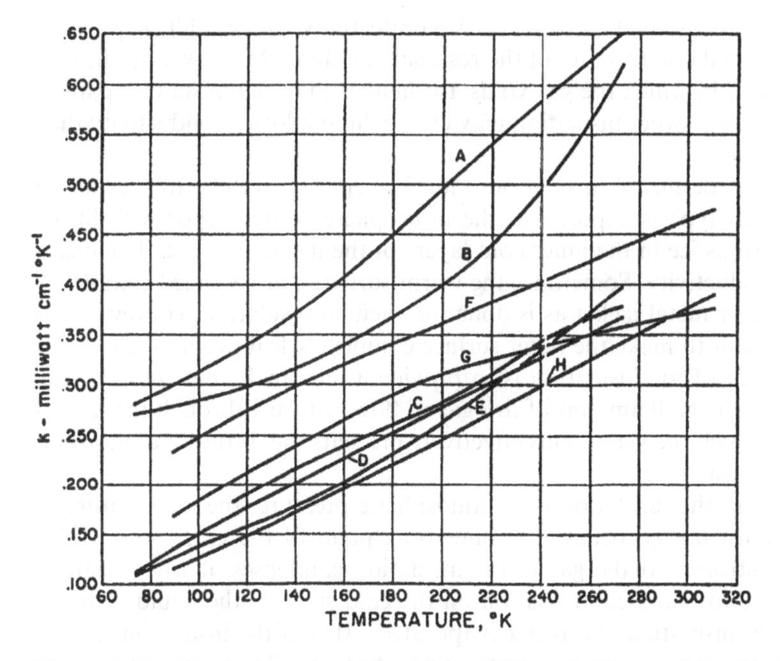


**Figure 7.13** Thermal conductivities of several permeable materials at low temperatures. A, vermiculite 10–14 mesh; B, sawdust; C, granulated cork, 50% 10–20 mesh, 50% less than 20 mesh; D, seaweed product (powder); E, crude cotton wadding; F, vegetable fiber board; G, corkboard; H, fiberglass board.

thermal conductivity is usually somewhat greater than that of the gas alone. There is one exception; very fine powders sometimes have such small distances between the solid particles that the mean free path of the gas is greater than the interstitial distance; hence its thermal conductivity is reduced as it would be by decreasing the pressure. Thus it is possible for a powder insulation to have a thermal conductivity less than that of the gas that fills the voids even when the gas is at atmospheric pressure.

The presence of the particulate solid also reduces radiation (often by about 5% of the overall conductivity) and inhibits gas conduction; thus, solid conduction and gas conduction through the voids serve as the predominant heat transfer mechanisms. The apparent thermal conductivity of gas-filled powders is given by

$$k_a = \left[\frac{V_r}{k_s} + \frac{1}{k_g/(1 - V_r) + 4\sigma T^3 d/V_r}\right]^{-1} \tag{7.20}$$

where $V_r$ is the ratio of the solid particulate volume to the total volume, $k_s$ the thermal conductivity of the solid particulate material, $k_g$ the thermal conductivity of the residual gas, $\sigma$ the Stefan-Boltzmann constant, $T$ the mean temperature of the insulation, and $d$ is the mean diameter of the fiber or powder.

At cryogenic temperatures, two assumptions can be made to simplify this expression: (1) The term containing $T^3$ is usually small relative to the term containing $k_g$ and normally can be neglected; and (2) the thermal conductivity of the solid material, $k_s$, is much greater than the thermal conductivity of the gas, $k_g$. Hence, Eq. (7.20) reduces to

$$k_a \simeq \frac{k_g}{1 - V_r}$$

This result reveals that the thermal conductivity of gas-filled powders approaches the thermal conductivity of the residual gas alone. Moreover, in the case of very fine powders, in which the gas voids are smaller than the mean free path of the gas, nonevacuated powder insulation may even exhibit a lower conductivity than the residual gas alone.

Needless to say, permeable insulations must be kept dry. If the warm surface of an insulation of this type is exposed to the atmosphere, water vapor will diffuse through and deposit as ice in the inner cold layers of the insulation, greatly increasing the thermal conductivity. Sometimes the warm surface can be sealed completely by providing an outer metal jacket as is done for vacuum insulation. However, it is not usually convenient to make the outer surface completely leakproof. A good way to keep moisture out of the insulation is to provide a few small vents in the outer boundary and allow a small amount of dry gas to flow outward through the insulation and escape through the vents. This effectively prevents the diffusion of moist air into the cold insulation.

Since the gas in the insulation is at atmospheric pressure, the temperature of the cold surface must not go below the condensing point of the gas, about 81.5 K for air, 77 K for nitrogen. If the gas in the insulation condenses, it will constitute a bad heat leak due to reflux-condensation on the cold surface, the liquid dripping off into the warmer insulation and being evaporated. Also, if the insulation is combustible, the condensation of air constitutes a real hazard. The condensate can be almost 50% oxygen, and combustible materials soaked in such a mixture are explosive. The most practical gas to use in gas-permeated insulation below 77 K is helium. However, the high thermal conductivity of helium (7.5 times that of nitrogen at

100 K) is a serious disadvantage. If helium were substituted for air in the voids of the materials listed in Fig. 7.13, one would expect that the resultant thermal conductivities would be larger by an amount approximately equal to the difference between the thermal conductivity of helium and that of air. This would mean that the conductivities of the better insulations would be raised by a factor of 5 or more. However, for a very fine powder such as silica aerogel or some grades of perlite, the ordinary thermal conductivity of helium may not apply. Since helium has a mean free path about three times that of air, its thermal conductivity would be reduced more by the small interstitial distances. Thus if it is necessary to use a permeable insulation at atmospheric pressure and at temperatures lower than 77 K, the best practical choice appears to be a very fine powder with helium as the interstitial gas. Of course, neon would be better than helium, but its cost would probably be prohibitive. Also neon could not be used below its boiling point (27 K) and so would not be applicable for insulating a liquid hydrogen vessel.

The final difficulty with powder insulations is that they tend to settle and pack due to vibrations, thermal contractions and expansions, creating higher than expected solid conduction in one area and leaving larger than expected voids in another. Consequently, care must be taken when initially packing the insulation space of a vessel with powder insulation so as to ensure the proper density of particles.

In most thermal insulation applications where vibration and shock are expected, fibers have distinct advantages because powders tend to compact whereas fibers do not. For cryogenic applications, submicrometer-diameter glass fibers have been widely used. Very similar considerations govern the use of both powder and fiber insulations. When fibers are used as spacers for multilayer insulations, to reduce the possibility of outgassing they are prepared without any binders or lubricants. Fibers with low vapor pressure binders have been used as supports to space an inner vessel from an outer shell. To ensure that multilayer insulations interposed between them are not subjected to compressive loads, such supports have been designed to take acceleration loads.

## 6. SOLID FOAMS

There is a class of insulating materials that have a cellular structure caused by the evolution of a large volume of gas during manufacture. When the cells in such a material are small and do not communicate with each other, the material has some properties that make it a useful insulator for certain low-temperature applications. Foams of this type have been made of polystyrene, polyurethanes (isocyanates), rubber, silicones, and other materials.

Since gas can penetrate such a foam only by diffusion through the cell walls, the material behaves as though it were completely impermeable when the exposure to a foreign gas lasts only a short time. Also, if the temperature of the cold side of the foam is so low that the gas in the nearby cells is condensed and has negligible vapor pressure, the thermal conductivity is greatly reduced by eliminating gaseous conduction in these regions. However, it should be remarked that the insulating value of such foams seldom approaches that of ordinary permeable materials such as powder because the solid conducting paths in the foam are continuous even though they may be tortuous. The foaming gas used in the manufacture of these foams is often carbon dioxide, which has a small vapor pressure at the temperature of liquid nitrogen. Thus when the foam is fresh, its thermal conductivity is reduced considerably by contact with liquid nitrogen. However, after the foam has been stored for several months, the foaming gas is largely replaced by air owing to slow diffusion, and the reduction of thermal conductivity by the condensation of the gas is of little consequence if the lowest temperature is that of liquid nitrogen. If the cold side is at the temperature of liquid hydrogen, the conductivity will be greatly lowered even if the cells of the foam contain air. However, if such a foam is left in an atmosphere of hydrogen or helium for a week or more, there will be another partial substitution by these gases and the thermal conductivity of the foam will be seriously increased.

Since solid foam insulations are almost impermeable to gases, they may be used to insulate regions below the condensing temperature of air or nitrogen. One of the applications is simply to hollow out a block or cylinder of foam and use it as an insulating vessel as shown in Fig. 7.14. Small vessels of this type have been quite useful. However, the logical extension of this idea—using a solid foam to insulate a large vessel made of metal—is attended with considerable difficulty. The principal cause of the trouble is the relatively high expansivity of the plastic foams.

In general, the thermal expansion coefficient of such foams is 2–5 times that of aluminum and 4–10 times that of steel. Therefore, it is necessary that the foam be designed and applied to accommodate for this difference in thermal expansion. It cannot be bonded directly to the metal surfaces because during cooldown from ambient to cryogenic temperatures the foam will shrink more than the metal and crack, leaving gaps. Water vapor and air normally enter these gaps and significantly increase the thermal conductivity of the insulation.

The dominant heat transfer mechanism is conduction through the interstitial gas, though there is also a small contribution from radiation. Thus, as with powder insulations, if the size of the voids in the foam is decreased so as to allow only free molecular gas conduction, the overall performance of the insulation can be significantly improved. Figure 7.15 shows the variation of the mean effective thermal conductivity in expanded polyurethane with cell size.

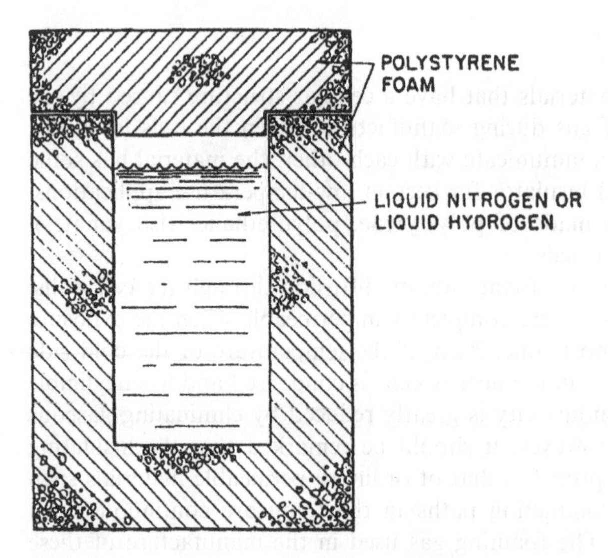


**Figure 7.14** Vessel made of plastic foam for cryogenic liquids.

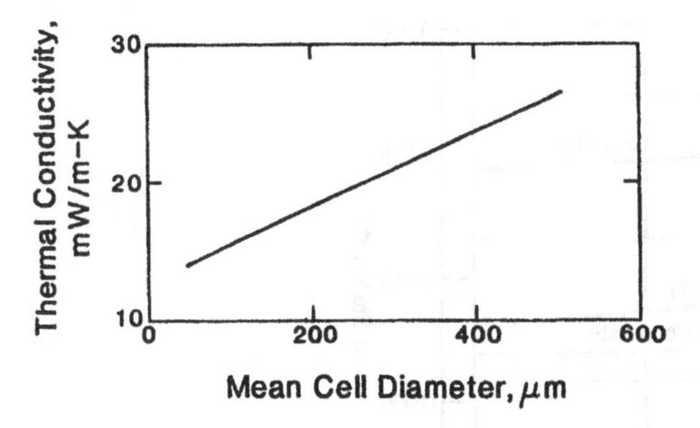


**Figure 7.15** Mean effective thermal conductivity of polyurethane foam as a function of cell size between temperatures of 300 and 77 K.

The thermal conductivities of some selected foams are shown in Table 7.13 (Kropschot, 1959,1963). In general, the thermal conductivity of a foam is determined by the interstitial gas contained in the cells plus internal radiation and a contribution due to solid conduction. Evacuation of most foams reduces the apparent thermal conductivity, as illustrated in Fig. 7.16, thus showing the partial open-cell nature of most of these materials (Kropschot, 1959,1963). Many freon-blown foams will be permeated by air over a period of time, which may increase their conductivity by as much as 30%. If hydrogen or helium is allowed to permeate the cells, the conductivity can be increased by a factor of 3 or 4. Glass and silica foams appear to be the only ones having fully closed cells. It should be noted that, even under evacuation, foam insulations have much higher conductivities than the other types of insulation previously discussed.

The tensile strength and shear strength of some foams are shown in Figs. 7.17 and 7.18 as a function of temperature.

Table 7.14 shows that a flexible polyether foam retained a degree of resilience even at liquid hydrogen temperature. The polyurethane foam became brittle and

**Table 7.13** Apparent Mean Thermal Conductivity of Some Selected Foams

| Foam | Density (g/cm$^3$) | Boundary temperature (K) | Test space pressure | Conductivity [μW/cm K)] |
|---|---|---|---|---|
| Polystyrene | 0.039 | 300,77 | 1 atm | 330 |
| | 0.046 | 300,77 | 1 atm | 260 |
| | 0.046 | 77,20 | $10^{-5}$ mm Hg | 81 |
| Epoxy resin | 0.080 | 300,77 | 1 atm | 330 |
| | 0.080 | 300,77 | $10^{-2}$ mm Hg | 168 |
| | 0.080 | 300,77 | $4 \times 10^{-3}$ mm Hg | 130 |
| Polyurethane | 0.08–0.14 | 300,77 | 1 atm | 330 |
| | | | $10^{-3}$ mm Hg | 120 |
| Rubber | 0.08 | 300,77 | 1 atm | 360 |
| Silica | 0.16 | 300,77 | 1 atm | 550 |
| Glass | 0.14 | 300,77 | 1 atm | 350 |

*Source*: Kropschot (1959,1963).

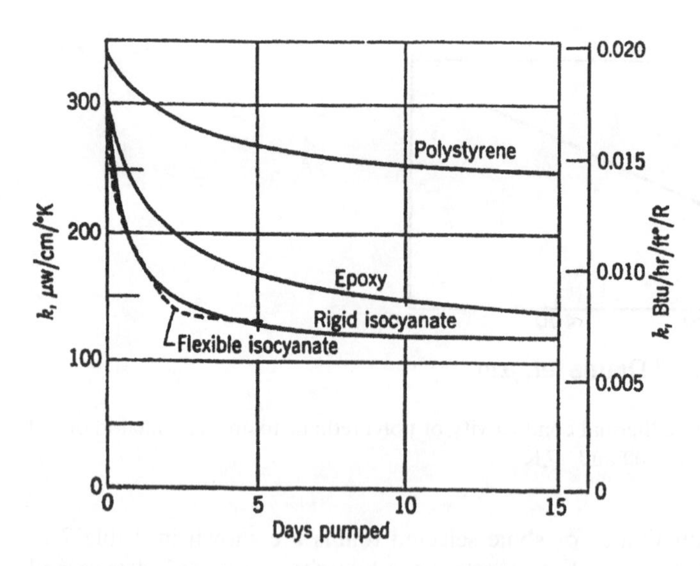


**Figure 7.16** Apparent mean thermal conductivity between surfaces at 300 and 76 K of some selected foams, as a function of time under evacuation.

yielded, as did the urethane foam. The epoxy foam had satisfactory properties at this temperature, although it would not withstand 50 psi at room temperature. These tests generally indicated that none of the tested foams would be fully satisfactory. The flexible foams would be too compressible to retain their thermal characteristics under pressure loadings typically encountered by internal insulation; similarly, if they were to be used as external insulations, the flexible foams would be too

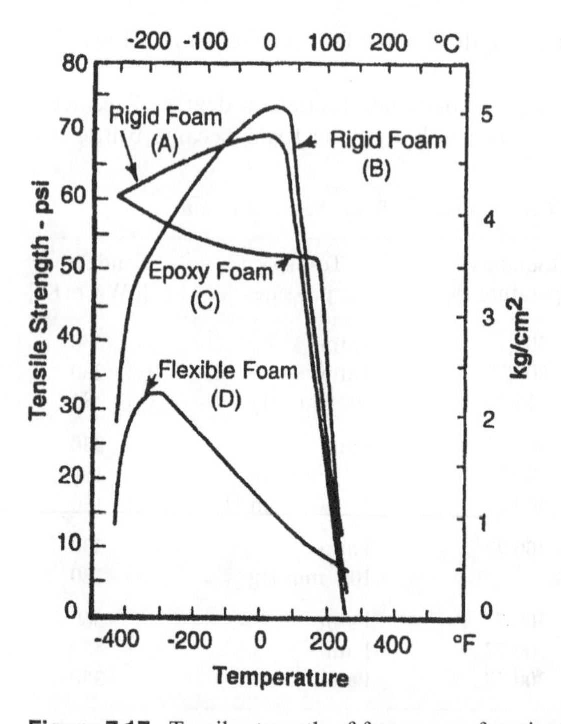


**Figure 7.17** Tensile strength of foams as a function of temperature.

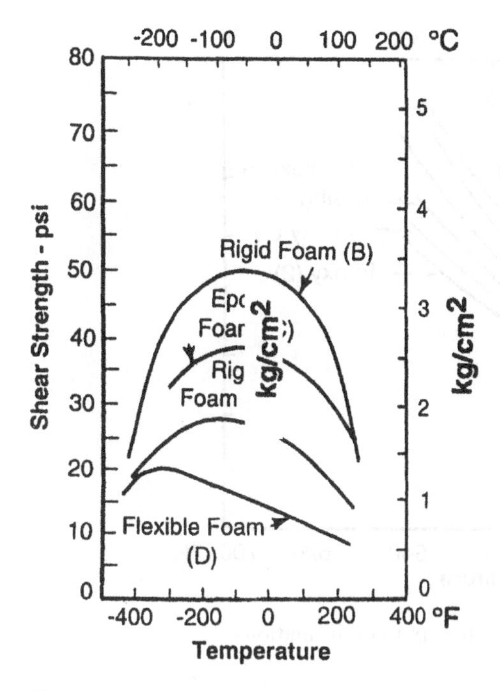


**Figure 7.18** Shear strength of foam as a function of temperature.

compressible to retain these thermal characteristics under most expected surface loadings. Although most rigid foams lack the necessary strength at either room or cryogenic temperatures, they are used in some applications.

## 6.1. Thermal Expansion

The thermal expansion properties of various foams are shown in Fig. 7.19. This figure shows that although the epoxy foams pose problems, they have the lowest thermal expansion coefficient. It also indicates that all of the foams have reasonably linear thermal expansion characteristics in the cryogenic range. The characteristics become nonlinear only at and above room temperature.

## 6.2. Thermal Conductivity

Typical thermal conductivity for carbon dioxide-blown urethane foams is about 0.23 Btu in./(h ft$^2$°F) [0.33 mW/(cm °C)]. For fluorocarbon-blown foams of

**Table 7.14** Load Compression Strength of Foams at −423°F (−253°C)

| | Yield | | | Compression(%) | |
|---|---|---|---|---|---|
| Foam | psi | kg/cm$^2$ | Set(%) | At 25 psi (1.7 kg/cm$^2$) | At 50 psi (3.4 kg/cm$^2$) |
| Rigid urethane | 28.0 | 1.9 | 68.7 | – | 30.4 |
| Epoxy | – | – | 2.8 | 4.0 | 6.8 |
| Rigid polyrethane | 25.0 | 1.7 | 67.4 | 10.9 | 82.5 |
| Flexible polyether | – | – | 7.6 | 82.6 | 86.9 |

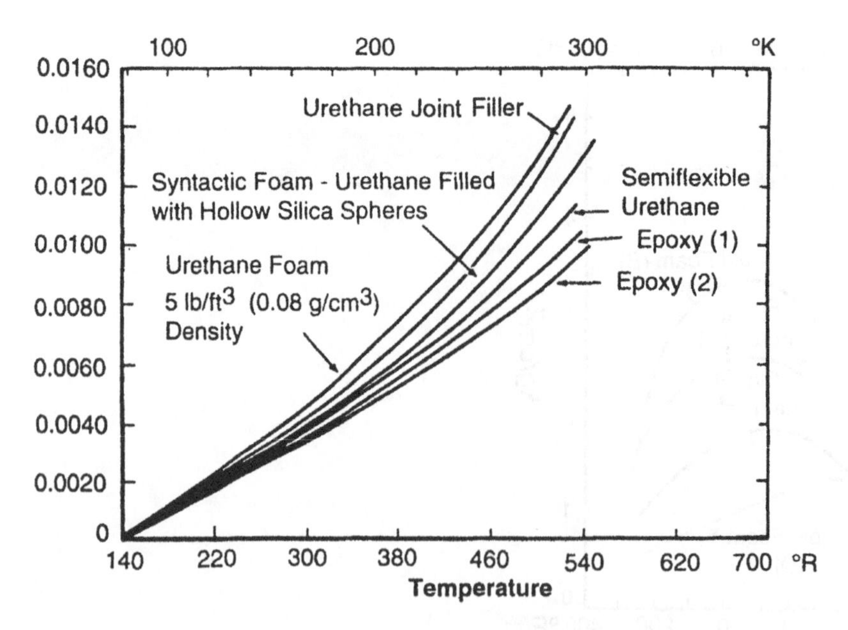


**Figure 7.19** Linear expansion vs. temperature for six foam insulations.

approximately 2 lb/ft$^3$ density, thermal conductivity is about 0.12 Btu in./(h ft$^2$ °F) [0.17 mW/(cm °C)]. Fluorocarbon-blown foams will maintain their low thermal conductivity only if the surfaces are sealed to prevent diffusion of the fluorocarbon out of the cells. In thin sections, the fluorocarbon will diffuse over a period of time, and the thermal conductivity will gradually increase to about 0.17 Btu in./(h ft$^2$ °F) [0.25 mW/(cm °C)]. For polystyrene foams, the room temperature thermal conductivity is in the range of 0.23–0.26 Btu in./(h ft$^2$ °F) [0.33–0.38 mW/(cm °C)].

The thermal conductivity of various foams has been measured over the temperature range of 75 to −423°F (21 to −253°C). Data were obtained on polystyrene foams, polyurethane foams, composite insulations of foam-filled honeycomb, and tetrafluoroethylene (TFE) sheet materials. These data indicate that for all of the foams tested the thermal conductivity at liquid hydrogen temperatures is in the range of 0.05–0.10 Btu in./(hft$^2$ °F) [0.07–1.4 mW/(cm °C)]. As room temperature is approached, these values increase and differences become more pronounced. The differences are attributable to the base resin and in large part to the blowing agent used in making the foam.

At liquid hydrogen temperatures, the gases in the cells liquefy or solidify, resulting in a partial vacuum in each cell. At this point, because convection is no longer a mechanism of heat transfer in the foam, the size and uniformity of the cells become the critical factors in determining thermal conductivity. In TFE sheet and in TFE sheet-TFE foam composite, the reduction in thermal conductivity achieved through the use of the foam is pronounced, even though the conductivity is above that of urethane and styrene foams.

Measurements were also made on the thermal conductivity of foams in the presence of helium gas and of air. At a mean temperature of about −385°F (−231°C), the conductivity was in the range of 0.2–0.3 Btu in./(h ft$^2$ °F) [0.29–0.43 mW/(cm °C)] in comparison with 0.05 Btu in./(hft$^2$ °F) [0.072 mW/(cm C)] in air. These data point out the increase in thermal conductivity caused by helium diffusion into evacuated or sealed insulations. In large space vehicle insulation systems, helium purging is done to reduce potential hazards and to prevent weight increase by atmospheric constituents leaking into the sealed insulation system and condensing or solidifying.

## 7. MULTILAYER INSULATION

Multilayer insulation (MLI) consists of highly reflecting radiation shields separated by spacers or insulators in high vacuum. Some forms have yielded the lowest thermal conductivity of any bulk cryogenic insulation developed to date. When properly applied, the insulation can have an apparent mean thermal conductivity between 300 and 20 K of 0.3–0.6 μW/(cm K). This insulation was first developed by Petersen (1958) and later was improved and perfected by many others.

The assembled insulation is placed perpendicular to the flow of heat. Each layer contains a thin, low-emissivity radiation shield, enabling the layer to reflect a large percentage of the radiation it receives from a warmer surface. The radiation shields are separated from each other to reduce the heat transferred from shield to shield by solid conduction. The gas in the space between the shields is removed to decrease the conduction by gas molecules. Each component of the insulation is designed specifically to perform a particular function:

Radiation shields: to attenuate radiation.
Spacers: to decrease solid conduction.
Evacuation: to decrease gas conduction.

The lowest values of thermal conductivity have been obtained by using layers of aluminum foil separated by fiberglass paper. However, many other materials can be used. Some typical choices are shown in Table 7.15.

For low-temperature applications, the radiation-reflecting shield is generally a 6 μm aluminum sheet; however, for greater strength and ease in application, it is often a thin (5–76 μm) plastic material such as polyethylene terephthalate (Mylar) or polyimide (Kapton) coated on one or both sides with a thin vapor-deposited layer ($\sim 1 \times 10^{-8}$ m) of high-reflectance metal such as aluminum. Spacer materials range from very coarse silk or nylon net to more continuous substances such as silica fiber felt, low-density foam, or fiberglass mat. Most common among these are glass fiber spacers, namely Dexiglas (71 μm nominal thickness) and Tissuglas (15 μm nominal thickness).

**Table 7.15** Choices of Reflecting and Insulating Materials for Fabrication of Multilayer Insulation

| Reflector | Insulator |
|---|---|
| Aluminum foil | Fiberglass paper |
| Copper foil | Fiberglass mat (bonded) |
| | Fiberglass (unbonded) |
| | Glass fabric |
| | Nylon net |
| | Finely divided powder |
| Aluminized Mylar | Aluminized Mylar |

**Table 7.16** Apparent Mean Thermal Conductivity of Eight Fiberglass Paper-Aluminum Foil Insulations

| Paper thickness (cm) | Sample thickness (cm) | Number of shields (per cm) | Density ($g/cm^3$) | Cold wall temperature[a] (K) | Thermal conductivity [μW/(cm K)] |
|---|---|---|---|---|---|
| 0.02 | 3.3 | 22 | 0.16 | 76 | 0.55 |
| | | | | 20 | 0.4 |
| 0.02 | 3.8 | 20 | 0.14 | 76 | 0.5 |
| | | | | 20 | 0.4 |
| 0.02 | 2.5 | 20 | 0.14 | 76 | 0.7 |
| | | | | 20 | 0.6 |
| 0.02 | 3.0 | 20 | 0.18 | 76 | 1.3 |
| 0.012 | 3.8 | 30 | 0.13 | 76 | 1.0 |
| 0.012 | 4.3 | 14 | 0.06 | 76 | 0.7 |
| 0.02 | 1.9 | 24 | 0.18 | 76 | 1.4 |
| 0.012 | 1.9 | 21 | 0.09 | 76 | 0.9 |
| | | | | 20 | 0.5 |

[a] Warm wall temperature, 300 K.
*Source*: Kropschot (1959,1963).

As mentioned, this assembly is placed perpendicular to the flow of heat. The overlapping reflecting shields attenuate radiation, and the spacer material decreases solid conduction between the shields. The entire system is then evacuated and baked out at an elevated temperature for an extended period of time to remove as much of the adsorbed residual gases as possible. To maintain the desired vacuum after sealing off the vacuum space, a getter material such as activated charcoal or molecular sieve (cooled by thermal contact with the inner vessel) is normally employed to adsorb gases, which will desorb with time from surfaces in the vacuum space. When the optimum density of multilayer insulation is properly applied, it can have an apparent thermal conductivity as low as $1.0 \times 10^{-5}$ W/(m K) between temperature boundaries of 20.3 and 300 K.

The results of tests of eight representative aluminum foil and glass fiber paper insulations are shown in Table 7.16.

## 7.1. Spacers

Multilayer insulations can be classified according to the type of spacers separating the radiation shields. Four kinds of spacers are discussed below; multiple-resistance, point contact, single-component, and composite.

### *7.1.1. Multiple-Resistance Spacers*

Two surfaces in contact, when not forced against each other, present a relatively high resistance to heat flow at the points of contact. Therefore, a material such as a fibrous mat arranged with fibers in a parallel array, in which the heat must pass from fiber to fiber to reach the next radiation shield, is an effective spacer. The finer the fibers, the greater the resistance per unit thickness the mat presents to the heat flow. Any fibers that are not in the plane perpendicular to the heat flow will act as thermal shorts across the mat.

Glass fibers (as used in glass wool or glass paper), quartz fibers, and plastic fibers (polyester mats and others) are being used as spacers in multilayer insulations. Fibrous mats are produced in various thicknesses ranging from 0.003 to 0.030 in. (75–750 μm). Fiber diameters range from 0.5 μm ($2 \times 10^{-5}$ in.) to several hundred micrometers.

The thermal performance of mats with fibers not bonded to each other is better than that of mats with bonded fibers, in which heat can flow across the bonded contact areas. However, the tensile strength of unbonded fiber mats is limited, and they are more difficult to handle. Fibers are either partially bonded or stitched to make the mat, thus increasing its tensile strength without greatly decreasing its thermal performance.

### *7.1.2. Point Contact Spacers*

Another method of separating neighboring radiation shields is to place particles along the surface of the shields. If the typical dimension of the individual particle is small compared to the distance between two particles along the shield, then the contacts between the shields can be considered to be point contacts. The concept is illustrated in Fig. 7.20 (Black and Glaser, 1966). In Fig. 7.20a, small spheres are placed between the shields. The distance from sphere to sphere is selected to be a maximum, with the upper limit determined by the sagging of the shields that cause direct contact between them. In Fig. 7.20b, point contacts between neighboring radiation shields are provided by embossing the spacer material.

Unattached spheres are impractical because the spheres will not maintain their positions. A grid of spheres may be approximated by a screen (Fig. 7.20c), whose crossover points (knots) are spheres formed of a thread that is only half the thickness of the knot and does not contact both radiation shields but keeps the spheres in position. Several insulations have been developed using screens made of nylon, silk, and vinyl-coated fiberglass as spacers. The thickness of the screens ranges from 0.003 to 0.025 in. (75–600 μm). Screen mesh sizes ranging from 1/16 to 1/4 in. have been investigated. Screens are dimensionally stable spacer material with tensile strength sufficient to simplify application. An additional benefit is derived from the open

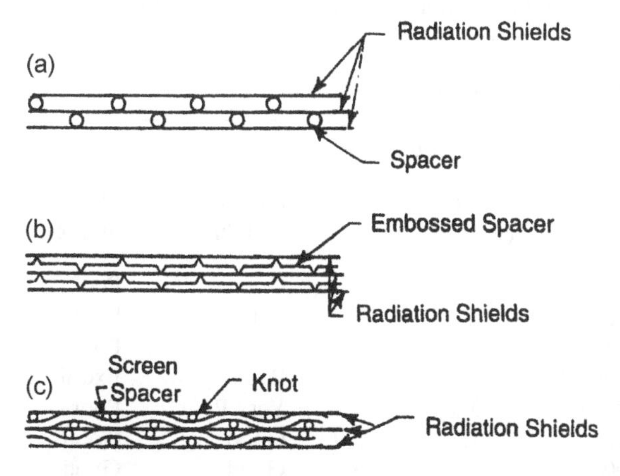


**Figure 7.20** Examples of point contact spacers: (a) spheres between shields, (b) point contacts between shields, and (c) screens between shields.

structure of a screen; that is, the evacuation of gas from a multilayer insulation with a screen spacer is greatly eased.

### 7.1.3. *Single-Component Multilayer Insulation*

Radiation shields can be spaced from each other without using any spacer material. Embossing or crinkling the shields produces random small-area contacts that create contact resistances high enough to reduce conductive heat transfer to a value comparable to the amount of heat transferred by radiation through the same assembly of radiation shields.

This method of spacing can only be used with radiation shields of low conductivity such as thin sheets of polyester film [0.00025 in. (6.3 μm) thickness] with a metal coating on only one side.

### 7.1.4. *Composite Spacers*

Some spacers consist of two or more materials with each material selected to perform a specific function. A composite spacer may consist of a very light, dimensionally stable, thin material of tensile properties that will permit easy application of the insulation. The thermal properties of this material are not of primary importance, because it is used only as a "carrier" to which a material of desired thermal properties is attached. The second material is not required to have any specific tensile strength or dimensional stability, only acceptable thermal properties. This concept permits a designer to select materials best suited to specific functions and the desired performance for the complete assembly.

### 7.1.5. *Spacer Materials*

Table 7.17 lists typical materials that can be used as spacers for multilayer insulation. Depending on the required performance, one or more layers of the spacer material may be applied between neighboring radiation shields.

To reduce the weight of the spacer, the material may be perforated, as long as the perforation is consistent with the conduction heat transfer limitations. Reducing the load-carrying areas of the spacer materials requires that the remaining areas carry increased compressive loads. A balance between the desire to reduce weight

**Table 7.17** Comparison of Typical Spacer materials

| Spacer material | Thickness (in.) | Weight per area lb/ft$^2$ ($\times 10^{-3}$) | Weight per area g/m$^2$ | Dimensional stability | Tensile strength |
|---|---|---|---|---|---|
| Foam | 0.020 | 3.3 | 16 | Good | Poor |
| Glass paper | 0.08 | 10.9 | 53 | Fair | Poor |
| | 0.04 | 6.1 | 30 | Fair | Poor |
| Crinkled polyester film | 0.00025 | 1.5 | 7.3 | Poor | Excellent |
| Glass fabric | 0.0015 | 3.8 | 19 | Very Poor | Poor |
| Silk screen | 0.003 | 1.2 | 5.9 | Good | Good |
| Nylon screen | 0.007 | 2.7 | 13 | Good | Good |
| Vinyl-coated fiberglass screen | 0.020 | 15.4 | 75 | Excellent | Excellent |

**Table 7.18** Apparent Mean Thermal Conductivity of Some Selected Multilayer Insulations of Aluminum Foil Spaced with Various Materials

| Sample thickness (cm) | Number of shields (per cm) | Density (g/cm$^3$) | Cold wall temperature[a] (K) | Thermal conductivity [μW/(cm K)] | Spacer material |
|---|---|---|---|---|---|
| 3.7 | 26 | 0.12 | 76 | 0.7 | Glass fiber mat |
| | | | 20 | 0.5 | |
| 2.5 | 24 | 0.09 | 76 | 2.3 | Nylon net |
| 1.5 | 76 | 0.76 | 76 | 5.2 | Glass fabric |
| 4.5 | 6 | 0.03 | 76 | 3.9 | Glass fiber (unbonded) |
| 2.2 | 6 | 0.03 | 76 | 3.0 | Glass fiber (unbonded) |

[a] Warm wall temperature, 300 K.

and the capability of the spacer material to carry loads without suffering a substantial increase in heat transfer has to be established for each spacer material.

Other low-conductivity spacer materials, such as glass fiber mat, nylon net, and glass fabric, have been used to separate aluminum foil shields. Some results are shown in Table 7.18. Table 7.19 shows the results using aluminized Mylar with no spacer material.

## 7.2. Radiation Shields

The primary requirement for a radiation shield is low emissivity. Silver, aluminum, and gold are low-emissivity materials that can be used either as coatings for radiation shields or to form thin foils. A comparison of the emissivity of metal-coated films is shown in Fig. 7.21. Aluminum and aluminum-coated plastic films are most frequently selected for the radiation shields because (1) the emissivity of aluminum is only slightly higher than that of clean silver, but, whereas silver tarnishes in air, aluminum forms a very thin layer of aluminum oxide that prevents further degradation of the surface, and (2) aluminum is inexpensive and readily available in various thicknesses of foil and as a coating on a variety of metallic and nonmetallic surfaces.

In applications where weight is at a premium, 0.00025 and 0.0005 in. thick (6–12 μm) aluminum foils are used for radiation shields even though these foils have very little tensile strength and are difficult to handle. Heavier foils (0.001–0.0005 in., 25–125 μm) find only limited application where the weight is of secondary importances to (1) ease of handling, (2) the tendency not to bridge between the support points, and (3) the maintenance of a given structural shape.

**Table 7.19** Apparent Mean Thermal Conductivity of Aluminized Mylar[a]

| Sample thickness (cm) | Number of layers (per cm) | Density (g/cm$^3$) | Cold wall temperature (K) | Thermal conductivity [μW/(cm K)] |
|---|---|---|---|---|
| 3.5 | 24 | 0.045 | 76 | 0.85 |
| 1.3 | 47 | 0.09 | 76 | 1.8 |

[a] Aluminum on one side only.

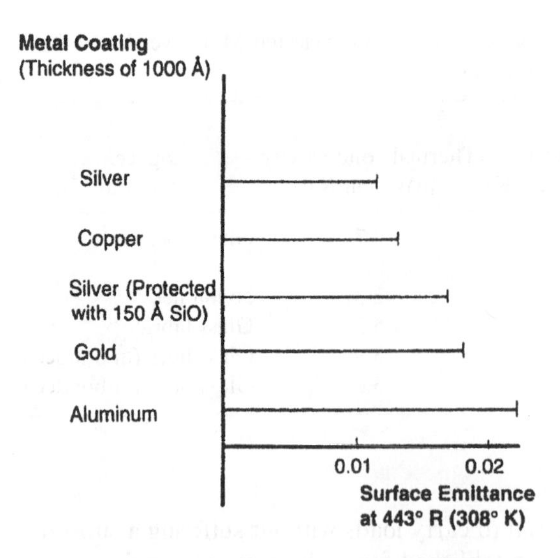


**Figure 7.21** Emissivity of metal-coated films.

Use of 0.00025–0.0005 in. thick metal-coated plastic films results in a further reduction in weight (plastic films are approximately one-half the weight of aluminum foils of equivalent thickness). In addition to the weight saving, 0.00025 in. polyester film offers a tear strength far superior to that of even 0.0005in. aluminum foil. Most polyester films, however, are limited to operating temperatures below 300°F (150°C), the point at which the film begins to deteriorate [although polyimide film can withstand temperatures as high as 750°F (400°C)]. Aluminum foils can be used in temperatures up to 1000°F (540°C).

The plastic film can be coated with metal on one or both sides, depending on the application. Single-component multilayer insulations are normally coated on one side to prevent metal-to-metal contact between neighboring radiation shields. The emissivity of an uncoated side is higher than that of a coated side (approximately 0.35 for 0.00025 in. thick film). More radiation shields are required to reduce the radiation heat transfer to the level that can be achieved by coating both sides.

Tests have shown that metal coatings less than 500Å thick are transparent to the radiation, while reflective properties of coatings more than 1000 Å thick change very little with the coating thickness (see Fig. 7.22). Because of the present state of

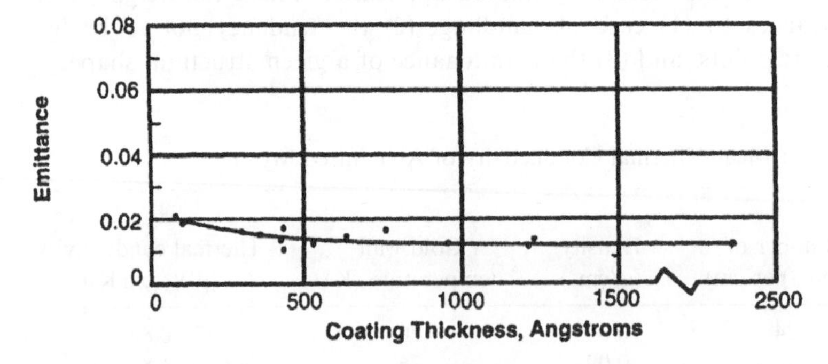


**Figure 7.22** Emissivity of polyester film coated with silver. (Silver coating was produced in 18 in. vacuum jar.)

the art, a metallic coating is formed on the plastic film by deposition of a vaporized pure metal in a vacuum. The process produces a clean, uniform, highly reflective metal layer on the film.

Aluminum vaporizes at a lower temperature than gold, making the aluminum deposition process easier to control. Plastic films with an aluminum deposit have been used for decorative purposes in industry for many years. As a result, aluminum-coated films are less expensive and of a better average quality than gold- or silver-coated films. (The amount of metal used in the deposition process is small, so the cost of the gold causes only a negligible increase in the price of the gold-coated film over that of aluminum-coated film.)

Checks of the thickness of metal deposits can be made in several ways. Two simple checks used for quality control are the measurement of electrical resistance and the weight of the deposited metal. (The weight of the deposited metal is determined by weighing the metal-coated plastic film before and after removal of the metal coating by a solvent.)

## 7.3. Typical Performance of Multilayer Insulation

Table 7.20 shows various combinations of spacers and radiation shields. It summarizes the performance of several multilayer insulations at zero compressive load and at loads of 15 psi. The insulations were tested between room temperature and the temperature of a boiling cryogen (liquid nitrogen or liquid hydrogen). To obtain the best possible performance, the tests were carried out under carefully controlled conditions.

The performance of any given insulation in an actual installation will be greatly affected by the following variables:

Applied compressive load.
Number of shields used in the sample.
Kind of gas filling the insulation and the pressure of that gas.
Size and number of perforations in the insulation to permit outgassing.
Temperatures of the warm and cold boundaries.

### *7.3.1. Effects of Number of Radiation Shields*

In theory, the heat flux passing through an uncompressed sample of multilayer insulation is inversely proportional to the sample thickness (i.e., number of shields), and therefore its thermal conductivity can be evaluated. Figure 7.23 shows that in the first approximation the heat flux is indeed inversely proportional to the number of shields. However, the experimentally obtained heat flux data for a sample with 40 shields are somewhat higher than predicted from the five-shield sample data. In Fig. 7.23, the solid line that passes through two experimental points for 5- and 10-shield samples predicts heat fluxes for 20- and 40-shield samples lower than were actually observed. This discrepancy can be explained by the compression exerted by the weight of the upper layers on the lower layers of the sample, which may cause the lower layers to perform less efficiently than the upper layers. If an accurate estimate of the heat flux is required, a correction factor for the effect of compression must be applied.

As noted earlier, radiant heat transfer is directly proportional to the emissivity of the shields between the warm and cold surfaces and inversely proportional to the

**Table 7.20** Thermal Conductivity of Typical Multilayer Insulation

| Material | Thickness (in.) | Properties at optimum density | | | | | Properties at 15 psi (1 kg/cm$^2$) compression | | | |
|---|---|---|---|---|---|---|---|---|---|---|
| | | Apparent thermal conductivity | | Density | | | Apparent thermal conductivity | | Density | |
| | | Btu in. (h ft °F) ($\times 10^{-4}$) | μW (cm °C) | lb/ft$^3$ | mg/cm$^3$ | Number of shields/in. | Btu in. (h ft$^2$ °F) ($\times 10^{-4}$) | μW (cm °C) | lb/ft$^3$ | mg/cm$^3$ |
| *Systems with 0.00025 in. thick aluminum-coated polyester radiation shields* | | | | | | | | | | |
| No spacer, crinkled shield aluminized on one side only | – | 2.0 | 0.29 | 1.4 | 22 | 67 | 410 | 59 | 29 | 470 |
| Polyurethane foam 2 lb/ft$^3$ (32 mg/cm$^3$) density | 0.020 | 1.0 | 0.14 | 1.4 | 22 | 151 | 68 | 10 | 3.0 | 48 |
| Polyurethane foam 2 lb/ft$^3$ density 11% support area | 0.020 | 0.8 | 0.12 | 0.83 | 13 | 20 | – | – | – | – |
| Polyurethane foam 2 lb/ft$^3$ with polyurethane foam grid $\frac{1}{4}$ in. wide on $1\frac{1}{2}$ in. centers | 0.020 | 2.4 | 0.35 | 1.3 | 21 | 9 | – | – | – | – |
| Two layers fiberglass cloth | 0.001 | 1.3 | 0.19 | 5.2 | 83 | 47 | 70 | 10 | 48 | 770 |
| One layer fiberglass cloth | 0.001 | 1.0 | 0.14 | 13.6 | 218 | 210 | 830 | 120 | 30 | 480 |
| $\frac{1}{8}$ in $\times$ $\frac{1}{8}$ in. mesh vinyl-coated fiberglass screen | 0.020 | 4.3 | 0.62 | 6.4 | 103 | 30 | – | – | – | – |
| silk screens (ea) | 0.003 | 3.0 | 0.43 | 1.4 | 22 | 30 | 42 | 6.1 | 12 | 190 |
| $3\frac{1}{16}$ in. $\times$ $\frac{1}{16}$ in. mesh silk screens (ea) | 0.003 | 2.5 | 0.36 | 1.4 | 22 | 22 | 18 | 2.6 | 16 | 260 |
| One $\frac{1}{16}$ in. $\times$ $\frac{1}{16}$ in. mesh silk screen with fiberglass paper | 0.008 | 3.0 | 0.43 | 6.0 | 96 | 19 | 58 | 8.4 | 15 | 240 |
| $\frac{1}{4}$ in. wide fiberglass strips | 0.008 | 2.0 | 0.29 | 1.9 | 31 | 22 | 41 | 5.9 | 11 | 180 |
| $\frac{1}{4}$ in. wide fiberglass strips | 0.004 | 4.2 | 0.60 | 1.7 | 27 | 20 | 41 | 5.9 | 13 | 210 |
| $\frac{1}{4}$ in. wide flexible foam strips | 0.100 | 5.5 | 0.79 | 0.74 | 12 | 9 | 180 | 26 | 9 | 140 |
| $\frac{1}{4}$ in. wide rigid foam | 0.100 | 3.6 | 0.52 | 0.62 | 10 | 10 | 130 | 19 | 4.4 | 71 |

| | | | | | | | | | | |
|---|---|---|---|---|---|---|---|---|---|---|
| Fiberglass paper 50% perforated | 0.008 | 2.0 | 0.29 | 3.7 | 59 | 32 | 87 | 13 | 20 | 320 |
| *Systems with 0.0005 in. thick soft aluminium radiation shields* | | | | | | | | | | |
| Three layers fiberglass cloth | 0.001 | 0.9 | 0.12 | 10 | 160 | 52 | 335 | 48 | 37 | 590 |
| Fiberglass mat | 0.014 | 1.0 | 0.14 | 5 | 80 | 22 | 44 | 6.4 | 13 | 210 |
| Fiberglass paper | 0.003 | 1.2 | 0.17 | 3 | 48 | 36 | 76 | 11 | 24 | 380 |
| *Systems with 0.002 in. thick H19 aluminium radiation shields* | | | | | | | | | | |
| Nylon screen | 0.007 | 0.43 | 0.062 | 15 | 240 | 40 | 160 | 23 | 35 | 560 |
| $\frac{1}{8}$ in × $\frac{1}{8}$ in. mesh vinyl-coated fiberglass screen | 0.020 | 1.6 | 0.23 | 16 | 260 | 30 | – | – | – | – |
| Impregnated fiberglass mat | 0.020 | 1.2 | 0.17 | 12 | 190 | 22 | 80 | 12 | 32 | 510 |
| Impregnated fiberglass mat with 89% perforation | 0.020 | 0.7 | 0.10 | 10 | 160 | 23 | – | – | – | – |
| Same | 0.080 | 3.3 | 0.48 | – | – | 7 | – | – | – | – |
| Fiberglass paper 50% perforated | 0.008 | 1.6 | 0.23 | 13 | 210 | 33 | 150 | 22 | 50 | 800 |
| *Systems with other radiation shields 0.0001 in. thick* | | | | | | | | | | |
| Silver-coated polyester film silk Screen with $\frac{1}{4}$ in. wide flexible Foam strips | 0.0005 0.003 0.100 | 5.4 | 0.78 | – | – | 8 | – | – | – | – |
| Gold-coated polyester film with 2 layers of silk screen per spacer | 0.00025 0.003 | 3.4 | 0.49 | 1.0 | 16 | 24 | – | – | – | – |

*Source*: Timmerhaus and Flynn (1989).

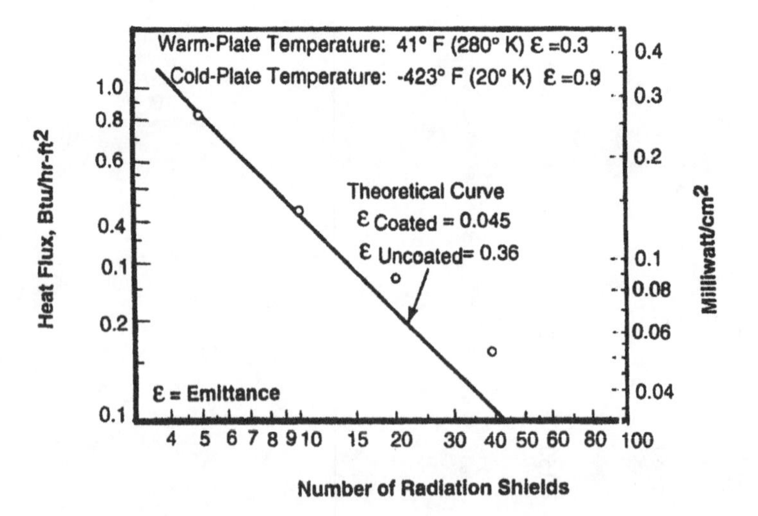


**Figure 7.23** Effect of number of radiation shields on net heat flux. (Radiation shields: crinkled aluminum polyester film; spacers: $\frac{1}{8} \times \frac{1}{8}$ in. mesh vinyl-coated fiberglass screen.)

number of these shields. An illustration of the effect of the number of shields on heat flux is shown in Fig. 7.24. Thus, by using many overlapping layers of low-emissivity material, radiation is significantly reduced. Further, the low conductivity of the fiberglass used as spacer material, the irregular size and geometry of the fibers, and the numerous discontinuities of the fibers all tend to reduce solid conduction.

A critical design variable for multilayer-insulated vessels is the most economical number of layers of insulation per unit thickness. If the insulation is compressed too tightly, the increase in solid conductivity outweighs the decrease in radiative heat. A typical dependency of the apparent thermal conductivity on layer density for MLI is presented in Fig. 7.25.

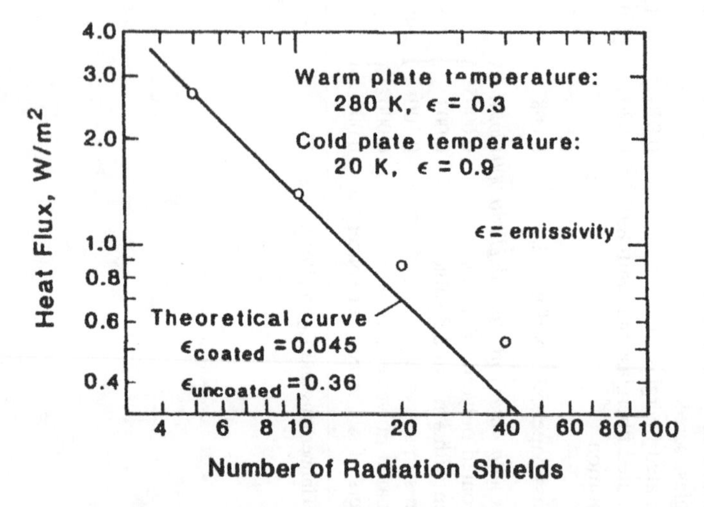


**Figure 7.24** Effect of number of radiation shields on net heat flux. (Shields: crinkled aluminum polyester film; spacers: 0.003 × 0.003 m mesh vinyl-coated fiberglass screen.) (From Black and Glaser, 1966.)

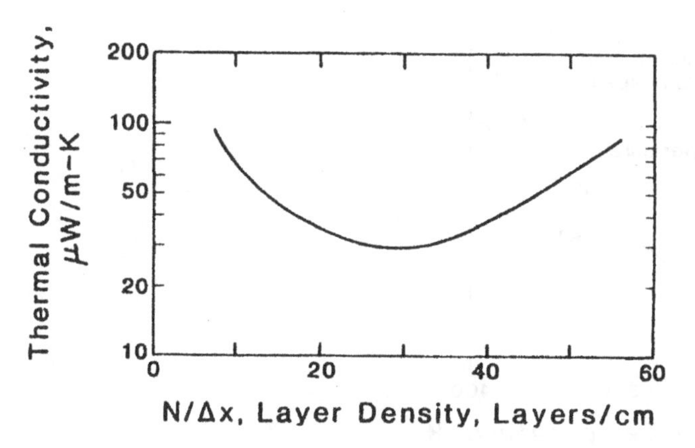


**Figure 7.25** Variation of thermal conductivity with layer density for a typical MLI with boundary temperatures of 294 and 78 K.

The bulk density of an MLI is a combination of the thickness and density of the reflective shields, the mass of the spacer material per unit area, and the number of layers per unit thickness, or

$$\rho_{\text{bulk}} = (m_s + \rho_r t_r (n/\Delta x) \tag{7.21}$$

where $\rho_{\text{bulk}}$ is the bulk density of the MLI, $m_s$ the mass of the spacer material, $\rho_r$ the density of the reflective shield material, $t_r$ the thickness of the radiation shields, and $n/\Delta x$ the number of insulation layers per unit thickness. Here one layer is defined as the combination of a reflecting shield and a spacer. Typical densities range from 32 to 320 kg/m$^3$

For highly evacuated MLI, on the order of $1.3 \times 10^{-10}$ MPa, heat is transferred primarily by radiation and solid conduction. The apparent thermal conductivity for MLI under these conditions can be obtained from

$$K_a = \frac{1}{n/\Delta x}\left[h_s + \left(\frac{\sigma e T_2^3}{2-e}\right)\left[1 + \left(\frac{T_1}{T_2}\right)^2\right]\left(1 + \frac{T_1}{T_2}\right)\right] \tag{7.22}$$

where $h_s$ is the solid conductance of the spacer material, $\sigma$ the Stefan–Boltzmann constant, $e$ the effective emissivity of the reflecting shields, and $T_1$ and $T_2$ are the temperatures of the cold and warm boundary surfaces, respectively. Equation (7.22) indicates that the thermal conductivity varies with the warm and cold boundary temperatures of the MLI. The curve in Fig. 7.26 shows that with a constant warm boundary temperature, a higher apparent thermal conductivity is observed as the cold boundary temperature is raised.

*Example 7.3.* Determine the mean apparent thermal conductivity of a multilayer insulation (a) between 294.4 and 20.4 K and (b) between 20.4 and 4.2 K if the insulation is constructed of 24 layers per centimeter of aluminum foil ($e = 0.05$) and fiberglass paper. The solid conductance of the spacer material may be assumed to be 0.0851 W/m$^2$ K).

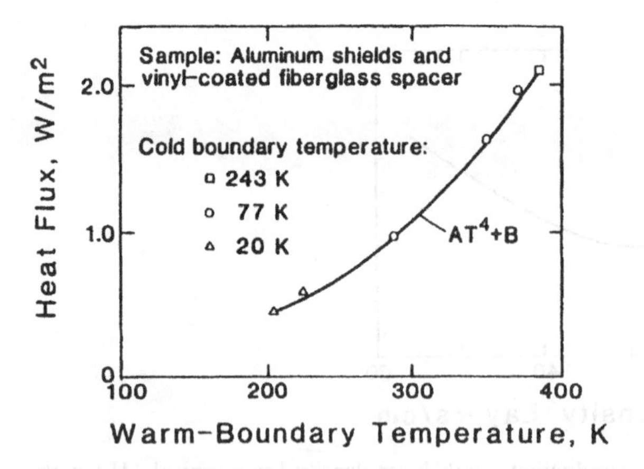


**Figure 7.26** Effect of warm boundary temperature on heat flux through MLI composed of aluminum shields and vinyl-coated fiberglass spacers.

The mean apparent thermal conductivity of the MLI may be obtained by using Eq. (7.22):

$$K_a = \frac{1}{(24)(100)}\left\{0.0851 + \frac{(5.669\times 10^{-8})(0.05)(294.4)^3}{(2-0.05)} \times\left[1+\left(\frac{20.4}{294.4}\right)^2\right]\left(1+\frac{20.4}{294.4}\right)\right\} = 5.20\times 10^{-5}\,\text{W/(m K)}$$
$$= 52\,\mu\text{W/(m K)} \tag{7.23a}$$

$$K_a = \frac{1}{2400}\left\{0.0851 + \frac{5.669\times 10^{-8}(0.05)(20.4)^3}{2-0.05} \times\left[1+\left(\frac{4.2}{20.4}\right)^2\right]\left(1+\frac{4.2}{20.4}\right)\right\} = 3.55\times 10^{-5}\,\text{W/(m K)}$$
$$= 35.5\,\mu\text{W/(m K)} \tag{7.23b}$$

Though useful, Eq. (7.22) does not take into account the effects of penetrations, the effects of compressive loads, or the method of attachment and how these may alter the effective thermal conductivity.

### 7.3.2. *Effects of Compressive Loads*

Compressive loads, either those caused by atmospheric pressure when a flexible outer skin is used to contain the insulation and permit evacuation or those developed during application of multilayer insulation, reduce overall insulating effectiveness. Even if the compression by the weight of the upper layers on the lower layers is disregarded, external forces (e.g., tension applied during wrapping of a multilayer insulation around a cylindrical object, thermal expansion or contraction of the insulation components with respect to the object, and localized loads in the vicinity of the object's supports) can compress the insulation.

These compressive loads may be in the range of 0.01–1 psi (0.0048–0.48 g/cm$^2$). Figure 7.27 shows the effects of compression on the heat flux through 16 different

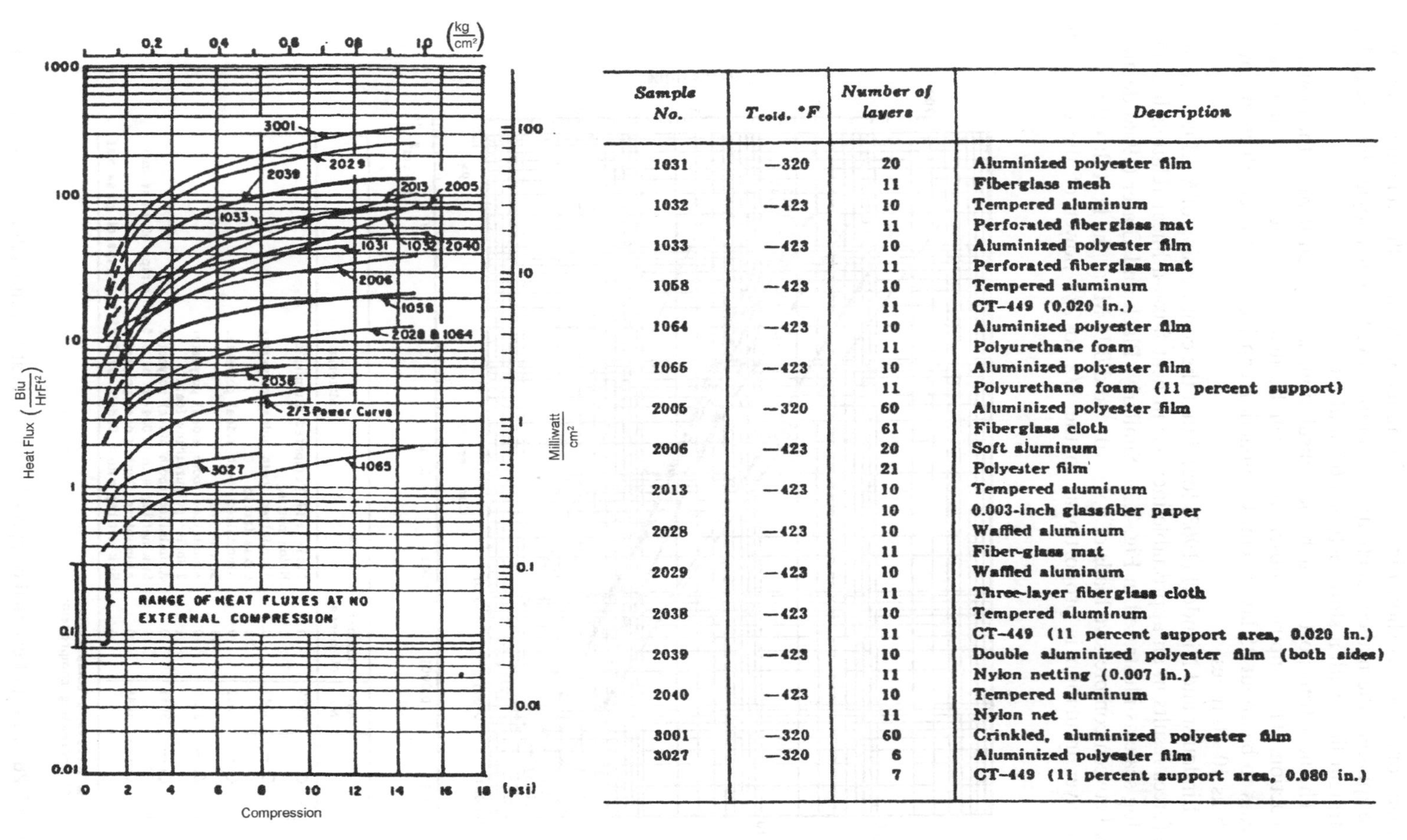


| Sample No. | $T_{cold}$, °F | Number of layers | Description |
|---|---|---|---|
| 1031 | −320 | 20 | Aluminized polyester film |
| | | 11 | Fiberglass mesh |
| 1032 | −423 | 10 | Tempered aluminum |
| | | 11 | Perforated fiberglass mat |
| 1033 | −423 | 10 | Aluminized polyester film |
| | | 11 | Perforated fiberglass mat |
| 1058 | −423 | 10 | Tempered aluminum |
| | | 11 | CT–449 (0.020 in.) |
| 1064 | −423 | 10 | Aluminized polyester film |
| | | 11 | Polyurethane foam |
| 1065 | −423 | 10 | Aluminized polyester film |
| | | 11 | Polyurethane foam (11 percent support) |
| 2005 | −320 | 60 | Aluminized polyester film |
| | | 61 | Fiberglass cloth |
| 2006 | −423 | 20 | Soft aluminum |
| | | 21 | Polyester film |
| 2013 | −423 | 10 | Tempered aluminum |
| | | 10 | 0.003-inch glassfiber paper |
| 2028 | −423 | 10 | Waffled aluminum |
| | | 11 | Fiber-glass mat |
| 2029 | −423 | 10 | Waffled aluminum |
| | | 11 | Three-layer fiberglass cloth |
| 2038 | −423 | 10 | Tempered aluminum |
| | | 11 | CT–449 (11 percent support area, 0.020 in.) |
| 2039 | −423 | 10 | Double aluminized polyester film (both sides) |
| | | 11 | Nylon netting (0.007 in.) |
| 2040 | −423 | 10 | Tempered aluminum |
| | | 11 | Nylon net |
| 3001 | −320 | 60 | Crinkled, aluminized polyester film |
| 3027 | −320 | 6 | Aluminized polyester film |
| | | 7 | CT–449 (11 percent support area, 0.080 in.) |

**Figure 7.27** Effect of mechanical loading on the heat flux through multilayer insulations.

multilayer insulations. When compression up to 2 psi (0.96 g/cm$^2$) is applied, the heat flux for the majority of the insulations is about 200 times greater than at the no-load condition. As shown in Fig. 7.28, plots of the heat flux (characteristic also of the apparent thermal conductivity) vs. compressive load on a logarithmic scale fall on straight lines with a slope between 0.5 and 0.67.

The heat transport through these insulations is very sensitive to the degree of compaction. The empirical observation can be made that the optimum packing appears to be about 25 layers/cm. Compaction can increase the conductivity by as much as 10–40 times.

Lindquist and Niendorf (1963) found that the compression of MLI using external forces results in an appreciable increase in heat flux—a linear relationship on a log–log scale as illustrated in Fig. 7.29. Similarly, Black and Glaser (1966) deduced that between compressive loads of $6.9 \times 10^{-5}$ and 0.10 MPa, the heat flux through a multilayer insulation is proportional to the 1/2 to 2/3 power of the externally

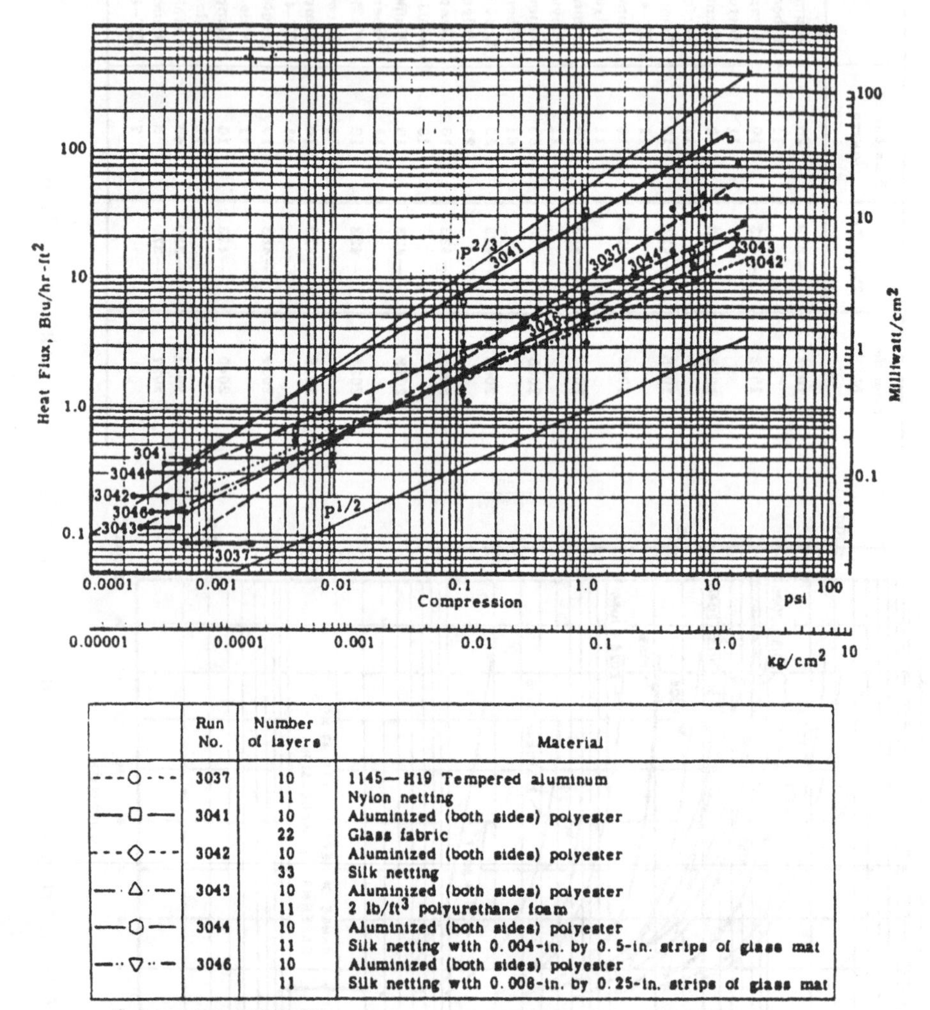


| | Run No. | Number of layers | Material |
|---|---|---|---|
| ---○--- | 3037 | 10 | 1145—H19 Tempered aluminum |
| | | 11 | Nylon netting |
| —□— | 3041 | 10 | Aluminized (both sides) polyester |
| | | 22 | Glass fabric |
| ---◇--- | 3042 | 10 | Aluminized (both sides) polyester |
| | | 33 | Silk netting |
| —·△·— | 3043 | 10 | Aluminized (both sides) polyester |
| | | 11 | 2 lb/ft$^3$ polyurethane foam |
| —○— | 3044 | 10 | Aluminized (both sides) polyester |
| | | 11 | Silk netting with 0.004-in. by 0.5-in. strips of glass mat |
| —··▽··— | 3046 | 10 | Aluminized (both sides) polyester |
| | | 11 | Silk netting with 0.008-in. by 0.25-in. strips of glass mat |

● Increasing compression.
○ Decreasing compression.

**Figure 7.28** Effect of external compression on the heat flux through multilayer insulations.

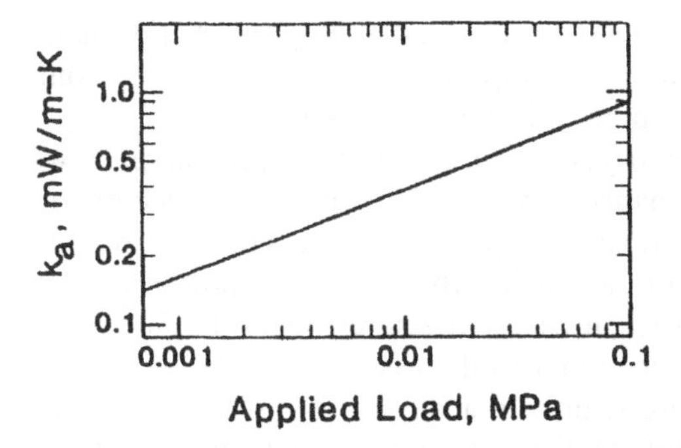


**Figure 7.29** Apparent thermal conductivity of MLI as a function of applied load on the insulation with cold and warm boundary temperatures of 77 and 300 K, respectively.

applied load. In other studies, it has been shown that once this compressive load is released, the MLI regains 90% of its original thickness and returns to nearly its original thermal conductivity. An advantage of crinkled aluminized Mylar is that it can withstand high acceleration loads without loss of insulation effectiveness. Generally, however, for large temperature differences such as those encountered in cryogenic systems, the overall heat flux is directly proportional to the fourth power of the warm boundary temperature.

### *7.3.3. Load-Supporting Multilayer Insulation*

Extensive work on the effects of compressive loads on multilayer insulations has indicated that it is feasible to develop MLI systems that can support light loads.

A prototype cryogenic container had insulation consisting of aluminized polyester film and fiberglass spacers. The fiberglass spacers with a silicone resin binder have a density of 9 lb/ft$^3$ (0.14 g/cm$^3$). Eighty percent of the spacer material was removed, and the rest was precompressed to achieve an installed density of 2.8 lb/ft$^3$ (0.04 g/cm$^3$). The insulation system was compressed between the inner container and an outer shroud so that the desired preload was obtained with the inner container at the cryogenic fluid temperature.

Small dynamic loads are carried in shear by the insulation around the circumference farthest from the axis of vibration. Larger dynamic loads are carried by sliding friction around the circumference away from the axis of motion and by compression at the ends of this axis. Coulomb damping in the insulation subjected to shear and hysteresis in the compressed insulation both provide damping to limit the relative motion between the two vessels.

Tests of the prototype container, which consisted of a 29 in. o.d., $\frac{1}{4}$ in. thick stainless steel inner vessel and a 31 in. i.d., 0.070 in. thick stainless steel outer vessel, indicated a heat leak corresponding to an apparent thermal conductivity of 0.0042 Btu in./(h ft$^2$ °F) [0.0060 m W/cm °C)].

### *7.3.4. Effects of Type of Gas and Gas Pressure*

The presence of residual gas inside fibrous and powder insulations decreases the thermal performance of a system. Gases of high thermal conductivity (e.g., helium or hydrogen cause more rapid performance deterioration than gases with low conductivity (e.g., nitrogen or air). These effects are even more pronounced for a multilayer insulation. The effect of gas pressure on the thermal conductivity of several multilayer insulations is shown in Fig. 7.30. For comparison purposes, data for one fibrous insulation (glass wool in air) are plotted in the same figure.

The thermal conductivity is related to gas pressure by an S-shaped curve. However, the effect of pressure on the performance of multilayer insulations is two orders of magnitude greater than the effect of pressure on powder or fibers (i.e., the performance of a multilayer insulation is 100 times higher than that of a powder, but a pressure 100 times lower is required to reach it). At pressures below $10^{-5}$ torr, the heat transferred by a gas is directly proportional to the gas pressure. However, the heat conducted by a gas at that pressure is only a small portion of the total heat transferred through the insulation; therefore, the apparent thermal conductivity of the multilayer insulation decreases only slowly at pressures below $10^{-5}$ torr.

At pressures of $10^{-5}$–$10^{-4}$ torr, the mean free path of the gas molecules approaches the distance the solid particles of the insulation. Beginning at these pressures, the apparent thermal conductivity of the insulation rapidly increases. For this reason, the multilayer insulation must be maintained at a pressure below $10^{-4}$ torr or it will not provide the desired insulation effectiveness.

When the gas pressure reaches atmospheric pressure, heat conduction by the gas becomes the dominant mode of heat transfer. For a warm boundary at room temperature, the radiation component becomes small in comparison to the gas conduction component. Therefore, the apparent thermal conductivity of a multilayer insulation approaches the conductivity of the interstitial gas. After atmospheric pressure has been reached, the conductivity of the gas remains nearly constant

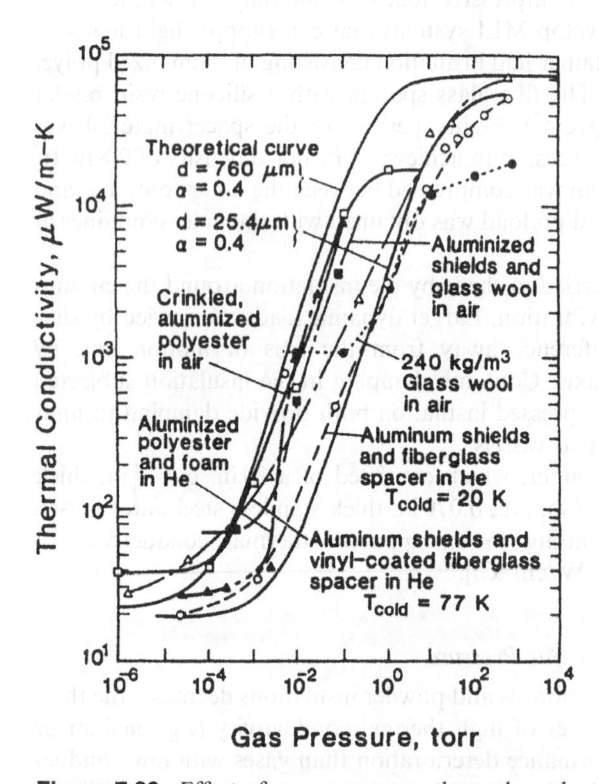


**Figure 7.30** Effect of gas pressure on thermal conductivity of MLI.

and independent of the pressure, as does the apparent thermal conductivity of a multilayer insulation.

### 7.3.5. *Effects of Perforations*

To ensure that multilayer insulations will operate at the desired low pressure, the interstitial gas has to be evacuated through the edges of the insulation. For an insulation with 100 shields per inch and spacers that effectively occupy 90% of the spaces between the shields, the maximum outgassing rate corresponds to the removal of one monomolecular layer from all exposed insulation surfaces every 2 weeks. Because much higher outgassing rates are likely to be encountered after the radiation shields and spacers have been exposed to atmospheric gases during manufacture, edge pumping by itself will not be adequate to ensure a low pressure within the insulation within a reasonable time.

Another difficulty with edge pumping is that at the exposed edges, the innermost radiation shields have to be brought into contact with a warmer outside boundary. This contact may permit radiation to bypass most of the shields and penetrate almost directly to the insulated object.

Broadside pumping of a multilayer insulation is made possible by perforating the radiation shields. If the perforations comprise 10% of the area of the shields, broadside pumping will increase the outgassing rate by three orders of magnitude compared to edge pumping and will produce a satisfactory low pressure in a short time. However, perforation of the radiation shields will lead to an increase in their effective emissivity; if an unperforated shield has an emittance of 0.05, perforating 10% of the shield area will increase the emittance to 0.25. Thus, a compromise must be reached between a sufficiently high pumping speed and a reduction in insulating effectiveness.

For the same amount of open area, optimum pumping and outgassing efficiency are obtained with many small holes rather than a few large holes. However, small holes cause a higher heat flux, as is observed from Fig. 7.31, which shows two theoretical limits, one for extremely small holes and one for large holes. Experimental data for holes 1/16, 1/8, and 1/4 in. in diameter fall as expected within these limits.

Figure 7.32 shows that perforations have a smaller effect on an insulation with crinkled, aluminized polyester shields than on one with aluminum shields, presumably because the crinkles introduce a more random distribution of the holes.

### 7.3.6. *Effects of the Boundary Temperature*

Table 7.21 and Fig. 7.33 illustrate the effect of the warm and cold boundary temperatures on the heat flux through multilayer insulations. Table 7.21 gives the heat flux through various insulations for which the cold boundary temperature was changed from −320 to −423°F (77 to 20 K). The heat flux through each sample is approximately the same for both temperature levels. Figure 7.33 shows heat flux vs. warm boundary temperature for 10 tempered aluminum shields spaced with 11 vinyl-coated fiberglass screen spacers. For large temperature differentials, heat flux is directly proportional to the fourth power of the warm boundary temperature.

The two curves of Fig. 7.34 show the apparent thermal conductivity vs. warm boundary temperature calculated from the heat flux data. The one that represents a cold boundary temperature of −423°F (20 K) indicates a lower thermal conductivity than the one representing a cold boundary temperature of −320°F (77 K). Thus,

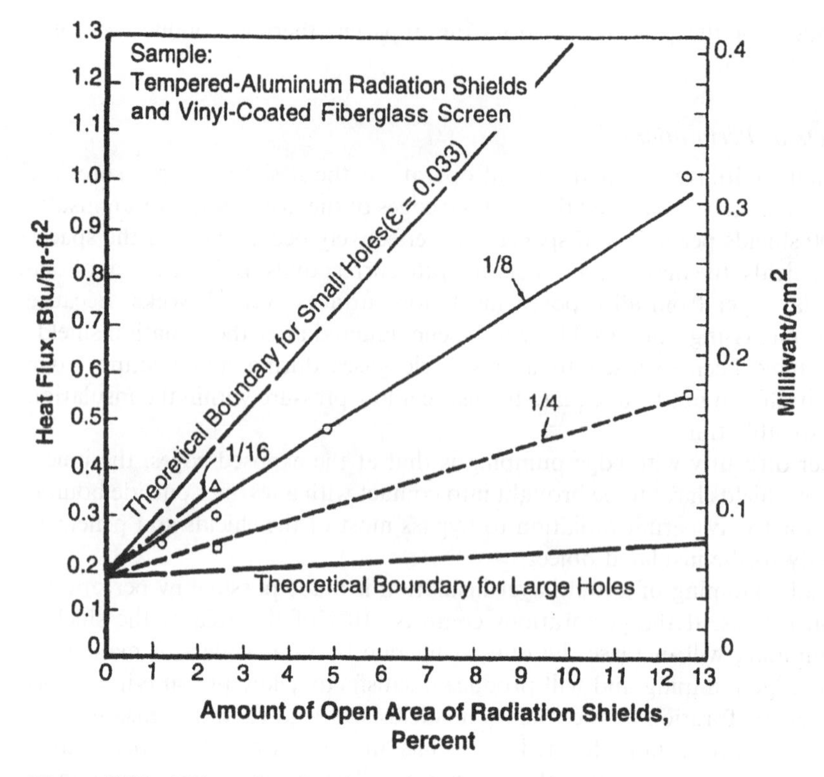


**Figure 7.31** Effect of perforations on heat flux (hole diameter 1/16, 1/8, and $\frac{1}{4}$in.).

when a warm boundary temperature is held constant, a higher thermal conductivity results from increasing the cold boundary temperature.

Apparent thermal conductivity is approximately proportional to the third power of the warm boundary temperature, as shown in Fig. 7.35, where the third

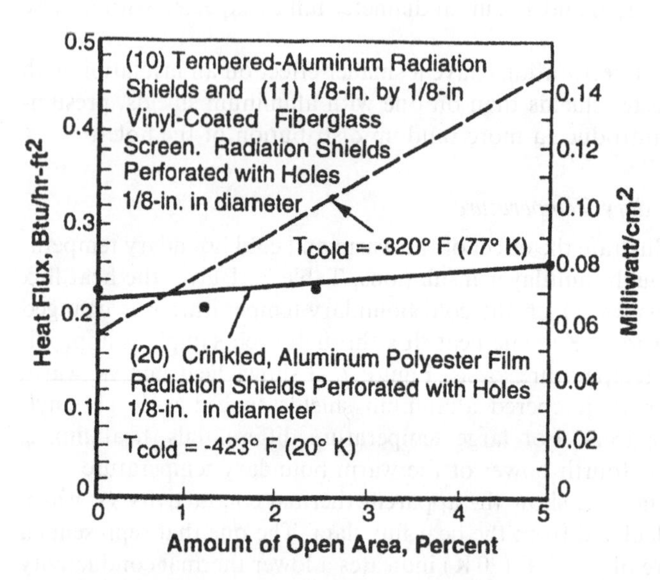


**Figure 7.32** Effect of perforations on the heat flux through a multilayer insulation.

**Table 7.21** The Effect of the Cold Boundary Temperature on the Heat Flux Through Multilayer Insulations

| | | | Boundary, cold | | Temperature, warm | | Density | | Heat flux | |
|---|---|---|---|---|---|---|---|---|---|---|
| Number of layers | Sample | Thickness (in.) | K | °F | K | °F | $g/cm^3$ | $lb/ft^3$ | $\mu W\ cm^2$ | $Btu\ h\ ft^2$ |
| 10 | Tempered aluminum foil | 0.300 | 77 | −320 | 290 | 61 | 0.21 | 13 | 73 | 0.23 |
| 11 | Fifty percent open fiberglass mat | | 20 | −423 | | | | | 63 | 0.20 |
| 20 | Aluminized polyester film, crinkled | 0.385 | 77 | −320 | 279 | 42 | 0.02 | 1.2 | 285 | 0.90 |
| | | | 20 | −423 | | | | | 276 | 0.87 |
| 95 | Aluminized polyester film | 0.800 | 77 | −320 | 286 | 55 | 0.31 | 20 | 133 | 0.42 |
| 96 | Glass fabric | | 20 | −423 | | | | | 136 | 0.43 |

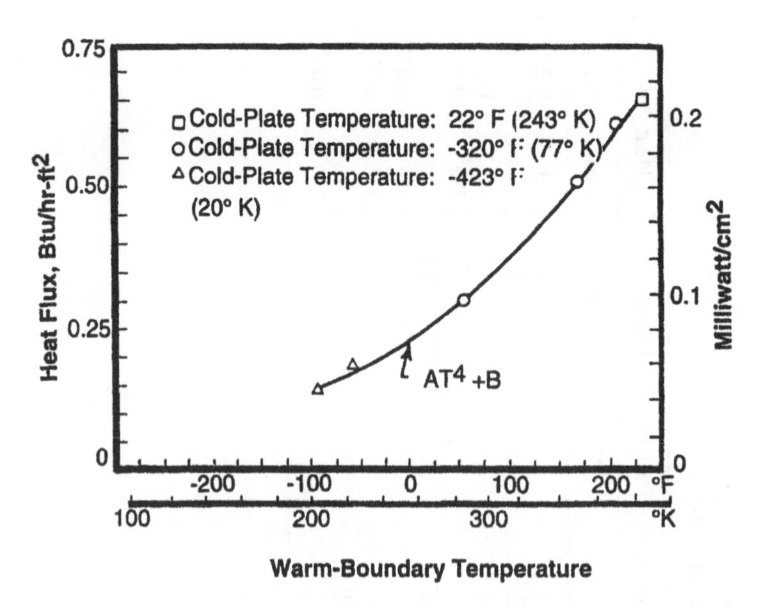


**Figure 7.33** Effect of warm boundary temperature on heat flux through a multilayer insulation. Cold plate temperature: ⟨□⟩ 22°F (243 K); ⟨O⟩ −320°F (77 K); ⟨△⟩ −423°F (20 K).

power of the warm boundary temperature is matched with experimental results for a multilayer insulation with crinkled, aluminized polyester film radiation shields.

### *7.3.7. Effects of Attachment Methods*

Analogous to compressive loads, methods of attachment can also alter the resistance of a multilayer insulation to thermal conductivity. Three basic techniques (Fig. 7.36), that were developed in the space program, have been used to attach MLI to a cryogenic vessel: button, tape, and shingle. The button system was developed to support the fragile fiberglass spacers with either aluminized Mylar or aluminum foil radiation shields. The buttons, made of very low conductivity material, add both a solid conduction element and a compressive load factor to the overall insulation conductivity.

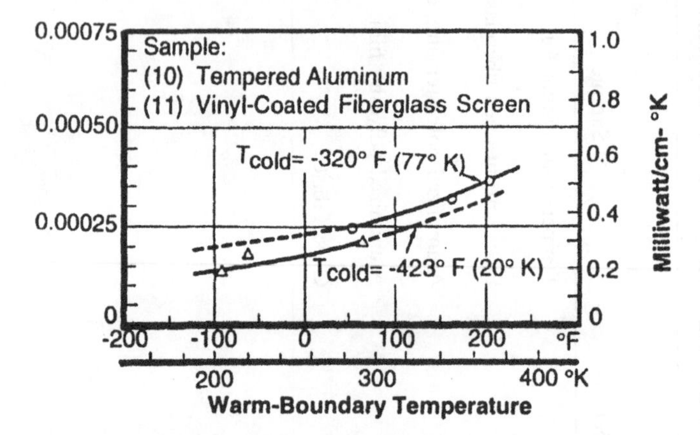


**Figure 7.34** Effect of boundary temperature on the mean apparent thermal conductivity of a multilayer insulation of tempered aluminum and vinyl-coated fiber-glass screen.

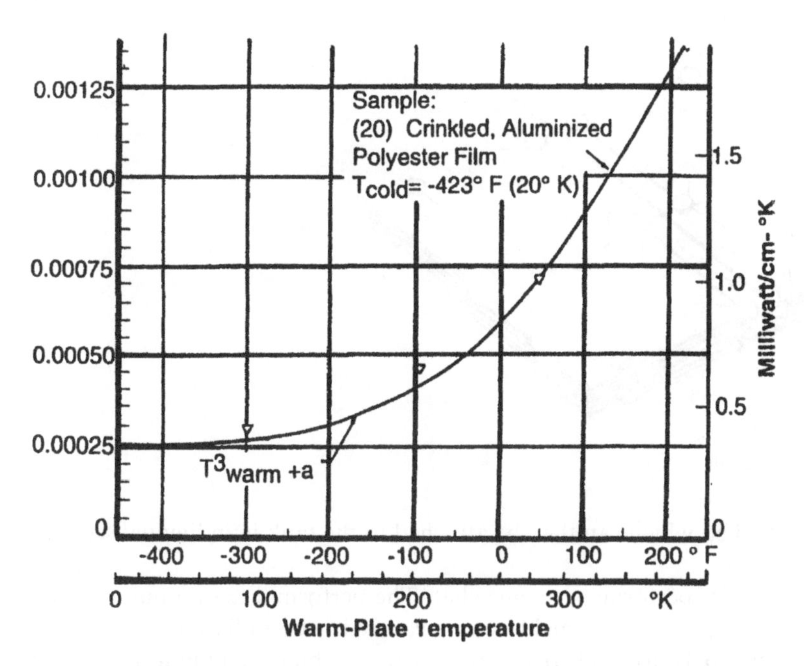


**Figure 7.35** Effect of boundary temperature on thermal conductivity of crinkled, aluminized polyester film (cold temperature 20 K).

On the other hand, the shingle and tape methods require adequate strength in the multilayers to allow attachment along one edge. Though the solid heat conduction resulting from the button attachments is eliminated, radiation tunneling between the shingles can increase the heat leak.

**(a) Pinned and Quilted Insulation.** Layers of a multilayer insulation can be pinned to the wall of the object to be insulated or they can be stitched or quilted

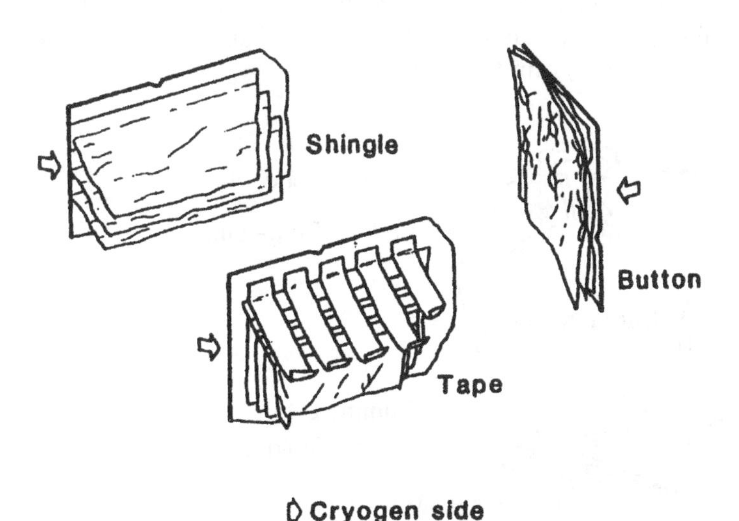


**Figure 7.36** Attachment techniques for multilayer insulation.

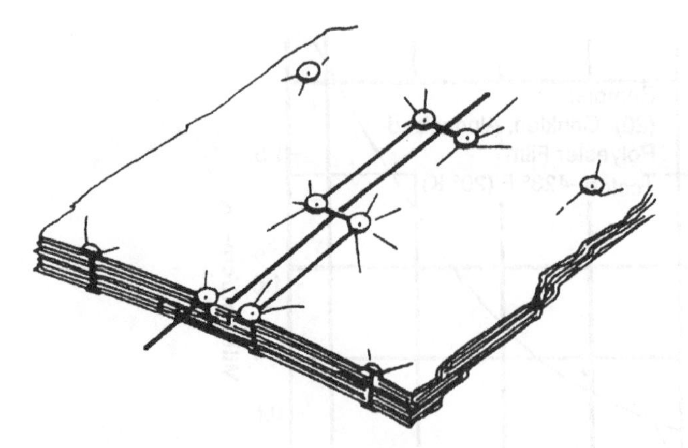

**Figure 7.37** Quilted insulation.

together into a blanket, which can then be attached to the tank in individual sections (Fig. 7.37).

Quilting degrades performance somewhat. The performance of a quilted insulation consisting of 10 radiation shields was measured. The radiation shields consisted of 0.00025 in. (6.3 μm) thick perforated polyester film, aluminum coated on both sides, and eleven 0.003 in. (76 μm) thick silk nettings [with $\frac{1}{4}$ in. diameter by 0.008 in. (200 μm) thick glass-paper disks sewn on both sides of silk netting on $1\frac{1}{2}$ in. centers with 0.006 in. (150 μm) diameter thread]. The quilting degraded the performance by 7%.

A quilted blanket insulation combined with a purgeable substrate system was tested on 1/2 (linear) scale model of a Saturn V Apollo landing craft module liquid hydrogen storage tank. The quilted blanket insulation system was chosen because it can function satisfactorily during prelaunch ground hold, launch, and prolonged space flight. The spherical tank, about 82 in. in diameter, was provided with fill, vent, pressurization, and drain lines.

The insulation system consists of several layers. First, a 0.4 in.-thick fiberglass substrate layer was placed over the tank and enclosed by a gastight bag. This layer can be purged with helium during prelaunch ground hold. (Figure 7.38 shows the arrangement of the helium purge system for the test tank.)

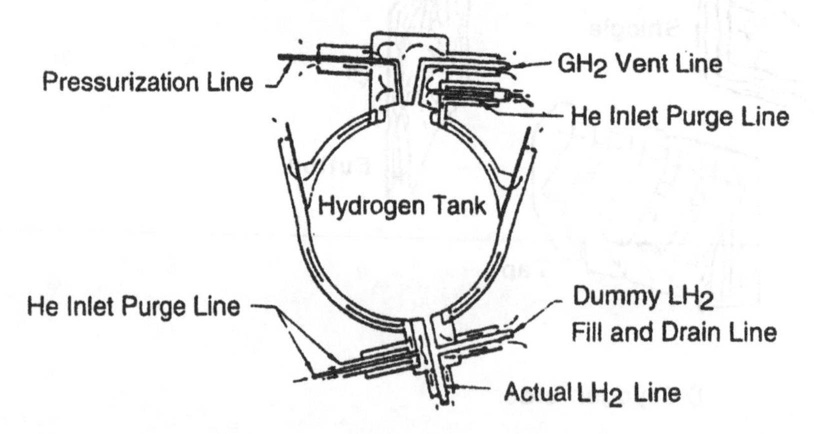


**Figure 7.38** Schematic of helium-purge substrate system.

The insulation consists of 30 radiation shields of double aluminized polyester film, 0.0025 in. (63 μm) thick, separated by 0.0028 in. (71 μm) glass-paper spacers, combined into blankets, each consisting of 10 shields and 9 spacers. The layers are held together with buttoned tie-throughs to form a quilted blanket. The blankets are in modular sections and completely cover the tank. Three blankets are placed over each other to build up the insulation system to the required number of layers. The separate blankets are held together with Velcro fasteners interspersed between the blankets at appropriate intervals. A polyester netting is placed over the outer blanket to provide additional strength to the system. The details of this system are shown in Fig. 7.39.

**(b) Bands.** When large objects are to be insulated with multilayer insulation, bands can be used to keep the insulation in place (Fig. 7.40). Bands apply compression to a relatively small portion of the total surface area of the insulation. For radiation shields that are poor heat conductors in the plane parallel to their surface (such as thin plastic shields), the adverse effect of compression on the thermal performance is localized.

**(c) Shingled Insulations.** The shingled multilayer insulations (Fig. 7.41) derive their name from their similarity to roofing shingles. One end of each shingle is attached to the tank wall (or substrate insulation) overlapping an adjacent shingle. The length of the shingle from the attachment point to the outer edge and the

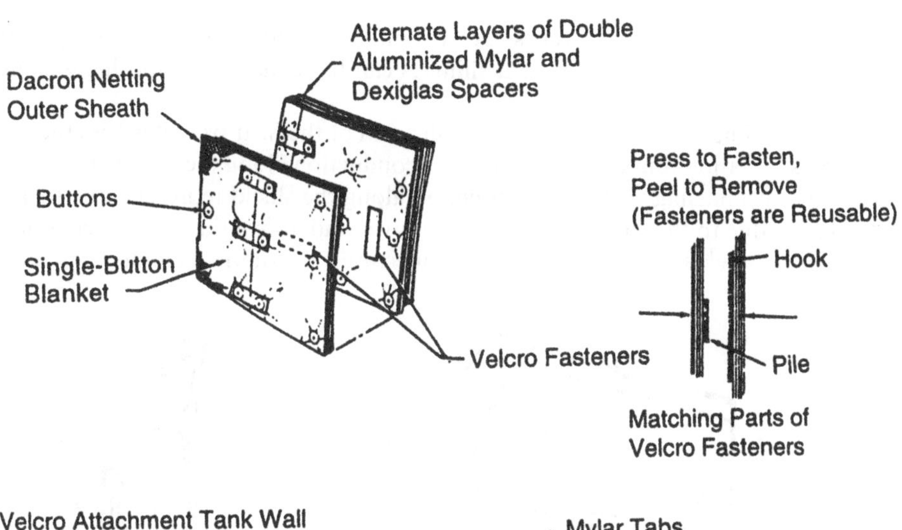


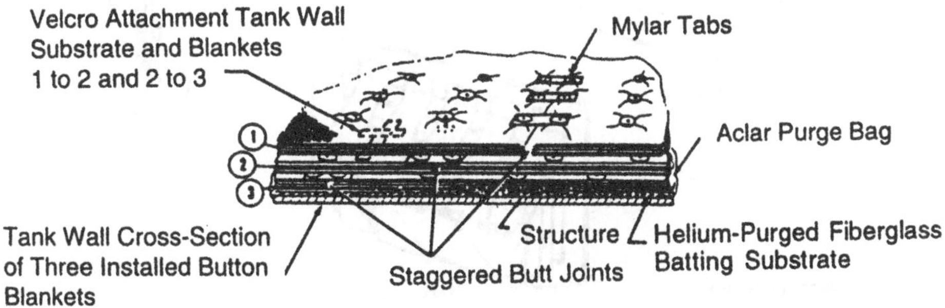


**Figure 7.39** Quilted blanket insulation.

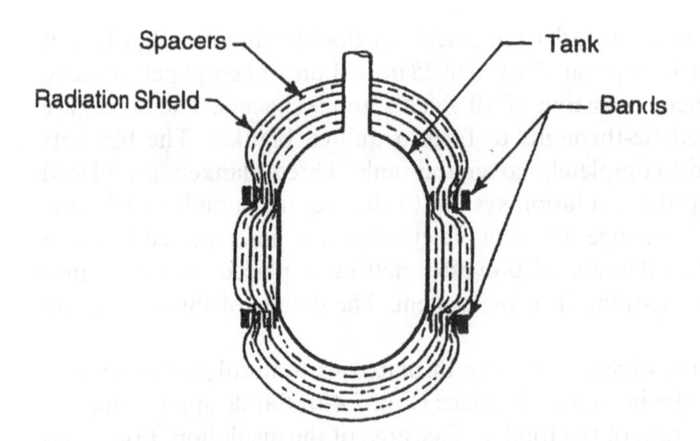


**Figure 7.40** A multilayer insulation kept in place by bands.

spacing between shingles at the bonded edge determine the number of layers that can be applied. Shingled insulation has the following advantages:

1. The multilayer insulation can be attached directly to the tank wall.
2. Gas trapped between the shingles can be rapidly vented.
3. Gas from any point source of gas leak, such as the tank surface at weld areas and manhole covers, will be trapped between two shingles and vented to space before it can permeate the outer layers.

Loss in insulation effectiveness due to solid conduction along the shingles and radiation tunneling between adjacent shingles can be reduced by choosing a shingle of suitable length.

The total heat flux is approximately the sum of the heat passed across the insulation plus radiation tunneling plus the solid conduction along the shield. The sum of the radiation tunneling and solid conduction along the shield represents the thermal degradation due to the shingle configuration over and above the thermal conductivity for the parallel layers. As shingle length increases, the thermal conductivity of the

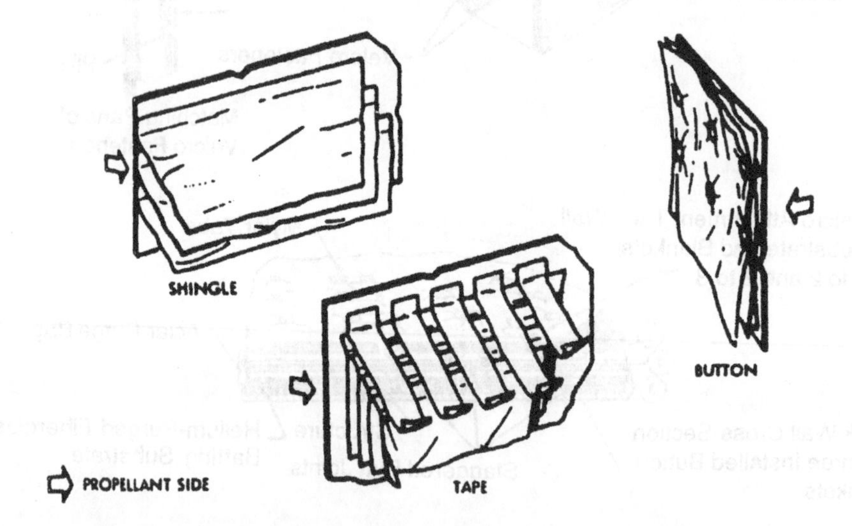


**Figure 7.41** Shingled multilayer insulation attachment methods.

shingled multilayer insulation asymptotically approaches a value equivalent to that of parallel multilayer insulation. For shingle lengths greater than 28 in., the thermal degradation is less than 40% of the thermal conductivity of parallel multilayer insulation.

Figure 7.42 shows the effects of gas leakage on the performance of shingled insulation. For comparison, the data for a multilayer system with parallel button radiation shields are shown on the same figure. The results indicate that the performance of the shingled insulation is nearly insensitive to gas leaks when the shingles are vented to the vacuum without permeating the outer layers; the performance of one shingle layer is degraded in the immediate vicinity of the gas leak.

The performance of shingled insulation was tested on a 4 in. diameter, 100 in. long calorimeter as well as on a 26 in. diameter test tank with hemispherical ends. The test included exposure to thermal shock, that is, rapid cooldown from ambient to 37°R (20 K) with no prechill, rapid pumpdown [0.2 lb/(in.$^2$ s)], vibration in the longitudinal and horizontal planes, acoustic load (to 158 dB), and acceleration up to 20 g. No structural damage to the shingles or significant change in insulation performance occurred during these tests.

Another characteristic of MLI is that the overall thermal conductivities parallel to the radiation shields are commonly between three and six orders of magnitude higher than those normal to the layers. Such a large disparity in the directional thermal resistance can present considerable problems in thermal insulation design for systems where structural supports and feedlines penetrate the insulation and provide convenient lateral heat leaks at the junctions. An advantage of aluminized Mylar over the aluminum–fiberglass combination is that it reduces this lateral heat conduction.

Due to the variety of degrading effects on MLI, such as the compressive loads, attachments, external penetrations, and high lateral conductivity, experimentally determined thermal conductivity values are generally considerably lower than those

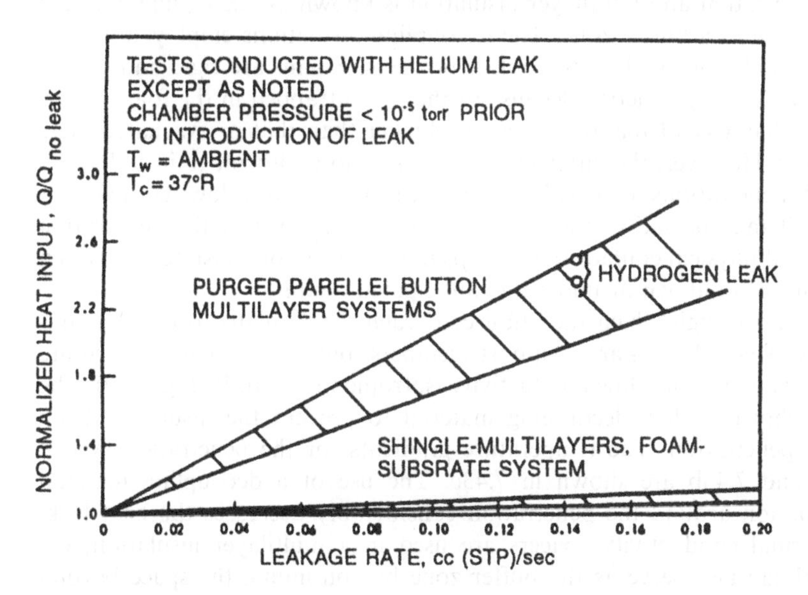


**Figure 7.42** Normalized heat input through MLI systems as a function of point source leakage rate.

actually observed in practice. Solutions to these practical problems generally take the form of improved reflecting shields with lower lateral thermal conductivity and efficient isotropic intermediary insulation systems around the external penetrations. Multilayer insulation, owing to its complicated structure, is not only more difficult to apply but also considerably more expensive than other insulation materials. Thus, the use of this insulation, though increasing rapidly, has been limited to those situations requiring high thermal efficiency. Recently, to simplify the installation process and also to reduce the material and installation costs, new insulation materials such as aluminized polyester nonwoven fabric and aluminized carbon-loaded paper have been successfully developed and tested.

### *7.3.8. Effect of Penetrations*

In any practical application, penetrations, such as fill lines and vent lines on a cryogenic tank or electrical leads on electronic equipment, are unavoidable.

Analyses describing the relationship between the penetration and highly anisotropic multilayer insulations have been carried out. Because of the important effect penetrations have on insulation effectiveness, considerable care must be taken in designing the system, selecting materials, and evaluating experimental data obtained on idealized systems.

**(a) Buffer Zones.** Multilayer insulations are highly anisotropic; that is, heat flow in a direction parallel to the layers is orders of magnitude greater than the heat flow perpendicular to the layers. This anisotropy has to be recognized in the design and construction of multilayer insulations whenever there is a penetration or a seam in the insulation. A simple case is illustrated in Fig. 7.43a. Here the multilayer insulation is in good thermal contact with a penetration. Heat flows not only directly through the penetration but also from the insulation layers into the penetration; this effectively short-circuits the insulation in this region. The isolation of a penetration (or any other short circuit) by additional, but isotropic, insulation in the space between the penetration and multilayer insulation is known as decoupling, and the space is referred to as a buffer zone. The decoupling insulations employed include urethane foams and evacuated glass wool or powders. Foams are rigid and can be produced with accurate geometry. Because of the gas entrapped in the foam pores, the thermal conductivity of foams is generally higher than that of evacuated glass wool or powders. However, the latter insulations are more difficult to install.

For small penetrations, especially those in a product with a low thermal conductivity area, it may be satisfactory to place the insulation near the penetration but prevent actual physical contact. These gaps in the insulation must neither widen nor narrow during cooldown or under any anticipated loading.

A large gap through which radiation can reach the insulated object is intolerable in terms of heat leak. Large support members, pipes, and manholes require that the gap be closed with a low-conductivity isotropic material. In Fig. 7.43b, the penetration is shown with a decoupling material to prevent the insulation from contacting the penetration. The temperature gradients for the penetration system in Fig. 7.43a and 7.43b are shown in 7.43c. The use of a decoupling material and axial insulation around the penetration considerably decreases the heat leak. When low thermal conductivity spacers are used in a multilayer insulation, the spacer material can also serve as the buffer zone by continuing the space beyond the radiation shields for a distance equal to about twice the thickness of the multilayer insulation.

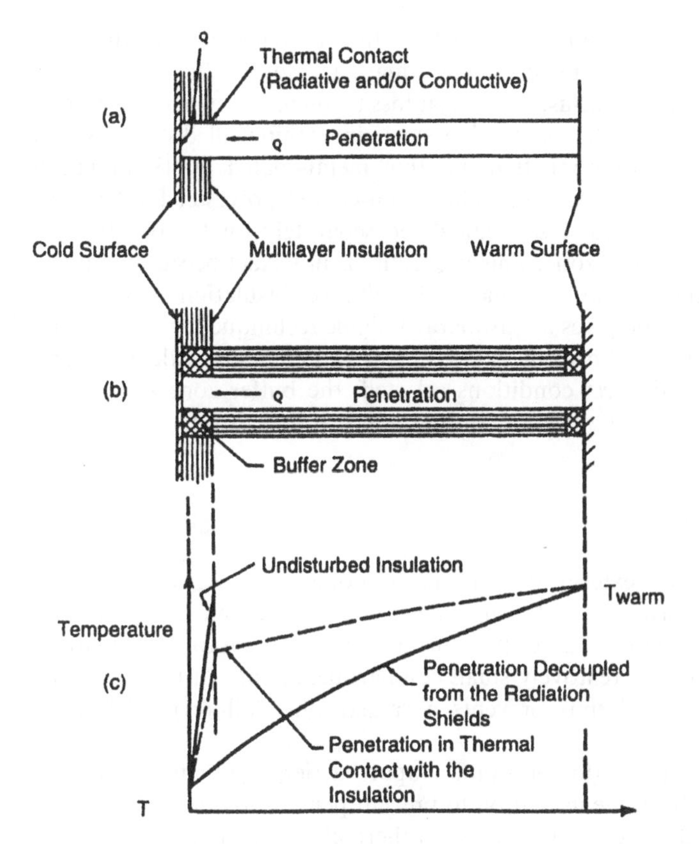


**Figure 7.43** Treatment of a penetration passing through multilayer insulation.

In the placement of multilayer insulations on tanks, seams of joints between insulating sections are often encountered. Instead of trying to match layer to layer (Fig. 7.44a), these seams are often decoupled with another (isotropic) insulation. For example, in Fig. 7.44b, such a decoupling is shown for a right angle turn in the cold wall.

The analyses of heat leaks through a buffer or decoupling zone are difficult and must be confirmed by actual tests. For the case shown in Fig. 7.44b, the heat flux due to the coupling section is $LK\Delta T$, where $L$ is the length of the joint, $K$ is the thermal

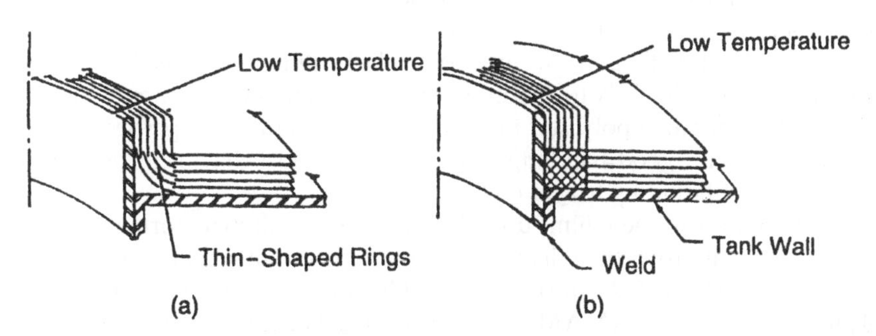


**Figure 7.44** Treatment of joint seams in the placement of multilayer insulation on tanks. (a) Layer-to-layer matching; (b) decoupling for a right angle turn.

conductivity of the decoupling insulation, and $\Delta T$ is the difference between the inner and outer temperatures. The dimensions of the square are not significant.

**(b) Metal-Coated Plastic Shields.** The heat loss through a system consisting of a small penetration such as a plastic or stainless steel pin or thin walled stainless steel tube through a multilayer insulation that has thin metal-coated plastic radiation shields [such as 0.00025 in. (6.3 μm) thick aluminum-coated polyester film] can be approximated by the sum of heat losses calculated separately for the penetration, multilayer insulation, and buffer zone. This procedure is not exact because no interaction between the penetration, buffer zone, and multilayer insulation is taken into account. However, the method gives a reasonably reliable technique for determining the effect of a penetration, as shown in Table 7.22, which compares calculated and experimental results. For the test conditions selected, the buffer zone around the penetration has an adverse effect: it adds to the total heat loss through the system.

## 8. VAPOR BARRIERS

For insulation applied to the exterior of a cryogenic propellant tank, vapor barriers are required to prevent permeation of atmospheric constituents into the insulation when the tank is chilled. Helium purges prevent this permeation into an insulation. Sealing off the insulation into relatively small compartments limits a failure to one portion of the insulation and thus prevents a catastrophic failure of the entire system.

For insulations on the interior of a tank, vapor barriers are required to separate the cryogenic liquid from the insulation to prevent penetration of the insulation by the liquid, which would lead to an increase in thermal conductivity and possible catastrophic failure of the insulation because of increased gas pressure created by the vaporizing liquid at the warm tank wall. Vapor barriers are even more important in filament-wound fiberglass tanks because of their tendency to be permeable to cryogenic liquids, particularly after repeated thermal cycling.

### 8.1. Vapor Barrier Materials

Various materials have been evaluated and used as vapor barriers for cryogenic insulation systems. The least permeable materials are metal foils; however, their high density, combined with handling difficulties, limits their use. A vapor barrier must conform readily and smoothly to a variety of contours. The small elongation of metal foils severely limits their ability to be formed and increases the permeability of foil-lined structures due to wrinkling and tearing.

The difficulties in handling metal foils, particularly aluminum foils, have been overcome by laminating the foils to thermoplastic films. The most common vapor barrier material is an oriented polyester film. Among the desirable properties of this film are its high strength, good handling characteristics, and behavior at temperatures as low as that of liquid hydrogen (−423°F; −253°C).

Lamination of foils to these films does not improve the ultimate formability of the foils, but it does improve their handling characteristics and limits damage to them. Laminates of 0.5 mil (12.7 μm) polyester film on each side of a 0.5–1 mil (12.7–25.4 μm) aluminum foil (MAM laminates) have been successfully used as vapor barriers. A double layer of aluminum foil (MAAM) has the lowest permeability of any available vapor barrier.

**Table 7.22** Comparison of Calculated and Experimental Heat Flows for Penetrated Multilayer Insulation

| Penetration | | | Heat flow | | | | | | | | | |
|---|---|---|---|---|---|---|---|---|---|---|---|---|
| | | | Penetration | | Buffer zone | | Insulation | | Total | | | |
| | | | | | | | | | Calculated | | Experimental | |
| Material | Description | Buffer zone diameter (in.) | Btu/h | mW | Btu/h | mW | Btu/h | mW | Btu/h | mW | Btu/h | mW |
| Stainless steel | Pin 0.020 in. | 0.25 | 0.10 | 29 | 0.013 | 4.0 | 0.017 | 5.0 | 0.13 | 38 | 0.12 | 3 |
| | | 0.04 | 0.11 | 32 | < 0.001 | < 0.3 | 0.017 | 5.0 | 0.13 | 37 | 0.12 | 3 |
| | | 1.88 | 0.18 | 53 | 0.24 | 70 | 0.016 | 4.7 | 0.44 | 128 | 0.47 | 13 |
| Nylon | Pin 0.032 in. | 0.04 | 0.006 | 2 | < 0.001 | < 0.3 | 0.046 | 13 | 0.052 | 20 | 0.063 | 1 |
| | | 0.28 | 0.006 | 2 | 0.003 | 1.0 | 0.046 | 13 | 0.055 | 16 | 0.042 | 1 |
| Stainless steel | Tube 0.25 in o.d., 0.001 in. thick | 0.28 | 0.25 | 73 | 0.016 | 5.0 | 0.046 | 13 | 0.31 | 91 | 0.32 | 9 |

Other thermoplastic films are useful as vapor barriers. Polyvinyl fluoride film has also been used as a vapor barrier because of its low permeability to gases, good strength, elongation characteristics, and availability in a readily bondable form. Other materials that have been considered and evaluated as vapor barriers include polyimide film, fluorinated ethylene propylene film, fiberglass cloth-reinforced plastics, chlorotrifluoroethylene polymers and copolymers, and polyethylene film.

## 8.2. Physical Properties of Vapor Barrier Materials

Polyester film, polyvinyl fluoride film, and polychlorotrifluoroethylene fluorohalocarbon films were tested for tensile and elongation properties at liquid nitrogen temperature (−320°F; −196°C). These tests indicated that of these three films polyester has both the highest tensile strength and highest elongation at failure at liquid nitrogen temperature. The polyester is a highly oriented film, and under some test conditions, it can be observed in both the longitudinal and transverse directions; therefore, it was tested in both directions.

Tensile tests were conducted on three vapor barrier materials at room temperature, +250°F,−320°F, and −423°F (+121°C,−196°C, and −253°C). The materials were 3 mil (76 μm) polyester film, a laminate of two layers of glass cloth and epoxy resin binder, and a laminate made of one layer of 1 mil (25.4 μm) aluminized polyester and one layer of glass cloth with epoxy resin binder. To determine whether there were any variations in properties, tests were made in both the longitudinal and transverse directions. The laminates increased in tensile strength as the temperature decreased to −423°F (−253°C). Most significantly, their elongation also increased, which is a desirable property. Measurements of thermal expansion of vapor barriers including 3 mil polyester, polyvinyl fluoride, and a laminate consisting of two layers of fiberglass cloth bonded to a 1 mil aluminized polyester film indicated that, because of the stabilizing effect of the glass fibers, the fiberglass laminate has by far the lowest thermal expansion. However, the polyester has only slightly higher linear expansion than the laminate, whereas the polyvinyl fluoride has a substantially higher linear expansion. To minimize thermal shear stresses in laminates subjected to cryogenic temperatures, these thermal expansion properties must be taken into consideration.

Data have been obtained on hydrogen permeation of various barrier materials. These measurements indicate the variation in hydrogen permeation of polyimide film, polyester film, polyvinyl fluoride film, polyethylene, chlorinated ethylene propylene, and two chlorotrifluoroethylene polymers over the range of +200°F to −100°F (93°C to −73°C).

## 8.3. Applications of Vapor Barriers

The primary internal application of insulation for liquid hydrogen tanks is in the Saturn S-IVB. In this system, the vapor barrier is a polyurethane resin reinforced with fiberglass cloth, which retains substantial resilience at liquid hydrogen temperature. Sealings of urethane foam with polyurethane resin did not decrease the permeability of the foam to helium.

In the Saturn S-II cryogenic tank insulation system, the outer vapor barrier is a double layer of polyamide plastic cloth laminated with epoxy resin, with a surface layer of polyvinyl fluoride film as the primary barrier to prevent air from permeating the insulation.

Several vapor barriers to prevent the permeation of propellants from filament-wound tanks have been evaluated. Fluorinated ethylene propylene and polyester tapes were bonded to the interior of small filament-wound tanks and evaluated by partially filling a tank with liquid hydrogen and then pressurizing it. Fluorinated ethylene propylene was found to be incompatible with the filament-wound structure, and failure occurred early in the test. The compatibility of polyester is limited because when the tank is pressurized the polyester liner will fail in tension before the burst pressure of the filament-wound structure is approached.

## 9. PROTECTIVE ENCLOSURES

To achieve minimum losses during the transportation and storage of cryogenic liquids, insulation must be protected from damage by handling during fabrication and during loading of the liquids. Water (in humid air or rain) must be prevented from penetrating and freezing on cold tank surfaces. Liquid hydrogen and liquid or solid air must not be present on a tank surface. Prior to launch, heat leaks into space vehicle propellant tanks have to be minimized to avoid excessive boiloff loss and reduce refilling requirements. The insulation must not be degraded by the significant thermal contraction of a tank during filling.

During the launch phase, the insulation system must be operable despite severe vibrations and buffeting, a high *g* loading, high surface temperatures caused by aerodynamic heating, localized radiant heat exchange from rocket engines, and a rapid reduction in ambient pressures.

In space, to prevent excessive heat leak, the propellant tank and insulation must be protected against solar and planetary radiation. Such hazards as the small, but finite, probability of collisions with micrometeoroids also have to be considered. Considering these various factors, an ideal protective enclosure should:

1. Form an integral part of an effective insulation system.
2. Withstand high temperatures.
3. Be sufficiently flexible to withstand thermal stresses during cooldown but be rigid enough to maintain integrity during the aerodynamic stresses of launch.
4. Be impermeable to water vapor (and, for liquid hydrogen tanks, air).
5. Allow pressure equalization.
6. Be of minimum weight.

Different requirements have led to different protective enclosure concepts depending on whether or not the enclosure is exposed to thermodynamic heating.

### 9.1. Protective Enclosure Not Exposed to Aerodynamic Heating

A protective enclosure for polyurethane foam or multilayer insulation that is not exposed to aerodynamic heating and forms a vapor barrier consists of two 0.5 mil thick polyester layers on either side of one or two 0.5–1 mil thick aluminum foils (MAM or MAAM). Thin lead sheets have also been used in place of aluminum. The polyester provides strength, and the aluminum acts as the vapor barrier. The barriers may leak if the seals between adjacent sections are not carefully made and handled. To allow for tank shrinkage during cooldown, constrictive wraps of fiberglass roving were applied so that the insulation is in a compressed state after application.

### 9.2. Protective Enclosure Exposed to Aerodynamic Heating

A protective enclosure satisfactory in the earth's environment is not necessarily suitable during the launch phase, where mechanical strength and resistance to high surface temperatures are necessary. Typical enclosures consist of a material that can withstand high temperature (e.g., fiberglass cloth) or can ablate. The enclosure permits the insulation to be evacuated in ground hold and sealed with a vapor-impermeable shroud. The enclosure can also provide solar thermal protection with applied low $\alpha/\varepsilon$ coatings. These coatings reduce the fraction of incident solar radiation that is absorbed. Metals and semiconductors generally have absorptivity/emissivity ($\alpha/\varepsilon$) values larger than unity, whereas nonconductors have $\alpha/\varepsilon$ values less than unity. One of the best low $\alpha/\varepsilon$ coatings is a calcium silicate pigment in a sodium silicate vehicle with an $\alpha/\varepsilon$ value of 0.17.

The enclosure can also incorporate solar radiation shields displaced from the cryogenic storage tank. These shields intercept and reradiate incident solar (or planetary) radiation. Such "shadow shields" are primarily useful in space applications where the tank can be oriented to accept the sun's radiation over its smallest projected area. These solar radiation shields are also useful as protection against micrometeorites. Meteoroid hazards can also be reduced by shielding, but the mass required is often considerable. Studies have indicated that even if foam-insulated cryogenic propellant tanks were to be punctured by meteoroids, the liquid escaping into space through holes in the foam would evaporate so rapidly that a solid plug of frozen liquid would form and effectively seal the leak.

When cryogenic liquid storage tanks are in the interior of a space vehicle, protection need not be provided against aerodynamic heating and pressures or solar and planetary radiation.

## 10. LIQUID AND VAPOR SHIELDS

### 10.1. Liquid Shields

A cheaper cryogenic liquid like nitrogen has at times been used as a thermal blanket to shield either expensive or hazardous fluids including liquid helium, hydrogen, or fluorine from ambient temperatures. In such a system, the inexpensive liquid nitrogen, in its own container, surrounds the main fluid, which in turn resides in a vacuum-jacketed inner vessel as shown in Fig. 7.45.

The benefits achieved by this liquid-shielding arrangement are obvious. Equation (7.2), the modified Stefan-Boltzmann relation for radiant heat transfer through a vacuum, shows that the reduction in the boiloff rate with the use of liquid nitrogen shielding is dependent on the difference between the fourth powers of the temperatures associated with the warm and cold surfaces. For instance, liquid helium at a temperature of 4.2 K would be subjected to an artificial ambient of 77.3 K provided by a liquid nitrogen shield instead of a possible ambient temperature of 300 K. Liquid fluorine, similarly, can be subcooled to 8.3 K by a liquid nitrogen shield, minimizing outgassing of its hazardous vapors. Studies have shown that by using a liquid nitrogen shield, approximately 99.6% of the total radiant heat influx can be intercepted by the less expensive liquid.

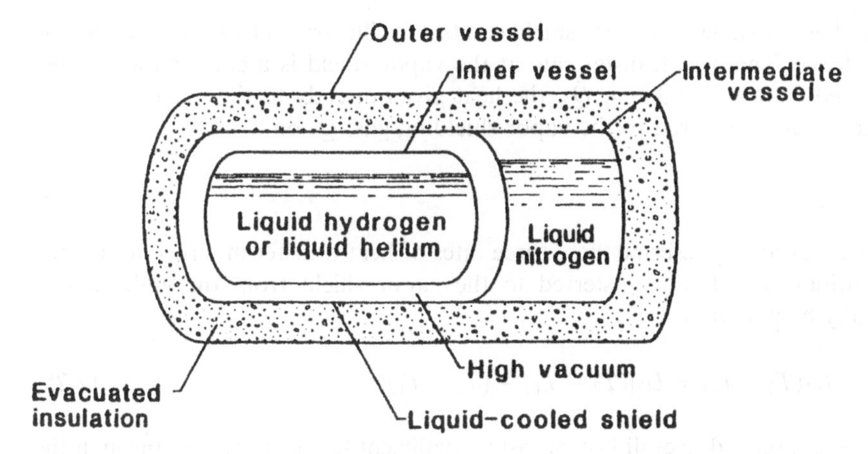


**Figure 7.45** Schematic of liquid-shielded vessel.

Due to the improvements in insulation quality, specifically the development of multilayer insulation, and to improved support and piping systems with lower conductivity, liquid shields have become obsolete. Disadvantages of the system include the inherent increase in bulk weight of the vessel, increase in fragility, and the need for liquid nitrogen. Further, as liquids such as helium are becoming less expensive, a savings in capital investment due to a cheaper vessel design generally outweighs the cost of their boiloff.

## 10.2. Vapor Shields

Unlike liquid shields, which use an inexpensive cooling fluid such as nitrogen, vapor shields exploit the cold vent gas liberated directly from the product liquid. This escaping vent gas is routed past an intermediate shield and sensibly absorbs some of the heat that would otherwise warm the contained liquid as shown in Fig. 7.46. This technique is used mainly for storing lower boiling liquids such as helium and hydrogen rather than nitrogen.

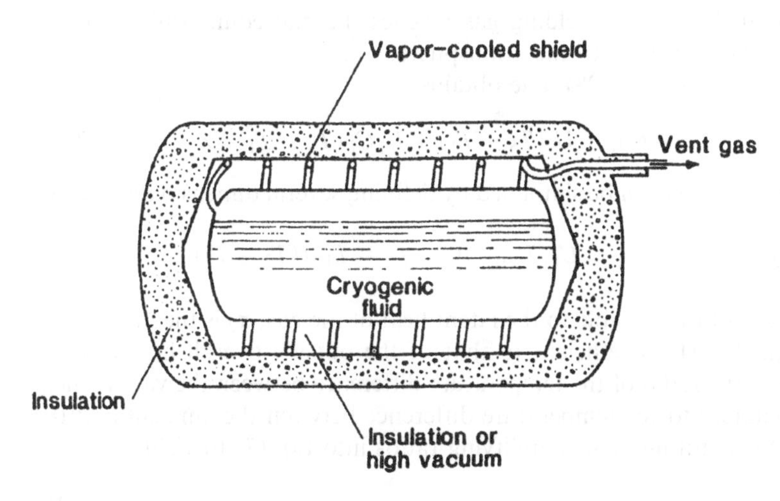


**Figure 7.46** Diagram of a vapor-shielded container.

The thermal value of a vapor shield can best be illustrated by means of a simple energy balance. The heat transfer rate at the vapor shield is a combination of the heat received from the ambient, $\dot{Q}_2$, the heat given directly to the inner container, $\dot{Q}_1$ and the heat transferred to the vapor shielding gas, $\dot{Q}_g$ or

$$\dot{Q}_2 = \dot{Q}_1 + \dot{Q}_g \tag{7.24}$$

It is necessary to describe these three rates of heat transfer in terms of measurable quantities. The heat transferred to the vapor shield from the ambient can theoretically be written as

$$\dot{Q}_2 = U_2(T_2 - T_s) = U_2[(T_2 - T_1) - (T_s - T_1)] \tag{7.25}$$

where $U_2$ is a modified overall heat transfer coefficient for heat transfer through the piping, supports, and insulation, which includes both the total area and path length involved in the heat transfer, namely,

$$U_2 = \sum\left(\frac{KA}{\Delta x}\right)_{\text{piping}} + \sum\left(\frac{KA}{\Delta x}\right)_{\text{supports}} + \sum\left(\frac{K_a A}{\Delta x}\right)_{\text{insulation}} \tag{7.26}$$

Additionally, in Eq. (7.25), $T_2$ is the ambient temperature $T_1$ the inner vessel temperature, and $T_s$ is the mean shield temperature.

Similarly, using the same definition for $U_1$ as was used for $U_2$, the heat transfer rate $\dot{Q}_1$ between the shield and the product container may be written as

$$\dot{Q}_1 = U_1(T_s - T_1) = \dot{m}_g \lambda \tag{7.27}$$

where $\dot{m}_g$ is the mass flow rate of the boiloff vapor and $\lambda$ the latent heat of the liquid product. The sensible heat absorbed by the gas while passing in contact with the vapor shield can be expressed as

$$\dot{Q}_g = \dot{m}_g C_P (T_s - T_1) \tag{7.28}$$

if one assumes that the vapor shielding gas reaches thermal equilibrium with the shield itself before being vented to the atmosphere.

By combining Eqs.(7.25–7.28), one obtains

$$U_2[(T_2 - T_1) - (T_S - T_1)] = U_1(T_S - T_1) + U_1 C_P (T_S - T_1)^2/\lambda \tag{7.29}$$

This cumbersome equation can be simplified by defining several dimensionless ratios:

$$\pi_1 = C_P(T_2 - T_1)/\lambda, \quad \pi_2 = U_1/U_2, \quad \theta = (T_S - T_1)/(T_2 - T_1) \tag{7.30}$$

where $\pi_1$ is the ratio of the sensible to the latent heat of the gas, $\pi_2$ is the ratio of the inside overall heat transfer coefficient to that of the outside overall heat transfer coefficient, and $\theta$ is the radio of the temperature difference between the vapor shield and the inner container to the temperature difference between the ambient and the inner container. Substituting these simplifying ratios into Eq. (7.30) yields

$$\pi_1 \pi_2 \theta^2 + (\pi_2 + 1)\theta - 1 = 0 \tag{7.31}$$

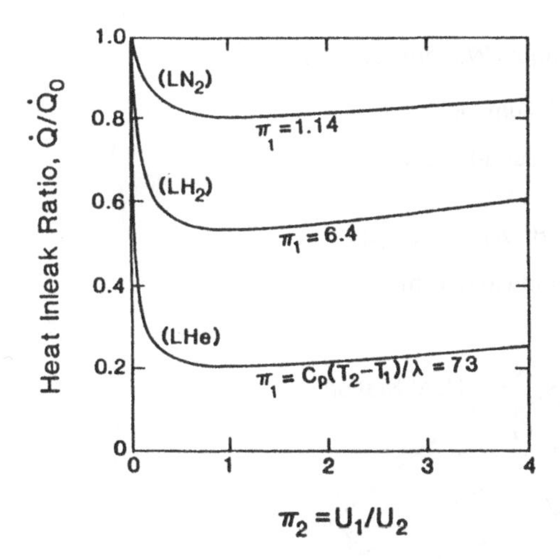


**Figure 7.47** Heat inleak ratio $Q/Q_0$ as a function of the overall heat transfer coefficient ratio $U_1/U_2$.

Solving for $\theta$ gives

$$\theta = \frac{\pi_2 + 1}{2\pi_1\pi_2}\left[\left(1 + \frac{4\pi_1\pi_2}{(\pi_2 + 1)^2}\right)^{1/2} - 1\right] \tag{7.32}$$

Now it is possible to quantify the value of a vapor shield. By assuming the mass flow rate to be 0 and making the same substitutions as before, an expression for $\theta$ can be derived that represents a system with no vapor shield, namely,

$$\theta_0 = 1/(\pi_2 + 1)$$

The ratio of heat inleak with vapor shielding to that without a vapor shield can now be expressed as the ratio of the $\theta$ parameters, or

$$\dot{Q}/\dot{Q}_0 = \theta/\theta_0$$

A plot of this heat inleak ratio for various values of $\pi_1$ is presented in Fig 7.47. As anticipated, the heat inleak is large when the fluid is helium and small when it is nitrogen. Thus, a vapor shield provides the greatest benefit with liquid helium systems; Fig. 7.47 shows that as much as 80% of the possible heat inleak can be intercepted by the vapor shield. Conversely, for a liquid nitrogen–contained vapor-shielded system, less than 20% of the possible heat inleak would typically be intercepted. The reason for this difference clearly is the higher sensible-to-latent heat ratio for helium than for nitrogen.

## 10.3. Use of Vapor Cooling

The design of a refrigerated support employing heat stations (Fig. 7.48) greatly reduces the conductive heat losses down the penetration. A plastic support tube filled with powder or fiber insulation carries the fill (withdrawal) and vent lines and cables that penetrate the insulation. Three equally spaced heat stations in the piping penetration pass transversely through it and are thermally bonded both to the support

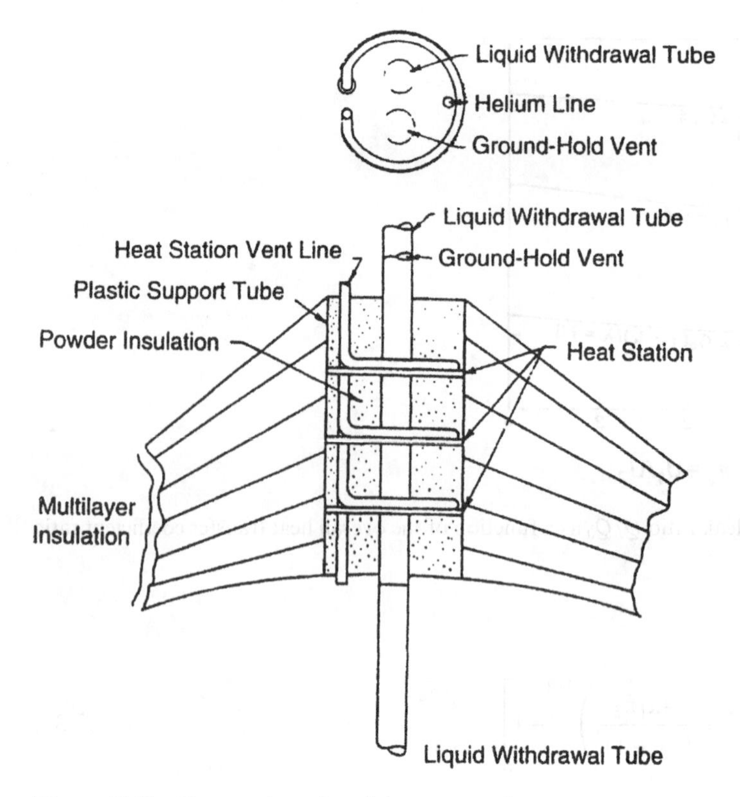


**Figure 7.48** Heat stations for piping penetration.

tube and to all of the pipes. The vent line is coiled and bonded to each of these heat stations so that a transverse isothermal surface is established at each heat station. This surface ensures that the specific heat available in the boiloff gases can be used to intercept conductive heat leaks down the pipes and reduce the total flux through this penetration. The multilayer insulation layers are attached to the structural support tube of the piping penetration at a point where the tube temperature matches the shield temperatures. Analysis of heat leak through the piping penetration as a function of a number of heat stations shows that when two heat stations are used the heat leak is reduced to 0.1, and for three heat stations to 0.05, of the heat leak with no heat station.

To reduce heat losses at points of attachment of multilayer insulation to vent tubes, the refrigeration potential of escaping boiloff gases can be used. To accomplish this, multiple shields attached to a vent tube are interleaved with multilayer insulation (Fig. 7.48).

The application of multiple shields to recover portions of the vapor refrigeration potential on a 100 L liquid helium storage tank indicated that heat leaks were reduced by a factor of about 18. The use of multiple shields reduces the liquid loss and the time required for cooldown.

## 11. COMPOSITE INSULATIONS

The optimum insulation system should combine maximum insulation effectiveness, minimum weight, and ease of fabrication. It would be desirable to use only one insulation material, but since no single insulation material has all the desirable physical and strength characteristics required in many applications, composite insulations have been developed. These composite insulations represent a compromise between thermal effectiveness, handling properties, and adequate service life. For cryogenic applications, composite insulations consist of a polyurethane foam; reinforcement for the foam to provide adequate compressive strength; adhesives for sealing and securing the foam to a tank; enclosures to prevent damage to the foam from mechanical contact, vibration, and aerodynamic heating; and vapor barriers to maintain a separation between the foam and atmospheric gases.

## 11.1. Honeycomb Foam Insulation

Several external insulation systems use honeycomb structures. Phenolic resin–reinforced fiberglass cloth honeycomb is most commonly used. The honeycomb is bonded to the tank with a suitable seal, and a vapor barrier is applied to the exterior of the honeycomb. In such a system, the cells in the honeycomb can be interconnected and evacuated through external pumping, interconnected and purged with helium, or sealed and allowed to reach a low pressure within the sealed cells when the tank is filled with a cryogenic liquid.

### *11.1.1. Foam-Filled Honeycomb*

The effectiveness of a honeycomb insulation can be considerably improved if its cells are filled with a low-density polyurethane foam. Several techniques have been used for preparing foam-filled honeycomb. For example, the liquid components can be foamed in place to fill the honeycomb cells. This system has not proved effective, however, because of large waste, nonuniform cell structure in the foam, and nonuniform filling of the honeycomb cells. The most common method of preparing foam-filled honeycomb is to press the rigid, phenolic resin-reinforced fiberglass honeycomb into a preformed sheet of low-density foam. The thin honeycomb walls cleanly cut through the foam so that it fills each cell. However, the foam is not sealed or bonded into the cells; it simply lies within them. This lack of bonding has certain disadvantages because atmospheric gases or helium can diffuse through the insulation. In this, as in other honeycomb insulations, the honeycomb provides compressive strength and allows the insulation to absorb stresses created by differential expansion between the insulating materials and the tank during chilldown. A sealed foam-filled system, however, is not completely reliable. Therefore, a helium-purged, externally sealed honeycomb system, was developed for the Saturn S-II insulation system.

### *11.1.2. The Saturn S-II Insulation System*

The S-II booster is the second stage of the Saturn C-5 launch vehicle. Overall, it is 33 ft in diameter and 85 ft long. The liquid hydrogen tank occupies the largest portion of the second stage's volume. The total insulated tank area is 6000 $ft^2$ exclusive of the bulkhead, which is common to the liquid hydrogen and liquid oxygen tanks. In addition to limiting heat flow during launch and preload, the insulation was designed to be self-snuffing when ignited, to prevent significant air condensation in the insulation, and to provide a means for identifying structural flaws in the tank during fabrication.

The insulation system consists of a foam-filled honeycomb having an overall thickness of 1.6 in. on the tank side wall and 0.5 in. on the bulkhead. The foam is an open-cell polyurethane of 2.2 lb/ft$^3$ (0.035 g/cm$^3$) density that is pressed into the honeycomb. This composite is bonded to the tank wall with an adhesive. The vapor barrier on the warm side of the insulation is made up of nylon-phenolic laminate impregnated with a 0.0015 in. (38.1 μm) seal coating of polyvinyl fluoride, as shown in Fig. 7.49. The cold side of the insulation is grooved to provide gas flow channels for helium-purging the insulation during prelaunch and for evaluating the insulation during launch. The weight of the composite 1.6 in. thick insulation structure is about 0.85 lb/ft$^2$ (4.15 kg/m$^2$).

The cylindrical section of the tank is made up of quadrant sections 27 ft wide by about 9 ft high that are welded into rings. The rings are welded together vertically. The insulation slab is applied to the quadrant sections except at the panel edges, where it is recessed from the weld areas to permit clearance for the weld tools and separation from the heat-affected zone. When the tank fabrication and hydrostatic tests have been completed, the gaps between insulation panels are filled with foam sections and bonded over with adhesives that cure at room temperature. The manufacturing process is depicted in Fig. 7.50.

Thermal conductivity values of 0.6–0.75 Btu in/(h ft$^2$ °F) [0.86–1.1 mW/(cm °C)] were obtained for the helium-purged insulation at a mean temperature of −200°F (−129°C). A comparison of the results is presented in Fig. 7.51. The measured foam conductivities are comparable to the thermal conductivity of helium.

Evaluation of purged honeycomb foam insulation indicated that the foam was not fully effective because in reducing thermal conductivity the composite insulation rapidly reached the thermal conductivity of the helium purge gas. To improve the insulation's effectiveness, the foam was cut to the proper size by pressing it into a slice of honeycomb. The foam was then punched out, and each piece of foam was sealed with a polyurethane resin and reinserted into the honeycomb cell. However, the thermal conductivity of the insulation containing the individually coated foam pieces was the same as if the pieces had not been coated, indicating that helium readily permeated the sealing resin. Composite insulation, which is first purged with helium and then evacuated, was also considered. However, if the vapor barrier is not leakproof, air can leak into the cold evacuated insulation and condense.

In addition to the basic hexagonal core honeycomb, modified honeycombs filled with rigid urethane foam have been evaluated for use in insulation systems.

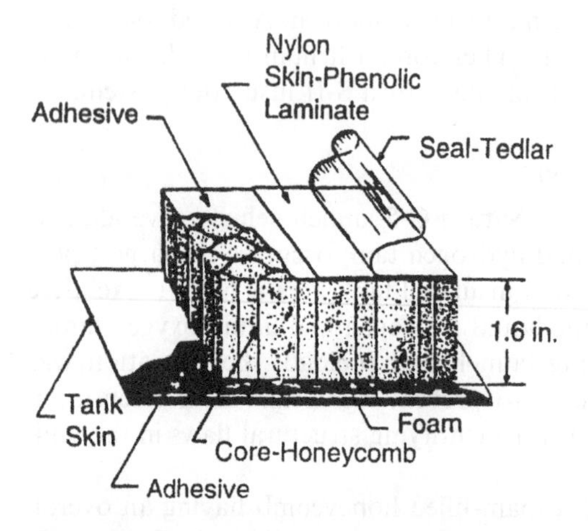


**Figure 7.49** Saturn S-II basic insulation configuration.

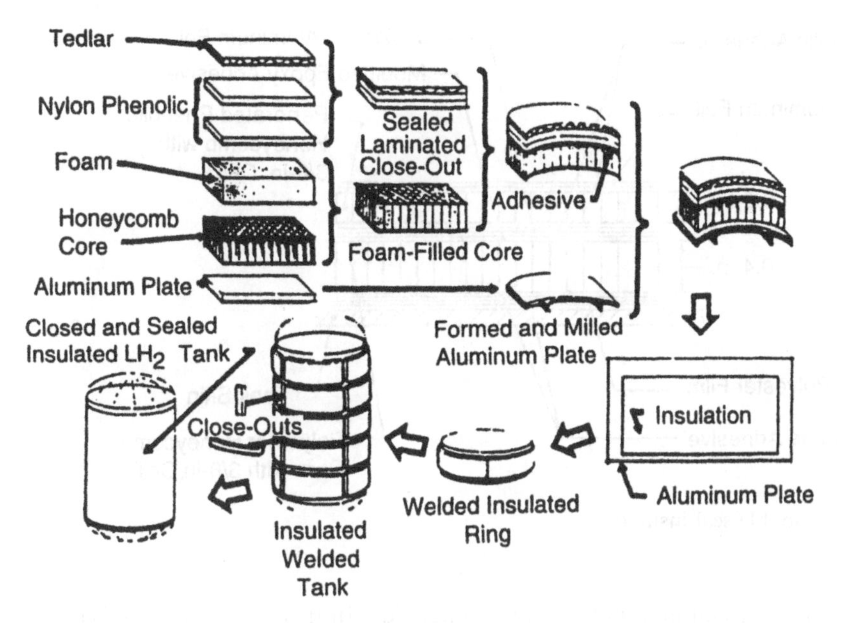


**Figure 7.50** Manufacturing process for Saturn S-II insulation system.

## 11.2. Sealed Evacuated Honeycomb Insulations

One concept considered for a cryogenic insulation was a sealed honeycomb in which the residual air in the individual cells was condensed to provide a vacuum in each cell. A removable bag of lightweight polyester film was to be placed around the exterior of the insulation to permit a helium purge and to prevent diffusion of air into the insulation in the event of any leaks in the outer seal. This outer bag was to fall away just prior to vehicle liftoff. This concept was not tried because no outer bag sufficiently reliable to withstand wind and mechanical loading and other ground hazards and capable of being removed rapidly from the vehicle was found.

A double-seal honeycomb insulation that incorporates a helium-purged layer and a dual layer of honeycomb has been developed. The layer of the double-seal insulation next to the liquid hydrogen tank is made of 3/8 in. cell polyester honeycomb to which is bonded a 0.002 in. (50.8 μm) polyester film on the tank side and

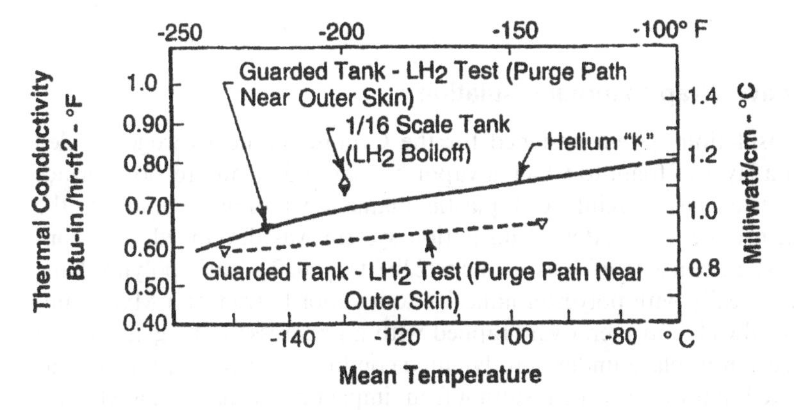


**Figure 7.51** Thermal conductivity of helium-purged insulation (1.5 in. thick insulation).

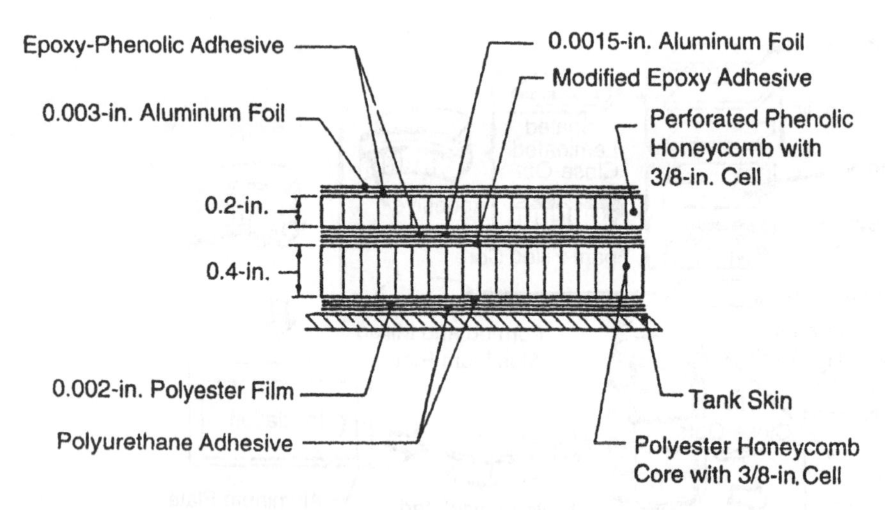


**Figure 7.52** Double seal insulation.

a 0.0015 in. (38.1 μm) aluminum foil on the outer side. In this layer, each cell is completely sealed; when the layer is chilled to liquid hydrogen temperature, each cell is evacuated by liquefaction and/or solidification of the gases within it. Thus, the layer provides a very effective thermal insulation. The outer layer of the insulation consists of a 3/8 in. cell perforated phenolic honeycomb, 0.2 in. thick, bonded to the 0.0015 in. aluminum foil on the outer side of the polyester honeycomb. The outer surface consists of a 0.003 in. (76.2 μm) aluminum foil bonded to the perforated phenolic honeycomb with an epoxy phenolic adhesive. In use, the perforated phenolic honeycomb is purged with helium under positive pressure to provide a complete encasement of the polyester honeycomb in a helium envelope. This prevents any possibility of air diffusing into the polyester honeycomb portion of the insulation. If the outer skin is punctured, helium will escape through the puncture. If both seals are punctured, the cell or cells of the polyester honeycomb in the punctured area will fill with helium, and helium will escape through the breach in the outer seal so that air is still excluded and the loss in insulation effectiveness is limited to those few cells of the polyester honeycomb that have become filled with helium. The double-seal insulation concept is shown in Fig. 7.52. This double-seal insulation weighs about 0.5 $lb/ft^2$ (2.44 $kg/m^2$) and has a conductance of 0.3 Btu in./(h $ft^2$ °F) [0.43 mW/(cm °C)].

### 11.3. Constrictive Wrap External Insulation

An alternative insulation system designed for the Centaur liquid hydrogen fueled stage consists of a layer of foam to which a vapor barrier and a compression surface winding are applied. Using reinforced plastic filament–winding techniques, the sealed foam constrictive wrap external insulation system was developed. The constrictive wrap system consists of 0.4 in. thick (2 $lb/ft^3$; 0.032 $g/cm^3$) polyurethane foam sealed hermetically with polyester adhesive to a vapor barrier of a Mylar–aluminum–Mylar sandwich and then overwrapped with a prestressed fiberglass roving to hold the insulation in place under aerodynamic conditions. Investigations of this insulation indicated that the hazards resulting from impacting insulation panels containing oxygen-rich liquefied air were negligible. The insulation effectiveness proved to be equal to that predicted for the system, and the installed weight was only 0.16 lb/ft$^2$ (078 kg/m$^2$).

## 11.4. Exterior Surface-Bonded Foam

The constrictive wrap prevents the cracking and delamination of foams subjected to low temperatures on the surface of cryogenic tanks. An alternative approach uses the excellent bonding strength of urethane foamed directly onto the tank surface. With appropriate temperature control [approximately 120°F (50°C)] at the tank wall, the foam can be made with a uniform density. The foam contains chopped strands of fiberglass that provide sufficient reinforcement to prevent it from cracking and delaminating when cold-shocked in liquid nitrogen, liquid hydrogen, or liquid helium.

The foam formulation is based on a quadral-triol mix combined with an isocyanate. It is fluorocarbon-blown and has a density of 4.0 lb/ft$^3$ (0.064 g/cm$^3$). When combined with the chopped fiberglass strands, the initial mix has sufficiently low viscosity to allow the foaming action to proceed to a uniform density. The chopped glass strands have a random orientation and provide strength to the foam in all directions.

Tests were performed with this insulation to measure thermal performance and to determine structural resistance to cold shock. The heat flux through the insulation was measured with liquid hydrogen and liquid nitrogen in the tank and with the tank exposed to the atmospheric environment. The measured heat flux ranged from 70 to 100 Btu/(h ft$^2$) (22–32 mW/cm$^2$). The average thermal conductivity was computed to be 0.13 Btu in./ (h ft$^2$°F) [0.19 mW/cm °C)]. The outer surface temperature ranged from −60°F to +62°F (−51°C to +17°C), which is above the condensation temperature of air.

Repeatedly filling the tank with liquid hydrogen and liquid nitrogen and spraying the outside of the insulation with liquid nitrogen while the tank was filled with liquid hydrogen did not impair the thermal performance of the insulation. It retained structural integrity except for minor liftoff of the vapor barrier.

Another application of external foam on an aluminum liquid hydrogen tank showed good results. The insulation was bonded to the tank with an intermediate layer of epoxy-coated fiberglass cloth. The adhesive qualities were best only when the aluminum surface had been precleaned and vapor-blasted.

## 11.5. Internally Insulated Fiberglass Cryogenic Storage Tank

Plastic foams and glass filament-wound structures can be combined to produce a light-weight liquid hydrogen tank. This approach was used on a tank of 40 gal capacity, 18 in. in diameter by 36 in. long, designed to operate at 100 psig. A layer of foam 1/2 in. thick is used as the mandrel for the filament-winding operation. A vapor barrier is placed on the interior of the foam structure to make the tank capable of containing gaseous and liquid hydrogen.

The foam layer of the tank is made of upper and lower dome sections and a cylindrical midsection. Each section is formed of polyurethane foam encased in an aluminum–Mylar–aluminum (AMA) laminate that serves as a vacuum-tight barrier. A glass cloth is placed between the foam and the AMA laminate to provide a path for gas flow when the foam is evacuated. The foam acts both as a thermal insulation and as a structural material for transmitting the internal pressure to the filament-wound shell structure. Construction details of the foam and vacuum-tight barrier in the vicinity of the joint between the cylinder and dome sections are shown in Fig. 7.53.

The structural shell is made up of two layers of longitudinal and four layers of circumferential wraps of glass filaments. The roving consists of 20-end S/HTS fiberglass impregnated with a low-temperature casing epoxy.

The liner for the tank was fabricated from AMA laminate. The domes were spin-formed from 0.005 in. (127 μm) thick laminate, and the cylinder section from 0.0025 in. (63 μm) thick laminate. The liner joints were sealed with an adhesive to maintain a hermetic seal at −423°F (−253°C) after curing at a temperature compatible with the foam insulator.

The vacuum-tightness of the enclosed foam was checked with a helium mass spectrometer both before and after the tank was cold-shocked with liquid nitrogen. The heat flow through the insulation was determined by monitoring the flow rate of vaporized hydrogen with the tank exposed to atmospheric conditions. To test its structural performance, the vessel was pressure-cycled while filled with cryogenic liquid.

When the tank was filled to about 75% of its volume, the heat leak was about 1800 Btu/h (530 W), that is, a 50% liquid loss per hour. A significant portion of the loss, about 1100 Btu (1160 KWh), was attributed to the conductive heat flow in the AMA laminate at the joint between the bottom dome and cylinder sections in the foam structure (Fig. 7.53). Depending on the environmental conditions, the heat flux in the cylindrical section was 80–110 Btu/(h ft$^2$) [25.2–34.6 mW/(cm$^2$)], corresponding to a thermal conductivity of the foam of about 0.1 Btu in./(h ft$^2$ °F) [0.144 mW/(cm °C)].

The liner of the vessel failed at an internal pressure of 68 psig after successful performance at lower pressures. Failure was attributed to overstressing of the liner material.

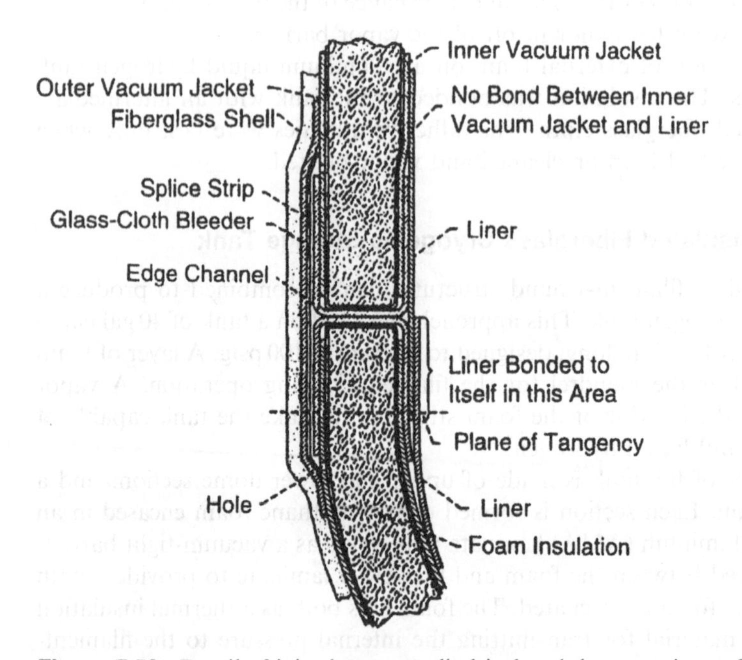


**Figure 7.53** Detail of joint between cylindrical and dome sections of tank foam layer.

### 11.6. Self-Sealing Tanks

Meteoroid particles impacting cryogenic propellant tanks will not only penetrate the tank wall but may also generate forces that could cause catastrophic failure of the tank. External thermal shielding and insulating systems for the tanks can be designed to incorporate an exterior meteoroid bumper together with a low-density energy-absorbing material between the bumper and the tank wall to provide meteoroid protection. Attention has been given to using structures similar to sealed exterior insulations as self-sealing shields for micrometeorite protection. If a cryogenic propellant tank is penetrated by a micrometeorite, the cryogenic propellant will solidify when it encounters the vacuum of space. If there is a porous medium around the penetration, the solidifying propellant will cling to this medium and plug the hole. Tests were made by filling a honeycomb with porous media such as fiberglass and an open-cell polyester foam. The most favorable self-sealing results were obtained with the open-cell polyester foam, which produced sealing of punctures as large as 0.050 in. in diameter.

## 12. OTHER MATERIALS

A large number of other materials have been screened as possible cryogenic insulations. Those that have proved useful are corkboard and balsa wood. Although balsa wood is a good nonporous insulator, it is brittle and difficult to fabricate into thin sections. Corkboard is readily available and relatively nonporous: it has a low conductivity but a high density, which limits its use for space vehicles. Corkboard tends to separate and crack near the tank walls during cooldown. Like balsa, it is difficult to fabricate in curved sections.

## 13. PLACEMENT OF INSULATION SYSTEMS

### 13.1. External Insulation

External insulation has been used for the liquid hydrogen storage tanks of large boosters such as the Saturn S-II and S-IV and for many smaller cryogenic fluid storage vessels developed for the Apollo program. External insulation is the most widely used because

1. Complex tank shapes can be insulated.
2. Insulation thickness can be built up in successive layers or in individual sections.
3. Insulation can be integrated with supports and pipes.
4. Sections of insulation can be more easily repaired.

External insulation requires that either an outer shell or a vaportight barrier be placed over it. After this process has been completed, however, the tank is quite fragile and must be handled with care to prevent damage to the insulation.

### 13.2. Internal Insulation

In some applications, primarily with large tanks, the insulation is placed inside the tank shell. The major advantages of such placement are that

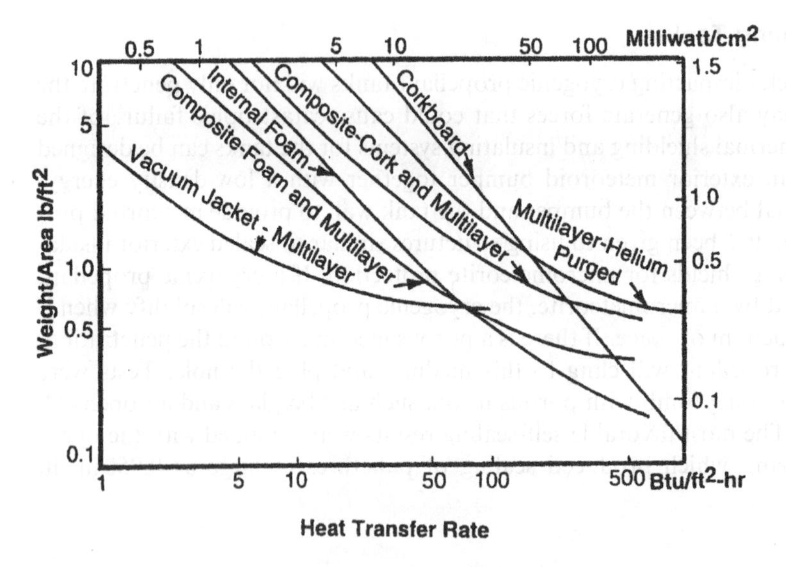


**Figure 7.54** Weight effectiveness of selected insulations.

1. The insulation does not need to withstand aerodynamic heating and pressures during vehicle launch.
2. Mechanical handling of the complete cryogenic tank is simpler than that of a tank with a fragile external insulation.
3. The insulation is not penetrated by external lines and projections.

The disadvantages of internal insulation are that

1. The insulation must be attached securely to the inner wall, and the cryogenic liquid must be prevented (by a vapor barrier) from leaking into the insulation. If such leakage occurs, the liquid will vaporize at or near the warm wall and drastically reduce the insulation's effectiveness.
2. Defective portions of the insulation are hard to locate, and repairs are difficult to make.
3. Any separation of the insulation from the wall caused by thermal contraction, vibration, or sloshing will result in an increased heat leak; loose sections of the insulation can plug the liquid supply lines.

As shown in Fig. 7.54, the weight of insulation materials that are to be used in space flight applications must also be optimized. For these applications, flexible materials are typically employed. Loose-fill materials, either evacuated or at atmospheric pressure, are used for storage, transportation, and transfer of cryogenic fluids, Rigid insulation materials are used for both space flight and terrestrial applications requiring mechanical strength.

## 14. ADHESIVES

Adhesives are an essential component of cryogenic insulations because no other means is satisfactory for joining the lightweight, and sometimes fragile, components of an insulation system into a structural entity. In addition, most insulation systems require sealing to prevent permeation of gases and/or liquids into or out of the insulation. Adhesives provide a combination of high strength, stress distribution, and sealing.

From a practical standpoint, two types of adhesives are used in the fabrication of composite insulation systems: heat-cured adhesives and ambient temperature cured adhesives. The heat-cured adhesives are used to achieve good properties at elevated temperatures in the shop assembly of components that may be exposed to such temperatures (e.g., aerodynamic heating). Ambient temperature cured adhesives are necessary to bond assembled structures where it would be impractical to apply heat to cure the adhesive.

Adhesives are used as either supported or unsupported films or as a liquid or semi-liquid mixture of two or more ingredients. Heat-cured adhesives are used in film form, whereas the ambient temperature cured systems are two-component semiliquid pastes. The film-type adhesives incorporate a fiberglass or other cloth carrier that not only lends handling strength to the adhesive film but also helps to reduce the coefficient of thermal expansion of the film and the resultant adhesive layer, thereby reducing thermal stresses in the completed bond. In both the heat-cured and ambient temperature cured adhesives, metal or metal oxide fillers can be added to the resin to modify the thermal expansion characteristics of the adhesives to more nearly match those of the substrate and thereby reduce thermal stresses in the completed bond during various environmental changes.

## 14.1. Adhesives for Cryogenic Application

Several basic types of adhesives are used for primary structural bonding in aircraft and space vehicles. These usually consist of a thermosetting phenolic or epoxy resin, which contributes high adhesion and thermosetting characteristics, and a modifier to reduce the brittleness of the thermosetting resin and thus provide a resilient bond unaffected by thermal and curing shrinkage stresses. Modifiers include nitrile rubber, vinyl acetal, vinyl butyral resins, and, more recently, polyamide resins.

In early work (before 1960), a number of structural adhesives were screened for bond strength with various types of substrates at temperatures ranging from room temperature down to −423°F (−253°C). This work clearly indicated the superiority of the then relatively new epoxy phenolic blends over the nitrile phenolic and vinyl phenolic adhesives that were the standard structural adhesives. Bond strengths of over 3000 psi (210 kg/cm$^2$) were achieved on aluminum lap shear joints at −423°F (−253°C) with the epoxy phenolics; all the other types of structural adhesives tested had bond strengths of about 2000 psi (140 kg/cm$^2$) or less.

About 1960, a new type of structural adhesive, based on epoxy polyamide resin blends, became available and was evaluated extensively for potential cryogenic use. The epoxy polyamide resin adhesives, available in film form, proved to have bond strength at liquid hydrogen temperature markedly superior to that of any other available adhesive. These adhesives showed bond strengths with stainless steel in lap shear in excess of 5000 psi (350 kg/cm$^2$), in comparison with approximately 3000 psi (210 kg/cm$^2$) for epoxy phenolic adhesives. Although nitrile phenolics have excellent bonded strength at temperatures in the range of −100°F (−73°C), their strength drops to around 2000 psi (140 kg/cm$^2$) at liquid hydrogen temperature. The bonded strength of epoxy polyamide resin adhesives, although slightly lower than that of the nitrile phenolics at room temperature and −100°F (−73°C), is slightly higher than that of the nitrile phenolics at −320°F (−196°C) and very much higher at −423°F (−253°C).

The epoxy polyamide and polyurethane adhesives are ambient temperature cured systems whose bond strengths are inherently lower than those of the heat-cured adhesives. However, the properties of these adhesives do not deteriorate drastically as the temperature is reduced from room temperature to −423°F, and, indeed, they have proven satisfactory for sealing and coating applications where high strength is not required but where the adhesive must not become brittle at −423°F.

For applications where good physical properties are required at both elevated temperatures (200–300°F; 93–150°C) and low temperature (−423°F; −253°C), the epoxy phenolic adhesives are used. For applications where high strength is not needed at temperatures above 150°F (66°C) but maximum strength is needed at cryogenic temperatures, the epoxy polyamide adhesives are used. For applications where an ambient temperature cured adhesive with good properties at cryogenic temperatures is required, epoxy polyamide or epoxy polyurethane adhesives can be used. In general, ambient temperature cured adhesives have poor strength at temperatures above 150°F (66°C).

Measurements have been made of the thermal expansion characteristics of several epoxy adhesives. The linear thermal expansion of various epoxy adhesives varies substantially. However, modifications in thermal expansion characteristics can be achieved through the incorporation of fillers.

### 14.2. Applications of Adhesives

In an internal insulation system, adhesives such as the polyurethane types are used for bonding the blocks of foam insulation into place, bonding a vapor barrier to the surface of the blocks, or actually forming a vapor barrier by impregnating a fiberglass cloth.

In external insulation systems, adhesives are required to bond the phenolic honeycomb or foam panels to the tank, to form the various seals that may be necessary in the intermediate layers, and to bond closeout panels between sections of shop-bonded insulation.

Several adhesives have been evaluated for bonding vapor barriers. Five adhesives were used to bond polyvinyl fluoride and polyester films to stainless steel substrates. Epoxy polyamide and epoxy both produced bond strengths of ~1200 psi (80 $kg/cm^3$) with polyvinyl fluoride at room temperature and at liquid nitrogen temperature, whereas the other adhesives produced lesser bond strengths. With the polyester vapor barrier, the epoxy polyamide again produced bond strengths of ~1200 psi (80 $kg/cm^2$) at room temperature and at liquid nitrogen temperature, whereas the epoxy had bond strength of ~300 psi (20 $kg/cm^2$) at room temperature and 1200 psi (80 $kg/cm^2$) at liquid nitrogen temperature. The epoxy polyamide appeared superior not only in bond strength but also from the standpoint of processing characteristics, because it can be cured at room temperature whereas epoxy curing requires an elevated temperature.

## 15. COMPARISON OF INSULATIONS

The boundary temperatures most frequently encountered in cryogenic design are room temperature, about 300 K; the oxygen boiling point, 90 K; the nitrogen boiling

**Table 7.23** Heat Transfer Between Parallel Surfaces Through Various Types of Insulation

| Insulation | Heat transfer through insulation having the indicated boundary temperatures[a] ($mW/cm^2$) | | |
|---|---|---|---|
| | 300 K, 77 K | 300 K, 20 K | 77 K, 20 K |
| High vacuum, $P = 10^{-6}$ mm Hg ($H_2$ residual). Emissivity 0.02 at each surface | 0.45 | 0.46 | 0.002 |
| Gases at atmospheric pressure (no convection). Fifteen cm between the warm and cold surfaces | | | |
| $H_2$ | 19.3 ($K = 1.30$) | 19.6 ($K = 1.05$) | 0.8 ($K = 0.22$) |
| He | 17.1 ($K = 1.15$) | 17.7 ($K = 0.95$) | 1.7 ($K = 0.45$) |
| Air or $N_2$ | 2.68 ($K = 0.18$) | – | – |
| Evacuated powder (expanded perlite with density of 5–6 $lb/ft^3$), 15 cm layer | 0.16 ($\bar{K} = 0.011$) | 0.13 ($\bar{K} = 0.007$) | 0.007 ($\bar{K} = 0.0002$) |
| Gas-filled powder (expanded perlite 5–6 $lb/ft^3$), 15 cm layer | | | |
| He | 18.7 ($\bar{K} = 1.26$) | 18.7 ($\bar{K} = 1.0$ est.) | 2.2 ($\bar{K} = 0.05$) |
| $N_2$ | 4.8 ($\bar{K} = 0.32$) | – | – |
| Polystyrene foam (2 $lb/ft^3$), 15 cm layer | 4.9 ($\bar{K} = 0.33$) | 5.1 ($\bar{K} = 0.27$) | 0.57 ($\bar{K} = 0.15$) |

[a] Temperatures are those of warm and cold surfaces.

$K$ is the apparent effective mean thermal conductivity for the temperature interval; $\bar{K}$ is the actual mean thermal conductivity. The units for $K$ and $\bar{K}$ are mW/(cm K).

**Table 7.24** Representative Apparent Thermal Conductivity Values for Various Insulations

| Insulation | Apparent thermal conductivity, $K_a$, between 77 and 300 K [mW/(m K)] |
|---|---|
| Pure gas at 0.101 MPa (180 K) | |
| $H_2$ | 104 |
| $N_2$ | 17 |
| Pure vacuum, $1.3 \times 10^{-10}$ MPa | 5 |
| Foam insulation | |
| Polystyrene foam (46 kg/m$^3$) | 26 |
| Polyurethane foam (11 kg/m$^3$) | 33 |
| Glass foam (140 kg/m$^3$) | 35 |
| Nonevacuated Powders | |
| Perlite (50 kg/m$^3$) | 26 |
| Perlite (210 kg/m$^3$) | 44 |
| Silica aerogel (80 kg/m$^3$) | 19 |
| Fiberglass (110 kg/m$^3$) | 25 |
| Evacuated powders and fibers($1.3 \times 10^{-7}$ MPa) | |
| Perlite (60–180 kg/m$^3$) | 1–2 |
| Silica aerogel (80 kg/m$^3$) | 1.7–2.1 |
| Fibreglass (50 kg/m$^3$) | 1.7 |
| Opacified powder insulations ($1.3 \times 10^{-7}$ MPa) | |
| 50/50 wt% Al/Santocel (160 kg/m$^3$) | $3.5 \times 10^{-1}$ |
| 50/50 wt% Cu/Santocel (180 kg/m$^3$) | $3.3 \times 10^{-1}$ |
| Multilayer insulations ($1.3 \times 10^{-10}$ MPa) | |
| Al foil and fiberglass | |
| (12–27 layers/cm) | $3.5\text{–}7 \times 10^{-2}$ |
| (30–60 layers/cm) | $1.7 \times 10^{-2}$ |
| Al foil and nylon net | |
| (31 layers/cm) | $3.5 \times 10^{-2}$ |
| Al crinkled, Mylar film | |
| (35 layers/cm) | $4.2 \times 10^{-2}$ |

*Source*: Timmerhaus and Flynn (1989).

point, 77 K; the hydrogen boiling point, 20 K; and the helium boiling point, 4.2 K. Three combinations most frequently encountered require somewhat different treatments in the selection of appropriate insulating procedures. These combinations are (1) warm surface at 300 K, cold surface at 77 K (or 90 K); (2) warm surface at 300 K, cold surface at 20 K; (3) warm surface at 77 K, cold surface at 20 K (or 4 K). Combination (1) is the condition existing in ordinary containers for liquid oxygen or nitrogen, combination (2) could refer to a container for liquid hydrogen that does not have a liquid nitrogen cooled shield, and combination (3) is encountered in vessels for liquid hydrogen or liquid helium that have a protective shield cooled with liquid nitrogen.

When high-vacuum insulation is used, heat is transferred from a surface at 300 K to one at 90 K or lower almost entirely by radiation. The radiant heat transfer is practically unaffected by reducing the temperature of the cold surface below 90 K because $300^4$ is so much greater than $90^4$. However, with boundaries at 77 K and 20 or 4 K, even a small a mount of residual gas (hydrogen or helium) will account for a large part of the total heat flow. If the separation between the walls that are at 300

**Table 7.25** Advantages and Disadvantages of Some Cryogenic Insulations

| Advantages | Disadvantages |
|---|---|
| (1) *Expanded perlite* | |
| Low thermal conductivity | Friable |
| Low density | Easily damaged during handling with increase in bulk density |
| Low cost | Flows with difficulty when crushed |
| Readily available | Dusty |
| Dimensional stability in static installation | Abrasive |
| Not flammable | Cannot be used to insulate moving parts |
| Not toxic | Maintenance may require complete removal of insulant |
| Not hygroscopic | Limited transferrability |
| Easy to install | Nonelastic in compression |
| Flows readily when new | Needs protection from atmosphere |
| Easily conveyed pneumatically | |
| Recoverable even after contact with water | |
| Does not cake in use | |
| Can be produced on site | |
| Useful in vacuum insulation | |
| Easy to evacuate | |
| Easy to install in fabricated vessel | |
| Typical trade names: Brelite, Cecaperl, Perlox, Pulvinsul, Ryolex | |
| (2) *Fiberglass* | |
| Low thermal conductivity | Expensive |
| Fine fibers comparatively easy to handle | Requires careful packing for best performance |
| Dimensionally stable | Difficult to install in vacuum-insulated vessels |
| Elastic in compression; allows for contraction and expansion | Needs protection from atmosphere |
| Not hygroscopic | |
| Recoverable after water contact | |
| Readily available | |
| (3) *Mineral or rock wool* | |
| Relatively inexpensive | More costly on a volume basis than expanded perlite because of high packing density |
| Readily available | Unpleasant to handle |
| Dimensionally stable | Requires experienced labor to install |
| Slight elastic compressibility | Requires careful specification to assume less than 2% organic matter in the wool and minimum sulfides |
| Not hygroscopic | Needs protection from atmosphere |
| Free from fine dust, can be used to insulate moving machinery | |
| Only problem area needs removal during maintenance | |
| Typical trade names: Banroc, Eldorite, Fluffex, Insulwool, Rocksil, Stillite | |

(*Continued*)

**Table 7.25** (*Continued*)

| Advantages | Disadvantages |
|---|---|
| (4) *Silica aerogel* | |
| Lightweight low thermal conductivity<br>Extremely free flowing<br>Dimensionally stable when dry<br>Does not shrink when larger particles are broken<br>Will not settle due to vibration after initial compaction<br>Not hygroscopic<br>Useful as the base for reflecting powders<br>Easy to evacuate in vacuum insulation<br>Trade names: Aerosil, Cab-O-Sil, Santocel | Price is approximately 10 times that of expanded perlite<br>After water contact, forms silica gel on drying with an increase in bulk density by a factor of 10 and decreased insulation properties<br>Needs protection from atmosphere |
| (5) *Basic magnesium carbonate* | |
| Relatively cheap<br>Readily available<br>Good insulating value<br>Dimensionally stable<br>Good performance in vacuum insulation | Slightly hygroscopic<br>Cakes easily; solidifies after exposure to moisture<br>Difficult to regenerate after water contact<br>Normally difficult to evacuate because of moisture content<br>Dissolves in acidic media<br>Needs protection from atmosphere |
| (6) *Vermiculite* | |
| Relatively inexpensive<br>Readily available<br>Lightweight<br>Dimensionally stable<br>Not hygroscopic<br>Not toxic<br>Good structural properties | Thermal conductivity 3–5 times that of expanded perlite<br>Large convective heat transfer rate<br>High solid-phase conduction<br>Needs protection from atmosphere |
| (7) *Glass foam* | |
| Strong<br>Readily available<br>Closed pores, not permeable to water<br>Useful as a support for inner vessel<br>Dimensionally stable<br>Not hygroscopic<br>Not flammable | Expensive<br>Needs skilled labor for proper installation<br>Needs vapor barrier between insulating blocks<br>Rather high conductivity |
| (8) *Foamed plastics* | |
| Lightweight<br>Low thermal conductivity<br>Relatively inexpensive<br>Readily available in many grades<br>Useful as support for inner vessel<br>Available with closed pores not permeable to water | Needs skilled labor for proper installation<br>Has large contraction and expansion<br>Upper temperature limit can be as low as 350 K<br>Forming materials generally flammable<br>Unusable in an oxygen-enriched atmosphere |

(*Continued*)

**Table 7.25** *(Continued)*

| Advantages | Disadvantages |
|---|---|
| Can be foamed in situ, simplifying installation | Requires nitrogen purge if used below 90 K |
| (9) *Fiberglass paper–aluminum foil combination* | |
| Very low thermal conductivity | Needs high vacuum for best performance |
| Easily applied to cylindrical shapes | Suppliers limited for special paper needed |
| Thermal conductivity increase with compression less than for other layer insulants | Paper relatively weak and easily torn |
| | Will not resist appreciable vibration |
| | Considerable skill needed to apply properly |
| Can be heated adequately for outgassing | Difficult to apply to complex geometries |
| Not flammable in oxygen | |
| (10) *Synthetic fiber net–aluminum foil combination* | |
| Very low thermal conductivity | Needs high vacuum for best performance |
| Organic fiber strong and easily handled | Difficult to apply to complex geometries |
| Dimensionally stable | Upper temperature limit of 425–475 K |
| Resists vibration | Application skill required for best performance |
| Easily applied to cylindrical and simple shapes | Flammable |
| Can be heated adequately for outgassing | Restricted to nonoxidizing materials |
| Organic nets give performance comparable to fiberglass but use fewer reflecting layers | |
| Easily obtained from many suppliers | |
| (11) *Crinkled aluminized Mylar* | |
| Only single layer needed | Does not resist compression |
| Very low thermal conductivity | Has greatest increase of conductivity with compression for layered insulation |
| Installation easier than for combination layered insulation | Needs high vacuum for best performance |
| Strong | Flammable |
| Easily handled | Restricted to nonoxidizing materials |
| Low lateral heat flow | Nonflammable material is available but more expensive |
| Easily evacuated; outgassing completed during aluminizing step | |

*Source*: Timmerhaus and Flynn (1989).

and 90 K or less is a few inches, the substitution of evacuated powder for high vacuum alone will usually improve the insulation by reducing the radiant heat transfer. Of course, this applies to surfaces having emissivities that are practically realizable, of the order of 1–2%. If the emissivities could be made low enough, the powder would not help. For the 77 K and 20 K boundary condition, the radiant heat transfer is so low that powder will increase the heat transfer because of its thermal conduction unless the separation between the surfaces is quite large.

An adequate thickness of evacuated powder will have less heat leak than that through high-vacuum insulation between room temperature and the hydrogen boiling point. However, in some cases—for example, liquid transfer lines that are used for only short periods of time—the additional heat capacity of the powder may consume more liquid than the additional heat leak when using high vacuum alone.

Permeable insulations, such as gas-filled powders, are at least an order of magnitude poorer than vacuum for the 300–77 K boundaries and are still worse for the 77–20 K conditions because the gases available for filling the voids have high thermal conductivities. The principal virtue of this type of insulation is that it does not require vacuum-tight boundaries and therefore can be lighter and less expensive than most evacuated insulations.

Solid foams are inferior to good gas-filled permeable insulation in the upper temperature region, but they usually improve greatly as the temperature is lowered to 20 K because the gas in the cells is condensed and the thermal conductivity of the solid constituting the cell walls is greatly reduced. One notable exception is a glass foam that contains hydrogen gas in the cells. Solid foams can also be used without vacuum-tight boundaries.

Table 7.23 summarizes representative behavior of the various kinds of insulation for the three sets of boundary temperatures most frequently encountered. The values of heat transfer listed are for parallel surfaces; in an actual case the appropriate area factor discussed earlier must be applied.

Table 7.24 lists some accepted thermal conductivity values for the more popular insulations between 77 and 300 K. Qualitative comparisons are shown in Table 7.25 (Timmerhaus and Flynn, 1989).

**Table 7.26** Heat Transfer Through 1.0 in. of Typical Cryogenic Insulations (μW/cm²)

| Boundary temperatures (K) | High vacuum ($E=0.02$) | Multiple layer ($p=0.1$ g/cm³) | Evacuated powder ($p=0.07$ g/cm³) |
|---|---|---|---|
| 300 and 76 | 910 | 70 | 260 |
| 300 and 20 | 920 | 55 | 220 |
| 300 and 4 | 920 | 47 | 170 |

**Table 7.27** Relative Cost of Insulation to Reduce Heat Flux to 100 μW/cm² [0.32 Btu/(h ft²)][a]

| | MicroCel | Perlite | Cab-O-Sil + 40% Al | Multilayer |
|---|---|---|---|---|
| $k$, $10^{-5}$ Btu ft/(h ft² °F) | 36 | 58 | 19 | 3.5 |
| $k$, μW/(cm K) | 6.3 | 10 | 3.3 | 0.6 |
| $t$ thickness, in. | 5.5 | 8.8 | 2.9 | 0.53 |
| Density, lb/ft³ | 19 | 12 | 5.5 | 5.6 |
| Material cost, $/ft³ | 2.38 | 0.65 | 5.12 | 12.80 |
| Installation cost, $/ft³ | 0.57 | 0.25 | 0.60 | 30.00 |
| Total cost, $/ft³ | 2.95 | 0.90 | 5.72 | 42.80 |
| Insulation cost, $/ft² | 1.34 | 0.66 | 1.39 | 1.90 |

[a] Boundary temperatures 300 and 77 K.
*Source*: Stoy (1960).

**Table 7.28** Range of Thermal Conductivity for Various Insulations (Unit Conversion Factors to the Common Engineering Systems)[a]

| Insulation | μW/(cm K) | Btu ft/(h ft$^2$ °F) | Btu in./(h ft$^2$ °F) |
|---|---|---|---|
| Foam | 300 | 0.0173 | 0.208 |
| | 200 | 0.0116 | 0.139 |
| | 100 | 0.00578 | 0.0693 |
| | 50 | 0.00288 | 0.0347 |
| Evacuated powder | 20 | 0.00116 | 0.0139 |
| | 15 | 0.000867 | 0.0104 |
| | 10 | 0.000578 | 0.00693 |
| | 5 | 0.000288 | 0.00347 |
| | 3 | 0.000173 | 0.00208 |
| | 2 | 0.000116 | 0.00139 |
| Multilayer | 1 | 0.0000578 | 0.000693 |
| | 0.3 | 0.0000173 | 0.000208 |

[a] 1W/(cm K) = 57.79 Btu ft/(h ft$^2$ °F) = 693.5 Btu in/(h ft$^2$ °F).

Table 7.26 shows the heat transport through 1 in. of typical (cryogenic) insulations. It is possible to reduce the heat flux to 100 μW/cm$^2$ by using a moderate thickness of multilayer insulation. Table 7.27 shows a comparison of different insulations as suggested by Stoy (1960) based on the cost required to do a specific job. To restrict heat flux to less than 100 μW/cm$^2$, one must use 5 in. of Micro-Cel, 8 in. of perlite, etc. Note that even though the material and installation costs of multilayer insulation are very high, it compares well with the other insulations because of its low thermal conductivity.

## 16. SUMMARY

Table 7.28 shows the range of apparent thermal conductivities for foam, evacuated powder, and multilayer insulations. The values given are apparent mean conductivities between room temperature and very low temperatures.

# 8

# Cryogenic Instrumentation

## 1. INTRODUCTION

Cryogenic engineering requires the measurement of both extensive and intensive properties of cryogenic liquids. Transducers are required for liquid level (quantity), both point and continuous systems, and mass rate systems. In addition, there must be transducers of pressure, temperature, density, and, occasionally, quality.

Cryogenic instrumentation may be regarded as a unique field of measurement requiring the development of new techniques. It should be considered a separate field of effort because of the increasingly higher accuracies required, the inherent remoteness of the measurements, and the peculiarities of the cryogenic fluids themselves. The latter consideration is among the strongest in setting cryogenic instrumentation apart.

### 1.1. Heat of Vaporization

In addition to the obvious characteristic of low boiling points, cryogenic fluids are characterized by extremely low heats of vaporization. We may expect from Trouton's rule, which states that the normal entropy of vaporization is a constant for all fluids, that the heat of vaporization decreases proportionately with the normal boiling point. This is indeed the case for oxygen, fluorine, and nitrogen. The situation for hydrogen is much worse than that which would be predicted by this rule, as hydrogen requires only about half the predicted energy. Liquid helium requires much less energy for vaporization.

The combination of low boiling point and low heat of vaporization increases the possibility that a cryogenic fluid will become a boiling, two-phase system. The influence this has on pumping, liquid density, and level determination is obvious. Any sensor adding energy to the system is in fact creating a vapor/liquid interface at the very point of measurement. An added consequence is that when the system is at equilibrium in the two-phase region, the measurement of both temperature and pressure is redundant because the system has only one degree of freedom.

The fact that two phases usually do exist means that two liquid levels often can be measured. The level within a stillwell, for example, will be lower than that of the surrounding bulk fluid. This results from the fact that the fluid within the stillwell can be a single-phase, more dense fluid, while the surrounding bulk fluid is often two-phase and hence less dense.

### 1.2. Expansivity

Liquid hydrogen has a rather large coefficient of thermal expansion compared to ordinary fluids. Also, the vapor pressure curve of hydrogen is rather steep. As a consequence, merely pressurizing a tank of liquid hydrogen from 1 atm to approximately 2 atm causes the liquid to warm to about 4 K and the level to rise by about 5%. Thus, the liquid level of a closed hydrogen tank will rise upon pressurization although the mass content actually remains the same.

### 1.3. Relative Density

Cryogenic vapors, being quite cold, have large densities compared to those of gases of our usual experience, and some cryogenic liquids are relatively light compared with ordinary fluids. Consequently, the liquid/vapor density ratio is often surprisingly low. The ratios for a few fluids are water, 1000/1; oxygen, 250/1; and hydrogen, 50/1 at the normal boiling point. Accordingly, an "empty" hydrogen propellant tank may still contain a considerable amount (up to 2%) of nonuseful (nonpumpable) mass. A total mass sensor, such as a capacitance type, will be subject to this error. The error may be only a few percent of the initial hydrogen propellant loaded but of course tends toward 100% as the tank empties.

### 1.4. Stratification

It is well established that cryogenic liquids experience thermal stratification. This condition tends to be a stable one, as the warmer, and hence less dense, fluid is at the top of the tank. This condition causes the ullage pressure and hence the tank wall thickness to be greater than would be necessary if stratification were avoided. Also, these systems may be described as anisotropic, or nonhomogeneous, which presents a sampling problem.

### 1.5. Molecular Simplicity

One advantage of instrumenting cryogenic systems is their molecular simplicity. The more complex molecular species have long since frozen out. As a result, we usually are able to deal with simple systems of a single component at high purity.

This fact of constant composition is very useful. For example, the difference in dielectric constant between a gas and a liquid is frequently used as a means of continuous liquid level indication. Capacitance measurements may give mass as well, as the dielectric constant is a simple function of density for some cryogenic systems.

### 1.6. Uniqueness of Cryogenic Instrumentation

Properties such as those discussed in the foregoing pages, which are peculiar to cryogenic fluids, demand a new philosophy of measurement that utilizes the characteristics of the cryogen in order to be most effective. It may be inappropriate to force conventional transducers into the low-temperature regions if the fundamental measurement coefficient becomes marginal. For these reasons, it is appropriate to regard cryogenic engineering instrumentation as a separate field of effort requiring the development of new measurement techniques and skills. Some of these techniques are discussed in the following sections.

## 2. STRAIN

The measurement of strain is commonly performed through the use of electrical stain gauges. Therefore, this discussion is limited to such gauges and excludes all purely mechanical methods for determining strain in materials and structures.

The strain gauge has importance as a transducer for two reasons. First, it is a basic transducer in its own right for the measurement of strain. Second, numerous other transducers, notably the pressure gauge and accelerometer, often employ a strain gauge as the electrical pickoff. Other transducers, such as some thermometers, sense strain as noise. Accordingly, some concepts for measuring strain or eliminating an unwanted strain signal are presented below.

A strain gauge is a transducer that is applied to the surface of a material to sense the strain of the material. The strain gauge is elongation-sensitive, that is, its electrical properties change in proportion to the elongation of the gauge. Strain elongation (of the gauge and the member on which it is mounted) is usually small as long as the applied stress does not exceed the elastic limit for the material. *Stress* is, by definition, the applied force $F$ divided by the cross-sectional area $A$ of the member:

$$\text{stress} = F/A$$

An applied stress produces a strain (dimensional change) in the material. The relationship between stress and the resulting displacement is defined as *Young's modulus* ($E$), where

$$E = \frac{\text{unit stress}}{\text{unit strain}}$$

Strain ($\varepsilon$), in the engineering usage, is defined as

$$\varepsilon = \frac{\text{change in length}}{\text{original length}} = \frac{\Delta L}{L}$$

Strain gauges indicate strain indirectly, that is, the change in length is measured in terms of a change of resistance. The method, although indirect, is precise; accuracies to 0.1% can be achieved. Coupled with this accuracy is great application flexibility.

When the test member is strained, so is the gauge. As the wire is strained, its electrical resistance changes. This resistance change is directly proportional to the strain in the wire, and the strain in the wire is directly proportional to the strain in the member.

Strain in the test member, $\varepsilon$, is defined as $\Delta L/L$. The unit resistance change that the strain produces is defined as $\Delta R/R$. The relation between the unit strain and the unit resistance change is defined as the *gauge factor* $G$. That is,

$$G = \frac{\Delta R/R}{\varepsilon} = \frac{\Delta R/R}{\Delta L/L}$$

The gauge factor $G$ is the conversion constant between strain and gauge resistance and depends on the type of material used for the strain gauge wire. Manufacturers maintain close control of wire composition, but the measured gauge factor, which is stated on each package, should be used in all calculations (see Table 8.1). The gauge factor is also an index of merit; the larger the gauge factor, the larger the change in resistance per unit strain.

**Table 8.1** Representative Gauge Factors and Temperature Coefficient of Resistance for Common Strain Gauge Materials

| Material | Gauge factor | Temperature coefficient of resistance (ohms/(ohm K)) × $10^{-4}$ |
|---|---|---|
| Advance™ | 2.1 | 1.0 |
| Chromel | 2.5 | – |
| Constantan | 2.0 | 0.1 |
| Copel | 2.4 | – |
| Isoelastic | 3.5 | 4.7 |
| Manganin™ | 0.47 | 0.1 |
| Monel | 1.9 | 20.0 |
| Nichrome™ | 2.5 | 4.0 |
| Nickel | −12.1 | 60.0 |
| Phosphor bronze | 1.9 | 20.0 |
| Platinum | 6.0 | 30.0 |
| Carbon | 20.0 | −5.0 |

## 2.1. Types of Strain Gauges

### *2.1.1. Bonded Strain Gauges*

The most common form of strain gauge consists of a short length of small-diameter (approximately 0.3 mm; 0.001 in.) wire of high electrical resistance. To keep the gauge length short, the wire runs the length of the gauge several times (see Fig. 8.1A). To simplify its mounting and to protect it, the wire is cemented between two thin pieces of paper. To apply the gauge, it can be cemented to the member to be tested; this is a bonded resistance wire strain gauge. Most gauges are now made by photo-etching thin foils, but the form and principle are the same.

### *2.1.2. Unbonded Strain Gauges*

In an unbonded strain gauge, the resistance wire is wrapped around pins on the test structure or affixed to pins on some type of motion-amplifying device that is in turn attached to the test structure. In this case, the wire is not cemented but is only mechanically held in place (see Fig. 8.1B).

### *2.1.3. Semiconductors*

Semiconductor strain gauges operate on the same piezoresistive effect that applies to metal strain gauges. This is the change in electrical resistivity of a material due to applied stress. The piezoresistive coefficients of germanium and silicon can be extremely high, with gauge factors up to 175, compared to 2–5 for metallic wires.

## 2.2. Low-Temperature Effects on Strain Measurements

Electrical strain measurements measure a change in resistivity. Ideally, this change is due only to the strain, but it may also be temperature-induced. As mentioned previously, the gauge factor is an index of merit with respect to strain.

It is nearly always true, however, that the greater the gauge factor the greater is the temperature coefficient of resistance (see Table 8.1). Hence, in using electrical strain gauges at low temperatures, it is necessary to compensate for the change in

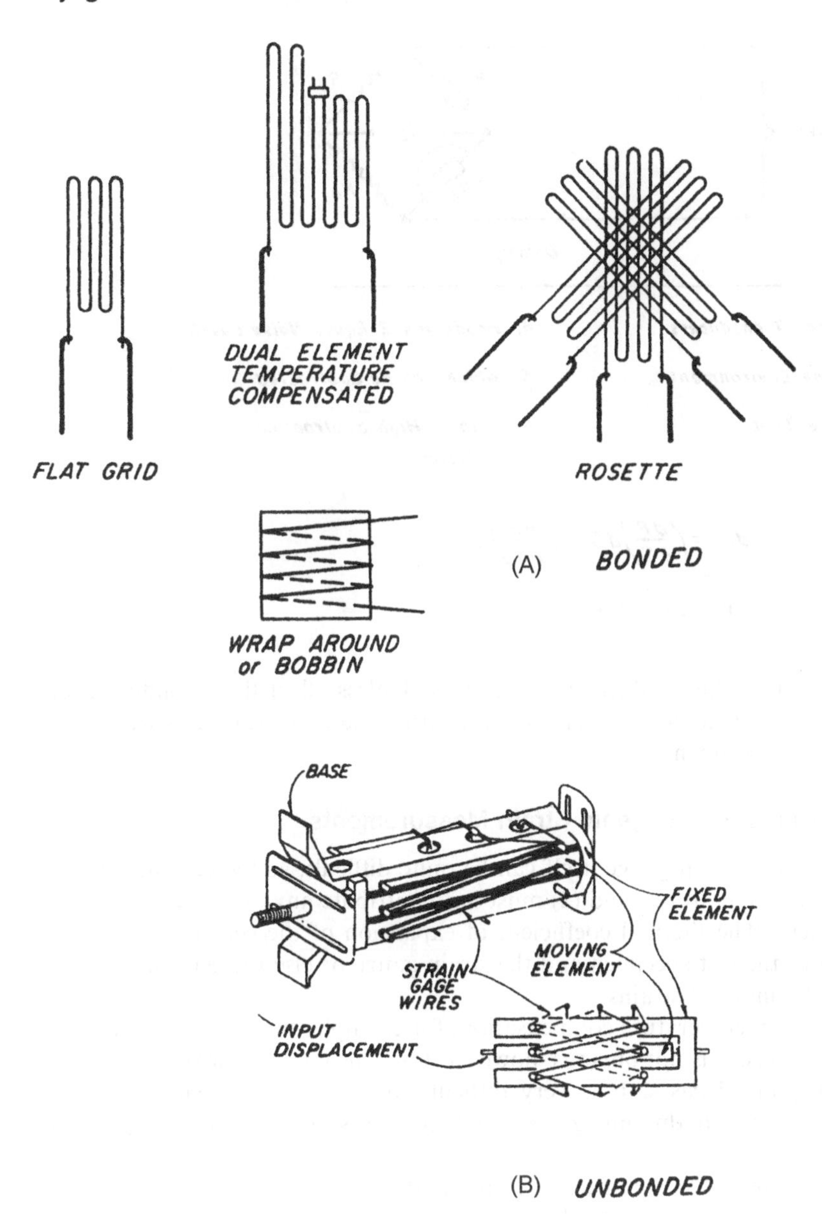


**Figure 8.1** Types of strain gauges.

resistivity that is caused by temperature changes. This is nearly always attempted by means of "common mode rejection," which is a cancelation of "noise" by subtraction (see Fig. 8.2). To accomplish this, two strain gauges are set into a Wheatstone bridge in such a way that their signals tend to cancel one another. One, referred to as the *active gauge,* experiences both the temperature and strain effects. The other gauge is referred to as the *dummy gauge* and experiences only the temperature effect. Thus, it is at least theoretically possible to cancel out the temperature effect so that only the strain effect is measured. For this to be true, however, the following conditions must be met: (1) there must be two identical strain gauges, (2) in the same environment, (3) at the same time, (4) producing two equal noise levels, (5) that can be

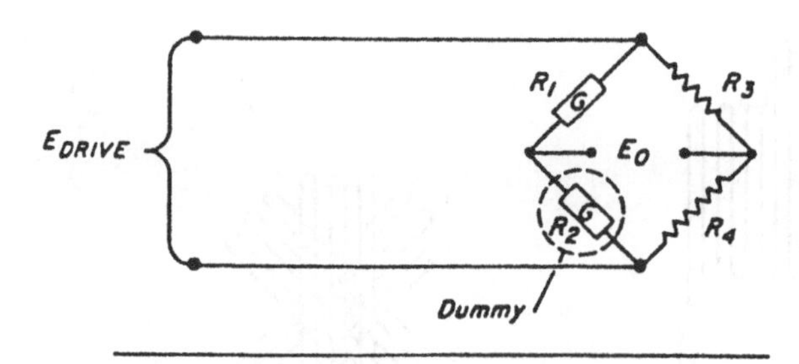


1. Two Identical Transducers
2. In the Same Environment
3. At the Same Time
4. Producing 2 Equal Noise Levels
5. Which Can Be Subtracted
6. With a High Subtraction Efficiency

$$dR = \left(\frac{\partial R}{\partial S}\right) dS + \left(\frac{\partial R}{\partial T}\right) dT$$

**Figure 8.2** Common mode rejection.

subtracted (6) with a high subtraction efficiency. Unless all of these conditions are met, one will, in fact, to some extent, be measuring the temperature as well as the strain of the test specimen.

## 2.3. Peculiarities of Cryogenic Strain Measurements

The bonding of the strain gauge itself can be quite difficult, as the cement (1) must hold at the low temperatures, and (2) must not contribute any extraneous effects to the measurement. The thermal coefficient of expansion of the cement must match carefully that of the test specimen and the strain gauge or else the cement itself will set up thermally induced strains.

The need for temperature compensation, for example, through common mode rejection, is obvious. It is not obvious, however, how this temperature compensation can be effected in all cases. It is very difficult to maintain thermal equilibrium between the active and dummy gauges and also very difficult to find perfectly matched gauges.

Usually in strain measurements we are faced with relatively low signals in the 25–50 mV range. This can cause problems because strain measurements in cryogenic systems are always remote, and extraneous signals can be induced in the cabling and connectors.

*Example 8.1.* Estimate the total apparent strain $\varepsilon_{\text{app}}$ that results solely from an applied temperature change $\Delta T$.

The apparent strain caused by $\Delta T$ will have two components: (1) that caused by the temperature coefficient of resistance $\gamma$ (Table 8.1) and (2) that caused by the difference in thermal contraction of the gauge, $\alpha_{\text{G}}$, and the thermal contraction coefficient of the substrate, $\alpha_{\text{S}}$.

The gauge factor $G$ was defined as

$$G = \frac{\Delta R/R}{\Delta L/L} = \frac{\Delta R/R}{\varepsilon} = \frac{\gamma \Delta T}{\varepsilon}$$

where $\varepsilon$ is strain and $\gamma$ is the temperature coefficient of resistance. Therefore, the apparent strain from the temperature coefficient of resistance is

$$\varepsilon_1 = \frac{\gamma}{G}\Delta T$$

The apparent strain caused by the difference in coefficients of thermal expansion is

$$\varepsilon_2 = (\alpha_S - \alpha_G)\Delta T$$

Thus, the total apparent strain caused by a temperature change alone is

$$\varepsilon_{app} = \varepsilon_1 + \varepsilon_2 = \left(\frac{\gamma}{G} + (\alpha_S - \alpha_G\right)\Delta T$$

Because both $\gamma$ and $\alpha$ are functions of temperature, $\varepsilon_{app}$ is a nonlinear function of temperature and can be a serious source of error.

## 3. DISPLACEMENT AND POSITION

Displacement and position transducers are intermediate between strain gauges and pressure transducers. They are, in fact, similar to both of these types of transducers. A displacement transducer, regardless of its type, always consists of three basic parts: (1) an input, (2) a link, and (3) an electrical conversion device (see Fig. 8.3). The input consists of the displacement and accompanying force. Next in the chain is a mechanical link that either transmits this displacement directly or can modify or enlarge it. The third element is an electrical conversion device, that is, some element that can convert the displacement or force into an electric signal. This latter element can take many forms. It can, for example, be a variable capacitor for which the plate separation is determined by the force and displacement. Variable potentiometers are also used, in which case the position of a wiper arm changes according to the input force and displacement. Various devices depending upon variable reluctance and variable inductance are used; for example, a phonograph needle. Piezoelectric and piezoresistive elements can also be used to perform the electrical conversion.

One of the more popular devices for cryogenic use is the linear variable differential transformer (LVDT) (see Fig. 8.4). The LVDT is a transformer whose core

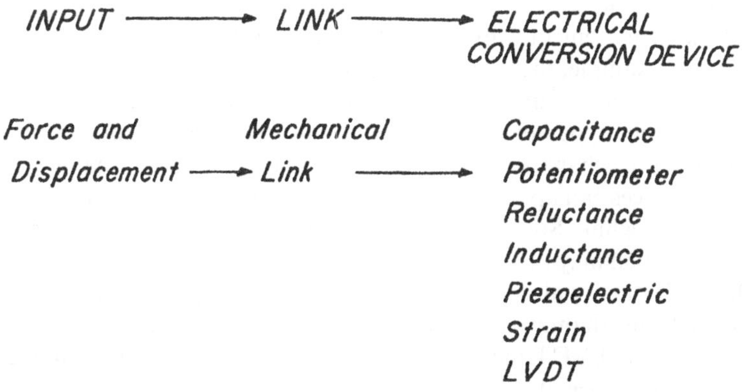

INPUT → LINK → ELECTRICAL CONVERSION DEVICE

Force and Displacement → Mechanical Link → Capacitance
Potentiometer
Reluctance
Inductance
Piezoelectric
Strain
LVDT

**Figure 8.3** Displacement and position transducer components.

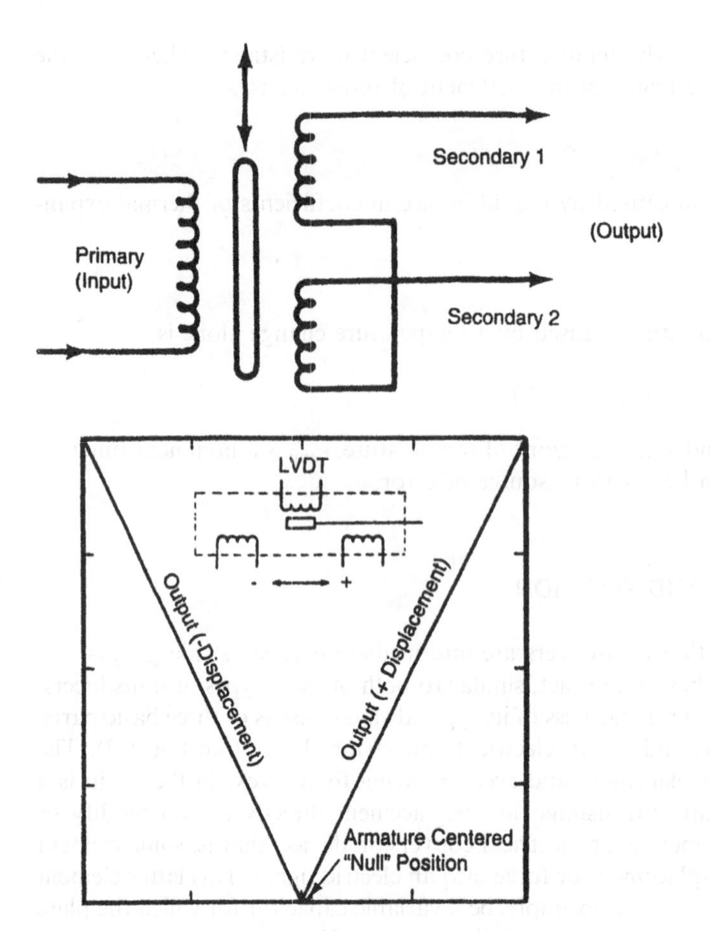


**Figure 8.4** Linear variable differential transformer.

moves in response to the displacement input. The primary is wound in an ordinary manner, but the secondary is split and wound in such a fashion that there will be no output when the core is exactly in the central position. As the core is displaced from the center, a voltage is generated across the terminals of the secondary that is directly proportional to the displacement of the core. The sign of the voltage is given by the direction of the motion of the core.

Regardless of the displacement transducer chosen, all suffer similar extraneous temperature effects. For example, there are temperature-induced changes in the electrical properties, such as changes in the resistivity of strain gauges, the change in dielectric of the capacitance type, and so on. All are also subject to temperature-induced changes in their mechanical properties. Temperature-induced changes in the dimensions of the devices can appear as an artificial displacement or as a change in the elastic modulus of the link. Some of these devices are also subject to freezing at very low temperatures, which may be due to moisture accumulating in the device and turning to ice. Or, if the materials of construction are not properly chosen, one dimension can change faster than another and cause a type of mechanical freezing. For example, the core in the LVDT may find itself frozen fast and unable to move.

Of all of the possible types of transducers, the LVDT is often preferred. This is so because it is a transformer; the change in the resistance of the windings is immaterial and, of course, there are no opportunities for any other changes in electrical properties to become significant. It does sometimes suffer from freeze-up problems, but care in the design and application can usually avoid these.

*Example 8.2.* A parallel-plate capacitor transducer is used to measure displacement. One plate of the capacitor is fixed and the other is free to move with the input displacement. The capacitor plates each have an area $A$ and are separated by a distance $d$. The dielectric constant of the medium separating the plates is $\varepsilon$. The output of the capacitor transducer is applied to a modified capacitance bridge that measures the fractional capacitance change $\Delta C/C$ rather than the absolute variation $\Delta C$. What is the nature of the response characteristic $\Delta C/C$ vs. $\Delta d/d$?

The capacitance between the terminals (neglecting fringe effects) is

$$C = k\varepsilon A/d$$

or, numerically,

$$C\mu\,\mu_F = 0.0885\,\varepsilon A/d \quad (d \text{ in cr.})$$

or

$$C\mu\,\mu_F = 0.225\,\varepsilon A/d \quad (\text{d in in.})$$

The response characteristic $C=f(d)$ is hyperbolic and only approximately linear over a small range of displacement.

The sensitivity of the capacitative transducer with variable separation is

$$S = \frac{\Delta C}{\Delta d} = \frac{\varepsilon A}{d^2}$$

or

$$S' = \frac{\Delta C}{C\Delta d} = \frac{1}{d}$$

and

$$S'' = \frac{\Delta C}{C} = \frac{\Delta d}{d}$$

Therefore, the response characteristic $\Delta C$ vs. $\Delta\, d/d$ is linear and is independent of the medium separating the capacitor plate

## 4. PRESSURE

Since pressure is, in fact, a normalized force (force per unit area), and since force cannot be measured without an accompanying displacement, we can expect that pressure transducers will look very much like displacement transducers. This is, in fact, the case and allows us to apply much of our knowledge regarding the behavior of displacement transducers to predicting the behavior of pressure transducers.

Pressure measurements in cryogenic systems have, for years, been made by simply running gauge lines from the point where the pressure measurement is desired to some convenient location at ambient temperature and attaching a suitable pressure-measuring device, such as the familiar Bourdon gauge. This system works quite well for many applications; however, there are disadvantages to this straightforward approach that may introduce problems in cryogenic systems. The two most important are (1) reduced frequency response and (2) thermal oscillations in the connecting gauge lines. In addition, heat leak and fatigue failure of gauge lines could become significant in some applications. Such problems associated with pressure measurement at cryogenic temperatures could be eliminated by installing pressure transducers at the point of measurement, thereby doing away with gauge lines. For these reasons, we consider in the following discussion only those pressure transducers that have an electrical output and can be flush-mounted directly on a transfer line or other cryogenic component under consideration.

### 4.1. Mechanical Description

It was noted that pressure transducers are very similar to displacement transducers in their general makeup. That is, all pressure transducers, regardless of their particular type, usually share at least three similarities. Three major subassemblies are common: (1) a force-summing device that converts the force or pressure to a displacement, (2) a mechanical link to transmit this displacement and perhaps amplify it, and (3) an electrical signal conversion device (see Fig. 8.5).

The force-summing device may take any of the several forms. It may be, for example, a diaphragm, whether flat, corrugated, or encapsulated. It may also be a bellows, or a Bourdon tube, whether circular or twisted, or simply a straight tube that is stressed under pressure.

The link may simply be a straight bar that transmits this displacement directly, or it may be an involved series of linkages that mechanically amplify the original displacement.

The electrical conversion device can also take several forms. It may be a variable capacitor whose plate spacing depends on the displacement and hence on the pressure. It may be a variable potentiometer in which the position of the wiper

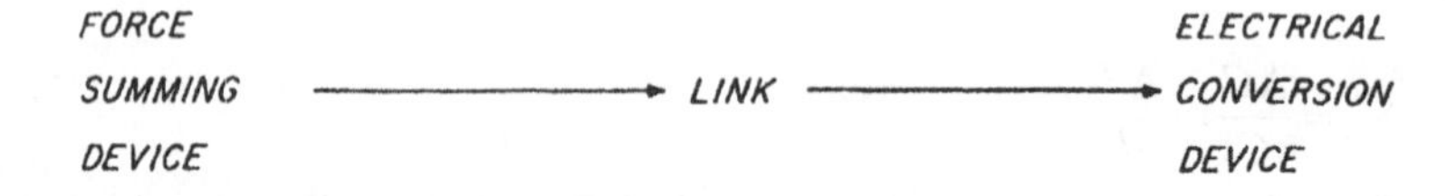


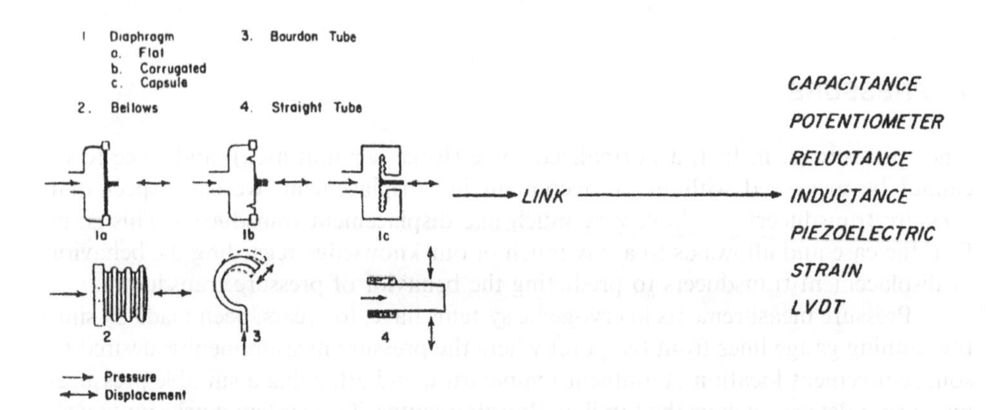


**Figure 8.5** Pressure transducer components.

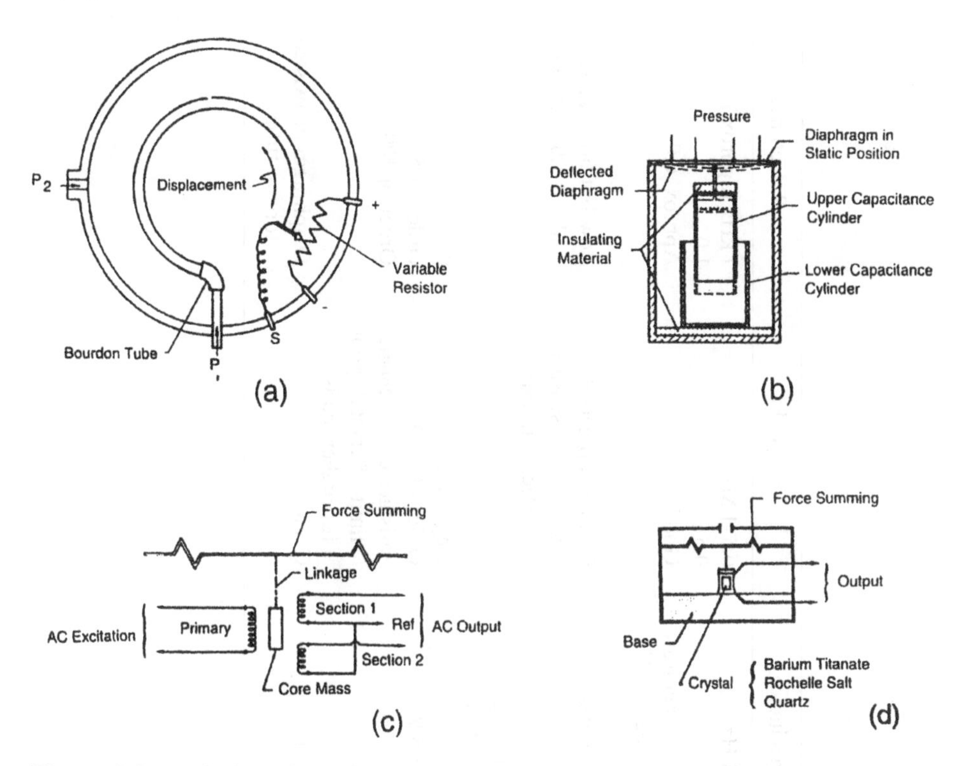


**Figure 8.6** Typical configurations of pressure transducers. (a) Bourdon tube; (b) capacitance; (c) LVDT; (d) piezoelectric.

arm is determined by the displacement and hence by the pressure. It may be any of several variable-reluctance or variable-inductance devices. It may also comprise a piezoelectric element, a strain gauge, or an LVDT.

Figure 8.6 illustrates a few of the combinations of force-summing, link, and electrical conversion devices that are embodied in various commercial pressure transducers.

Table 8.2 lists the general characteristics of some common pressure transducers as compiled by Arvidson and Brennan (1975). Their work is recommended as a comprehensive treatment of pressure measurement.

## 4.2. Steady-State Temperature Effects on Pressure Transducers

As with strain gauges and displacement devices, the output of a pressure transducer depends not only on the primary input, in this case pressure, but also on extraneous effects such as the effect of temperature on its various components. We examine these latter effects in the following paragraphs.

Figure 8.7 shows the steady-state temperature effects on two common cryogenic pressure transducers (Smelser, 1963). These data were obtained by simply calibrating the pressure transducer at three different fixed temperatures after a sufficient amount of time had elapsed that every component of the transducer was at the same temperature. There can be noted both changes in sensitivity and zero shifts.

It can be shown that the sensitivity shift is primarily due to the differences in the thermal expansivities of the several components that make up the transducer,

**Table 8.2** Pressure Transducer Characteristics

| Characteristic | Strain gauge | Capacitance | Crystal | Reluctance |
|---|---|---|---|---|
| Frequency response, Hz | To 40 kHz | To 500 kHz | 0.5–1 MHz | 1 kHz (approx.) |
| Nominal sensitivity, % full scale | 0.5 | 1.0 | 0.5 | 1.0 |
| Thermal stability | Excellent for compensated bridge winding | Approx. temperature drift, 0.025%/°F | Good | Approx. 0.02%/°F |
| Linearity, % | 0.1 | 1.0 | 0.5 | 1.0 |
| Response to vibration, noise, acceleration | Negligible for light-diaphragm and fully rigid types | Low but noticeable | Can be appreciable; also highly sensitive to electrical noise | Low but noticeable; about 1.0%/100 G |
| Drift, % | Negligible | 1.0 (excluding temperature) | | 2.0 |
| Hysteresis, % | Approx. 0.2 | 1.0 | Negligible | 0.5 |
| Open-circuit full-scale output | Order of 50 mV | Order of 5 V | Order of 1 mV | Order of 5 V |
| Overload, % | Generally about 100; to 500 for special types | Order of 100 | Satisfactory for elastic limit of crystal; poor for higher loads | Order of 100 |
| Typical pressure-sensing device | Tube or diaphragm | Diaphragm | Crystal | Diaphragm or twisted torsion tube |

*Source*: Arvidson and Brennan (1975).

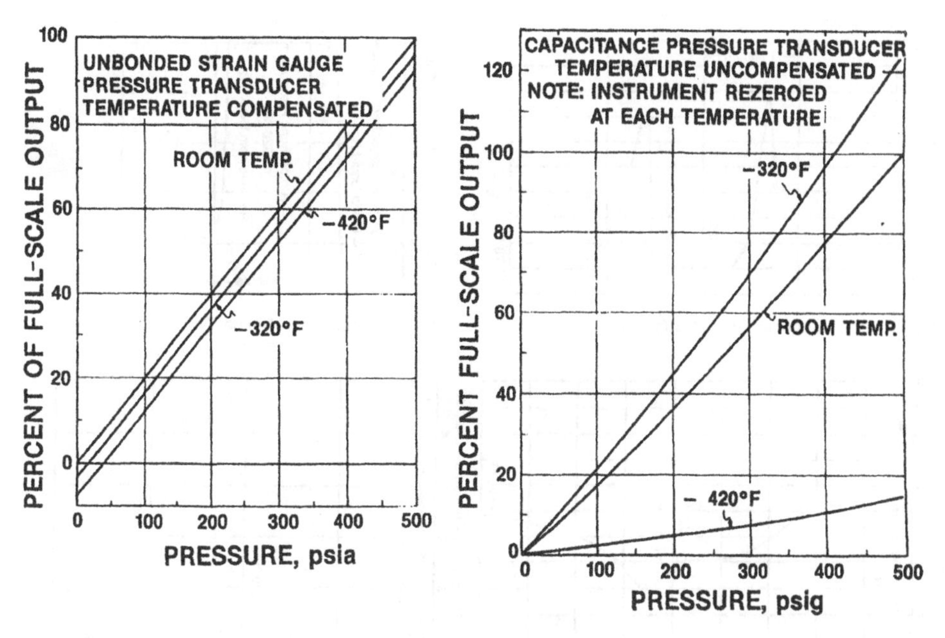


**Figure 8.7** Representative steady-state temperature effects on pressure transducers.

especially the change in the elastic modulus as a function of temperature. Likewise, the zero shift can be attributed to the differences in the thermal expansivities of the several components of the pressure transducers, but the more significant effect in this case is the change in the dimensions of the components due to changes in temperature. If the electrical conversion device is used in some form of a bridge network, additional zero shifts can occur owing to the inability of the bridge to perform adequate common mode rejection, as was discussed earlier. Figure 8.7 gives actual but illustrative results. It is not meant to imply that all pressure gauges of the type shown behave in exactly this fashion.

### 4.3. Thermal Shock Effects on Pressure Transducers

So far, we have been assuming that the pressure transducer is experiencing a uniform temperature throughout its many components. Figure 8.8 shows what happens when temperature gradients occur across a pressure transducer. These data were obtained by applying no pressure to the pressure transducer but subjecting it to a thermal shock by plunging it into liquid nitrogen. What we see, then, is purely the thermal shock or thermal gradient effect. It will be noted that the four pressure transducers illustrated would, in fact, be quite good thermometers.

To explain these results, consider, for example, a pressure transducer that is made up of a bellows, a link, and a linear resistor. Under conditions of thermal shock, the bellows will reach the new temperature environment at a different rate than the linear resistor. The difference in the expansivity of these materials varies dramatically over a wide temperature range. Therefore, the contributions of the expansivities of these components to the sensitivity coefficient are changed from the design condition, causing a sensitivity shift. As we have noted, a compensating resistor is often placed in the bridge to perform a common mode rejection to

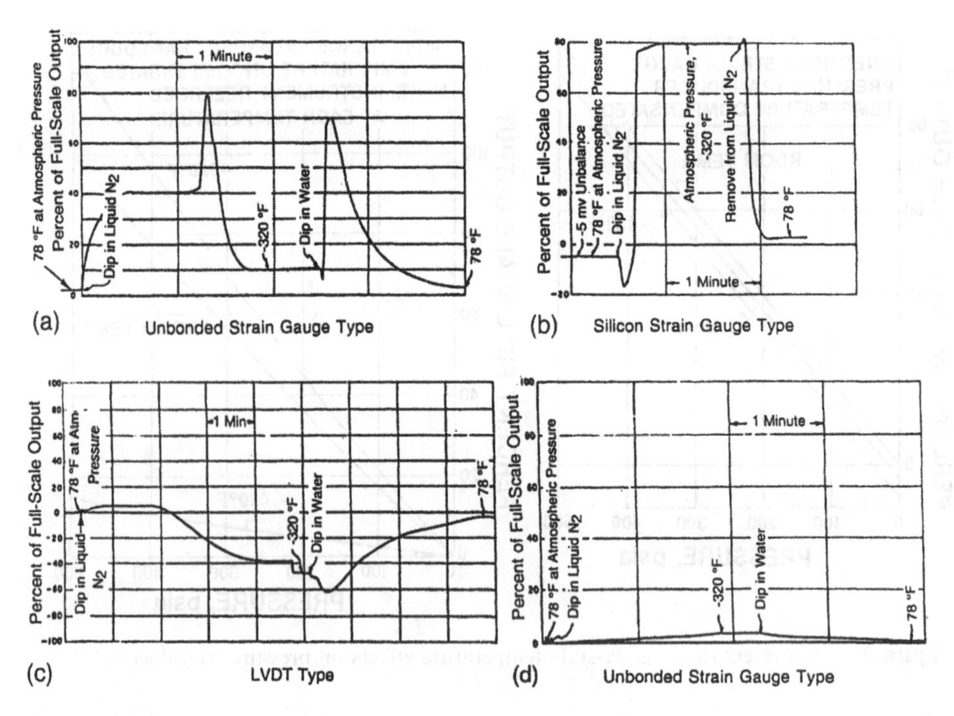


**Figure 8.8** Representative thermal shock effects on pressure transducers. (a,d) Unbonded strain gauge type; (b) silicon strain gauge type; (c) LVDT type.

compensate for the temperature dependency of the various components. Unfortunately, this compensating resistor is often some distance from the bellows, so that its temperature is, in fact, quite different from that of the bellows at any given time, and the conditions stated earlier for effective common mode rejection cannot be met. In fact, if temperature gradients are expected for a pressure transducer installation, it may be beneficial to remove the electrical compensation, as the error thus induced may far exceed the error due only to the temperature dependency of the elastic modulus.

The temperature gradient effect also causes zero shift. We can see from Fig. 8.8 that these zero-shift effects are dissipated with time as all of the members of the pressure transducer reach the same temperature. The small zero shift at thermal equilibrium is consistent with that shown on Fig. 8.7. The large output that occurs at several points during the process of reaching thermal equilibrium is caused by the mismatching of expansivities due to the temperature gradients that still remain.

## 4.4. Avoiding Temperature Effects on Pressure Measurements

Figure 8.9 is a record of the temperature shock of a potentiometer type of pressure transducer designed to minimize temperature effects. Zero shift on dipping the transducer into liquid nitrogen is less than 1%; the peaks on the record were caused by vibration used to make sure the wiper was not frozen to the variable resistor. This design (Fig. 8.10) uses a helically coiled Bourdon tube with a curved variable resistor placed perpendicular to its major axis. The wiper acts as a cantilever spring to absorb the differential contraction instead of being moved along the variable resistor. This

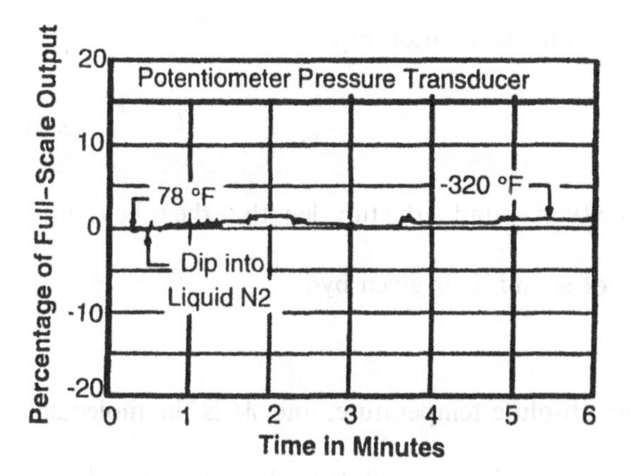


**Figure 8.9** Zero shift caused by thermal shocking of a helically coiled Bourdon-type actuated potentiometric pressure transducer.

configuration is limited to medium- and high-pressure ranges. Although the pressure transducer tested in Fig. 8.10 gave a good zero-shift result, there is room for improvement. This transducer is sensitive to vibration perpendicular to its major axis and to sensitivity changes due to Young's modulus temperature dependence.

The most common solution to temperature effects is to avoid extreme temperature environments. When the pressure instrumentation point is subjected to such temperatures, the standard procedure is to run tubing from the pressure source to a remote stable-temperature location. This is good procedure if the instrument engineer is interested in a frequency response on the order of 10 Hz. The tubing with attached transducer acts approximately like an organ pipe or a Helmholtz resonator. The resonant frequencies of an organ pipe closed at one end may be calculated from relationship

$$\omega = \frac{\nu}{4l} \tag{8.1}$$

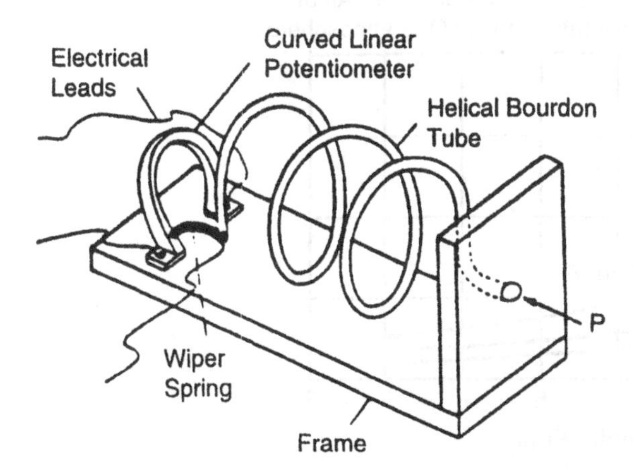


**Figure 8.10** Working diagram of a helically coiled Bourdon tube pressure transducer.

and those of a Helmholtz resonator from the relationship

$$\omega = \frac{\nu}{2\pi} \frac{\sqrt{\pi r^2}}{(1 + 1.7r)x} \tag{8.2}$$

where, $\omega$ is the frequency, $c$ the velocity of sound, $l$ the tube length, $r$ the tube radius, and $x$ is the volume of cavity.

For an ideal gas, the velocity of sound, $c$, is given by

$$c = \sqrt{RT/M}$$

where, $R$ is the gas constant, $T$ the absolute temperature, and $M$ is the molecular weight.

In an actual installation, the velocity of sound will depend on the gas or liquid used and its temperature and pressure. The state of the fluid will normally be variable, thus making a rigorous calculation of the resonant frequency difficult; however, Fig. 8.11, calculated for air at several temperatures, gives an indication of the resonant frequency as a function of tube length. The usable frequency range of the transducer and tube system may be from one-tenth to one-third of the resonant frequency, depending on the system damping. This is often on the order of 10 Hz. Tests with sinusoidal pressure generators show resonances occurring below those predicted by Eqs. (8.1) and (8.2). This subject is handled in more detail in the literature (Jones, 1963; Vaugh, 1954).

Another approach is to flush-mount one pressure transducer in the extreme environment and another at the end of a tube as discussed above. The object is to measure static pressure levels with the tube-mounted pressure transducer and dynamic levels with the flush-mounted pressure transducer on an ac device. The hope is that the temperature-shock contribution to dynamic response will be small or of low enough frequency to separate easily.

A better but more complicated solution for cryogenic application is shown in Fig. 8.12. A short thin-walled tube connects a cryogenic transfer line or tank to a pressure transducer. A heater and a thermostat are attached to the transducer,

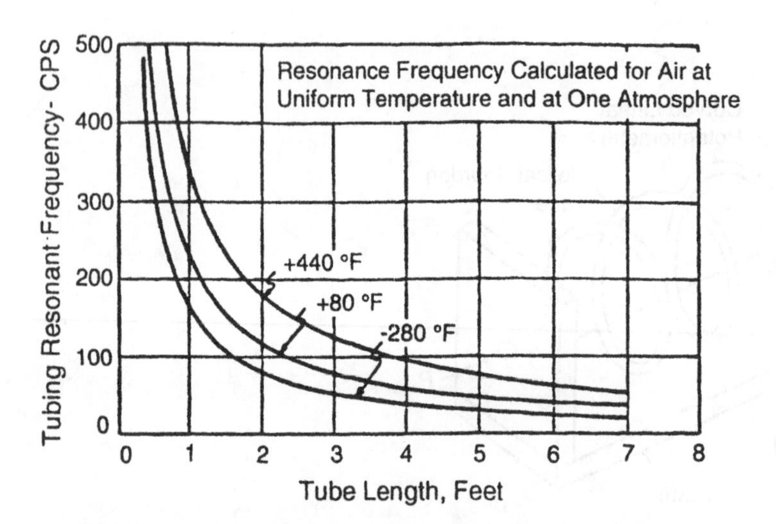


**Figure 8.11** Resonance frequency of a tube calculated as for an organ pipe.

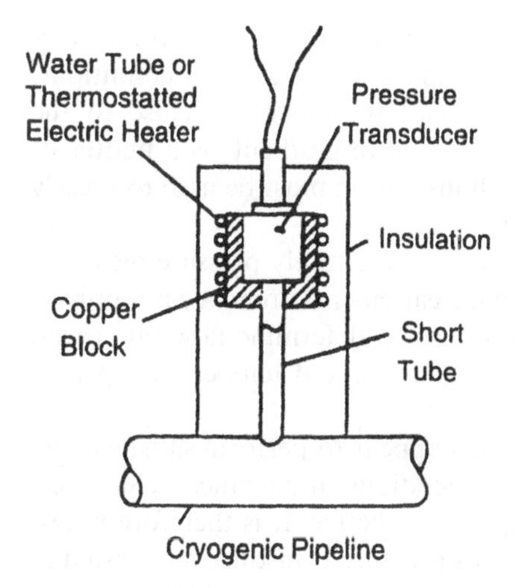


**Figure 8.12** Temperature-regulated pressure transducer, cryogenic installation.

regulating the temperature to perhaps 300 K. Insulation surrounds the transducer. Tube length is enough to reduce the power requirement to the heater to a few watts, hopefully without excessively reducing the frequency response. It is important that the heater and tube connection be arranged such that temperature gradients do not exist in the transducer. The transducer must be oriented such that the liquid cryogen does not flow into the diaphragm.

A similar cryogenic installation can be made without using the heater or thermostat. The purpose of this installation is to reduce temperature gradients that occur in the pressure transducer. The transducer will eventually approach the temperature of the cryogen in the tank or transfer line, but its thermal time constant has been greatly increased. Zero and sensitivity effects will still be experienced; however, the temperature gradient effects are reduced. It is important that the tube be sufficiently thin-walled and long enough to avoid excessive heat transfer. Test runs with several combinations of tube length and thickness show that a poor choice will increase the temperature gradient effects while a proper combination can reduce these effects by one-fourth or more.

### 4.5. Pressure Summary

Temperature effects on pressure transducers result in sensitivity changes and zero shifts. Sensitivity shifts are primarily proportional to the spring constant and thus the Young's modulus temperature dependency of the force-summing member, be it a diaphragm, a bellows, a Bourdon tube, or a capsule. Zero shifts are primarily due to dimensional instability caused by temperature changes of the pressure transducer components, including the case or frame. Temperature gradient effects are the combination of the above effects which the transducer component temperatures are varying. Proper design can reduce zero shift by locating the electrical sensing element such that dimensional changes are not detected as an input.

Sensitivity shifts can be electrically compensated, assuming that a resistor can be found that has the desired resistance–temperature properties. There is not much that can be done about temperature gradient effects except to avoid them.

Component material and compensating resistors that are carefully chosen for steady-state temperature conditions lose their relationships under transient conditions. However, it may be helpful to place the compensating resistor as close to the force-summing element as possible. Avoiding temperature gradients is a matter of heat transfer. Under cryogenic conditions, enough insulation must be used to greatly extend the thermal time constant of the system.

Of all the measurements made on cryogenic systems, surely pressure measurement must be one of the most common. Pressure measurements are made not only to determine the force per unit area in a system but also to determine flow rate (head meters), quantity differential pressure liquid-level gauges), and temperature (vapor pressure or gas thermometers).

Although some available pressure transducers appear to perform satisfactorily at low temperature, other types such as piezoelectric, diode, inductance, reluctance, and electro-kinetic devices may perform equally well or better. It is therefore necessary for the purpose of comparison that all types of pressure transducers be tested at low temperature and at various frequencies to determine their potential for cryogenic use. For further information the reader is referred to the comprehensive review of pressure transducers by Arvidson and Brennan (1975).

*Example 8.3.* If a diaphragm is used in a capacitance pressure transducer, it must not have a natural frequency or an appreciable harmonic in the range of pressure oscillations likely to be encountered. The following equation is used to calculate the natural frequency ($N_{\mathrm{f}}$) of a circular diaphragm:

$$N_{\mathrm{f}} = 0.4745\frac{t}{r^2}\left[\frac{E}{\rho(1-\nu^2)}\right]^{1/2}$$

where $t$ is the diaphragm thickness, cm; $r$ the diaphragm radius, cm; $\rho$ the material density, g/cm$^3$; $E$ the Young's modulus, g/cm$^2$; and $\nu$ is the Poisson's ratio.

Calculate the natural frequency for a mild steel diaphragm 2.54 cm in diameter and 0.0635 cm thick. For mild steel, $E = 2.0 \times 10^{12}$, $\rho = 7.8$, and $\sigma = 0.28$.

The natural frequency is

$$N_{\mathrm{f}} = 0.4745\frac{0.0635}{(1.27)^2}\left[\frac{2.0\times 10^{12}}{7.8(1-0.28^2)}\right]^{1/2}$$

$$N_{\mathrm{f}} = 9.853 \times 10^3\ \mathrm{Hz}$$

with higher harmonics occurring at $2N_{\mathrm{f}}$, $4N_{\mathrm{f}}$, etc.

## 5. FLOW

Flowmeters that find application in cryogenic service include (1) positive displacement type; (2) pressure drop type, such as the orifice and venturi; (3) turbine type; (4) momentum type, (5) vortex shedding, and (6) various miscellaneous types.

### 5.1. Positive Displacement Flowmeter

Most transfer of cryogenic liquids in commerce is performed by positive displacement meters. These devices have a long history of use in metering normal fluids and were simply extended into the cryogenic region. The following sections describe several positive displacement volumetric meters currently used in commerce.

### *5.1.1. Screw Impeller Volumetric Flowmeter*

The screw impeller volumetric flowmeter, in several configurations, is widely used in commerce for liquid oxygen, liquid nitrogen, and liquid argon. Many references from the literature cite performance, design, or usage information. Crawford (1963) discusses materials of construction and design, as does Close (1968, 1969). Fox (1967) refers to the screw impeller meter as a method of cryogenic flow measurement for custody transfer but gives no operational performance or design data. The most extensive evaluation was performed by Brennan et al. (1971), who provide performance information for several configurations on liquid nitrogen.

One common positive displacement flowmeter consists of a central screw impeller that meshes with two deeply grooved sealing screws, all selectively matched for close tolerances and fitted closely into a surrounding metal housing (see Fig. 8.13). The special contour of the screws provides a line seal where the rotors mesh. Liquid entering the meter causes rotation of the impellers, and the displacement obtained is transmitted through a gear system either to an electronic counter or to a totalizing register. Meters are provided with a cooldown mechanism composed of a priming line that allows circulation of liquid through the meter without causing operation of the impellers.

The flow rates employed with liquid oxygen, liquid nitrogen, and liquid argon are relatively low, approximately $3.2 \times 10^{-3}\,\mathrm{m^3/sec}$ (50 gal/min), and the meter is used primarily for trailer truck dispensing of fluids.

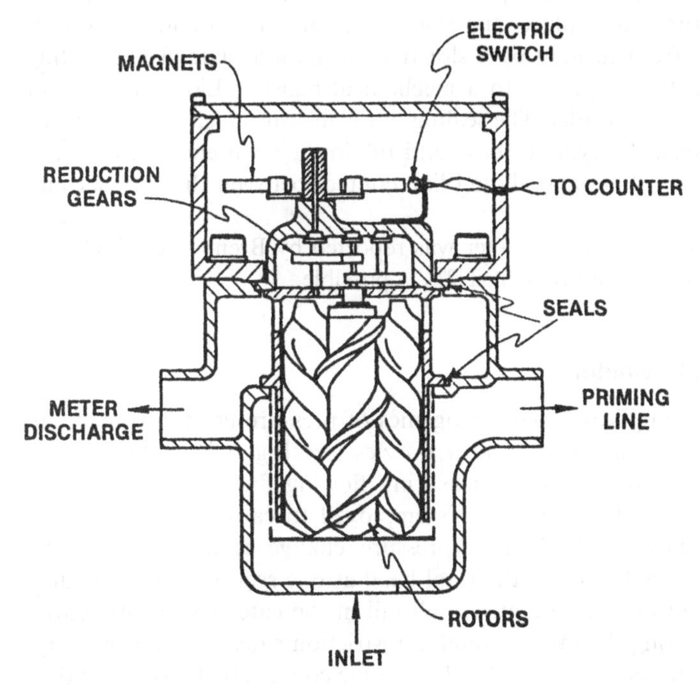


**Figure 8.13** Screw impeller meter.

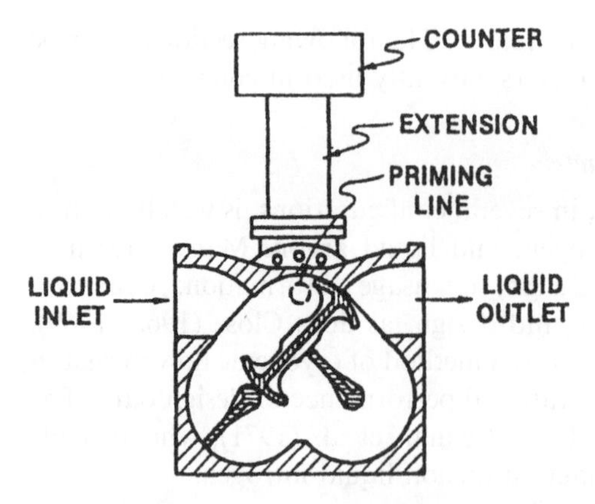


**Figure 8.14** Rotating vane meter.

*5.1.2. Rotating Vane Volumetric Flowmeter*

A schematic drawing of a rotating vane meter is shown in Fig. 8.14. The dumbbell-shaped element rotates in a clockwise mode while the two vanes rotate counterclockwise. This type of meter has only recently been adapted for cryogenic service. Because wear between the vanes and the housing may affect the accuracy of the measurement, more experience is needed before judging its suitability for cryogenic use.

*5.1.3. Oscillating Piston Volumetric Flowmeter*

An oscillating piston flowmeter is shown in Fig. 8.15. Liquid is admitted to the inlet and is first taken out the priming line to cool the meter to operating temperature. After priming, liquid flows through the piston assembly. It enters the piston assembly through the inlet port and displaces the piston horizontally around the vertical axis. The piston is kept from turning by a slot that rides on a plate. The resulting motion is an oscillation that is geared to a mechanical register. Liquid passes out the discharge port to the meter outlet. The sealing of the measuring chamber depends on sliding contact between the cylinder base and the lower open end of the piston and on a combination of rolling and sliding line contact between the cylinder and piston side walls.

The performance of this type of meter was reported by Brennan et al. (1971), who stated that precision on the order of $\pm 2\%$ is possible.

## 5.2. Pressure Drop Flowmeters

There have been various applications to cryogenics of pressure head flowmeters in the form of orifice plates, venturis, and flow nozzles (see Fig. 8.16). This type of meter is probably the oldest means of measuring flowing fluids. The distinctive feature of head meters is that a restriction is employed to cause a change in the static pressure of the flowing fluid. This pressure change is measured as the difference between the static head and the total head at one section of the channel.

Orifice, venturi, and flow nozzle meters all fall in the category of differential pressure meters. Accelerating the fluid through a restriction causes a pressure drop, which is what is actually measured. There has been some concern with using this type of meter with cryogenic fluids, because the pressure drop might cause cavitation

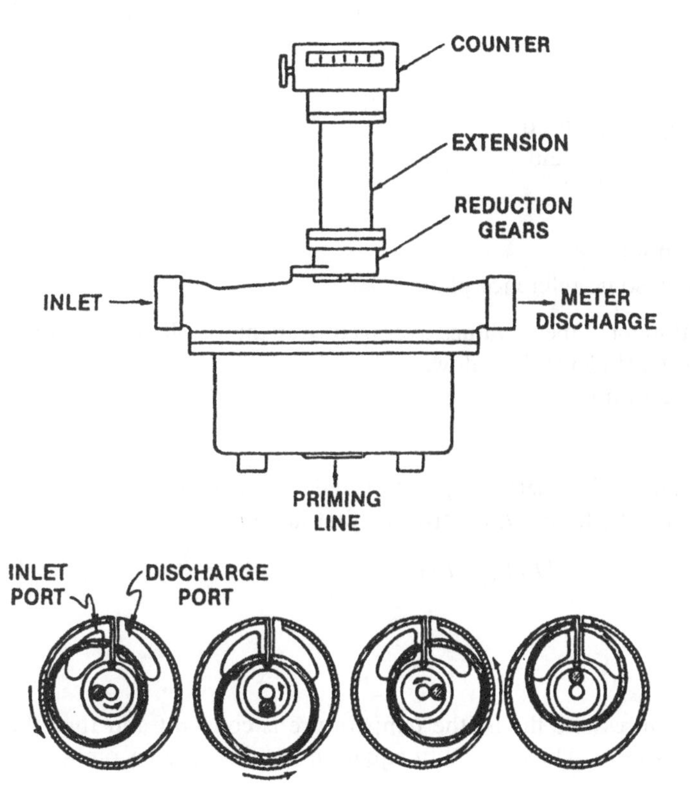


**Figure 8.15** Oscillating piston meter.

(flashing). However, it has been shown (see discussion below) that cavitation is not instantaneous, and indeed it is rather difficult to induce bubble formation in these differential pressure meters. Another concern centers on the transferability of experience with these meters on water and other normal fluids to cryogens. It has been shown experimentally that the vast amount of data developed on water for discharge coefficients vs. Reynolds number, etc., transfers directly to cryogenic systems. Water calibrations apply directly to liquid hydrogen, nitrogen, and oxygen with 1% accuracy. For these reasons, differential pressure meters are a good first choice for cryogenic flow measurements.

To develop the flow equations for differential pressure flowmeters, let

$A$ = area of first or upstream section of diameter $D$
$a$ = area of second or downstream section of diameter $d$

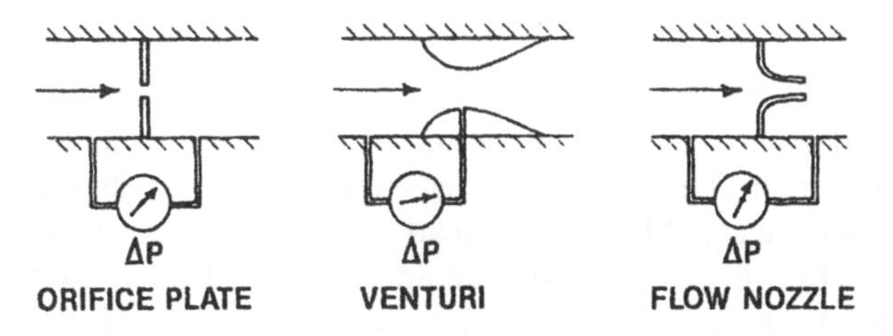


**Figure 8.16** Pressure drop flowmeters.

and let

$P$ = pressure
$U_i$ = internal energy of the fluid
$U_k$ = kinetic energy of the fluid
$v$ = velocity
$V$ = specific volume
$\rho$ = density of the fluid
$Z$ = elevation above some reference plane

These are the arithmetical mean values averaged over the whole section $A$ (or $a$) for a fully developed (turbulent) fluid flow.

The continuity equation is

$$Av_1\rho_1 = av_2\rho_2 \tag{8.3}$$

If the elevation change ($\Delta Z$) between points 1 and 2 is negligible and the heat flow between the pipe and the fluid, $Q$, is zero, the first law is

$$(U_{k2} + U_{i1}) - (U_{k1} + U_{i1}) = (P_1V_1 - P_2V_2) \tag{8.4}$$

### 5.2.1. *Liquids*

Treating liquids as incompressible fluids, the temperature is constant and thus the density is constant, $\rho_1 = \rho_2 = \rho$. The continuity equation for this case becomes

$$Av_1 = av_2 \tag{8.5}$$

$$v_1 = \frac{a}{A}v_2 \tag{8.6}$$

The energy equation reduces to

$$U_{k1} + P_1V_1 = U_{k2} + P_2V_2 \tag{8.7}$$

The kinetic energy per unit mass is

$$U_k = v^2/2g_c \tag{8.8}$$

and the general energy equation may be written

$$\frac{P_1}{\rho} + \frac{v_1^2}{2g_c} = \frac{P_2}{\rho} + \frac{v_2^2}{2g_c} \tag{8.9}$$

Using the value of $v_1$ from Eq. (8.5) in Eq. (8.9) and rearranging gives

$$v_2^2 = 2g_c\left(\frac{P_1 - P_2}{\rho}\right)\left(\frac{1}{1 - (a/A)^2}\right) \tag{8.10}$$

Two components of Eq. (8.10) deserve note. First, the quantity $(P_1 - P_2)/\rho$ is equal to the difference between the static pressures at $A$ and $a$ if measured by a column of the flowing liquid $h$ in height. This column is referred to generally as the "head" of the liquid and gives us the term, "differential head meter," which has been used in fluid-metering literature. (Recall, however, that a column of liquid when used as a measurement of pressure has the dimensions of force per unit area and not simply length.)

The second is $1/[1-(a/A)^2]$, the square root of which is called the "velocity of approach" factor, since it resulted from substituting the expression for $v_1$ into Eq. (8.9). If the areas are circular, the diameters are the dimensions measured and known; hence, $(a/A)^2$ may be and usually is replaced with $(d/D)^4 = \beta^4$.

To obtain the theoretical mass flow rate $\dot{m}_t$.

$$\dot{m}_t = \rho a v_2$$
$$= a\sqrt{2g_c\rho(p_1-p_2)}\sqrt{\frac{1}{1-\beta^4}} \quad (8.11)$$

and the theoretical volume rate of flow, $\dot{q}_t$, is

$$\dot{q}_t = a v_2$$

or

$$\dot{q}_t = a\sqrt{\frac{2g_c(P_1-P_2)}{\rho}}\sqrt{\frac{1}{1-\beta^4}}$$
$$= a\sqrt{2g_c h}\sqrt{\frac{1}{1-\beta^4}} \quad (8.12)$$

where

$$h = \frac{p_1-p_2}{\rho} \quad (8.13)$$

The subscript t indicates a theoretical value in contrast to actual rates of flow.

### *5.2.2. Compressible Fluids (Gases)*

The assumption that there is no transfer of heat between the fluid and the pipe, which implies no friction, permits the assumption that any change of state between sections $A$ and $a$ is a reversible isentropic (adiabatic) change for which

$$P_1V_1^{\gamma} = P_2V_2^{\gamma} = PV^{\gamma} = \text{a constant, } c' \quad (8.14)$$

If it is assumed that it is an ideal gas for which the general equation of state is

$$P/\rho = R/T \quad (8.15)$$

then the energy equation, Eq. (8.4), may be written in the form

$$P_1\rho_1 + \frac{v_1^2}{2g_c} + U_{i1} = \frac{p_2}{\rho_2} + \frac{v_2^2}{2g_c} + U_{i2} \quad (8.16)$$

or, applying the definition for enthalpy,

$$\frac{v_2^2}{2g_c} - \frac{v_1^2}{2g_c} = H_1 - H_2 \quad (8.17)$$

For any fluid

$$dH = T\,ds + V\,dP$$

Our assumption of adiabatic, reversible conditions (for which a correction will be made later) allows

$$ds = 0$$

and, therefore,

$$H_1 - H_2 = \int_{P_1}^{P_2} V\,dp \tag{8.18}$$

Using Eq. (8.14) in the form $V = c'/P^{1/\gamma}$ gives

$$H_1 - H_2 = c' \int_{P_1}^{P_2} P^{1/\gamma}\,dp \tag{8.19}$$

the integrand of which is

$$H_1 - H_2 = c' P_1^{(\gamma-1)/\gamma} \frac{\gamma}{\gamma - 1}(1 - r^{(\gamma-1)/\gamma}) \tag{8.20}$$

where $r$ is the ratio of pressures

$$P_2/P_1 \tag{8.21}$$

Since $c' = V_1 {P_1}^{1/\gamma}$, substituting Eq. (8.20) into Eq. (8.17) gives

$$\frac{v_2^2}{2g_c} - \frac{v_1^2}{2g_c} = P_1 V_1 \left(\frac{\gamma}{\gamma - 1}\right)(1 - r^{(\gamma-1)/\gamma}) \tag{8.22}$$

Since the mass rate of flow is the same at sections $A$ and $a$,

$$\dot{m} = A v_1 \rho_1 = a v_2 \rho_2 \tag{8.23}$$

Using Eq. (8.14) in the form $\rho_2/\rho_1 = r^{1/\gamma}$,

$$v_1 = v_2 \left(\frac{a}{A}\right) r^{1/\gamma} \tag{8.24}$$

Substituting the relation from Eq. (8.24) into (8.22) and solving for $v_2$ gives

$$v_2 = \left[\frac{\dfrac{2g_c p_1}{\rho_1}\left(\dfrac{\gamma}{\gamma - 1}\right)(1 - r^{(r-1)/\gamma})}{r - (a/A)^2 r^{2/\gamma}}\right]^{1/2} \tag{8.25}$$

Using this value of $v_2$ in Eq. (8.23) and noting that $\rho_2 = \rho_1 r^{1/\gamma}$ gives

$$\dot{m}_T = a \left[\frac{2g_c P_1 \rho_1 r^{2/\gamma}\left(\dfrac{\gamma}{\gamma - 1}\right)(1 - r^{(\gamma-1)/\gamma})}{1 - (a/A)^2 r^{2/\gamma}}\right]^{1/2} \tag{8.26}$$

This equation can be modified by using $P_1 = (P_1 - P_2)/(1 - r)$ and $\beta^2 = \mathrm{a/A}$ so that it can be written

$$\dot{m}_\mathrm{T} = a \left[\frac{2g_c (P_1 - P_2)\rho_1 r^{2/\gamma}\left(\dfrac{\gamma}{\gamma - 1}\right)\left(\dfrac{1 - r^{(\gamma-1)/\gamma}}{1 - r}\right)}{1 - \beta^4 r^{2/\gamma}}\right]^{1/2} \tag{8.27}$$

In contrast to Eq. (8.11), Eq. (8.27) is called the "theoretical adiabatic" equation for the mass rate of flow of an ideal compressible fluid across section a in terms of the initial pressure or the pressure difference and the density $\rho_1$. Eq. (8.27) can be written in the form

$$\dot{m}_T = a\left(\frac{2g_c\rho_1(p_1 - p_2)}{1-\beta^4}\right)^{1/2} \times \left[r^{2/\gamma}\left(\frac{\gamma}{\gamma-1}\right)\left(\frac{1-r^{(\gamma-1)/\gamma}}{1-r}\right)\left(\frac{1-\beta^4}{1-\beta^4 r^{2/\gamma}}\right)\right]^{1/2} \tag{8.28}$$

This amounts to using the hydraulic Eq. (8.11) modified by the expansion factor

$$Y = \left[r^{2/\gamma}\left(\frac{\gamma}{\gamma-1}\right)\left(\frac{1-r^{(\gamma-1)/\gamma}}{1-r}\right)\left(\frac{1-\beta^4}{1-\beta^4 r^{2/\gamma}}\right)\right]^{1/2} \tag{8.29}$$

The value of $Y$ depends upon the diameter ratio $\beta$, the pressure ratio $r$, and the ratio of specific heats $\gamma$. For routine computations it will be found convenient to prepare curves or tables from which the values may be read.

The values of $\rho_1$ in Eqs. (8.27 ) and (8.28 ) should be computed with the general equation of state for an actual gas, which, for this purpose, may be written

$$\rho_1 = P_1/Z_1 R_g T_1 \tag{8.30}$$

in which $Z_1$ is the compressibility factor for the particular gas being metered, corresponding to the conditions defined by $P_1$ and $T_1$.

### *5.2.3. Temperature Correction*

A temperature correction factor, $F_a$, is used to account for any change of the area of section $a$ of the primary element when the operating temperature differs appreciably from the ambient temperature at which the device was manufactured and measured. If the meter is to be used under temperature conditions within the range of ordinary atmospheric temperatures, any difference between the thermal expansion of the pipe and the primary element may be ignored, and the diameter ratio, $\beta$, may be considered to be unaffected by temperature. At cryogenic temperatures, the material for the throat liner of a venturi tube, a flow nozzle, or an orifice plate should have a coefficient of thermal expansion as close as possible to that of the pipe.

### *5.2.4. Discharge Coefficient*

The assumption of adiabatic-reversible (isentropic) flow is seldom the case. The actual rate of flow through a differential pressure meter is very seldom, if ever, exactly equal to the theoretical rate of flow indicated by the particular theoretical equation used. In general, the actual rate of flow is less than the indicated theoretical rate. Hence, to obtain the actual flow from the theoretical equation, an additional factor called the "discharge coefficient" must be introduced. This coefficient is represented by $C$ and is defined as

$$C = \frac{\text{actual rate of flow}}{\text{theoretical rate of flow}} \tag{8.31}$$

The rate of flow may be in terms of mass (or weight) or volume per unit of time. When volume is used, it is necessary that both the actual and the theoretical volumes be at (or be referred to) the same conditions of pressure and temperature. Thus, the actual rate of flow through a venturi tube, flow nozzle, or orifice, when the general hydraulic equation is used, will be

$$\dot{m} = \left(\frac{\pi d^2 F_a}{4}\right)\left(\frac{C}{\sqrt{1-\beta^4}}\right)[2g_c\rho_1(p_1 - p_2)]^{1/2}$$

or

$$q = \left(\frac{\pi d^2 F_a}{4}\right)\left(\frac{C}{\sqrt{1-\beta^4}}\right)\sqrt{2g_c h}$$

Using the adiabatic equation, Eq. (8.27 ), would require multiplying the right-hand side of Eqs. (8.38 ) or (8.39 ) by the expansion term, Eq. (8.29 ).

Values of $C$ will be different for each type of differential pressure flowmeter. Figure 8.17 gives typical values of the discharge coefficient for an orifice meter.

Differential pressure flowmeters are subject to all the issues concerning pressure measurements described in Sec. 4. Furthermore, the pressure taps must be located according to prescribed conventions if the discharge coefficients from one calibration are to apply to other situations. Three common designations for pressure taps related to their location are (1) corner taps, which are adjacent to the orifice plate, (2) flange taps, which are located 25.4 mm (1.00 in.) upstream and the same distance downstream of the orifice, and (3) $D$ and $D/2$ taps, which are located one pipe diameter upstream of the orifice and one-half pipe diameter downstream of the orifice.

For orifice meters having the $D$ and $D/2$ pressure tap configuration, the discharge coefficient is given by

$$C = 0.5959 + 0.0312\,\beta^{2.1} - 0.1840\,\beta^8 + 0.0390\,\beta^4/(1-\beta^4) - 0.01584\,\beta^3 + 91.71\,\beta^{2.5}N_{Re}^{-0.75}$$

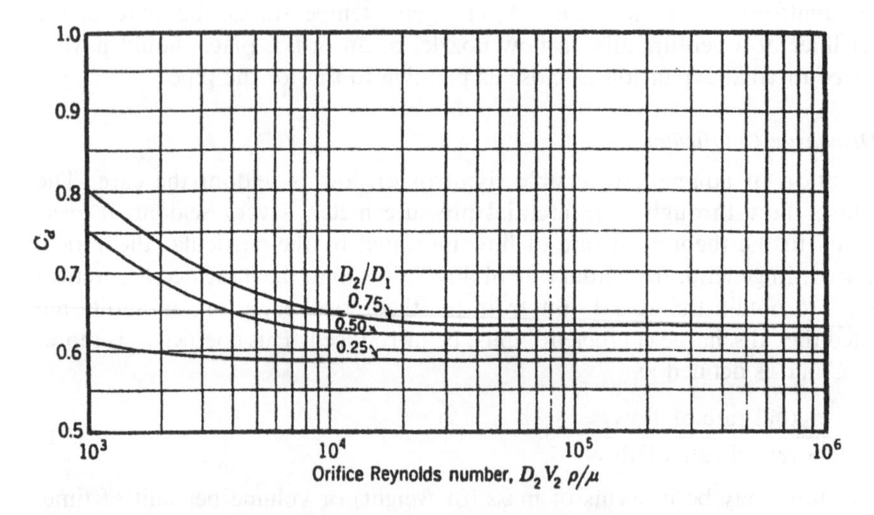


**Figure 8.17** Discharge coefficients for a sharp-edged orifice.

where $\beta = d/D$ is the orifice diameter ratio, $d$ the orifice diameter, $D$ the pipe inside diameter, $N_{\mathrm{Re}} = D_{\rho}G/\mu$ the pipe Reynolds number, and $G = \dot{m}/A$.The venturi meter consists of a conical reducer section and a straight throat section followed by a more gradual enlargement to the original tube diameter. The inlet cone angle is usually 20–22°, and the exit cone angle is about 5–7°. The throat diameter is on the order of one-half the tube diameter. Pressure taps are placed about one-half tube diameter upstream of the venturi entrance and at the center of the throat section.

The venturi flow equations are exactly the same as those developed above. No distinction was made during the thermodynamic analysis regarding the nature of the restriction causing the pressure drop. However, venturi discharge coefficients are nearly unity, as flow through a venturi more closely approximates the isentropic model assumed (see Fig. 8.18).

The best venturi discharge coefficients are obtained through actual calibrations. Figure 8.18 may be used to estimate $C$, and the following correlations are also useful:

1. For $3000 \leq N_{\mathrm{Re}} \leq 2 \times 10^5$,

$$C = \frac{\log N_{\mathrm{Re}}}{0.60 + 0.90 \log N_{\mathrm{Re}}}$$

2. For $N_{\mathrm{Re}} > 2 \times 10^5$,

$$C = 0.988$$

The Reynolds number is defined as

$$N_{\mathrm{Re}} = D\beta^2 (2g_c \rho \Delta p)^{1/2} / \mu$$

where $\beta$ is the ratio of throat diameter to pipe inside diameter $D$.

The appeal of these meters in cryogenics is more than simplicity and stems from the possibility of eliminating the necessity for calibration if proper design, application theory, and practice are followed. Design methods for square-edged

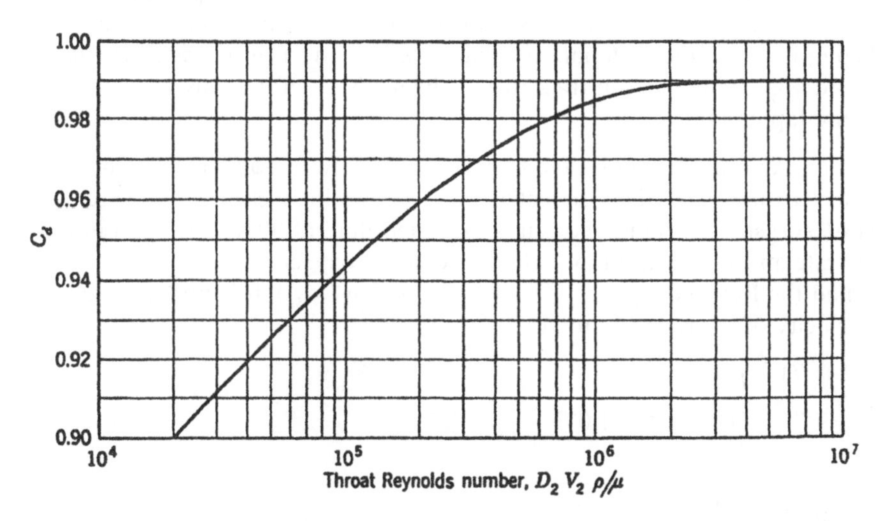


**Figure 8.18** Discharge coefficients for a venturi meter. The throat Reynolds number is based on the throat diameter and the theoretical fluid velocity at the throat cross section.

orifices, flow nozzles, and venturi tubes are provided by ASME (1971). Recommended practice for flange-mounted sharp-edged orifice plates can be found in publications of the ISA (1970) and ISO (1967). The application in general has been to follow these recommendations developed with water, correcting only for thermal contraction due to temperature change when operation is desired at cryogenic temperatures. Reynold's number increases caused by the low viscosity of cryogens and other differences between water and cryogens have been considered extensively.

There are potential disadvantages to the use of pressure drop meters in cryogenic service. First of all, the pressure drop flowmeter is strictly a volumetric meter, that is, it can of itself give no information regarding mass flow rate. Also, the pressure drop flowmeter follows a square root law such that the flow range is limited to about 4 – 1. It is also obvious that pressure measurements are required as the readout of these meters, and we have already discussed the severe difficulties of making precise pressure measurements at low temperatures. Further, it is the very nature of these devices that they generate a decrease in pressure by accelerating the fluid. If we are dealing with a saturated cryogen, then this sudden decrease in pressure may possibly lead to cavitation. Also, it is noteworthy that they are not suitable for determining two-phase flow.

An example of careful use is found in the work of Richards et al. (1960), who addressed the problem of potential cavitation. They made many attempts to obtain points off the calibration curve with the upstream static pressure lower than the vapor pressure. In spite of the fact that the downstream static pressure was as much as 3.3 kPa (10 in.Hg) below the vapor pressure, all of the points fell on the calibration curve within the accuracy of the experiments.

Richards et al. (1960) concluded, therefore, that it was probable that the time required for the nucleation and growth of bubbles in both nitrogen and hydrogen is so long that errors in the measurement of flow with sharp-edged orifices, arising from reduced downstream pressures, likely will not occur. However, it has been found that problems of cavitation are more severe with a venturi meter than with an orifice meter for the same flow rate and relative meter size (same $\beta$ ratio). To avoid the problem of cavitation in venturi meters, it is necessary to maintain the upstream pressure high enough that the vapor pressure is not reached at the throat of the venturi.

The estimated uncertainty in cryogenic flow measurement using head meters is in the range of $\pm$1–3%. This is composed of the uncertainty in bias shift caused by thermal contraction of the material, uncertainty in the effect of increased Reynolds number, and a large imprecision traceable to the methods of pressure measurement and pressure tap design.

*Example 8.4.* An orifice meter is used to measure the flow of liquid oxygen in a 5 cm (2 in.) diameter tube. The orifice diameter is 2.5 cm, and the measured pressure drop is 240 Pa (0.035 psi). The liquid oxygen is saturated fluid at 90 K (162°R). Find the mass flow rate and volumetric flow rate of the liquid oxygen.

To find the Reynolds number through the orifice ($N_{Re}$), it is first necessary to find the velocity of approach factor $C_a$,

$$C_a = 1/(1 - \beta^4)^{1/2}$$
$$= 1/[1 - (1/2)^4]^{1/2} = 1.033$$

At 162°R,

$$\rho = 70.8\,\text{lb}_m/\text{ft}^3$$
$$\mu = 0.455\,\text{lb}_m/(\text{ft hr})$$

The throat velocity is

$$v_2 = C_a[2g_c(P_1 - P_2)/\rho]^{1/2}$$
$$= 1.033[2(32.2)(0.035)(144)/70.8]^{1/2}$$
$$= 2.212\,\text{ft/sec}$$
$$N_{Re} = \frac{\rho v_2 d}{\mu} = \frac{(70.8)(2.212)(1)(1/12)(3600)}{0.455}$$
$$= 10.3 \times 10^4$$

From Fig. 8.17, $C_a = 0.62$. Then,

$$q = C_a A v_2 = (0.62)\frac{(\pi)(1)^2}{(4)(144)}(12.12)(60)$$
$$= 0.449\,\text{ft}^3/\text{min} = 2.12 \times 10^{-4}\,\text{m}^3/\text{sec}$$
$$m = \rho q = (70.8)(0.449) = 31.8\,\text{lb}_m/\text{min} = 0.240\,\text{kb/sec}$$

## 5.3. Turbine Type Meters

The turbine type of flowmeter is popular for measuring cryogenic fluid flow. It has had extensive cryogenic experience from which design improvements have resulted (Alspach and Flynn, 1965). To infer mass flow, usually the desired measurement, attempts have been made to couple the meter's volume flow measurement with fluid temperature and fluid pressure measurements. Then, by computation, an inference of mass flow is made. This inferred mass flow has not been highly accurate. This is understandable when one considers that three separate measurements, with their attendant inaccuracies, are combined in a computation that is then referred to fluid *PVT* data.

As a true volume flow measurement technique, the turbine flowmeter is an outstanding device that has excellent repeatability. To achieve this excellence, however, it must be used correctly. All too frequently the turbine flowmeter is misapplied. That is to say, it is used where the flow range is too great, it is used for flow rates lower than those for which it is designed, and it is used in two-phase fluid applications.

The turbine volumetric flowmeter is a simple mechanism that consists of a freely rotating bladed rotor supported by bearings inside a housing and an electrical transducer that senses rotor speed (see Fig. 8.19). Rotor speed is a direct function of flow velocity:

$$\omega = \frac{v \tan\alpha}{R} = \frac{Q}{A_h - A_r}\left(\frac{\tan\alpha}{R}\right)$$

where $\omega$ is the rotor angular velocity, $v$ the average flow velocity of the rotor blades, $\alpha$ the rotor blade angle with the pipe axis, $R$ the average radius of rotor blade (center of pressure), $Q$ the volumetric flow rate, $A_h$ the internal cross-sectional area of housing, and $A_r$ is the maximum cross-sectional area of rotor.

Deviations from this idealized mathematical expression can be expected to be due to (1) retarding forces on the rotor such as fluid drag, mechanical friction from rotary and thrust bearings, and transducer magnetic drag; (2) velocity profile variations; and (3) swirl of the incoming fluid stream.

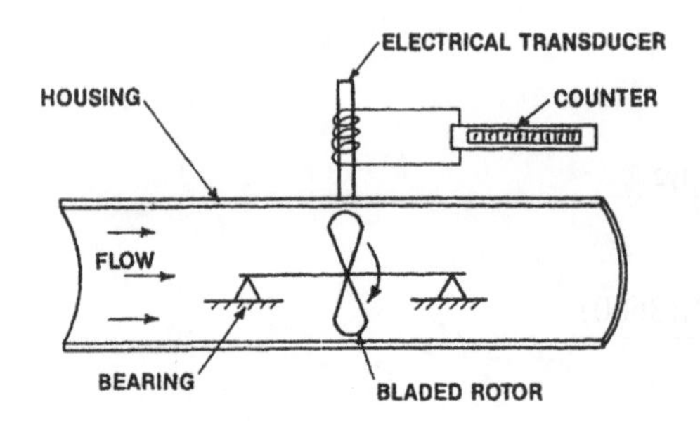


**Figure 8.19** Turbine flowmeter.

The turbine flowmeter is designed for use with a homogeneous fluid. Its use with a liquid/vapor fluid has resulted in measurement errors and, in a few cases, has also resulted in dangerous turbine runaway. In the cases of turbine runaway, the end result was catastrophic.

The advantages of the turbine flowmeter include:

1. small size—thus a minimum mass to cool down;
2. light weight;
3. versatility, 10:1 flow range;
4. linear calibration;
5. excellent repeatability;
6. digital output signal;
7. requires no shaft seals;
8. self-propelled; fluid flow supplies the driving force;
9. excellent response.

The disadvantages of the turbine flowmeter include:

1. enables volumetric measurement, not mass measurement;
2. designed for single-phase fluids;
3. overspeed will cause turbine runaway;
4. rotating mechanical component;
5. requires upstream flow straightening to eliminate fluid swirl effects;
6. attitude-sensitive; requires mounting in same position as when calibrated.

A variation of the turbine meter is the twin-turbine meter (see Fig. 8.20). In this case, two turbines of different blade pitch are coupled together by means of an elastic restraint. Because of the difference in pitch, the two turbines would tend to revolve at different speeds. However, this is prevented by the elastic coupling, so that they do revolve at the same speed but with a phase and between them. The magnitude of the phase angle is a measure of the flow rate and is said be, in fact, a measure of the mass flow rate.

## 5.4. Transverse Momentum Mass Flowmeters

Mass reaction or momentum flowmeters are of three types. In one type an impeller imparts a constant angular momentum to the fluid stream and the variable torque on a turbine that removes this momentum is measured. In a second type, an impeller is driven at either a constant torque, and the angular velocity of the impeller is

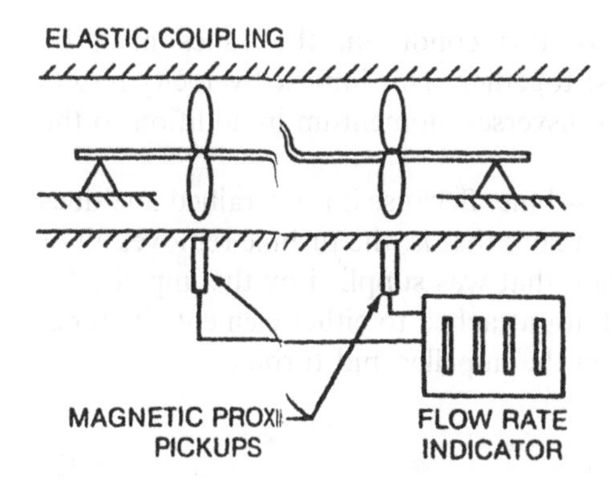


**Figure 8.20** Twin-turbine meter.

measured. In the third type, a loop of fluid is driven at either a constant angular speed or a constant oscillatory motion; then the mass reaction is measured.

All such meters are direct mass flowmeters because the reaction of their primary element is proportional to the momentum of the stream. Each imposes on the fluid stream a momentum (acceleration) transverse (at right angles) to the axial (longitudinal) flow of the stream. These transverse momentum flowmeters can be classified as (1) axial flow, (2) radial flow, and (3) gyroscopic. Since each of these classes is basic, several variations on mechanical design and primary element reaction detection have been devised. Some variations are discussed below for each class.

### *5.4.1. Axial Flow Transverse Momentum Mass Flowmeter*

Figure 8.21 shows a schematic of an axial flow transverse momentum flowmeter. The General Electric angular momentum mass flowmeter uses this design. Substantially, all fluid flow passes through both the impeller and the turbine. The impeller and the turbine are geometrically similar cylinders mounted in a cylindrical flow conduit on an axis coinciding with the conduit centerline. Each element is mounted on a separate shaft. Both the impeller and turbine are composed of several straight vanes located at the periphery of the elements and parallel to the centerline of the conduit. A means is provided for measuring the torque on the turbine shaft. If the impeller were locked (not rotating), the torque on the turbine shaft would be zero.

Now suppose the impeller is rotated at some angular speed $\omega$. As the fluid stream enters the impeller, it is set to rotating at an angular speed equal to the

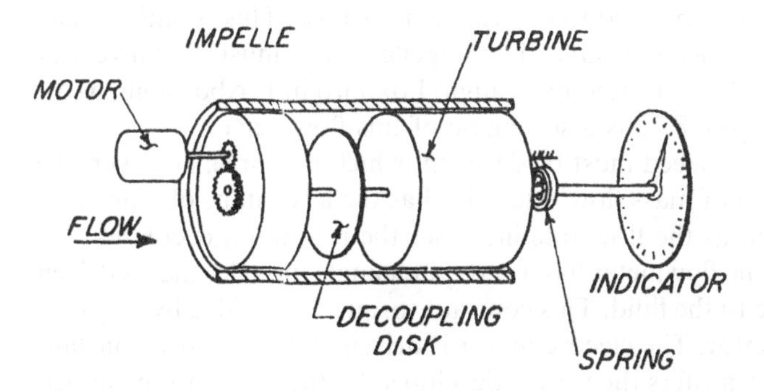


**Figure 8.21** Axial flow transverse momentum mass flowmeter.

angular speed of the impeller. To achieve this condition, the vanes must be sufficiently long and spaced sufficiently close together. This imposed velocity results in the fluid stream having an angular (or transverse) momentum in addition to the normal axial momentum of the stream.

Located adjacent to the impeller is the turbine. Because it is restrained and does not rotate, the turbine straightens the fluid stream; that is, the turbine removes all of the angular momentum from the fluid stream that was supplied by the impeller.

The decoupling disk is stationary and not attached to either element. Its function is to eliminate viscous coupling between the impeller and turbine.

The governing equation is

$$T = \omega K^2 \dot{m} \tag{8.32}$$

where $T$ is the torque, $\dot{m}$ is the mass flow rate, and $\omega$ is the speed of rotation. If the speed $\omega$ is constant and $K$ is a geometrical constant, then

$$T = C_1 \dot{m} \tag{8.33}$$

This equation shows that turbine shaft torque is directly proportional to mass flow rate. In the General Electric flowmeter this torque is detected by means of a calibrated spring that is attached to an electromechanical angle detector.

For cryogenic service, this flowmeter offers the following advantages (Alspach et al., 1966):

1. direct mass flow measurement;
2. applicable to two-phase fluids;
3. requires no flow straighteners;
4. good range;
5. linear calibration.

Disadvantages include:

1. moving parts in the fluid stream;
2. torque-measuring spring requires low-temperature calibration;
3. slow response time.

*5.4.2. Elbow Axial Flow Transverse Momentum Flowmeter*

If Eq. (8.33) is rewritten as

$$\omega = \frac{T}{K^2 \dot{m}} \tag{8.34}$$

then an alternative scheme may be used to determine mass flow. This equation illustrates that if torque is maintained constant, then impeller speed must vary inversely as the mass flow rate. This scheme is used by Waugh–Foxboro in Foxborough, MA., in their mass flowmeter. Figure 8.22 is a schematic of this flowmeter.

In this technique, motor speed must be detected, which is simple; however, the output is an inverse function of mass flow rate. This has the advantage of high resolution at low flow rates, but as the flow rate increases the resolution decreases.

The rotor portion of the flowmeter has a zero pitch angle impeller that is driven to impart a constant torque to the fluid. This constant torque is provided by a hysteresis drive and an electric motor. The electric motor is driven at essentially a constant speed. The hysteresis drive transfers the torque developed by the motor to the impeller. The hysteresis drive has the characteristic of transmitting a constant torque

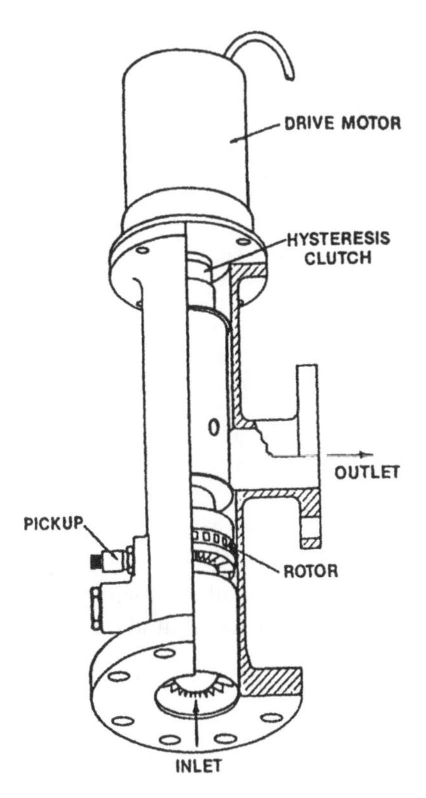


**Figure 8.22** Elbow axial flow transverse momentum mass flowmeter.

regardless of the operating slip speed. As a consequence, torque measurements are not required, thereby leaving only the measurement of impeller speed.

For cryogenic service, this flowmeter offers the following advantages (Alspach et al., 1966):

1. direct measurement of mass flow;
2. digital output signal;
3. does not require calibrated springs or other torque-sensing devices;
4. two-phase fluid capability.

Disadvantages include:

1. moving parts in the fluid stream;
2. will not measure zero flow;
3. nonlinear calibration (frequency vs. flow) ;
4. slow response time.

*5.4.3. Gyroscopic Transverse Momentum Mass Flowmeter*

The gyroscopic mass flowmeter is shown schematically in Fig. 8.23. It consists of a closed loop through which the fluid flows freely. Suitable bearings and seals at points 1 and 2 permit the closed loop to be rotated about the $x$ axis.

For any general gyroscope, the moment about the $y$ axis, $M_y$, can be expressed by

$$M_y = I_z \Omega \omega$$

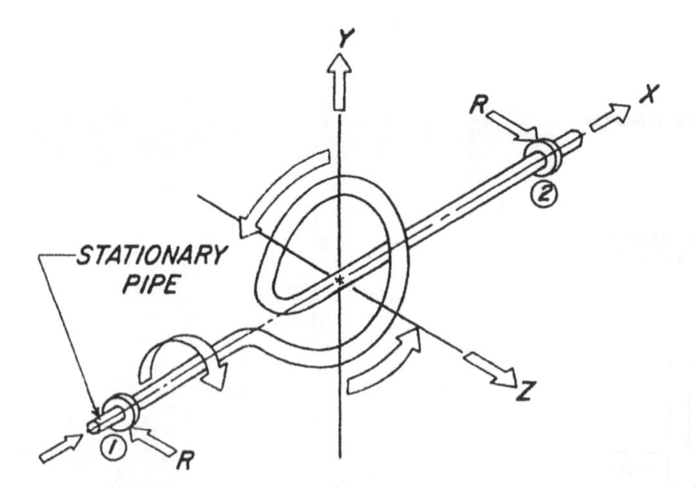


**Figure 8.23** Gyroscopic transverse momentum mass flowmeter.

where $\omega$ is the precessional velocity of the gyroscope, $\Omega$ is the angular velocity of the gyroscope rotor, $I_z$ is the polar moment of inertia of fluid in the conduit about the $z$ axis, and

$$I_z = \rho(2\pi r A)r^2 \quad \text{and} \quad V = \omega r$$

Then

$$M_y = (\rho V A)2\pi\omega r^2$$

The last equation now fits the familiar pattern of a moment proportional to the mass flow rate times the product of a constant speed and a geometrical constant, so that

$$M_y = C_3\dot{m}$$

In an actual flowmeter, therefore, it is obvious that the moment about the $y$ axis must be measured or, more realistically, the reaction $R$ at bearings 1 and 2 must be measured. Note that the magnitude of the reaction vector $R$ is directly proportional to mass flow rate $\dot{m}$ and that vector $R$ rotates at an angular speed equal to $\omega$.

This flowmeter is capable of measuring bidirectional flow. The flowmeter must be provided with rotating seals and suitable bearings.

The Decker Corporation, in a variation of this gyroscope flowmeter, has the angular precessional velocity applied about the $x$ axis as a small oscillatory angular velocity instead of a constant angular velocity. As a result, reaction $R$ is an oscillating value that can be measured. An advantage of this scheme is that rotating seals are not required. Instead, a flexible length of conduit or bellows can be attached upstream and downstream of the flowmeter.

In the Decker design, the loop is vibrated through a small angle of constant amplitude about an axis in the plane of the loop. This vibration results in an alternating gyro-coupled torque about the mutually orthogonal axes. The peak amplitude of this torque is directly proportional to mass flow rate. The torque acts against elastic restraints (torsion bars) to produce an alternating angular displacement about the torque axis. Velocity pickups, appropriately mounted, convert the alternating displacement to a proportional alternating electric signal. The peak amplitude signal is directly proportional to mass flow rate. This signal is then conditioned to provide a dc signal output linearly to mass flow rate.

For cryogenic service, this flowmeter offers the following advantages (Alspach et al., 1966) :

1. direct mass flow measurement;
2. two-phase fluid capability;
3. no moving parts in the flow stream;
4. requires no flow straighteners;
5. linear calibration;
6. bidirectional flow (for loading and unloading storage vessel).

Disadvantages include:

1. large size;
2. difficult to vacuum jacket;
3. complex mechanical design;
4. vibration-sensitive.

*5.4.4. Radial Flow Transverse Momentum Mass Flowmeter*

The radial flow transverse momentum flowmeter is commonly called the Coriolis mass flowmeter. Figure 8.24 is a schematic drawing of this flowmeter. The fluid stream enters the housing, which is driven at a constant angular velocity. The housing interior is composed of a flow-sensing element (impeller) and a downstream guide vane. Note that the housing, the impeller, and the turbine are attached and rotating at the same angular velocity. Both the impeller and the guide vanes are radial and of sufficient length and closely spaced so that the stream is set to rotating at the speed of the impeller. This causes the flow to leave the impeller and enter the guide vanes with zero tangential velocity.

The operation of this flowmeter is explained most simply by consideration of a special case where flow is confined to a straight radial channel of constant cross-sectional area (see Fig. 8.25). The fluid stream flowing down the tube will be

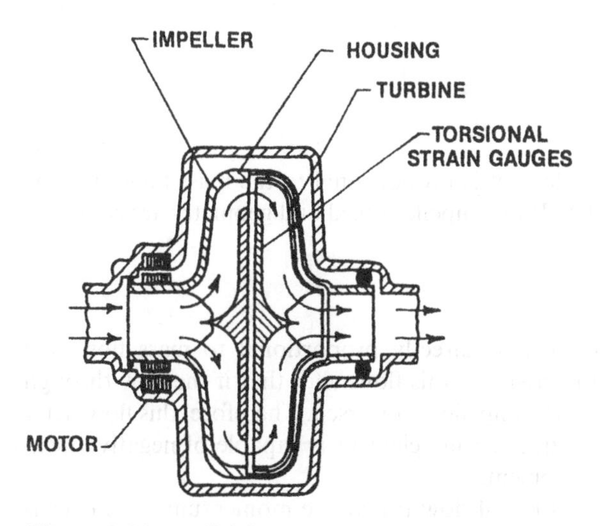


**Figure 8.24** Radial flow transverse momentum mass flowmeter.

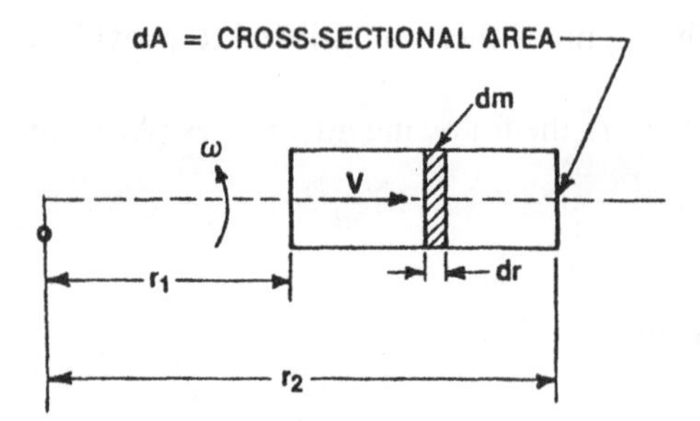


**Figure 8.25** Radial flow cross section.

subjected to a Coriolis acceleration as the tube is rotated about point O. This Coriolis acceleration is equal to

$$\alpha = 2v\omega$$

where $\alpha$ is the Coriolis acceleration, $v$ is the velocity of fluid mass down the tube, and $\omega$ is the angular velocity.

The moment created about O is

$$M_{\mathrm{O}} = \int r\alpha \, \mathrm{d}m$$

But

$$\mathrm{d}m = \rho A \mathrm{d}r$$

So

$$M_{\mathrm{O}} = \int_{r_1}^{r_2} 2v\omega A \, \mathrm{d}r = v\rho A\omega(r_2^2 - r_1^2)$$

Note that

$$v\rho A = \dot{m}$$

Therefore,

$$M_{\mathrm{O}} = \dot{m}\omega(r_2^2 - r_1^2)$$

Thus, the moment on the sensing element is proportional to the mass flow rate, the speed of the impeller, and geometry. If the impeller speed and geometry are constant, then

$$M_{\mathrm{O}} = C_2\dot{m}$$

This shows that the impeller moment is directly proportional to mass flow rate through the flowmeter. It is characteristic of this flowmeter that if the flow through the meter is reversed, the torque on the impeller is reversed. Therefore, this flowmeter is inherently bidirectional if the torque-sensing element is capable of negative measurement as well as positive measurement.

A particular advantage of this radial flow transverse momentum flowmeter is that the angular momentum supplied to the stream by the impeller is recovered by the guide vanes. Consequently, the power supplied to the flowmeter must only be sufficient to overcome bearing friction, seal friction, windage losses, and viscous losses at the meter entrance and exit.

A flowmeter of this type has been considered for cryogenic fluid applications by Space Sciences, Incorporated in Waltham, MA. Space Sciences proposed improved mounting or suspension of the impeller–turbine assembly and an improved method of torque measurement.

For cryogenic service, this flowmeter offers the following advantages (Alspach et al., 1966) :

1. direct measurement of mass flow;
2. two-phase fluid capability;
3. bidirectional flow (for loading and unloading a storage vessel) ;
4. good rangeability;
5. linear output signal;
6. low power requirement;
7. no flow straighteners.

Disadvantages include:

1. moving parts in the flow stream;
2. complex mechanical design;
3. no experience or evaluation in cryogenic fluid applications.

## 5.5. Vortex Shedding Meter

The vortex shedding meter is also a rate velocity meter, like the turbine meters, but in a distinct class by itself. The phenomenon upon which it is based is the Karman vortex trail, and its application to the measurement of flow of liquids and gases is fairly recent (Rodley, 1969), perhaps because only recently have sensors become available to detect the vortices.

The Karman vortex trail is used to explain certain phenomena associated with the flow around cylinders, ellipsoids, and flat plates. For flow around a circular cylinder, for instance, at Reynolds numbers above about 20, eddies break off alternately on either side of the cylinder in a periodic fashion. Behind the cylinder is a staggered, stable arrangement or trail of vortices. The alternate shedding produces a periodic force acting on the cylinder normal to the undisturbed flow. The force acts first in one direction and then in the opposite direction. Let $f$ represent the frequency of this vibration in cycles per unit time, $d$ the diameter of the cylinder, and $v$ the undisturbed velocity. Experiments have shown values of the dimensionless ratio $fd/v$ of between 0.18 and 0.27.

If the frequency of the vortex shedding approaches or equals the natural frequency of the elastic system consisting of the cylinder and its supports, the cylinder itself may have a small alternating displacement normal to the stream flow. (The vibration of some smoke-stacks, the vibration of some transmission lines, and the fatigue failure or progressive fracture of some transmission lines have been attributed to this resonance phenomenon.)

Such a meter was evaluated on liquid nitrogen (Brennan et al., 1974). It consisted of a bluff body located normal to the flow stream pipe that basically measured the average velocity of the flow passing through the meter (see Fig. 8.26). The

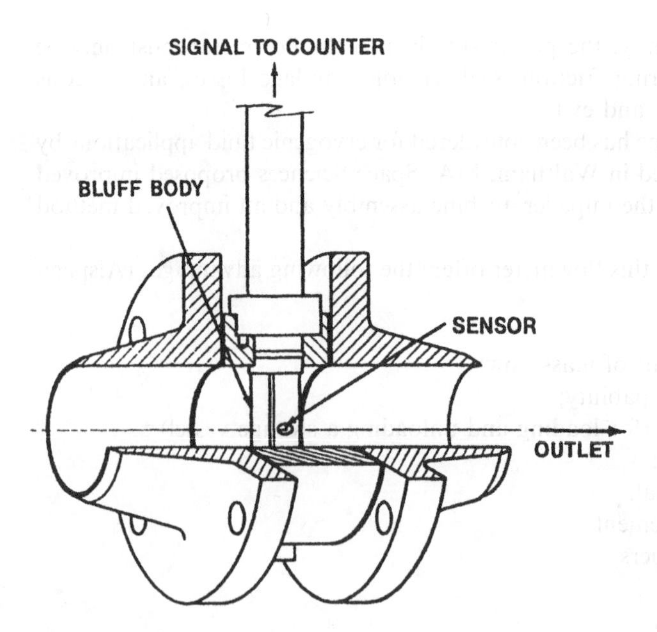


**Figure 8.26** Vortex-shedding flowmeter.

frequency $f$ of the vortex shedding was given approximately by

$$f = \frac{0.87v}{D} \text{ Hz}$$

where $v$ is the velocity in feet per second and $D$ is the nominal meter diameter in feet. The bluff body in this meter was in the shape of a modified delta with its base facing upstream.

The vortex sensors consist of electronically self-heated resistance elements whose temperatures and therefore resistances vary as a result of the velocity variations adjacent to the body. These velocity variations directly reflect the action of the vortices as they peel off from the downstream edge of the bluff body.

The precision of this meter was apparently quite good (approximately 0.5% 3 $\sigma$), but the range of linearity was found to be about 5:1 rather than the expected 10:1. Since these meters are relatively new to cryogenic service, one can reasonably hope that their performance will improve with further experience. They have the advantage of having no moving parts at low temperatures, and therefore they require minimum maintenance. The meter factor (pulses/gallon) depends only on the inside diameter of the pipe and the width across the bluff body face. Cryogenic evaluation is limited but shows promise.

## 5.6. Momentum–Capacitance Mass Flowmeter

The momentum–capacitance mass flowmeter is shown schematically in Fig. 8.27. The Bendix Corporation (Davenport, IA) employed this design in a cryogenic fluid mass flowmeter. This flowmeter has two sensing elements located in the fluid stream. One is a fluid-sampling densitometer, and the other senses fluid momentum. Both elements are open-mesh configurations offering a small impedance to flow. The densitometer element employs honeycomb plates in a parallel-plate configuration,

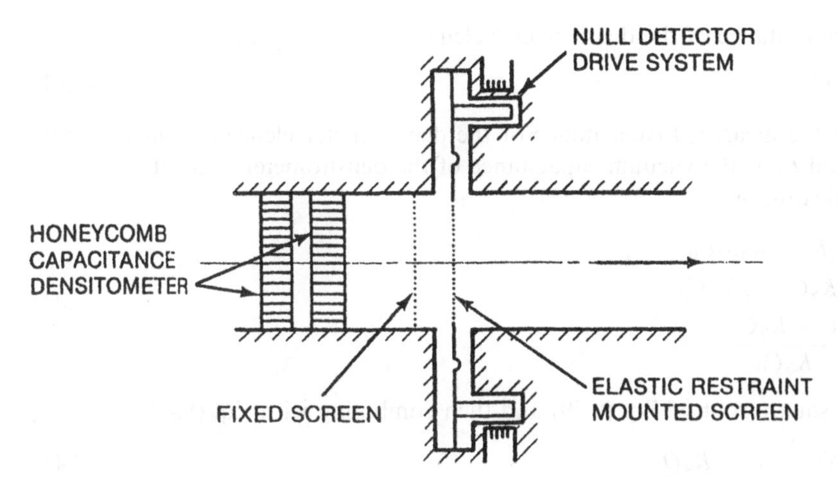


**Figure 8.27** Momentum–capacitance mass flowmeter.

while the momentum sensor employs round wire screens in a parallel-plate configuration.

The pressure drop across a wire screen can be expressed as

$$\Delta P = K_1 \rho v^2 \tag{8.35}$$

where $\Delta P$ is the pressure drop, $\rho$ the fluid density, $v$ the fluid velocity, and $K_1$ the constant when the Reynolds number is less than 1000 and there is a fixed solidity ratio for the screen.

The force exerted on a wire screen is

$$F_{\mathrm{F}} = A \Delta P \tag{8.36}$$

where $F_{\mathrm{F}}$ is the force on screen, $A$ the screen cross-sectional area, and $\Delta P$ is the pressure drop.

The balancing force of end-attracting electromagnets required to return the screen to the no-flow (null) position is

$$F_{\mathrm{E}} = K_2 I^2 \tag{8.37}$$

where $F_{\mathrm{E}}$ is the balancing force, $I$ the electromagnet current to hold screen at null position, and $K_2$ is the proportionality constant.

Therefore, with the screen held at the null position under fluid flow conditions,

$$\begin{aligned} F_{\mathrm{E}} &= F_F \\ K_2 I^2 &= A \Delta P \\ K_2 I^2 &= A K_1 \rho v^2 \end{aligned} \tag{8.38}$$

Solving for $I_1$,

$$I_1 = K_3 v \sqrt{A\rho} \tag{8.39}$$

where $K_3$ is a proportionality constant.

The Clausius–Mosotti relation of fluid density to fluid dielectric constant gives

$$\varepsilon = K_5 + K_6\, \rho \tag{8.40}$$

The capacitance of the densitometer element can be expressed as

$$C = \varepsilon C_0 \tag{8.41}$$

where, $C$ is the measured capacitance of the densitometer element, $\varepsilon$ the dielectric constant, and $C_0$ is the vacuum capacitance of the densitometer element.

By substitution,

$$\begin{aligned} C &= (K_5 + K_6\rho)C_0 \\ &= K_5C_0 + K_6\rho C_0 \\ \rho &= \frac{C - K_5C_0}{K_6C_0} \end{aligned} \tag{8.42}$$

Taking the square root of Eq. (8.39) and then combining with Eq. (8.42) gives

$$I_2 = K_3v, \quad I_2 = K_4Q \tag{8.43}$$

or

$$I_2 = K_7\dot{m} \tag{8.44}$$

For cryogenic service, this mass flowmeter offers the following advantages:

1. applicable to two-phase flow;
2. requires no flow straighteners;
3. no moving parts in fluid stream.

Disadvantages include:

1. inferential mass flow measurement;
2. slow response time;
3. screen and honeycomb open area may be insufficient to allow free passage of solid particles.

### 5.7. Electromagnetic Flowmeter

Faraday's law in induction, $E = uB$, where $E$ is the induced electric field, $u$ is the velocity of motion, and $B$ is the intensity of magnetic induction, is the basis for the electromagnetic flowmeter. In such a flowmeter, a uniform magnetic induction is established transverse to a flow pipe, and flow of the metered fluid generates a potential difference at suitable detecting electrodes.

Figure 8.28 is a schematic of the magnetoelectric flowmeter method showing the relative directions of the transverse electric field $E$, the flow velocity $u$, and the induced magnetic field $H$.

One electromagnetic flowmeter is based on the fact that a moving, polarized dielectric generates an effective magnetic moment ($M$) according to

$$M = Pu \tag{8.45}$$

where, $M$ is the induced magnetic moment per unit volume, $P$ the polarization in the dielectric medium, and $u$ is the velocity of the moving dielectric.

Equation (8.45) is analogous to the equation describing Faraday induction. For a linear dielectric, it can be written as

$$M = K_0(K - 1)Eu$$

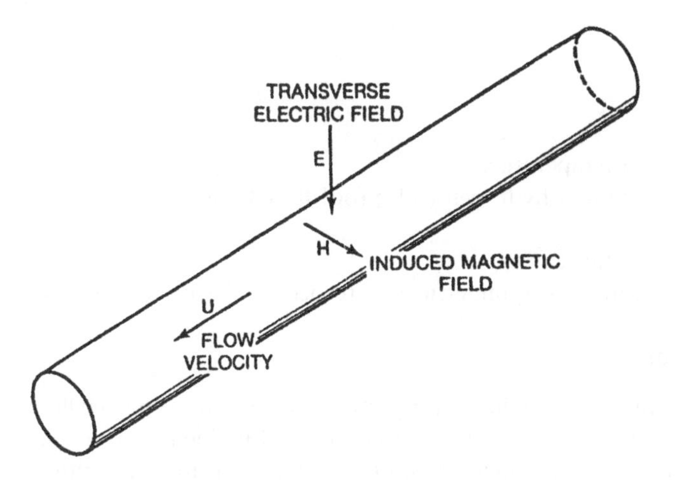


**Figure 8.28** Electromagnetic flowmeter.

where, $K_0$ is the permittivity of free space, $K$ the dielectric constant of flowing fluid, and $E$ the electric field that polarizes the dielectric. This flowmeter is dependent on the fact that it is operating in a dielectric regime, i.e., where conduction currents are negligible. The criterion that the flowmeter is operating in the dielectric regime is simply that the liquid's relaxation time must be large compared with the period alternation of the magnetic induction, a requirement that

$$\omega t > 1$$

where

$$t = K K_0 / \sigma$$

where $\omega$ is the angular frequency of magnetic induction and $\sigma$ is the electrical conductivity of the liquid.

Many cryogenic liquids, including hydrogen, are made up of homonuclear molecules, which have no permanent dipole moment, and so typically the dielectric constant is low, as is the electrical conductivity.

One mass flowmeter design is based on the expression

$$V_{\mathrm{M}} = \left(\frac{BMF_{\dot{m}}}{r}\right)\left(\frac{\sin\beta}{E(\beta, 1)}\right) \tag{8.46}$$

where $V_{\mathrm{M}}$ is the output voltage, $B$ the magnetic induction, $M$ the Clausius–Mosotti constant for the molecular species that makes up the metered fluid, $F_{\dot{m}}$ the mass flow rate, $r$ the interior radius of transducer tube, $\beta$ the semiangle subtended by the curvilinear detection electrodes, and $E(\beta, 1)$ is the tabulated elliptical integral.

Equation (8.46) shows that the output signal is proportional to the mass flow rate provided the Clausius–Mosotti constant is indeed constant for the fluid being measured and that the flow tube maintains structural integrity.

For cryogenic service, this mass flowmeter offers the following advantages:

1. no moving parts in the flow stream;
2. possible two-phase fluid capability;
3. direct mass flow measurement;

4. good time response;
5. linear calibration.

Disadvantages include:

1. zero point drift with temperature;
2. difficulty in maintaining a hydrogen leak-proof flow tube;
3. complex electronics;
4. still in development stage;
5. flow velocity profile must be symmetric with respect to the axis of the pipe.

### 5.8. Acoustic Flowmeter

Several attempts have been made to use the Doppler effect to measure cryogenic flow rates (see Fig. 8.29). So far these have not been entirely successful, due to extraneous acoustic signals and the necessity of achieving stable low-temperature transmitters and receivers.

### 5.9. Flowmetering Summary

The following conclusions may be drawn for cryogenic flow measurement. They were taken from the study by Mann (1974).

The general class of meters designated as positive displacement types are restricted in size and flow capacity merely by the bulk and inertia of the primary measurement elements. Precision is good (±0.3–0.6%), and units are generally rugged and dependable in service. Some work could be done on vapor elimination during cooldown for operation with cryogenic fluids.

According to Mann (1974), head flowmeters have been used extensively for gas measurement but have not shown great promise for cryogenic service. Precision of reported cryogenic meters is relatively poor (±1–3%) compared with other methods and even with ambient temperature gas service.

Turbine meters have the widest application from very small to intermediate flows of about 0.316 $m^3/s$ (5000 gal/min). There seems to be no practical limit in size. Expected precision is good—from ±0.5 to ±1.00% ($3\sigma$). Cooldown is a problem in

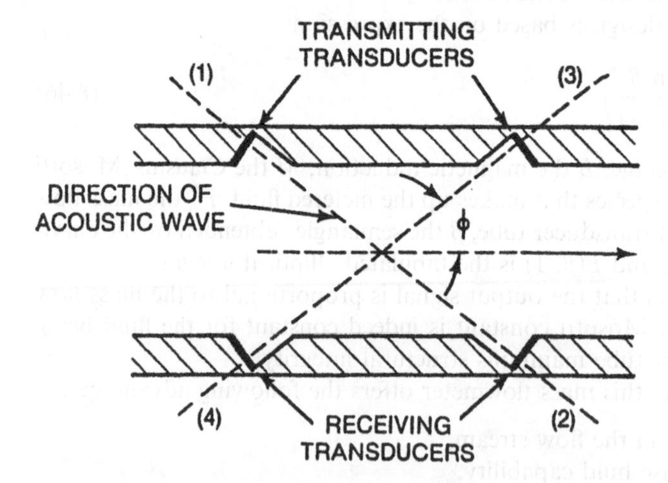


**Figure 8.29** Acoustic flowmeter.

that great care must be taken not to overspeed the turbine with gas flow. This can be accomplished by bypassing the meter during cooldown. Rangeability of individual meters is typically 10:1.

Momentum meters are a rather new entry into cryogenic flow measurement. The main advantage is a direct reading of mass flow. Precision ($3\sigma$) is good and may even be improved relative to volumetric meters when the density measurement uncertainty is added to the volumetric flow to give inferred mass flow.

Vortex shedding metering is even more recent than the above-cited momentum metering. It has the advantage of requiring no moving parts at the low temperatures and therefore requires minimum maintenance. The meter factor (pulses/gallon) depends only on the inside diameter of the pipe and the width across the bluff body face (Rodley, 1969). Cryogenic evaluation is limited at this time, but performance on liquid nitrogen and oxygen shows some deviations from ambient service. The linear flow range is decreased from 10:1 (specified by the manufacturer) to about 5:1 with liquid nitrogen (Mann, 1974).

There is no shortage of physical principles upon which to base a cryogenic flowmeter. Depending on the need, they range from simple pressure drop meters to sophisticated mass reaction meters. The need is thus not for more types of meters but rather stems from the problem common to every type. Namely, flow is a *derived*, not an intrinsic, quantity. As such, every flowmeter will require traceability to reference system to establish its credibility.

## 6. LIQUID LEVEL

Liquid level is but one link in a chain of measurements necessary to establish the contents of a container. Other links may include volume as a function of depth, density as a function of physical storage conditions, and sometimes discerning useful contents from total contents (e.g., pumpable liquid vs. liquid and dense gas). Fortunately, the actual discernment of the liquid/vapor interface (liquid level) is often the strongest link in this measurement process due to the significant progress made in the course of the space program. Liquid-level determinations are essential to propellant loading, management, and utilization and to other diagnostic or control functions such as engine cutoff. As a result of this intense interest, cryogenic liquid-level measurements can often be made with an accuracy and precision comparable to that of thermometry, for instance, and often with greater simplicity.

This period of development has led to many different physical embodiments of liquid-level sensors, and there are as many ways of classifying these devices as there are authors who write about them. One classification is by the principle of operation. This classification groups sensors according to the physics of the device, usually an impedance measurement (or mismatch) in some region of the electromagnetic spectrum. By this scheme, liquid-level sensors are grouped as acoustic, optical, thermal, magnetostrictive, etc. This classification is helpful if one is developing new instruments and needs to search the electromagnetic spectrum for a new type.

Another classification that includes the above is useful if one is interested in the outcome rather than the input. Here, sensors are grouped by whether the output is discrete or digital (point sensors) or a continuous analog of the measurand (continuous sensors). We shall use this classification, understanding from the outset that the principle of some point sensors can be stretched to make continuous sensors (e.g., capacitance measurements) and that no one type is intrinsically better than another for all cases. Special requirements of the particular application and engineering trade-offs will govern the choice.

## 6.1. Point Liquid-Level Sensors

All liquid-level sensors require a sharp property gradient at the liquid/vapor interface, for example, a sharp change in density. This may not always be the case if the fluid is stored near its critical point, for instance, or if stratification has built up after prolonged storage. Be that as it may, point level sensors are tuned to detect a sharp property difference and give an on–off signal. Several of the more commonly used types are described below.

### *6.1.1. Thermal*

The thermal or hot-wire type of point level sensor detects the large difference in heat transfer coefficient between liquid and vapor. Since the heat transfer coefficient is much greater in the liquid than it is in the vapor, one can expect, for a given power input, that the transducer will have a different temperature and hence different resistance in the liquid than in the vapor. It is this change in electrical resistance that is actually sensed in a bridge, and hence these devices can be made both simple and fast. Protection in the form of a stillwell is usually provided to prevent splashing or prevent drops falling on the hot wire from causing a false indication (see Fig. 8.30).

### *6.1.2. Capacitance*

The capacitive type of liquid-level sensor depends on the difference in dielectric constant between the liquid and vapor, which is essentially a function of the fact that the liquid is more dense than the vapor. This sensor is usually made in the form of a bull's-eye, with alternate rings forming the plates of the capacitor, and is installed

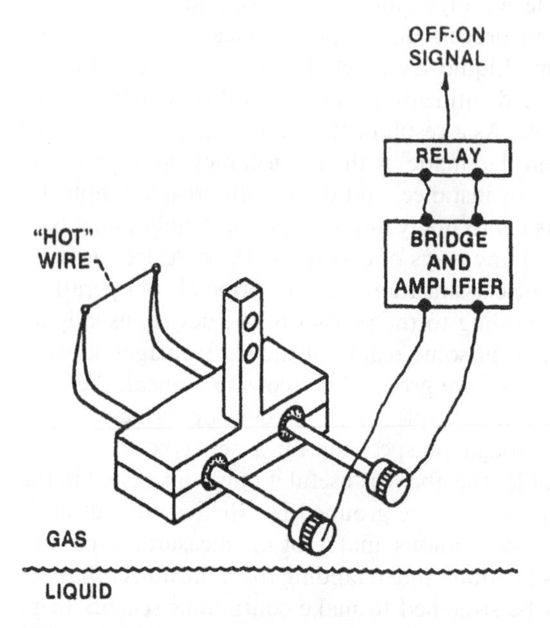


**Figure 8.30** Schematic of a point liquid-level sensor—thermal.

so that the plane of the rings is parallel to the liquid/vapor interface (see Fig. 8.31). Detection is accomplished much as with the hot-wire sensor, above, but in a capacitance bridge rather than a Wheatstone bridge.

### 6.1.3. *Optical*

The optical type of point liquid-level sensor senses the change in refractive index between the liquid and vapor, which, of course, is closely related to the dielectric constant and density. This type of transducer contains a light source and a light-sensitive cell that are isolated from one another but do communicate down a prism (see Fig. 8.32). The prism is cut in such a way as to have total internal reflection when in gas and yet let the light escape when in liquid. This is possible because there is a difference in the index of refraction of the liquid and the vapor and the critical angle for total internal reflection depends on the index of refraction. This type of transducer has a very high in–out signal ratio, for the light detector is almost totally illuminated or not.

### 6.1.4. *Acoustic*

There are also several acoustic or ultrasonic devices that depend on the fact that the daming of the vibrating member is greater in the liquid than in the vapor. Often, a magnetostrictive element is driven as a constant power and is more or less damped depending on whether it is in liquid or in vapor. The end of these sensors is about the size of a quarter, but this does not in any way indicate the limit of its resolution (see Fig. 8.33). These devices can be tuned to detect only a fraction of the total damping that would occur upon total immersion so that their sensitivity can be much less than that of the physical dimensions of the device. Peizoelectric versions of this acoustic type of liquid-level sensor are also available.

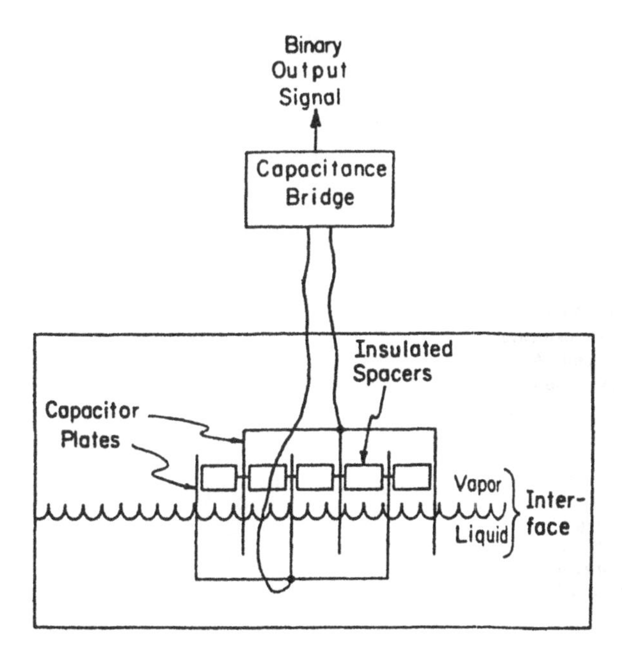


**Figure 8.31** Schematic of a point liquid-level sensor—capacitance.

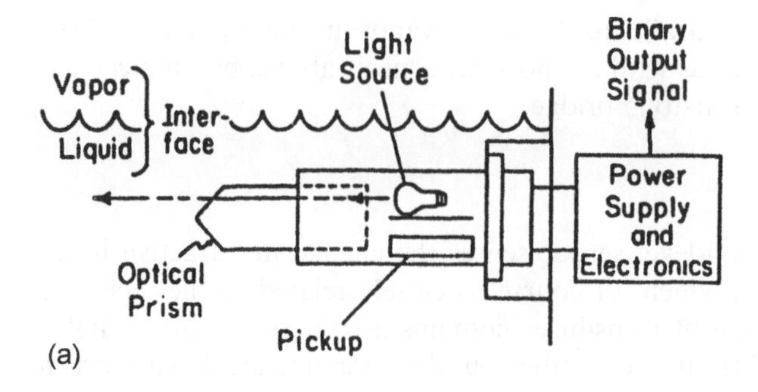


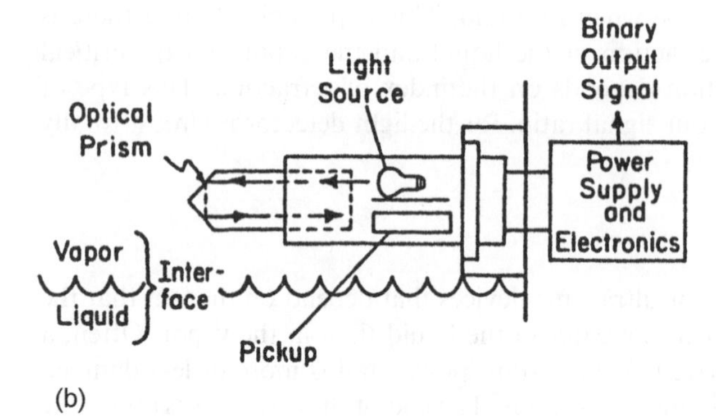


**Figure 8.32** Schematic of a point liquid-level sensor—optical.

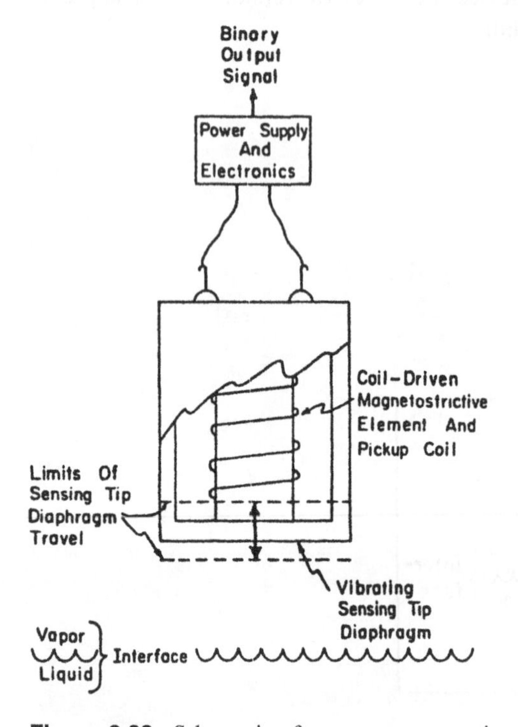


**Figure 8.33** Schematic of a magneto-acoustic sensor.

*6.1.5. Vibrating Paddle*

In the sensor shown in Fig. 8.34, a paddle is driven at mechanical frequencies, and a measure of the damping on the paddle indicates whether it is in liquid or vapor. This type of sensor has been used successfully for a number of years in liquid oxygen and nitrogen. It is not as successful in liquid hydrogen, however, because hydrogen is so much less dense and less viscous than the other fluids.

## 6.2. Continuous Liquid-Level Sensors

Figure 8.35 illustrates several types of continuous liquid-level sensors. Shown schematically are (a) a direct weighing scheme, (b) differential pressure, (c) capacitance, (d) acoustic, and (e) nuclear radiation attenuation.

Direct weighing schemes have indeed been used for hydrogen but are handicapped by the fact that the weight of the container is often large compared to the weight of the contents (see Fig. 8.35a).

Differential pressure measurements are simple to visualize but difficult to realize for liquid hydrogen. The signal is low due to the very low density of hydrogen. The noise can be high due to boiling and thermal oscillations at the lower pressure tap (see Fig. 8.35b).

Capacitance sensors (Fig. 8.35c) are widely used for the continuous measurement of the level of liquid hydrogen. They do not truly follow the interface, however, but rather are total content sensors. That is, the dielectric constant of the gas phase contributes to the output signal as well as that of the liquid phase. This can be a serious source of error in hydrogen as the density of the cold gas (and hence its dielectric constant) is significant compared to that of the liquid. This problem is especially

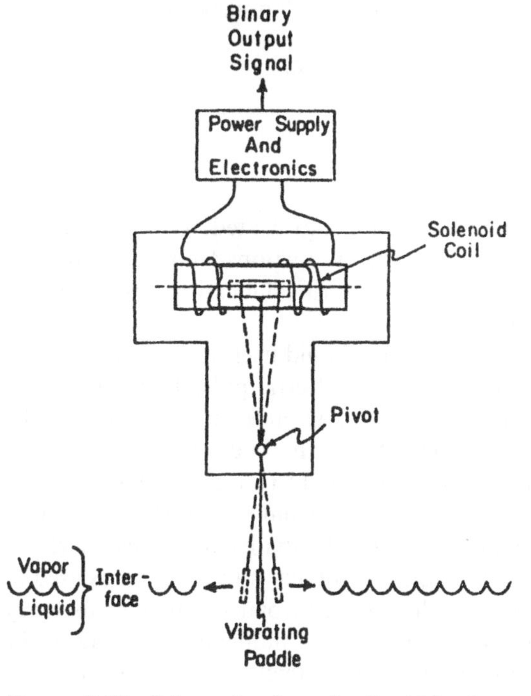


**Figure 8.34** Schematic of a point liquid-level sensor—vibrating paddle.

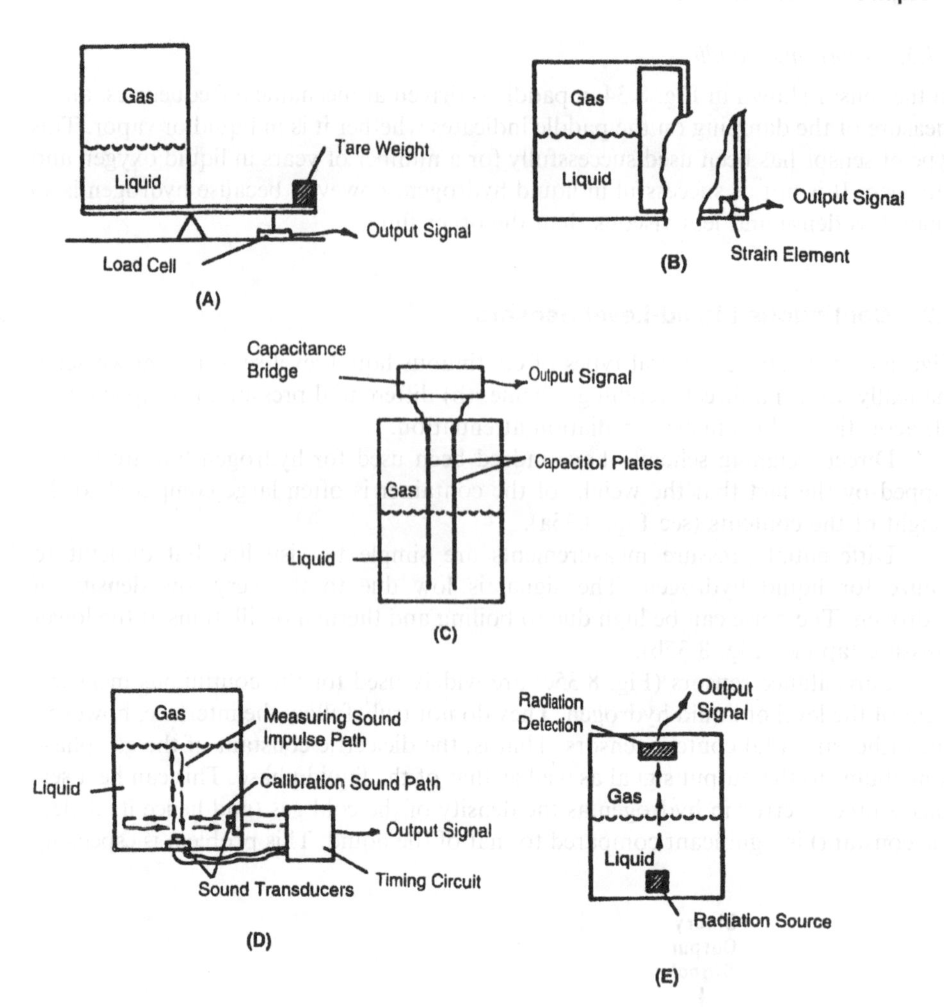


**Figure 8.35** Continuous liquid-level sensors.

troublesome in nearly empty tanks but is often compensated for by either (1) segmenting the capacitor and thus cutting out "dead" sections from the output as the level drops or (2) installing a few point sensors at critical locations and routinely calibrating the continuous sensor against them.

In the acoustic device shown in Fig. 8.35D, the liquid itself is used as the transmitting medium. In this case, a transmitter feeds an electric pulse to a transducer, where it is converted to an acoustic pulse traveling at sonic velocity to the liquid/vapor interface. There it is reflected back at the same speed to the transducer, where it is reconverted to an electric pulse and finally sent to a receiver. Knowing the velocity of sound in a given liquid, the pulse transit time from transmitter to interface to receiver becomes an indication of liquid level. To eliminate extraneous interfaces that may be caused by vapor bubbles in the liquid, a tubular stillwell often is used to isolate the measured fluid column from such physical disturbances. Bubbles also may be suppressed by slight pressurization of the vessel—e.g., momentary closing of the vent valve—immediately prior to any measurement. Sometimes an acoustic racetrack of known path length and variable time of flight is included to provide a measure of the density of the fluid. The product of the two measurements tends to give an indication of the mass of the tank.

Nuclear radiation attenuation (NRA) (Fig. 8.35E) has found some success in oxygen and nitrogen systems but comparatively little use for hydrogen. This is so because hydrogen is nearly transparent to the commonly available sources of nuclear radiation, while the tank walls are relatively "thick" to the same radiation. Hence the fundamental signal-to-noise ratio is inherently low. Hydrogen does absorb beta radiation rather well, but then windows of beryllium or equivalent light metals must be placed in the tank walls. On the whole, this technique seems too sophisticated for simple engineering measurements.

The radio-frequency cavity system consists of inserting a simple dipole antenna into the tank. In this system, microwave energy is introduced into a tank so as to illuminate it by setting up electromagnetic fields throughout the entire volume of the tank. The tank interior is a dielectric region completely surrounded by conducting walls. Such a system is called a cavity, and the resonant solutions are the normal theoretical modes of the cavity. Considerable development work has been done very recently on both uniform-density and nonuniform-density fluids, and good results have been obtained.

## 6.3. Performance Data

Figure 8.36 gives typical repeatability data for various types of liquid-level sensors. It is possible to determine the liquid level within milli-inches, and, as expected, the

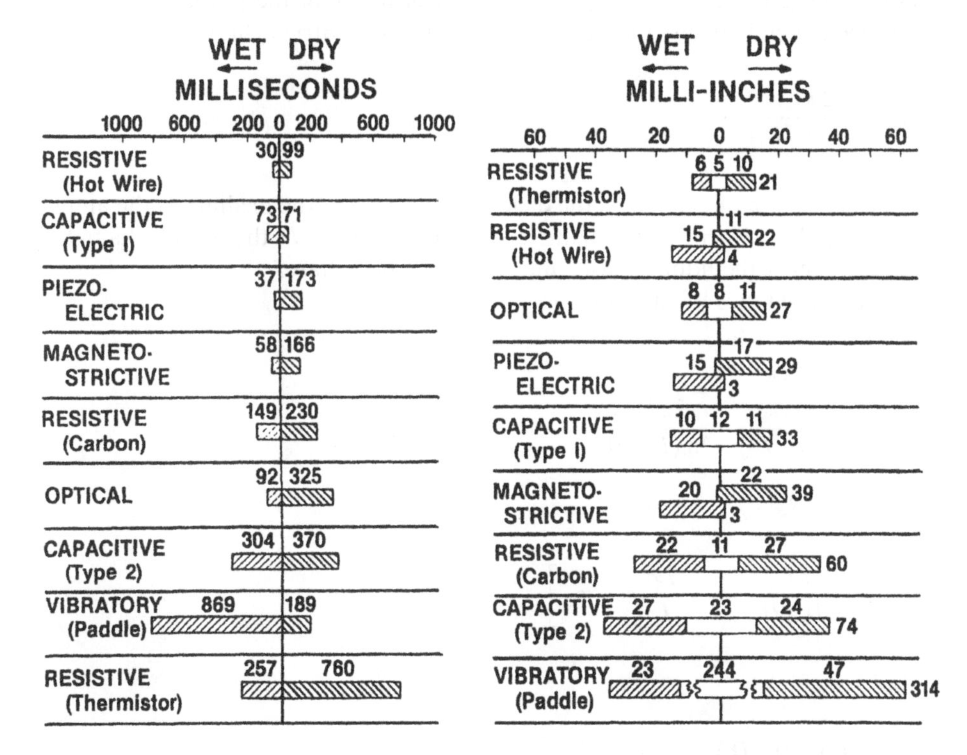


**Figure 8.36** Repeatability indication data for tests in liquid hydrogen by types of level sensors.

smaller devices usually tend to show the smallest precision. Also shown are typical time response data obtained from these various sensors. It is possible to obtain time response from milliseconds to a tenth of a second or so. Also, again, the smaller the device, the smaller the time response.

## 6.4. Liquid-Level Transducers—Summary

The data of Fig. 8.36 are merely indicative of the kind of performance one can expect from liquid-level transducers. The real performance depends on the skill of the cryogenic engineer in applying the instruments properly. He must exercise judgment in placing them in the proper locations and avoiding sloshing, avoiding vortexing, and avoiding having remaining liquid dropping down on them, which would cause erroneous readings. It should also be noted that it is not possible to make a comparative evaluation or to rank these devices. It is impossible to say that one is better than another, for this depends on the use to which they are put.

For a comprehensive review and exhaustive bibliography on this topic, see Roder (1974).

*Example 8.5.* A capacitance liquid-level gauge is used to determine the liquid level in a liquid nitrogen tank. The probe is made of concentric cylinders having the following characteristics: length 1.2 m (4 ft), inside tube diameter 7.6 cm (3 in.), and outside tube diameter 10.2 cm (4 in.). The pressure in the liquid nitrogen vessel is 101 kPa (1 atm). The measured capacitance is 310 pF. At these conditions, the dielectric constant of the liquid and vapor nitrogen are $\varepsilon_f = 1.439$ and $\varepsilon_g = 1.0020$, respectively. Determine the liquid level and the sensitivity of the gauge.

Neglecting end effects, the capacitance of the concentric cylinder capacitor is

$$C = \frac{2\pi L \varepsilon}{\ln(D_o/D_i)}$$

where $L$ is the length of cylinder, $\varepsilon$ the dielectric constant of the medium in the annulus of the cylinders, $D_o$ the inside diameter of outer cylinder, $D_i$ the outside diameter of inner cylinder, $f$ the liquid phase, and $g$ is the vapor phase.

Total capacitance of the gauge is

$$C = C_f + C_g$$
$$= \frac{2\pi L_f \varepsilon_f}{\ln(D_o/D_i)} + \frac{2\pi L_g \varepsilon_g}{\ln(D_o/D_i)}$$

Since $L = L_f + L_g$,

$$C = \frac{2\pi L}{\ln(D_o/D_i)}\left[\varepsilon_g + \frac{L_f}{L}(\varepsilon_f - \varepsilon_g)\right]$$

or

$$L_f = \frac{C \ln(D_o/D_i)}{2\pi(\varepsilon_f - \varepsilon_g)} - \frac{\varepsilon_g L}{\varepsilon_f - \varepsilon_g}$$

Using the data given,

$$L_f = \frac{310 \ln(4/3)}{2\pi(1.4319 - 1.0020)(0.3048)(10^3/36\pi)} - \frac{(1.0020)(4)}{1.4319 - 1.0020}$$
$$= 2.93 \text{ ft or } 0.892 \text{ m}$$

*Note*: Conversion factors used in the above equation are 0.3048 m/ft.
The factor $10^3/36\pi$ (pF/m) is the permittivity of free space, $\varepsilon_o$.
To find the sensitivity, differentiate the equation for $C$ vs. $L_f$,

$$\frac{dc}{dL_f} = \frac{2\pi(\varepsilon_f - \varepsilon_g)}{\ln(D_o/D_i)}$$
$$= \frac{2\pi(1.4319 - 1.0020)}{\ln(413)} = 9.39 \text{ pF/ft}$$

## 7. DENSITY

Density measurements are closely akin to liquid-level measurements because (1) both are often required simultaneously to establish the mass contents of a tank and (2) the same physical principle may often be used for either measurement, since, as noted, liquid level detectors sense the sharp density gradient at the liquid/vapor interface. Thus, the methods of density determination include the following techniques: direct weighing, buoyancy, differential pressure, capacitance, optical, acoustic, ultrasonic, momentum, rotating paddle, transverse momentum, nuclear radiation attenuation, and nuclear magnetic resonance. Each of the principles involved will be discussed along with its relative merits and shortcomings.

### 7.1. Direct Weighing

The direct weighing method of density determination measures the volume and weight of the mixture to give density. Although direct weighing is not considered a field type of procedure, it can serve as a primary calibration standard. Some of the advantages of this method are the simplicity of the equipment, repeatability, good frequency response, and the lack of dependence of pressure, temperature, and inhomogeneity effects of the fluid on the measurement. Disadvantages of direct weighing include relatively poor sensitivity, windage, and the fact that the entire container system must be weighted or nulled out. Tare weights are troublesome in hydrogen and helium systems.

### 7.2. Buoyant Force

Density of a fluid may be determined from the buoyant force of the fluid on a submerged plumb bob. This method is ideally suited for static laboratory use and involves relatively simple equipment. The drawbacks of the buoyancy method are slow response, poor sensitivity, and the need to make a remote reading of a mechanical displacement.

### 7.3. Differential Pressure

The differential pressure method measures the pressure of a vertical column of the fluid, which, along with the height of the column, gives the density. Advantages of this method are the relative simplicity of the equipment required, small component size, and the possibility of application as a field measurement.

This method has several disadvantages. It is dependent on two separate measurements: pressure differential and fluid liquid level. Errors in the accuracy of either of these will affect the overall accuracy of the method. Because of the extremely low density of liquid hydrogen, for instance, the accuracy, sensitivity, and hysteresis of the differential pressure measurement will be adversely affected.

### 7.4. Capacitance

The determination of density through capacitance measurements depends on the fact that the dielectric constant and density are related for simple fluids, such as hydrogen, by the well-known Clausius–Mossotti function,

$$P = \frac{\varepsilon - 1}{\varepsilon + 2}\left(\frac{1}{\rho}\right)$$

where $P$ is the macroscopic polarizability in units of induced dipole moment per unit mass per unit applied field, and $\varepsilon$ and $\rho$ are the dielectric constant and density in cgs units. Thus, a measure of the dielectric constant (capacitance measurement) is a direct measurement of the density. Disadvantages include the well-known stray signals and erroneous temperature effects of capacitance measurements.

### 7.5. Optical

The optical method uses the relationship between spectral absorption and fluid density. The density of hydrogen can be determined by measuring the amount of spectral absorption, caused by the fluid, of an incident light source,

$$\rho = \frac{1}{ul}\ln\frac{I_o}{I}$$

where $\rho$ is the density of fluid, $u$ the absorption coefficient, $l$ the light path length, $I$ the measured intensity after passing through length $l$, and $I_o$ is the incident intensity.

Using the infrared spectrum, this method can determine the density of liquid hydrogen, because a number of spectral absorption lines exist in the wavenumber range of 4100–5100 for these fluids.

Such a method should have good frequency response and adequate sensitivity and not be restricted to homogeneous fluids. However, the problem of providing optical windows may be overwhelming.

### 7.6. Acoustic

The acoustic method uses the fact that the velocity of propagation of a sound wave in a fluid is directly related to the fluid density:

$$c = \sqrt{B/\rho}$$

where $c$ is the velocity of propagation of a sound wave, $B$ the bulk modulus, a constant for a particular fluid, $\rho$ is the density of the fluid.

Solving the above equation for the fluid density yields

$$\rho = B/c^2 = Bt^2/l^2$$

where $l$ is the distance between the sound wave transmitter and, the sound wave receiver and, $t$ is the time required for the sound wave to travel distance $l$.

Thus, fluid density is measured by measuring the time required for the sound wave to travel from the sound wave transmitter, through the fluid, to the sound wave receiver.

This method has been used with some success for both static and dynamic fluid density measurements. Its advantages include a good frequency response and the fact that there are no moving parts located in the fluid. It also has several disadvantages, including a low signal-to-noise ratio, the fact that fluid turbulence and fluid nonhomogeneity will cause spurious echoes, and a large bulk size. Also, temperature effects that influence the fluid vessel geometry will adversely affect performance.

## 7.7. Ultrasonic

The ultrasonic method measures the impedance of a vibrating crystal in a fluid, which can be related to the fluid density. For instance, a piezoelectric crystal, in torsional vibration, immersed in a fluid has its electrical impedance related to the fluid density by

$$\rho = \frac{\pi}{f_0}\left(\frac{m}{s}\right)^2\left(\frac{(B_1R - B_2R_0)^2}{\mu}\right) \tag{8.47}$$

where $\rho$ is the density of the fluid, $f_0$ the resonant frequency, $m/s$ the mass-to-surface ratio, $R$ the impedance of the crystal at its natural frequency, $R_0$ the impedance of the crystal at its natural frequency in a vacuum, and $\mu$ the viscosity of fluid.

The terms $B_1$ and $B_2$ may be considered constant over a small fluid viscosity range. For density measurements where the fluid viscosity range is *appreciable*, a more general equation is applicable:

$$\rho = \frac{\pi}{f_0}\left(\frac{m}{s}\right)^2\frac{(\Delta f_{\text{total}} - \Delta f_0)^2}{\mu}$$

where $\Delta f_{\text{total}}$ is the bandwidth at the half-power point and $\Delta f_0$ is the bandwidth at the half-power point in a vacuum.

For most density-measuring applications the fluid viscosity range is not very great, so Eq. (8.47) is applicable. In this case, sufficient accuracy is generally achieved solely by measuring the impedance $R$ and neglecting the term $R_0$. With this, density is related to impedance by

$$\rho \,\alpha\, R^2$$

This ultrasonic method of density measurement has advantage that include excellent frequency response and sensitivity. Disadvantages include the delicate nature of the apparatus, the requirement of a quiescent fluid medium, and the viscosity dependence.

## 7.8. Momentum Methods

Density can be measured by several momentum methods. They density of the fluid is obtained by measuring fluid flow rate and fluid flow momentum. For instance, the rotating paddle method is based on the principle that an aerodynamic foil rotating through a fluid will experience a measurable drag that can be related to density. The rotating paddle serves a twofold purpose: density measurement and stirring of the mixture. The equipment required is relatively simple. Disadvantages include low frequency response and sensitivity and the presence of moving parts in the fluid.

The density of a fluid can also be measured by the angular momentum method, which measures the angular momentum of a rotating fluid and relates this momentum to the fluid's density. Like the rotating paddle method, this method measures the density and stirs the mixture. The equipment required could be relatively simple. Disadvantages of this method include the presence of bulky, moving parts in the fluid and the lack of any known investigation into the method's applicability to static fluid density measurement.

One particular liquid hydrogen momentum densitometer depends on the fact that the mass of any vibrating system is a primary factor in determining the dynamic characteristics of the system. If the system is designed so that the fluid flowing through it measurably affects the vibrating mass, a means of measuring fluid densities will have been provided.

The dynamometer, sensing the force exerted by the fluid passage, generates a signal ($E_{ac}$) in phase with the motion and proportional to the maximum force:

$$E_{ac} \sim F_{max} \sin \omega_f t$$

If $E_{ac}$ is rectified, the resulting signal, $E_{dc}$, is proportional to the force:

$$E_{dc} \sim F_{max}$$

The final relation for the fluid density takes the form

$$\rho = a + b' E_{dc}$$

where $a$ and $b'$ are experimental constants. This method can thus be used to determine fluid density with the fluid in either a static or dynamic condition.

Advantages of this method are in its frequency response repeatability, and linearity. It also has the advantages of being unaffected by fluid pressure effects, independent of fluid homogeneity, and compatible with many materials.

This type of densitometer has been evaluated in oxygen and nitrogen systems. Application to liquid hydrogen will require higher sensitivities than are presently available.

## 7.9. Nuclear Radiation Attenuation

The nuclear radiation method makes use of the adsorption of radiation energy by the fluid medium interposed between a radiation source and a radiation detector. Typically, for cryogenic liquids a beta or low-energy gamma radiation source is focused so as to impinge on a radiation detector placed on the opposite side of the fluid container. The fluid may be in either a static or a dynamic condition. The adsorption of a collimated radiation beam from the radiation source in the matter interposed between the radiation source and the radiation detector is given by

$$I = I_0 \exp(-VT)$$

where $I$ is the intensity at receiver, $I_0$ the intensity of source, $V$ the linear adsorption coefficient, and $T$ is the thickness of the adsorber. $V$, the linear adsorption coefficient, is related to the density of the matter interposed between the source and detector by the expression

$$V/\rho = \sigma$$

where $\rho$ is the apparent or combined density of the matter interposed between the source and detector. In this case, $\rho$ is the sum of the density of the fluid container, $\rho_c$, and that of the fluid, $\rho_f$. $\sigma$ is the mass adsorption coefficient, which for a particular fluid and container is a constant.

The fundamental adsorption equation may now be written in the form

$$I = I_0' \exp(-k\rho_f)$$

where

$$I_0' = I_0 \exp(-\rho_c \tau_c \sigma_c)$$

and

$$k = \sigma_f T_f$$

For a fluid container having a fixed geometry, the intensity of radiation received by the detecter is related to the density of the fluid inside the container by

$$\frac{dI}{I} = -k\rho_f$$

Thus, a linear change in fluid density causes a linear change in output signal.

This method has a distinct advantage in the fact that no moving parts are inside the fluid container. This method can be highly repeatable and have little hysteresis error.

Primary disadvantages, besides those inherent to any radiation measurement, include the need for windows transparent to the radiation (or long counting times) and the need for a fairly long adsorption path.

## 7.10. Nuclear Magnetic Resonance (NMR) Method

The NMR method utilizes the relationship of excess nuclei, when the fluid is in a magnetic saturated condition, to the fluid density. Assuming that the fluid storage vessel contains liquid hydrogen and the application of a steady-state magnetic field, the hydrogen protons may take two states, aligned parallel or antiparallel to the magnetic fields. The number of nuclei in excess of those that remain parallel (lower energy state) is referred to as the *excess number of spins*, $N_o$. At thermal equilibrium and in a steady-state magnetic field, the excess number of spins is given by the expression

$$N_o = \frac{NuH_o}{kT} \tag{8.48}$$

where $N$ is the total number of nuclei per $cm^3$, $u$ the magnetic moment, $H_o$ the steady-state magnetic field, $k$ the Boltzmann's constant, and $T$ is the temperature of fluid.

When an rf field is impressed on the above system, the excess nuclei will be reduced to a final steady-state value, $N_s$,

$$N_s = N_o\left(\frac{1}{1+2pT_1}\right) \tag{8.49}$$

Substituting in Eq. (8.48),

$$N_s = \frac{N}{T}\left[\frac{uH_o}{k}\left(\frac{1}{1+2pT_1}\right)\right] \tag{8.50}$$

where $T_1$ is the spin lattice relaxation time and $p$ the probability per unit time of a transition by the nuclei under the influence of the rf field.

From Eq. (8.50), the density of the fluid is proportional to the excess nuclei when the fluid is in the saturated condition and directly proportional to the fluid absolute temperature, or

$$p \,\alpha\, N_s T$$

With this method, a measure of the net transfer of energy supplied by the rf field to the nuclei is a measure of the fluid density.

The NMR method is known to exhibit good frequency response, sensitivity, and repeatability. Shortcomings of the system include the need to know fluid temperature and ortho–para hydrogen concentrations and the bulky and complex equipment required.

## 7.11. Summary of Density Determination

Several physical principles of hydrogen density measurement have been brought well forward by the space program. Notable are the capacitance, vibration, and nuclear radiation attenuation schemes. For a comprehensive review and exhaustive bibliography on this topic, see Roder (1974).

*Example 8.6.* It is desired to find the quality of flowing liquid nitrogen composed of saturated liquid and vapor nitrogen at 207 kPa (30 psia) using a concentric tube capacitor. The characteristics of this quality (density) gauge are

Length $L = 91$ cm (36 in.)
Inner diameter $D_i = 2.5$ cm (1.0 in.)
Outer diameter $D_o = 5.1$ cm (2.0 in.)
Measure capacitance $= 96$ pF

At the conditions given,

$$\rho_f = 777.7\,\text{kg/m}^3 (48.5\,\text{lb}_m/\text{ft}^3)$$
$$\rho_g = 8.92\,\text{kg/m}^3 (0.557\,\text{lb}_m/\text{ft}^3)$$
$$\varepsilon_f = 1.413$$
$$\varepsilon_g = 1.002$$

Neglecting end effects, the capacitance of a concentric tube capacitor is

$$C = \frac{2\pi L\varepsilon}{\ln(D_o/D_i)}$$

Hence the dielectric constant of the mixture in the annulus of the cylinders is

$$\begin{aligned}\varepsilon &= \frac{C \ln(D_o/D_i)}{2\pi L} \\ &= \frac{(96)\ln(2/1)}{(2\pi)(3)(0.3048)(10^3/36\pi)} = 1.310\end{aligned}$$

where 0.3048 is the m/ft conversion factor and $10^3/36\pi$ (pF/m) is $\varepsilon_0$, the permittivity of free space.

Assuming that the dielectric constant of the two-phase mixture is a linear function of the volume fractions of vapor and liquid,

$$\varepsilon = \frac{V_f}{V_f + V_g}\varepsilon_f + \frac{V_g}{V_f + V_g}\varepsilon_g$$

Then

$$\varepsilon = \varepsilon_f - \frac{V_g}{V_f + V_g}(\varepsilon_f - \varepsilon_g)$$

where $\varepsilon_f$ is the dielectric constant of saturated liquid, $\varepsilon_d$ the dielectric constant of saturated vapor, $V_f$ the volume of liquid in annulus of sensor, and $V_g$ the volume of vapor in annulus of sensor.

The quality, $X$, of the fluid mixture is

$$X = \frac{m_g}{m} = \frac{V_g \rho_g}{(V_f + V_g)\rho}$$

where $m$ is the total mass of fluid, $m_g$ the mass of vapor phase, $\rho_g$ the density of saturated vapor $= 1/V_g$, and $\rho$ is the density of mixture $= \rho_f - \dfrac{V_g(\rho_f - \rho_g)}{(V_f + V_g)}$ From the data given,

$$\frac{V_g}{V_f + V_g} = \frac{\varepsilon_f - \varepsilon}{\varepsilon_f - \varepsilon_g} = \frac{1.413 - 1.310}{1.413 - 1.002} = 0.251$$

and the density of the mixture is

$$\begin{aligned}\rho &= \rho_f - \frac{V_g(\rho_f - \rho_g)}{V_f + V_g} \\ &= 48.5\frac{\text{lb}_\text{m}}{\text{ft}^3} - (0.251)(48.5 - 0.557)\frac{\text{lb}_\text{m}}{\text{ft}^3} \\ &= 36.5\,\text{lb}_\text{m}/\text{ft}^3 = 585\,\text{kg/m}^3\end{aligned}$$

Therefore, the quality is

$$X = \frac{V_g}{V_f + V_g}\left(\frac{\rho_g}{\rho}\right) = (0.251)\left(\frac{0.557}{36.5}\right) = 0.00383$$

# 8. TEMPERATURE

In the field of cryogenics, as in many other areas of science and industry, the accurate measurement of temperature is a very critical matter. The measurement of temperature, however, is not as easy to accomplish as is the measurement of many of the other physical properties of a substance. Unlike properties such as volume or length, temperature cannot be measured directly. It must be measured in terms of another property. Physical properties that have been used so far include (1) pressure of a gas, (2) equilibrium pressure of a liquid with its vapor, (3) electrical resistance, (4) thermoelectric emf, (5) magnetic susceptibility, (6) volume of a liquid, (7) length of a solid, (8) refractive index, and (9) velocity of sound in a gas. In addition, there are thermometers that respond to a temperature-dependent phenomenon rather than to a physical property. Included in this category are the optical pyrometer and the "electrical noise" thermometer.

As pointed out by Scott (1959), the importance of choosing the thermometer or method most applicable to a specific situation is often not obvious. In some unusual circumstances, only one type of thermometer can be used. However, in most cases, this limitation does not exist, and the choice should be carefully considered with respect to (1) accuracy, (2) reproducibility, (3) sensitivity, (4) stability, (5) simplicity, (6) Joule heating effect, (7) heat conduction, (8) heat capacity, (9) convenience, (10) cost, and other characteristics. To meet these criteria, most engineering measurements are made with metallic resistance thermometers, nonmetallic resistance thermometers, or thermocouples. Each of these types is discussed in the sections below. Gas thermometers and vapor pressure thermometers find little practical engineering application and hence are mentioned only briefly in the following paragraphs.

## 8.1. Fluid Thermometry

### *8.1.1. Gas Thermometry*

If the equation of state—i.e., the function relating pressure, volume, and temperature—of a gas is known, the gas can be used to measure temperature. If one of the variables is held constant and a second is measured, the third can be calculated from the equation of state. In thermometry, it has been found most advantageous to hold the volume of a fixed amount of gas constant and measure the pressure. The temperature of the gas is then determined. When properly corrected, such a constant-volume gas temperature very closely approximates an ideal gas thermometer and is used to fix points on the absolute temperature scale.

One of the simplest forms of the constant-volume gas thermometer is the one used by Simon (1952) (see Fig. 8.37a). The bulb, which is at the low temperature to be measured, is connected by a fine capillary to a Bourdon-type pressure gauge maintained at room temperature.

If the gas thermometer contains $n$ moles of gas which, for the moment, we will assume to be an ideal gas, and we further assume that the mass of gas in the manometer and connecting capillary can be neglected, the pressure registered by the thermometer will be given by the ideal gas law,

$$P = \frac{nRT}{V} \tag{8.51}$$

where $n$ is the number of moles of gas, $R$ the universal gas constant, $T$ the Kelvin temperature, and $V$ the volume of the thermometer bulb. If, for bulb temperatures

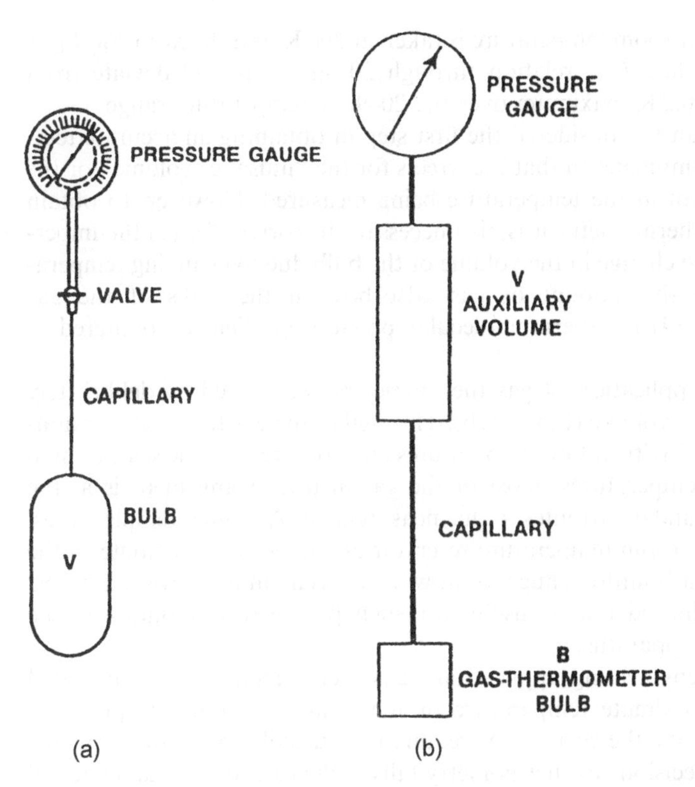


**Figure 8.37** Gas thermometers. (a) Constant-volume; (b) nonlinear.

$T_1$ and $T_2$, the corresponding pressures are determined to be $P_1$ and $P_2$, then

$$\frac{T_2}{T_1} = \frac{P_2}{P_1} \tag{8.52}$$

Thus, the gas thermometer determines ratios of absolute temperatures. If $T_1$ is chosen to be the defined ice point, 273.15 K, Eq. (8.52) gives $T_2$, the temperature being determined, on the Kelvin scale.

Of course, the above analysis is greatly oversimplified. Let us consider the case where the measuring bulb has a volume $V_B$ and the pressure spiral (Bourdon tube) in the measuring gauge has a volume $V_G$, which is considered constant. If the volume of the capillary is assumed to be negligible, and if the gas can be assumed to be ideal over the temperature range to be measured, then the temperature is given by the equation

$$T = \frac{PV_B T_R}{P_R(V_B + V_G) - PV_G} \tag{8.53}$$

where the subscript R denotes room conditions and $P$ is the pressure in consistent units. This simple device can be quite accurate at low temperatures because almost all of the gas will be in the bulb. At higher temperatures, however, a substantial amount of the gas will be in the volume $V_G$, and the thermometer will be rather insensitive to $T$. If the ratio $V_B/V_G$ is made sufficiently large, such a device can have an accuracy of $\pm 0.05$ K at temperatures below 30 K (White, 1959). In the same vein,

Scott (1960) noted that if room temperature is taken at 300 K and the ratio $V_B/V_G$ is equal to 20, a straight-line $P$–$T$ relation through 20 and 77 K will deviate from Eq. (8.53) by less than 0.2 K maximum over the 20–90 K temperature range.

Equation (8.53) can be considered the first step in obtaining an accurate temperature with a gas thermometer in that it corrects for the "nuisance volume" or the volume of gas that is not at the temperature being measured. However, to obtain accurate results in gas thermometry it is also necessary to correct for (1) the imperfection of the gas, (2) the change in the volume of the bulb due to changing temperature, (3) variations in the amount of gas adsorbed on the walls of the gas thermometer bulb, and (4) the thermomolecular pressure gradient encountered at extremely low pressures.

One engineering application of gas thermometers is achieved by deliberately increasing the extraneous volume (Fig. 8.37b). This makes the gas thermometer nonlinear, increasing its sensitivity at low temperatures and compressing the scale at high temperatures. At low temperatures, most of the gas in this arrangement is at the measuring temperature and contributes to the measurement. At higher temperatures, most of the gas is in the room temperature reservoir and does not contribute to the measurement. By using a Bourdon gauge to show the pressure in this device, a simple gas thermometer is achieved that is useful for such purposes as monitoring the cooldown of cryogenic apparatus.

Accordingly, for engineering applications, gas thermometry is recommended for indicating the approximate temperature or temperature trend. For precision temperature measurements, the corrections required are usually too bothersome for the process engineer. Precision gas thermometry falls in the category of fundamental thermometers rather than empirical thermometers. It is usually much too demanding for common use in the field.

*Example 8.7.* A constant-volume gas thermometer is filled with helium such that the standard conditions, $T_s$ and $P_s$ are:

$$T_s = 273.15\,\text{K}\,(491.67^\circ\text{R}) \quad \text{and} \quad P_s = 69\,\text{kPa}\,(10\,\text{psia})$$

Neglecting dead volume and imperfect gas corrections, find the temperature when the indicated pressure is 24 kPa (3.5 psia)

$$\begin{aligned} T &= P\left(\frac{T_s}{P_s}\right) \\ &= (3.5)\left(\frac{491.67}{10}\right) = 172.1^\circ\text{R} = 95.60\,\text{K} \end{aligned}$$

### *8.1.2. Vapor Pressure Thermometry*

The pressure exerted by a saturated vapor in equilibrium with its liquid is a very definite function of temperature and as such can be used to measure the temperature of the liquid. With a good pressure-measuring device, the vapor pressure thermometer is an excellent secondary standard because its temperature response depends upon a physical property of a pure compound or element. Many formulas have been derived that relate the vapor pressure of a liquid to its temperature. These are semiempirical in that the constants must be determined experimentally and change as new data become available. The most common forms of the vapor pressure equation are

$$\log P = A + \frac{B}{T} + CT$$

$$\log P = A + \frac{B}{C + T}$$

$$\log P = \frac{A}{T} + B \log T - CT + D$$

Here $A$, $B$, $C$, and $D$ are experimentally determined constants. Tables have been published, however, giving vapor pressure as a function of temperature, and it is usually easier to consult one of these tables than use the formulas. Fixed points and vapor pressure equations are given in Tables 8.3–8.5.

One of the great advantages of vapor pressure thermometers at low temperatures is their extreme sensitivity in the range over which they can be used. A temperature change of 1°R (1 K) will result in a sizable pressure change (see Table 8.6). This great sensitivity, however, also leads to the principal weakness of the vapor pressure thermometer. It is useful over only a very limited temperature range. These thermometers are most accurate in the area of the normal boiling temperature of the liquid used; hence, several regions are inaccessible to vapor pressure thermometry. For example, the range from 40 to 50 K is above the range of neon and below that of oxygen and nitrogen. Within its limits, however, this type of thermometer can be very accurate.

The apparatus for vapor pressure determinations is relatively simple. There are many variations of vapor pressure thermometers, but one described by Scott (1959), shown in Fig. 8.38, will suffice as a common example. Any type of accurate dial gauge or manometer can be used, and in the proper range great precision is obtained. For example, a mercury manometer that can be read to 0.1 mm will yield measurements that may be accurate to 0.0001 K at liquid helium temperature, whereas a sensitive, differential oil manometer may detect changes in temperature on the order of $10^{-5}$ K (Simon, 1952).

The materials commonly used for vapor pressure thermometry at low temperatures are oxygen, nitrogen, hydrogen, and helium. These characteristics should be noted:

1. For a given material, the sensitivity, $dp/dT$, increases rapidly with temperature.
2. The lower the boiling point of the material, the greater the sensitivity.

**Table 8.3** Fixed Points for Common Vapor Pressure Thermometer Fluids

| Fluid | Triple point (K) | Normal boiling point (K) | Critical point (K) |
|---|---|---|---|
| Oxygen | 54.351 | 90.180 | 154.576 |
| Nitrogen | 63.148 | 77.347 | 126.20 |
| Neon | 25.54 | 27.09 | 44.40 |
| Hydrogen | 13.803 | 20.268 | 32.976 |
| Helium-4 | 2.177[a] | 4.224 | 5.2014 |

[a] Lower lambda point.
*Source*: van Dijk, 1960; Johnson (1960).

**Table 8.4** Constants for Vapor Pressure Equations of Several Fluids

| Coefficient | Oxygen | Nitrogen | Neon | Hydrogen, 99.79% p-$H_2$ | Helium |
|---|---|---|---|---|---|
| $A_1$ | $-62.5967185$ | $0.8394409444 \times 10^4$ | 7.46116 | 1.772454 | $-3.9394635287$ |
| $A_2$ | 2.47450429 | $-0.1890045259 \times 10^4$ | $-106.090$ | $-4.436888 \times 10$ | $1.4127497598 \times 10^2$ |
| $A_3$ | $-4.68973315 \times 10^{-2}$ | $-0.7282229165 \times 10^1$ | $-3.56616 \times 10^{-2}$ | $2.055468 \times 10^{-2}$ | $-1.6407741565 \times 10^3$ |
| $A_4$ | $5.48202337 \times 10^{-4}$ | $0.1022850966 \times 10^{-1}$ | $4.11092 \times 10^{-4}$ | 2.000620 | $1.1974557102 \times 10^4$ |
| $A_5$ | $-4.09349868 \times 10^{-6}$ | $0.5556063825 \times 10^{-3}$ | | $-5.009708 \times 10$ | $-5.5283309818 \times 10^4$ |
| $A_6$ | $1.91471914 \times 10^{-8}$ | $-0.5944544662 \times 10^{-5}$ | | 1.0044 | $1.6621956504 \times 10^5$ |
| $A_7$ | $-5.13113688 \times 10^{-11}$ | $0.2715433932 \times 10^{-7}$ | | $1.748495 \times 10^2$ | $-3.2521282840 \times 10^5$ |
| $A_8$ | $6.02656934 \times 10^{-14}$ | $-0.4879535904 \times 10^{-10}$ | | $1.317 \times 10^{-3}$ | $3.9884322750 \times 10^5$ |
| $A_9$ | | $0.5095360824 \times 10^3$ | | $-5.926 \times 10^{-5}$ | $-2.7771806992 \times 10^5$ |
| $A_{10}$ | | | | $3.913 \times 10^{-6}$ | $8.3395204183 \times 10^4$ |

**Table 8.5** Vapor Pressure Equations for Several Fluids

| Substance | Equation | Units | Scale[a] |
|---|---|---|---|
| Oxygen | $\ln P = A_1 + A_2T + A_3T^2 + A_4T^3 + A_5T^4 + A_6T^5\ A_7T^6 + A_8T^7$ | $P$, atm; $T$, K | NBS-55 |
| | | | NBS-48 |
| Nitrogen | $\ln P = A_1/T + A_2 + A_3T + A_4(T_c - T)^{1.95} + A_5T^3 + A_6T^4 + A_7T^5 + A_8T^6 + A_9$ in $T$ | $P$, atm; $T$, $K$ | IPTS-68 |
| Neon | $\log P = A_1 + A_2/T_{55} + A_3T_{55} + A_4T^2{}_{55}$ | $P$, torr; $T$, K | NBS-55 |
| Hydrogen | (1) $13.803 < T < 21$ K; $\log P = A_1 + A_2/T_{55} + A_3T_{55}$ | $P$, atm; $T$,$K$ | NBS-55 |
| | (2) $20.268 < T < 29$ K; $\log P_a = (A_4 + A_5/T_{55} + A_6) + A_7T_{55}$ | Same | Same |
| | (3) 29 K $< T < 32.967$ K; $P = P_a + A_8(T_{55} - 29)^3 + A_9(T_{55} - 29)^5 + A_{10}(T_{55} - 29)^7$ | Same | Same |
| Helium | $\ln P = \sum_{i=1}^{10} A_iT^{2-i}$ | $P$, μmHg; $T$, $K$ | |

[a] $T_{58} = 1958$ helium vapor pressure–temperature scale.
See Appendix A for conversion among the temperature scales mentioned.

**Table 8.6** Characteristics of Some Vapor Pressure Thermometers

| | Pressure = 100 mmHg (13.3 kPa) | | | Pressure = 760 mmHg (101 kPa) | | | Pressure = 1000 mmHg (133.33 kPa) | | |
|---|---|---|---|---|---|---|---|---|---|
| | | $dP/dT$ | | | $dP/dT$ | | | $dP/dT$ | |
| Material | Temp. K | mmHg K | kPa K | Temp. K | mmHg K | kPa K | Temp. K | mmHg K | kPa K |
| Oxygen | 74.5 | 15.9 | (2.12) | 90.18 | 80.0 | (10.7) | 92.9 | 95.0 | (12.7) |
| Nitrogen | 63.5 | 17.6 | (2.35) | 77.35 | 88.0 | (11.7) | 79.9 | 110.0 | (14.66) |
| Parahydrogen | 15.0 | 50.0 | (6.66) | 20.3 | 225.0 | (29.99) | 21.2 | 277.0 | (36.92) |
| Helium | 2.63 | 180.0 | (23.99) | 4.2 | 720.0 | (95.98) | 4.52 | 867.0 | (115.6) |

*Source:* Scott (1960).

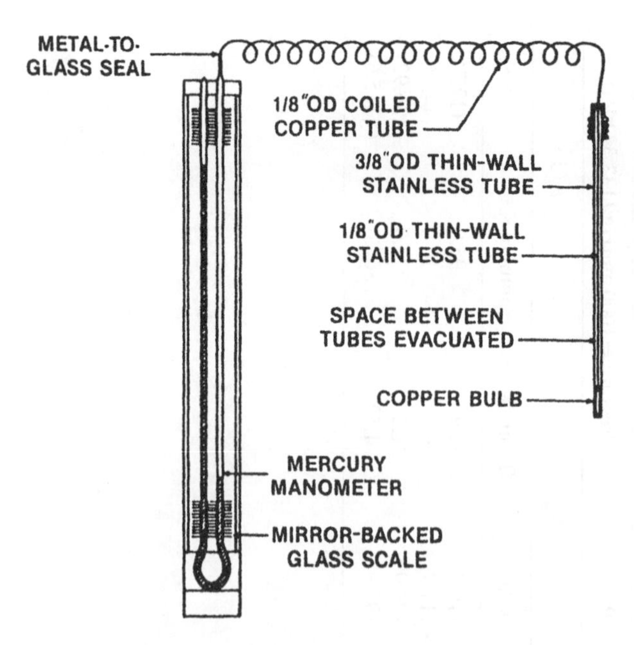


**Figure 8.38** Vapor pressure thermometer.

3. The temperature ranges are rather small. The interval 2–90 K cannot be covered completely by good vapor pressure thermometry because there are some regions in which no substance has a usable $dp/dT$.

Because of the hydrostatic head of the liquid, the temperature at equilibrium tends to increase with depth. However, it should not be assumed that this increase can be calculated according to the factor given in Table 8.6. Thus, the actual gradient may be smaller due to convection or greater due to superheating. For accurate work, such gradients are avoided by using a small vapor pressure bulb located where it is desired to determine the temperature.

It would seem obvious that the tubing between the bulb and the manometer should nowhere be colder than the bulb, or refluxing will occur and the pressure reading will be low. However, it has been argued in the case of helium that it is not possible to transfer heat from the vapor rapidly enough at such a cold spot to produce an appreciable effect and that a vacuum jacket to eliminate such a cold spot may itself cause errors by permitting heat from outside to reach the bulb. Experiments appear to support both sides of this question. However, it is agreed by all that radiation shielding of the bulb is necessary for accurate work.

Sample purity is also important. With hydrogen, the ortho/para ratio is the most important composition parameter. No common impurities are sufficiently soluble in liquid hydrogen or helium to affect the vapor pressure. Of the possible impurities in the vapors of these, only helium in hydrogen is at all likely to be a source of error. With commercial liquid nitrogen and oxygen, each is likely to be the most significant impurity in the other because of their common source in the atmosphere and their complete mutual solubility. Preparation by chemical methods or the use of high-purity cylinder gas is necessary to avoid such contamination.

As noted, these thermometers are most accurate in the area of the normal boiling point of the liquid used and hence can be very useful, albeit over a very limited temperature range.

Hydrogen is especially troublesome because of the dependence of vapor pressure upon ortho/para composition. It is doubtful that hydrogen vapor pressure thermometers will find much use except under carefully controlled conditions where high accuracy over a very narrow span is sought.

Advantages of vapor pressure thermometry are that it is sensitive, can have good time response, is not affected by magnetic fields, and needs no calibration. The primary disadvantage is that it can be used only between the triple point and critical point of the fill liquid.

## 8.2. Metallic Resistance Thermometry

Since the resistivity of pure metals varies with change in temperature, metals have been used as simple and reliable temperature-measuring devices. Many elements or compounds, however, are not suitable for use in low-temperature resistance thermometry because they lack one or more of the desirable properties of an ideal resistance thermometer. These properties include the following:

1. A resistivity that varies linearly with temperature to simplify interpolation.
2. High sensitivity.
3. High stability of resistance so that its calibration is retaiined over long periods of time and is not affected by thermal cycling.
4. Capability of being mechanically worked.

Although a number of metals are more or less suitable for resistance thermometry, platinum has come to occupy a predominant position, partly because of its excellent characteristics, such as chemical inertness and ease of fabrication, and partly because of custom. Certain desirable features such as ready availability in high purity and the existence of a large body of knowledge about its behavior have come into being as its use grew and have tended to perpetuate that use. Its sensitivity down to 20 K and its stability are excellent. Its principal disadvantages are (1) low resistivity, (2) insensitive low about 10 K, and (3) a variation in the form of the resistance–temperature relation from specimen to specimen below about 30 K.

The significance of the first item is that the excellent sensitivity in the liquid hydrogen region cannot be realized unless the resistor has a high ratio of length to cross-sectional area. If it is a wire, it must be long and thin, which results in a bulky and delicate resistor coil and complications in supporting and insulating it. If it is a film, nonreproducible behavior due to differential contraction stresses between the film and substrate is difficult to prevent.

The second disadvantage can be avoided only by using a different metal in which the lattice vibrations persist to lower temperatures, inasmuch as these are responsible for the temperature-dependent part of the resistivity.

According to Matthiessen's rule (White, 1959), the total resistivity of a pure metal is given by

$$R_T = R_r + R_{tv}$$

where $R_T$ is the total resistivity, $R_r$ is the residual resistivity, and $R_{tv}$ is the resistivity due to thermal vibrations. The residual resistivity is caused by static imperfections such as chemical or physical impurities. The resistivity due to thermal vibrations is proportional to the temperature down to temperatures in the vicinity of one-third of the Debye temperature $\theta_D$. Below this, $R_{tv}$ decreases more rapidly with temperature. Between $\theta_D/10$ and $\theta_D/50$, the resistivity due to thermal vibrations is proportional to the temperature to the *n*th power, where *n* is determined experimentally ($3 < n < 5$). The sensitivity is the first derivative of total resistivity with respect to the temperature, i.e., the slope of the $R_T$ vs. $T$ curve. For pure metals at very low temperatures, this slope is quite small, results in poor sensitivity. This is because $R_r > R_{tv}$, and it has been shown theoretically experimentally that Matthiessen's rule is not valid in that case.

It can be seen that a low value of the Debye parameter $\theta_D$ is required. Indium has been proposed and ought to have useful sensitivity almost down to its superconducting transition at 3.4 K. However, attention has been diverted from it by the advent of successful germanium thermometers, and the use of indium has undergone little or no development.

### 8.2.1. Platinum Resistance Thermometers

There are two basic designs of platinum resistance thermometers (PRTs) for engineering use: immersion probes and surface temperature sensors. The immersion probes (Fig. 8.39) feature a high–purity platinum wire encapsulated in ceramic or securely attached to a support frame. Features such as repeatability after thermal shocks, time response in different environments, interchangeability, and mechanical shock tolerance differ between companies and specific designs. The repeatability of the typical immersion sensor is usually certified to be about $\pm 0.1$ K at the ice point after several thermal cyclings to cryogenic temperatures. For most thermometers, this repeatability figure is conservative.

The time response is particularly difficult to assess in a general way because it depends critically on the design and on the method of testing. Flowing water, oil, or a cryogenic liquid are often used as the test medium. The time response of a capsule-type PRT may be stated as 2–7 s, which implies that in this time the sensor had reached the equilibrium temperature of the system (ignoring $I^2R$ heating). In the case of the industrial PRT, dynamic systems are frequently encountered. Convention has been to define the time response of a thermometer to be the time it takes the sensor to reach 63.2% of the temperature of a step function temperature change. With this definition of the time constant, a typical range of values for this type of resistor is 0.1–3 s.

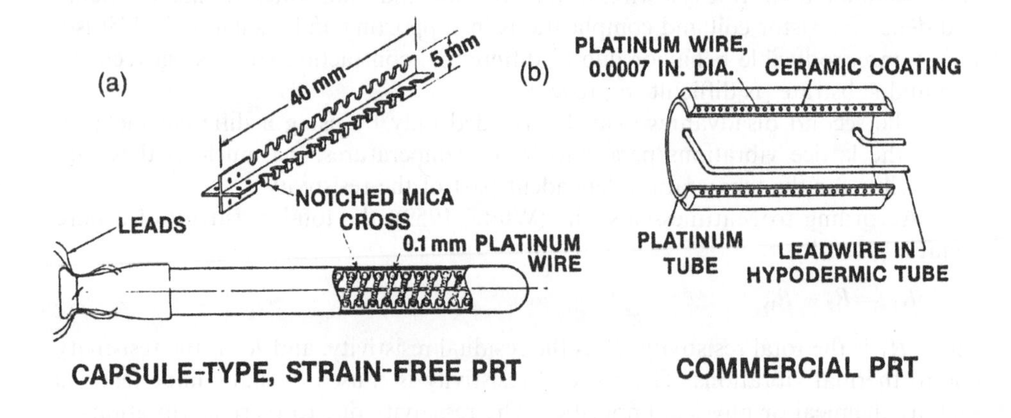


**Figure 8.39** Platinum resistance thermometers. (a) Capsule, strain–free; (b) commercial.

Interchangeability is measured in terms of the errors that result when more than one thermometer is used with a single $R$–$T$ relationship. This becomes a major concern in operations where control resistors must be replaceable without system interruption and where data reduction and calibration expense must be minimized. Interchangeability is ordinarily specified at a given temperature, i.e., the resistance of the thermometers will not vary more than a specified amount at a certain temperature.

Immersion–type sensors may generally be specified to have the same ice point resistance to within an equivalent of about $\pm 1.5$ K; surface sensors show slightly worse interchangeability, $\pm 4$–5 K; at 273 K. Some manufacturers provide different grades of interchangeability for particular models. Even after specifying a particular resistance value at a given temperature, the slope of $R$ vs. $T$, which depends on purity and strain, may vary from one thermometer to another.

Another type of industrial sensor is broadly known as a surface temperature sensor. Its geometry is such that it makes good thermal contact with surfaces of various shapes. Sensors are available for clamping around small tubing, fitting into milled slots, and clamping under bolt heads. The principal advantages of these thermometers are that they are small, typically $0.1 \times 0.5 \times 0.5$ in. with perhaps a factor of two variation in any dimension.

Both immersion and surface sensors are available in versions that have built-in bridge circuits. These bridge circuits allow adjustments to be made on individual sensors to increase their interchangeability. Two-, three-, and four-lead configurations are available in both surface and immersion sensors. Joule heating must be guarded against in all types of resistance thermometers by limitation of the measuring current.

The platinum resistance thermometer is the interpolating instrument for the International Practical Temperature Scale, IPTS-48 and, as revised, IPTS-68. IPTS-48 used the Callendar (1887) equation,

$$R_t = R_o(1 + At + Bt^2) \tag{8.54}$$

as the interpolating function above 0°C and the Callendar–VanDusen(1925) equation,

$$R_t = R_o[1 + At + Bt^2 + C(t - 10)t^3] \tag{8.55}$$

for interpolating between 0°C and 90 K. Below 90 K, IPTS-48 provided no adequate analytical relation.

Since resistance is measured, and the resulting temperature is desired, Eqs. (8.54) and (8.55) may be made explicit in temperature.

Callendar, $0 \le t \le 630.5$°C:

$$t(°C) = (R_t - R_o)/\alpha R_o + 0.01\partial \cdot t(0.01t - 1)$$

and the Callendar–VanDusen, $-182.97 \le t \le 0$°C:

$$t(°C) = \frac{(R_t - R_o)}{\alpha R_o} + 0.01\partial \cdot t(0.01t - 1) + \beta(0.01t)^3(0.01t - 1) \tag{8.56}$$

The constants $R_o$, $\alpha$ and $\partial$ in these equations are determined by resistance measurements at the ice point, steam point, and sulfur point. The oxygen point resistance allows $\beta$ (Eq. 8.56 ) to be determined.

There are many other resistance–temperature interpolating functions for platinum, including the so-called $Z$ function and its allies. In one embodiment, the universal function $Z_T$ is given by

$$Z_T = (R_T - R_1)/(R_2 + R_1)$$

This function is calculated using the existing platinum resistance thermometer (PRT) calibration: $R$ is the calibration resistance at $T$, and $R_1$ and $R_2$ are the calibration resistances at $T_1$ and $T_2$. This approximation applies when $T_1 < T < T_2$. Since $Z_T$ can be determined from an existing calibration table, the resistance $R_{Tx}$ of the uncalibrated PRT at temperature T is given by

$$R_{Tx} = R_{1x} + Z_T(R_{2x} - R_{1x})$$

where $R_{1x}$ and $R_{2x}$ are experimentally determined values of resistance at $T_1$ and $T_2$.

For industrial PRTs, this $Z$ gives temperature differences on the order of

$\pm 100\,\text{mK}$ for $T_1 = 4\,\text{K}$ $T_2 = 20\,\text{K}$

$\pm 20\,\text{mK}$ for $T_1 = 20\,\text{K}$ $T_2 = 77\,\text{K}$

$\pm 50\,\text{mK}$ for$T_1 = 77\,\text{K}$ $T_2 = 273\,\text{K}$

Compared to the $Z$ function, Callendar–VanDusen analytical relations require measurements at many fixed points. In case of the 13.81–90.188 K region as defined by IPTS-68, there are six fixed points; the reference function involves 20 coefficients and four polynomial equations. It is comforting note that the maximum difference between the IPTS-68 and, for example, the NBS-55 scale is only about 0.015 K.

As mentioned, the adoption IPTS-68 eliminated the need for the Callendar–VanDusen equation above 90 K. IPTS-68 provides analytical relationships that allow thermometers to be calibrated between 13.81 and 90.188 K. Since seven fixed points are required, the process is tedious at best Nonetheless, Eqs. (8.54) and (8.55) are often used for engineering purposes to represent the resistance–temperature relation of platinum resistance thermometers.

Table 8.7 lists several fixed points useful for calibrating low-temperature thermometers.

### *8.2.2. Other Metallic Resistance thermometers*

In addition to platinum, other metals can be used in resistance thermometers. Indium has been investigated because its lattice ations are still significant at lower temperatures. Indium thermometers are more sens than platinum at the lower temperatures, by about a factor of 10 in the 4.2–15 K region. If madc carefully enough, they exhibit a reproducibility as high as $\pm 0.01$ K over the entire 4.2–290 K range. Unfortunately, the reproducibility from one thermometer to another (even if made from the same spool of wire) is much worse. Also, due to a superconducting transition at 3.4 K, indium cannot be used below this point.

Several alloys that include magnetic transition metals have been proposed. One of these, platinum–cobalt, shows promise in the 3–30 K range (Fig. 8.40). Another, rhodium–0.5% iron, has been used in the tem ure range 0.4–20 K (Fig. 8.41). Selected sensors of this type are stable with respect to time and thermal cycling to within a few tenths of a millikelvin, which is better than can be obtained with germanium sensors. Also, their voltage sensitivity (volts per kelvin) is equal to or greater than that of platinum sensors below 20 K, so errors due to self-heating are less.

**Table 8.7** Temperature Fixed Points

| Fixed point | ITS-27(°C) | ITS-48(°C) | IPTS-48(°C) | IPTS-68 | | |
|---|---|---|---|---|---|---|
| | | | | °C | K | Uncertainty (K) |
| Equilibrium between solid, liquid, and vapor phases of equilibrium hydrogen (TP $e$-$H_2$) | | | | −259.34 | 13.81 | 0.01 |
| Equilibrium between the liquid and vapor phases of $e$-$H_2$·$P = 33{,}330.6\,N/m^2$ (25/76 standard atmosphere) | | | | −256.108 | 17.042 | 0.01 |
| Equilibrium between liquid and vapor phases of $e$-$H_2$ (NBP $e$-$H_2$) | | | | −252.87 | 20.28 | 0.01 |
| Equilibrium between liquid and vapor phases of neon/(NBP Ne) | | | | −246.048 | 27.102 | 0.01 |
| Equilibrium between solid, liquid and vapor phases of oxygen (TP $O_2$) | | | | −218.789 | 54.361 | 0.01 |
| Equilibrium between liquid and vapor phases of oxygen (BP $O_2$) | −182.97 | −182.970 | −182.97 | −182.962 | 90.188 | 0.01 |
| Equilibrium between solid and liquid phases of water (FP $H_2O$) | 0.000 | 0 | | | | |
| Equilibrium between solid, liquid, and vapor phases of water (TP $H_2O$) | | | 0.01 | 0.01 | 273.16 | Exact |
| Equilibrium between liquid and vapor phases of water (BP $H_2O$) | 100.000 | 100 | 100 | 100 | 373.15 | 0.005 |
| Equilibrium between solid and liquid phases of zinc (FP $Z_n$) | | | | 419.58 | 692.73 | 0.03 |
| Equilibrium between liquid and vapor phases of sulfur (BP S) | 444.60 | 444.600 | 444.6 | | | |
| Equilibrium between solid and liquid phases of silver (FP Ag) | 960.5 | 960.8 | 960.8 | 961.93 | 1235.08 | 0.2 |
| Equilibrium between solid and liquid phases of gold (FP Au) | 1063 | 1063.0 | 1063 | 1064.43 | 1337.58 | 0.2 |

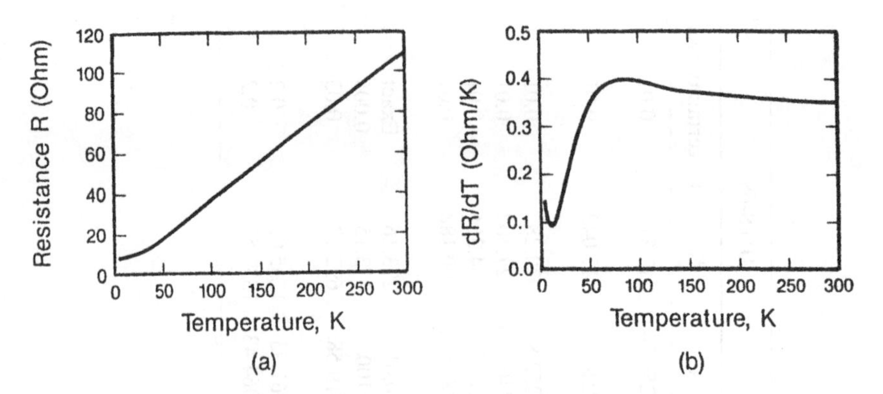


**Figure 8.40** Platinum–cobalt alloy resistance thermometer. (a) The reference *R–T* relation; (b) the first derivative of the reference *R–T* relation. (Courtesy of LakeShore Cryotronics Inc., Westerville, OH.)

Rhodium–iron metallic resistance thermometers are available commercially in configurations similar to the PRTs. These thermometers are produced in several sizes with overall lengths of 25–60 mm and diameters of 3.2–9 mm. Nominal ice temperature resistances are from 20 to 100 ohms. Although not as temperature-sensitive as semiconducting sensors ($T \leq 20$ K) or PRTs ($T \geq 20$ K), these metallic sensors do have certain advantages. Very important, the simple behavior of the resistivity allows for easier calibration and analytical representation compared to the semiconductors. The usable sensitivity over the entire temperature range ($T \leq 300$ K) offers a one-

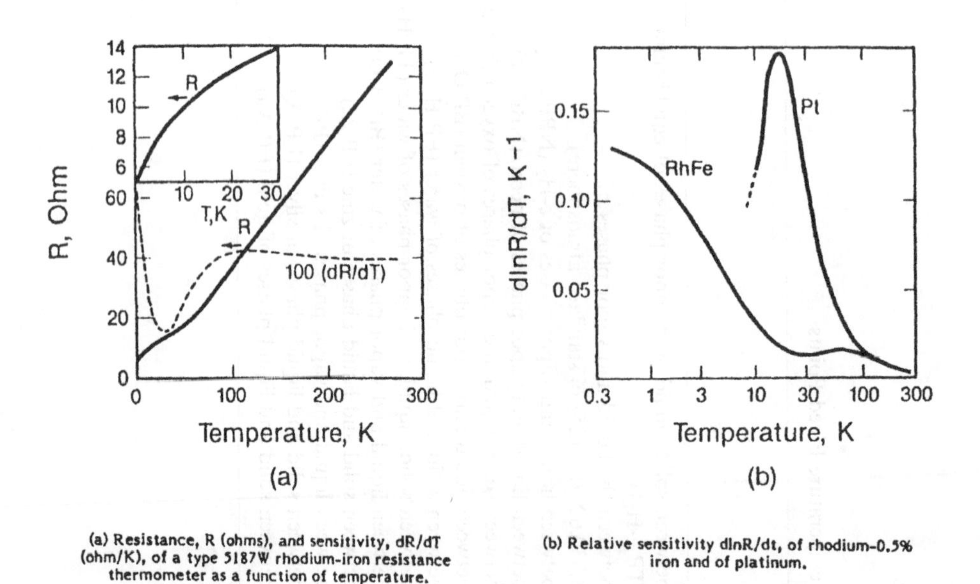


**Figure 8.41** Rhodium–iron resistance thermometer. (a) Resistance *R*(ohms) and sensitivity d*R*/d*T* (ohm/K) of a type 5187 W Rh–Fe resistance thermometer as a function of temperature; (b) relative sensitivity d(In *R*)/dt of Rh–0.5% Fe and of platinum. (Courtesy of LakeShore Cryotronics Inc., Westerville, OH.)

sensor capability that is extremely valuable. It appears that the chief disadvantage is lack of uniformity of resistive characteristics from thermometer to thermometer. This nonuniqueness is due to the difficulty in producing homogeneous wires from a single melt of Rh–Fe and, even more difficult, production of sufficiently identical alloys on a batch-to-batch basis. These concerns are currently overcome by the unique calibration of each thermometer and may diminish completely as experience with the Rh–Fe alloy increases.

For most purposes, platinum resistance thermometers remain the workhorses of metallic resistance thermometers. For wide-ranging measurements, $1\,K < T < 300\,K$, and at the lower temperatures, $T < 30\,K$, semiconductor and diode thermometers are the principal competitors to PRTs and are often to be preferred.

*Example 8.7.* A platinum resistance thermometer has a resistance of 100.0 ohms at $T = 273.16\,K$ (0°C). Find the corresponding temperature when the resistance is 50.0 ohms and the sensitivity at this point. The thermometer constants are

$$A = 3.946 \times 10^{-3} (^\circ C)^{-1}$$
$$B = -1.108 \times 10^{-6} (^\circ C)^{-2}$$
$$C = 3.33 \times 10^{-12} (^\circ C)^{-6}$$

and the interpolating function is

$$R_e = R_o[1 + AT + BT^2 + C(T - 100^\circ C)T^3]$$

where $T$ is in °C.

Most hand-held calculators have an iterating function that can conveniently be used if the above equation is rearranged in the form

$$\frac{R_e}{R_o} - 1 = AT + BT^2 + C(T - 100)T^3$$
$$\frac{50}{100} - 1 = -0.5 = AT + BT^2 + C(T - 100)T^3 = X$$

Using the iterating program for a value of $X = -0.5$ yields $T = -123^\circ C = 150\,K$.

The equation may also be solved by successive approximations. Truncate it,

$$R_e/R_o - 1 = AT_1$$

or

$$T_1 = (R_e/R_o - 1)/A$$

and solve directly for $T_1$. Use this $T_1$ in the next approximation,

$$R_e/R_o - 1 - AT_2 = BT_1^2 + C(T_1 - 100)T_1^3$$

And so on.

To find the sensitivity, differentiate the interpolating function,

$$\frac{dR_e}{dt} = R_o(A + 2BT + 4CT^3 - 300CT^2)$$

Substituting the appropriate values for $R_o$, $A$, $B$, and $C$ with the value obtained earlier for $T = -123°C$ yields

$$\frac{dR_e}{dt} = 0.4179\ \mathrm{ohm/°C}$$

## 8.3. Nonmetallic Resistance Thermometry

A number of semiconductors also have useful thermometric properties at low temperatures. A semiconductor has been defined as a material whose electric conductivity is much less than that of a metallic conductor but much greater than that of a typical insulator. There are, however, at least three distinct differences between semiconductors and pure metals as resistance thermometers. First, the sensitivity, $(1/R)(dR/dT)$, of a semiconducting thermometer in its useful range is usually much greater than that of a pure metal thermometer. Second, the temperature coefficient of resistivity of semiconductors is negative, whereas for pure metals it is positive; for example, the $dR/dT$ for a typical semiconductor is $-50 \times 10^{-3}\ K^{-1}$ at 273 K, whereas for platinum at the same temperature it is $4 \times 10^{-3}\ K^{-1}$. Thus the resistance of a semiconductor decreases with increasing temperature while the opposite is true for a pure metal. Third, as mentioned in the definition for a semiconductor, the resistivity of a semiconductor is usually several orders of magnitude greater than that of a pure metal. Consequently, a semiconducting thermometer element is usually short and of relatively large cross section, so its resistance is not too large for practical use.

The resistive properties of semiconducting materials are attributed to the existence of an energy gap between the valence band and the conduction band. At the absolute zero of temperature, the valence band is completely full, the conduction band is completely empty, and the semiconductor crystal behaves as an insulator. As the temperature increases to a certain level, sufficient thermal energy is added to allow some electrons to jump to the conduction band. Hence, the resistance of semiconductors increases as the temperature decreases, as described above, which is opposite to the behavior of metals.

The most promising semiconductors seem to be germanium, silicon, and carbon. The latter, though not strictly a semiconductor, is included in this group because of its similarity in behavior to semiconductors. Diode thermometers are included here for the same reason.

### *8.3.1. Germanium*

Of the three semiconductors mentioned, germanium has received by far the most attention, and germanium resistance thermometers are available from several commercial sources. The resistance element is usually a small single crystal. Inasmuch as the resistivity is high, the element can be short and thick. It is mounted strain-free in a protective capsule (see Fig. 8.42). Because of this combination of features, the germanium thermometer can be both rugged and reproducible.

Accurate temperature measurement in the temperature range 1.5–20 K can be done routinely using germanium resistance thermometers (GeRTs); thermometers

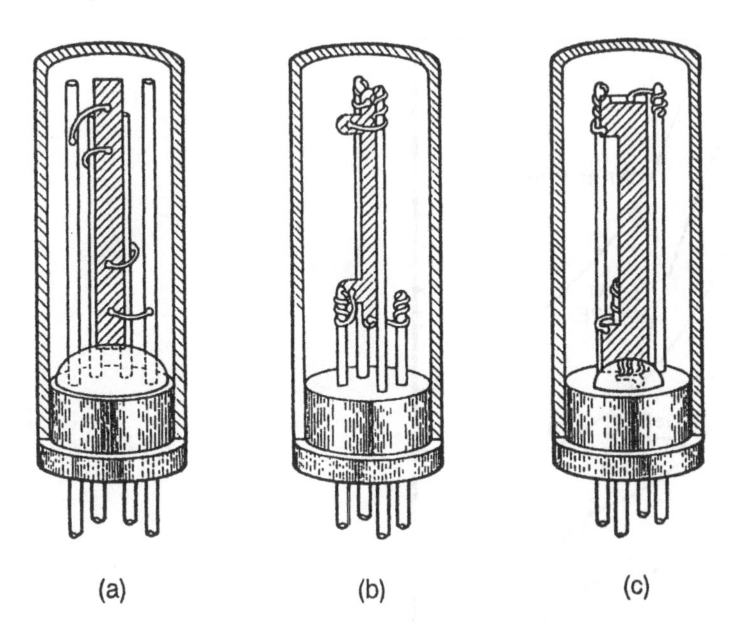


**Figure 8.42** Construction of three commercially available germanium resistance thermometers.

that allow temperatures up to 100 K to be measured with reduced sensitivity are available. Extensive development of these semiconducting devices has led to the most accurate and reproducible empirical thermometer yet devised for $T \leq 20$ K. For this reason, germanium resistance thermometers remain the standard sensor for accurate measurement in the 1–30 K range and may be used reliably both above and below that range. Typical temperature dependences of $R$ for n- and p-type GeRTs are shown in Fig. 8.43.

The reproducibility of the thermometers at 4 K is as good as the present ability to check them, that is, within a few tenths of a millidegree. Besley and Plumb established the reproducibility of 30 GeRTs at 20.28 K where the reduced sensitivity makes small resistance instabilities more significant. At this temperature and after 100 thermal cycles between 20 and 100 K, they found instabilities ranging from 0.1 to 20 mK, with 90% of the units being stable to better than 2 mK. Anyone who can tolerate errors of 0.1% ($\pm$ 0.05%) can be reasonably confident that a single sensor calibrated by the manufacturer will perform as promised.

These thermometers are small (typical size is approximately 1.3 cm long by 0.3 cm in diameter) and have small mass with a resulting time constant on the order of 0.1 sec. As seen in Fig. 8.42, the leads that actually contact the germanium crystal are very small. This results in the instruments burning out relatively easily if the measuring current is too high. Most models use 1 μA for $T < 2$ K, 10 μA for $2\text{ K} < T < 15\text{ K}$, 100μA for $15\text{ K} < T < 40\text{ K}$, and 1 mA for $40\text{ K} < T < 100\text{ K}$.

Joule heating must be guarded against in all types of resistance thermometers by limiting the measuring current. This is especially important when using germanium and its allies because of the very high resistance at low temperatures, on the order of $10^3$–$10^4$ ohms.

The major disadvantage associated with the use of germanium thermometers is the lack of a simple analytical representation. This is because conduction (in the $n$-type) changes gradually from impurity conduction at low temperatures to

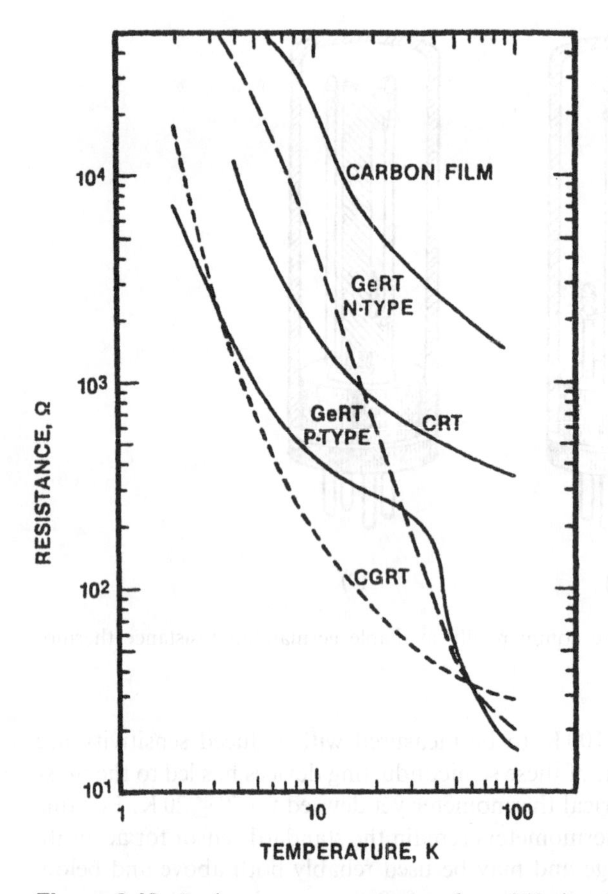


**Figure 8.43** Resistance vs. temperature for a 270 ohm (ambient temperature, 0.1 W) CRT; 1000/ohm (4.2 K) CGRT, n-type and p-type GeRTs, and a vacuum-deposited carbon-film resistor. (From Blakemore et al., 1970.)

free-electron conduction at higher temperatures, leading to a complex $R$–$T$ relationship. The most analytical forms commonly used with both n- and p-type thermometers are

$$\log R = \sum_{j=0}^{m} A_j (\log T)^j$$

and

$$\log T = \sum_{i=0}^{m} B_i (\log R)^i$$

Each thermometer must be calibrated by comparison at many points in the range of interest if the inherent reproducibility of the thermometers is to be utilized.

### 8.3.2. *Carbon*

The carbon resistance thermometer (CRT) has maintained its popularity in the field because it continues to provide high sensitivity at low temperatures in a very small, low-cost package. Carbon in one form or another is perhaps the most widely used of the high-resistivity materials for low-temperature thermometry.

The bulk of the use of carbon resistance thermometers is for temperatures at and below 20 K. Their use at higher temperatures is limited because of decreasing sensitivity, reported stability problems, and competition with metallic resistors. By far the most used form of the carbon thermometer is the standard radio type. In general, the 0.1–1 W resistors with room temperature resistances of 10–500 ohms are most used. The advantages of these resistors are that they are physically small, rugged, and cheap and are the least sensitive of the resistance thermometers to magnetic fields.

Preferences vary as to the exact type, size, and resistance of CRTs. Historically, the important brands have been Allen-Bradley, Ohmite, and Speer (more properly Airco Speer). The Airco Speer grade 1002 (1/2 W) resistors are no longer being manufactured and have not been since 1979. Also, Allen-Bradley has manufactured type BB (1/8 W), CB (1/4 W), and EB (1/2 W) resistors, the sizes commonly used in cryogenic thermometry. They are available through a number of direct distributor outlets.

The Allen-Bradley radio resistors in sizes from 0.1 to 1 W and resistance values from 10 to 500 ohms, remain the most popular choices above 1 K and are used occasionally below 1 K. The Speer radio resistors in various sizes and resistances are usually selected for the region below 1 K and may prove useful to as low as 0.02 K. A typical 1/2 W, 220 ohm resistor will measure roughly 1 kohm at 1 K, 20 kohm at 0.1 K, and 300 kohm at 0.015 K, the last at about $10^{-13}$ W dissipation. Usually 0.1–1 W resistors with room temperature resistances of 10–500 ohms are used.

The temperature dependence of $R$ for three types of carbon resistance thermometers is compared with that of two types of germanium resistance thermometers in Fig. 8.43 (Blakemore et al., 1970).

Clement and Quinnel (1952) found that commercial resistors of various types were reproducible to about 0.1% or $\pm 4$ mk when cycled between room temperature and liquid helium temperature in a vacuum environment.

Reproducibilities of the order discussed above are not possible if the CRT is exposed to moisture or is heated, for example with a soldering iron or by baking.

Clement and Quinnel (1952) investigated such thermometers extensively and determined that the 1 W size was the most suitable. The original interpolation equation that fits their data in the 2–20 K range was given as

$$\log R + \frac{K}{\log R} = A + \frac{B}{T}$$

and later revised to

$$\left(\frac{\log R}{T}\right)^{1/2} = A + B \log R$$

where $R$ is the resistance at temperature $T$, and $K$, $A$, and $B$ are experimentally determined constants. Temperatures calculated from this equation were found to be within $\pm 0.5\%$ of the measured temperature between 2 and 20 K for eight different resistors. As shown in Fig. 8.44, 2.7 and 10 ohm resistors were found to be suitable for temperature measurement below 1 K down to 0.3 K (Collier et al., 1973).

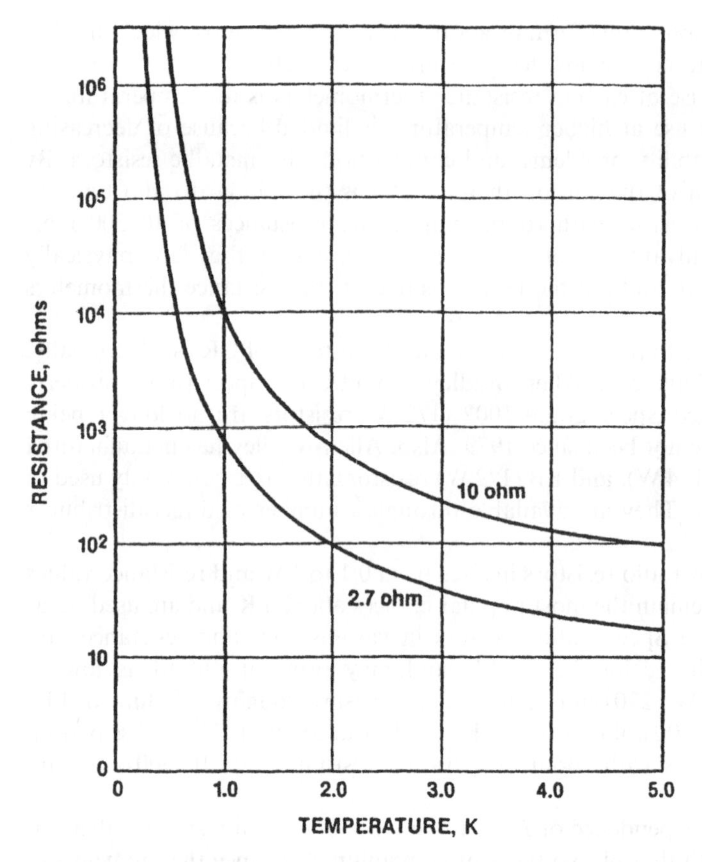


**Figure 8.44** Resistance–temperature curve for two Allen-Bradley carbon resistance thermometers.

This complex $R$–$T$ relation is characteristic of semiconductor thermometers, and the exact form is frequently revised to provide the best fit over the widest temperature range. Another form for $R$ vs. $T$ for a wide temperature range (4–296 K) is the empirical relationship,

$$\log R + \frac{D}{\log R} = A + \frac{B}{T}$$

Temperatures calculated from this equation agreed within $\pm 0.5$ K for a variety of resistors with different nominal room temperature resistances. This equation has been used with the coefficients $A = 4.478$, $B = 3.091$, and $D = 5.004$ to calculate $R_t / R_{296\,K}$ for a 0.1 W, 270 ohm CRT in the temperature range 4–296 K. The following differences were found between the calculated and experimental ratios: $\pm 6.8\%$ at 4.2 K, $\pm 2.2\%$ at 20.3 K, and $\pm 0.25\%$ at 77.4 K. These variations correspond to temperature uncertainties of $\pm 110$, $\pm 500$, and $\pm 500$ mK, respectively.

### *8.3.3. Carbon Glass*

A carbon glass resistance sensor can measure temperature between 1 and 300 K (Ricketson and Grinter, 1982). These thermometers are produced by treating porous alkali borosilicate glass with organic liquids. The resulting composite consists of carbon filaments in a glass matrix (Lawless, 1972). This sensor has a smoothly temperature-dependent resistance much like ordinary carbon resistors but is about twice as sensitive. Its magnetoresistance is much lower than that of germanium and is not orientation-dependent.

The carbon glass sensors now available are similar to commercially available germanium sensors in sensitivity, packaging, and price. Their construction is similar to that of precision germanium resistance thermometers (GeRTs)—the active resistor is mounted in a configuration that is as strain-free as possible and is encapsulated in a helium-filled metallic can. They are four-lead devices and are produced specifically for use as thermometers. The cylindrical encapsulated sensor has a 3.2 mm diameter and 11 mm length. The cost of carbon glass sensors is comparable to that of GeRTs and higher than that of the circuit carbon resistors discussed above.

A sensor with a nominal resistance of 1000 ohms at 4.2 K may have a resistance that varies from 120 kohms at 1.5 K to 10 ohms at 300 K. The sensitivity over this temperature range varies from 400 kohms/K to 0.015 ohm/K. Millikelvin reproducibility will certainly be available below 4.2 K but not necessarily above that temperature.

The temperature–resistance characteristics of the carbon glass resistance thermometers (CGRTs) are typified by the negative $\mathrm{d}R/\mathrm{d}T$ expected for semiconductors. The $R$–$T$ behavior is smooth and monotonic (Fig. 8.45), as is the slope $\mathrm{d}R/\mathrm{d}T$. This makes calibration and analytical representation easier than for GeRTs.

The equation

$$\log R = A_1 = A_2/T^p$$

has been used to represent CGRTs in the ranges 2–5, 2–10, 2–20, and 2–40 K with a standard deviation of less than 1% of the upper temperature of each interval. This equation was found to represent the CGRTs to within ±0.5 K up to 70 K. The sensors scaled well enough to allow the equation

$$R = AR_{jM}$$

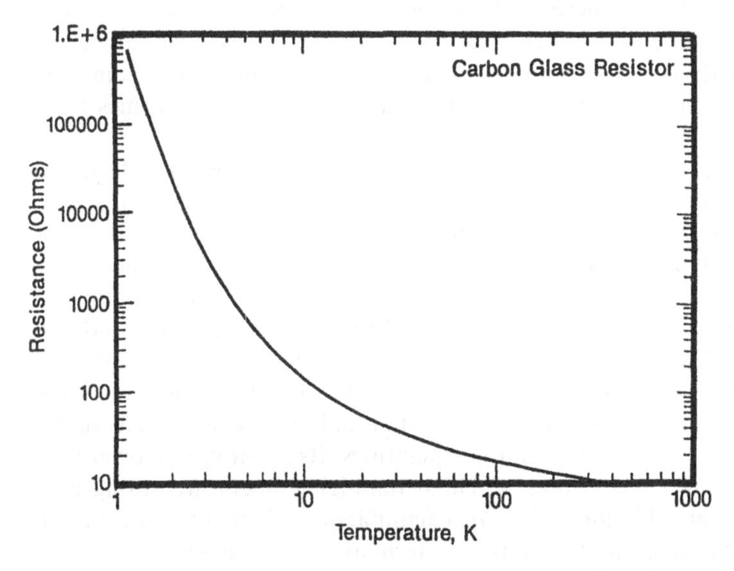


**Figure 8.45** Carbon glass resistance thermometer, typical $R$–$T$ curve. (Courtesy of Lake-Shore Cryotronics Inc., Westerville, OH.)

to be used within the uncertainties given above. Here $R_j$ is the temperature-dependent resistance of a calibrated CGRT and $M$ and $A$ are temperature-independent terms. This allows a CGRT to be calibrated by experimentally determining $R(T)$ at only two temperatures.

The remaining open questions regarding these sensors concern their long-term stability and repeatability. A group of six CGRTs were studied for repeatability after thermal cycling. Six sensors are a small sample, but those six were clearly not equal in their repeatability on rapid cycling from 300 to 20 K to the germanium sensors tested at the same time. Some speak of millikelvin errors, while others find errors of $\Delta T/T \sim 1\%$. Even though the repeatability question is not settled, it is clear that the carbon glass sensors have some advantages over resistors. They are hermetically sealed, so they do not require care to avoid vapor absorption. They do not drift after large temperature changes, no matter how sudden the change. No drift of resistance at constant temperature has been reported, so the sensors are quite stable.

In spite of their instability, CGRTs provide adequate precision for many users, particularly near 4 K, and usable sensitivity over a much broader temperature range than GeRTs. Their behavior in magnetic fields is also superior to that of germanium resistors. A low magnetoresistance, similar to that seen in an Allen-Bradley carbon resistance thermometer but with better reproducibility, makes this thermometer the preferred choice in many high-field applications (Swartz et al., 1975).

### *8.3.4. Thermometers for Radiation and Magnetic Field Environments*

Carbon glass has been used in magnetic fields since about 1976. It is a true four-electrical-contact, bulk element device. It is made up of carbon filaments in a porous glass. Its magnetic field behavior is well known and correction algorithms are available for field-induced errors. It is the most sensitive of the three in the range of 1.2–10 K. It is monotonic to room temperature, but is a factor of about 5 lower in sensitivity than Cernox at 300 K. It is not available to an interchangeability curve. It is moderately resistant to ionizing radiation, but the least resistant of the three. Its short-term stability during thermal shock cycles to 4.2 K is good (10 cycles), but its mean time to failure under these conditions is the shortest of the three. Its long-term stability under slow thermal cycles is good and is the best documented of the three types of sensors. It is manufactured at LakeShore Cryotronics from ultrapure materials, and thus its future availability is secure.

Cernox has been available since 1992. It is a thin film ceramic, two-electrical-contact device on a sapphire substrate. It is the only one of the three that can be used above room temperature, and is monotonic, in one model, down to 300 mK. A wider range of sensitivities and packages is available in this sensor than in any of the others. Assuming the same instrumentation is used for the two, its temperature resolution will be a factor of five or ten better than that of Ruox. Its magnetic field-induced errors are, in general, lower than those of carbon glass, but vary in a complicated way with temperature and sensitivity. It is more resistant to ionizing radiation (very good, in fact) and is the most stable and shock resistant of the three. It is not available as an interchangeable sensor except in limited quantities. Its sensitivity is lower than that of carbon glass at low temperatures, and higher at high temperatures. Long-term stability data are still being developed. It is also manufactured from ultrapure materials at LakeShore Cryotronics, and thus its long-term availability is secure.

Ruox sensors are thick film resistor chips with two electrical contacts, and are available in only one package. They are the lowest overall sensitivity of the three, and only one model is monotonic up to room temperature. Two models are useable to 0.05 K and are the only choice at those temperatures for low magnetic field error sensors. They cannot be used above room temperature because upon temperature cycling, they develop microcracks which heal at higher temperatures. They are thermally shocked 50 times into liquid nitrogen during production to precondition them and increase their stability in use. Their short-term stability in thermal shock tests is less than the other two, and their long-term stability, when it is established, will undoubtedly prove to depend more on their history than is true of the others. Their radiation resistance is very good—better than that of carbon glass and a little worse than that of Cernox. The magnetic field errors are, in general, superior to those of carbon glass in two of the models, and depend more regularly on temperature than those of Cernox. The field errors of certain Cernox models in limited temperature ranges are lower, however. The greatest strength of the Ruox sensors is that all models have an inter-changeability curve with tolerance bands, and thus may not have to be calibrated.

### *8.3.5. Diodes*

The forward voltage of a semiconductor junction at a constant current is a well-defined function of temperature and has been widely used for over 40 years to measure temperature in the range 1–400 K (Rao, 1982). The forward voltage–temperature characteristics (VTC) of germanium, gallium arsenide, and silicon junctions have been extensively investigated over the last two decades or so. Commercially available temperature sensors have been primarily fabricated from either silicon or gallium arsenide *p–n* junctions. The silicon diode is currently the most widely used sensor, as it has proven to be the most stable and reproducible. Gallium arsenide, although less stable and predictable, is preferred for magnetic field use because of its low magnetic field dependence at low temperatures. Gallium aluminum arsenide diode temperature sensors are an alternative for gallium arsenide, offering better matching characteristics and improved sensitivity (Krause and Swinehart, 1985).

The *p–n* junction sensors share many of the resistance thermometer (RT) characteristics, being most often current-biased with voltage readout. They are equal to, or better than, RTs for many practical applications. Diodes are more sensitive and more nearly linear over more of their usable range than carbon, germanium, and platinum RTs. Their biggest advantage is unquestionably the wide temperature range over which reasonable resolution is maintained, perhaps 10 mK over the 2–300 K range.

The forward voltage reading across a diode is current-dependent (Fig. 8.46), which requires that calibration and operation be at identical currents. However, this current dependency is not very strong. For example, a current tolerance of $10\,\mu a \pm 0.05\%$ will limit equivalent temperature errors to less than $\pm 6$ mK for silicon diodes. This tolerance is not very difficult to obtain with present-day electronics, and current sources are available that can quite easily and inexpensively meet this specification. On the other hand, the diode voltage is quite sensitive to temperature variations, making excellent temperature resolution achievable with moderately priced voltmeters (Figs. 8.46 and 8.47).

The voltage–temperature characteristic shown in Fig. 8.47 is typical of both germanium and silicon diodes. It is made up of two distinct regions with different slopes, namely an exponential extrinsic low-temperature region ($T < 20$ K) and a

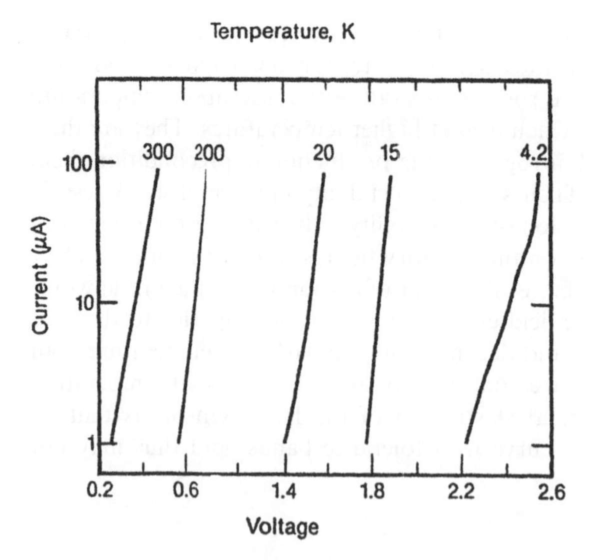


**Figure 8.46** Diode thermometer. Dependence of forward voltage on current and temperature. (Courtesy LakeShore Cryotronics Inc., Westerville, OH.)

linear intrinsic high-temperature region ($T > 20\,K$). Extrapolation of the linear intrinsic junction characteristics to absolute zero gives the intrinsic bandgap energy of the semiconductor. A distinct change in slope is observed for silicon junctions at 1.1 V and for germanium junctions at 0.72 V. The temperature the change in slope takes place is current-dependent and in the range 12–25 K; calculated on the basis of conventional $p$–$n$ junction theory.

Diode sensors that are designed specifically for use as wide-range thermometers are routinely available. These sensors are pretested for stability and provide well-characterized $V(T)$ behavior. Accurate and traceable calibrations are available on an individual basis, or interchangeable sensors conforming to standardized calibration curves can be purchased. In addition, these sensors are available in a wide variety of specially designed packages, making them adaptable for almost any situation. Small,

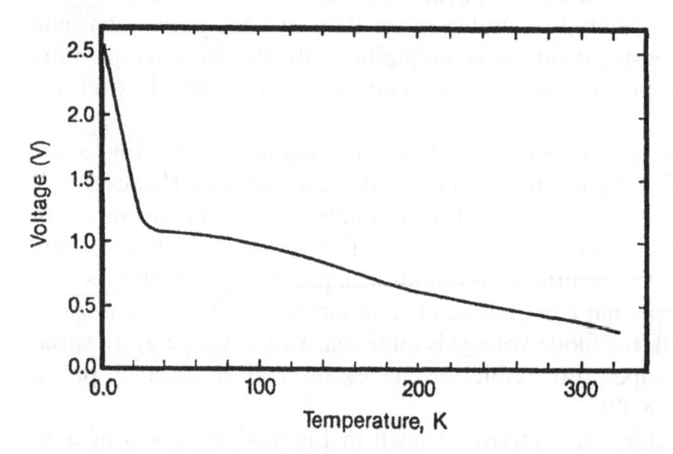


**Figure 8.47** Voltage–temperature curve for a typical silicon diode temperature sensor at a constant current of 107 μA. (Courtesy of LakeShore Cryotronics Inc., Westerville, OH.)

pinhead-sized epoxy encapsulations are useful when size and mass considerations are important. Cylindrical geometries are used for inserting into drilled holes, while more elaborate screw-in mounts and bolt-on bobbins can make installation and replacement extremely easy. Nonmagnetic packages are also available for magnetic field work (Krause and Swinehart, 1985).

When using a diode or other high-resistance semiconductor thermometer below about 10 K, the possibility of self-heating should be considered. This situation occurs when the diode warms slightly above the surrounding environment because of its relatively high power dissipation. Typical power dissipation at 4.2 K is on the order of tens of microwatts for a 10 μA operating current. Although this appears small to someone inexperienced in cryogenics, it is still several orders of magnitude greater than the power dissipated in a germanium or carbon glass thermometer. The obvious solution seems to be to drop the operating current, for example, to 1 μA. However, this raises the static impedance of the diode into the megohm range and requires tighter constraints on the measurement instrumentation to achieve comparable temperature measurement accuracies. To minimize any self-heating effects, the diode must be placed in good thermal contact with the sample being measured. Improper installation techniques can result in temperature errors approaching a kelvin.

As indicated above, silicon sensors have become the standard in diode thermometry, and an extensive amount of data exists for them. Typical long-term stability is on the order of ±50 mK, while short-term stability can be as low as a few millikelvin. This stability makes the diode competitive with industrial grade platinum thermometers with the added benefit of being usable to as low as 1 K. The upper temperature limit for commercially available diode sensors is around 400 K. This limit is determined by the properties of silicon, the metallurgy of the contact areas, and the construction materials and techniques used in device packaging.

In a forward-biased *p–n* junction, the forward current is related to the junction voltage as

$$I_f = I_s \exp(qV/nkT) \tag{8.57}$$

where $q$ is the electronic charge, $n$ is a constant, $k$ is Boltzmann's constant, and $T$ is the absolute temperature. The value of $n$ is 2 for the dominant generation–recombination current at high injection levels. $I_s$ is the intrinsic carrier concentration,

$$I_s = AqDn_i/L \tag{8.58}$$

where $A$ is the junction area, $D$ and $L$ are the diffusion constant and length of the carriers, and $n_i$ is the intrinsic carrier concentration,

$$n_i = BT^3 \exp(-qV_g/kT) \tag{8.59}$$

where $B$ is a constant (related to junction geometry) and $qV_g$ is the bandgap energy of the semiconductor at absolute zero. Solving Eqs. (8.57)–(8.59) for $V$ gives

$$V = V_g - CT(-\log I_f + E \log T) \tag{8.60}$$

where $C$ and $E$ are constants and $V_g$ is the bandgap voltage of the intrinsic semiconductor at absolute zero (1.16 for Si).

Equation (8.60 ) represents a practical linear dependence of $V$ on $T$, which can be expressed as

$$V = V_g - m(I_f)T \tag{8.61}$$

This equation not only emphasizes the dependence of the slope on the forward current but also indicates the possibility of matching the slopes (interchangeability) of different junctions.

The Chebyshev polynomial is useful for the *R–T* interpolation formula for diode thermometers (as well as other semiconductor resistance thermometers). The Chebyshev polynomial has two desirable properties not associated with a power polynomial:

1. The fitting and solution of a Chebyshev polynomial can be made to high orders without the large computer power needed for power polynomials. (The IPTS-68 scale below the ice point is defined in terms of a power polynomial; to derive the dependent variable to seven places, each coefficient is defined to 16 places.)
2. A user may truncate the coefficients and delete any order to suit the desired accuracy.

The Chebyshev polynomial is defined as

$$Y = \frac{A_0}{2} + \sum_{n=1}^{n=N} A_n \cos[n \arccos(BX + C)] \tag{8.62}$$

where $A_0, A_1, A_2, \ldots, A_N$ are the coefficients, $B$ and $C$ are constants chosen so that $BX + C$ lies between 1 and –1, and $X$ and $Y$ are the independent and dependent variables, respectively.

As a temperature–resistance interpolating function, the Chebyshev equation takes the form

$$T(x) = \sum_{n=0} a_n t_n(x) \tag{8.63}$$

where $T(x)$ represents the temperature in kelvin, $t_n(x)$ is a Chebyshev polynomial, and $a_n$ represents the Chebyshev coefficients. The parameter $x$ is a normalized variable given by

$$x = [(Z - Z_L) - (Z_U - Z)]/(Z_U - Z_L) \tag{8.64}$$

For diodes, $Z$ is simply the voltage $V$. For resistors, either $Z$ is the resistance $R$, or $Z = \log R$. $Z_L$ and $Z_U$ designate the lower and upper limits of the variable $Z$ over the fit range.

The Chebyshev polynomials can be generated from the recursion relation

$$t_{n+1}(x) = 2x t_n(x) - t_{n-1}(x), \quad t_0(x) = 1, \quad t_1(x) = x \tag{8.65}$$

Alternatively, these polynomials are given by

$$T_n(x) = \cos(n \arccos x) \tag{8.66}$$

The use of Chebyshev polynomials is no more complicated than the use of the regular power series, and they offer significant advantages in the actual fitting process. The first step is to transform the measured variable, either $R$ or $V$, into the normalized variable using Eq. (8.64 ). Equation (8.63 ) is then used in combination with Eq. (8.65 ) or (8.66 ) to calculate the temperature. Programs 1 and 2 give sample BASIC subroutines that will take the parameter $Z$ and return the temperature $T$ calculated from the Chebyshev fits. The subroutines assume that the values $Z_L$ and $Z_U$ have been input along with the degree of the fit, *N*degree. The Chebyshev coefficients are also assumed to be in an array *A*(0), *A*(1), ... , *A*(*N*degree).

*Program 1.* BASIC subroutine for evaluating the temperature *T* from the Chebyshev series using Eqs. (8.63)–(8.65). An array Tc(*N*degree) should be dimensioned.

```
100  REM Evaluation of Chebyshev series
110  X=[(Z-ZL)-(ZU-Z)]/(ZU-ZL)
120  Tc(0)=1
130  Tc(1)=X
140  T=A(0)+A(1)*X
150  FOR I=2 to Ndegree
160  Tc(I)=2*X*Tc(I-1)-Tc(I-2)
170  T=T+A(I)*Tc(I)
180  NEXT I
190  RETURN
```

*Program 2.* BASIC subroutine for evaluating the temperature *T* from the Chebyshev series using Eqs. (8.63) and (8.66). ACS denotes the arccosine function. Doublprecision calculations are recommended.

```
100  REM Evaluation of Chebyshev series
110  X=[(Z-ZL)-(ZU-Z)]/(ZU-ZL)
120  T=0
130  FOR I=0 to Ndegree
140  T=T+A(I)*COS(I*ACS(X) )
150  NEXT I
160  RETURN
```

An interesting and useful property of the Chebyshev fits is evident in the form of the Chebyshev polynomial given in Eq. (8.66). The cosine function means that $t_n(x) \leq 1$, and so no term in Eq. (8.63) will be greater than the absolute value of the coefficient. This property makes it easy to determine the contribution of each term to the temperature calculation and where to truncate the series if the full accuracy of the fit is not required.

There are several significant measurement advantages in using diodes instead of resistance thermometry or thermocouples. As the signal level is on the order of a volt and sensitivities are quite large, thermal emf's are not a significant source of temperature uncertainty. Also, because the diode gives an absolute measurement, reference baths or reference junctions are not required. Constant-current operation avoids the necessity of a variable current source as in the case of resistance thermometry, and the wide range of the diode sensor eliminates the need for multiple sensors. These reasons, combined with the ability to select diodes that follow standardized calibration curves, make the diode an excellent choice for use in temperature controllers and other instrumentation.

Cernox™ is another type of resistance temperature sensor which offers advantages over the more bulky type of RTD's. Cernox™ is a ceramic oxynitride and is the trademark of LakeShore Cryotronics, Inc. The smaller package size of these thin film sensors (Fig. 8.48) makes them useful in a broader range of experimental mounting schemes, and they are also available in a chip form. Additionally, they have been proven very stable over repeated thermal cycling and under extended exposure to ionizing radiation. They are easily mounted in packages designed for excellent heat

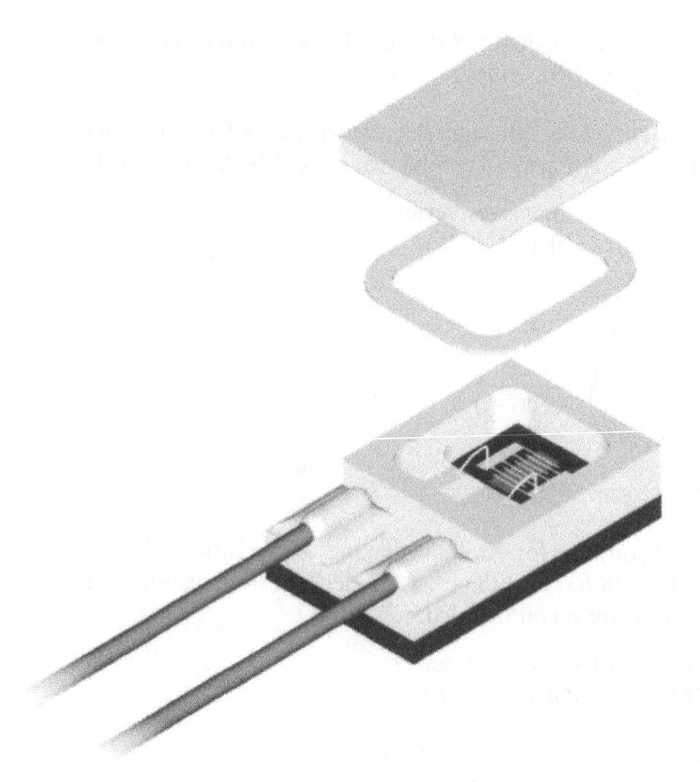

**Figure 8.48** Cernox$^{TM}$ resistance temperature sensor construction. The base of sapphire is chosen because of its very high thermal conductivity at low temperatures. LakeShore Cryotronics, Westerville OH, www.lakeshore.com.

transfer, yielding a characteristic thermal response time much faster than possible with bulk devices requiring strain-free mounting.

Figure 8.48 shows the configuration of a diode temperature sensor suitable for the range 1.4 – 500 K. Such a wide range is simply not possible for platinum or germanium detectors. As mentioned above, diode sensors are also interchangeable, unlike platiunum or germanium. For many applications, these sensors do not require individual calibration. The figure shows the sapphire mounting base. Sapphire is chosen because of its very high thermal conductivity at low temperatures. (See Chapter 5 for the discussion of thermal conductivity of dielectrics at low temperatures.)

Figure 8.49 shows a complete system for using diodes as well as other transducers as temperature sensors. The system shown has excellent resolution, accuracy and stability for demanding temperature measurement and control applications to as low as 300 mK. It supports diodes, negative temperature coefficient (NTC) and positive temperature coefficient (PTD) resistance temperature sensors (RTD). It supports two or eight additional diode/RTD sensor inputs, one capacitance sensor or two thermocouples.

### *8.3.6. Thermistors*

Thermistors (TMs) (thermally sensitive resistors) are essentially resistors made up of metal oxides. Frequently used materials are nickel, manganese, and cobalt oxides. The temperature–resistance for this type of resistor has a negative slope much like

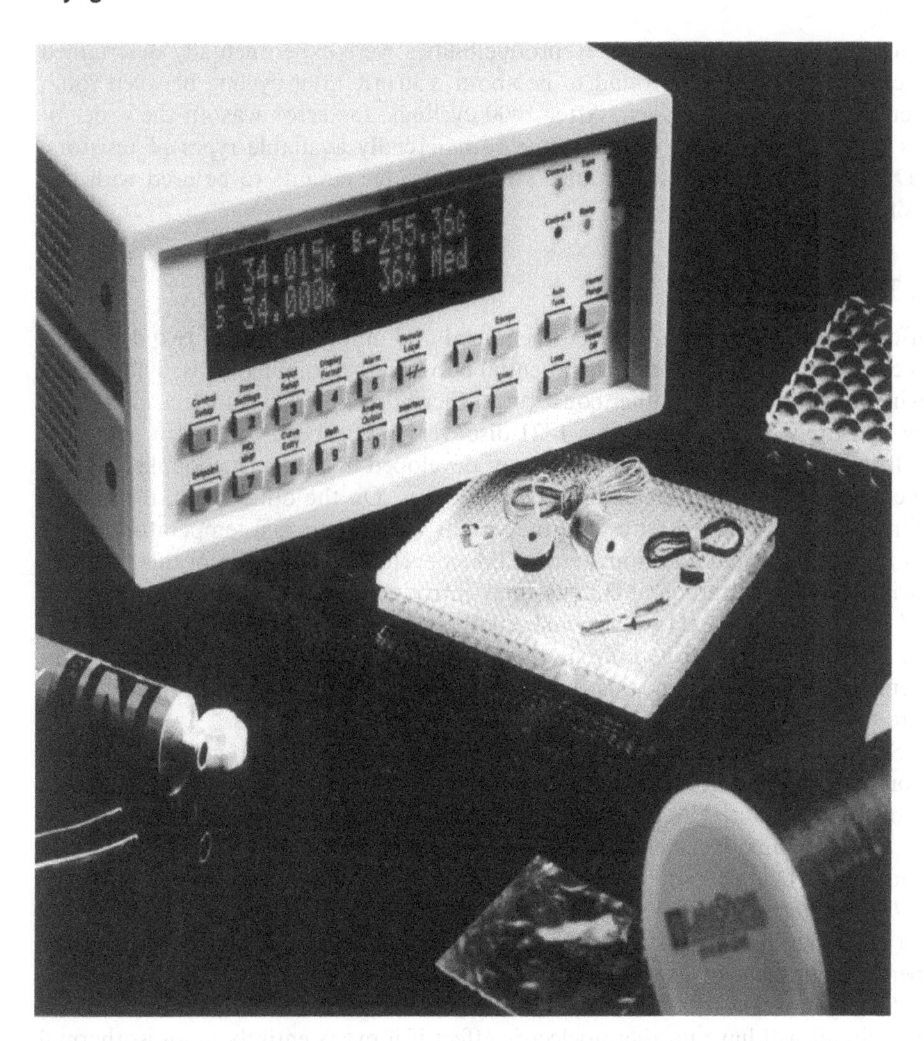


**Figure 8.49** A complete system for low–temperature measurement and control using diodes as well as other transducers as temperature sensors. LakeShore Cryotronics, Westerville OH, www.lakeshore.com.

the carbon or germanium resistors. The fact that thermistors are popular in measurement and control circuits is attested to by the number of companies selling them ("Measurements and Data 7,"1973). Some reasons for their increasing use are that (1) they are small, which tends to make the time response significantly less than 1 sec; (2) they are typically high-resistance units, which reduces the overall effect of lead resistances; and (3) their temperature–resistance characteristics are dependent on materials and procedures that allow thermometers to be developed that are particularly sensitive in limited ranges of temperature.

There are, of course, disadvantages also. Items 2 and 3 above may be considered disadvantages as well as advantages. No one thermistor is usable over a wide range of temperatures because its resistance becomes exceedingly high. The analytical representation of the resistance vs. temperature characteristics are represented by

$$\rho = A \exp \frac{B}{T}$$

for short ranges of temperature. Reproducibilities were experimentally determined by Sachse (1962) and were found to be about ±30 mK after cycling between room temperature and liquid oxygen. After 1000 cyclings, the error was on the order of tenths of a degree. Sachse discusses the commercially available types of resistors, and Droms (1962) presents a detailed study of bridge circuits to be used with this type of resistor.

## 8.4. Thermocouples

A difference of temperature between the joined and unjoined ends of two dissimilar wires causes a difference of electric potential between the unjoined ends. The ends of the wire connected to each other constitute a thermocouple. Thermoelectric thermometry is entirely dependent on this 1821 discovery by Seebeck.

If the wires are homogeneous, the emf developed in the thermocouple circuit is independent of the length or diameter of the wires. On the other hand, if inhomogeneities are present in the wires and they occur where there is a temperature gradient in the wire, then an additional emf will be developed. Variation of this parasitic emf with temperature gradient changes may introduce a serious error into the reading of the thermocouple. However, it should be emphasized that no emf is produced by an intermediate metal as long as the junctions are kept in regions of constant temperature. It follows from this that great care must be exercised in the selection of wire for thermocouples to reduce the effect of inhomogeneities.

Sign conventions have been adopted for the direction of current flow. The convention is that if current flows from material A to material B at the cooler of the two junctions, then material A is thermoelectrically positive with respect to material B. There are also three basic empirical laws that govern the use of thermocouples as thermometers (Dike, 1954). They may be stated as follows.

*Law of the Homogeneous Circuit.* No current will flow in a thermoelectric circuit if materials A and B are identical regardless of the magnitude of the temperature gradient or its distribution along the wires.

*Law of Intermediate Materials.* A third material introduced into the thermoelectric circuit will have no thermoelectric effect if it exists entirely in an isothermal region.

*Law of Intermediate Temperatures.* If a thermocouple pair generates a voltage $E_1$ when its junctions are at $T_1$ and $T_2$ and a voltage $E_2$ when its junctions are at $T_2$ and $T_3$, then it will generate the voltage $E_{12} = E_1 + E_2$ when its junctions are at $T_1$ and $T_3$.

Application of the second law assures us that a feedthrough or disconnect (being an intermediate material) will produce no thermoelectric effect if it is isothermal.

Various thermocouple combinations used at low temperatures are tabulated below. A graphical representation of the thermoelectric power is given in Fig. 8.50.

Copper vs. constantan (60% Cu and 40% Ni)
Gold–cobalt (Au + 2.11 at% Co) vs. copper
Gold–coblat vs. normal silver (Ag + 0.37 at% Au)
Iron vs. constantan
Chromel P (90% Ni and 10% Cr) vs. alumel [95% Ni and 5% (Al, Si, Mn)]

Thermocouple types as defined by the ASTM follow.

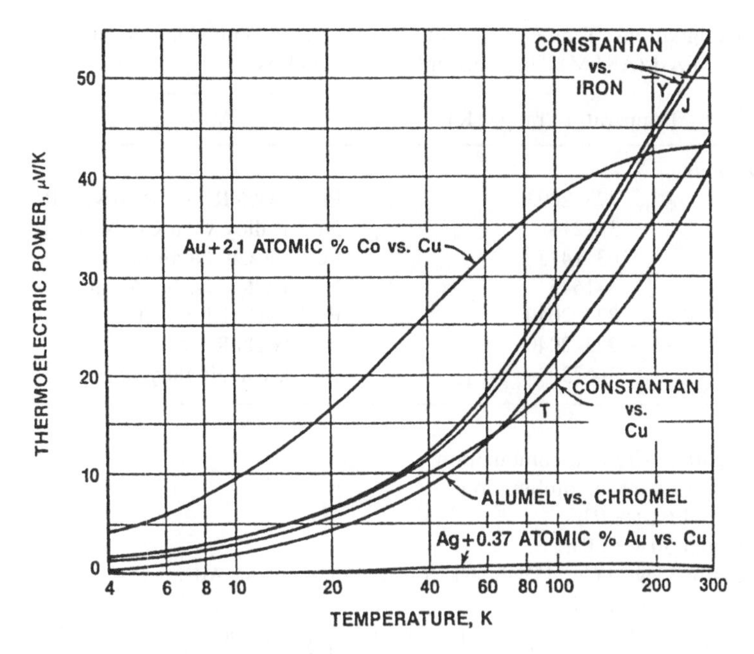


**Figure 8.50** Thermoelectric power as a function of temperature for various thermocouple combinations.

### *8.4.1. Type E Thermocouples*

The ASTM designation type E indicates a thermocouple pair consisting of a Ni–Cr alloy and a Cu–Ni alloy (the common trade names and nominal compositions for these alloys are given in Table 8.8) (Reed and Clark, 1983). This type of thermocouple has the highest Seebeck coefficient $S$ of the three ASTM standard thermocouple types commonly used at low temperatures, types E, K, and T. Also, both elements of this thermocouple have low thermal conductivity, reasonable homogeneity, and corrosion resistance in moist atmospheres. This type is the best thermocouple to use for temperatures down to about 40 K.

### *8.4.2. Type K Thermocouples*

The ASTM designation type K indicates a thermocouple pair consisting of a Ni–Cr alloy and a Ni–Al alloy (the common trade names and nominal compositions for these alloys are given in Table 8.8). The sensitivity is only about half that of the type E combination at 20 K (4.1 μV/K compared with 8.5 μV/k). The negative element is a bit more homogeneous than the EN element. Both materials have low thermal conductivity and are corrosion-resistant in moist atmospheres.

Type K thermocouples are recommended by the ASTM for continuous use at temperatures within the range 3–1533 K in inert atmospheres.

### *8.4.3. Type T Thermocouples*

The ASTM type T designation indicates a thermocouple pair consisting of Cu and Cu–Ni alloy (the common trade names and nominal composition for these alloys are given in Table 8.8). Type T is one of the older, more popular combinations

**Table 8.8** Material Designations, Temperature Ranges, and Single-Wire Component Designations for Standardized (ASTM) Thermocouple Combinations

| Type designation | Temperature range (K) | Materials |
|---|---|---|
| *Thermocouple combinations* | | |
| B | 273–2093 | Pt–30 wt% Rh vs. Pt–6 wt% Rh |
| E | 3–1273 | Ni–Cr alloy vs. a Cu–Ni alloy |
| J | 63–1473 | Fe vs. a Cu–Ni alloy |
| K | 3–1645 | Ni–Cr alloy vs. Ni–Al alloy |
| R | 223–2040 | Pt–13 wt% Rh vs. Pt |
| S | 223–2040 | Pt–10 wt% Rh vs. Pt |
| T | 3–673 | Cu vs. Cu–Ni alloy |

| *Single-wire materials* | |
|---|---|
| N | The negative wire in a combination |
| P | The positive wire in combination |
| BN | Pt-nominal 6 wt% Rh |
| BP | Pt-nominal 30 wt% Rh |
| EN or TN | Nominally 55 wt% Cu, 45 wt% Ni, constantan; Cupron, Advance, ThermoKanthal JN |
| EP or KP | Nominally 90 wt% Ni, 10 wt% Cr: Chromel, Tophel, ThermoKanthal KN |
| JN | A Cu–Ni alloy similar to, but not always interchangeable with, EN and TN |
| JP | Nominally 99.5 wt% Fe: ThermoKanthal JP |
| KN | Nominally 95 wt% Ni, 2 wt% Al, 2 wt% Mn, 1 wt% Si: Alumel, Nial, T-2, ThermoKanthal KN |
| RN and SN | High-purity platinum |
| RP | Pt–13 wt% Rh |
| SP | Pt–10 wt% Rh |
| TP | Copper, usually electrolytic tough pitch |

*Source:* Reed and Clark (1983).

and is the only one of the standardized types for which limits of error below 273.15 K have been established.

Type T thermocouples are recommended by the ASTM for use in the temperature range 89–644 K in vacuum or in oxidizing, reducing, or inert atmospheres.

### 8.4.4. *Gold–Iron Alloy Thermocouples*

The increasing use of liquid hydrogen and liquid helium has created a demand for specialized thermometry below 25 K. Ordinary thermocouple combinations are only marginally acceptable because of their low sensitivity in this range. Dilute alloys of noble metals and transition metals, however, have relatively high temperature sensitivity below 25 K; Au–2.1 at% Co is perhaps the best known of this type. Unfortunately, this alloy is a supersaturated solid solution. Powell et al. (1961) found that the thermoelectric power decreases with time when the alloy is stored at room temperature. This gradual change is probably due to diffusion of cobalt atoms to grain boundaries. Dilute alloys of iron in gold are metallurgically stable and have extremely useful thermoelectric properties at very low temperatures. A differential thermocouple made with Au–0.02, 0.03, or 0.07 at% Fe as the negative element and copper, "normal" silver (Ag–0.37 at% Au), or KP (see Table 8.8) as the positive element provides a usable sensitivity even below 4 K.

### 8.4.5. *Thermocouple Reference Data*

Reference data for thermocouples of type E, K, and T over the entire usable temperature range are given by Powell et al. (1974). Data for KP wire vs. Au–0.07 at% Fe; KP wire vs. Au–0.02 at% Cu. TP wire vs. Au–0.07 at% Fe, Cu wire vs. Au–0.02 at% Fe, normal silver wire (Ag–0.37 at% Au) vs. Au–0.07 at% Fe, and normal silver wire vs. Au–0.02 at% Fe are given by Sparks and Powell (1973). The low-temperature tables are based on data from the thermocouple thermometry program at the National Bureau of Standards at Boulder, Colorado.

The Seebeck voltage (μV) as a function of temperature (K) is represented by

$$V = \sum_{i=1}^{n} A_i T^i \tag{8.67}$$

where $n$ represents the number of coefficients for the particular combination and $T < 280$ K. The power series coefficients, $A_i$, used to represent these combinations, $V = f(T)$, are given in Powell et al. (1974) with a 0 K reference temperature. The experimental range for the data is $5 < T < 280$ K. See Table 8.9.

Coefficients representing $T = f(V)$ are often needed and are also given by Powell et al. (1974) for types E, K, and T in reduced temperature ranges. These power series coefficients are given for second-, third-, and fourth-order equations, with maximum uncertainties indicated for each order and temperature range.

For engineering use, Eq. (8.67 ) and its inverse may be reduced to

$$e = a_1 t + a_2 t^2 + a_3 t^3 + a_4 t^4 \tag{8.68}$$

or

$$t = b_1 e + b_2 e^2 + b_3 e^3 + b_4 e^4 \tag{8.69}$$

where $t$ is the difference in temperature and $e$ is the thermocouple output emf. Values of the coefficient $a$ and $b$ for three commonly used cryogenic thermocouples are given in Table 8.10.

Thermocouples in general are very simple and inexpensive to make. They can be easily installed in complex apparatus and have low heat capacities. They are very responsive to temperature changes, are fairly accurate over large temperature ranges, and can be read at a convenient location. The inherently low output and parasitic emf's pose the principal problems in their use.

## 8.5. Other Thermometers

### 8.5.1. *Capacitance Thermometers*

Glass ceramic ($SrTiO_3$) thermometers are available commercially and are very useful for temperature measurements in high magnetic fields (Rubin, 1970; Rubin et al., 1982a,b). The operating range for these devices is from $T < 1$ K to above room temperature. The sensitivity at 4.2 K is ~350 pF/K at 5 kHz and 7.5 mV (rms). The capacitance–temperature relationship for a particular sensor is shown in Fig. 8.51. The peak near 70 K may be caused by a paramagnetic-to-antiferroelectric phase transition. The capacitance–temperature relationship, $T \lesssim K$, is accurately represented by a simple fourth-order series expansion. The temperature region below the knee in the capacitance–temperature curve is of particular importance because of the good sensitivity and essentially zero magnetic field dependence.

**Table 8.9** Power Series Coefficients for Representation of Thermoelectric Voltage, $V(T)$, in the Range 0–280 K with a 0 K Reference Temperature

| | Thermocouple type | | | |
|---|---|---|---|---|
| Power series coefficient | T | E | K | KP vs. Au–0.07 at% Fe |
| $A_1$ | $-3.9974007864 \times 10^{-1}$ | $-2.0344697205 \times 10^{-1}$ | $2.4061140104 \times 10^{-1}$ | $6.9864426367$ |
| $A_2$ | $2.6329515981 \times 10^{-1}$ | $3.0220985715 \times 10^{-1}$ | $7.3438313272 \times 10^{-2}$ | $9.0607276605 \times 10^{-1}$ |
| $A_3$ | $-9.6491216443 \times 10^{-3}$ | $-5.7844373965 \times 10^{-3}$ | $1.2873437647 \times 10^{-3}$ | $-4.3469694773 \times 10^{-2}$ |
| $A_4$ | $3.8973308068 \times 10^{-4}$ | $1.7879650162 \times 10^{-4}$ | $-2.2622572598 \times 10^{-5}$ | $1.2468246660 \times 10^{-3}$ |
| $A_5$ | $-9.8186150331 \times 10^{-6}$ | $-3.6597667313 \times 10^{-6}$ | $2.1765238991 \times 10^{-7}$ | $-2.3500537590 \times 10^{-5}$ |
| $A_6$ | $1.6059280063 \times 10^{-7}$ | $4.9073685405 \times 10^{-8}$ | $-1.3304091711 \times 10^{-9}$ | $3.0837610415 \times 10^{-7}$ |
| $A_7$ | $-1.7932074012 \times 10^{-9}$ | $-4.4751468891 \times 10^{-10}$ | $5.2493539029 \times 10^{-12}$ | $-2.9032251684 \times 10^{-9}$ |
| $A_8$ | $1.4080710479 \times 10^{-11}$ | $2.8331235582 \times 10^{-12}$ | $-1.2997123230 \times 10^{-14}$ | $1.9881512159 \times 10^{-11}$ |
| $A_9$ | $-7.8671373053 \times 10^{-14}$ | $-1.2476596612 \times 10^{-14}$ | $1.8403309812 \times 10^{-17}$ | $-9.9174829612 \times 10^{-14}$ |
| $A_{10}$ | $3.1144995156 \times 10^{-16}$ | $3.7536769066 \times 10^{-17}$ | $-1.1382797374 \times 10^{-20}$ | $3.5645229362 \times 10^{-16}$ |
| $A_{11}$ | $-8.5433550766 \times 10^{-19}$ | $-7.3627479508 \times 10^{-20}$ | | $-8.9864698504 \times 10^{-19}$ |
| $A_{12}$ | $1.5448411036 \times 10^{-21}$ | $8.4898427718 \times 10^{-23}$ | | $1.5071673023 \times 10^{-21}$ |
| $A_{13}$ | $-1.6565456476 \times 10^{-24}$ | $-4.3671808488 \times 10^{-26}$ | | $-1.5093916059 \times 10^{-24}$ |
| $A_{14}$ | $7.9795893156 \times 10^{-28}$ | | | $6.8264293980 \times 10^{-28}$ |

*Source*: Powell et al. (1974).

**Table 8.10** Values of $a$ and $b$ in the Simplified Thermocouple emf-T Interpolation Formulas, Eq. (8.68) and (8.69)[a]

| Coefficient | Copper–constantan,[b] type T | Chromel–Alumel,[b] type K | Chromel–Au/0.03 Fe[c] |
|---|---|---|---|
| $a_1$ | $-3.87706 \times 10^{-2}$ | $-3.94841 \times 10\text{-}2$ | $-1.53129 \times 10^{-2}$ |
| $a_2$ | $-4.56877 \times 10^{-5}$ | $-2.83938 \times {}^{10^{-5}}$ | $-7.87084 \times 10^{-5}$ |
| $a_3$ | $4.35205 \times 10^{-8}$ | $1.13868 \times 10^{-7}$ | $-9.79295 \times 10^{-7}$ |
| $a_4$ | $1.51931 \times 10^{-11}$ | $2.57457 \times 10^{-11}$ | $-1.59091 \times 10^{-9}$ |
| $b_1$ | −25.47783 | −24.90286 | −64.79915 |
| $b_2$ | −1.45756 | −1.33496 | −28.39826 |
| $b_3$ | 0.29713 | 0.37298 | 24.75019 |
| $b_4$ | −0.06419 | −0.06882 | −2.34307 |

[a] t = temperature minus reference temperature; $e$ = absolute value of emf. $t$ is in °C or K: $e$ is in mV.
[b] The reference junction for type K and type T thermocouples is 0°C (32°F).
[c] The reference junction for the Chromel–Au/0.03 Fe thermocouple is 77.3 K (139.1°R).

The bulk of the work on capacitance thermometers since 1969 has been for two types of materials: $SrTiO_3$ and doped alkali halide crystals. The former have been studied since 1970 both above and below the liquid helium range. Commercially available units of this type are compact in size, and equations have been developed to fit the $C$ vs. $T$ relationship. Unfortunately, aging effects (i.e., drifts with time) and slight calibration irreproducibilities with thermal cycling limit their usefulness as a reliable secondary standard thermometer for experiments in any temperature range (Rubin et al., 1982a).

The aging or drift problem associated with capacitance thermometers may be due to relaxation of the glass matrix and will occur whenever this matrix is perturbed. Perturbation is caused by thermal expansion ($T > 77$ K), by electrostrictive coupling of the $SrTiO_3$ microcrystals in the glass matrix following a voltage pulse,

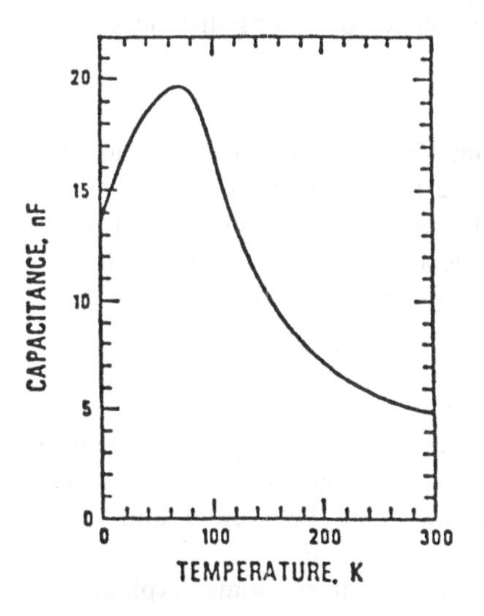


**Figure 8.51** Capacitance as a function of temperature for a glass ceramic capacitance thermometer. The measurement frequency was 5 kHz, and the rms amplitude was 7.5 mV. (Courtesy of LakeShore Cryotronics Inc., Westerville, OH.)

or by crossing the ~65 K phase transition temperature. The aging brought about by any of these occurrences may cause a change in capacitance equivalent to ~15 mK over a period of 30 min. The aging phenomenon and such things as stray lead capacitances and frequency and amplitude dependence limit the use of this type of sensor as an absolute thermometer at the present time.

Capacitance thermometers are nonetheless widely used and well-characterized sensors. Although not suitable at the present time for absolute temperature measurements, they are valuable as control devices in magnetic field environments.

### *8.5.2. Pyroelectric Thermometers*

Pyroelectric thermometers utilize the temperature dependence of spontaneous electrical polarization that occurs in some anisotropic solids. The thermometers can be used to detect extremely small temperature changes (on the order of a few tenths of a microkelvin) and rates of temperature change (on the order of hundredths of μK/sec) in the temperature range 4–500 K.

### *8.5.3. Resonance Thermometers*

Two resonance phenomena, nuclear quadrupole resonance (NQR) and nuclear magnetic resonance (NMR), can be used to measure low temperatures. NQR provides usable temperature sensitivity in the range from ~20 to 500 K; accuracy of ~1 mK is possible near 77 K. Accuracies on the order of 1 mK can be achieved using an NMR thermometer at temperatures above 50 K.

### *8.5.4. Inductance Thermometers*

Inductance thermometers make use of the temperature dependence of the magnetic properties of core materials. When an inductor is incorporated into a tunnel diode oscillator circuit, the temperature dependence of the core is reflected in the oscillator frequency. Inductors with cores of specially treated alloys can be used in the temperature range 4–300 K with a resolution of better than 10 mK. Standard broadcasting techniques can be used to transmit the frequency over long distances.

### *8.5.5. Quartz Thermometers*

Development of quartz crystal frequency thermometers continues. These devices use the small but highly reproducible variation of the natural vibration frequency of appropriately cut piezoelectric quartz samples. The sensors are compact and useful over a wide cryogenic range (4–400 K) with an accuracy within a few hundredths of a kelvin. There are some problems with hysteresis and aging, however.

## **8.6. Thermometry Summary**

Table 8.11 summarizes some representative characteristics of the best known types of cryogenic thermometers. Figure 8.52 gives a graphic summary of the approximate useful range of many common cryogenic thermometers between 4 and 300 K (Sparks, 1983).

The terms “reproducibility” and “accuracy” require some explanation. *Reproducibility* means the variability observed in repeating a given measurement using the best present-day laboratory techniques. Changes produced on thermal cycling of the thermometer to and from ambient are included in this parameter.

**Table 8.11** Some Approximate Characteristics of the Most Widely Used Classes of Thermometers

| Type | Range (K) | Best reproducibility (mK) | Best accuracy[a] (mK) | Response time (sec) | Relative size[b] |
|---|---|---|---|---|---|
| Resistance thermometers | | | | | |
| Platinum | ≥30 | 0.1–1 | 10 | 0.3–1.3 | 2, 3 |
| Rhodium–iron | 0.5–300 | 0.5 | 10 | 0.3–1.3 | 2 |
| Carbon | 1–30 | 1–10 | 1–10 | 0.1–10 | 2 |
| Carbon glass | 1–300 | 0.75 | 1 | 0.1–10 | 2 |
| Germanium | ≤30 | 0.5 | 1–10 | 0.1–10 | 2 |
| Diode (Se/GaAs) 1.4–400 | ±25 | 10 | 0.01–0.001 | 1, 2 | |
| Thermocouples | | | | | |
| Gold–cobalt vs. copper | 4–300 | 10–100 | 10 | 1 or less[c] | 1 |
| Constantan vs. copper | 20–600 | 10–100 | 10 | 1 or less[c] | 1 |
| Vapor pressure | | | | | |
| Helium | 1–5 | 0.1–1 | 0.2 | 0.1–100 | 4 |
| Hydrogen | 14–33 | 1 | 2 | 0.1–100 | 4 |
| Nitrogen | 63–126 | 1–10 | 2 | 0.1–100 | 4 |
| Oxygen | 54–155 | 1–10 | 2 | 0.1–100 | 4 |

[a] Including nonreproducibility, calibration errors, and temperature scale uncertainty.
[b] From 1 (smallest) to 4 (largest).
[c] Bare wire.

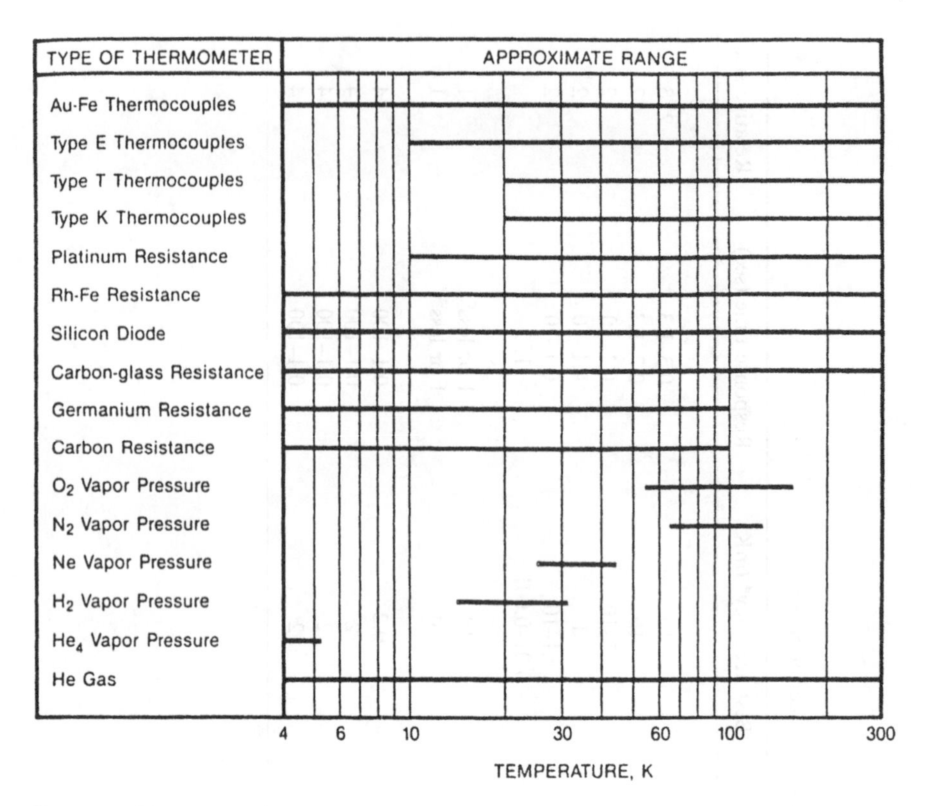


**Figure 8.52** Approximate range of use of several thermometers between 4 and 300 K.

*Accuracy* means the significance with which the thermometer can indicate the absolute thermodynamic temperature. This includes errors of calibration and errors due to nonreproducibility, the former usually being much more significant.

The approximate numbers given for these quantities represent good current practice. It may be possible to do better by exercising extreme care. On the other hand, in most engineering measurements a lower order of accuracy is permissible, and this may allow relaxation of certain requirements such as strain-free mounting of resistors, homogeneity of thermocouple materials, purity of vapor pressure fluid, and sensitivity and accuracy of associated instruments.

For temperatures above about 20 K, the metallic resistance thermometers are more sensitive than the nonmetallic resistance thermometers. Temperatures above 20 K can be measured routinely with an industrial platinum resistance thermometer with an accuracy of better than 100 mK and time responses somewhat better than 1 sec. Accuracy at the millikelvin level requires a precision capsule type of platinum resistance thermometer and careful calibration.

Carbon thermometers are generally used for low-temperature measurements ($T < 80\,\text{K}$) when accuracies of $\pm 0.1\,\text{K}$ or $\pm 1\%$ of the absolute temperature are needed. Millikelvin accuracy is attainable using germanium resistance thermometers at temperatures below 20 K. The primary drawback to germanium thermometers is that no simple analytical representation is available for the resistance–temperature characteristics, even for a given class of doped germanium crystals. A many-point comparison calibration is required if all the inherent stability of the resistor are to be utilized.

Thermocouples of type E, K, T, and KP vs. Au–0.02, 0.07 at % Fe can be used very well in the liquid oxygen range. If a gold–iron alloy is to be used above about 20 K, the positive thermoelement should be type KP. Use of the KP material allows the gold–iron alloys to be used from below the normal boiling point of helium up to room temperature. These couples have an extremely linear sensitivity above 20 K. Type E is recommended for general use when temperatures are not below 20 K (Sparks, 1974).

Vapor pressure and gas thermometry offer sensitive methods of temperature measurement with the advantage that no calibration is necessary. Further advantages are that these transducers are not sensitive to magnetic or electric fields. In the case of vapor pressure thermometers, the time response may be made comparable to that of the resistance thermometers.

A sensible recommendation about which type to use cannot be made without knowing the specific measurement requirements. The generalization most likely to hold true, however, is the following: for crude temperature indications and monitoring trends, a gas thermometer will probably be suitable; for more accurate work, vapor pressure or resistance thermometry is recommended; for engineering application in this latter area, the need is for inexpensive devices that do not call for unique calibration.

Outstanding reviews of cryogenic temperature measurements have been published by Sparks (1974, 1983) and by Rubin et al. (1982a,b) and are highly recommended.

# 9

# Cryogenic Equipment and Cryogenic Systems Analysis

## 1. INTRODUCTION

### 1.1. Design Pitfalls

What is it about cryogenics that makes cryogenic equipment different from the equipment of ordinary chemical or mechanical engineering? The most common pitfall in cryogenic equipment design and selection is assuming that cryogenic fluids and materials behave much like those of our ordinary experience. They do not. Indeed, the whole point of treating cryogenic engineering as if it were a separate discipline of engineering is that there are special properties of cryogenic fluids and materials that demand special techniques. We are not better than any other engineer, just different. Our work is not more or less challenging than his (or hers), just different. I do not think much of articles proclaiming how tough we have it, how smart we must be to do cryogenics. We do have to be different, however.

What is it about cryogenic engineering that is different to such an extent that might cause a well-trained engineer difficulty? This idea came forcefully to mind while I was reviewing some pressure drop calculations. The engineer used the ordinary Fanning equations to calculate a pressure drop for flowing liquid nitrogen. What he did not consider was that the liquid nitrogen was flowing in an uninsulated line at nearly atmospheric pressure and would therefore be a boiling two-phase mixture of liquid and vapor. Secondly, the pressure drop in question occurred during cooldown, a transient condition that can last several minutes and during which the fluid can even flow backward in the transfer line and often does. His calculations were precise but totally inaccurate.

Listed below are a few design pitfalls that can and do occur in cryogenic equipment selection.

1. Cryogenic fluids have unexpectedly low heats of vaporization. There is a rule of thumb from physical chemistry that says that the normal heat of vaporization is proportional to the normal boiling point in absolute units. As mentioned earlier in Chapter 3, it is called Trouton's rule and says that the entropy of vaporization at the NBP is a constant, and hence the heat of vaporization divided by the normal boiling point is a constant. We can therefore expect that the heat of vaporization of liquid nitrogen (NBP 77 K) is about one-fourth of a liquid boiling at room temperature (300 K), and that liquid hydrogen (NBP 20 K) would have a heat of vaporization one-fifteenth that of a room temperature boiling fluid. Hence you can expect that it does not require much heat leak for liquid hydrogen to boil away.

However, the truth is even much worse than that. Both hydrogen and helium have a built-in energy called the zero (temperature) point energy as a result of their being quantum fluids (see Chapter 3). This built-in energy means that it takes even less external energy to boil (or melt) these fluids (or solids) than given by the above rule, which is low enough already. In fact, the heat of fusion (there is a Trouton's rule for any phase change) is about 20% less for solid hydrogen than the already low number predicted above. The heat of vaporization of helium is reduced by about 50% over the expected value. Cryogenic fluids do not require much energy to boil.

2. Accordingly, cryogenic fluids are always boiling two-phase systems unless deliberate steps are taken to the contrary. The boiling point is low, almost by the definition of cryogenics. This means that there is a high $\Delta T$ compared to our room temperature experience and that heat is always leaking into the cryogenic system. The heat of vaporization is correspondingly low, as seen above. This combination of high $\Delta T$ and low heat of vaporization means that the normal state for cryogenic fluids is always a two-phase boiling system. This normally occurring boiling two-phase system can be temporarily suspended by increasing the pressure above the NBP pressure.

3. Because cryogenic fluids are nearly always boiling two-phase systems, the net positive suction head (NPSH) is nearly always 0. The initial stage of a cryogenic pump is more fan-like than compressor-like.

4. Because cryogenic fluids are normally two-phase systems, the pressure depends upon the temperature, and the temperature depends upon the pressure. It is not possible to arbitrarily set the pressure and temperature independently of one another. One is always a dependent variable of the other. This situation can be temporarily upset by pressurization; see (2) above.

5. Because cryogenic fluids are boiling, they must be treated as potentially high-pressure systems. They must be vented or afforded adequate pressure relief devices. There are no exceptions. Cryogenics is always potentially a high-pressure game. (See Chapter 11.)

6. Cryogenic fluids have a relatively high coefficient of thermal expansion, compared to fluids of our normal experience. A closed cryogenic tank will gradually warm due to the relentless heat leak inward. See (2) above. If not vented, the liquid will warm and its density will decrease. This means that the liquid level will rise over time, even though the mass in the tank has not changed. The tank will appear to get more full, as measured by any liquid level detector. This effect can cause a lot of consternation, as an operator may wonder where the leak of liquid into the tank is. This is the reason that an ullage space must be provided above any stored cryogenic fluid. See Chapter 11.

7. Cryogenic fluids are not necessarily homogenous. Another consequence of (6) above is that the warmer liquid, being less dense, rises to the top of the tank. This is a stable configuration. Since the pressure in the tank is controlled by the vapor pressure of the top layer only, the tank will tend to reach venting pressure much earlier than anticipated. Another consequence is that it is difficult to describe the density or temperature of the fluid in the tank. Since both are anisotropic, where shall we sample for the representative density or temperature?

8. There is no such thing as an "empty" cryogenic tank. Because the vapors are cold, a hydrogen tank that has been emptied of its liquid will still contain about 2% by mass of the original contents. This is product that cannot be removed by pumping but must stay behind. This is especially vexing in the cryogenic space

propulsion business where the "empty" hydrogen fuel tank can still contain up to 2% of its original mass.

9. All construction materials are not weaker at low temperatures. The first pitfall here results from confusing strength with ductility. All crystalline solids become stronger (as measured by yield strength or ultimate tensile strength) as the temperature decreases (see Chapter 4). An added safety factor is built in just by using the room temperature values for design, and thus over designing. Some materials do become brittle at low temperatures, and some retain their ductility down to absolute zero. Is this always a source of concern? Not necessarily. It depends upon the service condition. We have been making cryogenic apparatus out of glass since the time of Dewar, and there is no material more brittle than glass. If the container is just going to sit there, experiencing steady state one-dimensional loads, there is not necessarily a harm in using a brittle material. If, however, one must plan on the fork-lift truck backing into the cryogenic apparatus, and in such a case, a brittle material will be less than satisfactory.

10. Scaling, that is making something bigger from a smaller model, is tricky. Once a group of otherwise intelligent people bought a 1-L dewar type vessel at a coffee shop. Honestly. They filled it with $LN_2$ and noted the boil-off was 1 liter per day. They then proceeded to make a 10-L dewar of the same construction and expected the boil-off rate to be constant at 1 L per day. They had been sold on the idea of the NER (normal evaporation rate) being the appropriate scaling factor. In fact, what works best is to use the percent-of-contents-per-day as the scaling factor. The 1-L coffee thermos had an NER of one but 100% of contents per day. The 10-L scale up would boil away 100% of its contents per day, or 10 L per day. It did, much to their chagrin.

11. Refrigeration by isentropic expansion is not necessarily more efficient than expansion through a valve. When the expansion occurs below the critical point, there is hardly any difference between the performance of a real expansion engine and the expansion valve. This has to do with the shape of the isentropic lines and the isentropic lines in the two-phase region. See Chapter 3.

12. Superinsulation is not always the best insulation. Superinsulation, or multilayer insulation (MLI), has the lowest thermal conductivity of any insulation known (Chapter 7). However, whether it is better or not depends upon the application. In fact, the thermal performance of one insulation can be made as good as another, if we can vary the thickness. One-half inch of real MLI as actually applied has about the same thermal performance of 6 inches of closed cell foam (foam ice chest). The whole business of "better and best" anything depends very much on intended use.

13. Cryogenic fluids make poor paperweights. They are not noted for their density, except for oxygen. Liquid oxygen has a specific gravity of about 1.13, or 113% times the density of water. That is why we like to make and store oxygen as a liquid. Liquid nitrogen has a specific gravity of about 0.81, and the specific gravity of liquid hydrogen is only 0.07. Liquid hydrogen does not pack well. There is more hydrogen in a gallon of water than there is in a gallon of pure liquid hydrogen. The coffee-jug-dewar crowd mentioned above also tried to make a liquid nitrogen shipper bottom heavy by making the inner container in the shape of an Erlenmeyer flask, a conical laboratory flask with a narrow neck and a flat, broad bottom. That might have been useful had the liquid nitrogen had the density of water but it does not. They would have been better off putting a brick in the bottom of the dewar if bottom weighting were really important.

14. Liquid nitrogen is inert. Well, maybe, sometimes. The first pitfall is that liquid nitrogen, like every cryogen, represents a potential high-pressure gas hazard. Worse, liquid nitrogen, when in contact with a bare uninsulated pipe, will condense enriched air from the atmosphere. See Chapters 3 and 11. The composition of the enriched air is 50% oxygen, or 100 proof as the distillers say. Fifty percent oxygen is definitely a fire and explosion hazard. There are documented cases of liquid nitrogen causing 50% oxygen enriched air to condense onto flammable substances (say, a greasy floor) with disastrous results, including death. Beware of inert liquid nitrogen.

## 1.2. Commonly Used Cryogenic Equipment

Historically, the first concern of cryogenics was the liquefaction of gases, and still today gas liquefaction is employed in nearly all processes requiring quite low temperatures.

The fundamental requirement for gas liquefaction is a refrigerative process capable of removing heat from the gas until the liquefaction temperature is reached. In the quest for better liquefaction processes, the principal goals are (1) the development of a refrigerator with the highest possible efficiency and (2) the design of equipment which will not waste the refrigeration that has been produced. These goals are common to all cryogenic processes.

The production of low temperatures requires an assortment of highly specialized pieces of equipment including compressors, expanders, heat exchangers, pumps, transfer lines, and storage tanks. See Fig. 9.1. As a general rule, the design of such equipment follows methods applicable to ambient temperature design. However, pervading every aspect of this design must be a thorough understanding of the temperature effect on the properties of the fluids being handled and the materials of construction being selected. See the section on "Design Pitfalls".

These design considerations become particularly apparent in heat-transfer design where the downward trend in temperature is accompanied by a downward trend in many of the properties such as specific volume, heat capacity, and thermal conductivity which enter as variables in a Nusselt-type heat-transfer relation. With a decrease in temperature, the refrigeration cost increases, making it imperative to recover as economically as possible the refrigeration effect in effluent streams by heat exchange with input streams. Since irreversible effects become magnified with a decrease in temperature, lower temperature differences and lower entropy increases are in order for heat exchangers.

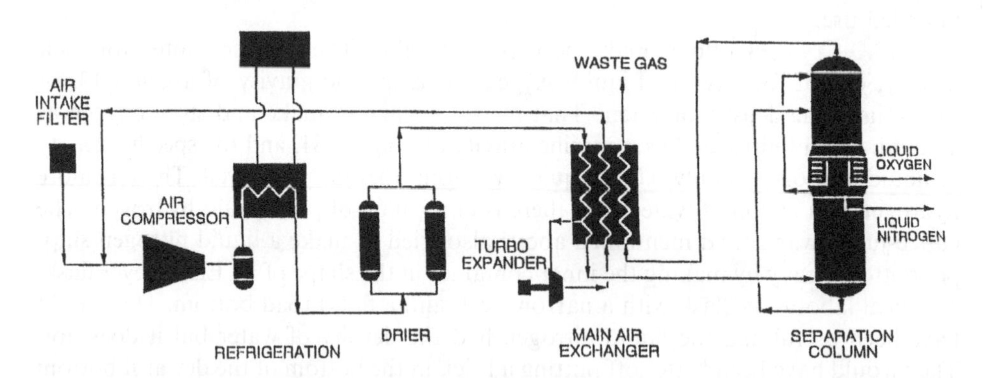


**Figure 9.1** Typical cryogenic process flow diagram.

**Table 9.1** Cryogenic Process Equipment

- Low-temperature process systems utilize
  - Compressors
  - Expanders
  - Heat exchangers
  - Pumps
  - Transfer systems
    - Piping; fluid control
- Irreversible effects are magnified with decreasing temperatures
- Thermal effects may lead to undesirable mechanical stresses
- Material mismatches can be hazardous
  - Potential leak sources, line rupture; others

Similar considerations must be given to materials of construction since these usually contract as temperature decreases and can result in unwanted stresses on expanders, pumps, piping, etc. Some materials suffer a great loss of impact resistance and fatigue strength below a characteristic temperature. Operation below such temperatures requires the choice of other, perhaps more exotic, materials of construction. Additionally, combinations of materials may be hazardous, especially at interface conditions such as at flanges where reduction to the low working temperature may relieve the gasket pressure and cause leaks. Likewise, welds of dissimilar interfaces are another source of stress and can also result in potential hazards. See Table 9.1.

Table 9.2 organizes the equipment shown in Fig. 9.1 in the order normally used in a cryogenic process. Table 9.2 also shows that the equipment basic to the liquefaction of gases is basic to every other cryogenic process. For very high performance processes involving hydrogen and helium, the performance of storage and transfer systems becomes critical, and these components are further broken down in Table 9.2.

Thus the equipment required for the liquefaction of gases (left column, Table 9.2) constitutes not only a logical but also a very practical introduction to cryogenic equipment. In the following sections, we shall follow the outline of Table 9.2.

**Table 9.2** Cryogenic Equipment

| | |
|---|---|
| Process equipment for air separation and liquefaction plants | Storage systems for *higher performance* applications |
| Compressors | Dewar systems |
| Expanders | Supports |
| Joule–Thompson valves | Penetrations |
| Heat exchangers | Structural elements |
| Storage systems | Insulation/vapor cooled shields |
| Transfer systems | Safety devices |
| Transportation | Vacuum technology |
| | Transfer systems for cryogenic systems |
| | Flow control devices |
| | Piping/flex lines |
| | Vaporizers |
| | Filters/miscellaneous |

## 2. COMPRESSORS

In the production of industrial gases such as oxygen, nitrogen, and helium, or natural-gas liquefaction, compression power accounts for more than 80% of the total energy requirements. Thus careful selection of the compressor and its drive are essential if cost and maintenance of the cryogenic facility are to be minimized. Essential to any evaluation are (1) cost of equipment and installation, (2) energy or fuel costs, and (3) cost of maintenance. Finally, the effect of plant site, available fuel source, and its reliability, existing facilities, and electrical power structure will ultimately determine the final selection.

### 2.1. Compressor Fundamentals

The principles discussed below apply (only) to the most common use of compressors, namely the compression of fluids in a steady state flow process, which was first described in Chapter 2.0, Basics.

*2.1.1. Compression Paths*

There are three basic compression paths that may be treated thermodynamically: reversible adiabatic, reversible isothermal, and reversible polytropic. The real world of irreversible compressors is handled by assigning an efficiency to the real process. We first predict performance by use of one of the reversible processes above. Then we describe the real world approach to the reversible process by an efficiency, based on experience. The efficiency issue is discussed in Sec. 2 of this chapter.

**(a) The reversible Adiabatic Compression (Isentropic) process**

The energy balance for a steady state flow system is

$$\Delta H + \frac{\Delta u^2}{2g_c} + \frac{g}{g_c}\Delta Z = Q - W_s$$

Applied to a reversible adiabatic compression ($S_{in} = S_{out}$), where $Q = 0$, and where gravitational and velocity effects are negligible, the equation becomes

$W_s = \Delta H$, where $W_s$ is the shaft, or compressor, work

assuming that kinetic and potential energy changes are negligible, as above.

A reversible adiabatic compression path is the path that requires the *maximum* work. This is not good for compressors but is exactly what is desired for expanders (Sec. 3 of this chapter).

1. *Ideal gas isentropic compression*

If the gas is ideal the work may be calculated from

$$W_s = \frac{RT_1\gamma}{\gamma - 1}\left[1 - \left[\frac{P_2}{P_1}\right]^{(\gamma-1)/\gamma}\right]$$

where $\gamma = C_P/C_V$.

Note that the work is proportional to the pressure ratio, not the pressure difference. This is the idea behind the Linde Dual Pressure Process (Sec. 6.0).

2. *Real fluid isentropic compression*

There are essentially three methods of evaluating $\Delta H$ for real fluids: equations of state, thermodynamic diagrams, and generalized correlations.

*Equations of state:* An equation of state combined with heat capacities ($C_p$) can be used to evaluate $\Delta H$. The thermodynamic relations (see Chapter 3)

$$\mathrm{d}H = C_P\,\mathrm{d}T + \left[V - T(\partial V/\partial T)_P\right]\mathrm{d}P, \qquad \mathrm{d}S = \frac{C_P}{T}\mathrm{d}T - (\partial V/\partial T)_P\,\mathrm{d}P$$

are used in the following manner.

First the outlet temperature, $T_2$, is calculated by integrating the equation for d$S$, remembering that a reversible adiabatic process is isentropic, so that $\mathrm{d}S=0$. The equation of state is used to evaluate $[-(\partial V/\partial T)_P\,\mathrm{d}P]$ and the ideal gas heat capacities are used for $C_P\,\mathrm{d}T$. After determining $T_2$, the enthalpy difference between outlet and inlet conditions is determined by integration of the d$H$ relation. This technique may be used with either pure fluids or mixtures.

*Thermodynamic diagrams:* If a thermodynamic chart ($T$ vs. $S$, $H$ vs. $S$, etc.) is available for the fluid, the work is calculated in the following manner. Since $\Delta S=0$, a line of constant entropy is traced on the thermodynamic chart from $P_1$, $T_1$ to the intersection with $P_2$, thus establishing $T_2$. The value of $\Delta H$ is then read directly from the chart. Since $T$–$S$ charts are based on real properties, with no assumptions, this method is to be preferred.

*Generalized correlations:* When dealing with mixtures or pure fluids for which specific thermodynamic charts are unavailable, generalized correlations of thermodynamic properties may be used. The procedure is illustrated in the *GPSA Engineering Data Book* (GSPA, 2001).

**(b) Reversible Isothermal Compression**

Inlet conditions $T_1$, $P_1$
Outlet conditions $T_1$, $P_2$ (since $T_1=T_2$)
$Q$: heat rejected to maintain isothermal process
$W_s$: shaft work done in compression from $P_1$ $T_1$ $P_2$

The energy balance for this situation is

$$W_s = Q - \Delta H = T_1\Delta S - \Delta H$$

assuming that potential and kinetic energy effects are negligible.

A reversible, isothermal compression path is the path that requires the *minimum* work.

1. Ideal gas

For the ideal gas, the work is

$$W_s = RT_1 \ln(P_2/P_1)$$

Note that, as in the adiabatic compression, the work is proportional to the *pressure ratio* rather than the pressure difference.

2. *Real fluid*

As in the adiabatic compression, there are essentially three methods for evaluating the work: equations of state, thermodynamic diagrams, and generalized correlations.

*Equations of state:* The thermodynamic relations for d$H$ and d$S$ presented above in adiabatic compression may be used with ideal gas heat capacities ($C_P$) and an equation of state to evaluate $\Delta H$ and $\Delta S$.

*Thermodynamic diagrams:* The values of $\Delta H$ and $\Delta S$ may be read directly from the available charts ($T$ vs. $S$, $H$ vs. $S$, etc.).

*Generalized correlations:* Refer to the discussion under adiabatic compression described in the *GPSA Engineering Data Book* (GPSA, 2001).

**(c) Reversible Polytropic Compression**

$T_1 \neq T_2$

$S_1 \neq S_2$

$Q$ and $W_s$ as before

Reversible adiabatic and reversible isothermal compressions described above represent the two extremes. A reversible polytropic process is one intermediate between the isothermal and adiabatic situations and is an attempt to model real behavior more closely. In a polytropic process, heat is transferred, with the result that the process is not isothermal. A number of possible polytropic paths may be used, but probably the most common is

$$PV^n = \text{constant}$$

For an ideal gas, $n = \gamma = C_P/C_V$, $\frac{7}{5}$ for a diatomic gas, $\frac{5}{3}$ for a monatomic gas. Note how the compression of He, Ar, Ne is fundamentally different from the compression of $H_2$ and $N_2$, for instance. Why?*

For real gases, the exponent, $n$, which is assumed constant for the path, is obtained experimentally by evaluating $PV^n$ at the initial and final conditions:

$$n = \frac{\ln(P_2/P_1)}{\ln(V_2/V_1)}.$$

Note that to obtain a value for $n$, it is necessary to know both the outlet pressure and temperature. (Knowing $P_2$ and $T_2$, $V_2$ may be obtained from the known physical properties of the fluid, or calculated from an equation of state.) The work for a reversible polytropic process is calculated from

$$w_s = P_1 V_1 \left(\frac{n}{n-1}\right)\left[1 - \left[\frac{P_2}{P_1}\right]^{(n-1)/n}\right]$$

### 2.1.2. *Multistaging*

Compression processes are often performed in several stages rather than a single operation for three reasons. First, compression processes take place so rapidly that they approach adiabatic behavior (maximum work) rather than isothermal behavior (minimum work). Allowing the compression to take place in several independent stages, and cooling the fluid between stages, allows a closer approach to isothermal operation and thus results in substantial savings in compressor work requirements. Second, for high compression ratios, the outlet gas temperature will be unacceptably high. For example, the adiabatic compression of an ideal diatomic gas ($C_P/C_V = 1.4$) at 80°F results in an outlet temperature of 595°F for a pressure ratio of 10:1. The mechanical design limit for outlet temperature is usually 350–400°F (GSPA). Thus high-pressure ratios demand staging with intercooling. Third, mechanical considerations make it difficult to design a cylinder large enough to take in substantial

*Because He, Ar, and Ne are monatomic gases ($n = 5/3 = 1.67$) and $H_2$ and $N_2$ are diatomic gases ($n = 7/5 = 1.4$).

volumes of the lower pressure gas and yet strong enough to withstand the high outlet pressure resulting from a large compression ratio.

It is not possible to give an exact rule for determining the number of stages for a compression process, but some general guidelines are available. A common guide states that the maximum pressure ratio per stage for a reciprocating compressor is usually 3:1 to 4:1. Furthermore valve design in reciprocating machines may limit the pressure rise per stage to 1000 psi or less. In large multistage machines, the pressure ratio per stage will usually be between 3:1 and 5:1, with somewhat higher ratios for single stage machines. The Gas Processors Suppliers Association (GPSA, 2001) states that the gas discharge temperature normally controls the compression ratio, and they make the following recommendations for outlet temperatures of reciprocating compressors (GPSA, 2001):

> The maximum ratio of compression permissible in one stage is usually limited by the discharge temperature. When handling gases containing oxygen, which could support combustion, there is always a possibility of fire and explosion because of the oil vapors present.
>
> To reduce carbonization of the oil and the danger of fires, a safe operating limit may be considered to be approximately 300°F. Where no oxygen is present in the gas stream, temperatures of 350–400°F may be considered as the maximum, even though mechanical or process requirements usually dictate a somewhat lower figure.
>
> Packing life may be significantly shortened by the dual requirement to seal both high pressure and high temperature gases. For this reason, at higher discharge pressures, a temperature closer to 250°F or 275°F may be the practical limit.
>
> In summary, the use of 300°F maximum would be a good average. Recognition of the above variables is, however, still useful.

It can be demonstrated that the minimum work requirement for reversible, adiabatic compression of an ideal gas having a constant $\gamma$ is obtained when the work, or pressure ratio, for each stage is the same. The same principle, equal work per stage, also applies for non-adiabatic compressions of non-ideal gases, but of course the pressure ratios for the stages will not be the same.

#### 2.1.3. *Efficiencies*

There are a number of different efficiencies that may be defined and used.

Adiabatic efficiencies are normally used when discussing reciprocating compressors, while polytropic efficiencies are generally used for axial and centrifugal machines. For quick estimates, common practice is to use an overall value of 0.75, to 0.8. The efficiency of compressors is definitely downward with size. This decreasing efficiency with size is a definite problem for small-scale cryocoolers, where the compressor efficiency may be only 30% or less.

### 2.2. Compressors

A general classification of compressors is shown below.

The approximate ranges of application for the various types are given in Table 9.3.

Note that the application ranges are approximate and that the figure and table, from different sources, are not in perfect agreement.

In these notes, consideration will be given only to the more common compressors.

**Table 9.3** Compressor Maximum Capabilities

| | Speed range, rpm | Maximum capacity, cfm | Maximum pressure, psig | Maximum, bhp |
|---|---|---|---|---|
| Reciprocating | 100–1800 | 10,000 | 60,000 | 26,000 |
| Rotary vane | 600–2200 | 6000 | 400 | 860 |
| Helical lobe (screw) | 1800–12,000 | 20,000 | 250 | 6000 |
| Straight lobe | 600–1800 | 30,000 | 25 | 2000 |
| Liquid piston | 570–3500 | 13,000 | 90 | 400 |
| Centrifugals | | | | |
| Sectionalized | 1000–20,000 | 20,000 | 10 | 600 |
| Horizontally split | 1000–20,000 | 650,000 | 1000 | 35,000 |
| Vertically split | 1000–10,000 | 250,000 | 5000 | 20,000 |
| Axial | 1000–10,000 | 1,000,000 | 500 | 100,000 |

### *2.2.1. Reciprocating Compressors*

The trademark of reciprocating compressors is their adaptability to a wide range of volumes and pressures with high efficiency. Some of the largest units for cryogenic gas production range up to 15,000 brake horse-power (bhp) and use the balance opposed machine concept in multistage designs with synchronous motor drive. When designed for multistage, multiservice operation, these units incorporate manual or automatic, fixed or variable, volume clearance pockets and externally actuated unloading devices where required. Balanced opposed units not only minimize vibrations, resulting in smaller foundations, but also allow compact installation of coolers and piping, still further increasing savings.

Air compressors for constant speed service normally utilize piston-type suction valve loaders for low-pressure-type lubricated machines. Non-lubricated units require diaphragm-operated unloaders. Medium pressure compressors for argon, hydrogen, etc., generally use this type also. The trend toward non-lubricating machines has led to piston designs using glass-filled Teflon rider rings and piston rings, with cooled packings for the piston rods.

Rotative speeds of larger units have increased to 277 r/min, and with piston speeds for air service of up to 850 ft/min. The larger compressors with provision for multiple services reduce the number of motors, or drivers, and minimize the accessory equipment, resulting in lower maintenance cost.

Compressors for oxygen service are characteristically operated at lower piston speeds, on the order of 650 ft/min. Maintenance of these machines requires rigid control of cleaning procedures and inspection of parts to insure absence of oil in the working cylinder and valve assemblies. See Sec. 2 of this chapter.

Selection of the driver for a reciprocating compressor is dependent upon the economics of a particular plant site. If steam-generating equipment is in use and an adequate supply is assured, then turbine-geared drive is feasible. Where low power rates are available from local utilities, motor drive should be the first selection. For low-speed reciprocating compressors, synchronous motor drives offer high efficiency and power factor, low starting kVA, and lower overall cost. Recent new brushless designs employing semiconductor devices with built-in automatic excitation simplify construction and motor control equipment. These are particularly suitable for hazardous areas. Gas engine prime movers are used where gaseous fuels are still in good supply and at reasonable cost. However, because of the rising costs of energy, consideration is being given to the turbocharged tripower engine which makes changeover to liquid fuels or conversion to spark ignition possible.

Engine drivers that are of the variable speed type can generally control the speed with a 100% to 50% variation. Such control results in high efficiency since compressor fluid friction losses decrease at the lower r/min.

Reciprocating compressors are classed as constant-volume, variable-pressure machines for low to medium flow and covering the entire range of discharge pressures. Reciprocating machines, operating at constant speed, will compress *approximately* the same inlet acfm (actual cubic feet per minute) regardless of inlet conditions, gas composition, or discharge pressure. The inlet acfm depends only (approximately) on the compressor speed (rpm).

Reciprocating machines have the disadvantages, when compared to the continuous flow centrifugal or axial units, of higher maintenance and more vibration due to the unbalanced moving parts. The pulsating flow of the reciprocating units may also be a serious disadvantage in some applications.

The capacity of a fixed-speed machine can be controlled by

1. external recycle of gas from the outlet to the inlet,
2. cylinder unloaders,
3. cylinder clearance pockets, and
4. a combination of the above.

### 2.2.2. *Centrifugal Compressors*

Advanced technology in centrifugal compressor design has resulted in improved high-speed compression equipment in capacities to over 60,000 $ft^3$ min in a single unit. Because of their high efficiency, better reliability, and design upgrading, centrifugal compressors have become accepted for low-pressure cryogenic processes.

The separately driven centrifugal compressor for process refrigeration is adapted to low-pressure cryogenic systems because it can be directly coupled to the steam turbine, is less critical from the standpoint of foundation design criteria, and lends itself to gas turbine or combined cycle application. Isentropic efficiencies of 80–85% are usually obtained. Separate lubrication packages can provide filtration and cooling of lubricant. Safety devices can also be provided to monitor the vibration, low oil pressure, and water temperature.

Axial flow multistage units, with impellers mounted on a common shaft, offer a compact high-pressure ratio design with isentropic efficiencies of 88–93%. In these units, diaphragms separate the individual stages, controlling direction and velocity of the gas into the next stage. Guide vanes are fitted at the entry to each stage impeller and to the compressor inlet to obtain flow or pressure control. Thus these units can operate at a greater turndown than the centrifugal units. Special seals are provided at the coupling end to eliminate gas leakage. These can be of the labyrinth, contact seal, or pressure oil film type.

Condensing turbines, topping turbines, electric motors, and gas turbines are generally used as conventional drivers for the multistage centrifugal compressor. A number of geared engine drives have also been successfully applied. Where low-cost power is attainable or no need for process heat exists, electric motor drive will normally be the choice because of costs. Here, again, because of greater demands for energy, standby capability needs to be investigated. Totally enclosed air–water cooled motor enclosures are favored for large centrifugal compressor electric drives of 5000 hp and up. For still larger high-speed machines, consideration should be given to pressurized housings with special shaft sealing and inert gas cooling.

The gas turbine unit is often a modified aircraft jet engine coupled to a power turbine which, in turn, drives the compressor through a high-speed-type coupling. Combined cycles have been suggested using waste exhaust gas heat from the turbine to provide steam for the helper turbine. Thermodynamic efficiencies exceeding 70% are predicted for combustion gas turbines when their power and exhaust are correlated with process requirements, either for direct use or for heat recovery boilers.

### 2.2.3. *Axial-flow Compressors*

These machines are used for high-flow and medium to high-pressure applications.

The centrifugal compressor is less efficient than the axial type, due primarily to the irregular flow path in the centrifugal type, but axial flow machines have a narrower operating range than the centrifugal type.

Centrifugal compressors approximately follow the well-known "fan laws".

$$\frac{N_1}{N_2} = \frac{Q_1}{Q_2} = \frac{\sqrt{H_1}}{\sqrt{H_2}}$$

where $N$ = speed, $Q$ = volumetric capacity, and $H$ = head. Compressor control is thus best achieved by varying the speed through the use of a variable speed driver. Steam and gas turbines are suited to variable speed operation and thus are often coupled with centrifugal compressors.

If a constant speed drive is used with the centrifugal, then control is achieved by:

1. inlet guide vanes,
2. throttling of suction pressure, and
3. throttling of discharge pressure.

There are two limiting flow conditions associated with *centrifugal* machines: *surge* (pumping) and *choked flow.* Surge occurs when the flow falls to such a low value (about 50–75% of the rated value) that compressor operation becomes unstable, causing excessive vibration and possible overheating. Choked flow occurs when sonic velocity is reached at the exit of the compressor wheel, thus preventing further increases in flow rate.

## 3. PUMPS

### 3.1. Introduction

Pumps are, of course, very similar to compressors with one great difference. Compressors move gases, while pumps move liquids. Gases are single phase and stay that way. Liquids, on the other hand, are subject to vaporization change from liquid to vapor with an accompanying density change of 600 or so to 1. Herein lies the problem faced by pumps and the reason they differ from compressors.

The job of the first stage of a pump (inlet impeller) is to suck the liquid into the inlet by causing a partial vacuum at that point. We know that if the pressure on a liquid at its boiling point is reduced, the liquid boils, or cavitates, forming a two-phase mixture. The greater the suction, the more the fluid cavitates. Hence the fluid is not drawn into the first stage as hoped but merely forms two phases, and the pump loses suction, the flow rate goes to 0.

The second issue is that cavitation may not occur at the inlet of the pump but further along. Whenever the pressure in a liquid drops below the vapor pressure corresponding to its temperature, the liquid will vaporize, as described above. When this happens in an operating pump, the vapor bubbles may be carried along to a point of higher pressure, where they suddenly collapse. This phenomenon is also known as cavitation. Cavitation in a pump should be avoided, as it is accompanied by metal removal, vibration, reduced flow, reduction of efficiency, and noise. When the absolute suction pressure is low, cavitation may occur at the inlet (as above) and mechanical damage may result in the pump suction and on the impeller vanes and inlet edges.

One solution is to maintain a net positive suction head (NPSH). To avoid the cavitation phenomena, it is necessary to maintain a required NPSH, which is the *equivalent total head of liquid* at the pump less the vapor pressure $p$. NPSH is a measure of how far away the liquid is from its saturated condition at the operating temperature. If the operating conditions at the pump inlet are those of the vapor pressure curve, then the NPSH is 0. Either the pump will lose prime (all vapor at the inlet) or cavitation will occur and the collapsing bubbles may cause material damage to the pump. NPSH values depend upon the liquid being pumped and the design of the pump. Thus, pump manufacturers publish a required NPSH ($NPSH_r$) for a specific pump model. As a practical matter, the NPSH required to avoid cavitation will be somewhat greater than this theoretical value (Daney, 1988; Edmonds and Hord, 1969).

Positive NPSH can be obtained by increasing the "head" or height of liquid above the pump inlet, a case that requires the bulk storage tank to be thermally stratified, as it usually is. If the tank is somehow well stirred, then the contents will be at the saturated vapor conditions everywhere, and the NPSH will be 0. For the usual condition where thermal stratification exits, the NPSH will be equal to the height of liquid above the pump inlet. Many cryogenic pumps are often installed submerged at the bottom of the storage tank to insure the greater NPSH. If the pump were installed somewhere outside the tank, whereas the head might be the same, the inleak to the transfer line will degrade the apparent head by an unknown amount.

Precooling the liquid to below its saturation temperature is another way of gaining NPSH.

The other solution is to design a pump that can operate with a zero NPSH. This is accomplished by treating the fluid as if it were a dense gas. In this case, the inlet impeller resembles a fan. This design technique is most often employed for liquid hydrogen and liquid helium.

These are critical differences between compressors and pumps. Otherwise, the hardware is much the same. As is the case with compressors, pumps fall into two main categories: *positive displacement pumps* and *centrifugal pumps.*

### 3.2. Positive Displacement Pumps

*Positive displacement pumps* are preferred for high pressure rises and low flow rates. Available piston pumps give pressure rises of over 700 bar. *Centrifugal pumps* are favored for lower head rises and higher flow rates, although pumps with capacities as low as 0.1 L/sec are reported (DiPirro and Kittel, 1988). Reciprocating piston pumps are regularly used as adjuncts to air liquefaction and separation columns.

The pumps, operating directly upon the high-density liquid, force the liquid oxygen or nitrogen through a heat exchanger where it is vaporized, still at high pressure, and is delivered to high-pressure gas cylinders. This method is not only more economical of power than the old method of pumping the gas, but since the part of the pump handling the liquid is non-lubricated, the gas is delivered to the cylinders without contamination by lubricating oil or water.

Positive displacement (reciprocating) pumps are used in applications requiring a very high pressure (head) rise such as filling high-pressure gas cylinders from a liquid cryogen supply. These pumps have also been used to circulate liquid or supercritical helium in cooling loops where pressure rises of a few bar are required.

Figure 9.2 is an example of a high-pressure commercial pump that, depending on the bore and stroke, delivers from 0.04 L/sec at 700 bar to 1.9 L/sec at 70 bar. These single-action pumps used for filling gas cylinders place the piston rod in compression. Consequently, the rod diameter is large to prevent buckling, and the heat leak is high, which is unimportant in this application.

Ideal characteristics of a positive displacement pump are that it (1) be lightweight to allow for mobile applications, (2) be compact with a fast cool-down, (3) uses no oil for lubrication to assure contamination-free pumping, and (4) requires a very low net positive suction head (NPSH)to minimize cavitation.

### 3.3. Centrifugal Pumps

Conversely, for high liquid flow rates up to 3.8 $m^3$/sec and low pressure, a centrifugal pump is used. These pumps, which are commonly constructed of aluminum, stainless steel, or brass, should be easy to maintain, have a high efficiency, have a thermal barrier to isolate the bearings and lubricant from the cryogenic temperatures, as well as have a low required NPSH and a low thermal mass.

Some of the problems associated with the centrifugal or turbine pump are (1) cavitation, (2) pump bearings, and (3) pump seals. Frequently the centrifugal pump is run submerged in the fluid being pumped to provide the required net positive suction head to prevent cavitation. Cooling and lubrication of the bearing are provided by the cryogenic fluid in which the pump is submerged. The critical component of ball bearings used at cryogenic temperatures in pumps is the retaining ring. One of the most satisfactory materials for the retaining ring is a fabric-reinforced phenolic or glass-filled polytetrafluoroethylene. The sealing problem can be alleviated, as usual, by locating the seal at ambient temperatures instead of cryogenic temperatures if possible. See Fig. 9.3 for an illustration of a wide range of cryogenic liquid centrifugal pumps.

Centrifugal pumps, as illustrated by the small experimental liquid helium pump (Ludtke and Daney, 1988) in Fig. 9.4, are simple devices with one moving assembly (motor shaft, ball bearings, and impeller in this case). This pump consists of (1) a stationary, two-piece *pump housing*, which includes a *vaneless diffuser* for efficient recovery of fluid kinetic energy; (2) a rotating, radial flow *impeller*, which imparts pressure and kinetic energy to the fluid; (3) an *inducer*, which is an axial flow, low-head preimpeller that suppresses cavitation in the impeller by boosting the pressure at the impeller inlet; and (4) a *drive motor* and *drive shaft* and bearings that power the impeller. The motor in this submersible pump is immersed in the liquid cryogen, so motor losses are transferred to the cryogen, which cools the motor and bearings. Bleed ports in the pump housing control the coolant flow. The

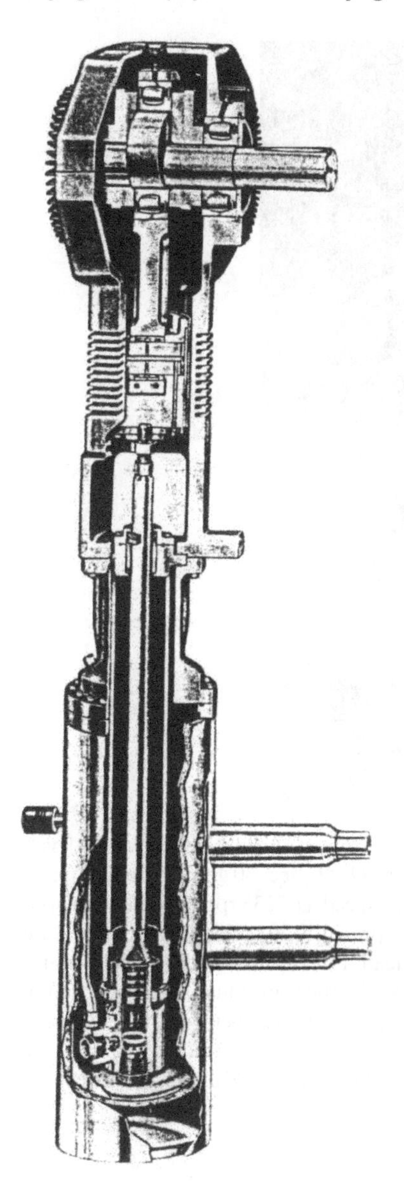

**Figure 9.2** Small-bore commercial high-pressure pump. (Courtesy CVI Inc.).

submersible design offers the possibility of higher rotational speeds because the drive shaft is a short, cantilevered extension of the motor shaft. The overall pump efficiency of the submersible design tends to be lower because all the motor losses are adsorbed at the operating temperature.

The pump shown in Fig. 9.4 was a pump typically used for He I circulation and modified to handle He II. The major modification involved the installation of a new state-of-the-art screw inducer since the propeller inducer in the He I pump provided unsatisfactory performance at low suction heads with He II. This screw inducer has three blades with the blade pitch and root diameter increasing from inlet to exit. The leading edges of the blades are faired back and sharply honed. A special adapter mates the screw inducer to the impeller (shown separately in the figure). A vaneless

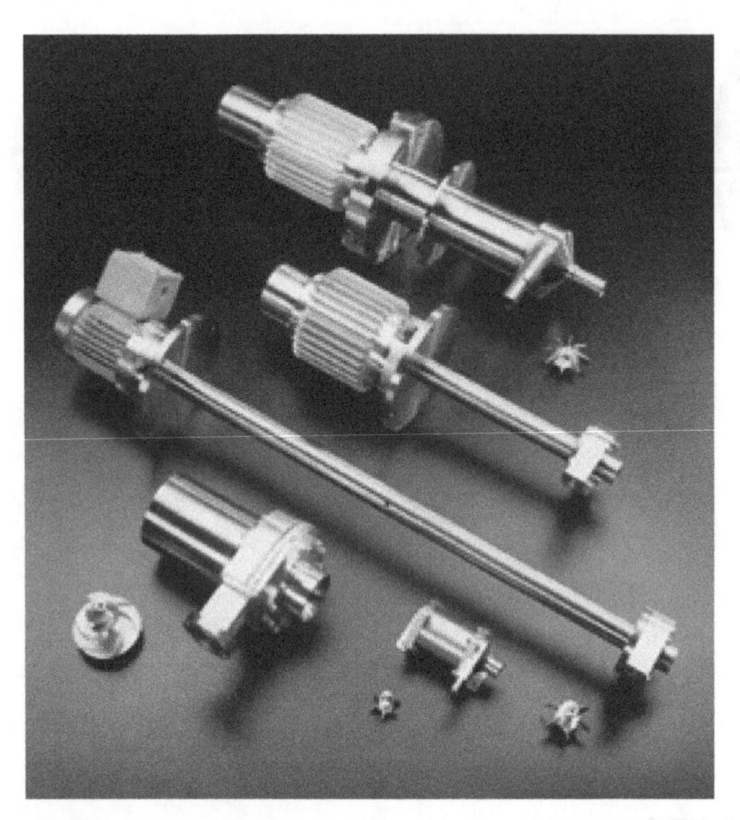

**Figure 9.3** Cryogenic liquid pumps from top to bottom: Supercritical helium pump, develops a head of 61 m (over 200 ft) at a flow rate of 250 g/sec (1.9 L/sec, 30 gpm). The inlet pressure is 57.3 psia, the inlet temperature is 4.5 K and the speed is 7235 rpm. Liquid nitrogen pump, head of 33.5 m, flow 0.1 L/sec (1.6 gpm), inlet pressure 203 psia, inlet temperature 77.6 K, and speed 6633 rpm. Liquid nitrogen pump, head 15.7 m, flow 0.1 L/sec (1.6 gpm), inlet pressure 15.5 psia, inlet temperature 77.6 K and speed 4500 rpm. Slush hydrogen pump, head 48.8 m, flow 6.3 L/sec (100 gpm), inlet pressure 1.6 psia, inlet temperature 13.7 K, speed 6700 rpm. Liquid nitrogen, the smallest pump in this series, head 0.9 m, flow 0.05 L/sec (0.8 gpm), inlet pressure 20 psia, inlet temperature 77.0 K, speed 1620 rpm. Courtesy of Barber-Nichols, Arvada Colorado, barber-nichols.com.

diffuser on the periphery of the impeller handles the flow from the impeller. A bearing coolant flow path is provided in which part of the flow from the high-pressure side of the pump is diverted between the back hub of the impeller and the pump housing, then through the motor ball bearings, and finally out through the motor housing.

This pump has shown that He II obeys the pump scaling laws in spite of the unique characteristics of the superfluid component. One major concern in using centrifugal pumps for transfer of He II in space applications is that a net positive suction head (NPSH) is required. The fountain effect pump described by DiPirro and Kittel (1988) for this application appears to be simpler and more reliable than the centrifugal pump because it has no moving parts.

Figure 9.5 shows a very low heat leak pump designed especially for supercritical helium.

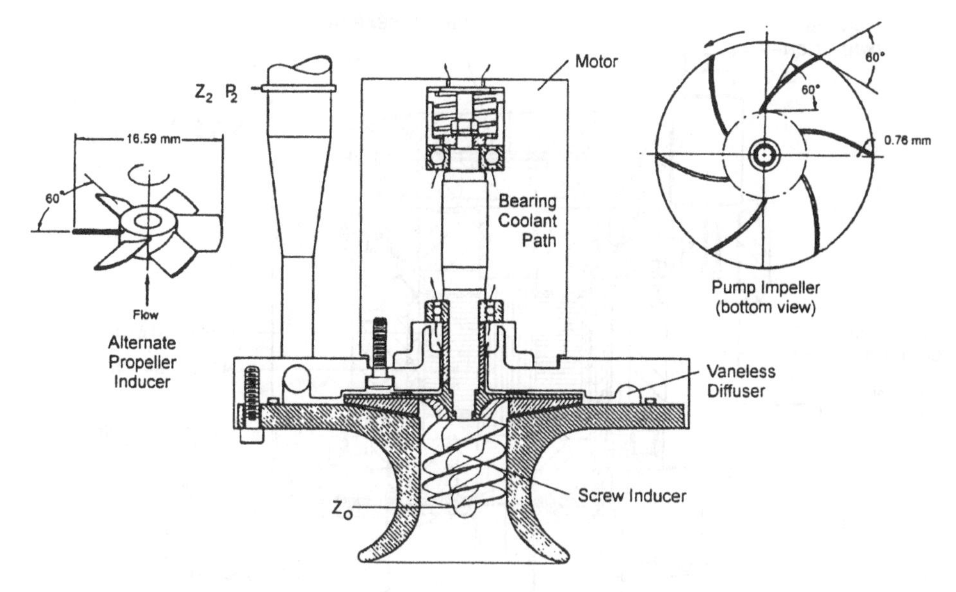


**Figure 9.4** Small centrifugal pump for liquid helium with submersible motor (Ludtke and Daney, 1988).

**Figure 9.5** Low heat leak supercritical helium pump (pictured without pump housing). This low heat leak, supercritical helium pump was developed to provide supercritical helium for use in supercolliders. The pump operates at 7380 rpm, it moves 112.43 L/min (19.7 gpm), the fluid temperature is 4.8 K, and the head rise is 61 m (200). Barber-Nichols, Arvada Colorado.

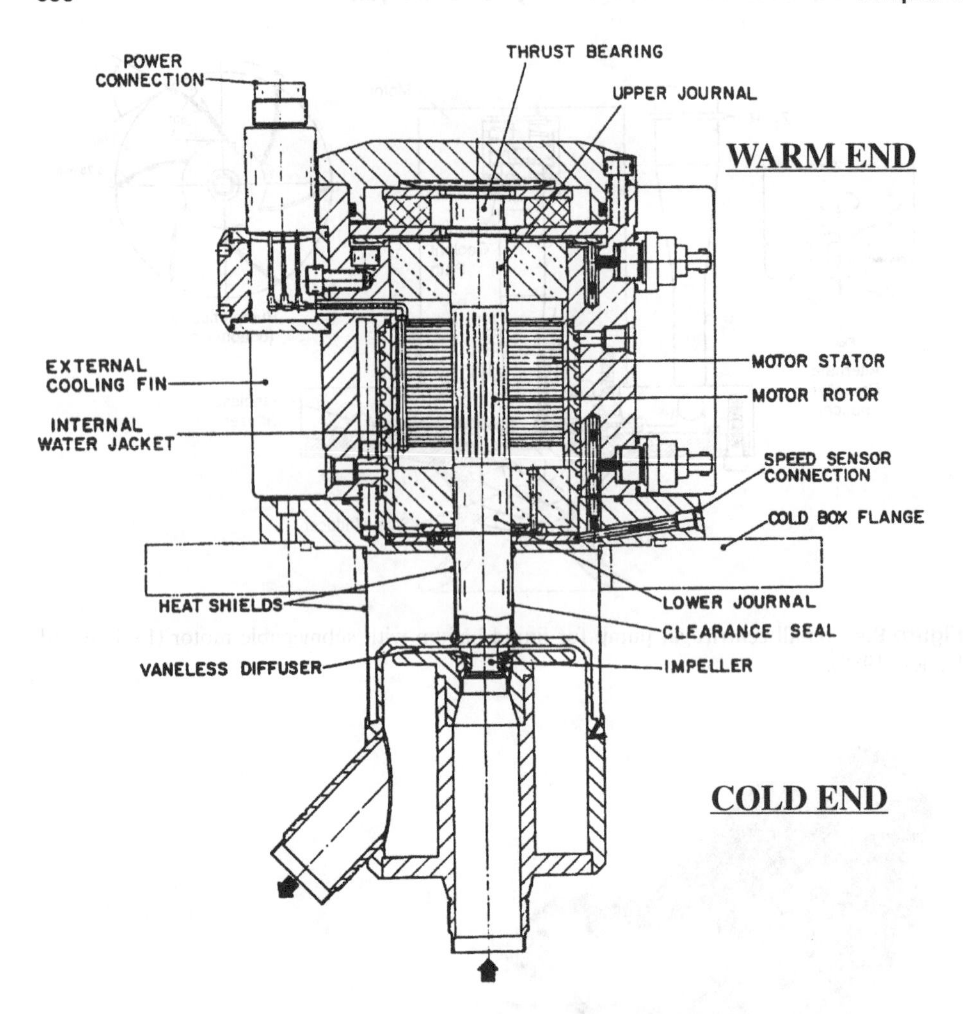


**Figure 9.6** High-speed helium pump (Jasinski et al., 1988).

Warm drive pumps, illustrated in Figs. 9.6 and 9.7, remove the motor losses to ambient temperature but have conduction heat leak down the drive shaft and shaft housing. The drive shaft must be designed so that it does not operate near its critical speed yet has a relatively small heat leak. The pumps illustrated are designed with hermetic housings, and thus there are no dynamic seals in the assembly. This arrangement eliminates (1) the small but inevitable leakage that escapes through a dynamic seal during normal operation, (2) the possibility of massive leaks that can occur with seal failure, and (3) air leaking into the system during vacuum purging for charging of the system with pure gas because it will hold a hard vacuum (which a dynamic seal typically will not). This characteristic is especially important with flammable gases or liquids.

Conventional centrifugal pumps use a *full emission* design, but an unconventional *partial emission* design offers advantages in high-head, low-flow applications. Figure 9.8 compares impeller and diffuser flow passages for the two designs. The conventional full-emission design uses backward-curved blades on the impeller and a collector scroll that allows flow from all impeller channels to pass continually

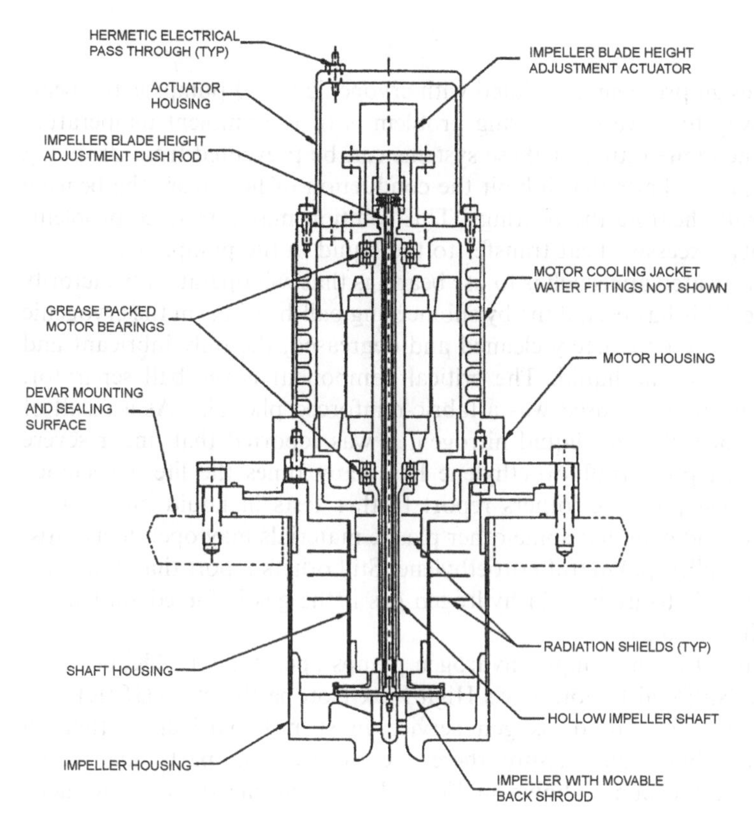


**Figure 9.7** Liquid helium circulating pump.

to the pump discharge. The partial emission design uses straight radial blades on the impeller and a diffuser that only allows flow from a small sector of the impeller channels to pass to the pump discharge. In some applications where the performance of a conventional full emission pump would be unacceptably poor, a partial emission pump can be used in lieu of a fixed displacement pump.

**Figure 9.8** Comparison of full emission and partial emission flow paths.

### 3.4. Bearings

One mechanical design problem associated with cryogenic liquid pumps is the bearing. An obvious way to solve the bearing problem is to use ambient temperature bearing systems; the temperature of these systems can be prevented from dropping too low by having long shafts that inhibit the conduction of heat from the bearing region, or by actually heating the bearings. This solution may introduce problems with shaft flexibility, excessive heat transfer to the liquid in the pump, etc.

A more sophisticated approach is to use bearings that will operate satisfactorily at low temperature. This has been done by submerging ball-bearings in the cryogenic liquid. The bearings are completely cleaned and degreased, the only lubricant and coolant being the cryogenic liquid. The critical component is the ball separator; the most satisfactory material used was a fabric-reinforced phenolic. As a result of an extensive series of tests with liquid nitrogen, it was reported that under severe loading, a glass-filled polytetrafluoroethylene has many times the life expectancy of the aforementioned phenolic; others report further tests in liquid nitrogen, at 9200 rev/min, which indicate that some other plastic materials may operate as satisfactorily as the glass-filled polytetrafluoroethylene. Still others report that ball-bearings will operate satisfactorily in cold hydrogen gas if the gas is forced sufficiently rapidly through the bearings.

Figures 9.9 and 9.10 show liquid hydrogen pumps operating at 20 K, designed for the Oak Ridge National Laboratory's High Flux Isotope Reactor (HFIR).

Pumps designed for liquid oxygen service must pay particular attention to materials compatibility and assure there are no rubbing parts that could ignite. Cleanliness of the pump shown in Fig. 9.11 is guaranteed by its hermetic construction.

## 4. EXPANSION ENGINES

### 4.1. Introduction

Expansion engines or turbines are the devices used to permit gases to expand and perform work, thereby effecting a very large reduction in temperature of the gas which does the work. Ideally, such an expansion would occur at constant entropy, for isentropic expansion provides the greatest cooling for a given pressure ratio of expansion. Expansion of gases by performing work is a very effective means of refrigeration.

Refrigeration by the expansion of compressed air was developed during the 19th century to make ice and store and transport frozen meats. Indeed, in 1899, more ships having refrigeration facilities used air expansion engines than any other means of refrigeration. See Chapter 1. Improvements in the vapor compression refrigeration systems, using ammonia, carbon dioxide, or other fluids soon relegated air refrigeration to the technical scrap yard. Today, one of the few applications of air refrigeration is the cooling of aircraft cabins.

As the air expansion engine was giving way to vapor compression refrigerants for the frozen meat industry, it was employed by Claude to provide refrigeration for the liquefaction of air. This was first accomplished in 1902. Since that time, all large and many small air separation plants have used expansion engines or turbines. The reciprocating and rotary expansion machines will be discussed in detail in later sections of this chapter.

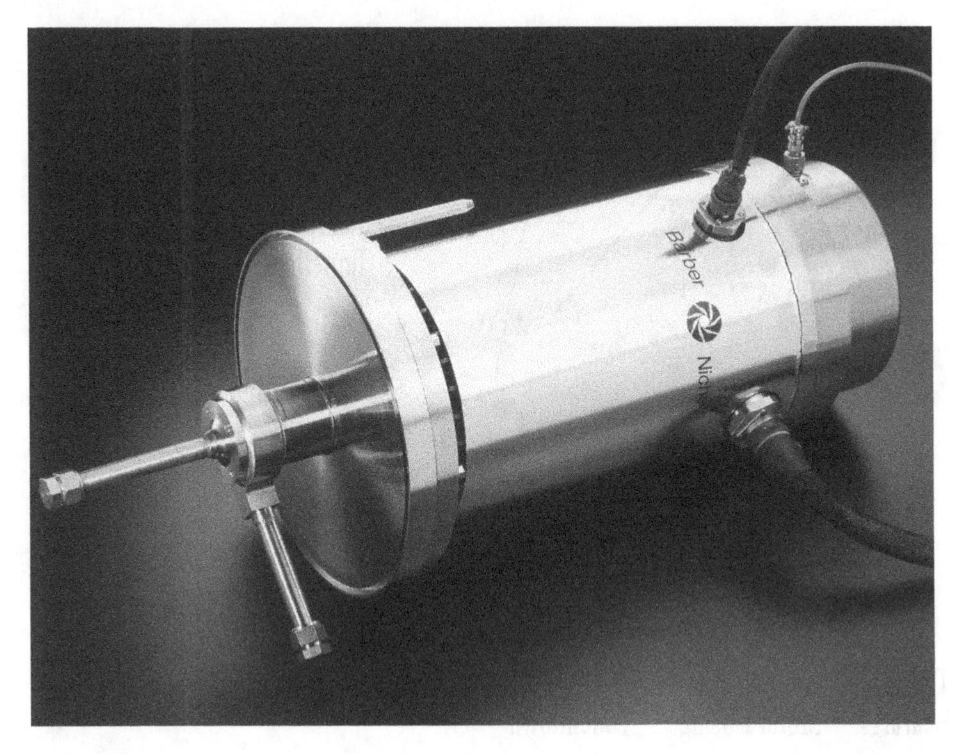


**Figure 9.9** Supercritical hydrogen circulator. This supercritical hydrogen circulator is designed for use at the Oak Ridge National Laboratory's High Flux Isotope Reactor (HFIR). The supercritical hydrogen circulator aids in investigating how materials are assembled at the atomic level. The circulator starts operating at room temperature and cools a hydrogen loop down to 20 K. To avoid the vapor-to-liquid phase change, the system operates at a supercritical pressure of 15 bar (14.8 atm, 217.6 psia). This wide temperature (and corresponding density) range requires wide speed variations to maintain the correct moderator cooling flow. Therefore, the circulator was designed to operate continuously between 60,000 and 15,000 rpm. Magnetic bearings were selected to make the wide speed range and high speed operation practical and reliable. The resulting design provides a high temperature (288 K) flow rate of 4.15 L/min (approximately 1 gpm) with a 0.27 bar (4 psia) pressure rise and a low temperature (20 K) flow rate of 1.02 L/min (0.27 gpm) with a 1.0 bar (14.5 psia) pressure rise. Barber-Nichols, Arvada, Colorado.

## 4.2. Expansion Engine General Design Considerations

In the design of any piece of equipment, a decision must be made regarding the relative significance of such diverse factors as efficiency, cost, useful life, flexibility, appearance, ease of operation and maintenance, size, and weight. For example, the design of a household vacuum cleaner might emphasize cost, appearance, and ease of operation. The importance of these parameters in the design of expansion machines for gas liquefaction or cryogenic refrigeration systems is much less than such factors as efficiency and reliability. An attempt will be made to evaluate design priorities for expansion devices.

*Primary* design considerations for expansion machines are as follows:

a. *Efficiency*—The overall performance of a cryogenic system is affected directly by the efficiency of the expansion engine. In superior designs, the efficiency ranges from 80% to 90%.

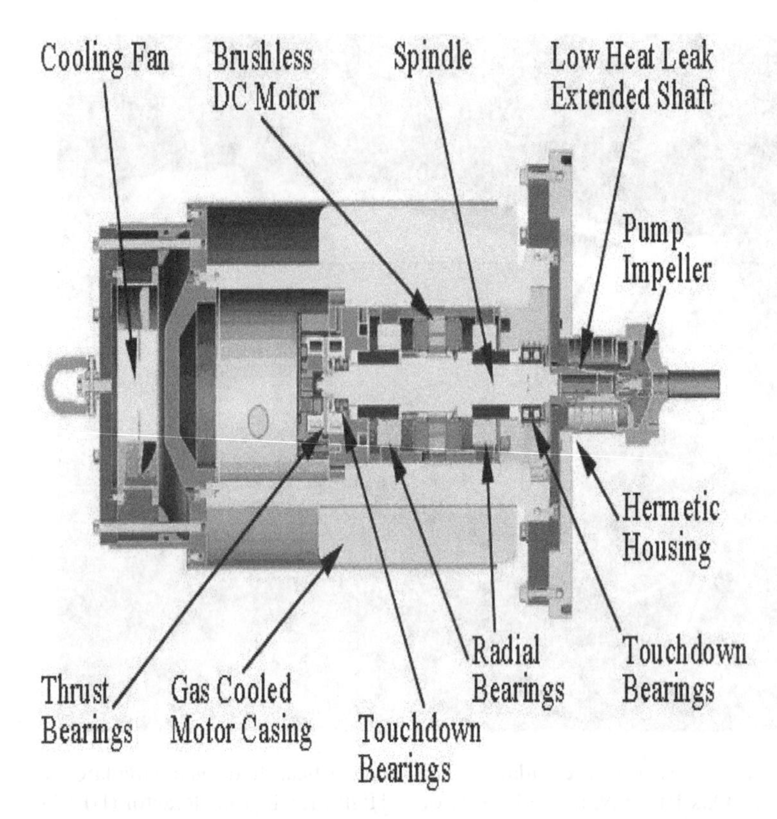


**Figure 9.10** Drawing of supercritical hydrogen circulator. See also Fig. 9.9 Barber-Nichols, Arvada, Colorado.

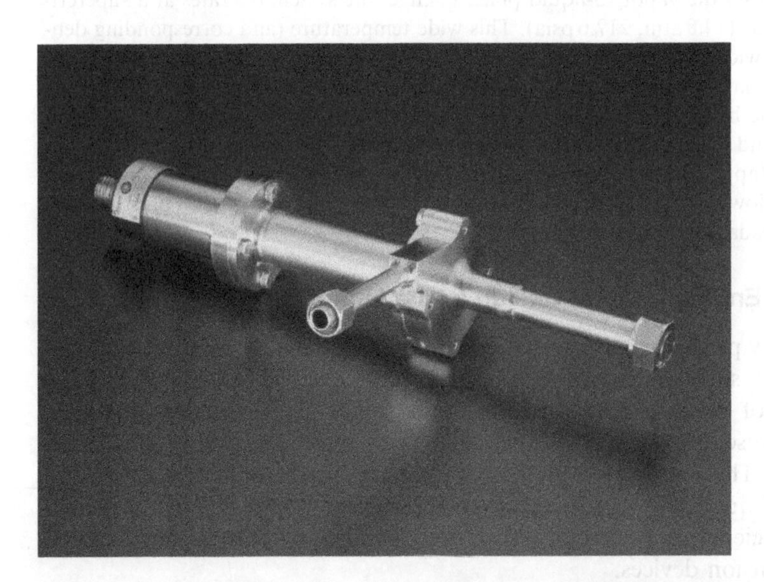

**Figure 9.11** Hermetic liquid oxygen pump. This lightweight, hermetically sealed liquid oxygen pump operates at 4100 rpm. It moves 9.46 L/min (2.5 gpm), and the fluid temperatures can vary from 78 to 133 K. The purpose was to complement NASA's development of non-toxic, liquid oxygen/alcohol thrusters to replace the toxic hydrazine thrusters that are currently in use. Barber-Nichols, Arvada Colorado.

e. *Reliability*—With the exception of certain miniature (cryocooler) refrigeration systems, most cryogenic processes operate for extended periods of time. Some air separation plants have operated for years without shutdown. The expansion engine is a vital component of such plants and must be able to run continuously as well as efficiently. Frequently a spare expansion machine is installed to permit periodic maintenance of the mechanical equipment without stopping the overall process. A spare machine greatly improves the reliability factor.
f. *Gas contamination*—Any contamination of the gas being expanded must be avoided to prevent fouling of the heat exchangers and other low-temperature equipment. The primary source of such contamination is lubricating oil. Not only will oil foul equipment but it can create a hazardous condition if it enters the distillation column of an air separation plant.

*Secondary* considerations in the design of expansion machines include the following:

a. *Initial cost*—Although cost is always a design consideration, it is of secondary importance in most applications of expansion machines. Usually the cost of the expansion machine is a small fraction of the total equipment cost for a cryogenic system. For example, the expansion turbine for a 200 ton/day oxygen plant represents about $25,000 of the total plant cost. Thus a 10% change in the cost of the turbine is completely lost in the total figure.
b. *Power recovery*—The recovery of useful power is of less importance than the extraction of the power from the expanding gas. However, the shaft work must be dissipated in some manner outside the cold portion of the cryogenic system. If the power exceeds 100 hp, it is usually economically worthwhile to recover it. For smaller units, the power is dissipated in a brake or may be incidentally recovered in a speed control device.
c. *Size and weight*—The size and weight of expansion machines are of importance only in those cases of specialized airborne mounted refrigeration systems.
d. *Flexibility*—Most expansion machines require little flexibility in terms of capacity since the overall cryogenic refrigeration system is usually designed for and operated at maximum capacity at all times. It is essential for the designer to know the complete operating requirements in order to determine the need for flexibility.

*Little or no* consideration in expansion machine design is given to such items as:

a. Design which is readily adaptable to mass production techniques.
b. Appearance, other than clean engineering design.
c. Salvage value after the useful life of the machine.
d. Resale value or use for an unknown future application.

### 4.3. Reciprocating Expanders

Generally, reciprocating expansion engines are used when the inlet pressure and pressure ratio are high and when the volume is low. The inlet pressure to expansion engines used in air separation plants varies between 600 and 3000 lb/in.$^2$ gauge and the isentropic efficiencies achieved are from 80% to 87%.

The design features of expander engines include rigid, guided cam-actuated valve gears, renewable hardened valve seat, helical steel or air-springs, and special valve packings that eliminate leakage. Cylinders are normally steel forgings effectively insulated from the rest of the structure. Removable cylinder liners and floating piston design offer wear resistance and good alignment in operation. Piston rider rings serve as guides for the piston. Non-metallic rings are used for non-lubricated service. Both horizontal and vertical design, and one- and two-cylinder versions, have been used successfully.

There are five principal sources of inefficiency for expansion engines. These include: (1) Valve losses, (2) Incomplete expansion, (3) Heat leakage, (4) Non-ideal expansion, and (5) Clearance losses. Generally, each one of these losses contributes between 2% and 6% to the inefficiency of expansion engines.

Reciprocating expander controls are designed to adjust to changing flow rates manually or automatically. A variable cutoff uses a double cam actuator for regulation of inlet valve timing providing maximum valve lift at all cutoff points. Flow control is therefore completely variable from zero to maximum cutoff at constant expander speed. Variable cutoff at reduced capacity improves thermal efficiency at low throughputs with some compensation for increased plant losses. However, variable speed control has been limited to small size expanders.

Reciprocating expanders, in normal operation, should not accept liquid in any form during the expansion cycle. However, the reciprocating engine can tolerate some liquid for short periods of time providing none of the gas constituents freeze out in the expander cylinder and cause serious mechanical problems. If design conditions selected indicate possibilities of entering the liquid range on expansion during normal operation, then inlet pressure and temperature must be revised or thermal efficiency modified.

The choice of expander design depends on process and plant factors such as (1) expander sizing for best thermal efficiency consistent with plant cost and power charges, (2) method of energy absorption—whether for primary gas compressor, induction generator, or power brake, (3) cost of equipment, and (4) type of control, either constant or variable speed, manual, automatic, or semiautomatic.

The reciprocating expansion-engine refrigerator was introduced by Claude. See Chapter 6. In principle, it is simply a reciprocating piston-and-cylinder engine similar to the steam engine which had been in use for over a century. The most difficult practical problem is that of providing lubrication that will remain effective at the low operating temperatures. Claude's first solution was the use of light hydrocarbons which remained fluid at the engine temperatures. More recently engines have been designed with pistons or piston rings made of special materials which operate unlubricated with little friction and little wear. Expansion engines used in the helium separation plants of the US Bureau of Mines use plastic rings (Micarta®). The great cryogenic pioneer of the United States, Sam Collins, invented a liquid nitrogen generator which utilized an air expansion engine with a Micarta® piston sleeve operating in a chromium-plated bronze cylinder. The Collins Helium Cryostat used expansion engines having cylinder and pistons of nitrided steel. It seems that in such engines the gas itself is the lubricant, a very thin layer of gas being always between the piston and cylinder during operation. Collins stated that for air expansion machines, the chromium-plated-bronze–plastic combination was preferable to the nitrided steel cylinder–piston assemblies. Not only is the former easier to manufacture but it is also

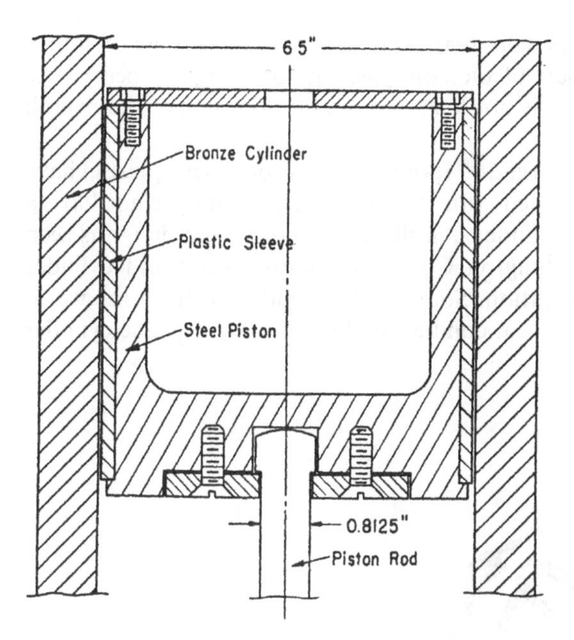


**Figure 9.12** Piston and cylinder of Collins' air expansion engine.

less vulnerable to seizure from solid foreign matter. The efficiency of the Collins air engine is approximately 85% and was not diminished after several thousand hours of operation. Figure 9.12 shows the piston and cylinder of the Collins air engine. Collins also constructed expansion engines with pistons which present a surface of laminated leather disks to contact the walls of the cylinder.

*4.3.1. Non-Lubricating Reciprocating Expansion Engines*

Early reciprocating expansion engines, such as developed by Claude (above), were adaptations of steam engines. Generally they were single cylinder, single-acting vertical machines operating at fairly low speeds. Claude was not able to solve the problem of lubricating and sealing the piston with the use of a fluid. He found it expedient to use a leather cup as a piston seal. The pressure inside of the cup forces the leather against the cylinder wall and effects a reasonably good gas seal. Although oil impregnated leather has a certain amount of self-lubrication, there is a significant frictional drag which decreases the overall efficiency of expansion. Fortunately, at very low temperatures, the leather of the cup becomes hard and glassy, and if the cylinder is smooth and hard, the friction is minimized. Leather cups are still in use today in some older designs of air expansion engines.

An example of a different non-lubricated piston and cylinder arrangement was shown in Fig. 9.12. This design is by Collins and has been used in a liquid nitrogen generator for a number of years. The bronze cylinder is plated with chromium to give a very hard-wearing surface. The steel piston is an example of the successful use of a brittle material at low temperature. It is fitted with a sleeve of Micarta®, a laminated plastic whose layers of fabric are impregnated with a thermosetting resin, or nylon. The original annular gap between the cylinder and sleeve was about 0.002 in. In this design, the gas pressure does not force the sleeve against the cylinder and wear is slight. After 10,000 hr of service, the sleeve wear was several thousandths of an inch.

The leakage past the piston increases with wear. This results in a throttling or Joule–Thompson type of expansion of part of the cold gas. The blow-by is piped to the expansion engine discharge line so it is not lost, although it does reduce the overall efficiency of expansion.

The self-lubricating qualities of Teflon® (poly-tetra-fluoroethylene plastic) have made it possible to have a piston sealed with expandable rings and yet not have a fluid lubricant introduced into the cylinder. An illustration of such a design is seen in Fig. 9.13, courtesy of the Worthington Corp. The entire piston and cylinder assembly becomes quite cold in operation. Some heat is conducted from the warm cross-head along the piston rod. The heat conducted along the frame to the cylinder

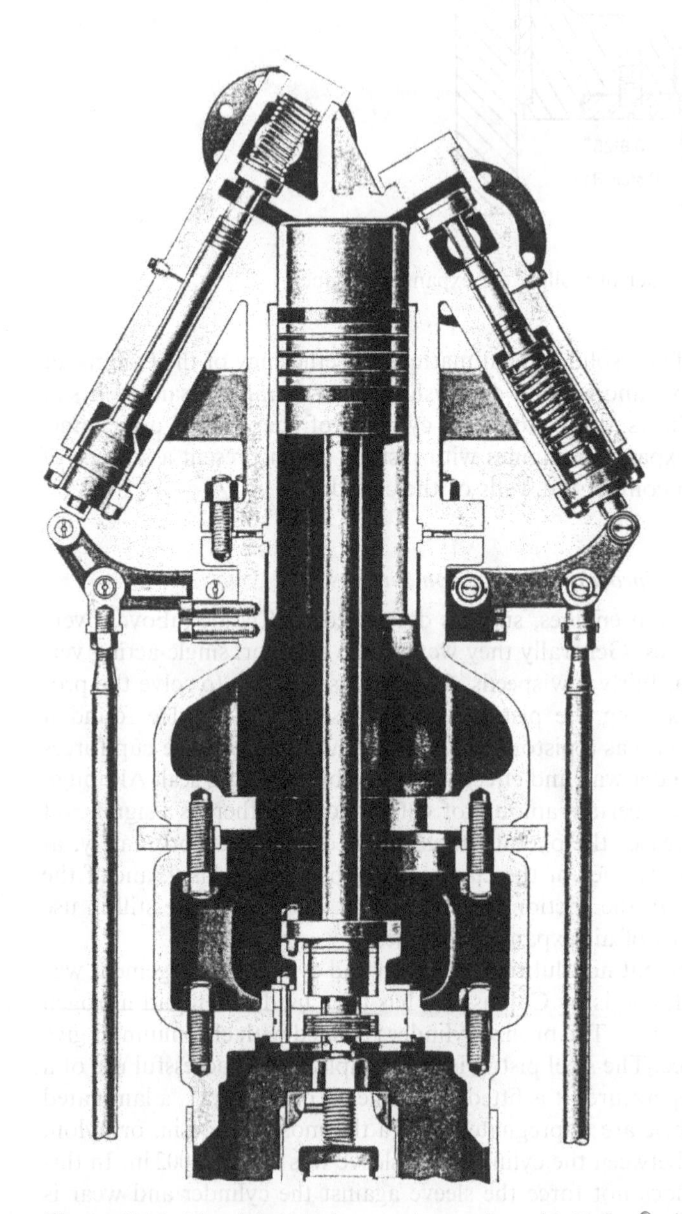

**Figure 9.13** Non-lubricated expansion engine with Teflon® piston rings.

is greatly reduced by the insertion of a non-metallic heat dam. Micarta® is suitable for this purpose.

Because of the flexibility of piston rings, a side load cannot be taken up by the sealing rings. As can be seen in the illustration, solid rider rings are fitted on the piston to accommodate side thrust. These rings do wear but the blow-by does not increase as long as the expandable rings are sealing against the cylinder.

Pure Teflon® has a large coefficient of thermal expansion and a tendency to flow under load. A filler material, such as carbon or bronze, improves performance by increasing the modulus of elasticity, thermal conductivity, and wear resistance. As Teflon® rings wear in, they coat the mating surface with fine particles of resin. For optimum transfer of the resin, without excessive wear, the cylinder wall should have a finish between 12 and 30 $\mu$in. The number of rings per piston varies with the pressure ranges, from a minimum of two rings for 300 psi differential to five rings for 3000 psi differential pressure.

#### *4.3.2. Oil Lubricated Reciprocating Engines*

In 1912 Heylandt patented an engine which was especially suitable for oil lubrication and very high pressure ratios, such as an inlet pressure of 3000 psi and an exhaust pressure of 100 psia. The essence of the design is a very long piston, with the head end in contact with the cold expanding gas and with piston rings, oil lubricated, located well downward toward the crank end of the piston. Conventional cast iron piston rings and cylinder lubricants can be used because the rings operate in a warm environment. Sometimes the crank end of the cylinder is equipped with a water jacket for keeping this portion warm. Usually this is unnecessary because there is enough frictional heat to keep the area warm. With the very high inlet pressures, the inlet temperature is also high, about −20°F to +40°F. Exhaust temperatures are very low, although they are maintained well above the saturation temperature to avoid the possibility of liquefying in the cylinder. If such liquefaction does occur, the engine efficiency suffers because the liquid droplets deposit on the cylinder walls and are evaporated, thus producing a refrigeration loss.

The general characteristics of reciprocating expanders are shown in Fig. 9.14.

### 4.4. Expansion Turbines

#### *4.4.1. Development*

In 1934, Linde published an article on the use of expansion turbines for cooling. In 1934, Kapitza published a thorough theoretical study of the application of the expansion turbine to the liquefaction of air. His analysis and subsequent development produced a radial-inflow, or centripetal, turbine having an efficiency of over 80% of theoretical. One of the arguments he presented in favor of the radial-inflow turbine for air liquefaction was that air, at the temperature and pressure existing in such an expansion turbine, had characteristics more like water than like steam, so the turbine should have some attributes similar to those of the very efficient modern water turbines. Kapitza's success was followed by further development, e.g., that described by Swearingen, so that today the radial-inflow expansion turbine is in regular use for air liquefaction.

Kapitza's turbine consisted of an overhung wheel, very much like a centrifugal pump, with the two ball bearings supporting the shaft in the warm region. This general arrangement has been used on almost all low-temperature expansion turbines.

- Units offer high efficiency
  - 80% to 87%
- Offers high pressure ratios
  - Low flows
- Care must be exercised not to condense liquid during expansion cycle
- Sources of inefficiency:
  - Valve losses
  - Incomplete and non-ideal expansion
  - Heat leaks
  - Clearance losses
- Problems with reciprocating expanders
  - High Maintenance
  - Valve problems
  - Not usable with condensibles

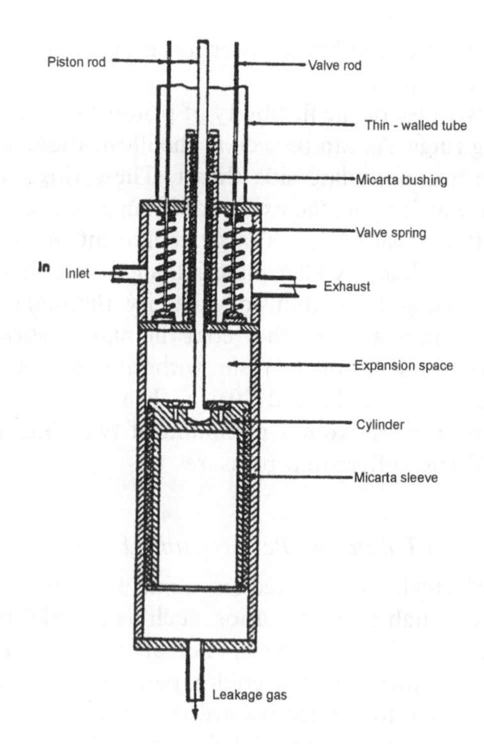


**Figure 9.14** Reciprocating expanders.

The monel wheel of 8 cm diameter rotated at 40,000 rpm and developed up to 10 hp at an efficiency of about 80%. The exhaust temperature of –305°F was the saturation temperature at the discharge pressure of 23 psia. It is interesting to note that Kapitza used a flexible, or De Laval, shaft, and he was able to get above the critical speed only by developing a special damping device. He points out that such damping fortunately occurs in the oil film of journal bearings but needs to be introduced when using ball bearings.

The high efficiency achieved by Kapitza through the radial reaction design and the high rotative speed was a significant improvement over the earlier design. Almost all subsequent low-temperature expansion turbines have been based on these principles.

The first American expansion turbine was developed in 1942 by the Elliott Company under the sponsorship of the National Defense Research Committee. The purpose of the turbine was to provide refrigeration for mobile oxygen generators. For this purpose, the low-temperature turbine, with its features of light weight, simplicity, power recovery, and high efficiency, is ideally suited.

The Elliott turbine followed the Kapitza lead in utilizing the radial-reaction principle and high rotative speeds. It consisted of a $6\frac{7}{8}$ in. diameter aluminum wheel rotating at 22,000 rpm, developing about 40 hp at an efficiency of more than 80%. The bearings consisted of a plain journal near the wheel and a ball bearing, carrying both radial and thrust loads, near the warm end of the shaft. The journal bearing also acted as a shaft seal. The shaft was a stiff-shaft design with a lowest critical speed of 32,000 rpm. This prevented any stability problems which Kapitza encountered, although it is probable that the dampening action of the journal would have rendered these harmless. The power developed by the Elliott turbine was absorbed in an induction generator, driven through suitable gearing.

Herbie Sixsmith, of Creare, Inc., has developed a miniature turbine in connection with a small air liquefaction plant. This turbine, with a $\frac{9}{16}$ in. diameter wheel and a 2 in. long shaft, rotates at 240,000 rpm and develops slightly over 1 hp. The shaft is supported on air-lubricated journal and thrust bearings, using a portion of the air to be expanded. This miniature turbine demonstrates that extremely small flow quantities can be handled by rotating machinery with efficiencies of about 60%. This leads to the conclusion that very high pressure air, with correspondingly small volumes, may be expanded by turbines rather than reciprocating expansion engines, at a significant saving in weight of machinery and capital cost. A liquid oxygen plant, employing air pressures up to 3000 psig, could incorporate a miniature turbine to expand air from 3000 to 600 psig and a conventional radial-reaction turbine to expand from 600 to 75 psig, with overall efficiencies approaching that of reciprocators. Sixsmith has pioneered micro-turbines and produced models spinning at 500,000 rpm.

### 4.4.2. *Analysis of a Centripetal Turbine*

Figure 9.15 is a diagram of a centripetal turbine: a cross-section perpendicular to the axis. Gas enters at the inlet, is accelerated and directed by the turbine nozzles, and then enters the rotor passages with a high tangential velocity and a small radial velocity. As the gas passes through the rotor, its kinetic energy is transmitted to the rotor and it leaves with negligible velocity. Also, during its passage through the rotor, the gas is subjected to a strong centrifugal force; so it experiences an additional expansion as it traverses the diminishing centrifugal force field. The energy of this expansion also is transmitted to the rotor.

Let $p$, $v$, $u$, and $T$ designate the pressure, molar volume, molar internal energy, and absolute temperature, respectively, of the gas. Let the subscript 1 refer to the entering high-pressure gas, subscript 2 to the gas as it enters the rotor, and subscript 3 to the exhaust gas as it leaves the center of the rotor. Assume no losses. Assume also that this is a "zero angle" turbine—that the gas enters the rotor almost tangentially. This, of course, is an idealized case; if the inlet angle were really zero there would be zero radial velocity and no gas would pass through the turbine. However, the angle can be such that the radial velocity is as much as 10% of the tangential

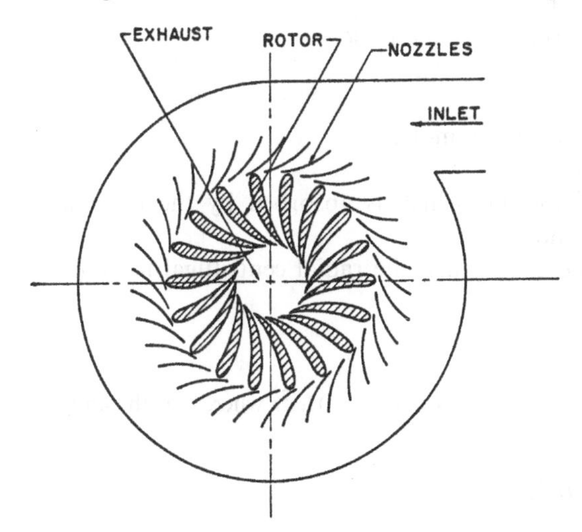


**Figure 9.15** Schematic diagram of a radial-flow or centripetal turbine. This is a cross-section through the housing, nozzles, and blades, perpendicular to the axis of the shaft.

velocity without seriously departing from the characteristics of the zero-angle turbine. The tangential component of velocity will, even in this practical case, still be 99% of the total velocity.

For one mole of gas passing through the nozzles, the specific energy entering the nozzles can be equated to that leaving.

Entering:

(1) $u_1$ = internal energy entering with one mole, and
(2) $p_1v_1$ = work done to inject one mole. (See Chapter 2.)

Leaving:

(3) $u_2$ = internal energy leaving with one mole,
(4) $p_2v_2$ = work done to expel one mole, and
(5) $MV^2/2$ = kinetic energy of one mole, where $M$ is the mole weight of the gas and $V$ its velocity as it leaves the nozzles.

Then

$$u_1 + p_1v_1 = u_2 + p_2v_2 + MV^2/2 \tag{9.1}$$

or since $h = u + pv$,

$$h_1 - h_2 = MV^2/2 \tag{9.2}$$

and, since $\Delta h = C_p\Delta T$,

$$C_p(T_1 - T_2) = MV^2/2 \tag{9.3}$$

where $h = u + pv$ is the molar enthalpy of the gas and $C_p$ is the molar specific heat at constant pressure.

The energy relations of the gas passing through the rotor can be treated in a similar fashion.

Entering:

(1) $u_2$ = internal energy entering with one mole,
(2) $p_2v_2$ = work to inject one mole, and
(3) $MV^2/2$ = kinetic energy entering with one mole.

Leaving:

(4) $u_3$ = internal energy leaving with one mole,
(5) $p_3v_3$ = work to expel one mole,
(6) $MV^2/2$ = kinetic energy of the gas which is absorbed by the rotor and transmitted to the shaft, and
(7) $MV^2/2$* = work done by the gas against the radial centrifugal force (also transmitted to the shaft).

---

*The centrifugal force on a mole at radial distance $r$ is $MV^2/r = M\omega^2 r$ where $\omega$ is the anuglar velocity of the rotor. Then the work

$$W = \int f\,dx = \int_{r_1}^{r_2} M\omega^2 r\,dr = \frac{1}{2}M\omega^2(r_2^2 - r_1^2)$$

where $r_1$ and $r_2$ are the inner and outer radii of the rotor blades. When $r_1 << r_2$, $W \cong M\omega^2 r_2^2 = MV^2/2$.

Then $u_2 + p_2v_2 + MV^2/2 = u_3 + p_3v_3MV^2$, so

$$h_2 - h_3 = MV^2/2$$

or

$$C_p(T_2 - T_3) = MV^2/2 \tag{9.4}$$

Thus it is seen that the enthalpy change that occurs in the nozzles is the same as that accompanying the expansion in the rotor, and the total energy absorbed from one mole of gas is $C_p(T_1 - T_3) = MV^2$.

It should be noted that this centripetal turbine is similar to a reaction turbine; however, high velocities of the gas relative to the rotor passages are avoided because the exit ends of the rotor passages are near the axis and hence they are moving at a velocity much less than the peripheral velocity. In a reaction turbine, the gas enters the moving blades with little relative velocity and is discharged in a backwards direction with a relative velocity equal to the forward velocity of the blades, so that its absolute exhaust velocity is practically zero.

The characteristic behavior of a loss-free reaction turbine having sonic* gas velocity at the nozzles ($V = V_c$) can be computed from well-known thermodynamic relations. These relations are:

$$pv^{\gamma} = \text{constant} \tag{9.5}$$

where $\gamma = C_p/C_v$, the ratio of the specific heat at constant pressure to that at constant volume;

$$pv = RT \tag{9.6}$$

$$V_c^2 = \frac{\gamma p_2 v_2}{M} \tag{9.7}$$

($V_c$ is the velocity of sound) and Eqs. (9.3) and (9.4)

$$C_p(T_1 - T_2) = C_p(T_2 - T_3) = \frac{MV_c^2}{2} \tag{9.8}$$

From (9.5)

$$\frac{p_1}{p_2} = \left(\frac{v_2}{v_1}\right)^{\gamma} \tag{9.9}$$

From (9.6)

$$\frac{v_1}{v_2} = \frac{p_2}{p_1}\frac{T_1}{T_2} \tag{9.10}$$

Substituting (9.10)in (9.9) we have

$$\frac{p_1}{p_2} = \left[\frac{T_1}{T_2}\right]^{\gamma/(\gamma-1)} \tag{9.11}$$

*Sonic velocity is chosen because it is the upper limit for normal design. Supersonic design introduces additional complexities and probable losses.

Similarly

$$\frac{p_1}{p_3} = \left[\frac{T_1}{T_3}\right]^{\gamma/(\gamma-1)} \tag{9.12}$$

From Eqs. (9.6), (9.7), and (9.8)

$$\frac{T_1}{T_2} - 1 = 1 - \frac{T_3}{T_2} = \frac{\gamma R}{2C_p} \tag{9.13}$$

$$\frac{T_1}{T_3} = \frac{1 + \gamma R/2C_p}{1 - \gamma R/2C_p} \tag{9.14}$$

For monatomic gases $\gamma = \frac{5}{3}$ and $C_p = \frac{5}{2}R$, so

$$\frac{T_1}{T_3} = 2 \quad \text{and} \quad \frac{p_1}{p_3} = 2^{\gamma/(\gamma-1)} = 5.64$$

For diatomic gases $\gamma = \frac{7}{5}$ and $C_p = \frac{7}{2}R$, so

$$\frac{T_1}{T_3} = 1.5 \quad \text{and} \quad \frac{p_1}{p_3} = (1.5)^{\gamma/(\gamma-1)} = 4.13$$

As a practical example, consider a nitrogen expansion turbine operating at inlet conditions $T_1 = 120\,\text{K}$, $p_1 = 6\,\text{atm}$. Then $T_3 = 80\,\text{K}$ and $p_3 = 1.45\,\text{atm}$.

From Eq. (9.13)

$$\frac{T_1}{T_2} = 1 + \frac{\gamma R}{2C_P} = \frac{6}{5}$$

so $T_2 = 100\,\text{K}$. From Eq. (9.8)

$$V_c^2 = \frac{2C_p}{M}(T_1 - T_2)$$

and

$$V_c = 2.04 \times 10^4 \text{cm/sec}$$

$V_c$ is also the peripheral speed of the rotor.

This is a fair approximation of what may be expected of such a turbine. Of course the performance of an actual turbine is somewhat different because the losses tend to increase the pressure drop and reduce the temperature drop. Since excessive speed is a disadvantage of the expansion turbine, it should be noted that the speed may be reduced by (1) using a heavy gas, (2) operating at a low temperature. The lower temperature may offset the disadvantage of a lighter gas. For example, a sonic expansion turbine using helium at an inlet temperature of 15 K would have a somewhat lower peripheral speed than the nitrogen turbine just analyzed. Of course, another way to reduce speed is to provide a multistage turbine in which several rotors and sets of nozzles share the pressure drop in a series arrangement.

This discussion is intended only to give some idea of the physics and thermodynamics of expansion turbines. It is greatly oversimplified because a sophisticated turbine analysis is very complex. For example, a great variety of inlet and exist angles and blade shapes are possible which still yield good efficiencies. Also the velocity at which the gas leaves the rotor blades can be quite large, provided the accompanying pressure is sufficiently low; the excess kinetic energy is utilized in a diffuser which raises the pressure up to the exit pressure of the turbine in a process which approaches an isentropic one.

Note that these equations tell us that, for a given $P_1$ and $P_3$, the lower the inlet to the expander, the less the cooling effect achieved.

Using a temperature-entropy diagram for $N_2$, for $T_1 = 200$, $P_1 = 10\,\text{atm}$, $P_3 = 2.34\,\text{atm}$, and $T_3$ will be approximately 130 K.

*Example.* Isentropic expansion of an ideal gas

$$\frac{T_2}{T_1} = \left(\frac{P_2}{P_1}\right)^{(\gamma-1)/\gamma}$$

assume $(P_2/P_1) = \frac{1}{4}$, $\gamma = 1.40$ (average value for diatomic gas)

$$\frac{T_2}{T_1} = (1/4)^{(1.4-1)/1.4} = 0.673$$

or $T_2 = 0.673T_1$

| $T_1$ (K) | $T_2$ (K) | $\Delta T$ | $T_1/T_2$ |
|---|---|---|---|
| 300 | 202 | 98 | ~1.5 |
| 250 | 168 | 82 | ~1.5 |
| 200 | 135 | 65 | ≅1.5 |
| 150 | 101 | 49 | ≅1.5 |
| 100 | 67 | 33 | ≅1.49 |

As the inlet temperature to the expander is lowered, the $\Delta T$ is reduced. The *J–T* effect is the opposite, i.e., the lower the inlet temperature the greater the $\Delta T$. See Fig. 9.16 for the general characteristics of centrifugal expanders.

## 5. VALVES

### 5.1. Valves for Cryogenic Liquids

Some desirable characteristics of a satisfactory valve for cryogenic service are (1) low heat leak, (2) reliable operation at the required temperature, (3) small heat capacity, (4) small resistance to flow, (5) simplicity and economy of construction, and (6) adaptability for insertion into ordinary vacuum-insulated lines. See Table 9.4. The following discussion is devoted to descriptions of some of the valves that have been used in vacuum-insulated transfer lines and some comments on their characteristics.

A design frequently employed consists of extending the stem of an ordinary commercial valve so as to reduce heat leak and surrounding the assembly in the vacuum enclosure which is a part of the insulating vacuum of the line. Such a valve is illustrated in Fig. 9.17. A valve of this type is probably the most economical in first cost because the principal parts are stock items, commercially available. It is common practice to replace the closing disk, usually made of fiber, with a disk of Teflon® which gives a better seal at low temperatures. There is some heat leak through the extended stem but this can be made small by employing thin-wall alloy tubing for this part of the stem. The valve may have an objectionably high heat capacity

- High reliability and efficiency
  - Used in large tonnage liquefaction plants
- Two types radial and axial
- Can handle liquid formation within turbines
- Major function is to provide expansion with recovery of resulting work
- As is all rotating machinery; uses journal and thrust bearings
  - Cryogenic environments
- Most cryogenic expanders are of the radial type
  - Special blade designs to accommodate condensing flow

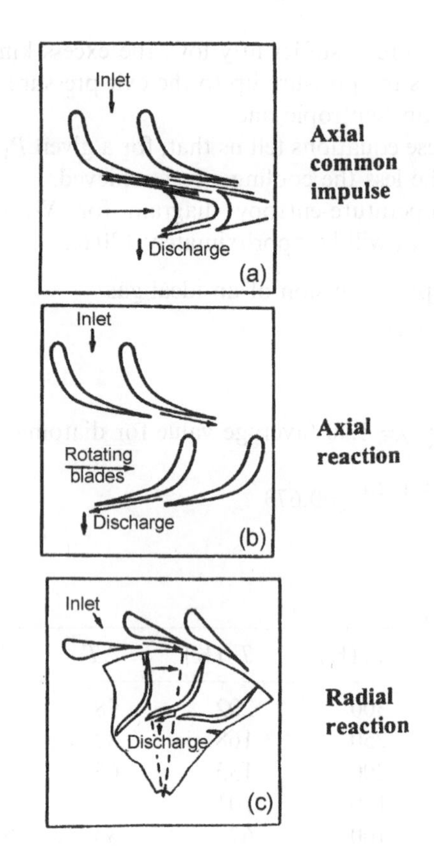


**Figure 9.16** General characteristics of centrifugal expanders.

because the commercial valve body was intended for use at room temperature, hence there was no incentive to minimize heat capacity.

Figure 9.18 illustrates a valve in which metallic heat conduction from the exterior is eliminated except while the valve is being adjusted. In this valve, the actuating stem can be withdrawn after the valve is turned so that there is no metallic contact. Valves of a somewhat similar principle have been devised in which the actuating

**Table 9.4** Cryogenic Flow Control Devices

Flow control devices are generally components which initiate, stop, alter or regulate the flow of cryogenic fluids.

- Valves
- Check valves
- Regulators

Characteristics to evaluate in component selection include

- Low heat leak
- Reliability at low temperature
- Small heat capacity
- Small resistance to flow
- Simplicity of construction and maintenance
- Cost

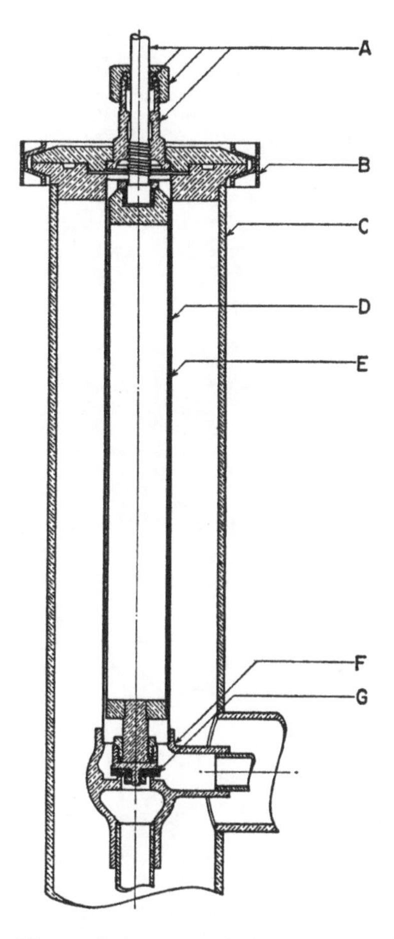


**Figure 9.17** Extended stem valve. A, commercial valve stem and packing for gas seal; B, Marman V-band coupling; C, outer wall of vacuum jacket; D, valve stem; E, fluid line and inner wall of vacuum jacket; F, commercial valve; G, closing element.

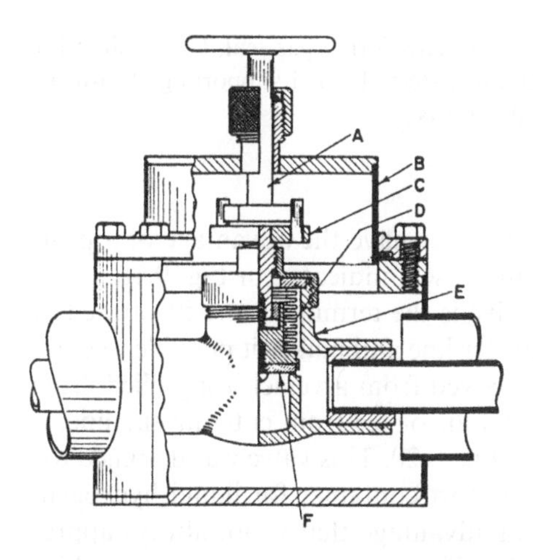


**Figure 9.18** Broken-stem valve. A, valve stem; B, outer wall of vacuum jacket; C, valve wheel; D, bellows seal between liquid line and vacuum jacket; E, modified commercial valve; F, closing element.

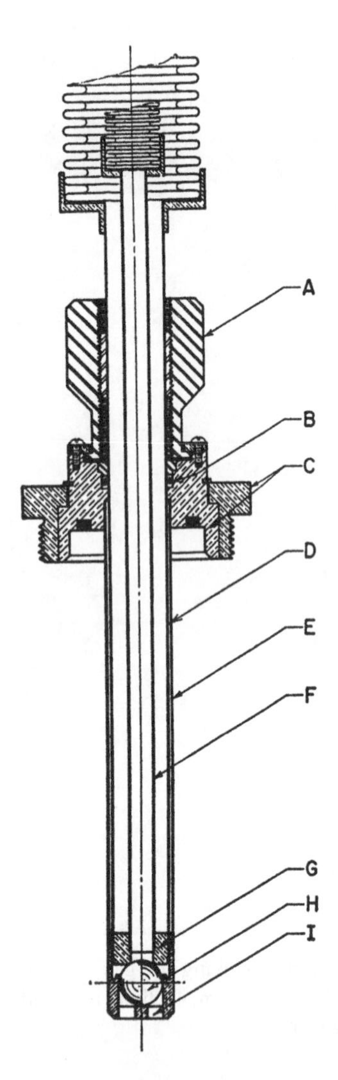


**Figure 9.19** Terminal valve. A, valve handle; B, neoprene O-ring gas seal; C, coupling for transfer line; D, outer wall of vacuum jacket and valve stem; E, ball supporting sheath; F, liquid line; G, valve seat; H, closing element; I, valve ports.

device operates through a metal bellows. These latter have the advantage of a positive, soldered, vacuum seal rather than the O-ring seal indicated in Fig. 9.18.

A special-purpose but still very useful valve is the terminal valve illustrated in Fig. 9.19. This valve is used at the end of a transfer line and makes it possible to close the transfer line at the very end before it is removed from a vessel being filled. This end closure prevents contaminants such as solid air or ice from entering the line.

A somewhat similar valve is illustrated in Fig. 9.20. This valve was designed for use in the fill-and-empty fixture of a large transport container for liquid hydrogen. Both this valve and the preceding one have an advantage that is not always appreciated. The stainless stell ball closing member is self-centering so that minor machining errors do not prevent tight closure. It has been found desirable to prepare the seat in the shape of a sharp edged cylinder of deformable metal (e.g., brass); then

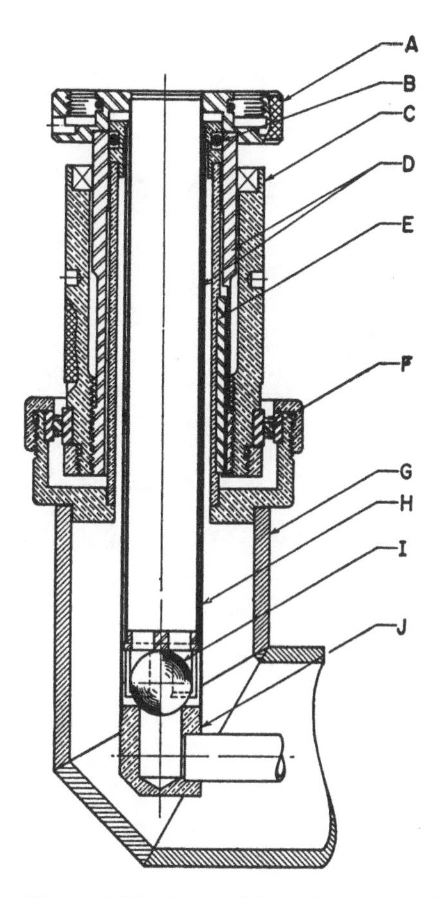


**Figure 9.20** Dewar inlet valve. A, coupling for transfer line; B, neoprene O-ring gas seal; C, valve handle; D, valve stem; E, key to hold valve stem from turning; F, bearing; G, outer wall of vacuum jacket; H, fluid line and inner wall of vacuum jacket; I, closing element; J, valve seat.

when the truly spherical ball is pressed against the seat the deformation provides an accurate match and a good seal.

A valve that has been found quite convenient for controlling the transfer of liquid hydrogen is the helium-operated valve shown in Fig. 9.21. In this valve gaseous helium admitted through a small, poorly conducting tube into the interior of the metal bellows provides the force required to close the valve. The principal advantage of this valve is that it is readily controlled from a remote location.

Figures 9.22 and 9.23 illustrate typical cryogenic valves. Thin-walled stainless steel valve stems and vacuum jackets minimize the heat leak to the cryogen. Enclosing the valve stem in a stainless steel sheath allows an all-welded construction, which assures integrity of the vacuum insulation and makes replacement of the valve plug easy. The annular space between the valve stem and its sheath is typically 0.1 mm or less to prevent convection and its associated heat leak. Valve plugs are typically fabricated from fluorocarbons such as Kel-F®. Either an O-ring or packing seals the cryogen gas space. Figure 9.23 shows a cross-section of a typical cryogenic valve. A family of cryogenic globe valves is shown in Fig. 9.24. Figure 9.25 details a vacuum-jacketed valve. Figure 9.26 shows how large these valves can be.

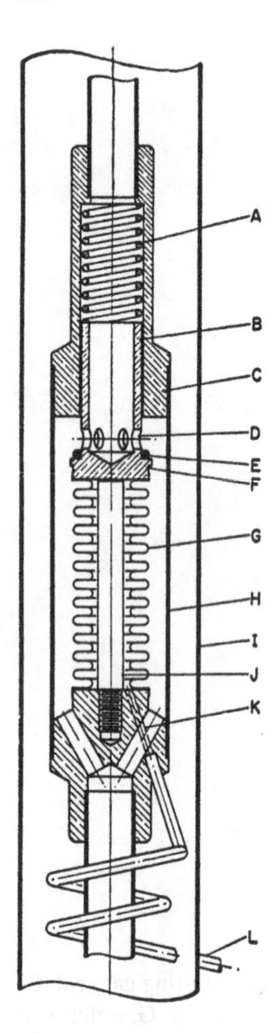


**Figure 9.21** Helium-operated valve. A, spring which opens valve; B, valve guide; C, valve body; D, valve ports; E, Teflon® O ring; F, part which forms one side of O-ring seal; G, operating bellows; H, fluid lines; I, outer wall of vacuum jacket; J, valve stop; K, helium inlet port; L, stainless steel helium line.

## 5.2. Joule–Thomson Valves

The principal function of a Joule–Thomson valve is to obtain isenthalpic cooling of the gas flowing through the valve. A typical expansion valve is shown in Fig. 9.27. To reduce sealing problems in the valve and minimize leakage through the valve-stem packing, the high-pressure stream is always directed into the lower compartment of the valve away from the valve seat. Positive control on the valve position is recommended over spring-returned valves because impurities can cause the valve seat to stick. The valve stem is generally long enough so that the stem sealing is at ambient temperature. The walls of the tubes are purposely thin and hollow to reduce heat transfer. Galling of the valve stem threads can be prevented by either using a solid lubricant or brass as the female-threaded member. Another function of the expansion valve is to separate the liquid formed after expansion from the vapor. Liquid that is entrained in the vapor can be separated by swirling the mixture in a centrifuge-type action. This can be accomplished by curving the outlet tubes.

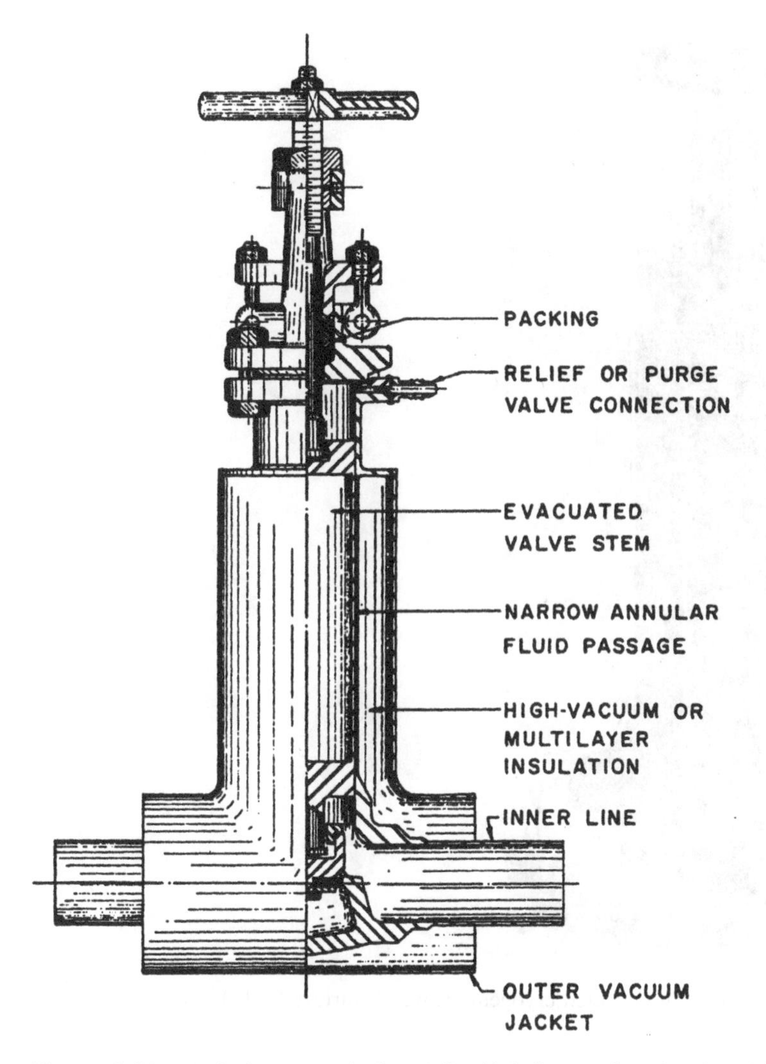


**Figure 9.22** Typical vacuum-jacketed liquid helium valve showing thin-walled evacuated stem design with warm-end connection on helium gas space. (From Ref. [3].)

Joule–Thomson valves, besides being widely used in refrigeration processes, now are offering an attractive alternative to the turboexpander for small-scale light hydrocarbon (LPG) recovery applications, provided that ethane recovery is limited to below 30%. In a Joule–Thomson plant, the expander would be eliminated, but to obtain the same level of recovery as in a turboexpander plant, the compressor would have to be scaled up. Another advantage attributed to a Joule–Thomson plant besides overall simplicity of operation is the ability to operate under more widely varying inlet gas flow rates.

## 6. HEAT EXCHANGERS

### 6.1. The Thermodynamics of Heat Exchangers

The counterflow heat exchanger is one of the most important devices used in cryogenics. It constituted the final touch that permitted the continuous-flow liquefaction

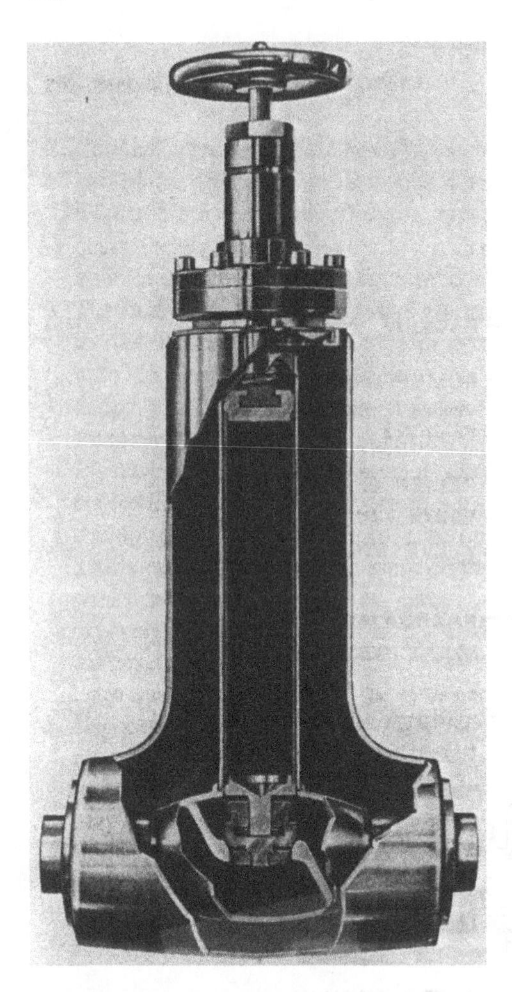

**Figure 9.23** Typical vacuum-jacketed cryogenic valve. (Courtesy CVI, Inc.)

of the "permanent" gases. It now exists in a multitude of forms, but its purpose is always to transfer heat from one fluid stream to another and thus in cryogenic processes, to conserve "cold" by using the outgoing cold fluid to cool the incoming warm stream. If this can be done with negligible temperature differences and no appreciable resistance to flow, the process approaches thermodynamic reversibility. Thus the designer of a heat exchanger tries to provide large surface for heat flow and yet tries to avoid excessive pressure drop. These two objectives are actually somewhat opposed, because increasing the surface area, and thus enhancing the opportunity for heat exchange of a channel of given cross-section, also increases its resistance to flow. There is another practical consideration in heat-exchanger design; the heat capacity should be kept small. This reduces the time required to reach a steady thermal state.

One of the more important aspects of any low-temperature heat-exchanger design is that of efficient heat transfer. This point is aptly demonstrated by considering the effect of heat-exchanger effectiveness on the liquid yield for a simple nitrogen Joule–Thomson liquefaction process with lower and upper operating pressures of 1 and 100 atm, respectively. It can be shown (see below) that the liquid yield under these conditions will be 0 for an exchanger that has an effectiveness below 90%.

**Figure 9.24** A family of cryogenic globe valves. Cryogenic globe valves are available from $\frac{1}{2}$ in. OD to 12 in. NPS and available as manual, remote on/off, remote flow control, and check valves. Valve materials are 304L or 300-series stainless steel. Bonnets are bronze alloys. Valve bonnets have permanently cast or marked flow arrows when vacuum jacketed. The preferred butt welds are "true butt weld" without closed-end cavities (such as socket welds, which cannot be easily cleaned for oxygen service). Seal disk material for the valves shown here is Neoflon® PCTFE—400H (formerly Kel-F). Bonnet (Viton®, fluorocarbon rubber, O-ring) seals remain warm when the valve is in cryogenic service. Stem packing utilizes Viton® O-rings on the static and dynamic areas of the sealing surfaces. These vacuum-jacketed valves are serviceable without breaking vacuums. Stem packing in manual valves is replaceable without removing the valves from service. PHPK Technologies Inc., Westerville, OH, www.phpk.com.

*Example 9.1.* In a simple Linde liquefier (see Fig. 6.22, Chapter 6), nitrogen gas enters the compressor at 1 atm and 60°F and is compressed isothermally to 100 atm. Determine the minimum allowable effectiveness of the heat exchanger (i.e., determine the heat-exchanger effectiveness for which the liquid yield is 0.

*Solution:* Heat-exchanger effectiveness is defined as the ratio of the actual heat transfer in an exchanger to the maximum heat transfer possible. An energy balance around the cold heat exchanger and liquid pot provides the mass liquefied. (Refer to Fig. 6.22 and Section D of Chapter 6 for the notation used, remembering that, since this is a liquefier, $\dot{m}_f$ will be withdrawn from the liquid pot.)

$$h_2 = \dot{m}_f h_f + (1 - \dot{m}_f) h_1 \tag{9.15}$$

**VACUUM JACKETED VALVE**

**Item:**
Valve, extended stem, vacuum jacketed, cryogenic uses.

**Service:**
Cryogenic liquids and gases; helium, hydrogen, nitrogen, natural gas (Methane), and oxygen (with special cleaning and packaging).

**Operating Ranges:**
Temperature: –425°F to +300°F.
Maximum operating pressure: 400 psig (ambient).
Proof pressure: 600 psig (ambient).

**Tests:**
Valve: Mass spectrometer to $1 \times 10^{-9}$cc G He/sec.
Seat: Bubble tight with 100 psig G He across the seat at –320°F.

**Materials:**
Metal components are predominantly 300 series stainless steel. All soft goods are Teflon, Kel-F or Viton.

**Options:**
"Cold box" jacketing, purge ports, pneumatic and electric actuators, positioners, limit switches and position indicators available (please consult factory).

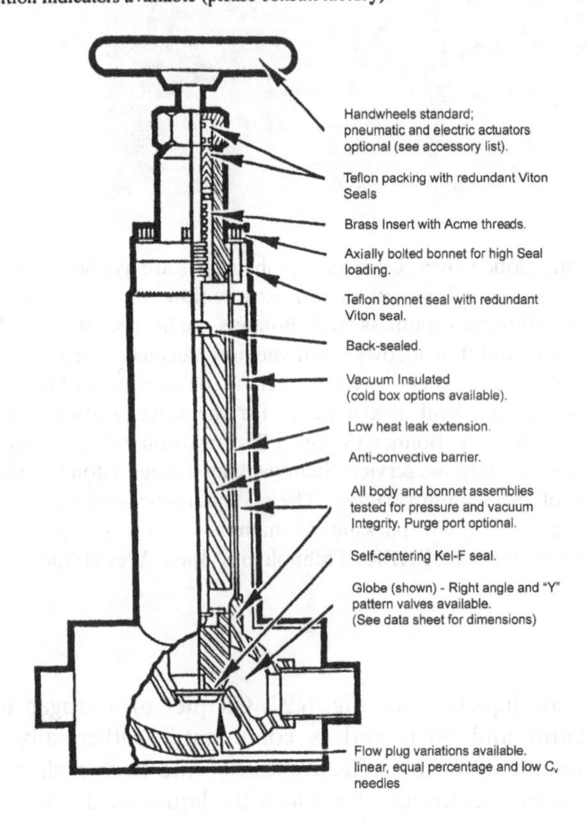
Handwheels standard; pneumatic and electric actuators optional (see accessory list).
Teflon packing with redundant Viton Seals
Brass Insert with Acme threads.
Axially bolted bonnet for high Seal loading.
Teflon bonnet seal with redundant Viton seal.
Back-sealed.
Vacuum Insulated (cold box options available).
Low heat leak extension.
Anti-convective barrier.
All body and bonnet assemblies tested for pressure and vacuum Integrity. Purge port optional.
Self-centering Kel-F seal.
Globe (shown) - Right angle and "Y" pattern valves available. (See data sheet for dimensions)
Flow plug variations available. linear, equal percentage and low Cv needles

**Figure 9.25** Examination of a cryogenic valve.

Solving for $\dot{m}_f$ results in

$$\dot{m}_f = \frac{h_1 - h_2}{h_1 - h_f} \tag{9.16}$$

(A) (B)

**Figure 9.26** Large cryogenic valves. (A) An 8 in. NPS liquid oxygen valve for a NASA Stennis test stand. The fellow beside it is actually the photographer. He was the only person that day who had a tie on. (B) The same valve installed at NASA Stennis. PHPK Technologies Inc., Westerville, OH, www.phpk.com.

For zero liquid yield, $\dot{m}_f = 0$; therefore

$$\frac{h_1 - h_2}{h_1 - h_f} = 0 \tag{9.17}$$

and from this $h_1 = h_2$.

Thermodynamic values for the ideal condition (zero temperature difference between points 1 and 2) are as follows:

$$h_2 = 2680\,\text{cal/g mol}(60°\text{F}, 100\,\text{atm})$$
$$h_1' = 2830\,\text{cal/g mol}(60°\text{F}, 100\,\text{atm})$$
$$h_f = 1320\,\text{cal/g mol}(60°\text{F}, 100\,\text{atm})$$

Heat-exchanger effectiveness under such conditions is then

$$\varepsilon = \frac{h_1 - h_f}{h_1' - h_f} = \frac{h_2 - h_f}{h_1' - h_f} = \frac{2680 - 1320}{2930 - 130} = 0.90$$

See Table 9.4.

The importance of minimizing temperature differences between streams exchanging heat (as shown in Example 9.1) and achieving small pressure drops in

- Principal function is to obtain isenthalpic cooling of gas
- Joule-Thomson plant advantages
  - Overall simplicity
  - Can operate with more widely varying gas flow rates
- Offers alternative to turboexpander for smaller light hydrocarbon recovery applications
  - Eliminate expander
  - Compressor scale up to obtain same level of recovery

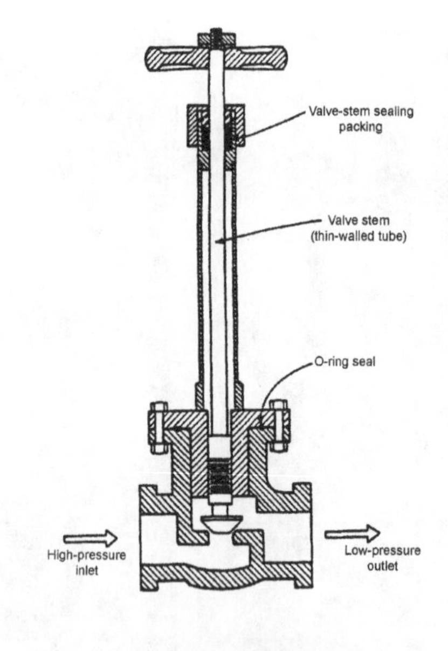


- Lee ViscoJet
  - Designed for small flows
  - Multiple flow paths to reduce pressure in flowing fluid
    - Helps prevent clogging since path size is larger than single orifice

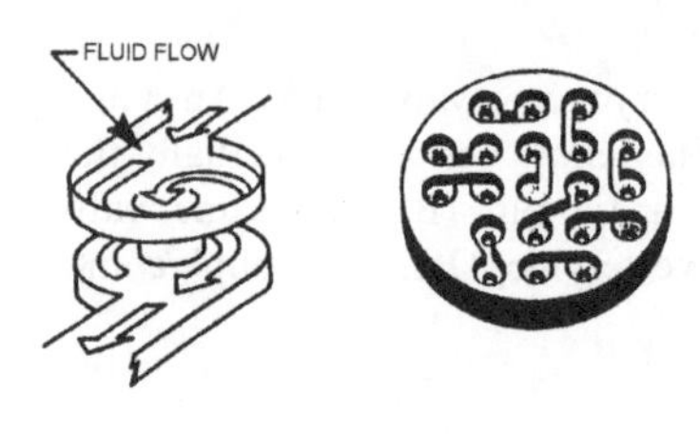


- General Pneumatics Cryostat
  - Designed for high pressure gas feed
  - Self-cooling
  - Clog-resistant due to multiple path approach

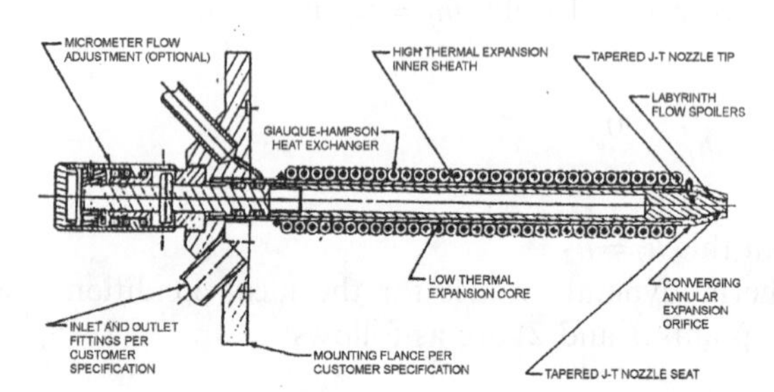


**Figure 9.27** Typical Joule–Thomson valves.

each stream, consistent with reasonable sizes and hence costs, is clear when one realizes that these are all irreversible effects, which, in turn, require increased work input.

The thermodynamic measure of the general quality of either a low-temperature piece of equipment, such as a heat exchanger, or an entire process, is its reversibility. The second law, or more precisely the entropy increase, is an effective guide to the degree of irreversibility associated with such an item or process. However, to obtain a clearer picture of what these entropy increases mean, it has become convenient to relate such an analysis to the additional work that is required to overcome these

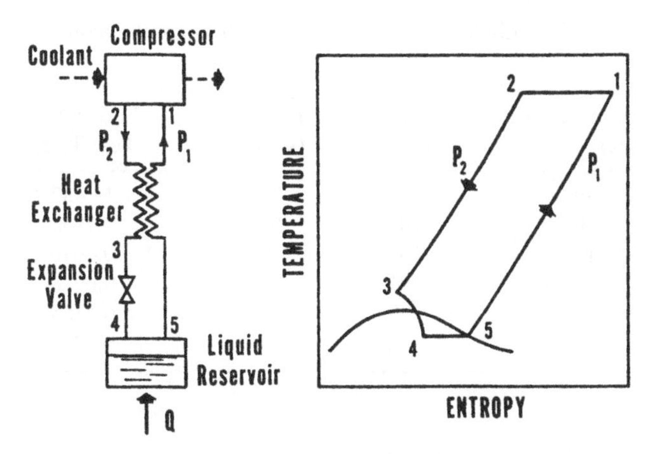


**Figure 9.28** Refrigerator using simple Linde cycle.

irreversibilities. The fundamental equation for such an analysis is

$$W = W_{\text{rev}} + T_0 \sum \dot{m}\,\Delta s \tag{9.18}$$

where the total work is the sum of the reversible work plus a summation of the losses in availability for various steps in the analysis.

Application of this method to an actual process can best be demonstrated by a numerical example. Consider a simple nitrogen liquefaction process using the same concept as the refrigerator shown in Fig. 9.28. Conditions for this process are indicated in Table 9.5. Assuming, for simplicity, that there is no heat leak to the process, the fraction of nitrogen liquefied is 0.0198 as calculated from

$$y = \frac{h_1 - h_2}{h_1 - h_f} \tag{9.19}$$

The total work involved in the compressor, assuming three-stage adiabatic compression, 75% compression efficiency, and an initial temperature of 300 K, is 16,400 kcal for each 2 kg of liquid nitrogen produced.

For the exchanger, the losses are given by

$$T_0 \sum m\Delta s = 300[99(1.0482 - 0.7155) - 101(0.7158 - 0.4646)] = 2270\,\text{kcal}$$

**Table 9.5** Heat Exchangers

Critical component in any low-temperature liquefaction and refrigeration system
Heat-exchanger effectiveness has significant influence on process results
  Consider the liquid yield of a simple Joule–Thomson liquefaction process
  With $N_2$, $P_1 = 1.0$ atm, $P_u = 100$ atm psia; liquid yield $= 0$ if heat-exchanger effectiveness is $< 0.90$
Well-established principles of mechanics and thermodynamics at room temperature hold true for cryogens
  Similar heat-transfer correlations for simple low-temperature designs
  Need for higher effectiveness resulted in very efficient, complex heat exchangers

**Table 9.6** Process Conditions for Simple Linde Cycle (Location Numbers are the Same as for Fig. 9.19)

| Location | Pressure (atm) | Temperature (K) | Mass (kg) | Enthalpy (kcal kg) | Entropy (kcal/kg K) |
|---|---|---|---|---|---|
| 1 | 1 | 290 | 99 | 108.0 | 1.0482 |
| 2 | 100 | 300 | 101 | 106.0 | 0.7158 |
| 3 | 100 | 158 | 101 | 52.8 | 0.4646 |
| 4 | 1 | 77.3 | 101 | 52.8 | 0.6918 |
| 5 | 1 | 77.3 | 99 | 54.6 | 0.7155 |
| Liquid | 1 | 77.3 | 2 | 7.0 | 0.0999 |

For the throttling valve, the losses are evaluated as

$$T_0 \sum m\Delta s = 300[101(0.6918 - 0.4646)] = 6880\,\text{kcal}$$

Since temperature approaches in the intercoolers for the compressor have not been specified, it is more convenient to calculate the work involved for the reversible isothermal compression and assume that the difference between the latter and the total work is the loss associated with the compression step (Table 9.7). The reversible work forthis compression (9.20) is obtained from the expression

$$-W/\dot{m}_f = T_0(s_1 - s_f) - (h_1 - h_2) \tag{9.20}$$

which gives for this particular process 9890 kcal. Therefore, the loss in the compression step is (16,400 – 9890) or 6510 kcal. Table 9.7 presents a summary of the results.

The sum of the losses plus the reversible work closely approximates the total compression work evaluated earlier and shows at a glance where the greatest gains in efficiency can be made. Since all low-temperature processes encounter heat leaks, this can be included in the calculations by evaluating the fraction liquefied from Eq. (9.19 )and recalculating the work per mass of liquid produced. This calculation will show that even small heat leaks can contribute significantly to the losses and thus greatly increase the work of compression.

Numerous analyses and comparisons of refrigeration and liquefaction cycles are available in the literature. Great care must be exercised in accepting these comparisons since it is quite difficult to put all processes on a strictly comparable basis.

Thus, we see that the total work for a process is equal to the sum of the reversible work plus the work to overcome all of the irreversibilities. The latter quantity was given as a summation term in Eq. (9.18) and when multiplied by the heat-sink temperature $T_0$ gives the total loss in availability due to all the irreversible effects

**Table 9.7** Losses Associated with Simple Linde Cycle

| Location of Losses | Losses (kcal) | % |
|---|---|---|
| Loss in heat exchanger | 2270 | 14.5 |
| Loss in throttling valve | 6880 | 43.9 |
| Loss in compressor | 6510 | 41.6 |
| | 15,660 | 100.0 |

**Table 9.8** Design Criteria for Low-Temperature Heat Exchangers

| |
|---|
| Small temperature differences between inlet and outlet streams (high effectiveness) |
| Large fluid surface area-to-volume ratio to minimize heat leak |
| High heat transfer to reduce surface area |
| Low mass to minimize start-up time (cool-down) |
| Multichannel capability to minimize the number of exchangers |
| High-pressure capability to provide design flexibility |
| Low pressure drop to minimize compression requirements |
| High reliability with minimal maintenance to reduce shutdowns |

in a process or piece of equipment. Even though these losses are smaller for the heat-exchange process than the ones associated with the compression step, they are nevertheless sizable and have to be made up by the expenditure of expensive mechanical energy. For example, in a simple gaseous oxygen process, these losses may account for as much as 13% of the total needed for the cycle. (The loss due to the temperature difference between two streams in a heat exchanger becomes even more significant as the absolute temperature is lowered.)

Most cryogens, with the exception of helium II, behave as "classical" fluids. As a result, it has been possible to predict their behavior by using well-established principles of mechanics and thermodynamics applicable to many room-temperature fluids. However, the requirements imposed by the need to operate more efficiently at low temperatures make the use of simple exchangers impractical in many cryogenic applications. In fact, some of the important advances in cryogenic technology are directly related to the development of rather complex but very efficient types of heat exchangers. Some of the criteria that have guided the development of these units for low-temperature service are (1) a small temperature difference at the cold end of the exchanger to enhance efficiency, (2) a large surface area-to-volume ratio to minimize heat leak, (3) a high heat transfer to reduce surface area, (4) a low mass to minimize start-up time, (5) multichannel capability to minimize the number of exchangers, (6) high-pressure capability to provide design flexibility, (7) a low or reasonable pressure drop to minimize compression requirements, and (8) minimum maintenance to minimize shutdowns. See Table 9.8.

The criteria of minimizing the temperature difference at the cold end of the exchanger are not without their problems, particularly if the specific heat of the cold fluid increases with increasing temperature as demonstrated by hydrogen. In such a case, the temperature difference between the warm and cold streams is considerably smaller in the middle of the heat exchanger than at each end of the exchanger, as shown in Fig. 9.29. The difference at the cold end wastes refrigeration. Attempts to design the heat exchanger with a smaller temperature difference run the risk of violating the second law of thermodynamics. The heat exchanger itself could not operate under these conditions, and the exit temperatures would adjust themselves so that the appropriate direction of heat flow would be maintained within the heat exchanger.

The problem in cryogenic heat exchangers is generally alleviated by adjusting the mass flow rates of the key stream into the heat exchanger. In other words, the capacity rate is adjusted by controlling the mass flow rates to offset the change in specific heats. Problems of this nature can be avoided by making enthalpy balances in incremental steps from one end of the exchanger to the other.

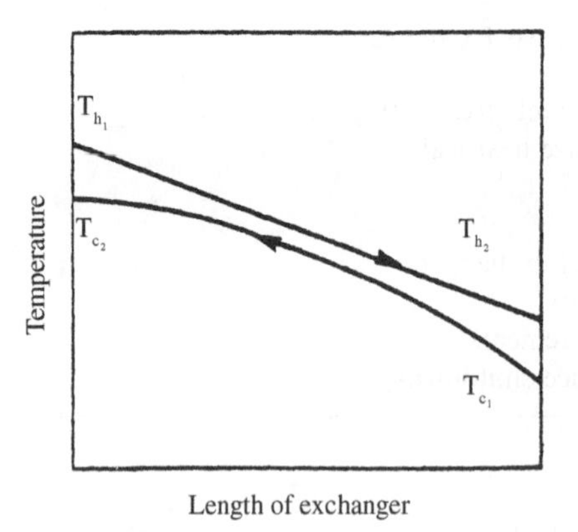


**Figure 9.29** Temperature profile in countercurrent heat exchanger with cold fluid demonstration more rapid increase in specific heat with increasing temperature.

## 6.2. Types of Low-Temperature Heat Exchangers

It may be seen that there exists the possibility of almost infinite variation in heat-exchanger design. The form chosen depends upon several factors, among which are (1) the importance of avoiding pressure drop, (2) the need for low heat capacity, (3) the cost, (4) the pressures to be accommodated, and (5) the dimensions of the space available for the exchanger.

These criteria are often conflicting: (1) the need for high surface area vs. the requirement of low pressure drop, and (2) high lateral heat transfer between fluid streams vs. low heat leak longitudinally from top to bottom. It is not surprising that many design attempts have been made to accommodate conflicting requirements. Some types are discussed below.

### *6.2.1. Coiled Tube Heat Exchangers*

In the form used by Hampson the exchanger consisted of small copper tubes wound in a closely spaced coil. The high-pressure gas stream flowed in the tubes and the low-pressure stream filtered back outside the tubes through the interstices between the turns. Figure 9.30 shows a modern version of the Hampson heat exchanger, a design of W.F. Giauque of the University of California. Multiple-tube exchangers of rather similar design have been used in very large commercial installations. It is important in exchangers of this type to make the tube spacing quite uniform; otherwise, the low-pressure return gas stream will tend to channel, that is, it will have a preferred path and will not be equally distributed over a cross-section of the exchanger. This precaution is particularly important when the diameter of the exchanger is large.

Figure 9.31 shows the heat exchanger designed by S.C. Collins. It consists of several coaxial copper tubes, each closely wrapped with an edge-wound helix of copper ribbon. The helix is soft-soldered to the outside of the tube it surrounds and to the inside of the next outer tube. The helix thus greatly extends the surface for heat transfer and at the same time provides a lateral path for heat conduction. This heat-exchanger tubing was manufactured by the Joy Manufacturing Company of Michigan City, Indiana, and was available with as many as three annular channels. The channels chosen for the ingoing and outgoing gas flows can be selected

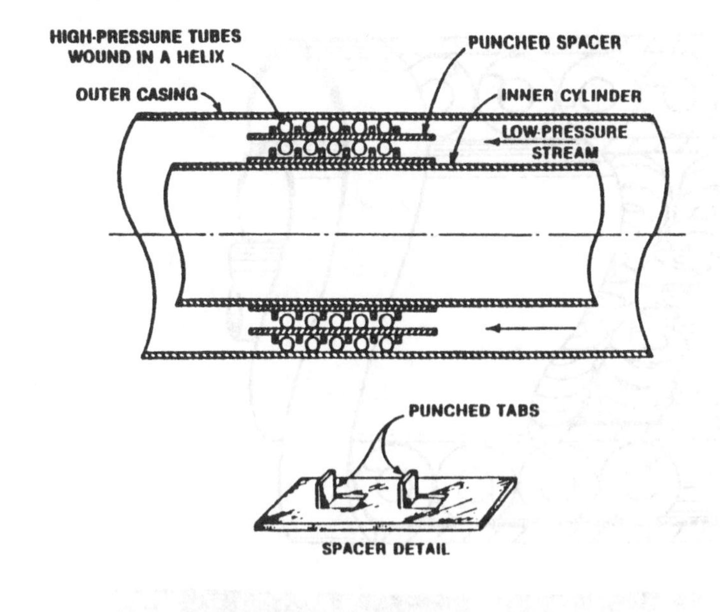


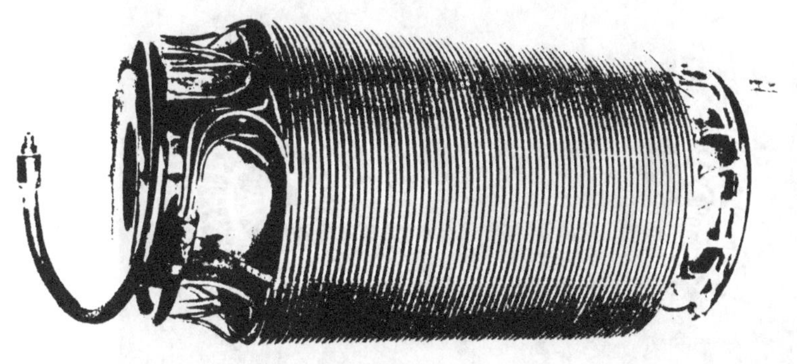

- Classical heat exchanger for large-scale liquefaction systems
- Consists of carefully spaced helixes of small-diameter tubes through which the high-pressure stream flows
  - Low pressure stream flows across the outside of the small tubes
- Tube spacing is critical for uniform flow
  - Utilizes punched spacer strips
- Heat transfer correlations are available for this type of heat exchanger

**Figure 9.30** Giauque–Hampson heat exchanger.

and combined to accommodate gas streams of different pressure and volumetric rate of flow.

A heat exchanger of good efficiency, low heat capacity, and relatively simple construction was described by Parkinson. It consists of a number of helices (of small diameter and small pitch) of high-pressure copper tubing. These are wound as a multiple-thread helix on a central, poorly conducting thin-wall tube and are surrounded by a close-fitting sheat of low thermal conductivity as shown in Fig. 9.32. The low-pressure counterflowing stream returns through the annular space bounded by the

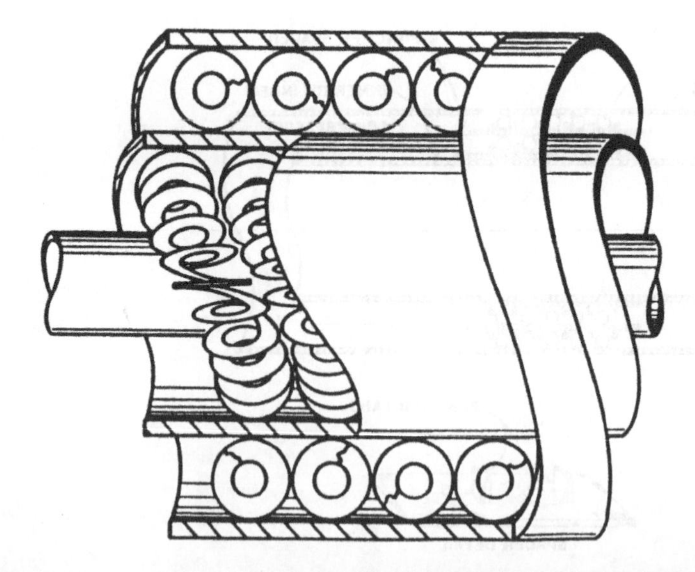

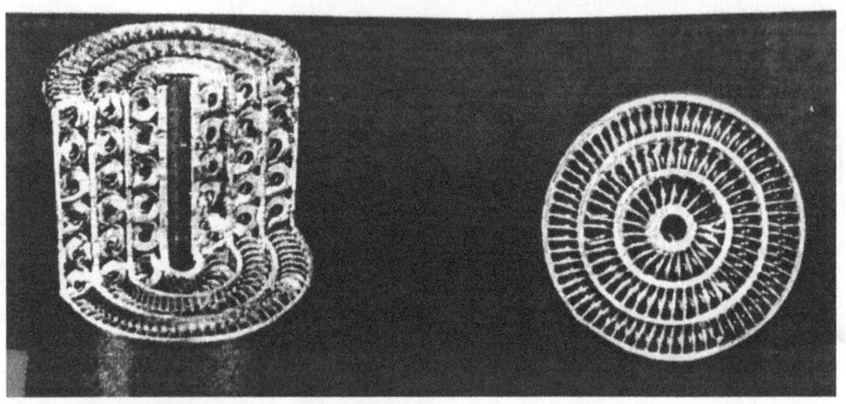

- More efficient version of concentric tube
- Consists of several concentric copper tubes with an edge wound copper helix wrapped in annular space
  - Copper is soft-soldered to both sides of annular
  - Performs like a fin (extending heat transfer surface)
- Number of tubes range from 4 to 10
- Warmer high-pressure stream flows in the innermost tube

**Figure 9.31** Collins type extended surface heat exchangers.

central tube and the outer sheath and makes good contact with the high-pressure helices.

*6.2.2. Concentric Tube Heat Exchangers*

Heat exchangers have been made consisting of seven high-pressure tubes spaced inside the tube which constitutes the return passage of the low-pressure gas as shown in cross-section in Fig. 9.33. It will be noted that in this design all the surfaces contribute to heat exchange except the large-diameter outer tube. This will serve

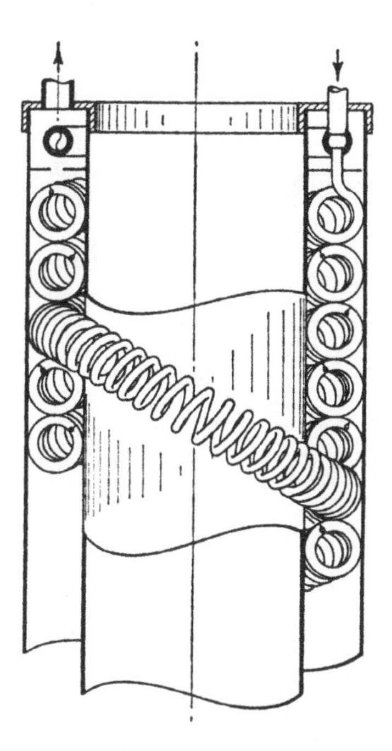

**Figure 9.32** Parkinson's heat exchanger.

to illustrate a desirable objective in heat-exchanger design. Since all the surfaces bounding either gas stream cause frictional loss and thus contribute to the pressure drop, it is desirable to have them contribute also to heat exchange between the two streams, in order to have the most favorable ratio of heat-exchange efficiency to pressure drop. In the Hampson heat exchanger, the thermally useless surface is the outer case. In the Collins exchanger, the conducting helix provides thermal paths which make all the surfaces useful in heat exchange.

### *6.2.3. Plate–Fin Exchangers*

These types of heat exchangers normally consist of heat-exchange surfaces obtained by stacking alternate layers of corrugated, high-uniformity, die-formed aluminum sheets (fins) between flat aluminum separator plates to form individual flow passages. Each layer is closed at the edge with solid aluminum bars of appropriate shape and size. Figure 9.34 illustrates by exploded view the elements of one layer, in relative position, before being joined by a brazing operation to yield an integral rigid structure with a series of fluid flow passages. The latter normally have integral welded headers. Several sections may be connected together to form one large exchanger. The main advantages of this type of exchanger are that it is (1) compact (about nine times as much surface area per unit volume as conventional shell and tube exchangers), yet permits wide design flexibility; (2) involves minimum weight; and (3) allows design pressures to $875\,\text{lb/in.}^2$ gauge from $-452°\text{F}$ to $150°\text{F}$.

Various flow patterns of brazed plate–fin exchangers (Fig. 9.35) can be developed to provide multipass or multistream arrangements by incorporating suitable internal seals, distributors, and external headers. In the simple cross-flow layout, the corrugations extend throughout the full length of each set of passages with

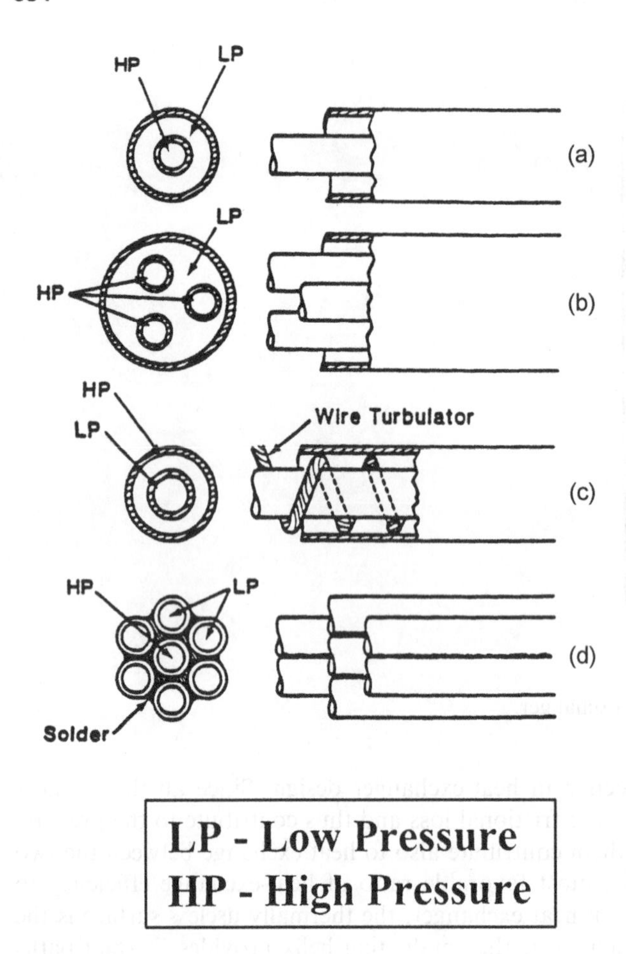
HP
LP
(a)
LP
HP
(b)
HP
LP
Wire Turbulator
(c)
HP
LP
(d)
Solder
LP - Low Pressure
HP - High Pressure
• Widely used in small-scale operations
• Simple design consists of small inner tube and an annulus (a)
• Multi-tube heat exchangers are used in high pressure large-flow systems (b)
• Heat transfer coefficient enhanced by inserting wire spacer in annulus (c)
• Bundle heat exchanger has all tubes in good thermal contact with each other (d)

**Figure 9.33** Concentric tube heat exchangers.

no internal distributors. This arrangement is suitable when the effective mean temperature difference obtained in cross-flow is not too far below the logarithmic mean temperature difference calculated for countercurrent flow. Such a condition is encountered in liquefiers where there is little change in temperature on the condensing side and the large flow of low-pressure gas on the warming side requires a large cross-section and short passage length. The multipass cross-flow arrangement can be considered as comprising several cross-flow sections, assembled in counterformation, to provide effective mean temperature difference approaches that more closely approximate those obtained with countercurrent operation. Construction of this type is often used for gas–gas and gas–liquid applications. Countercurrent construction is normally selected when very high thermal efficiencies (95–95%) of heat exchange are required.

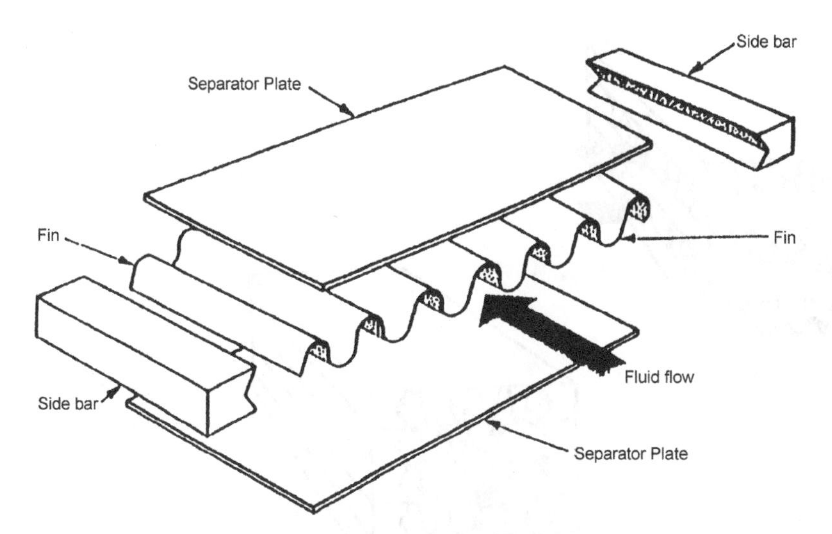


- Constructed by stacking alternating layers of corrugated aluminum sheets between flat aluminum separator plates
  - Edges sealed with solid aluminum bars
  - Brazing connects the elements to form rigid structure
  - Flow passages have integral welded headers
- Plate-fin heat exchangers features
  - Approximately 9 times as compact as shell-tube type
  - Can withstand high pressures

**Figure 9.34** Plate–fin heat exchanger.

Design of the brazed aluminum plate–fin exchanger resolves itself into selecting a geometry and surface arrangement to give a product $UA$ of the right magnitude to satisfy the equation

$$Q = UA\,\Delta T_{\mathrm{c}} \tag{9.21}$$

The overall $UA$ of a plate–fin exchanger can be related to the individual $hA$ values for the individual component fluid streams by

$$UA = \left[\sum (hA)_{\mathrm{h}} \sum (hA)_{\mathrm{c}}\right] / \left[\sum (hA)_{\mathrm{h}} + \sum (\mathrm{hA})_{\mathrm{c}}\right] \tag{9.22}$$

where the subscripts h and c refer to the warm and cold streams, respectively.

For single-phase flow, the heat-transfer coefficient is generally expressed in

$$h = j_1 C_p G (\mathrm{Pr})^{-2/3} \tag{9.23}$$

where $j_1$ separates the effects of the fluid properties on the heat-transfer coefficient and permits correlation as a function of the Reynolds number. Because of the many varied shapes for the fins, it is usually necessary to test each configuration individually to determine both the heat-transfer and frictional characteristics for a specific surface. Such information, in the form of curves relating $j_1$ as a function of the Reynolds number, is available from the heat-exchanger manufacturers of the various surface configurations.

Information on two-phase heat-transfer coefficients in plate–fin heat exchangers is essentially non-existent, and no general correlations are available for the prediction of these coefficients. In view of this deficiency, coefficients for condensing,

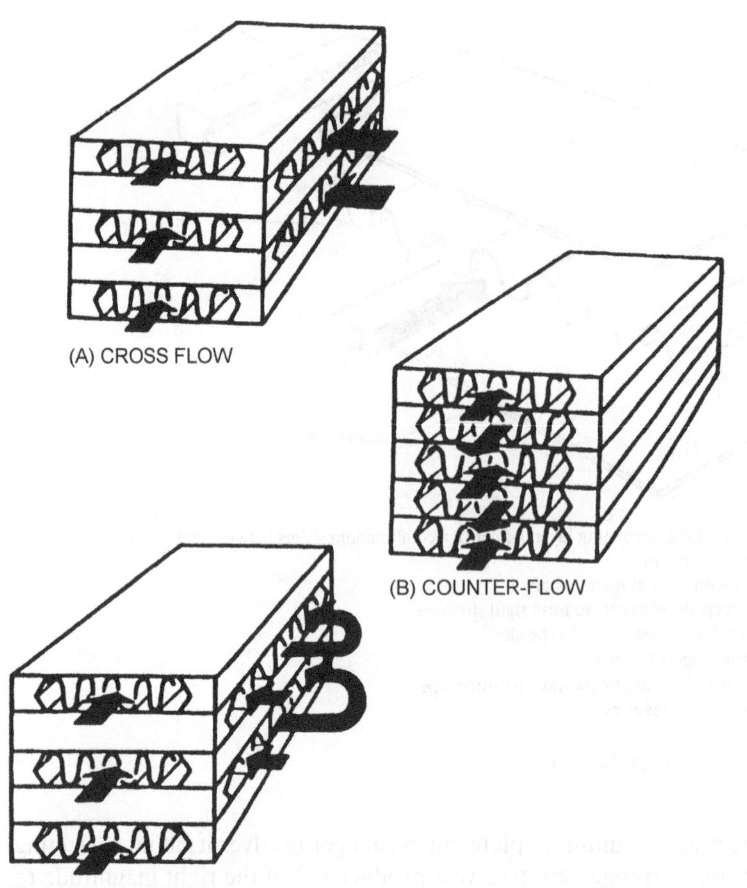


- Many different types of flow patterns are possible
  - Internal seals, distributors and external headers provide for either multipass or multistream arrangements
- For simple cross-flow arrangements (a)
- Counter-flow arrangement (b)
- Multipass cross-flow arrangement (c)

**Figure 9.35** Plate–fin heat exchangers flow patterns.

boiling, and multicomponent systems are approximated using predictions commonly adopted for single tubes. For example, the relations developed for two-phase condensing flow in tubes have been used with reasonable success to determine heat-transfer coefficients under similar operating conditions in individual channels of plate–fin exchangers.

Pressure drops for single-phase flow are normally expressed in terms of a friction factor as

$$\Delta P = 4f_1 \frac{LG^2}{D_c 2\rho g_c} \tag{9.24}$$

Values of the factor are conveniently plotted as a function of the Reynolds number and are obtainable from the manufacturers for their various fin geometries.

Proprietary methods for determining the pressure drop in the inlet and exit nozzles, distributors, and headers are also obtainable from the manufacturers.

Pressure drop data for two-phase flow are just as scarce as data for the heat-transfer coefficients. Consequently, the pressure drop is usually assumed to be equal to that produced for all-gas flows. An excellent source of information is the manufacturer of the specific configuration.

### *6.2.4. Reversing Exchangers*

Continuous operation of low-temperature processes requires that impurities in feed streams be removed almost completely prior to cooling the streams to very low temperatures. The removal of impurities is necessary because their accumulation in certain parts of the system creates operational difficulties or constitutes potential hazards. Thus there is an advantage in carrying out the necessary purification steps in the heat exchangers themselves. One method for accomplishing this is by using reversing heat exchangers.

A typical arrangement of a reversing exchanger for an air separation plant is shown in Fig. 9.36. Channels *A* and *B* constitute the two main reversing streams. Operation of such an exchanger is characterized by the cyclical changeover of one of these streams from one channel to the other. The reversal normally is accomplished by pneumatically operated valves on the warm end and by check valves on the cold end of the exchanger. The warm-end valves are actuated by a timing device which is set to a period such that the pressure drop in the feed channel is prevented from increasing beyond a certain value because of the accumulation of impurities.

- Uninterrupted operation at low temperatures requires removal of all impurities in gas streams
- Accumulations of impurities on cold surfaces can cause operational difficulties/hazards
- Typical arrangement is illustrated
  - Channels A & B are two main reversing streams
  - Reversal operation accomplished by pneumatically operated valves on warm end and check valves on cold end
  - When reversal waste stream reevaporates the deposited impurities and removes them from system

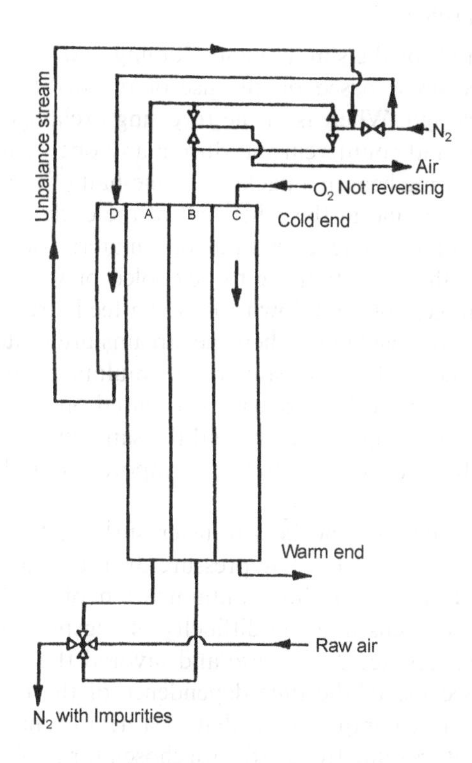


**Figure 9.36** Reversing (purging) heat exchangers.

Feed enters the warm end of the exchanger and as it is progressively cooled, impurities are deposited on the cold surface of the exchanger. When the flows are reversed, the waste stream reevaporates the deposited impurities and removes them from the system. Pressure differences of the two streams, which in turn affect the saturation concentrations of impurities in those streams, permit impurities to be deposited during the warming period and reevaporated during the cooling period. Deviations from ideality at temperatures below those encountered for air separation plants unfortunately work against this pressure effect.

Proper functioning of the reversing exchanger will depend on the relationship between the pressures and temperatures of the two streams. Since pressures are normally fixed by other considerations, the purification function of the exchanger is usually controlled by the proper selection of temperature differences throughout the exchanger. These differences must be such that at every point in the exchanger where reevaporation takes place, the vapor pressure of the impurity must be greater than the partial pressure of the impurity in the scavenging stream. Thus a set of critical values for the temperature differences exist depending on the pressures and temperatures of the two streams. Since ideal equilibrium concentrations can never be attained in an exchanger of finite length, allowances must be made for an exit concentration in the scavenging stream well below the equilibrium one. Generally a value close to 85% of equilibrium is chosen.

The design of a multistream exchanger is necessarily more complicated when two of the streams are subject to flow reversals. Personally, this author (TMF) would rely on the manufacturer's recommendations.

### *6.2.5. Regenerators*

Another method for the simultaneous cooling and purification of gases in low-temperature processes is based on the use of regenerators first suggested by Fränkl nearly 50 years ago. Whereas in the reversing exchanger the flows of the two fluids are continuous and countercurrent during any one period, the regenerator operates periodically, storing heat in a packing in one-half of the cycle and then giving up the stored heat to the fluid in the other half of the cycle. Such an exchanger consists of two identical columns packed with a solid material such as metal ribbon or stones, through which the gases that are to be cooled or warmed flow.

In the process of cool-down the warm feed stream deposits impurities on the cold surface of the packing. When the streams are switched, the impurities are reevaporated in the cold gas stream while simultaneously cooling the packing. Thus the purifying action of the regenerator is based on the same principles as presented earlier for the reversing exchanger and the same limiting critical temperature differences need to be observed if complete reevaporation of the impurities is to take place. See Fig. 9.37.

The cheapness of the heat-transfer surface, the large surface area per unit volume of packing, and the low pressure drop are the principal advantages of the regenerator. However, the intercontamination of fluid streams by mixing due to periodic flow reversals and the difficulty of regenerator design to handle three or more fluids has restricted their use and favored the adoption of brazed aluminum exchangers. Because of the time dependence of the temperatures in a regenerator, its analysis is more complex than that of a heat exchanger.

Regenerators quite frequently are chosen for applications where the heat-transfer effectiveness, defined as $Q_{\text{actual}}/Q_{\text{ideal}}$, must approach values of 0.98 to 0.99. It is

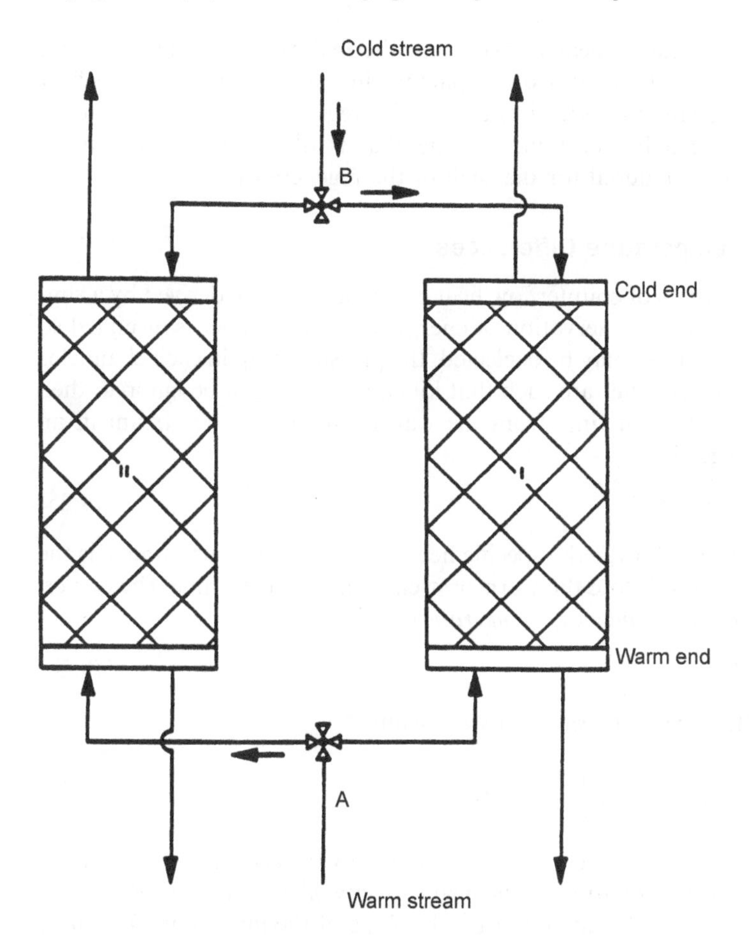


- Type of reversing heat exchanger
- Provides simultaneous cooling and purification of gases
- Does not operate continuously like the reversing exchanger
  - Periodically store heat in a packing during first half of cycle
  - Gives up stored heat to fluid in second half of cycle
- Consists of two identical columns
  - Packing with a solid material
  - Good heat capacity such as metal ribbon

**Figure 9.37** Regenerator for purification.

clear that a high regenerator effectiveness requires a high heat capacity per unit volume and a large surface area per unit volume.

Another variable that must be recognized is the effect of condensibles in the feed gas. For example, an air separation plant processing a 5-atm, 20°C feed saturated with water vapor will require an increase in regenerator length of 20% over that required to handle a dry gas feed. Since switchover losses in regenerators can approach 3% of the feed, the optimum regenerator volume must be determined by an economic analysis involving volume of unit, switching time, and capital cost of shell plus packing.

As with all other low-temperature exchangers, there are certain precautions that must be observed in designing regenerators which are in addition to those required to obtain the required heat transfer and thermal capacity characteristics. For example, the thermal conductivity of the packing in the direction of the gas flow should be small, otherwise the effectiveness of the regenerator can be reduced. Generally this is remedied by selecting packing that is discontinuous and using low thermal-conductivity material for the shell of the regenerator.

### 6.3. Determining Temperature Differences

Some general characteristics of counterflow heat exchangers can be derived by a simple application of the law of conservation of energy. If the exchanger is insulated so that losses to the surroundings may be neglected, the pressure drop in each channel is negligible, and the flow velocities are such that kinetic energies can be ignored, then the heat $dq$ transferred in unit time from the warmer to the cooler stream in an element of the exchanger is

$$dq = m'c'dT' = m''c''dT'' \tag{9.25}$$

where $m$ is the mass rate of flow, $c$ the specific heat at constant pressure, and $T$ is the temperature. Single primes denote the warmer stream and double primes the cooler. If the *specific heats are independent of temperature*

$$m'c'T' = m''c''T'' + d \tag{9.26}$$

The temperature differences between the two streams is

$$\Delta T = T' - T'' = \left(1 - \frac{m'c'}{m''c''}\right)T' + d/m''c'' \tag{9.27}$$

where $d$ is the constant of integration to be determined by the conditions existing for a specific application of this equation. If the heat capacity $m'\ c'$ of the warmer stream is greater than that of the cooler stream $m''c''$, the slope of the line $\Delta T$ vs. $T'$ will be negative; that is, $\Delta T$ will diminish as the temperature increases. This is the condition existing in the Hampson air liquefier. Since part of the air is condensed and removed as liquid, the heat capacity of the cold returning stream is less than that of the incoming stream. Thus in the Hampson liquefier the $\Delta T$ at the warm end can be made negligible if the heat exchanger is sufficiently large. However, if $m''c''$ is greater than $m'c'$, $\Delta T$ will increase at higher temperatures, and if $m''\ c'' = m'c'$, $\Delta T$ will not change.

When the specific heat of either stream varies, it is better to use the specific enthalpy, $h$ of the gas as given in temperature-entropy or Mollier diagrams. Thus, since $c\,dT = dh$, instead of Eq. (9.25) we can write

$$dq = m'\,dh' + m\,dh \tag{9.28}$$

and

$$m'h' = m''h'' + d \tag{9.29}$$

or

$$m'\Delta h' = m''\Delta h'' \tag{9.29a}$$

This relation is valid for any variation of specific heats and can be used to make a more detailed analysis of temperature differences along a counterflow heat exchanger when the specific heats of the fluid streams are changing.

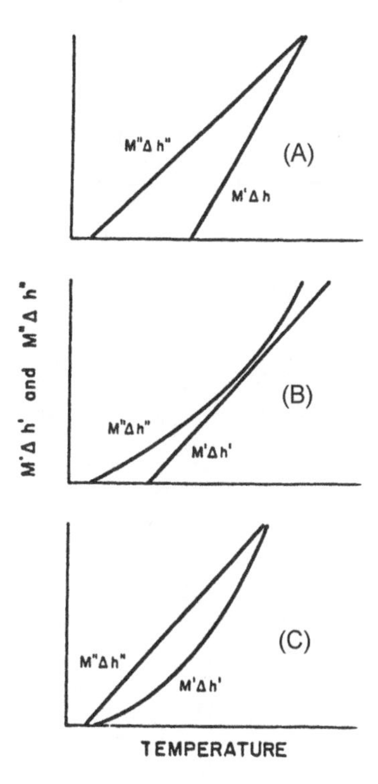


**Figure 9.38** Three types of temperature distribution along a counterflow heat exchanger.

It is recommended that the student of heat-exchanger design becomes thoroughly familiar with this concept because it is fundamental and is often the limiting factor in heat-exchanger efficiency, and the controlling consideration in selecting a liquefaction circuit. The usefulness of Eq. (9.29) or its equivalent, (9.29a), can be illustrated graphically by plotting both $m'\Delta h'$ and $m''\Delta h''$ as ordinate with temperature as abscissa, as shown in Fig. 9.38A. Figure 9.38A illustrates the conditions usually encountered in the final exchanger of a Joule–Thomson liquefier, having a returning cold stream of much smaller mass rate of flow than that of the incoming warm stream. Figure 9.38B illustrates a difficult heat-exchanger situation. Here the specific heat of one of the streams has a temperature dependence such that large temperature differences are required at both ends of the exchanger in order to maintain a temperature difference in the center of the proper sign to transfer heat. By switching the gases of Fig. 9.38B so that the cold stream consists of the gas which formerly occupied the warm channel, and vice versa, the very favorable situation illustrated by Fig. 9.38C will result. Here the temperature difference is larger in the middle of the heat exchanger than it is at the ends.

Such peculiar temperature difference distributions are common at low temperatures, where $C_p$ of gases varies significantly with temperature. This is a major difference between low temperature heat-exchanger design and that of ordinary temperatures. See Fig. 9.39 for other low-temperature examples.

## 6.4. Basic Heat-Exchanger Design Approaches

There are two basic approaches to heat-exchanger design: (1) the effectiveness-NTU approach and (2) the log-mean-temperature-difference (LMTD) approach. See

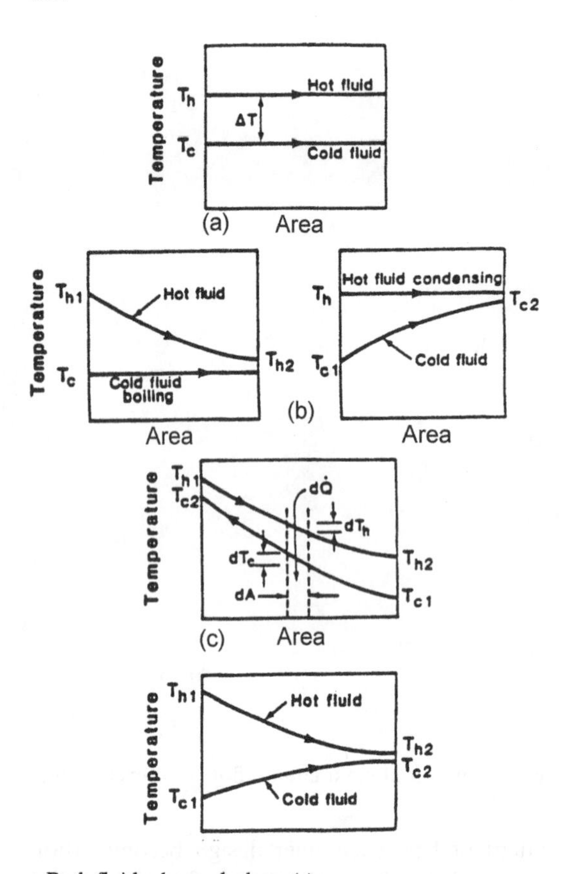


- Both fluids changed phase (a)
  - Constant temperature when hot fluid is condensed at $T_h$ and cold fluid is evaporated at $T_c$
- One fluid changes phase (b)
- Heat transferred without phase change
  - Counterflow (c)–fluids flow in opposite directions
  - Parallel flow (d)–fluids flowing in same direction
- Major advantage if counterflow exchanger
  - Outlet temperature of cold fluid can be greater than outlet temperature of hot fluid

**Figure 9.39** Low-temperature heat-exchanger temperature distributions.

Table 9.9. The LMTD approach is used most frequently when all the required mass flows are known and the size of the exchanger is to be determined. The effectiveness-NTU approach is used more often when the inlet temperatures and the flow rates are known in a given exchanger and the outlet temperatures are to be found. Both methods may be used in either case, but the LMTD method often requires an iterative solution for the second situation.

The rate of heat transfer may be represented by the relationship

$$\dot{Q} = UA\Delta T_m \tag{9.30}$$

where $\Delta T_m$ is the suitable mean temperature difference across the heat exchanger. In the log-mean-temperature-difference approach for counterflow or parallel flow

**Table 9.9** Basic Heat-Exchanger Design Approaches

| |
|---|
| Two basic approaches to heat-exchanger design |
| Effectiveness-NTU |
| Log-mean-temperature-difference (LMTD) |
| LMTD approach used when all required mass flows are known and size of exchanger is to be determined |
| Effectiveness-NTU approach used when inlet temperatures and flow rates are known in a given exchanger and outlet temperatures are unknown |
| Interchangeable approaches |
| LMTD would be iterative solution |

exchangers, $\Delta T_m$ is defined as

$$\Delta T_m = \frac{\Delta T_{min} - \Delta T_{min}}{\ln(\Delta T_{max}/\Delta T_{max})} = \text{LMTD} \tag{9.31}$$

where $\Delta T_{max}$ is the largest local temperature difference and $\Delta T_{min}$ is the smallest local temperature difference between the two fluid streams at the inlet and outlet of the exchanger.

Typical temperature distributions encountered in a heat exchanger are shown in Fig. 9.39. The temperature difference remains constant between the two fluids only when the hot fluid is condensing on one side at $T_h$ while the cold fluid is evaporating on the other side of the exchanger surface at $T_c$, as shown in Fig. 9.39a. The condensor-reboiler of a Linde double-column rectification system exhibits such behavior. In Fig. 9.39b, one fluid is changing phase due to heat transfer with the other fluid, whose temperature increases or decreases while passing through the exchanger. When heat is transferred from one fluid to another without a phase change, the situations shown in Fig. 9.39c and d occur. Figure 9.39c shows the fluids flowing in opposite directions, commonly referred to as counterflow, while Fig. 9.39d shows the fluids flowing in the same direction, designated as parallel flow. In the case of counterflow, the temperature difference between the two streams can remain fairly constant throughout the length of the heat exchanger. In the case of parallel flow, the temperature difference decreases from a maximum at the inlet to a minimum at the outlet.

A major advantage in the use of counterflow exchangers over parallel flow exchangers is that the outlet temperature of the cold fluid in the former can be much greater than the outlet temperature of the hot fluid. This is not possible with parallel flow. The large temperature changes necessary in cryogenic systems cannot be obtained with parallel flow exchangers.

The heat-transfer analysis of a heat exchanger in which the temperatures of both fluids are changing may be carried out for the counterflow case by referring to Fig. 9.39c. The inlet temperatures of the hot and cold streams are designated at $T_{h1}$ and $T_{c1}$, respectively, while the outlet temperatures are designated at $T_{h2}$ and $T_{c1}$, respectively. By considering an incremental area as shown, the incremental heat-transfer rate may be written in three different ways:

$$\mathrm{d}\dot{Q} = -C_h\,\mathrm{d}T_h \quad \text{(for the warm fluid)} \tag{9.32}$$

$$\mathrm{d}\dot{Q} = -C_c\,\mathrm{d}T_c \quad \text{(for the cold fluid)} \tag{9.33}$$

$$\mathrm{d}\dot{Q} = U\,\mathrm{d}A(T_h - T_c) \quad \text{(between the fluids)} \tag{9.34}$$

where $C_h$ is defined as $\dot{m}_h C_{ph}$, the capacity rate of the hot fluid, and $C_c$ is defined as $\dot{m}_c C_{pc}$ the capacity rate of the cold fluid. By manipulating Eqs. (9.32) and (9.33), the differential temperature changes may be determined by

$$dT_h - dT_c = d(T_h - T_c) = -d\dot{Q}\left[\frac{1}{C_h} - \frac{1}{C_c}\right] \tag{9.35}$$

Substituting Eq. (9.34) in for $d\dot{Q}$, and setting up the integration, gives

$$\int_{T_{h1}-T_{c2}}^{T_{h2}-T_{c1}} \frac{d(T_h - T_c)}{T_h - T_c} = -U\left[\frac{1}{C_h} - \frac{1}{C_c}\right]\int_0^A dA \tag{9.36}$$

Upon integration,

$$\frac{T_{h2} - T_{c1}}{T_{h1} - T_{c2}} = \exp\left[-UA\left(\frac{1}{C_h} - \frac{1}{C_c}\right)\right] = \exp\left[-\frac{UA}{C_h}\left(1 - \frac{C_h}{C_c}\right)\right] \tag{9.37}$$

If the hot fluid is assumed to have the smaller capacity rate, then $C_h = C_{min}$ and $C_c = C_{max}$, and the argument of the exponential is always negative. Equation (9.37) may be rewritten as

$$\frac{T_{h2} - T_{c1}}{T_{h1} - T_{c2}} = \exp\left[\frac{-UA}{C_{min}}\left(1 - \frac{C_{min}}{C_{max}}\right)\right] \tag{9.38}$$

If the cold stream had been assumed as the one with the minimum capacity rate, the same form of Eq. (9.38) would have been obtained, provided the proper sign conventions for Eqs. (9.32), (9.33) and (9.34) had been used. Typically, the cold stream has the smaller capacity rate in cryogenic systems. The quantity $(UA/C_{min})$ is defined as the number of transfer units or NTU, and the quantity of $C_{min}/C_{max}$ is referred to as the capacity-rate ratio.

The effectiveness of a heat exchanger is defined as

$$\varepsilon = \frac{\text{heat actually transferred}}{\text{maximum have available for transfer}} \tag{9.39}$$

The maximum temperature change occurs in the fluid with the minimum capacity rate. In order to satisfy the energy balance, the heat lost by the hot fluid is equal to the heat gained by the cold fluid, or

$$(\dot{m}C_p)_c(T_{c2} - T_{c1}) = -(\dot{m}C_p)_h(T_{h2} - T_{h1}) \tag{9.40}$$

This can also be expressed in the form

$$C_{min}\Delta T_{max} = -C_{max}\Delta T_{min} \tag{9.41}$$

Theoretically, the temperature of the exit cold fluid could increase to that of the inlet hot fluid or the temperature of the exit hot fluid could decrease to that of the inlet cold fluid. Therefore, the maximum temperature change is $(T_{h1} - T_{c1})$, and the maximum possible energy transfer if $C_{min}(T_{h1} - T_{c1})$ The effectiveness may now be expressed in reference to the hot or cold fluid, respectively, as

$$\varepsilon = \frac{C_h(T_{h1} - T_{h2})}{C_{min}(T_{h1} - T_{c1})} = \frac{C_c(T_{c2} - T_{c1})}{C_{min}(T_{h1} - T_{c1})} \tag{9.42}$$

The left side of Eq. (9.37) may now be expressed in terms of the inlet and exit streams of the warm and cold streams by

$$\frac{T_{h2} - T_{c1}}{T_{h1} - T_{c2}} = \frac{1 - [(T_{h1} - T_{h2})/(T_{h1} - T_{c1})]}{1 - (C_h/C_c)[(T_{h1} - T_{h2})/(T_{h1} - T_{c1})]} \tag{9.43}$$

Since the hot fluid was assumed to be the minimum fluid in the derivation of Eq. (9.37), Eq. (9.42) reduces to

$$\varepsilon = (T_{h1} - T_{h2})/(T_{h1} - T_{c1}) \tag{9.44}$$

By substitution, Eq. (9.43) becomes

$$(T_{h2} - T_{c1})/(T_{h1} - T_{c2}) = [1 - (C_{min}/C_{max})\varepsilon] \tag{9.45}$$

Equations (9.38) and (9.45) may now be combined to solve for the effectiveness-NTU expression for counterflow heat exchange:

$$\varepsilon = \frac{1 - \exp[-\mathrm{NTU}(1 - C_{min}/C_{max})]}{1 - (C_{min}/C_{max})\exp[-\mathrm{NTU}(1 - C_{min}/C_{max})]} \tag{9.46}$$

Equation (9.46) is the expression for the heat-transfer effectiveness in the case of unbalanced flow, as in a liquefier. For the case of balanced flow, as in a refrigerator, with $C_{min}/C_{max} = 1$, Eq. (9.46) reduces to

$$\varepsilon = \mathrm{NTU}/(1 + \mathrm{NTU}) \tag{9.47}$$

If a phase change occurs on one side of the heat-exchanger surface, as in boiling or condensation, $C_{max} = \infty$. Therefore, $C_{min}/C_{max} = 0$, and Eq. (9.46) reduces to

$$\varepsilon = 1 - \exp(-\mathrm{NTU}) \tag{9.48}$$

A similar analysis can be carried out for other flow arrangements as well. For the case of parallel flow, the effectiveness equation derived as

$$\varepsilon = \frac{1 - \exp[-\mathrm{NTU}(1 + C_{min}/C_{max})]}{1 + (C_{min}/C_{max})} \tag{9.49}$$

For balanced parallel flow,

$$\varepsilon = 1/2[1 - \exp(-2\mathrm{NTU})] \tag{9.50}$$

As can be seen from Eq. (9.50), the maximum effectiveness for a parallel-flow heat exchanger is only 50%, whereas the limit for the counterflow situation is 100%. The effectiveness for condensation or boiling on one side of the heat-exchanger surface for parallel flow is the same as that for counter-flow, as indicated by

Eq. (9.48), since the flow arrangement does not affect heat transfer at the exchanger surface.

## 7. STORAGE

### 7.1. Some Common Factors in Storage Systems

The need for the storage of cryogenic fluids is obvious. There are many factors involved in the economics of these problems, and additional factors such as logistics and reliability are also involved.

The most important single factor is insulation, because the economics involved are very much related to the rate of evaporation, and this is determined by the effectiveness of the insulation. The matter of insulation in some detail was covered in Chapter 7. In this chapter, we will consider the major technical factors which enter into the storage of cryogenic fluids.

A large part of the discussion will consist of descriptions of actual containers. Some special techniques dealing with precooling containers and utilizing the refrigerative value of the escaping cold vapor will also be discussed.

The invention of the vacuum-insulated vessel for liquefied gases by James Dewar in 1892 was a break-through in the field of thermal insulation that has not yet been matched by further developments. All of the advances since Dewar's time have been improvements on Dewar's original concept, usually means of reducing radiant heat transfer by attaining surfaces of higher reflectivity or by interposing shields which reflect or intercept the radiant energy.

At its best the vacuum is so good that heat transfer by residual gas is almost negligible. Accordingly, the designer is mainly concerned with reducing the heat transferred by mechanical supports and that which is conveyed across the insulating vacuum by radiation. The heat transferred by the supports is a problem involving the mechanical strength required for a specific application and has no general solution. If there is no restriction on the dimensions of the vessel, the heat conducted through the supports can usually be made quite small simply by using supports of low thermal conductivity and making their lengths great enough. Even in restricted space the clever designer can usually find a way to provide a long thermal path through the solid supports. On the other hand, the heat transferred by radiation is little affected by the thickness of the insulating space. In fact, the insulation is somewhat better for a thin vacuum space which approaches the ideal configuration of parallel surfaces. However, heat transfer by radiation can be reduced by interposing thermal shields between the warm and cold surfaces of the vessel.

One of the most effective shields for vessels containing liquid hydrogen or helium is a surface cooled with liquid nitrogen. By interposing in the insulating vacuum of a helium or hydrogen container a surface cooled to 77 K, the radiant heat transfer is reduced by a factor of about 250 under that emanating from a surface at 300 K (room temperature). Nearly all the heat radiated from the room temperature surface is intercepted and causes the evaporation of relatively inexpensive liquid nitrogen.

Although the refrigerated shield is very effective, a simple non-refrigerated "floating" shield will cause a substantial reduction in energy transfer by radiation. The floating shield is an opaque, highly reflecting surface, geometrically similar to the surface of the liquid container and suspended with the minimum of thermal contact approximately midway between the inner and outer walls of the vacuum space. If the shield has the same reflectivity as that of the boundary walls and contains no openings or irregularities, the radiant heat transfer to the liquid being stored will be reduced by about one-half. Additional shields will further reduce the heat transferred by radiation. However, the design of mechanical supports which will hold the shields in place without providing objectionable heat leaks becomes a formidable task as the number of shields is increased. Also, openings and irregularities which are needed to accommodate the piping and supports for the inner container greatly reduce the effectiveness of this type of multiple shielding.

Fortunately a method has been discovered which does not complicate the mechanical design and yet provides a barrier to thermal radiation comparable with multiple shields. This method consists of filling the insulating vacuum space with a fine light powder such as perlite, silica aerogel, carbon black, or diatomaceous earth. If the thickness of the insulating space is sufficient, filling with powder will greatly reduce the radiant heat transfer. Moreover, when powder is used, the vacuum requirement is quite modest; the minimum heat transfer is very closely approached when the interstitial gas pressure is reduced to 10 μm of mercury in an insulating powder (e.g., perlite or silica aerogel) with one boundary at room temperature, 300 K, and the other at the temperature of liquid nitrogen, 77 K. A layer of evacuated powder several centimeters thick will effectively stop radiation and yet not provide an objectionable amount of thermal conduction through the touching particles of powder. When the space between the walls must be small, of the order of 1 cm or less, the thermal conduction through the powder is objectionable. In this case, the most effective insulation is a high vacuum bounded by highly reflecting walls.

Since the principal function of the powder is to stop radiation, it is useful only when radiation constitutes an important heat leak. If an insulating vacuum is bounded by highly reflecting surfaces such as clean annealed copper, silver, or aluminum, and the surfaces are at 77 and 20 K, respectively, adding powder may actually increase heat transfer by providing paths of solid conduction. The radiant heat transfer between the surfaces of these temperatures is so small that its elimination by adding powder is of little consequence. Radiation from a surface at 77 K is only about 0.004 of that from an identical surface at 300 K, since it is proportional to the fourth power of the absolute temperature. Of course the thickness of the insulating space must be considered in such a comparison. It is probably more correct to say that for boundaries at 77 and 20 K, a rather large thickness of powder would be required to compete with high-vacuum insulation.

It appears then that evacuated powder insulation is very desirable if the warm boundary is at ordinary room temperature and the insulating space is thick enough (10 cm or more) that solid conduction through the powder is not serious.

Of course, evacuated powder insulation is not new. It has been used for many years in large industrial containers for liquid oxygen and liquid nitrogen. It is particularly suitable for large containers. Evacuated powder appears attractive as the insulation to use in large liquid-hydrogen vessels. Evacuated powder has been used in conjunction with a liquid-nitrogen-cooled shield for some vessels holding 750 L of liquid hydrogen, and some liquid-hydrogen vessels of 42,000,000 gallons capacity have been constructed which utilize only powder insulation (no refrigerated shield). See Table 9.10.

**Table 9.10** Cryogenic Storage Systems

| |
|---|
| Cryogens require specialized storage containers to maintain any kind of shelf life |
| (system boil-off time > minutes) |
| Commonly known as "Dewars"! |
| Storage containers range in size from 1 L to 42,000,000 gallons |
| Economics and *safety* determine best container design |
| Rate loss (heat leak) |
| Weight |
| Materials of construction (system compatibility) |
| Insulation system (dependent upon life requirements) |
| Specialized requirements (dynamic environments, etc.) |
| Many standard designs are available through a variety of suppliers and manufacturers |

## 7.2. Dependence of Rate of Evaporation upon Size and Shape of Vessel

It is quite easy to formulate for liquefied gas containers some general relationships between the size and shape of the vessel and the rate of evaporation of the liquid it contains. First, since the heat leak into the container increases as the surface area increases, the most favorable shape is the sphere (the greatest ratio of volume to surface). The sphere also has some advantages of mechanical strength, but in large sizes, it may be expensive to fabricate. The next most favorable common shape is the square cylinder in which the length equals the diameter. The practical approach to this shape is the cylinder with standard commercial dished or elliptical heads. This design has good mechanical strength and is readily fabricated. It is stronger and has a somewhat larger ratio of volume to surface than the cylinder with the flat ends. Also the ratio of surface to volume of the cylinder with dished ends is only about 10% greater than that of the sphere. However, in many cases, particularly vessels which are to be transported by rail or highway, the maximum diameter is fixed.

Thus most of the designs for large insulated tanks for storing and transporting liquefied gases can be assigned to two categories: (1) *fixed shape* and (2) *fixed diameter.* It is of interest to see how the performance of such containers varies with the size of the vessel. The effectiveness of a container for liquefied gas is commonly given as the fraction (or percent) of the total contents which evaporates in one day. (See Sec. 1.1.10, "Scaling", of this chapter.)

### 7.2.1. *Vacuum-Insulated Vessels*

For vacuum-insulated containers of *fixed shape*—that is, containers of geometric similarity but of different sizes—the heat leak by radiation and residual-gas conduction is proportional to the surface area of the inner container—that is, proportional to $L^2$, where $L$ is a linear dimension of the container. The heat leak through the supports may also be considered proportional to $L^2$, since they will probably have a length nearly proportional to the linear dimension and a total cross-sectional area proportional to the volume (and weight) of the inner container. Then the total rate of heat leak is

$$W = CL^2 = C_1V^{2/3}$$

where $C_1$ is a constant and $V$ is the volume of the inner container. The fractional rate of evaporation is

$$\frac{W}{L_{\mathrm{v}}\rho V} = \frac{C_1 V^{2/3}}{L_{\mathrm{v}}\rho V} = \frac{C_1}{L_{\mathrm{v}}\rho V^{1/3}}$$

where $L_{\mathrm{v}}$ is the heat of vaporization of the liquid and $\rho$ its density.

### *7.2.2. Vessels with Porous Insulation*

For containers insulated with porous materials, e.g., evacuated powders, the situation is somewhat different because for similar geometry, the thickness, and therefore the effectiveness of the insulation, is proportional to the linear dimension of the container. The relation between the size and the performance of such vessels is not simple unless the heat conducted by the supports can be neglected. In this case, the total heat leak is proportional to the linear dimension; so the fractional evaporation is proportional to $1/V^{2/3}$. Thus it is seen that for vessels utilizing this type of insulation, large size is even more advantageous than it is for high-vacuum-insulated vessels.

### *7.2.3. Vessels of Fixed Diameter and Different Lengths*

The container for rail or road transportation, in which the maximum diameter cannot exceed a certain value but the length may be varied, can be conveniently analyzed on the assumption that the ends are spherical; so its shortest length and minimum volume are those of a sphere having the permitted diameter. In this case, since the thickness of the insulation is greatly restricted, it will be assumed that the thickness is constant. This means that the heat leak through the insulation will be proportional to the surface area both for high-vacuum insulation and for porous insulation. It should be emphasized that this does not imply that the effectiveness of the various insulations is the same. It only means that in this circumstance the heat leak for unit area is constant for a given type of insulation and is not a function of the total area. Then if we again ignore the heat leak through the supports

$$W = C[4\pi r^2 + 2\pi r(L - 2r)]$$

where $C$ is a constant, $r$ the maximum permissible radius, and $L$ is the length of the container.

The fractional evaporation is

$$\frac{W}{L_{\mathrm{v}}\rho V} = \frac{2CL}{L_{\mathrm{v}}\rho r(L - (2/3)r)}$$

As one would expect, the fractional evaporation changes very little with length, decreasing by only one-third when the length of the container is changed from $2r$ to infinity.

### *7.2.4. Comparison of Evaporation Rate vs. Size for Different Vessels*

Figure 9.40 shows graphically how the fractional rate of evaporation depends upon the volume of the container for the three conditions just discussed. The three curves should be compared with each other in regard to trend, not in regard to magnitude. The actual rates of evaporation can be determined only by computing the heat leaks for the particular insulation being used.

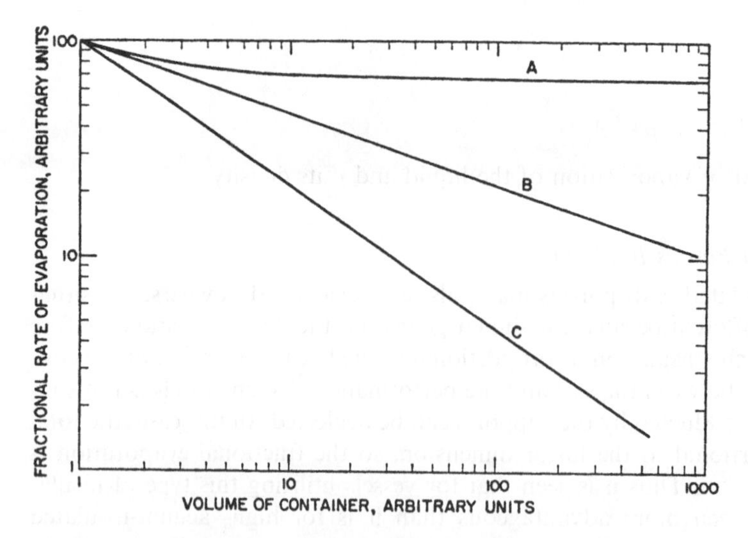


**Figure 9.40** Dependence of evaporation rate upon size of the vessel. Curve A represents vessels of fixed diameter and various lengths; B, high-vacuum-insulated containers or containers having constant insulating thickness and similar geometry; and C, powder-insulated vessels of similar geometry throughout, including the feature that the insulation thickness bears a ratio to the other linear dimensions of the vessel.

Most low-temperature research laboratories use silvered-glass dewar flasks for small quantities of cryogenic liquids. These are available in a wide variety of shapes and sizes. Two of the commonly used types are shown in Fig. 9.41. The cylindrical shape is often used as a part of an experimental assembly; sometimes parts are left unsilvered, providing transparent windows or slits for visual observation of the interior. Since glass can be heated to a high temperature while being evacuated, adsorbed and dissolved gases can be thoroughly removed. Accordingly, a sealed glass dewar

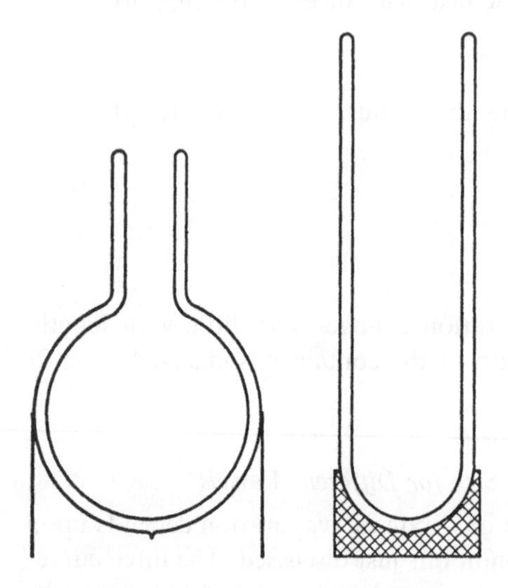

**Figure 9.41** Two types of glass dewars. For best insulation, the surfaces facing the vacuum space are silvered. For some experiments, the vessel is left unsilvered, or a strip or window is left unsilvered for visual observation.

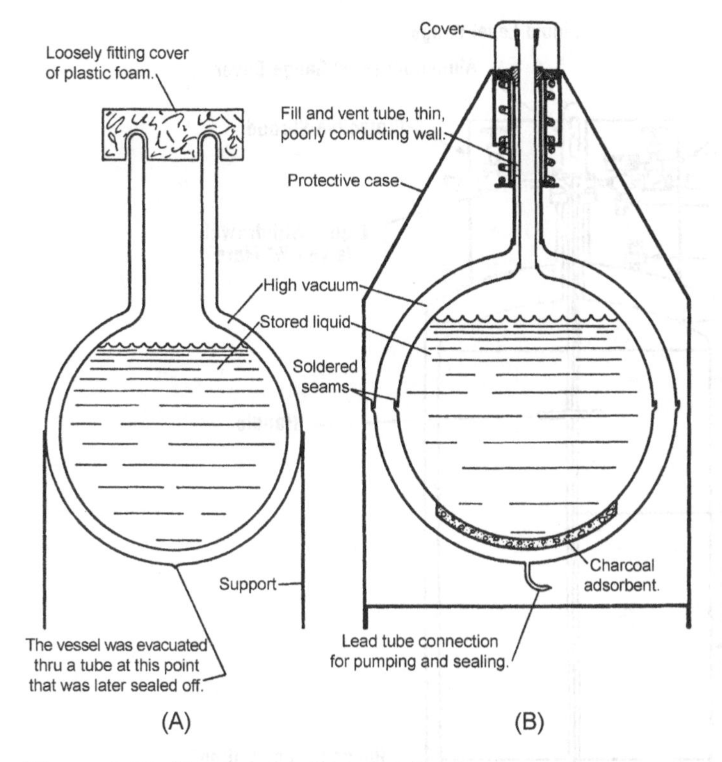


**Figure 9.42** Two types of vacuum-insulated containers for liquefied gases. Vessel A is Pyrex® glass. Surfaces of the glass facing the vacuum space are silvered to reduce heat transfer by radiation. Vessel B is metal. Spheres are copper, with surfaces facing the vacuum space cleaned to achieve the high intrinsic reflectivity of copper.

will maintain a good vacuum for many years. Cylindrical metal laboratory dewars also are available commercially with inside diameters ranging from 3 or 4 to 17 in. and lengths up to several feet. In large diameters, the manufacturer is usually willing to make the length to the customer's specifications.

For the storage and transportation of somewhat larger quantities of liquid, 15L or more, the vessel commonly used is the metal dewar flask illustrated in Fig. 9.42. Vessels of this type are rather rugged, but should not be subjected to severe shocks. With proper handling they will maintain a good insulating vacuum for several years (Fig. 9.43). In time, however, the gas given up by the metal accumulates in the vacuum space to a degree that the charcoal adsorbent cannot maintain a sufficiently good vacuum, and it is necessary to recondition the container by heating and pumping. Containers of this type are manufactured in sizes up to 1000 L capacity. See Figs. 9.44–9.48. The larger containers are usually of the form of vertical cylinders with spherical ends, so that the diameter does not become excessive.

F44–F48

### 7.3. Liquid-Nitrogen-Shielded Dewars

Figure 9.45 illustrates a type of very successful commercial container for liquid helium. Vessels of this type are manufactured in sizes ranging from 15 to 100 L. The rate of evaporation of the stored liquid is very small; in the larger sizes the helium loss is less than 1% of the capacity per day. Many laboratories take

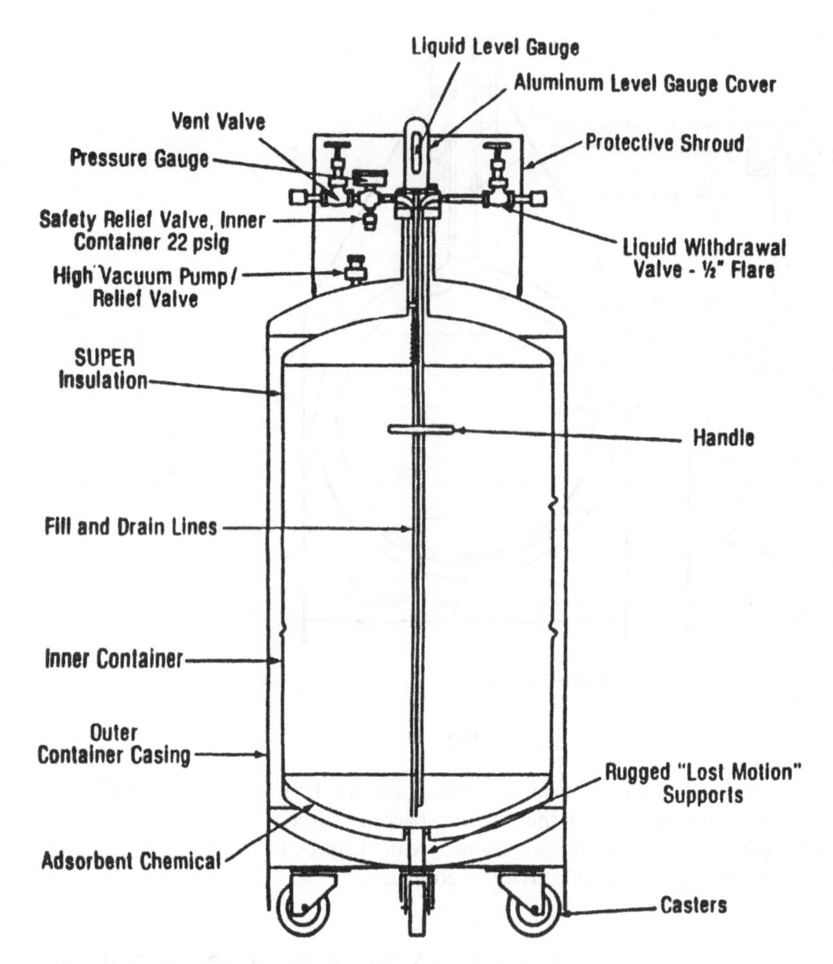


**Figure 9.43** Commercial cryogenic storage system.

advantage of this high-quality storage and maintain a continuous supply of liquid helium and liquid hydrogen for their research projects. By careful handling such vessels can be shipped.

Liquid-nitrogen-shielded dewars (Fig. 9.46) have very high performance. However, they have largely been replaced by the ML1 insulated dewars of Fig. 9.45.

## 7.4. A Large Powder-Insulated Transport Dewar for Liquid Hydrogen

Current designs of transport dewars for liquid hydrogen dispense with refrigerated shields and achieve an acceptable evaporation rate with evacuated powder insulation alone. This is made possible because of the large volume of the container and the room to accommodate a substantial thickness of powder insulation. The advantage of large size in a powder-insulated dewar was discussed at the beginning of this chapter (see Fig. 9.30). Figure 9.47 is an axial cross-sectional diagram of the vessel as designed by the Beech Aircraft Corporation in consultation with the Los Alamos Scientific Laboratory and the National Bureau of Standards. The evacuated perlite insulation is approximately 12 in. thick. A feature that should be mentioned is the collection pipe, which shortens the time required for evacuation by decreasing the

**Figure 9.44** A sampling of standard portable atmospheric dewars. All are on wheels to facilitate distribution, and are for liquid use only (not high-pressure gas, for instance). Cryofab, Kenilworth, NJ.

distance that the gas must travel through the powder. The powder offers such a great resistance to the flow of low pressure gas that it usually takes several days to obtain a satisfactory vacuum in a large powder-insulated vessel. The filter keeps powder out of the vacuum pump.

The internal volume of this container is 6000 L. In normal use, it carries 5400 L of liquid, leaving a vapor space of 600 L. The loss rate is approximately 1.5% of the rated capacity in 24 hr. It is mounted on a trailer for highway transportation.

Figure 9.48 illustrates some of the features that most dewar vessels have in common. An oxygen dewar is shown in Fig. 9.49.

## 7.5. Cryogenically Cooled Shields

It has been noted that losses of liquid while filling a helium dewar can be substantially reduced by causing the vapor to furnish the major part of the refrigeration needed to cool the container. The cold vapor has 80 times the cooling capacity of the vaporizing liquid helium alone. After the vessel has been filled, the vapor resulting from normal evaporation can be used to absorb some of the heat leaking into the vessel. In most vessels, part of this refrigeration is utilized more or less accidentally in absorbing some of the heat which flows by solid conduction down the vent pipe. (Fig. 9.50). An analysis of this process will be given later.

### *7.5.1. Vapor-Refrigerated Shields*

A way to make good use of the refrigeration available in the escaping vapor is to have it cool a shield suspended between the inner liquid container and the outer

**Figure 9.45** Liquid helium portable dewar. Used in large-scale liquid helium distribution. Cryofab, Kenilworth, NJ.

warm wall of the vessel. This device can be applied to high-vacuum-insulated vessels as well as vessels utilizing evacuated powder insulation. The openings in the shield required to admit piping and supports or re-entrant irregularities that may be produced by attaching piping to one side of a shield will have little effect when powder insulation is used; but when the shield is in an empty space, these openings and irregularities may somewhat reduce the effective overall reflectivity of the shield. Moreover, in a vessel having sufficient room between the walls to accommodate a shield, the insulation thickness may be great enough to show a substantial improvement when powder is used. Because of the simpler construction and the fact that the ideal is more nearly approached, it is believed that the vapor-cooled shield is more aptly applied to a vessel with evacuated powder insulation.

Some commercial powder-insulated vessels make partial use of the cold escaping vapor by having it flow through several turns of pipe which lie between the inner

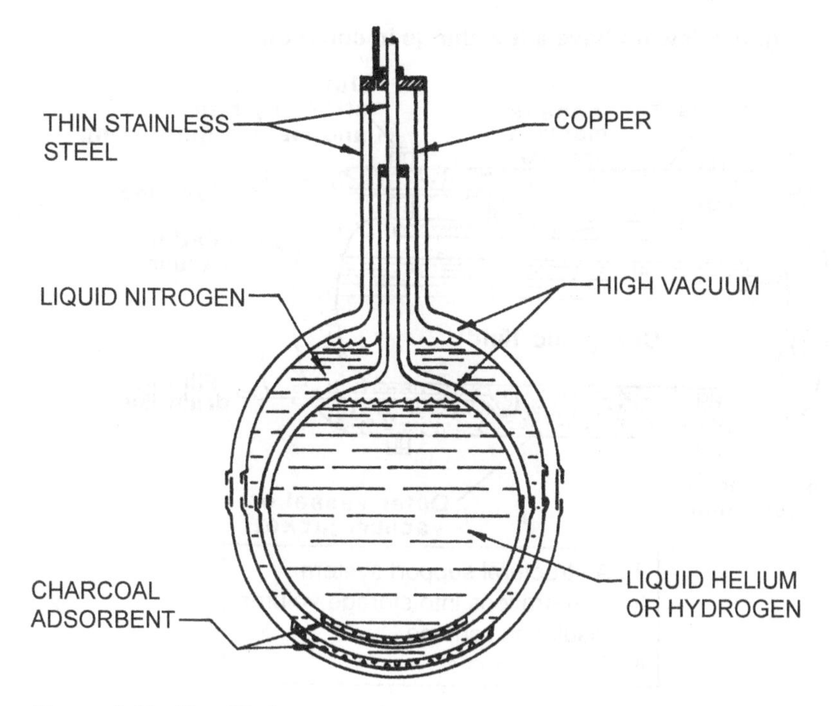


**Figure 9.46** Simplified cross-section of a commercial dewar for the low-loss storage and transportation of liquid hydrogen and helium.

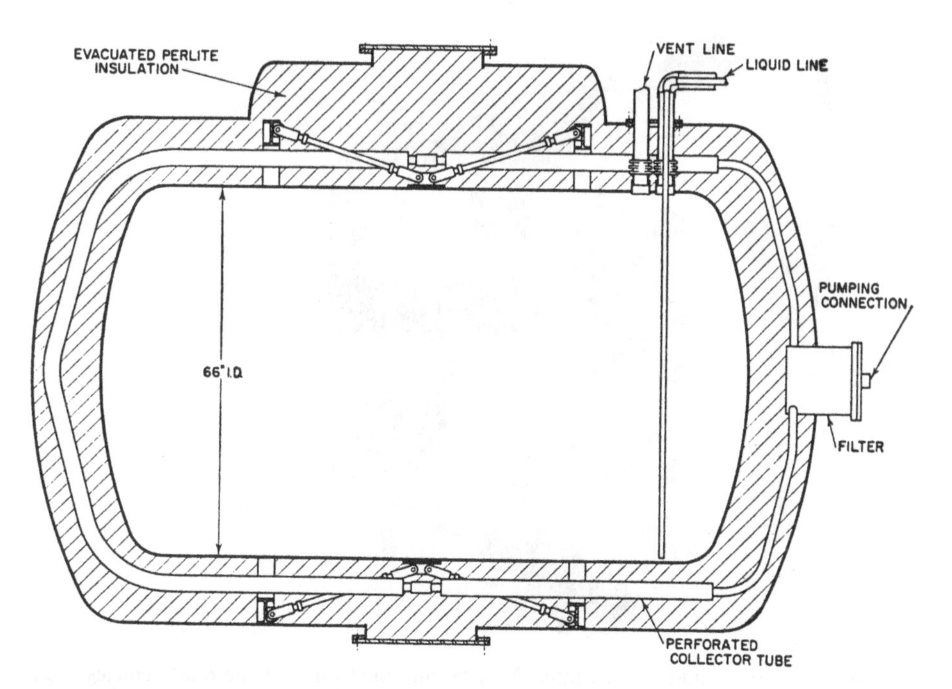


**Figure 9.47** Transport vessel for liquid hydrogen with evacuated powder insulation. (Courtesy, Beech Aircraft Corporation.)

• All cryogenic dewars have a few things in common :

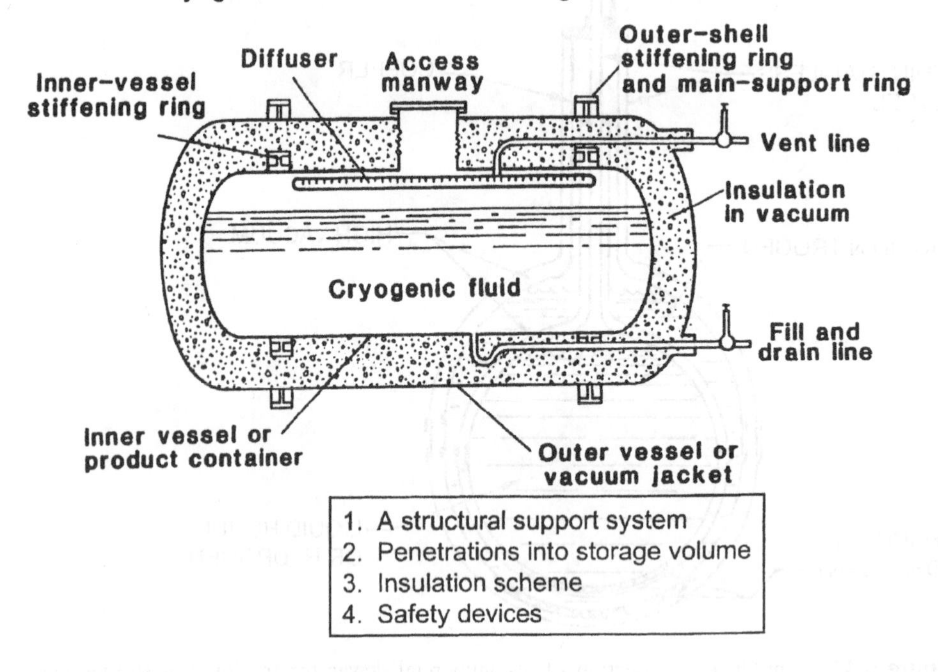


**Figure 9.48** Dewar system commonality.

**Figure 9.49** Portable liquid oxygen tank. May be mounted on over the road vehicles such as a van or pickup truck. A principal use is the delivery of liquid oxygen to homes to refill respiratory equipment DOT certified. Cryofab, Kenilworth, NJ.

and outer walls. It is believed, however, that the available refrigeration can be used more effectively if it is made to cool a shield which is located at the optimum position between the inner and outer walls of the insulating space.

The following analysis of Scott (1959) presents the general method of determining the optimum location of a vapor-cooled shield and the expected gain in performance. The results are applied to the specific case of spherical containers for liquid hydrogen and helium with evacuated perlite insulation.

With evacuated powder insulation filling the entire insulating space the rate at which heat reaches the liquid is

$$W_1 = \frac{4\pi k_1 r_1 r_2 (T_2 - T_1)}{r_2 - r_1} = \dot{m} L_v \tag{9.51}$$

where $k_1$ is the mean thermal conductivity of the powder in the temperature range $T_1$ to $T_2$, $r_1$ the radius of the liquid container, $r_2$ that of the vapor-cooled shield, $\dot{m}$ the mass rate of evaporation, and $L_v$ is the heat of vaporization of unit mass of the liquid.

The rate at which heat reaches the shield is

$$W_2 = \frac{4\pi k_2 r_2 r_3 (T_3 - T_2)}{r_3 - r_2} = \dot{m} L_v + \dot{m} \Delta h \tag{9.52}$$

where $r_3$ is the radius of the outer boundary of the insulation, $k_2$ the mean thermal conductivity of the powder in the temperature range $T_2$ to $T_3$ and $\Delta h$ is the change of specific enthalpy of the vapor as it is warmed from $T_1$ to $T_2$.

Dividing Eq. (9.51) by Eq. (9.52) gives

$$\frac{L_v}{L_v + \Delta h} = \frac{k_1}{k_2} \frac{(r_3 - r_2)}{(r_2 - r_1)} \frac{r_1}{r_3} \frac{(T_2 - T_1)}{(T_3 - T_2)}$$

Let $r_3 = a r_1$ and $r_2 = b r_1$; then

$$\frac{a(b-1)}{a-b} = \frac{k_1}{k_2} \frac{L_v + \Delta h}{L_v} \frac{(T_2 - T_1)}{(T_3 - T_2)} = g$$

From Eq. (9.51)

$$\frac{L_v \dot{m}}{4\pi r_1} = k_1 \frac{b}{b-1} (T_2 - T_1) \tag{9.53}$$

Since $b$ is a function of $a$ and $g$ only, and $g$ and $k$ are functions of $T_2$ only, $\dot{m}/r_1$ is a function of $a$ and $T_2$ only.

For each value of $a$, there is an optimum value of $T_2$ and a corresponding optimum value of $b$ which will make $\dot{m}/r_1$ a minimum. Since for a given insulation $k_1$, $k_2$ and $g$ depend only upon the properties of the liquid and vapor, the fixed boundary temperatures $T_1$ and $T_3$, and the variable temperature $T_2$, we can compute values of $g$ for a series of temperatures $T_2$ in the range $T_1 < T_2 < T_3$ and use these $g$'s for all subsequent computations dealing with the particular liquid being stored.

Then for a given value of $a$, values of $\dot{m}/r_1$ can be computed and the minimum can be determined by graphical means. By plotting $\dot{m}/r_1$ vs. the position of the shield, $(b-1)/(a-1)$, the location of minimum will also yield the optimum value of $b$, the important design parameter. It will be noted also that for this vessel, the rate of evaporation is proportional to the first power of the radius or to (volume) $^{1/3}$, if heat conduction through supports is neglected. Thus this design offers the same premium for large size as that shown in curve C of Fig. 9.31.

Figure 9.50 shows the results of a representative computation for a liquid-hydrogen container, values of $a$ ranging from 1.05 to 1.50. It is seen that at the optimum position of the shield, about 35% of the distance from the cold wall to the warm wall, the shielded vessel will have an evaporation rate of about 38% of the unshielded vessel. It is seen also that the optimum location of the shield is not critical; the curve of $\dot{m}/m^*$ has a very flat minimum. It was found that the effectiveness of the shield $\dot{m}/\dot{m}^*$ at its optimum location is practically the same for values of $a$ ranging from 1.05 to 1.5.

It will be noted that the shield is somewhat warmer than liquid nitrogen. The rate of evaporation will be lower if a liquid-nitrogen-cooled shield is used. The decision about the use of this method rather than liquid-nitrogen shielding will depend upon the relative costs of liquid nitrogen and liquid hydrogen and the cost of keeping the liquid-nitrogen shield replenished. Also, the complete design study will include heat leaks from piping and supports which cannot be resolved here because they are peculiar to the specific application and mechanical design.

It is remarkable also that the temperature of the shield and its optimum locations, that is its distance from the inner container divided by the distance between the inner and outer walls, $(b-1)/(a-1)$, are practically independent of the value of $a$. This is a consequence of the specific heat of the vapor. As more complete data on the thermal conductivity of powder insulations become available, the optimum position should be recomputed.

To gain an idea of the performance that might be realized in a practical vessel, the results of these computations were applied to a spherical container for liquid hydrogen having an inner shell 2 m in outside diameter and an outer shell 3 m in inside diameter. This will have a capacity of slightly more than 4000 L. The evaporation rate resulting from heat leak through evacuated perlite insulation was found to be sufficient to evaporate 25 L or 0.62% of the contents per day. A vapor-cooled shield reduces this to 0.24% per day.

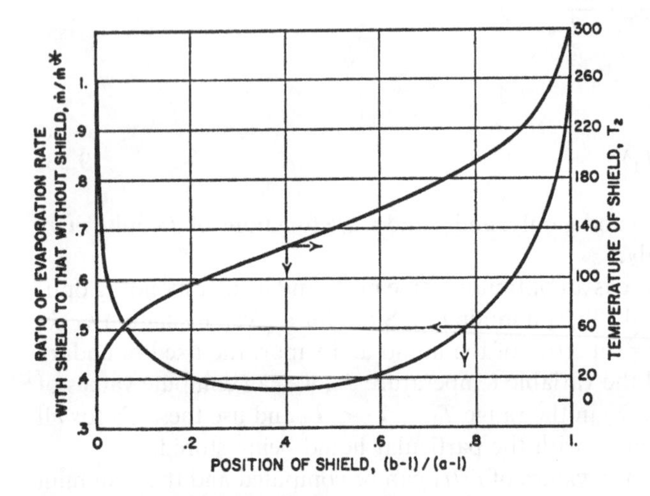


**Figure 9.50** Optimum position and temperature of a vapor-cooled shield in a powder-insulated vessel for liquid hydrogen.

*7.5.2. Containers for Liquid Helium*

It is quite apparent that the effectiveness of the vapor-cooled shield depends upon the ratio of the heat of vaporization of the liquid being stored to the specific heat of the vapor. Accordingly the application of this device to the storage of liquid helium is very attractive because of the favorable ratio between the specific heat of the vapor and the heat of vaporization.

The results of computations for $a=1.5$ fully justified this optimism: it was learned that the vapor-cooled shield of a powder-insulated container for liquid helium at its optimum location will have a temperature of 62 K, substantially lower than that of a liquid-nitrogen-cooled shield. The optimum position of such a shield was found to be $(b-1)/(a-1)=0.25$. The designer should again be cautioned against using this value after more data on the thermal conductivity become available. Measurements now in progress will undoubtedly yield more accurate data for this application.

Applying these results to the 4000-L spherical container described above, it was found that the heat leak through the insulation would cause liquid helium to evaporate at a rate of 7.7% per day from the unshielded vessel and 0.7% per day from the shielded vessel. See Fig. 9.51.

*7.5.3. Combining High-Vacuum and Evacuated Powder Insulation*

A logical refinement of the foregoing design is to use a high vacuum in the space between the coldest wall and the vapor-cooled shield, and powder insulation in the outer space. If the surfaces bounding the vacuum space have a low emissivity, this arrangement will result in a considerable improvement, an improvement gained at the expense of greater construction costs. The heat leak to the inner vessel, exclusive of solid conduction, can be computed by a method rather similar to that used earlier. In this case, however, the thickness of the high-vacuum space should be made as small as practical, so we can idealize the problem by assuming that the vacuum space has negligible thickness ($r_1=r_2$). Then the rate at which heat is transmitted to the inner container by radiation from a shield at the temperature $T_2$ is

$$\sigma\frac{e}{2-e}4\pi r_1^2(T_2^4-T_1^4)=\dot{m}L_v \tag{9.54}$$

where $\sigma$ is the Stefan–Boltzmann radiation constant, $e$ is the emissivity of the surfaces facing the vacuum space, $r_1$ is the radius of the inner container and $T_2$ and $T_1$ are the temperatures of the shield and inner container, respectively, $m$ is the mass rate of evaporation of the liquid being stored and $L_v$ is its heat of vaporization. The heat which reaches the shield by conduction through the powder is

$$\frac{k_2 4\pi r_1 r_3(T_3-T_2)}{(r_3-r_1)}=\dot{m}L_v+\dot{m}\Delta h \tag{9.55}$$

where $k_2$ is the mean thermal conductivity of the powder, $r_3$ is the radius of the outer shell at the temperature $T_3$, and $\Delta h$ is the change in enthalpy of the vapor as it is warmed from $T_1$ to $T_2$. Dividing Eq. (9.54) by (9.55) and letting $a=r_3/r_1$ gives

$$\frac{L_v}{L_v+\Delta h}=\frac{\sigma e r_1(a-1)(T_2^4-T_1^4)}{(2-e)k_2 a(T_3-T_2)} \tag{9.56}$$

**Figure 9.51** Novel liquid helium dewar. Features integral liquid transfer eliminating the need for standard lances, which must be inserted into the top of all other dewars every time liquid is transferred. Dewars requiring the top insertion of lances for the liquid transfer operation typically need a room with a minimum 12-foot ceiling rather than the standard 8-foot ceiling required here. The dewar is insulated with multilayered super-insulation and multiple vapor barriers for high storage efficiency. The dewars are also fitted with dual safety relief valves, a road valve and vent valve. Technifab Products Inc., Brazil, IN.

### *7.5.4. Comparison of Heat Leaks*

Table 9.11 shows the rates of heat transfer in terms of loss of liquid hydrogen and liquid helium in percent per day for several different types of insulation for a vessel of 4000 L capacity. These results do not take account of heat transfer by conduction through the supports nor heat conduction through the residual gas in a high vacuum.

A very obvious extension is the use of multiple shields. Since the cold vapor has a substantial refrigerative value after it leaves the shield discussed here, it is apparent that it could be utilized to cool additional shields.

Another source of refrigeration that could be utilized to cool the shield of a container for liquid parahydrogen is the heat absorbed by para-to-ortho conversion. If the proper catalyst is put into the vent tubes which cool the shield the endothermic conversion to a higher ortho concentration will provide extra refrigeration. The cooling produced by converting hydrogen of 99.8% para (the equilibrium concentration

**Table 9.11** A Comparison of the Heat Leaks Through Various Insulations for a 4000-L Spherical Dewar[a]

| Insulation | Evaporation rate caused by the heat flow through the insulation[b] | |
|---|---|---|
| | Liquid hydrogen, percent per day | Liquid helium, percent per day. |
| High vacuum[c] | 4.0 | 49.0 |
| Evacuated perlite | 0.62 | 7.7 |
| Evacuated perlite plus vapor-cooled shield | 0.24 (shield temperature = 124 K) | 0.70 (shield temperature = 62 K) |
| Evacuated perlite plus high vacuum[c] plus vapor-cooled shield | 0.13 (shield temperature = 129 K) | 0.33 (shield temperature = 86 K) |
| High vacuum[c] plus liquid $N_2$ shield (77 K) | 0.017 | 0.21 |

[a] The outside diameter of the inner vessel is 2 m and, when powder insulation is used, the outer shell has an inside diameter of 3 m. For the cases in which no powder is used, the diameter of the outer shell will be smaller. The emissivity of the surfaces facing the high-vacuum spaces is taken as 0.02.

[b] Excluding conduction by solid supports and assuming no conduction by residual gas.

[c] These are idealized conditions. The vacuum space is assumed to be negligibly thin and so good that heat conducted by residual gas can be ignored.

at the normal boiling temperature) to 38.6% para (the equilibrium concentration of 100 K) is 44% of the cooling produced by warming the gas from the hydrogen boiling point to 100 K.

Finally a word should be said about the storage of liquid oxygen and nitrogen. Since the ratio of the heat of vaporization to the change in enthalpy of the vapor upon being warmed to room temperature is much greater for these materials than it is for hydrogen and helium, it should be expected that the value of a vapor-cooled shield will be correspondingly less. However, the principles just discussed are equally valid for this application and should be given adequate consideration in the design of vessels for the long-term storage of liquid oxygen and nitrogen. See Fig. 9.52.

### *7.5.5. Vapor-Cooled Vent Tube*

In those cases in which heat conduction through the supports or fill and vent tubes is a substantial part of the total, it may be worthwhile to use the escaping vapor to intercept part of the heat. This is an important consideration in the vent tubes of some dewars. The escaping vapor maintains good thermal contact with the wall of the tube and absorbs a great deal of the heat that would otherwise enter and cause evaporation of the liquid being stored. The maximum possible saving from this process can be computed upon the assumption that the heat transfer between the vapor and the tube is perfect—that both have the same temperature at each level. Also it is assumed that there is no lateral conduction or radiation to or from the tube. Since, in the temperature regions that are of interest, the thermal conductivity of the tube or support is a function of temperature, the practical solution of the problem can be simplified by assuming that this function is linear, that

- **Use of either cryogenic solids, liquids or vapors may be use to shield other, more thermally susceptible, cryogen**
  - **A"cheaper" one protecting a more expensive one!**
  - **Liquid shields are mostly extinct due to insulation improvement, solid cooled shields are in use within the aerospace industry**
- **Vapor shields use cold vent gas liberated directly from the cryogen to be protected**
  - **Vent gas is routed past an intermediate shield**
  - **Absorbs some of the heat that would otherwise warm the liquid**

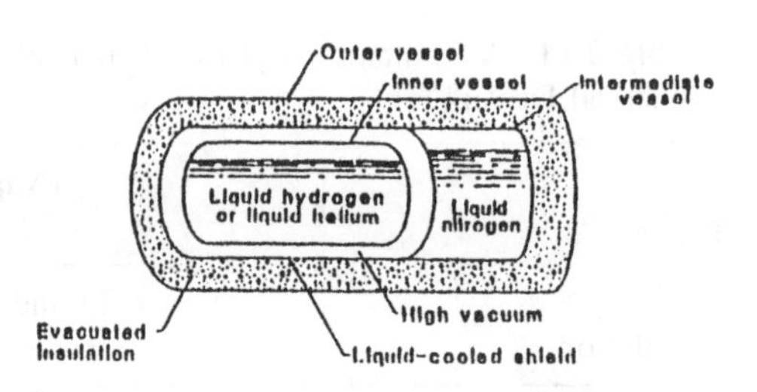


Liquid-shielded vessel

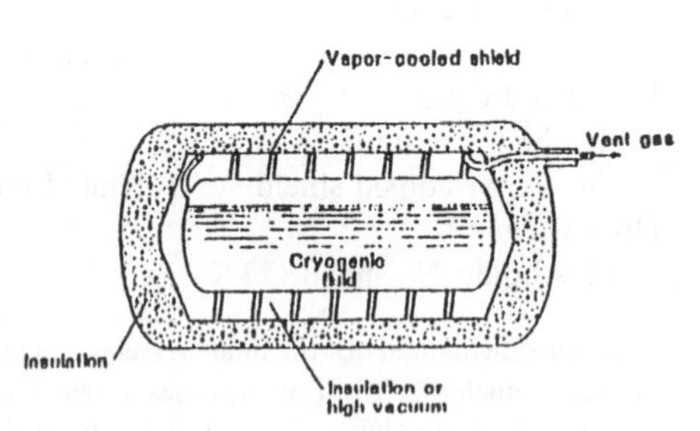


Vapor-shielded container

**Figure 9.52** Cryogenically cooled thermal shields.

$$k = k_0 + a(T - T_0)$$

where $k$ is the thermal conductivity at the variable temperature $T$, $k_0$ is the thermal conductivity at the cold end which has the temperature $T_0$, and $a$ is a constant. then

$$W = A[k_0 + a(T - T_0)](\mathrm{d}T/\mathrm{d}x) \tag{9.57}$$

which is the equation for heat conduction where $W$ is the heat current at any point in the tube, $A$ is the cross-sectional area of the tube, and $x$ is the distance from the cold end. Now the issuing vapor will absorb some of the heat traveling down the tube, so that

$$W = W_0 + \dot{m}C_p(T - T_0) \tag{9.58}$$

where $W_0$ is the heat current that reaches the cold end of the tube, $\dot{m}$ is the mass rate of flow of the issuing vapor and $C_p$ is its specific heat.

Upon combining Eq. (9.57) and (9.58) and integrating, it is found that

$$L/A = \left[\frac{1}{\dot{m}C_p}\right]\left[a\Delta T + \left(k_0 - W_0\frac{a}{\dot{m}C_p}\right)\ln\frac{W_0 = \dot{m}C_p\Delta T}{W_0}\right] \tag{9.59}$$

where $L$ is the length of the tube and $T$ is the temperature difference between the warm and cold ends.

Thee thermal conductivities of alloys commonly used for vent tubes of metal dewars are rather complex functions of temperature. However, Eq. (9.59) can be employed by dividing the total temperature interval into smaller intervals in which the thermal conductivity can be considered to be a linear function of temperature.

Then Eq. (9.59) is applied to the lowest temperature interval, $\Delta T_1 = T_2 - T_1$, and the dependence of $W_0$ upon $L_1/A$ for various values of $\dot{m}$ is determined; $L_1$ is the length of tube lying in the lowest temperature interval. Then values of heat current $W'_0$ at the bottom of the next higher temperature interval are determined by the relation

$$W'_0 = W_0 + \dot{m}C_p\Delta T_1$$

These values of $W'_0$ are then used in Eq. (9.59) to compute values of $L_2/A$ for the second temperature interval and the process is repeated as required. Finally total values of $L/A$ are obtained by adding the individual values, $L_n/A$.

### *7.5.6. Conclusions*

The foregoing analyses show quite clearly that important benefits can be realized by utilizing the refrigerative value of the vapor which escapes from an insulated vessel for liquid hydrogen or helium. Table 9.12 summarizes the foregoing information.

## 7.6. Support Systems

The heat transfer through rods and cables which support the inner structure, as well as through fill and vent tubes, and any instrumentation connections are important considerations in the design of containers and transport vessels. The structural numbers must have sufficient strength and rigidity for the given application but must also be so designed that the heat transfer along these members is a minimum. The heat

**Table 9.12** Vapor Cooled Shields (VCS)

| |
|---|
| Basic features |
|   Aluminum sheet cooled by vapor vented from dewar |
|   Shield prevents portion of heat from reaching dewar |
| Types of shields |
|   Boiler shield used to intercept heat and vaporize liquid |
|   Vapor shield absorbs heat by warming vapor |
|   Multiple shields improve effectiveness |
| Special features |
|   Helium requires several shields because $c_p$ is large and $h_{fg}$ is small |
|   Hydrogen requires one shield because $c_p$ is large and $h_{fg}$ is large |
|   Nitrogen does not require shields because $c_p$ is small and $h_{fg}$ is large |
| Solid cryogen cooled shields by intercepting incoming heat with shield attached to solid cryogen dewar/tank and removes via sublimation of solid to vapor |
| Vapor cooling can lower cryogen loss rate considerably, depending on ratio of latent heat-to-sensible heat and relative levels of internal heat load and parasitic heating |
|     Factor of 20 improvement possible with helium |
|       Three or four vapor cooled shields (VCS) typical |
|     Factor of 5 achievable with hydrogen |
|       Two or three VCSs typical |
|     Factor of 2 achievable with heavier cryogens, like nitrogen, oxygen, methane, ammonia, neon |
|       One VCS typical |
| VCS is a thin aluminum shield with vent tube mechanically (thermally) attached |

- **Supports must carry load of inner tank and fluid to outer tank**
- **High strength, low conductivity materials are preferred**
  - **-Steel cables or chains**
  - **-Insulating blocks or rings**
  - **-Fiberglass struts or straps**
  - **-Load-bearing insulation**

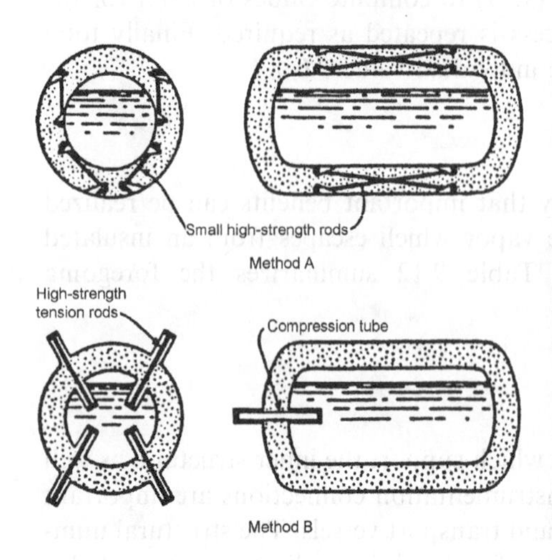


| Material | Strength/ Conductivity Ratio |
|---|---|
| Teflon | 13.6 |
| Nylon | 48.4 |
| Mylar | 286.6 |
| Dacron fibers | 630.7 |
| Kel F oriented fibers | 542.8 |
| Glass fibers | 183.6 |
| 304 stainless steel | 50.5 |
| 316 stainless steel | 66.1 |
| 347 stainless steel | 71.1 |
| 1100-H16 aluminum | 0.52 |
| 2024-0 aluminum | 0.86 |
| 5056-0 aluminum | 1.20 |
| K Monel (45% cold drawn) | 34.4 |
| Hastelloy C (annealed) | 30.7 |
| Inconel (cold drawn) | 23.2 |

**One example of support rods**

**Figure 9.53** Commercial cryogenic inner tank storage tank supports.

transfer along these members is given by the relation

$$Q = \frac{A}{L}\int_{T_1}^{T_2} k\,\mathrm{d}T$$

From this, it is evident that the heat transfer will be a minimum if the cross-sectional area is small and the length large. This immediately suggests that the supporting members be in tension rather than compression, since buckling is not a problem in tension but is often the limiting design factor in compression, particularly if one is attempting to decrease the area and increase the length. When space limitations are imposed, techniques such as shown in Fig. 9.53 may be used.

It is also evident that one should use materials which have a low thermal conductivity. Fortunately there are several materials, such as stainless steel, which have high strength, which permit the use of small cross-sectional areas for a given load and low thermal conductivity. See Table 9.13.

## 8. TRANSFER OF LIQUEFIED GASES

The transfer of small amounts of liquid nitrogen, oxygen, or hydrogen through short tubes is a common laboratory procedure and usually presents little difficulty. Since the line is short, evaporation losses and pressure drop are not troublesome. However, for large-scale operations, long lines and rapid transfer are often required. In

**Table 9.13** Support Systems for Cryogen Storage

| |
|---|
| Main design features |
| Long and thin to reduce conductive heat leak (long thermal path) |
| Stiff enough to keep natural frequency above that of source of vibration |
| Low conductive materials such as fiberglass |
| Flexible enough to accommodate thermal contraction of tank |
| Frequently used designs |
| Fiberglass straps in tension from outer tanks girth ring to head of tank |
| Stainless steel rods in tension |
| Fiberglass tubes in tension or compression from outer tanks girth ring to head of tank |
| Fiberglass or low conductivity metallic neck ring (with or without liner) |
| Special techniques/features |
| Load bearing insulations |
| Bumpers used to limit tank motion under large loads |
| Thermal short to vapor cooled shields to intercept heat leak along a support |
| Multistage support to reduce heat leaks (fiberglass or composite straps or concentric tubes) |

such cases, the size of the line and the effectiveness of the insulation become very important. This chapter deals with the behavior of cryogenic liquids during transfer and the design of transfer equipment. Some special attention will be devoted to the transfer of liquid helium, because losses of liquid helium during transfer may be serious even in the relatively short transfer lines used in the laboratory.

## 8.1. Two-Phase Flow

Heat flow into a pipe carrying a liquefied gas has two objectionable consequences; first, it wastes liquid by causing evaporation, and second, the vapor thus formed seriously reduces the carrying capacity of the line. The mixture of vapor and liquid, having a lower density than that of the pure liquid, must have a greater velocity in order to maintain a given mass rate of flow of liquid when rapid delivery is desired. The stream velocity in the two-phase region constitutes a serious limitation. The maximum velocity of flow of a fluid in a pipe is equal to the velocity of sound in the fluid. The velocity of sound in cryogenic liquids is quite high: for liquid oxygen, nitrogen and hydrogen of the order of 1000 m per second, and for liquid helium about 200 m per sec. However, in a two-phase mixture of liquid and vapor, the velocity of sound is low because of the high adiabatic compressibility resulting from the presence of vapor.

Two-phase flow has been studied by many and is extremely complex. Since this focuses on "what's different" at low temperatures, the reader is advised to consult the classic publications on two-phase flow. Ordinary two-phase flow is very complex involving slug flow, surging, independent flow rates of the two phases if separated, acoustic oscillations, and even backward flow.

For cryogenic fluids, the complexities of ordinary two-phase flow are further compounded by the fact that there is mass transition, usually from the liquid phase to the vapor phase. This mass transition depends upon heat influx into the pipe and upon pressure changes. Moreover there is considerable evidence that non-equilibrium between liquid and vapor phases during transfer is a common condition. As a consequence of these and other complexities, there is not at present a good analysis of the problem of two-phase, single component flow. The reader is strongly advised to seek out the most recent and comprehensive studies in the appropriate volumes of *Advances in Cryogenic Engineering* (Kluwer Press, formerly Plenum Press). Here, we shall address just the general problem and try to advise of common pitfalls.

There are three different kinds of two-phase flow: (1) a rather homogeneous mixture of vapor bubbles in the liquid; (2) "slug" flow consisting of alternate regions of pure liquid and pure vapor, each filling the pipe (a condition most common in transfer lines of small diameter); and (3) annular flow, wherein the liquid flows along the annular region next to the wall of the tube and the vapor, moving at a much higher velocity, occupies the central core of the tube.

Fortunately, when high-speed transfer of cryogenic liquid is required, it is usually feasible to control the conditions so that two-phase flow is avoided; all of the fluid in the line is kept in the liquid state so that the ordinary relations between pressure drop, flow velocity, properties of the liquid, and dimensions of the pipe-line are applicable and high-velocity transfer is readily accomplished. Accordingly, *it is much more profitable for the cryogenic engineer to devise means of avoiding two-phase flow* than to attempt to design equipment that will accommodate two-phase flow. Of course, two-phase flow is often unavoidable, for example during the cool-down of a pipeline; therefore, a solution of the two-phase condition is necessary for the completely reliable design and predictable performance of cryogenic transfer equipment.

## 8.2. Cool-Down

When a cryogenic liquid (somewhat supercooled by pressurization or pumping) is started through a typical transfer line initially at room temperature, at first all the liquid which enters is quickly evaporated and nearly all of the line will contain only gas. As the process continues, a length of the upstream end of the line will be cooled below the saturation temperature and this portion will contain the pure liquid phase. Downstream of this section, there will be a region in which both liquid and vapor will be present, and in the remainder of the line, only gas will be flowing. As the line is further cooled by the evaporating liquid and by the resulting cold vapor, the liquid phase will persist farther along the line until finally liquid will be discharged at the exit. This behavior is readily observed in a glass "dewar" siphon (unsilvered) used to transfer liquid hydrogen in the laboratory. It has also been found economical of liquid to watch the start-up behavior of such a siphon and insert the delivery end into apparatus requiring refilling only after liquid hydrogen is seen to wet the entire length of the siphon. If the warm vapor which precedes the liquid is discharged into the receiver, it evaporates some of the liquid already present. This evaporation can be largely avoided by making sure that the siphon is ready to discharge liquid before it is inserted into equipment which already contains liquid hydrogen. With liquid helium this precaution is *much more essential* because of the very much lower heat of vaporization of helium.

In a long transfer line, the vapor may be warmed nearly to the initial temperature of the line during a part of the cool-down process. This, of course, will be economical of liquid because heat is extracted from the line by both evaporating the liquid and warming the vapor. The minimum amount of liquid required to cool a long line therefore can be estimated by equating the decrease in enthalpy of the line to the total increase in the enthalpy of the fluid as it is vaporized and the vapor warmed to room temperature. Such an estimate will be on the optimistic side because it takes no account of the heat leak to the line during cool-down, the frictional energy dissipation or the final cooling of the exit region while cold vapor is discharged. Obviously there is merit in keeping the mass (and therefore the heat capacity) of the line small.

In the initial stages of the cool-down of a long transfer line, a large part of the resistance to flow results from the high-velocity, low-density gas which evaporates from and precedes the liquid during its passage through the line. The warm line may quickly raise the temperature of the gas nearly to the ambient temperature, and at this condition its density may be only about $\frac{1}{1000}$ the density of the liquid. Accordingly the *velocity of the gas will be about 1000 times the velocity of the liquid,* since the mass rates of flow are necessarily the same. The flow resistance associated with the high velocity greatly increases the time required for cooling the transfer line.

In several installations the cool-down of long liquid transfer lines is expedited by providing vapor taps at intervals. In this way, the pressure is relieved and the liquid front moves more rapidly, cooling the line. The line is cooled most rapidly by leaving a given vapor tap open until liquid arrives. However, this may be wasteful of liquid because the cold vapor preceding the liquid is not completely utilized in refrigerating the line; hence more of the cooling just be done by the evaporating liquid.

Even in short lines, the consequences of cool-down may be very important when liquid helium is being delivered to add to liquid already present in the receiving vessel. If the warm line is permitted to deliver warm gaseous helium to the receiver during cool-down, a very serious evaporation of liquid may result. When one considers the fact that the heat of vaporization of a unit mass of helium is sufficient to cool an equal mass of gaseous helium only 4 K, it is apparent that a little warm helium gas can cause the evaporation of a large quantity of liquid already present in the receiver. The provision of a vapor tap near the exit end of the transfer line can greatly reduce this loss. A transfer line described later in this chapter employs such a vapor tap.

## 8.3. The Case of Zero Delivery

This heading is used to emphasize a condition that can arise if a line is designed without taking account of the effect of heat leak. If the line is perfectly insulated and liquid is continuously introduced into one end, it is only a matter of time before liquid will appear at the discharge end. However, in practical lines, there is a heat influx which evaporates liquid; and if the maximum rate of flow is insufficient, it is quite possible that liquid will *never reach the exit.* Sometimes such a line can be induced to deliver liquid by using vapor bleeding taps as described earlier to effect the initial cool-down. After the line is cold, the greatly increased velocity of flow may cause liquid to be discharged even after part of the liquid has been evaporated by the heat influx.

In some cases, it may be profitable to provide occasional gas–liquid separators along the transfer line and allow most of the vapor formed in the line to escape continuously at these places during transfer. This will not avoid two-phase flow, but it will reduce the ratio of the volume of vapor to that of liquid and therefore speed up the rate of delivery of liquid.

## 8.4. Transfer Through Uninsulated Lines

Liquid air, oxygen, and nitrogen are often transferred through uninsulated metal pipes or tubes. The adjective "uninsulated" may be slightly misleading because even with a line made of a metal with high thermal conductivity, such as copper, there are two mechanisms which automatically provide some resistance to the flow of heat and thus supply a little insulation to the liquid being transferred. These are the so-called *film coefficients*, the coefficient of heat transfer from the ambient air to the outer surface of the tube carrying the cold liquid, and the coefficient of heat transfer from the inner surface of the tube to the liquid flowing within.

In many operations involving only short-time, short-distance transfers of liquid air, oxygen, or nitrogen, the uninsulated line may be economically the most desirable. The cost of insulating the line may exceed the resulting saving. The choice is not always easy to make. For example, when transferring liquid oxygen or nitrogen from a tank truck or railway car to stationary storage through an uninsulated line, two-phase flow is usually present and the transfer may take several hours. Since the flow rate can be substantially increased by insulating the line, thereby reducing the amount of vapor in the line, insulation may pay because of the time and labor saved, even though the saving of liquid may be inconsequential. This is another case in which a reliable solution of the two-phase, single-component, fluid-flow problem would be useful.

Some NASA test stands use uninsulated lines for liquid hydrogen, which, at first glance, seems unthinkable. Several analyses of such uninsulated line hydrogen transfer have been made, all with approximately the same conclusion. The duration of flow for rocket engine tests is short, measured in seconds. Hence, insulation has little or no effect on the economics of this part of the flow schedule. Cool-down of the massive transfer lines ($mC_p\Delta T$) is huge by comparison and would not benefit at all from any degree of insulation. Hence, no degree of insulation has any economic benefit in this particular type of cool-down. The justification for insulation is based on safety, namely preventing the condensation of enriched air on the outside of the transfer line. The condensation of atmospheric air on the outside of such a line will provide an essentially infinite source of heat for the evaporation of liquid hydrogen. The effectiveness of this heat source may be better appreciated by pointing out that the formation of a unit volume of liquid air will cause the evaporation of more than 10 volumes of liquid hydrogen. The transfer of liquid helium without insulation should not even be considered because, since it has only $\frac{1}{12}$ the heat of vaporization of hydrogen, on a volumetric basis, even the best of insulation is none too good.

## 8.5. Transfer Lines Insulated with Porous Materials

Lines carrying liquid oxygen or nitrogen may be insulated with porous materials. For liquid nitrogen, some extra precautions should be taken because the temperature is such that fractional condensation of air can occur on the nitrogen-cooled surfaces. The air so condensed will be rich in oxygen and will constitute a fire and explosion hazard if the adjacent insulation is combustible. Several fatal explosions have been attributed to a similar circumstance. This danger can be avoided by using non-combustible insulation or by providing a purge flow of gaseous nitrogen through the insulation so as to keep out atmospheric air. The gaseous nitrogen purge is preferred because there will be no condensation as long as the nitrogen pressure is below saturation. The use of non-combustible porous insulators such as Fiberglas®, diatomaceous earth, expanded perlite, vermiculite, etc., without the gaseous nitrogen purge is quite feasible, although the insulating value may be found to be far below that computed because air may condense on the cold line, flow to a warm region, be evaporated and again return to the cold line to be recondensed, thus constituting an additional heat-transfer process. Also when insulation of this type communicates with the atmosphere for long periods of time while it is cold, atmospheric moisture will collect and spoil its insulating properties. It is necessary to prevent the accumulation of moisture in the insulation, and when the line is at a temperature below 83 K, air should be excluded. Thus this type of insulation may be suitable for insulating lines carrying liquid oxygen; and with proper precautions, liquid nitrogen lines may be so insulated.

However, difficulties are almost certain to arise if this type of insulation is used for lines carrying liquid hydrogen. The cold surface can condense air, and the only practical purge gases to prevent this are hydrogen and helium, both of which have very high intrinsic thermal conductivities and therefore will greatly diminish the insulating properties of the porous insulation. Solid impermeable foams such as polystyrene foam may be used if the seams are adequately sealed, but permeable insulations such as powders or fibrous materials are not suitable if they are air-filled. Of course, if the porous insulation is surrounded by a perfectly vacuum-tight enclosure, condensation of the air by the cold surface will produce a good vacuum in the interstices of the insulation and the result will be "evacuated powder" insulation.

The NASA Transonics Flow Facility at Langley, Virginia, is cooled to $LN_2$ temperatures to increase the reynolds number without increasing the pressure. This enormous facility is insulated as described above to prevent enriched air condensation. The external tank of the Columbia shuttle was insulated with a special sprayon foam mainly to prevent the freezing of air and water on this tank. Alas, the foam cracked due to the differences in thermal contraction between the metal tank substrate and the polymeric foam, with well-known disastrous results.

## 8.6. Vacuum-Insulated Transfer Lines

For the efficient transfer of liquid hydrogen and helium, and in many cases also liquid oxygen or nitrogen, vacuum-insulated transfer lines appear to be the most appropriate.

There are a number of versatile transfer system consisting of sets of standard vacuum-insulated sections in the form of straight runs, ells, valves, etc., each provided with the appropriate vacuum-insulated couplings so that an assemblage of such components can be readily arranged to make connection between any two points. Transfer lines of this general form are readily available. See Figs. 9.54–9.58. Flexible vacuum-insulated lines for liquid hydrogen consisting of coaxial metal hose are also available. See Fig. 9.59.

## 8.7. Design Considerations

The formulae given earlier for the heat transfer across the space between long coaxial cylinders apply very precisely to vacuum-insulated transfer lines. However, there are two items which introduce uncertainties into the computation of heat leak for a practical line. First is the problem of heat flow through the spacers or supports separating the inner, liquid-carrying pipe from the outer pipe which provides the exterior surface of the evacuated annulus. The spacers, commonly constructed of a poorly conducting organic plastic, make rather indeterminate thermal contact with both the inner and outer pipe; hence the heat leak cannot be computed. It was once observed that a transfer line in which the vacuum had deteriorated developed cold

- **Minimizes heat leak**
- **Eliminates condensation of air and water**
- **Available in standard sizes to 6"**
  - **Larger are special orders**

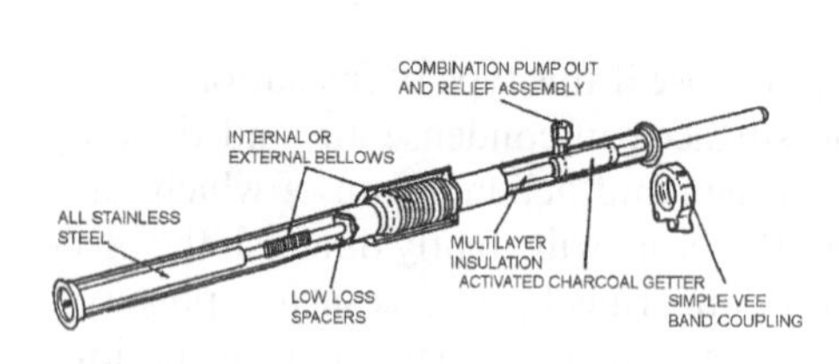


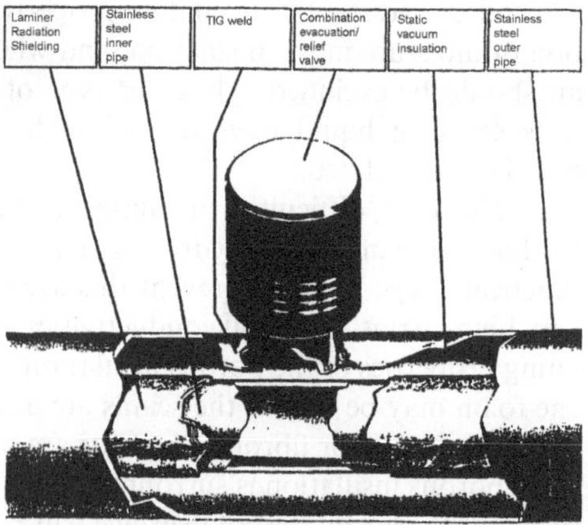


**Figure 9.54** Vacuum-jacketed transfer piping.

(A) (B)

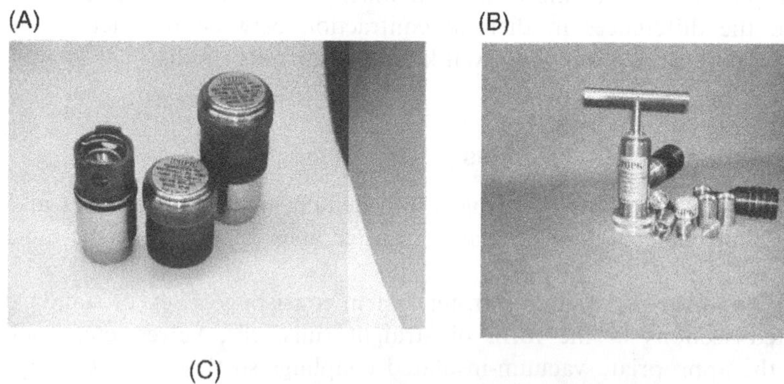

(C)

**Figure 9.55** Combination vacuum evacuation and over-pressurization valves. It is crucial that vacuum-insulated transfer lines have a pressure relief device in the event there is leakage of cryogenic fluid into the vacuum space. See Chapter 11. The valves shown in (A) relieve at 15 psig. (B) shows such valves with a removable operator. Valves shown in (C) are all stainless steel with dual Viton® O-rings, suitable for severe outdoor weather conditions. PHPK Technologies Inc., Westerville, OH, www.phpk.com.

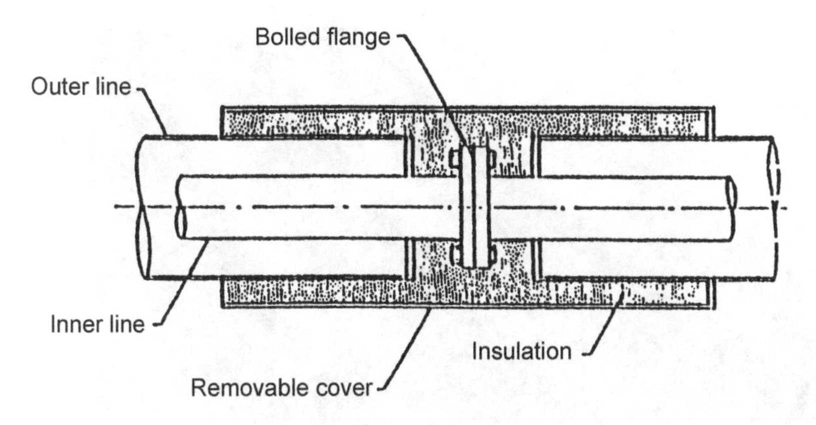


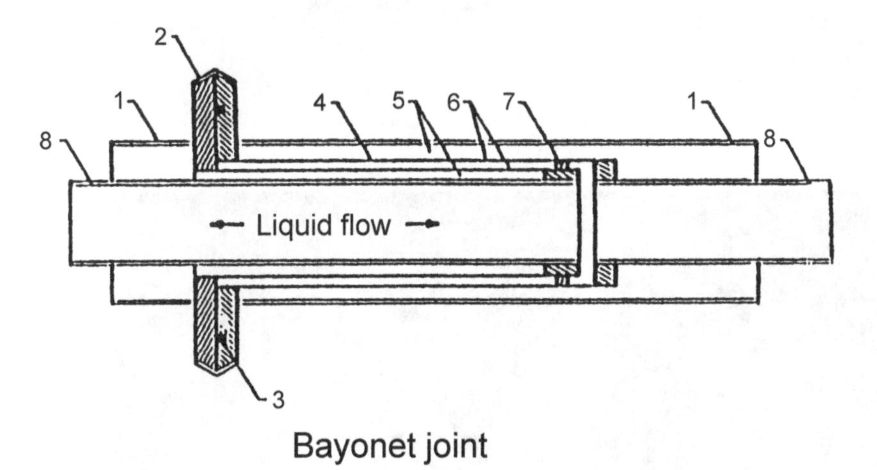


**Figure 9.56** Vacuum-insulated line joints (bayonets). 1, outer line, vacuum shell; 2, line coupling; 3, warm temperature O-ring seal; 4, static gas leg; 5, vacuum insulation space; 6, vacuum barriers; 7, additional liquid seal; and 8, liquid line.

spots on the outside at each spacer position. When the vacuum was good there was no external indication of heat conduction through the spacers.

### *8.7.1. Spacer Designs*

Spacer designs frequently used are shown in Fig. 9.60. The central hole fits the inner tube and may make rather good thermal contact. The corners of the square spacers make poor thermal connection with the outer tube.

### *8.7.2. Emissivity of the Inner Line*

The second uncertainty in computing heat leaks for practical lines arises from the difficulty of ensuring low emissivity of the lines after assembly. Transfer lines are usually assembled by soldering with a torch, and unless considerable care is exer-

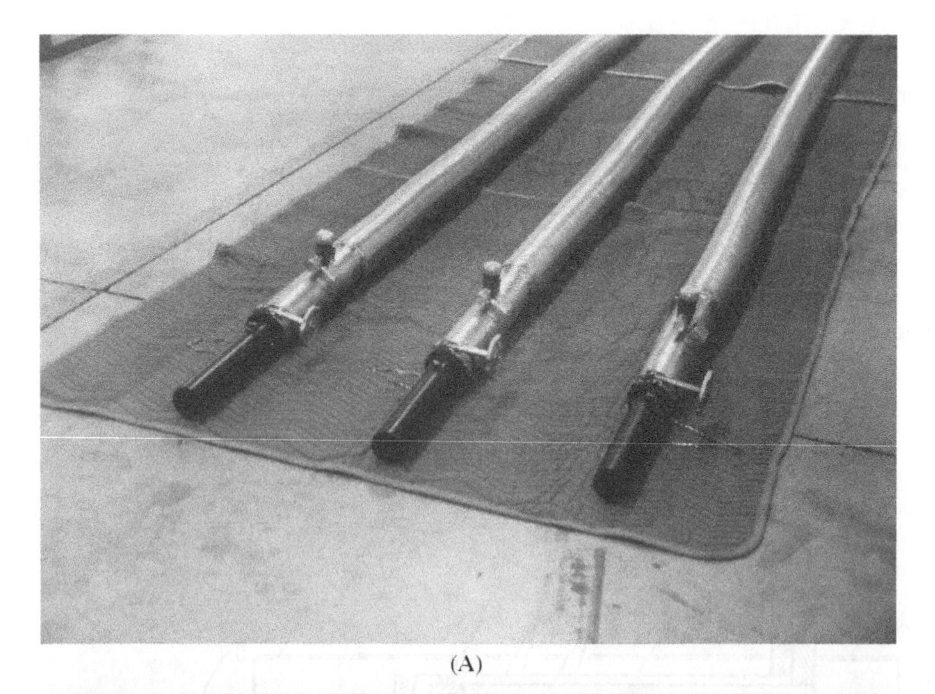

(A)

(B)

**Figure 9.57** Cryogenic couplings (bayonets). Are commercially available from $\frac{1}{2}$ in. to 14 in. (B). The couplings shown are manufactured from 304/304L-grade stainless steel with special close tolerance design considerations to eliminate sweating and frosting. The bayonet design eliminates cavities, such as socket welds, which could trap contaminants, an important feature in oxygen systems (see Chapter 11). PHPK Technologies Inc., Westerville, OH, www.phpk.com.

**Figure 9.58** Low heat-leak bayonet connections for rigid vacuum-jacketed pipe. These bayonet connections are constructed of type 304 stainless steel schedule 5 welded or seamless pipe ASTM A 312. Standard inside pipe diameter available range in size from $\frac{1}{2}$ to 2 in. Compte systems of rigid VJ pipe and connections are available to make hundreds of feet of piping. The low heat leak allows such systems to pay for themselves in cryogen savings in less than 18 months. Technifab Inc., Brazil, IN.

**Figure 9.59** Flexible vacuum-insulated transfer lines incorporating remote actuated valves for liquid helium distribution. Cryofab, Kenilworth, NJ.

- **Internal fluid must be supported within the vacuum jacket**
  - **Structurally sound**
  - **Thermally sound**
- **Spacer materials**
  - **Plexiglass**
  - **Fluorocarbon plastic**
  - **Stainless steel**
- **Heat transfer through minimized by small contact area**
- **Square or triangle spacers effective for small lines**
- **Roller or ball spacers used for heavier lines**
- **All systems must handle thermal transients**

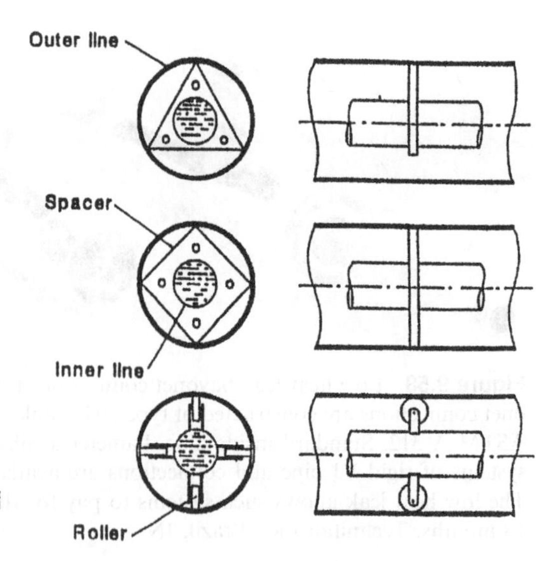


**Figure 9.60** Line spacer designs.

cised, it is quite likely that oxidation, soldering flux, or carbon deposited from the torch flame will greatly deteriorate the surfaces. (Washing with a 10% ammonium persulfate solution after the assembly is almost finished will clean the copper, removing oxide and other corrosion.) Some of the spacers may be overheated during assembly and the resulting produces of decomposition may coat the line and spoil its reflectivity. A spacer material that is favored by many cryogenic workers is Teflon®, polytetrafluoroethylene. This plastic is a good insulator, is chemically very stable, has a high decomposition temperature, and does not outgas excessively.

If the insulating space is thick enough, more than 2 cm, evacuated powder may be better than high vacuum alone. There is an increase in the amount of liquid consumed during cool-down because the powder near the inner line must be cooled. On the other hand, the fact that the powder behaves as an adsorbent for residual gas, plus the fact that an extremely good vacuum is not required, strongly favors evacuated powder insulation. Perhaps the greatest objection to powder is the structural-assembly problem. It is not easy to fill the annular space with powder so completely that subsequent shaking during use will not cause the formation of voids. These voids can be serious heat leaks because a small amount of residual gas can conduct heat readily where there is no powder. The pressure required for high-vacuum insulation is two or three orders of magnitude less than that required for evacuated powder insulation.

### *8.7.3. Thermal Contraction*

Thermal contraction complicates the design of vacuum-insulated lines, because the inner pipe is cooled while the outer remains at room temperature. A copper tube 25 ft long will contract approximately 1 in. upon being cooled from room temperature to liquid hydrogen temperature. There are several ways of designing lines to allow for this contraction: (1) the inner tube may be provided with metal bellows inserts to that these can lengthen as the rest of the rube shortens; (2) the outer tube can have the bellows inserts and be supported in rollers or flexible supports so that it can follow the length changes of the inner tube without much restraint; and (3) for short U-shaped lines, sufficient space may be allowed between the spacers and the inner or outer tube that the contraction will not subject the assembly to undue strain. There is a definite advantage in having the bellows in the outer rather than the inner tubes, because metal bellows constitute a potential vacuum leak. Such leaks are easier to find and repair if they are in the outside tube.

#### *8.7.4. Material Choice*

The materials used for transfer lines depend a great deal upon the intended service. Copper is often used for the inner tube because it has a low emissivity and is easy to assemble. The outer tube may very well be copper also except for the ends, which should have low thermal conduction where they provide a heat path from room temperature to the temperature of the cryogenic liquid being transferred. When repeated batch transfers are to be made with the transfer line left in place and warming up between transfers, the ends of the inner line also should be made of poorly conducting material.

The other metals often used are stainless steel and Monel®. These latter have three very desirable characteristics: (1) they are poor conductors of heat, (2) they are available as thin-wall tubing, and (3) they are strong and resist accidental deformation. The thin wall reduces both the longitudinal heat leak and the heat capacity of the inner line. Several of the designs referenced earlier employ Monel® or stainless steel. The principal objection to these alloys is their relatively high emissivity. The radiant heat transfer across a vacuum space bounded by stainless steel or Monel® walls will be several times larger than that across a space bounded by clean copper walls. Seamless flexible tubing can be used to fabricate the terminals of vacuum-insulated lines. This makes it convenient to insert the terminal into the filling connection of the receiver.

Flexible tubing has a very high effective emissivity because the convolutions present recesses that behave as black bodies; incident radiation is almost totally absorbed. In some designs radiant heat transfer has been greatly reduced by wrapping the inner flexible tube with aluminum foil. Just a few layers of MLI do a world of good.

### 8.8. Transfer Line Issues

My friend David Daney made major contributions to the theory and practice of transferring cryogenic liquids both at his career at the NBS Cryogenic Engineering Laboratory and also later at the Los Alamos National Laboratory. Daney (1998) points out some specific design considerations for cryogenic transfer lines:

1. *Optimization of the heat leak to the transfer line.* The quality of transfer line insulation is primarily an economic problem of minimizing the sum of the capital cost (of the insulation system) and the operations cost (incurred by the heat leak, taking into account the duty cycle). Large-diameter or low-duty-cycle LNG and $LN_2$ lines are often insulated with glass or plastic foam, whereas liquid helium and liquid or slush hydrogen are vacuum insulated with MLI—sometimes with a vapor-cooled or liquid-nitrogen-cooled shield. In general, the lower the temperature, the better the insulation required, because the refrigeration is more expensive.

2. *Cavitation and two-phase flow.* When cryogens are transferred with little subcooling, two-phase flow may occur at positions where the pressure falls below saturation owing to the combined effects of pressure drop and heat transfer. Cavitation may occur locally in the low-pressure regions of flowmeters and valves with the vapor bubbles collapsing in the pressure recovery regions of these devices.
3. *Thermal-acoustic oscillations.* Helium and slush hydrogen systems are susceptible to thermal-acoustic oscillations in pressure taps, valve stems, and other conduits with large length-to-diameter ratios that traverse between these cryogens and ambient temperature. The oscillations can have heat pumping rates on the order of watts; their frequency is typically tens of hertz.
4. *Pressure collapse and low-frequency pressure oscillations.* Slush hydrogen and subcooled liquid cryogen flow systems are susceptible to pressure collapse in relief valve standpipes, valve stems, and so forth, in which a temperature-stratified layer of liquid maintains the subcooling pressure on the flow system. If flow turbulence disturbs the stratification, the pressure collapses as vapor condenses on the slush or subcooled cryogen. The cryogen then surges into the warmer regions of the pipe, where it flashes, causing a pressure pulse. The resulting low-frequency oscillations can pump large quantities of heat into the cryogen.
5. *Negative gauge pressures.* Systems at negative gauge pressure, such as slush hydrogen and pumped helium, are susceptible to contamination by air and water vapor that enter through leaky valve stems, relief valves, and pipe joints. These leaks may plug relief valve lines with ice and solid air. Air, if it reaches the cryogen, frequently condenses in submicron-size particles, giving hydrogen and helium a milky appearance. Because these crystals have settling times of hours or longer, they are easily transported throughout a system. They tend to collect in stagnant zones, such as downstream of valve seats, where they may block the flow. In hydrogen systems, accumulated oxygen crystals represent a potential safety hazard. Special precautions such as the use of welded-bellows valve packing and avoidance of compression fittings (welded or soldered joints are preferred) can reduce these problems.
6. *Thermal contraction.* Contraction of stainless steel from ambient to cryogenic temperatures is about 0.003 in./in., which on a 100-ft length of pipe is 3.6 in. Bellows placed periodically along straight pipe runs are used to eliminate thermal cool-down stresses. On lengths up to several feet, thermal contraction can be accommodated with bends on smaller-diameter lines.
7. *Trapped line segments.* Segments of cryogenic piping that can be isolated between valves must be provided with relief valves to prevent rupturing the line by trapped cryogens as they warm up. The isometric pressure rise experienced by liquid nitrogen warming to 300 K, for example, is about 2900 bar.

These are serious issues peculiar to cryogenics and should be constantly in your mind.

# 10

# Natural Gas Processing and Liquefied Natural Gas

## 1. INTRODUCTION

The processing of natural gas and the liquefaction of natural gas have quite different objectives but share many of the same processing techniques and unit operations. For this latter reason, the two topics are considered together in this chapter. Initially, however, we consider some of the differences.

The major objectives of natural gas processing are (1) purification (removal of $CO_2$, $H_2S$, $H_2O$, etc.), (2) hydrocarbon recovery [liquefied petroleum gas (LPG), ethane], (3) upgrading ($N_2$ rejection), and (4) helium recovery. Liquefaction or partial liquefaction of the natural gas feedstock can be used to meet some of these objectives.

The objective of producing liquefied natural gas (LNG) is the huge reduction in volume, a factor of 650, decreasing the size and cost of storage and transportation containers. The ability to store and transport natural gas in containers has become more important as shortages are met by imports from Africa and the Middle East—sources that cannot be tapped via a pipeline. Even in areas where gas is piped, the flow may not be great enough during peak periods to satisfy all users on the pipeline. In these cases, liquefied natural gas (LNG) can be made on low-use days and vaporized on peak days to meet the demand.

These two uses of LNG, shipping and storage, lead to two types of plants, baseload and peakshaving. Baseload plants are located near gas fields to produce LNG for shipment. In baseload plants, approximately two-thirds of the capital investment is for liquefaction equipment, with the remainder distributed among shipping terminals, storage, and purification costs. Peakshaving facilities, on the other hand, act as reservoirs, have lower liquefaction rates than baseload plants, and are located near the points of consumption. Compared to the baseload plants, almost half the capital investment for peakshaving plants is for storage, with the balance invested in liquefaction, vaporization, and purification equipment. Thus, baseload plants are liquefaction-intensive, and peakshaving plants are storage-intensive. The unit operations are much the same for both.

In both natural gas processing and LNG production, it must be recalled that natural gas is not a pure substance but is a mixture that may vary in composition depending on the source. Typical composition ranges for domestic and foreign gases are given in Table 10.1 (Kirk et al., 1980). This variation in composition will affect

**Table 10.1** Composition of Some Natural Gases

| Components (mol%) | Alberta (Canada) | BuHasa (Abu Dhabi) | Western Colorado | SW Kansas | Pincher Creek, Alberta, Canada | Typical pipeline, USA and Canada |
|---|---|---|---|---|---|---|
| He | – | – | – | 0.45 | – | – |
| $N_2$ | 3.2 | 0.30 | 26.10 | 14.65 | 0.40 | 20.6 |
| $CO_2$ | 1.7 | 4.48 | 42.66 | – | 6.30 | 0.35 |
| $H_2S$ | 3.3 | 0.02 | – | – | 10.50 | – |
| $C_1$ | 77.1 | 67.48 | 29.98 | 72.89 | 74.69 | 90.33 |
| $C_2$ | 6.6 | 11.54 | 0.55 | 6.27 | 3.26 | 5.19 |
| $C_3$ | 3.1 | 8.20 | 0.28 | 3.74 | 1.25 | 1.40 |
| i-$C_4$ | 0.7 | 1.48 | 0.06 | 0.33 | 0.35 | 0.22 |
| n-$C_4$ | 1.3 | 3.08 | 0.15 | 1.05 | 0.46 | 0.31 |
| i-$C_5$ | 0.5 | 0.93 | 0.05 | 0.20 | 0.21 | 0.07 |
| n-$C_5$ | 0.5 | 1.07 | 0.06 | 0.21 | 0.22 | 0.05 |
| $C_6$ | 0.7 | 0.83 | 0.06 | 0.10 | 0.45 | 0.02 |
| $C_7$+ | 1.3 | 0.59 | 0.05 | 0.11 | 1.91 | |

*Source*: Kirk et al. (1980).

the processing or liquefaction cycle, so it must always be considered in any natural gas process.

Olefins, carbohydrates, and carbon monoxide are seldom found in natural gas. Argon, hydrogen, and mercury are reported to occur very infrequently, but mercury can be especially troublesome when it does occur. If hydrogen sulfide is present, then COS, $CS_2$, $CH_3SH$, and $C_2H_5SH$ may also appear. Water is nearly always present.

Because natural gas is a mixture, special correlations have been developed to estimate the properties of the mixture from the properties of the pure components alone. For instance, Haynes (1977) presented a method for estimating the density of LNG. The Klosek–McKinley method was modified and recommended as the most accurate, simple technique. To use this method, the molar density of the mixture, $\rho_{\text{mix}}$, is determined from

$$\frac{1}{\rho_{\text{mix}}} = \sum_i \frac{x_i}{\rho_i} - \left(k_1 + \frac{(k_2 - k_1)x_{N_2}}{0.0425}\right) x_{CH_4}$$

where $x_i$ is the mole fraction of component $i$, and $\rho_i$ is the molar density of component $i$ from Table 10.2. The mixture's average molecular weight is found from

$$M_{\text{mix}} = \sum x_i M_i$$

with $M_i$ the molecular weight of component $i$ from Table 10.2.

$k_1$ and $k_2$ are functions of temperature, and $M_{\text{mix}}$ as shown in Tables 10.3 and 10.4. The mass density of LNG is then given by

$$\rho_{\text{mix}}(\text{mass density}) = \rho(\text{molar density})\, M_{\text{mix}}$$

The answer is in g/L or kg/m$^3$ (liquid). To convert to lb/cu ft, multiply by 0.0624.

*Example 10.1.* An LNG mixture has the following composition: methane 92 mol%, ethane 5.0 mol%, and propane 3.0 mol%. Find its density at 110 K (−260°F).

| Component | $M_i$ | $x_i$ | $M_ix_i$ | $\rho_i$ | $x_i/\rho_i$ |
|---|---|---|---|---|---|
| Methane | 16.04 | 0.92 | 14.76 | 26.50 | 0.0347 |
| Ethane | 30.07 | 0.05 | 1.50 | 20.97 | 0.0024 |
| Propane | 44.09 | 0.03 | 1.32 | 16.07 | 0.0019 |
| | | | 17.58 | | 0.0390 |

From Tables 10.3 and 10.4, with $T = 110$ K and $M_{mix} = 17.58$, $k_1 = 0.30 \times 10^{-3}$ and $k_2 = 0.48 \times 10^{-3}$. Then,

$$1/\rho_{mix} = 0.0390 - [0.30 \times 10^{-3} + (0.48 - 0.30)(10^{-3})(0.0)/0.0425](0.92) = 0.0387$$
$$\rho_{mix} = 25.82 \text{mol/L}$$
$$= (25.82)(17.58) = 454.0 \text{kg/m}^3$$
$$= (454.0)(0.0624) = 28.33 \text{lb/cuft}$$

**Table 10.2** Densities of Saturated Liquid of the Pure Components (mol/L)

| $T$(K) | $CH_4$ | $C_2H_6$ | $C_3H_8$ | $n$-$C_4H_{10}$ | $i$-$C_4H_{10}$ | $N_2$ |
|---|---|---|---|---|---|---|
| 90 | 28.216 | 21.781 | 16.540 | 13.386 | 13.143 | 26.636 |
| 92 | 28.051 | 21.629 | 16.493 | 13.353 | 13.111 | 26.260 |
| 94 | 27.886 | 21.356 | 16.446 | 13.320 | 13.078 | 25.873 |
| 96 | 27.718 | 21.484 | 16.400 | 13.287 | 13.045 | 25.475 |
| 98 | 27.550 | 21.411 | 16.353 | 13.255 | 13.012 | 25.064 |
| 100 | 27.379 | 21.339 | 16.307 | 13.222 | 12.979 | 24.639 |
| 102 | 27.207 | 21.266 | 16.260 | 13.190 | 12.946 | 24.198 |
| 104 | 27.033 | 21.193 | 16.213 | 13.157 | 12.913 | 23.737 |
| 106 | 26.858 | 21.120 | 16.167 | 13.124 | 12.880 | 23.255 |
| 108 | 26.680 | 21.047 | 16.120 | 13.092 | 12.848 | 22.746 |
| 110 | 26.501 | 20.974 | 16.074 | 13.059 | 12.815 | 22.207 |
| 112 | 26.319 | 20.901 | 16.028 | 13.027 | 12.782 | 21.631 |
| 114 | 26.135 | 20.827 | 15.981 | 12.994 | 12.749 | 21.008 |
| 116 | 25.950 | 20.754 | 15.935 | 12.962 | 12.716 | 20.334 |
| 118 | 25.762 | 20.680 | 15.888 | 12.929 | 12.683 | 19.652 |
| 120 | 25.571 | 20.606 | 15.842 | 12.897 | 12.650 | 18.970 |
| 122 | 25.378 | 20.532 | 15.795 | 12.864 | 12.617 | 18.288 |
| 124 | 25.183 | 20.458 | 15.749 | 12.832 | 12.585 | 17.607 |
| 126 | 24.984 | 20.384 | 15.703 | 12.799 | 12.552 | 16.925 |
| 128 | 24.783 | 20.309 | 15.656 | 12.767 | 12.519 | 16.243 |
| 130 | 24.579 | 20.234 | 15.610 | 12.735 | 12.486 | 15.561 |
| 132 | 24.372 | 20.159 | 15.563 | 12.702 | 12.453 | 14.879 |
| 134 | 24.161 | 20.084 | 15.517 | 12.670 | 12.420 | 14.197 |
| 136 | 23.946 | 20.009 | 15.470 | 12.637 | 12.387 | 13.516 |
| 138 | 23.728 | 19.933 | 15.424 | 12.605 | 12.354 | 12.834 |
| 140 | 23.506 | 19.858 | 15.377 | 12.573 | 12.321 | 12.152 |
| 142 | 23.280 | 19.781 | 15.331 | 12.540 | 12.288 | 11.470 |
| 144 | 23.048 | 19.705 | 15.284 | 12.508 | 12.255 | 10.788 |
| 146 | 22.812 | 19.629 | 15.237 | 12.475 | 12.222 | 10.106 |
| 148 | 22.571 | 19.522 | 15.191 | 12.443 | 12.189 | 9.425 |
| 150 | 22.324 | 19.475 | 15.144 | 12.411 | 12.156 | 8.743 |

*Source*: LNG Information Book (1981).

**Table 10.3** Values of $k_1 \times 10^3$ for LNG Density Correlation[a]

| | Average molecular weight of mixture, $M_{mix}$ | | | | | | | | | | | | | | |
|---|---|---|---|---|---|---|---|---|---|---|---|---|---|---|---|
| $T$(K) | 16 | 17 | 18 | 19 | 20 | 21 | 22 | 23 | 24 | 25 | 26 | 27 | 28 | 29 | 30 |
| 90 | −0.005 | 0.12 | 0.22 | 0.34 | 0.43 | 0.51 | 0.59 | 0.66 | 0.75 | 0.82 | 0.88 | 0.93 | 0.98 | 1.03 | 1.08 |
| 95 | −0.006 | 0.14 | 0.26 | 0.38 | 0.49 | 0.58 | 0.67 | 0.75 | 0.85 | 0.92 | 0.99 | 1.05 | 1.10 | 1.15 | 1.20 |
| 100 | 0.0007 | 0.15 | 0.30 | 0.42 | 0.54 | 0.65 | 0.75 | 0.86 | 0.96 | 1.04 | 1.11 | 1.17 | 1.23 | 1.28 | 1.34 |
| 105 | −0.007 | 0.17 | 0.34 | 0.46 | 0.62 | 0.73 | 0.84 | 0.96 | 1.08 | 1.16 | 1.24 | 1.31 | 1.37 | 1.43 | 1.49 |
| 110 | −0.008 | 0.19 | 0.38 | 0.54 | 0.70 | 0.82 | 0.93 | 1.06 | 1.21 | 1.30 | 1.39 | 1.47 | 1.53 | 1.60 | 1.67 |
| 115 | −0.009 | 0.22 | 0.42 | 0.61 | 0.79 | 0.93 | 1.06 | 1.20 | 1.36 | 1.47 | 1.56 | 1.65 | 1.71 | 1.79 | 1.86 |
| 120 | −0.010 | 0.25 | 0.50 | 0.70 | 0.90 | 1.04 | 1.23 | 1.40 | 1.54 | 1.65 | 1.75 | 1.85 | 1.92 | 2.00 | 2.08 |
| 125 | −0.013 | 0.30 | 0.59 | 0.79 | 1.02 | 1.18 | 1.38 | 1.60 | 1.73 | 1.86 | 1.97 | 2.08 | 2.15 | 2.24 | 2.33 |
| 130 | −0.015 | 0.35 | 0.70 | 0.92 | 1.14 | 1.36 | 1.60 | 1.73 | 1.96 | 2.09 | 2.22 | 2.34 | 2.42 | 2.52 | 2.62 |
| 135 | −0.017 | 0.40 | 0.80 | 1.06 | 1.39 | 1.51 | 1.77 | 1.90 | 2.23 | 2.38 | 2.52 | 2.64 | 2.74 | 2.85 | 2.95 |

[a] Multiply by $10^{-3}$ to obtain value of $k_1$.
*Source*: LNG Information Book (1981).

If LNG is to be produced or the natural gas processing train involves liquefaction, then contaminants such as $H_2O$, $CO_2$, and the mercaptans must be reduced to very low levels to prevent them from freezing and possibly plugging the process lines. Table 10.5 gives allowable impurity levels in the LNG product for baseload plants. These values should be contrasted with that of the feedstock shown in Table 10.1 to appreciate the difficulty of the purification task.

In an LNG operation, these minor components are regarded as impurities to be removed and discarded (purification). In natural gas processing, the same components may be removed as a valuable by-product (separation).

The following section discusses some of the separation and purification techniques common to both natural gas processing and LNG production.

## 2. PURIFICATION

The principal objective of the following discussion is to identify some of the most important industrial scale purification and recovery processes. A *purification process*

**Table 10.4** Values of $k_2 \times 10^3$ for LNG Density Correlation

| | Average molecular weight of mixture, $M_{mix}$ | | | | | | | | | | | | | | |
|---|---|---|---|---|---|---|---|---|---|---|---|---|---|---|---|
| $T$(K) | 16 | 17 | 18 | 19 | 20 | 21 | 22 | 23 | 24 | 25 | 26 | 27 | 28 | 29 | 30 |
| 90 | −0.004 | 0.10 | 0.22 | 0.35 | 0.50 | 0.60 | 0.69 | 0.78 | 0.86 | 0.95 | 1.03 | 1.12 | 1.20 | 1.27 | 1.35 |
| 95 | −0.005 | 0.12 | 0.28 | 0.43 | 0.59 | 0.71 | 0.83 | 0.94 | 1.05 | 1.14 | 1.22 | 1.28 | 1.37 | 1.45 | 1.54 |
| 100 | −0.007 | 0.16 | 0.34 | 0.49 | 0.64 | 0.79 | 0.94 | 1.08 | 1.17 | 1.27 | 1.37 | 1.47 | 1.57 | 1.67 | 1.77 |
| 105 | −0.010 | 0.24 | 0.42 | 0.61 | 0.75 | 0.91 | 1.05 | 1.19 | 1.33 | 1.45 | 1.58 | 1.69 | 1.81 | 1.92 | 2.03 |
| 110 | −0.015 | 0.32 | 0.59 | 0.77 | 0.92 | 1.07 | 1.22 | 1.37 | 1.52 | 1.71 | 1.83 | 1.97 | 2.10 | 2.23 | 2.36 |
| 115 | −0.024 | 0.55 | 0.72 | 0.95 | 1.15 | 1.22 | 1.30 | 1.45 | 1.65 | 2.00 | 2.17 | 2.32 | 2.47 | 2.63 | 2.79 |
| 120 | −0.032 | 0.75 | 0.91 | 1.23 | 1.43 | 1.63 | 1.85 | 2.08 | 2.30 | 2.45 | 2.60 | 2.77 | 2.95 | 3.13 | 3.32 |
| 125 | −0.043 | 1.00 | 1.13 | 1.48 | 1.73 | 1.98 | 2.23 | 2.48 | 2.75 | 2.90 | 3.10 | 3.30 | 3.52 | 3.74 | 3.96 |
| 130 | −0.058 | 1.34 | 1.46 | 1.92 | 2.20 | 2.42 | 2.68 | 3.00 | 3.32 | 3.52 | 3.71 | 3.95 | 4.20 | 4.46 | 4.74 |
| 135 | −0.075 | 1.75 | 2.00 | 2.40 | 2.60 | 3.00 | 3.40 | 3.77 | 3.99 | 4.23 | 4.47 | 4.76 | 5.05 | 5.36 | 5.69 |

*Source*: LNG Information Book (1981).

**Table 10.5** Typical Maximum Allowable Impurity Levels in Natural Gas

| Pipeline gas | | Liquefied natural gas[a] | | |
|---|---|---|---|---|
| Impurity | Maximum allowable level | Impurity | Limit set by[b] | Allowable level in LNG product |
| Water | Varies from 6 to 7 lb/$10^6$ scf in USA to 4 lb/$10^6$ scf in Canada (Requirements are more stringent for pipelines at low temperatures and pressures greater than 1000 psi.) | Water | A | 0.5 ppm |
| Hydrocarbon dew point | Usually + 15°F at 850 psia | | | |
| $H_2S$ | 1/4 grain/100 scf (4 ppm) | $H_2S$ | C | 3.5 mg/$Nm^3$ (or 0.25 grains/100 scf) |
| | | $CO_2$ | B | 50–125 ppm |
| | | Organic sulfur (mercaptans, disulfides, thioethers, $CS_2$, COS,[c] etc.) | C | 150 mg total sulfur per $Nm^3$ or regasified gas (0.5 grains/100 scf) |
| | | Nitrogen | C | 0.5–1.5 mol% |
| Total sulfur | 1 grain/100 scf | Aromatics | A or B | 1.10 ppm v/v |

1 ppm = 1 $cm^3/m^3$; $Nm^3$ = a cubic meter at normal (STP) conditions.

[a] Allowable level of impurities in LNG product for baseload LNG plants (Chiu, 1978).

[b] A, quantity allowed to accumulate at the cold end of plant above the solubility limit without restricting production; B, solubility limit; C, product specifications.

[c] Hydrolysis of COS, $COS + H_2O \rightarrow H_2S + CO_2$, in gas will add to the concentration of $H_2S$.

*Source*: Chiu (1978).

is defined as one where the material removed from the gas stream is discarded (e.g., water), and a *recovery process* is one where the material removed has economic value (e.g., helium or ethane).

Impurities must be removed not only to obtain a desired product purity but also, particularly in cryogenic processes, to prevent the deposition of solids that would block passages and soon make the process inoperative, to remove an explosion hazard, to prevent the blanketing of heat transfer surfaces by noncondensible gases, and sometimes to prevent corrosion.

The most common impurities that must be removed are water vapor, hydrogen sulfide, and occasionally nitrogen or carbon dioxide. The water vapor is removed to prevent hydrate formation, the hydrogen sulfide because it is corrosive and toxic, and the removal of carbon dioxide and nitrogen is occasionally required to raise the heating value of the gas to contract specifications. It is sometimes necessary to consider the removal of carbonyl sulfide (COS), carbon disulfide ($CS_2$), and mercaptans because these materials appear in small quantities in some gases. The removal of trace amounts of mercury is also required in a few instances.

Since the nature of the removal process depends on the concentration and type of impurity, separation processes are presented here under the heading of a particular impurity. In general, small concentrations are expressed in ppm (molar parts per million, the same as ppm by volume in the case of gases at low pressure, where they approach ideality) and larger concentrations as percent, also molar. It is understood that all the source gases to be considered will contain varying amounts of water vapor up to a maximum corresponding to saturation at the given pressure and temperature, but composition is invariably expressed on a dry basis.

The methods used to remove specific impurities could be divided into just two classes: external and internal. External methods are processes that are carried out at near-ambient temperatures or ahead of the "cold box," and internal methods are the purification steps that take place after the feed gas has entered the low-temperature system.

Another very arbitrary but sometimes useful classification is on the basis of the volatility of the impurity. For example, one might consider two classes: (1) those used to remove high-boiling impurities such as many hydrocarbons, organic sulfur compounds such as carbon disulfide, carbonyl sulfide, mercaptans, water, and sulfur dioxide and (2) those that remove low-boiling impurities like nitrogen, carbon monoxide, argon, and methane, with perhaps an intermediate class for methods that remove carbon dioxide, hydrogen sulfide, and ammonia.

A more specific and perhaps more useful classification is the following:

1. condensation to a liquid by increase of pressure at constant temperature;
2. condensation to liquid or solid by cooling at constant pressure;
3. absorption by a liquid;
    a. purely physical;
    b. with chemical reaction;
4. adsorption by a solid;
    a. physical;
    b. with chemical reaction;
5. chemical reaction;
6. permeation through a membrane;
7. mechanical separation.

The absorption by a liquid, adsorption by a solid, and chemical reaction methods are most commonly used in large-scale natural gas processing and LNG production. Examples of each of these three methods, applied to specific impurities, are discussed below.

## 2.1. Water Removal—Dehydration

The quantities and types of impurities present in most domestic LNG peakshaving plants, while variable, do not vary greatly. Natural gas for peakshaving LNG facilities typically contains 48–112 kg $H_2O/10^6\,m^3$ gas (3–7 lb $H_2O/10^6$ scf). In most instances, pretreatment in these plants involves primarily the removal of carbon dioxide and water vapor from natural gas that has previously been processed for some degree of hydrocarbon recovery.

Natural feed to baseload plants can contain as much as 1600 kg $H_2O/10^6\,m^3$ gas (100 lb $H_2O/10^6$ scf). This gas must be dehydrated to a minimum practical amount of water to prevent plugging by hydrates in the downstream liquefaction system. In practice, dew points ranging from 200 to 172 K (−100 to −150°F) have proved satisfactory for LNG plants operating at 1380–4830 kPA (200–700 psig).

The removal of water vapor is accomplished either by absorption in a liquid or by adsorption on a solid. The absorption processes generally use diethylene glycol or triethylene glycol as the drying liquid in a countercurrent absorber (see Fig. 10.1). In this system, wet natural gas enters the bottom of the absorber, flows countercurrent to the glycol solution, and exits at the top as a dry gas. The lean (very low water content) glycol enters the top of the absorber, removes the water from the natural gas, and exits at the bottom as a rich (high water content) solution. The glycol is filtered to remove any solids and then flows through a surge tank heat exchanger and into the stripping column, where the water is removed by distillation. The lean glycol is recirculated, and the water is discarded (Campbell, 1975).

Adsorption on a solid is more commonly used to produce very low water content in the natural gas. Although dehydration may be the principal objective, the acid gases ($CO_2$ and $H_2S$) can also be removed by solid adsorption. Hence, acid gas removal is discussed briefly here along with the major subject of dehydration.

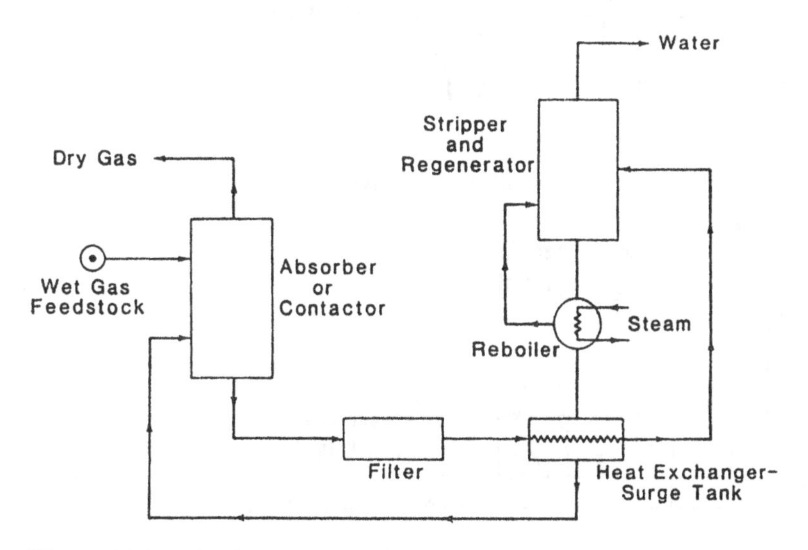


**Figure 10.1** Glycol water absorption process.

Of the commonly available desiccants, molecular sieves show the greatest affinity for water due to the nature of their structure. Also, they are the only commercially proven adsorbent that can effectively remove both acid gases ($H_2S$ and $CO_2$) and water at or near ambient conditions. Using molecular sieves for complete prepurification in LNG production was first demonstrated in the 28,320 $m^3$ ($1 \times 10^6$ scf) peakshaving plant built by Wisconsin Natural Gas at Oak Creek, Wisconsin, in 1964 (Naeve, 1968). Most of the peakshaving plants built in the United States since 1965 use molecular sieves for both water and carbon dioxide removal. Over 75% of all cryogenic hydrocarbon recovery plants rely on molecular sieves for dehydration.

Although sieves have gained acceptance throughout the liquefied natural gas industry, they have disadvantages, which include

1. high capital costs, limitation of maximum economical throughput;
2. danger of adsorbent contamination and undesirable coadsorption;
3. need for source of clean, dry reactivation gas;
4. possibility of adsorbent dusting.

A flow sheet of a typical molecular sieve dehydrator appears in Fig. 10.2. After physical separation of entrained solids and liquids, the inlet gas is simply passed through a tower containing the solid adsorbent. When the molecular sieve approaches saturation, the inlet stream is switched to a second tower, while the adsorbent in the first is regenerated by flowing heated, dry gas counter to the direction of the stream that was being dried. After leaving the tower, the warm, moist regeneration gas is cooled, and much of the water is condensed, separated, and removed from the system. The regeneration gas is then either mixed with the wet inlet gas to the adsorbing tower (closed-cycle operation) or returned to a lower pressure distribution line (open-cycle operation). Once dry, the regenerated tower must be cooled by a flow of cool, dry gas before being placed back in service.

Because of the great affinity of molecular sieves for water, regeneration temperatures in the range of 550–600°F are required for proper regeneration. Proper

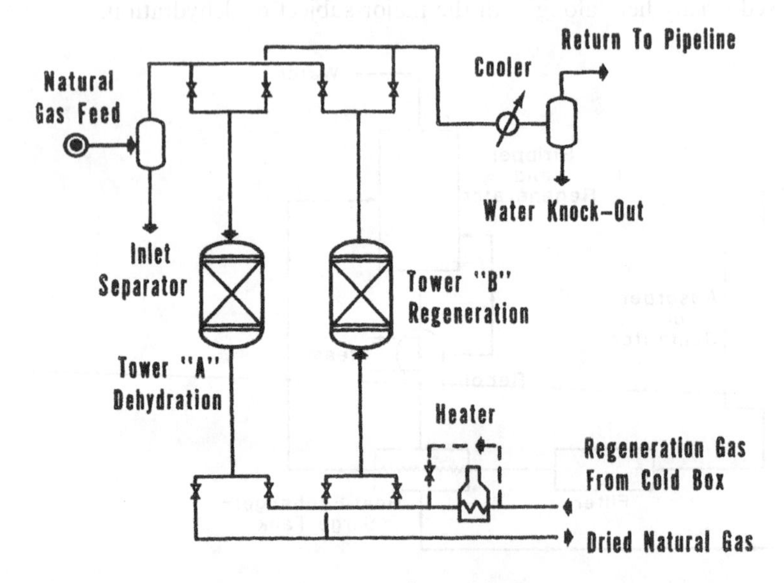


**Figure 10.2** Molecular sieve natural gas dehydrator.

reactivation also requires a source of clean, dry gas, which is generally available with domestic peakshaving plants because such plants seldom liquefy all the available pipeline gas. A quantity of purified feed gas equal to 30–50% of the initial feed stream is used for reactivation, then either returned to the pipeline or used as on-site fuel gas.

Inlet gas streams containing unusually high concentrations of water may be partially dried by using glycol dehydration (described above) followed by final dry desiccant dehydration. In most instances, however, a single molecular sieve adsorption system can be designed to handle the entire water load.

Dehydration to dewpoints of 200 K (–100–105°F) is usually accomplished through the use of molecular sieve synthetic zeolite adsorbents (Schools, 1900). A dew point of 200 K (–100°F) corresponds to approximately 1 ppm water. Other dry desiccants such as activated alumina, activated bauxite, and silica gel have been used in some instances.

It is interesting to calculate the amount of water and carbon dioxide that must be removed from a natural gas stream to prevent freeze plugging of the liquefier components at low temperatures.

A gas saturated with water vapor at 101.3 kPa (1 atm) absolute and 311 K (100°F) contains 6.45% water, which is the most one would normally encounter. At 273 K and 101.3 kPa (32°F and 1 atm), the water content is reduced to 0.60%. The water content of a saturated gas is inversely proportional to pressure, assuming an ideal gas. Thus, a gas saturated at 300 K and 20.26 MPa (80°F and 200 atm) would contain only 0.0172% water. Due to the departure from ideality, a saturated gas at high pressure will contain considerably more than the ideal concentration. Some data on the humidity of compressed air over the ranges 367–228 K (200 to –50°F) and 101.3 kPa to 101.3 MPa (1–1000 atm) are available. Table 10.6 presents values of the enhancement factor $E$ (defined as the ratio of the actual molar concentration to the ideal molar concentration or $P{\cdot}y/P_s$) that were calculated from these data. $P$ is the total pressure, $y$ the mole fraction of water vapor in the gas, and $P_s$ the vapor pressure of water at the temperature of interest. Thus, at high pressure, the assumption of an ideal gas can be greatly in error, especially at low temperatures. Unfortunately, little or no data are available on the system methane–water.

Many of the contamination limits for LNG plants are based on the considerable experience obtained with air plants. Accordingly, it is interesting to note that in a $9.07 \times 10^4$ kg/day (100 ton/day) oxygen plant, assuming 90% recovery, about 196 kg (432 lb) of carbon dioxide would be carried into the cold box every day if none of it were removed ahead of the cold box. Even at only 1 ppm carbon dioxide

**Table 10.6** Enhancement Factors for the Water–Air System

| Temperature | | Pressure | | |
|---|---|---|---|---|
| K | °F | MPa | atm | Enhancement factor |
| 297 | 75 | 2.026 | 20 | 1.02 |
| 297 | 75 | 20.26 | 200 | 1.65 |
| 256 | 0 | 2.026 | 20 | 1.07 |
| 256 | 0 | 20.26 | 200 | 2.30 |
| 228 | −50 | 2.026 | 20 | 1.26 |
| 228 | −50 | 20.26 | 200 | 4.12 |

in the air after a purification, the amount carried in is 0.62 kg (1.37 lb) per day, which would cause serious plugging in a short time if it were deposited as a solid in some restricted space. Air with a dew point of 200 K at 689 kPa (–100°F at 100 psia) entering an oxygen plant of the same size would carry into the cold box 60 g (0.13 lb) of water per day. Most of these solids would be deposited in the exchangers, and if the exchangers were of the reversing or regenerative type the deposits would be reevaporated into the returning cold streams. If the temperature leaving an exchanger were 100 K, the vapor pressure of water would be so low that no appreciable amount of water as vapor would be carried into the column even in a year's time. In the case of carbon dioxide at 100 K and 689 kPa (100 psia), the equilibrium amount in the vapor would be about 0.033 ppm, and for the 100 ton oxygen plant it would amount to about 21 g (0.046 lb) per day carried into the columns. Some water and carbon dioxide probably passes through the exchangers as snow, but much is filtered out and does not deposit in the column. In any case, it eventually is necessary to warm up the plant for defrosting, but with good purification systems this should be necessary only about once a year.

*Example 10.2.* An LNG plant dehydrates the natural gas feed to a dew point of 203 K at 689 kPa (–94.6°F at 100 psia). The plant processes $9.07 \times 10^4$ kg/day (100 tons/day) of methane. How much water is carried by the dehydrated gas into the cold box?

The vapor pressure of ice at 203 K (–70°C) is 0.2586 Pa (0.00194 mmHg). Assuming ideality, the mole fraction of water in the gas is

$$0.2586\,\text{Pa}/689\,\text{kPa} = 3.75 \times 10^{-7}$$

For a natural gas plant processing $9.07 \times 10^4$ kg/day (100 US tons/day), the entering moles of methane (MW = 16) are

$$(9.07 \times 10^{-4}\,\text{kg/day})/16 = 5.67 \times 10^{-3}\,\text{kg mol/day}$$

The number of moles of water remaining after dehydration is

$$\left(5.67 \times 10^{-3}\,\frac{\text{kg mol}}{\text{day}}\right)(3.75 \times 10^{-7}) = 2.13 \times 10^{-3}\,\text{kg/day}$$

The mass of water per day carried into the plant is then

$$(2.13 \times 10^{-3}\,\text{kg mol})\left(\frac{18\,\text{kg water}}{\text{kg mol}}\right) = 0.038\,\text{kg water}$$

$$= 38\,\text{g water}$$

$$= 0.08\,\text{lb}_\text{m}\,\text{water}$$

## 2.2. Removal of Carbon Dioxide and Hydrogen Sulfide

Inlet natural gas typically contains 0.5–2.0 vol % $CO_2$, but the limits on the permissible concentrations of carbon dioxide entering the liquefaction unit are not as well defined.

In processing natural gas at low temperatures, a much higher concentration of carbon dioxide can be tolerated than in the case of air separation plants, because solid $CO_2$ is sufficiently soluble in hydrocarbons that solid deposits are minimized. However, experience in the operation of air separation plants and other cryogenic processing plants has shown that local freeze-out of impurities such as $CO_2$ can occur at concentrations well below the solubility limit. This is due to factors such as transient shifts in temperatures and pressures, local cold spots on equipment surfaces, unknown composition effects, and limitations in methods of analysis (Schoofs, 1966). For this reason, the $CO_2$ content of the feed gas subject to the minimum operating temperature is usually kept below 50 ppm.

Hydrogen sulfide is removed for two reasons: to eliminate an undesirable component in the gas stream and (if economically justified) to recover the $H_2S$ as elemental sulfur. The processes that remove $H_2S$ generally also remove $CO_2$, and thus these two operations are considered together here.

The $H_2S$ and $CO_2$ removal operations may be roughly grouped into three categories:

1. processes using a reversible chemical reaction;
2. processes using physical absorption;
3. processes employing fixed beds of solids.

The processes involving chemical reaction may be divided into two subgroups: amine reactions and carbonate reactions. The amine reactions use monoethanolamine (MEA; 2-aminoethanol; $NH_2C_2H_4OH$), diethanolamine [DEA; 2,2′-dihydroxyamine; 2,2′-iminodiethanol; $NH(C_2H_4OH)_2$], and diglycolamine [$O(H_2CONH_2)_2$]. The amine and similar purification processes are based on a chemical reaction between the active ingredient of the solution and the impurity to be removed. Each amine has at least one hydroxyl group, which reduces the vapor pressure and increases water solubility, and one amino group, which provides the alkalinity to cause the absorption of the acid gas.

Monoethanolamine is the most commonly used amine for purification because of its low price, high reactivity, excellent stability, and ease of regeneration (Kohl and Reisenfeld, 1959). Some disadvantages of the use of MEA are its low selectivity for $H_2S$ or $CO_2$, irreversible reactions with carbonyl sulfide (COS) and carbon disulfide ($CS_2$), high utility cost, low effective mercaptan removal, and greater vaporization losses. (A mercaptan is the chemical combination of a hydrocarbon and a sulfur acid, R–SH.) Diethanolamine has been used to remove COS and $CS_2$. It has lower circulation rates, vaporization losses, and heat requirements than MEA. Diglycolamine has the ability to remove carbonyl sulfide, carbon disulfide, and to some extent mercaptans. Diglycolamine also partially dehydrates the gas stream with lower utility costs and circulation rates and a lower freezing point than MEA.

The general reactions are

$$2R - NH_2 + H_2S \underset{116^\circ C}{\overset{38^\circ C}{\rightleftharpoons}} (R - NH_2)_2 \cdot H_2S \text{ (water-soluble salt)}$$

$$2R - NH_2 + CO_2 + H_2O \underset{159^\circ C}{\overset{49^\circ C}{\rightleftharpoons}} (R - NH_2)_2 \cdot H_2CO_3 \text{ (water-soluble salt)}$$

The purification and regeneration processes are based on the fact that at low temperatures the salt formation is favored, while at high temperatures the salt decomposes. Generally, monoethanolamine is used, but if significant amounts of COS are present, diethanolamine must be used because COS forms an unregenerable complex with monoethanolamine (Kirk et al., 1980).

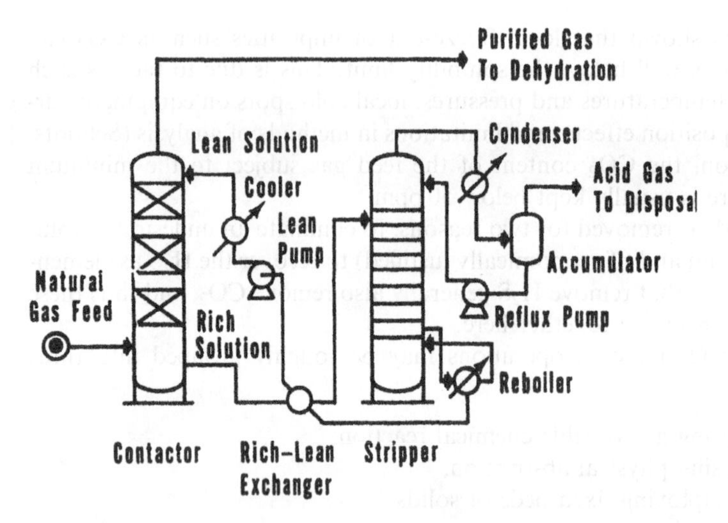


**Figure 10.3** Amine system for natural gas $CO_2$ removal.

The carbonate reactions involve sodium or potassium carbonate in the following reversible reactions:

$$K_2CO_3 + CO_2 + H_2O \rightleftharpoons 2KHCO_3$$
$$K_2CO_3 + H_2S \rightleftharpoons KHCO_3 + KHS$$
$$COS + H_2O \rightleftharpoons CO_2 + H_2S$$

The amine process and the molecular sieve process are most widely used for $CO_2$ removal. The amine process involves absorption of "acid gases" ($CO_2$ and $H_2S$ simultaneously) by a lean aqueous organic amine solution, usually monoethanolamine (MEA).

Amine solutions, in general, are more selective to $H_2S$ absorption than to $CO_2$ absorption and can easily yield a treated gas containing less than $5 \times 10^{-5}\,kg/m^3$ (0.25 grains $H_2S$/100 scf). With sufficient amine recirculation rates, $CO_2$ in the treated gas can be reduced to less than 25 ppm. Several variations are possible if $CO_2$ removal is critical. These include pretreatment of the natural gas with a hot carbonate solution if large quantities of $CO_2$ are present in the feed gas and secondary treatment of partially purified gas with caustic to remove the last traces of $CO_2$.

A simplified flow diagram of an amine system appears in Figure 10.3. Feed gas is passed counterflow to the liquid in a tray or packed tower. The absorption process is a chemical reaction between the acid gases, water, and amine in which amine carbonates, bicarbonates, and hydrosulfides are formed. The rich amine solution is stripped of the acid gases in a distillation column by the addition of heat, which reverses the chemical reaction.

## 2.3. Combined Water and Carbon Dioxide Removal

The low dew points required for natural gas liquefaction and related low-temperature natural gas processing can easily be obtained by adsorption on a suitable desiccant such as alumina, silica gel, or molecular sieve. Helium recovery plants built in the early 1960s provided the first large-scale use of fixed bed adsorbers. In addition to their ability to achieve very low dew points, fixed bed adsorbers have the following advantages compared to glycol and similar adsorption approaches:

1. system is dry and clean;
2. no inventory of liquid chemicals required;
3. simpler operation with less equipment, well-suited to automatic or semiautomatic operations;
4. generally lower operating costs;
5. no carryover of chemicals or solvents into system.

The molecular sieve process for $CO_2$ removal has been used extensively for peakshaving LNG production. Unlike the amine process described earlier, which requires subsequent dehydration of the natural gas before liquefaction, the molecular sieve process may simultaneously remove both water and carbon dioxide in a single unit. The process consists of essentially the same equipment and involves about the same operation as ordinary dry desiccant dehydration systems. A simplified flow diagram of a molecular sieve process appears in Fig. 10.4.

Incoming gas flows downward through one tower filled with molecular sieve. Water is removed in the upper section of the adsorbent bed, and carbon dioxide is removed in the lower section. Effluent natural gas typically contains less than 20 ppm by volume $CO_2$ and less than 1 ppm by volume $H_2O$. The adsorber towers are designed for relatively short cycle times of 2 hr or less, with automatic switching valves for unattended operation.

Depending on such factors as the particular cycle and the location of the liquefaction plants, regeneration of the second tower may be affected at essentially the same pressure as adsorption (e.g., pipeline pressure) or at some lower pressure (e.g., the distribution system pressure). Regeneration gas passes through a heater and into the regenerating tower in an upflow direction to remove (desorb) both

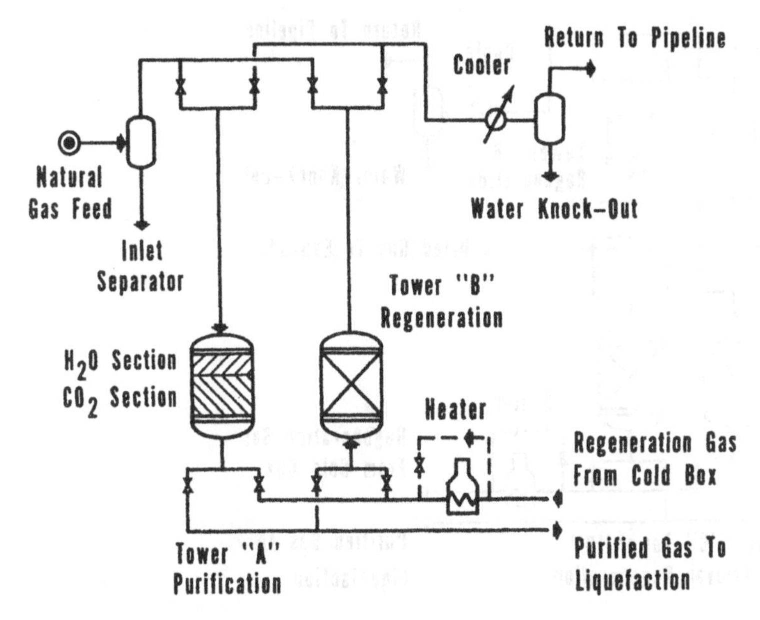


**Figure 10.4** Molecular sieve system for combined natural gas dehydration and $CO_2$ removal.

$CO_2$ and $H_2O$. Exit gas passes through a cooler–knockout combination and then back into the pipeline. After sufficient heating and desorption, the inlet gas heater is bypassed and the bed is cooled.

### 2.4. Two-Stage Adsorption and Purification

A dual adsorption and prepurification process is possible in plants that use the expander liquefaction cycles, as these cycles require more dry gas than is used for liquefaction. With these systems, only a small portion of the initial feed gas stream reaches the liquid methane temperature level and need be treated for $CO_2$ removal. One set of adsorbers containing one type of molecular sieve and operating at about 278 K (40°F) treats all the gas entering the plant for removal of moisture and hydrogen sulfide. That portion of feed gas that is to be liquefied is further treated for $CO_2$ removal in a second set of vessels containing another type of molecular sieve. Thus, a four-tower molecular sieve system is used—two towers (dryers) for removing water from the total inlet natural gas stream and two towers (purifiers) for removing carbon dioxide from the portion of gas being liquefied. Figure 10.5 shows such a four-tower process.

The four-tower, two-stage system has two advantages.

1. The first-stage adsorbers can be efficiently designed for water and $H_2S$ removal, taking advantage of the high adsorption capacity of the adsorbent for these impurities.
2. Second-stage adsorbers can be designed specifically for the higher concentrations and lower adsorptive capacity of $CO_2$. The adsorber thus handles a smaller quantity of feed gas, thereby reducing overall system requirements.

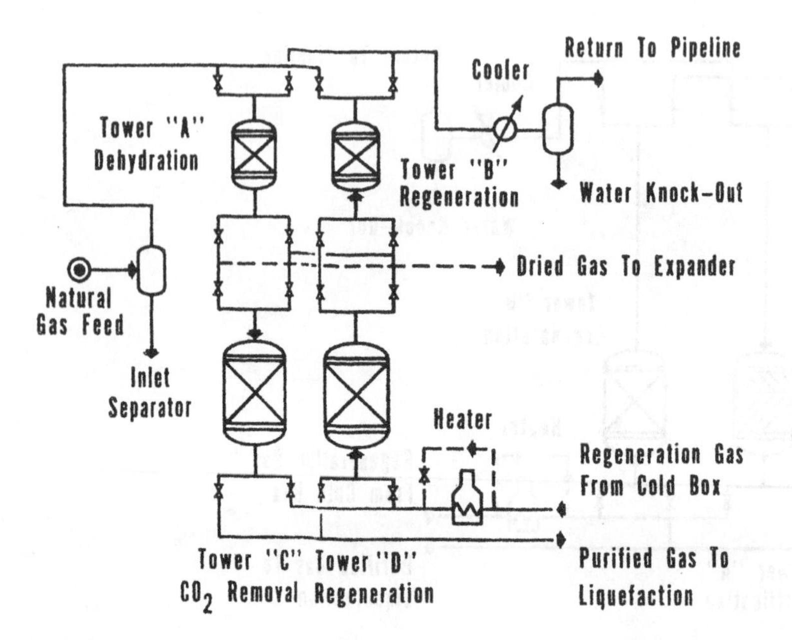


**Figure 10.5** Four-tower molecular sieve system for natural gas dehydration and $CO_2$ removal.

Equipment and utility costs can be substantially reduced, and efficiency is increased, because the dryers and purifiers can be designed for unequal time cycles. Long cycles of 12–24 hr are practical in gas drying because the molecular sieve material has a high affinity for water and because water content normally is under $1.12 \times 10^{-4}$ kg/m$^3$ (7 lb per million cubic feet). The purifier cycle must be shorter because a gas containing 0.5% $CO_2$ has 9.3 g/m$^3$ (580 lb $CO_2/10^6$ scf). Unequal time cycles are advantageous, then, because each stage can be tailored to the most efficient operation. Also, the unequal cycles help minimize problems associated with sulfur removal.

Regeneration of the first-stage adsorber is accomplished using heated reactivation gas from the second-stage adsorbers. Since molecular sieves adsorb to some degree all of the organic sulfur compounds as well as $H_2S$, care must be taken in the disposal of regeneration gas if any of these compounds are present in the feed gas. Simple dilution of organic sulfur compounds (which are normally present in only trace quantities) in the relatively large amounts of regeneration gas is usually sufficient for most disposal applications. However, since the concentration of impurities being desorbed tends to reach a maximum at some point during the desorption cycle, peak concentrations of these impurities can be returned to the distribution system. Proper sequencing of the purge cycle can help overcome this problem.

Caution must be exercised in the manner in which the regeneration gas stream is reinjected into the transmission or distribution line. First, large quantities of regeneration gas must be disposed of. This can be done only if the main receiving the regeneration gas has a favorable flow of its own. Second, the regeneration gas does not sweep out the adsorbed $CO_2$ uniformly during the regeneration part of the cycle, resulting in periods of time of significant duration when gas of poor quality is being returned to the mains. Third, $H_2S$ and odorant mercaptans are also removed in the molecular sieve vessels, and these come off during regeneration in peaks that can cause high odor levels in the mains. The regeneration gas may be used as fuel for a boiler or prime mover to avoid these problems.

In the majority of cases of gas purification, the impurities removed have no value and are simply discarded. In some cases, they may be valuable by-products. Carbon dioxide, when present in relatively high concentration, may be used to synthesize urea or condensed to a liquid or solid and sold as such. Hydrogen sulfide and other sulfur compounds may be converted to sulfur. This not only produces a useful chemical but at the same time avoids pollution of the atmosphere. Offgases high in hydrocarbons frequently are used as fuel in the plant where they originated.

## 2.5. Mercury Removal

Mercury concentrations ranging from 0.001 to 180 mg/m$^3$ exist in all natural gas streams (Nm$^3$ = cubic meter at normal STP conditions) (Phannensteil et al., 1975, 1976). Mercury removal by adsorption and oxidation and by cooling, condensing, and phase separation has been ruled out by developmental work. The use of sulfur-impregnated activated carbon seems promising. Laboratory tests indicate that 15% mercury adsorptive capacity (weight of mercury per weight of carbon) is possible with this method. Since most of the mercury in natural gas occurs in the elemental form and no problems in natural gas processing plants have been caused by the presence of either organic or inorganic mercury compounds, removal methods focus on the removal of elemental mercury rather than mercury-containing compounds (Chiu, 1978).

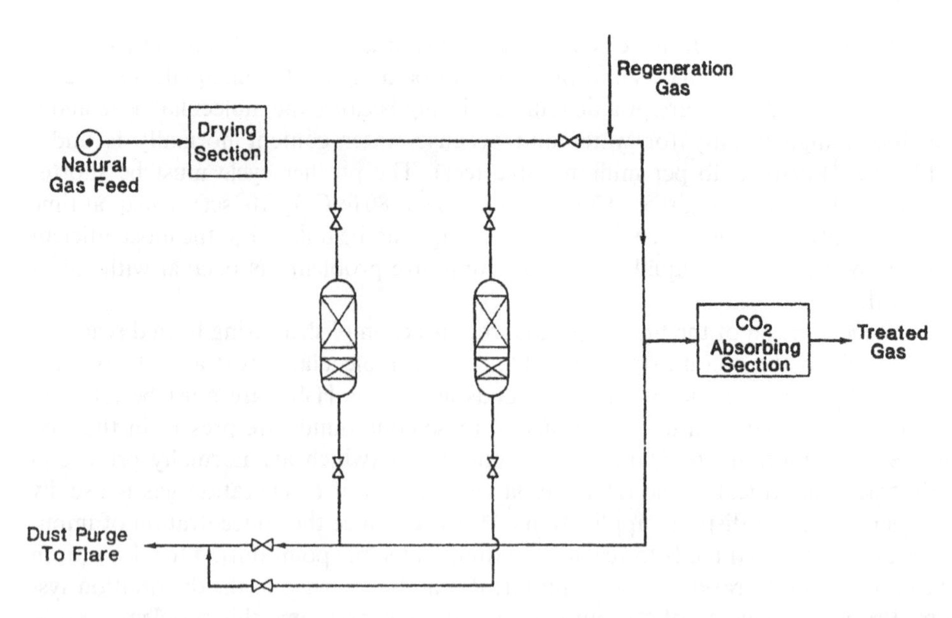


**Figure 10.6** Mercury removal.

One liquefaction facility was shut down because of mercury-related corrosion. Laboratory tests and operating experience indicate that the simultaneous presence of water, foreign ions, and mercury is required for significant aluminum corrosion. A type of molecular sieve for mercury removal from natural gas has been developed by one manufacturer. Mercury removal using activated carbon or metal salt filters at the inlet to an LNG plant has also been considered (Khenat and Hasni, 1977).

One commercial process started up in 1982 by the Institut Francais du Petrole uses a train of parallel catalytic reactors placed between the entering natural gas drying section and the exiting $CO_2$-absorbing section (see Fig. 10.6). The available pressure drop determines the number of reactors required, although normally one reactor will suffice (Khenat and Hasni, 1977). Mercury removal in this process is based on the high reactivity of mercury with sulfur and its compounds; less than $10\,\text{ng/m}^3$ is obtained with a high space velocity. The catalyst lifetime is estimated to be 5 years, and the presence of condensable compounds, such as $C_{5+}$ hydrocarbons and water, does not affect mercury removal.

## 3. HYDROCARBON RECOVERY

### 3.1. Liquefied Petroleum Gas Recovery

The object of liquefied petroleum gas (LPG) recovery is to separate valuable components such as propane, butane, and natural gasoline from natural gas. If the lean gas specifications allow $CO_2$, this process may be used when a high level of $CO_2$ (greater than 2%) is present. A schematic is shown in Fig. 10.7.

Associated or feedstock gas is chilled once it has been compressed and has undergone glycol dehydration. It then flows into a de-ethanizing absorber/stripper where it comes in contact with a presaturated refrigerated lean oil. To comply with the propane and ethane content specifications, the rich oil is partially stripped in the

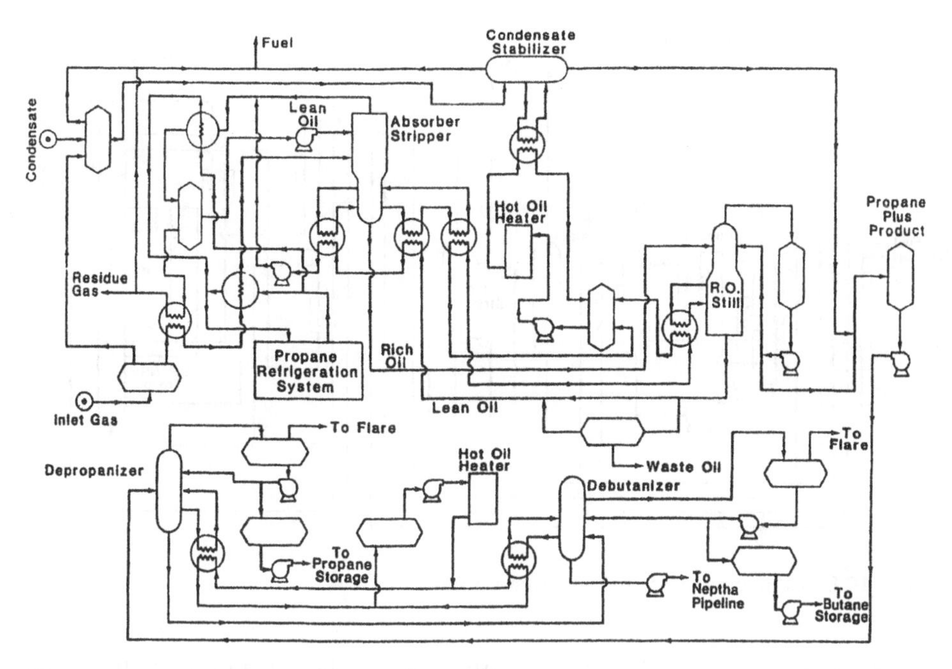


**Figure 10.7** LPG recovery.

lower half of the column. This rich oil is then sent to the rich oil (R.O.) still, where the overhead product contains propane, butane, and natural gasoline, and the lean oil comes off for recycle as a bottom product. Two-column fractionation of the overhead product yields propane overhead in the first column and butane overhead and natural gasoline as a bottom product in the second column. An auxiliary propane cycle provides the necessary refrigeration for this process, using air or water for cooling and condensation.

Assuming a negligible vapor pressure at the top tray of the absorber/stripper, the composition of the lean oil may be fixed by limiting the maximum rich oil still base temperature to 561 K. Typically, 90% of the propane and nearly 100% of the butane and heavier hydrocarbons can be recovered economically with this process. Equilibrium losses of lean oil in the absorber/stripper are made up by recovery of heavier components (heptane plus) from the feed gas.

## 3.2. Ethane Recovery and Nitrogen Rejection

The process of enhanced oil recovery (EOR) uses nitrogen or flue gas injection to increase hydrocarbon yields. The process shown in Fig. 10.8 is used to recover the hydrocarbons by separating them from the injected gases. This process is designed to handle feeds with low to high nitrogen content while ethane recovery remains fairly constant at 75%. The feed is pretreated by drying and reducing the $CO_2$ level to 5 ppm by volume.

An expander-driven compressor raises the feed gas pressure from 6.3 to 7.6 MPa, and the high-pressure (HP) stripper removes condensed liquids. The exhaust pressure of the expander is 2.8 Mpa. At the base of the low-pressure (LP) stripper, two low-nitrogen streams are produced, and these are demethanized to produce LPG. The remaining $N_2$–$CH_4$ mixture is cooled below 189 K and enters the

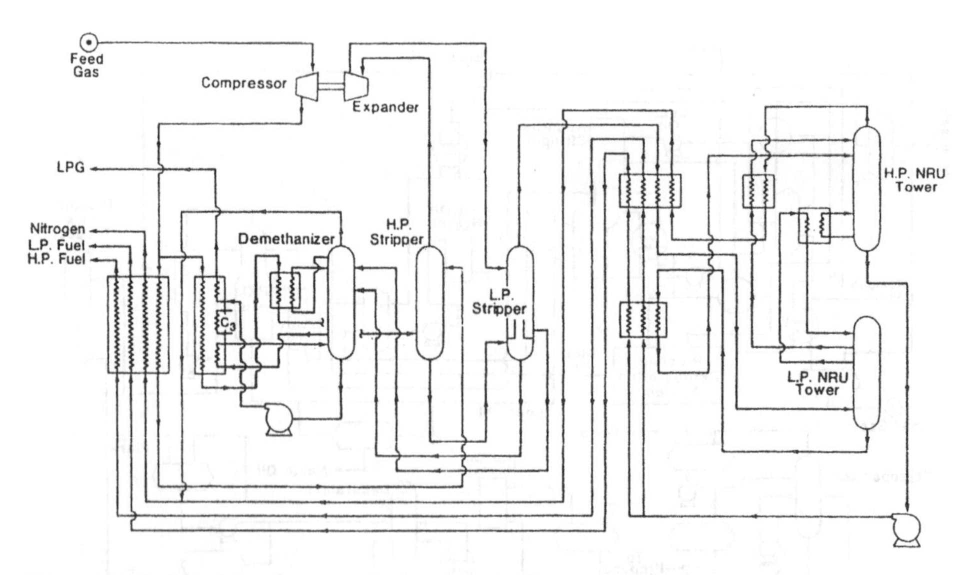


**Figure 10.8** Ethane recovery and nitrogen rejection.

nitrogen rejection unit (NRU). Once the stream enters the base of the LP tower, it is partially condensed, and two liquid streams are produced to feed the LP tower. The common reboiler/condenser links the towers, and inert components such as hydrogen or helium are purged from the top of the HP tower. The nitrogen product is warmed against the feed, and the feed temperatures are controlled and varied by revaporizing the bottoms liquid methane stream at different pressures as the nitrogen content changes.

### 3.3. Use of Natural Gas Liquids Turboexpander in Hydrocarbon Recovery

A process that uses a natural gas liquids (NGL) turboexpander may be used with hydrocarbon gases that contain no lube oil, no sulfur compounds, limited $CO_2$, and no foaming agents. Molecular sieves are used to dehydrate the feed gas to less than 1 ppm water. The operational setup of this process varies with respect to optimum cryogenic temperature, expander, and heat exchanger arrangements depending on the throughput, pressure, and composition of the feed and the desired recovery of ethane and propane. The pressure of the feed gas typically ranges from 0.52 to 20.8 MPa (60–3000 psig), with precompression up to 7.0 MPa (7000 psig) necessary. A schematic of such a process is shown in Fig. 10.9.

The demethanizer reboiler and side reboiler are heated with part of the warm feed, which is then cooled (with external refrigeration if necessary, depending on the hydrocarbon content of the feed gas). This portion of the feed joins the remainder of the feed, which has been cooled in a heat exchanger. Condensed liquids from the separator are fed to the demethanizer, and the vapor phase is fed to an expander. The feed pressure is reduced nearly isentropically to the demethanizer pressure and temperature, between 172.2 and 200 K (–150 and –100°F), while delivering shaft work to the compressor for partial recompression of the sales gas. Final recompression, often to 7.0 MPa (1000 psig), is carried out using turbine- or engine-driven compressors.

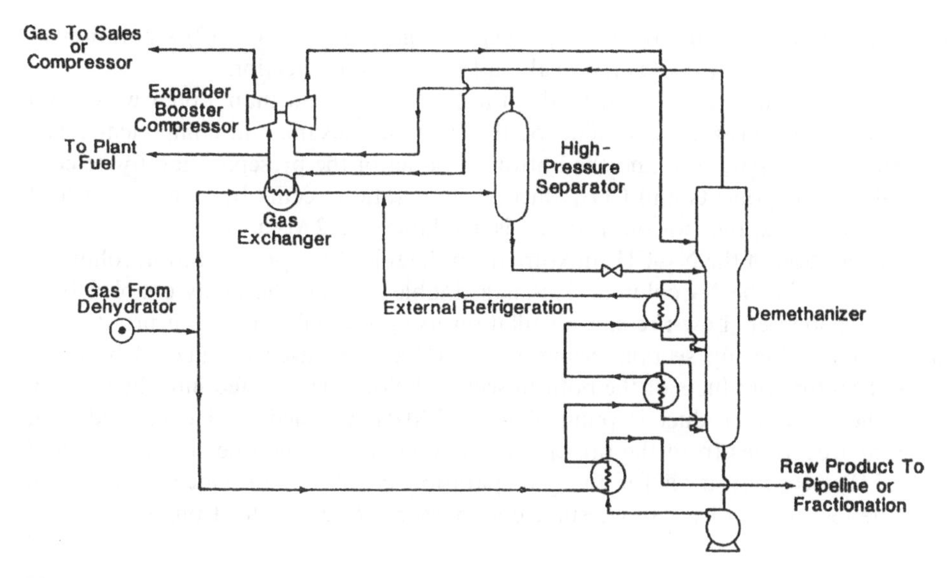


**Figure 10.9** Natural gas liquids (NGL) turboexpander.

The recovery of ethane may be as high as 98%, with considerably higher recovery for propane. Heavier hydrocarbons are essentially completely removed. If desired, ethane may be rejected into sales gas by adding a partial condenser and trays above the demethanized feed tray. Over 200 plants currently use this process.

## 4. CRYOGENIC UPGRADING OF NATURAL GAS

### 4.1. Nitrogen Removal

The process for nitrogen removal shown in Fig. 10.10 produces sales gas fractions and very pure (99.5%) nitrogen from natural gas ranging from 5 to 40 vol% nitrogen.

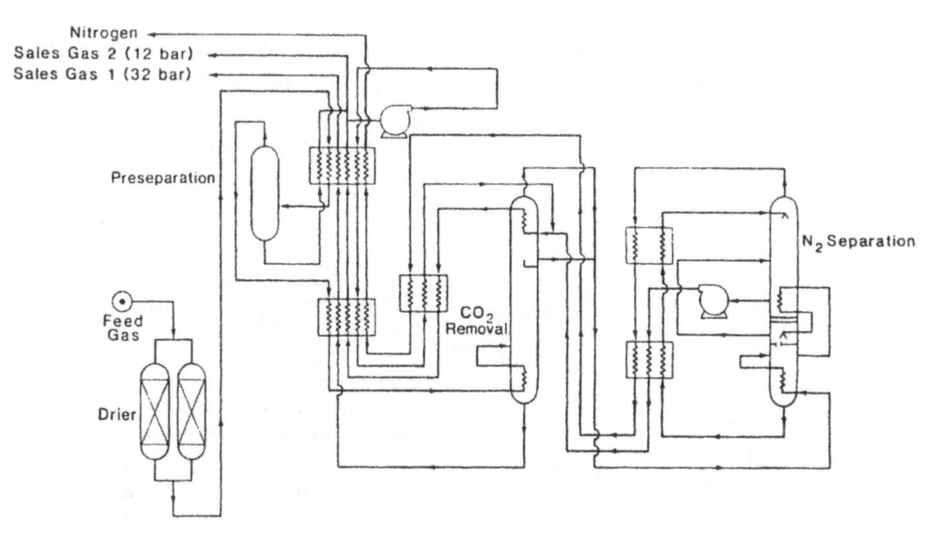


**Figure 10.10** Nitrogen removal.

There are four major treatment steps: (1) feed gas dehydration, (2) separation of heavy hydrocarbons, (3) $CO_2$ removal, and (4) nitrogen rejection.

Activated alumina is used to dry the feed gas to less than 1 ppm water and heavy hydrocarbons are separated, totally removing hexane and components. The next step is preseparation, and the bottom product of the preseparation tower contains all $C_2$ components and $CO_2$. Once revaporization occurs by transferring heat from the feed gas, this bottom product is available at 3.2 MPa.

Separation of the $N_2$–$CH_4$ mixture from the top of the preseparation column is accomplished in the $N_2$ column. This is a double column thermally coupled by a reboiler/condenser. Low-pressure distillation using the reflux nitrogen produced is carried out in the top section. Separation products are used to subcool both the top and bottom products of the bottom section before they are fed into the top section. The bottom product is pumped to 1.25 MPa, warmed, and combined with heavy hydrocarbons from the preseparation column, and the sales gas is delivered at 1.2 MPa. Refrigeration for the preseparation column is provided by an open methane loop driven by a single-stage compressor acting as a heat pump.

## 4.2. Cryogenic Upgrading of Petrochemical Gases

Two major feedstocks for the cryogenic upgrading process are refinery/petrochemical vent gases and ammonia purge gas, which are rich in hydrogen and contain various other constituents as well. Recovering hydrogen from these sources is much cheaper than producing it by the stream reforming of hydrocarbons. Recovery rates are 90–99% or more hydrogen together with LPG fractions and nitrogen. Argon plus rare gases are also recovered when economically attractive. A schematic of this process is shown in Fig. 10.11.

### 4.2.1. *Refinery/Petrochemical Vent Gases*

Approximately 90–95 mol% hydrogen can be recovered from these gases by cryogenic upgrading, rejecting methane-rich gas into the fuel gas header and $C_{3+}$ components into the LPG recovery system. Toluene or pyrolysis gasoline hydrodealkylation units integrated with a cryogenic upgrading unit can maximize hydrogen utilization by processing recycled gas together with available refinery gas streams that are rich in $H_2$. Processes combining cryogenic and PSA (pressure swing adsorption) units are often used when there is a requirement for both high purity and high recovery.

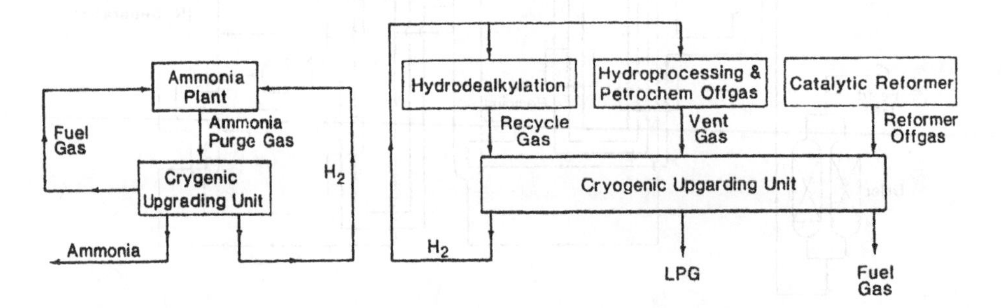


**Figure 10.11** Cryogenic upgrading of petrochemical gases.

*4.2.2. Ammonia Purge Gas*

Steam reformer-based ammonia plants are increasing their use of cryogenic upgrading units. Purge gas from the ammonia converter is processed to separate 90 mol% hydrogen for recycle back into the ammonia converter. This process can either save energy or increase ammonia production. Processing of the combined purge gas from several ammonia plants on the same site may be accomplished by integrating standard purge gas recovery units into a new or existing ammonia plant. Liquid argon for the industrial gas market, together with nitrogen for inerting, can be separated from ammonia purge gas upon further processing. If necessary, hydrogen of higher purity can be supplied for downstream petrochemical applications.

*4.2.3. Cryogenic Upgrading Units*

To remove any high freezing point components such as $NH_3$, $CO_2$, $H_2S$, $H_2O$, and aromatic compounds from the feed gas, cryogenic upgrading units usually require a pretreatment section that contains wet scrubbing or molecular sieve and activated carbon units to accomplish the absorption or adsorption required. These units are often skid-mounted and fully automatic.

The entire cryogenic section is contained within an insulated enclosure (cold box) and shop-fabricated. All welded construction is maximized to avoid leaks, and the complete cold box is shipped to the site to minimize site construction work. Refrigeration for the cryogenic $H_2$ recovery process is provided by expansion of the rejected condensable gas into the low-pressure fuel gas header. Nearly, 30 of these cryogenic upgrading units are in use commercially today.

## 4.3. Cryogenic Upgrading of Synthesis Gas

The cryogenic upgrading of synthesis gas, shown in Fig. 10.12, is used to recover high purity hydrogen, carbon monoxide, and adjusted blends from synthesis gas (syngas). The hydrogen recovered has a purity of $\geq$99.9% with less than 10 ppm

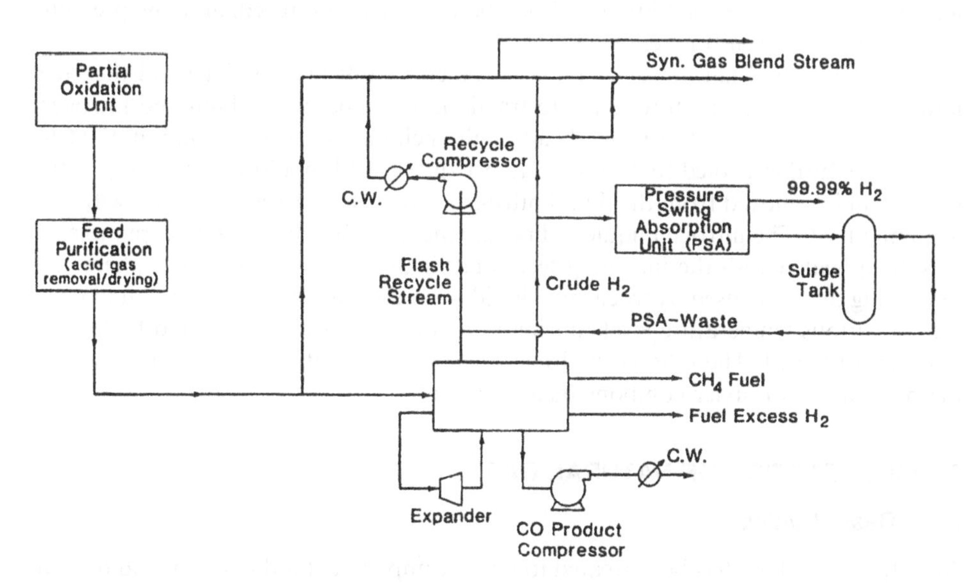


**Figure 10.12** Cryogenic upgrading of synthesis gas.

CO. A carbon monoxide product of 98.5–99.5% purity is available. Syngas blends are supplied at $H_2/CO$ ratios of 1.1–2.5.

The syngas unit consists of two components—a cryogenic processing unit (cold box) and a pressure swing adsorption (PSA) unit. Pure CO is separated from the synthesis gas in the cold box. Crude hydrogen from the cold box is purified in the PSA unit. The purified syngas is cooled in the cold box to condense bulk methane and CO in the feed. Partial condensation is integrated with phase separation of the liquids formed at various temperature levels.

The resultant hydrogen-rich vapor fraction is reheated and expanded to generate refrigeration, which is the driving force for the heat transfer process. Expander design and configurations vary, depending on total high purity hydrogen requirements and related use pressures. Upon further reheating in the cold box, a portion of the hydrogen-rich stream is returned as crude hydrogen (94–98% purity) to be used for blending and PSA feed. Excess hydrogen from the CO-rich liquid is rejected by throttling condensed liquid phases and carrying out phase separation. These fractions are processed further in a distillation column to reject the methane from the overhead product. The overhead CO is rewarmed in the cold box and directed to product compression. A methane-rich stream is rejected from the cold box as a second fuel stream.

Part of the crude hydrogen is fed to the PSA unit, which uses a proprietary process employing four adsorbent beds to provide a continuous and constant hydrogen product flow. The crude hydrogen stream is purified to greater than 99.99% purity with less than 10 ppm carbon monoxide. Over 40 cryogenic units and 100 PSA units are in service for hydrogen purification of synthesis gas and similar streams.

## 5. HELIUM EXTRACTION, NITROGEN REJECTION, AND HYDROCARBON RECOVERY

The process depicted in Fig. 10.13 can handle various feed compositions at various temperatures and pressures while recovering about 96% of the helium present in the feed with a purity of 85%. Liquid hydrocarbons are also produced, and low-pressure nitrogen (99.5% purity) is removed.

A compressed natural gas stream is cooled to 169.4 K (–155°F) and then fed into a nitrogen–methane tower, where fractionation separates helium and nitrogen from the methane. The tower overhead of the helium–nitrogen mixture at 120.6 K (–243°F) is further cooled to 82.2 K (–312°F) and flashed to a lower pressure, where the helium is removed from the liquid nitrogen. The helium vapor is heated with the incoming feed. The nitrogen liquid is first expanded to help cool the heat pump system, then heated with the incoming feed, and finally vented to the atmosphere. By separating the condensed hydrocarbon liquid from the chilled feed vapor–liquid mixture at various temperature levels during feed cooling, a variety of liquid hydrocarbons are obtained. These liquid hydrocarbons are then fed to a deethanizer, where the propane and heavier components are recovered.

## 6. LIQUEFACTION OF NATURAL GAS

### 6.1. Basic Cycles

After the natural gas has been treated to remove impurities (and nitrogen and helium recovery is complete, if desired), the next step is liquefaction. There are three basic

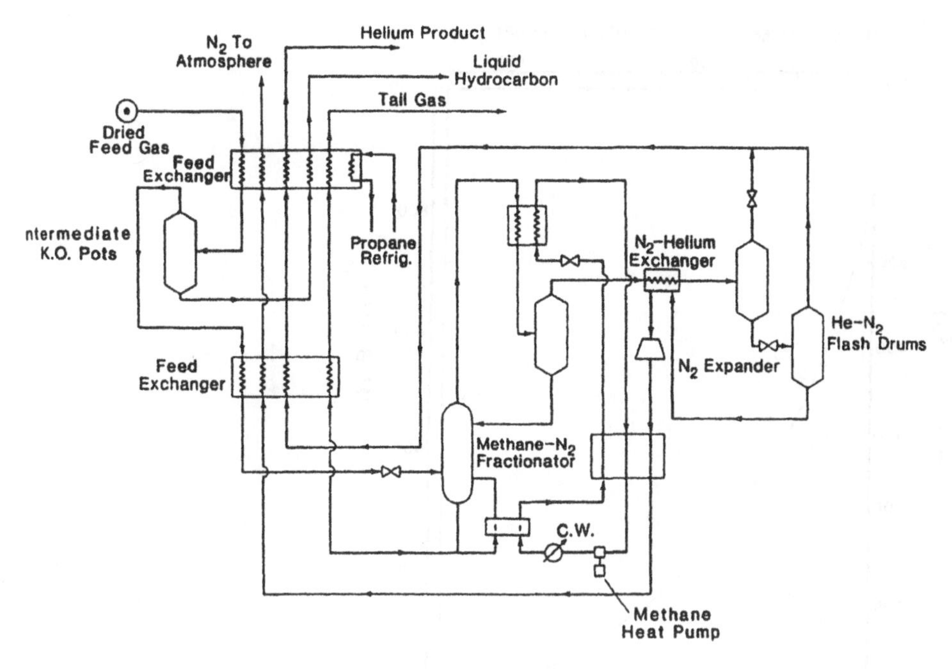


**Figure 10.13** Helium extraction, nitrogen rejection, and hydrocarbon recovery.

liquefaction cycles: (1) the classical cascade cycle, (2) the expander cycle, and (3) the mixed-refrigerant cascade cycle.

### *6.1.1. The Classical Cascade Cycle*

A process flowsheet for the classical cascade cycle used in the liquefaction of natural gas is shown in Fig. 10.14. Typically, a propane–ethylene–methane cascade

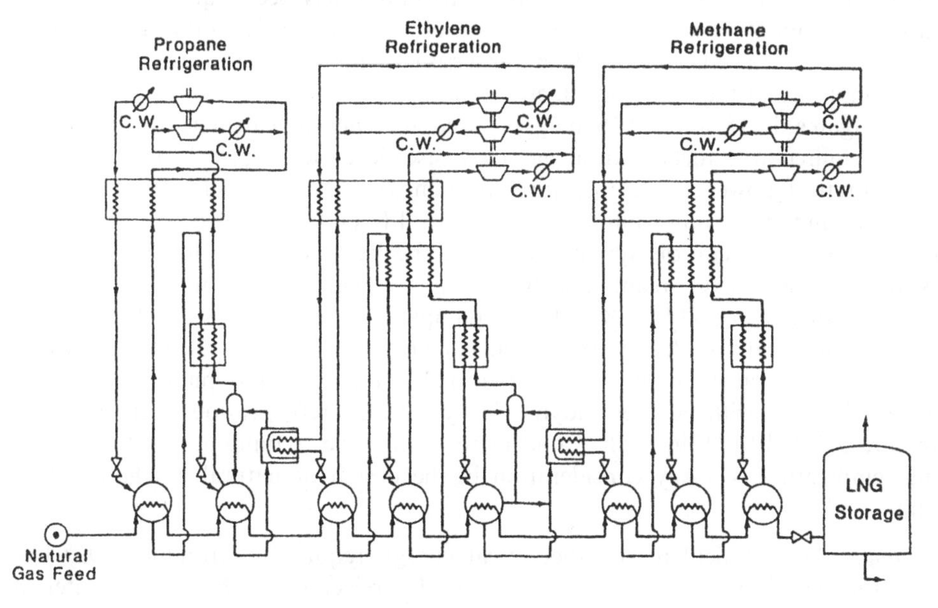


**Figure 10.14** Cascade process flowsheet.

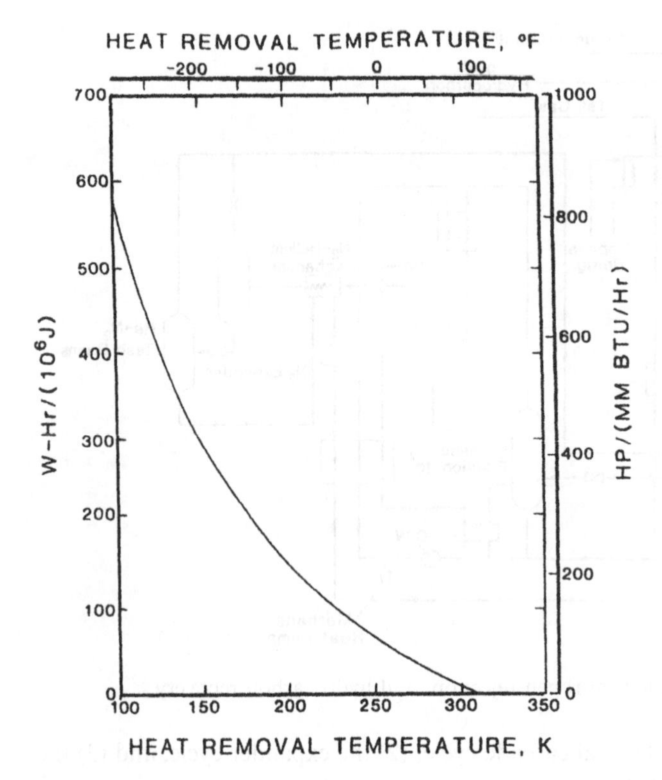


**Figure 10.15** Theoretical power requirements for an ideal refrigerator with heat rejection at 311 K (100°F).

combination is used with the natural gas being cooled through the vaporization of each component in turn. Each of these refrigerants has been liquefied previously in a conventional refrigeration cycle. By vaporizing each refrigerant at two or three pressure levels, the efficiency of the natural gas cooling process is increased, but so is the complexity of the entire liquefaction cycle.

Figures 10.15 and 10.16 may be used together to calculate the minimum work required to liquefy natural gas in a cascade cycle. Figure 10.15 is a graph of theoretical power vs. temperature for removing 292.9 kW ($1 \times 10^6$ Btu/hr) at various temperatures and pumping it up to 311 K (100°F) with a thermodynamically perfect refrigeration system. Rejecting the heat at 311 K (100°F) would be typical for an LNG plant rejecting heat to seawater. Removing 292.9 kW ($1 \times 10^6$ Btu/hr) at 211 K (–80°F) and pumping it up to 311 K (100°F) with a thermodynamically perfect refrigeration system would require 138.7 kW (186 hp). Removing the same quantity of heat at 111 K (–260°F) and pumping it up to 311 K (100°F) with a thermodynamically perfect refrigeration system would require 527.2 kW (707 hp). The rapid increase in power requirement with decreasing temperature is a key consideration in the design of natural gas liquefaction units.

*Example 10.3.* Calculate the theoretical energy required to produce 292.9 kW ($10^6$ Btu/hr) of refrigeration at 200°R (–260°F), rejecting the heat to seawater at 560°R (100°F). Compare the calculated value with that found from Fig. 10.15.

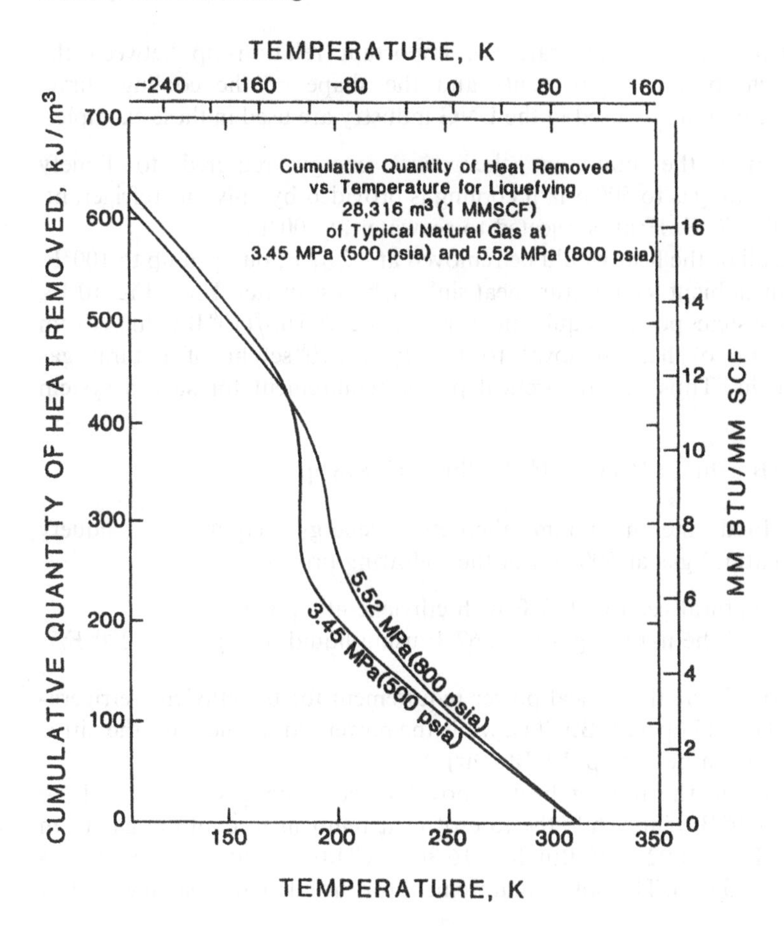


**Figure 10.16** Cumulative quantity of heat removed from natural gas.

The minimum energy is given by the Carnot cycle equation:

$$\frac{W}{Q} = \frac{Q_h - Q_l}{Q_l} = \frac{T_h - T_l}{T_l}$$

$$= \frac{560 - 200}{200} = 1.8$$

$$W = (1.8)\left(10^6 \frac{\text{Btu}}{\text{hr}}\right)\left(0.2929 \frac{\text{W hr}}{\text{Btu}}\right)\left(\frac{1}{745.7}\frac{\text{hp}}{\text{W}}\right)$$

$$= 707\,\text{hp}$$

This value is plotted in Fig. 10.15. This is the method used to derive that figure.

Figure 10.16 is a graph in the units commonly used by the LNG industry of cumulative quantity of heat removed vs. temperature for liquefying $1 \times 10^6$ scf of typical natural gas at 500 and 800 psia (see Table 9.8 for conversion factors). The shape of the cooling curve is dependent on the composition of the gas and the pressure. This cooling curve defines the work that must be performed to liquefy the natural gas.

Figures 10.15 and 10.16 can be used to calculate the theoretical power requirement for liquefying typical natural gas at 500 psia with various liquefaction systems.

The following two examples illustrate the important relationship between the refrigeration system power requirements and the shape of the cooling curve. Measurement units commonly used in the LNG industry are used in these examples.

*Example 10.4.* Find the minimum theoretical energy required to liquefy $1 \times 10^6\,\text{ft}^3/\text{h}$ of natural gas to 500 psia if cooling is provided by only one refrigerant, liquid nitrogen at –270°F. Heat is rejected to seawater at 100°F.

In this case, all of the heat would be removed at −270°F, pumped up to 100°F, then rejected to an ambient temperature heat sink such as seawater. From Fig. 10.15, the refrigeration system power requirement would be $765\,\text{hp}/(10^6\,\text{Btu/hr})$. From Fig. 10.16, the rate of heat removal to liquefy $1 \times 10^6\,\text{scf/hr}$ of natural gas is $16.86 \times 10^6\,\text{Btu/hr}$. Thus, the theoretical power requirement for such a system would be

$$765\,\text{hp}/(10^6\,\text{Btu/hr}) \times 16.86 \times 10^6\,\text{Btu/hr} = 12,898\,\text{hp}$$

*Example 10.5.* Find the minimum theoretical energy required to liquefy $1 \times 10^6\,\text{ft}^3/\text{hr}$ of natural gas at 500 psia in the following process:

1. cool the natural gas to −145°F with ethylene at −150°F;
2. further cool the natural gas to –267°F using liquid nitrogen at –270°F.

From Fig. 10.15, the theoretical power requirement for the ethylene refrigeration system would be $317\,\text{hp}/(10^6\,\text{Btu/hr})$, while the power requirement for the nitrogen system would again be $765\,\text{hp}/10^6\,\text{Btu/hr})$.

From Fig. 10.16, the rate of heat removal to cool the gas from 100°F to −145°F is $11.94 \times 10^6\,\text{Btu/hr}$, while the rate of heat removal to cool the gas from −45°F to −267°F is $4.92 \times 10^6\,\text{Btu/hr}$ ($16.86 \times 10^6\,\text{Btu/hr}$ at −267°F minus $11.94 \times \text{Btu/hr}$ at −155°F). The power requirement for the ethylene system would be

$$317\,\text{hp}/(10^6\,\text{Btu/hr}) \times 11.94 \times 10^6\,\text{Btu/hr} = 3785\,\text{hp}$$

The power requirement for the nitrogen system would be

$$765\,\text{hp}/(10^6\,\text{Btu/hr}) \times 4.92 \times 10^6\,\text{Btu/hr} = 3765\,\text{hp}$$

The total power requirement for the system would be

$$3785 + 3764 = 7549\,\text{hp}$$

This is considerably less than the 12,898 hp for the system that used nitrogen alone.

These two examples demonstrate that the greater the number of refrigerations stages, the lower the power requirement. The thermodynamically perfect liquefaction unit would have an infinite number of refrigeration stages, each removing an infinitesimally small quantity of heat over an infinitesimally small temperature range. Such a system for liquefying $1 \times 10^6\,\text{scf/hr}$ of natural gas at 500 psia would require 4533 hp. Practical liquefaction systems use 2.5–4 times as much power as a thermodynamically perfect system.

Figure 10.16 also shows cooling curves for liquefying $1 \times 10^6$ typical natural gas at 800 psia. The characteristics of the gas at the higher pressure more heat to be removed at higher temperatures, thus decreasing the power requirement. The theoretical horsepower for liquefying $1 \times 10^6\,\text{scf/hr}$ at 800 psia is 4061 hp compared to 4533 hp at 500 psia. Thus, natural gas feed pressure is an important consideration in LNG plant design.

### 6.1.2. *The Expander Cycle*

The expander cycle uses the cooling effect obtained by expansion of the natural gas to finally liquefy a portion of the natural gas. Expansion turbines are used in this process, with liquefaction occurring upon expansion through a Joule–Thomson valve. (See Fig. 10.17.) A modification of this cycle involves obtaining refrigeration by turboexpansion of both the gas being liquefied and the flow-by gas (the portion of the original gas stream not to be liquefied) (see Fig. 10.18).

An alternative approach uses nitrogen separated from the feed gas as the low-temperature refrigerant, as shown in Fig. 10.19. The nitrogen circulats in an auxiliary refrigeration loop, with nitrogen added as makeup when necessary. The nitrogen column removes nitrogen and some methane from the feed stream, and these two components are further separated inthe methane column.

### 6.1.3. *The Mixed-Refrigerant Cascade Cycle*

The multicomponent mixed-refrigerant cascade cycle is based on the same principle as the classical cascade cycle, but instead of three separate refrigeration loops complete with their own compressors, a single refrigeration loop with a single compressor is used (Fig. 10.20) (Kidnay, 1972). If the composition of the feed changes, the composition of the mixed refrigerant must also be changed to reduce thermodynamic

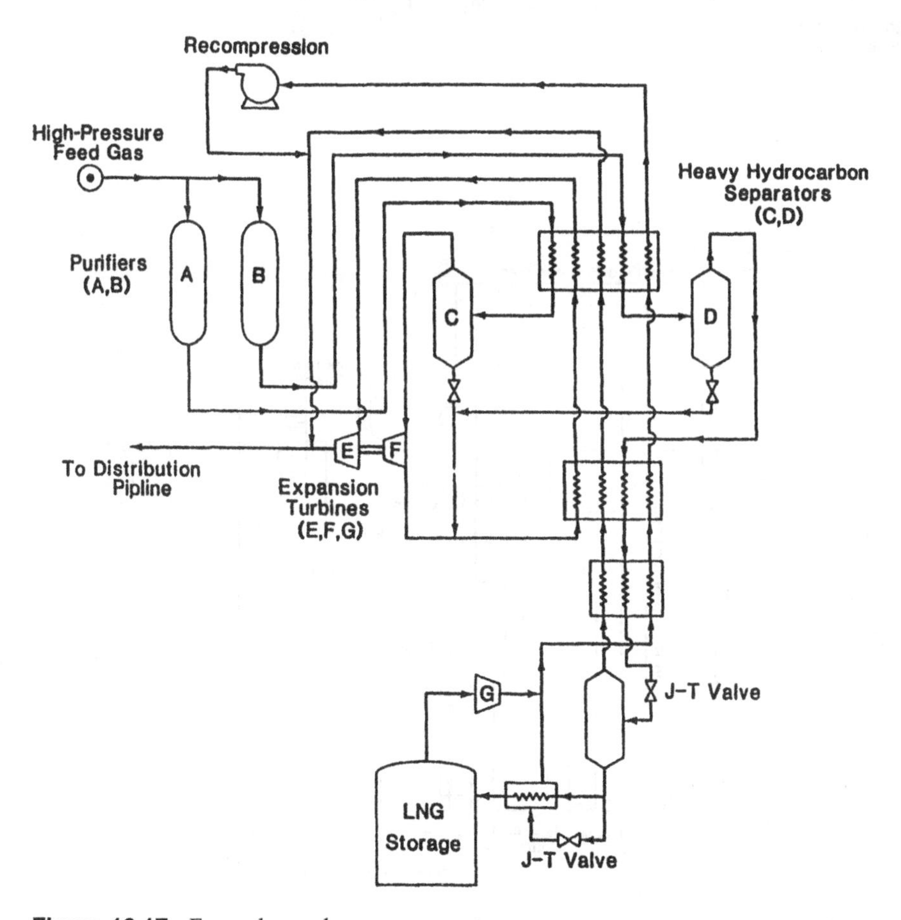


**Figure 10.17** Expander cycle.

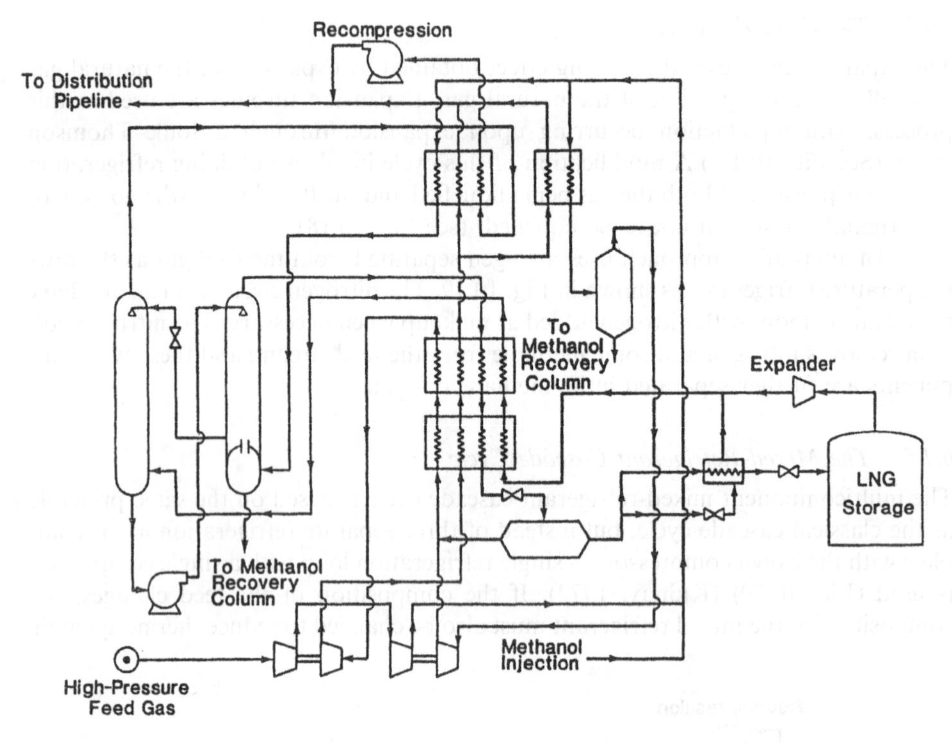


**Figure 10.18** Double expander cycle.

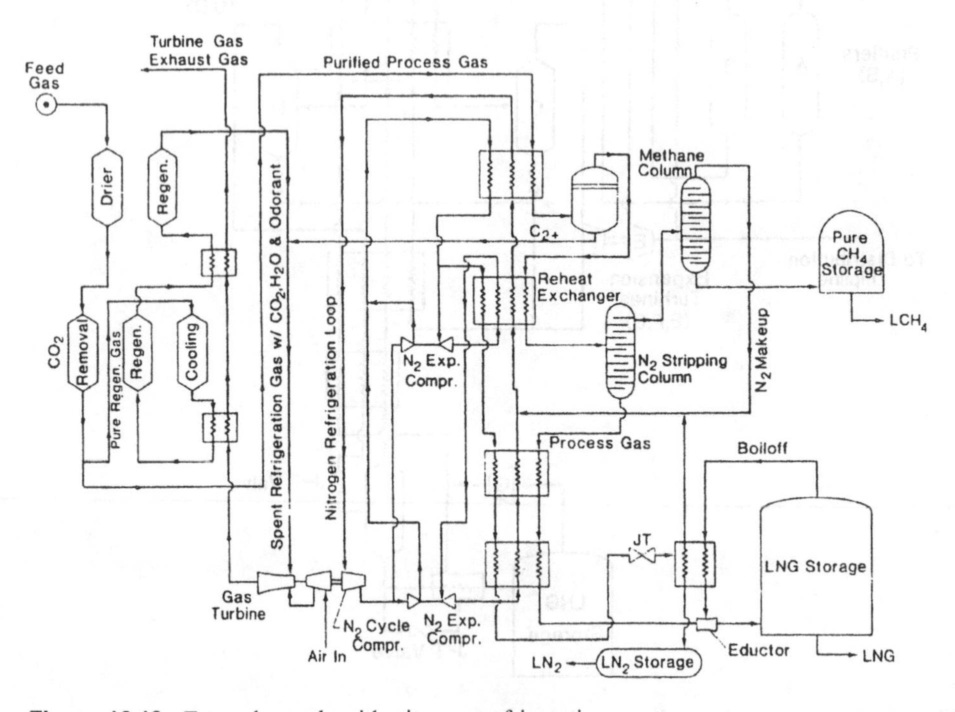


**Figure 10.19** Expander cycle with nitrogen refrigeration.

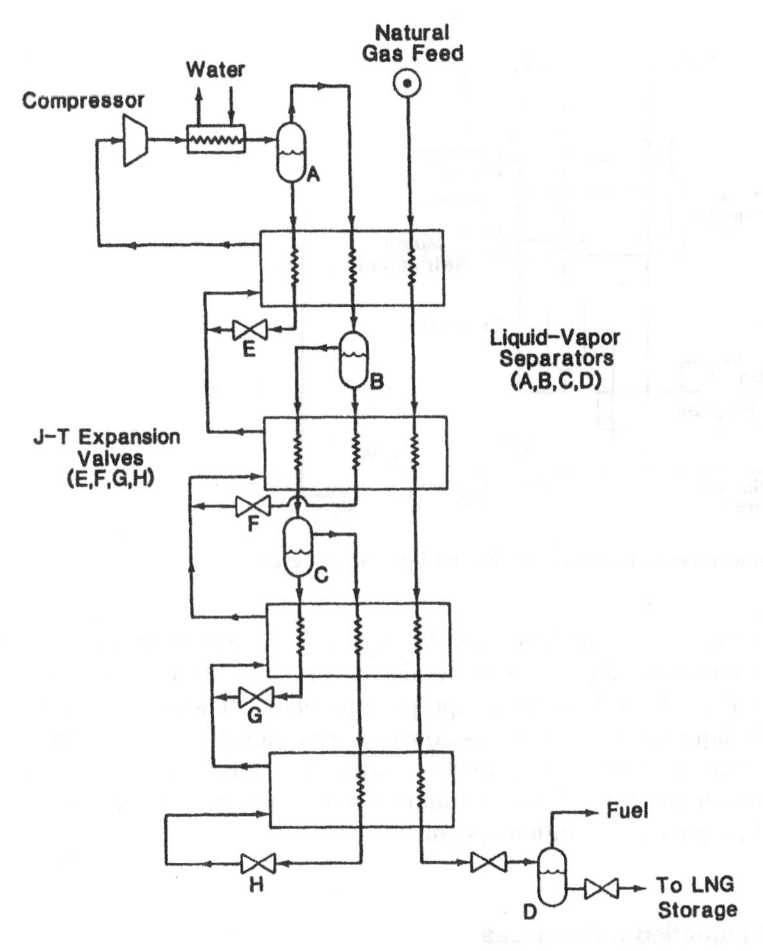


**Figure 10.20** Closed-cycle mixed-refrigerant cascade.

irreversibilities. The mixed refrigerant is repeatedly condensed, vaporized, separated, and expand more thorough knowledge of the thermodynamic properties of gaseous mixtures and more sophisticated design methods are required with the mixed-refrigerant cycle than with the classical cascade or expander cycles. A major disadvantage of the mixed-refrigerant cycle is that problems occur in handling two-phase multicomponent mixtures in the heat exchangers.

## 6.2. Commercially Used Multicomponent Refrigerant (MCR) Liquefaction Cycle

The mixed-refrigerant cycle is commonly used in commercial liquefaction processes today for both baseload and peakshaving operations. Figure 10.21 illustrates a typical industrial process using this technique. After water and acid gases are removed, the natural gas is cooled with successive stages of propane to about 238.9 K (–30°F). The cooled gas enters the MCR heat exchanger, where it is liquefied with a multicomponent refrigerant consisting of nitrogen, methane, ethane, and propane. Nitrogen is obtained from the air, while the other three components are obtained from the natural gas feed. The liquefied natural gas is then sent to storage.

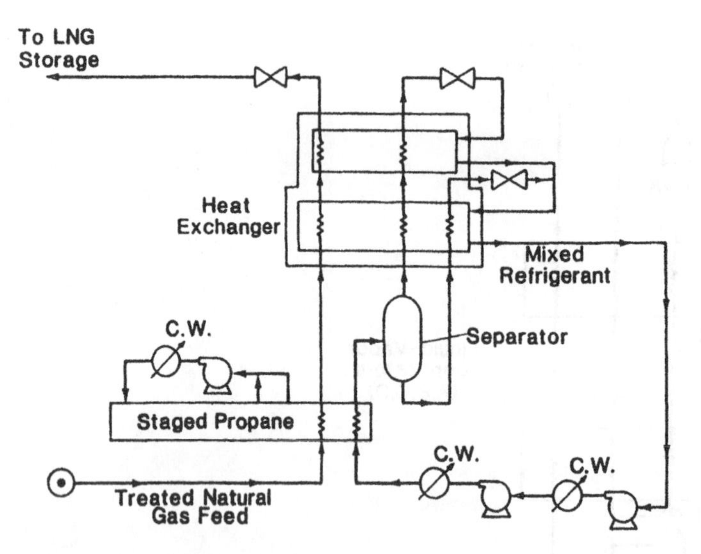


**Figure 10.21** Multicomponent refrigerant (MCR) liquefaction cycle.

The mixed refrigerant stream leaves the MCR heat exchanger at about 236 K (−35°F) and is compressed and cooled with cooling water in two stages to 4.2 MPa (600 psi) and 238.9 K (−30°F). Evaporating propane partially condenses the mixed refrigerant, and the liquid and vapor phases flow to a separator. Both streams then enter the MCR heat exchanger to provide continuous cooling, as shown in Fig. 10.21. Cooling for the natural gas feed and the mixed refrigerant is provided by a conventional propane refrigeration system.

### 6.3. Storage of Liquefied Natural Gas

Natural gas stored as a liquid occupies approximately 1/650 of its gaseous volume under standard conditions. Both baseload and peakshaving operations require storage of LNG, although their requirements differ somewhat.

In the international transport of LNG, the need for large storage depots at both the loading and market ports is evident. In such baseload operations, the storage facility is essentially a surge tank smoothing out the nonuniform baseload demand and fuel delivery operations. In this case, it is not necessary to preserve the LNG in the tank for long periods. Since it is passed on to distribution within a short period of time, relatively higher heat influx resulting in greater boil off rates can be tolerated.

Domestic use of LNG is primarily concerned with peakshaving, the practice of providing large-volume storage near metropolitan areas to meet the winter peak loads. In this case, LNG is stored for relatively long periods of time and used (i.e., vaporized and distributed) during only a few days of the winter. Consequently, heat influx to the storage tank must be held to a minimum.

Whether the LNG plant is to be used for baseload operations or peakshaving, a large portion of the cost goes toward construction and operation of the storage facilities. Two main types of LNG storage in general use are (1) conventional above-ground, double-walled metal tanks and (2) prestressed concrete tanks, above or below ground. Two other techniques, frozen earth storage and mined caverns, have rarely been used. Figure 10.22 diagrams four of the most common storage techniques.

### *6.3.1. Cryogenic In-Ground (CIG) Storage*

A typical sketch of a frozen earth storage container is shown in Fig. 10.23. The cavity is initially cooled by spraying LNG in the vapor space. The roof reaches its steady-state temperature rapidly, but due to the low thermal conductivity of the frozen earth it may take several years for the surrounding soil to reach its steady-state temperature. The final effective thermal conductivity attained by the frozen earth will depend strongly on the type of geologic formation and the moisture content of the wall.

There were four early LNG plants that incorporated CIG units for storage. The first is located in Arzew, Algeria, and is owned by SONATRACH, the Algerian national petroleum company. It is 37 m (122 ft) in diameter and 36 m (119 ft) deep and has a capacity of $24 \times 10^6\,m^3$ gas ($850 \times 10^6$ cu ft) of natural gas. The next to be constructed was a $28 \times 10^6\,m^3$ gas ($1000 \times 10^6$ cu ft) unit connected with a peakshaving facility owned by the Transcontinental Gas Pipe Line Corporation. This plant was located near Carlstadt, New Jersey. It was 35 m (115 ft) in diameter and extended 50 m (165 ft) below existing grade.

Next in line was the construction of the Tennessee Gas Pipeline Company's peakshaving facility near Hopkinton, Massachusetts. In this plant, two $42.5 \times 10^6\,m^3$ gas ($1.5 \times 10^9$ cu ft) units were constructed. Each of these units was about 41 m (135 ft) in diameter and about 52 m (172 ft) deep.

Finally, four frozen-ground storage units were built at the Canvey Island receiving facilities in England for the British Gas Council, now the British Gas Corporation. Each of these is about 40 m (130 ft) in diameter and about 39 m (128 ft) deep and provides approximately $28 \times 10^6\,m^3$ ($1 \times 10^9$ cu ft) of gas storage per unit.

Of the four plants mentioned that incorporated cryogenic in-ground (CIG) storage units, only two still have CIG units in service—the one at Arzew, Algeria, and the one at Canvey Island, England. The two plants in the United States, those near Carlstadt, New Jersey, and near Hopkinton, Massachusetts, were abandoned in favor of other forms of storage because of failure of the CIG units to perform satisfactorily as peakshaving storage units. It is significant that both CIG units in the United States were intended for peakshaving application, whereas those in Algeria and England were for baseload plants. Because of the lack of published operating data, it is not possible to assess differences in performance between these four CIG applications. What is acceptable for baseload operations may not be acceptable for peakshaving.

### *6.3.2. Prestressed Concrete Tanks*

Hundreds of prestressed concrete tanks and reservoirs have been built for many uses, including the storage of liquid oxygen (Closner, 1968), which is both heavier and colder than LNG. These latter tanks have been in continuous use since 1968.

A concrete tank is considered to be prestressed when it is designed and constructed so that the ring force of a completely filled tank does not cause any circumferential tensile stresses in the concrete. This condition is created by applying high-tensile wires stressed on the order of 1034 MPa (150,000 psi) around the concrete core. In certain cases, vertical pre-stressing may be employed to control the magnitude of, or eliminate, vertical bending stresses in the concrete wall.

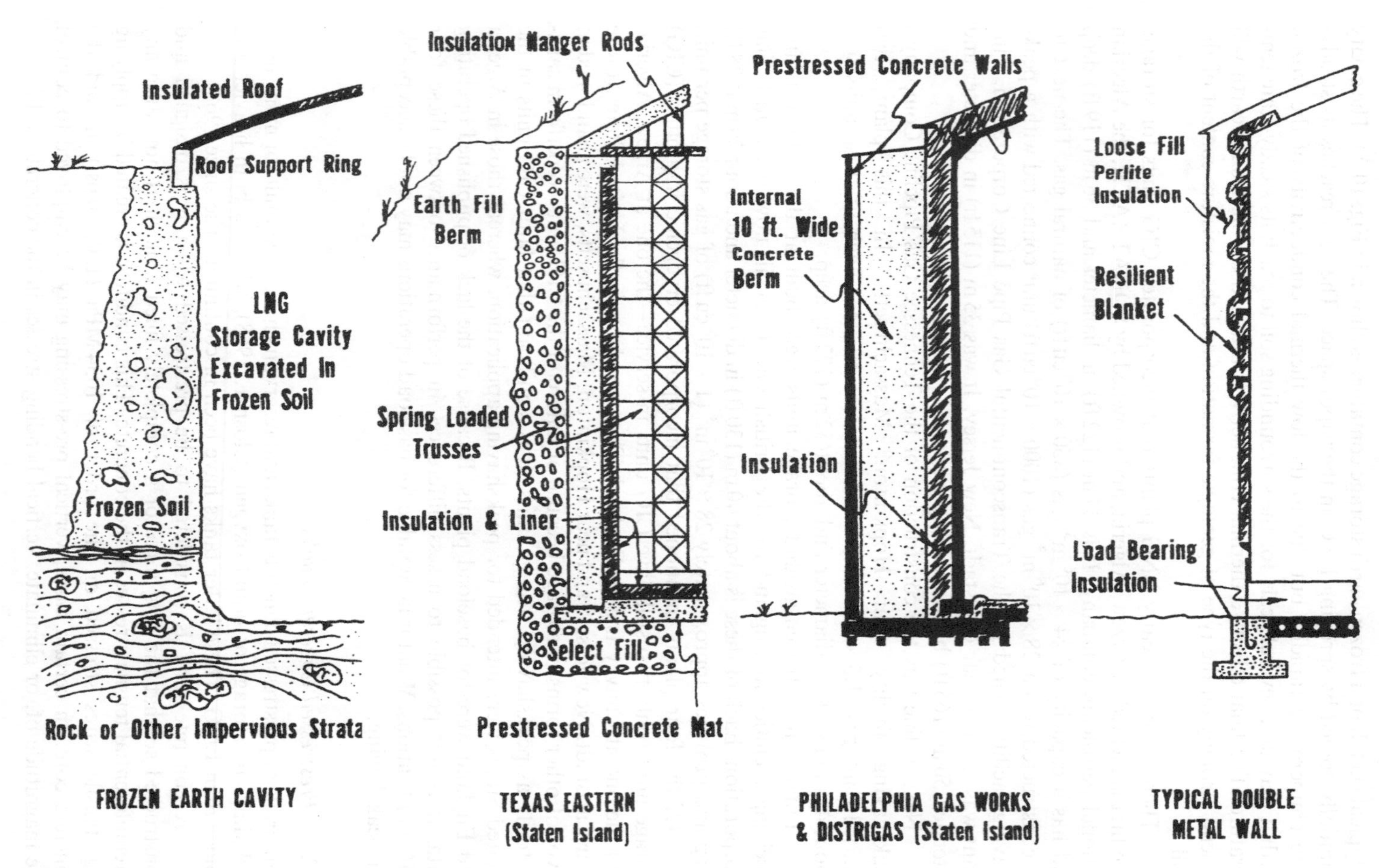


Figure 10.22 Common LNG storage techniques.

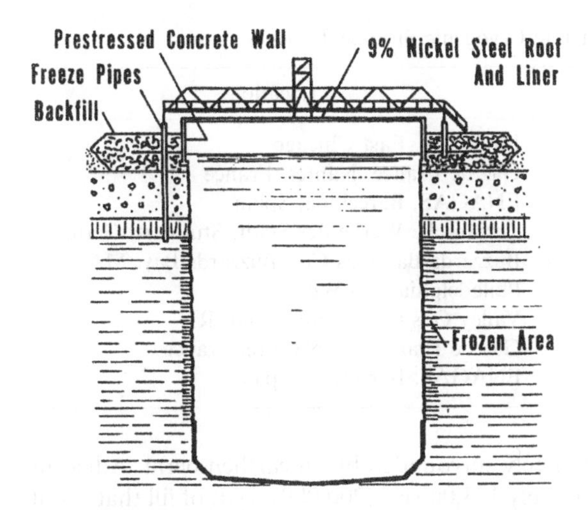


**Figure 10.23** Cryogenic in-ground storage.

The Portland Cement Association (Monfore and Lentz, 1962) and others have shown that concrete cured at ordinary temperatures can be used to form tanks for holding cryogenic fluids. The mechanical properties of concrete are not significantly impaired at low temperatures. The compressive strength of moist concrete at 117–172 K is almost triple that of concrete at room temperature.

Prestressed concrete tanks may be constructed above ground, at grade, below ground, or partially below ground, depending on site conditions or other factors. For tank sites below ground, excavation and control of water follow standard construction procedures. Above-ground tanks may be of either the single-shell or dual-shell type, depending primarily on insulation requirements. A $5.7 \times 10^7 \, m^3$ ($2 \times 10^9$ cu ft) prestressed concrete LNG tank was constructed for the Texas Eastern Transmission Corporation on Staten Island, New York, and placed in operation in early 1970. This storage tank was approximately 81.7 m (268 ft) in diameter and allowed for a liquid depth of 18.6 m (61 ft) (see Fig. 10.24). The tank was constructed

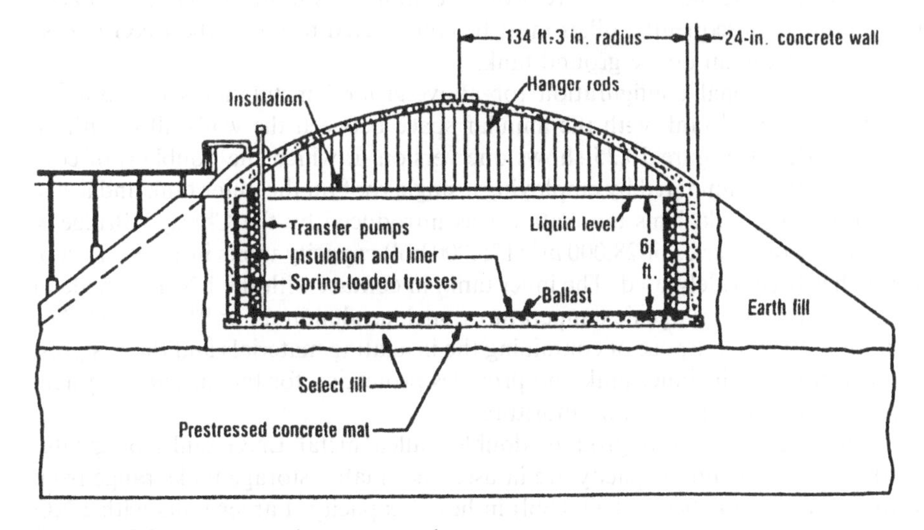


**Figure 10.24** Prestressed concrete tank.

**Table 10.7** Typical Prestressed Concrete Cryogenic Storage Tanks

| Service | Location |
|---|---|
| Liquid oxygen | Linde Co., East Chicago |
| 2000 m$^3$ (12,500 bbl) | Gaz de France, Nantes, France |
| 40,000 m$^3$ (252,000 bbl) | ENAGAS, Barcelona, Spain |
| 30,000 m$^3$ (180,000 bbl) | Technische Werke der Stadt, Stuttgart, Germany |
| 8000 m$^3$ (50,000 bbl) | Buzzards Bay Gas Co., Buzzards Bay, MA |
| 92,500 m$^3$ (582,000 bbl) | Philadelphia Gas Works |
| 4000 m$^3$ (25,000 bbl) | Valley Gas Co., Cumberland, RI |
| 120,000 m$^3$ (755,000 bbl) | Gaz de France, St. Nazaire, France |
| 80,000 m$^3$ (500,000 bbl) | ENAGAS, Barcelona, Spain |

with its base at grade but was completely surrounded by an earthen berm. Adjacent to the tank wall and floor, approximately 153,000 m$^3$ (200,000 cu yd) of fill that is not susceptible to frost heave was used to eliminate possible damage from subsequent soil freezing. As an additional precaution against frost heave, 8169 m (26,800 ft) of heating cable was installed 1.8 m (6 ft) below the concrete floor.

Several other cryogenic projects using prestressed concrete are listed in Table 10.7.

### *6.3.3. Above-Ground Metal Tanks*

Storage of LNG in above-ground metal tanks has been the most widely accepted and used storage method for both baseload and peakshaving uses. The apparent advantages of above-ground storage are that

1. The technology is more fully developed than that of in-ground storage.
2. There are no geographic limitations for above-ground storage containers.
3. Both mined caverns and frozen earth cavities have higher heat leaks and much larger cooldown times than the standard above-ground tank (Kidnay, 1972).

On the negative side, the above-ground container is more susceptible to accidental rupture. Consequently, dikes must be constructed to retain the LNG in case there is a spill from an above-ground tank.

The conventional configuration for above-ground metal tanks is a double-walled, flat-bottomed tank with the annular space between the walls filled with an insulating material. Figure 10.25 shows one version in which the double-roof construction has been eliminated by employing a single roof on the outer tank and a suspended insulating deck. This innovation was introduced by the Chicago Bridge & Iron Company on a series of 28,000 m$^3$ (175,000 bbl) capacity tanks in 1968 and subsequently has been widely used. The inner tank, in contact with the LNG, is made of materials suitable for cryogenic temperatures, such as aluminum or 9% Ni steel. The outer tank serves the purpose of containing the insulating materials and the gas pressure that surrounds the inner tank and provides protection for the insulation system against fire, impact, weather, and moisture.

Several hundred above-ground, double-walled metal LNG tanks of greater than 7000 m$^3$ (45,000 bbl) capacity are in use. The smaller storage tanks range from 7000 to 100,000 m$^3$ (45,000–630,000 bbl) in liquid capacity. Larger units with LNG capacities up to 200,000 m$^3$ (1,250,000 bbl) are also in use.

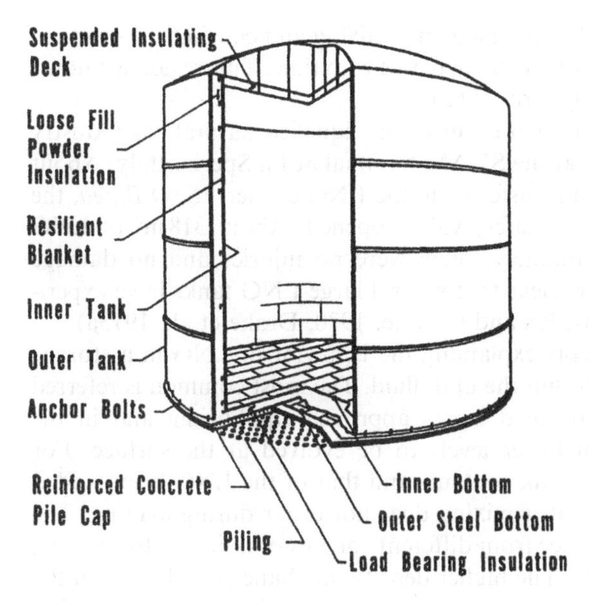


**Figure 10.25** Above-ground metal tank for LNG.

There are a number of acceptable materials for LNG tank construction. The material for constructing inner tanks must be one that retains its strength and ductility through the range of temperatures possible in the tank. The materials must be particularly notch-tough for the important and more highly stressed components. Aluminum, stainless steels, and 9% Ni alloy steel have been proven to remain notch-tough beyond the very low temperatures of LNG and are indicated as acceptable materials by the codes. Where material exhibits a transition from ductile to brittle behavior as the temperature decreases, acceptance is based on rigid tests that the material would normally be expected to pass. Stainless and 9% Ni steel are thoroughly tested by means of the Charpy V-notch test.

Aluminum and 9% Ni steel are the most extensively used metals for large LNG tanks. Because of its high cost, stainless steel is generally used only for small shop-constructed vessels and especially for LNG plant piping and heat exchangers.

Experience with large cryogenic above-ground metal tanks is not limited to LNG tanks. Over the past 25 years, owing to space exploration and the rapidly expanding liquefied industrial gas field, over 600 similar types of tanks have been constructed and placed in service for the storage of liquid oxygen, nitrogen, argon, hydrogen, and ethylene. Storage of these other cryogenic fluids has been accomplished not only in double-walled, flat-bottomed tanks, but also in double-walled spheres, spheres suspended in cylindrical vessels, cylindrical tanks suspended within cylindrical tanks, and other configurations. Selection of tank configuration is generally made on the basis of the requirements for storage capacity, operating pressure, external pressure, and other service conditions.

### 6.4. Stratification

A major problem in the storage of LNG and other cryogenic fluids is stratification. Stratification is caused by the free convective flow of heated liquid along the side walls of a tank and into the upper regions near the liquid–vapor interface. Here it flows toward the center of the tank, dispersing and mixing and causing a downward motion of heated liquid, the depth of which increases with time. This region is known as the thermal stratification layer (Clark, 1965).

Stratification led to an unexpected event in the liquefied natural gas industry that occurred on August 21, 1971, at the SNAM terminal at La Spezia, Italy. About 18 hr after the completion of cargo transfer from the LNG carrier *ESSO Brega*, the tank pressure suddenly rose and the safety valves opened. About 318 $m^3$ of LNG vaporized and was released. Fortunately, there were no injuries and no damage was done (Sarsten, 1972). It is now clear that several large LNG tanks have experienced such "rollover" incidents (Bellus and Gineste, 1970; Drake et al., 1973a).

There are at least two concepts explaining the behavior of LNG in a storage tank where there is stratification within the bulk fluid. The most common is referred to as rollover, wherein density-stratified layers approach equilibrium and in the process allow stored energy from lower levels to be evolved at the surface. For instance, if the composition of the tank contents and that of the LNG being added are different, stratification will result if mixing does not occur during loading. The different compositions can arise either from different sources of liquid or from aging (weathering) of the tank's contents. The higher density, methane-poor layers on the bottom are warmer, and the vaporization is suppressed by the lower density, cooler, methane-rich layers on top. Mixing between layers is slow, and only the top layer is in thermal equilibrium with the vapor space. As lower layers warm, the density differences decline until the densities of the layers are about equal; then the layers rapidly mix, hence the term rollover. Similar phenomena can be studied in salt solutions. Now as the warmer layer or layers reach the topmost layer, the suppressed vaporization is released, and rapid flashing occurs. The danger lies in overpressurizing the tank or in a large vapor cloud issuing from the safety valves.

The second concept is referred to as thermal overfill—a disturbance disrupts a top layer of liquid, causing the tank pressure to change in correspondence with the bulk (warmer) liquid temperature.

Either explanation results in excessive release of vapor that could cause physical damage to the storage tank if there is insufficient venting capacity. The two explanations have some common technical elements, and both are supported by physical evidence.

### *6.4.1. Common Principles*

It has been shown that if the Rayleigh number of a tank of fluid is above about 2000, natural circulation within the tank makes stratification impossible. The Rayleigh number is defined as

$$\mathrm{Ra} = \frac{g_c \alpha \, \Delta T \, h^3}{\eta \, K/\rho C_P}$$

where, $g$ is the gravitational constant, $\alpha$ the thermal expansion coefficient, $\Delta T$ the temperature difference between fluid at top and bottom, $h$ the height of LNG, $\eta$ the kinematic viscosity, $\rho$ the density, $C_P$ the specific heat, and $K$ the thermal conductivity.

The Rayleigh number for a full tank of LNG is typically on the order of $1 \times 10^{15}$, which is many orders of magnitude greater than required for stratification. Thus, natural convection currents can occur with ease in LNG and thereby maintain the liquid at a uniform temperature.

*Example 10.6.* A liquid natural gas tank is filled to a depth of 5.03 m (16.5 ft). The properties of the contents are as follows:

| | |
|---|---|
| Viscosity, $\eta$ | $0.00267\ cm^2/s$ |
| Thermal conductivity, $K$ | 1.86 mW/(cm K) |
| Specific heat, $C_p$ | 3.50 J/(g K) |
| Density, $\rho$ | $0.4226\ g/cm^3$ |
| Thermal expansion coefficient | $0.00115\ K^{-1}$ |
| Acceleration and gravity, $g$ | $9.807\ m/s^2$ |

If the temperature difference between the top and bottom of the tank is 4.44 K (8°F), is natural circulation sufficiently well developed to prevent stratification?

If Ra > 2000, natural circulation will prevent stratification. Using the above data, the Rayleigh number can be calculated as follows:

$$\mathrm{Ra} = \left(980.7\frac{\mathrm{cm}}{\mathrm{sec}^2}\right)\left(\frac{0.00115}{\mathrm{K}}\right)(4.44\ \mathrm{K})(503\ \mathrm{cm})^3 \times \left(0.4226\frac{\mathrm{g}}{\mathrm{cm}^3}\right)\left(3.50\frac{\mathrm{J}}{\mathrm{g\ K}}\right)\left(\frac{1\ \mathrm{sec}}{0.00267\ \mathrm{cm}^2}\right) \times \left(\frac{1}{0.00186}\frac{\mathrm{cm\ K}}{\mathrm{W}}\right) = 1.89 \times 10^{14}$$

Therefore, natural circulation is well developed.

The heat coming into the tank causes boundary layer heating at the walls. The boundary layer increases in velocity and width as it flows up the wall. Near the top of the wall, the boundary layer has been calculated to be several centimeters thick and have a velocity of 0.06–1.2 m/sec (0.2–0.4 ft/sec)—well within the turbulent flow region.

The fluid in the moving boundary layer is slightly warmer than the bulk fluid as it approaches the liquid surface because it has gained heat from the wall. The average warm-up is on the order of 0.6 K (0.1°F). No vapor is generated before the fluid reaches the surface because the hydrostatic head increases its boiling point more than the fluid temperature increases. There is no visible boiling even at the surface because the temperature driving force is too small to form bubbles. Part of the warm liquid flashes or vaporizes when it reaches the surface and returns to the equilibrium temperature set by the tank pressure level.

The natural convection circulation is quite large and induces displacement of the entire tank contents in 10–20 hr. Such a time increment is very short compared to the weathering process (see below). Once a tank of LNG is well mixed, it would not be expected to naturally stratify. However, when stratification is artificially imposed by the filling process, total mixing can be inhibited.

### 6.4.2. *Weathering*

The change of composition and density of LNG due to weathering is strongly influenced by the initial nitrogen content. Since nitrogen is the most volatile LNG component, it will preferentially vaporize before methane and heavier hydrocarbons. If the nitrogen composition is substantial, the weathered LNG density will decrease with time compared to the original density. If the nitrogen composition is low, as is most often the case, the weathered LNG density will increase due to vaporization of methane. It is important, therefore, to know the composition of both tank and feed LNG prior to tank filling.

Since liquid density differences between layers are key to stratification and subsequent rollover behavior, it is important to have a clear understanding of the influence of liquid composition and temperature on LNG density. The units common to the LNG industry (US Customary) are used in this discussion. Table 10.8 gives the conversion factors most often used.

For LNG mixtures in equilibrium with atmospheric pressure, the liquid temperature is a function of composition. Pure methane has a molecular weight of 16, a higher heating value (HHV) of about 1010 Btu/scf, a density of about 26.5 lb/cu ft, ft, and a boiling point of −259°F at atmospheric pressure. If LNG mixtures containing heavier hydrocarbans—ethane, propane, etc.—are compared to pure liquid methane, the larger the proportion of heavies, the greater the increase in HHV, molecular weight, density, and saturation temperature. To give a rough idea of the magnitudes of the increases, a 100 Btu/scf increase in HHV corresponds approximately to a 0.6 increase in molecular weight, a 2 lb/cu ft increase in liquid density, and a 2°F increase in atmospheric boiling point. If the liquid is stored above atmospheric pressure, its temperature is further elevated by about 1.5°F per 1 psi increase in pressure. Each degree of superheat produces about a 0.2% expansion of the fluid (Drake, 1976).

Density variations in a stratified LNG storage tank can be described if heat and mass balances are made for each stratum (Drake et al., 1973b). The upper layer "weathers" by losing methane (and nitrogen, if present). The heavier constituents (ethane, propane, etc.) have a very low vapor pressure, sothey preferentially remain in the liquid phase. Consequently, the LNG becomes denser (with greater HHV) as some of the methane boils off. An approximateequation for estimating density change by weathering (neglecting $N_2$) is

$$\Delta\rho = 0.017\left(\frac{Q}{m\gamma}\right)(\mathrm{HHV} - 1013)$$

where $\Delta\rho$ is the density change, lb/cu ft, $Q$ the net heat input to upper layer over period of interest (Btu), $m$ the total pound-moles of liquid in layer, $\gamma$ the heat of vaporization of methane, 3519 Btu/lb mol, and HHV the higher heating value of liquid (Btu/scf).

*Note*: 1013 = HHV of pure methane in Btu/scf (assumed boiloff composition); 0.017 = approximate slope, in (lb/cu ft)/(Btu/scf)

**Table 10.8** LNG Conversion Factors

| Item | $lb_m$ | kg | $ft^3$ (liq) | $m^3$ (liq) | 1 $m^3$ (gas) | HHV (Btu) | HHV (kJ) |
|---|---|---|---|---|---|---|---|
| 1 cubic foot (liq) | 26.38 | 11.97 | 1 | 0.02832 | 17.59 | $0.6302 \times 10^6$ | $0.6647 \times 10^6$ |
| 1 MMscf (gas)[a] | 42474 | 19266 | 1610 | 45.59 | 28,316 | $1014.6 \times 10^6$ | $1070.2 \times 10^6$ |
| 1 US gallon (liq) | 3.526 | 1.599 | 0.1337 | 0.003786 | 2.351 | $0.08423 \times 10^6$ | $0.08885 \times 10^6$ |
| 1 API barrel (liq) | 148.1 | 67.18 | 5.614 | 0.1590 | 98.73 | $3.538 \times 10^6$ | $3.7319 \times 10^6$ |
| 1 cubic meter (liq) | 931.6 | 422.6 | 35.31 | 1 | 621.1 | $22.25 \times 10^6$ | $23.47 \times 10^6$ |
| 1 $m^3$ (gas) | 1.499 | 0.6801 | 0.0569 | 0.00161 | 1 | $35.82 \times 10^3$ | $37.78 \times 10^3$ |

[a] Dry basis, 16.696 psia, 60°F.

For a given rate of heat input to (or removal from) an LNG stratum, the density change due to thermal expansion alone is

$$\Delta\rho = 0.06\left(\frac{Q}{mC_p}\right)$$

with $C_P$ = Btu/(lb mol F).

*Note*: The coefficient of thermal expansion of methane is about 0.002/°F. A typical LNG density is assumed as 30 lb/cu ft, giving a 0.006 coefficient.

### 6.4.3. Rollover

*Rollover* is a term that has been used to describe a phenomenon in which a bottom portion of the liquid in a cryogenic storage tank is superheated as a result of heat leak and then spontaneously migrates to the surface accompanied by the evolution of a large quantity of vapor. There is no evidence to suggest that density stratification can occur in nearly pure fluids.

The addition of LNG of different densities to partially filled LNG tanks can lead to the temporary formation of stratified layers. When these layers of different densities exist, the less dense upper layer can convect normally and release heat by flashing into the tank vapor space. However, if the convective boundary layer in the dense lower layer is unable to penetrate the upper layer because buoyancy-driven convective forces are too weak to drive the denser fluid up through the lighter overlying layer, then the lower layer will develop an internal convective pattern. The two patterns remain separate until heat and mass transfer processes bring upper and lower layer densities close enough to each other to allow rapid mixing and the release of the deferred boiloff from the lower portion of the tank. These changes in vaporization rates may be small and insignificant for many modes of operation. Under some conditions, however, the increases in vaporization rates may be large and lead to overpressurization of storage tanks.

Analytical work on the rollover concept indicates that density differences associated with HHV differences of only a few Btu/scf are sufficient to induce rollover. (Rollover with an associated pressure increase of 1 psi can be created with an HHV difference of 4 Btu/scf, a density difference of ~0.1 lb/cu ft.) A change in LNG composition has a greater effect on density than does a change in fluid temperature.

A small dense layer near the bottom of a tank (the heel) does not generate a serious problem; that is, tank pressure is not greatly affected by rollover. Conversely, a thin light layer at the top of the tank provides the most intense rollover potential.

A large initial temperature difference between tank and feed LNG leads to a large increase in boiloff rate after only a short period of time, while a small temperature difference delays the effect for a much longer time period.

Factors determining the time until densities equalize in a two-layer stratified LNG system are:

1. composition change due to boiloff from the upper layer;
2. heat and mass transfer between layers;
3. heat leak into the bottom layer.

Boiloff gas does not have the same composition as upper layer LNG unless the liquid is pure methane. If LNG contains both methane and some heavier hydrocarbons, the boiloff gas is essentially pure methane. Consequently, LNG becomes more concentrated in heavy hydrocarbons as a result of preferential methane loss. This tends to make the upper layer denser over a period of time and drives the system toward density equalization. The process can be retarded if significant $N_2$ is present in the LNG, since $N_2$ boils off preferentially to methane and its loss tends to decrease LNG density. Neglecting the effect of $N_2$ in computing the time for density equalization conservatively underestimates the rollover incubation period.

The heavy layer (again neglecting $N_2$) will be both warmer and more concentrated in heavier hydrocarbons than the upper layer. Heat transfer from this layer to the upper layer accelerates the boiloff rate and therefore increases the upper layer density. Mass transfer between layers is expected to be a slower process than heat transfer, but such effects also move the system toward density equalization as methane diffuses to the lower layer and heavier species counterdiffuse to the upper layer.

Finally, heat leak from portions of the tank contacting the bottom layer will accumulate energy in that layer. If such heat addition is larger than the heat loss across the interface between strata, the heavy layer temperature will gradually increase and layer density will decrease by thermal expansion. If less heat is added than is transferred between layers, the bottom layer will cool, tending to increase density, stabilize stratification, and delay the occurrence of rollover.

### 6.4.4. *Thermal Overfill*

Several cases of sudden uncontrolled venting have been reported in LNG storage tanks at import terminals. In each situation, a warmer and more dense liquid had been filled into the bottom of the storage tank. (There have been instances where sudden venting has occurred in peakshaving plants following top filling of the storage tank.) This type of venting is most often attributed to a rollover occurring in the storage tank. However, some investigators feel that the available data do not necessarily support the rollover concept, but rather "thermal overfill" in the presence of a surface layer phenomenon.

As discussed previously, the mode of heat removal from a storage tank is by evaporation at the free surface. Since the gas and the liquid are in equilibrium at the gas/liquid interface, this process requires the transfer of energy from the slightly warmer bulk fluid to the interface. Most of the resistance to this transfer occurs in a very thin fluid layer at the top surface. The vapor withdrawal rate, then, establishes a surface layer on the liquid that equilibrium temperature colder than the temperature of the underlying bulk liquid. Tank gauge pressure depends only on the liquid surface temperature. This is generally referred to as the layer or top layer phenomenon.

The presence of the surface layer is observed as a sudden decrease in barometric pressure. The vapor withdrawal rate increases in order to cool the top layer sufficiently to maintain a guage pressure less than the vent setting. It is not necessary to vent an amount of vapor sufficient to cool the entire body of liquid to a temperature corresponding to the lower barometer. It is estimated that only a fraction of 1% of full tank contents need be vaporized to restore equilibrium.

The term "thermal overfill" has been used to describe the situation in which a tank has been filled with a liquid whose saturation pressure exceeds the maximum operating pressure of the tank. When an LNG tank is thermally overfilled with a warm and more dense liquid, the surface layer phenomenon establishes a liquid surface temperature corresponding to the operating pressure of the tank rather than the higher vapor pressure of the underlying liquid. Eventually, however, refrigeration must be provided for this underlying warm LNG layer. One mechanism by which this refrigeration may be provided is increased evaporation at the liquid surface.

Reports indicate that if the top layer is disturbed, the evaporation rate will substantially increase depending on the amount of thermal overfill present in the warm underlying liquid. This disturbance in the top layer may be produced by mechanical agitation of the surface layer or by abrupt changes in the operating pressure of the tank. Tank pressure changes can be caused by the malfunction of a relief vent or by the sudden interruption or substantial reduction in the rate of vapor withdrawal from the tank.

### *6.4.5. Stratification*

Regardless of the conceptual explanation, if an LNG tank is filled with liquids of moderately different differensities and/or temperatures, the probability of stratification is high unless precautions are taken in tank design and terminal operating procedures to ensure proper mixing between the two fluids.

Stratification in LNG tanks can be detected by

1. Measuring the vertical temperature (and density) distribution of the liquid within the tank. Temperature differences of 6–7°F have been reported in a tank prior to rollover.
2. Monitoring density differences between feed and tank liquids. A large heating value difference between feed and tank LNG can be an early warning of probable stratification.
3. Monitoring tank vaporization rates. Stratification usually causes an initial depression of boiloff rates, but barometric pressure also has a strong influence.

### *6.4.6. Summary of Stratification*

The rollover concept suggests that the filling procedures should be modified to accommodate the density difference between the incoming fill liquid and the tank heel. The fill lines should be situated appropriately to induce the mixing of tank contents. Thus, a more dense incoming liquid should be top-filled and a less dense incoming liquid should be bottom-filled. It should also be recalled that it has been observed that a fill liquid at a higher temperature is likely to induce an earlier sudden evolution of vapor. This latter point would encourage equilibrating the temperature of the incoming liquid to be in saturation with the tank operating pressure.

Physical mixing of liquids is a problem because of the large size of most LNG storage tanks. Mixing by recirculation pumps often requires days and may take too long to prevent rollover. It also adds heat to the tank, increasing the total vapor load. However, this method can be used if the time required for removing and mixing the lower layer is less than that required for density equalization.

Another approach uses the momentum of the feed stream to mix the inlet fluid jet with the stored LNG. In general, jet mixing is useful when pumping pressures and flow quantities produce large, high-velocity jets. This situation is found in ship terminals where unloading rates can be high. On the other hand, good results will not be obtained by jet mixing from a truck unloading operation or from a peakshaving liquefier because of the small flow quantities relative to tank volume.

Jet penetration into the liquid increases as fluid velocity increases in the inlet pipe. A theoretical analysis (Drake et al., 1973b) shows that the jet penetration distance Z for cases where momentum and buoyancy forces are opposed is

$$Z = 1.4 \left(\frac{\rho_L}{\Delta\rho}\right)^{1/2} \left(\frac{v}{\sqrt{g\,D}}\right) D$$

where $\rho_L$ is the liquid density (lb/cu ft), $\Delta\rho$ the difference in density between jet fluid and stagnant fluid (lb/cu ft), $v$ the jet entrance velocity (ft/s), $g$ the gravitational acceleration (ft/s$^2$), and $D$ the jet entrance diameter (ft).

However, experimental studies indicate that mixing may be more thorough than predicted by this analysis. Nevertheless, in some experiments stratification occurred in the early stage of mixing and prevented the jet from functioning as designed. Additional work may be needed to design mixing jets for LNG tanks without imposing a very high degree of conservatism, as the problem involves interaction of inertial and buoyancy effects.

Proponents of the thermal overfill concept suggest only that the top surface layer should be continually agitated during filling to force the bulk liquid temperature to be in equilibrium with the tank gauge pressure. Such agitation could be produced by top filling the tank for both dense and light incoming liquids.

The filling procedures required to accommodate the two conceptual explanations would therefore seem to be in conflict. However, a recent innovation is being employed in the design of both peakshaving facilities and import terminals that would appear to be of promise to alleviate the spontaneous evolution of vapor by either the rollover or thermal overfill phenomenon. These recent designs use an open-topped, bottom-fill standpipe in the LNG storage tank in addition to top-fill capability. The beneficial aspects of the standpipe structure are that it allows bottom filling of a less dense liquid and at the same time allows the incoming liquid to come into equilibrium with the tank ullage pressure and temperature, avoiding the introduction of higher thermal energy liquid into the stored LNG heel. Flash gas resulting from pressure reduction after the unloading and transfer process is removed in the vapor space of the storage tank. The additional top-fill capability would be used for filling a tank with a heavier incoming liquid automatically equilibrated with the ullage pressure.

For the high circulation rates associated with ship unloading, the open-topped stand-pipe configuration also serves to reduce fluid dynamic reaction forces on the incoming bottom fill line support structure. Typical piping sizes for a terminal facility are a 60 cm (24 in.) diameter fill line emptying into a 27 m (108 in.) diameter standpipe. Peakshaving facilities have considerably smaller standpipes, ranging down to 10 cm (4 in.) in diameter. The standpipe is supported by the tank wall as well as the bottom inner shell.

It may also be interesting to note that some top fill line configurations have induced significant tank vibrations due to two-phase slug flow and its impact on the side wall of the storage tank. The fluctuating fluid forces have produced vibrations up to $\pm$0.6 cm (0.24 in.) in amplitude at approximately 2.5 Hz. In the original configuration, the 60 cm (24 in.) fill line terminated in a vertical section approximately 150 cm (5 ft) from the tank wall. The tank vibration has apparently been reduced to an insignificant level by modifying the fill line to terminate with a 45 cm (18 in.) diameter 45 in. diverter elbow directing the fluid toward the center of the tank.

### 6.5. Marine Transport of LNG

Three methods are available for transporting LNG—trucks, pipelines, and marine tankers. The most widely used method involves marine tankers. Ffooks and Montagu (1967) present a detailed discussion of the evolution of LNG tanker design, and they list several basic design criteria:

1. The low density of LNG and the requirement for separate water ballast containment require a large hull with low draft and high freeboard.
2. The low temperature of LNG requires the use of special and expensive alloys in tank construction. For free-standing tanks, only aluminum or 9% Ni steel is suitable, whereas for membrane tanks, stainless steel or Invar may be used.
3. Due to the large thermal cycling possible in the storage tanks, special supporting arrangements are necessary for free-standing tanks, and flexibility for membranes is necessary in membrane designs.
4. Because the hull of the vessel is carbon steel, good thermal insulation is required between the tanks and the hull. In addition, for membrane tanks, the insulation must be capable of supporting the full weight of the cargo.
5. Cargo-handling equipment must also be carefully designed to account for thermal expansion and contraction.

The design of the tankers of the Alaska-to-Japan LNG project incorporated these principles, as described by Emery and others. Figure 10.26 is a basic sketch of one of these vessels, along with a cutaway view of the insulation. A 14.9 MW (20,000 hp) steam turbine powers the ship, using either boiloff LNG or fuel oil for fuel, and the tanker has a surface speed of 17 knots. The delivered volume is estimated to be about 95% of the loaded quantity, based on an estimated cargo boiloff rate of 0.25% of the cargo per day.

There are two safety problems that are peculiar to LNG marine transport in addition to the dangers of handling flammable materials. These stem from the low-temperature condition of the product. The first problem concerns materials of construction. Thermal expansion and contraction problems during the normal thermal cycling of the tanks and cargo-handling equipment may be minimized by carefully choosing construction materials. The low temperatures cause many materials, such as carbon steel, to become dangerously brittle and unsatisfactory for normal use. The second problem concerns the so-called LNG–water "explosion" that may occur with a maritime accident, when large quantities of LNG are rapidly dumped

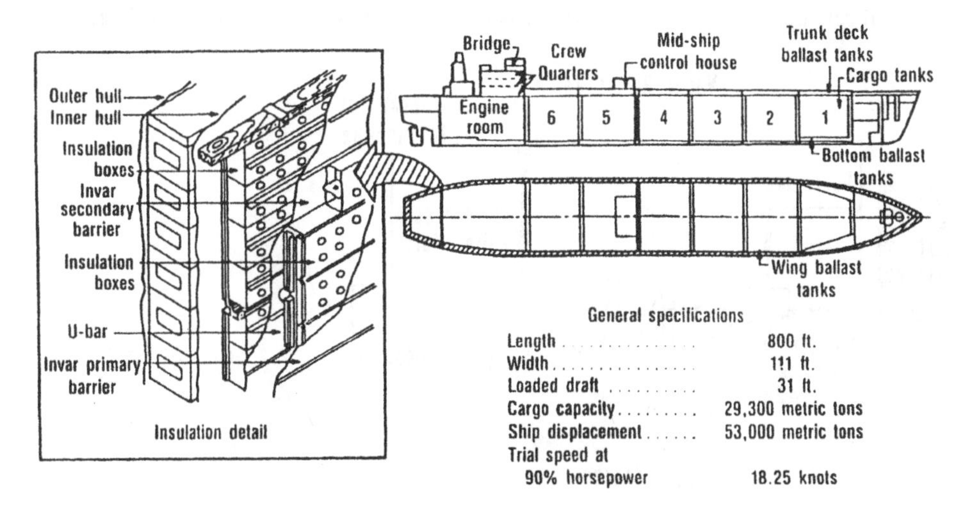


**Figure 10.26** LNG tanker for Alaska-Japan service.

on the ocean. Nakanishi and Reid (1971) suggest an explanation based on the rapid, often violent, boiling that occurs due to the extreme temperature difference between the LNG and the water. Because of its unique character, this phenomenon has received much attention (Nakanishi and Reid, 1971; Burgess et al., 1970; Barber et al., 1971).

## 6.6. Revaporization and Cold Utilization of LNG

Regasification of the stored liquid natural gas is the final step in the operation of an LNG facility. The regasification or vaporization is accomplished by the additional heat from ambient air, ambient water, and integral-fired or remote-fired vaporizers. The cost of the regasification system generally represents only a small fraction of the cost of the storage plant. However, reliability of the system is most important because failure or breakdown would defeat the purpose of the facility.

The regasification section of a peakshaving plant is designed for only a few days of operation during the year to meet the extreme peak loads. To obtain adequate reliability, total sendout capacity may be divided into several independent parallel systems, each capable of handling all or a substantial fraction of the total demand. Designs for the regasification section of a baseload plant are similar to those for a peakshaving plant, but the capacities of these facilities may be much higher, and many more spare items of equipment may be required to achieve adequate reliability.

Figure 10.27 is a simplified flow diagram of a typical LNG regasification system. Liquid is pumped from the storage container to the vaporizer. The pump discharge pressure must be high enough to provide the desired gas pressure for entry into the transmission or distribution system. Heat is added to vaporize the high-pressure LNG and to superheat the gas. Gas leaving the vaporizer must usually be odorized because the liquefaction process removes most or all of the odorant originally in the gas.

### 6.6.1. *Cold Utilization*

One of the most striking characteristics of LNG is the vast amount of cold, or potential refrigeration, available, which is now almost completely wasted. Liquefied

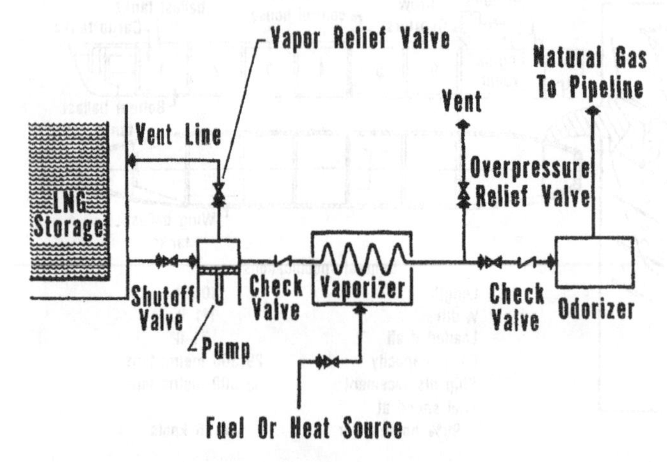


**Figure 10.27** LNG gasification system.

natural gas at 112.2 K, when referenced to a 289 K datum plane, has approximately the same potential thermodynamic credit as low-pressure steam at about 433.3 K. Accordingly, the problem of utilizing the cold from LNG vaporization has been examined for years, but very little has been done in this field of energy conservation. There are two general approaches advanced for utilizing the cold in LNG vaporization: (1) the extraction of work from a power cycle that uses the LNG cold as a heat sink and (2) use of the cold as a source of refrigeration.

Only four of the more than 100 large LNG facilities in the world use the cold in LNG, and none are in the United States. These four commercial facilities are (1) an LNG receiving facility in Barcelona, Spain, which uses an LNG cryogenic process to recover heavy hydrocarbons from Libyan LNG (Rigola, 1970); (2) a similar operation at La Spezia, Italy (Picutti, 1970); (3) a Japanese facility that produces liquid oxygen and liquid nitrogen and affords warehouse refrigeration for frozen food as well (Kataoka et al., 1970); and (4) an LNG-assisted air separation plant in France (Grenier, 1971).

At the La Spezia LNG terminal, refrigeration is used to separate ethane and heavy liquids by fractionation and for cooling requirements in the subsequent liquid reforming plant. Fractionation of the rich Libyan gas is also performed in the ENAGAS terminal at Barcelona. The process employed consumes a great percentage of the LNG cold as a refrigeration medium in the fractionation plant and in subcooling the LPG for refrigerated atmospheric storage. One further cold recovery step has been proposed and considered by several terminal operators, that of producing ethylene by using previously extracted ethane as a feedstock.

In Japan, Tokyo Gas has supplied Tokyo Electric with chilled water, taking the discharge from open-rack seawater vaporizers to supplement the condensing water requirements for a steam turbine power plant. Tokyo Gas is recovering refrigeration from the LNG stream for use in a refrigerated warehouse near the Negishi Works. An intermediate refrigerant is circulated between LNG plant heat exchangers and the unit coolers in the warehouse. For this installation, a 60% power savings has been realized. Available LNG refrigeration can also be used for the reliquefaction of boiloff gas from tank and piping heat leak. At British Gas's Canvey Island terminal, excess boiloff gas is recondensed against send-out LNG flows. Reliquefaction of boiloff gas may be more economical than compression to pipeline sendout pressures and where the low-pressure gas cannot be used as plant fuel.

There are two locations in which LNG cold recovery systems have been installed to provide refrigeration for air separation units and for production of liquid oxygen and nitrogen. An 18,160 kg/hr (480 ton/day) liquid oxygen/nitrogen plant has been operated since 1971 outside the Tokyo Gas Negishi Works. A power savings of 38% has been reported for liquid oxygen production in this plant. Gas de France and L'Air Liquide installed a similar plant in conjunction with the Fos-sur-Mer LNG import terminal. Cold in the LNG is used to lower the temperature of cooling water to the "oxyton" plant. The cooling water is used for compressor cooling and for condensing steam from the turbine drivers.

# 11

# Safety with Cryogenic Systems

## 1. INTRODUCTION

Experience has shown that cryogenic fluids can be used safely in industrial environments as well as sophisticated laboratories, provided that all facilities are properly designed and maintained and personnel handling of these fluids are adequately trained and supervised. The general safety aspects of handling cryogenic fluids can be divided into four main categories.

1. physiological (personnel exposure);
2. suitability of materials and construction;
3. explosions and flammability;
4. excessive pressure.

We shall discuss each of these categories in turn, followed by some special considerations for hydrogen and for oxygen. We shall conclude this section with a few general safety principles and a safely checklist.

## 2. PHYSIOLOGICAL HAZARDS

### 2.1. Frostbite

Severe cold "burns" (frostbite) can be inflicted when the human body comes into contact with surfaces cooled by cryogenic liquids. Damage to the skin or tissue is similar to that inflicted by an ordinary burn. Because the body is composed mainly of water, the low temperature effectively freezes the tissues, damaging or destroying them. The severity of the frostbite depends on the contact area and the contact time; prolonged contact results in deeper burns. Frostbite is accompanied by stinging sensations and pain similar to those of ordinary burns. The ordinary reaction is to withdraw that portion of the body that is in contact with the cold surface. Severe frostbite is seldom sustained if fast withdrawal is possible.

Protective clothing can and should be used to insulate the body from these low temperatures and prevent frostbite (See Fig. 11.1). The choice of appropriate protective clothing depends upon the work environment and is quite flexible. For example, in areas where construction or overhead work is common, a hard hat should be included in the safety clothing. Normally, safety goggles, gloves, and boots are required for transfer of cryogenic fluids. Face shields are sometimes substituted for goggles, but goggles are usually preferable because of the possibility of

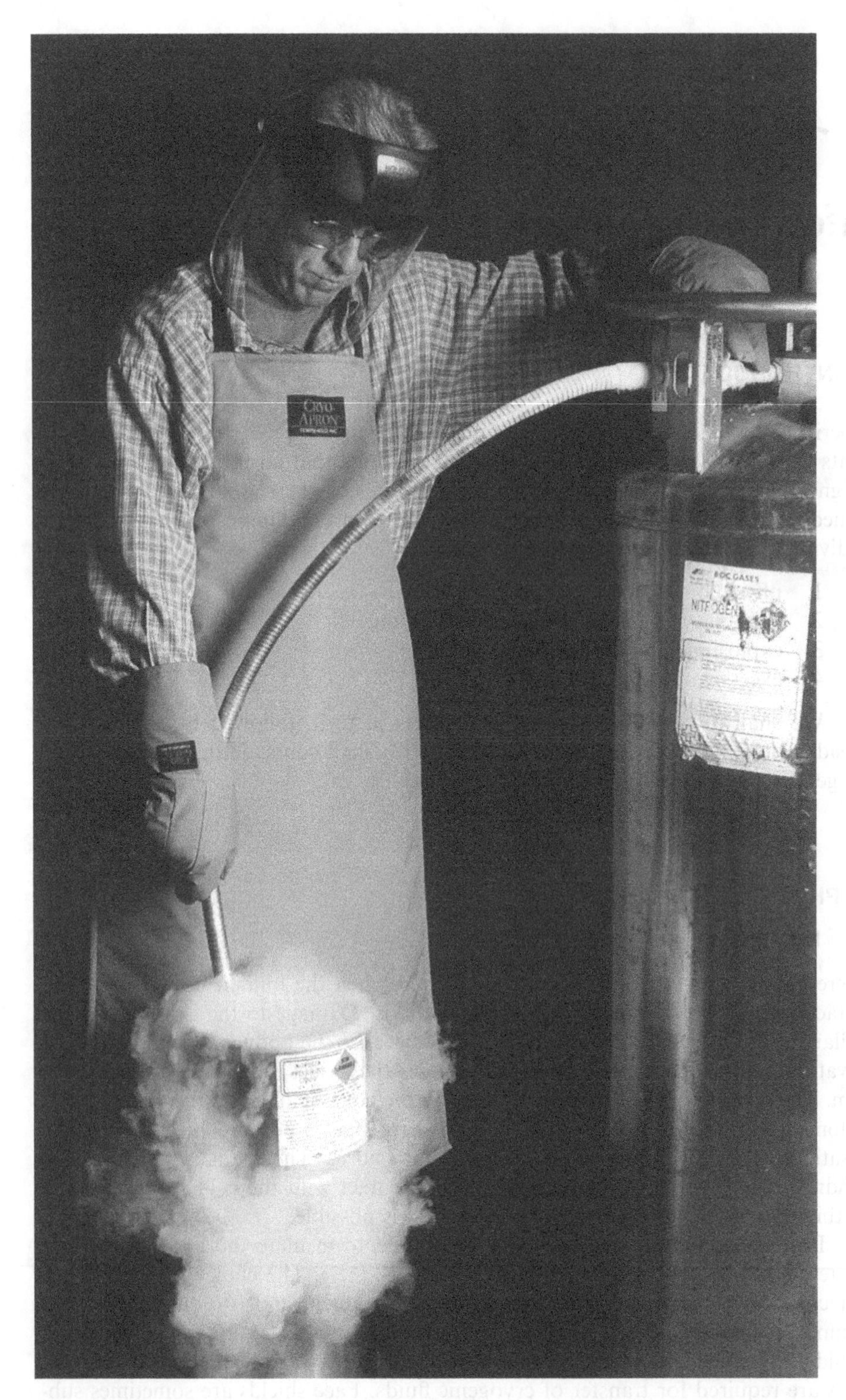


**Figure 11.1** Cryo-gloves, Cryo-apron, Face shield, basic protection.

**Table 11.1** Protective Clothing

- Eye and body protection are required for all workers handling cryogens
- Safety glasses or goggles are required at all times
- Face shields are required when operating systems under pressure, and when:
  Connecting and disconnecting lines
  Venting, unless vent system releases all gases away from personnel
- Loose-fitting, grease-free leather gloves should be worn when handling cryogens
- Adequate footwear should be provided, with trousers outside boots or work shoes
  Open or porous shoes are not permitted in the work area
- Exposed skin should be avoided when working with cryogens to avoid contacting cold surfaces
- All clothing should be free of grease and oils
  Clothing worn for hydrogen service should minimize static electricity buildup
  Clothing worn for oxygen service should minimize flammability
- Any clothing which is splashed or soaked with hydrogen or oxygen should be removed

getting splashes under the shields. Both asbestos and leather gloves are used, but the leather gloves must be washed in a suitable solvent to remove tanning oils before they are used for handling liquid oxygen. What types of gloves to use has been the subject of considerable debate; some prefer tightly fitted cuffs up inside the sleeves, while others prefer the gauntlet-style cuff. The gauntlet style permits temporary entrapment of liquid in the cuff but may be quickly removed should the glove fingers become frozen. Large spills, capable of filling the cuffs, are not usually encountered, but cold glove fingers resulting from handling cold equipment are rather common; therefore the gauntlet style of glove is generally the best. Close-fitting boots may be used to prevent catching and trapping liquid cryogens in shoes, which could result in frozen feet. Cuffless coveralls or trousers that cover high-top shoes are frequently substituted for boots. See Table 11.1. White coveralls usually are worn when handling liquid oxygen so that it is easy to tell when they are dirty; the concern here is the incompatibility of oxygen with dirty materials, a subject discussed later in this chapter.

Regardless of the particular style of clothing selected, the idea is to prevent direct contact between the skin and a low-temperature fluid or cold surface. Should exposure occur and a burn be inflicted, the only first aid treatment is to liberally flood the affected area with lukewarm water. Do not rub or massage the affected parts under any condition, as additional tissue damage may be incurred. In less severe cases, the affected area can be identified by its whitish appearance (like dead skin). Frostbite should receive medical attention. See Table 11.2.

## 2.2. Nitrogen Asphyxiation

Personnel should work in groups of two or more when handling liquid cryogens. If required to work in confined spaces with questionable or uncertain ventilation, the area should be continually monitored with an oxygen-analyzing device to sound an alarm if the oxygen content of the air gets too low or too high (16–25% is acceptable). Airline respirators or a self-contained breathing apparatus should be used in oxygen-deficient atmospheres.

**Table 11.2** Physiological Personnel Exposure

- Frostbite
  - Very brief contact with a small amount of cryogenic fluid will probably not do a great deal of damage
  - Prolonged contact (above a second or two) with a cryogenic fluid or even momentary contact with a cold surface can cause painful "burns"
  - Shut off the flow (if possible) or escape the area (or both)
- Treatment of frozen body tissue
  - In the field, wrap the affected area loosely and transport to medical treatment facility
  - Immerse affected tissues in lukewarm water between 105°F and 110°F
  - Do not remove frozen clothing except in a slow, careful manner
  - Do not massage or rub affected part(s)
  - Do not apply snow or ice
  - Do not use safety showers or eyewash fountains
  - Do not apply ointment

Because nitrogen is a colorless,* odorless, inert gas, personnel must be aware of the associated respiratory and asphyxiation hazards. Whenever dilution of the oxygen content of the atmosphere occurs, due to spillage or leakage of nitrogen, there is danger of nitrogen asphyxiation. One is most likely to encounter such atmospheres when entering (1) large vessels that have contained gaseous or liquid nitrogen or (2) laboratories, rooms, or compartments where spillage of liquiud or gaseous nitrogen has occurred. In either case, a good air purge permits entry without hazard. Tanks should be fresh air purged prior to entry, and the forced ventilation should be continued while personnel occupy the tank. In addition, a strong rope should be attached to each person entering the vessel so that teammates may physically retract them from the tank if they lose consciousness. One should never be allowed to enter or remain inside a tank without a companion stationed outside to man the lifeline. The companion should continuously check on the reactions of the person within the tank. Table 11.3 contains some life-saving reminders. Obviously, proper ventilation is of prime importance in situations where the atmosphere may contain too much nitrogen and not enough oxygen gas. A person who becomes groggy or loses consciousness while working with nitrogen systems should be taken to a well-ventilated area immediately. Apply artificial respiration or mouth-to-mouth resuscitation if breathing has stopped; summon a physician. In general, the oxygen content of air should not be allowed to fall below 16%; Table 11.4 lists the physiological effects of lower concentrations of oxygen in air.

### 2.3. Effects of Pure Oxygen

Oxygen gas is also colorless, odorless, and nontoxic but, unlike nitrogen, produces exhilarating effects when breathed. Nevertheless, lung damage can occur if the oxygen concentration in the air exceeds 60%, and prolonged exposure to an atmosphere of pure oxygen may cause bronchitis, pneumonia, or lung collapse. Perhaps the greatest threat of oxygen-enriched air lies in its increased flammability and explosion

*The white vapor cloud, visible upon exposure of cold nitrogen gas or liquid to the atmosphere, is water vapor being condensed from the air.

**Table 11.3** Asphyxiation and Toxicity

- Cryogenic fluids released or spilled in a confined area alter the air composition by displacement, introducing an asphyxiating risk Oxygen concentrations as low as 13 vol% can be tolerated by healthy subjects at 1 atm total pressure
- Hazards may exist in "empty" tanks, spill areas, or purged areas
- Always work in pairs with one team member outside the work area
  A worker entering a vessel should wear a harness-type safety belt with an attached lifeline
  The line should be tended by a watcher outside the vessel where the watcher can be in constant touch with the worker throughout the operation
- Use portable oxygen monitors to determine the adequacy of the local atmosphere

**Table 11.4** Respiratory Hazards of Nitrogen-Enriched Atmosphere

| Percent oxygen in air | Physiological reactions |
|---|---|
| 12–14 | Respirations deeper, pulse up, coordination poor |
| 10–12 | Respiration fast and shallow, giddiness, poor judgment, lips blue |
| 8–10 | Nausea, vomiting, unconsciousness, ashen face |
| 6–8 | 8 min, 100% fatal; 6 min, 50% fatal; 4–5 min, all recover with treatment |
| 4 | Coma 40 sec, convulsions, cessation of respiration, death |
| 0 | Death in 10 sec |

**Table 11.5** Physiological Effects of Pure Oxygen

- Lung damage can occur if oxygen concentration exceeds 60%
- Prolonged exposure to pure oxygen may cause:
  Bronchitis
  Pneumonia
  Lung collapse
- Prevention of oxygen exposure
  Locate oxygen detectors near the source
  Watch for fog or mists
- If oxygen exposure occurs:
  Move affected personnel to clear area
  Seek medical attention if exposure warrants

hazards, as discussed later in this chapter. Housekeeping (tidiness and cleanliness) also becomes important in systems using oxygen because of the increased possibility of fires and explosions. See Table 11.5.

## 3. SUITABILITY OF MATERIALS AND CONSTRUCTION TECHNIQUES

### 3.1. Brittle Fracture

Construction materials for noncryogenic service are usually chosen on the basis of tensile strength, fatigue life, weight, cost, ease of fabrication, corrosion resistance, and similar factors. When dealing with low temperatures we must also consider the ductility of the material. Low temperature has the effect of making many materials brittle or less ductile. The *ductility* of a material is a measure of its ability to be stretched, to bend, or to absorb impact. Some materials become brittle at low temperatures but can still absorb considerable impact, while others get brittle and lose their impact strength. Materials that fail under impact at low temperature are said to fail due to "brittle fracture". Ordinary iron and steel (commonly referred to as "carbon steel") are candidates for brittle fracture.

Whether or not a material may fail due to brittle fracture at low temperatures has been the subject of considerable investigation. As a result of this study there are two excellent standards by which the impact strength of a material may be judged. The Naval Research Laboratory has developed a drop–weight test to determine the nil ductility transition (NDT) temperature of a material specimen. The NDT temperature varies with material, thickness, hardness, and other properties; and brittle fracture may occur below this temperature with very little stretching or bending. For brittle fracture to occur, the stresses in the material must exceed a certain critical value and the temperature of the material must be below the NDT temperature. Another standard for impact strength is the Charpy Impact Test. The Charpy tester also uses a drop weight to measure the ability of a specimen to withstand an impact load. The Charpy standard is used most extensively because smaller specimens can be tested at a lower cost. Test results show that brittle fracture of materials may be directly related to the Charpy impact value for that material. Such tests have led to the development of codes to specify the minimum allowable impact value for various materials. The ASME Boiler and Pressure Vessel Code for Unfired Pressure Vessels, Sec. VIII (1968), supplies these data. The code does not require impact testing of low–alloy and carbon steels used at temperatures above −20°F but permits their use at temperatures below −20°F only if they pass the minimum requirements for impact resistance. The NDT temperature range for most carbon steels is about −100 to −10°F. The impact resistances of various materials at low temperatures are shown in Fig. 11.2 all of the materials shown, except carbon steel and (50–50) soft solder, are satisfactory for low-temperature service. For cryogenic service, type 18-8 stainless steels, aluminum, copper, and brass are acceptable materials; low–alloy and carbon steels are not. Materials such as Monel, Inconel, and 9% nickel steel may also be used with some sacrifice in cost, ease of fabrication, or impact resistance.

Brittle fracture can occur very rapidly, resulting in almost instantaneous failure. This failure can cause shrapnel damage if the equipment is under pressure. Also, release of a fluid such as oxygen can result in fire or explosions. Figure 11.3 shows a carbon steel pipe that failed due to brittle fracture; the pipe was accidentally cooled below its brittle transition temperature while pressurized internally. It is interesting to note that some carbon steels have brittle transition temperatures above −10°F and that many World War II ship failures were attributed to brittle fracture. In passing, it should be pointed out that Pyrex glass is frequently used in cryogenic research apparatus. Glass has an "elephant's memory," recalling all abuse, scratches, stresses, etc., and consequently can fail without warning. Glass pipe and glass dewars under internal or external pressure can be a severe shrapnel hazard and should be adequately shielded for personnel protection. Clear plastic shields are often used where fluid visibility is required. Table 11.6 gives the definitions of commonly used metals.

### 3.2. Thermal Stress

Low-temperature equipment also can fail due to thermal stresses. Thermal stresses are caused by thermal contraction of the materials. Where two dissimilar materials

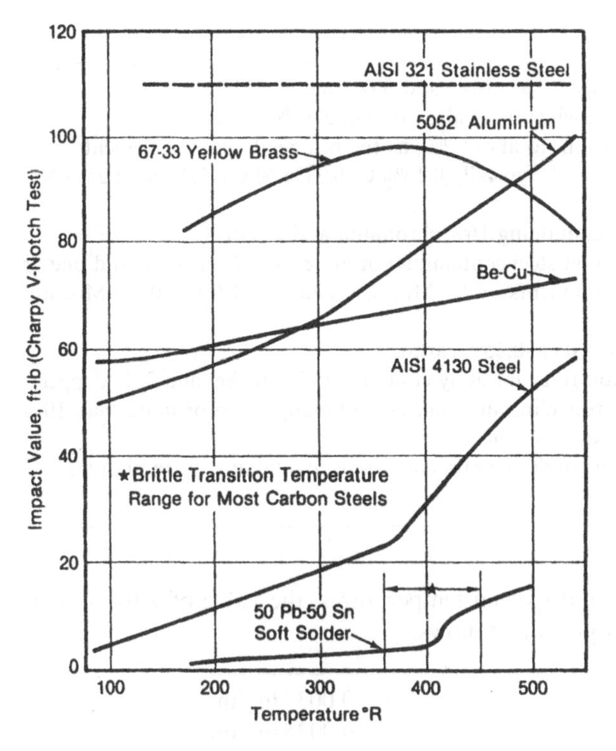


**Figure 11.2** Impact resistance of various metals.

are joined, the materials contract by different amounts when cooled and consequently may pull apart if not adequately restrained. The change in length of a material when cooled may be calculated as

$$\Delta l = \alpha l_0$$

where $\Delta l$ is the change in length, $\alpha$ the overall coefficient of thermal contraction (or expansion), and $l_0$ is the initial or room temperature length. Between room tempera-

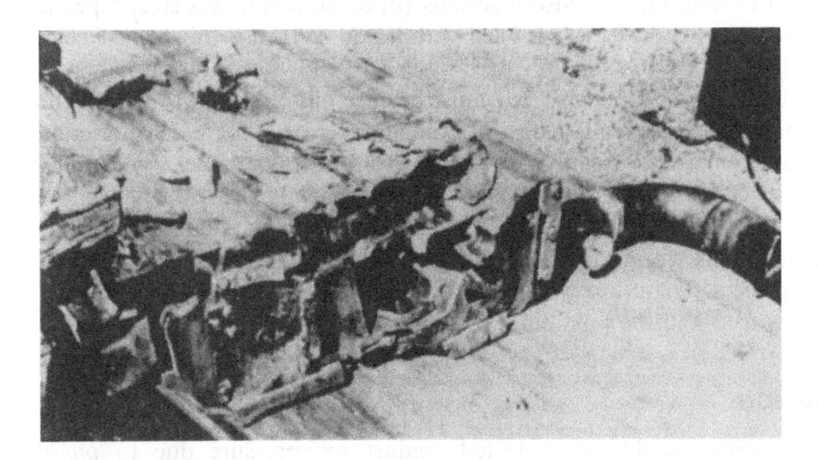

**Figure 11.3** Brittle fracture of a carbon steel pipe.

**Table 11.6** Metal Definitions

*Brass*—A copper–zinc alloy, typically 67% copper and 33% zinc
*Bronze*—An alloy of copper and tin which may also contain Zn, Ni, and Pb
*Carbon Steel*—Steel containing carbon, to about 2%, as the principal alloying element
*Hy-80*—A high strength, low alloy steel, typically 0.15% C, 0.31% Mn, 0.22% Si, 2.25% Ni, 1.4% Cr, 0.31% Mo
*Inconel*—Trade name for an alloy containing 16% chromium and 7% iron
*Low-alloy Steel*—A hardenable carbon steel containing not more than 1% carbon and one or more of the following alloyed components: $<2\%$ Mn, $<4\%$ Ni, $<2\%$ Cr, $<0.6\%$ Mo, and $<0.2\%$ V
*Mild Steel*—Carbon steel containing 0.05–0.25% carbon
*Monel*—Trade name for a corrosion resistant alloy containing 67% nickel and 30% copper
*Stainless Steel*—A large group of iron–chromium alloys containing up to or more than 10% chromium plus Ni, Si, Mo, W, Ni, and others
*Steel*—An iron base alloy, malleable under certain conditions, containing up to about 2% carbon

ture and liquid nitrogen and liquid oxygen temperatures, the value of $\alpha$ for several common structural materials is given as follows:

| | |
|---|---|
| Aluminum | $\alpha = 0.0039$ in./in. |
| Brass | $\alpha = 0.0035$ in./in. |
| Copper | $\alpha = 0.0030$ in./in. |
| 18–8 Stainless steel | $\alpha = 0.0028$ in./in. |

An aluminum tube that is 10 in. long at room temperature, when cooled to liquid nitrogen temperature would shrink 0.0039 in./in. $\times$ 10 in. $= 0.039$ in. In solder joints, the solder must be strong enough to resist stresses caused by differential contraction where two dissimilar metals are joined. Soft solder (50–50) is generally not used for this job in low-temperature service where the joint is subjected to cyclic duty. The soft solder loses its ductility at low temperature as evidenced by its loss of impact resistance as shown in Fig. 11.2.

Contraction in long pipes is also a serious problem; a stainless steel pipeline 100 ft long would contract 3.4 in. when filled with liquid oxygen or nitrogen. Provisions must be made for this change in length by using bellows, expansion joints, or flexible hose. Pipe anchors, supports, etc., must be carefully designed to permit the contraction to take place without tearing the pipeline apart. Thermal contraction is not always bad; sometimes it can be put to good use. For example, when trying to make a seal with concentric mating parts of dissimilar metals, the seal will be more effective if the material with the larger $\alpha$ is placed outside of the material with the smaller $\alpha$. The outer part will shrink more than the inner part and make a tighter seal between the two pieces. The primary hazard of failure due to thermal contraction is spillage of the cryogen and the possibility of fire or explosion.

### 3.3. Overpressure

All cryogenic systems should be protected against overpressure due to phase change zfrom liquid to gas. Noncryogenic systems should also be protected against overpressure that may arise as a result of fire or some other unforeseen incident. Relief valves and burst disks are commonly used to relieve piping systems at a pressure near the design pressure of the equipment. Such relief should be provided between valves, on tanks, and at all points of possible (though perhaps unintentional) pressure rise in a piping system. Gas lines can reach rather high pressures during a fire if they are not adequately relieved. Systems containing liquid cryogens can reach bursting pressure very rapidly, if not relieved, simply by trapping the liquid in an enclosure. The rate of pressure rise depends on the rate of heat transfer into the liquid. In uninsulated systems the liquid is vaporized rapidly, and pressure in the closed system rises very fast. If a vessel containing liquid nitrogen or liquid oxygen is sealed off at atmospheric pressure without relief protection, the pressure in the container will reach the values shown in Fig. 11.4. The more liquid there is in the tank when it is sealed off, the greater will be the resulting pressure. Figure 11.5 shows how high the pressure in the tank can rise when all of the liquid is vaporized (737 psia for oxygen and 493 psia for nitrogen); however, once the liquid is vaporized, the resultant cold fluid warms to room temperature and increases the pressure in the container even more. Figure 11.5 is applicable only to those cases where the liquid has sufficient room to expand inside the tank. If the tank is 100% full, the lock-up (or seal-off) pressure will exceed the values shown in the figure while the liquid phase still exists. So it is evident that conversion of a cryogenic liquid to warm gas causes excessive pressures unless pressure relief is provided.

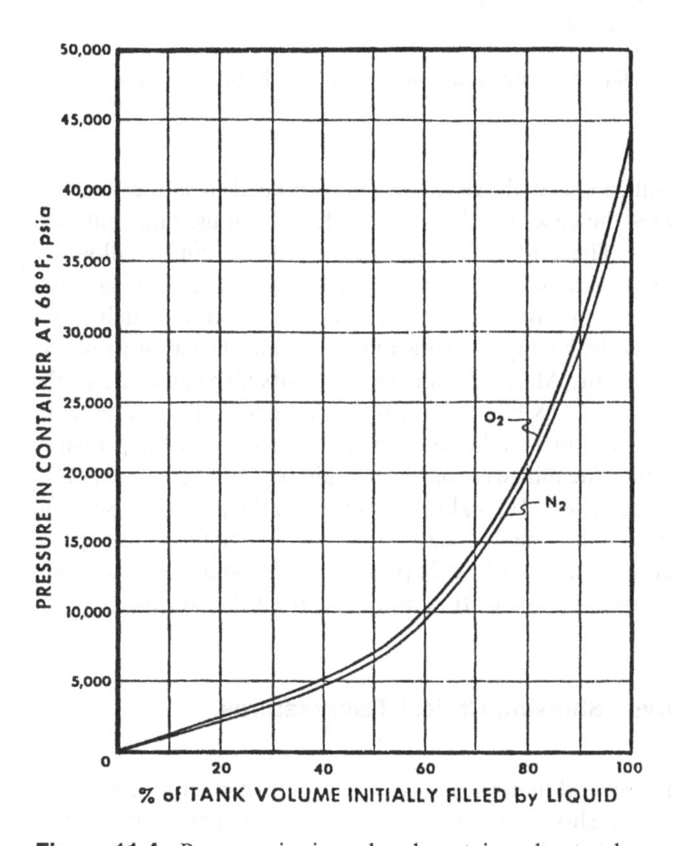


**Figure 11.4** Pressure rise in a closed container due to phase change.

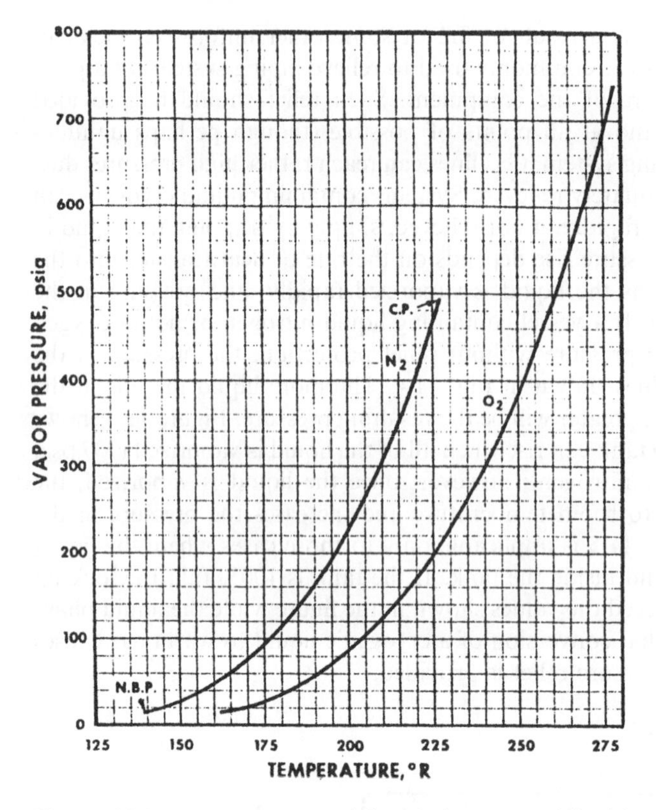


**Figure 11.5** Vapor pressure of liquid nitrogen and liquid oxygen. N.B.P. = normalboiling point; C.P. = critical point.

Ullage space, a vapor-filled space above the liquid, is crucial to safe cryogenic storage. Figure 11.6 illustrates the case of a 1000 gallon LOX storage tank initially 70% full, or an ullage of 30%. The tank is "locked down", no venting is allowed, to assure that all the contents remain available for their intended use. This tank has a heat leak equivalent to boiling away 1% of the contents per day, which is an industry norm. At first, this heat leak simply warms the LOX, and the vapor pressure rises at a modest rate of 0.36 psi/hr. Alas, warming the contents also cause the LOX to expand, so that after 620 hr, the LOX has expanded to completely fill the tank, and the ullage is now zero %. This is shown as the sharp increase in slope of the pressure vs. time shown in Fig. 11.6. Since the ullage is gone, the liquid has nowhere to expand, and now the rate of pressure increase is 21 psi/hr. For this case, the pressure rise is an unalarming rise of 0.36 psi/hr for the first 25+ days. Suddenly, when the ullage space disappears, the pressure rise is an astounding 21 psi/hr. One would have to make haste to avoid a catastrophic over-pressurization and rupture of the storage tank.

### 3.4. Vapor Pressure Curves Showing Critical Temperatures and Pressures

The vapor pressure curves for several materials—hydrogen, nitrogen, argon, oxygen, carbon dioxide, and water—are shown in Fig. 11.7. The critical temperature, the temperature above which the liquid phase cannot remain in stable existence,

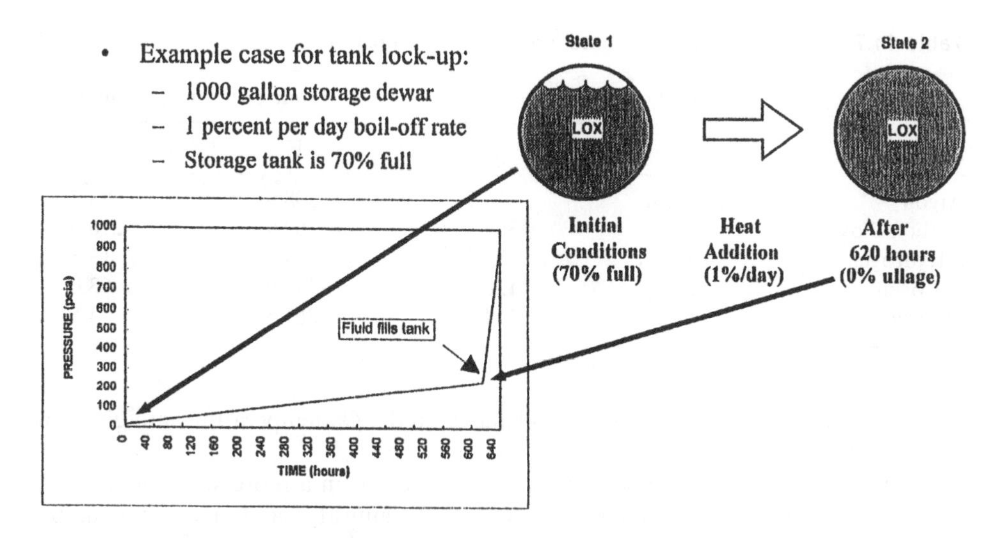

Example case for tank lock-up:
1000 gallon storage dewar
1 percent per day boil-off rate
Storage tank is 70% full
State 1
State 2
LOX
LOX
Initial Conditions (70% full)
Heat Addition (1%/day)
After 620 hours (0% ullage)
PRESSURE (psia)
Fluid fills tank
TIME (hours)
Parasitic heat leak expands the liquid in the tank
Pressure rises slowly, 0.36 psi/hour
Once liquid fills the entire tank, pressure rises rapidly, 21 psi/hour

**Figure 11.6** Storage tank self-pressurization.

is shown as a circle at the upper limit of the vapor pressure curve. The triangles represent the triple point, the temperature at which the liquid, solid, and gaseous phases of the component all exist in equilibrium. The liquid and vapor phases exist in the temperature range from the triple point to the critical; only the solid and vapor phases exist at temperatures below the triple point. The critical temperatures and

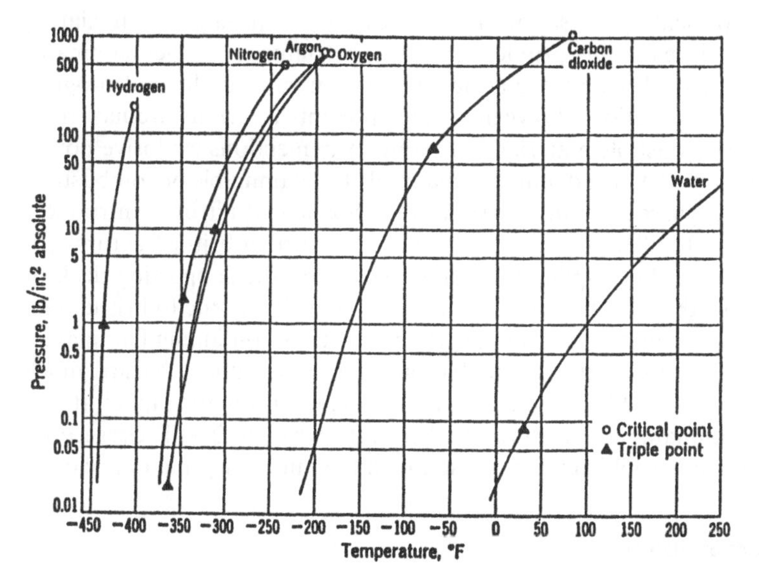

Hydrogen
Nitrogen
Argon
Oxygen
Carbon dioxide
Water
o Critical point
▲ Triple point
Pressure, lb/in.$^2$ absolute
Temperature, °F

**Figure 11.7** Vapor pressures of hydrogen, nitrogen, argon, oxygen, $CO_2$, and water. (Courtesy of Air Products and Chemicals, Inc., Allentown, PA.)

**Table 11.7** Critical Temperatures and Triple-Point Temperatures

| | Critical temperatures | | Triple point temperatures | |
|---|---|---|---|---|
| | °F | K | °F | K |
| Argon | −188.4 | 150.7 | −308.9 | 83.7 |
| Carbon dioxide | 87.98 | 304.3 | −69.9 | 216.0 |
| *p*-Hydrogen | −400.4 | 32.9 | −434.8 | 13.8 |
| Nitrogen | −232.7 | 126.1 | −346.0 | 63.0 |
| Oxygen | −181.2 | 154.7 | −361.5 | 54.5 |

triple point temperatures for the four elements and $CO_2$ temperatures are given in Table 11.7.

Overpressure in cryogenic systems can also occur in a more subtle way. Vent lines without appropriate rain traps can collect rainwater and freeze closed. So can exhaust tubes on relief valves and burst disks. Small-necked open-mouth dewars can collect moisture from the air and freeze closed. In this event the ice plug can be melted or loosened by inserting a warm copper rod. Entrapment of cold liquids or gases can occur when water or other condensables are frozen in some portion of the cold system. Should this entrapment occur in the wrong place, the relief valve or burst disk may be isolated and afford no protection. Such a situation usually arises from improper operating procedures and emphasizes the importance of good operating practices.

### 3.5. Secondary Accidents

In making an accident or safety analysis, it is always wise to consider the possibility of a bad situation getting worse. For example, one of the failures mentioned above (brittle fracture, contraction, overpressure, etc.) may release sizable quantities of cryogenic liquids, presenting a severe fire or explosion hazard, asphyxiation possibilities, further brittle fracture problems, or shrapnel damage to other flammable or explosive materials. In this way, the accident situation can rapidly and progressively become much more serious. Oxygen spillage presents a serious fire hazard in the presence of any combustible material and creates a tremendous explosive potential when it becomes mixed with a flammable material. The flammable or combustible material may be in the gaseous, liquid, or solid state. Spillage of nitrogen introduces respiratory hazards, and spillage of any cryogen can cause further brittle fracture problems; e.g., when a vessel fails, spilling a liquid cryogen on the supporting deck or structure, it is possible that the deck or support structure will fail due to brittle fracture. Spillage or drippage of cryogens on a carbon steel tank containing flammables can result in brittle fracture of the flammable tank and a secondary hazard much greater than the initial one. Likewise, an initial accident can fling shrapnel about, causing damage to flammable containers and other explosive materials. These secondary accidents can be catastrophic in comparison with the initial or primary accident.

### 3.6. Oxygen Compatibility

It is of vital importance to select materials and methods of construction compatible with the chemical properties of oxygen. Oxygen reacts readily with almost all materials known under conditions favorable to combustion. This reaction may take place in the form of fire or an explosion. To avoid this hazard, materials must be carefully selected and meticulously assembled, and ignition sources must be effectively eliminated. There is little need to discriminate between the gas and liquid phases when dealing with the chemical compatibility of oxygen. Compatible materials are materials that will not explode when simultaneously exposed to liquid oxygen and an impact hammer or drop weight. One compatibility test requires that a $\frac{1}{2}$in. diameter sample of the test material be submerged in liquid oxygen and subjected to 72 ft lb of impact energy. If 20 samples of a material can be tested in this fashion without incident, the material is accepted for oxygen service and is "compatible" with oxygen. Materials failing this test are not to be used in oxygen service and are said to be "noncompatible". Any material that passes the compatibility test in liquid oxygen is automatically qualified for gaseous oxygen service.

Most structural materials recommended for avoidance of brittle fracture are compatible with oxygen; aluminum, copper and its alloys, and stainless steel are acceptable materials. The most commonly used seal and gasket materials are Kel-F, Teflon, and asbestos. Metallic gaskets may be made of annealed aluminum or copper. In general, all petroleum-derived lubricants and silicone-based greases and fluids are noncompatible with oxygen. Many of the fluorinated and chlorinated fluids and greases are compatible, but the chlorofluorocarbon lubricants such as Fluorolube, Kel-F, and Halocarbon should not be used in aluminum systems. The fully fluorinated greases and oils are compatible with oxygen in all of the accepted metal systems. Teflon tape is often used as lubricant and sealant for threaded connections. Where moving parts come into contact with oxygen, the oxygen itself or Teflon bearings usually provide the necessary lubrication. The only lubricant recommended for use in vacuum pumps that may pump oxygen gas is tricresyl phosphate (TCP). TCP is toxic and should be handled accordingly. Insulation materials for liquid oxygen storage tanks are chosen for their low flammability and insulation qualities. The most widely used insulation materials are granular types like perlite, silica gel, and diatomaceous earth; multiplayer radiation shield insulations also are used. Liquid oxygen transfer lines should not be insulated with wrap-on or slip-on materials such as cork or plastic foams. These materials are either noncompatible or can collect contaminants that are noncompatible with liquid oxygen. If the transfer line must be insulated, it should be vacuum-jacketed with a suitable metal. Some cellular glass insulations are compatible with liquid oxygen and may be used for low-cost, temporary installations; however, care must be taken to avoid contaminating this applied insulation with greases, oils, mastics, and other noncompatible debris.

Obviously, cleanliness is of utmost importance in handling oxygen. One must be continuously aware of possible sources of contamination in oxygen systems. A casual fingerprint or human hair on a part exposed to liquid oxygen results in an impact-sensitive fuel that can explode. Remember that *clean* compatible materials are the only materials that can be allowed to come into contact with gaseous or liquid oxygen. Valves, pipes, vessels, etc., used for oxygen service must be completely free of grease, oils, and other contaminants. Solvents suitable for cleaning oxygen equipment are trichloroethylene, trichloroethane, and methylene chloride. All of these solvents must be used with adequate ventilation and skin protection. Trichloroethylene is probably the most widely used solvent and is suitable for hot vapor, immersion, or scrub action cleaning. Hot water and an appropriate nonhydrocarbon detergent such as trisodium phosphate may also be used to degrease and clean equipment for oxygen service. The equipment must be thoroughly flushed and clean for oxygen service. It must be thoroughly flushed and dried when water is used. Parts cleaned for oxygen storage should be protected against subsequent contamination by plastic bags, plastic plugs, etc. A dust-tight seal must be made to ensure that the parts stay clean.

In the fabrication and construction of oxygen equipment, it is necessary to design and assemble parts so that cracks, recesses, pores, etc., are effectively eliminated. All surfaces contacting the fluid should be smooth, nonporous, clean, and compatible with oxygen. Pockets, cracks, and other irregularities are difficult to clean properly and act as collection sites for foreign particles or contaminants traveling with the oxygen stream. Gaskets should be designed and fitted to avoid recesses insofar as possible. Acceptable methods of joining materials or lengths of pipeline are welding, silver soldering, and the use of flanges, threaded joints, and flared joints. Soft solder is *not* acceptable and should *not* be used for oxygen service. In addition to being susceptible to brittle fracture in low-temperature cyclical service, soft solder is impact-sensitive in liquid oxygen. Soft solders melt at low temperatures and contain tin or lead. The silver solders melt at temperatures above 1000°F, contain no lead or tin, and are compatible with oxygen. Joints that cannot be disassembled for cleaning and equipment with dimensions and voids that make cleaning by immersion difficult are best cleaned in an ultrasonic cavitation cleaning device using an appropriate solvent. Hot trichloroethylene degreasing units are also effective in difficult cleaning situations. Threaded pipe fittings, though acceptable for oxygen service, should be avoided wherever possible because the threads that are not engaged on the female part of the fitting can collect debris. Sealing the threads while maintaining liquid oxygen compatibility is also a problem. Figures 11.8 and 11.9 show the essential features of good and bad joining techniques. The examples given in these figures are intended to illustrate how crevices can be avoided by proper construction methods; other joint configurations may be stronger. The ASME Boiler and Pressure Vessel Code for

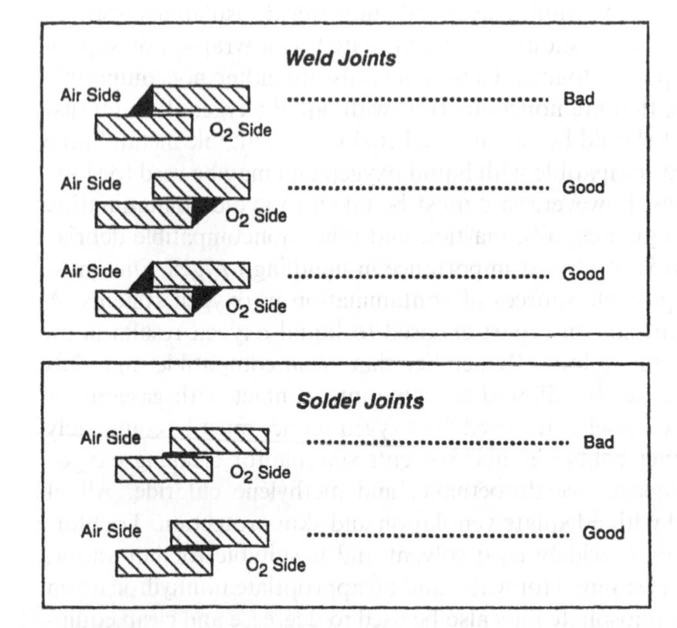


**Figure 11.8** Construction details for equipment in oxygen service—lap joints.

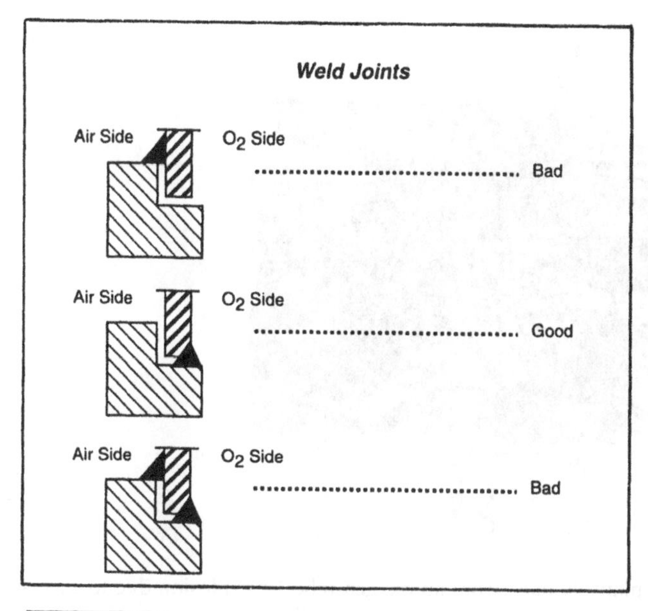


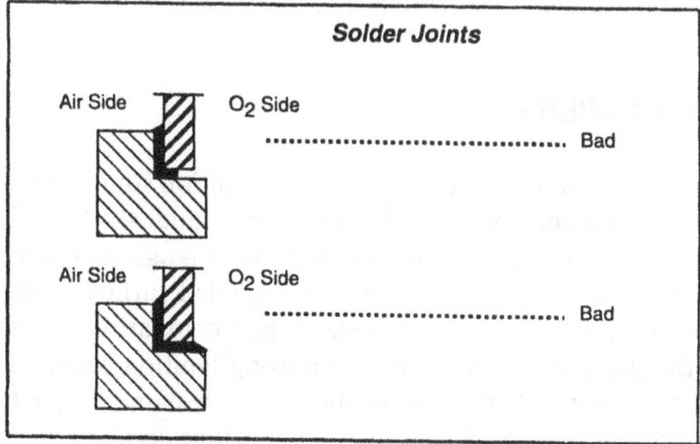


**Figure 11.9** Construction details for equipment in oxygen service—slip joints.

Unfired Pressure Vessels, Sec. VIII, is helpful in the design of weld joints and in related structural design details.

Contamination of liquid oxygen can also result from the use of contaminated pressurizing gas to transfer liquid out of a storage dewar. In oxygen liquefaction (air separation) plants, care must be taken to avoid contamination from compressor lubricants and from the atmosphere itself. The air taken into the liquefier must be free of contamination. Dust and suspended particles can be removed by mechanical filtration and electrostatic precipitators, but hazardous concentrations of hydrocarbons must be avoided. The fresh air intake of such plants should be monitored to detect hydrocarbons unless pure intake air can be guaranteed; see Fig. 11.10 for one case in point. Oxygen liquefaction and storage facilities should be monitored on a periodic basis to determine the accumulation of acetylene and total hydrocarbons. Certain inspection routines at periodic intervals are also necessary to ensure that oxygen equipment has not become contaminated.

**Figure 11.10** Air separation plant damage from an oxygen reboiler explosion due to excessive atmospheric hydrocarbon contamination at the compressor inlet.

## 4. EXPLOSIONS AND FLAMMABILITY

In order to have a fire or an explosion an oxidant, a fuel, and an ignition source must exist in combination. Where nitrogen and oxygen fluids are being handled, the oxidizer will be oxygen. The oxygen may be present due to spillage or leakage, or it may occur as a result of the condensation of air on liquid nitrogen-cooled surfaces. The fuel may be almost any noncompatible material or flammable gas; compatible materials can also act as fuels in the presence of extreme heat (strong ignition sources). The ignition source may be a mechanical or electrostatic spark, a flame, impact, or heat from kinetic effects, friction, chemical reaction, etc. Certain combinations of oxygen, fuel, and ignition source will always result in fire or explosion.

The order of magnitude of flammability and detonability limits for fuel–oxidant gaseous mixtures of hydrogen and methane is shown in Table 11.8. Selected properties of hydrogen, methane, gasoline, and oxygen are shown in Table 11.9.

### 4.1. Flammability and Detonability Data

Table 11.10 portrays the order of magnitude of flammability and detonability limits for fuel–oxidant gaseous mixtures (NFPA, Vol. V, 1962–1963; NFPA, Vol. 1, 1962).

**Table 11.8** Flammability and Detonability Limits of Hydrogen and Methane Gas

| Mixture | Flammability limit (mol%) | Detonability limit (mol%) |
|---|---|---|
| $H_2$–air | 4–75 | 20–65 |
| $H_2$–$O_2$ | 4–95 | 15–90 |
| $CH_4$–air | 5–15 | 6–14 |

**Table 11.9** Selected Properties of Hydrogen, Methane, Gasoline, and Oxygen

| Property | Hydrogen | Methane | Gasoline | Oxygen |
|---|---|---|---|---|
| Molecular weight | 2.016 | 16.043 | ~107.0 | 31.9988 |
| Triple point pressure, atm | 0.0695 | 0.1159 | – | 0.0015 |
| Triple point temperature, K | 13.803 | 90.680 | 180–220 | 54.35 |
| Normal boiling point (NBP) temperature, K | 20.268 | 111.632 | 310–478 | 90.18 |
| Critical pressure, atm | 12.759 | 45.387 | 24.5–27 | 49.76 |
| Critical temperature, K | 32.976 | 190.56 | 540–569 | 154.58 |
| Density at critical point, $g/cm^3$ | 0.0314 | 0.1604 | 0.23 | 0.44 |
| Density of liquid at triple point, $g/cm^3$ | 0.0770 | 0.4516 | – | 1.306 |
| Density at solid triple point, $g/cm^3$ | 0.0865 | 0.4872 | – | 1.359 |
| Density of vapor at triple point, $g/m^3$ | 125.597 | 251.53 | – | 10.77 |
| Density of liquid at NBP, $g/cm^3$ | 0.0708 | 0.4226 | ~0.70 | 1.1408 |
| Density of vapor at NBP, $g/cm^3$ | 0.00134 | 0.00182 | ~0.0045 | 0.00447 |
| Density of gas at NTP, $g/cm^3$ | 83.764 | 651.19 | ~4,400 | 1331.15 |
| Density ratio: NBP liquid to NTP gas | 845.0 | 649.0 | 156.0 | 674.3 |
| Heat of fusion, J/g | 58.23 | 58.47 | 161 | 13.90 |
| Heat of vaporization, J/g | 445.59 | 509.88 | 309 | 242.55 |
| Heat of sublimation, J/g | 507.39 | 602.44 | – | 256.47 |
| Heat of combustion (low), kJ/g | 119.93 | 50.02 | 44.5 | N/A |
| Heat of combustion (high), kJ/g | 141.86 | 55.53 | 48 | N/A |
| Specific heat ($C_P$) of NTP gas, J/(g K) | 14.89 | 2.22 | 1.62 | 0.919 |
| Specific heat ($C_P$) of NBP liquid, J/(g K) | 9.69 | 3.50 | 2.20 | 1.696 |
| Specific heat ratio ($C_P/C_V$) of NTP gas | 1.383 | 1.308 | 1.05 | 1.40 |
| Specific heat ratio ($C_P/C_V$) of NBP liquid | 1.688 | 1.676 | – | 1.832 |
| Viscosity of NTP gas, g/(cm s) | 0.0000881 | 0.000110 | 0.000052 | |
| Viscosity of NBP liquid, g/(cm s) | 0.000132 | 0.001130 | 0.002 | |
| Thermal conductivity of NTP gas, mW/(cm K) | 1.914 | 0.330 | 0.112 | 0.2575 |
| Thermal conductivity of NBP liquid, mW/(cm K) | 0.99 | 1.86 | 1.31 | 1.515 |
| Surface tension of NBP liquid, N/m | 0.00193 | 0.01294 | 0.0122 | 0.01320 |
| Dielectric constant of NTP gas | 1.00026 | 1.00079 | 1.0035 | 1.00049 |
| Dielectric constant of NBP liquid | 1.233 | 1.6227 | 1.93 | 1.4870 |

(*Continued*)

**Table 11.9** (*Continued*)

| Property | Hydrogen | Methane | Gasoline | Oxygen |
|---|---|---|---|---|
| Index of refraction of NTP gas | 1.00012 | 1.0004 | 1.0017 | 1.00025 |
| Index of refraction of NBP liquid | 1.110 | 1.2739 | 1.39 | 1.219 |
| Compressibility factor (Z) in NTP gas | 1.0006 | 1.0243 | 1.0069 | 0.9992 |
| Compressibility factor (Z) in NBP liquid | 0.01712 | 0.004145 | 0.00643 | 0.00379 |
| Limits of flammability in air, vol% | 4.0–75.0 | 5.3–15.0 | 1.0–7.6 | N/A |
| Limits of detonability in air, vol% | 18.3–59.0 | 6.3–13.5 | 1.1–3.3 | N/A |
| Stoichiometric composition in air, vol% | 29.53 | 9.48 | 1.76 | N/A |
| Minimum energy for ignition in air, mJ | 0.02 | 0.29 | 0.24 | N/A |
| Autoignition temperature, K | 858 | 813 | 501 to 744 | N/A |
| Hot air jet ignition temperature, K | 943 | 1493 | 1313 | N/A |
| Flame temperature in air, K | 2318 | 2148 | 2470 | N/A |
| Percentage of thermal energy radiated from flame to surroundings, % | 17–25 | 23–33 | 30–42 | N/A |
| Burning velocity in NTP air, cm/sec | 265–325 | 37–45 | 37–43 | N/A |
| Detonation velocity in NTP air, km/sec | 1.48–2.15 | 1.39–1.64 | 1.4–1.7 | N/A |
| Diffusion coefficient in NTP air, $cm^2$/sec | 0.61 | 0.16 | 0.05 | N/A |
| Diffusion velocity in NTP air, cm/sec | <2.00 | <0.51 | <0.17 | N/A |
| Buoyant velocity in NTP air, m/sec | 1.2–9 | 0.8–6 | Nonbuoyant | N/A |
| Maximum experimental safe gap in NTP air, cm | 0.008 | 0.12 | 0.07 | N/A |
| Quenching gap in NTP air, cm | 0.064 | 0.203 | 0.2 | N/A |
| Detonation induction distance in NTP air | $L/D \approx 100$ | – | – | N/A |
| Limiting oxygen index, vol% | 5.0 | 12.1 | 11.6 | N/A |
| Vaporization rate (steady state) of liquid | 2.5–5.0 | 0.05–0.5 | 0.005–0.02 | N/A |
| Burning rate of spilled liquid pools, cm/min | 3.0–6.6 | 0.3–1.2 | 0.2–0.9 | |
| Flash point, K | Gaseous | Gaseous | ≈230 | N/A |
| Energy of explosion (g TNT)/(g fuel) | ≈24 | ≈11 | ≈10 | N/A |
| Energy of explosion (g TNT)/$cm^3$ NBP liquid fuel) | 1.71 | 4.56 | 7.04 | N/A |
| Energy of explosion, (kg TNT)/($m^3$ NTP gaseous fuel) | 2.02 | 7.03 | 44.22 | N/A |
| Toxicity | Nontoxic (asphyxiant) | Nontoxic (asphyxiant) | Slight (asphyxiant) | |

NBP–normal boiling point; NTP = 1 atm and 20°C (293.15 K); TNT = symmetrical trinitrotoluene. Thermophysical properties listed are those of parahydrogen. Gasoline property values are the arithmetic average of normal heptane and octane in those cases where "gasoline" values could not be found (unless otherwise noted). Also, atm × 0.101325 = MPa.

**Table 11.10** Flammability and Detonability Limits of Some Gases

| | Lower limit (mol%) | | Upper limit (mol%) | |
|---|---|---|---|---|
| Mixture | Flammability | Detonability | Detonability | Flammability |
| $H_2$–$O_2$ | 4.65 | 15.0 | 90.0 | 94.0 |
| $H_2$–air | 4.0 | 18.3 | 59.0 | 74.0 |
| CO–$O_2$ (moist) | 15.5 | 38.0 | 90.0 | 93.9 |
| CO–air | 12.5 | – | – | 74.2 |
| $CH_4$–$O_2$ | 5.4 | – | – | 59.2 |
| $CH_4$–air | 5.0 | – | – | 15.0 |
| $C_2H_2$–air | 2.5 | 4.2 | 50.0 | 80.0 |
| $C_4H_{10}$ (ether–$O_2$ | 2.1 | 2.6 | 40.0 | 82.0 |
| $C_4H_{10}O$–air | 1.85 | 2.8 | 4.5 | 36.5 |
| $C_2H_6$–$O_2$ | 4.10 | – | – | 50.5 |
| $C_2H_6$–air | 3.0 | – | – | 12.5 |

Keep in mind in studying the information in this table that almost any flammable mixture will, under favorable conditions of confinement, support the propagation of an explosive flame. A somewhat more favorable mixture of the fuel and oxidant and proper confinement may also support a detonation.

The lower flammability limit, expressed as mole percent of the flammable material, varies quite substantially from one material to another. The lower limit for hydrogen is 4.6% in oxygen, 4.0% in air. Carbon monoxide's lower limit is 15% in oxygen, 12% in air; the lower limit for methane is about 5% in either oxygen or air. The flammability limit for the higher molecular weight hydrocarbons is somewhat lower.

A flame is not supported and will automatically extinguish itself at fuel concentrations below those listed for the lower flammability limit. An upper flammability limit also exists, as shown in the last column of Table 11.10. At fuel concentrations higher than those listed in this last column, the flame again will automatically extinguish itself. Between these two flammable fuel concentrations, a narrower range of fuel concentration exists in which the character of burning may change into an explosion or detonation. For the hydrogen–oxygen system, the detonability limits are seen to be within the flammability limits.

When a detonable mixture is weakened by dilution with either the oxidant or the fuel or an inert gas or any heat sink, a limit of detonability will be found. Further dilution will lead to the limit of inflammability, and beyond this a flame of any type will not be supported. Similar lower and upper explosion and detonability limits may be sought for any fuel–oxidant combination, whether it is a homogeneous gas-phase mixture, a homogeneous liquid phase, or a mist or dust of one material in the other.

### *4.1.1. The Dilutent Role*

A gas, liquid, or solid dilutent located in the advancing flame front will, because of its heat capacity, absorb some of the heat of reaction. It will therefore reduce the resultant flame temperature, the combustion products temperature, the rate of the chemical reactions, and the associated flame velocity.

#### 4.1.2. Flame and Detonation Velocities

Before proceeding with this discussion, we must define some basic terms common to flammability and explosion studies. A *flammable* material or mixture is simply one that is easily set on fire and will burn readily. An *explosion* is the result of a sudden release of energy. Ordinarily this involves the rapid combustion of material and is accompanied by explosive blast waves and a loud noise. The air blast waves cause a temporary overpressure. The bursting of a steam boiler, cryogenic tank, or pipeline is also called an explosion even though combustion may not be involved. An explosion, then, is a sudden release of energy that is dissipated in the formation of shock waves, acceleration of shrapnel, thermal radiation, etc. When the explosion is due to a chemical reaction, the concept of rapid burning is convenient.

In the study of burning rates of flammable gas–air mixtures, two more terms are commonly used. A *deflagration* is an explosion resulting from subsonic flame speed, and a *detonation* is an explosion resulting from supersonic flame speed. Thus, in combustible gas–oxidizer mixtures, the flame speed is the yardstick used to measure the flammability and explosive potential of the mixture. A slow-burning flammable mixture results in lower overpressures and less shrapnel hazard-deflagration. A fast-burning mixture causes higher overpressures and a greater shrapnel hazard-detonation. A deflagration may develop into a detonation in a closed system due to the influence of the confining walls. The distance in a tunnel, pipeline, or room that it takes to progress from a deflagration to a detonation is called the *induction distance*. A detonation may occur in open or closed systems if the ignition source has sufficient energy. Matches, sparks, electrostatic sparks, and open flames are considered weak ignition sources; strong ignition sources are blasting caps, dynamite, high voltage capacity shorts, lightning, and other explosive charges. In condensed phase mixtures such as liquid oxygen–oil, detonation may be initiated by any source that produces a shock, e.g., spark discharge, detonator caps, or impact.

When a fuel–oxidant mixture of a composition most favorable for high-speed combustion is weakened by dilution, it will first lose its ability to detonate, then it will lose its ability to burn explosively, and finally, it will not maintain its combustion temperature and a flame front will not be supported.

Several examples of combusion front velocities are noted in Table 11.11. The first two items in the table are for methane burning in air and in oxygen. With the gases being fed to the burner at atmospheric pressure and ambient temperature, a 10% methane–90% air mixture burns at 1.15 ft/sec. The hotter mixture, 35%

**Table 11.11** Flame and Detonation Velocities at 1 Atm and 70°F (294 K)

| Mixture | Velocity (ft/sec) |
|---|---|
| Methane 10% + air 90% | 1.15 |
| Methane 35% + oxygen 65% | 10.8 |
| Hydrogen 43% + air 57% | 8.7 |
| Hydrogen 70% + oxygen 30% | 29.1 |
| Methane 33.3% + oxygen 66.7% | 7040.0 |
| Hydrogen 66.7% + oxygen 33.3% | 9270.0 |
| Coal dust + air[a] | 500–890 |

[a] Concentration, 17 g/ft$^3$; particle size, 1–150 μm.

methane–65% oxygen, burns at 10.8 ft/sec. Essentially, the same mixture will detonate at the supersonic velocity of 7040 ft/sec. Hydrogen in a 43% mixture with air burns at 8.7 ft/sec, and a richer mixture of 70% hydrogen in oxygen burns at 29.1 ft/sec. Such a mixture in a vessel or pipe of sufficient size will pass over from relatively slow combustion to a detonation velocity of 9270 ft/sec. If combustible mixtures are handled, conditions must be carefully controlled to avoid those in which slow combustion can pass over to an explosive or detonating reaction.

#### *4.1.3. Flammability Limits—Methane–Oxygen–Nitrogen Mixtures*

The fuel–oxidant–dilutent situation can be further analyzed by studying of the methane–oxygen–nitrogen gaseous system shown in Fig. 11.11. This may be considered as qualitatively typical of any system. Methane can represent any fuel, oxygen can represent any oxidant, and nitrogen can represent any dilutent. Furthermore, the corners of this diagram may qualitatively be used to represent the oxidant, the fuel, or the dilutent in any state—solid, liquid, or gas.

Each of the corners of the equilateral triangle represents the pure component indicated at that corner, with the concentration of that component decreasing linearly as one moves away from the particular apex to the opposite side. The lower and upper explosive limits of methane in oxygen are indicated along the base of the diagram, and the lower and upper limits of methane in air are indicated along the heavier line that extends between the point marked air and the methane corner. All other flammable mixtures of methane with oxygen and nitrogen lie within the cross-hatched area.

Gas mixtures with compositions that lie outside the shaded envelope will not burn or explode. They do not release enough heat upon reaction of the methane with the oxygen to raise the temperature to a level at which combustion is supported. The fuel in the mixture would be consumed if it were passed into a flame but will not support its own combustion without preheating or an external source of energy. The flame temperatures are at a minimum around the edge of the envelope. As one proceeds inward within the cross-hatched area, higher flame temperatures are obtained and detonability conditions may be reached.

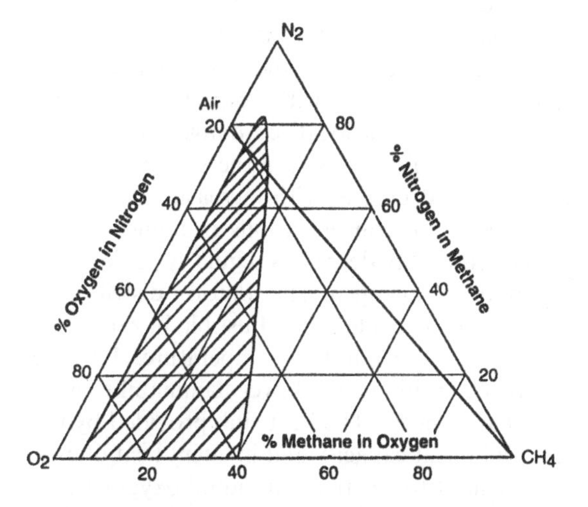


**Figure 11.11** $CH_4$–$O_2$–$N_2$ flammable limits. (Courtesy of Air Products and Chemicals, Inc., Allentown, PA.)

## 4.2. Solid Dilutents—Flame Arresters

A flame arrester is inserted in a line or vessel to prevent the propagation of a flame front. Such a device functions by absorbing energy from the flame and thus lowering its temperature below that at which combustion is supported. The minimum ignition energy varies somewhat from system to system. The ignition energy for a hydrogen–air mixture is an order of magnitude lower than that for a methane–air mixture. The minimum ignition energy is also a function of the quenching distance. It is this latter relationship that is important in flame arrester behavior.

Lewis and Von Elbe (1961) tabulated a great number of data for quenching distances and minimum ignition energies for the hydrocarbon–air systems. An empirical relationship that has been partially substantiated theoretically exists between these two quantities. The form of their relationship is

$$D = \text{constant} \times \sqrt{E}$$

The quenching diameter in millimeters is about 25 times the square root of minimum ignition energy in millijoules.

The effectiveness of a flame arrester depends on its ability to dissipate heat at a rate greater than the rate at which it is generated. The ratio of the gaseous volume of reactants to the surface area of the heat acceptor is therefore important. This is shown in Fig. 11.12. The particle diameter of a bed of spheres that will quench a methane–oxygen flame is plotted against the methane concentration with void fraction as a parameter. It is seen that at the closest packing in spheres ($\varepsilon = 0.259$), a flame containing less than 22% or more than 34% methane will not propagate in a bed of 2 mm diameter spheres. If the void fraction in the same bed of spheres is increased to 0.400, the range of flammability limits is increased to 15–43%.

## 4.3. Solubility of Hydrocarbons in Liquid Oxygen

For any combustible, there is some level of concentration in liquid oxygen above which the mixture is hazardous and below which it is not. For highly soluble materials, this level is the liquid-phase flammability limit; for less soluble materials, the dangerous concentration level is, for practical purposes, the saturated solution concentration. Even though the total quantity of fuel with respect to the total quantity of oxygen is below the flammability limit, a localized dangerous mixture may be produced by localized concentrations of fuel exceeding the solubility level.

The solubility of saturated hydrocarbons decreases with increasing molecular weight (see Fig. 11.13). Methane is soluble at all concentrations in liquid oxygen at the normal boiling point of oxygen. The solubility decreases regularly to 76 mol% at −320.5°F (77 K), the normal boiling point of nitrogen. The uniform distribution of dissolved methane in liquid oxygen will not provide a combustible hazard until sufficient fuel is present to heat the mass to the kindling temperature. For the standard temperature and pressure gaseous conditions, the lower flammability limit for methane in oxygen is slightly higher than 5 mol%. The heat sink provided by the low temperature of the liquid oxygen, both in heat capacity and in latent heat of evaporation, results in the lower flammability limit for methane in liquid oxygen being increased to a little above 10 mol%. In the absence of data for the liquid system, it is conservative to adopt the gas-phase flammability limits for the liquid system.

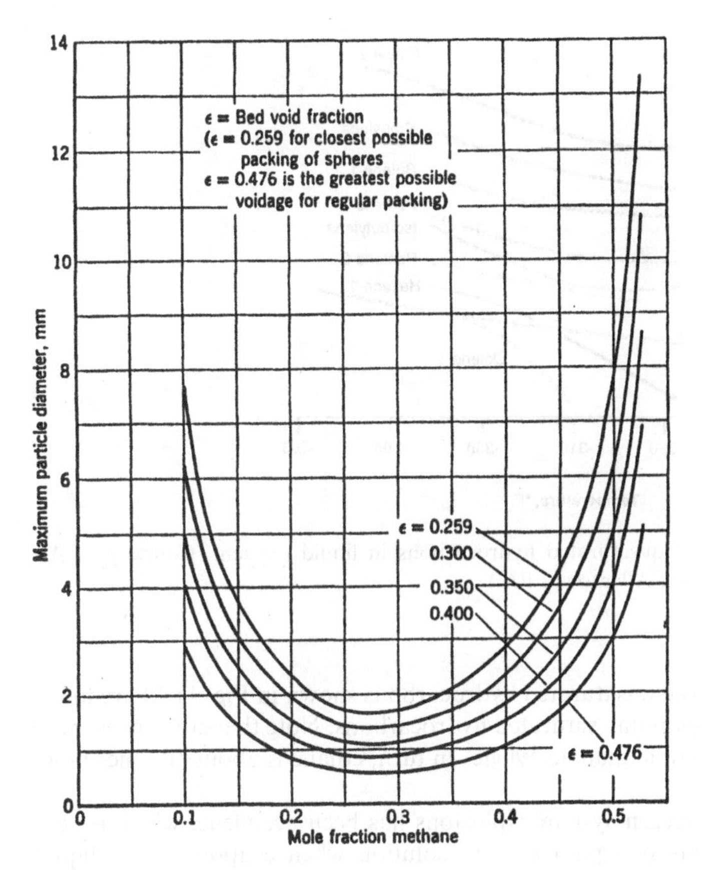


**Figure 11.12** Quenching methane–oxygen flame with a bed of spheres. Particle diameter vs. methane concentration. (Courtesy of Air Products and Chemicals, Inc., Allentown, PA.)

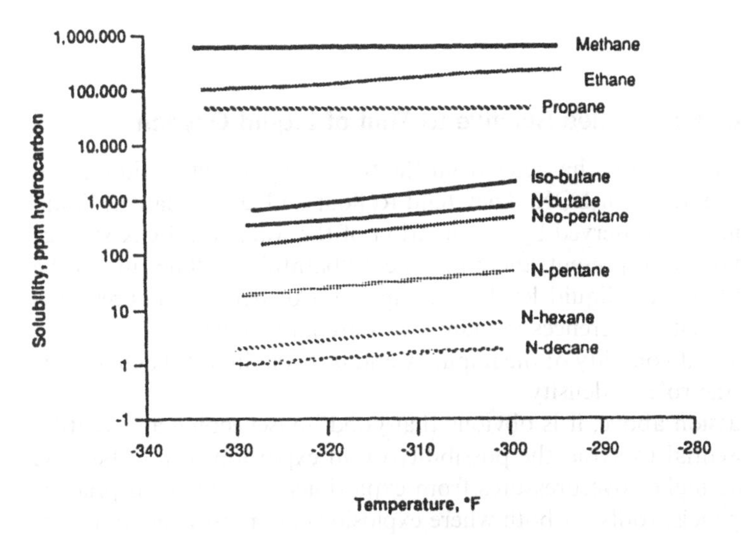


**Figure 11.13** Solubility of saturated hydrocarbons in liquid oxygen. (Courtesy of Air Products and Chemicals, Inc., Allentown, PA.)

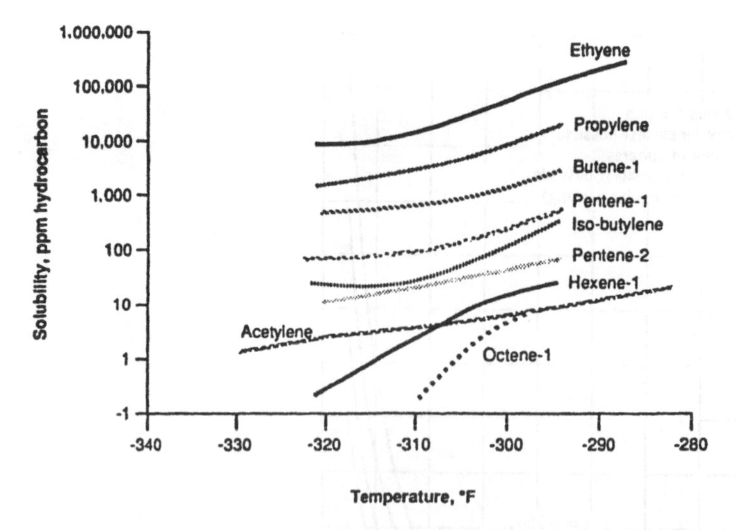


**Figure 11.14** Solubility of unsaturated hydrocarbons in liquid oxygen. (Courtesy of Air Products and Chemicals, Inc., Allentown, PA.)

The solubilities of the unsaturated hydrocarbons shown in Fig. 11.14 are lower than those for the corresponding saturated hydrocarbons. Note that ethylene is more than 1000 times more soluble than acetylene. In turn, ethane is about 10 times more soluble than ethylene.

The fuel in most oxygen system explosions has been acetylene, which is very sparingly soluble and has precipitated from solution when evaporation of liquid oxygen was taking place.

It can be generalized that if either the explosive limit or the solubility limit is exceeded for a fuel in a liquid oxidant, a hazardous condition exists. As either of these limits is approached at any point in a cryogenic system, corrective action must be taken.

## 4.4. Density of Solid Impurities Relative to That of Liquid Oxygen

Most impurities of a dangerous character from the fuel viewpoint in liquid oxygen (LOX) handling are lighter than LOX. They tend to float and accumulate at liquid level surfaces. This may be observed by examination of the walls in a LOX storage tank. If impurities have been present, there may be a "bathtub" telltale ring in the tank approximating the usual liquid level. A comparison of densities is presented in Table 11.12. The density differences in many cases are large. One must also keep in mind the solubility and volatility of the impurity and its crystalline bulk characteristics in considering the role of density.

From the discussion above, it is obvious that good housekeeping and ventilation practices are essential to avoid the possibility of an explosion or fire. Because confined spaces invite higher overpressures from explosions, it is accepted practice to provide blowout panels, roofs, or both where explosions are most likely to occur. For the same reason, vessels, pipes, and plumbing systems are sometimes provided with blowout plugs, burst disks, etc.

**Table 11.12** Density of Potential Impurities in Liquid Oxygen

| Substance | Density at −297°F (90 K)(g/cm$^3$) |
|---|---|
| Acetylene | 0.625 |
| Ethylene | 0.727 |
| *n*-Butane | 0.770 |
| Propylene | 0.775 |
| *n*-Hexane | 0.795 |
| Methyl alcohol | 0.815 |
| Toluene | 0.960 |
| Ammonia | 1.01 |
| Aniline | 1.03 |
| Oxygen | 1.15 |
| Carbon dioxide | 1.58 |
| Carbonyl sulfide | 1.65 |

## 4.5. Flammability Hazards Associated with Liquid Nitrogen

Nitrogen is an inert gas and will not support combustion; however, there are some subtle means by which a flammable or explosive hazard may develop. Cold traps or open-mouth dewars containing liquid nitrogen can condense air and cause oxygen enrichment of the liquid nitrogen. The composition of air as it condenses into the liquid nitrogen container is about 50% oxygen and 50% nitrogen. This condensation takes place at about 148°R. As the liquid nitrogen evaporates, the liquid oxygen content steadily increases so that the last portion of liquid to evaporate may be half oxygen. The nitrogen container must then be handled as if it contained liquid oxygen; fire, compatibility, and explosive hazards all apply to this oxygen-enriched liquid nitrogen. Because air condenses at temperatures below 148°R and liquid nitrogen boils at 140°R (at 1 atm), uninsulated pipelines will condense air. This 50-50 condensate can drop on combustible materials, causing an extreme fire hazard or explosive situation. The oxygen-rich air condensate can saturate clothing, rags, wood, asphalt pavement, etc., and cause the same problems as those associated with the handling and spillage of liquid oxygen. Again, good housekeeping is mandatory. Splash shields and condensate gutters can be used to divert condensed air to safe areas.

## 4.6. Flammability Hazards Associated with Oxygen

As mentioned earlier, spillage or leakage of gaseous or liquid oxygen can create a severe fire or explosion hazard. Clothing and other porous combustible materials will burn rapidly and intensely when saturated with oxygen gas. When liquid oxygen is spilled on clothing, it may require as long as 1 hr for the clothes to be fully aired. Therefore, no sparks or flames should be allowed in or near areas where oxygen is being handled. Clothes that have not been fully aired should not be worn by or near anyone on a "cigarette break". Tightly woven white cotton coveralls should be worn when handling liquid oxygen. The tight weave will minimize oxygen penetration and saturation of the cloth, and the white color will advertise the need for laundry service. Cotton generates less static electricity than other clothing materials, lowering the possibility of electrostatic sparks, and cotton also is a reasonably slow-burning cloth. Although electrostatic sparks are considered to be relatively weak ignition sources and may not be capable of igniting oxygen-soaked clothing, they are proven sources of ignition for oxygen-enriched combustible vapors and gases. All combustible materials burn much more rapidly in the presence of oxygen-enriched atmospheres; thus, such atmospheres should be avoided.

There have been many serious accidents involving oxygen-soaked clothing because clothing can absorb or otherwise entrap and contain a relatively high concentration of oxygen for a substantial period of time.

One such incident involved a man who was wearing prescribed protective safety clothing and working in an oxygen-rich area. After withdrawing to a safe smoking area, his clothing ignited simultaneously with the lighting of a cigarette.

In another case, a patient in a semiprivate room awoke to see the oxygen tent on the other bed afire. He gave the alarm, but the patient in the tent could not be rescued in time to save his life. Extensive investigation failed to reveal positively the source of ignition, but a cigarette butt was found on the bed table and the remains of a book of matches in the victim's bed.

At a different hospital, an oxygen-powered respirator was delivering oxygen to a patient through a mouthpiece or mask at an alternating (intermittent) cycle rate in an intensive care room containing two patients and a special duty nurse. The doctor instructed the nurse in the operation of the respirator, but the respirator failed to operate properly. The oxygen therapist was called and was adjusting the respirator. The nurse decided to move a floor lamp from behind the respirator closer to the patient. (The nurse could not recall whether the light was on or even plugged in.) At the moment she moved the lamp she heard a sputtering noise, and upon turning saw a multitude of sparks fly from the main body (which contains the various pressure and flow controls) of the respirator. The light in the room failed, but in the darkness sparks, smoke, and fumes were very noticeable. The patients were removed from the room to protect them from the thick, acrid smoke.

The cause of the fire was traced to defective wiring in the lamp, which allowed one side of the circuit to ground to the lamp base, which in the process of being moved, contacted the respirator, which in turn was in contact with a combination electrocardioscope and pacemaker with a three-wire cord. Apparently, current was delivered from the lamp, through the respirator and the pacemaker, to the ground, and in passing through the respirator caused ignition at a point where high electrical resistance existed (probably coiled stainless steel wire used to prevent plastic tubing from collapsing). The patient was burned but survived.

A fire occurred in the Two Man Space Environment Simulator at Brooks Air Force base, Texas, on January 31, 1967. An animal experiment was under way in the chamber to investigate the hematopoietic effects of exposure to 100% oxygen. Environmental conditions in the chamber at the time of the fire were approximately 100% oxygen at 380 mmHg (7.35 psia; 50.7 kPa). The simulator was built and equipped with materials of low combustibility, but large quantities of paper, inorganic litter, and other highly combustible materials were brought into the chamber for use in the animal experiments. The chamber occupants were wearing combustible clothing. A portable electric light with an ordinary two-wire cord had been brought into the chamber. Shorting of this cord on the metal floor of the chamber is believed to have been the ignition source. The animals' fur caught fire, and their movements helped to spread the fire. The two airmen occupants of the chamber at the time of the fire were fatally burned even though the chamber was repressurized with air and opened within approximately 30 sec after initiation of the fire. The fire extinguishing equipment in the chamber at the time of the fire consisted of two small portable, manually operated carbon dioxide extinguishers, neither of which was used although one overheated and discharged through the pressure relief valve.

On January 27, 1967, three astronauts died as a result of a fire in an Apollo spacecraft command module on the launch pad at Kennedy Space Center, Florida. The atmosphere in the spacecraft was 100% oxygen at approximately 16 psia (110 kPa). The origin or ignition source of the fire was not definitely determined in spite of an extremely intensive investigation. The most probable source of ignition was thought to be an anomaly associated with the spacecraft wiring. Communication cables (telephone wires) had been laid temporarily on the floor, leading out through the hatch. They may have been abraded by opening and closing the hatch door. The extent of damage to the vehicle prevented final determination. The fire was propagated through the spacecraft by materials that were not considered significantly flammable in an air atmosphere but were very flammable in the 100% oxygen, 16 psia (110 kPa) atmosphere.

An accident occurred in a chamber with an internal atmosphere of 100% oxygen at 5 psia (34 kPa). Four men in the chamber were taking part in experiments. A light bulb in the ceiling fixture burned out. One man climbed up to replace the bulb. After the bulb was replaced, he heard a "sound like the arcing of a short circuit". A small flame [about $\frac{1}{2}$ in. (12 mm) long] was seen coming from an insulated wire in the fixture. The composition of this insulation is still not known. The subject requested water but was told to snuff the fire out with a towel. The towel caught on fire and blazed so vigorously that it set the man's clothing afire. An "asbestos fire blanket" was used to snuff out the clothing fire, but it too burst into flames. The asbestos blanket reportedly had an organic filler or coating that "kept the asbestos from flaking off." The clothing of the other subjects who were using the blanket also caught on fire. All four men received second-degree burns. Carbon dioxide was used to extinguish the fire after personnel were evacuated from the chamber. It was stated that the blanket and towel had been "saturated with oxygen for 17 days and burned much more vigorously than would be expected under sea level conditions". *Note:* Long storage in oxygen does significantly enhance the combustibility of textile materials. An interesting aspect of this case is the fact that burning insulation dripped from the light fixture onto a bunk. One crewman tried to snuff out the resulting fire, and "his skin caught on fire". The burns on his hands were "severe" and required hospital treatment for 11 or 12 days. The cabin was being continuously vented, and no analysis of the vapors was being performed at that time of the accident.

*Note:* Laboratory experiments suggest that human skin is difficult to ignite in low-pressure oxygen. It will burn readily, however, in the presence of other more easily ignitable combustibles, such as grease or molten plastics, which can act as localized ignition sources.

Leakage or spillage of liquid oxygen may also cause impact sensitivity. Asphalt pavement saturated with liquid oxygen has been detonated by dropped tools and pedestrian traffic. So have graveled and cinder-filled areas, which often unsuspectingly become accumulators of combustible materials—spilled oil, grease, etc. Wooden floors, oil-soaked rags, solvent, dirt, etc., are noncompatible with liquid oxygen and can be detonated by impact or other means when soaked or drenched with liquid oxygen. Perhaps the best solution to the multitude of problems and hazards associated with the use of asphalt, gravel, cinders, wood, and even dirt lies in the application of concrete—crack-free, with the pores sealed against spills and kept clean.

A noncompatible and explosive condition exists whenever a liquid fuel such as gasoline or kerosene is mixed with liquid oxygen. It may be formed by spillage, leakage, or inadvertent (process) mixing. Should a mixture occur, it should not be moved. The area should be cleared of personnel and the liquid oxygen allowed to evaporate, eliminating the hazard. If it is necessary to speed up the evaporation process, use a low-flow water spray or water deluge to increase the heat transfer rate. Under no circumstances should a possibly detonable mixture be transferred or transported. Accidental mixtures in liquid oxygen storage tanks may be warmed by partially destroying the insulating vacuum with an inert gas. Contaminated oxygen gas may be vented to the atmosphere, taking care to vent the contaminated vessels very slowly—low velocities, and slow valve opening and closing. In liquefaction plants, transport trailers, and storage vessels where the total hydrocarbon accumulation exceeds established minimum permissible concentrations, the contaminated liquid may be slowly transferred or drained out of the system and evaporated in a safe (preferably remote) location.

Leakage or spillage of oxygen in an area that is partially or totally confined presents an immediate explosion hazard. If a flammable gas is present in sufficient quantity, a spark or flame is all that is needed to set off a devastating explosion. The confinement afforded by walls, floors, and ceilings increases the resultant overpressure.

Care should be taken to avoid sparks, flame, and other ignition sources in areas where oxygen is handled. "NO SMOKING" signs should be conspicuously posted. The use of arcing motors, electrical hand tools, torches, welding equipment, etc., should be prohibited or restricted to known "safe periods". To minimize the possibility of electrostatic sparks, oxygen equipment should be electrically grounded in those installations where flammable vapor–oxygen mixtures can occur. Use of cotton clothing and prohibition of synthetic, wool, and silk clothing will minimize the possibility of electrostatic spark. The merits of "sparkproof" tools have been widely discussed. These tools can cause sparks and are easily deformed in use; nevertheless, they produce lower temperature sparks and require considerable energy to produce a spark. Therefore, sparkproof tools offer some margin of protection in those situations where flammable vapor–oxygen mixtures may occur. Personnel should realize the limitations of such tools and not be lulled into a false sense of security. Aluminum-painted steel surfaces are another source of sparks; a hard object with sufficient impact energy can generate a spark from the aluminum–iron mixture on the surface.

A Titan missile in a silo was being defueled when a leak in the liquid oxygen line was detected. The liquid oxygen infiltrated the adjacent equipment silo through a utility tunnel. It is thought that a spark from some of the equipment ignited combustibles in the oxygen-enriched atmosphere, causing a fire in the equipment silo and a subsequent explosion of fuel in the missile silo. Fortunately, all workers were safely evacuated before the explosion, although some were injured by smoke inhalation. Estimated loss: in excess of $7 million.

Electrical fixtures and equipment in or near oxygen storage, liquefaction, and handling areas should be designed to prevent open sparks. Electrical motors should be totally enclosed, oxygen concentrations should be held to about normal atmospheric composition (21%), and equipment should be purged continuously with an inert gas in those cases where high concentrations of oxygen are unavoidable. In an oxygen system, the equipment itself is the fuel, and an inert purge is the only certain way to exclude the oxygen. The electrical equipment should be located where it is not exposed to low temperatures. Adequate ventilation is the best preventive measure to avoid trouble with such equipment. Where oxygen and combustible gases and vapors are handled, electrical equipment should either be purged with an inert gas or conform to Article 500 of the National Electrical Code. This code classifies the electrical hazard according to the fuel present. Where significant oxygen enrichment of the atmosphere is likely to occur, the inert gas purge technique is the only foolproof, absolute protection against electrical hazard.

The storage of oxygen is regulated by the National Fire Protection Association (NFPA) in circular NFPA-566 (1962). Under certain circumstances, liquid oxygen storage tanks should be diked to retain the contents of the tanks in the event of rupture. A clean, well-kept dike will prevent liquid oxygen from draining into areas that may constitute an extreme fire or explosion hazard. In some installations, diking may also be desirable to prevent structural damage.

In oxygen flow systems an explosion may occur inside the equipment. The explosion may be due to contaminated oxygen, incompatibility of seal or structural materials, or violent combustion of a compatible material. The latter may be caused by an ignition mechanism that elevates the temperature of the oxygen-compatible material to its ignition temperature. Stainless steel, for example, if locally heated to its ignition temperature will be rapidly consumed in the presence of pure oxygen. The ignition temperature may be reached locally by supplying energy in the form of impact, friction, electrical discharge spark, or adiabatic compression. Impact energy can be delivered internally to a system in a number of ways: rapidly closing valves, rough handling, or loose particles (rust, weld slag, dirt, etc.) transported in the flow stream. Friction can ignite pumps and other rotating machinery where there are moving parts and in the transport of loose particles. Electricity can cause ignition in heater instrumentation and control circuits. Sparking is usually caused by electrical failure or loose debris striking carbon steel pipelines and fittings.

### 4.7. Adiabatic Compression as Ignition Source

As indicated above, adiabatic compression of the gas in high-pressure gas systems brought about by rapidly opening or closing a valve can cause equipment to ignite. The rapid actuation of a valve results in compression waves of sufficient energy to cause an appreciable local increase in the temperature of the gas. Figure 11.15 indicates the temperatures that can be attained due to adiabatic compression of oxygen gas. When a pipeline or other component of any oxygen system starts burning, the first thing to do is shut off the oxygen source. If the oxygen source cannot be isolated, the system piping and adjacent structures will burn until all of the available oxygen is consumed.

Figure 11.16 illustrates the results of a fire involving the sudden release of a large quantity of liquid oxygen. The incident took place at a missile launching site, where it is common practice to store enough liquid oxygen to run through a complete launching sequence. The liquid oxygen is transported to the site in insulated trailers having a capacity of several thousand gallons. A typical installation might store 14,000–30,000 gal. In this particular system, a liquid oxygen meter was included to measure the rate of flow during certain operations. At the time of the accident, two liquid oxygen trailers were just starting to unload. Faulty operation was noted by the man in charge, but before corrective action could be taken, a section of pipe burned out, releasing about 10,000 gal of liquid oxygen. This resulted in a substantial conflagration. The rubber tires, lubricating oils, gasoline, and even the metal comprising the tractor trailers became fuel for the fire. The bursting of the trailer tanks added more oxygen to the fire. The fire continued until the excess oxygen was consumed. The inside tank of the trailer was made of stainless steel, and the outside tank

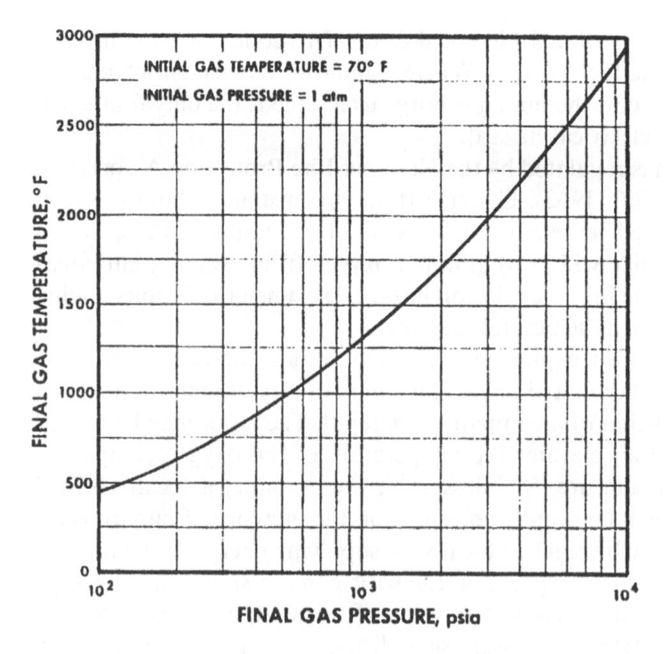


**Figure 11.15** Temperature rise due to adiabatic compression of oxygen gas.

of carbon steel, with vacuum insulation between. The carbon steel burned, but the stainless steel did not. It has been deduced that under the particular chain of circumstances, the liquid meter became exposed to a high-velocity reverse flow of gaseous oxygen, causing it to rotate like a turbine at sufficient speed to disintegrate and produce ignition.

High-pressure gas storage cylinders are especially vulnerable to damage from internal combustion; contamination, adiabatic compression, and sparks due to steel

**Figure 11.16** Damage sustained by liquid oxygen transport trailer in an oxygen-fed fire.

**Figure 11.17** High-pressure gaseous oxygen storage. (Courtesy of Air Products and Chemicals, Inc., Allentown, PA.)

slag are prime ignition sources in high-pressure gas lines. Ignition of the gas line will cause combustion of the line to progress toward the storage cylinders. If the oxygen flow is not stopped, the line will burn into the end of the cylinder, releasing oxygen at a greater rate. This situation can create a virtual rocket effect, "launching" the storage cylinder unless it is well fastened down.

Figure 11.17 shows a rather conventional high-pressure gaseous cascade oxygen storage system associated with a missile test project. The cylinders are connected into a common manifold, downstream of which is the main isolation or shutoff valve. A few feet downstream of this valve is a pressure-reducing regulator that controls the pressure to the requirements of each experiment. An accident occurred during the manipulation of valves in this system. Just prior to the accident, the cylinder bank had been recharged to 2000 lb pressure, during which sequence the shutoff valve was closed. The storage system and the operating procedure had been used repeatedly without incident. On this particular occasion, when the operator opened the shutoff valve, combustion took place. The combustion spread in both directions inside the stainless steel piping, apparently using the metal as fuel, until it reached the cylinders. The ends of the cylinders burned out, and several of them took off like missiles. One of these cylinders landed in a parking lot 300 ft away and damaged several automobiles (see Fig. 11.18).

Figure 11.19 shows what the burned-out cylinders looked like after they were recovered and stacked in their initial (prefire) positions.

It is not known what really caused the ignition. One could speculate that some foreign matter or oil might have been introduced into the system during the loading operation that immediately preceded the incident, although there is no evidence whatsoever to support this theory. Another speculation might be, as with the filter incident previously discussed, that friction caused by the opening of the shutoff valve initiated combustion. There might have been some hard foreign matter that created an excess of friction at this point. This theory is supported by the condition of the shutoff valve, which indicated that ignition originated in

**Figure 11.18** High-pressure oxygen gas cylinder damage to cars in a parking lot 300 ft from storage bank.

this area. Another theory that might be considered included the possibility that a rapid adiabatic compression wave was set up between the shutoff valve and the upstreamside of the pressure-reducing valve at the moment of opening. Such an effect would relate to the rate of opening. This theory is somewhat supported by

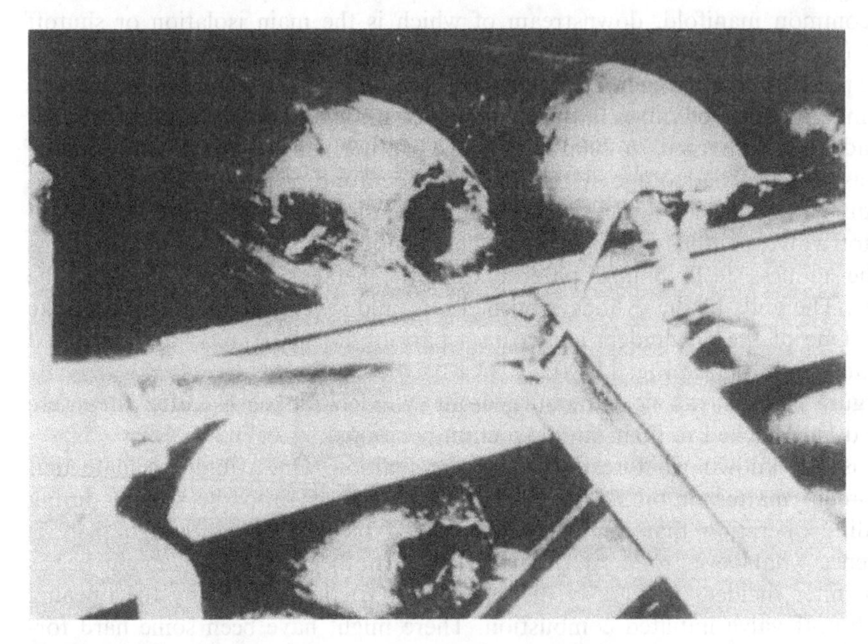

**Figure 11.19** Oxygen storage bank, re-stacked in prefire positions, following pipeline fire and cylinder flight.

the evidence of combustion inside the pressure-reducing valve but not by the observed condition in the shutoff valve.

In the missile industry, other serious accidents that have taken place simultaneously with the quick opening of shutoff valves against high-pressure oxygen have been explained in this manner. In a series of mockup tests, the possibility of generating compression wave-front temperatures approximating the ignition temperatures of stainless steel was demonstrated. The presence of any nonmetallic material with a lower ignition temperature would increase the chances of ignition.

In general, when handling high-pressure gaseous oxygen, quick changes of pressure should be avoided. The equipment should be designed and selected to minimize friction at the time of opening or closing of valves. Large, high-pressure isolation valves should be bypassed by small equalizing valves that can be opened prior to opening the main valve. Any rapid change in pressures in a high-pressure gas system should be avoided to the maximum extent possible consistent with the objectives of the system.

Because of this hazard in oxygen gas lines, carbon steel pipes should be avoided wherever possible.

For reasons of economy, carbon steel pipes are frequently used with the following precautions. Use stainless steel tees, elbows, and fireback sections, and keep the gas velocities below those given by the relation

$$v = 100\sqrt{\frac{165}{P}}$$

In this equation, $v$ is the velocity in ft/sec and $P$ the operating pressure in psia (and is not to exceed 1000 psia). So if the operating pressure is 165 psia, the maximum allowable velocity for oxygen gas is 100 ft/sec. The stainless steel tees and elbows are used to direct steel particles and debris around flow corners without causing a spark. The stainless steel firebreak sections are about 3 ft long and are spaced at intervals of 150 ft or less. They are installed because of stainless steel's greater resistance to burning relative to that of carbon steel. There are no velocity or pressure restrictions on the handling of oxygen gas in copper, brass, or stainless steel pipes. Pipes made of these materials are simply sized to do the job. When carbon steel lines are used (only at pressures below 1000 psia), every effort should be made to clean and deburr them prior to introduction of oxygen gas. A high-velocity blowdown of the system, using nitrogen gas at velocities far in excess of those anticipated with oxygen gas, is a good way of tearing off potentially troublesome weld slag, etc.

Whenever a fire occurs as a result of oxygen spillage, leakage, pipeline rupture, or similar event, the first thing to do is isolate the oxygen source if at all possible. Water deluges and fire hoses may be used to keep structures cool and put out fires in combustible materials. Where oil or electrical fires persist, appropriate $CO_2$ or dry chemical extinguishers should be used. In other words, shut off the oxygen feeder source, then put the fire out according to accepted rules for fighting fires in combustible and flammable materials. *Caution:* Do not play a stream of water on a potentially impact-sensitive material such as an oxygen–kerosene mixture or an oxygen-soaked pavement or wooden deck. Standard firefighting and safety equipment—e.g., safety showers, fire blankets, first-aid kits, fire hydrants and hoses, $CO_2$ extinguishers, class B firefighting agents for fuels—should be on hand and in good working condition.

## 5. EXCESSIVE PRESSURE GAS

The usual procedures and safeguards for handling industrial gases must be observed with gases under excessive pressure. Avoid entrapment between two valves, provide relief devices on all storage containers, avoid physical damage to the container, protect gas cylinder valves, anchor gas cylinders securely, and store cylinders in a safe area. Small gas cylinders should always be chained, strapped, or otherwise anchored to guard against valve damage due to falling or bumping. Large gas cylinders should also be well secured to avoid the possibility of unscheduled and undesirable flight, as illustrated by Fig. 11.19. Whether the gas storage vessel becomes airborne as a result of a sheared valve (any gas) or a burned-out pipeline (oxygen) is of little significance. The propelling mechanism is the thrust created by the escaping gas. Considerable force is exerted on a vessel that is rapidly emptying itself of gas. The thrust is calculated approximately by the equation

$$\frac{F}{A} = P - P_a$$

where $F$ is the thrust (lb), $A$ the area for gas flow (in.$^2$), $P$ the initial cylinder gas pressure (psia), and $P_a$ is the atmospheric pressure (psia). This relation is valid for all gases at or near room temperature and expresses the thrust to be expected when a gas vessel first ruptures. Figure 11.20 graphically illustrates the thrust of gas escaping from a ruptured vessel or broken line. Note that vent lines can experience similar reaction forces and should be securely anchored. The force exerted on a 2000 psia gas cylinder caused by breaking off a 1 in. diameter valve would be calculated as

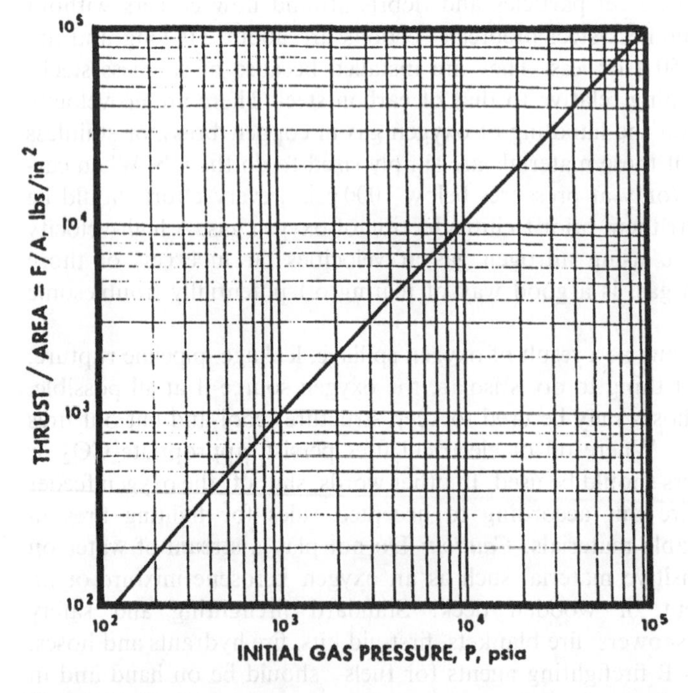


**Figure 11.20** Initial thrust exerted on a ruptured vessel or broken pipeline by escaping gas.

follows (see Fig. 11.20):

$$A = \frac{\pi D^2}{4} = \frac{\pi(1)^2}{4} = 0.785 \text{ in.}^2$$

$$\frac{F}{A} = 2000 \frac{\text{lb}_f}{\text{in.}^2}$$

$$F = \left(2000 \frac{\text{lb}_f}{\text{in.}^2}\right) A = \left(2000 \frac{\text{lb}_f}{\text{in.}^2}\right)(0.785 \text{ in.}^2) = 1570 \text{ lb}_f$$

If the gas cylinder weighs 155 lb, it will be accelerated at a rate of about 10 *g* unless it is tied down securely. Obviously, broken or loose lines can experience similar thrusts when gas flows at high velocities. Loose or broken lines present an additional hazard—"hose-whip". The reaction of escaping gas causes the line to whip and thrash around with considerable energy and velocity, and extensive personnel and equipment damage can result.

The thrust exerted by gas escaping from an open line is proportional to the gas flow rate and to the gas velocity. Sonic flow from a $\frac{1}{2}$ in. opening with an upstream nitrogen gas pressure of 4000 psi will result in a force of about 600 lb. If the line is loose and this force can be exerted to move it, the acceleration of the line can be very great. If a line section with a mass of 50 lb is being subjected to this force of 600 lb, the resultant 12 *g* acceleration can cause very rapid and dangerous whipping of the line section. As the line moves, the sonic choking section may move along the line as a result of movement of the restriction through kinking. This will cause rapid changes in the physical location of the energy being released, with the associated rapid and unpredictable line movement.

A *final word of caution* in the handling of high-velocity gas jets: high-velocity streams of gas issuing from a small orifice or bleed port have been known to penetrate the skin of a man, stripping it from the flesh and inflating it like a balloon. Exposure of the skin to this lancing effect of gas jets is to be avoided.

### 5.1. What is Excessive Pressure?

It is convenient to speak of the TNT equivalent energy of explosions. If one views an energy release from a distance or inspects the damage incurred, it is quite difficult to determine the energy source, i.e., whether the explosive material was TNT, flammable vapors, or a ruptured vessel. Knowledge of the potential energy of high-pressure containers is essential to personnel safety.

Rupture of high-pressure vessels used to store compressed gases, low-temperature liquefied gases, and ambient temperature liquids can be considered dangerous under certain conditions; consequently, the question of whether an experimental test item should be pneumostatically or hydrostatically proof-tested frequently arises. Pneumostatic tests should never be substituted for hydrostatic tests, where personnel safety is concerned, because the stored energy in gas-filled systems is much greater than that of water-filled systems. The potential energy of gas- and water-filled vessels is shown graphically in Figs. 11.21 and 11.22. Comparison of the two graphs at the same burst pressure shows that the gas-filled containers are much more hazardous. The potential energy of vessels or dewars containing cryogenic liquids may be greater than the energy available if the tank were filled

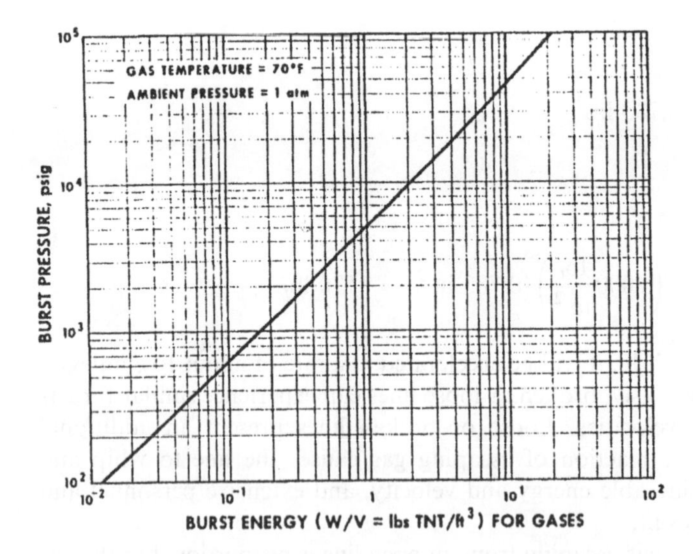


**Figure 11.21** Equivalent TNT burst energy for gas-filled vessels.

to the same pressure with room temperature gas. In tanks containing cryogenic liquids, failure at temperatures near or somewhat below the critical temperature can result in explosive potentials three or four times greater than those predicted by Fig. 11.21. Failure of cryogenic liquid-filled systems at temperatures well above the critical temperature or very near the normal boiling point temperature will not produce as much burst energy as room temperature gas at the same burst pressure

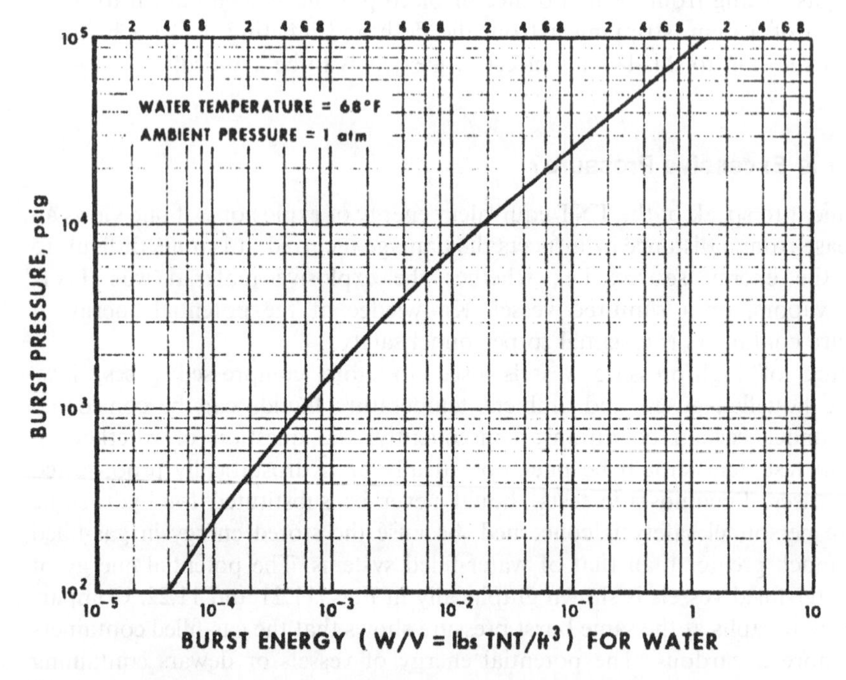


**Figure 11.22** Equivalent TNT burst energy for water-filled vessels.

(Fig. 11.21). To be on the safe side, burst energies taken from Fig. 11.21 should be multiplied by 4 when dealing with liquid nitrogen or liquid oxygen-filled systems. These graphs are not applicable to steam boilers and assume that the pressure surrounding the bursting vessel is atmospheric and that the temperature of the fluid is at or near room temperature. Secondary energy releases attributable to chemical or combustion reactions are excluded; e.g., oxygen may react with portions of the tank walls. With an understanding of the above, then, the TNT equivalent obtained from Fig. 11.21 or 11.22 may be used in conjunction with Table 11.13 to evaluate personnel hazard due to overpressure. Figure 11.21 illustrates a very important point: the energy stored in a pressurized tank depends on the quantity of gas contained in the tank. A low-pressure, large-volume vessel can be just as dangerous as a high-pressure, small-volume vessel. For example, a 50 cu ft gas-filled tank, failing at 400 psig, has a burst energy of 3 lb of TNT. This number is obtained by entering Fig. 11.22 at a burst pressure of 400 psig and reading

$$\frac{W}{V} = 0.06 \text{ lb TNT/ft}^3 \text{ tank volume}$$

Then

$$W = \left(0.06 \frac{\text{lb}}{\text{ft}^3}\right)(50 \text{ ft}^3) = 3 \text{ lb TNT}$$

To complete our example, find the burst energy of a 1 ft$^3$ gas-filled tank failing at 10,000 psig. The burst energy of this vessel is 2 lb of TNT—1 lb less than the burst energy of the 50 ft$^3$, 400 psig vessel. So it is shown that only "excessive" pressure is needed to constitute a hazard, not necessarily "high" pressure.

Blowout panels and plugs are designed to relieve the explosion overpressure rapidly and prevent damage to operating personnel and equipment. The blowout relief devices must have minimum inertia to provide the maximum protection. Properly designed and located barricades afford protection against shrapnel and overpressure. The latter usually requires rather massive structures. In deciding where to place protective barriers, one must consider the problem of confinement

**Table 11.13** Overpressure from 1 lb Hemispherical Charge of TNT—Surface Explosion in Open Air[a]

| Distance (ft) | Overpressure (psig) | $t_1$(msec) | $t_2$ (msec) |
|---|---|---|---|
| 2 | 320.0 | 0.2 | 0.8 |
| 4 | 70.0 | 0.8 | 1.6 |
| 6 | 28.0 | 1.8 | 2.4 |
| 10 | 9.6 | 4.3 | 3.2 |
| 20 | 3.0 | 12.0 | – |
| 40 | 1.2 | 29.0 | – |
| 100 | 0.35 | 82.0 | – |
| 200 | 0.13 | 169.0 | – |
| 400 | 0.05 | 346.0 | – |

[a] $t_1$ = blast wave arrival time; $t_2$ = time required for the overpressure to decay back to zero.

**Table 11.14** Structural and Biological Response to Overpressure and Shrapnel

| Blast overpressure or shrapnel momentum | Structural or biological response to blast effects |
|---|---|
| 0.5–1 psig | Glass windows shatter |
| 1–2 psig | Corrugated asbestos siding shatters; corrugated steel or aluminum paneling buckles |
| 2–3 psig | Nonreinforced concrete or cinderblock walls (8–12 in. thick) shatter |
| 5 psig | Eardrums rupture |
| 7–8 psig | Nonreinforced brick walls (8–12 in. thick) shear and fail |
| 15 psig | Lung damage occurs |
| 115 ft/sec for a 0.35 oz glass projectile | Projectile penetrates abdomen |
| 10 ft/sec for a 10 lb masonry projectile | Projectile penetrates abdomen |
| 10 ft/sec for a 160 lb man | Skull fractures from impact |

discussed above; i.e., barricading can result in sufficient confinement to cause increased overpressure in the event of an explosion. Ordnance handbooks deal with barricade design for shrapnel and overpressures. It should be noted that shrapnel shields made of materials with a large modulus of elasticity (Young's modulus) are the most effective; for example, steel is better than copper, and copper is better than aluminum in resisting shrapnel penetration. The opposite is true for shrapnel projectiles themselves. A projective having a given size and momentum will penetrate deeper if it has a lower density; aluminum will penetrate deeper than copper, and copper deeper than steel. Recall that momentum is mass × velocity and a constant momentum and size requires a higher velocity for the lower density projectiles.

Explosions that create shrapnel and overpressure hazards are rated in terms of the amount of energy that is released. The energy release may be given directly in energy units such as the British thermal unit, but it is commonly expressed as an equivalent quantity of TNT (symmetrical trinitrotoluene). Because the explosive strength of TNT is well known and reproducible, it is a good standard for rating explosive hazards and the explosive potential of various substances. Table 11.13 indicates the overpressure to be expected from a 1 lb TNT charge exploded in air. The times required for the blast wave to reach a particular station and for the pressure at that station to recede back to atmospheric pressure are also given. By the use of appropriate scaling laws, the data in Table 11.14 may be used to prepare a charge indicating overpressure for any distance or explosive charge. This plot is shown on Fig. 11.23. To use this figure, simply enter the chart with the appropriate distance and TNT charge. For example, to estimate the overpressure at 100 ft from an explosive charge having a TNT equivalent of 10 lb, in Fig. 11.23 connect 10 lb TNT and 100 ft to obtain an overpressure of 1 psig. Both structures and personnel are subject to damage from shrapnel and overpressure as indicated in Table 11.14. It should be noted that the overpressures shown on Fig. 11.23 can be amplified up to eightfold when the blast wave is reflected from a surface. Most petroleum products mixed with liquid oxygen have an explosive potential of about 1–2 lb of TNT per pound of detonable mixture. Consequently,

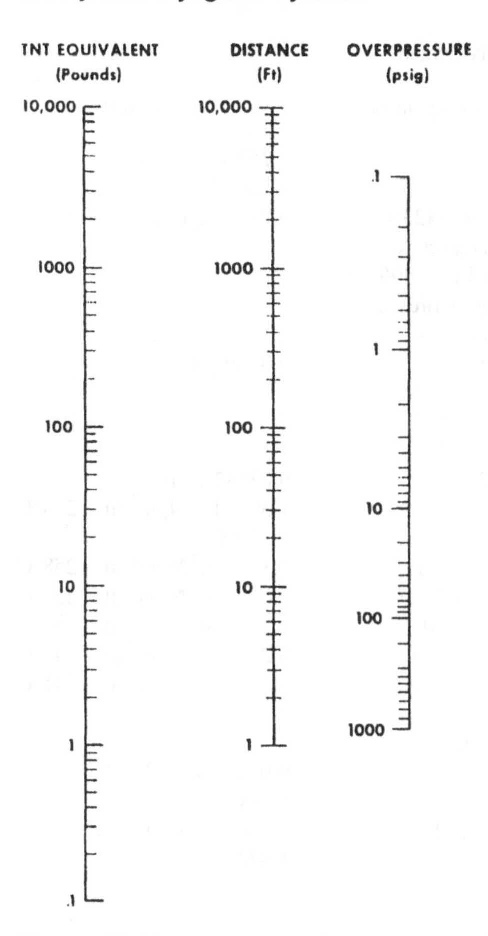


**Figure 11.23** Nomograph for estimating overpressures resulting from open-air surface explosion of a hemispherical TNT charge.

Fig. 11.23 may be used directly to estimate side-on overpressures for known quantities of mixtures, e.g., 1 lb of kerosene–liquid oxygen may be treated as 1 lb of TNT. To obtain this TNT equivalent, the weight of oxygen consumed in the reaction must be about three times the weight of fuel burned. This means that 1 lb of kerosene properly mixed with 3 lb of liquid oxygen is equivalent to 4 lb of TNT. Other mixture ratios result in lower TNT equivalents.

Where extensive fires occur without explosion, cooling water is frequently used to prevent structural damage or failure.

## 6. SPECIAL CONSIDERATIONS FOR HYDROGEN

High-purity liquid hydrogen is a transparent, colorless, odorless liquid. When observable, it is usually boiling vigorously because of its low boiling point, and when exposed to the atmosphere it creates a voluminous cloud of condensed water vapor.

The physical properties of equilibrium hydrogen are given in American Customary English and metric units in Table 11.15.

**Table 11.15** Physical Properties of Liquid Hydrogen

| Property | Engineering units | Metric units |
|---|---|---|
| Boiling point | −423°F | −253°C |
| Freezing point | −435°F | −259°C |
| Liquid density | 0.59 lb/gal at −423°F<br>Density decreases to 0.45 lb/gal at −405°F where vapor pressure is 120 psia | 0.07077 g/cc at −253°C |
| Specific gravity—vapor (relative to standard temperature and air pressure) | 0.07 at $\frac{32}{32}$°F | 0.07 at $\frac{0}{0}$°C |
| Critical density | 0.262 lb/gal | 0.03142 $g/cm^3$ |
| Critical pressure | 188 psia | $1.30 \times 10^4$ $N/m^2$ at −258°C |
| Critical temperature | −400°F | −240°C |
| Vapor Pressure | 1.9 pisa at −433°F<br>14.7 psia at −423°F<br>23.7 psia at −420°F<br>120 psia at −405°F<br>162 psia at −402°F | $1.31 \times 10^4$ $N/m^2$ at −258°C<br>$1.01 \times 10^5$ $N/m^2$ at −253°C<br>$1.63 \times 10^5$ $N/m^2$ at −251°C<br>$8.27 \times 10^5$ $N/m^2$ at −243°C<br>$1.12 \times 10^6$ $N/m^2$ at −241°C |
| Coefficient of viscosity | | |
| Kinematice | 0.20 cS at −423°F | – |
| Absolute | – | 0.014 cP at −253°C |
| Autoignition temperature in air | 1075°F | 579°C |
| Flammability range in air | 4–75 vol % at 68°F | 3.2–60 $g/m^3$ at 20°C |
| Volume expansion of liquid, from NBP to gas at 70°F and 1 atm | 1:840 | 1:840 |
| Vapor density at −423°F | 0.083 $lb/ft^3$ (1.02 times heavier than air at 32°F) | |
| Gas density at 32°F | 0.0056 $lb/ft^3$ (14.5 times lighter than air at 32°F) | |

$N/m^2 = 7.50 \times 10^{-3}$ mmHg $= 9.87 \times 10^{-6}$ std atm.

## 6.1. Fire and Explosion Hazards

A fire hazard always exists when hydrogen gas is present. Hydrogen–air mixtures containing as little as 4% or as much as 75% hydrogen by volume are readily ignited. Hydrogen–oxygen mixtures are flammable over the range of 4–94% hydrogen by volume. When no impurities are present, hydrogen burns in air with an invisible flame. In enclosed spaces, all personnel must be evacuated when the atmosphere reaches 20% of the LEL (lower explosive limit—same as lower limit of flammability). See Fig. 11.24.

Unconfined hydrogen–air mixtures generally burn rapidly but without detonation when ignited by heat, spark, or flame. Since hydrogen diffuses rapidly in air, it will not form persistent flammable mixtures when the liquid evaporates in open, unconfined areas. However, in confined areas or when ignition of the hydrogen–air mixture is caused by a shock source equivalent to a blasting cap or small explosive charge, the mixture can detonate.

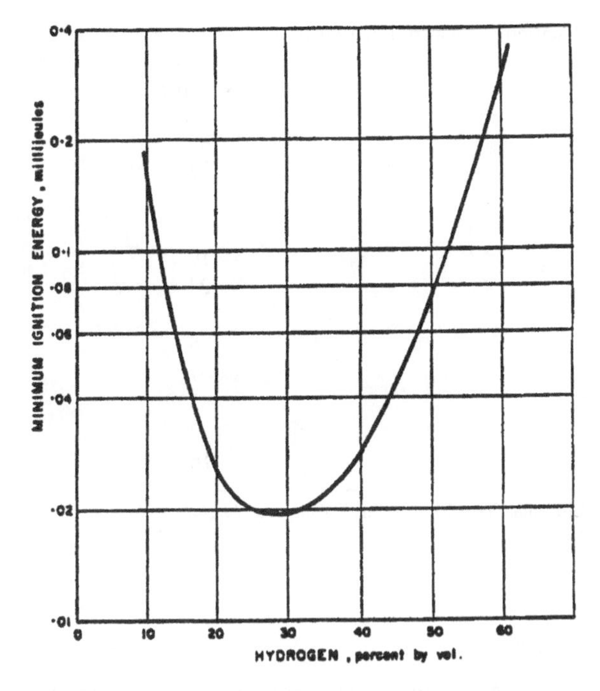


**Figure 11.24** Minimum ignition energy for hydrogen–air mixtures.

### *6.1.1. Industrial and Aerospace Accidents*

Industrial and aerospace accidents from the use of hydrogen have occurred. Tables 11.16–11.18 list types of accidents. Analyses of these accidents indicate the following factors are of primary importance in causing system failures:

1. mechanical failure of the containment vessel, piping, or auxiliary components (brittle failure, hydrogen embrittlement, or freeze-up);
2. reaction to the fluid with a contaminant (such as air in a hydrogen system);
3. failure of a safety device to operate properly;
4. operational error.

**Table 11.16** Hydrogen Accidents—Industrial

| Category | Number of incidents | Percentage total accidents |
|---|---|---|
| Undetected leaks | 32 | 22 |
| Hydrogen–oxygen off-gassing explosions | 25 | 17 |
| Piping and pressure vessel ruptures | 21 | 14 |
| Inadequate inert gas purging | 12 | 8 |
| Vent and exhaust system incidents | 10 | 7 |
| Hydrogen–chlorine incidents | 10 | 7 |
| Others | 35 | 25 |
| Total | 145 | 100 |

Zalosh, R.G., and T.P. Short. *Comparative Analysis of Hydrogen Fires and Explosion Incidents.* C00-4442-2, Factory Mutual Research Corp., Norwood, MA (1978).
Zalosh, R.G., and T.P. Short. *Compilation and Analysis of Hydrogen Accident Reports.* C00-4442-4, Factory Mutual Research Corp., Norwood, MA (1978).

**Table 11.17** Hydrogen Accidents—Ammonia Plants

| Classification | Number |
|---|---|
| Gaskets | 46 |
| Equipment flanges | 23 |
| Piping flanges | 16 |
| Valve flanges | 7 |
| Valve packing | 10 |
| Oil leaks | 24 |
| Transfer header | 9 |
| Auxiliary boiler | 8 |
| Primary reformer | 7 |
| Cooling tower | 3 |
| Electrical | 2 |
| Miscellaneous | 16 |

Williams, G.P. "Causes of Ammonia Plant Shutdowns." *Chemical Engineering Progress*, Vol. 74(9), (1978).

**Table 11.18** Hydrogen Accidents—Aerospace

| Description | Accidents involving release of hydrogen | Percentage of total accidents |
|---|---|---|
| Accidents involving release of liquid or gaseous hydrogen | 87 | 81 |
| Location of hydrogen release: | | |
| To atmosphere | 71[a] | 66 |
| To enclosures (piping, containers, etc.) | 26[a] | 24 |
| Ignition of hydrogen releases: | | |
| To atmosphere | 44 | 41 |
| To enclosures | 24 | 22 |

[a] Hydrogen was released to both location in 10 accidents.
Ordin, P.M. "A Review of Hydrogen Accidents and Incidents in NASA Operations," In *9th Intersociety Energy Conversion Engineering Conference*. New York: ASME (1974).

### *6.1.2. Ignition*

Because the ignition energy is so low (20 micro Joules), fires and explosions have occurred in various components of hydrogen systems as a result of a variety of ignition sources. Ignition sources have included mechanical sparks from rapidly closing valves, electrostatic discharges in ungrounded particulate filters, sparks from electrical equipment, welding and cutting operations, catalyst particles, and lightning strikes near the vent stack. Tables 11.19 and 11.20 list additional ignition sources.

### *6.1.3. Fire and Explosions*

A potential fire hazard always exists when hydrogen is present. $GH_s$ diffuses rapidly with air turbulence increasing the rate of $GH_2$ dispersion. Evaporation can rapidly occur in an $LH_2$ spill, resulting in a flammable mixture forming over a considerable distance. Although ignition sources may not be present at the leak or spill location,

**Table 11.19** Potential Ignition Sources

| Thermal ignition | Electrical ignition |
|---|---|
| Personnel smoking | Electrical short circuits, sparks, and arcs |
| Open flames | Metal fracture |
| Shock waves from tank rupture | Static electricity (two-phase flow) |
| Fragments from bursting vessels | Static electricity (flow with solid particles) |
| Heating of high-velocity jets | Lightning |
| Welding | Generation of electrical charge by equipment operations |
| Explosive charges | |
| Friction and galling | |
| Resonance ignition (repeated shock waves in a flow system) | |
| Mechanical impact | |
| Tensile rupture | |
| Mechanical vibration | |
| Exhaust from thermal combustion engine | |

Ordin, P.M. "Safety," Chapter 1, *Liquid Cryogens, Vol 1, Theory and Equipment*. K.D. Williamson, Jr. and F.J. Edeskuty, eds., CRC Press, Boca Raton, FL (1983).

fire could occur if the movement of the flammable mixture causes it to reach an ignition source.

Observation alone is not a reliable technique for detecting pure hydrogen–air fires or assessing their severity. A fire resulted from an accident in which a small leak developed. The equipment was shut down and the flame appeared to diminish; however, molten metal drippings from the equipment indicated a more severe fire was in progress. See Table 11.21.

A *deflagration* could result if a mixture within flammability limits is ignited at a single point. A *detonation* could occur if the $GH_2$–air mixture is within detonability limits and an appropriate energy source is available. A deflagration could transform into a detonation if there is confinement or a mechanism for flame acceleration. Flash fires or *boiling liquid expanding vapor explosions* (BLEVE) could occur when an external source of thermal energy is heating $LH_2$ or sluch hydrogen ($SLH_2$) and there is a path to the surroundings.

**Table 11.20** Safety Procedures for Hydrogen—Ignition Sources

- Eliminate all ignition sources when possible but assume ignition sources are *always* present
- Use explosion-proof or purged electrical equipment
  Purge with nitrogen or fresh air
- Electrical ignition sources should be eliminated by grounding and proper choice of equipment
  Avoid synthetic fabrics—use cotton
  Ground all equipment and transfer lines
  Use conductive floors if possible
- Thermal sources should be eliminated by proper procedures and controls
  No smoking
  Use spark-proof tools
  No objects hotter than 871°F (80% of $H_2$ ignition temperature)
  Exclude vehicles and machinery from hydrogen containers by at least 50 ft

**Table 11.21** Hydrogen Hazards

- Colorless, odorless gas not detectable by human senses
- Extremely wide flammable range
  4–75% in air
  4–94% in oxygen
- Low ignition temperature and energy
  Invisible sparks can ignite hydrogen–air mixtures
- Hydrogen leaks often ignite and burn with an invisible flame
  Burns can result from skin contact with hot surfaces
- Air contamination in liquid hydrogen can form an unstable mixture which detonates
- Cold vapor clouds will travel at ground level until hydrogen becomes buoyant and rises

### *6.1.4. Leaks*

Leaks can occur within a system or to the surroundings. Hazards can arise by air or contaminants leaking into a cold hydrogen system, such as by cryo-pumping. Hydrogen leaks generally originate from valves, flanges, diaphragms, gaskets, and various types of seals and fittings. The leaks usually are undetected because of the absence of a continuous hydrogen monitor in the area. An example of this condition was a large sphere partitioned by a neoprene diaphragm with hydrogen stored under the diaphragm and air above it. An explosion-proof fan was placed on top of the sphere to provide a slight positive pressure on the diaphragm. A violent explosion occurred in the sphere after the plant was shut down. Hydrogen leaked past the diaphragm when the fan was turned off. Ignition was attributed to an electrostatic discharge caused by motion of the diaphragm or a source associated with the explosion-proof fan. The explosion could have been avoided by using an inert gas instead of air across the diaphragm or monitoring the hydrogen concentration in the upper hemisphere.

For leaks involving $LH_2$ vaporization of cold vapor hydrogen to the atmosphere may provide a warning because moisture condenses and forms a fog. Undetected hydrogen leaks can lead to fires and explosions.

### *6.1.5. Hydrogen Dispersion*

The buoyancy of hydrogen tends to limit the horizontal spread of combustible mixtures from a hydrogen spill. Although saturated hydrogen is heavier than air, it quickly becomes lighter than air, making the cloud positively buoyant. Although condensing moisture is an indication of cold hydrogen, the fog shape does not give an accurate description of the hydrogen cloud location. The use of dikes or barricades around hydrogen storage facilities should be carefully planned because it is preferred to disperse any leaked or spilled $LH_2$ as rapidly as possible. Dikes or berms generally should not be used unless their purpose is to limit or contain the spread of a liquid spill because of nearby buildings, ignition sources, etc.

### *6.1.6. Vent and Exhaust System*

Vent and exhaust system accidents are attributed to inadequate ventilation and the inadvertent entry of air into the vent. Backflow of air can be prevented with suitable vent stack designs, provision of makeup air (or adequate supply of inert gas as the situation demands), check valves, or molecular seals.

**Table 11.22** Safety Procedures for Hydrogen—Purging Equipment

- Equipment in liquid hydrogen service must be purged before and after use for storage or servicing
- Vacuum purge is most effective but not always feasible
  Evacuating to 5 torr (0.1 psia) reduces hydrogen concentration to less than 1 ppm in 3 cycles
- Pressure purging can require large gas volumes and may not remove trapped gas pockets
  Pressurizing to 50 psig four times reduces hydrogen content to less than 0.3% (if well mixed)
- Flowing purge works well for well-mixed systems such as transfer lines but will not remove trapped gas in buoyant pockets such as gauge lines

#### *6.1.7. Purging*

Pipes and vessels should be purged with an inert gas before and after using hydrogen in the equipment. Nitrogen may be used if the temperature of the system is above 80 K (−316°F). whereas helium should be used if the temperature is below 80 K (−316°F). Alternatively, a $GH_2$ purge may be used to warm the system to 80 K (−316°F) and then switch to a nitrogen purge if the system is below 80 K (−316°F); however, some condensation of the $GH_2$ may occur if the system contains $LH_2$. Residual pockets of hydrogen or the purge gas will remain in the enclosure if the purging rate, duration, or extent of mixing is too low. Reliable gas concentration measurements should be obtained at a number of different locations within the system for suitable purges. See Table 11.22.

#### *6.1.8. Condensation of Air*

An uninsulated line containing $LH_2$ or cold hydrogen gas, such as a vent line, can be sufficiently cold (less than 90 K [−298°F] at 101.3 kPa [14.7 psia]) to condense air on the outside of the pipe. The condensed air, which can be enriched in oxygen to about 50%, must not be allowed to contact sensitive material or equipment. Materials not suitable for low temperatures, such as carbon steel, can become embrittled and fail if condensed enriched air drips onto them. Condensed air must not be permitted to drip onto combustible materials such as tar and asphalt.

### 6.2. Safety Procedures

A written procedure should be used for all operations involving propellants. All liquid hydrogen handling operations must be performed by two or more persons. All sources of ignition must be prohibited in areas where liquid hydrogen is handled and stored. Hydrogen vapors should be vented or flared at locations remote from sources of ignition.

Liquid hydrogen containers and systems should be kept under positive pressure to prevent entry of air, thus avoiding potential hazards of explosion and of plugging constricted areas such as valves and orifices with frozen air.

Special procedures and storage precautions must be taken with systems containing slush hydrogen under subatmospheric pressure. Any leak will be *into* the system and will be near impossible to detect. A rule of thumb for hydrogen that has saved many lives is "maintain positive pressure and exclude air". Clearly, this is impossible with subatmospheric systems such as slush hydrogen.

### 6.2.1. *Hydrogen Leak Detection*

Gaseous hydrogen is colorless and odorless and normally not detectable by human senses. Therefore, means must be provided to detect the presence of hydrogen in all areas where leaks, spills, or hazardous accumulations may occur.

Well-placed, reliable hydrogen detectors are imperative for safe installation. Continuous automatic sampling equipment with sample points strategically located should be provided as needed. The sampling equipment shall be calibrated to provide for short response times and detection of at least 25% of the LFL. A portable hydrogen detector should be available.

Tables 11.23–11.25 give information on hydrogen gas detectors. Table 11.23 shows a list of typical $GH_2$ detectors. Table 11.24 shows a survey and analysis of commercially available hydrogen sensors. Table 11.25 shows sensitivity limits of hydrogen detectors.

Detection and alarm at 1% by volume hydrogen concentrations in air, equivalent to 25% of the LFL, is required for enclosed areas in which $GH_2$ buildup is possible (29 CFR 1910.106 1996). Detection and alarm at 0.4% by volume hydrogen concentrations in air (equivalent to 10% of the LFL) is required for permit-required confined spaces (29 CFR 1910.146 1996). See Table 11.26.

### 6.2.2. *Detection Technologies*

*Bubble testing* is one of the simplest methods of leak detection; however, it is not a continuous monitoring system, needs to be applied directly on the source of the leak,

**Table 11.23** Typical Hydrogen Gas Detectors

| Type of detector | Description |
|---|---|
| Catalytic | A palladium and/or platinum catalyst is used to facilitate the combustion of hydrogen with oxygen. A sensing element detects the heat of combustion |
| Electrochemical | Liquid or solid electrolytes surrounding a sensing electrode and a counter electrode. Reaction with hydrogen product produces a current. The hydrogen gas must flow through a gas permeable membrane to reach the electrolyte |
| Semiconducting oxide | Hydrogen gas reacts with chemisorbed oxygen in a semiconductor material, such as tin oxide, and changes the resistance of the material |
| Thermal conductivity | The rate of heat conduction from a heat source into the surrounding environment is dependent on the thermal conductivity of that environment |
| Mass spectrometer | The gas is ionized and then accelerated through an electric field along a curved path. The amount of curvature induced by the electric field is dependent on the mass of the particle and is used to separate the particles by mass. A detector is placed in the path of the desired gas to be measured |
| Sonic | Leaking gas can produce acoustical emissions in the range of 30–100 kHz, with 40 kHz being the most common |
| Optical | The differences in the refractive index of various gases can be used for detection in sensors using optical interferometry |
| Glow plugs | Glow plugs are not a true gas detection technique. When a combustible gas mixture exists, the glow plug ignites the mixture and then the fire is detected with heat sensors |

**Table 11.24** Commercially Available Hydrogen Sensors

| Conditions | Catalytic combustion sensors (CC) | | | Electrochemical sensors (BC) | | | Semi conducting oxide sensors (SO) | | | Thermal conductivity detectors (TC) | | |
|---|---|---|---|---|---|---|---|---|---|---|---|---|
| | D | RT | PR | D | RT | PR | D | RT | PR | D | RT | PR |
| *In air* | | | | | | | | | | | | |
| 293 K/200 ppm | + | 0 | − | + | + | + | + | + | + | + | 0 | + |
| 293 K/2% $H_2$ | + | 0 | − | + | + | + | + | + | + | + | 0 | + |
| 293 K/100% $H_2$ | − | − | − | + | + | + | + | + | + | + | 0 | + |
| 77 K/200 ppm | + | ? | − | − | − | − | + | ? | ? | + | ? | ? |
| 77 K/2% $H_2$ | + | ? | − | − | − | − | + | ? | ? | + | ? | ? |
| 77 K/100% $H_2$ | − | − | − | − | − | − | + | ? | ? | + | ? | ? |
| *In helium* | | | | | | | | | | | | |
| 293 K/200 ppm | − | − | − | + | + | + | − | − | − | − | − | − |
| 293 K/2% $H_2$ | − | − | − | + | + | + | − | − | − | − | − | − |
| 293 K/100% $H_2$ | − | − | − | + | + | + | − | − | − | − | − | − |
| 77 K/2% ppm | − | − | − | − | − | − | − | − | − | − | − | − |
| 77 K/2% $H_2$ | − | − | − | − | − | − | − | − | − | − | − | − |
| 77 K/100% $H_2$ | − | − | − | − | − | − | − | − | − | − | − | − |
| *In nitrogen* | | | | | | | | | | | | |
| 293 K/200 ppm | − | − | − | + | + | + | − | − | − | + | 0 | + |
| 293 K/2%$H_2$ | − | − | − | + | + | + | − | − | − | + | 0 | + |
| 293 K/100% $H_2$ | − | − | − | + | + | + | − | − | − | + | 0 | + |
| 77 K/200 ppm | − | − | − | − | − | − | − | − | − | + | ? | ? |
| 77 K/2% $H_2$ | − | − | − | − | − | − | − | − | − | + | ? | ? |
| 77 K/100% $H_2$ | − | − | − | − | − | − | − | − | − | + | ? | ? |
| *In vacuum* | | | | | | | | | | | | |
| 293 K/200 ppm | − | − | − | − | − | − | − | − | − | + | 0 | + |
| 293 K/2% $H_2$ | − | − | − | − | − | − | − | − | − | + | 0 | + |
| 293 K/100% $H_2$ | − | − | − | − | − | − | − | − | − | + | 0 | + |
| 77 K/200 ppm | − | − | − | − | − | − | − | − | − | + | ? | ? |
| 77 K/2% $H_2$ | − | − | − | − | − | − | − | − | − | + | ? | ? |
| 77 K/100% $H_2$ | − | − | − | − | − | − | − | − | − | + | ? | ? |

Each type of sensor is evaluated for detection, response time, and power requirements according to the following system.
Detection (D): (a) detects (+); (b) inoperative (−).
Response time (RT): (a) less than 3 sec (+); (b) less than 45 sec (0); (c) greater than 45 sec (−).
Power requirements (PR): (a) $<1$ W (+); (b) $>1$ W (−).
? Indicates unknown or not available.

can only be used with inert gases at low pressure, and is limited to temperatures above freezing. Bubble solutions can detect very small leaks but do not measure concentrations.

*Catalytic combustion sensors* detect hydrogen gas by sensing the heat generated by the combustion of hydrogen and oxygen on the surface of a catalytic metal such as palladium or platinum. The sensors work well for detection of hydrogen in the 0–4% by volume (0–100% LFL) in air but do not operate in inert environments or 100% by volume hydrogen. Sampling systems can be designed to mix air in with the sample before exposure to the catalytic sensor for operation in inert

**Table 11.25** Sensitivity Limits of Hydrogen Detectors

| | Minimum detection limits, average values | | | | | |
|---|---|---|---|---|---|---|
| | In air | | | In nitrogen | | |
| Principle | atm-cc/sec | % Hydrogen | %LEL | atm-cc/sec | % Hydrogen | Distance (ft) |
| Catalytic combustion | 8.0 | 0.02 | 0.5 | 80 | 0.2[a] | 2000[b] |
| Bubble testing | $1 \times 10^{-4}$ | NA | NA | $1 \times 10^{-4}$ | NA | NA |
| Sonic-ultrasonic | $1 \times 10^{-2}$ | NA | NA | $1 \times 10^{-2}$ | NA | NA |
| Thermal conductivity | $1 \times 10^{-3}$ | $5 \times 10^{-4}$ | 0.01 | $1 \times 10^{-3}$ | $5 \times 10^{-4}$ | – |
| Gas density | $1 \times 10^{-2}$ | $5 \times 10^{-3}$ | 0.1 | $1 \times 10^{-2}$ | $5 \times 10^{-3}$ | NA |
| Hydrogen tapes | 0.25 | 1.5 | 35 | – | – | (d) |
| Scott-Draeger tubes | – | 0.5 | 13 | – | – | NA |
| Electrochemical | – | 0.05 | 1.2 | – | 0.05 | 1000[b] |
| Optical interferometer | – | 0.2 | 5 | – | 0.2 | NA |

[a] Only one commercial catalytic instrument has claimed to detect hydrogen in nitrogen.
[b] The sensing head is remote from readout.
Dashes indicate information is not available.
*Soruce:* Rosen, B., V.H. Dayan, and R.L. Proffitt. "Hydrogen Leak and Fire Detection—A Survey". NASA SP-5092, Washington, DC (1970).

environments; however, this results in a longer response time. Catalytic sensors such as sintered bronze utilize a heated filament and need to be enclosed in flame arrestors to prevent the sensors from becoming ignition sources. Also, catalytic sensors are not hydrogen specific and will respond to other combustible gases such as methane.

*Thermal conductivity sensors* work well in stable environments with minimal temperature variations and a constant background gas. Thermal conductivity sensors work well in background gases that have a thermal conductivity that varies

**Table 11.26** Safety Procedures for Hydrogen—Leak Detection

- A reliable leak detection system
  - Identifies all possible leak sources
  - Detects 1 ± 0.25% $H_2$ in air or 1–10% $H_2$ in an inert gas (by volume)
  - Responds in less than 1 sec
  - Sets off alarms for out-of-limit conditions
- Detection and monitoring systems should include
  - Manually portable gas detectors
  - Periodic detector recalibration
  - Assurance that leakage gas will pass detectors
  - Thorough coverage of all potential leak sources
- Many detector types and accuracies are available
  - Catalytic combustion provides ±1% FS with a response time of 2 sec
  - Mass spectrometers are the most sensitive but flow conductance can limit response time
  - Thermal conductivity detectors are 1–5% accuracy with variable response times
  - Changes in background gas (He, $N_2$, Air) can affect accuracy or offsets and may require recalibration

significantly from hydrogen, such as air or nitrogen. However, they do not work in helium backgrounds; a significant drawback for $LH_2$ systems because helium purges are often required because helium does not solidify at $LH_2$ temperatures. Thermal conductivity sensors can go from 0.02% to 100% by volume hydrogen detection.

*Electrochemical sensors* typically utilize a liquid electrolyte and require a gas-permeable membrane for the hydrogen to reach the electrolyte. The sensors are of low power and can operate from 0.02% to 100% by volume hydrogen. Exposure of the membrane to cryogenic or time-varying temperatures greatly affects the gas diffusion and can make the sensor unreliable.

*Semiconducting oxide sensors* rely on surface effects with a minimum oxygen concentration present and do not work in inert environments. Semiconducting oxide sensor can operate at lower powers than catalytic sensors, but performance at lower temperatures is degraded. Semiconducting oxide sensors are relatively new and not as common as the older catalytic-based systems.

*Mass spectrometers* are extremely sensitive (1 ppm), very specific to the gas being detected, linear over a wide dynamic range, and provide continuous monitoring. The complexity and high cost of mass spectrometers requires skilled operators and the use of sampling systems to monitor multiple locations with one instrument. The use of long sample lines can significantly reduce the response time.

*Gas chromatographs* are similar to mass spectrometers in their sensitivity and accuracy; however, measurement times are extremely slow. Gas chromatographs typically are used in a laboratory to analyze the gas collected in the field with sample bottles.

*Ultrasonic leak detection* can be used when hydrogen specificity is not required and there is minimal background interference noise. Ultrasonic systems typically are used to pinpoint the source of a leak and cannot measure whether a combustible mixture is present.

*$LH_2$ leaks* can be detected through loss of vacuum and by observing the formation of frost, the formation of solid air, or a decrease in outer wall temperature on vacuum-jacketed equipment.

*Glow plugs* and heat sensors are a less common technique used for rapid leak detection. The glow plugs ignite any combustible mixture present, and a heat sensor detects the fire and provides rapid shutdown of the process. The theory is it is better to burn the hydrogen gas rather than letting a combustible mixture accumulate, and the facility can shut down before the fire can do significant damage if a fire is started.

### *6.2.3. Hydrogen Fire Detection*

Because hydrogen fires themselves are nearly invisible in daylight, special imaging systems are required for determining the size and location of a flame. Table 11.27 gives information on hydrogen fire detection.

The solar-transmission-emission-detector sensitivity trade off follows:

1. Solar. The radiation from the sun can overpower the hydrogen flame emission, resulting in an invisible flame during the day, in the visible spectrum.
2. Transmission. A large percentage of the radiation emitted from a hydrogen fire originates from the hot water molecule. Emission peaks occur at the same location water (humidity) in the atmosphere absorbs radiation.
3. Emission. Hydrogen fires tend to emit over a broad range and are not characterized by extreme peaks such as the $4.3 \times 10^{-6}$-m($1.4 \times 10^{-5}$-ft) peak for hydrocarbon fires.

**Table 11.27** Typical Hydrogen Fire Detectors

| Type of detector | Description |
|---|---|
| Temperature sensor | Thermocouple or resistance temperature device (RTD) that detects the heat of the fire |
| Heat sensitive cable | Wire or fiber-optic-based cable that changed resistance or optical properties if any portion of the cable is exposed to high temperatures or is burned through |
| Optical | Ultraviolet, mid/near-infrared, and thermal infrared detectors for the detection of radiation emitted by the hydrogen flame. Infrared detectors must be optimized for hydrogen flame emissions that are not the same as hydrocarbon fires |
| Broadband imaging | Thermal or mid-infrared imaging systems effectively image hydrogen flames but require an operator to interpret the image for detection of fire |
| Narrow band imaging | Band-pass filters centered around the 950, 1100, and 1400 nm peaks can produce adequate images with low-cost silicon CCD cameras, image converter tubes, and vidicon systems. The filters must be carefully selected to block unwanted solar background while optimizing the imaging band for atmospheric transmission of the hydrogen fire radiation |
| Brooms/dust | Putting flammable objects or dust particles into a hydrogen flame will cause the flame to emit in the visible spectrum. Corn straw brooms, dirt, and dry fire extinguishers have been used for this purpose |

4. Detector sensitivity. Different types of detectors are sensitive to different parts of the spectrum. A higher sensitivity detector at a smaller emission peak may outperform a less sensitive detector at a larger emission peak.

### *6.2.4. Detection Technologies*

*Thermal fire detectors* classified as rate-of-temperature-rise detectors and overheat detectors have been manufactured for many years and are reliable. Thermal detectors need to be located at or very near the site of a fire.

*Optical sensors* for detecting hydrogen fires fall into two spectral regions: ultraviolet (UV) and infrared (IR). UV systems are extremely sensitive; however, they are susceptible to false alarms and can be blinded in foggy conditions. Infrared systems typically are designed for hydrocarbon fires and are not very sensitive to hydrogen fires.

*Imaging systems* mainly are available in the thermal IR region and do not provide continuous monitoring with alarm capability. The user is required to determine if the image being viewed is a flame. UV imaging systems require special optics and are very expensive. Low-cost systems, using low-light silicon charge coupled device (CCD) video technology with filters centered on the 940- and 1100-nm emission peaks have been used at some facilities.

A *broom* has been used for locating small hydrogen fires. The intent is a dry corn straw or sage grass broom easily ignites as it passes through a flame. A dry fire extinguisher or throwing dust into the air also causes the flame to emit visible

**Table 11.28** Safety Procedures for Hydrogen—Fire Detection

- Flames are nearly invisible in daylight
  Visibility is improved by moisture and impurities in the air
  Flames are readily visible in dark or subdued light
- Large fires are readily detectable due to convective "heat ripples" and thermal radiation
- Several detection systems can (and should) be employed
  Conventional or optical thermal sensors detect temperature rise
  Closed circuit infrared or ultraviolet cameras can monitor locations remotely
  Intumescent paints char, swell, and smell at 200°C
  Infrared binoculars are useful for manual detection at safe distances

radiation. This technique should be used with care in windy, outdoor environments in which the light hydrogen flame can easily be blown around. See Table 11.28.

## 6.3. Equipment

### *6.3.1. Pipes and Fittings*

Pipes and fittings must be of approved materials and construction and be hydrostatically tested at prescribed pressures. Welded and welded flanged connections are recommended. Eutectic brazing procedures are satisfactory with stainless steel. Threaded connections must be avoided, especially where directly exposed to liquid hydrogen temperatures. Liquid hydrogen lines must be insulated to prevent condensation of air or excessive heat transfer. All lines in which liquid hydrogen may be trapped between closed valves must be equipped with safety relief devices.

Facility and transfer piping systems shall include safeguards in accordance with ANSI/ASME B31.3 (1996) for protection from accidental damage and for the protection of people and property against harmful consequences of vessel, piping, and equipment failures. Barriers should be considered for operator protection particularly from metal parts associated with pump failures. Within a process area, hydrogen transport piping shall be treated similar to hydrogen storage in that all such piping shall be isolated by an exclusion zone in which access is restricted and certain types of operations are prohibited while hydrogen is present in the piping system.

The best practice for piping for $GH_2$ or $LH_2$ is that it should not be buried. Piping should be placed in open trenches with removable grating if placed below ground.

Hydrogen lines should not be located beneath electric power transmission lines. Electric wiring systems permitted above hydrogen lines should comply with the appropriate code (NFPA 50B 1994).

Sufficient grounding connections should be provided to prevent any measurable static charge from accumulating on any component. Each flange should have bonding straps in addition to metal fasteners, which are primarily structural.

Joints in piping and tubing should be made by welding or brazing. Mechanical joints, such as flanges, should only be used for ease of installation and maintenance and other similar considerations.

Provisions should be made for the expansion and contraction of piping connected to a pressure vessel to limit forces and moments transmitted to the pressure vessel, by providing substantial anchorage at suitable points, so there shall be no undue strain to the pressure vessel. See Table 11.29.

**Table 11.29** Liquid-Hydrogen Piping Systems

- All piping materials and designs should meet applicable ASME and ASTM codes and should have a 2-hr resistance rating
- Piping systems should be protected from accidental damage and for the protection of people and property in the event of a failure
- Piping should not be buried; below-grade installations should use open trenches with metal grates
- Piping systems should have sufficient flexibility to prevent piping failures or leaks caused by thermal expansion or contraction.
- Each section of liquid-hydrogen pipe should be considered as a pressure vessel and provided adequate relief paths
- Piping should not be painted white so that frost build-up can be readily detected
- Uninsulated piping is not recommended, but if used, should be installed away from asphalt and combustible surfaces due to the possible condensation of enriched liquid air

### 6.3.2. *Joints and Connections*

*Joints* in piping and tubing may be made for $GH_2$ by welding or brazing or using flanged, threaded, socket, or compression fittings. Joints in piping and tubing for $LH_2$ preferably shall be made by welding or brazing; flanged, threaded, socket, or suitable compression fittings may be used. Brazing materials shall have a melting point above 811 K (1000°F) (NFPA 50A 1994 and NFPA 50B 1994).

*Welding* is the first preference for all hydrogen systems, and all forms of welding can be used (ANSI/ASME B31.1 1995 and ANSI B31.3 1996). The type of weld to be used generally is determined by factors other than the system is for hydrogen use.

*Threaded joints* with a suitable thread seal are acceptable for use in $GH_2$ systems but are to be avoided in $LH_2$ systems. Consideration should be given to back-welding threaded joints inside buildings for $GH_2$. Bayonet joints should be used for $LH_2$ transfer operations.

*Soft solder joints* are not permissible in hydrogen systems. Soft solder has a low melting point and will quickly fail in fire, potentially releasing more hydrogen. Also, soft solders containing tin may become crumbly and lose all strength at cryogenic temperatures (CFR 29 1910.103). Soft solder may be used as appropriate in nonhydrogen portions of the system such as a vacuum jacket.

### 6.3.3. *Flexible Hoses*

Flexible hoses pressurized to greater than 1.14 MPa (165 psia) should be restrained at intervals not to exceed 1.83 m (6 ft) and should have an approved restraint device such as the Kellems hose containment grips attached across each union or hose splice and at each end of the hose.

### 6.3.4. *Gaskets*

Gaskets may be made of materials suitable for the particular application.

Ferrous alloys, except for the austenitic nickel–chromium alloys, lose their ductility when subjected to the low temperatures of liquid hydrogen and become too brittle for this service. The *metals suitable* for this service are

| | |
|---|---|
| Stainless steel (300 and other austenitic series) | Brass |
| | Monel |
| Copper | Aluminum |
| Bronze | Everdur |

*Nonmetallic materials* must also be selected to withstand the low temperature of liquid hydrogen. Nonmetals found suitable for this service include

| |
|---|
| Polyester fiber (Dacron or equivalent) |
| Tetrafluoroethylene (TFE, Halon TFE, Teflon, or equivalent) |
| Unplasticized chlorotrifluoroethylene (Kel-F, Halon CTF, or equivalent) |
| Asbestos impregnated with TFE |
| Mylar or equivalent |
| Nylon |

See Table 11.30.

Type 321 stainless steel O-rings with a coating such as Teflon® or silver should be used in stainless flanges with stainless bolts. Likewise, Teflon®-coated aluminum O-rings should be used in aluminum flanges with aluminum bolting. Using similar materials avoids the leakage possibility from unequal contraction or corrosion of dissimilar metals.

Surface finishes in the O-ring groove and contact area should be in accordance with the manufacturer's specifications. All machine or grind marks should be concentric.

### *6.3.5. Gaskets, Seals, and Valve Seats*

Kel-F® (polytrifluorochloroethylene) or Teflon® (polytetrafluoroethylene) can be used in $LH_2$ systems for the following sealing applications:

Valve seats may be Kel-F® or modified Teflon® (Fluorogreen® is preferred). Gland packing or seal only if it is maintained near ambient temperatures as in an extended bonnet of a shutoff valve. The contraction or shrinkage of Teflon® when cooled from ambient to cryogenic temperatures allows leakage.

### *6.3.6. Valves*

Valves to be used with liquid hydrogen must conform to particular specifications. Extended stem, globe-type, or gate-type valves are recommended, but plug-type or ball-type valves may also be used. It must be possible to purge the valves efficiently. Also, they must have an adequate packing design to provide good sealing and to prevent plugging or air condensation. See Table 11.31.

A *shut off valve* shall be located in liquid product withdrawal lines as close to the container as practical. On containers of over 7.6 $m^3$ (2000 gal) capacity, the shut-off valve shall be a remote control type with no connections, flanges, or other appurtenances (other than a welded manual shutoff valve) allowed in the piping

**Table 11.30** Recommended Materials for $H_2$ Service

| Component | Liquid hydrogen service | Gaseous hydrogen service |
|---|---|---|
| Valves | Forged 304 stainless steel or brass body with extended bonnet | Conventional |
| Fittings | Stainless-steel bayonet type for vacuum jackets | Conventional |
| O-rings | Stainless steel (or Kel-F) | Conventional |
| Gaskets | Soft aluminum, lead, or annealed copper between serrated flanges; Kel-F, Teflon | Encapsulated asbestos, mylar, lead, annealed copper, Teflon |
| Hoses | Flexible 316 stainless steel | High pressure |
| Rupture disk assembly | 304 or 304L stainless steel flanges | Forged steel flanges |
| Piping | 304 or 304L stainless steel | Uncoated wrought steel or any 300-series stainless steel |
| Dewars | 304 or 304L stainless steel | N/A |
| Tapes | Teflon | – |
| Threat compound | DC-4 (threaded connections not recommended) | – |
| Lubricant for O-rings | Dupont Krytox 240AC, Fluoramics OXY-8, Dow Corning FS-3452, Dow Corning DC-33, Bray Oil Braycote 601, General Electric Versilube, Houghton Cosmolube 5100 | Lubricants for liquid service suitable for gaseous environment |

**Table 11.31** Valves and Components for Liquid Hydrogen Service

- All valves in liquid hydrogen service should be specifically rated and tested for liquid hydrogen
  Typically, vacuum-jacketed, extended-stem construction is employed
  Many "cryogenic" valves are not rated for liquid hydrogen service
- Any valve actuators (pneumatic, solenoid, etc.) must conform with NFPA and other electrical requirements for hydrogen service
  Pneumatic actuators are often used for remotely controlled hydrogen operations
  Solenoids are often not rated for hydrogen service
  Manual valves (hand wheels) are acceptable provided they meet the automatic control criteria
  All vessels must have an automatic shutoff valve (see LeRC Hydrogen 3.6.5 for exceptions)
- Components in liquid hydrogen service must be compatible with hydrogen and the extremely low temperatures
  Manufacturers test capability in liquid hydrogen is often limited
  Liquid helium testing can provide required temperature ratings
  Since complex components can contain many materials and electronics, thermal contraction can easily create malfunctions

between the shutoff valve and its connection to the inner container (29 CFR 1910.103 1996).

Vessels used as components of a test facility should have remote operating fail-safe shutoff valves. A manual over-ride must be provided for use if the power fails.

*Rupture disks or relief valves* must be installed in all enclosures that contain liquid or can trap liquids or cold vapors for protection against the hazards associated with ruptures. Rupture disks or relief valves may not be necessary if the liquid or the cold vapor trapped between two valves can be relieved through one of the valves at a pressure less than the design operating pressure.

A rupture disk or relief valve shall be installed in every section of a line where liquid or cold gas can be trapped. This condition exists most often between two valves in series. A rupture disk or relief valve may not be required if at least one of the valves will, by its design, relieve safely at a pressure less than the design pressure of the liquid line. This procedure is most appropriate in situations where rupturing of the disk could create a serious hazard.

Rupture disks are designed to break at a specific pressure and temperature and tensile strength of the disk; therefore, the pressure at which a disk will rupture is directly affected by the temperature. The disks should be located to limit temperature variations. Rupture disks are susceptible to failure from cyclic loading. Rupture disks have finite lifetimes and routine replacement should be planned.

*Pressure-relief devices discharge vents* should not be connected to a common line when feasible. The effect of the back pressure that may develop when valves operate should be considered when discharge lines are long or outlets of two or more valves having different relief set pressures are connected to a common line. Discharges directly to the atmosphere should not impinge on other piping or equipment and should be directed away from platforms and other areas used by personnel because the discharge gas may ignite and burn. Reactions on the piping system because of actuation of pressure-relief devices should be considered, and adequate strength should be provided to withstand the reactions. See Fig. 11.25 and Table 11.32.

- **Relief valves and burst disks are often used in combination**
- **Each relief device must be sized to handle the relief requirements of the entire vessel**
- **Redundant relief valves are preferred for some test articles and flight hardware because they re-seal after relieving pressure**
- **Relief device sizing is given in CGA Publication S-1.1**

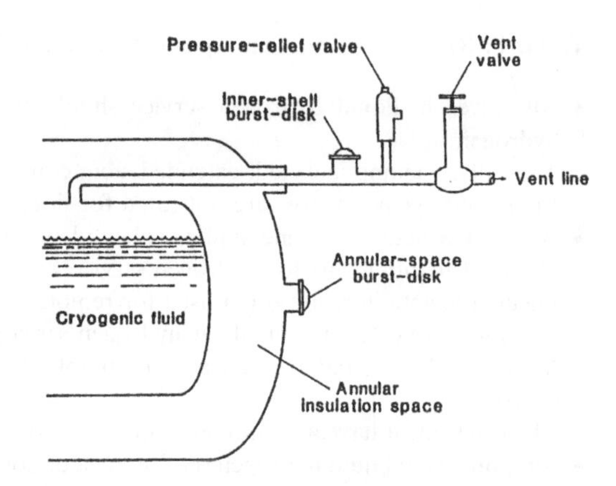


**Figure 11.25** Redundant relief devices are required on all cryogenic vessels.

### *6.3.7. Pumps and Hoses*

Pumps and shaft seals must be specifically designed and qualified by test for use with liquid hydrogen. Hoses must be of a proper design and engineered specifically for liquid hydrogen service.

### *6.3.8. Pressure Gauges*

Liquid hydrogen systems must be monitored with approved types of pressure gauges, as required. Clean standard gauges are acceptable. To minimize operator reading errors, all pressure gauges used for the same purpose should have identical scales.

### *6.3.9. Venting Systems and Safety Relief*

In general, vessels that contain liquids (or gases with low boiling points) should be individually relieved. However, when two or more vessels are connected by piping without valves, the group may be treated as a whole. Relief valves should be connected into the vapor space. The relief valve setting should be such as to protect

**Table 11.32** Rupture Disks and Relief Valves

- Install rupture disks in a warm zone to avoid temperature fluctuations
- Premature failure of rupture disks can occur from fatigue and creep
- Rupture disks must be provided with a "catch" for broken fragments
- Rupture disks may be installed upstream of a relief valve if:
  Disk pressure is less than MAWP
  Disk failure or leakage detection is in place
  Disk failure cannot interfere with relief valve operation
  Sizing is correct
- Rupture disks can be installed downstream of relief valves under approved conditions
- Supplemental relief devices should be installed to protect against excessive pressures created by exposure to fires
  These devices should be set at 110% of MAWP
  Flow capacity should limit vessel pressure to 90% of test pressure
- Disk failure can create a hazardous condition; replacement requires emptying and purging the tank

the vessel or vessels involved and should be sized for operating emergencies or exposure to fire.

## 6.4. Building Areas

Sprinklers or deluge systems and fire hoses with stream-to-spray nozzles should be installed for critical areas where large quantities of liquid hydrogen are handled. Because hydrogen burns with a colorless flame, there should also be a fire detection and alarm system.

Enclosures, if used, must be of noncombustible materials with a ridge-type roof well ventilated at the roof peak.

At least two access roads should be provided to transfer and storage sites, and they should be wide enough at each site to give adequate space for turning motor vehicles.

### *6.4.1. Diking and Drainage*

Major storage facilities may be surrounded by a dike high enough to contain 110% of the capacity of the storage vessels and constructed so as to allow diversional ditching to a prepared area. All storage sites must be properly drained to prevent damage in other areas or mixing of drainage with noncompatible materials. Under no conditions must drainage be into an area where it might mix with an oxidizer. The drainage must not block established routes of personnel exit.

### *6.4.2. Ventilation*

If it is possible for liquid hydrogen to evaporate into a confined area, the area must be adequately ventilated to prevent the accumulation of an excessive concentration of gaseous hydrogen in the atmosphere. Buildings must be ventilated at the highest point so that hydrogen gas cannot accumulate. Depending on the building's location and type of construction, fans of an approved type may be required. A combustible gas detection and alarm system should be arranged to analyze gas samples from all critical locations.

Ventilation shall be established before hydrogen enters the system involved and continued until the system is purged. Ventilation should not be shut off as a function of an emergency shutdown procedure. Suspended ceilings and inverted pockets shall be avoided or adequately ventilated.

The normal air exchange should be about 1 $ft^3$/min of air per square foot of solid floor in the space (NFPA 59A, 1966).

Oxygen concentrations below 19.5% by volume require an air-line respirator in occupied spaces.

Ventilation rates should be sufficient to dilute hydrogen leaks to 25% of the LFL, about 1% by volume for $GH_2$ (29 CFR 1910.106 1996).

A more stringent $GH_2$ concentration limit of 10% of the LFL or about 0.4% by volume is required for permit-required confined space (29 CFR 1910.146 1996).

Inlet openings shall be located at floor level in exterior walls. Outlet openings shall be located at the high point of the room in exterior walls or roof. Inlet and outlet openings shall have a minimum total area of 0.003 $m^2/m^3$ (0.001 $ft^2/ft^3$) of room volume (29 CFR 1910.103 1996).

Hydrogen containers in buildings shall have their safety relief devices vented, without obstruction, to the outdoors at the minimum elevation to ensure area safety. Vents shall be located at least 15.2 m (50 ft) from air intakes (29 CFR 1910.103 1996).

### *6.4.3. Electrical Equipment*

The electrical installation in hazardous areas in transfer and storage locations must conform to the National Electrical Code for Class I Group B, Division 1 or Division 2, as applicable. All lines and equipment, fixed or movable, should be grounded so that static electricity will not accumulate.

All electrical equipment for outdoor locations within 4.6 m (15 ft), separate buildings, and special rooms shall be in accordance with NFPA 70 (1993) for Class I, Division 2 locations for $GH_2$ systems (29 CFR 1910.103 1969 and NFPA 50A 1994).

Electrical wiring and equipment located within 0.9 m (3 ft) of a point where connections are regularly made and disconnected shall be in accordance with NFPA 70 (1993) for Class I, Group B, Division 1 locations for $LH_2$ systems. Electrical wiring and equipment located within 7.6 m (25 ft) of a point where connections are regularly made and disconnected or within 7.6 m (25 ft) of a $LH_2$ storage container, shall be in accordance with NFPA 70 (1993) for Class I, Group B, Division 2 for distances greater than 0.9 m (3 ft).

When equipment approved for Class I, Group B atmospheres is not commercially available, the equipment may be one of the following (29 CFR 1910.103 1996):

1. purged or ventilated in accordance with NFPA 496 (1982);
2. intrinsically safe;
3. approved for Class I, Group C atmospheres.

### *6.4.4. Housekeeping*

Tanks, piping, and equipment used in liquid hydrogen service must be kept clean and free of grease and oil. Surrounding areas must be kept free of grease, oil, oily waste, and other combustible materials. Supervisory personnel should make periodic inspections to ensure good housekeeping practices (see also Table 11.33).

## 6.5. Liquid Hydrogen Systems and Equipment Cleaning

Liquid hydrogen systems should be (1) clean of any surface film, oxidant, grease, oil, or other lubricants; (2) free of particulate matter such as rust, dirt, scale, or weld spatter that will obstruct flow passages; and (3) dry and free of water and of any liquid with a boiling point higher than that of hydrogen. Since many special problems and circumstances are encountered when preparing systems for use, only general guideline procedures are given here.

### *6.5.1. Components—Valves, Fittings, Tubing, Regulators*

**a. Disassembly.** Cryogenic units are disassembled into their component parts, except for plastic inserts that might be damaged by removal. Other metal parts are separated from stainless steel parts to prevent marring.

**b. Solvent Degreasing.** Exceedingly greasy or dirty parts are wiped free of loose dirt and grease. The parts are degreased with perchloroethylene vapor or inhibited methyl chloroform for 30 min. Remove all parts containing plastics and clean

**Table 11.33** Summary of Liquefied Hydrogen Spill Data

| Size of spill (gal) | Surface | Vaporization | Dispersal | Ignition | Radiation |
|---|---|---|---|---|---|
| *AD 324/94*[a] | | | | | |
| 0.18 | Paraffin wax | Yes | No | No | No |
| 1.8 | Macadam | No | Visible cloud rise only | No | No |
| 1.8 | Gravel | No | Visible cloud rise only | No | No |
| 14.4 | Gravel | No | No | Yes | Flame size only |
| 15.4 | Steel | No | No | Yes | Flame size only |
| 0.8 | Macadam | No | Vertical concentrations in quiescent atmosphere | No | No |
| 2.0–23.6 | Gravel | No | No | Yes | Flame size and radiant heat flux |
| *BM-RI-5707*[b] | | | | | |
| 23.5–44.4 | Sand | Yes | No | No | No |
| 35.2 | Bank Gravel | Yes | No | No | No |
| 25.0–50.0 | Crushed rock | Yes | No | No | No |
| 1.3 | Sand | Average | Observations of visible vaporization time cloud | Yes | Fireball size only |
| 32.0 | Sand | Average | Observations of visible vaporization time cloud | No | Fireball size only |
| 500.0 | Sand | Average | Observations of visible vaporization time cloud | No | Fireball size and radiant heat flux |
| 5000.0 | Sand | No | No | Yes[c] | Fireball size and radiant heat flux |

[a] A.D. Little, Inc., "An Investigation of Hazards Associated with the Storage and Handling of Liquid Hydrogen". AD 324/94, A. D. Little, Inc., Cambridge, MA, March 1960.
[b] Zebetakis, M.G., and D.S. Burgess, "Research on the Hazards Associated with the Production and Handling of Liquid Hydrogen". BM-RI-5707, Bureau of Mines Report of Investigation (1961).
[c] Inadvertently.

separately at about 12°F (49°C). Certain plastics may be cleaned with solvents at 120°F without danger of solvent absorption, but in general it is preferable to avoid the possibility of solvent contamination. Rinse degreased parts well with alcohol and then with water.

**c. Detergent Cleaning.** Place stainless steel parts in a 4% detergent solution for 30 min, with the temperature controlled at 120°F. The temperature may be raised if no plastic parts are present, but it should never be over 150°F (65.5°C). Rinse parts several times with water. Clean aluminum parts in a similar manner with a 4% solution of aluminum cleaner for 30 min and then rinse them thoroughly with water. Handle all finished parts with clean gloves.

**d. Acid Pickling.** Place stainless steel parts in a bath of 40–50% nitric acid for at least 1 hr. Castings and rough finished parts should remain in the nitric acid for a longer time. After pickling, remove parts and wash them thoroughly with water several times. Rinse parts thoroughly with distilled or deionized water or steam them clean.

**e. Final Treatment.** For metal parts, continue with the following steps:

1. steam parts clean;
2. blow parts absolutely dry with nitrogen gas;
3. package each small part in a plastic bag and close the bag securely until the part is to be used. Cover the opening and the clean areas of each large part with plastic film and tape until the part is to be used.

*6.5.2. Plastic Parts—O-Rings, Gaskets, etc*

Clean plastic parts as follows:

1. clean parts with a 4% detergent solution for 30 min at 120°F;
2. rinse parts with distilled water several times;
3. blow parts dry with nitrogen gas;
4. package parts in plastic bags until they are to be used.

*6.5.3. Stainless Steel Tanks*

Stainless steel tanks should be prepared for liquid hydrogen service as follows:

**a. Cleaning.** First prepare the tank for cleaning.

1. Inspect tank for rust, dirt, scale, etc.
2. Remove rust and scale mechanically or with a nitric acid–hydrofluoric acid mixture.

The actual cleaning comprises the following four steps:

1. degrease the tank with inhibited methyl chloroform;
2. rinse the tank thoroughly with alcohol;
3. rinse the tank thoroughly with water or steam clean;
4. fill the tank partially with 4% detergent solution, heat it to 150°F, and maintain it at that temperature for 30 min. Rinse it with potable water.

**b. Welds.**

1. Inspect all welds inside the tank. If welds are blackened, add a nitric acid–hydrofluoric acid mixture (enough to cover the welds), leave for 30 min, drain, and rinse with potable water.
2. Inspect welds again. If they are still black, repeat step 1. If they are clean, rinse tank thoroughly with water or steam it clean.

**c. Final Treatment.**

1. Fill tank completely and wash it thoroughly with particle-free deionized or distilled water.
2. Blow tank dry with nitrogen, and cover all openings with plastic film. Plastic caps or plugs should not be used where there is a possibility of the plastic material shearing off and contaminating the system.

**d. Aluminum and Aluminum Alloy Tanks.** Proceed as follows to clean an aluminum or aluminum alloy tank.

1. inspect the inside of the tank and remove burrs, grease, dirt, scale, etc;
2. degrease the tank with inhibited methyl chloroform for 30 min;
3. rinse it with alcohol;
4. add 4% aluminum cleaning solution and leave for 20 min at room temperature before decanting. Revolve tank thoroughly so that the solution covers the entire tank or scrub the tank walls;
5. wash the tank thoroughly with water or steam it clean;
6. blow the tank dry with nitrogen, and cover all openings with plastic film.

## 6.6. Transfer Operations

Operating procedures for liquid hydrogen transfer are defined by the cognizant authority or by the manufacturer of the cryogenic equipment. All operating personnel must be thoroughly instructed before operating the equipment. All valves, pumps, switches, etc., must be identified and tagged. In-line filters are recommended for use on all transfer operations in order to scavenge residual solidified contaminants. A written operating procedure must be used at each operational site.

### *6.6.1. Initial Use of the Liquid Hydrogen Transfer System*

If possible, before initial filling or use with liquid hydrogen, the system should be precooled and purged with liquid nitrogen. (Consult container manufacturer for maximum weight of liquid nitrogen that may be used.) This precooling will remove air and minimize flash-off when the system is filled with liquid hydrogen. If, for structural or operational reasons, the system cannot be precooled with liquid nitrogen, the system may be purged with gaseous nitrogen to remove all air. Before adding liquid hydrogen, all nitrogen used for precooling or purging must be removed by using hydrogen gas to displace the nitrogen. (Nitrogen will freeze in liquid hydrogen.) Only flash-off or vaporized gas from liquid hydrogen should be used, and this must be warmer than 300°F. The impurities in most cylinder hydrogen make it unsuitable for purging or pressurizing liquid hydrogen containers.

### *6.6.2. Removal from Service*

When liquid hydrogen equipment or containers are removed from service, before air is admitted they must be purged with inert gas to prevent the formation of flammable mixtures. Containers that have been in service for long periods of time may have an accumulation of solid air in the vacuum space. It is recommended that the warmup proceed slowly and the vacuum be monitored for any pressure rise above design limitations.

Nitrogen inerting of rooms and buildings housing transfer equipment has been practiced successfully at test facilities.

## 6.7. Emergency Procedures

### *6.7.1. Spills, Leaks, and Decontamination*

The principal danger from a spill or leak is fire. Unconfined hydrogen gas does not detonate, it burns, as dramatically shown in the Hindenberg accident. Proper ventilation of storage and transfer areas and the avoidance of sources of ignition will help reduce the danger of fire. There are no effective means of decontamination other than rapid vaporization and dilution with air.

If a large spill occurs, evacuate an area several hundred feet in radius around the source of the spill. If a leak occurs in an enclosed area, exercise care to eliminate potential sources of ignition and to adequately ventilate the area before entering it.

Experiments have been conducted on the dispersion of flammable clouds resulting from large $LH_2$ spills. Such data are necessary for evaluating the safety of hydrogen systems and separation distances from buildings and roadways and developing satisfactory ignition and spill controls.

Table 11.33 summarizes initial data obtained in experiments performed by A. D. Little, Inc. (1961) and by Zebetakis and Burgess (1961). Gas evolution rates measured with a gas meter were compared with calculated vaporization rates. Most of the A.D. Little, Inc. (1961) vaporization data were obtained from tests in a 0.6-m (2 ft) diameter, vacuum-insulated cylinder. Data for spills on sand and bank gravel exhibited an initial vaporization rate of 2.12–2.96 mm/s (5–7 in./min), decreasing to an apparently constant liquid regression rate of about 0.635 mm/s (1.5 in./min). See Fig. 11.26.

Experiments (Fig. 11.26) provide basic information regarding the physical phenomena governing the dispersion of flammable clouds resulting from large $LH_2$ spills. The experiments consisted of ground spills of $LH_2$ as large as 5.7 $m^3$ (1500 gal), with spill durations of approximately 35 sec. Instrumented towers downwind of the spill site gathered data on the temperature, hydrogen concentration, and turbulence levels as the hydrogen vapor cloud drifted downwind. Visual phenomena were recorded by motion picture and still cameras. Results of the experiments indicate, for rapid spills, thermal and momentum-induced turbulences cause the cloud to disperse to safe concentration levels and to become positively buoyant long before

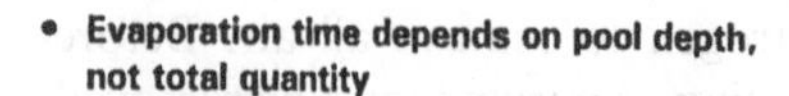

• Evaporation time depends on pool depth, not total quantity


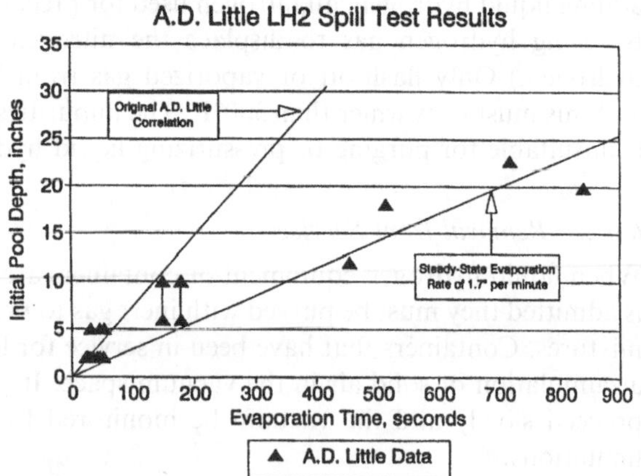


**Figure 11.26** Liquid hydrogen spills.

mixing from normal atmospheric turbulence becomes a major factor. On the basis of the $LH_2$ spill quantities, rates, and modes reported and the limited data analyses conducted, the following conclusions were drawn:

1. Rapid $LH_2$ spills that might occur from a storage facility rupture are characterized by a brief period of ground-level flammable cloud travel during which the violent turbulence generated by the momentum of the spill, the quick phase change from a liquid to a vapor, and the thermal instability of the cloud cause the hydrogen vapors to mix quickly with air, disperse to nonflammable concentration, warm up, and become positively buoyant. The ground-level cloud travels approximately 50–100 m (164–328 ft) and then rises at 0.5–1.0 m/sec (1.64–3.28 ft/sec).
2. Prolonged, gentle spills or spills that might occur from an $LH_2$ pipeline rupture are characterized by prolonged ground-level cloud travel. Ground-level cloud travel is prolonged by low spill or momentum-induced cloud turbulence and suspected to be aggravated by long-term ground cooling; the major heat transfer mechanism for determining the vaporization rate.
3. The experiments show that the use of dikes around $LH_2$ storage facilities probably prolong ground-level flammable cloud travel and that it *may be preferable not to use dikes* and take advantage of the dispersion mechanisms provided by spill and vaporization-induced turbulence.

### *6.7.2. Firefighting*

*Note*: Hydrogen fires normally are not extinguished until the supply of hydrogen has been shut off because of the danger of reignition and explosion.

The most effective way to control a hydrogen fire is to shut off the supply of hydrogen. Fires from hydrogen gas can also be controlled effectively with very high concentrations of the common extinguishing agents such as water, $CO_2$, and steam. Equipment should be designed for effective control and isolation in case of failure. It should be remembered that if hydrogen flames resulting from a leak are extinguished but the leak is not plugged, then hydrogen will continue to leak and form a cloud of combustible gas that may explode if ignited. *Remember*: The outer limits of a hydrogen flame or fire cannot generally be seen. If possible, use large quantities of water, preferably in the form of spray, to cool adjacent exposures. See Table 11.34.

Employees, other than trained professional firefighters, trained volunteers, or emergency response personnel, shall not fight fires except in cases in which the fire is incipient in nature (NHB 1700.1 1993).

The only positive way of handling a hydrogen fire is to let it burn under control until the hydrogen flow can be stopped. A hazardous combustible mixture starts forming at once if the hydrogen fire is extinguished and the hydrogen flow is not stopped. It is very possible that the mixture will ignite with an explosion to cause more damage and restart the fire.

Although the hydrogen fire should not be extinguished until the hydrogen flow can be stopped, water sprays, etc., shall be used to extinguish any secondary fire and prevent the spread of the fire. The hydrogen-containing equipment should be kept cool by water sprays to decrease the rate of hydrogen leakage and prevent further heat damage.

Carbon dioxide may be used in the presence of hydrogen fires. Although some toxic carbon monoxide may be produced in the flame, it will not be a large amount.

**Table 11.34** Liquid Hydrogen Fires and Firefighting

- Liquid hydrogen fires begin with a "flash hot-gas ball" and then continue to burn rapidly until extinguished
- Uncontained hydrogen fires will not detonate and are usually allowed to burn under control until the source of hydrogen is exhausted or shut off
- Hydrogen fires in contained areas (buildings and other enclosures) can result in detonation
- Extinguishing a hydrogen fire without removing the hydrogen source usually results in reignition or explosion
- Water is typically used to cool surrounding structures rather than extinguishing the hydrogen fire
  Water sprayed on liquid hydrogen fires can cause vents to freeze which increases the likelihood of explosions
  Secondary fires caused by hydrogen are treated by carbon dioxide, dry chemical, water, or Halon systems

Anyone breathing in the hot flame gases will be affected in any case, regardless of the presence of carbon monoxide. The carbon monoxide will be down to tolerable levels by the time the flame gases are diluted with fresh air and reach breathable temperatures. Confined spaces shall be well ventilated and verified as safe before they are entered, unless the appropriate protective apparatus are being used. Dry chemicals are better than carbon dioxide because they make the flames visible.

Remotely controlled water-spray equipment, if it has been installed, should be used instead of hoses to cool equipment and reduce the spread of fire. If it is necessary to use hoses, personnel using them shall stay behind protective structures.

### *6.7.3. First Aid*

*General.* Remove persons from high concentrations of hydrogen gas (oxygen-deficient atmosphere) to a well-ventilated area. If breathing has stopped, apply artificial respiration and then call for medical aid. Keep all sources of ignition away. Treat parts of the body that have been frozen by contact with liquid hydrogen by soaking them in *lukewarm* water. *Extensive frostbite requires prompt medical attention.*

Direct physical contact with $LH_2$, cold vapor, or cold equipment can cause serious tissue damage. Momentary contact with a small amount of the liquid may not pose as great a danger of a burn because a protective film may form. Danger of freezing occurs when large amounts are spilled and exposure is extensive. Cardiac malfunctions are likely when the internal body temperature drops to 300 K (80.2°F), and death may result when the internal body temperature drops to 298 K (76.4°F).

*Medical assistance* should be obtained as soon as possible. Treatment of truly frozen tissue requires medical supervision because incorrect first aid practices invariably aggravate the injury.

It is safest in the field to do nothing other than to protect the involved area with a loose cover. Transport the injured person to a medical facility if directed.

Attempts to administer first aid are often harmful. The following are important things to remember:

1. frozen gloves, shoes, or clothing shall not be removed except in a slow, careful manner, because the skin may be pulled off. Unremoved clothing can easily be put into a warm water bath;
2. the affected part shall not be massaged;

3. The affected part shall not be exposed to temperatures higher than 318 K(112°F). Exposure to warm temperatures superimposes a burn, and gravely damages already injured tissues;
4. water outside the temperature range of 311–318 K (100–112°F) shall not be applied to expose the affected part;
5. safety showers, eyewash fountains, or other sources of water shall not be used, because the temperature will almost certainly be therapeutically incorrect and aggravate the injury. Safety showers shall be tagged, NOT TO BE USED FOR TREATMENT OF CRYOGENIC BURNS;
6. snow or ice shall not be applied;
7. ointments shall not be applied.

### 6.8. Disposal of Liquid Hydrogen

Small quantities of liquid hydrogen may be disposed of outdoors in isolated, well-ventilated areas away from sources of ignition by (1) pouring the liquid hydrogen on the ground and allowing rapid evaporation and diffusion into the air to take place or (2) vaporizing the hydrogen through a vent stack of suitable height. Large quantities are best disposed of in a burn pond, where the hydrogen is vented at a number of points through an underwater piping system and burning takes place at the surface of the water. Large quantities may also be disposed of by burning at an isolated flare stack. See Table 11.35.

### 6.9. Key Documents for the Design of Hydrogen Facilities Include

1. AF 127-100 Explosive Safety Standards: revised August 3, 1990.
2. DOD Contractors Safety Manual for Ammunition and Explosives, March 1986.
3. NASA Safety Standard NSS 1740.16 "Safety Standard for Hydrogen and Hydrogen Systems," February 1977. (Originally available from NASA Center for Aerospace Information (CASI) Tel. 301-621-0390. Other sources for this publication include the National Technical Information Center (NTIS) Tel. 703-485-4650; NASA Special Publications Tel. 202-554-4380, and Main NASA publication office Tel. 202-554-4380.
4. Compressed Gas Association (CGA), 1235 Jefferson Davis Highway, Arlington, VA 22202-3269, Tel. 703-979-4341, Fax 703-979-0134.
5. National Fire Protection Association (NEPA), 1 Battery March Park, PO 9101, Quincy, MA 02269-9904, Tel. 1-800-344-3555.
6. American Society for Testing and Materials (ASTM), 1916 Race St., Philadelphia, PA 19103, Document Center Tel. 415-591-7600.

## 7. SPECIAL CONSIDERATIONS FOR OXYGEN

LOX has many meanings. LOX can be found in almost any good delicatessen. LOX also stands for Liquid Oxygen eXplosive, which is a mixture of charcoal, or any carbonaceous fuel, and liquid oxygen. Up until about 1960, this LOX was the most widely used explosive in the United States mining industry. In that year alone, 10 million pounds per year were used. This LOX was made by filling canvas tubes 4–6 in. in diameter (depending upon the size of the blast hole) with charcoal or

**Table 11.35** Hydrogen Disposal by Venting or Flaring

- Unburned hydrogen can be routinely vented without incident
  Nominal flowrates are 0.25–0.50 lb/sec for a single vent 16 ft above roof
  Multiple vents across prevailing winds spaced 16 ft apart can be used
  Over-the-road trailers regularly vent hydrogen in remote areas
- Larger quantities of hydrogen are disposed by burning in a flare stack or burn pond
  Burn ponds do not have well-established design criteria
  Flare-stack design is well established based on flowrates, prevailing winds, and thermal radiation protection
  Hydrogen flare stacks should be at least 200 ft from work areas

Design level

| (K), Btu/hr ft$^2$ | Conditions |
|---|---|
| 500 | Value of K at design flare release at any location where personnel are continuously exposed |
| 1500 | Heat intensity in areas where emergency actions lasting several minutes may be required by personnel without shielding but with appropriate clothing |
| 2000 | Heat intensity in areas where emergency actions lasting up to 1 min may be required by personnel without shielding but with appropriate clothing |
| 3000 | Value of K at design flare release at any location to which people have access, for example, at grade below the flare or on a service platform of a nearby tower. Exposure must be limited to a few seconds, sufficient for escape only |
| 5000 | Heat intensity on structures and in areas where operators are not likely to be performing duties and where shelter from radiant heat is available, for example, behind equipment |

API 521, Sec. 4, Table 3

| Radiation level, Btu/hr ft$^2$ | Time to pain threshold, sec |
|---|---|
| 550 | 60 |
| 740 | 40 |
| 920 | 30 |
| 1500 | 16 |
| 2200 | 9 |
| 3000 | 6 |
| 3700 | 4 |
| 6300 | 2 |

powdered coal and soaking the tube in liquid oxygen for several hours. The carbon/liquid oxygen tube was then placed in the blast hole and tamped into position. This LOX has the impact sensitivity three times that of nitroglycerine, which was abandoned as a mining or construction explosive for that very reason.

This LOX was largely discontinued in the USA as a mining explosive after about 1960 due to the obvious safety hazards. This LOX, however, is still the explosive agent of choice in many countries, including the nation of India. A rich literature exists on its manufacture and use.

This little story illustrates two things: (1) the name LOX is not unambiguous, and for this reason shall be avoided in this text; and (2) liquid oxygen by any name should give one pause, as it is intrinsically an explosive. In my opinion, oxygen is much more hazardous than is hydrogen because hydrogen has only one reaction (with air), whereas oxygen reacts with almost anything under the right (or wrong) conditions.

### 7.1. Basic Principles of Oxygen Safe Use

Although oxygen itself is chemically stable, is not shock-sensitive, will not decompose, and is not flammable, its use involves a degree of risk that shall never be overlooked. This risk is that oxygen is a strong oxidizer that vigorously supports combustion. Oxygen is reactive at ambient conditions, and its reactivity increases with increasing pressure, temperature, and concentration. Most materials, both metals and nonmetals, are flammable in high-pressure oxygen; therefore, systems must be designed to reduce or eliminate ignition hazards. The following principles apply to all oxygen systems:

1. Materials that are highly reactive in oxygen must be avoided.
2. Materials that are less reactive, but are still flammable, can be used if protected from ignition sources, such as:

- mechanical impact; screening required;
- abrasion—particle impact;
- reasonance from repeated shock waves;
- adiabatic compression;
- fracture;
    - diaphragm rupture in titanium;
    - fresh metal surface;
- rubbing or rotating friction;
- electric;
- stupidity—smoking or open flames;
- all ignition hazards are exacerbated by contamination.

3. Oxygen systems shall be kept clean because organic compound contamination, such as hydrocarbon oil, can ignite easily and provide a kindling chain to ignite surrounding materials. Contamination can also consist of particles that could ignite or cause ignition when impacting other parts of the system.
4. Some of the most important features for safe oxygen systems include leak prevention, adequate ventilation, elimination of or minimizing the severity of ignition sources, proper material selection, good housekeeping, suitable design of system components, system cleanliness, and proper system operation. The necessity of maintaining system cleanliness and using ignition- and combustion-resistant materials cannot be overemphasized. In this respect, oxygen safety is very similar to hydrogen safety.
5. Safety systems including at least two barriers or safeguards shall be provided under normal and emergency conditions so that at least two simultaneous

**Table 11.36** Oxygen Hazards

- Oxygen is one of the five strongest oxidizers that include
  Oxygen
  Fluorine
  Chlorine
  Bromine
  Iodine
- Oxygen reactivity is the principal concern
  Forms oxidation films
  Ignites spontaneously
  Burns with most metals and nonmetals
- Oxygen is more dangerous than fluorine
  Fluorine is too hypergolic to mix
  Fluorine reacts on contact at combination rate
- Oxygen mixes passively and is easily ignited
  Fuel is ubiquitous—clothing, insulation, metals, asphalt paving
- Liquid oxygen represents a frostbite hazard
  Boils at 90.18 (–297°F)

undesired events must occur before any possibility arises of personnel injury or loss of life, or major equipment or property damage.

6. Liquid oxygen is a frostbite hazard. See Table 11.36 for a list of oxygen hazards.

## 7.2. Igniton Mechanisms and Sources

Potential ignition mechanisms and ignition sources include:

1. Particle impact. Heat is generated from the transfer of kinetic, thermal, or chemical energy when small particles moving at high velocity strike a component. This heat, which is adequate to ignite the particle, may be caused by the exposure of unoxidized metal surfaces or the release of mechanical strain energy. The heat from the burning particle ignites the component. *Example:* High-velocity particles from assembly-generated contaminants striking a valve body just downstream of the control element of the valve have caused particle impact ignition.
2. Mechanical impact. Heat is generated from the transfer of kinetic energy when an object having a relatively large mass or momentum strikes a component. The heat and mechanical interaction between the objects is sufficient to cause ignition of the impacted component. *Example:* the poppet of a solenoid-operated valve striking the seat has caused mechanical impact ignition. Aluminum, tin, lead, and titanium alloys have been ignited experimentally in this way, but iron, nickel, cobalt, and copper alloys have not.
3. Pneumatic impact. Heat is generated from the conversion of mechanical work when a gas is compressed from a low to a high pressure. Pneumatic impact is an effective ignition mechanism with polymers but less so with metals. *Example:* high-pressure oxygen released into a dead-end tube or pipe compresses the residual oxygen in the tube ahead and causes pneumatic impact.

4. Promoted ignition. A source of heat input occurs (perhaps caused by a kindling chain) that acts to start the nearby materials burning. *Example:* the ignition of contaminants (oil or debris) combusts, releasing heat that ignites adjacent components, thus causing promoted ignition. Or a polymer valve seat can ignite and combust, igniting the valve stem.
5. Galling and friction. Heat is generated by the rubbing of two parts together. The heat and interaction of the two parts, along with the resulting destruction of protective oxide surfaces or coatings, causes the parts to ignite. *Example:* the rub of a centrifugal compressor rotor against its casing causes galling and friction.
6. Resonance. Acoustic oscillations within resonant cavities cause a rapid temperature rise. This rise is more rapid and reaches higher values if particles are present or gas velocities are high. *Example:* a gas flow into a tee and out of a branch port can form a resonant chamber at the remaining closed port.
   Results of studies with several types of tee configurations indicated that temperature increases caused by resonance heating would be sufficient to ignite both aluminum and stainless steel tubes. Tests with aluminum and stainless steel particles added to the resonance cavity indicated that ignition and combustion would occur at lower temperatures. Some of the tests with 400-series stainless steel resulted in ignition, but ignition appeared to depend more on system pressures and system design. Cleanliness is crucial.
7. Electrical arcing. Electrical arcing can occur from motor brushes, electrical power supplies, lighting, etc. Electrical arcs can be very effective ignition sources for any flammable material. *Example:* an insulated electrical heater element can experience a short circuit and are through its sheath to the oxygen gas causing an ignition.
8. Figure 11.15 showed the temperature ruse caused by the *adiabatic compression* of oxygen. In Fig. 11.27 the dashed line shows the theoretical adiabatic compression temperature of Fig. 11.15 that can arise in a dead-end fitting when the pressure increases suddenly (near adiabatic, since heat transfer depends upon time). The solid lines of Fig. 11.27 show how the ignition temperature of several materials reduces with increase in oxygen pressure. For example, if a 3000 psi cylinder of oxygen is suddenly opened to a dead-end space (such as to a pressure gauge), the temperature of that dead end will exceed 1000°C. This is well above the autoignition temperature of many common materials at a pressure of 3000 psi. In short, a fire should result nearly every time we do this. What keeps us out of such trouble? The valve opening is not instantaneous and hence the temperature spike is not the theoretical adiabatic compression temperature. Message: (1) sometimes we get lucky; (2) no sudden moves when dealing with high-pressure oxygen.

### 7.3. Materials Selection for Oxygen Service

Factors affecting the ignition of solid materials include material composition and purity; size, shape, and condition of the sample; characteristics of oxide layers; phase; testing apparatus; ignition source; gas pressure; and gas composition. Ignition temperatures are provided in Tables 11.37 and 11.38 for several solid materials in air

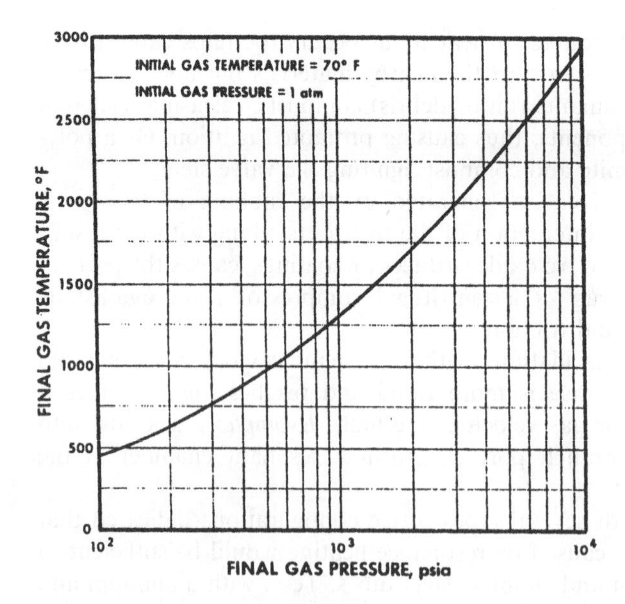


**Figure 11.27** Metal ignition temperatures compared with adiabatic compression temperatures.

and oxygen. The ignition process depends on the geometry and operating conditions; therefore, caution must be taken in interpreting the results of any ignition experiment and in generalizing ignition data.

Care must be exercised in applying ignition temperature data, especially for metals, to actual components. Ignition temperatures are not inherent materials properties but are dependent upon the items listed above. When applying ignition temperature data, it must be ensured that the ignition temperature data were obtained in a manner similar to the end-use application. Failure to do this can result in erroneous materials selection decisions. For example, the ignition temperatures of aluminum in oxygen vary from 680°C (1255°F), which is the melting point of aluminum, to 1747°C (3176°F), which is the melting point of $Al_2O_3$. The ignition temperature obtained depends on whether or not the oxide is protective during the ignition process.

With these caveats in mind, see Tables 11.39–11.42 for recommended materials for oxygen service.

### 7.4. Design Principles for Oxygen

We suggest a holistic approach to assuring the best possible safety in cryogenic systems. These four steps work well:

1. What are the properties of the cryogen that cause concern?
   For hydrogen, the property that concerns us the most is its very low ignition energy. It is safe to assume that every leak or spill of hydrogen will ignite but not necessarily detonate. Detonation of hydrogen requires that the hydrogen be contained. Hydrogen has only one reaction, with oxygen (air), but it is a doozy. A life saving rule with hydrogen has always been "maintain positive pressure and exclude oxygen".

**Table 11.37** Ignition Temperatures of Some Solids at 1 atm Oxygen Pressure

| | Ignition Temperature | |
|---|---|---|
| Metal | (K) | (°F) |
| Mild Steel | 4248–4410 | 2240–2330 |
| W | 4302–4446 | 2270–2350 |
| Ta | 4284–4428 | 2260–2340 |
| Ti Alloys | | |
| RC-70 | 5400–5544 | 2880–2960 |
| RS-70 | 5418–5508 | 2890–2940 |
| RS-110-A | 5364–5454 | 2860–2910 |
| RS-110-BX | 5346–5472 | 2850–2920 |
| Stainless Steel | | |
| 430 | 4644–4698 | 2460–2490 |
| Berylco® 10 | 3366–3384 | 1750–1760 |
| Mg | 2323 | 1171 |
| Mg Alloys | | |
| 20% Al | 1900 | 936 |
| 70% Zn | 2023 | 1004 |
| 25% N | 1897 | 934 |
| 20% S | 2194 | 1099 |
| 63% Al | 1767 | 862 |
| Fe | 3286 | 1706 |
| Sr | 2606 | 1328 |
| Ca | 2055 | 1022 |
| Th | 1893 | 932 |
| Ba | 840 | 347 |
| Mo | 2736 | 1400 |
| U | 1310 | 608 |
| Ce | 1310 | 60 |

For oxygen, the problem is just the opposite. Oxygen reacts with almost everything if the conditions are right but requires a respectable ignition temperature. See Tables 11.39ff. Thus, the principal issue with oxygen is materials selection, scrupulous cleanliness, and avoiding ignition.

2. Given these physical properties of the cryogen, what is the best way to design-out problems and design-in safety? For instance, with oxygen, what is the safest material to use?
3. Having minimized all the risks possible with a safety conscious design, what is the safest way to operate the system? For instance, with hydrogen, eliminate hydrogen leaks and sources of ignition. With oxygen, eliminate the nonobvious sources of ignition, such as adiabatic compression (see Fig. 11.27).
4. Having done all the above, what to do when things go wrong? What are the emergency procedures? Clearly, it is best to plan emergency procedures beforehand. There is no time during an emergency to think.

Safe use of oxygen requires the control of potential ignition energy mechanisms within oxygen systems by judiciously selecting ignition-resistant materials and

**Table 11.38** Minimum Ignition Temperatures of Some Nonmetals in Oxygen at 1-Atmosphere Pressure

| Material | Minimum ignition temperature,°C |
|---|---|
| *Selected plastics and elastomers* | |
| Polyimide (Vespel) | 500 |
| Chlorotrifluoroethylene (TFE) | 500 |
| Tetrafluoroethylene (TFE) | 500 |
| Duroid 5600 | 468 |
| Duroid 5813 | 463 |
| Duroid 5870 | 452 |
| Duroid 5650 | 444 |
| Rulon A | 463 |
| Rulon B | 460 |
| Rulon C | 458 |
| Graphite asbestos | 460 |
| Vinylidene fluoride/hexafluoropropylene (Viton A/Fluorel) | 435 |
| Polymethylmethacrylate (Plexiglas) | 430 |
| TFE (carbon filled) | 420 |
| Fluorinated ethylene propylene copolymer (FEP) | 412 |
| Polyvinylalcohol | 400 |
| Silicone rubber | 390 |
| Kel-F 5500 | 340–352 |
| Kel-F 3700 | 332–341 |
| Carbon | 330 |
| Flurosilicone | 320 |
| Polyvinychloride | 390 |
| Nylon[a] | 345 |
| Mylar[a] | 320 |
| Butyl rubber[a] | 275 |
| Neoprene[a] | 250–270 |
| Polyethylene[a] | 160–220 |
| Natural rubber[a] | 150 |
| *Selected lubricants* | |
| TFE lubricant | 505 |
| CTFE grease | 500 |
| Halocarbon grease 25-10 | 431 |
| Halocarbon oil 13-21 | 427 |
| Mano pipe and joint | 422–430 |
| Oxyweld 64 | 410 |
| Silicone grease | 370 |
| *Selected adhesives* | |
| Fused TFE | 465 |
| Epoxy (3 M EC 1469) | 380 |
| Oxyseal (2) | 347–360 |
| Epoxy (Armstrong A-6) | 310 |
| *Selected fabrics* | |
| Nomex | 490 |
| Orlon | 430 |
| Nylon | 360 |
| Dacron | 310–390 |

[a] Not recommended for oxygen—included for comparison only.

**Table 11.39** Recommended Materials for Oxygen Service

| Application | Material low-moderate-pressure | High pressure-near sonic vel |
|---|---|---|
| Component bodies | Nickel alloy steel | Monel |
| | Stainless steel | Inconel 718 |
| Tubing and fittings | Copper, stainless steel, steel aluminum and aluminum alloys | Monel |
| | | Inconel 718 |
| Internal parts | Stainless steel | Monel |
| | | Inconel 718 |
| | | Beryllium copper |
| Springs | Stainless steel | Beryllium copper |
| | | Elgiloy |
| | | Monel |
| Valve seats | Stainless steel | Gold or silver |
| | Monel | Plated over Monel |
| | Inconel | Or Inconel 718 |
| Valve balls | Stainless steel | Sapphire |
| | Tungsten carbide | |
| Lubricants | Everlube 812 | Batch/lot tested |
| | Microseal 100-1 and 200-1 | Braycote 3L-38RP |
| | Triolube 1175 | Batch/lot tested |
| | Krytox 240AB and 240AC | Everlube 812 |
| | Braycote 3L-38RP | Krytox 24DAD |
| O-seals and backup | TFE, Halon TFE | Batch/lot tested |
| | Teflon | Viton |
| | Kel F | Batch/lot tested |
| | Viton | Teflon |
| Pressure vessels | Nickel steel | Inconel 718 |
| | Stainless steel | |
| | Steel | |
| | Aluminum alloys | |

Operating pressure ranges as they are used in this section are defined as follows:

| | |
|---|---|
| Low pressure | 0–500 psia |
| Medium pressure | 500–2000 psia |
| High pressure | 2000–7500 psia |
| Very high pressure | 7500–10,000 psia |
| Ultra-high pressure | 10,000–15,000+ psia |

system designs, maintaining scrupulously clean systems, and using appropriate operational procedures.

The primary consideration for resolving oxygen hazards shall be the ELIMINATION BY DESIGN. See Table 11.43.

The following design principles shall be adopted to achieve maximum oxygen safety.

**Inherent Safety.** Oxygen systems and operations shall have a high degree of built-in safety. The selection of ignition and combustion resistant materials at the maximum expected operating conditions and the suitable design of components and systems is essential. Oxygen safe systems must include special designs for pre-

**Table 11.40** Recommended Metals for Oxygen Service

- Nickel and nickel alloys
  - Nickel
  - Hastelloy B
  - Inconel-X
  - K-Monel
- Stainless steel types
  - 304, 316, 304L
  - 310, 321
  - 304ELC
- Copper and copper alloys
  - Copper
  - Cupro-nickel
- Naval and admiralty brass

venting leaks, eliminating ignition sources, and maintaining a clean system, avoiding cavitation, and preventing resonant vibration. See Table 11.44.

**Two Lines of Defense.** In addition to the inherent safety features, at least two failure-resistant, independent barriers shall be provided to prevent a given failure from mushrooming into a disaster. The desire is that at least two undesirable, independent events must occur simultaneously under either normal or emergency conditions before there is a potential danger to personnel or major damage to equipment and property.

**Fail-safe Design.** The equipment, power, and other system services shall be designed and verified for safe performance in the normal and maximum designed operational regimes. Any failures shall cause the system to revert to conditions that are safest for personnel and cause the least property damage. Redundant components shall be incorporated into the design to prevent shutdowns.

**Automatic Safety Devices.** System safety valves, flow regulation, and equipment safety features shall be installed to automatically control hazards.

**Caution and Warning Systems.** Warning systems shall be part of oxygen systems design to monitor those parameters of the storage, handling, and use of oxygen that may endanger personnel and cause property damage. Warning systems shall consist of sensors to detect abnormal conditions, measure malfunctions, and indicate

**Table 11.41** Prohibited Metals in Oxygen Service

- Cadmium is not allowed because of its toxicity and vapor pressure
- Titanium and its alloys
  - Impact sensitive
  - Unsafe in liquid oxygen at any pressure
  - Unsafe above 30 psi with air or gaseous oxygen
- Magnesium and its alloys
  - Corrodes easily
  - Reacts with halogenated lubricants
- Mercury
  - Accelerated stress cracking of aluminum and titanium
- Veryllium is too toxic for use

**Table 11.42** Recommended Nonmetals for Oxygen Service

- Nonmetals have lower ignition temperatures, thermal conductivities, and heat capacities
- Acceptable nonmetals include
  Tetrafluoroethylene polymer (TFE, Halon TFE, Teflon, or equivalent)
  Unplasticized chlorotrifluoroethylene polymer (Kel F, Halon CTF, or equivalent)
  Fluoro-silicone rubbers (Viton type), of a NASA tested compound
- Acceptable lubricants include
  Kryotox (Dupont)
  Tribolube 16 (Aerospace lubricants)
- The nonmetallic materials used should be based on data presented in the NASA Publication JSC 02681, *Non-Metallic Materials Design Guidelines and Test Data Handbook*, Latest Edition

incipient failures. Data transmission systems for caution and warning systems shall have sufficient redundancy to prevent any single-point failure from disabling an entire system.

An excellent reference source for oxygen design principles is: *Standard Guide for Designing Systems for Oxygen,* ASTM G88-84. 1984.

**Table 11.43** Design Approach for Oxygen

| Factors causing fires | Design approach |
|---|---|
| • Temperature<br>Reduces added energy margin<br>Sustains combustion | • Avoid elevated temperatures<br>Location<br>Dissipation<br>Monitor and shutdown |
| • Pressure<br>Ignition temperatures decreases<br>Rate of propagation increases | • Avoid elevated pressure<br>• Reduce pressure near supply point |
| • Contamination<br>Initial<br>Introduced<br>Generated | • Design for cleanliness<br>Easy to clean<br>Easy to maintain<br>Avoid dead ends<br>• Use adequate filters<br>Maintain them<br>Avoid friable, ignitable materials |
| • Particle impact<br>Low thermal diffusivity<br>Self-generate | • Avoid these conditions<br>Filters<br>High gas velocities<br>Throttling valves<br>Local impingement<br>High pressurization rates |
| • Heat of compression in dead ends | • Avoid high rates of compression<br>Valves, line restrictors |
| • Friction and galling | • Avoid rubbing<br>• Provide adequate clearances |

**Table 11.44** Design Principles for Oxygen

- Control ignition energy mechanisms:
  Select ignition-resistant materials and designs
  Maintain scrupulously clean systems
  Use appropriate operating procedures
- Inherent safety design considerations:
  Prevent leaks
  Eliminate ignition sources
  Maintain cleanliness
  Avoid cavitation, resonance
- Two lines of defense in all systems:
  Include two failure resistant, independent barriers
- Fail safe designs automatically control hazards
- Thermal design considerations:
  Start-up conditioning
  Insulate against external condensation
  Avoid lock-up

## 7.5. Equipment for Oxygen Service

### *7.5.1. Pipes and Fittings*

Lines, tubing, pipes, ducting, and their associated hardware are subject to two conditions or circumstances that are of special concern in oxygen systems: (1) unaccounted for overstress and high-velocity flow of gaseous oxygen; (2) their length to diameter ratio allows very high leverage for excessive bending stresses in addition to stresses from internal pressure, thermal expansion–contraction, water-hammer, and cyclic stresses from vibration or chugging. Lines are often assembled without regard to slight misalignments (because it is so easy to do) and a prestress result which is additive to the working stress. These stresses are generally concentrated at a discontinuity such as a weld joint, connector, valve, elbow, or material flaw.

High flow rates of gaseous oxygen in lines can impale entrained solid particles into an elbow or flow system component with sufficient energy to cause erosion or even ignition in sensitive materials.

*Materials* for pipe, tubing, or ducting for liquid oxygen systems are selected to meet the chemical compatibility requirements of the use environment as well as the required structural ductility and notch sensitivity characteristics at cryogenic temperatures.

Examples of preferred materials, based on the preceding criteria, and general experience for oxygen plumbing systems are presented in Table 11.45 for a range of pressures and temperatures.

Cryogenic cycling tends to loosen bolted connections; therefore, flange bolt tensions must be checked after 1 or 2 cycles and then again before each use interval. Many compression-type fittings and mechanical clamping techniques to hold the fitting together are commercially available. The design and choice of sealing materials in many of these fittings are suitable for use in oxygen systems.

**Line rupture.** A gas at ambient temperature escaping from the open end of a line or vessel yields a specific impulse of approximately 50 sec, which is a considerable amount when it is realized that no chemical reaction is taking place. For example, a 1-in. line charged to 200 psi can develop approximately 175 pounds of thrust (see Table 11.46). Therefore, should a rupture occur in a line, flex hose, or bellows,

**Table 11.45** Materials for Lines for Oxygen Systems

| Material | Remarks |
|---|---|
| Aluminum 1114, 2024-0, 3003, 5052-0, 6061-0, 6061-T6, 6061-T4 | Aluminum alloys used mainly as tubing with tube connectors; pressure limits determined by type of connector; not recommended for very high pressures (over 7500 psi) |
| Monel 400, K500 | Tubing and pipe stock good for very high pressures and welded connections |
| Inconel 600, 718 | Tubing and pipe stock good for very high pressures and welded connections |
| Stainless steel 304, 316, 317, 321, 347 | Tubing and pipe stock good for very high pressures and welded connections |
| Some 400 series stainless AM-350 Rene 41 | Tubing and pipe stock good for very high pressures and welded connections |
| Carbon steel | Welded joints or other; not usable below –30°F |
| Copper, copper alloys | Brazed joints or tube fitting connectors; not recommended for very high pressures |

the unrestrained sections may cause severe damage or injury. Restraints, fasteners, and braces should be used to prevent this type of occurrence especially in high-pressure systems (over 2000 psia). From this example, the hazards of accidentally uncoupling an "empty" line that in reality is under pressure becomes obvious.

A useful flared coupling is the machined flare where the flare and the ferrule, or sleeve, is one machined part; however, the tube must be welded to the fitting.

Threaded couplings for use in oxygen systems should be designed so that the threaded nuts, sleeves, and collars which provide the compression force may be lubricated without any portion of the seal interface being contaminated with lubricant; that is, the parts that effect the seal should not require lubricant for proper torque.

Thermal distortion of the connector will occur when the flow of liquid oxygen begins and ends. Repeated cycling of such distortion can cause sealing pressure relaxation and increase the leakage rate. To avoid such leaks, retorquing should be part of the standard operating procedures for systems whose connectors are subjected to such temperature cycling. The best practice is to use a connector which incorporates a replaceable seal and to always replace the seal (gasket) when reassembling the coupling.

**Table 11.46** Open Line Thrust Force

To calculate the force acting on line opening, select applicable diameter and multiply right-hand column by the source pressure (psi).

| Diameter opening, in. | Calculated force factor for each psi of source pressure | Diameter opening, in. | Calculated force factor for each psi of source pressure |
|---|---|---|---|
| 1/8 | 0.18506 | 5/8 | 0.5777 |
| 1/4 | 0.2832 | 3/4 | 0.6759 |
| 3/8 | 0.3814 | 7/8 | 0.7741 |
| 1/2 | 0.4796 | 1 | 0.8723 |

**Table 11.47** Bellows Material for Oxygen Use

| |
|---|
| *Metal* |
| 304 Stainless steel |
| 301 Stainless steel |
| 347 Stainless steel |
| A-286 |
| Inconel 718 |
| Inconel X |
| Hastelloy-C |
| Aluminum 6061-T6 |
| Aluminum 7075 |
| Beryllium copper |
| *Nonmetal* |
| TFE |
| CTFE |

B-nuts and sleeves of certain alloys develop stress corrosion cracks when exposed to seacoast atmosphere. B-nuts and/or sleeves manufactured of 2014, 2017, 2024, and 7075 aluminum alloys in the T4 or T6 tempers or type 303 stainless steel are not recommended for use for extended periods in a salt atmosphere.

Bellows materials suitable for oxygen use are given in Table 11.47. The materials listed are primarily for formed bellows. Materials used most for electrodeposited bellows are nickel and copper.

There are a multitude of *coupling and threaded connector* configurations available from many manufacturers incorporating various seal designs. A few of the more commonly used are discussed here.

The AN (Air Force/Navy Aeronautical), MS (Military Standard), and MC (NASA Marshall Center) connectors are the most commonly used flared tube connectors which can be made to government or industry specifications. These connectors are basically the same except for their sealing interface specifications.

Standard AN's (including MS and MC) are excellent for low-pressure requirements; for higher pressures, up to 3000 psia for larger sizes and 7500 psia for smaller sizes, the addition of conical seals is recommended. At the higher pressures, the possibility of coupling failure exists because of the tube slipping out of the fitting. This is more prevalent with aluminum or copper alloy fittings than with steel. A "rule of thumb" has evolved that recommends using these types of couplings be limited to 2 in. in diameter for steel tubing and $\frac{3}{4}$ in. for aluminum or copper tubing at pressures in the range of 1500 psia. However, this is not a valid limitation because of the many variables in coupling designs.

For *flexible hose* requirements, wire-reinforced stainless steel corrugated sections are generally used. Flexible metal ducting up to 18 in. in diameter is available for cryogenic liquid up to several atmospheres. The pressure drop through convoluted flexible tubing can be 10–15 times the pressure losses of straight wall tubing which suggests minimal use. Nonmetal materials are not generally used since they do not meet the flexibility requirements for use with low-temperature gaseous or liquid oxygen. The brittle point of most oxygen compatible polymeric materials is above the temperature of liquid oxygen (–297°F). The TFE, CTFE, and FET fluorocarbon polymers have brittle points lower than liquid oxygen temperature and could be used as a liner for flexible hose with liquid oxygen. However, this is not recommended because of the reduced cycle life at these temperatures.

Where flexible sections are necessary, *bellows* with conventional convoluted shape are recommended for liquid oxygen use. Welded bellows are more prone to overpressure damage and early fatigue failure than formed bellows, and they are generally more difficult to clean and keep clean. Overdeflection due to mishandling has also contributed to some failures. Minor flaws have a tendency to propagate rapidly unless low magnitude stresses are maintained. Mechanical restraint linkage must be used to limit overflexure of the bellows beyond acceptable stress limits.

### *7.5.2. Gaskets and Seals*

An example of the types of polymers preferred for use in liquid or gaseous oxygen systems is given in Table 11.48. Except for the silicone rubber, these materials meet NASA specifications for compatibility with oxygen. Their physical properties vary enough to meet almost any sealing requirement. Viton was being used as a seal material; however, the Viton O-rings exposed to a 7500 psia oxygen pressure failed because of oxygen absorption. When depressurized, the material was fractured from the pressure of the absorbed gas. A modified Viton, identified as D-80, has been developed for use with oxygen at pressures up to 3000 psia.

### *7.5.3. Valves*

The major problems with valves are their inherent tendency to leak, trap, and be affected by contaminants (see Table 11.49). Leakage through valve seats and

**Table 11.48** Preferred Seal Materials for Oxygen

| Trade name | Chemical name or description |
|---|---|
| (a) *Nonmetals* | |
| Teflon TFE | Polytetrafluoroethylene |
| Halon TFE, G-80 | Polytetrafluoroethylene |
| Rulon A | Polytetrafluoroethylene with $MoS_2$ filler |
| Nickel-filled Teflon | Polytetrafluoroethylene with nickel-powder filler |
| 15% glass-filled Teflon | Polytetrafluoroethylene with glass fibers |
| Teflon FEP | Fluorinated ethylene propylene |
| Kel-F 81 | Polychlorotrifluoroethylene (CTFE) |
| Kel-F 82 | Copolymer of CTFE and 3 mol% vinylidene fluoride |
| Plaskon 2400 | Polychlorotrifluoroethylene (CTFE) |
| Halon TVS (now called Plaskon 2200) | Polychlorotrifluoroethylene(CTFE) |
| Kynar | Vinylidene fluoride |
| Viton A | Copolymer of vinylidene fluoride and hexafluoropropylene |
| Fluorel | Copolymer of vinylidene fluoride and hexafluoropropylene |
| Silastic | Silicone rubber |

| (b) *Metals* | | | |
|---|---|---|---|
| Aluminum | Copper alloys | Lead | Rhodiuma[a] |
| Aluminum alloys | Golda | Monel | Silver[a] |
| Chromium[a] | Inconel[a] | Nickel | Stainless steel |
| Copper | Indium[a] | Platinum[a] | Tin[a] |

[a] Usually restricted to use as a surface plating material.

**Table 11.49** Valves for Oxygen Service

- Inherent tendency to leak, trap, and become contaminated
- Wear, thermal cycling, erosion, pressure spikes
- Cause flow discontinuity by design
- Provide a trap for particulates, liquid
- Avoid self-cleaning, turbulence
- Materials for valve construction
  High melting point
  High ignition temperature
  High thermal conductivity
  Low heat of reaction
- Recommended for high pressure service (>7500 psia)
  Copper
  Brass
  Monel
  17-7 and 17-4 PH Steels
  300 and 400 series stainless steels
  Hastelloy
  Avoid aluminum

through actuator seals is aggravated by wear, thermal cycling, erosive flow, water hammer, high pressures, and pressure spikes. The problem of sealing to prevent leakage through valve seats and valve element actuators (or valve stems) is one of the more challenging parts of cryogenic system design.

Most of the design requirements for valves for oxygen service are the same as those for other cryogenic liquid or ambient temperature gaseous systems. However, in oxygen service the consequences of a design weakness, a material incompatibility, or a contaminant can be more severe than for inert fluids or for most fuels.

**Design Criteria.** Valves create a discontinuity in the flow line both fluid dynamically and structurally. The flow passage in many valve designs incorporate recesses, pockets, and crevices to accommodate the valve element actuating mechanism and the valve element supporting structure. These valve mechanisms cause stagnation zones where particulate matter may accumulate and create a hazard in oxygen systems. Also, in liquid systems, the element may trap liquid in a closed cavity; unless the cavity is vented, an overpressure condition is created when the liquid vaporizes. Therefore, the primary design criterion is to avoid valve design features that create stagnation zones or crevices where particulate matter can accumulate.

**Materials Selection for Valves.** A list of materials that have been used for oxygen valves in major rocket engine propellant systems is given in Table 11.50. It should be noted in Table 11.50 that Mylar has been used successfully as a secondary seal material in several operational valves even though it has been shown to react with oxygen. Therefore, its use elsewhere in the valve or in valves of different design operating above 1500 psi should be reviewed carefully and validated by tests which cover all possible service conditions.

For very high and ultra-high pressures (over 7500 psi), the preferred materials are metals and alloys which have high melting points, high ignition temperatures, a relatively high thermal conductivity, and a low heat of reaction with oxygen. Copper, brass, Monel, 17-7 and 17-4 PH steels, 400 and 300 series stainless steel, and Hestelloy, in this approximate order, best meet these requirements. Aluminum and

**Table 11.50** Oxygen Valve Materials for Rocket Systems

| Vehicle and engine | Valve application | Fluid media | Valve element | Valve element material | Seal or seat material | Valve housing-material | Dynamic seal type | Secondary seal material | Line size, in. | Operating pressure, Dsi | Flow rate lb/sec | Actuator type | Response time, sec |
|---|---|---|---|---|---|---|---|---|---|---|---|---|---|
| Atlas Ma-S | Booster main oxidizer value | $LO_2$ | Butterfly | 7015-T6 forging | CTFE | TENS-50 aluminum | Lip seal | Mylar[a] | 4.0 | 762 | 424.5 | Pneumatic piston | 400 |
| | Sustainer gas generator | $LO_2$ | Blade | 17-7 CRES | CTFE | 356-T6 permanent mold casting | race seal | Kel-F | 0.75 | 570 | 7.27 | Pneumatic piston | 100 |
| | Vernier propellant valve | LOX | Poppet | 2024-T4 | CTFE | 7075-T6 | Lip seal | Kel-F | – | – | – | | – |
| Saturn S-111 H-1 | Main oxidizer value | $LO_2$ | Butterfly | 431 CRES | CTFE | 356-T6 permanent mold casting | Lip seal | Mylar[a] | 4.00 | 975 | 545 | Hydraulic piston | 333 |
| | Gas generator (oxidizer) | $LO_2$ | Poppet | A-286 CRES | FEP Teflon | 347 CRES | Bellows | A286 CRES | 0.609 | 940 | – | hydraulic piston | – |
| Saturn S-IC | Main oxidizer | $LO_2$ | Poppet | 5061-T6 aluminum forging | TFE Teflon | 6151-T6 aluminum forging | Vented redundant lip seals | Mylar[a] | 8.0 | 1600 | 2000 | Hydraulic piston | Goo |
| F-1 | Gas generator | $LO_2$ | Ball | Two piece welded A-286 | CTFE | 7075-T73 die forging/ TENS-50 permanent mold | Vented redundant lip seals | Kel-F | 1.5 | 1600 | 121.5 | hydraulic piston | – |

*(Continued)*

**Table 11.50** *(Continued)*

| Vehicle and engine | Valve application | Fluid media | Valve element | Valve element material | Seal or seat material | Valve housing-material | Dynamic seal type | Secondary seal material | Line size, in. | Operating pressure, Dsi | Flow rate lb/sec | Actuator type | Response time, sec |
|---|---|---|---|---|---|---|---|---|---|---|---|---|---|
| | | | | | | casting | | | | | | | |
| | Oxidizer prevalve | 102 | Ball(visor) | – | – | – | – | – | 17.0 | 115 | 3970 | Pneumatic piston | 500 |
| Saturn S-11 | Main oxidizer Valve | $LO_2$ | Butterfly | 431 CRES | CTFE | TENS-50 permanent mold casting | Lip seal | Mylar[a] | 4.0 | 911 | 386 | Pneumatic piston | 50 |
| | Gas generator (oxidizer) | $LO_2$ | Popped | A-286 | FEP Teflon | 6061-T8 | Bellows | A-286 | 0.531 | 1010 | 3.29 | Pneumatic bellows | 3.0 |
| | Propellant utilization | $LO_2$ | Rotating sleeve | 440C | None | TENS-50 aluminum casting | Lip seal | Mylar[a] | 2.0 | 1310 | 1 do 120 | Electric motor | Throttling |
| | Prevalve | $LO_2$ | Butterfly | A-206 | CTFE | 6061-T6 | V-Seal | Teflon-coated inconel ×750 | 8.0 | 132 | 386 | Pneumatic bellows | 1000 |
| | Propellant Tank vent and relief | $LO_2$ | Poppet | 7075-T73 | CTFE | 356-T | RACO and flat gaskets | Teflon with CRES and Eel-F and Mylar | – | – | – | – | – |

[a] Mylar generally nod recommended for use in liquid oxygen valve seals because it has produced positive reactions in LOX impact sensitivity tests add high-impact energy levels.

its alloys may be used, but caution should be exercised when using aluminum in gaseous oxygen in the very high pressure range (>7500 psia) because of its low melting point and very high heat of reaction. Therefore, Monel for strength requirements and copper or brass for nonstructural requirements would provide the greatest resistance to ignition at high pressures and temperatures.

**Metallic Seats.** Metal seats are generally preferred for very high pressure applications because they are more resistant to damage from pressure stresses and erosive flow. Generally, the valve element in the plug or poppet-type valves is of a harder material than the seat material to prevent galling and contact surface deterioration from the stress of open-close cycling. But this is not critical when the contacting surfaces are not caused to rub or slide under pressure during closure and where closure pressure is within the elastic limit of the valve element–valve seat combination.

The design objective in metal-to-metal valve element–valve seat surfaces is complete contact between the surfaces since effective valve closure depends on close conformity of the mating seat and element surfaces. This is difficult to achieve, even with optically flat surfaces and assuming perfect alignment because of the surface asperities.

**Nonmetal seats.** Polymeric valve seat materials are not specifically recommended for pressures above 2000 psi or flow velocities above 500–700 km per hour (455–638 ft/sec). However, with proper protection, if the seat material is recessed in a high thermal conductivity metal, polymeric materials may be used after proof tests substantiate the design. This technique reduces the possibility of ignition and prevents "cold-flow" relaxation of the seat.

While several nonmetallic materials have been used successfully in liquid and gaseous oxygen systems, tetrafluoroethylene (TFE), fluorinated ethylenepropylene copolymer (FEP), and chlorotrifluoroethylene (CTFE) filled with glass or metal fibers are preferred for most applications. For oxygen systems at pressures above 2000 psi, such nonmetallic seats should be qualified by proof tests which simulate the use cycle and environment.

### *7.5.4. Filters*

As noted earlier, solid particles in oxygen systems can cause abrasion and impact ignition of sensitive materials and clogging of close clearances. Generally, the oxygen as supplied by the vendor has been filtered to remove particles greater than 10 μm nominal and 40 μm; absolute. This is adequate for most oxygen system requirements provided that no additional contaminant is introduced by subsequent handling.

Particulate contamination is most critical in high-pressure gaseous oxygen systems for two reasons: first, gaseous systems are generally at room temperature or higher, which reduces the energy requirement to cause ignition, and second, gaseous systems can easily accelerate particles to sonic or even supersonic velocities. This results in very high impact energies. Consequently, filters may be required, especially in high pressure gaseous oxygen systems.

Where it is possible to provide contaminant-controlled oxygen to a contaminant-free system, system filters may not be required. Where filters are required to guarantee particle control, they should be installed in the section of the oxygen system which has a low flow velocity. This provides for a minimum $\Delta P$, lower potential for high-velocity particle impact, and minimal scrubbing action of the fluid through the filter.

Unless test data permit otherwise, the following criteria are suggested:

1. Filter to a particle size, that is, 10% less than the smallest orifice, pore, or passage to remain unplugged.
2. Remove by filtration particles that exceed 10 times the asperity height of the land surfaces where all metal valve surfaces are used in a shutoff valve application.
3. Solid particulates should be limited as follows:
   a. 4000–1000 μin. for low pressure gaseous oxygen systems (500–2000 psia);
   b. 1000–250 μin. for medium pressure (2000–7500 psia);
   c. 250–100 μin. for high pressure (7500–10,000 psia);
   d. 100–40 μin. for pressure above 10,000 psia.

Oxygen filters require special design and maintenance considerations. Usually, filters are changed when the pressure loss across the element becomes too great. With oxygen filters it must be rememebered that the unit is filtering and *storing* solid contaminants. If the contaminants include a high percentage of hydrocarbons, which may be increasing with each use, it should also be assumed that the filter could be sensitive to shock and ignition sources.

Filter isolating valves should be installed in the line so that the pressure in the filter can be relieved and the filter element removed for cleaning. This may mean two filters in parallel where the flow of oxygen cannot be interrupted while the filter element is being cleaned. Bypass lines to permit removal of the filter element for cleaning arc not recommended unless a high degree of control can be exercised over the use of the bypass line.

Since pressure losses across the filter increase with the filter fineness and residue accumulation, system tolerance for such pressure loss must be provided. The standard procedure for monitoring the filter condition is by $\Delta P$ measurement.

Because of the pressure drop across a filter, filters may not be tolerable in liquid systems. However, components may be protected from damage due to large particulates by a screen which introduces only a slight pressure drop.

When conical wire mesh screens are used, the cone must point upstream so that the screen will be self-cleaning by washing debris to the wall of the pipe. This orientation will result in collecting particles in a location which will result in the least possible pressure drop across the screen.

**Limiting Velocities.** A high velocity flow of oxygen can effect ignition and increase the burning rates of materials. Also, solid particles entrained in a high-velocity flow stream can cause surface erosion, and particle impact can cause, or be the source of, ignition.

Ignition of oxygen piping from impact and abrasion of entrained particulate matter has been a frequent occurrence in gaseous oxygen systems especially at room temperatures or higher. This is especially true where a high degree of cleanliness has not been maintained. Those organizations familiar with the problem maintain that the required cleanliness limits the flow velocity in the piping. In most cases, the oxygen flow velocity limits are set in consideration of past experience, the pipe material, and operating pressure and temperature. These vary with the organization. Typical values are listed in Table 11.51.

Actual tests have confirmed what has been observed from experience: exceeding a velocity limit when entrained particulates are present results in ignition of the

**Table 11.51** Typical Velocity Limits for Entrained Particulates in Oxygen with Carbon Steel Pipe

| | | Maximum flow velocity | |
|---|---|---|---|
| Pressure, atm | Temperature, °C | ft/sec | m/sec |
| 10 | 175 | 150 | 46 |
| 10 | 150 | 175 | 53 |
| 10 | 120 | 200 | 61 |
| 20 | 175 | 75 | 23 |
| 20 | 150 | 85 | 26 |
| 20 | 120 | 100 | 30 |
| 20 | 65 | 130 | 40 |
| 40 | 120 | 50 | 15 |
| 40 | 100 | 60 | 18 |
| 40 | 65 | 65 | 20 |
| 60 | 100 | 30 | 9 |
| 60 | 65 | 35 | 10–11 |

flow duct, and bends, valves, and other flow system obstructions or deviations are the most critical ignition sites.

In summary, the primary source of concern under high velocity flow conditions is the entrainment of particulates and their subsequent impingement on a surface such as at bends in piping. Whether material erosion or ignition is due to entrained particulate impact and abrasion, to the erosive effects of the fluid flow, or to both, the fact remains that the effects exist and they increase with pressure and flow velocities.

Until a more quantitative limit can be established, the following practices are recommended:

1. Where practicable, avoid sonic velocity in gases and cavitation in liquid oxygen.
2. If possible, avoid the use of nonmetals at locations within the system where sonic flow or cavitation can occur.
3. Maintain fluid system cleanliness and limit entrained particulates. See Table 11.52.

### *7.5.5. Oxygen Pumps*

The most commonly used pump for pumping liquid or gaseous oxygen is the centrifugal pump, which is used almost exclusively in rocket engine systems. Centrifugal pumps are also widely used in ground applications for liquid oxygen pumping and

**Table 11.52** Filter Guideline

- Control velocity
- 100-mesh (150 μm) may not be sufficient
- Monel, bronze, or nickel is preferred
- How bad is it? A 1 mg particle of 2024 aluminum caused ignition when it hit a 21–6-9 steel target at 510°F and 4000 psia
- Develop and maintain a good filter cleaning plan
- Point the filter cone upstream

as high flow rate gaseous oxygen compressors. Where very high pressures must be attained (usually at low flow rates), positive displacement piston-type pumps may be used; however, by means of staging, high pressures may be obtained from centrifugal pumps. Axial flow pumps tend to be impractical for pumping oxygen because of the relatively high density of the liquid and the complexity caused by bearing location and lubrication requirements for pumping gaseous oxygen. Diaphragm pumps can be designed to pump either liquid or gaseous oxygen, but practicality would generally dictate using a centrifugal or piston pump. This is because of problems such as diaphragm flexure fatigue, gaseous compression heating, and materials compatibility to meet the internal pump conditions. To achieve high-pressure liquid oxygen at very high flow rates, centrifugal pumps with one or more stages would be preferable. Most very high pressure oxygen requirements are for gaseous breathing oxygen storage purposes such as for life support systems. The easiest way to generate gaseous oxygen pressures up to 10,000 psi or higher is to pump liquid oxygen to that pressure through a vaporizer. Since this does not generally require high liquid flow rates, the positive displacement piston pump is the most appropriate.

### 7.6. Cleaning

System cleanliness is critical in oxygen components and systems because contaminants may cause functional anomalies or ignition. Components used in oxygen systems should always be reasonably clean before initial assembly to ensure contaminants do not damage the hardware. After initial mockup assembly, oxygen systems must be disassembled and thoroughly cleaned, reassembled, leak tested, and purged with clean, oil-free, filtered, dry, $GN_2$, or helium before they are wetted with oxygen.

Cleaning should ensure the removal of contaminants that could cause mechanical malfunctions, system failures, fires, or explosions. Effective cleaning (Table 11.53) will:

1. remove particles, films, greases, oils, and other unwanted matter;
2. prevent loose scale, rust, dirt, mill scale, weld spatter, and weld flux deposited on moving and stationary parts from interfering with the component function and clogging flow passages;

**Table 11.53** Effective Cleaning

1. Clean surfaces of films, greases, oils, and other organic materials
2. Prevent functional interference with components and the clogging of flow passages
3. Remove combustible contaminants
4. Avoid the accumulation of finely divided contaminants

- How is effective cleaning accomplished:
  Integrate with construction operations
  Plan for cleaning during all phases of development
- The cleaning process consists of several steps
  Solvent degreasing with methyl chloroform
  Detergent cleaning
  Acid pickling of stainless steel parts
  Final treatment consisting of steam and blow dry
  Seal in polychlorotrifluoro ethylene
- Good recordkeeping is required throughout the cleaning process

3. reduce the concentration of finely divided contaminants, which are more easily ignited than bulk material.

Oxygen systems and components should be thoroughly cleaned in accordance with established procedures (ASTM G 93 1985; Bankaitis and Scueller, 1972; CGA G-4.1 1987; MIL-STD-1246B (latest revision); NASA JHB 5322B 1994; SSC 79-0010 1989; KSC-C-123G 1994; MSFC-PROC-1831 1990; MSFC-PROC-1832 1990). Additional highly recommended requirements are given in:

1. NASA Reference Publication 1113, *Design Guide for High Pressure Oxygen Systems*, 1983, Chapters 6, 7, and 8.
2. NASA SP-3072, *Cleaning Requirements, Procedures, and Verification Techniques*, ASRDI Oxygen Technology Survey, Volume 2.
3. MSFC-SPEC-164A (October 1970), *Cleanliness of Components for Use in Oxygen, Fuel, and Pneumatic Systems;* contains acceptable methods of cleaning pipe, tubing, and flexhose.

Note: Under The Clean Air Act Amendments of 1990 and the UN Montreal Protocol, the use of chlorofluorocarbons (CFC), 1,1,1-trichloroethane, and other ozone-depleting substances are illegal. Alternative cleaning solvents are being developed and investigated and will probably come into use in the near future. These include the new hydrochlorofluorocarbons (HCFC), deionized (DI) water, and isopropyl alcohol (IPA). The effects of these changes on the cleaning of oxygen systems are being assessed.

Whenever possible, oxygen-system cleaning should begin by disassembling all components to their individual parts. In situ cleaning of systems and flow cleaning of components are generally ineffective.

### *7.6.1. Cleaning Procedures*

No single cleaning procedure will meet all cleanliness requirements. Visual cleanliness is not a sufficient criterion when dealing with oxygen systems, because of the hazards associated with contaminants invisible to the naked eye. General oxygen-system cleaning procedures are discussed in ASTM G 93 (1985).

Items to be cleaned should be completely disassembled before cleaning when possible. Piping systems should be cleaned and inspected before assembly. Components or parts that could be damaged during cleaning should be cleaned separately.

Preparing components for oxygen service includes degreasing, disassembling and examining, hydrostatic testing (if necessary), precleaning, inspection, precision cleaning, reassembly, functional testing, and packaging.

**Degreasing.** Degrease metal parts with a degreasing agent by immersing, spraying, or vapor-rinsing the part until all surfaces have been thoroughly flushed or wetted.

Note: This step is required only for heavily oil- or grease-contaminated items. Alkaline cleaners used to preclean metallic parts and detergents used to preclean nonmetallic parts will effectively remove small amounts of grease and oil.

**Disassembling and Examining.** Components should be disassembled and their parts grouped according to the method of cleaning.

**Precleaning.** Various commercially available cleaning solutions can be used in conjunction with ultrasonics to remove firmly attached contaminants. Commonly used cleaning solutions include alkaline solutions, acid solutions, mild alkaline liquid detergents, and rust and scale removers.

The cleaning solutions used depend on the material to be cleaned. Stainless steels (300 series), Monel® alloys, Inconel® alloys, and Teflon® are usually cleaned in an alkaline solution and then in an acid solution. Carbon steel is cleaned by a rust and scale remover, if required, and then in an alkaline solution. In severe cases of rust or corrosion, carbon steel may be sand or glass-bead blasted. Copper and brass are cleaned in alkaline solution, then acid pickled. Aluminum and nonmetals are cleaned in liquid detergent.

**Chemical Cleaning.** Chemical cleaning may be a single-step or multistep process, depending upon the material involved (Table 11.54).

Note: Parts should be handled only with approved, clean gloves from this point on.

- Detergent cleaning. Mild detergent is usually used to clean nonmetallic and aluminum parts. Spray and/or immerse the items in a detergent solution for a specified period of time.
- Alkaline cleaning. Corrosion-resistant metals and Teflon® are usually cleaned with an alkaline cleaning agent. Spray, soak, and/or immerse items in a solution of alkaline cleaner for a specified period of time.
- Acid cleaning. Place nonwelded stainless steel parts or other acid-resistant metal parts in an acidic cleaning solution and allow them to soak for a specified period of time.
- Rinsing. After the detergent, alkaline, or acid cleaning, thoroughly spray, rinse, or immerse parts in DI, distilled, filtered water to remove all the cleaning agent.
- Acid pickling. Acid pickling is used to remove welding discoloration and slag. Place newly welded stainless steel parts in a pickling bath, typically 3–5% hydrofluoric and 15–20% nitric acid solution, for about 3–5 min.

**Inspecting.** The parts shall be visually inspected under a strong white light for contaminants, including lint fibers. Visual inspection will detect particulate matter larger than 50 μm as well as moisture, oils, and greases. See Table 11.55.

### *7.6.2. Clean Assembly*

Even the best-designed oxygen systems can contain hazardous ignition sources if they are fabricated or assembled incorrectly. Recommended techniques for clean assembly and inspection to verify correct fabrication are described below.

**Table 11.54** Oxygen System Cleaning

| |
|---|
| • Precleaning—removal of gross foreign materials by brushing, vacuuming, sand blasting, etc. |
| • Steam or hot water with detergent—removal of dirt, oil, and loose scale |
| • Caustic cleaning and rinse—removal of heavy or tenacious surface contaminants |
| • Acid cleaning—removal of oxides and other contaminants |
| • Solvent washing (ultrasonic)—removal of organics, oils, greases, and other contaminants |
| • Vapor degreasing—removal of soluble organic materials by condensation of solvent vapors |
| • Mechanical cleaning—removal of rust, scale, paint, weld slag, loose particles, and general crud |

**Table 11.55** Cleanliness Levels

- Visually clean
  No corrosion, scale, grease, or chips
  No UV fluorescence
- Wipe-test clean
  Lint free wiper such as filter paper or white gloves
  No particulates larger than 2000 μm
  No fluorescence under UV
- Free of hydrocarbons and/or particulates
  Residue from solvent extraction is $< 50\,\text{mg/ft}^2$
- Surface particle count
- Clean by particle count and NVR (nonvolatile residue)
  NVR $< 1\,\text{mg/ft}^2$

**Careful assembly.** Careful assembly is extremely important for high-pressure oxygen systems because contaminants generated during assembly are a potential source of readily ignitable material. Elimination of all contaminants is the goal. Careful assembly procedures can minimize the quantity of contaminants remaining in a system and, thus, the potential for contaminant ignition.

**Maintaining Cleanliness During Assembly.** Procedures for system and component assembly or reassembly after cleaning must be stringently controlled to ensure that the required cleanliness levels are not compromised. All components requiring reassembly (such as valves, regulators, and filters) shall be reassembled in a clean room or flow bench.

Assembly or reassembly of systems should be accomplished in a manner that minimizes system contamination. Components should be kept in clean bags until immediately before assembly.

**Burrs.** Removal of burrs and sharp edges is of critical importance in high-pressure oxygen systems. Burr removal in small-diameter internal passageways at the intersection of cross drills is a common problem.

**Lubricants.** Lubricants shall be used whenever they are required to reduce abrasion and damage to seals during assembly and to enhance the operational sealing or sliding of parts. Lubricants should be applied lightly, and excess lubricant should be removed to prevent future migration.

Hydrocarbon-based lubricants must not be used in high-pressure or LOX systems because they can easily ignite; the incorrect use of hydrocarbon-based lubricants is a common cause of oxygen system fires. The best lubricants for compatibility with high-pressure oxygen are highly fluorinated materials. Even the best lubricants can react with oxygen when system design limits on temperature, pressure, or pressure rise rates are exceeded.

#### *7.6.3. Record Keeping*

Record keeping is especially critical for equipment in oxygen service. At a minimum, labels must specify clean level and what cleaning specification was used.

### 7.7. Emergency Procedures

#### *7.7.1. General*

**S-W-I-M.** *Stop* the source of oxygen if that can be done safely and quickly. *Warn* everyone in the vicinity. *Isolate* the hazardous area, attempt to control the hazard using installed safety systems and preplanned procedure. *Move* it. Evacuate the vicinity of the hazard. Only personnel trained in specific rescue techniques should engage in these techniques. All other personnel should stay clear. Persons on fire in an oxygen-enriched atmosphere cannot be rescued by a person entering the area to pull them out, as the rescuer will most probably catch fire also.

### *7.7.2. Leaks and Spills*

Plan the facility for spills. Consider the prevalent wind direction and provide a means of disposal. See Table 11.56.

The primary danger from oxygen leaks and spills is a fire or explosion caused by combustible materials in the presence of a high concentration of oxygen. Oxygen-enriched environments greatly increase the rate of combustion of flammable materials.

*Gaseous oxygen* leaks can result in oxygen-enriched environments, especially in confined spaces. Impingement of gaseous oxygen onto an organic material such as grease can cause a fire. When leaks are detected, the source of the oxygen should be halted or disconnected.

*Liquid oxygen* spills and leaks cause oxygen enrichment of the immediate vicinity as the liquid vaporizes. When a spill or leak is detected, the source of the supply should be immediately halted or disconnected. Any equipment inherently heat- or spark-producing should be turned off or disconnected. Affected areas should be completely roped off or otherwise controlled to limit personnel movement. The equipment or piping should be thoroughly vented and warmed before repair of the leak is attempted.

Liquid oxygen spills on pavements such as asphalt have resulted in impact-sensitive conditions that caused explosions from traffic or dropped items. The same condition can occur from liquid oxygen leakage onto concrete that is contaminated with oil, grease, or other organic materials. The affected areas should be completely roped off or otherwise controlled to limit vehicle and personnel movement. Electrical sources should be turned off or disconnected. No attempt should be made to hose off the affected area, and the area should not be cleared for access until the oxygen-rich cold materials are adequately warmed and absorbed oxygen has evaporated.

**Table 11.58** Venting and Disposal of Oxygen

- Liquid oxygen should not be dumped on the ground for disposal
- Contained vaporization should direct the stream of an open-ended vaporizer
  Clean gravel heat sink
- Liquid oxygen contaminated with fuel should be isolated
  Evacuate the area
  Inert the system before returning
- Gaseous oxygen should be disposed by planned venting
  Interconnections of vent discharges is discouraged
- Safe distances for venting are highly variable
  Direction, amount, wind, and surroundings
- Plan for a total loss of insulation in vent system design
- Use corrosion-resistant materials in vent construction

#### *7.7.3. Overpressurization*

If liquid oxygen is trapped in a closed system and allowed to warm, extreme pressures can overpressurize the system. Pressure relief of some kind must be provided where trapping might occur. Moreover, relief and vent systems must be sized to accommodate the flow so that excessive backpressures will not occur. Cryogenic liquid storage vessels are protected from overpressurization by a series of pressure relief devices. These relief devices are designed to protect the inner vessel and the vacuum-insulated portion of the tank from failures caused by inner and outer shell damage, overfilling, and heat load from insulation damage or from a fire.

*Frost* appearing on the outer wall of an insulated cryogenic vessel indicates vessel insulation loss. Frost appearance is only a clue to the type of insulation loss. This insulation loss could be caused by a movement of the insulation in the annular area of the tank, by loss of vacuum in the annular area, or by inner vessel failure. Personnel should listen and watch for indication of pressure-relief device actuation. Constant relief actuation is an indication that a major problem has occurred. Continued pressure rise while the relief device is actuated indicates a major system malfunction. If constant relief device actuation is occurring, immediately evacuate the area and physically rope off and control the area if this can be performed safely. Do not apply water, as this would only act as a heat source to the much colder oxygen and aggravate the boiloff.

#### *7.7.4. Cold Injury*

Direct physical contact with LOX, cold vapor, or cold equipment can cause serious tissue damage. Medical assistance should be obtained as soon as possible for any cold injury. First aid procedures should be administered by trained medical professionals and are beyond the scope of this handbook. However, proper immediate bystander response should be as follows:

1. If it is safe to do so, remove the patient from the source of the cold.
2. In the event of limb-size or smaller cryogenic exposure, appropriate response may include an attempt to rapidly warm the affected area with moist heat from a shower, eyewash or warm water bath, not to exceed 38.9°C (102°F). Do not allow a heavy stream of water to impinge directly on frozen skin. In some cases, it is safest to do nothing other than cover the involved area until professional medical help is available.
3. Massive full-body cryogenic exposures present significant additional concerns, but removal of the victim from the exposure atmosphere and keeping the victim's airway open are important. Loosely wrapping the victim in a blanket until the arrival of the ambulance team is also advised.
4. Some important *do nots*:
    - *Do not* remove frozen gloves, shoes, or clothing. Salvageable skin may be pulled off inadvertently.
    - *Do not* massage the affected part.
    - *Do not* expose the affected part to temperatures higher than 44°C (112°F), such as a heater or a fire. This superimposes a burn and further damages already injured tissues.
    - *Do not* apply snow or ice.
    - *Do not* apply ointments.

- *Do not* allow any smoking, open flames, or other hazardous conditions near the victim.
- *Do not* apply fire blankets to cover personnel whose clothing is saturated with oxygen as the fire blanket may also burn.

*7.7.5. Safety Clothing*

Protective clothing and equipment should be included in personnel protective measures.

**Hand and Foot Protection.** Gloves for work around cryogenic systems must have good insulating quality. They should be designed for quick removal in case liquid oxygen gets inside. Because this is also a danger with footwear, shoes should have high tops and pant legs should be worn outside and over the shoe tops. The shoes should be of leather.

**Head, Face and Body Protection.** Personnel handling liquid oxygen shall wear head and face protection. A face shield or a hood with a face shield shall be worn. If liquid oxygen is being handled in an open system, an apron of impermeable material should be worn.

**Flammability of Clothing.** Oxygen will saturate clothing, rendering it extremely flammable. Clothing described as flame resistant or flame retardant in air may be flammable in an oxygen-enriched atmosphere.

**Oxygen Vapors on Clothing.** Any clothing that has been splashed or soaked with oxygen vapors shall be removed until it is completely free of the gas.

**Exposure to Oxygen-rich Atmospheres.** Personnel exposed to high oxygen atmospheres should leave the area and avoid all sources of ignition for at least 20 min or longer, until the oxygen in their clothing dissipates. The safest course is to change clothing.

*7.7.6. Fire-Fighting Techniques*

When fighting a fire involving oxygen-enriched atmospheres, the first step should be to shut off the oxygen supply and, if possible, to shut off and remove fuel sources. Combustible materials must be cooled below their ignition temperatures to stop the fire. Water is an effective extinguishing agent for fires involving oxygen-enriched atmospheres.

In some cases, when the oxygen supply cannot be shut off, the fire may burn so vigorously that containment and control is more prudent than trying to put out the fire.

If fuel and liquid oxygen are mixed but not burning, quickly isolate the area from ignition sources, evacuate personnel, and allow the oxygen to evaporate. Mixtures of fuel and liquid oxygen are an extreme explosion hazard.

If a fire is supported by liquid oxygen flowing into large quantities of fuel, shut off the oxygen flow. After the excess oxygen is depleted, put out the fire with an extinguishing agent recommended for the particular fuel.

If a fire is supported by fuel flowing into large quantities of liquid oxygen, shut off the fuel flow and allow the fire to burn out. If other combustible material in the area is burning, water streams or fogs may be used to control the fires.

If large pools of oxygen and water-soluble fuels, such as hydrazine or alcohol, are burning, use water to dilute the fuel and reduce the fire's intensity.

## 8. GENERAL SAFETY PRINCIPLES

Personnel safety cannot be guaranteed; however, a high level of safety can be achieved if operating personnel have the proper attitude, understanding, and training. Safety regulations must be conscientiously practiced and rigidly enforced. It is the painful truth that many of these rules have been written because of the death or suffering of those who did not know or chose to ignore vitally important safety principles.

The best assurance of personnel safety lies in the safety education of the people themselves. If they can be made aware of potential hazards and means of protecting their own lives, most of them will respond in a responsible fashion. Responsibility for the safety of one's self and others cannot be obtained solely with a set of written regulations. Responsibility is secured on an individual basis, in varying degrees, and is the framework for all safety education. There would be little need for safety rules if everyone were extremely responsible and knowledgeable. Unfortunately, this is not always the case. A lack of maturity on the part of an individual, or a new or unfamiliar job assignment, work in a manner contrary to the possession of such responsibility and knowledge. Safety rules, then, become a primary tool in securing safety-conscious, well-trained personnel. In many instances, safety education is conducted on a haphazard basis and taken seriously only when required by top management. It is not uncommon for safety procedures to evolve following a serious accident that has resulted in injury or death. The safety of personnel can be almost ensured only when there is a thorough understanding of potential hazards, correct procedures and equipment are used, and the equipment is in good working condition. Unsafe equipment should be repaired and unsafe conditions corrected as soon as their existence is known. A safety officer should be available for consultation and for the enforcement of proper safety policies and practices. Experience has shown that safety committees and safety officers are a necessary part of any organization dedicated to the safety of its personnel.

Even normally safe jobs can be hazardous when the job is not approached properly. Activities involving explosives, extreme temperatures, high pressures, flammable liquids, and similar hazards should be undertaken only by knowledgeable persons. Any activity can be considered only as safe as its weakest link—an action taken by a person who is not fully aware of the magnitude and varieties of hazards involved. Knowledge of a job situation and appropriate safety equipment are vital to successful completion of the job and should be included in the job plans. There is no substitute for knowledge and planning. Good plant design and system layout are essential for safe operation. Operating personnel must thoroughly understand the equipment, the system layout, and how the system functions. All of the normal, abnormal, manual, and automatic operations and actions that can occur must be analyzed to determine potential safety hazards. Sequential and individual events must be considered in making these determinations. Analysis of potential hazards provides an opportunity to isolate the cause and take the necessary precautions to minimize any possibility of an accident stemming from this cause.

Corrective action for personnel protection must be planned for those cases where an accident can occur; this usually involves planning for emergencies. Emergency planning requires that personnel know beforehand what to do and what not to do and where to go and where not to go in various types of emergency situations. Personnel should be trained in the location and use of fire alarms, safety showers, fire blankets, firefighting equipment, and first aid kits. Accidents are effectively reduced or perhaps eliminated when planning is based on good judgment. Good judgment grows from a responsible worker attitude, experience, and training and is best exercised with appropriate support from management.

The six points given below describe an excellent safety philosophy, adapted from a major US chemical producer. It is outstanding in the way it engages both management and all levels of employees.

## 8.1. Safety Philosophy

1. All injuries can be prevented. The key word here is "all." This is a realistic goal and not just a theoretical objective. A supervisor with responsibility for the well-being of employees cannot be effective without fully accepting this principle.
2. Management, which includes all levels through the first-line supervisor, has the responsibility for preventing personal injuries. Since the line organization has the responsibility for every operational activity of the company, each supervisor must accept his share of the responsibility for the safety of the employees.
3. It is possible to safeguard all operating exposures that may result in injuries. It is preferable, of course, to eliminate the sources of danger. However, where this is not reasonable or practical, supervision must resort to such measures as the use of guards, safety devices, and protective clothing. No matter what the exposure, an effective safeguard can be provided.
4. It is necessary to train all employees to work safely. They must understand that it is to their advantage, as well as the company's to work safety and that they have a definite responsibility to do so. Adequate training of the employees is a responsibility of supervision.
5. It is good business to prevent personal injuries on the job and off the job. In addition to causing personal suffering, injuries cost money and reduce efficiency.
6. Safety is a condition of employment. Employees are expected to accept their responsibility for safety when they join the company—the safety of themselves and of the people and facilities with which they work.

Management, which includes you, is responsible for setting standards and guidelines for managing safety. These standards apply to several areas (see Table 11.57):

- Safety rules and procedures;
- Work practices and conditions;
- Emergency procedures;
- Operating procedures;
- Safe facility maintenance;
- Housekeeping.

In addition to establishing standards and rules, you must communicate your standards and those of your site; all your employees must understand the safety standards and know what level of safety performance you expect of them.

Your attitude to the safety rules of your workplace is the key to effective enforcement of those rules. You must demand of yourself full adherence to your site's and your own safety rules; you can accept no less from anyone you supervise.

**Table 11.57** Safety Reference for Observers

| Inadequate personal protective equipment | Incorrect positions of people | Observed actions of people |
|---|---|---|
| Eyes and face | Striking against | Adjusting personal protective equipment |
| Ears | Struck by | |
| Head | Caught between | Changing position |
| Hands and arms | Falling | Rearranging job |
| Feet and legs | Temperature extremes | Stopping job |
| Respiratory system | Electrical current | Attaching grounds |
| | Inhaling | Hiding, dodging |
| Trunk | Absorbing | Changing tools |
| **Tools and equipment** | **Procedures** | **Housekeeping** |
| Right for job? | Is standard practice adequate for job | Standards established? |
| Used correctly? | | Standards understood? |
| In safe condition? | Is standard practice established? | Obstruction in passageways? |
| Seat belts in use? | | |
| Barricades or warning lights | Is standard practice being maintained? | Disorganized tools, materials? |
| Chocks/restraints properly used? | Work permit proper? Fire watch adequate? | Obstructions by stairs/platforms? |
| | Gas test (if needed) satisfactory? | Obstructions by emergency equipment |

This philosophy emphasizes management. When it comes to safety, *everyone is a manager*, for he or she is the ultimate manager of his or her own actions.

## 8.2. The Role of the Safety Department

It is *your* job as supervisor to ensure that the employees in your group or area work as safely as possible. You should use the Safety Department as a resource, not as a department that can do your job for you.

Your safety officers can assist you in the following ways (see Table 11.58):

- By contributing their expertise. They are the safety professionals in your company.
- By asking probing questions.
- By seeing contributing causes that you might not recognize.

The safety officers can also clarify questions on safety by asking outside sources for any necessary extrainformation.

## 8.3. Safety Reviews

"Nature Punishes Any Neglect of Prudence"—Emerson

Emerson might well have added: "And any act of arrogance." As good as you are, you will never know it all or think of everything.

Accordingly, safety assessment should be integrated into the overall facility design review of major projects. Each design review phase will address and evaluate the safety aspects of the project consonant with its level and maturity. The DoD and NASA have a well-developed system of reviews keyed to the progress of the program. They are:

1. Concept Design Review (10 Percent Review). Purpose and design performance criteria should be established. Proposed and selected design approaches and basic technologies should be delineated sufficiently to indicate the type and magnitude of the principal potential hazards. Applicable

**Table 11.58** Checklist of General Safety Principles

1. A safe worker is a responsible, safety-conscious person
2. A safe worker thoroughly understands the system and equipment with which he or she works
3. A safe worker is fully aware of the accepted safety practices and regulations
4. All equipment is in good order
5. A qualified safety officer is available with authority to take correction action
6. The equipment and system have been properly designed, and good operating procedures have been prepared
7. The equipment and system have been analyzed for potential hazards, and emergency plans have been drafted
8. Good judgment has been exercised in the system and equipment design and in the formulation of operating procedures, safety rules, and emergency actions

design codes, safety factors, and safety criteria should be specified. A preliminary hazards analysis shall be started. Appropriate safety tasks should be planned and become the foundation for safety efforts during the system design, manufacture, test, and operations.

2. Preliminary Design Review (30/60 Percent Review). Stress calculations for critical structures shall show that design codes, safety factors, and safety criteria have been met. The preliminary hazards analysis shall be completed; system/subsystem hazards analyses should be under way.
3. Critical Design Review (90 Percent Review). The design shall be reviewed for conformance to design codes, required safety factors, and other safety criteria. Proposed construction methods and arrangements shall make clear that construction hazards will be effectively controlled. Procurement documents, such as a statement of work (SOW) shall specify appropriate safety requirements. The system/subsystem hazards analyses shall be completed and close-out actions shall be proceeding. An operational hazards analysis shall be under way.
4. Design Certification Review (100 Percent Review). All project documentation (drawings, SOWs, specifications) should be complete, reviewed, and approved. All hazards analyses shall be complete, including close-out actions. Actions from previous reviews should be verified as complete.
5. Ongoing System/Subsystem Hazards Analyses. Hazards and operational analyses shall be continued during operations and testing.
6. Other Reviews. Other reviews that may be conducted consist of the following:

   a. Test Readiness Review. Operational procedures, along with instrumentation and control systems, shall be evaluated for their capacity to provide the required safety. Equipment performance should be verified by analysis or certification testing. It may be necessary to develop special procedures to counter hazardous conditions.
   b. Emergency Procedures Review. The safety of personnel at or near cryogenic systems should be carefully reviewed and emergency procedures developed in the earliest planning and design stages. Advance planning for a variety of emergencies such as fires and explosions should be undertaken so the first priority is to reduce any risk to life.
   c. Operational Readiness Review. An Operational Readiness Review may be required for any major facility change. Hydrogen hazards should be reviewed for compliance.

## 9. SAFETY CHECKLIST

### 9.1. Job Preparation

1. Be safety-conscious.
2. Be sure all equipment is in good working order (including safety equipment).
3. Conspicuously post warning signs—"NO SMOKING", "FLAMMABLE LIQUIDS", etc.
4. Report unsafe equipment and operating practices to the appropriate safety officer.

5. Do not tackle a job unless you have thorough knowledge of the hazards involved. Always work in groups of two or more.
6. Plan the job and determine what safety equipment is required. Use good judgment.
7. Know what steps to take in an emergency. Know your escape routes. Know first aid for yourself and others.
8. Know your firefighting equipment and how to use it.

## 9.2. Doing the Job

1. Wear protective clothing. Use goggles or face shield, gauntlet leather gloves, boots or high-top shoes covered by cuffless, clean, tightly woven, white cotton coveralls. Avoid skin contact with cryogenic surfaces. Use cotton clothing to minimize electrostatic spark hazards.
2. Use adequate ventilation. Keep the oxygen concentration in the air between 16% and 25%. Use auxiliary breathing equipment where the oxygen level is below 16%. Do not enter any tank without a lifeline and have a companion stationed outside.
3. Mind your housekeeping. Be tidy and clean.
4. In designing a system, make provisions for thermal contraction and system pressure relief.
5. Know and use materials and construction methods that are compatible with oxygen service. *Cleanliness* is imperative. Use proper solvents and cleaning techniques.
6. Avoid spillage, leakage, and drippage of oxygen. Do not permit clothes or other combustible materials to become saturated with oxygen. Avoid all ignition sources in areas that may contain flammable vapors and/or excessive oxygen concentrations. Do not permit liquid oxygen to come in contact with any fuel or noncompatible material. Take special care to avoid impact energy sources in liquid oxygen systems. Avoid liquid air drippage from liquid nitrogen systems.
7. Provide blowout panels, burst disks, etc., as required to guard against explosion overpressure, and provide protective barriers where needed to guard against fire, shrapnel, and overpressure.
8. Open and close valves on oxygen gas cylinders *very slowly*. Keep oxygen gas flow velocities low. Use stainless steel fittings and firebreak sections in carbon steel pipelines carrying oxygen gas.
9. Monitor for contamination. Monitor all liquid oxygen storage and liquefaction plants regularly to determine the contamination level. Keep fresh air intake of air separation plants free of all contaminants by continuously monitoring the intake air. Inspect oxygen equipment on a routine basis for evidence of accumulated contaminants.
10. Treat all gas-filled and cryogenic liquid-filled vessels as potential bombs—they are! Securely anchor all gas cylinders in a safe position. Guard against accidental damage to the valves or plumbing on these containers. Never pneumostatically pressure-test a vessel when you can hydrostat it.
11. Securely fasten all relief and vent pipes, loose lines, and similar equipment to prevent “hose-whip”. Do not expose the skin to high-velocity gas jets.

13. Keep vent lines open. Avoid entrapment and freezing of moisture in vent pipes, relief valve exhaust tubes, burst disk exhausts, small-mouth dewars, etc. Failure of these devices nullifies system pressure-relief protection.
14. Construct dikes around liquid oxygen storage tanks if tank rupture is likely to result in liquid oxygen draining into areas where a structural, fire, or explosion hazard may develop.
15. Use nonsparking tools wherever possible where flammable vapors and oxygen gas may mix in air. However, remember that sparkproof tools *can* cause sparks.
16. Use an inert gas purge for all electrical equipment where a substantial oxygen-enriched atmosphere is certain. A guaranteed fresh air purge may be substituted for the inert purge if necessary. Where flammable gases exist, the equipment must conform to Article 500 of the National Electrical Code (1962–63). Where oxygen-enriched air may occur occasionally, electrical equipment should be suitably protected to prevent open sparking and the possible ignition of combustible materials in the vicinity of the equipment. Where flammable vapor–oxygen mixtures may occur, ground all oxygen systems to minimize electrostatic spark hazards.

# Cited References and Suggested Reading

## CHAPTER 1: SUGGESTED READING


Anderson, O. E. Jr. (1953). *Refrigeration in America; A History of a New Technology and Its Impact*. Princeton University Press.

Barron, R. F. (1985). *Cryogenic Systems*. 2nd ed. New York: Oxford University Press.

Burke, J. (1978). *Connections*. Boston: Little, Brown and Company.

Creuzbauer, R. (1899). *Power Magazine*. May, p. 18.

Hepworth, T. C. (1899). Liquid air. *Chambers J.* Sixth Series, 2:361.

Johnson, V. J. (1960). *ACE*. Vol. 5. New York: Plenum Press, p. 566.

McClintock, M. (1964). *Cryogenics*. New York: Reinhold Publishing Corp.

Morton, H. (1899a). Engineering fallacies—liquid air as a motive power. *Cassier's Mag.* 16:29.

Morton, H. (1899b). *Stevens Inst. Indicator* 16:113. [Reprinted in (1899). *Sci. Amer.* 80:245].

Remson, A. (1899). Liquid air. *Sci. Am.*, June 10, 1899, p. 372.

Scott, R. B. (1959). *Cryogenic Engineering*. Princeton, NJ: D. Van Nostrand Co., Inc. [Prepared for the Atomic Energy Commision.]

Scurlock, R. G. ed. (1992). *History and Origins of Cryogenics. Part II: Cryogenic Research.* Oxford: Clarendon Press, 1992.

Sloane, T. O. (1899). *Liquid Air and the Liquification of Gases: Theory, History, Biography, Practical Applications, Manufacture*. London: Sampson Low, Marston and Co.

Thomson, E. (1899). The possibilities of liquid air power. *Eng. Mag.* 17:197.

Timmerhaus, K. D. (1982). *ACE*. 27. New York: Plenum Press, p. 1.

Tripler, C. E. (1899). Liquid air—newest wonder of science. *The Cosmopolitan*. 25:199. [This is the only known account of Tripler's demonstrations written himself.]

Wilshire, H. G. (1899). *Sci. Am.* 81:118.


## CHAPTER 2: SUGGESTED READING


Abbott, M. M., Van Ness, H. C. (1972). *Schaum's Outline of Theory and Problems of Thermodynamics*. New York: McGraw-Hill.

Banchero, J. T., Barker, G. E., Boll, R. H. (1951). Heat transfer characteristics of boiling oxygen, fluorine, and hydrazine, Proj. M834. Engineering Research Inst., Univ. of Michigan.

Bell (1968). K. J. Adv. Cryo. heat transfer. In: Chemical Engineering Progress Symposium Series. Vol. 64, No. 87.

Brentari, E. G., Giarratano, P. J., Smith, R. V. (1965). Boiling heat transfer for oxygen, nitrogen, hydrogen, and helium. US Nat. Bur. Stand. Tech. Note No. 317, Sept.

Drayer, D. E., Timmerhaus, K. D. (1962). An experimental investigation of the individual boiling and condensing heat transfer coefficients for hydrogen. *Adv. Cryo. Eng* 7: 401–412.

Flynn, T. M., Draper, J. W., Ross, J. J. (1962). The nucleate and film boiling curve of liquid nitrogen at one atmosphere. *Adv. Cryo. Eng.* 7:539–545.

Frederking, T. H. K. (1959). Film boiling of helium I and other liquefied gases on single wires. *AIChE J.* 5(3):403–406.

Krishnaprakas, C. K., et al. (2000). Heat transfer correlations for multilayer insulation systems. *Cryogenics* 40(7):431–435.

Lewis, G. N., Randall, M., Pitzer, K. S., Brewer, L. (1961). *Thermodynamics*. 2nd ed. New York: McGraw-Hill.

Lyon, D. N. (1964). Peak nucleate boiling heat fluxes and nucleate boiling heat transfer coefficients for liquid $N_2$, liquid $O_2$ and their mixtures in pool boiling at atmospheric pressure. *Intl J. Heat Mass Transfer* 7(10):1097.

Lyon, D. N. (1965). Peak nucleate boiling fluxes and nucleate boiling heat transfer coefficients in saturated liquid helium between the $\lambda$ and critical temperatures. *Adv. Cryo. Eng.* Vol. 10.

Ramirez, J. A. (2000). Measurement of heat transfer coefficients in high NTU regenerative heat exchangers. *Adv. Cryog. Eng.* 45:403–410.

Sandler, S. I. (1977). *Chemical and Engineering Thermodynamics*. New York: Wiley.

Schmidt, A. F. (1972). *ASRDI Oxygen Technology Survey*. Vol. I. *Heat Transfer and Fluid Dynamics*. Tech. Rept. NASA SP-3076. Vol. III. Jan, 177 pages.

Seader, J. D. (1965). *Boiling Heat Transfer for Cryogenics*. NASA-CR-243. 171 pages.

Sliepcevich, C. M., Hashemi, H. T., Colver, C. P. (1968). Heat transfer problems in LNG technology. In: Chemical Engineering Progress Symposium Series. Vol. 64, No. 87, pp. 120–126.

Smith, J. M., Van Ness, H. C. (1975). *Introduction to Chemical Engineering Thermodynamics*. 3rd ed. New York: McGraw-Hill.

Waynert, J. et al. (1990a). Transient heat transfer characteristics of liquid hydrogen, including freezing. In: Proceedings of the AIAA Second International Aerospace Planes Conference AIAA-90-5213.

Weil, L. (1951). Heat transfer coefficients of boiling liquified gases. In: Proceedings of the Eighth International Congress on Refrigeration. London, IC: 9193947, 181 pages.

Williamson, K. D., et al. (1968) *Studies of Forced Convection Heat Transfer to Cryogenic Fluids*. In: Chemical Engineering Progress Symposium Series. Vol. 64, No. 87, pp. 103–110.

Zemansky, M. W. (1957). *Heat and Thermodynamics*. 4th ed. New York: McGraw-Hill.

Zuber, N., Fried, E. (1962). Two-phase flow and boiling heat transfer to cryogenic liquids. *ARS J.* 32:1332–1341.

## CHAPTER 2: REFERENCES

Brentari, E. G., Giarratano, P. J., Smith, R. V. (1965). Boiling heat transfer for oxygen, nitrogen, hydrogen, and helium. US Nat. Bur. Stand. Tech. Note No. 317, Sep.

Drayer, D. E., Timmerhaus, K. D. (1962). An experimental investigation of the individual boiling and condensing heat transfer coefficients for hydrogen. *Adv. Cryo. Eng.* 7:401–412.

Flynn, T. M., Draper, J. W., Ross, J. J. (1962). The nucleate and film boiling curve of liquid nitrogen at one atmosphere. *Adv. Cryo. Eng.* 7:539–545.

Frederking, T. H. K. (1959). Film boiling of helium I and other liquefied gases on single wires. *AIChE J.* 5(3):403–406.

Lewis, G. N., Randall, M., Pitzer, K. S., Brewer, L. (1961). *Thermodynamics*. 2nd ed. New york: McGraw-Hill.

Lyon, D. N. (1964). Peak nucleate boiling heat fluxes and nucleate boiling heat transfer coefficients for liquid $N_2$, liquid $O_2$ and their mixtures in pool boiling at atmospheric pressure. *Intl J. Heat Mass Transfer* 7(10):1097.

Lyon, D. N. (1965). Peak nucleate boiling fluxes and nucleate boiling heat transfer coefficients in saturated liquid helium between the $\lambda$ and critical temperatures. *Adv. Cryo. Eng.* 10.

Sandler, S. I. (1977). *Chemical and Engineering Thermodynamics*. New York: Wiley.

Smith, J. M., Van Ness, H. C. (1975). *Introduction to Chemical Engineering Thermodynamics*. 3rd ed. New York: McGraw-Hill.

Weil, L. (1951). Heat transfer coefficients of boiling liquified gases. In: Proceedings of the Eighth International Congress on Refrigeration. London.

Zemansky, M. W. (1957). *Heat and Thermodynamics*. 4th ed. New York: McGraw-Hill.

Zuber, N., Fried, E. (1962). Two-phase flow and boiling heat transfer to cryogenic liquids. *ARS J.* 32:1332–1341.

## CHAPTER 3: SUGGESTED READING

Arp, V. (1998). A Summary of fluid properties including near-critical behavior. In: Proceedings of the International Conference on Cryogenics and Refrigeration, held April 21–24, International Academic Publishers, pp. 387–392.

Gershman, R., Sherman, A. I. (1979). *Fluid Properties Handbook*. NBS No. 48307, R-2732, 176–364.

Mann, D. B. (1962a). The thermodynamic properties of helium from 3 to 300 K between 0.5 and 100 Atmospheres. NBS Technical Note No. 154, PB 172217. US Government Printing Office, pp. 95.

Mann, D. B. (1962b). The thermodynamic properties of helium from 6 to 540°R between 10 and 1500 psia. NBS Technical Note No. 154A, PB 182435. US Government Printing Office. 89 pages.

McCarty, R. D. (1973). Thermodynamic properties of helium-4 from 2 to 1500 K with pressures to $10^8$ Pa. *J. Phys. Chem. Ref. Data* 2(4):923–1041.

McCarty, R. D., Weber, L. A. (1972). Thermophysical properties of para hydrogen from the freezing liquid line to 5000 R for pressures to 10,000 psia. NBS Technical Note No. 617. US Government Printing Office, 169 pages.

Roder, H. M., Weber, L. A., Goodwin, R. D. (1965). Thermodynamic and related properties of parahydrogen from the triple point to 100 K at pressures to 240 Atmospheres. NBS Monograph 94, 110 pages.

Storvick, T. S., Sandler, S. I. (1977). Phase equilibria and fluid properties in the chemical industry estimation and correlation. In: ACS Symposium Series, No. 60, p. 550.

Strobridge, T. R. (1962a). The thermodynamic properties of nitrogen from 64 to 300 K between 0.1 and 200 Atmospheres. NBS Technical Note No. 129, PB 161630. US Government Printing Office, 85 pages.

Strobridge, T. R. (1962b). The thermodynamic properties of nitrogen from 114 to 540°R between 0.1 and 3000 psia. Supplement A (British units) NBS Technical Note No. 129A. US Government Printing Office, 85 pages.

Weber, L. A. (1975). Thermodynamic and related properties of parahydrogen from the triple point to 300 K at pressures to 1000 bar. NASA Spec. Publ. No. SP-3088 and NBS Interagency Rep. No. NBSI R 74–374, NASA-SP-3088, 105 pages.

Zudkevitch, D., Gray, R. D. Jr. (1975). Impact of fluid properties on the design of equipment for handling LNG. *Adv. Cryog. Eng.* 20:103–123.

## CHAPTER 3: REFERENCES

Arp, V., Kropschot, R. H. (1962). Helium. In: Vance, R. W., ed. *Applied Cryogenic Engineering*. New York: John Wiley and Sons, Inc.

Baly, E. C. C., Donnan, F. G. (1902). *J. Chem. Soc.* 81:907.

Berman, R., Mate, C. R. (1958). *Phil. Mag.* 8(3):461–469.

Brickwedde, F. G., van Dijk, H., Durieux, M., Clement, J. R., Logan, J. K. (1960). *J. Res. Natl. Bur. Stand.* 64A:1.

Din, F. (1956). *Thermodynamic Functions of Gases*. 2nd ed. London: Butterworths.

Elverum, G. W., Doescher, R. N. (1952). *J. Chem. Phys.* 20:1834–1836.

Franck, E. U., Stober, (1952). *Zeits für Naturforschung* 7A:822.

Frank, E. U., Wicke, E. (1951). *A. Elektrochem* 55:636–643.

Fricke, E. F. (1948). Report No. F-5028 101 ATI 121 150. Farmingdale, NY: Republic Aviation Corp.

Furukawa, G. T., McCoskey, R. E. (1953). NACA Technical Note 2969.

Gerold, E (1951). *Ann. Physik* 65:82–96.

Grenier, C. (1951). *Phys. Rev.* 83(3):589–603.

Hilsenrath, J., et al. (1955). NBS Circular 564. Washington, DC: US Government Printing Office, p. 424.

Hoge, H. J. (1950). *J. Res. Natl. Bur. Stand.* 44:321–345.

Hoge, H. J., King, G. K. (1950). *NACA Tables of Thermal Properties of Gases*. Table 11.50. Washington, DC: National Bureau of Standards. US Government Printing Office.

Hu, J. H., White, D., Jophnston, H. L. (1953). *J. Am. Chem. Soc.* 75:5642–5645.

Jacobsen, R. T., Stewart, R. B., McCarty, R. D., Hanley, H. J. M. (1973). Thermophysical properties of nitrogen from the fusion line to 3500 R (1944 K) for pressures to 150,000 psia ($10342 \times 10^5$ N/m$^2$). NBS Technical Note 648. Washington, DC: US Government Printing Office.

Jarry, R. L., Miller, H. C. (1956). *J. Am. Chem. Soc.* 78:1552–1553.

Jelatis, J. G. (1948). *J. Appl. Phys.* 19:419–425.

(1960). Johnson, V. J., ed. *A Compendium of the Properties of Materials at Low Temperatures (Phase I)*. Part 1, Properties of fluids. WADD Technical Report 60–56, 874 pages.

Johnson, H. L., McCloskey, K. E. (1940). *J. Phy. Chem.* 44(9):1038–1058.

Keesom, W. H. (1922). *Onnes-Festschrift*. pp. 89–163.

Keesom, W. H. (1923). *Physik Ber* 4:613.

Kestin, J., Wang, H. E. (1958). *Trans. ASME* 80:11–17.

Lounasmaa, O. U. (1958). Thesis submitted for the degree of Doctor of Philosophy. England: University of Oxford.

Mathias, E., Crommelin, C. A. (1924). In: Proceedings of the Fourth International Congress of Refrigeration 1, pp. 89–106a.

Mathias, E., Onnes, H. K. (1911). *Commun. Phys. Lab. No. 117*. The Netherlands: University of Leiden.

McCarty, R. D. (1972). Thermophysical properties of helium-4 from 2 to 1500 K with pressures to 1000 atmospheres. NBS Technical Note 631. Washington, DC: US Government Printing Office.

McCarty, R. D., Roder, H. M. (1981). Selected properties of hydrogen (engineering design data). NBS Monograph 168. Washington, DC: US Government Printing Office.

McLennan, J. C., Jacobsen, R. C., Wilhelm, J. O. (1930). *Trans. Roy. Soc. Can.* 24(3):37–46.

Millar, R. W., Sullivan, J. D. (1928). Bureau of Mines Technical Paper 424.

Mills, R. L., Grilley, E. R. (1955). *Phys. Rev.* 99(2):480–486.

Roder, H. M., Weber, L. A. (1972). Thermophysical properties. *ASRDI Oxygen Technology Survey*. Vol. I. NASA SP-3071.

Rossini, F. D., et al. (1952). NBS Circular 500. Washington, DC: US Government Printing Office.

Rudenko, N. S. (1939). *J. Exptl. Theoret. Phys.* (USSR) 9:1078.

Rudenko, N. S., Shubnikow, L. V. (1934). *Physik. Z. Sowjetunion* (USSR), 6:470–477.
Schmidt, H. W., Forney, D . E. (1975). Oxygen systems engineering review. *ASRDI Oxygen Technology Survey*. Vol. IX. NASA SP-3090.
Simon, F., Lange, F. (1923). *Z. Physik* 15:312.
Van Itterbeek, A. (1955). *Progress in Low Temperature Physics*. Vol. 1. Amsterdam, The Netherlands: North Holland Publishing Company.
Verschaffelt, J. E. (1929). *Commun. Kammerlingh Onnes Lab.* Supplement No. 64d. The Netherlands: University of Leiden.
Wooley, H. W., Scott, R. B., Brickwedde, F. G. (1948). *J. Res. Natl. Bur. Stand.* 41:379.

## CHAPTER 4: SUGGESTED READING

Ho, C. Y., Li, H. H. (1986). Computerized comprehensive numerical data system on the thermophysical and other properties of materials established at Cindas/Purdue University. *Int. J. Thermophy's.* 7(4):949–962.
Hust, J. G., Kirby, R. K. (1978). Standard reference materials for thermophysical properties. *Adv. Cryo. Eng.* Vol. 24, Plenum Press, pp. 232–239.
Vasiliev, L. L. et al. (1977). Thermophysical properties of composite polymer materials in cryogenic techniques. In: International Cryogenic Materials Conference, No. 18.

## CHAPTER 4: REFERENCES

ASME Boiler and Pressure Vessel Code (1983). *Sec. VIII, Rules for Construction of Pressure Vessels.* New York: American Society of Mechanical Engineers.
Barrett, C. S. (1957). *Trans. Am. Soc. Metal* 49:53.
(1978). In: Baumeister, T., ed. *Mark's Standard Handbook for Mechanical Engineers*. 8th ed. New York: McGraw-Hill.
Brick, R. M. (1953). *Behavior of Metals at Low Temperatures*. Cleveland, OH: American Society for Metals (ASM).
Campbell, J. E. (1974). *Handbook on Materials for Superconducting Machinery* (MCIC-HB-04). Columbus, OH: Metals and Ceramics Information Center, Battelle Memorial Institute.
Corruccini, R. J. (1957). *Chem. Eng. Prog.* 53:397.
Cottrell, A. H. (1953). *Dislocations and Plastic Flow in Crystals*. London: Oxford University Press.
Durham, T. F., McClintock, R. M., Reed, R. P. (1962). *Cryogenic Materials Data Handbook* (PB 171809), Air Force Contract AF04 (647)-59-3, NBS Cryogenic Engineering Laboratory. Office of Technical Services. Washington, DC: US Department of Commerce.
Fisher, J. C., ed. (1957). *Dislocations and Mechanical Properties of Crystals*. New York: John Wiley and Sons, Inc., 634 pages.
Johnson, V. J., ed. (1960). *A Compendium of the Properties of Materials at Low Temperatures (Phase I)*. Part II, Properties of solids, WADD Technical Report 60–56.
Jones, G. O. (1956). *Glase*. New York: John Wiley and Sons, Inc.
Kropschot, R. H., Mikesell, R. P. (1957). *J. Appl. Phys.* 28:610.
Read, R. P. and Clark, A.F. (1983). *Materials at Low Temperatures*. Metals Park, OH: American Society for Metals, 590 pages.
McClintock, R. M., Gibbons, H. P. (1960). Mechanical properties of structural materials at low temperatures. NBS Monograph 13, Washington, DC: US Government Printing Office.
Quarrel, A. G. (1959). *The Structure of Metals*. New York: Interscience Publishers, Inc.

Teed, P. L. (1950). *The Properties of Metallic Materials at Low Temperatures.* New York: John Wiley and Sons, Inc.
Van Buren, H. G. (1960). *Imperfections in Crystals.* New York: Interscience Publishers, Inc.
Weitzel, D. H., Robbins, R. F., Copp, G. R., Bjorklund, W. R. (1978). Elastomers for static seals at cryogenic temperatures. *Advances in Cryogenic Engineering.* 6th ed. New York: Plenum Press.
Weitzel, D. H., Robbins, R. F., Copp, G. R., Bjorklund, W. R. (1960). *Rev. Sci. Instrum.* 31:1350.

## CHAPTER 5: SUGGESTED READING

Dillard, D. S. (1968). Thermal transport properties of selected solids at low temperatures. In: Chemical Engineering Progress Symposium Series. Vol. 64. p. 87.
Dillard, D. S. (1979). Thermal transport properties of selected solids at low temperatures. In: *AICHE Symposium on Advances in Cryogenic Heat Transfer.*
Johnson, V. J., Diller, D. E. (1970). *Thermodynamic and Transport Properties of Fluids and Selected Solids for Cryogenic Applications.* Rept. No. 9782. Boulder, CO: National Bureau of Standards, 57 pages.
(1959). In: Touloukian, Y.S. ed. Thermodynamic and transport properties of gases, liquids and solids. In: *ASME Symposium on Thermal Properties,* 472 pages.

## CHAPTER 5: REFERENCES

Bardeen, J., Cooper, L. N., Schrieffer, J. R. (1957). Theory of superconductivity. *Phys. Rev.* 108:1175–1204.
Corruccini, R. J., Gniewek, J. J. (1960). Specific heats and enthalpies of technical solids at low temperatures. NBS Monograph 21. Washington, DC: US Government Printing Office.
Corruccini, R. J., Gniewek, J. J. (1961). Thermal expansion of technical solids at low temperatures. NBS Monograph 29. Washington, DC: US Government Printing Office.
Crittenden, E. C., Jr. (1963). Properties of solids and liquids. *Cryogenic Technology.* New York: John Wiley and Sons, Inc.
Dillard, D. S., Timmerhaus, K. D. (1968). *Chem. Eng. Prog.*, Symp. Ser. Vol. 64, No. 87, p. 1.
File, J., Mills, R. G. (1963). *Phys. Rev. Lett.* 10:93.
Fulk, M. M. (1959). *Progress in Cryogenics.* I. London: Heywood, 65.
Gopal, E. S. R. (1966). *Specific Heats at Low Temperatures.* New York: Plenum Press.
Gruneisen, E. (1926). The state of a solid body. Trans. from *Handbuch der Physik.* Vol. 10.
Johnson, V. J., ed. (1960). *A Compendium of the Properties of Materials at Low Temperatures (Phase I).* Part II, Properties of Solids. WADD Technical Report 60–56.
Landoldt-Bornstein (1961). *Zahlenwerte and Funktionen.* Vol. 2. Pt. 4, 6th ed. Berlin: Springer.
Lynton, E. A. (1969). *Superconductivity.* New York: John Wiley and Sons, Inc.
Onnes, K. H. (1911). *The Resistance of Pure Mercury at Helium Temperature.* Leiden Commun. No. 120b. The Netherlands: University of Leiden.
Powell, R. L. (1963). *American Institute of Physics Handbook.* 2nd ed. New York: McGraw-Hill, pp. 4–76.
Powell, R. W., Ho, C. Y., and Liley, P. E. (1966). Thermal conductivity of selected materials. NSRDS-NBS. 8th ed. Washington, DC: US Government Printing Office.
Schmidt, A. F. (1968). *Technical Manual of Oxygen/Nitrogen Systems.* NAVAIR 06-30-501.
Silsbee, F. B., Jr. (1916). Electric conduction in metals at low temperatures. *J. Wash. Acad. Sci.* 6.

Touloukian, Y. S., ed. (1964). Metallic elements and their alloys. *Thermophysical Properties Research Center Data Book*. I. Purdue University Press.

White, G. K. (1959). *Experimental Techniques in Low Temperature Physics*. Oxford, England: Clarendon Press.

Zemansky, M. W. (1957). *Heat and Thermodynamics*. New York: McGraw-Hill.

## CHAPTER 6: SUGGESTED READING

### Refrigeration and Liquefaction

Gschneidner, Jr., K. A., et al. (1994). *New magnetic refrigeration materials for the liquefaction of hydrogen*. Advances in Cryogenic Engineering. Vol. 39. London: Plenum Press, pp. 1457–1465.

Ishizuka, M., et al. (1975). New type screw compressor for helium refrigerators and liquefiers. *Cryo. Eng. (Tokyo)* 10(4):134–139.

Kakimi, Y., et al. (1997). Pulse-tube refrigerator and nitrogen liquefier with active buffer system. Availability: approved for public release; distribution is unlimited. *Cryocoolers*. Vol. 9. London: Plenum Press, pp. 247–253.

Kanazawa, M. (1993). Refrigeration by a small liquefaction plant. *Cryo. Eng. (Japan)* 28:9–16.

Matsumoto, K., et al. (1996). Improvement of the performance of helium liquefaction system on dilution refrigerator using Gm Precooled Jt expansion refrigerator with magnetic regenerator. *Cryo. Eng. (Japan)* 31(4):86–92.

Nagao, M., et al. (1989). Helium liquefaction by Gifford-mcmahon cycle cryogenic refrigerator. *Cryo. Eng. (Japan)* 24(4):34–39.

Narinsky, G. B., et al. (1995). Influence of thermodynamic parameters on main characteristics of helium liquefaction and refrigeration plant. *Cryogenics* 35(8): 483–487.

Ohira, K., Nakamichi, K., Furimoto, H. (2000). Experimental study on magnetic refrigeration for the liquefaction of hydrogen. *Advances in Cryogenic Engineering*. Vol. 45, Kluwer Academic/Plenum Publishers, pp. 1747–1754.

### Cryocoolers

Collins, S. A., Paduano, J. D. (1994). Multi-axis vibration cancellation for Stirling cryocoolers. *Proc. SPIE* 2227:145–155.

Curran, D. G. T., et al. (2000). Cryocooler state of the art for space-borne applications. Advances in Cryogenic Engineering. Vol. 45. Kluwer Academic/Plenum Publishers, pp. 585–594.

Gao, C. M., et al. (2000). Study on a pulse tube cryocooler using gas mixture as its working fluid. *Cryogenics* 40(7):475–480.

Gong, M. Q., et al. (2000). Prediction of transport for multicomponent cryogenic mixtures used in J–T cryocoolers. *Advances in Cryogenic Engineering*. Vol. 45. Kluwer Academic/Plenum Publishers, pp. 1159–1165.

Harvey, J. P., Kirkconnell, C. S., Desai, P. V. (2000). Regenerator performance evaluation in a pulse tube cryocooler. *Advances in Cryogenic Engineering*. Vol. 45, Kluwer Academic/ Plenum Publishers, pp. 373–381.

Levenduski, R. L., Scarlotti, R. (1994). *Development of a cryocooler for space applications*. Proc. SPIE 2227, 109–126.

Luo, E. C., et al. (2000). *The research and development of cryogenic mixed-refrigerant Joule–Thomson cryocoolers*. Advances in Cryogenic Engineering. Vol. 45. Kluwer Academic/Plenum Publishers, pp. 299–306.

Tomlinson Jr., B. J., et al. (2000). *Air force research laboratory cryocooler characterization status and lessons learned*. Advances in Cryogenic Engineering. Vol. 45. Kluwer Academic/Plenum Publishers, pp. 595–601.

Tomlinson Jr., B. J., et al. (2000). *Air force research laboratory spacecraft cryocooler endurance evaluation.* Advances in Cryogenic Engineering. Vol. 45. Kluwer Academic/Plenum Publishers, pp. 609–616.

Unger, R. Z., Wood, J. G. (2000). Performance comparison of M77 Stirling cryocooler and proposed pulse tube cryocooler. *Advances in Cryogenic Engineering*. Vol. 45, Kluwer Academic/Plenum Publishers, pp. 539–544.

## CHAPTER 6: REFERENCES

Barclay, J. A., Steyert, W. A., Zrudsky, D. R. (1979). Proceedings of the XVth International Congress of Refrigeration.

Brown, G. V. (1976). *J. Appl. Phys.* 47(8):3673–3680.

Brown, G. V. (1979). *Phys. Today*, June 18.

Gifford, W. E., Longsworth, R. C. (1964). Pulse tube refrigeration progress. *International Advances in Cryogenic Engineering*. Vol. 10. Pt. 2. New York: Plenum Press (reprinted from Proceedings of the 1964 Cryogenic Engineering Conference. Philadelphia, PA).

Kirk, A. G. (1874). *Min. Proc. Inst. Civ. Eng.* 37:244–315.

Kohler, J. W. H. (1956). Gas refrigerating machines. *Advances in Cryogenic Engineering*. Vol. 2. New York: Plenum Press, (1960) (reprinted from Proceedings of the 1956 Cryogenic Engineering Conference. NBS, Boulder, CO).

Kohler, J. W. H. (1959). Refrigeration below −100°C. *Advances in Cryogenic Engineering*. Vol. 5. New York: Plenum Press (1960) [reprinted from Proceedings of the 1959 Cryogenic Engineering Conference. University of California, Berkeley].

London, H. (1951). Proceedings of the International Conference on Low Temperature Physics. Oxford, England, p. 157.

Longsworth, R. C. (1967). An experimental investigation of pulse tube refrigeration heat pumping rates. *Advances in Cryogenic Engineering*. Vol. 12. New York: Plenum Press.

Mann, D. B. (1962). The thermodynamic properties of helium from 3 to 300 K between 0.5 and 100 atmospheres. NBS Technical Note 154. Washington DC: US Government Printing Office.

Merkli, P., Thomann, H. (1975). Thermoacoustic effects in a resonance tube. *J. Fluid Mech.* 70:161.

Mikulin, E. I., Tarason, A. A., Shkrebyonock, M. P. (1984). Low-temperature expansion of pulse tubes. *Advances in Cryogenic Engineering*. 29. New York: Plenum Press.

Pratt, W. P., Jr., Rosenblum, S. S., Steyert, W. A., Barclay, J. A. (1997). A continuous demagnetization refrigerator operating near 2 K and a study of magnetic refrigerants. *Cryogenics* 17:689.

Radebaugh, R. (1977). Refrigeration Fundamentals: A view toward new refrigeration systems, Applications of closed-cycle cryocoolоers to small superconducting applications. NBS Special Publications 508. Washington, DC: US Government Printing Office.

Radebaugh, R. (1986). A comparison of three types of pulse tube refrigerators: new methods for reaching 60 K. *Advances in Cryogenic Engineering*. 31. New York: Plenum Press.

Radebaugh, R. (1990). A review of pulse tube refrigeration. *Advances in Cryogenic Engineering*. 35. New York: Plenum Press.

Sato, T. 1968. Some experiments on the reciprocating expander for large He liquefiers. Proceedings of the 1968 Cryogenic Engineering Conference. Tokyo, Japan Heywood-Temple. London: pp. 217–218.

Sloane, T. O. (1900). *Liquid Air and the Liquefaction of Gases: Theory, History, Biography, Practical Applications, Manufacture.* 2nd ed. New York: Henley Press.

Solvay, E. (1887). Deutches Reichspatent. No. 39, 280.

Steyert, W. A. (1978a). Applications of closed-cycle cryocoolers to small superconducting devices. *J. Appl. Phys.* 49:1227.

Steyert, W. A. (1978b). *J. Appl. Phys.* 49:1216.

Strobridge, T. R. (1974). Cryogenic Refrigerators—an Updated Summary. NBS Technical Note 655. Washington, DC: US Government Printing Office.

Van Geuns, J. R. (1966). *Phillips Res. Rep. Suppl.* No. 6, Eindhoven, Netherlands.

Walker, G. (1983). *Cryocoolers*. Vol. I. New York: Plenum Press.

Wheatley, J., Hafler, T., Swift, G. W., Migliori, A. (1983). An intrinsically irreversible thermoacoustic heat engine. *J. Acoust. Soc. Am.* 74:153.

Wheatley, J., Hafler, T., Swift, G. W., Migliori, A. (1985). Understanding some simple phenomena in thermoacoustics with applications to acoustical heat engines. *Am. J. Phys.* 53:147.

Zemansky, M. W. (1957). *Heat and Thermodynamics*. 4th ed. New York: McGraw-Hill.

## CHAPTER 7: SUGGESTED READING

Augustynowicz, S. D., Fesmire, J. E. (2000). *Cryogenic insulation system for soft vacuum. Advances on Cryogenic Engineering*. Vol. 45, Kluwer Academic/Plenum Publishers, pp. 1691–1698.

Blevins, E., Sharpe, J. (1995). Water blown urethane insulation for use in cryogenic environments. NTIS N95-31765.

Celik, E., Schwartz, J. (1999). Evaluation of adhesion strength of sol–gel ceramic insulation for Hts magnets. *IEEE Trans. Appl. Superconductivity* 9(2):1916–1919.

Heaney, J. B. (1998). Efficiency of aluminized Mylar insulation at cryogenic temperatures. *Proc. SPIE* 35:150–157.

Kamiya, S., et al. (2000). Thermal test of the insulation structure for Lh2 tank by using the large experimental apparatus. *Cryogenics* 40(11):737–748.

Kosaki, M., et al. (1998). Solid insulation and its deterioration. *Cryogenics* 38(11): 1095–1104.

Krishnaprakas, C. K., et al. (2000). Heat transfer correlations for multilayer insulation systems. *Cryogenics* 40(7):431–435.

Kumar, A. S., et al. (2000). Thermal performance of multilayer insulation down to 4.2 *Advances in Cryogenic Engineering*. Vol. 45. Kluwer Academic/Plenum Publishers, pp. 1675–1682.

## CHAPTER 7: REFERENCES

Black, I. A., Glaser, P. E. (1966). *Advances in Cryogenic Engineering*. Vol. 11. New York: Plenum Press, 26.

Corruccini, R. J. (1957). *Chem. Eng. Prog.* 53: 6:262–267; 7:342–346; 8:397–402.

Corruccini, R. J. (1958). Gaseous heat conduction at low pressure and temperature. *Vacuum* 7:8, 19–29.

Fulk, M. M. (1959). Evacuated powder insulation for low temperatures. *Progress in Cryogenics*. Vol. 1. London: Heywood and Co.

Hunter, B. J., Kropschot, R. H., Schrodt, J. E., Fulk, M. M. (1960). Metal powder additives in evacuated powder insulation. *Advances in Cryogenic Engineering*. Vol. 5. New York: Plenum Press.

Knudsen, M. (1910). *Ann. D. Physik* 31:205–229; 32:809–842; 33:1435.

Knudsen, M. (1930). *Radiometer Pressure and Coefficient of Accommodation*. Copenhagen, Denmark: A. F. Hast and Son.

Kropschot, R. H. (1959). *ASHRAE J.* 1:48–54.

Kropschot, R. H. (1963). Insulation technology. In: Vance, R. W., ed. *Cryogenic Technology*. New York: John Wiley and Sons, Inc.

Lindquist, C. R., Niendorf, L. R. (1963). *Advances in Cryogenic Engineering*. Vol. 8. New York: Plenum Press, p. 398.

Partington, J. R. (1952). *Advanced Treatise on Physical Chemistry*. Vol. I. New York: Longmans, Green and Co., Inc.

Petersen, P. (1958). Swedish technical research council report no. 706 (1951). *Sartryck Ur TVF* 29(4):151.
Stoy, S. T. (1960). Cryogenic insulation development. *Advances in Cryogenic Engineering*. Vol. 5. New York: Plenum Press.
Timmerhaus, K. D., Flynn, T. M. (1989). *Cryogenic Process Engineering*. New York: Plenum Press.

## CHAPTER 8: CRYOGENIC INSTRUMENTATION SUGGESTED READING

### Thermometers

Balle, Ch. (2000). Influence of thermal cycling on cryogenic thermometers. *Advances in Cryogenic Engineering*. Vol. 45, Kluwer Academic/Plenum Publishers, pp. 1817–1823.
Phillips, R. W. (2000). Approximating the resistance–temperature relationship of platinum resistance thermometers from 20 K to 273 K. *Advances in Cryogenic Engineering*, Kluwer Academic/Plenum Publishers, pp. 1809–1815.
Safrata, R. S., et al. (1980). Deuterized cerium lanthanum magnesium nitrate as a magnetic thermometer. *J. Low Temp. Phys.* 41(3/4):405–407.
Soulen, Jr., et al. (1980). A self-calibrating rhodium–iron resistive squid thermometer for the range below 0.5 K. *J. Low Temp. Phys.* 40(5/6):553–569.

### Liquid level

Haraguchi, K. (1992). Measurement of flow-rate, liquid level and stress–strains. *Cryo. Eng. (Japan)* 27(5):10–15.
Karunanithi, R., et al. (2000). Development of discrete array type liquid level indicator for cryogenic fluids. *Advances in Cryogenic Engineering*. Vol. 45. Kluwer Academic/Plenum Publishers, pp.1803–1808.

## CHAPTER 8: REFERENCES

Alspach, W. J., Flynn, T. M. (1965). Considerations when using turbine type flowmeters in cryogenic service. *Advances in Cryogenic Engineering.* Vol. 10. New York: Plenum Press, pp. 246–252 (reprinted from Proceedings of the 10th Cryogenic Engineering Conference. Philadelphia, PA, August 1964).
Alspach, W. J., Miller, C. E., Flynn, T. M. (1966). Mass flowmeters in cryogenic service. Flow Measurement Symposium Proceedings. ASME Flow Measurement Conference, Pittsburgh, PA, September.
Arvidson, J. M., Brennan, J. A. (1975). Pressure measurement. ASRDI oxygen technology survey. Vol. VIII. NASA SP-3092.
ASME (1971). *Fluid Meters, Their Theory and Applications.* 6th ed. New York: ASME.
Blakemore, J. S., Winstel, J., Edwards, R. V. (1970). Computer fitting of germanium thermometer characteristics. *Rev. Sci. Instrum.* 41(6):835–842.
Brennan, J. A., Dean, J. W., Mann, D. B., Kneebone, C. H. (1971). An evaluation of positive displacement cryogenic volumetric flowmeters. NBS Technical Note 605. Washington, DC: US Government Printing Office.
Brennan, J. A., Stokes, R. W., Kneebone, C. H., Mann, D. B. (1974). An evaluation of selected angular momentum, vortex shedding and orifice cryogenic flowmeters. NBS Technical Note 650. Washington, DC: US Government Printing Office.
Callendar, H. L. (1887). On the practical measurement of temperature. *Phil. Trans. Roy. Soc. London* 178:160–230.

Clement, J. R., Quinnel, E. H. (1952). *Rev. Sci. Instrum.* 23:213–216.

Close, D. L. (1968). Flow measurement of cryogenic fluids. *Instrum. Control Syst.* 41(2):109–114.

Close, D. L. (1969). Cryogenic flowmeters. *Cryogenics and Industrial Gases.* August.

Collier, R. S., Sparks, L. L., Strobridge, T. R. (1973). Carbon thin film thermometry. NBS Information Report 74-355. Boulder, CO: National Bureau of Standards.

Crawford, R. B. (1963). A broad look at cryogenic flow measurement. *Instrum. Soc. Am. J.* 10(6):65–72.

Dike, P. H. (1954). *Thermoelectric Thermometry.* 1st ed. Philadelphia, PA: Leeds and Northrup Company.

Droms, C. R. (1962). Thermistors for temperature measurements. *Temperature—Its Measurement and Control in Science and Industry 3.* Part 2. New York: Reinhold Publishing Corporation, pp. 339–346.

Fox, A. (1967). Cryogenic flow measurement for custody transfer. ASME Winter Annual Meeting. Pittsburgh, PA, November 12–17. National Cylinder Gas Division, Chemetron Corp. Chicago: ILL.

ISA (1970). *Standard Practices for Instrumentation.* Pittsburgh, PA: Instrument Society of America.

ISO Recommendation R-541 (1967). *Measurement of Fluid Flow by Means of Orifice Plates and Nozzles.*

Johnson, V. J., ed. (1960). A compendium of the properties of materials at low temperature, Phase I. Part I: Properties of Fluids. WADD Technical Report 60-56.

Jones, H. B. (1963). Transient pressure measuring methods. Aerospace Engineering Report No. 595a, Princeton, NJ: Princeton University.

Krause, J. K., Swinehart, P. R. (1985). Reliable wide range divide thermometry. *Advances in Cryogenic Engineering* (reprinted from Proceedings of the 1985 Cryogenic Engineering Conference).

Lawless, W. N. (1972). Thermometric properties of carbon-impregnated porouos glass at low temperatures. *Rev. Sct. Instrum.* 43(12):1743–1747.

Mann, D. B. (1974). Flow measurement instrumentation. ASRDI oxygen technology survey. Vol. VI. NASA SP-3084.

(1973). *Measurements and Data 7* No. 2, pp. 101–112.

Powell, R. L., Bunch, M. D., Caywood, L. P. (1961). *Advances in Cryogenic Engineering.* Vol. 6. New York: Plenum Press, 537.

Powell, R. L., Hall, W. J., Hyink, C. H., Sparks, L. L., Burns, G. W., Scroger, M. G., Plumb, H. H. (1974). Thermocouple reference tables based on IPTS-68. NBS Monograph 125. Boulder, CO: National Bureau of Standards.

Rao, M. G. (1982). Semiconductor junctions as cryogenic temperature sensors. *Temp. Amer. Inst. Phys.* 5:1205–1211.

(1983). In: Reed, R. P., Clark, A. F., eds. *Materials at Low Temperatures.* Metals Park, OH: American Society of Metals.

Richards, R. J., Jacobs, R. B., Pestalozzi, W. J. (1960). Measurement of the flow of liquefied gases with sharp-edged orifices. *Advances in Cryogenic Engineering.* Vol. 4. New York: Plenum Press, pp. 272–285 (reprinted from Proceedings of the 4th Cryogenic Engineering Conference. August 1958).

Ricketson, B. W., Grinter, R. (1982). Carbon-glass sensors: reproducibility and polynomial fitting of temperature vs. resistance. *Temp. Amer. Inst. Phys.* pp. 845–851.

Roder, H. M. (1974). Density and liquid level measurement instrumentation for the cryogenic fluids oxygen, hydrogen and nitrogen. ASDRI oxygen technology survey. Vol. V. NASA SP-3083.

Rodley, A. E. (1969). Vortex shedding flowmeter. *Measurements and Data 18.* November–December.

Rubin, L. G. (1970). Cryogenic thermometry: a review of recent progress. *Cryogenics,* Feb. 1970, 14–22.

Rubin, L. G., Brandt, B. L., Sample, H. H. (1982a). Cryogenic thermometry: a review of recent progress II. *Cryogenics* 22(10):491–503.

Rubin, L. G., Brandt, B. L., Sample, H. H. (1982b). Cryogenic thermometry: a review of recent progress II. *Temp. Amer. Inst. Phys.* Vol. 5, Part II, pp. 1333–1344.

Sachse, H. B. (1962). Measurement of low temperatures with thermistors. *Temperature—Its Measurement and Control in Science and Industry 3.* Pt. 2. New York: Reinhold Publishing Corporation. pp. 347–353.

Scott, R. B. (1959). *Cryogenic Engineering*. Princeton, NJ: D. Van Nostrand Company.

Scott, R. B. (1960). Proceedings of the 6th ISA National Flight Test Instrumentation Symposium. San Diego, CA, May 2–5.

Simon, F. E. (1952). *Low Temperature Physics, Four Lectures*. New York: Academic Press.

Smelser, P. (1963). Pressure measurement in cryogenic systems. *Advances in Cryogenic Engineering*. Vol. 8. New York: Plenum Press, pp. 378–386.

Sparks, L. L. (1974). Low temperature measurement. ASRDI oxygen technology survey. Vol. IV. NASA SP-3073.

Sparks, L. L. (1983). Temperature, strain, and magnetic field measurements. *Materials at Low Temperature*. Metals Park, OH: American Society for Metals.

Sparks, L. L., Powell, R. L. (1973). Low temperature thermocouples KP 'Normal' silver, copper versus Au–0.02 at.% Fe and Au–0.07 at.% Fe. NBS Technical Note 76A. Washington, DC: US Government Printing Office, pp. 263–283.

Swartz, J. M., Gaines, J. R., Rubin, L. G. (1975). Magneto-resistance of carbon-glass thermometers at liquid helium temperatures. *Rev. Sci. Instrum.* 46(9):1177–1178.

van Dijk, H. (1960). *Progress in Cryogenics*. Vol. 2. New York: Academic Press.

Van Dusen, M. J. (1925). Platinum-resistance thermometry at low temperatures. *J. Am. Chem. Soc.* 47:326.

Vaugh, H. (1954). The response characteristics of airplane and missile pressure measuring systems. Sandia Corporation, 174-54-51.

White, G. K. (1959). *Experimental Techniques in Low Temperature Physics*. Oxford, England: Oxford University Press.

## CHAPTER 9: REFERENCES

Barron, R.F. (1999). Cryogenic Heat Transfer. Philadelphia, PA: Taylor & Francis Publishers.

Daney, D. E. (1988). Cavitation in flowing superfluid helium. In: Proceedings of space cryogenics. Workshop. Plemum, MA, Butterworth & Co., Ltd., p. 132.

DiPirro, M. J., Kittel, P. (1988). The Superfluid on-orbit transfer (SHOOT) flight. *Adv. Cryo. Eng.* 33:893.

Edmonds, D. K., Hord, J. (1969). Cavitation in liquid cyrogenics. *Adv. Cryo. Eng.* 14:274.

(2004). *GPSA Engineering Data Book*. 12th ed. Gas Processors Suppliers Association, Tulsa OK: 74145, gpa@gasprocessors.com.

Jasinski, T., Stacy, W. D., Honkonen, S. C., Sixsmith, H. (1986). A generic pump/compressor design for circulation of cryogenic fluids. *Adv. Cryo. Eng.* 31:991.

Ludtke, P. R., Daney, D. E. (1988). Cavitation characteristics of a small centrifugal pump. *Cryogenics* 28:96.

Scott, R. B. (1959). *Cryogenic Engineering*. NY: Van Nostrand (the classic in the field, never-out of date).

## CHAPTER 9: SUGGESTED READING

### Design

Swartz, E. T. (1995). Efficient cryogenic design, a system approach. *J. Low Temp. Phys.* 101(1/2).

## Compressors

Asakura, H., et al. (2000). Performance of 80K turbo compressor system without LN2 cooling for high reliable and efficient helium refrigerator. *Advances in Cryogenic Engineering*. Vol. 45. Kluwer Academic/Plenum Publishers.

Decker, L., et al. (1997). A cryogenic axial–centrifugal compressor for superfluid helium refrigeration. In: Sixteenth International Cryogenic Engineering Conference/International Cryogenic Materials Conference. Elsevier Science.

Stehrenberger, W. (1988). Cold compressor development. In: Twelfth International Cryogenic Engineering Conference (ICEC 12), held July 12–15, 1988, in Southampton, UK. Proceedings. Butterworth and Company.

Yamamura, H., et al. (1994). Measurements of impurity concentration for a helium compressor. *Cryo. Eng.* (Japan) Vol. 29, No. 1.

## Expanders

Claudet, G., Verdier, J. (1972). Simplified cryogenic reciprocating expansion Engine. In: Proceedings: ICEC4, IPC Business Press Ltd.

Danilov, I. B., et al. (1972). Two-stage expansion engine with differential piston. In: Proceedings: ICEC4. IPC Business Press Ltd.

Kaneko, J., et al. (1982). Performance of reciprocating expansion engine with electronic control valves. In: Proceedings of the Ninth International Cryogenic Engineering Conference, May 11–14, 1982, Kobe, Japan; Butterworth and Company.

Kobayashi, S. (1990). Experimental investigations on the reciprocating expansion engine. *Cryogenics* 30: September Supplement.

Patton, G., et al. (1982). Hydraulically controlled helium expansion engine. *Advances in Cryogenic Engineering*. Vol. 27. Plenum Press.

Von Bredow, H., Vogelhuber, W. W. (1971). Cryogenic expansion engine. US Patent 3,574,998 (April 13, 1971).

## Valves

Haycock, R. (1996). Remote controlled, stepper-motor-activated cryogenic valve: design, development and testing. *Proc. SPIE* 2814.

Haycock, R. (1996). Remote controlled, stepper-motor-activated cryogenic valve: design, development and testing. *AIAA Paper* 97-3315.

Hobbs, W., Kaufman, W. (1979). *Design, Development and Testing of Non-modulating Pressure Control Valves.* Vol. III. US Dept of Commerce, NIST, Cryogenic Technologies Group, Cryogenic Data Center.

Mills, G. L. (1991). Design and development of a leak tight helium II valve with low thermal impact. *Advances in Cryogenic Engineering*. Vol. 37, Plenum Press.

Ratts, E. B., Smith, Jr., J. L., Iwasa, Y. (1994). Design of a cold magnetically-actuated exhaust valve. *Advances in Cryogenic Engineering*. Vol. 39, Plenum Press.

Struzik, L. P. (1979). Design and testing of fire safe cryogenic Valves. US Dept of Commerce, NIST, Cryogenic Technologies Group, Cryogenic Data Center.

## Heat Exchangers

Alexeev, A. et al. (2000). Study of behavior in the heat exchanger of a mixed gas Joule–Thomson cooler. *Advances in Cryogenic Engineering*. Vol. 45. Kluwer Academic/Plenum Publishers.

Anthony, M. L., Greene, W. D. (1997). Analytical model of an existing propellant densification unit heat exchanger. AIAA/ASME/SAE/ASEE Joint Propulsion Conference and Exhibit, July 6–9.

Boyko, V., Siemensmeyer, K. (2001). Pt black powder as a heat exchanger at ultralow temperature. *J. Low Temp. Phys.* 122(3/4).

Darve, Ch. et al. (2000). A He II heat exchanger test unit designed for the LHC interaction region magnets. *Advances in Cryogenic Engineering*. Vol. 45. Kluwer Academic/Plenum Publishers.

Das, S. K., Roetzel, W. (1998). Second law analysis of a plate heat exchanger with an axial dispersive wave. *Cryogenics.* Vol. 38, No.8.

Liang, J. et al. (2000). Test of recuperative pulse tube refrigerator with simplified perforated plate heat exchanger. *Advances in Cryogenic Engineering*. Vol. 45. Kluwer Academic/ Plenum Publishers.

Pradeep Narayanan, S., Venkatarathnam, G. (1999). Performance of a counterflow heat exchanger with heat loss through the wall at the cold end. *Cryogenics.*Vol. 39, No. 1.

Venkatarathnam, G., Narayanan, S. P. (1999). Performance of a counter flow heat exchanger with longitudinal heat conduction through the wall separating the fluid streams from the environment. *Cryogenics.*Vol. 39, No. 10.

Yuen, W. W., Hsu, I. C. (1999). An experimental study and numerical simulation of two-phase flow of cryogenic fluids through micro-channel heat exchanger. *Cryocoolers*. Vol. 10 Kluwer Academic/Plenum Publishers.

Zhang, X. S. et al. (1998). The heat transfer characteristics of plate-fin heat exchanger in the field of refrigeration and air-conditioning. In: International Conference on Cryogenics and Refrigeration. International Academic Publishers.

## Transfer Lines

Haruyama, T. et al. (1996). Pressure drop of two-phase helium flowing in a large solenoidal magnet cooling path and a long transfer line. *Cryogenics*. Vol. 36, No. 6.

Hasan, A. R. et al. (2000). Modeling of cryogenic transfer line cool down. *Advances in Cryogenic Engineering*. Vol. 45. Kluwer Academic/Plenum Publishers.

Hosoyama, K. et al. (2000). Development of a high performance transfer line system. *Advances in Cryogenic Engineering*. Vol. 45. Kluwer Academic/Plenum Publishers.

Ng, Y. S., Lee, J. G. (1989). Analysis of dewar and transfer line cooldown in superfluid helium on-orbig transfer flight experiment (Shoot). Cryogenic Optical Systems and Instruments III Conference. *Proc. SPIE* 973.

## Cooldown

Bonney, G. E., Stubbs, D. M. (1994). Design fundamentals of rapid cooldown JT cryostats and sensors. *Proc. SPIE* 2227.

Commander, J. C., Schwartz, M. H. (1966). Cooldown of large-diameter liquid hydrogen and oxygen lines. Aerojet General Corporation Report 8800-54, NASA CR-54809.

## Two-Phase flow

Bernard, R. et al. (1997). Thermohydraulic behaviour of He II in stratified co-current two-phase flow. In: Sixteenth International Cryogenic Engineering Conference/Interna-International Cryogenic Materials Conference, May 20–24, 1996, in Kitakyushu, Japan; Proceedings, Elsevier Science.

Rousset, B. et al. (2000). He II two phase flow in an inclinable 22 m long line. *Advances in Cryogenic Engineering.* Vol. 45. Kluwer Academic/Plenum Publishers.

Preclik, D. (1992). Two-phase flow in the cooling circuit of a cryogenic rocket engine. American Institute of Aeronautics and Astronautics Library, AIAA Technical Information Service, 555 W. 57th St., Suite 1200, New York, NY 10019.

Wang, J. (1998). Analysis of the fluctuation characteristics of two-phase flow in exchanger of refrigerator. In: International Conference on Cryogenics and Refrigeration, April 21–24, 1998, Zhejian University, Hangzhou, China; Proceedings, International Academic Publishers.

Wang, J. et al. (1989). The void fraction measurement in the two phase flow of helium (4.2 K). In: International Conference on Cryogenics and Refrigeration, May 22–26, 1989, at Zhejian University, Hangzhou, China; Proceedings. International Academic Publishers.

Yuen, W. W., Hsu, I. C. (1999). An experimental study and numerical simulation of two-phase flow of cryogenic fluids through micro-channel heat exchanger. *Cryocoolers.* Vol. 10. Kluwer Academic/Plenum Publishers.

## Bearings

Bosson, R. et al. (1999). High performance cryogenic ball bearings demonstration. Presented at the 35th AIAA/ASME/SAE/ASEE Joint Propulsion Conference and Exhibit, June 20–24.

Ohta, T. et al. (1999). LH2 Turbopump test with hydrostatic bearing. Presented at the 35th AIAA/ASME/SAE/ASEE Joint Propulsion Conference and Exhibit. June 20–24.

Soyars, W. M., Fuerst, J. D. (2000). Fermilab cold compressor bearing lifetime improvements. *Advances in Cryogenic Engineering*. Vol. 45. Kluwer Academic/Plenum Publishers.

## Seals

Carlile, J. A. et al. (1993). Preliminary experimental results for a cryogenic brush seal configuration. Presented at the AIAA/SAE/ASEE Joint Propulsion Conference and Exhibit, 29th, held in Monterey, CA on June 28–30.

Haycock, R. H. et al. (1990). A compact indium seal for cryogenic optical windows. *Proc. SPIE* 1340.

Hendricks, R. C. et al. (1990). Brush seal configurations for cryogenic and hot gas applications. In: Advanced Earth-to-Orbit Propulsion Technology Conference held at NASA George C. Marshall Space Flight Center, Huntsville, AL on May 15–17.

Hendricks, R. C. et al. (1992). Development of advanced seals for space propulsion turbomachinery. NASA-TM-105659; E-7024.

Liu, L. Q., Zhang, L. (2000a). Study on a new type of sealing—regeneration-labyrinth sealing for displacer in cryocoolers: Part I—theoretical study. *Cryogenics.* Vol. 40, No. 2.

Oike, M. et al. (1995). Characteristics of a shaft seal system for the LE-7 liquid oxygen turbopump. Presented at the AIAA/ASME/SAE/ASEE Joint Propulsion Conference and Exhibit, 31st, held in San Diego, CA on July 10–12.

Proctor, M. P. (1993). Brush seals for cryogenic applications. In: Annual Propulsion Engineering Research Center Symposium, 5th, held at the Pennsylvania State University, University Park on September 8–9.

## CHAPTER 10: SUGGESTED READING

Arkharov, A. M. et al. (1998). Measurements of void fraction and flow rate of LNG flow. *Adv. Cryo. Eng.* 43:795–802.

Jurns, J. M. et al. (1998). Testing of a buried LNG tank. *Adv. Cryo. Eng.* 43:1215–1221.

Konishi, H., Teramoto, K. (1997). Natural gas gathering, liquefaction and ship transportation for thermal power generation. *Cryo. Eng. (Japan)* 32(3):6–12.

Marrucho, I. M., et al. (1994). An improved extended-corresponding-states theory for natural gas mixtures. *Int. J. Thermophys.* 15(6):1261–1269.

Shi, Y., Gu, A. (1998). The enthalpy and entropy in the LNG process. In: Proceedings of the International Conference on Cryogenics and Refrigeration. International Academic Publishers, pp. 111–114.

Suprunova, Z. A., Seriogin, V. E. (1998). Regime of uniform heating of liquid phase during static storage of liquefied natural gas. *Cryo. Eng.* 43:1223–1228.

Yoshimuta, T. (1997). LNG facilities for power plants availability: approved for public release. *Cryo. Eng. (Japan)* 32(3):13–22.

## CHAPTER 10: REFERENCES

Barber, H. W., Reed, E. E., Sharp, H. R. (1971). *Cryo. Ind. Gases.* Sept./Oct.

Bellus, M., Gineste, M. (1970). Proceedings of LNG-2. Session III, Paper 2, Paris (October).

Burgess, D. S., Murphy, J. N., Zabetakis, M. G. (1970). Hazards associated with the spillage of liquefied natural gas on water. US Bureau of Mines Report 7448.

Campbell, J. M. (1975). *Gas Conditioning and Processing 2.* Campbell Petroleum Series. J. M. Campbell and Co.

Chiu, C. H. (1978). Evaluate separation for LNG plants. Hydrocarbon Processing, September, pp. 266–272.

Clark, J. A. (1965). A review of pressurization, stratification, and interfacial phenomena. *International Advances in Cryogenic Engineering.* New York: Plenum Press.

Closner, J. J. (1968). Prestressed concrete storage tanks for LNG distribution. In: Conference Proceedings. American Gas Association, May 8.

Drake, E. M. (1976). LNG rollover-update. *Hydrocarbon Processing.* January.

Drake, E. M., Geist, J. M., Smith, K. A. (1973a). *Hydrocarbon Processing.* 87, March.

Drake, E. M., Geist, J. M., Smith, K. A. (1973b). Prevent LNG rollover. *Hydrocarbon Processing.* March.

Ffooks, R. C., Montagu, H. E. (1967). *Cryogenics.* December.

(1982). *Gas Processing Handbook. Hydrocarbon Processing.* Vol. 61, No. 4, p.86–165.

Grenier, M. R. (1971). La Centrale d'Oxygene de Fos-sur-Mer. In: XIII Congress of IIR. Washington, D.C.

Haynes, M. (1977). Densities of LNG for custody transfer. In: Proceedings of the Fifth International Conference on Liquified Natural Gas. Düsseldorf, Germany.

Kataoka, H., Fujisawa, S., Inoue, A. (1970). Utilization of LNG cold for the production of liquid oxygen and liquid nitrogen. In: Proceedings of the Second International Conference and Exhibition on LNG. Paris.

Khenat, B., Hasni, T. (1977). The first years of operation of the LNG plant with a discussion of mercury chemical corrosion of aluminum cryogenic exchangers. In: Proceedings of the Fifth International Conference on Liquefied Natural Gas. Düsseldorf, Germany.

Kidnay, A. J. (1972). Liquefied natural gas. Mineral Industries Bulletin. Colorado School of Mines, 15, No. 2, March.

(1980). In: Kirk, R. E., Othmer, D. F., Grayson, M., Eckroth, D., eds. *Encyclopedia of Chemical Technology.* Vol. I and II. 3rd ed. New York: John Wiley and Sons, Inc.

Kohl, A. L., Riesenfield, F. C. (1959). Today's processes for gas purification. *Chem. Eng.* 129–130.

*LNG Information Book* (1981). American Gas Association, Arlington, VA.

Monfore, G. E., Lentz, A. E. (1962). Physical properties of concrete at very low temperatures. Portland Cement Association Research Bulletin No. 145, Chicago, ILL.

Naeve, L. K. (1968). Operating experience of a small LNG plant. In: Proceedings of the First International Conference on LNG. Chicago, ILL, April 7–12.

Nakanishi, E., Reid, R. C. (1971). *Chem. Eng. Prog.* 67(12):36–41.

Phannenstiel, L. L., McKinley, C., Sorenson, J. C. (1975). Mercury in natural gas. In: Proceedings of the XIV International Congress of Refrigeration. Moscow.

Phannenstiel, L. L., McKinley, C., Sorenson, J. C. (1976). Mercury in natural gas. In: Operating Section Proceedings. American Gas Association, T202 to T204.
Picutti, E. (1970). The LNG terminal at La Spezia. In: Proceedings of the Second International Conference and Exhibition on LNG. Paris.
Rigola, M. (1970). Recovery of cold in a heavy LNG. In: Proceedings of the Second International Conference and Exhibition on LNG. Paris.
Sarsten, J. A. (1972). *Pipeline Gas J.* 199:37–39.
Schoofs, R. J. (1966). Natural gas clean-up prior to liquefaction. *Gas.*
Wenzel, L. A. (1975). LNG peakshaving plants—a comparison of cycles. *Advances in Cryogenic Engineering*. Vol. 20. New York: Plenum Press, p. 93.

## CHAPTER 11: SUGGESTED READING

Alcorta, J. J. (1998). Cryogenic safety in the United States Antarctic Program. *Advances in Cryogenic Engineering*. Vol. 43. Plenum Press, pp. 1041–1045.
Cassidy, K. (1993). Risk assessment and the safety of large cryogenic systems and plant in the UK and Europe. *Cryogenics* 33(8):755–76.
Stanek, R., Kilmer, J. (1993). Evolution of cryogenic safety at Fermilab. *Cryogenics* 33(8):809–812.
Webster, T. J. (1982). Proceedings of the Ninth International Cryogenic Engineering Conference. Kobe, Japan: Butterworth and Co.

## CHAPTER 11: REFERENCES

American Institute of Chemical Engineers. *Safety in Air and Ammonia Plants* Vol. 1–10. CEP Technical Manuals.
(1968). *ASME Boiler and Pressure Vessel Code for Unfired Pressure Vessels*, Sec. VIII. New York: American Society of Mechanical Engineers.
Bankaitis, H., Scueller, C. F. (1972). ASRDI oxygen technology survey. Vol. II. Clearing requirements, procedures, and verification techniques. NASA SP-3072.
Burgoyne, J. H. (1965). Explosion and fire hazards associated with the use of low-temperature industrial fluids. *Trans. Insit. Chem. Eng. London* 185:CE 7–10.
(1961). *General Safety Precautions for Missle Liquid Propellants.* AFT. O. 11C-1-6 (October).
Hauser, R. L., Rumpel, W. F. (1963). Reactions of organic materials with liquid oxygen. *Advances in Cryogenic Engineering*. Vol. 8. New York: Plenum Press, pp. 242–250.
Hauser, R. L., Sykes, G. E., Rumpel, W. F. (1961). Mechanically initiated reactions of organic materials in missle oxidizers. ASD TR 61-324, June 1960–June 1961.
Jackson, J. D., Boyd, W. K., Miller, P. D. (1963). Reactivity of Metals with Liquid and Gaseous Oxygen. Battelle Memorial Institute, Defense Metals Information Center. Columbus, OH: DMIC Memo. No. 163 (January).
Key, C. F. (1966). *Compatibility of Materials with Liquid Oxygen.* NASA TMX-53533. Huntsville, AL: NASA George C. Marshall Space Flight Center.
Key, C. F., Gayle, J. B. (1966). Effect of liquid nitrogen dilution on LOX impact sensitivity. *J. Spacecraft Rockets* 3(2):274–276.
Key, C. F., Riehl, W. A. (1963). *Compatibility of Material with Liquid Oxygen*. Huntsville, AL: NASA George C. Marshall Space Flight Center. Internal Report MTP-P & VE-M-63-14 (December).
Lapin, A. (1972). Liquid and gaseous oxygen safety review: Final Report. Vols. I–IV. NASA CR-120922.
Lewis, B., Von Elbe, G. (1961). *Combustion, Flames and Explosions of Gases.* 2nd ed. New York: Academic Press.

McCamy, C. S. (1957). Survey of hazards of handling liquid oxygen. *Indust. Eng. Chem.* 49(9):81A–82A.
McKinley, C. (1962). Safety aspects of cryogenic systems. *Applied Cryogenic Engineering*. New York: John Wiley and Sons, Inc.
McKinley, C., Himmelberger, F. (1957). Oxygen plant safety principles. *Chem. Eng. Prog.* 53(3):112–121.
NASA Kennedy Space Center Safety Office (1965). *Liquid Propellants Safety Handbook*. Cocoa Beach, FL: NASA Report SP-4-44-s, (April).
(1962–63). *National Fire Codes* Vol. V. Electrical, National Fire Protection Association (NFPA), containing Articles 500–503 and 510–517 of the National Electrical Code, Boston, MA.
Neary, R. M. (1961). Safe handling of cryogenic fluids, Paper presented at Chemical Section, National Safety Congress. Chicago, IL (October).
NFPA (1962). Flammable liquids and gases, bulk oxygen systems. Vol. 1. NFPA Circular No. 566.
Pelouch, J. J., Jr. (1974). *ASRDI Oxygen Technology Survey*. Vol. VII. Characteristics of metals that influence system safety. NASA SP-3077.
Reynales, C. H. (1958). *Compatibility of Materials with Oxygen*. Douglas Aircraft Company Report D81-444 (October).
Reynales, C. H. (1959). Safety aspects in the design and operation of oxygen systems. Douglas Aircraft Company Eng. Paper No. 713 (January).
Reynales, C. H. (1961). Selection of lubricants and thread compounds for oxygen missle systems. *Advances in Cryogenic Engineering*. Vol. 6. New York: Plenum Press, pp. 117–129.
Schmidt, H. W., Forney, D. E. (1975). *ASRDI Oxygen Technology Survey*. Vol. IX. Oxygen systems engineering review. NASA SP-3090.
Spencer, E. W. (1964). Cryogenic safety. *J. Chem. Educ.* p. 41.
(1966). Union Carbide Corp. Precautions and safe practices for handling liquefied atmospheric gasses. Publication F-9888. Linde Division.
Van Dolah, R. W. (1969). Fire and explosion hazards of cryogenic liquids. *Applications of Cryogenic Technology*. Los Angeles: Tinon-Brown, Inc.
Van Dolah, R. W., Zabetaklis, M. G., Burgess, D. S., Scott, G. S. (1961). Review of fire and explosion hazards of flight vehicle combustibles. ASD TR 61–278, April.
Zabetaklis, M. G. (1967). *Safety with Cryogenic Fluids*. New York: Plenum Press.
Zebetakis, M. G., D. S. Burgess (1961). Research on the hazards associated with the production and handling of liquid hydrogen. BM-RI-5707, Bureau of Mines Report of Investigation.
Zenner, G. H. (1960). Safety engineering as applied to the handling of liquified atmospheric gasses. *Advances in Cryogenic Engineering*. Vol. I. New York: Plenum Press, pp. 291–295.

## APPENDIX: SUGGESTED READING

American Petroleum Institute (1976). Conversion of operational and process measurement units to the metric (SI) System. API 2564 (March).
American Society for Testing and Materials. Standard for metric practice. ASTM E 380–76, Philadelphia, PA.
Rossini, F. D. (1974). *Fundamental Measures and Constants for Science and Technology*. Cleveland, OH: CRC Press.
(1971). SI Unit Conversion Factors for Chemical Engineers. *BCP Process Technol.* 16(9): 829–832.

# Index